The Complete A level maths

Orlando Gough

Heinemann Educational Books

Acknowledgements

To Peter Templeton, my students, the Lansdowne photocopier, my friends—Joanna in particular, Alan Daykin, Stephen Ashton: thank you very much.

Heinemann Educational Books Ltd
22 Bedford Square, London WC1B 3HH

LONDON EDINBURGH MELBOURNE AUCKLAND
SINGAPORE KUALA LUMPUR NEW DELHI
IBADAN NAIROBI JOHANNESBURG
PORTSMOUTH (NH) KINGSTON

ISBN 0 435 51345 1

© Orlando Gough 1987
First published 1987

Typeset by J. W. Arrowsmith Ltd, Bristol
Printed and bound in Great Britain by
Butler and Tanner Ltd, Frome

Preface

The Complete A Level Maths covers, in one volume, the requirements of the pure-with-mechanics syllabuses for all the examining boards. I have tried to write a book which, while similar in content to existing texts, is more practical for use in the classroom.

I feel the exercises are the most important part of any A level maths textbook, since the exam tests problem-solving ability and the development of this ability depends mainly on practice. Therefore the majority of this book is devoted to a comprehensive set of graded exercises – far more than in other texts. The questions range in difficulty from the trivial to the almost impossible. They contain almost all the creative input – the insights, the conclusions, the jokes – and much of the book's educational thrust. Students are invited to prove theorems; they are led deeper into the subject and shown applications; they are asked to think about what they are doing and comment on it.

I have used the following symbols to make it easier for teachers and students to find their way around each group of problems:

- •   These problems are suitable for use in class as **worked examples**. Some students could do them without help, others need to be shown the method.
- :   These problems are suitable for **class discussion**.
- →   These problems, taken alone, investigate most of the work covered in the section concerned. Solving them successfully indicates a good grasp of the subject. They form the **central core** of problems.
- +   These are more **difficult problems**.
- R   This set of problems forms a useful **revision course**.

I was also concerned that students solving problems outside the classroom should not get stuck for want of a helping hand. For many problems I have therefore indicated a relevant worked example. (The symbol ⑤ alongside a question refers the student back to worked example 5 in that section).

I have provided a fairly brief text, enabling teachers to choose how to present a particular topic. Furthermore, I have tried, by careful referencing within the text, to ensure that the order in which I have presented the topics does not dictate the order in which they are taught.

At the end of every chapter, there is a **Miscellaneous Exercise**, and at the end of the book there is a selection of **A Level Exam Questions**. I would like to thank the following for permission to reproduce questions:

The Associated Examining Board (AEB)
University of Cambridge Local Examinations Syndicate (C) and (SMP)
University of London School Examinations Board (L)
University of Oxford Delegacy of Local Examinations (O)
Oxford and Cambridge Schools Examinations Board (O & C)
Joint Matriculation Board (JMB).

A letter following a Board's initials refers to the syllabus. All answers have been provided by myself and not the Examination Boards who accept no responsibility for the accuracy of those answers given.

Orlando Gough
July 1987

Contents

Preface

1 Introduction 1
1.1 Algebra 1
1.1:1 Logical symbols 1
1.1:2 Sets 3
1.1:3 The real numbers 6
1.1:4 Algebraic expressions: equations, identities and inequalities 9
1.1:5 Expressions involving square roots 13
1.1:6 Indices 15
1.1:7 Logarithms 17
1.2 Functions and graphs 20
1.2:1 The coordinate system 20
1.2:2 The straight line 25
1.2:3 Functions 28
1.2:4 Graphs 35
1.2:5 Limits and continuity 40
1.2:6 Asymptotes 43
1.2:7 Graphical solution of equations and inequalities 46
1.2:8 Implicit functions 49
1.3 Differentiation 50
1.3:1 Gradient of a curve 50
1.3:2 Basic derivatives 53
1.3:3 Equation of the tangent and the normal to a curve 56
1.3:4 Higher derivatives 57
1.3:5 Stationary points and points of inflexion of a curve 59
1.3:6 Function of a function rule 63
1.3:7 Product and quotient rules 66
1.3:8 Implicit function rule 68
1.3:9 Small changes 70
Miscellaneous exercise 1 72

2 Algebraic, exponential and logarithmic functions 74
2.1 Quadratic functions 74
2.1:1 Quadratic equations 74
2.1:2 Quadratic functions 76
2.1:3 Quadratic inequalities 79
2.1:4 Roots of a quadratic equation 82
2.2 Polynomial functions 85
2.2:1 Multiplication and division 85
2.2:2 The remainder theorem 86
2.2:3 The factor theorem; solution of polynomial equations 87
2.2:4 Graphs of polynomial functions 88
2.3 Rational functions 91
2.3:1 Graphs of rational functions 91
2.3:2 Inequalities involving rational functions 94
2.3:3 Partial fractions 96
2.4 Modulus functions 99
2.4:1 Graphs of modulus functions 99
2.4:2 Equations and inequalities 100
2.5 Exponential and logarithmic functions 102
2.5:1 Power functions 102
2.5:2 The exponential function e^x 104
2.5:3 Logarithmic functions 106
2.5:4 The logarithmic function $\ln x$ 106
2.5:5 Logarithmic differentiation 109
2.6 Approximate solution of equations 110
2.6:1 Curve sketching 110
2.6:2 Iterative methods 112
2.7 Reduction of the relationship between two variables to linear form 115
Miscellaneous exercise 2 119

3 Trigonometry 123
3.1 Trigonometric functions 123
3.1:1 Definitions of $\sin\theta$, $\cos\theta$ and $\tan\theta$ 123
3.1:2 Graphs of $\sin\theta$ and $\cos\theta$ 125
3.1:3 Graphs of $\tan\theta$, $\cot\theta$, $\sec\theta$ and $\csc\theta$ 129
3.2 Trigonometric identities and equations 131
3.2:1 The equations $\sin\theta = k$, $\cos\theta = k$ and $\tan\theta = k$ 131
3.2:2 The identity $\sin^2\theta + \cos^2\theta = 1$ 136
3.2:3 Compound angle formulae 139
3.2:4 Double angle formulae 141
3.2:5 t formulae 143
3.2:6 The form $a\sin\theta + b\cos\theta \equiv R\sin(\theta + \alpha)$ 145
3.2:7 Factor formulae 147
3.3 Solution of triangles 149
3.3:1 Sine and cosine formulae 149
3.3:2 Area of a triangle 152

3.4	**Derivatives of trigonometric functions**	**153**
3.4:1	Radians, arcs and sectors	153
3.4:2	Small angles	156
3.4:3	Derivatives of trigonometric functions	159
3.5	**Inverse trigonometric functions**	**162**
3.5:1	Definitions and graphs	162
3.5:2	Derivatives of inverse trigonometric functions	164
Miscellaneous exercise 3		165

4	**Integration and differential equations**	**168**
4.1	**Integration**	**168**
4.1:1	The reverse of differentiation	168
4.1:2	Basic integrals	169
4.1:3	Area under a curve	173
4.1:4	Integral as the limit of a sum; volume of a solid of revolution	180
4.2	**Methods of integration**	**184**
4.2:1	Integration by substitution	184
4.2:2	Integrand containing a function and its derivative	188
4.2:3	Substitutions of the type $x = a \sin \theta$	191
4.2:4	The substitution $t = \tan \frac{1}{2} x$	194
4.2:5	Integration of rational functions	195
4.2:6	Integration by parts	197
4.3	**Approximate integration**	**200**
4.4	**Differential equations**	**203**
4.4:1	Formation of differential equations	203
4.4:2	First order differential equations with separable variables	205
4.4:3	Natural occurrence of differential equations	207
Miscellaneous exercise 4		210

5	**Series and probability**	**214**
5.1	**Sequences and series**	**214**
5.1:1	Sequences	214
5.1:2	Series	216
5.1:3	Arithmetic progressions	216
5.1:4	Geometric progressions	220
5.1:5	Convergence of series	223
5.1:6	Convergence of geometric series	225
5.1:7	Method of differences	229
5.2	**Method of induction**	**234**
5.3	**Permutations and combinations**	**236**
5.3:1	Permutations of objects which are all different	236
5.3:2	Combinations of objects which are all different	241
5.3:3	Permutations and combinations of objects which are not all different	246

5.4	**Binomial expansions**	**249**
5.4:1	Binomial theorem	249
5.4:2	Binomial series	252
5.5	**Probability**	**256**
5.5:1	Sample space, probability	256
5.5:2	$p(A')$, $p(A \cup B)$	262
5.5:3	Mutually exclusive and independent events	264
5.5:4	Conditional probability	270
Miscellaneous exercise 5		274

6	**Vectors, coordinate geometry and complex numbers**	**280**
6.1	**Vectors**	**280**
6.1:1	Displacements	280
6.1:2	Vectors	280
6.1:3	Use of vectors for geometry	287
6.2	**Coordinate geometry in two dimensions**	**291**
6.2:1	Components of a vector; vectors in a plane	291
6.2:2	Position vector of a point in a plane	296
6.2:3	Equation of a line: parametric form	299
6.2:4	Equation of a line: scalar product form	301
6.2:5	Loci	305
6.2:6	The circle	306
6.2:7	The parabola	313
6.2:8	Tangents	315
6.3	**Parameters**	**317**
6.3:1	Graphs	317
6.3:2	Tangent and normal to a curve given in parametric form	320
6.3:3	Secondary loci	324
6.3:4	Area under a curve given in parametric form	327
6.4	**Coordinate geometry in three dimensions**	**328**
6.4:1	Vectors in three-dimensional space	328
6.4:2	Position vector of a point in space	331
6.4:3	Equations of a line	332
6.4:4	Equation of a plane: parametric form	336
6.4:5	Equation of a plane: scalar product form	338
6.5	**Complex numbers**	**343**
6.5:1	Definition of a complex number	343
6.5:2	Simple operation on a complex number	345
6.5:3	Complex roots of polynomial equations	349
6.5:4	Argand diagram	352
6.5:5	Addition and subtraction in the Argand diagram	353
6.5:6	Modulus and argument of a complex number	355
6.5:7	The polar form of a complex number	357
6.5:8	Multiplication and division of complex numbers in polar form; De Moivre's theorem	358

Miscellaneous exercise 6 — 362

7 Kinematics of a particle — 367

Introduction — 367
7.1 Motion of a particle in a straight line — 368
7.1:1 Displacement, velocity and acceleration — 368
7.1:2 Use of integration — 372
7.1:3 Constant acceleration — 375
7.1:4 Free vertical motion under gravity — 378
7.1:5 Use of differential equations — 380
7.1:6 Collisions — 384
7.2 Motion of a particle in a plane — 386
7.2:1 Use of differentiation — 386
7.2:2 Use of integration — 391
7.2:3 Relative motion — 394
7.3 Projectiles — 401
7.3:1 Motion of a projectile — 401
7.3:2 Properties of the flight of a projectile — 406
7.4 Angular velocity and circular motion — 411
7.4:1 Angular velocity — 411
7.4:2 Circular motion — 414
Miscellaneous exercise 7 — 415

8 Forces on a particle — 420

8.1 Newton's laws — 420
8.1:1 Newton's first law — 420
8.1:2 Newton's second law — 425
8.1:3 Types of force — 427
8.1:4 Newton's third law — 430
8.1:5 Hooke's law — 434
8.2 Particles in equilibrium — 435
8.2:1 Equilibrium problems — 435
8.2:2 Equilibrium problems (friction) — 439
8.2:3 Equilibrium problems (elastic strings) — 443
8.2:4 Connected particles — 445
8.3 Motion of a particle — 448
8.3:1 Dynamics problems — 448
8.3:2 Dynamics problems (friction) — 451
8.3:3 Use of constant acceleration equations — 455
8.3:4 Connected particles — 457
8.3:5 Circular motion — 465
8.4 Work and energy; power — 470
8.4:1 Work done by a force — 470
8.4:2 Energy — 474
8.4:3 Work Energy Principle — 476
8.4:4 Power — 483
8.5 Momentum — 488
8.5:1 Impulse Momentum Principle — 488
8.5:2 Impulsive forces — 491
8.5:3 Direct elastic impact — 496
Miscellaneous exercise 8 — 500

9 Forces on a rigid body — 507

9.1 Systems of coplanar forces — 507
9.1:1 Turning effect of a force — 507
9.1:2 Couples — 511
9.1:3 Equivalent systems of forces — 512
9.1:4 Systems reducing to a single resultant force — 512
9.1:5 Systems reducing to a couple — 519
9.1:6 Systems in equilibrium — 522
9.2 Centres of gravity — 526
9.2:1 Centre of gravity, centre of mass, centroid — 526
9.2:2 Centre of gravity of a set of particles in a plane — 528
9.2:3 Centre of gravity of a composite body and a remainder — 530
9.2:4 Use of integration — 535
9.3 Rigid bodies in equilibrium — 541
9.3:1 Equilibrium problems — 541
9.3:2 Three-force equilibrium problems — 548
9.3:3 Suspension — 549
9.3:4 Toppling and sliding — 552
Miscellaneous exercise 9 — 556

Questions from examination papers — 560

Answers — 568
Index — 599

1 Introduction

1.1 Algebra

1.1:1 Logical symbols

The symbol $\Rightarrow$ means **implies**.

If p and q are statements, $p \Rightarrow q$ means 'p implies q', or 'if p is true, then q is true'. For example:

(i) $x = 2 \Rightarrow x^2 = 4$
(ii) $2x - y = 5 \Rightarrow y = 2x - 5$
(iii) There is life on every planet $\Rightarrow$ There is life on Mars.

The symbol $\not\Rightarrow$ means **does not imply**. For example:

(iv) $x = 2 \not\Rightarrow x^2 = 5$
(v) $x^2 = 4 \not\Rightarrow x = 2$
If $x^2 = 4$, then x might equal 2, but it might equal -2; it does not *necessarily* equal 2.
(vi) Some newspapers have left-wing tendencies $\not\Rightarrow$ The *Daily Telegraph* has left-wing tendencies.

If $p \Rightarrow q$ and $q \Rightarrow p$ we write: $p \Leftrightarrow q$. This means: 'if p is true then q is true, and if q is true then p is true'. We say that p and q are **logically equivalent**. For example:

(vii) $x = 3 \Leftrightarrow 2x = 6$
(viii) The sum of the interior angles of the polygon P is $360° \Leftrightarrow P$ is a quadrilateral.

Note that we are at present only interested in deciding whether or not one statement follows from another, and not whether each statement is, in itself, true. In example (i), we said that the statement '$x = 2$' implies the statement '$x^2 = 4$' without even asking: 'Does $x = 2$?' Similarly, in example (iii), it doesn't matter what exactly we mean by life on another planet, nor whether the first statement is true.

$\sim p$ is the **negation** of p. For example:

(ix) $p: x = 2$
       $\sim p: x \neq 2$
(x) p: All policemen are gentle, law-abiding citizens.
      $\sim p$: Not all policemen are gentle, law-abiding citizens.

Exercise 1.1:1

→ **1** Consider the following pairs of statements. Does $p \Rightarrow q$? Does $q \Rightarrow p$?
Which of the pairs of statements are logically equivalent?

   (i) p: Mandy lives in London.
       q: Mandy lives in England.
   (ii) p: $x = 4$
        q: $3x - 2 = 10$
   (iii) p: $x = 4$
         q: $3x - 2 = 11$
   (iv) $ABCD$ is a square.
        p: The length of AB is 3 cm.
        q: The area of $ABCD$ is 9 cm^2.
   (v) $ABCD$ is a rectangle.
        p: $AB = 5$ cm, $BC = 3$ cm.
        q: The area of $ABCD$ is 15 cm^2.
   (vi) $ABCD$ is a quadrilateral.
        p: $AB = 5$ cm, $BC = 3$ cm.
        q: The area of $ABCD$ is 15 cm^2.
   (vii) p: $x^2 - 3x - 2 = 0$
         q: $x = 1$
   (viii) p: x is a parallelogram.
          q: x is a square.
   (ix) p: x is a rectangle.
        q: x is a rhombus.
   (x) p: x is a banana.
       q: x is a fruit.
   (xi) p: x is a rhinoceros.
        q: x is a fruit.

2 (i) Let p be the statement: All pop-stars are drug addicts.
Which of these statements are $\sim p$?

   q_1: Not all pop-stars are drug addicts.
   q_2: No pop-star is a drug addict.
   q_3: There is a pop-star who is not a drug addict.

   (ii) Let p be the statement: Some cats like Pal.
   Which of these statements are $\sim p$?

   q_1: Some cats do not like Pal.
   q_2: No cat likes Pal.
   q_3: There is a cat that does not like Pal.

   (iii) Write down the statement $\sim p$ when p is the statement
   (a) All cows eat grass.
   (b) All politicians are dangerous criminals.
   (c) Anyone who chooses to study A level Maths should have his (or her) head examined.
   (d) Some Englishmen make good lovers.
   (e) Some people think that the Queen is the most beautiful woman in the world.
   (f) There is more nutritional value in a Corn Flakes packet than in its contents.

3 Consider the pairs of statements in question 1. In each case, write down $\sim p$ and $\sim q$.
Does $\sim p \Rightarrow \sim q$? Does $\sim q \Rightarrow \sim p$?
For which pairs is it true to say that $p \Rightarrow \sim q$? What does this mean? In these cases, does $q \Rightarrow \sim p$?

4 Invent *two* pairs of statements p and q for which:
   (i) $p \Leftrightarrow q$
   (ii) $p \Rightarrow q$ but $q \not\Rightarrow p$
   (iii) $\sim p \Rightarrow q$

5 Which of these statements are true, and which are false?
   (i) $x = 2 \Rightarrow x^2 = 4$
   (ii) $x = 2 \Rightarrow x^2 = 5$
   (iii) $x^2 = 4 \Rightarrow x = 2$
   (iv) $x^2 = 4 \Leftrightarrow x = 2$
   (v) $x \neq 2 \Rightarrow x^2 \neq 4$
   (vi) $x = 2 \Rightarrow x^2 \neq 5$
   (vii) $x^2 \neq 4 \Rightarrow x \neq 2$

6 Which of these statements are true, and which are false?
   (i) (a) I have broken the law $\Rightarrow$ I am in prison.
       (b) I am in prison $\Rightarrow$ I have broken the law.
       (c) I am in prison $\Rightarrow$ I am not taking the dog for a walk.
       (d) I am not taking the dog for a walk $\Rightarrow$ I am in prison.
   (ii) (a) My mother is French $\Rightarrow$ I am French.
        (b) I am French $\Rightarrow$ My mother is French.
        (c) My mother is an idiot $\Rightarrow$ I am an idiot.
        (d) My mother is not an idiot $\Rightarrow$ I am not an idiot.

→ **7** Decide whether each of the following general statements is true or false. Give examples to illustrate your answers:
   (i) $(p \Rightarrow q) \Rightarrow (q \Rightarrow p)$
   (ii) $(p \Rightarrow q) \Rightarrow (\sim p \Rightarrow \sim q)$
   (iii) $(p \Rightarrow q) \Rightarrow (\sim q \Rightarrow \sim p)$
   (iv) $(p \Rightarrow \sim q) \Rightarrow (q \Rightarrow \sim p)$

What is $\sim \sim q$? So what can you say about statements (iii) and (iv)?

8 Which of the following pairs of statements are logically equivalent?
   (i) p: If John is in love, he doesn't go to school.
       q: If John is at school, he is not in love.
   (ii) p: If I go on the tube today, I will get mugged.
        q: If I do not go on the tube today, I will not get mugged.

• → **9** Most results in Mathematics are proved by **deduction**.
Suppose we are trying to prove a statement q.
We start from a statement p that we know is true (either a statement that we have proved before, or a

statement that we are confident is true anyway, e.g. 1 = 1). Then we write a sequence of statements, starting with p and ending with q, each of which implies the next one: $p \Rightarrow a \Rightarrow b \Rightarrow c \Rightarrow d \Rightarrow q$; a series of logical steps leading from p to q.

Use this method to prove that the angles opposite the equal sides of an isosceles triangle are equal, stating your assumptions and indicating the steps in your proof. (Look up the proof if you need to: the point is to be aware of the logical method.)

• → **10** Suppose we are trying to disprove some general statement, e.g. the statement q: Politicians are liars.

All we have to do (all!) is to find one politician z who is *not* a liar. z is called a **counterexample** to q.

Describe the nature of a counterexample to each of the following statements:

(i) All quadrilaterals are rectangles.
(ii) Women enjoy housework.
(iii) I help an old lady across the road every day.
(iv) Mathematicians wear glasses and have trouble talking to people at parties.
(v) Guinness is good for you.

• → **11** We will occasionally use a method of proof known as 'proof by contradiction'.

Suppose we are trying to prove a statement q; we assume that q is *not* true and show that this assumption leads to a contradiction (sometimes written $※$, i.e. $(\sim q \Rightarrow ※) \Rightarrow q$).

Use this method to show that the tangent to a circle is perpendicular to the radius.

+ : **12** A card has two statements p and q written on it, one on each side.

p: The statement on the other side of this card is true.
q: The statement on the other side of this card is false.

Does $p \Rightarrow q$? Does $q \Rightarrow p$? Does $p \Rightarrow \sim q$? Does $q \Rightarrow \sim p$? What do you conclude?

1.1:2 Sets

A **set** is a collection of objects, e.g. the set of all the planets in the Solar System. The objects are called the **elements** of the set. If x is an element of the set A, we write $x \in A$.

We specify a set either by listing its elements, or by stating properties which characterize its elements, e.g.

$$A = \{2, 4, 6, 8\} = \{\text{even numbers between 2 and 8}\}$$

$2 \in A$, $6 \in A$, $7 \notin A$.

The number of elements in the set A is written $n(A)$.

A **subset** A of a set B is a set containing only elements of B: $x \in A \Rightarrow x \in B$ (fig. 1.1:1). We write $A \subset B$, or $B \supset A$, e.g. $\{2, 4, 6, 8\} \subset \{1, 2, 4, 5, 6, 8, 9\}$.

A diagram like fig. 1.1:1, illustrating the relationship between two (or more) sets, is called a **Venn Diagram**.

Note: any set B is a subset of itself. If $A \subset B$ and $A \neq B$, A is called a **proper subset** of B.

The set that contains all the elements that we are interested in is called the **universal set** $\mathscr{E}$. Then all sets that we consider are subsets of $\mathscr{E}$. For example, if $\mathscr{E} = \{1, 2, 3, 4, 5, 6, 7, 8, 9\}$ then $A = \{2, 4, 6, 8\}$ is a subset of $\mathscr{E}$.

We can write $A = \{x \in \mathscr{E} : x \text{ is even}\}$. The colon means 'such that'.

The **empty set** $\varnothing$ is the set with no elements. Note that $\varnothing$ is a subset of every other set.

The **complement** A' of a set A is the set of elements of $\mathscr{E}$ which are not in A (fig. 1.1:2).

$$A' = \{x \in \mathscr{E} : x \notin A\}$$

For example, when $\mathscr{E} = \{1, 2, 3, 4, 5, 6, 7, 8, 9\}$ and $A = \{2, 4, 6, 8\}$, $A' = \{1, 3, 5, 7, 9\} = \{x \in \mathscr{E} : x \text{ is odd}\}$.

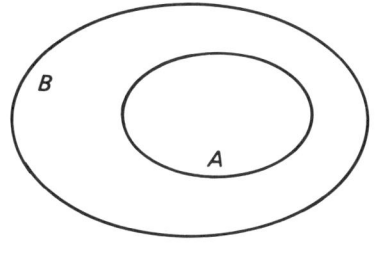

Fig. 1.1:1

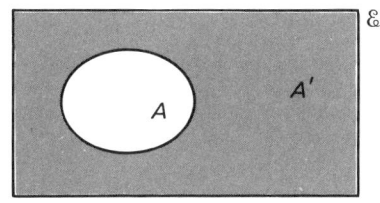

Fig. 1.1:2

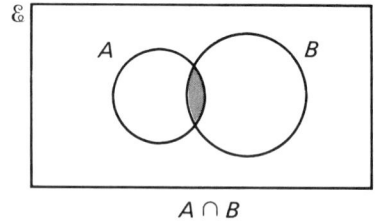

Fig. 1.1:3a

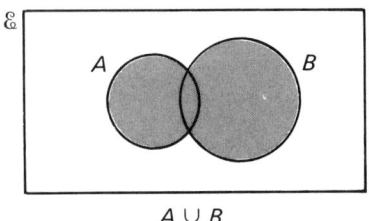

Fig. 1.1:3b

The **intersection** $A \cap B$ of two sets A and B is the set of elements which are both in A and in B (fig. 1.1:**3a**):

$$A \cap B = \{x \in \mathscr{E} : x \in A \text{ and } x \in B\}$$

The **union** $A \cup B$ of two sets A and B is the set of elements which are in A or in B or in both (fig. 1.1:**3b**):

$$A \cup B = \{x \in \mathscr{E} : x \in A \text{ or } x \in B\}$$

For example, if $A = \{2, 4, 6, 8\}$ and $B = \{2, 5, 6, 9\}$, $A \cap B = \{2, 6\}$ and $A \cup B = \{2, 4, 5, 6, 8, 9\}$.

(To remember the meanings of the symbols $\cap$ and $\cup$, think i∩tersection and ∪nion.)

Exercise 1.1:2

→ 1 A, B, C, D, and E are sets: $A = \{a\}$, $B = \{b, c\}$, $C = \{a, b, c\}$, $D = \{c, d\}$, $E = \{a, b, d\}$.
Which of these statements is true and which false?

(i) $A \subset C$ (ii) $B \subset D$ (iii) $B \subset C$
(iv) $D \subset E$ (v) $C \subset E$ (vi) $E \subset A$

2 A, B and C are sets: $A = \{1, 2, 4, 5\}$, $B = \{3, 4, 5\}$, $C = \{2, 5\}$.
Which of these statements is true and which false?

(i) $A \subset B$ (ii) $B \subset C$ (iii) $C \subset A$

Draw a Venn diagram illustrating the relationship between the sets.

3 Let A, B, C, D and E be the sets of all trapezia, all parallelograms, all rectangles, all rhombi and all squares respectively.
Determine which of these sets are subsets of any of the others. Draw a Venn diagram illustrating the relationship between the sets.

4 (i) Write down all the subsets of A (including A and $\varnothing$) when:
(a) $A = \{a, b, c\}$ (b) $A = \{a, b, c, d\}$

(ii) If $n(A) = k$, how many subsets has A?

5 If $\mathscr{E} = \{1, 2, 3, 4, 5, 6, 7, 8, 9, 10\}$, write down the elements of these subsets of $\mathscr{E}$:

(i) $A = \{x \in \mathscr{E} : x \text{ is even}\}$
(ii) $B = \{x \in \mathscr{E} : x \text{ is divisible by 3}\}$
(iii) $C = \{x \in \mathscr{E} : \sqrt{x} \text{ is a whole number}\}$
(iv) $D = \{x \in \mathscr{E} : x - 6 \in \mathscr{E}\}$
(v) $E = \{x \in \mathscr{E} : x^2 \in \mathscr{E}\}$
(vi) $F = \{x \in \mathscr{E} : x \text{ is divisible by 11}\}$

Write down $n(A)$, $n(B)$, $n(C)$, $n(D)$, $n(E)$, $n(F)$.
Write down A', B', C', D', E', F'.

6 (i) Let A be the set $\{1, 3, 4\}$. Find A' if:
(a) $\mathscr{E} = \{1, 2, 3, 4, 5\}$; (b) $\mathscr{E} = \{1, 2, 3, 4\}$;
(c) $\mathscr{E} = \{\text{whole numbers less than 7}\}$;
(d) $\mathscr{E} = \{\text{everything}\}$.
(You are advised not to spend too long on (d).)

(ii) Identify the sets (a) $\mathscr{E}'$; (b) $\varnothing'$; (c) $(A')'$.

→ 7 Let $\mathscr{E} = \{a, b, c, d, e\}$, $A = \{a, b, d\}$ and $B = \{b, d, e\}$. Find:

(i) $A \cup B$ (ii) $A \cap B$ (iii) A' (iv) B'
(v) $A' \cup B$ (vi) $A \cup B'$ (vii) $A' \cap B'$
(viii) $(A \cap B)'$ (ix) $(A \cup B)'$

Draw a Venn diagram showing the sets $\mathscr{E}$, A and B (and their elements).

8 Repeat question 7 with $A = \{3, 4, 5, 7, 9\}$, $B = \{2, 4, 6, 7, 8\}$ and $\mathscr{E} = \{1, 2, 3, 4, 5, 6, 7, 8, 9, 10\}$.

→ 9 Let $\mathscr{E} = \{a, b, c, d, e, f, g\}$, $A = \{a, b, c, d, e\}$, $B = \{a, c, e, g\}$ and $C = \{b, e, f, g\}$. Find:

(i) $(A \cup B) \cup C$ (ii) $(A \cap B) \cap C$
(iii) $(A' \cup B') \cup C'$ (iv) $(A' \cap B') \cap C'$
(v) $(A \cup C) \cap B$ (vi) $(A' \cup B') \cap C'$
(vii) $(A \cap C) \cup B$ (viii) $(A' \cup B') \cap C'$

Draw a Venn diagram showing the sets $\mathscr{E}$, A, B and C (and their elements).

10 Let $\mathscr{E} = \{1, 2, 3, 4, 5, 6\}$, $A = \{1, 2, 5, 6\}$, $B = \{1, 5\}$ and $C = \{2, 4, 6\}$. Show these sets, and their elements, on a Venn diagram.

(i) What can you say about (a) A and B; (b) B and C?

(ii) Find: (a) $(A \cap B') \cap C'$; (b) $(A' \cup C) \cap B$.

→ **11** A and B are subsets of a universal set $\mathscr{E}$. Sketch on Venn diagrams the sets:

(i) $A \cap B'$ (sometimes written $A \backslash B$)
(ii) $A' \cap B$ (iii) $A \cup B'$ (iv) $A' \cup B$
(v) $A' \cap B'$ (vi) $(A \cap B) \cup (A' \cap B')$
(vii) $(A \cap B) \cup A'$ (viii) $(A \cap B) \cup B'$

12 De Morgan's laws state that:

(i) $(A \cup B)' = A' \cap B'$
(ii) $(A \cap B)' = A' \cup B'$

Illustrate these laws by drawing appropriate Venn diagrams.

(It is difficult, and laborious, to *prove* results in set theory, like trying to explain to someone how to ride a bicycle, so we shall be content simply to verify the results.)

13 A is a subset of a universal set $\mathscr{E}$. Simplify:

(i) $A \cap \mathscr{E}$ (ii) $A \cap \varnothing$ (iii) $A \cup \mathscr{E}$ (iv) $A \cup \varnothing$
(v) $A \cap A'$ (vi) $A \cup A'$ (vii) $n(A) + n(A')$

14 Two sets A and B are **mutually exclusive** if they have no element in common, i.e. if $A \cap B = \varnothing$.

(i) If $\mathscr{E} = \{1, 2, 3, 4, 5, 6, 7, 8, 9, 10\}$, write down two mutually exclusive subsets of $\mathscr{E}$. Draw a Venn diagram showing $\mathscr{E}$ and these subsets.

(ii) Which of the following pairs of sets are mutually exclusive?
(a) A and A'
(b) $A \cap B$ and $A \cup B$
(c) $A \cap B$ and $(A \cup B)'$
(d) $(A \cap B)'$ and $A \cup B$
(Assume that A and B are not themselves mutually exclusive.)

(iii) Show that if A and B are mutually exclusive, then: (a) $A' \cap B = B$ (b) $A \subset B'$

15 A and B are subsets of a universal set $\mathscr{E}$. Show that

(i) $n(A \cup B) = n(A) + n(B) - n(A \cap B)$
(ii) $n(A' \cap B') = n(A') + n(B') - n(A' \cup B')$

What can you say if A and B are mutually exclusive?

16 A, B and C are subsets of a universal set $\mathscr{E}$. Sketch on Venn diagrams the sets:

(i) $(A \cap B) \cup C$ (ii) $(A \cup B) \cap C$
(iii) $(A' \cap B) \cap C$ (iv) $(A \cup B') \cup C'$
(v) $(A' \cap B') \cup C$ (vi) $(A' \cup B') \cap C'$

17 A, B and C are three sets.

(i) Illustrate by drawing Venn diagrams the laws
(a) $A \cup (B \cup C) = (A \cup B) \cup C$
(b) $A \cap (B \cap C) = (A \cap B) \cap C$

These laws may seem hardly worth mentioning, but they imply that we can write unambiguously $A \cup B \cup C$ and $A \cap B \cap C$.

(ii) Show by drawing Venn diagrams that:
(a) $A \cap (B \cup C) = (A \cap B) \cup (A \cap C)$
(b) $A \cup (B \cap C) = (A \cup B) \cap (A \cup C)$
Verify these results when $A = \{1, 2, 4\}$, $B = \{2, 3, 4, 5\}$, $C = \{1, 3, 4\}$.

18 A, B and C are subsets of a universal set $\mathscr{E}$. Sketch on Venn diagrams the sets:

(i) $A' \cap B' \cap C'$ (ii) $A' \cup B' \cup C'$
(iii) $(A' \cup B' \cup C')'$ (iv) $(A' \cap B' \cap C')'$

19 Let $\mathscr{E} = \{\text{students in your class}\}$,
$A = \{\text{students under 170 cm (5ft 7in)}\}$,
$B = \{\text{students who own dogs}\}$,
$C = \{\text{students who are in love}\}$.

Make a list of the students in each set (the elements), and find the elements of:

(i) $A \cap B$ (ii) $B \cap C$
(iii) $C \cap A$ (iv) $A \cap B \cap C$

What do you conclude, if anything?

→ **20** Write down symbols representing the shaded sets:

(i)

(v)

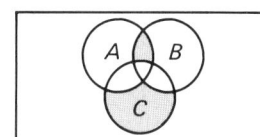

(ii)

(vi)

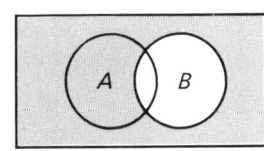
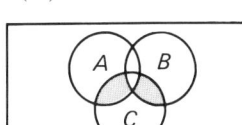

(iii)

(vii)

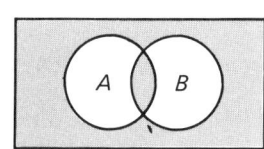
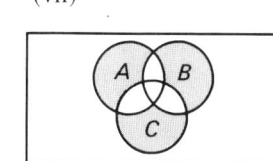

(iv)

(viii)

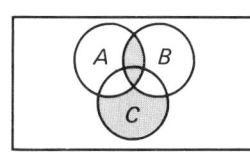
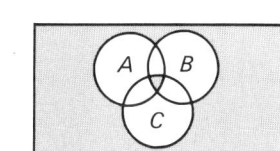

21 Simplify:

(i) $(A' \cap B')'$ (ii) $(A' \cup B')'$
(iii) $A \cap (A' \cup B)$ (iv) $(A \cap B) \cap (A' \cup B')$
(v) $(A \cup B)' \cup B'$ (vi) $(A \cup B) \cap (A \cup B')$

22 Simplify:

(i) $(A' \cap B' \cap C')$ (ii) $(A' \cap B' \cap C')'$
(iii) $(A' \cup B' \cup C')'$ (iv) $(A \cup B \cup C')'$
(v) $(A \cup B) \cap (A \cup B' \cap C)$
(vi) $(A \cup B) \cap (A \cup C) \cap (B \cup C) \cap C'$

23 A team of scientists is making a search of a region of land 2 km square in an attempt to locate possible mineral deposits. They perform the search in two stages. The first stage is to fly over the entire region five times in a helicopter equipped with detection devices; a circle of radius 100 m is centred about each point at which a positive signal is recorded. The second stage is to make a detailed examination of those parts of the region which are within at least three of the circles.

Explain the reasons for this procedure.

The results are shown in the diagram on the right. Copy the diagram and hatch the parts of the region which are subjected to the second stage of the search. Cross-hatch the parts which you think are most likely to yield success at the second stage of the search.

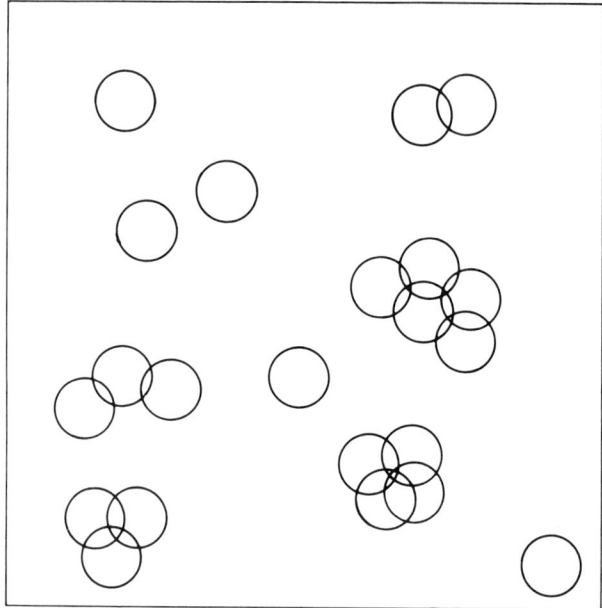

1.1:3 The real numbers

The set of **whole numbers** or **natural numbers** is:

$$\mathbb{N} = \{0, 1, 2, \ldots\}$$

The set of **integers** is:

$$\mathbb{Z} = \{\ldots -3, -2, -1, 0, 1, 2, 3, \ldots\}$$

$\mathbb{Z}$ stands for *Zahl*, the German word for number.

The set of **rational numbers** (i.e. fractions) is:

$$\mathbb{Q} = \left\{\frac{m}{n} : m \in \mathbb{Z}, n \in \mathbb{Z}, n \neq 0\right\}$$

The restriction $n \neq 0$ is necessary because $m/0$ is meaningless: you cannot divide a number into no parts. Do not make the mistake of thinking that, for example, $3/0 = \infty$. Infinity is not a number, it's an idea. (See section 1.2:6).

> **Theorem 1:** Any rational number can be expressed as a decimal which either terminates or recurs and vice versa. For example:
>
> $$\tfrac{5}{7} = 0.714\,285\,714\,285\,71\ldots = 0.\dot{7}14\,28\dot{5}$$
>
> $$0.325 = \tfrac{325}{400} = \tfrac{13}{40}$$

Proof
See Section 5.1:4.

Suppose we draw a straight line and choose one point on it to represent 0, another to represent 1. Then all the rational numbers can be represented by points on the line, as in fig. 1.1:4.

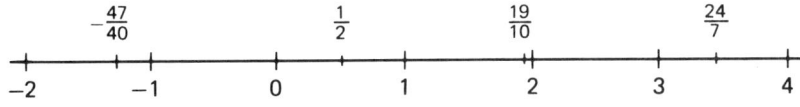

Fig. 1.1:4

Are there any other points on the line? Consider the point representing the number $\sqrt{2}$ (fig. 1.1:5). $\sqrt{2}$ is not a rational number (see question 4). $\sqrt{x}$ means the *positive* square root of x; e.g. $\sqrt{16} = 4$, $x^2 = 3 \Rightarrow x = \pm\sqrt{3}$).

In fact, there are many other numbers which are not rational, e.g. $\sqrt{3}$, $\sqrt[3]{2}$, $\sqrt[5]{17}$, π.

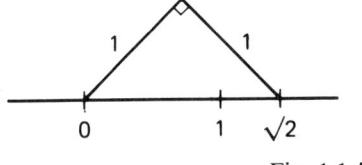

Fig. 1.1:5

The set of all numbers represented by points on the line is called the set $\mathbb{R}$ of **real numbers**. The line is called the **real line**.

The set of all numbers that are real but not rational is called the set $\mathbb{J}$ of **irrational numbers**: $\sqrt{2} \in \mathbb{J}$, $\pi \in \mathbb{J}$.

(Note: the sets of *positive* integers, *positive* rational numbers and *positive* real numbers are denoted by $\mathbb{Z}^+$, $\mathbb{Q}^+$ and $\mathbb{R}^+$ respectively.)

> **Theorem 2:** Any irrational number can be expressed as a decimal which neither terminates nor recurs, and vice versa. For example:
>
> $$\sqrt{2} = 1.414\ 213\ 5\ldots$$
> $$\pi = 3.141\ 592\ 6\ldots$$

Proof
Theorem 2 is logically equivalent to Theorem 1 (see question 7).

Meaning of inequality signs
$x > y$ means 'x is greater than y', i.e. x is to the right of y on the real number line.
$x < y$ means 'x is less than y', i.e. x is to the left of y on the real number line.
For example $9 > -4$, $-23 < -10$.

Worked example 1 Represent on the real line the sets
(i) $A = \{x: x > 3\}$ (ii) $B = \{x: 2 \leq x \leq 5\}$ (iii) $C = \{x: x \leq -3 \text{ or } x > 1\}$

(i) A:

(ii)  B:

The dots indicate that $2 \in B$, $5 \in B$.

(iii) 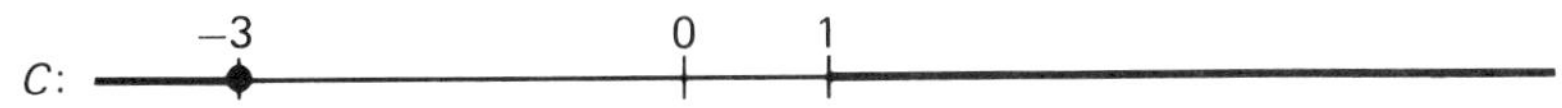 C: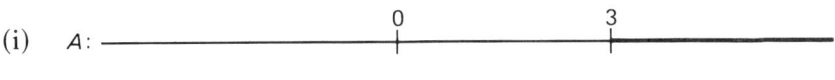

Note that $-3 \in C$, $1 \notin C$.

8   *The real numbers*

Useful notation: the use of round and square brackets saves a lot of writing:

$$[a, b] = \{x: a \leqslant x \leqslant b\}$$
$$(a, b) = \{x: a < x < b\}$$

For example:

$$(-3, 5) \cup [6, 10] = \{x: -3 < x < 5 \text{ or } 6 \leqslant x \leqslant 10\}$$
$$(3, \infty) = \{x: x > 3\}$$

Exercise 1.1:3

1 (i) Express each of the following decimals as a rational number, cancelled to its lowest terms:
(a) 0.85; (b) 0.625; (c) 0.475; (d) 3.75; (e) 2.16.

(ii) Express these fractions as decimals:
(a) $\frac{5}{4}$; (b) $\frac{13}{8}$; (c) $\frac{87}{40}$.

2 (i) Express $0.\dot{5}$ as a rational number. (Hint: let $x = 0.\dot{5}$, write down $10x$, hence find $9x$.)

(ii) Express $2.\dot{4}\dot{5}$ as a rational number. (Hint: let $x = 2.\dot{4}\dot{5}$; write down $100x$, hence find $99x$. Your answer should be cancelled down to its lowest terms.)

3 (i) Express each of the following decimals as a rational number, cancelled to its lowest terms:
(a) $0.\dot{2}$; (b) $0.\dot{7}$; (c) $5.\dot{2}\dot{7}$; (d) $0.2\dot{8}$; (e) $1.37\dot{8}$; (f) $4.5\dot{7}\dot{6}$.

(ii) Express each of the following fractions as a (recurring) decimal:
(a) $\frac{5}{3}$; (b) $\frac{4}{9}$; (c) $\frac{8}{7}$; (d) $\frac{48}{37}$.

→ • 4 Show that $\sqrt{2}$ is an irrational number. (Hint: use the method of 'proof by contradiction' (see 1.1:1 Q11), i.e. assume that $\sqrt{2}$ is a rational number, in particular that $\sqrt{2} = p/q$ where p and q are integers with no common factors. By squaring both sides of this equation, show that p is even; hence show that q is even. Explain how this contradicts the original assumption.)

5 Using the method of question 4, prove that the following are irrational:
(i) $\sqrt{3}$;   (ii) $\sqrt{5}$.
Where does the proof break down when you try to prove that $\sqrt{4}$ is irrational?

6 Show that if n is an integer and $\sqrt{n}$ is rational, then $\sqrt{n}$ is an integer.

7 What exactly does 'and vice versa' mean in the contexts of:
(i) Theorem 1;   (ii) Theorem 2?
Explain why Theorems 1 and 2 are logically equivalent.

→ 8 (i) How many real square roots has:
(a) a positive real number;
(b) a negative real number?

(ii) How many real cube roots has:
(a) a positive real number;
(b) a negative real number?
Give examples. (Section 6.5 introduces another set of numbers, called **complex numbers**, which includes all real numbers and also non-real or **imaginary numbers** such as $\sqrt{-1}$. In this wider context, every number has two square roots, three cube roots, and so on.)

9 Draw a Venn diagram illustrating the relationship between the sets $\mathbb{N}$, $\mathbb{Z}$, $\mathbb{Q}$, $\mathbb{J}$ and $\mathbb{R}$.

① → 10 Represent the following sets on the real line:
(i) $\{x: x > 4\}$ (ii) $\{x: x \geqslant -1\}$
(iii) $\{x: x < 2\}$ (iv) $\{x: 3 < x < 7\}$
(v) $\{x: -2 \leqslant x \leqslant 1\}$ (vi) $\{x: 2 \leqslant x < 4\}$
(vii) $\{x: x < 0 \text{ or } x > 2\}$
(viii) $\{x: x \leqslant -4 \text{ or } x \geqslant -1\}$
(ix) $\{x: x < -3 \text{ or } 0 \leqslant x \leqslant 2 \text{ or } x > 3\}$

① 11 Identify the elements of these sets:
(i) $\{x \in \mathbb{Z}: 2 \leqslant x \leqslant 8\}$
(ii) $\{x \in \mathbb{Z}: -2 \leqslant x \leqslant 3\}$
(iii) $\{x \in \mathbb{Z}: -2 < x < 3\}$
(iv) $\{x \in \mathbb{Z}^+: -2 < x \leqslant 3\}$

① 12 Represent the following sets on the real line:
(i) $(2, 4)$ (ii) $[2, 4]$ (iii) $(-3, \infty)$
(iv) $(-\infty, 4)$ (v) $(1, 2) \cup (5, \infty)$
(vi) $(-\infty, -1) \cup [-\frac{1}{2}, \frac{1}{2}] \cup (1, \infty)$

① → 13 If $A = (-1, 2)$, $B = (0, 5)$ and $C = (3, 6)$, find:
(i) $A \cap B$ (ii) $A \cup B$ (iii) $B \cap C$
(iv) $B \cup C$ (v) $A \cup B \cup C$

① 14 If $\mathscr{E} = \mathbb{R}$, $A = [-2, 3]$ and $B = [-4, 1]$, find:
(i) A' (ii) B' (iii) $A \cap B$
(iv) $A' \cap B'$ (v) $A \cap B'$ (vi) $A' \cap B$

① **15** Give meanings to $(a, b]$ and $[a, b)$.

(i) Represent the following sets on the real line:
(a) $[4, 7)$; (b) $(-2, 1]$; (c) $(-\infty, 3] \cup [5, \infty)$;
(d) $[0, 1) \cup (1, 3]$.

(ii) If $\mathscr{E} = \mathbb{R}$, $A = [-5, 4)$ and $B = (2, 6]$, find:
(a) A'; (b) B'; (c) $A \cap B$; (d) $A' \cap B'$; (e) $A \cup B$;
(f) $A' \cup B'$.

① → **16** If $\mathscr{E} = \mathbb{R}$, $A = (2, 5)$ and $B = [4, 6)$, find:
(i) $A \cap B$ (ii) $A \cup B$ (iii) $A' \cap B$ (iv) $A \cap B'$

① **17** If $\mathscr{E} = \mathbb{R}$, $A = (2, 5)$, $B = [4, 6)$ and $C = (1, 3]$, find:
(i) $A \cap (B \cup C)$ (ii) $A \cup B \cup C$
(iii) $A' \cap B' \cap C'$ (iv) $A \cap B' \cap C'$

1.1:4 Algebraic expressions: equations, identities and inequalities

Consider these expressions:

$$2x^2 + 3 = 11 \tag{1}$$

$$(x+1)^2 = x^2 + 2x + 1 \tag{2}$$

$$x - 2 > 1 \tag{3}$$

Expression (1): $2x^2 + 3 = 11$ is an **equation**. (You knew?) It is true only when x takes certain values, in fact -2 and $+2$, since $2(-2)^2 + 3 = 11$ and $2(2)^2 + 3 = 11$, whereas, for example, $2(5)^2 + 3 \neq 11$.

The values -2 and $+2$ are called the **roots** or **solutions** of the equation (fig. 1.1:6).

Fig. 1.1:6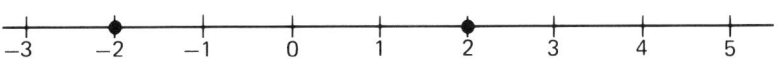

Expression (2): $(x+1)^2 = x^2 + 2x + 1$ is an **identity**. It is true for *all* values of x.

To distinguish between identities and equations we (often) write identities with a three-line equal sign (identity sign):

$$(x+1)^2 \equiv x^2 + 2x + 1$$

Expression (3): $x - 2 > 1$ is an **inequality**. It is true only when x is in a certain set of values, in fact when x is any number greater than 3, since, for example $25 - 2 > 1$ whereas $-13 - 2 \not> 1$. So the solution of the inequality (fig. 1.1:7) is the set $\{x \in \mathbb{R}: x > 3\}$ or $(3, \infty)$.

Fig. 1.1:7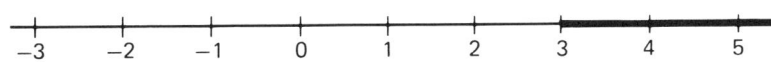

Proof of identities

Worked example 2 Prove that $2x^2 + 3x - 20 \equiv (2x - 5)(x + 4)$.

$$\text{RHS} \equiv (2x - 5)(x + 4) \equiv (2x - 5)x + (2x - 5)4$$
$$\equiv 2x^2 - 5x + 8x - 20$$
$$\equiv 2x^2 + 3x - 20 \equiv \text{LHS}$$

Do not make the mistake of assuming the answer:

q:
$$2x^2 + 3x - 20 \equiv (2x - 5)(x + 4)$$
$$2x^2 + 3x - 20 \equiv 2x^2 - 5x + 8x - 20$$

p:
$$0 \equiv 0$$

Here we have shown that $q \Rightarrow p$, whereas what we need to show is that $p \Rightarrow q$.

Solution of equations

Finding the roots of an equation is called **solving** the equation. We solve an equation by carrying out simple operations on each side of the equation:
    adding the same number to each side;
    multiplying each side by the same number;
    taking the square root of each side;
    grilling each side for five minutes.

Worked example 3   Solve $2x^2+3=11$.

Add -3 to each side: $2x^2=8$
Multiply each side by $\frac{1}{2}$: $x^2=4$
Take the square root of each side: $x=2$ or -2

We can also use an identity to simplify one side of an equation, as in the following example.

Worked example 4   Solve $x^2+2x-4=0$.

Add 5 to each side:

$$x^2+2x+1=5$$

Simplify LHS using the identity $x^2+2x+1 \equiv (x+1)^2$:

$$(x+1)^2=5$$

Take the square root of each side:

$$x+1=\sqrt{5} \text{ or } -\sqrt{5}$$

Add -1 to each side:

$$x=-1+\sqrt{5} \text{ or } -1-\sqrt{5}$$

Note: an equation of the form $ax^2+bx+c=0$ (a, b, c are constants) is called a **quadratic** equation (see section 2.1:1).

Solution of inequalities

We solve an inequality in the same way as we solve an equation, i.e. by adding the same number to each side, by multiplying each side by the same number, etc., with one crucial difference—that *if we multiply each side of an inequality by a negative number we must reverse the inequality.*
    Notice that, for example, $3>2$ but (multiplying by -2) $-6<-4$.

Worked example 5   Solve the inequalities:
(i) $3x+2<8$ (ii) $-5x-2>13$

(i) $3x+2<8$
     $+(-2)$ $3x<6$
     $\times \frac{1}{3}$ $x<2$

(ii) $-5x-2>13$
     $+2$ $-5x>15$
     $\times(-\frac{1}{5})$ $x<-3$ (note reversed inequality)

Exercise 1.1:4

1 Which of the following are equations, and which identities?

(i) $x^2 + 2 = 3$
(ii) $x^2 - 16 = (x-4)(x+4)$
(iii) $x^2 + 9 = (x+3)^2$
(iv) $(x^2)^3 = x^6$
(v) $x^2 - 6x + 10 = (x-3)^2 + 1$
(vi) $x^2 - 6x + 10 = (2x-3)^2 + 1$
(vii) $\dfrac{1}{x+1} + \dfrac{1}{x-1} = \dfrac{x}{x^2-1}$
(viii) $\dfrac{1}{x+1} + \dfrac{1}{x-1} = \dfrac{2}{x^2-1}$

② **2** Prove the identities:

(i) $x^2 - 9 \equiv (x+3)(x-3)$
→ (ii) $6x^2 - 13x - 5 \equiv (3x+1)(2x-5)$
→ (iii) $x^3 + 8 \equiv (x+2)(x^2 - 2x + 4)$
→ (iv) $\dfrac{1}{x+3} - \dfrac{2}{3x-1} \equiv \dfrac{x-7}{(x+3)(3x-1)}$
(v) $\dfrac{4-x^2}{2-x} \equiv x + 2$

② **3** Factorize:

(i) $x^2 - 3x + 2$ (ii) $x^2 + x - 12$
(iii) $10 + 3x - x^2$ (iv) $3x^2 + 4x + 1$
(v) $6x^2 + 7x - 5$ (vi) $15x^2 + 49x + 24$
(vii) $3x^3 - 6x^2$ (viii) $x^3 + 4x^2 - 5x$

② **4** (i) Factorize $a^2 - b^2$, where $a, b \in \mathbb{R}$. Is there a similar factorization for $a^2 + b^2$? Factorize, as far as possible, $a^4 - b^4$.

(ii) Factorize:
(a) $x^2 - 1$; (b) $1 - x^2$; (c) $4 - (x+3)^2$.

(iii) Evaluate:
(a) $101^2 - 99^2$
(b) $101^2 - 102^2 + 103^2 - 104^2 + 105^2 - 106^2$

② → **5** (i) Show that, for $a, b \in \mathbb{R}$,
●
$$a^3 + b^3 \equiv (a+b)(a^2 - ab + b^2) \quad (4)$$

(ii) Deduce a similar factorization for $a^3 - b^3$. (Hint: put $-b$ instead of b in identity (4); if you are worried about this process see Q17.)

(iii) Hence factorize $x^3 + 1$. (Hint: put $a = x$, $b = 1$ in identity (4).) Also factorize $x^3 - 1$.

(iv) Hence factorize $x^6 - 1$.

(v) Factorize:
(a) $x^3 + 27$; (b) $8x^3 - 1$; (c) $2\sqrt{2}x^3 + 3\sqrt{3}$.

② **6** (i) Under what circumstances are the following equations true?
(a) $a^2 + b^2 = (a+b)^2$; (b) $a^3 + b^3 = (a+b)^3$.

(ii) Show that
$$1 - x^2 \equiv (1-x)(1+x)$$
and
$$1 - x^3 \equiv (1-x)(1 + x + x^2)$$

Deduce similar expressions for $1 - x^4$ and $1 - x^5$, and prove them.

(iii) Put $-x$ instead of x in the four identities of part (ii) and write down the resulting identities.

(iv) Simplify:
(a) $1 + (\tfrac{1}{2}) + (\tfrac{1}{2})^2 + (\tfrac{1}{2})^3 + (\tfrac{1}{2})^4 + (\tfrac{1}{2})^5$
(b) $1 - (\tfrac{1}{2}) + (\tfrac{1}{2})^2 - (\tfrac{1}{2})^3 + (\tfrac{1}{2})^4 - (\tfrac{1}{2})^5$

②: **7** Suppose that $a, b \in \mathbb{R}$. Note that
$$(a+b)^1 \equiv a + b$$

(i) Show that
$$(a+b)^2 \equiv a^2 + 2ab + b^2 \quad (5)$$
and
$$(a+b)^3 \equiv a^3 + 3a^2b + 3ab^2 + b^3$$

(ii) Deduce similar expressions for $(a+b)^4$ and $(a+b)^5$.

(iii) Write down the coefficients (i.e. the numerical factors of the terms) in these expansions in a triangular array, beginning

```
        1   1
      1   2   1
    1   3   3   1
```

This array is called Pascal's Triangle.

(iv) Given one row of the array, is it possible to deduce the next row? How?

+ (v) Explain why the method works.

(vi) Without having a nervous breakdown, write down expansions of $(a+b)^6$, $(a+b)^7$ and $(a+b)^8$.

(vii) Write down expansions of
(a) $(1+x)^3$ (Put $a = 1$, $b = x$ in identity (5).)
(b) $(1-x)^3$ (Put $a = 1$, $b = -x$ in identity (5).)
(c) $(1+x)^4$
(d) $(2-x)^4$

(viii) Without using a calculator or tables, find:
(a) $(1.01)^3$; (b) $(3.1)^4$.

8 Solve the following equations. Explain what operation you have carried out at each stage.

(i) $\frac{1}{2}x - 3 = 2$
(ii) $6x + 5 = 20$
(iii) $x^2 + 3 = 12$
→ (iv) $4x^2 - 7 = 13$

Check that your answers do actually satisfy the original equation.

9 Solve the equations:

(i) $x(x+3) = 0$
(ii) $(2x+5)(x-2) = 0$
→ (iii) $(x-1)(x-3)(x-5) = 0$

Explain the reasoning you have used.

10 Factorize the LHS of each of these equations and hence solve it:

(i) $x^2 - 3x + 2 = 0$
(ii) $5x^2 - 4x = 0$
→ (iii) $3x^2 - 11x - 4 = 0$
(iv) $15x^2 + 49x + 24 = 0$
(v) $2x^3 - x^2 = 0$
→ (vi) $x^3 - 9x = 0$

11 Using the identity $(x-3)^2 \equiv x^2 - 6x + 9$, solve the equation

$$x^2 - 6x + 7 = 0$$

Explain what operation you have carried out at each stage.

12 Solve the equations:

(i) $x^2 + 4x + 1 = 0$
(ii) $2x^2 + 12x + 1 = 0$
(iii) $x^2 - x - 1 = 0$

• → **13** Suppose that on the real line the variable number x is represented by the point X, the number 4 by the point A. Explain the meaning of the equation

$$|x - 4| = 3$$

Hence find the two values of x satisfying the equation.

14 Solve the equations:

(i) $|x-3| = 2$ (ii) $|x+4| = 1$ (iii) $|2x-3| = 7$

15 (i) What is wrong with this solution?

$$x + 2 = 5 \quad (6)$$
$$\Rightarrow (x+2)^2 = 25$$
$$\Rightarrow x^2 + 4x + 4 = 25$$
$$\Rightarrow x^2 + 4x - 21 = 0$$
$$\Rightarrow (x-3)(x+7) = 0$$
$$\Rightarrow x = 3 \text{ or } x = -7$$

(ii) By squaring both sides of the equation, solve:

$$\sqrt{(x-1)} = 3 \quad (7)$$

Explain why squaring both sides of equation (6) leads to an extra solution, whereas squaring both sides of equation (7) leads to the correct solution only.

16 Solve the equation:

(i) $\sqrt{(x+4)} = 5$
(ii) $\sqrt{(9-2x)} = 7$
(iii) $\sqrt{(x^2-9)} = 4$

• → **17** This question highlights an important difference between equations and identities.

(i) Consider the equation:

$$2x^2 + 3 = 11 \quad (8)$$

Write $3x$ instead of x in equation (8) and simplify the resulting equation (9). Does equation (9) have the same solutions as equation (8)?

(ii) Consider the identity:

$$(x+1)^2 \equiv x^2 + 2x + 1 \quad (10)$$

Write $3x$ instead of x in equation (10) and simplify the resulting equation (11). Is equation (11) an identity? Write x^2 instead of x in equation (10) and simplify the resulting equation (12). Is equation (12) an identity?

18 Solve these inequalities, explaining what operation you have carried out at each stage. Illustrate your solution set on the real line.

(i) $x - 5 > 2$
(ii) $x + 1 < -4$
(iii) $3x + 7 \geqslant 11$
(iv) $1 - x > 4$
(v) $-\frac{1}{2}x - 1 \leqslant \frac{3}{2}$
→ (vi) $-7x + 3 > -1$
(vii) $1 < 2x + 3 < 7$
→ (viii) $-1 \leqslant 4 - \frac{1}{3}x \leqslant 1$

19 (i) If $a, b \in \mathbb{R}$ and $ab > 0$, what can you say about the signs of a and b?

(ii) Solve the inequalities:
(a) $(x-1)(x-3) > 0$; (b) $(x-1)(x-3) < 0$.

In each case, illustrate your solution set on the real line (see section 1.1:3).

20 Solve the inequalities:

(i) $(x+4)(x-2) > 0$
(ii) $(2x-1)(3x+5) < 0$
(iii) $x^2 - 6x - 7 < 0$
(iv) $2x^2 - 1 > 5x + 2$
(v) $(x-\alpha)(x-\beta) > 0$ $(\alpha < \beta)$

⑤ **21** Solve the inequalities:
  (i) $|x-4|>3$
  (ii) $|2x-5|<1$ (see question 13)

⑤ **22** Solve the inequalities:
  (i) $\sqrt{(x-1)}>3$
  (ii) $\sqrt{(3x+1)}>4$ (see question 15(ii))

:+ **23** (i) Explain exactly what the difference is between an equation and its solution, e.g. between the equation $4x+3=11$ and its solution $x=2$ (which is, after all, also an equation). What exactly does 'solving an equation' mean? Similarly, explain the difference between an inequality and its solution, e.g. between the inequality $x-2>1$ and its solution $x>3$.

(ii) Explain the difference between the *kind* of solution you expect of an equation and of an inequality.
Discuss the solutions of:
(a) $x-2=7$ and $x-2<7$
(b) $x^2-2=7$ and $x^2-2<7$
(c) $(x-2)^2(x-3)^2=0$ and $(x-2)^2(x-3)^2 \geq 0$

(iii) Discuss the inequalities:
(a) $x^2+6>0$; (b) $x^2+6<0$.

1.1:5 Expressions involving square roots

Theorem 3: If a and b are positive real numbers, then
$$\sqrt{(ab)} = \sqrt{a}\sqrt{b} \tag{13}$$

Proof
$$(\text{LHS})^2 = ab$$
$$= (\text{RHS})^2$$
$$\Rightarrow \quad \text{LHS} = \pm\text{RHS}$$

But both sides are positive (since $\sqrt{}$ means *positive* square root); so LHS = RHS.

Worked example 6 Simplify:

(i) $\sqrt{450}$ (ii) $\dfrac{25}{\sqrt{5}}$ (iii) $\dfrac{1}{\sqrt{3}-\sqrt{2}}$

(i) $450 = 5^2 \cdot 3^2 \cdot 2$
By equation (13), $\sqrt{450} = \sqrt{5^2}\sqrt{3^2}\sqrt{2}$
$= 5 \cdot 3 \cdot \sqrt{2} = 15\sqrt{2}$

(ii) $\dfrac{25}{\sqrt{5}} = \dfrac{25}{\sqrt{5}} \cdot \dfrac{\sqrt{5}}{\sqrt{5}}$
$= \dfrac{25\sqrt{5}}{5} = 5\sqrt{5}$

(iii) $\dfrac{1}{\sqrt{3}-\sqrt{2}} = \dfrac{1}{\sqrt{3}-\sqrt{2}} \cdot \dfrac{\sqrt{3}+\sqrt{2}}{\sqrt{3}+\sqrt{2}}$
$= \dfrac{\sqrt{3}+\sqrt{2}}{3-2} = \sqrt{3}+\sqrt{2}$

Note: These three expressions are called **surds**. A surd is an irrational number represented in one of these forms using the $\sqrt{}$ sign.

14 Expressions involving square roots

Exercise 1.1:5

⑥ **1** Simplify:
  (i) $\sqrt{12}$ (ii) $\sqrt{18}$ (iii) $\sqrt{20}$
  (iv) $\sqrt{27}$ (v) $\sqrt{32}$ (vi) $\sqrt{45}$
  (vii) $\sqrt{50}$ → (viii) $\sqrt{72}$ (ix) $\sqrt{96}$
  (x) $\sqrt{216}$ (xi) $\sqrt{1210}$ (xii) $\sqrt{18\,900}$
  → (xiii) $\sqrt{(a^4 b^7 c)}$

⑥ → **2** Simplify:
  (i) $\sqrt{8}$ (ii) $\sqrt{80}$ (iii) $\sqrt{800}$ (iv) $\sqrt{8000}$

⑥ **3** Simplify:
  (i) $\sqrt{8}+\sqrt{18}$
  (ii) $\sqrt{72}-\sqrt{50}+\sqrt{18}-\sqrt{32}$
  (iii) $\sqrt{45}-\sqrt{20}$

⑥ **4** Simplify:
  (i) $\dfrac{1}{\sqrt{3}}$ (ii) $\dfrac{2}{\sqrt{2}}$ (iii) $\dfrac{14}{\sqrt{7}}$
  (iv) $\dfrac{1}{2\sqrt{2}}$ (v) $\dfrac{20}{\sqrt{8}}$ → (vi) $\dfrac{9}{4\sqrt{6}}$
  (vii) $\sqrt{\left(\dfrac{5}{3}\right)}$ (viii) $\dfrac{a}{\sqrt{a}}$ (ix) $\dfrac{a^2 b}{\sqrt{(ab)}}$

⑥ **5** Simplify:
  → (i) $\sqrt{\left(\dfrac{81}{8}\right)}$ (ii) $\sqrt{\left(\dfrac{27}{50}\right)}$ (iii) $\sqrt{\left(\dfrac{125}{24}\right)}$
  (iv) $\sqrt{\left(\dfrac{98}{243}\right)}$ (v) $\sqrt{\left(\dfrac{a^3 c}{b^5}\right)}$

⑥ **6** Simplify:
  → (i) $(\sqrt{3}+1)(\sqrt{3}-1)$
  (ii) $(\sqrt{5}-3)(\sqrt{5}-2)$
  (iii) $(3\sqrt{2}-2)(3+2\sqrt{2})$
  (iv) $(\sqrt{5}-\sqrt{2})^2$
  (v) $(\sqrt{a}+\sqrt{b})(\sqrt{a}-\sqrt{b})$
  → (vi) $(\sqrt{(x+1)}+\sqrt{x})(\sqrt{(x+1)}-\sqrt{x})$
  (vii) $(\sqrt{a}+\sqrt{b})^2$
  → (viii) $(\sqrt{3}+1)^3$
  (ix) $(\sqrt{3}-\sqrt{2})^3$

⑥ **7** Simplify:
  (i) $\dfrac{1}{3-\sqrt{2}}$ (ii) $\dfrac{1}{\sqrt{2}+1}$
  (iii) $\dfrac{1}{2-\sqrt{3}}$ (iv) $\dfrac{55}{2\sqrt{5}-3}$
  (v) $\dfrac{1}{\sqrt{5}-\sqrt{2}}$ (vi) $\dfrac{1}{3\sqrt{2}-2\sqrt{3}}$
  → (vii) $\dfrac{1}{\sqrt{a}-\sqrt{b}}$

⑥ **8** Simplify:
  (i) $\dfrac{\sqrt{2}-1}{3-\sqrt{2}}$
  (ii) $\dfrac{\sqrt{8}-\sqrt{7}}{\sqrt{8}+\sqrt{7}}$
  (iii) $\dfrac{\sqrt{2}+2\sqrt{5}}{\sqrt{5}-\sqrt{2}}$

⑥ **9** Simplify:
$$\dfrac{1}{2+\sqrt{3}}+\dfrac{1}{2-\sqrt{3}}$$

⑥ **10** Given that $\sqrt{5}\approx 2.236$, find, without using a calculator, $\dfrac{1}{\sqrt{5}}$ and $\dfrac{1}{\sqrt{5}+1}$ to 3 dec. pls.

⑥ **11** Given that $\sqrt{3}\approx 1.732$, find, without using a calculator,
$$\dfrac{\sqrt{3}+1}{\sqrt{3}-1} \quad \text{and} \quad \dfrac{\sqrt{3}-1}{\sqrt{3}+1}$$
to 3 dec. pls.

⑥ **12** Show that, when $x>1$,
  (i) $\dfrac{1}{\sqrt{x}-\sqrt{(x-1)}}\equiv \sqrt{x}+\sqrt{(x-1)}$
  (ii) $\sqrt{(x-1)}+\dfrac{1}{\sqrt{(x-1)}}\equiv \dfrac{x}{\sqrt{(x-1)}}$
  → (iii) $\dfrac{1}{\sqrt{(x-1)}-1}-\dfrac{1}{\sqrt{(x-1)}+1}=\dfrac{2}{x}$

  Why is the restriction $x>1$ necessary?

13 Find the positive square root of $4+2\sqrt{3}$. (Hint: let the square root be $\sqrt{a}+\sqrt{b}$. Then $(\sqrt{a}+\sqrt{b})^2 = 4+2\sqrt{3}$.)

14 Find the two square roots of $7-2\sqrt{10}$.

• → **15** By squaring both sides, solve the equation
$$\sqrt{(x+5)}=5-\sqrt{x}$$
Check that your answer satisfies the equation.

16 Solve the equation
$$\sqrt{x}+\sqrt{(x+9)}=9$$

- **17** Solve the equation

$$\sqrt{(2x+5)} - \sqrt{(x-1)} = 2$$

(Hint: square both sides; collect terms and square both sides again; you should find two values for x; explain why one of these values is in fact the solution of the equation $\sqrt{(2x+5)} + \sqrt{(x-1)} = 2$.)

18 Solve the equations:
(i) $\sqrt{(3x+1)} - \sqrt{(x+4)} = 1$
→ (ii) $2\sqrt{x} - \sqrt{(x+5)} = 1$

1.1:6 Indices

Definition If a is any number and n is a positive integer, then

$$a^n = \underbrace{a \times a \times a \times a \times \cdots \times a}_{n \text{ times}} \qquad (14)$$

a is called the **base**, n the **index**. For example,

$$3^4 = 3 \times 3 \times 3 \times 3$$

From this definition we deduce three laws of indices:

$$a^m \times a^n \equiv a^{m+n} \qquad (15)$$
$$a^m \div a^n \equiv a^{m-n} \quad (m > n) \qquad (16)$$
$$(a^m)^n \equiv a^{mn} \qquad (17)$$

To give a sensible meaning to fractional and negative indices, we begin by assuming that laws (15), (16) and (17) are true whatever the values of m and n.

Theorem 4:

$$a^{p/q} \equiv \sqrt[q]{a^p} \equiv (\sqrt[q]{a})^p \qquad (18)$$
$$a^0 \equiv 1 \qquad (19)$$
$$a^{-n} \equiv \frac{1}{a^n} \qquad (20)$$

Proof
Equation (18): First, consider $a^{1/q}$. By law (17),

$$(a^{1/q})^q = a, \text{ so } a^{1/q} = \sqrt[q]{a}$$

Now look at $a^{p/q}$. By law (17),

$$(a^{p/q})^q = a^p, \text{ so } a^{p/q} = \sqrt[q]{a^p}$$

Also by law (17),

$$(a^{1/q})^p = a^{p/q}, \text{ so } a^{p/q} = (\sqrt[q]{a})^p$$

Equation (19): Put $m = n$ in law (16)

$$a^0 = a^n \div a^n = 1$$

Equation (20): Put $m = 0$ in law (16)

$$a^{-n} = a^0 \div a^n = \frac{1}{a^n}$$

Worked example 7 Simplify:

(i) $(125)^{2/3}$ (ii) $\left(\dfrac{25}{16}\right)^{-3/2}$

(i) $\quad (125)^{2/3} = \sqrt[3]{(125^2)} = \sqrt[3]{15\,625} = 25$

or $\quad (125)^{2/3} = \{\sqrt[3]{125}\}^2 = 5^2 = 25$

Notice that the second method is much simpler.

(ii) $\quad \left(\dfrac{25}{16}\right)^{-3/2} = \dfrac{1}{\left(\dfrac{25}{16}\right)^{3/2}} = \dfrac{1}{\left(\dfrac{125}{64}\right)} = \dfrac{64}{125}$

Worked example 8 Solve the equation:
$$2^{2x} - 12 \cdot 2^x + 32 = 0$$

Put $2^x = y$, so that $2^{2x} = y^2$; the equation then becomes a quadratic equation in y.

$$\begin{aligned} & y^2 - 12y + 32 = 0 \\ \Rightarrow \quad & (y-8)(y-4) = 0 \\ \text{So} \quad & y = 8 \quad \text{or} \quad y = 4 \\ \text{i.e.} \quad & 2^x = 8 \quad \text{or} \quad 2^x = 4 \\ \Rightarrow \quad & x = 3 \quad \text{or} \quad x = 2 \end{aligned}$$

Exercise 1.1:6

⑦ **1** Evaluate:
 (i) $2^0, 5^0$
 (ii) $2^{-3}, 4^{-2}, 3^{-4}, 2^{-7}$
 (iii) $4^{1/2}, 256^{1/2}, 8^{1/3}, 64^{1/3}, 1000^{1/3}, 16^{1/4}$
 (iv) $(\frac{1}{100})^{1/2}, (\frac{25}{49})^{1/2}, (12\frac{1}{4})^{1/2}$
 (v) $4^{3/2}, 9^{3/2}, 8^{2/3}, 8^{5/3}, 27^{4/3}, 64^{2/3}, (\frac{8}{27})^{4/3}$
 (vi) $4^{-1/2}, 49^{-1/2}, 4^{-3/2}, 8^{-1/3}, 27^{-2/3}, 125^{-4/3}$
→ (vii) $(\frac{121}{25})^{-1/2}, (\frac{1}{16})^{-3/2}, (\frac{27}{8})^{-2/3}, (\frac{16}{81})^{-3/4}$
 (viii) $\dfrac{4^{-5/2}}{8^{-2/3}}, \dfrac{9^{1/3} \cdot 3^{1/3}}{6}, 5^{2/3} \cdot 5^{4/3}, 8^{1/2} \cdot 8^{1/6}$

⑦ **2** Simplify:
 (i) $\dfrac{x^{1/2} + x^{-1/2}}{x^{1/2}}$
→ (ii) $\dfrac{(1+x)^{1/2} - 2x(1+x)^{-1/2}}{(1+x)^{1/2}}$
 (iii) $\dfrac{(1+x)^{1/3} - x(1+x)^{-2/3}}{(1+x)^{4/3}}$

 (iv) $\dfrac{(1+x^2)^{1/2} - x^2(1+x^2)^{-1/2}}{x^2}$
 (v) $\dfrac{(1-x)^{1/2}(1+x)^{-1/2} + (1-x)^{-1/2}(1+x)^{1/2}}{(1+x)(1-x)}$

3 Solve the equations:
• (i) $3^{2x} = 27$
 (ii) $25^x = 125$
→ (iii) $64^{x/3} x = 128$
 (iv) $(\frac{1}{9})^x = 3$
 (v) $(\sqrt{2})^{3x} = \frac{1}{64}$

• **4** Show that $8^{x-2} \equiv 2^{3x-6}$.
 Hence solve the equation $2^{2x-1} = 8^{x-2}$.

5 Solve the equations:
 (i) $3^{3x} - 1 = 9^{x+1}$
 (ii) $25^{x-1} = 125^{x-3}$
 (iii) $4^{2x-1} = (\frac{1}{8})^{1-x}$

6 Solve the equations:
   (i) $2^{x^2} = 4^{(x+4)}$
   (ii) $9^{(x^2-1)} = 27^{(x-1)}$
   (iii) $16^{(x^2-4)} = 32^{(2x-2)}$

⑧ **7** Solve the equations:
   (i) $5^{2x} - 6 \cdot 5^x + 5 = 0$
   (ii) $2^{2x} - 10 \cdot 2^x + 16 = 0$
→ (iii) $3^{2x} - 4 \cdot 3^{x+1} + 27 = 0$
   (iv) $3^{2x+1} - 10 \cdot 3^x + 3 = 0$
   (v) $3^{2x+1} - 28 \cdot 3^x + 9 = 0$
   (vi) $2 \cdot 4^{2x} - 5 \cdot 4^x + 2 = 0$

⑧ **8** Show that the equation $2^{2x} - 7 \cdot 2^x - 8 = 0$ has only one real root.

⑧ **9** Solve the equations:
   (i) $4^{2x} - 7 \cdot 4^x - 8 = 0$
   (ii) $4 \cdot 3^{2x+1} + 17 \cdot 3^x - 7 = 0$
   (iii) $5^{2x+2} + 49 \cdot 5^x - 2 = 0$

: **10** Explain what you think $5^{\sqrt{2}}$ means.

1.1:7 Logarithms

What do we mean by $\log_a x$?

Definition	$y = \log_a x \Leftrightarrow a^y = x$	(21)

For example,
$$10^2 = 1000 \Leftrightarrow \log_{10} 100 = 2$$
$$2^3 = 8 \Leftrightarrow \log_2 8 = 3$$
$$27^{1/3} = 3 \Leftrightarrow \log_{27} 3 = \tfrac{1}{3}$$

Notice that a logarithm is an index.

Theorem 5:	If a, x and y are positive real numbers, then	
	$\log_a x + \log_a y \equiv \log_a xy$	(22)
	$\log_a x - \log_a y \equiv \log_a (x/y)$	(23)
	$n \log_a x \equiv \log_a x^n$	(24)

Proof
See question 3.

Worked example 9   Simplify: $2 \log_{10} 25 - 3 \log_{10} 5 + \log_{10} 20$.

$$2 \log_{10} 25 - 3 \log_{10} 5 + \log_{10} 20$$
$$= \log_{10} 25^2 - \log_{10} 5^3 + \log_{10} 20 \quad \text{by equation (24)}$$
$$= \log_{10} \left(\frac{25^2}{5^3}\right) + \log_{10} 20 \quad \text{by equation (23)}$$
$$= \log_{10} \left(\frac{25^2}{5^3} \cdot 20\right) \quad \text{by equation (22)}$$
$$= \log_{10} 100$$
$$= 2$$

Note that we often write '$\log_{10}$' as 'lg'. For example:

$$\lg 1000 = 3, \qquad \lg \tfrac{1}{100} = -2$$

18 Logarithms

Theorem 6 (change of base formula):

$$\log_a b = \frac{\log_c b}{\log_c a} \qquad (25)$$

In particular, putting $c = b$,

$$\log_a b = \frac{1}{\log_b a} \qquad (26)$$

Proof

Let $x = \log_a b$, i.e. $a^x = b$. Then

$$\log_c(a^x) = \log_c b$$

By equation (24)

$$x \log_c a = \log_c b$$

$$x = \frac{\log_c b}{\log_c a}$$

Worked example 10   Find $\log_3 7$.

By equation (25)

$$\log_3 7 = \frac{\lg 7}{\lg 3}$$

$$\approx \frac{0.8451}{0.4771} \approx 1.771$$

Worked example 11   Solve the equation: $\log_3 x - 2\log_x 3 = 1$.

Let $\log_3 x = y$; then by equation (26), $\log_x 3 = 1/y$. So the equation becomes:

$$y - \frac{2}{y} = 1$$

i.e. $\qquad\qquad y^2 - y - 2 = 0$

$\Rightarrow \qquad\qquad (y-2)(y+1) = 0$

So $\qquad\qquad y = 2 \quad\text{or}\quad y = -1$

i.e. $\qquad\qquad \log_3 x = 2 \quad\text{or}\quad \log_3 x = -1$

$\Rightarrow \qquad\qquad x = 3^2 = 9 \quad\text{or}\quad x = 3^{-1} = \frac{1}{3}$

Exercise 1.1:7

1 (i) Express in logarithmic notation:
(a) $2^4 = 16$; (b) $25^{-1/2} = 0.2$; (c) $10^{1.3} = 20$; (d) $5^0 = 1$.

(ii) Express in index notation:
(a) $\log_2 128 = 7$; (b) $\log_{81} 3 = 0.25$; (c) $\log_4 5 = 1.16$; (d) $\log_5 4 = 0.86$.

2 Evaluate:
(i) $\log_5 25$, $\log_3 27$, $\log_2 2$, $\log_7 7$, $\log_2 1$
(ii) $\log_{10}(\frac{1}{10})$, $\log_7(\frac{1}{49})$, $\log_2(\frac{1}{32})$
(iii) $\log_4 2$, $\log_8 2$, $\log_4 8$, $\log_3 3\sqrt{3}$, $\log_{27} 9$, $\log_9 \sqrt{3}$, $\log_4 8\sqrt{2}$
(iv) $\log_{1/5} 25$, $\log_{\sqrt{3}} 9$

3 (i) Prove that $\log_a x + \log_a y = \log_a xy$.
(Hint: put $\log_a x = m$ (so that $x = a^m$) and $\log_a y = n$, and use equation (15).)

(ii) Similarly, using equation (16), prove that:
$$\log_a x - \log_a y = \log_a \left(\frac{x}{y}\right)$$

(iii) Prove that $n \log_a x = \log_a x^n$.

Explain why these results hold only when a, x and y are positive.

4 Simplify:
(i) (a) $\log_6 3 + \log_6 2$
(b) $\log_2 24 - \log_2 3$
(c) $\log_2 15 - \log_2 60$
(d) $\log_3 36 - \log_3 12$

(ii) (a) $\log_5 80 - 2 \log_5 4$
(b) $\log_5 4 + 2 \log_5 3 - 2 \log_5 6$
(c) $\log_a x^3 - \log_a x^2$
(d) $\log_a (x-1) - \log_a (x^2 - 1)$

5 (i) Show that $2 - 2 \lg 5 = \lg 4$.

(ii) Simplify:
(a) $\frac{1}{3} \lg 27 - 1$; (b) $\lg 45 - 2 \lg 3 + 1$.

6 Given that $\lg 2 \approx 0.3010$ and $\lg 3 \approx 0.4771$, find to 3 sig. figs. and without using a calculator, the values of:

(i) $\lg 4$ (ii) $\lg 36$ (iii) $\lg 5$

7 (i) Find $\log_5 5^2$, $\log_2 2^7$, $3^{\log_3 27}$, $4^{\log_4 2}$.
(ii) Simplify: $\log_a a^x$, $a^{\log_a x}$
(iii) Simplify: $\log_a 1$, $\log_a a^2$, $\log_a (1/a)$, $\log_a (1/a^3)$, $\log_a \sqrt{a}$, $\log_a a^2\sqrt{a}$.

8 Write in terms of $\log_a x$: $\log_a x^2$, $\log_a (1/x)$, $\log_a \sqrt{x}$, $\log_a (1/x\sqrt{x})$.

9 Express in terms of $\log_a x$, $\log_a y$ and $\log_a z$:
(i) $\log_a xyz$ (ii) $\log_a (xy/z)$
(iii) $\log_a (x^2/y^3 z)$ (iv) $\log_a \sqrt{(x/y)}$
(v) $\log_a ax$ (vi) $\log_a (xy/a^2)$

10 Simplify:
(i) $\log_a 8 / \log_a 2$ (ii) $\log_a 125 / \log_a 25$
(iii) $\log_a 49 / \log_a 343$ (iv) $\log_a x^2 / \log_a x^6$
(v) $\log_a x^y / \log_a x^2$

11 Find:
(i) $\log_3 4$ (ii) $\log_7 5$ (iii) $\log_6 10$
(iv) $\log_{0.4} 3$ (v) $\log_{0.2} 0.08$

12 (i) Show that $\log_2 12 = \log_4 144$.
Hence evaluate $\log_2 12 - \log_4 9$.

(ii) Evaluate $\log_3 63 - \log_9 49$.

13 (i) If a, b and c are positive numbers, simplify:
(a) $(\log_a b)(\log_c a)$; (b) $(\log_a b)(\log_b a)$.

(ii) Evaluate:
(a) $(\log_5 4)(\log_4 125)$; (b) $(\log_3 49)(\log_7 27)$;
(c) $\log_{10} 2 \log_2 10$; (d) $\log_3 4 / \log_{27} 16$.

(iii) Simplify:
(a) $(\log_a b^3)(\log_b a^2)$; (b) $(\log_a b^x)(\log_b a^y)$.

14 (i) If a and b are positive numbers other than 1, show that
$$\log_a b + \log_{1/a} b = 0$$

(ii) Evaluate $\log_5 10 / \log_{0.2} 10$

(iii) Evaluate $\log_a x / \log_{1/a} x$. Is it possible to simplify $\log_a x / \log_b x$?

15 (i) If a, b and c are positive numbers other than 1, show that
$$\log_b a \, \log_c b \, \log_a c = 1$$

(ii) Evaluate $\log_{10} 25 \cdot \log_2 10 \cdot \log_5 4$.

16 Using tables or a calculator, solve these equations, correct to 2 dec. pls.:

(i) $5^x = 7$
(ii) $6^x = 4$
(iii) $3^x = 0.2$
(iv) $4^{2x} = 50$
(v) $2^{3x-2} = 5$
(vi) $5^{x^2} = 12$

17 Using a calculator or tables, solve these equations (see worked example 8):

(i) $3^{2x} - 5 \cdot 3^x + 6 = 0$
R (ii) $2^{(2x+1)} - 13 \cdot 2^x + 20 = 0$
(iii) $7^{2x} - 2 \cdot 7^{(x+1)} + 40 = 0$
(iv) $2^{(2x+2)} + 2^x - 3 = 0$

18 Solve these equations, correct to 3 dec. pls.:

(i) $2^x = 3^{x-1}$
(ii) $10^{x-3} = 2^{10+x}$
(iii) $3^{x+1} = 4^{(2x-1)}$
(iv) $5^x = 2^{2+2x}$

19 Solve the equation:
$$\lg x^3 - 2 \lg x^2 + 2 \lg x + \lg \sqrt{x} = 3$$

20 Solve the equations:
(i) $\log_x 16 = -4$
(ii) $\log_x 8 = 1.5$
(iii) $\log_x 10 = 5$
(iv) $\log_x 7 = 0.9$
(v) $\log_x 2 = -0.4$

20 The coordinate system

21 Solve the equation:
$$\log_x 24 - 3 \log_x 4 + 2 \log_x 3 + 3 = 0$$

→ **22** (i) Show that $\log_3 (1+x) \equiv \log_9 (1+x)^2$.
Hence solve the equation:
$$\log_3 (1+x) = \log_9 (6x - 2)$$

(ii) Solve the equations:
(a) $\log_5 (2-x) = \log_{25} (5-4x)$
(b) $\log_2 x = \log_4 (x+6)$ (Only one of the solutions you obtain is valid—why?)

(iii) Show that $\log_2 x = \log_4 x^2$. Hence solve the equations:
(a) $\log_4 x - \log_2 x = 6$
(b) $\log_2 x - \log_4 (x+4) = \frac{1}{2}$

⑪ **23** Solve the equations:
(i) $\log_{10} x = 4 \log_x 10$
(ii) $6 \log_x 8 + 6 \log_8 x = 13$
(iii) $\log_2 x + 4 \log_x 2 = 5$
R (iv) $4 \log_x 3 - \log_3 x = 3$
(v) $\log_{10} x + \log_x 100 = 3$

• **24** If $\lg x + \lg y = 2$, show that $xy = 100$.
Hence solve the simultaneous equations:
$$\lg x + \lg y = 2, \quad x + y = 25$$

25 Solve the simultaneous equations:
$$\lg x - 3 \lg y = 1, \quad xy = 160$$

→ **26** Show that $2 \log_{a^2} xy \equiv \log_a x + \log_a y$.
Hence solve the simultaneous equations:
(i) $\log_{16} xy = 7$, $\log_4 x / \log_4 y = -8$
(ii) $\log_9 xy = \frac{1}{2}$, $\log_3 x \log_3 y = -6$

• → **27** By writing $\log_x y = u$, show that, if
$$\log_x y + \log_y x = \frac{5}{2}$$
then $\log_x y = \frac{1}{2}$ or 2.
Hence solve the simultaneous equations:
$$\log_x y + \log_y x = \frac{5}{2}, \quad xy = 64$$

28 Solve the simultaneous equations:
(i) $\log_y x = 3$, $xy = 16$
R (ii) $\log_y x = \frac{3}{2}$, $\log_2 x + \log_2 y = 5$
(iii) $\log_y x = 2$, $12 \log_x y = 5y - x$

29 Find y in terms of x when:
• (i) $\lg y = 3 + 2 \lg x$
→ (ii) $2 \log_3 y = 4 - \log_3 x$
(iii) $\log_2 y = \log_4 x$
(iv) $\log_x y + \log_y x^2 = 3$, $x \neq y$

1.2 Functions and graphs

1.2:1 The coordinate system

As a convenient way of specifying the position of any point on a plane, we mark a particular point O on the plane and a pair of perpendicular axes Ox and Oy through O (fig. 1.2:**1**).

The position of any point P on the plane can be specified by an ordered pair (x, y), as in fig. 1.2:**2**.

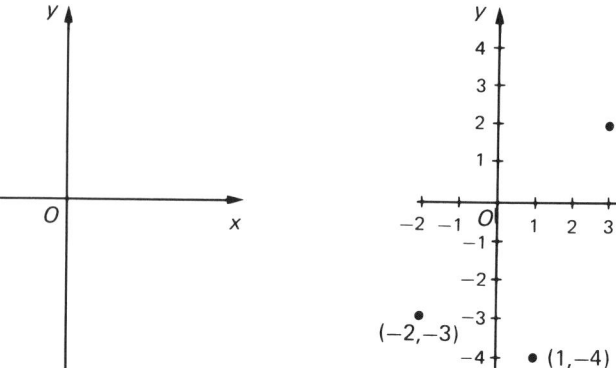

Fig. 1.2:**1**

Fig. 1.2:**2**

Introduction 21

O is called the **origin**, Ox the **x-axis** and Oy the **y-axis**. x and y are called the **cartesian coordinates** of P. (Cartesian coordinates are named after René Descartes (1596-1650), French mathematician and philosopher.) x is the x-coordinate, or **abscissa** of P. y is the y-coordinate, or **ordinate** of P.

Theorem 1: If A and B are the points (x_1, y_1) and (x_2, y_2),

(i) the distance AB is $\sqrt{\{(x_2-x_1)^2+(y_2-y_1)^2\}}$ **(1)**

(ii) the midpoint of AB is $(\tfrac{1}{2}(x_1+x_2), \tfrac{1}{2}(y_1+y_2))$ **(2)**

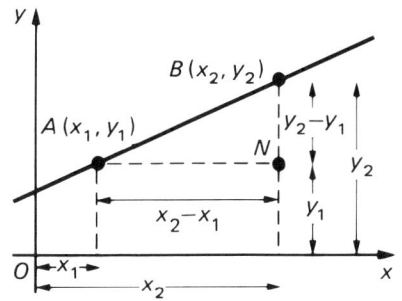

Fig. 1.2:**3**

Proof

Statement (1): $AN = x_2 - x_1$, $BN = y_2 - y_1$ (fig. 1.2:3). By Pythagoras' Theorem,

$$AB^2 = AN^2 + BN^2 = (x_2-x_1)^2 + (y_2-y_1)^2$$

i.e. $\quad AB = \sqrt{\{(x_2-x_1)^2+(y_2-y_1)^2\}}$

Statement (2): $AN = x_2 - x_1$ (fig. 1.2:4), since $AM = \tfrac{1}{2}AB$, $AH = \tfrac{1}{2}AN$ (similar triangles), i.e.

$$AH = \tfrac{1}{2}(x_2 - x_1)$$

So x-coordinate of $M = x_1 + \tfrac{1}{2}(x_2 - x_1) = \tfrac{1}{2}(x_1 + x_2)$.
Similarly, y-coordinate of $M = \tfrac{1}{2}(y_1 + y_2)$.

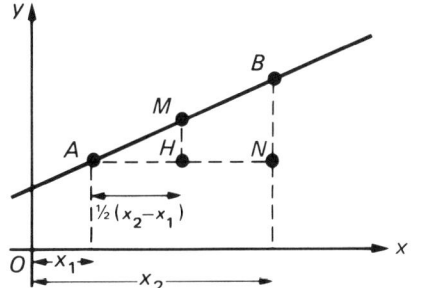

Fig. 1.2:**4**

Worked example 1 A and B are the points $(-2, 1)$ and $(4, 3)$ in fig, 1.2:**5**. Find (i) the distance AB; (ii) the coordinates of the midpoint of AB.

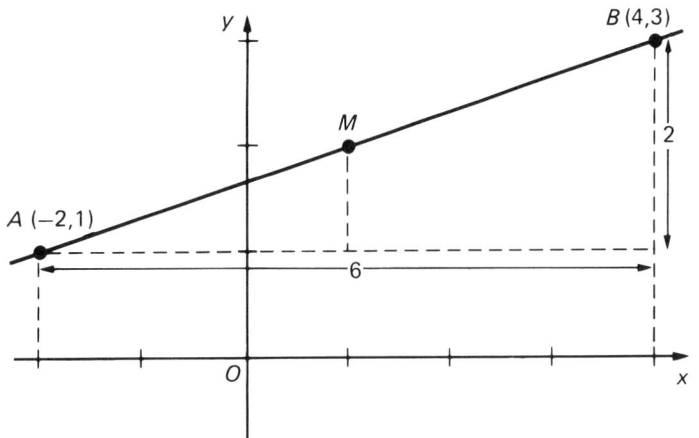

Fig. 1.2:**5**

(i) $AB^2 = \{4-(-2)\}^2 + \{3-1\}^2 = 6^2 + 2^2 = 40$.

So $\quad AB = \sqrt{40} = 2\sqrt{10}$

(ii)   Midpoint of $AB = \left(\dfrac{-2+4}{2}, \dfrac{1+3}{2}\right) = (1, 2)$.

22 *The coordinate system*

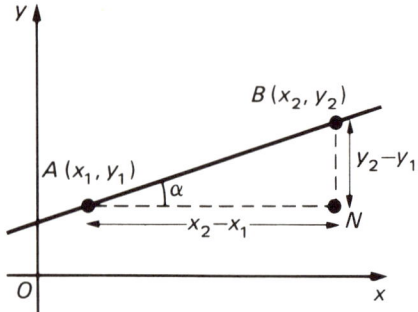

Fig. 1.2:6

Definition The **gradient** of the straight line through the points $A(x_1, y_1)$ and $B(x_2, y_2)$ is

$$\frac{y_2 - y_1}{x_2 - x_1}, \text{ i.e. } \frac{\Delta y}{\Delta x}$$

where Δy is the increase in y and Δx is the increase in x. (See fig. 1.2:6.)

Worked example 2 Find the gradient of the line through the points: (i) $A(2, 1)$ and $B(5, 5)$; (ii) $C(1, -2)$ and $D(-5, 1)$.

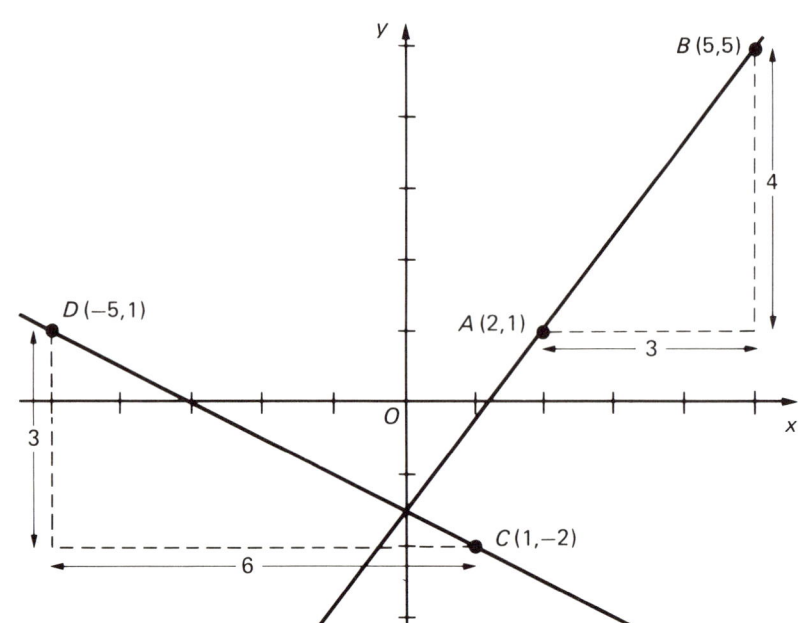

Fig. 1.2:7

From fig. 1.2:7:

(i) gradient of $AB = \dfrac{5-1}{5-2} = \dfrac{4}{3}$.

(ii) gradient of $CD = \dfrac{1-(-2)}{-5-1} = \dfrac{3}{-6} = -\dfrac{1}{2}$.

Note: (i) the gradient is independent of the positions of A and B on the line (see question 13).
(ii) The gradient of $AB = BN/AN = \tan \alpha$, where α is the angle between the line and the positive direction of Ox; in fact this is true whether the gradient is positive (fig. 1.2:8) or negative (fig.1.2:9).

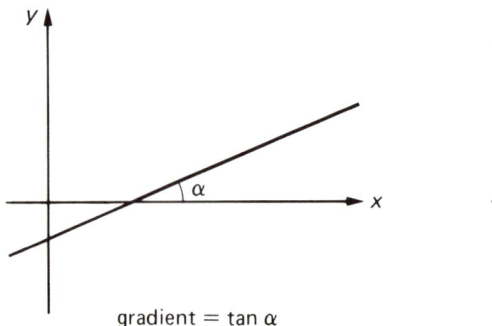
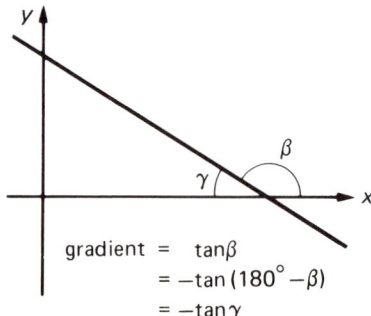

Fig. 1.2:8 Positive gradient
Fig. 1.2:9 Negative gradient

Theorem 2
(i) If two lines are parallel, they have the same gradient. (3)

(ii) If two lines are perpendicular, the product of their gradients is -1. (4)

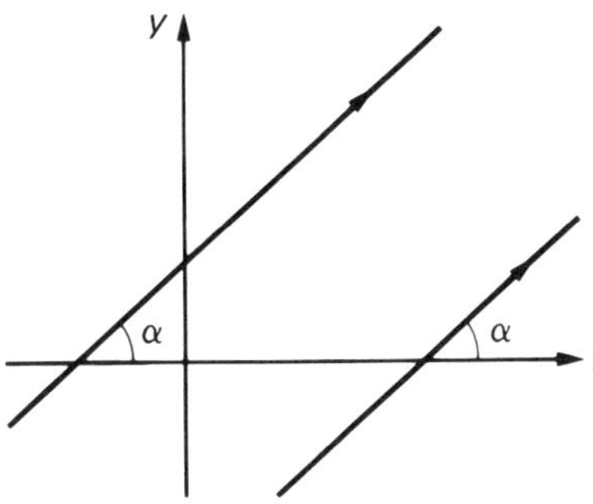

Fig. 1.2:**10**

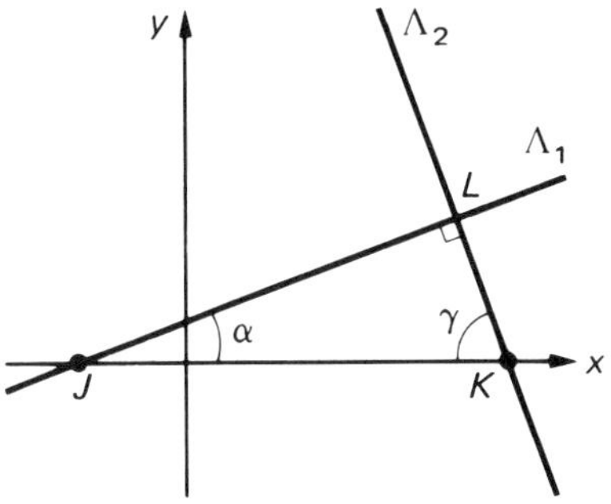

Fig. 1.2:**11**

Proof

Statement (3): The gradient of each line is $\tan \alpha$ (fig. 1.2:**10**).

Statement (4): Gradient of Λ_1, $m_1 = \tan \alpha = LK/JL$ (fig. 1.2:**11**).
Gradient of Λ_2, $m_2 = -\tan \gamma = -JL/LK$.

So $$m_1 m_2 = \left(\frac{LK}{JL}\right) \cdot \left(\frac{-JL}{LK}\right) = -1$$

Exercise 1.2:1

→ 1 Mark a point O and perpendicular axes Ox and Oy on a diagram. Mark in the points $(2, 0)$, $(-5, 0)$, $(0, 1)$, $(0, -3)$, $(2, 1)$, $(-5, -3)$, $(4, 2)$, $(4, -3)$, $(-\frac{5}{2}, \frac{1}{2})$ and $(-\frac{5}{2}, -3)$.

① 2 Sketch a diagram showing the points A and B, and find the distance AB:

(i) $A(0, 0)$, $B(4, 6)$
(ii) $A(3, 1)$, $B(9, 9)$
(iii) $A(1, 0)$, $B(3, 4)$
(iv) $A(2, 3)$, $B(8, 5)$
→ (v) $A(5, 1)$, $B(2, 4)$
(vi) $A(-3, -2)$, $B(4, 3)$
(vii) $A(-2, 10)$, $B(0, 0)$
(viii) $A(1, -2)$, $B(-3, 7)$
(ix) $A(-9, 1)$, $B(6, -7)$
(x) $A(-2, 3)$, $B(4, 3)$
→ (xi) $A(0, 0)$, $B(a, b)$
(xii) $A(a, b)$, $B(3a, 5b)$

① 3 A, B and C are the points $(-1, 2)$, $(6, 1)$ and $(4, 7)$. Show that $\triangle ABC$ is isosceles. Is it equilateral?

① 4 A, B, C and D are the points $(4, -1)$, $(5, 2)$, $(2, 3)$ and $(1, 0)$. Show that $ABCD$ is a square. (**Note:** you need to do more than simply show that $AB = BC = CD = DA$.)

① 5 A, B and C are the points $(-3, 3)$, $(-1, -1)$ and $(5, 2)$. Show that $AB^2 + AC^2 = BC^2$. What can you say about $B\hat{A}C$?

① → 6 Find the midpoint of AB for each of the pairs of points in question 2.

① 7 A and B are the points $(-3, 4)$ and $(7, 6)$. Find the length of the line from the origin to the midpoint of AB.

① → 8 A, B and C are the points $(-5, 7)$, $(1, -2)$ and $(5, 4)$. D is the midpoint of BC.

(i) Find the length of AD (which is called a **median** of $\triangle ABC$).
(ii) Verify that $AB^2 + AC^2 = 2(AD^2 + BD^2)$.

① 9 A is the point $(-1, 2)$, M the point $(3, 5)$. M is the midpoint of AB. Find the coordinates of B.

① 10 A is the point $(4, 1)$, B the point $(2, 2)$ and C is the point on BA produced such that $BC = 2AB$. Find the coordinates of C.

② → 11 Find the gradient of AB for each of the pairs of points in question 2.

12 If A and B are the points $(3, -1)$ and $(3, 4)$, what can you say about the gradient of AB?

13 Suppose that we are trying to find the gradient of a straight line passing through the points A, B, A' and B'. By using statement (3), show that you get the same answer whether you work with the pair of points A, B or the pair A', B'.

14 A and B are the points $(1, -3)$ and $(-4, 2)$. Show that the gradient of $AB = -\tan \gamma$, where γ is the angle between AB and the *negative* direction of the x-axis. What is the angle γ?

15 Write down the gradients of the lines which are inclined at the following angles to the positive direction of the x-axis:

(i) $0°$ (ii) $45°$ (iii) $135°$ (iv) $60°$ (v) $120°$

(Note: $\tan 60° = \sqrt{3}$.)

16 A, B and C are the points $(4, 2)$, $(-3, -3)$ and $(5, 1)$. Show that OA is parallel to BC.

17 A, B, C and D are the points $(8, 1)$, $(-1, 4)$, $(0, -2)$ and $(-3, -1)$. Is AB parallel to CD? Is AD parallel to BC?

18 P is the point (a, b). (Think of P as being in the first quadrant.) When OP is rotated anticlockwise through $90°$, P lands on the point P'. What are the coordinates of P'?

Write down the gradients of OP and OP' and show that their product is -1.

19 A and B are the points $(1, k)$ and $(-k, 1)$. Show that $\widehat{AOB} = 90°$. Sketch a diagram to illustrate your answer.

20 A, B and C are the points $(3, -1)$, $(8, 1)$ and $(-1, 9)$. Show that $\widehat{ABC} = 90°$.

21 Is AB parallel to CD, perpendicular to CD, or neither?

(i) $A(0, 0)$, $B(3, 2)$, $C(-1, 1)$, $D(7, 5)$
(ii) $A(-2, 0)$, $B(6, 2)$, $C(3, 5)$, $D(4, 1)$
(iii) $A(0, 4)$, $B(3, 5)$, $C(2, 1)$, $D(5, 2)$

22 Is AB parallel to CD, perpendicular to CD, or neither?

(i) $A(1, 2)$, $B(-4, 7)$, $C(-4, -2)$, $D(-1, 1)$
(ii) $A(-1, 0)$, $B(4, 4)$, $C(2, 1)$, $D(7, -5)$
(iii) $A(3, -2)$, $B(-5, 2)$, $C(0, -1)$, $D(-2, 0)$

23 A, B and C are the points $(4, 1)$, $(6, 4)$ and $(2, 3)$. Show that OA is parallel to CB and OC is parallel to AB. What do you deduce about the quadrilateral $OABC$? Check that $OA = CB$ and $OC = AB$.

24 A, B, C and D are the points $(-3, -1)$, $(3, 0)$, $(-1, 3)$ and $(-7, 2)$. Show that $ABCD$ is a parallelogram.

Check that the midpoints of AC and BD coincide, i.e. that AC and BD bisect each other.

R 25 A, B and C are the points $(-1, 4)$, $(5, 2)$ and $(1, 6)$. Find the coordinates of the midpoints L and M of AB and AC. Show that LM is parallel to BC, and that $LM = \tfrac{1}{2} BC$.

26 A, B, C and D are the points $(5, -1)$, $(3, 2)$, $(-4, 3)$ and $(2, -2)$. Show that $ABCD$ is *not* a parallelogram. Find the coordinates of the midpoints P, Q, R and S of AB, BC, CD and DA. Show that $PQRS$ is a parallelogram.

27 Find the area of $\triangle ABC$ when A, B and C are the points:

(i) $A(1, 2)$, $B(5, 2)$, $C(1, 4)$
(ii) $A(-3, 1)$, $B(3, 1)$, $C(3, -6)$
(iii) $A(1, 0)$, $B(4, 0)$, $C(2, 3)$
(iv) $A(-2, -3)$, $B(-2, 5)$, $C(2, 1)$

28 Show that $\widehat{BAC} = 90°$ and hence find the area of $\triangle ABC$ when A, B and C are the points:

(i) $A(1, 0)$, $B(5, 3)$, $C(-5, 8)$
(ii) $A(0, 3)$, $B(-1, 5)$, $C(-8, -1)$
(iii) $A(1, 2)$, $B(5, 3)$, $C(-3, 9)$
(iv) $A(-3, 1)$, $B(-1, 7)$, $C(6, -2)$

29 For the following sets of points A, B and C, show that $AB = AC$, find the coordinates of the midpoint M of BC, verify that AM is perpendicular to BC, find the distance AM, and hence find the area of $\triangle ABC$.

(i) $A(0, 4)$, $B(-1, -3)$, $C(5, -1)$
(ii) $A(-4, 1)$, $B(5, 3)$, $C(3, -5)$

30 A, B, C and D are the points $(1, -3)$, $(7, 1)$, $(3, 7)$ and $(-3, 3)$. Show that $AB = BC = DC = DA$ and that $AC = BD$.

What kind of quadrilateral is $ABCD$? Check that:

(i) AB is perpendicular to BC;
(ii) AC is perpendicular to BD;
(iii) AC and BD bisect each other.

31 A, B, C and D are the points $(-1, 1)$, $(4, 6)$, $(-3, 7)$ and $(-8, 2)$. Show that:

(i) AC and BD bisect each other;
(ii) AC is perpendicular to BD;
(iii) AB is not perpendicular to BC.

What kind of quadrilateral is $ABCD$? Find its area.

32 A, B and C are the points $(2, 0)$, $(0, 3)$ and $(x, 7)$, and $\widehat{ABC} = 90°$. Find x.

33 A, B and C are the points $(-4, 1)$, $(2, 4)$ and $(x, 0)$, and $\widehat{ABC} = 90°$. Find x.

R 34 A, B and C are the points $(-5, -2)$, $(3, 4)$ and $(0, y)$, and $\widehat{ABC} = 90°$. Find y and the area of $\triangle ABC$.

- **35** A, B and C are the points $(-2, 1)$, $(4, 3)$ and $(7, 4)$. Find the gradients of AB and AC. What can you say about A, B and C?

→ **36** If the points $A(-2, 5)$, $B(2, 1)$ and $C(x, 0)$ are collinear (i.e. if A, B and C lie in a straight line), find x.

37 If the points $A(5, -2)$, $B(-5, 7)$ and $C(-1, y)$ are collinear, find y.

→ **38** If the points $A(1, -3)$, $B(4, 2)$ and $P(x, y)$ are collinear, show that $5x - 3y - 14 = 0$.
  Draw a diagram showing the possible positions of P. Show that the coordinates of A and B themselves satisfy the equation.

39 If the points $A(3, -2)$, $B(1, 5)$ and $P(x, y)$ are collinear, show that $7x + 2y = 17$.
  Draw a diagram showing the possible positions of P. Show that the coordinates of A and B themselves satisfy the equation.

• → **40** The point $(x, 0)$ is equidistant from the points $(-4, 2)$ and $(-1, 6)$. Find x.

R 41 Find the coordinates of the point on the y-axis which is equidistant from $(-5, 3)$ and $(2, 4)$.

→ **42** (i) Prove that $C(6, 4)$ is on the perpendicular bisector Λ of the line joining $A(-1, 3)$ and $B(1, -1)$ (i.e. the line perpendicular to AB through the midpoint of AB).

  (ii) If $D(x, -2)$ is on Λ, find x. Show that $ACBD$ is a parallelogram.

  (iii) If $P(x, y)$ is on Λ, show that $x - 2y + 2 = 0$. Check that the coordinates of C and D satisfy this equation.

43 A long straight avenue, bordered on either side by parks, runs along the x-axis. Brenda, the Queen of Cartesia, rides down the avenue in a Ford Cortina, starting at the point $(-10, 0)$ and moving at 10 m s^{-1} (1 unit = 10 m). At the moment she sets out, a man with a bomb in his pocket, who intends to rid the country of the drudgery of monarchy, sets out from the point $(11, 12)$. He runs at 5 m s^{-1}, intercepts the Royal Cortina, lobs his bomb delicately onto Brenda's lap and disappears into the park. At what point does the interception occur?
  In fact the bomb fails to detonate and her popularity is increased further...

1.2:2 The straight line

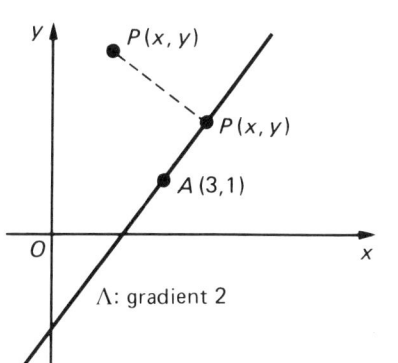

Fig. 1.2:12

Consider the straight line Λ of gradient 2 passing through the point $A(3, 1)$ (fig. 1.2:12).
  Let P be the point (x, y). Then gradient of $AP = (y-1)/(x-3)$.
  If P lies on Λ, gradient of AP = gradient of Λ, i.e.

$$(y-1)/(x-3) = 2$$
$$\Rightarrow \quad y = 2x - 5 \tag{5}$$

The equation (5) gives a condition that $P(x, y)$ lies on Λ, since for any point (x, y) on Λ, equation (5) is satisfied and for any point (x, y) *not* on Λ, equation (5) is *not* satisfied.
  Equation (5) is called the **cartesian equation** of Λ.

Worked example 3 Find the cartesian equation of the straight line through the points $A(-1, 2)$ and $B(4, -1)$ (fig. 1.2:13).

Gradient of $AB = \dfrac{-1-2}{4-(-1)} = -\dfrac{3}{5}$. Let P be the point (x, y). Then

$$\text{gradient of } AP = \frac{y-2}{x-(-1)} = \frac{y-2}{x+1}$$

So if P lies on AB, $(y-2)/(x+1) = -\tfrac{3}{5}$, i.e.

$$y = -\tfrac{3}{5}x + \tfrac{7}{5} \tag{6}$$
$$\Rightarrow \quad 5y = -3x + 7$$

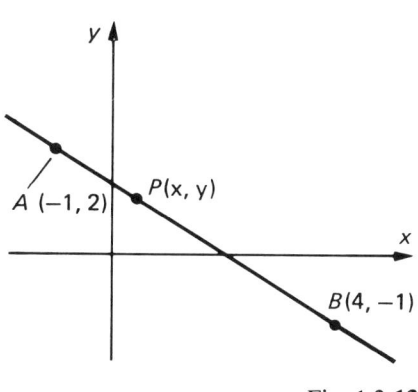

Fig. 1.2:13 So

$$3x + 5y = 7 \tag{7}$$

26 *The straight line*

Notice that the equation of a straight line can be written in either of the forms:

$$y = mx + c, \quad Ax + By = C \qquad (8), (9)$$

corresponding to equation (6) and equation (7) respectively.

Consider the straight line through $A(0, c)$ with gradient m. Its cartesian equation is $\dfrac{y-c}{x-0} = m$, i.e. $y = mx + c$

c is called the **y-intercept** of the line. So,

the equation of the straight line with gradient m and y-intercept c is $y = mx + c$ \hfill (10)

Worked example 4   Find the point of intersection of the lines $\Lambda_1: x - 3y = 1$ and $\Lambda_2: x + 2y = 6$ (fig. 1.2:14).

Suppose $P(x, y)$ lies on Λ_1 and Λ_2. Then:

$$x - 3y = 1, \quad x + 2y = 6 \qquad (11), (12)$$

Taking equation (11) from equation (12): $5y = 5 \Rightarrow y = 1$
From equation (11) or equation (12): $x = 4$
So the point of intersection of Λ_1 and Λ_2 is $(4, 1)$.

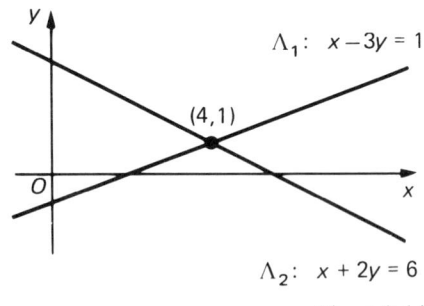

Fig. 1.2:14

Exercise 1.2:2

1 Find the cartesian equation of the line with given gradient passing through the given point.

(i) 3, (2, 0)
→ (ii) ¼, (5, 3)
→ (iii) −2, (1, 4)
(iv) −⅔, (0, 0)
(v) b/a, $(a, 2b)$
→ (vi) 0, (2, −5)
→ (vii) m, (0, 0)
→ (viii) m, (0, c)
(ix) m, (x_1, y_1)

Write the equation of each line in the form $y = mx + c$ and the form $Ax + By + C = 0$.

2 (i) Find the cartesian equation of the line with gradient $-\frac{2}{3}$ passing through the point $(-6, 4)$. Draw a diagram to illustrate your answer.

(ii) Find the cartesian equation of the line with gradient m passing through the point (a, am).

3 Find the cartesian equation of the line with gradient $\frac{1}{3}$ passing through the point:

(i) (0, 0) (ii) (3, 0) (iii) (−2, 5)

Sketch these lines on the same pair of axes.

4 Find the equations of the line which passes through the point (2, −5) and makes angle

(i) 45° (ii) 135°

with the positive direction of the x-axis. Sketch these lines on a diagram.

5 Consider worked example 3 again. Write down the gradient of BP, where P is the point (x, y). Using the fact that if P lies on AB, the gradient of $BP =$ the gradient of AB, find the cartesian equation of AB. Check that your answer agrees with the one found in the text.

→ **6** Consider 1.2:1 Q2 *again*. For each part, find the cartesian equation of the line passing through each pair of points A and B.

7 Find the cartesian equation of the line passing through the points:

(i) (2, 4) and (2, −3)   (ii) (−5, 1) and (−5, 7)

8 Find the gradient of the line with cartesian equation:

(i) $y = 3x + 5$
(ii) $4y = 5x - 2$
→ (iii) $x - 3y - 2 = 0$
(iv) $6x + 3y - 1 = 0$
(v) $y = 2$
(vi) $Ax + By + C = 0$
(vii) $\frac{1}{4}x + \frac{1}{6}y = 1$
(viii) $x/a + y/b = 1$

9 Show that the lines $4x - 2y + 3 = 0$ and $6x - 3y + 5 = 0$ are parallel.
   Show that the lines $2x - 3y + 5 = 0$ and $6x + 4y - 7 = 0$ are perpendicular.

→ **10** Which of the following pairs of lines are parallel, which perpendicular and which neither?

   (i) $y = 3x + 2$, $y = \frac{1}{3}x + 2$
   (ii) $y = x - 4$, $y = -x + 5$
   (iii) $y = -\frac{1}{2}x - 7$, $3x + 6y - 2 = 0$

11 Which of the following pairs of lines are parallel, which perpendicular and which neither?

   (i) $4x - 5y + 3 = 0$, $4x + 5y - 1 = 0$
   (ii) $y = 2x + 1$, $x + 2y + 1 = 0$
   (iii) $8x + 12y - 5 = 0$, $3x - 2y + 5 = 0$

12 (i) Show that the lines $Ax + By + C = 0$ and $kA + kBy + C' = 0$ are parallel.
   (ii) Show that the lines $Ax + By + C = 0$ and $Bx - Ay + C' = 0$ are perpendicular.

Questions 13-16: Find the cartesian equation of the line which passes through the point A and is

(i) parallel   (ii) perpendicular
to the line Λ.

• **13** $A(2, 1)$, $\Lambda: y = x + 5$

→ **14** $A(0, -5)$, $\Lambda: x + 3y - 4 = 0$

15 $A(-1, 4)$, $\Lambda: 5x - 2y = 0$

16 $A(0, 0)$, $\Lambda: 4x + y + 7 = 0$

17 A and B are the points $(5, 0)$ and $(-1, 4)$. Find the equation of the line parallel to the line AB through:

   (i) O (ii) $C(-4, 0)$

Sketch a diagram showing the three lines.

R 18 A and B are the points $(5, -4)$ and $(-1, -3)$. Find the equation of the line perpendicular to the line AB through:

   (i) O (ii) $C(4, 1)$

Sketch a diagram showing the three lines.

19 Find the coordinates of the midpoint of AB and hence find the equation of the perpendicular bisector of AB, when A and B are the points:

• (i) $(4, 0)$, $(0, 6)$
→ (ii) $(-3, -2)$, $(3, 4)$
   (iii) $(1, -3)$, $(7, 1)$
   (iv) $(-2, 1)$, $(1, 4)$
   (v) (a, b), $(-5a, 3b)$

20 Find the gradient of each of these lines and the coordinates of the points where it crosses the axes; hence sketch the line.

   (i) $y = x - 2$ (ii) $y = 2x - 3$
   (iii) $y = \frac{1}{4}x + 1$ (iv) $y = -3x - 6$
   (v) $x + 2y + 2 = 0$ → (vi) $3x - 4y - 5 = 0$
   (vii) $\frac{1}{4}x + \frac{1}{5}y = 1$ (viii) $x + 3y = 0$

④ **21** Find the coordinates of the point of intersection of the lines:

   (i) $y = 4x - 5$, $y = \frac{1}{2}x + 2$
   → (ii) $3x + y - 1 = 0$, $x + 2y - 7 = 0$
   (iii) $y = \frac{1}{3}x - 2$, $y = -x + 6$
   (iv) $x - 4y = 5$, $x + 5y = 4$
   → (v) $4x + 3y - 3 = 0$, $6x + 5y - 1 = 0$

④→ **22** Attempt to find the point of intersection of the lines $6x - 2y + 5 = 0$ and $y = 3x + 1$.
   What goes wrong? Why?

④ **23** Find, where possible, the point of intersection of the lines AB and CD when A, B, C and D are the points:

   → (i) $A(\frac{1}{2}, 0)$, $B(-2, 2)$, $C(5, -3)$, $D(-3, 1)$
   (ii) $A(5, 3)$, $B(3, -3)$, $C(1, -3)$, $D(6, 2)$
   (iii) $A(4, 4)$, $B(-2, 1)$, $C(6, -2)$, $D(2, -4)$

④ **24** Sketch the triangle formed by the given lines, and find the coordinates of its vertices.

   → (i) $y + 1 = 0$, $y = x + 1$, $y = -x + 7$
   (ii) $2x - y + 2 = 0$, $4x + y + 14 = 0$, $2x + 5y + 2 = 0$
   (iii) $5x - 7y + 8 = 0$, $2x + y + 12 = 0$, $7x - 6y + 15 = 0$

④ • **25** Find the equation of the perpendicular from the point $A(3, 2)$ to the line $\Lambda: 3x + 4y - 2 = 0$.
   Find the coordinates of the point of intersection of the two lines. Hence find the distance from A to Λ.

26 Find the distance from the point A to the line Λ:

   (i) $A(4, 5)$, $\Lambda: y = -x + 1$
   → (ii) $A(6, -2)$, $\Lambda: 3x - y + 4 = 0$
   (iii) $A(1, 1)$, $\Lambda: 2x + 4y - 3 = 0$
   (iv) $A(0, 0)$, $\Lambda: 4x - 3y - 5 = 0$

④ **27** Find the area of the triangle bounded by the lines:

   (i) $3x + y - 7 = 0$, $x - 3y + 1 = 0$, $x + y + 7 = 0$
   R (ii) $x - 2y - 1 = 0$, $3x - 2y + 5 = 0$, $x + 2y - 9 = 0$

④→ **28** A, B and C are the points $(6, 2)$, $(1, -1)$ and $(4, 8)$.
• Find the coordinates of the point D of intersection of BC and the line through A parallel to the x-axis.
   Find the area of $\triangle ADC$ and of $\triangle ADB$. Hence find the area of $\triangle ABC$.

④ **29** Find the area of $\triangle ABC$ when A, B and C are the points:

   (i) $(5, 0)$, $(-3, -4)$, $(2, 6)$
   (ii) $(3, 1)$, $(7, -3)$, $(-3, 2)$
   (iii) $(0, 0)$, $(-3, 12)$, $(7, 4)$

④ **30** A barren landscape. A windmill O. Stretching across the landscape a straight fence Λ, the frontier between two countries—Blueland to the north-west and Redland to the south-east. At equal intervals along Λ, outposts manned by armed guards. Relative to axes Ox and Oy due east and due north respectively, the equation of Λ is $x - 2y + 4 = 0$. The coordinates of two adjacent outposts A and B are $(p, 1)$ and $(q, 3)$ respectively. (1 unit = 100 m.)

Find p and q. Write down the coordinates of a general outpost on Λ.

A marathon runner is defecting from Redland to Blueland. He intends to run, taking a straight course, to a point on the frontier between A and B, climb over the fence, and keep running in the same direction. What is the cartesian equation of the line along which he should best run? (Assume that the closer he is to an observation post, the more likely he is to be spotted.) How close does he then come to the windmill (which unknown to him, is a secret observation post)?

He has organised all the greatest marathon runners in Redland to defect simultaneously. Each runner will run along the best possible straight course between a pair of observation posts. Write down the cartesian equation of the line along which the general defector runs.

1.2:3 Functions

Let D and T be two sets.

> **Definition** A **function** or **mapping** $f: D \to T$
>
> is a process which assigns to each element $x \in D$ a *unique* element $f(x) \in T$.

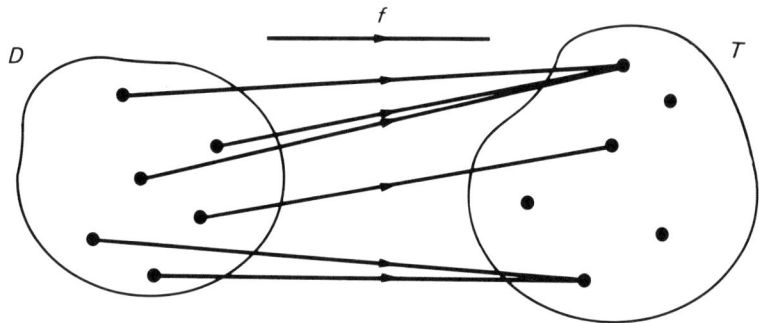

Fig. 1.2:**15**

D is called the **domain** of f. T is called the **codomain** or **target** of f.

For each element $x \in D$ there is exactly *one* element $f(x)$ (fig. 1.2:**15**). We say that x is **mapped** into $f(x)$ by f. $f(x)$ is called the **image** of x under f.

x and $f(x)$ can both vary, but whereas x varies freely, the value of $f(x)$ depends on the value of x; so we call x the **independent variable** and $f(x)$ the **dependent variable**.

The **range** R of f consists of all the elements in T which are images of elements in D (fig. 1.2:**16**): $R = \{y \in T : y = f(x), x \in D\}$

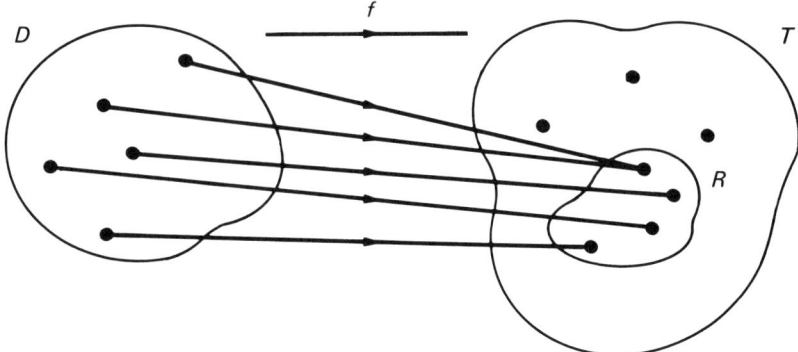

Fig. 1.2:**16**

Introduction 29

The elements $x \in D$ which map into an element $y \in R$ are called the **pre-images** of y under f, i.e. x is a pre-image of y under f if $y = f(x)$.

Worked example 5 A and B are the sets $A = \{a, b, c, d, e\}$, $B = \{p, q, r, s, t\}$. The function $f: A \to B$ is defined by:

$$f(a) = p, f(b) = q, f(c) = s, f(d) = s, f(e) = p$$

(i) What is the range of f?
(ii) What are the pre-images of p under f?

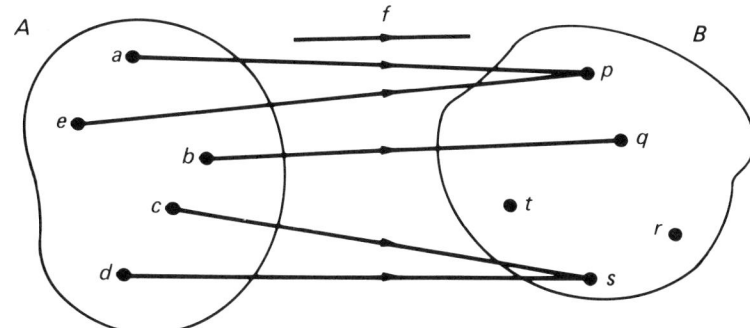

Fig. 1.2:**17**

(i) The elements in B which are images of elements in A are p, q and s (fig. 1.2:**17**). So the range of f is $\{p, q, s\}$.
(ii) $f(a) = p$ and $f(e) = p$, i.e. the elements a and e are mapped into p, so the pre-images of p under f are a and e.

Worked example 6 The function $f: [-3, 3] \to \mathbb{R}$ is defined by $f(x) = x^2$.

(i) Find the image of 1.5 under f;
(ii) Find the range of f;
(iii) Find the pre-images of 4 under f.

The function is shown in fig. 1.2:**18**.

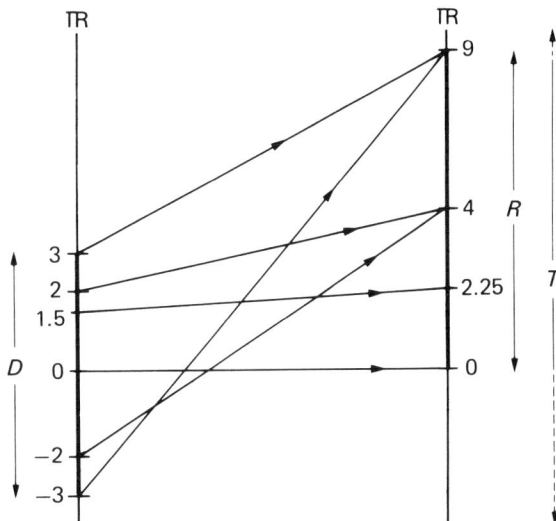

Fig. 1.2:**18**

(i) $f(1.5) = (1.5)^2 = 2.25$
(ii) $-3 \leqslant x \leqslant 3 \Rightarrow 0 \leqslant x^2 \leqslant 9$, i.e. $0 \leqslant f(x) \leqslant 9$. So the range of f is $[0, 9]$.
(iii) $f(x) = 4 \Rightarrow x^2 = 4 \Rightarrow x = \pm 2$. The pre-images of 4 under f are $+2$ and -2.

Composition of functions

If $f: D \to T$ and $g: D' \to T'$ are two functions, the composition $g \circ f$ of f and g is defined by

$$(g \circ f)(x) = g(f(x)) \quad x \in D \qquad (13)$$

Note that $g \circ f$ means 'do f first, then g' (fig. 1.2:**19**).

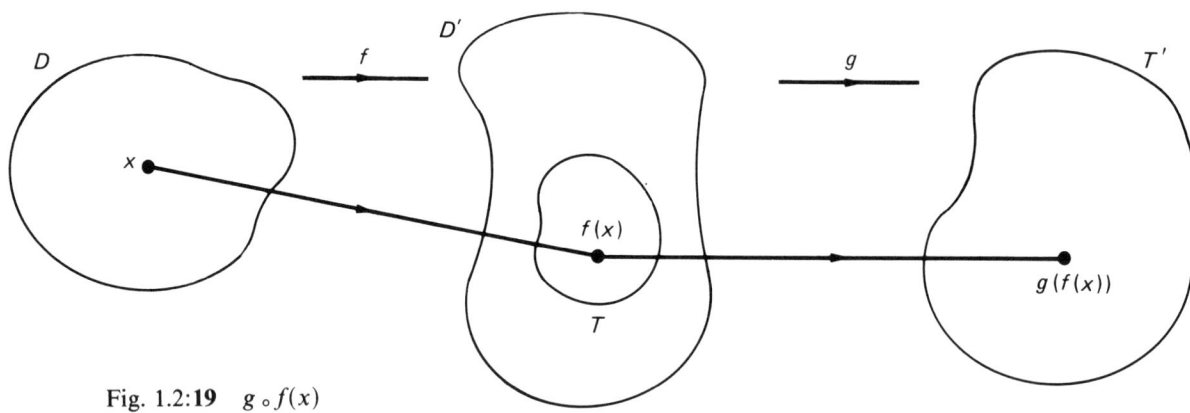

Fig. 1.2:**19** $g \circ f(x)$

Worked example 7 f and g are the functions $f(x) = x^2$ and $g(y) = \sqrt{(3y+2)}$. Find: (i) $(g \circ f)(x)$ (ii) $(f \circ g)(y)$.

(i) $(g \circ f)(x) = g(f(x)) = g(x^2) = \sqrt{(3x^2+2)}$
(ii) $(f \circ g)(y) = f(g(y)) = f(\sqrt{(3y+2)}) = \{\sqrt{(3y+2)}\}^2 = 3y+2$

Note that $(f \circ g)(x) = 3x+2$ (see question 38).
  Note also that in this case (and in fact generally),

$$g \circ f \neq f \circ g$$

Inverse functions

Suppose that the function $f: D \to T$ is **one–one**, i.e. each element in T has exactly one pre-image in D (fig. 1.2:**20**).

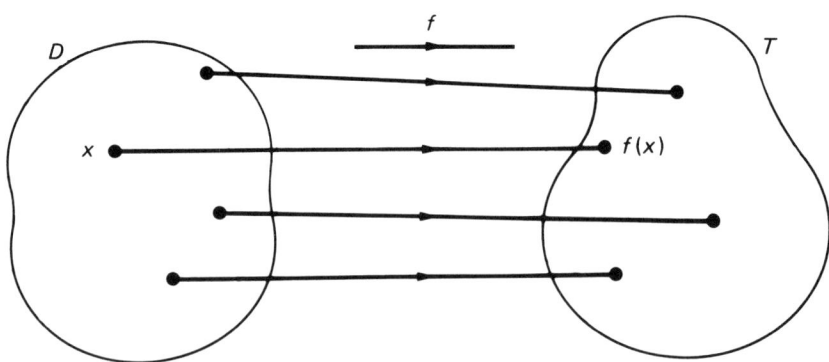

Fig. 1.2:**20**

By reversing the arrows (fig. 1.2:**21**) we can define a function $g: T \to D$ with the property that

$$g(f(x)) = x \quad \forall x \in D$$

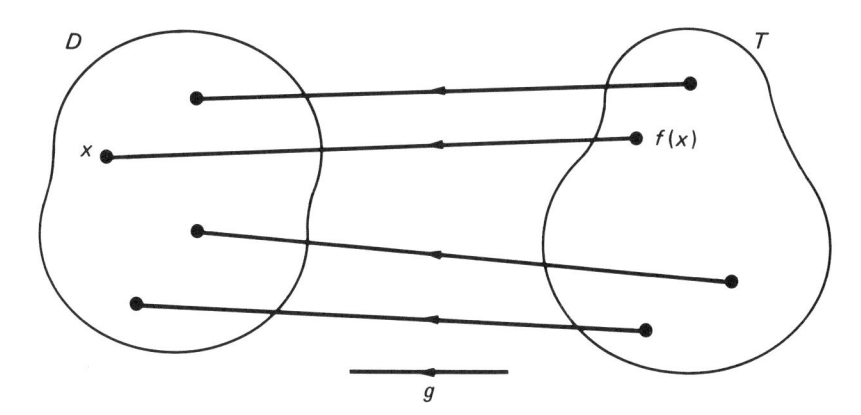

Fig. 1.2:**21**

Note also that

$$f(g(y)) = y \quad \forall y \in T$$

The function g is called the **inverse** of f and is written f^{-1}.

$$y = f(x) \Leftrightarrow x = f^{-1}(y) \tag{14}$$

Worked example 8 Find the inverse f^{-1} of the function f defined by $f(x) = x/(x+3)$.

$$y = f(x) \Rightarrow y = \frac{x}{x+3} \Rightarrow (x+3)y = x \Rightarrow x - xy = 3y$$

$$\Rightarrow x(1-y) = 3y \Rightarrow x = \frac{3y}{1-y}$$

So by statement (14), $f^{-1}(y) = 3y/(1-y)$.
Note that $f^{-1}(x) = 3x/(1-x)$ (see question 38).

Exercise 1.2:3

Questions 1-5: If $A = \{a, b, c, d, e\}$, $B = \{p, q, r, s, t\}$, draw a diagram (see fig. 1.2:**17**) showing the effect of $f: A \to B$. What is the range of f? What (if any) are the pre-images of q?

1. $f(a) = p, f(b) = q, f(c) = p, f(d) = r, f(e) = q$.

2. $f(a) = p, f(b) = p, f(c) = q, f(d) = q, f(e) = q$.

3. $f(a) = p, f(b) = q, f(c) = r, f(d) = s, f(e) = t$.

4. $f(a) = q, f(b) = t, f(c) = p, f(d) = r, f(e) = s$.

5. $f(a) = r, f(b) = t, f(c) = t, f(d) = p, f(e) = r$.

Questions 6-11: Draw a diagram showing the effect of f. What is the range of f?

6. $A = \{\text{apple, banana, courgette, leek}\}$
   $B = \{\text{arm, leg, ear, elbow, mouth}\}$
   $f: A \to B$ is defined by $f(\text{apple}) = \text{mouth}, f(\text{banana}) = \text{ear}, f(\text{courgette}) = \text{leg}, f(\text{leek}) = \text{ear}$.

7. $A = \{\text{Alan, Brian, Colin, David}\}$
   $B = \{\text{Alice, Babs, Carol, Dotty, Edith}\}$
   $f: A \to B$ is defined by $f(\text{Alan}) = \text{Babs}, f(\text{Brian}) = \text{Babs}, f(\text{Colin}) = \text{Dotty}, f(\text{David}) = \text{Dotty}$.

8. $A = \{a, b, c, d, e\}$
   $f: A \to A$ is defined by $f(a) = b, f(b) = d, f(c) = d, f(d) = a, f(e) = b$.

9. $A = \{a, b, c, d, e\}$
   $f: A \to A$ is defined by $f(a) = b, f(b) = c, f(c) = d, f(d) = e, f(e) = a$.

10. $A = \{0, 1, 2, 3\}$, $B = \{0, 1, 2, 3, 4, 5, 6\}$
    $f: A \to B$ is defined by $f(x) = 2x$.

11. $A = \{0, 1, 2, 3\}$, $B = \{-2, -1, 0, 1, 2, 3, 4, 5, 6, 7, 8\}$
    $f: A \to B$ is defined by $f(x) = 3x - 2$.

12 Draw a diagram showing the effect of f on a few elements of its domain:

(i) $f:\{\text{polygons}\}\to\mathbb{R}$, $f(x)=$ number of sides of x
(ii) $f:\{\text{people}\}\to\mathbb{R}$, $f(x)=$ age of x (in years)
(iii) $f:\{\text{words}\}\to\{\text{letters of the alphabet}\}$, $f(x)=$ initial letter of x
(iv) $f:\{\text{films}\}\to\{\text{people}\}$, $f(x)=$ director of x
(v) $f:[0,1]\to\mathbb{R}$, $f(x)=3x$
(vi) $f:[0,2]\to\mathbb{R}$, $f(x)=x^2$

13 Explain why each of the following relations is *not* a function:

(i) $f:\{\text{people}\}\to\{\text{people}\}$, $f(x)=$ daughter of x
(ii) $f:\{1,2,3,4,5\}\to\{1,2,3,4,5\}$, $f(x)=$ numbers less than x
(iii) $f:\mathbb{Q}\to\mathbb{Z}$, $f(m/n)=n$

14 The function $f:[-1,4]\to\mathbb{R}$ is defined by $f(x)=3x-2$.

(i) What is the domain of f?
(ii) Find $f(0)$, $f(\tfrac{2}{3})$, $f(1)$, $f(\tfrac{4}{3})$, $f(3)$.
(iii) Find $f(-1)$ and $f(4)$. What is the range of f?
(iv) Draw a diagram (see worked example 6) illustrating f.
(v) Find the pre-images under f of 1, 0, 7, $\tfrac{1}{2}$.

15 The function $f:\mathbb{R}\to\mathbb{R}$ is defined by $f(x)=4x+1$.

(i) Find $f(0)$, $f(1)$, $f(-2)$.
(ii) Find the pre-image under f of -7.
(iii) Show that if $f(x)=y$ then $x=\tfrac{1}{4}(y-1)$.

Hence *write down* the pre-images under f of 0, 5, -15.

16 The function $f:[-4,4]\to\mathbb{R}$ is defined by $f(x)=3x^2+2$.

(i) Find the image under f of 0, 2, -2, 0.75, -0.75.
(ii) Find the pre-image of the number 2. Show that the number 29 has two pre-images—what are they? Explain why the number -10 has no pre-image.
(iii) What is the range of f?
(iv) Draw a diagram illustrating f.

17 The function $f:[0,2]\to\mathbb{R}$ is defined by $f(x)=x(x-2)$.

(i) Find $f(0)$, $f(1)$, $f(2)$.
(ii) Show that $f(x)$ takes only negative values.
(iii) Show that $f(x)\equiv(x-1)^2-1$. Hence show that the range of f is $[-1,0]$.
(iv) Draw a diagram illustrating f.
(v) Find the pre-images under f of 0, $-\tfrac{3}{4}$.

18 The function $f:\mathbb{R}\to\mathbb{R}$ is defined by $f(x)=x^2+x$. Find the pre-image under f of:

(i) 0   (ii) 2   (iii) 6   (iv) 12.

19 Find:

(i) $f(0)$, $f(1)$, $f(-1)$, $f(2)$
(ii) the pre-images of 0

when the function $f:\mathbb{R}\to\mathbb{R}$ is defined by:
(a) $f(x)=x^2-4x$; (b) $f(x)=x^2-4$;
(c) $f(x)=x^3-4x^2$; (d) $f(x)=x^3-4x$.

20 The function $f:\mathbb{R}^+\to\mathbb{R}$ is defined by $f(x)=(x+4)/(x+2)$. Find:

(i) $f(0)$, $f(4)$, $f(5)$
(ii) the pre-images under f of 2, $\tfrac{3}{2}$, $\tfrac{5}{4}$

21 The function $f:\mathbb{R}^+\to\mathbb{R}$ is defined by $f(x)=(x-2)/(x+1)$. Show that if $f(x)=y$, then $x=(y+2)/(1-y)$. Hence *write down* the pre-images under f of 0, $\tfrac{1}{2}$, -2, -5.

22 Write down the domain and find the range of f:

(i) $f:(0,1)\to\mathbb{R}$, $f(x)=2x$
(ii) $f:[0,1]\to\mathbb{R}$, $f(x)=3x+2$
(iii) $f:[-1,2]\to\mathbb{R}$, $f(x)=-4x+1$
(iv) $f:[-6,3]\to\mathbb{R}$, $f(x)=-\tfrac{1}{3}x$
(v) $f:\mathbb{R}\to\mathbb{R}$, $f(x)=3x+5$

23 Find the range of the function:

(i) $f_1:[0,1]\to\mathbb{R}$, $f_1(x)=x^2$
(ii) $f_2:[-5,5]\to\mathbb{R}$, $f_2(x)=x^2$
(iii) $f_3:[-2,3]\to\mathbb{R}$, $f_3(x)=x^2$
(iv) $f_4:[4,6]\to\mathbb{R}$, $f_4(x)=x^2$

24 Find the range of the function $f:\mathbb{R}\to\mathbb{R}$:

(i) $f(x)=x^2$ (ii) $f(x)=x^2+2$
(iii) $f(x)=3x^2-4$ (iv) $f(x)=-x^2+1$
(v) $f(x)=-\tfrac{1}{2}x^3-3$

25 Show that $x^2+2x+10\equiv(x+1)^2+9$. Hence find the range of the function $f:\mathbb{R}\to\mathbb{R}$ defined by $f(x)=x^2+2x+10$.

26 Show that $x^2-6x+5\equiv(x-3)^2-4$. Hence find the range of the function:

(i) $f_1:\mathbb{R}\to\mathbb{R}$, $f_1(x)=x^2-6x+5$
(ii) $f_2:[1,5]\to\mathbb{R}$, $f_2(x)=x^2-6x+5$
(iii) $f_3:[-1,5]\to\mathbb{R}$, $f_3(x)=x^2-6x+5$

27 Find the range of f:

(i) $f:[3,4]\to\mathbb{R}$, $f(x)=\dfrac{2}{x}$

(ii) $f:[3,4]\to\mathbb{R}$, $f(x)=\dfrac{5}{x+2}$

(iii) $f:[3,4]\to\mathbb{R}$, $f(x)=\dfrac{x+1}{x-5}$

28 Find the range of f:

(i) $f_1:\mathbb{Z}\to\mathbb{R}$, $f_1(n)=(-1)^n$
(ii) $f_2:\mathbb{Z}\to\mathbb{R}$, $f_2(n)=3n$
(iii) $f_3:\mathbb{R}^+\to\mathbb{R}$, $f_3(x)=[x]$, where $[x]$ is the largest integer less than or equal to x.

29 Let $f: D \to T$ be a function with range R.

If $R = T$, i.e. if all the elements in T are images of elements in D, we say that f is **surjective** or **onto**.

If each element in R has exactly one pre-image, we say that f is **injective** or **into**.

If f is onto and into, we say that f is **bijective** or **one–one**.

(i) Consider again the functions of questions 6-11. State whether each function is onto, into or both (one-one).

(ii) Which of these functions is onto, which into, which one-one?
(a) $f_1: [0, 1] \to [2, 5]$, $f_1(x) = 3x + 2$
(b) $f_2: [0, 1] \to \mathbb{R}$, $f_2(x) = 3x + 2$
(c) $f_3: \mathbb{R}^+ \to \mathbb{R}$, $f_3(x) = x^2$
(d) $f_4: \mathbb{R} \to \mathbb{R}$, $f_4(x) = x^2$

(iii) Draw a diagram illustrating a function which is:
(a) onto but not into;
(b) into but not onto;
(c) one-one.
Give two examples of each of these types of function.

30 We shall be considering primarily functions which take real values, i.e. those whose codomain is $\mathbb{R}$ and whose domain is $\mathbb{R}$ or a subset of $\mathbb{R}$. We often do not bother to specify the domain D of a function f; this suggests that D is the largest possible subset of $\mathbb{R}$ on which f can be defined, e.g.

$$f(x) = 5x + 2 \qquad D = \mathbb{R}$$
$$f(x) = \sqrt{x} \qquad D = \mathbb{R}^+$$
$$f(x) = \frac{1}{x-3} \qquad D = \{x \in \mathbb{R}: x \neq 3\}$$

Write down the largest possible subset of $\mathbb{R}$ on which f can be defined:

(i) (a) $f(x) = \sqrt{(x-2)}$
(b) $f(x) = \sqrt{(-x)}$
(c) $f(x) = \sqrt{(3-x)}$
(d) $f(x) = \sqrt{(1-x^2)}$
(e) $f(x) = \sqrt{(x^2-9)}$
(f) $f(x) = \sqrt{(4x^2-5)}$
(g) $f(x) = \sqrt{\{(x-1)(x+2)\}}$
(h) $f(x) = \sqrt{\{x(3-x)\}}$

(ii) (a) $f(x) = 1/(x-4)$
(b) $f(x) = 1/(x+3)$
(c) $f(x) = 1/(3x-5)$
(d) $f(x) = 1/\{(x-1)(x-2)\}$
(e) $f(x) = 1/\{(2x+5)(5x+4)\}$
(f) $f(x) = 1/\sqrt{(x-1)}$

31 Write down the largest subset D of $\mathbb{R}$ on which f can be defined; with D as domain find the range of f:

(i) (a) $f(x) = 3x - 2$
(b) $f(x) = 3x^2$
(c) $f(x) = 5x^2 - 8$
(d) $f(x) = \sqrt{x}$
(e) $f(x) = \sqrt{(-x)}$
(f) $f(x) = \sqrt{(4-x)}$
(g) $f(x) = \sqrt{(4-x^2)}$
(h) $f(x) = \sqrt{(9x^2-16)}$
(i) $f(x) = \sqrt{\{(x-3)(1-x)\}}$
(j) $f(x) = \sqrt{\{x(3x-2)\}}$

(ii) (a) $f(x) = 1/x$
(b) $f(x) = 1/x^2$
(c) $f(x) = 1/(x-1)$
(d) $f(x) = 1/(3x+2)$

32 Write down the largest subset of $\mathbb{R}$ on which f can be defined when

$$f(x) = \frac{3x+2}{2x+1}$$

Show that $f(x)$ cannot take the value $\frac{3}{2}$. What is the range of f?

33 Explain why each of the following relations is *not* a function:

(i) $f_1: \mathbb{R} \to \mathbb{R}$, $f_1(x) = 1/(x-5)$
(ii) $f_2: \mathbb{R} \to \mathbb{R}$, $f_2(x) = (1+2x)/(1+5x)$
(iii) $f_3: \mathbb{R} \to \mathbb{R}$, $f_3(x) = \sqrt{(x+2)}$

Suggest domains A, B and C such that $f_1: A \to \mathbb{R}$, $f_2: B \to \mathbb{R}$ and $f_3: C \to \mathbb{R}$ are functions. What are the ranges of the functions?

34 Explain why the relation $f: \mathbb{R} \to \mathbb{R}$ defined by $\{f(x)\}^2 = x$ is not actually a function. Would f be a function if:

(i) its domain were restricted to $\mathbb{R}^+$?
(ii) its domain and range were restricted to $\mathbb{R}^+$?

35 Let $f(x) = x/(x+1)$.

(i) Find $f(0)$, $f(1)$, $f(-2)$.
(ii) What can you say about $f(-1)$? Suggest a domain for f.
(iii) Find the pre-image of $\frac{3}{4}$.
(iv) Show that if $f(x) = y$, then $x = y/(1-y)$.

Hence *write down* the pre-images of 2, 5, −3. Explain why the number 1 has no pre-image. What is the range of f?

36 Let $f(x) = x^2/(x+4)$.

(i) Find $f(0)$, $f(1)$, $f(-2)$.
(ii) Suggest a domain for f.
(iii) Find the pre-images under f of 0 and 2.

37 Let $f(x) = \sqrt{(x^2-9)}$.

(i) Find $f(3)$, $f(-3)$, $f(9)$, $f(-9)$.
(ii) Suggest a domain for f.
(iii) Find the pre-images under f of 4 and 6.

38 We will often talk about 'the function $f(x) = x^2$' —or even simply 'the function x^2'. This is convenient but confusing. It is not $f(x)$ that is the function, $f(x)$ is a number. It is f itself that is the function, i.e. the process that maps x onto $f(x)$.

So 'the function $f(x) = x^2$' is shorthand for 'the function f with the property that for any real number x, $f(x) = x^2$'.

Notice also that we can use any other letter, such as t, instead of x and talk about 'the function $f(t) = t(t-3)$'. x (or t or whatever) is used only to tell us what f does.

Write down $f(4), f(t), f(2x), f(-4x), f(x^2), f(\sqrt{x}), f(3x+1), f(s+t)$ when:

(i) $f(x) = 3x + 2$
(ii) $f(x) = x^2$
(iii) $f(x) = x(x+1)$
(iv) $f(x) = x^3$
→ (v) $f(x) = (x+2)/(x-1)$

39 If $f(x) = 3x + 1$, solve the equations:

(i) $f(x) = x$
(ii) $f(x) = f(2x+3)$
(iii) $f(x^2) = x + 3$

40 If $f(x) = x^3$, solve the equations:

(i) $f(x) = x$
(ii) $f(2x) + x^2 = 0$
(iii) $f(x+2) = 0$

41 If $f(x) = x$, x is said to be **invariant** under f. Find the numbers invariant under f when:

(i) $f(x) = 3 - \tfrac{1}{2}x$
(ii) $f(x) = x^2$
(iii) $f(x) = x^2 - 6$
(iv) $f(x) = 4x^3$
(v) $f(x) = (x+3)/(x-1)$

• → **42** If two variables x and y are related by an equation of the form $y = f(x)$, we say that **y is a function of x**.

The period T (seconds) of a pendulum, i.e. the time it takes to swing back and forth once, is a function of its length l (metres):

$$T = 2\sqrt{l}$$

Let us write $f(l) = 2\sqrt{l}$; so $T = f(l)$.

(i) What is the meaning of $f(9)$? Evaluate $f(9)$.
(ii) What is the meaning of the equation $f(l) = 2.5$? Solve the equation.
(iii) What is the domain of f?

43 What is the relation between the area A of a circle and its radius r? Note that A is a function of r. Let us write $A = f(r)$.

(i) What is the meaning of $f(2)$? Evaluate $f(2)$.
(ii) What is the meaning of the equation $f(r) = 20$? Solve the equation.
(iii) Does $f(-2)$ have a meaning? What is the domain of f?

44 When a ball is thrown vertically upwards from ground level with speed 20 m s^{-1}, its height h m above the ground is a function of the time t s for which it has been in the air:

$$h = 20t - 5t^2$$

(i) Find the height of the ball above the ground after $\tfrac{1}{2}$ s, 1 s, $\tfrac{3}{2}$ s, 2 s, $\tfrac{5}{2}$ s, 3 s, $\tfrac{7}{2}$ s, 4 s.
(ii) Deduce the greatest height reached by the ball.
(iii) Find the times at which the ball is at $\tfrac{160}{9}$ m above the ground.
(iv) The formula only applies when $0 \leqslant t \leqslant 4$. Why?

Write $h = f(t)$ where $f(t) = 20t - 5t^2$. Interpret the results of (i) to (iv) as statements about f.

⑦ **45** Find $(g \circ f)(x)$ and $(f \circ g)(y)$ when:

→ (i) $f(x) = x^2$, $g(y) = \sqrt{(2y+3)}$
(ii) $f(x) = 5x - 2$, $g(y) = 3y + 4$
(iii) $f(x) = x - 3$, $g(y) = (2y+1)(2y+5)$
(iv) $f(x) = x^4$, $g(y) = \sqrt{(y+1)}$
→ (v) $f(x) = (x+3)/(x-2)$, $g(y) = y - 3$
R (vi) $f(x) = x/(x+1)$, $g(y) = y/(1-y)$

In each case find, where possible, $(g \circ f)(1)$ and $(f \circ g)(1)$.

⑦ **46** Find $(g \circ f)(x)$ when $g(y) = 1 + y^2$ and:

(i) $f(x) = 3x$ (ii) $f(x) = x + 1$
(iii) $f(x) = \sqrt{(x+1)}$

In each case find $(g \circ f)(0)$ and $(g \circ f)(1)$.

⑦ **47** (i) $f: D \to T$ and $g: D' \to T'$ are functions. If $g \circ f$ is defined what can you say about the sets T and D'? If $f \circ g$ is defined what can you say about T' and D?

What are the domain and target of (a) $g \circ f$; (b) $f \circ g$?

(ii) Explain why $g \circ f$ is undefined, and find a suitable restricted domain A for f such that $g \circ f : A \to \mathbb{R}$ is defined:
(a) $f: \mathbb{R} \to \mathbb{R}; f(x) = x^2$
  $g: (-\infty, 1] \to \mathbb{R}; g(x) = \sqrt{(1-x)}$
(b) $f: \mathbb{R} \to \mathbb{R}; f(x) = x + 3$
  $g: \mathbb{R}^+ \to \mathbb{R}; g(x) = \sqrt{x}$
(c) $f: (-\infty, 4] \to \mathbb{R}; f(x) = \sqrt{(4-x)}$
  $g: [-3, 3] \to \mathbb{R}; g(x) = \sqrt{(9-x^2)}$

⑦ **48** f and g are the functions:

$$f: \{x \in \mathbb{R} : x \neq 0\} \to \mathbb{R}; f(x) = 1 + x - 6/x$$
$$g: \{x \in \mathbb{R} : x \neq 0\} \to \mathbb{R}; g(x) = 1/x$$

Find a and b such that $g \circ f$ is defined on $\{x \in \mathbb{R} : x \neq 0$ or a or $b\}$.

⑦ **49** Find the range of $g \circ f$:

(i) $f: [1, 4] \to \mathbb{R}; f(x) = 8/x$
  $g: \mathbb{R} \to \mathbb{R}; g(x) = x^2 - 1$
(ii) $f: [2, 3] \to \mathbb{R}; f(x) = x^2$
  $g: [2, \infty) \to \mathbb{R}; g(x) = \sqrt{(x-2)}$

⑦ 50 (i) If $f(x) = x^2$, $g(x) = 1/(1+x)$, $h(x) = x-1$, find $(g \circ f)(x)$ and $(h \circ g)(x)$.
Show that

$$[h \circ (g \circ f)](x) = [(h \circ g) \circ f](x)$$

This implies that we can write unambiguously $(h \circ g \circ f)(x)$ (c.f. 1.1:2 Q17(i)).

(ii) Find $(h \circ g \circ f)(x)$ when

$$f(x) = 3x, \ g(x) = 1 + x^2, \ h(x) = \sqrt{x}.$$

Do you expect $(f \circ g \circ h)(x)$ to equal

$$(h \circ g \circ f)(x)?$$

Find $(f \circ g \circ h)(x)$.

⑦ 51 Find $(f \circ f)(x)$ and $(f \circ f \circ f)(x)$ when:
(i) $f(x) = x + 1$ (ii) $f(x) = 3x$
(iii) $f(x) = 4 - 5x$ (iv) $f(x) = x^2$
(v) $f(x) = 2x^2 - 1$ (vi) $f(x) = x^3$
(vii) $f(x) = 1/x$ (viii) $f(x) = x/(x-1)$

⑦ → 52 The **sum** $f + g$ of two functions f and g is defined by

$$(f + g)(x) = f(x) + g(x)$$

The **product** fg of two functions f and g is defined by

$$fg(x) = f(x)g(x)$$

(i) Find $(f + g)(x)$ and $fg(x)$ when:
(a) $f(x) = x^2$, $g(x) = x - x^2$
(b) $f(x) = 1/(x-1)$, $g(x) = (x-1)/(x+1)$

(ii) Give two examples to show that in general

$$(g \circ f)(x) \neq g(x)f(x)$$

In fact, can you think of two functions f and g for which $g \circ f = f \circ g$?

⑦ 53 Find $(g \circ f)(x)$ and $(f \circ g)(x)$ when:
(i) $f(x) = 4x + 1$, $g(x) = \frac{1}{4}(x - 1)$

(ii) $f(x) = \dfrac{x}{x+1}$, $g(x) = \dfrac{x}{1-x}$
(C.f. questions 15(iii) and 35(iv).)

⑧ 54 Find x in terms of y when $y = f(x)$ and hence find $f^{-1}(x)$ when:

(i) (a) $f(x) = 2x$ (b) $f(x) = \frac{1}{3}x - 2$
    (c) $f(x) = 7x + 4$ → (d) $f(x) = 5 - \frac{3}{4}x$
→ (ii) (a) $f(x) = x^3$ (b) $f(x) = (x-1)^3$

(iii) (a) $f(x) = \dfrac{1}{x}$ (b) $f(x) = \dfrac{5}{x}$

    (c) $f(x) = \dfrac{3}{x-2}$ (d) $f(x) = \dfrac{1}{1+5x}$

    (e) $f(x) = \dfrac{x}{x+4}$ → (f) $f(x) = \dfrac{x-2}{x+1}$

R   (g) $f(x) = \dfrac{2x-5}{7x+4}$

⑧ → 55 (i) Show, by drawing a diagram, that if a function $f: D \to T$ is onto but not into (see question 29), we can define f^{-1} by restricting f to a domain $D' \subset D$ on which f is into.

(ii) Show that the function $f: \mathbb{R} \to \mathbb{R}$ defined by $f(x) = x^2$ has an inverse if its domain is restricted to the set $\mathbb{R}^+$.
What is the domain of f^{-1}?

⑧ 56 Find a subset A of R such that the function f has an inverse f^{-1}. Find f^{-1}. What is the domain of f^{-1}?
(i) $f(x) = x^2 + 1$
(ii) $f(x) = (x + 3)^2$
(iii) $f(x) = 1/(x^2 + 1)$

⑧ 57 Show that the function $f(x) = x$ is its own inverse. Give two other examples of functions which are their own inverses.

⑧ 58 Find $(g \circ f)^{-1}$ and $f^{-1} \circ g^{-1}$ when:
(i) $f(x) = 2x$, $g(x) = \sqrt{(1+x)}$
→ (ii) $f(x) = 2x - 3$, $g(x) = x/(x+2)$
What do you notice about $(g \circ f)^{-1}$ and $f^{-1} \circ g^{-1}$?

1.2:4 Graphs

The set Π of points in the xy plane is $\{P(x, y): x \in \mathbb{R}, y \in \mathbb{R}\}$.

The general point $P(x, y)$ can take any position in the plane. x can be any real number, y can be any real number, and there is, as yet, no connection between x and y. We say in this case that P has **two degrees of freedom**.

36 Graphs

> **Definition** The **graph** of a function $f: \mathbb{R} \to \mathbb{R}$ is the set G_f of points $(x, f(x))$.
> G_f is a subset of Π. It is precisely those points (x, y) of Π for which $y = f(x)$, i.e.
> $$G_f = \{(x, y): x \in \mathbb{R}, y = f(x)\} \qquad (15)$$

The points in G_f have only **one degree of freedom**. x can be any real number, but for any given value of x, y is completely restricted; it has to be $f(x)$.

For every point (x, y) on G_f, $y = f(x)$ (fig. 1.2:22). For every point (x, y) not on G_f, $y \neq f(x)$. We can think of the equation $y = f(x)$ as being a restriction on the position of a general point $P(x, y)$.

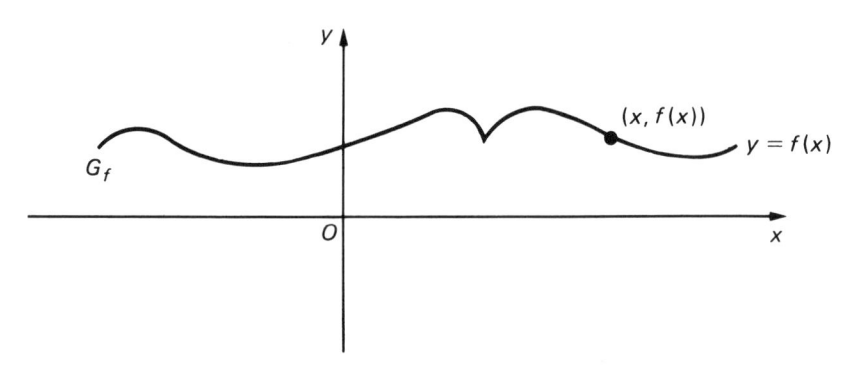

Fig. 1.2:22

The graph G_f of a function $f: \mathbb{R} \to \mathbb{R}$ is in general a curve or a straight line. The equation $y = f(x)$ is the **equation of the curve**. We can refer to 'the curve $y = f(x)$' (though not to 'the function $y = f(x)$' since it is f that is the function).

Worked example 9 Sketch the curve $y = f(x)$ where $f(x) = (x+1)(2x-5)$.

x	-2	-1	0	1	2	3	4
$f(x)$	9	0	-5	-6	-3	4	12

Note: the curve (fig. 1.2:23) cuts the y-axis where $x = 0$, i.e. at the point $C(0, -5)$. The curve cuts the x-axis where $f(x) = 0$, i.e. where

$$(x+1)(2x-5) = 0 \Rightarrow x = -1 \text{ or } \tfrac{5}{2},$$

i.e. at the points $A(-1, 0)$ and $B(\tfrac{5}{2}, 0)$.

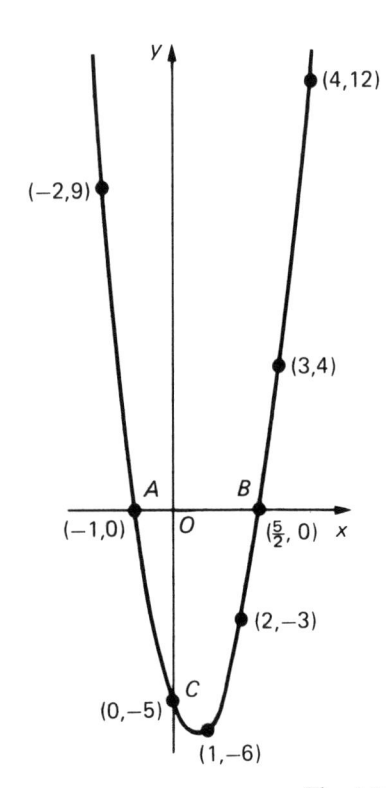

Fig. 1.2:23

Exercise 1.2:4

⑨ **1** Make a table showing integer values of x between -6 and $+6$, and the corresponding values of $f(x)$; plot these points $(x, f(x))$ on the xy plane and hence sketch the graph of f.

    (i) $f(x) = x - 3$ (ii) $f(x) = -\frac{1}{4}x + 1$
→ (iii) $f(x) = x^2 + 3$ (iv) $f(x) = 4 - x^2$
→ (v) $f(x) = (x+3)(x-1)$ (vi) $f(x) = (x+1)^2$
→ (vii) $f(x) = (x-1)(x+2)(x-5)$
    (viii) $f(x) = (x+4)(x+1)x(x-2)$

State the coordinates of: (a) the point at which the graph cuts the y-axis; (b) the point(s) at which the graph cuts the x-axis.

⑨ → **2** Sketch graphs of the functions:

    (i) $f(x) = x$ (ii) $f(x) = x^2$
    (iii) $f(x) = x^3$ (iv) $f(x) = x^4$

⑨ **3** A function of the type $f(x) = ax + b$, where a and b are constants, is called a **linear function**.

Sketch graphs of the functions:

    (i) $f(x) = x - 2$ (ii) $f(x) = 2x - 3$
    (iii) $f(x) = \frac{1}{4}x + 1$ (iv) $f(x) = -\frac{2}{5}x + \frac{4}{5}$

Explain the description 'linear' function (see 1.2:2 Q20).

⑨ **4** A function of the type $f(x) = ax^2 + bx + c$, where a, b and c are constants, is called a **quadratic function**. Notice (worked example 9 and question 1(iii), (iv)) that the general shape of the graph of a quadratic function is always the same. The shape is called a **parabola** (see section 6.2:7).

Find $f(0)$, solve the equation $f(x) = 0$ and hence sketch graphs of the functions:

    (i) $f(x) = x(x-3)$ (ii) $f(x)$
    (iii) $f(x) = x(4-x)$ (iv) $f(x) = x^2 + 3x + 2$
→ (v) $f(x) = (2x+5)(x-1)$ (vi) $f(x) = 3x^2 - 10x + 8$

⑨ **5** (i) Look at 1.2:3 Q43 again. Sketch a graph of A against r ($r \geqslant 0$).

(ii) Look at 1.2:3 Q44 again. Sketch a graph of h against t ($0 \leqslant t \leqslant 4$).

⑨ **6** A function of the type $f(x) = ax^3 + bx^2 + cx + d$, where a, b, c and d are constants, is called a **cubic function**.

Find $f(0)$, solve the equation $f(x) = 0$ and hence sketch graphs of the functions:

→ (i) $f(x) = (x-1)(x-2)(x-3)$
    (ii) $f(x) = (x+3)(x-1)(2-x)$
    (iii) $f(x) = x^3 - x$

⑨ **7** Sketch graphs of the functions:

    (i) $f: [0, 1) \to \mathbb{R}; f(x) = 3x + 2$
    (ii) $f: [0, 1] \to \mathbb{R}; f(x) = \dfrac{1}{x+1}$
    (iii) $f: [0, 1] \to \mathbb{R}; f(x) = \dfrac{x-3}{x-2}$
    (iv) $f: [1, 5] \to \mathbb{R}; f(x) = \dfrac{1}{x}$

⑨ **8** Sketch graphs of the functions:

    (i) $f: \mathbb{Z} \to \mathbb{R}; f(n) = n + 2$
    (ii) $f: \mathbb{Z} \to \mathbb{R}; f(n) = n^2$
    (iii) $f: \mathbb{Z}^+ \to \mathbb{R}; f(n) = 1/n$
    (iv) $f: \mathbb{Z} \to \mathbb{R}; f(n) = (-1)^n$

⑨ → **9** Sketch the graph of the function $f: \mathbb{R}^+ \to \mathbb{R}$ defined by $f(x) = \sqrt{x}$.

⑨ **10** Write down the largest possible subset D of $\mathbb{R}$ on which f can be defined (see 1.2:3 Q30), and sketch the graph of f for values of $x \in D$.

    (i) (a) $f(x) = \sqrt{(x-1)}$ → (b) $f(x) = \sqrt{(x+2)}$
         (c) $f(x) = \sqrt{(2x-5)}$ (d) $f(x) = \sqrt{(1-x)}$
    (ii) (a) $f(x) = \sqrt{\{(x-2)(x+3)\}}$
         (b) $f(x) = \sqrt{(1-x^2)}$
         (c) $f(x) = \sqrt{\{(1-x)(2+x)\}}$
         (d) $f(x) = x + \sqrt{\{(x-1)(x-2)\}}$

⑨ **11** Sketch a graph of the function $f: \mathbb{R} \to \mathbb{R}$ defined by:

    (i) $f(x) = \begin{cases} x-1 & \text{when } x < 0 \\ x+1 & \text{when } x \geqslant 0 \end{cases}$

    (ii) $f(x) = \begin{cases} 3x-2 & \text{when } x \leqslant 2 \\ \frac{1}{2}x+5 & \text{when } x > 2 \end{cases}$

→ (iii) $f(x) = \begin{cases} x^2 & \text{when } x < 0 \\ x(2-x) & \text{when } x \geqslant 0 \end{cases}$

    (iv) $f(x) = \begin{cases} (x+1)^2 & \text{when } x \leqslant -1 \\ 0 & -1 < x < 1 \\ (x-1)^2 & \text{when } x \geqslant 1 \end{cases}$

In each case write down the range of f.

Questions 12–16: Sketch the graphs of the given functions on the same axes.

⑨ **12** (i) $f(x) = \frac{1}{2}x, f(x) = \frac{1}{2}x + 5,$
         $f(x) = \frac{1}{2}x + 1, f(x) = \frac{1}{2}x - 4$
→ (ii) $f(x) = x^3, f(x) = x^3 + 3, f(x) = x^3 - 2$

⑨ **13** (i) $f(x) = 2x, f(x) = 2(x-1), f(x) = 2(x+3)$
→ (ii) $f(x) = x^2, f(x) = (x-4)^2, f(x) = (x+2)^2$
    (iii) $f(x) = 3x^3 + 1, f(x) = 3(x-1)^3 + 1,$
          $f(x) = 3(x+1)^3 + 1$

14 (i) $f(x) = -\frac{1}{2}x + 3$ and $f(x) = \frac{1}{2}x - 3$
(ii) $f(x) = x^2$ and $f(x) = -x^2$
(iii) $f(x) = x^2 - 9$ and $f(x) = 9 - x^2$

15 (i) $f(x) = 2x + 4$ and $f(x) = -2x + 4$
(ii) $f(x) = (3-x)^2$ and $f(x) = (3+x)^2$
(iii) $f(x) = x^2 + x - 6$ and $f(x) = x^2 - x - 6$

16 (i) $f(x) = x, f(x) = 2x, f(x) = \frac{1}{2}x$
(ii) $f(x) = x^2, f(x) = 2x^2, f(x) = \frac{1}{2}x^2$
(iii) $f(x) = x^3, f(x) = 2x^3, f(x) = \frac{1}{2}x^3$

Questions 17-19: Sketch on the same axes the graphs of:

(i) $f(x)$ and $-f(x)$
(ii) $f(x), f(x) - \alpha, f(x) + \alpha$
(iii) $f(x), \alpha f(x), (1/\alpha)f(x)$
(iv) $f(x)$ and $f(-x)$
(v) $f(x), f(x - \alpha), f(x + \alpha)$
(vi) $f(x), f(\alpha x), f((1/\alpha)x)$

where α is a positive constant, for the functions f whose graph is given.

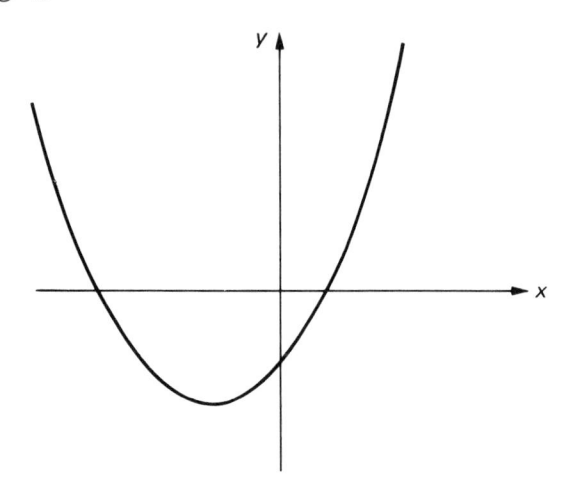

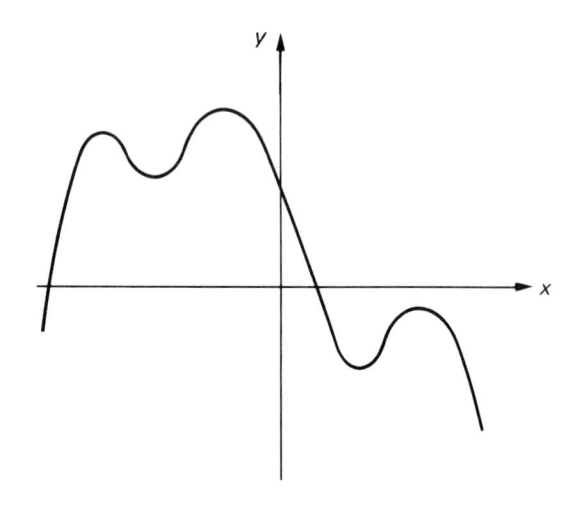

17

18

19

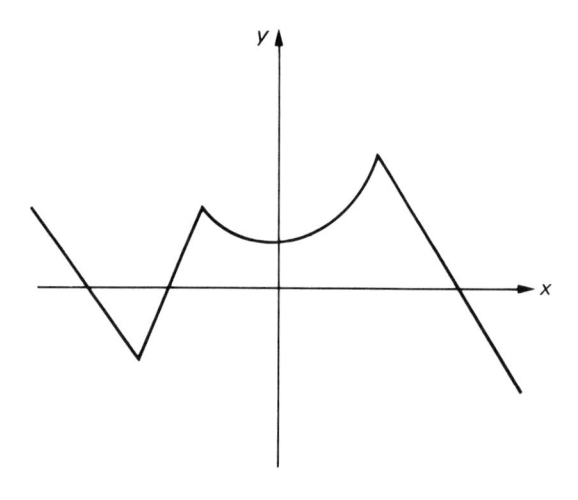

20 Sketch on the same axes the graphs of the functions:

(i) (a) $f(x) = (x+5)(x-2), f(x) = (x+4)(x-3)$,
$f(x) = (x+7)x$
(b) $f(x) = (2x+1)(x+3), f(x) = (2x-1)(x+2)$,
$f(x) = (2x+3)(x+4)$
(ii) $f(x) = (x+1)(x-3), f(x) = (2x+1)(2x-3)$,
$f(x) = (3x+1)(3x-3)$

21 A function $f: \mathbb{R} \to \mathbb{R}$ is **even** if $f(-x) \equiv f(x)$, and **odd** if $f(-x) \equiv -f(x)$. **(16)**

(i) Give three examples of an even function and three examples of an odd function.

(ii) Why do you think they are called 'even' and 'odd' functions?

(iii) Which of these curves are the graphs of even functions, which of odd, and which of neither?

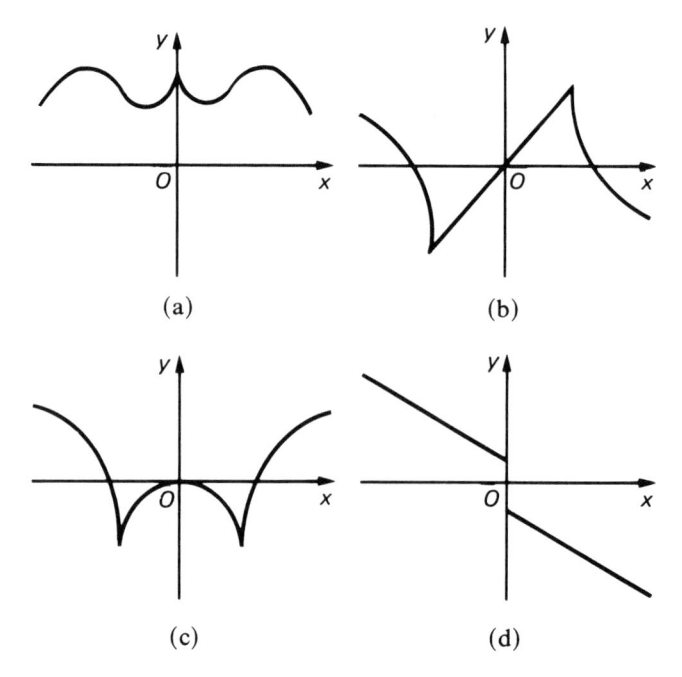

(a)

(b)

(c)

(d)

(e)   (f)

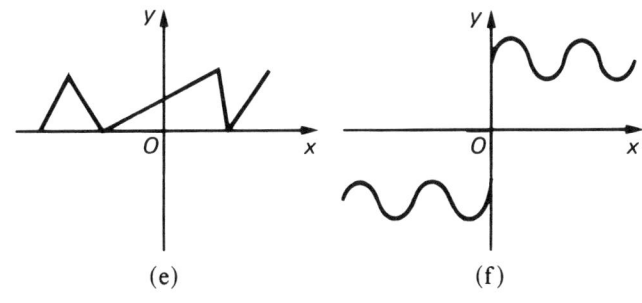

(iv) If f is an even function, G_f has an axis of symmetry. What is it?

If f is an odd function, does G_f have an axis of symmetry? Show that if f is an odd function, then G_f passes through the origin.

(v) Write down three even functions and three odd functions, and sketch a graph of each.

22 Sketch graphs of the following functions. Which are even, which odd and which neither?

(i) $f(x) = \frac{1}{3}x$ (ii) $f(x) = \frac{1}{3}x + 1$;
(iii) $f(x) = (x+1)(x-2)$ (iv) $f(x) = 4 - x^2$
(v) $f(x) = \begin{cases} (x+1)(x-2) & x \geq 0 \\ (x-1)(x+2) & x < 0 \end{cases}$
(vi) $f(x) = (x+1)x(x-1)$

23 Which of these functions are even, which odd and which neither?

(i) $f(x) = x^5 + x^3$
(ii) $f(x) = 4x^4 - 5$
(iii) $f(x) = x^3 + x^2$
(iv) $f(x) = 4x^7 + 1$
(v) $f(x) = x^2/(x^2+1)$
(vi) $f(x) = x/(x^2+1)$
(vii) $f(x) = \sqrt{(x^2+1)}$
(viii) $f(x) = x^{1/3}$

24 Decide whether the functions $f + g$, fg and $g \circ f$ are even, odd or neither, when:

(i) f and g are both even;
(ii) f and g are both odd;
(iii) f is odd and g is even.

Give examples.

Do even and odd functions behave in any way like even and odd numbers?

→ **25** A function f is called **periodic** if there is a real number l such that $f(x+l) \equiv f(x)$. l is called the **period** of f.

Sketch the graph of the function $f(x) = x - [x]$ (see 1.2:3 Q28(iii)). What is the period of this function?

26 The function f is defined by:

$$f(x) = x^2 \quad -1 \leq x < 1$$

and f is periodic with period 2.

Sketch a graph of $f(x)$ for values of x between -3 and $+5$.

27 The function f is defined by:

$$f(x) = \begin{cases} x & 0 < x \leq 1 \\ x(2-x) & 1 < x \leq 3 \end{cases}$$

and f is periodic with period 3.

Sketch a graph of $f(x)$ for values of x between -4 and $+8$.

R 28 The function f is defined by:

$$f(x) = \begin{cases} 4x+2 & -1 < x \leq 0 \\ 2-2x^2 & 0 < x \leq 2 \\ 2x-10 & 2 < x < 4 \end{cases}$$

and f is periodic with period 5.

Sketch a graph of $f(x)$ for values of x between -6 and $+9$.

→ • **29** Show that the point (k, h) is the mirror image of the point (h, k) in the line $y = x$.

Show that if the point (h, k) lies on the graph G_f of a function, then (k, h) lies on $G_{f^{-1}}$.

Hence show that $G_{f^{-1}}$ is the mirror image of G_f in the line $y = x$. Sketch on the same axes the graph of an imaginary function f, the line $y = x$ and the graph of the f^{-1}.

30 Sketch on the same axes the graphs of f and f^{-1} when:

(i) $f(x) = 5x$
(ii) $f(x) = 3x + 4$
(iii) $f(x) = -2x + 1$
(iv) $f(x) = x^3$
(v) $f(x) = x^2 \quad (x > 0)$
(vi) $f(x) = \frac{1}{2}x^2 \quad (x > 0)$
(vii) $f(x) = x(x-2) \quad (x > 1)$
→ (viii) $f(x) = \begin{cases} 2x+1 & x < -1 \\ x & -1 \leq x \leq 1 \\ 2x-1 & x > 1 \end{cases}$
R (ix) $f(x) = \begin{cases} 1+x^2 & x < 0 \\ 1-x^2 & 0 \leq x \leq 1 \\ 1-x & x > 1 \end{cases}$

31 Suppose that $D \subset R$ and that $f: D \to \mathbb{R}$ is a function.

(i) (a) Explain why, whatever f is,

$$x_1 = x_2 \Rightarrow f(x_1) = f(x_2) \quad x_1 x_2 \in D$$

(b) What kind of function does f have to be in order that

$$f(x_1) = f(x_2) \Rightarrow x_1 = x_2? \quad (17)$$

Draw graphs of two functions for which statement (17) is true, and two functions for which it is false.

(ii) f is an **increasing** function if $x_1 > x_2 \Rightarrow f(x_1) > f(x_2)$, a **decreasing** function if $x_1 > x_2 \Rightarrow f(x_1) < f(x_2)$. Show that

$$f^{-1} \text{ exists} \Leftrightarrow f \text{ is increasing or decreasing}$$

Which of the following functions are increasing?
(a) $f(x) = x$; (b) $f(x) = x^2$; (c) $f(x) = \sqrt{(x-2)}$.
Sketch these graphs.

32 Sketch on the same axes the graphs of f and g, and find the coordinates of their point(s) of intersection.

(i) $f(x) = x - 1$, $g(x) = -\frac{1}{3}x + 3$
(ii) $f(x) = 4x - 5$, $g(x) = \frac{1}{2}x + 2$
(iii) $f(x) = 2x + 1$, $g(x) = \frac{1}{2}x - 5$
• (iv) $f(x) = x^2$, $g(x) = x$
→ (v) $f(x) = x(x+1)$, $g(x) = 6$
(vi) $f(x) = (x-1)(x+2)$, $g(x) = x + 7$
(vii) $f(x) = (x-4)(x+3)$, $g(x) = 3(x-5)$
→ (viii) $f(x) = 9 - x^2$, $g(x) = 3x - 1$
(ix) $f(x) = (x+2)(x+5)$, $g(x) = x(x-3)$
R (x) $f(x) = 5 - x^2$, $g(x) = x(x-3)$
(xi) $f(x) = \sqrt{(x-1)}$, $g(x) = 3$

33 When an island is inhabited by several species of a certain organism, the number n of species of the organism is thought to be a function of the area A km³ of the island:

$$n = kA^\gamma \quad (A \in \mathbb{R}^+)$$

where k and γ are constants.
Sketch on the same axes rough graphs of n against A in the cases:
(i) $\gamma = 0.5$ (ii) $\gamma = 1$ (iii) $\gamma = 1.5$

34 When a parachutist is descending his (or her) acceleration a m s⁻² at any instant is given by

$$a = 10 - \tfrac{2}{5}v^2 \quad (0 \leq v \leq 5)$$

where v m s⁻¹ is his (or her) speed at that instant.
Sketch a graph of a against v.
The speed 5 m s⁻¹ is called the **terminal speed** of the parachutist. Why do you think this is?

35 The number T tourists visiting the island of Shnaxos in any particular year is a function of the number H of hotels on the island:

$$T = 50H^{5/4}.$$

Sketch graphs of T against H, and T/H against H, for $0 \leq H \leq 100$.

Does building more hotels result in the hotels being fuller, or emptier? Attempt to explain the mentality of the tourists.

1.2:5 Limits and continuity

Worked example 10 Consider the function $f: \mathbb{R} \to \mathbb{R}$ defined by

$$f(x) = \begin{cases} x - 1 & x < 0 \\ 0 & x = 0 \\ x + 1 & x > 0 \end{cases}$$

shown in fig. 1.2:24. Notice that

(i) $f(0) = 0$; but
(ii) if we approach the y-axis ($x = 0$) along the graph from positive values of x, $f(x)$ tends to 1; and
(iii) if we approach the y-axis along the graph from negative values of x, $f(x)$ tends to -1.

We write:

$$\text{as } x \to 0^+, f(x) \to 1 \quad \text{or} \quad \lim_{x \to 0^+} f(x) = 1$$

$$\text{as } x \to 0^-, f(x) \to -1 \quad \text{or} \quad \lim_{x \to 0^-} f(x) = -1$$

Fig. 1.2:24   The statement '$\lim_{x \to 0^+} f(x) = k$' is read as 'the limit, as x tends to zero from above, of $f(x)$, is k'.

In general, the statement:

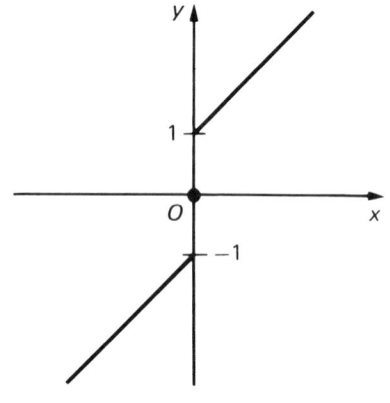

$$\text{as } x \to a^+, f(x) \to k \quad \text{or} \quad \lim_{x \to a^+} f(x) = k$$

means that as we approach the line $x = a$ from the right along the graph of f, the value of $f(x)$ tends to k; and the statement:

$$\text{as } x \to a^-, f(x) \to k' \quad \text{or} \quad \lim_{x \to a^-} f(x) = k'$$

means that as we approach the line $x = a$ from the left along the graph of f, the value of $f(x)$ tends to k'.

This is shown in fig. 1.2:**25**.

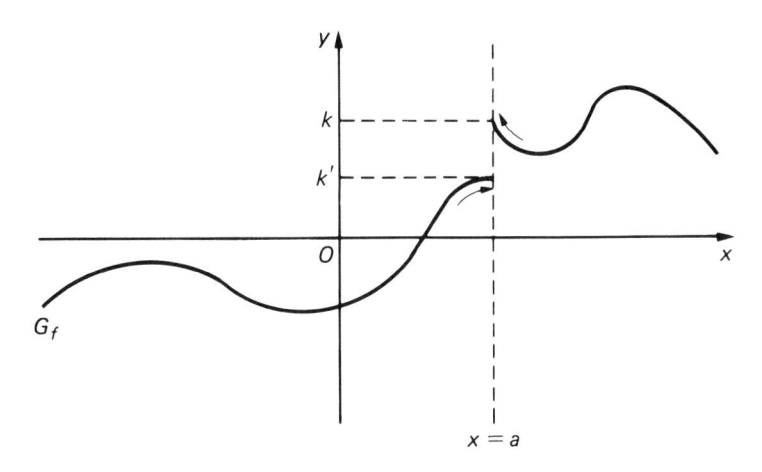

Fig. 1.2:**25**

Definition A function f is **continuous at $x = a$** if

$$\lim_{x \to a^-} f(x) = f(a) = \lim_{x \to a^+} f(x)$$

---

i.e. if you can draw the graph of f near $x = a$ without taking your pen off the paper (fig. 1.2:**26**). (Note that this definition of continuity applies only if the function f has a graph, i.e. if f takes real values. For more general functions we need a more complicated definition, based on the idea of the neighbourhood of a point.)

Definition A function is **discontinuous at $x = a$** (or has a **discontinuity** at $x = a$) if it is not continuous at $x = a$ (fig. 1.2:**27**).

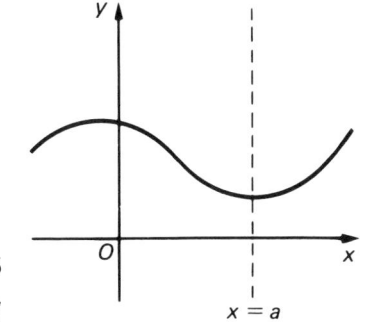

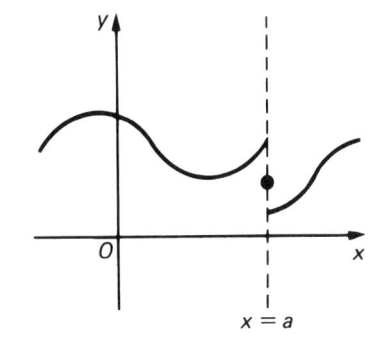

Fig. 1.2:**26**

Fig. 1.2:**27**

Definition A function f is **continuous** if it is continuous for all values of x.

Limits and continuity

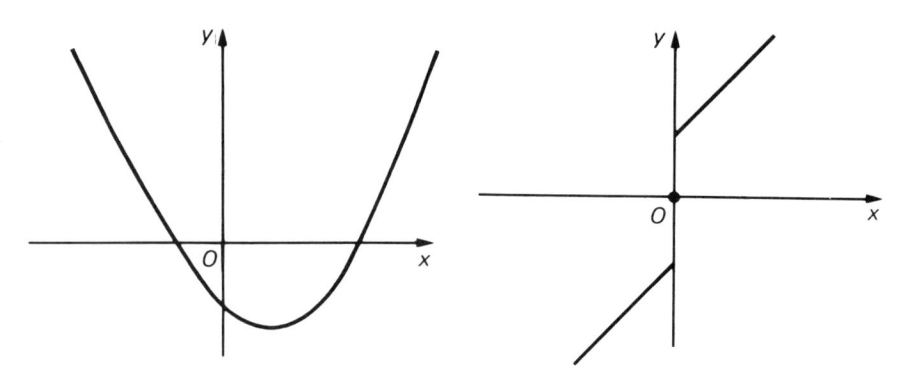

Fig. 1.2:28

Fig. 1.2:29

Fig. 1.2:**28** shows a continuous function, and fig. 1.2:**29** shows a discontinuous function.

Exercise 1.2:5

1 Sketch a graph of f. Write down: (i) $f(0)$ (ii) $\lim_{x\to 0^+} f(x)$ (iii) $\lim_{x\to 0^+} f(x)$

Is f continuous or discontinuous at $x=0$?

(a) $f(x) = \begin{cases} x & x<0 \\ 1 & x=0 \\ 2x+3 & x>0 \end{cases}$

(b) $f(x) = \begin{cases} x-1 & x\leq 0 \\ x^2 & x>0 \end{cases}$

→ (c) $f(x) = \begin{cases} 1-x & x<0 \\ 0 & x=0 \\ 1+x^2 & x>0 \end{cases}$

(d) $f(x) = \begin{cases} 4+x^2 & x<0 \\ 4-x & x\geq 0 \end{cases}$

2 (i) $f(x) = \begin{cases} x-2 & x<1 \\ 0 & x=1 \\ x & x>1 \end{cases}$

Write down $\lim_{x\to 1^-} f(x)$, $f(1)$, $\lim_{x\to 1^+} f(x)$. Is f continuous at $x=1$? Sketch a graph of f.

(ii) $g(x) = \begin{cases} 2x & x\leq 1 \\ x+1 & 1<x\leq 2 \\ -x-1 & x>2 \end{cases}$

Is g continuous at (a) $x=1$; (b) $x=2$? Sketch a graph of g.

3 Sketch a graph of f. Is f continuous?

(i) $f(x) = \begin{cases} x & x<0 \\ x^2 & x\geq 0 \end{cases}$

(ii) $f(x) = \begin{cases} 2x-1 & x<0 \\ 2x+1 & x\geq 0 \end{cases}$

(iii) $f(x) = \begin{cases} \frac{1}{2}x & x\leq 2 \\ x-1 & x>2 \end{cases}$

→ (iv) $f(x) = \begin{cases} x^2 & x\leq 2 \\ 4-x^2 & x>2 \end{cases}$

→ (v) $f(x) = \begin{cases} -x-1 & x<-1 \\ 1-x^2 & -1\leq x\leq 1 \\ x-1 & x>1 \end{cases}$

(vi) $f(x) = \begin{cases} x(1-x) & x<1 \\ (x-3)(1-x) & x\geq 1 \end{cases}$

4 The function $f: \mathbb{R} \to \mathbb{R}$ is defined by:

$$f(x) = \begin{cases} x+1 & -1\leq x<0 \\ 1-x^2 & 0\leq x<1 \end{cases}$$

and f is periodic with period 2 (see 1.2:4 Q25). Sketch a graph of f. Is f continuous?

R 5 The function $f: \mathbb{R} \to \mathbb{R}$ is defined by:

$$f(x) = \begin{cases} x(2-x) & 0\leq x<2 \\ 1-\frac{1}{2}x & 2\leq x<3 \\ \frac{1}{2}x-2 & 3\leq x<5 \end{cases}$$

and f is periodic with period 5. Sketch a graph of f. Is f continuous?

6 A man is deeply in love. At 8 o'clock one morning he sets out from his home and walks to his girlfriend's home, 30 miles away across difficult terrain. He arrives at 6 o'clock in the evening. At 8 o'clock the next morning he walks back home by the same route, arriving at 6 o'clock in the evening. Prove that he reaches a certain place at the same time on both days.

1.2:6 Asymptotes

Worked example 11 Consider the function f defined by

$$f(x) = \frac{1}{x-2}$$

Look at the values of $f(x)$ as $x \to 2^+$ and $x \to 2^-$.

x	3	2.5	2.1	2.05	2.01	2.001
$f(x)$	1	2	10	20	100	1000

As $x \to 2^+$, $f(x)$ gets very large. We write:
$$\text{as } x \to 2^+, f(x) \to \infty \quad \text{or} \quad \lim_{x \to 2^+} f(x) = \infty$$

x	1	1.5	1.9	1.95	1.99	1.999
$f(x)$	-1	-2	-10	-20	-100	-1000

As $x \to 2^-$, $f(x)$ gets very large and negative. We write:
$$\text{as } x \to 2^-, f(x) \to -\infty \quad \text{or} \quad \lim_{x \to 2^-} f(x) = -\infty$$

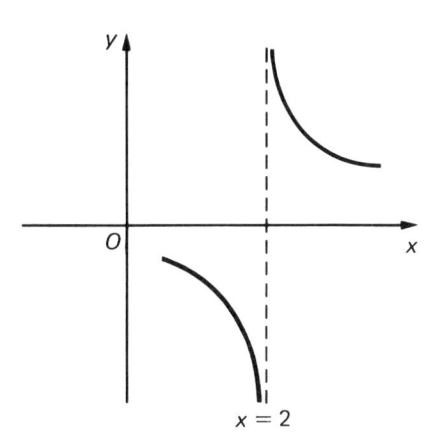

Fig. 1.2:**30** The graph of f in the neighbourhood of $x = 2$ is shown in fig. 1.2:**30**. These is no value of $f(x)$ corresponding to $x = 2$. As $x \to 2$, the curve gets closer and closer to the line $x = 2$ (without ever actually crossing it).

The line $x = 2$ is called an **asymptote** (in particular an **x-asymptote**) to the curve. In general,

---
if $\lim_{x \to a^+} f(x) = \pm\infty$, the line $x = a$ is an (x-) asymptote to the graph of f
---

Note: if $x = a$ is an asymptote to the graph of f, then f is discontinuous at $x = a$ (figs. 1.2:**31a** and **b**).

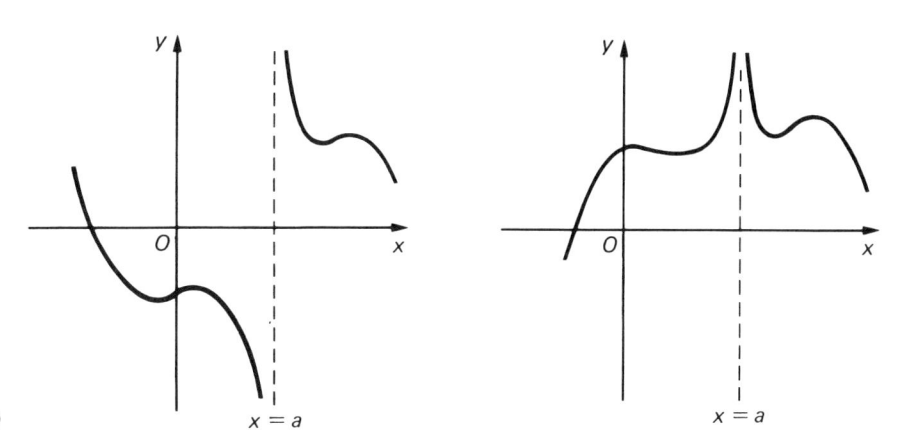

Figs. 1.2:**31a** and **b**

Worked example 12 Consider the function f defined by $f(x) = \dfrac{2x-1}{x+1}$.

Look at the values of $f(x)$ as x gets very large and positive ($x \to \infty$) and as x gets very large and negative ($x \to -\infty$). As $x \to \infty$, $f(x) \to 2$, i.e. $\lim_{x \to \infty} f(x) = 2$.

x	10	100	1000	10 000
$f(x)$	1.7273	1.9703	1.9970	1.9997

44 Asymptotes

As $x \to -\infty$, $f(x) \to 2$, i.e. $\lim_{x \to -\infty} f(x) = 2$.

x	-10	-100	-1000	$-10\,000$
$f(x)$	2.3333	2.0303	2.0030	2.0003

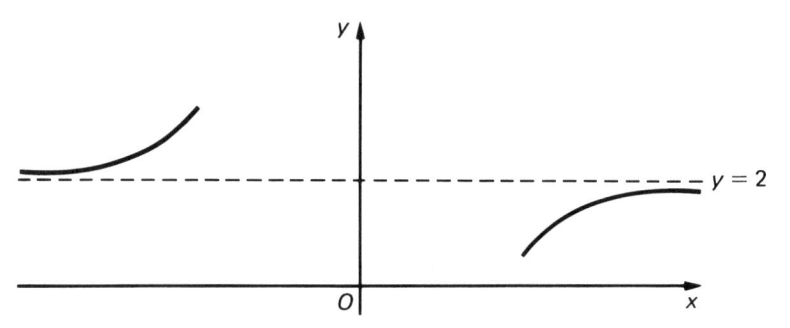

Fig. 1.2:32

So in the neighbourhood of $y = 2$ the graph looks like fig. 1.2:32. As $x \to \pm\infty$, the curve gets closer and closer to the line $y = 2$. The line $y = 2$ is an asymptote (in particular a *y*-**asymptote**) to the curve. In general,

> if $\lim_{x \to \pm\infty} f(x) = k$, the line $y = k$ is a (*y*-) asymptote to the graph of f (figs. 1.2:**33a** and **b**).

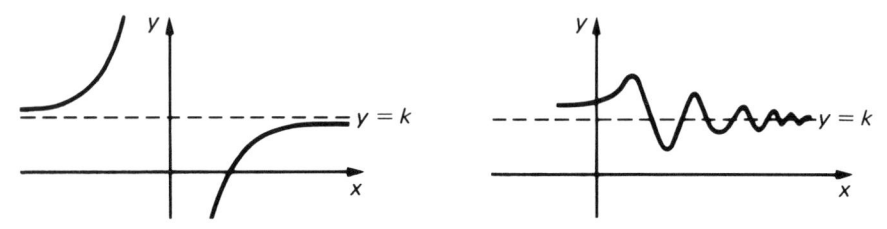

Fig. 1.2:33

Worked example 13 Sketch the curve $y = f(x)$ if $f(x) = (x+1)/(x-3)$

(i) When $x = 0$, $f(x) = -\frac{1}{3}$, so the curve cuts the *y*-axis at the point $(0, -\frac{1}{3})$.
(ii) When $f(x) = 0$, $x = -1$, so the curve cuts the *x*-axis at the point $(-1, 0)$.
(iii) The *x*-asymptote is the line $x = 3$.
Note: as $x \to 3^+$, $f(x) \to +\infty$ and as $x \to 3^-$, $f(x) \to -\infty$
(iv) As $x \to \infty$, $f(x) \to 1$; so the line $y = 1$ is a *y*-asymptote of the curve. The curve is shown in fig. 1.2:**34**.

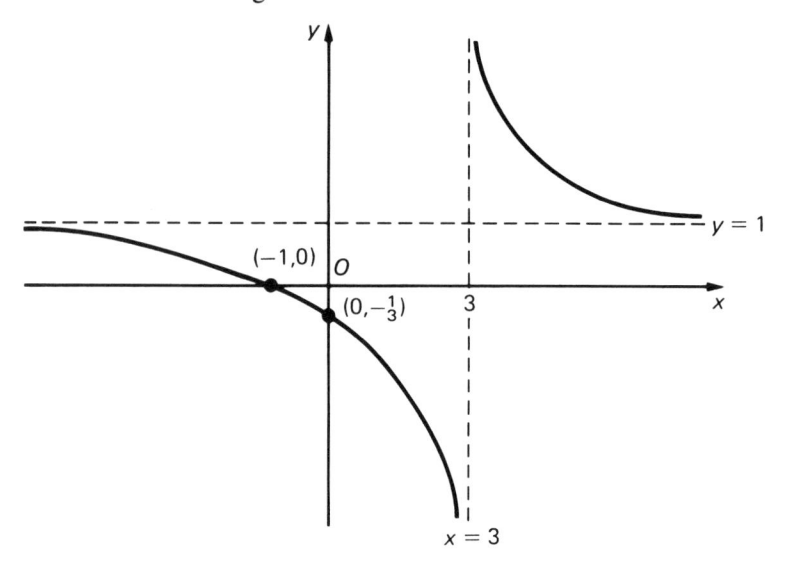

Fig. 1.2:34

Exercise 1.2:6

⑪ → **1** The function f is defined by $f(x) = 1/(x-1)$. Explain why $f(1)$ is undefined.

Complete this table:

x	0.5	0.9	0.95	0.99	0.999
$f(x)$					

x	1.001	1.01	1.05	1.1	1.5	2
$f(x)$						

Sketch the graph of f for values of x between 0 and 2.

⑪ **2** Find the x-asymptotes of the following curves, and sketch the curve in the neighbourhood of each asymptote.

(i) $y = \dfrac{1}{x-4}$

(ii) $y = \dfrac{1}{x+3}$

(iii) $y = \dfrac{2}{3x-5}$

→ (iv) $y = \dfrac{x+1}{x-2}$

→ (v) $y = \dfrac{1}{(x-1)(x+2)}$

⑫ → **3** The function f is defined by
$$f(x) = (5x+3)/(2x-1)$$

Complete this table:

x	$-10\,000$	-1000	-100	-10
$f(x)$				

x	10	100	1000	10 000
$f(x)$				

Find the value of k for which
$$\lim_{x \to -\infty} f(x) = \lim_{x \to \infty} f(x) = k$$

Sketch the graph of f in the neighbourhood of $y = k$.

⑫ **4** Find the y-asymptotes of the following curves, and sketch the curve in the neighbourhood of its y-asymptote.

(i) $y = \dfrac{x+2}{x-1}$

→ (ii) $y = \dfrac{3x+2}{4x-1}$

(iii) $y = \dfrac{2-5x}{x+2}$

(iv) $y = \dfrac{4-4x}{1-2x}$

(v) $y = \dfrac{1}{3x+2}$

(vi) $y = \dfrac{x}{x^2+2}$

⑫ **5** Write down the y-asymptote of the curve $y = f(x)$.

→ (i) $f(x) = \dfrac{6x-5}{2x-1}$

(ii) $f(x) = \dfrac{1-4x}{3x}$

(iii) $f(x) = \dfrac{x}{3x^2-1}$

→ (iv) $f(x) = \dfrac{ax+b}{px+q}$ $(p \neq 0)$

⑬ **6** Find the coordinates of the points, if any, where the curve $y = f(x)$ crosses the axes, find the x-asymptote to the curve, investigate the behaviour of the curve in the neighbourhood of its x-asymptote(s), find the y-asymptote to the curve, and hence sketch the curve.

(i) $f(x) = \dfrac{1}{x-2}$

(ii) $f(x) = \dfrac{1}{2x+3}$

(iii) $f(x) = \dfrac{x+4}{x-2}$

→ (iv) $f(x) = \dfrac{2x-5}{3x+5}$

(v) $f(x) = \dfrac{x+3}{x}$

(vi) $f(x) = \dfrac{1}{x^2}$

⑬ **7** Sketch a graph of the function $f:(-1,1) \to \mathbb{R}$ defined by $f(x) = 1/\sqrt{(1-x^2)}$.

⑬ **8** Sketch a graph of the function $f(x) = 1/x$. Explain why the graph is symmetrical about the line $y = x$ (see 1.2:4 Q29).

⑬ **9** Sketch on the same axes the graphs of f and f^{-1} when:

(i) $f(x) = \dfrac{1}{x-3}$

→ (ii) $f(x) = \dfrac{x+2}{x-2}$

R (iii) $f(x) = \dfrac{x}{x-1}$

46 *Graphical solution of equations and inequalities*

⑬ **10** Sketch on the same axes the graphs of f and g, and find the coordinate of their point(s) of intersection, when:

(i) $f(x) = \dfrac{x}{x+3}$, $g(x) = -\tfrac{1}{2}$

(ii) $f(x) = \dfrac{4}{x-3}$, $g(x) = x-3$

(iii) $g(x) = \dfrac{2}{x+1}$, $g(x) = x$

(iv) $f(x) = \dfrac{1}{x-1}$, $g(x) = x+1$

(v) $f(x) = \dfrac{3x}{x+2}$, $g(x) = 2x-1$

(vi) $f(x) = \dfrac{x}{x+3}$, $g(x) = \dfrac{x+1}{x-1}$

11 We are also interested in the behaviour, as $x \to +\infty$ and $x \to -\infty$, of functions which do *not* have asymptotes.

Investigate the behaviour of $f(x)$ as $x \to \pm\infty$, when:

(i) $f(x) = 3x - 2$
(ii) $f(x) = 2 - 3x$
(iii) $f(x) = x^2$
→ (iv) $f(x) = 3x - x^2$
(v) $f(x) = x^3 - x$
→ (vi) $f(x) = (1-x)(2-x)(3-x)$

Illustrate your answer by sketching the graph of f.

1.2:7 Graphical solution of equations and inequalities

Type A
Suppose that the curve $y = f(x)$ is that in fig. 1.2:35.

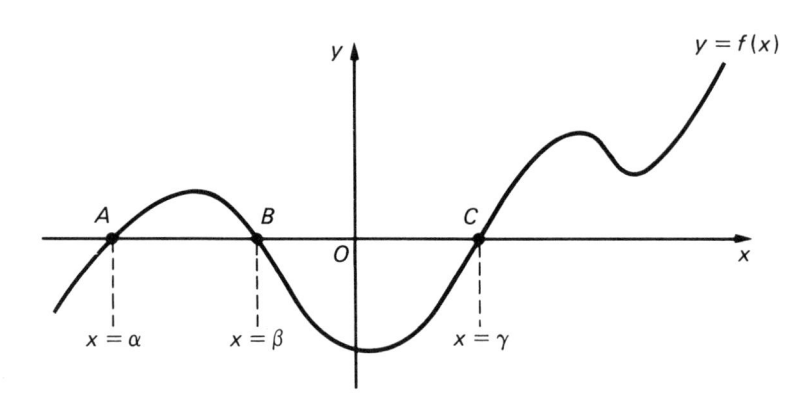

Fig. 1.2:35

(i) *The equation $f(x) = 0$:* The solutions of $f(x) = 0$ correspond to the points of intersection of $y = f(x)$ and $y = 0$ (the x-axis), i.e. $x = \alpha$, β and γ.

(ii) *The inequalities $f(x) > 0$, $f(x) < 0$:* The crucial points are those at which $f(x)$ changes sign, i.e. A, B and C.

$f(x) > 0$ (curve *above* x-axis): $\alpha < x < \beta$ or $x > \gamma$ (fig. 1.2:**36a**).

$f(x) < 0$ (curve *below* x-axis): $x < \alpha$ or $\beta < x < \gamma$ (fig. 1.2:**36b**).

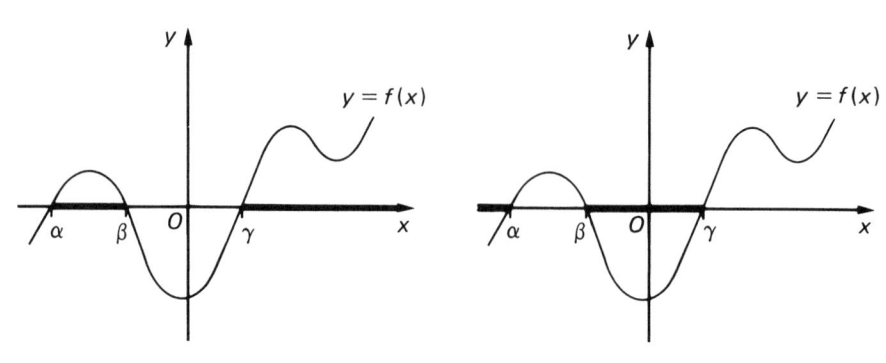

Fig. 1.2:**36a** $f(x) > 0$

Fig. 1.2:**36b** $f(x) < 0$

Introduction 47

Worked example 14 Sketch the curve $y = f(x)$ where $f(x) = (x-1)(x+2)$. Find the values of x for which

(i) $f(x) = 0$ (ii) $f(x) > 0$ (iii) $f(x) < 0$

The curve is shown in fig. 1.2:**37**.

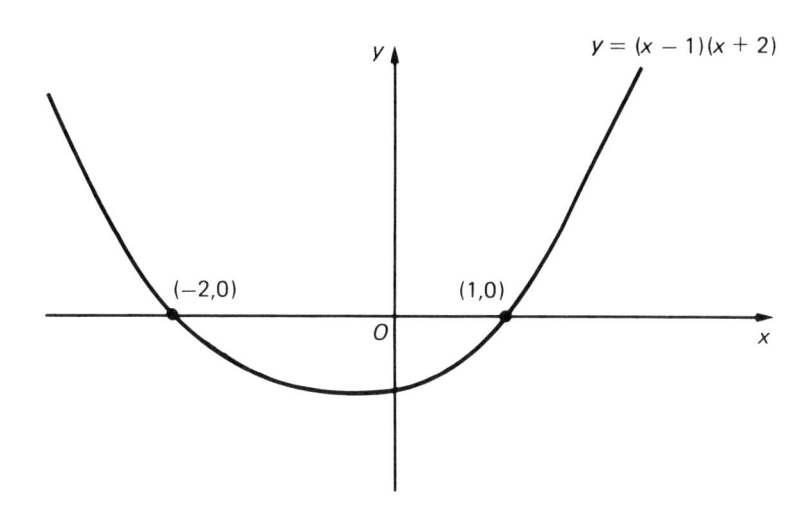

Fig. 1.2:**37**

(i) $f(x) = 0 \Rightarrow (x-1)(x+2) = 0 \Rightarrow x = -2$ or $x = 1$.
(ii) $f(x) > 0$ (curve above x-axis): $x < -2$ or $x > 1$ (fig. 1.2:**38a**).
(iii) $f(x) < 0$ (curve below x-axis): $-2 < x < 1$ (fig. 1.2:**38b**).

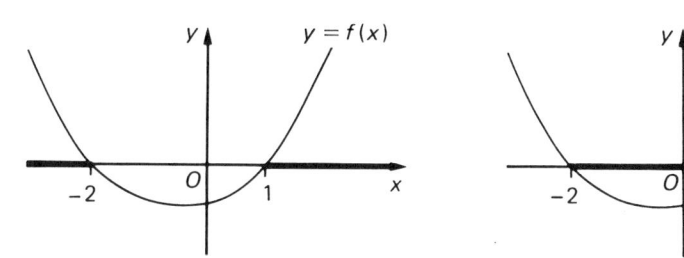

Fig. 1.2:**38a** $f(x) > 0$
Fig. 1.2:**38b** $f(x) < 0$

Type B

Suppose that the curves $y = f(x)$, $y = g(x)$ are those in fig. 1.2:**39**.

(i) *The equation $f(x) = g(x)$:* The solutions of $f(x) = g(x)$ correspond to the points A, B, C and D of intersection of $y = f(x)$ and $y = g(x)$, i.e. $x = \alpha, \beta, \gamma$ and δ. (Note: $f(\alpha) = g(\alpha)$, etc.)
(ii) *The inequalities $f(x) > g(x)$, $f(x) < g(x)$:* The crucial points are A, B, C and D.

$f(x) > g(x)$ (thin line above thick line): $\alpha < x < \beta$ or $\gamma < x < \delta$.
$f(x) < g(x)$ (thin line below thick line): $x < \alpha$ or $\beta < x < \gamma$ or $x > \delta$.

(Note: Type A is just a particular case of Type B, with $g(x) = 0$.)

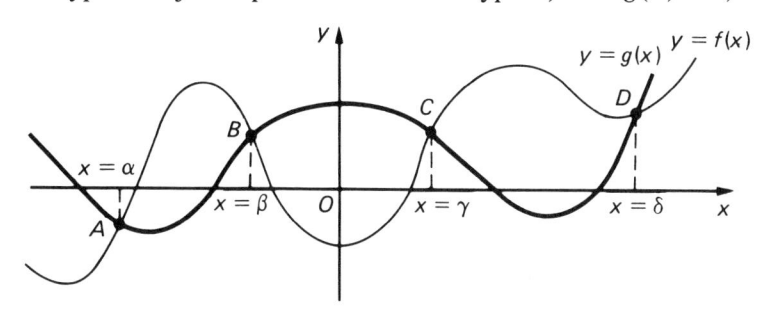

Fig. 1.2:**39**

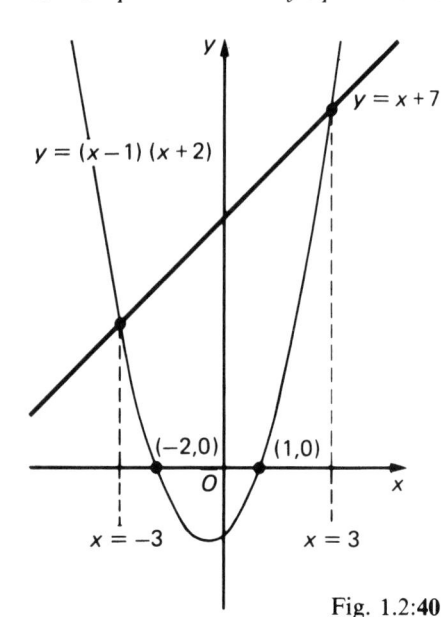

48 *Graphical solution of equations and inequalities*

Worked example 15 Sketch on the same axes the curve $y=(x-1)(x+2)$ and the line $y=x+7$.

Find the values of x for which:

(i) $(x-1)(x+2)=x+7$ (ii) $(x-1)(x+2)>x+7$
(iii) $(x-1)(x+2)<x+7$

The curve is shown in fig. 1.2:**40**.

(i) $(x-1)(x+2)=x+7 \Rightarrow x^2+x-2=x+7 \Rightarrow x^2=9 \Rightarrow x=3$ or $x-3$.
(ii) $(x-1)(x+2)>x+7$ (thin line above thick line): $x<-3$ or $x>3$.
(iii) $(x-1)(x+2)<x+7$ (thin line below thick line): $-3<x<3$.

(See 1.2:4 Q32(vi).)

Fig. 1.2:**40**

Exercise 1.2:7

Questions 1-14: Sketch the curve $y=f(x)$. Find the value(s) of x for which:

(i) $f(x)=0$ (ii) $f(x)>0$ (iii) $f(x)<0$

⑭ **1** $f(x)=x+4$

⑭ **2** $f(x)=-2x+5$

⑭ **3** $f(x)=(x+3)(x-2)$

⑭ **4** $f(x)=(2x-7)(x-1)$

⑭→ **5** $f(x)=(x+2)(4-3x)$

⑭ **6** $f(x)=x(3x+5)$

⑭ **7** $f(x)=9-4x^2$

⑭ **8** $f(x)=(x-1)(x-3)(x-5)$

⑭→ **9** $f(x)=(x+3)(2x-1)(5-3x)$

⑭ R **10** $f(x)=x^3-4x$

⑭ **11** $f(x)=\dfrac{x-3}{x}$

⑭ **12** $f(x)=\dfrac{x+1}{x-2}$

⑭ **13** $f(x)=\begin{cases}-3x+6 & x\leq 4\\ x-10 & x>4\end{cases}$

⑭ **14** $f(x)=\begin{cases}-4x-8 & x<-1\\ -\tfrac{1}{2}x+1 & -1\leq x\leq 3\\ \tfrac{1}{4}x-\tfrac{5}{4} & x>3\end{cases}$

⑭ **15** Solve the inequality $(x+1)(2x-5)>0$:

(i) analytically, by the method of 1.1:4: Q19;
(ii) graphically, by sketching the curve

$$y=(x+1)(2x-5)$$

Questions 16-26: Sketch on the same axes the curves $y=f(x)$ and $y=g(x)$. Find the value(s) of x for which:

(i) $f(x)=g(x)$
(ii) $f(x)>g(x)$
(iii) $f(x)<g(x)$

⑮ **16** $f(x)=x$, $g(x)=-\tfrac{1}{2}x+3$

⑮ **17** $f(x)=2x+1$, $g(x)=3x-2$

⑮ **18** $f(x)=(x+1)(x-2)$, $g(x)=4$

⑮→ **19** $f(x)=(x+3)(x-1)$, $g(x)=3x+3$

⑮ **20** $f(x)=5-x^2$, $g(x)=4x$

⑮ **21** $f(x)=x^2+4$, $g(x)=(x+2)(4-x)$

⑮ R **22** $f(x)=x(x+1)$, $g(x)=(x-2)(x-3)$

⑮ **23** $f(x)=\dfrac{3}{x+5}$, $g(x)=\tfrac{1}{2}x$

⑮ **24** $f(x)=\dfrac{x}{x-3}$, $g(x)=3x-8$

⑮ **25** $f(x)=\dfrac{2x-3}{x-1}$, $g(x)=3-\dfrac{x}{4}$

⑮ **26** $f(x) = \dfrac{x+1}{x-2}$, $g(x) = \dfrac{4x-6}{x}$

⑮ **27** By writing the inequality in the form $f(x) > 0$ and sketching the curve $y = f(x)$, find the set of values of x for which:

(i) $(x+3)(x-1) > 3x+3$ (ii) $x/(x-3) < 3x-8$

(c.f. questions 19 and 24).

⑮ **28** Solve the inequality $x^2 - 3x < 12 - 2x$:

(i) analytically, by the method of 1.1:4 Q19;

(ii) graphically, by sketching the curve $y = x^2 - 3x$ and the line $y = 12 - 2x$ and finding their points of intersection;

(iii) graphically, by first expressing the inequality in the form $f(x) < 0$ and then sketching the curve $y = f(x)$;

(iv) pictorially, by first sending the function $x^2 - 3x$ on a postcard to a person chosen at random from the phone book and then....

29 By considering the intersection of the curve $y = f(x)$ and the line $y = k$, discuss the number of solutions of the equation $f(x) = k$ when f is

(i) one-one   (ii) onto but not into
(iii) not onto

(see 1.2:3 Q29).

1.2:8 Implicit functions

We say that y is an **implicit function** of x if there exists an equation relating x and y, but y *cannot* be expressed explicitly in terms of x. For example,

$$y^2 = x \qquad x^2 + y^2 = 9 \qquad xy^3 - x\sqrt{(y-1)} + y = 0$$

Note: If y is an implicit function of x, y is unlikely to be a function of x in the normal sense. For example, $y^2 = x \Rightarrow y = \pm\sqrt{x}$; the relation f defined by $f(x) = \pm\sqrt{x}$ is not a function (see 1.2:3 Q34).

Worked example 16 Sketch the curves: (i) $y^2 = x$; (ii) $x^2 + y^2 = 9$.

(i)

x	0	1	4	9
y	0	± 1	± 2	± 3

The curve (fig. 1.2:**41**) is a parabola (see section 6.2:7).

(ii)

x	0	± 1	± 2	± 3
y	± 3	$\pm\sqrt{8}$	$\pm\sqrt{5}$	0

The curve (fig. 1.2:**42**) is a circle.

The equation $x^2 + y^2 = 9$ expresses the fact that, if $P(x, y)$ is a general point on the curve, $OP^2 = 9$, i.e. $OP = 3$, so the curve is a circle centre O, radius 3.

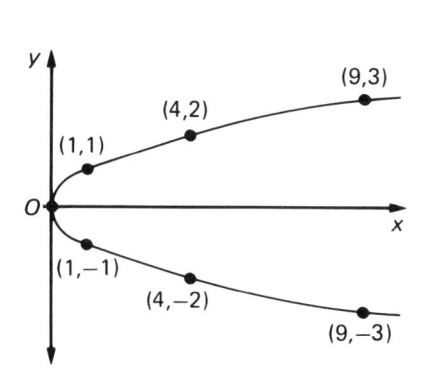

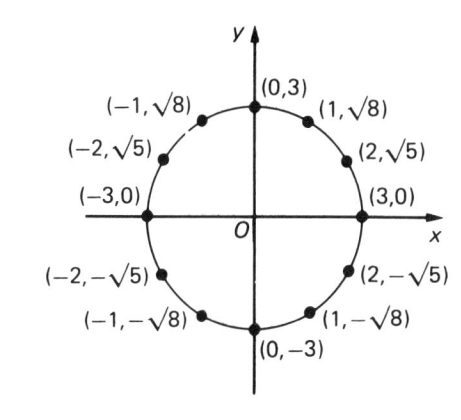

Fig. 1.2:**41**

Fig. 1.2:**42**

Exercise 1.2:8

(16) → **1** Sketch the curves:
(i) $y^2 = 4x$ (ii) $x^2 + y^2 = 25$

(16) → **2** Sketch on the same axes the curves $y = x^2$ and $y^2 = x$.

(16) **3** Sketch on the same axes the circles:
(i) $x^2 + y^2 = 4$ (ii) $x^2 + y^2 = 16$ (iii) $x^2 + y^2 = 36$

(16) **4** Find the possible values of y when $x = -4, -3, -2, -1, 0, 1, 2, 3, 4$, and hence sketch the curve defined by the equation:
(i) $y^2 = x - 2$
(ii) $x^2/16 + y^2/9 = 1$ (an *ellipse*)
→ (iii) $2x^2 + y^2 = 8$
→ (iv) $y^2 = x^3$ (a *semi-cubical parabola*)
(v) $y^3 = 8x$

(16) **R 5** Sketch on the same axes the curves $y^3 = x$ and $y = x^3$.

(16) **6** A is the point $(3, 0)$ and P the variable point (x, y). Show that if the distance $AP = 2$,

$$(x - 3)^2 + y^2 = 4$$

Sketch the curve represented by this equation.

(16) **7** Sketch the circles:
(i) $x^2 + (y - 1)^2 = 9$
(ii) $(x - 4)^2 + (y - 3)^2 = 4$

(16) **8** Sketch the pair of lines represented by the equations:
(i) $4x^2 - y^2 = 0$ (ii) $x^2 - 3xy + 2y^2 = 0$

(Hint: factorize the LHS.)

1.3 Differentiation

1.3:1 Gradient of a curve

A line joining two points P and Q on a curve is called a **chord** of the curve.

The line which touches the curve at P is called the **tangent** to the curve at P (fig. 1.3:1).

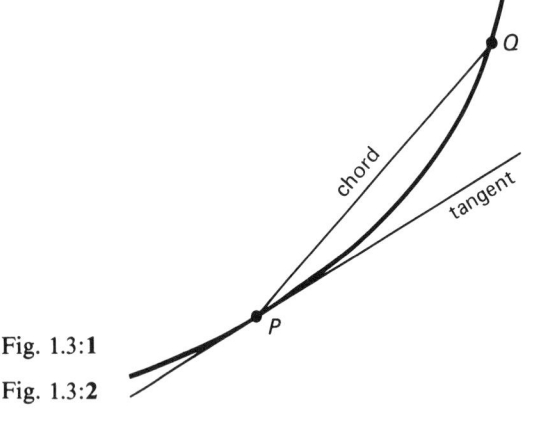

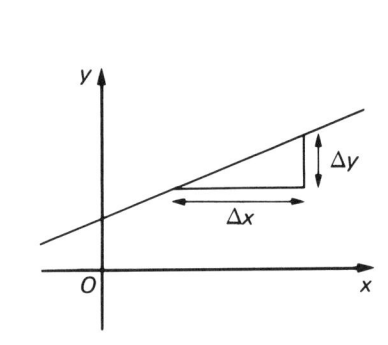

Fig. 1.3:1

Fig. 1.3:2

We have defined the gradient of a straight line in the xy plane as $\Delta y / \Delta x$ (see 1.2:1), i.e. the rate at which y increases compared with x, or the rate of increase of y with respect to x (fig. 1.3:2).

We would like to give a similar meaning to 'the gradient of a curve'.

Definition The **gradient of a curve** in the xy plane is the rate of increase of y with respect to x.

Notice that the gradient of a curve changes as we move along the curve.

Introduction 51

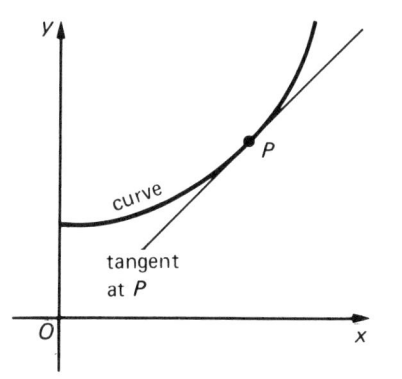

Consider the gradient of a curve at a point P on the curve (fig. 1.3:3). The tangent to the curve at P is the continuation of the curve in a straight line; so the rate at which y increases with respect to x is unchanged as we move off the curve and onto the tangent, i.e.

> The gradient of a curve at a point P on the curve equals the gradient of the tangent to the curve at P.

Suppose P is the point $(x, f(x))$ on the curve $y = f(x)$. We shall find the gradient—call it $f'(x)$—of the tangent at P in terms of $f(x)$.

Fig. 1.3:3

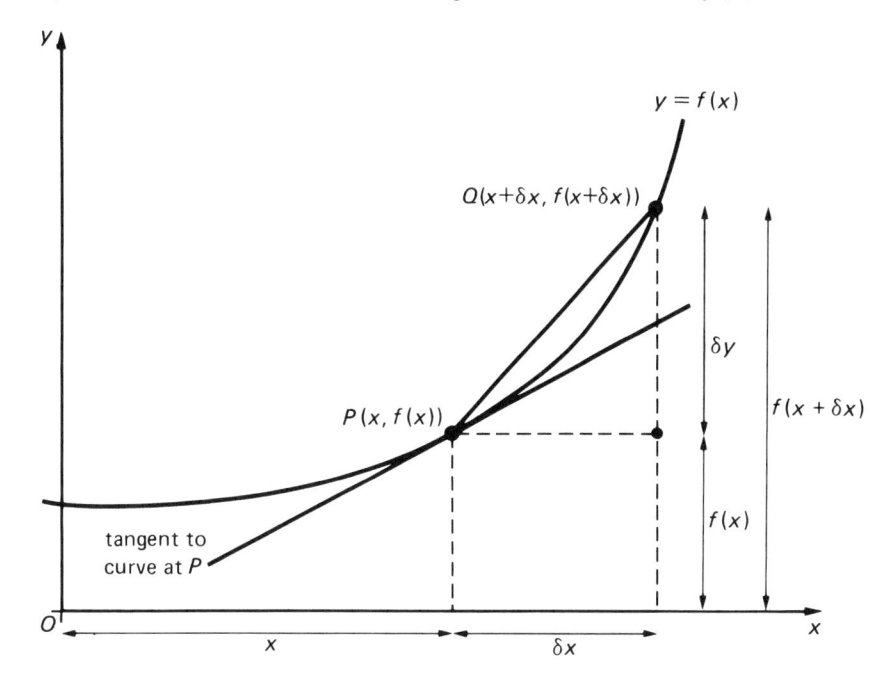

Fig. 1.3:4

Let Q be a point near P on the curve (fig. 1.3:4). Using δx ('delta-ex') to denote a small increase in x, suppose that the x-coordinate of Q is $x + \delta x$. Then, since Q is on the curve, its y-coordinate is $f(x + \delta x)$.

$$\text{gradient of } PQ = \frac{\delta y}{\delta x} = \frac{f(x + \delta x) - f(x)}{\delta x} \tag{1}$$

As $\delta x \to 0$, $PQ \to$ tangent at P, so the gradient of $PQ \to$ the gradient of the tangent at P, i.e.

$$\frac{\delta y}{\delta x} \to f'(x)$$

In other words,

$$f'(x) = \text{the limit, as } \delta x \to 0, \text{ of } \frac{\delta y}{\delta x}$$

We can express $f'(x)$ as $\lim_{\delta x \to 0} \frac{\delta y}{\delta x}$, i.e.

$$f'(x) = \lim_{\delta x \to 0} \frac{\delta y}{\delta x} = \lim_{\delta x \to 0} \frac{f(x + \delta x) - f(x)}{\delta x}$$

(from equation (1)).

52 Gradient of a curve

We write $\lim\limits_{\delta x \to 0} \dfrac{\delta y}{\delta x}$ as $\dfrac{dy}{dx}$. So, if $y = f(x)$,

$$\frac{dy}{dx} = f'(x) = \lim_{\delta x \to 0} \frac{f(x + \delta x) - f(x)}{\delta x} \qquad (1a)$$

Note: dy and dx do *not* have any meaning in themselves; in particular we cannot think of dy/dx as meaning $dy \div dx$.

Worked example 1 Find $f'(x)$ when: (i) $f(x) = x^2$; (ii) $f(x) = 1/x$.

(i) $f(x) = x^2$. By equation (1a),

$$f'(x) = \lim_{\delta x \to 0} \frac{(x + \delta x)^2 - x^2}{\delta x}$$

$$= \lim_{\delta x \to 0} \frac{2x\delta x + (\delta x)^2}{\delta x}$$

$$= \lim_{\delta x \to 0} 2x + \delta x$$

$$= 2x$$

(ii) $f(x) = 1/x$. By equation (1a),

$$f'(x) = \lim_{\delta x \to 0} \frac{\dfrac{1}{x + \delta x} - \dfrac{1}{x}}{\delta x}$$

$$= \lim_{\delta x \to 0} \frac{x - (x + \delta x)}{(x + \delta x)x(\delta x)}$$

$$= \lim_{\delta x \to 0} \frac{-1}{x(x + \delta x)}$$

$$= -\frac{1}{x^2}$$

The process of finding $f'(x)$ given $f(x)$ is called **differentiation** with respect to (abbreviated to w.r.t.) x.

$f'(x)$ is called the **derivative** of $f(x)$ w.r.t. x, e.g. $2x$ is the derivative of x^2 w.r.t. x.

The symbol $\dfrac{d}{dx}$ means 'the derivative, w.r.t. x, of'; e.g.

$$\frac{d}{dx}(y) = \frac{dy}{dx}, \qquad \frac{d}{dx}\{f(x)\} = f'(x), \qquad \frac{d}{dx}\left(\frac{1}{x}\right) = \left(-\frac{1}{x^2}\right).$$

Note that the derivative of a function is itself a function.

Exercise 1.3:1

: 1 Sketch the curve $y = x^2$ for values of x between 0 and 2. Mark on the curve the points A, B, C, D, E, F and G with x-coordinates 1, 1.01, 1.05, 1.1, 1.2, 1.5 and 2.

Find the y-coordinates of these points.
Find the gradients of the chords AG, AF, AE, AD, AC and AB. Hence estimate the gradient of the tangent to the curve at A.

2 By a similar method to the one you used in question 1, estimate the gradient of the tangent to:

(i) the curve $y = x^2$ at the point $(3, 9)$;
(ii) the curve $y = x^3$ at the point $(1, 1)$;
(iii) the curve $y = x(3-x)$ at the point $(2, 2)$.

: 3 Let $P(x, x^2)$ and $P'(x+h, (x+h)^2)$ be points on the curve $y = x^2$.

(i) Show that the gradient of the chord PP' is $2x + h$.

(ii) By considering the gradient of this chord as h gets smaller and smaller, estimate the gradient of the tangent at P.

4 By considering the gradient of the chord PP' where P and P' are the points on the curve $y = x(3-x)$ with x-coordinates x and $x+h$, find the gradient of the tangent to the curve at P.

① **5** Using equation (1a), find $f'(x)$ when:

(i) $f(x) = 5x^2$
→ (ii) $f(x) = 2x - x^2$
(iii) $f(x) = 1/3x$
→ (iv) $f(x) = 1/x^2$

① → **6** (i) Expand $(x + \delta x)^3$ in ascending powers of δx. (Use Pascal's Triangle—see 1.1:4 Q7.) Hence differentiate x^3.

(ii) Similarly, differentiate x^4.

① • **7** Use equation (1a) to show that, if n is a positive integer,

$$\frac{d}{dx}(x^n) = nx^{n-1}$$

(Write down the first two terms only in the expansion of $(x + \delta x)^n$ in ascending powers of δx. Explain why two terms are enough.)
See also 1.3:8 Q14.

1.3:2 Basic derivatives

$$\boxed{\text{If } \alpha \text{ is any real number,} \qquad \frac{d}{dx}(x^\alpha) = \alpha x^{\alpha - 1}} \qquad (2)$$

(see 1.3:1 Q7). For example

$$\frac{d}{dx}(x^{-5}) = -5x^{-6}, \qquad \frac{d}{dx}(x^{8/3}) = \frac{8}{3}x^{5/3}$$

Theorem 1

(i) If a is a constant $\dfrac{d}{dx}\{af(x)\} = af'(x)$ \hfill (3)

(ii) $\qquad \dfrac{d}{dx}\{f(x) + g(x)\} = f'(x) + g'(x)$ \hfill (4)

Proof
This can be proved using equation 1(a).

Worked example 2 Find $\dfrac{d}{dx}\left(2x^4 + \dfrac{3}{x^2}\right)$.

$$\frac{d}{dx}\left(2x^4 + \frac{3}{x^2}\right) = \frac{d}{dx}(2x^4) + \frac{d}{dx}\left(\frac{3}{x^2}\right) \quad \text{by equation (4)}$$

$$= 2\frac{d}{dx}(x^4) + 3\frac{d}{dx}\left(\frac{1}{x^2}\right) \quad \text{by equation (3)}$$

$$= 8x^3 - \frac{6}{x^3}$$

Exercise 1.3:2

Questions 1-3: Using equation (2), differentiate with respect to x.

→ **1** (i) x^7 (ii) $x^{5/2}$
(iii) $\sqrt[3]{x}$ (iv) $x\sqrt{x}$
(v) $\dfrac{1}{x^3}$ (vi) $\dfrac{1}{\sqrt{x}}$
→ (vii) $\dfrac{1}{x\sqrt{x}}$ (viii) $\dfrac{1}{\sqrt[3]{x}}$

② **2** (i) $5x^{12}$ (ii) $9x^{4/3}$
(iii) $7x$ (iv) 4
(v) $7x+4$ (vi) $\frac{1}{2}x^4+\frac{1}{6}x^3$
→ (vii) $2\sqrt{x}+\dfrac{4}{\sqrt{x}}$ (viii) $\dfrac{3}{x^3}-\dfrac{1}{3x^6}$

② • **3** (i) $(1+x^2)^2$ (ii) $x^3(x^2+2x)$
(iii) $\dfrac{10x^5+3x^4}{2x^2}$ (iv) $\dfrac{2x^3+3}{x^2}$
→ (v) $\left(\dfrac{3x+1}{x}\right)^2$ (vi) $\dfrac{1+5x}{\sqrt{x}}$

② → **4** Explain why, if $y = $ constant, $\dfrac{dy}{dx} = 0$. Interpret this result geometrically.
Differentiate $y = mx+c$ (m and c constant) with respect to x. Interpret this result geometrically.

② **5** Using equation (1a), show that:
(i) $\dfrac{d}{dx}\{af(x)\} = af'(x)$
(ii) $\dfrac{d}{dx}\{f(x)+g(x)\} = f'(x)+g'(x)$

② : **6** If $f(x) = x^4$, $g(x) = x^3$, find:
(i) $f'(x)$ (ii) $g'(x)$
(iii) $\dfrac{d}{dx}\{f(x)g(x)\}$ (iv) $\dfrac{d}{dx}\left\{\dfrac{f(x)}{g(x)}\right\}$

Repeat this for $f(x) = x+1$, $g(x) = \sqrt{x}$.
Investigate the truth of the statements:
(a) $\dfrac{d}{dx}\{f(x)g(x)\} = f'(x)g'(x)$
(b) $\dfrac{d}{dx}\left\{\dfrac{f(x)}{g(x)}\right\} = \dfrac{f'(x)}{g'(x)}$

7 Sketch (on separate axes) graphs of $f(x)$ and $f'(x)$ when:
• (i) $f(x) = x^2$
(ii) $f(x) = x^2 - 3x + 2$
→ (iii) $f(x) = x^3 - 9x$

In each case, write down $f'(-2), f'(-1), f'(0), f'(1), f'(2)$.

Questions 8-11: Copy the given graph. Suppose it is the graph of the function f. Sketch a rough graph of the function f'.

→ **8**

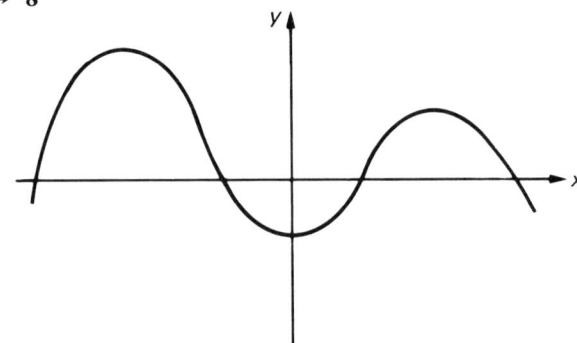

9

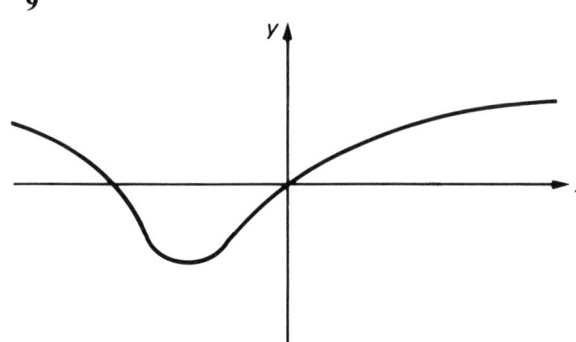

10

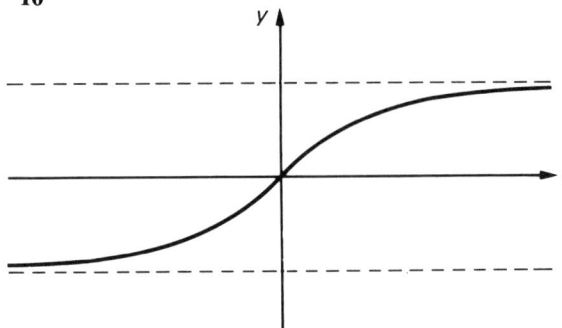

11

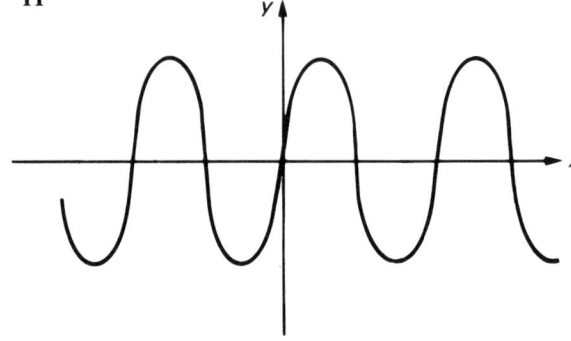

12 Sketch graphs of $f(x)$ and $f'(x)$ in each of the following cases. Is $f'(x)$ continuous at $x=0$? (See section 1.2:5.) Does it make sense to talk about 'the gradient of the tangent to the curve $y=f(x)$ at the point where $x=0$'?

(i) $f(x) = \begin{cases} x & x \geq 0 \\ -x & x < 0 \end{cases}$ (ii) $f(x) = \begin{cases} x & x \geq 0 \\ x^2 & x < 0 \end{cases}$

(iii) $f(x) = \begin{cases} x & x \geq 0 \\ x(x+1) & x < 0 \end{cases}$

R 13 $f(x) = \begin{cases} -x^2 & x < -1 \\ x^3 & -1 \leq x < 1 \\ x^2 & x > 1 \end{cases}$

Is $f'(x)$ continuous at:

(i) $x = -1$ (ii) $x = 0$ (iii) $x = 1$?

→ **14** $f(x) = \begin{cases} (x+3)(1-x) & x < -1 \\ 1 & -1 \leq x \leq 1 \\ x^2 - 2x + 2 & x > 1 \end{cases}$

Find $f'(-2), f'(-1), f'(0), f'(1), f'(2)$.

Questions 15-17: Copy the given graph. Suppose it is the graph of the function f. Sketch a rough graph of the function f'.

15

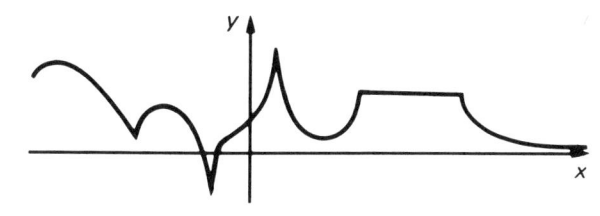

→ **16**

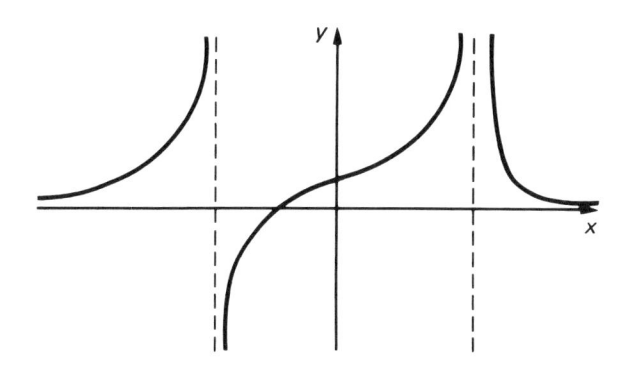

17

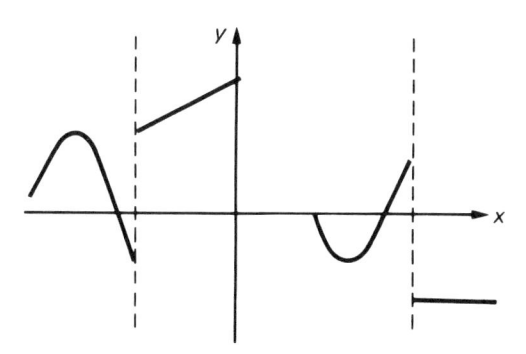

18 (i) If $s = 10t - 5t^2$, find ds/dt.
(ii) If $T = 2\pi \sqrt{(l/g)}$ (g constant), find dT/dl.
(iii) If $A = \theta + 2/\theta^2$, find $dA/d\theta$.

19 The weight W g of a female mouse after t weeks is given approximately (for $t \leq 8$) by

$$W = 1 + \tfrac{1}{4}(t+2)^2$$

(i) Find the rate of growth of the mouse after:
(a) t weeks; (b) 1 week; (c) 2 weeks; (d) 8 weeks.

(ii) The **relative growth rate** is defined to be the growth rate divided by the weight.
Find the relative growth rate of the mouse after:
(a) t weeks; (b) 1 week; (c) 2 weeks; (d) 8 weeks.

20 The capital £C of a gambler t months from the time when he lays his first bet is given by

$$C = 1000(8 - 9t + 6t^2 - t^3)$$

(i) Find C when $t = 0, 1, 2, 3, 4, 5$. Sketch a graph of C against t for $0 \leq t \leq 5$.

(ii) Find the rate of increase of his capital after:
(a) 2 months; (b) 3 months; (c) 4 months.

(iii) Find the times at which he has maximum capital.

(iv) What are the time intervals during which his capital is:
(a) increasing, (b) decreasing?

(v) What happens to him eventually?

Is there a moral to be drawn? Have you drawn it?

1.3:3 Equation of the tangent and the normal to a curve

The normal to a curve at a particular point A on the curve is the line through A perpendicular to the tangent at A (fig. 1.3:5).

The gradient of the tangent to the curve $y = f(x)$ at a general point $P(x, f(x))$ is $f'(x)$. So

(i) the gradient of the tangent to the curve $y = f(x)$ at a particular point $A(a, f(a))$ is $f'(a)$,   **(5)**

(ii) the gradient of the normal to the curve at A is $-1/f'(a)$ **(6)**
(See 1.2:1 Theorem 2.)

tangent at A: gradient $= f'(a)$

normal at A gradient $= -1/f'(a)$

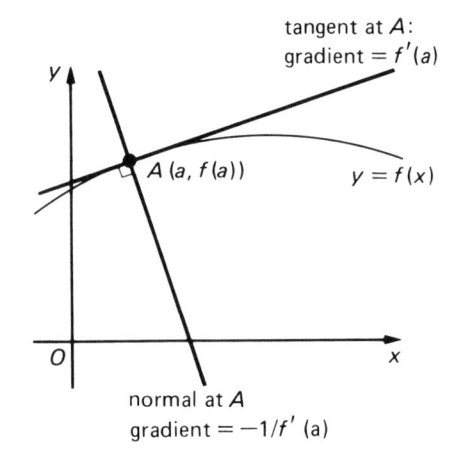

Fig. 1.3:5

Worked example 3 Find the equations of (i) the tangent and (ii) the normal to the curve $y = x^3$ at the point $A(2, 8)$.

$$f(x) = x^3 \Rightarrow f'(x) = 3x^2$$

(i) Gradient of tangent at $A = f'(2) = 3 \cdot (2)^2 = 12$.
The tangent passes through $A(2, 8)$ so its equation is

$$\frac{y-8}{x-2} = 12$$

i.e. $y = 12x - 16$ (see section 1.2:2).

(ii) Gradient of normal at $A = -1/f'(2) = -\frac{1}{12}$.
The normal passes through $A(2, 8)$ so its equation is

$$\frac{y-8}{x-2} = -\frac{1}{12}$$

i.e. $y = -\frac{1}{12}x + \frac{49}{6}$.

The tangent and normal to $y = x^3$ at $A(2, 8)$ are shown in fig. 1.3:6.

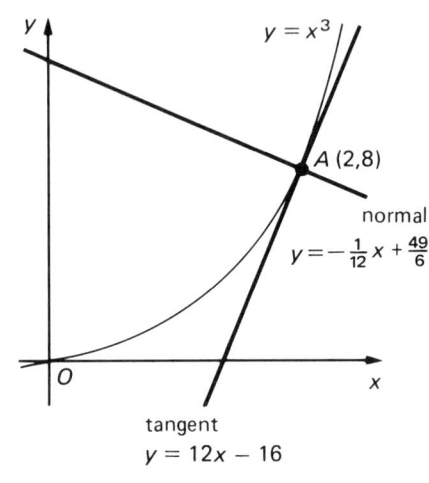

Fig. 1.3:6

Exercise 1.3:3

1 Find the gradients of the tangent and the normal to each of the following curves at the given point:

(i) $y = 4 - x + x^2$, $(3, 10)$
(ii) $y = 6\sqrt{x}$, $(4, 12)$
→ (iii) $y = 3x^2 - 1/x$, $(-1, 4)$
(iv) $y = (x-1)(x+2)(x-3)$, $(1, 0)$
(v) $y = x^2/4 + 4/x$, $(2, 3)$

Interpret your answer to part (v).

2 Find the gradient of the curve $y = 3x^2$ at the points:

(i) $(-2, 12)$ (ii) $(-1, 3)$ (iii) $(0, 0)$
(iv) $(1, 3)$ (v) $(2, 12)$

Illustrate your answers on a sketch of the curve.

3 Find the coordinates of the point(s) on the curve with the given gradient:

→ (i) $y = 5x^2 - 3x$, 7
(ii) $y = x^3$, $\frac{3}{4}$
(iii) $y = x^3 - x^2 - 5x + 1$, -1
(iv) $y = x - (4/x^2)$, -7
→ (v) $y = x^3 - 3x^2 + 5$, 0

4 Find the coordinates of the points on the curve $y = x^2 - 2x - 8$ with gradient:

(i) 4 (ii) 2 (iii) 0 (iv) -2 (v) -4

Illustrate your answers on a sketch of the curve.

③ **5** Find the equation of the tangent to the curve $y = x^2$ at the point $(2, 4)$.
(What is wrong with this solution?

$$y = x^2 \Rightarrow \frac{dy}{dx} = 2x \Rightarrow \text{gradient of tangent} = 2x$$

Tangent passes through $(2, 4)$ so equation of tangent is $(y-4)/(x-2) = 2x$, i.e. $y = 2x^2 - 4$.)

③ **6** Find the equations of the tangent and normal to the curve at the point corresponding to the given value of x:

(i) $y = 3x^2$, $x = 1$
→ (ii) $y = 9x - x^3$, $x = -1$
(iii) $y = x + 1/x$, $x = 2$
(iv) $y = 5\sqrt{x}$, $x = 4$
(v) $y = \dfrac{9}{\sqrt{x}}$, $x = 9$
(vi) $y = (x-1)/x^2$, $x = 1$

③ **7** Find the equations of the tangent and normal to the curve $y = 4x - x^2$ at the point $(1, 3)$. Illustrate your answer by sketching the curve, the tangent and the normal, marking in the points where they cut the y-axis.

③ **8** The tangent and normal to the curve $y = 4\sqrt{x}$ at the point $A(9, 12)$ cut the x-axis at T and N. Find the area of $\triangle ATN$.

③ **9** The curve $y = x^2 - 5x + 4$ cuts the x-axis at A and B. The tangents to the curve at A and B meet at T, and the normals to the curve at A and B meet at N. Find the area of the quadrilateral $ATBN$.

• **10** Find the gradient of the tangent to the curve $y = 2x^2$ at the general point $(x, 2x^2)$.
Find the coordinates of the point on the curve whose tangent has gradient 12.
Find the equation of this tangent.

11 Find the equation of the tangent of gradient 2 to the curve $y = x^2 - 4x$.

→ **12** Find the equations of the tangents of gradient 4 to the curve $y = x^3 + 3x^2 - 5x + 6$.

R **13** Find the equations of the tangents to the curve $y = x^3 - 11x$ which are parallel to the line $y = x$.

14 Find the equation of the tangent to the curve $y = x + x^2$ at the point $(a, a + a^2)$.
Find the values of a for which this line passes through the point $C(2, -3)$. Hence find the equations of the tangents from C to the curve.
Illustrate your answer by sketching a diagram.

15 Find the coordinates of the two points on the curve $y = 1 - x^2$ whose tangents pass through the point $(-1, 4)$.
Illustrate your answer by sketching a diagram.

16 Find the points of intersection of the curve $y = x(3 - x)$ with the lines:

(i) $y = x - 3$ (ii) $y = x$ (iii) $y = x + 1$

Illustrate your answers by sketching the curve.
Verify by differentiation that the line $y = x + 1$ is a tangent to the curve.

Questions 17–20: Find the equation of the normal to the curve at the given point. Where does the normal meet the curve again?

• **17** $y = x^2$, $(2, 4)$

18 $y = \frac{1}{2}x^2 - 1$, $(-2, 1)$

19 y

R **20** $y = x + 4/x$, $(1, 5)$

21 Find the value of k for which the curves $y = x^2$ and $y = k/x$ intersect at right angles (i.e. the tangent to one curve at their point of intersection is the normal to the other).

1.3:4 Higher derivatives

Notice that if $f(x)$ is a function of x, then so is $f'(x)$. So we can sensibly differentiate $f'(x)$ itself to obtain a new function which we call $f''(x)$, i.e.

$$\frac{d}{dx}\{f'(x)\} = f''(x) \qquad (7)$$

Similarly $\dfrac{d}{dx}\{f''(x)\} = f'''(x)$, etc.

Note: $f'(x)$ is the gradient of the curve $y = f(x)$ at a general point; in the same way $f''(x)$ is the gradient of the curve $y = f'(x)$ at a general point.

58   *Higher derivatives*

We write

$$\frac{d}{dx}\left(\frac{dy}{dx}\right) \quad \text{as} \quad \frac{d^2y}{dx^2}$$

$$\frac{d}{dx}\left(\frac{d^2y}{dx^2}\right) \quad \text{as} \quad \frac{d^3y}{dx^3} \qquad (8)$$

etc.

So if $y = f(x)$, $\frac{dy}{dx} = f'(x)$, $\frac{d^2y}{dx^2} = f''(x)$, $\frac{d^3y}{dx^3} = f'''(x)$, etc., e.g. if $y = 2x^4 - 5x^2$, $\frac{dy}{dx} = 8x^3 - 10x$, $\frac{d^2y}{dx^2} = 24x^2 - 10$, $\frac{d^3y}{dx^3} = 48x$.

$\frac{d^n y}{dx^n}$ is called the **differential coefficient** of y of order n. As before, $d^n y$ and dx^n do not have any meaning by themselves (in particular, the ns, $n = 1, 2, \ldots$, are not indices).

Exercise 1.3:4

1 Find $f'(x)$ and f'' when:

(i) $f(x) = x^5$
(ii) $f(x) = 1/x^2$
(iii) $f(x) = x^{4/3}$
→ (iv) $f(x) = x^4 - 3x^3 + 4x$
(v) $f(x) = 2x\sqrt{x} + 3\sqrt{x}$
(vi) $f(x) = x^2 - 1/\sqrt{x}$

2 If $y = 3x^5 - x^4 + 2x^2 + 8x - 20$, find $d^n y/dx^n$, $n = 1, 2, 3, 4, 5$. What can you say about $d^n y/dx^n$, $n \geq 6$?

3 Sketch graphs of $f(x)$, $f'(x)$, $f''(x)$ when:

• (i) $f(x) = 6x - x^2$
(ii) $f(x) = x^3$
→ (iii) $f(x) = x^3 - 3x^2$
(iv) $f(x) = x^4 - 3x^2$

: **4** Describe with the aid of sketches the behaviour of the graph of $f(x)$ when:

(i) $f'(x) > 0$ (ii) $f'(x) < 0$ (iii) $f'(x) = 0$
(iv) $f''(x) > 0$ (v) $f''(x) < 0$ (vi) $f''(x) = 0$
(vii) $f'(x) = 0$ and $f''(x) = 0$

Questions 5 and 6: Copy the given graph. Suppose that it is the graph of the function f. Sketch rough graphs of f' and f''.

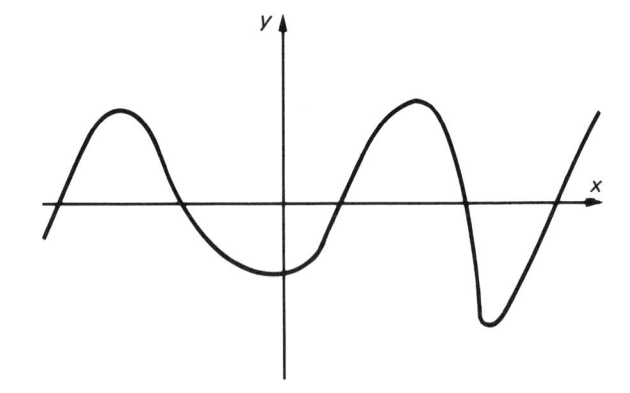

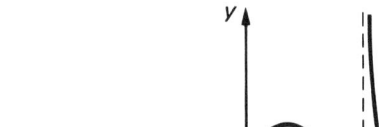

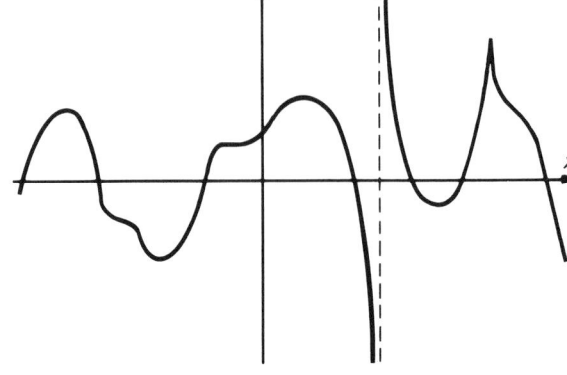

→ **5**

6

1.3:5 Stationary points and points of inflexion of a curve

A **stationary point** of the curve $y = f(x)$ is a point at which the tangent to the curve is parallel to the x-axis, i.e. the gradient of the curve is zero:

$$f'(x) = 0 \qquad (9)$$

In fig. 1.3:7, B, D and F are the stationary points of the curve.

To find the coordinates of the stationary points of the curve $y = f(x)$, we solve equation (9). For each solution $x = \lambda$, the point $P(\lambda, f(\lambda))$ is a stationary point.

Each value $f(\lambda)$ is called a **stationary value** of $f(x)$.

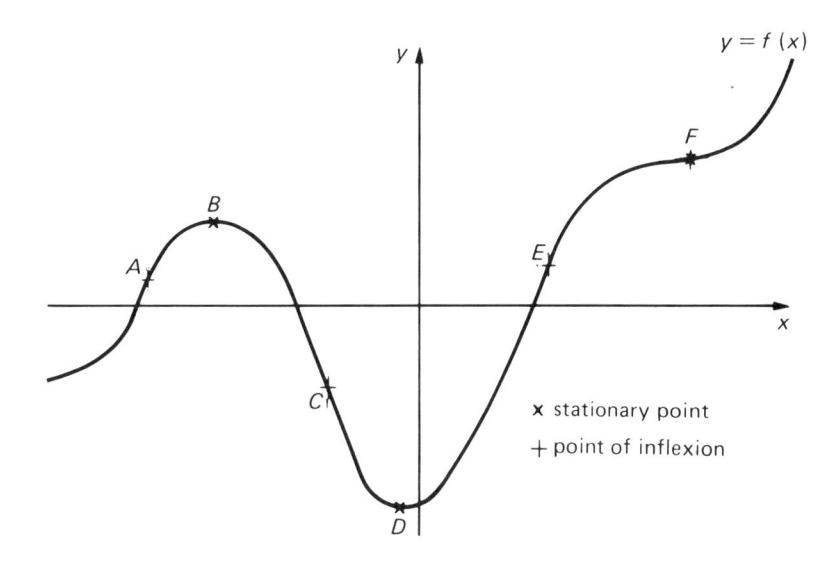

Fig. 1.3:7

Worked example 4 Find the stationary values of the function

$$f(x) = x^4 - 4x^3$$

$f'(x) = 4x^3 - 12x^2$. At a stationary point $f'(x) = 0$.
$f'(x) = 0 \Rightarrow 4x^3 - 12x^2 = 0 \Rightarrow 4x^2(x - 3) = 0 \Rightarrow x = 0$ or 3.
So the stationary values of $f(x)$ are $f(0) = 0$ and $f(3) = -27$.

Suppose the point P is a stationary point of the curve $y = f(x)$.
 (i) If the gradient of the curve is decreasing as the curve passes through P, then P is a **(local) maximum**.
 (ii) If the gradient of the curve is increasing as the curve passes through P, P is a **(local) minimum**.
 (iii) If the gradient of the curve on each side of P is of the same sign, P is a **point of inflexion**.

Local maxima and local minima are called **turning points** of the curve. In fig. 1.3:7, B and D are turning points—B is a local maximum and D is a local minimum. (Why are they called 'local' maxima and minima? See question 7.)

How can we tell whether a stationary point $P(\lambda, f(\lambda))$ is a maximum, a minimum or a point of inflexion?
 (i) If $f''(\lambda) < 0$, then, as the curve $y = f(x)$ passes through P, the gradient $f'(x)$ is decreasing (from $+$ to $-$); so P is a maximum (fig. 1.3:**8a**).
 (ii) If $f''(\lambda) > 0$, then, as the curve passes through P, $f'(x)$ is increasing (from $-$ to $+$); so P is a minimum (fig. 1.3:**8b**).

60 *Stationary points and points of inflexion of a curve*

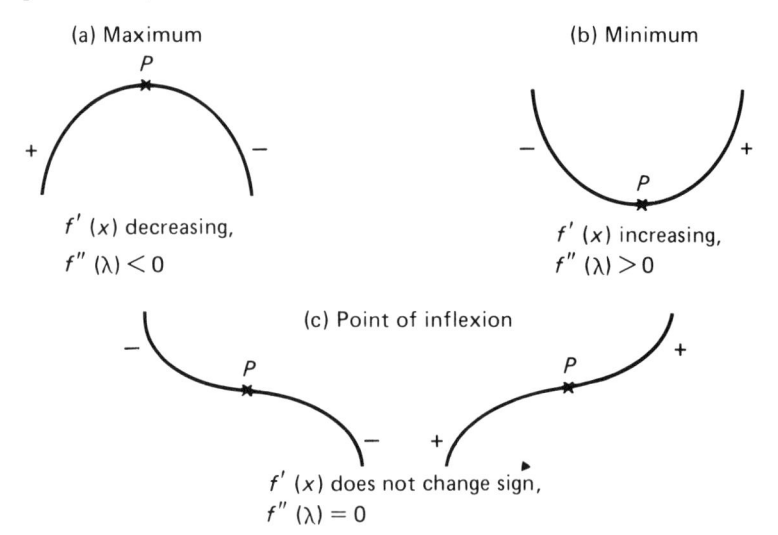

(iii) If $f''(\lambda) = 0$, then in general, $f'(x)$ has the same sign on both sides of P; so P is a point of inflexion (fig. 1.3:**8c**). (Why 'in general'? See question 8.)

Fig. 1.3:**8**

(a) Maximum

$f'(x)$ decreasing, $f''(\lambda) < 0$

(b) Minimum

$f'(x)$ increasing, $f''(\lambda) > 0$

(c) Point of inflexion

$f'(x)$ does not change sign, $f''(\lambda) = 0$

Worked example 5 Distinguish between the stationary points of the curve $y = x^4 - 4x^3$.

The stationary points of the curve are $(0, 0)$ and $(3, -27)$ (see worked example 4).

$$f''(x) = 12x^2 - 24x$$

$f''(0) = 0$, so $(0, 0)$ is a point of inflexion. $f''(3) = 36 > 0$, so $(3, -27)$ is a minimum. The curve is shown in fig. 1.3:**9**.

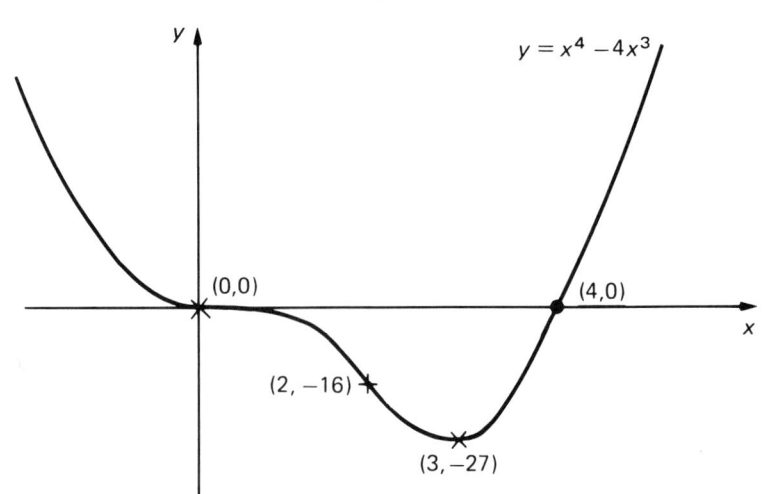

Fig. 1.3:**9**

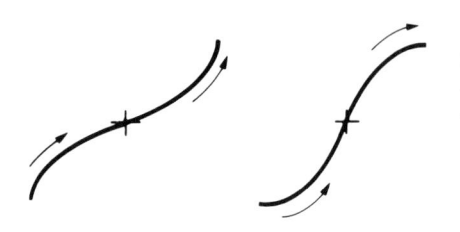

Fig. 1.3:**10**

A **point of inflexion** of the curve $y = f(x)$ is any point where the sense in which the curve is turning changes (from clockwise to anticlockwise or vice-versa). At a point of inflexion, the gradient $f'(x)$ has a stationary value (in fig. 1.3:**10**, $f'(x)$ is maximum at each point of inflexion); i.e.

$$f''(x) = 0 \qquad (10)$$

In fig. 1.3:**7**, A, C, E and F are the points of inflexion of the curve. Note that F is both a stationary point and a point of inflexion.

To find the coordinates of the points of inflexion of the curve $y = f(x)$, we solve equation (10). For each solution $x = \mu$, the point $Q(\mu, f(\mu))$ is, in general, a point of inflexion. (Why 'in general'? See question 8.)

Worked example 6 Find the points of inflexion of the curve $y = x^4 - 4x^3$.

$f''(x) = 12x^2 - 24x$. At a point of inflexion $f''(x) = 0$.
$f''(x) = 0 \Rightarrow 12x^2 - 24x = 0 \Rightarrow 12x(x-2) = 0 \Rightarrow x = 0$ or $x = 2$.
So the points of inflexion of the curve (fig. 1.3:9) are $(0, 0)$, which is (as we found in worked example 4) also a stationary point, and $(2, -16)$.

Exercise 1.3:5

1 Find the stationary value(s) of $f(x)$ when:

(i) $f(x) = x^2 - 6x$
(ii) $f(x) = 5 - 4x - x^2$
(iii) $f(x) = 4x - 3x^3$
(iv) $f(x) = 2x^3 - 3x^2 - 12x - 7$
(v) $f(x) = x + 1/x$
(vi) $f(x) = (x-1)^3$
(vii) $f(x) = x^4 - 4x^3 - 8x^2$

In each case write down the coordinates of the stationary points of the curve $y = f(x)$.

2 For each of the curves in question 1, find which of the stationary points are maxima, which minima and which points of inflexion.

3 Find the coordinates of the stationary points of the curve and distinguish between them. Hence sketch the curve.

(i) $y = x^3 - x^2$
(ii) $y = x^3 - 24x$
(iii) $y = x^3 - 5x^2 + 3x + 6$
R (iv) $y = x^3 - 6x^2 + 12x + 2$
(v) $y = x(6-x)^2$
(vi) $y = 3x^4 + 4x^3$

4 Consider the functions $f(x) = 3x^2 + 2x + 5$ and $g(x) = x^3 - 4x^2 - 3x + 6$. Show that there is one value of x for which $f(x)$ and $g(x)$ both have a stationary value.
Sketch on the same axes the graphs of $f(x)$ and $g(x)$.

5 Find the turning point of the curve $y = 2x + 27/x^2$ and show that it is a minimum.

6 (i) Find the turning point of the curve $y = (x^3 + 2)/x$.
(ii) Show that the gradient of the curve is negative for all negative values of x.
(iii) Sketch the curve.

7 Sketch the graphs of:
(i) a discontinuous function
(ii) a continuous function which has a 'maximum' value *less than* a 'minimum' value

(For this reason maxima and minima are called, strictly, **local maxima** and **local minima**.)
Find the stationary values of the function $f(x) = x^3 - 3x + 7$. What is the greatest possible value of $f(x)$? What is the least possible value?

8 Consider a function $g(x)$ for which there is a value of x, say a, such that $g'(a) = g''(a) = g'''(a) = 0$.
By sketching possible graphs of $g(x)$ and $g'(x)$ in the neighbourhood of $x = a$, show that the turning point $(a, g(a))$ is not necessarily a point of inflexion of the graph of $g(x)$.

This suggests that if $f'(a) = f''(a) = 0$, the turning point $(a, f(a))$ is a point of inflexion of the curve $y = f(x)$, *unless* $f'''(a) = 0$ also, in which case it may be a maximum, a minimum, or a point of inflexion. (To find which, consider the sign of $f'(x)$ on either side of $x = a$.)

Discuss the turning points of the curves:

(i) $y = x^4$ (ii) $y = 4x^5 - 5x^4$ (iii) $y = x^5$

Illustrate your answers by sketching the curves.

9 Find (all) the points of inflexion of the following curves:

(i) $y = x^3$
(ii) $y = x^3 - 6x^2 + 5$
(iii) $y = x^4 - 4x^3 - 18x^2 + 7x - 5$
(iv) $y = x^4 - 6x^3 + 12x^2 - 8x$

Are any of these points of inflexion also turning points?

10 Find the turning points and the points of inflexion, if any, of these curves and hence sketch them:

(i) (a) $y = 3x^4 - 4x^3$
(b) $y = 4x^3 - 3x^2 - 6x$

(ii) (a) $y = x + 9/x$
R (b) $y = x^2 + 16/x$

(The curves of (ii) both have an asymptote $x = 0$.)

11 The point $P(x, x+2)$ lies on the line $y = x + 2$; the point A has coordinates $(3, 1)$. Express AP^2 in terms of x. Hence find the position of P for which AP^2 is least.

Show that with *P* in this position,

(i) $AP = 2\sqrt{2}$ (ii) *AP* is perpendicular to the line.

12 Find the shortest distance from:

(i) the origin to the line $y = 2x - 1$
(ii) the point $(3, -1)$ to the line $y = 2x - 1$.

13 (i) Find the positions of the points *P* on the curve $y = x^2$ which are closest to the point $A(0, 1)$.
Verify that for each of these positions *AP* is perpendicular to the tangent to the curve at *P*. Illustrate your result on a sketch.

(ii) Find the shortest distance to the curve from the point $(6, 3)$.

• **14** A square sheet of cardboard of side 1.2 m has four equal square portions removed at the corners and the sides are then turned up to form an open rectangular box.
Show that when the depth of the box is *x* m, its volume $V = x(1.2 - 2x)^2$ m³. Hence show by differentiation that the maximum possible volume occurs when $x = 0.2$. What is the maximum possible volume?
Explain why the formula $V = x(1.2 - 2x)^2$ is meaningful only when $0 \leq x \leq 0.6$. Sketch a graph of *V* against *x* in this range.

→ **15** A square sheet of cardboard of side *a* m has four equal square sections removed at the corners and the sides are then turned up to form an open rectangular box. Show that when the box has maximum volume its depth is $\frac{1}{6}a$ m.

16 A rectangular sheet of metal 8 cm × 5 cm has four equal square sections removed at the corners and the sides are then turned up to form an open rectangular box. Find its maximum possible volume.

17 A farmer makes a rectangular sheepfold with 80 m of electric fence.

(i) Find the maximum possible area of the sheepfold.

(ii) Find the maximum possible area of the sheepfold if one side is provided by a long hedge already in existence.

18 The headmaster of a primary school is overenthusiastic about the enforcement of discipline. He wants to construct a rectangular enclosure in the playground in which pupils who have misbehaved will be displayed for ridicule. He buys 20 straight 3 m lengths of iron fencing, high enough to prevent escape. What is the largest possible area of the enclosure if:

(i) it is to be free-standing?

(ii) one of the sides is provided by a wall of the school?

He decides eventually to make a free-standing enclosure of the largest possible area, using the 20 lengths of railing (without restricting the shape). What shape should the enclosure be? (Use common sense rather than calculus.) One day, one of the wrongdoers decides to tunnel out through the tarmac, using a road drill which he has smuggled into school ...

19 An open box with a square base is to have a volume of $4a^3$ m³. What should the dimensions of the box be if the surface area is to be as small as possible?

R **20** A cylinder of height *x* fits exactly into a sphere of radius *a*. Show that its volume *V* is given by

$$V = \tfrac{1}{4}\pi x(4a^2 - x^2)$$

Show that the volume is maximum when $x = \tfrac{2}{3}a\sqrt{3}$ and that then the cylinder occupies $\tfrac{1}{3}\sqrt{3}$ of the volume of the sphere.

→ **21** A cylinder of height *x* fits exactly into a cone of height *h* and base radius *r*. Show that its volume *V* is given by

$$V = \pi r^2 x \left(1 - \frac{x}{h}\right)^2$$

Show that the maximum volume for the cylinder occurs when $x = \tfrac{1}{3}h$.
Show also that the maximum curved surface area for the cylinder occurs when $x = \tfrac{1}{2}h$.

22 Show that the surface area *A*, the volume *V* and the radius *r* of a cylinder are connected by the equation

$$A = \frac{2V}{r} + 2\pi r^2$$

(i) Find the minimum possible surface area of a cylinder if its volume is to be 100π cm³.

(ii) A cylindrical block of ice is melting at a rate proportional to its surface area. Find the ratio of length to radius if a cylindrical block of given volume is to melt as slowly as possible.

23 A car travelling at *v* km h⁻¹ uses fuel at a rate of $(6 + 0.0001\, v^3)$ l h⁻¹. Find the total amount of fuel used on a journey of 1000 km at a steady speed of *v* km h⁻¹.
Hence find:

(i) the value of *v* which gives the greatest economy of fuel consumption;
(ii) the amount of fuel then used.

Is the answer to (i) affected by the length of the journey?

24 The profit £x on the sale of a certain car and the time t h of its manufacture are related by the formula

$$x = 20\left\{200 - \frac{250}{t} - t^2\right\} \quad (t > 3)$$

Find the maximum possible profit.

25 The cost £$c(x)$ of producing an output of x units of a certain product is given by

$$c(x) = a + bx^2$$

(a and b constants, $a > 0$, $b > 0$). The selling price of each unit is £p. Write down the profit £$P(x)$ on the sale of x units.

Sketch an approximate graph of $P(x)$ against x. Find the value of x which ensures the maximum possible profit.

26 An object travels from the point (1, 4) directly to a point P on the x-axis and from P directly to the point (7, 8). Find the distance of P from the origin if the total distance travelled by the object is a minimum.

27 A fountain consists of several jets of water issuing from a nozzle O at ground level. Referred to axes Ox and Oy (horizontal and vertically upward), the equation of the path of one of the jets is

$$y = \tfrac{4}{3}x - \tfrac{4}{45}x^2 \quad (1 \text{ unit} = 1 \text{ m})$$

Find the greatest height above ground level of the water in this jet.

28 The cross-section of a slice of bread is in the shape of a rectangle surmounted by a semicircle. Find the ratio of the height of the rectangle to its width which ensures that the slice has the least possible amount of crust, for any given cross-section area.

1.3:6 Function of a function rule

To differentiate the composition of two functions f and g (see section 1.2:3) we proceed as follows.

> **Theorem 2:** If y is a function of u and u is itself a function of x, then
>
> $$\frac{dy}{dx} = \frac{dy}{du} \cdot \frac{du}{dx} \tag{11}$$
>
> Suppose $y = g(u)$ and $u = f(x)$, so that $y = g\{f(x)\}$; then
>
> $$\frac{dy}{dx} = g'(u) \cdot f'(x) = g'\{f(x)\} \cdot f'(x)$$
>
> i.e. $$\frac{d}{dx}\{g\{f(x)\}\} = g'\{f(x)\} \cdot f'(x) \tag{12}$$

Note that $g'(u)$ means 'the derivative of the function g *with respect to u*'.

Proof
$$\frac{dy}{du} \cdot \frac{du}{dx} = \left\{\lim_{\delta u \to 0} \frac{\delta y}{\delta u}\right\} \cdot \left\{\lim_{\delta x \to 0} \frac{\delta u}{\delta x}\right\}$$

$$= \lim_{\delta x \to 0} \frac{\delta y}{\delta u} \cdot \frac{\delta u}{\delta x}$$

$$= \lim_{\delta x \to 0} \frac{\delta y}{\delta x}$$

$$= \frac{dy}{dx} \quad \text{(but see question 5)}$$

You may have thought that this theorem didn't need proving, since

$$\frac{dy}{du} \cdot \frac{du}{dx} = \frac{dy}{\cancel{du}} \cdot \frac{\cancel{du}}{dx} = \frac{dy}{dx}$$

But remember: dy, du and dx do not mean anything by themselves; so the cancellation is an invalid step. Think rather of the theorem as a justification for writing

$$\lim_{\delta x \to 0} \frac{\delta y}{\delta x} \quad \text{as} \quad \frac{dy}{dx}.$$

Worked example 7 Find dy/dx when $y = \sqrt{(1+x^2)}$.

Let $u = 1 + x^2$; so $du/dx = 2x$.

$$y = \sqrt{u} = u^{1/2} \Rightarrow \frac{dy}{du} = \tfrac{1}{2} u^{-1/2} = \frac{1}{2\sqrt{u}}$$

By equation (11),

$$\frac{dy}{dx} = \frac{1}{2\sqrt{u}} \cdot 2x = \frac{x}{\sqrt{u}}$$

$$= \frac{x}{\sqrt{(1+x^2)}}$$

Worked example 8 A spherical balloon is being blown up so that its volume increases at a rate of 50 cm³ s⁻¹. When the radius is 12 cm, at what rate is it increasing?

If at time t the volume of the balloon is V and its radius is r, then $V = \tfrac{4}{3}\pi r^3$. Note that both V and r are functions of t. By equation (11),

$$\frac{dV}{dt} = \frac{dV}{dr} \cdot \frac{dr}{dt}$$

$$= 4\pi r^2 \cdot \frac{dr}{dt}$$

$dV/dt = 50$ cm³ s⁻¹, $r = 12$ cm; so

$$50 = 4\pi \cdot 12^2 \frac{dr}{dt} \Rightarrow \frac{dr}{dt} = \frac{50}{4\pi \cdot 12^2}$$

$$\approx 0.025 \text{ cm s}^{-1}$$

Exercise 1.3:6

⑦ • **1** Find dy/dx when:

(i) $y = (3x^2 + 1)^4$
(Let $u = 3x^2 + 1$. Show that $dy/du = 4u^3$ and find du/dx. Hence, using equation (11), find dy/dx.)

(ii) $y = \sqrt{(x^3 + 4)}$
(Let $u = x^3 + 4$. Show that $dy/du = 1/2\sqrt{u}$ and find du/dx. Hence, using equation (11), find dy/dx.)

(iv) $y = \dfrac{1}{3x^2 - 1}$

(v) $y = \dfrac{1}{(x^2 + 2)^3}$

(vi) $y = \dfrac{1}{(1 - \sqrt{x})^2}$

→ (vii) $y = \sqrt{(1 + x^2)}$
(viii) $y = \sqrt{(1 + 3x)}$
(ix) $y = \sqrt[3]{(1 - x^2)}$

(x) $y = \dfrac{1}{\sqrt{(1-x)}}$

(xi) $y = \dfrac{1}{\sqrt{(3 + 2x^3)}}$

(xii) $y = (\sqrt{x} + 1)^3$

⑦ **2** Find dy/dx when:

(i) $y = (1 - x^2)^6$
(ii) $y = (x^2 - x)^5$
→ (iii) $y = \left(x + \dfrac{1}{x}\right)^3$

⑦ **3** Differentiate with respect to x:
(i) $(ax+b)^n$ (ii) $(ax^2+b)^n$ (iii) $(a\sqrt{x}+b)^n$
where a, b and n are constants.

⑦ **4** Differentiate with respect to x:
(i) $(1+x^2)^2$ (ii) $(1+x^2)^3$
(a) by expanding the expression first;
(b) using the function of a function rule.
Check that your answers agree with each other.

⑦: **5** What assumptions are made in the proof of Theorem 2?

⑦ **6** The function of a function rule is sometimes called the **chain rule**. Why do you think this is?
Formulate a rule for differentiating with respect to x functions of the type $h[g\{f(x)\}]$—(see 1.2:3 Q50).
Differentiate $\sqrt{[x+\sqrt{(1+x)}]}$ with respect to x.

⑦ → **7** For
(i) $y = x^2 \ (x>0)$ (ii) $y = (x+1)^3$
(a) express x as a function of y;
(b) find $\dfrac{dx}{dy}$ $\left(\text{is } \dfrac{dx}{dy} = \dfrac{1}{dy/dx}?\right)$
(c) find $\dfrac{d^2x}{dy^2}$ $\left(\text{is } \dfrac{d^2x}{dy^2} = \dfrac{1}{d^2y/dx^2}?\right)$

⑦ **8** Find the maximum and minimum values of the function $y = x + \sqrt{(1-x^2)}$.

⑦ R **9** A man walks from one corner A of a square ploughed field $ABCD$ of side a m to the opposite corner C. He first walks x m along the edge AB of the field and then walks directly towards C. He walks at 2.5 m s^{-1} if he keeps to the edge of the field and at 1.5 m s^{-1} when he walks on the ploughed land.
Find in terms of a and x how long he takes.
Show that if he takes the least possible time, $x = \tfrac{1}{4}a$.

⑦ **10** A canal runs east–west between straight banks. A woman is standing on the south bank of the canal. She sees her daughter floundering around in the canal, 60 m east, 30 m north of where she is standing. If the woman can run at $\sqrt{10} \text{ m s}^{-1}$ and swim at 1 m s^{-1}, find the least time in which she can reach her daughter.

⑦ **11** A, B, C and D are four towns situated at the corners of a flat square desert of side $25\sqrt{2}$ km. An explorer P, who is searching for his companion Q, sets out from A with a pair of binoculars and walks towards C with speed 4 km h^{-1}. At the same moment Q, who is at B, sets out with a pair of binoculars and a beautifully furled umbrella to search for P, and walks towards D with speed 3 km h^{-1}.
Show that their distance apart after t hours ($t \leq 12.5$) is $5\sqrt{(50-14t+t^2)}$ km.
They will spot each other if their distance apart is less than 4 km. Is there a happy ending?

⑧ → **12** A spherical balloon is blown up so that its radius increases at a constant rate of 0.01 cm s^{-1}. Find the rate of increase in the volume when the radius is 5 cm.

⑧ **13** A spherical balloon is blown up so that its volume increases at a constant rate of $0.05 \text{ m}^3 \text{ s}^{-1}$. Find the rate of increase of the radius when the volume is 0.008 m^3.

⑧ **14** A spherical balloon is blown up so that its radius increases at a constant rate of 0.4 cm s^{-1}. Find the rate of increase of the surface area when the radius is 20 cm.

⑧ **15** A spherical balloon is blown up so that its volume increases at a constant rate of $8 \text{ cm}^3 \text{ s}^{-1}$. Find the rate of increase of the surface area when the radius is 10 cm.

⑧ **16** A spherical balloon is blown up so that its volume increases at a constant rate. Show that the rate of increase of its surface area is inversely proportional to its radius.

⑧ **17** The great chef Chabrol makes an ice-sculpture in the shape of the Taj Mahal. Each part of the sculpture melts at a rate proportional to its surface area. Assuming that the dome is approximately a sphere, show that the rate of decrease in its radius is constant.
If the dome is half-melted (i.e. its volume is reduced by half) after 2 hours, find the time at which it is completely melted.

⑧ **18** A circular ripple spreads across a pond. If the radius of the ripple increases at 0.02 m s^{-1}, at what rate is the area increasing when the radius is 3 m?
If the area continues to increase at this rate, at what rate will the radius be increasing when it is 5 m?

⑧ **19** (i) The side of a cube is increasing at 0.1 m s^{-1}. Find the rate of increase of the volume when the length of a side is 2 m.

(ii) The surface area of a cube is increasing at $0.1 \text{ m}^2 \text{ s}^{-1}$. Find the rate of increase of the volume when the length of a side is 2 m.

⑧ **20** A street lamp is at a height of 5 m. A man of height 2 m is running away from it. Find the length of his shadow when he is at a distance x m from the lamp.
He runs at 4 m s^{-1}. Find the rate of increase in the length of his shadow.

⑧ **21** The top of a ladder 5 m long is at a height 4 m up a wall and is slipping down the wall at a rate of 0.1 m s^{-1}. At what rate is the foot of the ladder sliding along the ground? (Hint: consider the ladder in a general position when the top is at a height x m and the foot is at a distance y m from the wall; find y in terms of x and hence find dy/dt in terms of dx/dt.)

66   *Product and quotient rules*

⑧ **22** A ladder 4 m long rests against a wall. The top of the ladder is at a height 2 m up the wall. A builder is pushing the foot of the ladder towards the wall at a rate of 0.05 m s^{-1}. At what rate is the top of the ladder sliding up the wall?

⑧ **23** A flask in the shape of a cone of height 20 cm and radius 8 cm is held vertex downwards. Show that when the depth of water in the flask is x cm, the volume of the water is $\frac{4}{75}\pi x^3 \text{ cm}^3$. (Hint: use similar triangles to find the radius of the surface of the water.)
   The water leaks out from the vertex at the rate of $2 \text{ cm}^3 \text{ s}^{-1}$. Find the rate of change in the depth of the water when the depth is 10 cm.

⑧ → **24** A flask in the shape of a cone of height 15 cm and radius 10 cm is held vertex downwards. Water flows into the flask at a rate of $4 \text{ cm}^3 \text{ s}^{-1}$. Find the rate of change of the depth of the water when

   (i) the depth is 10 cm;
   (ii) the flask is half-full (by volume) (hint: find the depth of the water);
   (iii) the flask is full.

⑧ **25** The body of a cocktail glass is in the shape of a cone of height 4 cm and radius 3 cm. The barman at the Ritz is pouring an 'Exocet', one of the most dangerous cocktails ever invented, into this glass at a rate of 15 cm s^{-1}. Find the rate of change of the depth of the liquid in the glass when the glass is:

   (i) half-full (by volume);
   (ii) full. (Assume that the barman is distracted by a beautiful young man, and continues pouring at the same rate until after the glass is full.)

⑧ **26** Coal falls from a chute at a rate of $0.5 \text{ m}^3 \text{ s}^{-1}$, making a conical pile of semi-vertical angle 60°. Find the rate of increase in the depth of the pile when the area of the base of the pile is:

   (i) 5 m^2 (ii) 10 m^2

⑧ **27** A device for watering houseplants while the owner is away consists of a conical funnel of height 30 cm and semi-vertical angle 30° with an outlet at the vertex. The funnel is suspended vertex downwards over the plant, the outlet is shut and the funnel is filled with water. The outlet is then opened once every 24 hours for 10 seconds by a time-switch. When the outlet is open the water streams out at a rate of $30\pi \text{ cm}^3 \text{ s}^{-1}$.
   Show that when the funnel is half-empty (by volume) the rate of decrease in the height of the water is approximately 0.02 cm s^{-1}.
   For how long are the plants watered?

⑧ **28** A flask in the shape of a cone is fixed with its vertex pointing downwards. The cone is filling with liquid at a rate which at any instant is proportional to the surface area of the liquid at that instant. Show that the rate of increase in the depth of the liquid is constant.

Questions 29 and 30: When the depth of water in an hemispherical bowl of radius r is x, the volume of the water is $\frac{1}{3}\pi x^2(3r-x)$.

⑧ **29** A hemispherical bowl of radius 12 cm is initially full of water. The water runs out of a hole in the bottom of the bowl at a rate of $96\pi \text{ cm}^3 \text{ s}^{-1}$. When the depth of the water is x cm, show that the depth is decreasing at a rate of $48/x(24-x) \text{ cm s}^{-1}$.
   Find the rate at which the depth is decreasing when:

   (i) the bowl is full;
   (ii) the depth is 6 cm.

⑧ **30** A hemispherical bowl of radius a cm is being filled with water at a rate of $\frac{1}{12}\pi a^3 \text{ cm}^3 \text{ s}^{-1}$. Find the rate at which the level is rising when the depth of water is $\frac{1}{3}a$ cm.

1.3:7 Product and quotient rules

Theorem 3:   If $u = uv$, where u and v are functions of x, then

$$\frac{dy}{dx} = u\frac{dv}{dx} + v\frac{du}{dx} \tag{13}$$

Suppose that $u = f(x)$, $v = g(x)$; then

$$\frac{dy}{dx} = f(x)g'(x) + g(x)f'(x)$$

i.e. $$\frac{d}{dx}\{f(x)g(x)\} = f(x)g'(x) + g(x)f'(x) \tag{14}$$

Proof
See question 7.

Worked example 9 Find dy/dx when $y = x^2\sqrt{(1+x^2)}$.

Let $u = x^2$, so $du/dx = 2x$; and $v = \sqrt{(1+x^2)}$, so $dv/dx = x/\sqrt{(1+x^2)}$ (see worked example 7). Then $y = uv$, so by equation (13):

$$\frac{dy}{dx} = x^2 \cdot \frac{x}{\sqrt{(1+x^2)}} + \sqrt{(1+x^2)} \cdot 2x$$

$$= \frac{x^2 + 2x(1+x^2)}{\sqrt{(1+x^2)}} = \frac{2x + 3x^3}{\sqrt{(1+x^2)}}$$

Theorem 4: If $y = u/v$ where u and v are functions of x, then

$$\frac{dy}{dx} = \frac{v\dfrac{du}{dx} - u\dfrac{dv}{dx}}{v^2} \tag{15}$$

i.e.

$$\frac{d}{dx}\left\{\frac{f(x)}{g(x)}\right\} = \frac{g(x)f'(x) - f(x)g'(x)}{\{g(x)\}^2} \tag{16}$$

Proof
See question 18.

Worked example 10 Find dy/dx when $y = x^3/1+x^2$.

Let $u = x^3$, so $du/dx = 3x^2$; and let $v = 1+x^2$, so $dv/dx = 2x$; then $y = u/v$, so by equation (15):

$$\frac{dy}{dx} = \frac{3x^2(1+x^2) - x^3 \cdot 2x}{(1+x^2)^2}$$

$$= \frac{x^2(3+x^2)}{(1+x^2)^2}$$

Exercise 1.3:7

⑨ **1** Differentiate with respect to x, simplifying your answers carefully:

(i) $x(1+x)^3$
(ii) $x^2(3x+1)^4$
(iii) $(x+1)^2(x-2)^3$
(iv) $(x+1)^2(x^2-5)$
→ (v) $x\sqrt{(2x^3+1)}$
(vi) $\dfrac{x^2}{\sqrt{(1+x^2)}}$
(vii) $\dfrac{x-1}{x+1}$
(viii) $\dfrac{\sqrt{x}}{1-x^2}$

⑨ **2** Differentiate with respect to x:

(i) $x^2(1+x^2)$ (ii) $x^2(1+x^2)^3$

(a) by first multiplying out;
(b) using the product rule.
Check that your answers agree.

⑨ **3** Write down a rule for finding dy/dx when $y = uvw$ where u, v and w are functions of x.

Differentiate $\dfrac{x\sqrt{(1+x)}}{1+2x}$ with respect to x.

⑨ **4** Find the equations of the tangent and normal to the curve $y = x\sqrt{(1+x)}$ at the point $(3, 6)$.

⑨ **5** Find the coordinates of the turning points of the curves: (i) $y = x(x-2)^2$ (ii) $y = (x-1)^2(x+3)^2$. Sketch these curves.

⑨ **6** Find the maximum and minimum values of the functions:
→ (i) $f(x) = x^2\sqrt{(1-x)}$
R (ii) $f(x) = (x^2-1)\sqrt{(1+x)}$

- **7** Using 1.3:1 definition (1a) of the derivative, write down expressions for

  (i) $\dfrac{d}{dx}\{f(x)g(x)\}$ (ii) $f(x)g'(x)+g(x)f'(x)$

  and show that they are equal.

  Describe the assumptions you have made in your proof.

(10) **8** Differentiate with respect to x, simplifying your answers carefully:

→ (i) $\dfrac{x}{2x^2+1}$ (ii) $\dfrac{x-1}{x+1}$

(iii) $\dfrac{1-x^2}{1+x^2}$ (iv) $\dfrac{x}{(3x+1)^2}$

(v) $\dfrac{(x+1)^2}{(x-4)^2}$ (vi) $\dfrac{1+\sqrt{x}}{1-\sqrt{x}}$

→ (vii) $\sqrt{\left(\dfrac{1-x}{1+x}\right)}$ (viii) $\sqrt{\left(\dfrac{x}{2-x}\right)}$

(ix) $\dfrac{x}{\sqrt{(a^2-x^2)}}$ (a is a constant)

(10) **9** Differentiate: (i) $\dfrac{x}{x^2+1}$ (ii) $\dfrac{x+1}{2x-1}$

(a) using the quotient rule;
(b) by expressing the functions as products, and using the product rule.

(10) → **10** If $y = x/\sqrt{(1+x^2)}$, show that

$$(1+x^2)\dfrac{d^2y}{dx^2}+3x\dfrac{dy}{dx}=0$$

(10) **R 11** If $y = x^2/(1+x^2)$ show that:

(i) $(1+x^2)\dfrac{dy}{dx}+2xy=2x$

(ii) $(1+x^2)\dfrac{d^2y}{dx^2}+4x\dfrac{dy}{dx}+2y=2$

(10) **12** Show that $\dfrac{d}{dx}\left\{\dfrac{1}{1-x}\right\} = \dfrac{d}{dx}\left\{\dfrac{x}{1-x}\right\}$

Explain why.

(10) **13** Find the equations of the tangent and normal to the curve $y = x^3/(5x-2)$ at the point $(2, 1)$.

(10) **14** Find the equations of the tangent and normal to the curve $y = (2x-7)/(x+1)$ at the point $(2, -1)$.

Show that the gradient of the curve is always positive.

Sketch the curve, showing its asymptotes (see 1.2:6 WE13).

(10) **15** If $f(x) = x/(1+x^2)$, find $f'(x)$.

Find the maximum and minimum values of $f(x)$.

Show that as $x \to \pm\infty$, $f(x) \to 0$.

Sketch the graph of $f(x)$.

(10) **16** Find the coordinates of the stationary points of the curves:

(i) $y = \dfrac{x^2}{x-1}$ (ii) $y = \dfrac{(x-3)^2}{(x+2)^2}$ (iii) $y = \dfrac{x^3}{1-x^2}$

Explain why in part (ii) $x = -2$ does not give a stationary point.

(10) **17** The rate $P(I)$ of photosynthesis of a plant is given by $P(I) = \dfrac{I}{a+bI}$ $I \geqslant 0$; a, b constants where I is the light intensity.

(i) Find $P'(I)$ (which is called the photochemical efficiency of the plant).
(ii) Find $P'(0)$ and $\lim_{I\to\infty} P'(I)$.
(iii) Show that $P'(I) = \dfrac{1}{a}\{1-bP(I)\}^2$.

Sketch a graph of $P'(I)$ against $P(I)$ when $a = b = 1$.

1.3:8 Implicit function rule

We would like a rule for differentiating implicit functions such as

$$xy^2 + y^{1/2} - 5x = 0$$

(see section 1.2:8). In particular, we would like to be able to differentiate a function of y with respect to x.

We use the function of a function rule (1.3:6 equation (11)):

$$\dfrac{d}{dx}g(y) = \dfrac{dg}{dy} \cdot \dfrac{dy}{dx} \qquad (17)$$

e.g. $\dfrac{d}{dx}(y^3) = 3y^2 \dfrac{dy}{dx}$

Worked example 11 Find dy/dx when $2x^2 - 3y^2 = 2x$.

$$2x^2 - 3y^2 = 2x \Rightarrow 4x - 6y\frac{dy}{dx} = 2 \quad \text{(by equation (17))}$$

So
$$\frac{dy}{dx} = \frac{4x - 2}{6y} = \frac{2x - 1}{3y}$$

Note that dy/dx is a function of x and y.

Exercise 1.3:8

1 Find dy/dx in terms of x and y when:

(i) $x^2 + y^2 = 4$
(ii) $4x^2 + 9y^2 = 36x$
(iii) $x^3 + y^3 = 5y$
(iv) $x^3 - y^3 - 4x^2 + 3y = 11x + 4$
(v) $y + \sqrt{y} = x$

2 Use the product rule to differentiate $x^2 y^3$ with respect to x.
Hence find dy/dx when $x^2 y^3 + y^2 = 4$.

3 Find dy/dx when:

(i) $x\sqrt{y} = 6$
(ii) $3x^2 - 4x^2 y = 7$
(iii) $x^3 + xy^2 = 5xy$
(iv) $3(x-y)^2 = 2xy + 1$.

4 Calculate the gradients of the following curves at the given points:

(i) $x^2 - y^2 = 3$; $(2, 1)$
(ii) $x^2 y = 2$; $(-1, 2)$
(iii) $x^2 + 3xy + 2y^2 = 5$; $(1, -4)$
(iv) $x^3 + 3y^3 = 3$; $(3, -2)$

5 Show that the curve $x^3 + y^3 = 2xy$ passes through the point $P(1, 1)$. Find its gradient at P, and hence find the equation of the tangent to the curve at P.

6 Find the coordinates of the two points on the curve $x^2 - y^2 = 7$ at which $x = 4$. Find the equations of the tangents to the curve at these points.

7 Find the point of intersection of the tangents to the curve $2x^2 + 3y^2 = 14$ at the points on the curve where $x = 1$.

8 (i) Find the equations of the tangents to the curve $3x^2 + y^2 = 4y$ at the points on the curve where $x = 1$.

(ii) At what points on the curve is the tangent parallel to the x-axis?

9 Find the equation of the tangent to the curve $y^2 = 4x$ which is parallel to the line $y = 3x + 5$.
What are the coordinates of the point of contact? Find the equation of the normal at this point.

10 If $x^2 + y^2 = y$, show that $\dfrac{dy}{dx} = \dfrac{2x}{1 - 2y}$ \hfill (18)

Using the quotient rule; show that

$$\frac{d^2 y}{dx^2} = \frac{2(1 - 2y) - 4x\dfrac{dy}{dx}}{(1 - 2y)^2}$$

and hence, using equation (18), find $d^2 y/dx^2$ in terms of x and y.
Is there an alternative way of finding $d^2 y/dx^2$?

11 Find dy/dx and $d^2 y/dx^2$ in terms of x and y when:

(i) $x^2 - y^2 = 1$
(ii) $3x^2 - 2y^2 = 2x$
R (iii) $x^3 - y^3 = y$

12 (i) If $x^2 + y^2 = 10y$, find dy/dx and $d^2 y/dx^2$ at the point $(3, 1)$.

(ii) $x^2 + xy - y^2 = 1$, find dy/dx and $d^2 y/dx^2$ at the point $(1, 1)$.

13 Show that the curve $x^3 + y^3 = 3xy$ has a turning point at $(2^{1/3}, 2^{2/3})$.
Find whether this gives a maximum or minimum value of y.

14 (i) If $y = x^{m/n}$ (m and n positive integers) find dy/dx in terms of x, by writing $y^n = x^m$ and using the implicit function rule.

(ii) If $y = x^{-m/n}$, find dy/dx in terms of x, by writing $x^m y^n = 1$ and using the implicit function rule.
(Note: This proves that $d/dx (x^\alpha) = \alpha x^{\alpha - 1}$ for all $\alpha \in \mathbb{Q}$.)

1.3:9 Small changes

Theorem 5: Suppose that y is a function of x. Then if we make a small change δx in x, the corresponding change δy in y is given by:

$$\delta y \approx \frac{dy}{dx}\delta x \qquad (19)$$

Proof From fig. 1.3:**11**:

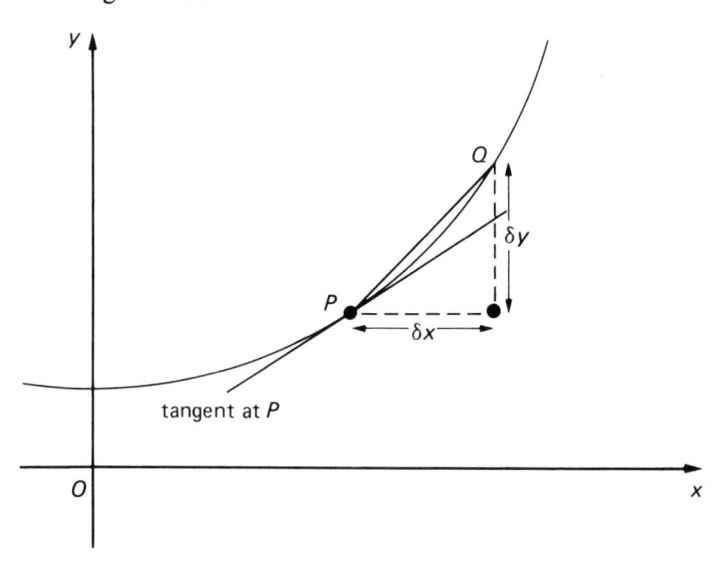

Fig. 1.3:**11**

$\dfrac{\delta y}{\delta x}$ = gradient of $PQ \approx$ gradient of tangent at $P = \dfrac{dy}{dx}$. So $\delta y \approx \dfrac{dy}{dx}\delta x$.

Worked example 12 Find an approximation for $\sqrt{(9.01)}$.

Let $y = \sqrt{x}$; so $\dfrac{dy}{dx} = \dfrac{1}{2\sqrt{x}}$. When $y = 3$, $x = 3$, $\dfrac{dy}{dx} = \dfrac{1}{6}$.

Suppose that we now make a small change 0.01 in x, so that $x = 9.01$. By approximation (19), δy, the corresponding change in y, is given by:

$$\delta y \approx \frac{dy}{dx} \cdot \delta x = \frac{1}{6} \cdot 0.01 \approx 0.0017$$

i.e. when $x = 9.01$, $y = 3.0017$ so $\sqrt{(9.01)} \approx 3.0017$

Worked example 13 An error of 3% is made in measuring the radius r of a sphere. What is the resulting percentage error in the calculation of its volume V?

$$V = \tfrac{4}{3}\pi r^3 \Rightarrow \frac{dV}{dr} = 4\pi r^2$$

Error in measuring r is 3%, i.e. 3% of the value of r; so $\delta r = \tfrac{3}{100}r$

By approximation (19), the resulting error in calculating V, is

$$\frac{\delta V}{V} \approx \frac{\dfrac{dV}{dr} \cdot \delta r}{V} = \frac{4\pi r^2 \cdot \dfrac{3r}{100}}{\tfrac{4}{3}\pi r^3} = \frac{9}{100} = 9\%$$

Exercise 1.3:9

(12) **1** Show that if $y = \sqrt{x}$, then $y = 2$ and $dy/dx = \frac{1}{4}$ when $x = 4$.

Find approximate values for:

(i) $\sqrt{(4.01)}$
(ii) $\sqrt{(3.99)}$

(12) **2** Find approximate values for:

(i) $\sqrt{(16.02)}$
→ (ii) $\sqrt{(24.9)}$
(iii) $\sqrt[3]{(8.01)}$
(iv) $\sqrt[3]{213}$
(v) $33^{3/5}$

(12) **3** If $y = 4x^3$ find approximately the change in y caused by a change of 0.01 in x when:

(i) $x = -1$ (ii) $x = 0$ (iii) $x = 1$ (iv) $x = 2$
(v) $x = 10$

(12) → **4** A spherical balloon is being blown up. Calculate the approximate increase in its surface area when the radius increases from:

(i) 10 cm to 10.1 cm
(ii) 20 cm to 20.2 cm

(12) R **5** A hemispherical bowl of radius 20 cm contains water to a depth of 10 cm. Find the decrease in the depth of the water if 100 cm³ of the water evaporates. (The volume of a cup of height x of a sphere of radius r is $\frac{1}{3}\pi x^2(3r-x)$.)

(12) **6** Calculate the approximate percentage change in y caused by a 3% change in x when

(i) $y = x^2$
(ii) $y = \sqrt{x}$
(iii) $y = 1/\sqrt{x}$
(iv) $y = 1/x^2$

(13) → **7** Calculate the approximate percentage change in y caused by a $p\%$ change in x when:

(i) $y = x^\alpha$
(ii) $y = kx^\alpha$

(13) → **8** An error of 2% is made in measuring the radius of a sphere. What is the resulting percentage error in the calculation of its surface area?

(13) R **9** The period T of a pendulum of length l is given by

$$T = 2\pi \sqrt{\left(\frac{l}{g}\right)}$$

where g is the acceleration due to gravity.

(i) Find the percentage increase in the period caused by a 1% increase in the length of a pendulum.

(ii) Find the percentage increase in the length of a pendulum which would cause a 5% increase in the period.

(13) **10** The time T a planet takes to revolve about the Sun and the average distance r from the planet to the Sun are related by the equation $T = kr^{3/2}$. If the Earth's distance from the Sun were increased by 1%, how much longer would a year become?

(13) **11** If the volume of a soap bubble increases by 3%, find the approximate percentage increase in its radius.

(13) **12** A cube of metal is heated, and each side expands by 2%. Find the percentage increases in its surface area and its volume.

(13) **13** When a wire is stretched, the volume remains constant. If the length is increased by $p\%$, find the percentage decrease in the wire's diameter.

(13) → **14** For a certain gas at a constant temperature, the pressure P and the volume V obey the law

$$PV = k \quad (k \text{ a constant})$$

Find:

(i) the percentage change in pressure due to a 2% increase in volume;

(ii) the percentage change in volume due to a 2% increase in pressure

(13) **15** $y = 2 + k/x$ (k is a constant). Find the percentage error in y due to a 3% error in x when:

(i) $y = 0.5$ (ii) $y = 1$ (iii) $y = 1.5$

16 A group of sociologists has made a survey of unemployment in a certain London borough.

(i) The group claims that the number U of unemployed people in the borough at any time is related to the rate r of inflation by the formula

$$U^3 = k/r^2 \quad (k \text{ a constant})$$

Using this formula, find the percentage change in U corresponding to a 6% increase in r.

(ii) The group also claims that the number C of burglaries committed per week in the borough is related to U by the formula

$$C = 100 + k'U^2 \quad (k' \text{ a constant})$$

Find the percentage increase in C corresponding to a 3% decrease in r when $C = 150$.

Miscellaneous exercise 1

1 Solve the equations:
  (i) $9 \log_8 x = 6 + 8 \log_x 8$
  (ii) $(\frac{1}{4})^x = 8^{1-2x}$
  (iii) $2^{2x+3} - 33(2^x) + 4 = 0$

2 (i) Solve the the equations:
    (a) $\sqrt{x} + \sqrt{(2x-2)} = 7$
    (b) $\log_2 x + \log_3 x = 1$
  (ii) Given that $\log 4 \approx 0.602\,06$, $\log 6 \approx 0.778\,15$, find, without using a calculator,
    (a) $\log 12$; (b) $\log 25$; (c) $\log 30$.
  Give your answers to 4 sig. figs.

3 The functions $f: \mathbb{R}^+ \to \mathbb{R}$ and $g: \mathbb{R}^+ \to \mathbb{R}$ are defined by $f(x) = 4x^2 - 1$, $g(x) = \sqrt{x}$. Find $(f \circ g)(x)$ and $(g \circ f)(x)$. Sketch on the same axes, for $x \in \mathbb{R}^+$, the graphs of $y = f(x)$ and $y = g(x)$. Solve the equations:
  (i) $f(x) = (f \circ g)(x)$ (ii) $f(x) = (g \circ f)(x)$

4 The function $f: [-1, 3] \to \mathbb{R}$ is defined by
$$f(x) = \begin{cases} 3x & -1 \leq x \leq 1 \\ 4 - x^2 & 1 < x \leq 3 \end{cases}$$
Sketch a graph of f and find the range of f. Explain why f has no inverse. Show that if f is restricted to the domain $(1, 3]$, f does have an inverse, and find it.

5 The functions $f: [0, 1] \to \mathbb{R}$ and $g: \mathbb{R}^+ \to \mathbb{R}$ are defined by $f(x) = 1/(1+x^2)$, $g(x) = (2+x)/x$. Find $(g \circ f)(x)$. What is the domain of $g \circ f$? What is the range of $g \circ f$? Sketch the graph of $g \circ f$. Find its inverse.

6 The functions f and g are defined by
$$f(x) = \frac{x}{x-1}, \quad g(x) = x^2$$
  (i) Find $(f \circ g)(x)$ and $(g \circ f)(x)$. Which of the functions $f, g, f \circ g$ and $g \circ f$ are even?
  Find $\lim_{x \to \infty} f(x)$ and $\lim_{x \to \infty} (f \circ g)(x)$.
  (ii) Find $f^{-1}(x)$. Explain why the function $f \circ g$ has no inverse.

7 D is the set $\{x: x \in \mathbb{R}, x \neq 3\}$. The function $f: D \to \mathbb{R}$ is defined by $f(x) = (2x-1)/(x-3)$. Sketch the graph of f. Find:
  (i) $(f \circ f)(x)$ (ii) $f^{-1}(x)$
What is the domain of the function f^{-1}? Sketch its graph.

8 Given that $f(x) = 2x - 2/x - 2$,
  (i) find $(f \circ f)(x)$;
  (ii) show that the function f is its own inverse;
  (iii) sketch on the same axes the graphs of $y = f(x)$ and $y = x$; what is the relationship between the two graphs?
  (iv) solve the equation $x = f(x)$.

9 The function $f: [-6, 9) \to \mathbb{R}$ is periodic with period 5, and
$$f(x) = \begin{cases} -x & -1 \leq x < 0 \\ \frac{1}{2}x^2 & 0 \leq x < 2 \\ 4/x & 2 \leq x < 4 \end{cases}$$
Sketch a graph of f. Is f continuous? Find:
  (i) $f'(3)$ (ii) $f'(\frac{9}{2})$ (iii) $f'(-4)$

10 The function $f: \mathbb{R} \to \mathbb{R}$ is periodic with period 4, and
$$f(x) = \begin{cases} x^2 & 0 \leq x < 1 \\ 2-x & 1 \leq x < 2 \end{cases}$$
Sketch the graph of $f(x)$ for $-4 \leq x \leq 6$, given that f is:
  (i) an even function;
  (ii) an odd function;
  (iii) $f(x) = 0, -2 \leq x < 0$.

11 The function $f: \mathbb{R} \to \mathbb{R}$ is defined by
$$f(x) = \begin{cases} \frac{1}{4}x^3 & x \leq 2 \\ 3x - 4 & x > 2 \end{cases}$$
Is f continuous? Is f' continuous? Is f'' continuous?

12 (i) Sketch the curve $y = x(4 - x^2)$. Find the coordinates of the points of intersection of the curve and the x-axis. Hence solve the inequality $x(4 - x^2) > 0$.
  (ii) Sketch on the same axes the curve $y = x(4-x)$ and the line $y = 3x - 2$. Find the coordinates of the points of intersection of the curve and the line. Hence solve the inequality
$$x(4-x) > 3x - 2$$

13 (i) Find dy/dx when:
    (a) $y^2(1+2x)/(1-2x)$; (b) $x^2y^3 - 2xy^2 = 3y$.
  (ii) If $x^2 - 2y^2 = 2x$, find:
    (a) dy/dx; (b) d^2y/dx^2
  at the point $(4, 2)$.

14 (i) Find the coordinates of the turning points of the curve $y = x^3 - 3x$. Sketch the curve. Show that the tangent to the curve at the origin does not meet the curve again.

(ii) Find the coordinates of the turning points of the curve $y = x + 4/x$. Sketch the curve. Show that the normal to the curve at the point $(1, 5)$ meets the curve again at the point $(6, 6\frac{2}{3})$.

15 Find the turning points and the points of inflexion of the curve $y = x^3 - 6x^2 + 9x$. Sketch the curve.

The curve meets the line $y = x$ at O, A and B, where A is between O and B. Find the coordinates of A and B. Find the area of the triangle bounded by the line, the x-axis, and the tangent to the curve at B.

16 A cylindrical soft drinks can, closed at both ends, is to contain $\frac{1}{4}$ litre of a fluid which rots the teeth, fouls the breath and diseases the mind. If the surface area of the can is, for reasons of economy, to be as small as possible, find its height and base radius. (1 litre = 1000 cm^3.)

17 Air is pumped into a spherical balloon at the rate of 0.05 m^3 s^{-1}. Find the rate of increase of the surface area of the balloon when its radius is 0.2 m.

18 (i) Air is pumped into a spherical balloon at a rate which is proportional to its surface area. Show that its radius is increasing at a constant rate.

(ii) The top of a ladder of length l m is sliding down a wall at a rate of r m s^{-1}. Find the rate at which the bottom of the ladder is sliding away from the wall when the top of the ladder is b m from the ground.

19 A flask in the shape of a cone of semi-vertical angle 30° is suspended with its axis vertical and its vertex downwards. Water drips into the flask at the rate of 0.15 cm^3 s^{-1}. Find the rate of increase in the height h cm of the water when $h = 2$.

20 The attractive force F between two objects of masses m_1 and m_2, when they are a distance r apart, is given by

$$F = \frac{Gm_1m_2}{r^2}$$

(G is a constant.) Calculate the percentage change in F caused by a $\frac{1}{2}\%$ change in r.

21 The demand D for a certain product and its price P are related by the formula $D^2 = k/p^3$, where k is a constant. Find the percentage decrease in the demand for the product if its price increases by 2%.

22 The happiness H of a man is a function of the volume V of Guinness that he has drunk during the day:

$$H = 12V - V^2 - 20, \ V \geq 0$$

where V is measured in pints, H in IEU (international ecstasy units).

(i) What is the most sensible amount of Guinness for him to drink every day?

(ii) What change in happiness results from drinking an extra $\frac{1}{4}$ pint of Guinness when he has already drunk:
(a) 3 pints; (b) 20 pints?

(iii) How much Guinness must he drink to reach Total Despair (-500 IEU)?

2 Algebraic, exponential & logarithmic functions

2.1 Quadratic functions

2.1:1 Quadratic equations

> A **quadratic equation** is an equation of the form $ax^2 + bx + c = 0$, where a, b and c are constants, $a \neq 0$. (Why the restriction $a \neq 0$?)

A quadratic equation is satisfied by two values of x, which are called the solutions, or roots of the equation. Given a particular quadratic equation, we can always solve it, i.e. find its roots. Some quadratic equations can be solved directly by factorization, e.g. $2x^2 + 3x - 5 = 0 \Rightarrow (2x+5)(x-1) = 0 \Rightarrow x = 1$ or $-\frac{5}{2}$. But *any* quadratic equation can be solved by **completing the square** as in worked example 1.

Worked example 1 Solve the equation $x^2 - 3x + 1 = 0$.

We first express the LHS in the form $(x+p)^2 + q$.

$$x^2 - 3x + 1 \equiv (x^2 - 3x) + 1$$
$$\equiv \{(x - \tfrac{3}{2})^2 - \tfrac{9}{4}\} + 1$$
$$\equiv (x - \tfrac{3}{2})^2 - \tfrac{5}{4} \quad (p = -\tfrac{3}{2}, q = -\tfrac{5}{4})$$

Since $x^2 - 3x + 1 = 0$,

$$(x - \tfrac{3}{2})^2 - \tfrac{5}{4} = 0$$
$$\Rightarrow \qquad (x - \tfrac{3}{2})^2 = \tfrac{5}{4}$$
$$\Rightarrow \qquad x - \tfrac{3}{2} = \pm\tfrac{1}{2}\sqrt{5}$$
$$\Rightarrow \qquad x = \tfrac{3}{2} \pm \tfrac{1}{2}\sqrt{5}$$

(C.f. 1.1:4 WE4.)

> **Theorem 1:** The roots of the equation $ax^2 + bx + c = 0$ ($a \neq 0$) are
> $$x = \frac{-b \pm \sqrt{(b^2 - 4ac)}}{2a} \qquad (1)$$

Proof
See question 10.

$\Delta = b^2 - 4ac$ is called the **discriminant** of the expression $ax^2 + bx + c$.

> If $\Delta > 0$, the equation has two distinct real roots (2)

Algebraic, exponential and logarithmic functions 75

e.g. $x^2 - 4x + 3 = 0 \Rightarrow x = 1$ or 3; $\Delta = (-4)^2 - 4.3 = 4 > 0$

$$\boxed{\text{If } \Delta = 0, \text{ the equation has two equal real roots} \qquad (3)}$$

e.g. $x^2 - 4x + 4 = 0 \Rightarrow x = 2$; $\Delta = (-4)^2 - 4.4 = 0$

$$\boxed{\text{If } \Delta < 0, \text{ the equation has two non-real roots} \qquad (4)}$$

e.g. $x^2 - 4x + 5 = 0 \Rightarrow x = 2 \pm \sqrt{-1}$; $\Delta = (-4)^2 - 4.5 = -4 < 0$. (In fact, in this case the roots are **complex numbers**—see section 6.5:1.)

Worked example 2 Find the sets of values of k for which the equation $(k-2)x^2 + 2kx + (k+3) = 0$ has (a) distinct real roots; (b) equal roots; (c) non-real roots.

$$\Delta = (2k)^2 - 4(k-2)(k+3) = -4k + 24$$

(a) Distinct real roots: $\Delta > 0 \Rightarrow -4k + 24 > 0 \Rightarrow k < 6$
(b) Equal roots: $\Delta = 0 \Rightarrow -4k + 24 = 0 \Rightarrow k = 6$
(c) Non-real roots: $\Delta < 0 \Rightarrow -4k + 24 < 0 \Rightarrow k > 6$

Note that when $k = 6$, the equation is $4x^2 + 12x + 9 = 0 \Rightarrow (2x+3)^2 = 0 \Rightarrow x = -\frac{3}{2}$.

Exercise 2.1:1

1 Factorize the expression on the LHS and hence solve the equation:

(i) $x^2 + 2x - 3 = 0$
(ii) $4 - 3x - x^2 = 0$
(iii) $2x^2 + 5x - 7 = 0$
(iv) $20x^2 + 33x - 36 = 0$
(v) $4x^2 - 9 = 0$
(vi) $3x^2 - 5 = 0$

In each case check that the two values of x you have found really do satisfy the equation.

•• **2** (i) Solve the equation $6x^2 + 10x = 0$.
(What is wrong with this solution:
$6x^2 + 10x = 0 \Rightarrow 3x^2 + 5x = 0 \Rightarrow 3x + 5 = 0 \Rightarrow x = -\frac{5}{3}$?)

(ii) Solve the equations:
(a) $5x^2 + x = 0$ (b) $9x^2 - 12x = 0$

•• **3** (i) Solve the equation $3x^2 - x = 1$.
(What is wrong with this solution
$3x^2 - x = 1 \Rightarrow x(3x-1) = 1 \Rightarrow x = 1$ or $3x - 1 = 1$

$\Rightarrow x = 1$ or $\frac{2}{3}$?)

(ii) Solve the equation $9x^2 - 12x = 5$.

4 Factorize the expression on the LHS and hence solve the equation:

(i) $x^2 - 4x + 4 = 0$
(ii) $x^2 - 2px + p^2 = 0$ (p constant)
(iii) $9x^2 + 24x + 16 = 0$

① → **5** Show that $x^2 - 4x - 21 \equiv (x-2)^2 - 25$. Hence solve the equation $x^2 - 4x - 21 = 0$.
Check your answer by factorizing the expression $x^2 - 4x - 21$.

① **6** Show that $x^2 + \frac{5}{2}x + \frac{1}{2} \equiv (x + \frac{5}{4})^2 - \frac{17}{16}$.
Hence solve the equation $2x^2 + 5x + 1 = 0$.

① **7** Show that $x^2 - \frac{2}{3}x - \frac{2}{3} \equiv (x - \frac{1}{3})^2 - \frac{7}{9}$.
Hence solve the equation $3x^2 - 2x - 2 = 0$.

① **8** Complete the square on the LHS and hence solve the equation:

(i) $x^2 + 4x - 5 = 0$
(ii) $x^2 - 2x - 8 = 0$
(iii) $x^2 - 6x - 7 = 0$
→ (iv) $2x^2 + 5x + 3 = 0$
(v) $3x^2 + 14x - 5 = 0$
(vi) $6x^2 - x - 2 = 0$

In each case check your answers by factorizing the LHS.

① **9** Complete the square on the LHS and hence solve the equation:

(i) $x^2 + 2x - 1 = 0$
(ii) $x^2 - 4x + 1 = 0$
(iii) $x^2 + 8x - 5 = 0$
(iv) $x^2 + x - 1 = 0$
→ (v) $x^2 + 3x + 1 = 0$
(vi) $2x^2 - 3x - 1 = 0$
(vii) $3x^2 + 7x + 1 = 0$
(viii) $5x^2 + 5x + 1 = 0$
→ (ix) $x^2 - 2px + q = 0$
(x) $x^2 + px + p = 0$

76 *Quadratic functions*

① → **10** By completing the square on the LHS, solve the equation $ax^2 + bx + c = 0$.

11 Using the result of Theorem 1, solve the equations:
- (i) $x^2 - 2x - 2 = 0$
- → (ii) $x^2 + 5x + 5 = 0$
- (iii) $x^2 - x - 3 = 0$
- → (iv) $5x^2 + 3x - 1 = 0$
- (v) $4x^2 = 8x + 1$

→ **12** Using the result of Theorem 1, solve the equations:
- (i) $x^2 - 4x + 2 = 0$
- (ii) $x^2 - 4x + 3 = 0$
- (iii) $x^2 - 4x + 4 = 0$

being careful to express the answers in the simplest possible form. In each case check that the two values of x you have found really do satisfy the equation.

13 Solve the simultaneous equations:
- (i) $x - y = 1$, $xy = 12$
- (ii) $x + y = 1$, $1/x + 1/y = -1$

14 A and B are two points. P is the point on AB which divides AB so that $AP/PB = PB/AB = k$. Show that $k^2 + k - 1 = 0$ and hence find k. (This division of the line is called the Golden Section).

15 Find the discriminant of $f(x)$ and hence find the nature (distinct real/equal real/non-real) of the roots of the equation $f(x) = 0$.
Check that your answers to (a) and (b) are correct, by factorizing $f(x)$.
- (i) (a) $f(x) = x^2 - 6x + 8$
  - (b) $f(x) = x^2 - 6x + 9$
  - (c) $f(x) = x^2 - 6x + 10$
- → (ii) (a) $f(x) = 4x - x^2 - 1$
  - (b) $f(x) = 4x - x^2 - 4$
  - (c) $f(x) = 4x - x^2 - 7$
- (iii) (a) $f(x) = 4x^2 + 7x + 3$
  - (b) $f(x) = 16x^2 + 56x + 49$
  - (c) $f(x) = 4x^2 + 7x + 4$

- (iv) (a) $f(x) = x^2 - 4$
  - (b) $f(x) = x^2$
  - (c) $f(x) = x^2 + 4$

② **16** For what value(s) of k does the equation $f(x) = 0$ have equal roots?
- (i) $f(x) = x^2 - 16x + k$
- (ii) $f(x) = kx^2 - 16x + k$
- (iii) $f(x) = x^2 + 2kx + (2 - k)$
- → (iv) $f(x) = x^2 + (k + 5)x + (5k + 4)$
- (v) $f(x) = 3x^2 + (4 - k)x - (1 + 2k)$

② **17** (i) Show that the equation $x^2 + kx + k^2 = 0$ cannot have real roots, except when $k = 0$.

(ii) Show that the equation $x^2 - px - q^2 = 0$ always has real roots, whatever the values of p and q.

(iii) Show that the equation $p^2x^2 + 2pqx + q^2 = 0$ always has equal roots. Is this the only type of equation with equal roots?

② **18** Find the set of values of k for which the equation $f(x) = 0$ has real roots:
- (i) $f(x) = kx^2 + 8x + 2$
- (ii) $f(x) = 3x^2 + 5x + (k - 3)$
- → (iii) $f(x) = (k + 2)x^2 + 2kx + (k + 1)$
- R (iv) $f(x) = 2kx^2 + (4k - 1)x + (2k - 3)$

② **19** Find the set of values of k for which the equation $x^2 - 2x + k = 0$ has real roots. Find the roots of the equation in the cases $k = -3$, $k = 1$. What happens when $k = 3$?

② **20** (i) What kind of roots has the equation

$$(3k + 2)x^2 - (6k + 1)x + (3k - 1) = 0?$$

(ii) Find the set of values of k for which the equation

$$(3k + 2)x^2 - (6k + 1)x + (3k + 1) = 0$$

has real roots.

2.1:2 Quadratic functions

> A **quadratic function** is a function of the form $ax^2 + bx + c$, where a, b and c are constants, $a \neq 0$.

Consider the graph $y = ax^2 + bx + c$ of a quadratic function $ax^2 + bx + c$. The roots are α and β.
(i) When $x = 0$, $y = c$.

Algebraic, exponential and logarithmic functions 77

(ii) When $y = 0$, $ax^2 + bx + c = 0$. There are then three possibilities:

(a) $\Delta > 0$: the equation has 2 distinct real roots, so the graph cuts the x-axis in 2 points.
(b) $\Delta = 0$: the equation has 2 equal roots, so the graph cuts the x-axis in one point, i.e. it touches the x-axis.
(c) $\Delta < 0$: the equation has 2 non-real roots, so the graph does not cut the x-axis (since a graph shows only real values of x and y).

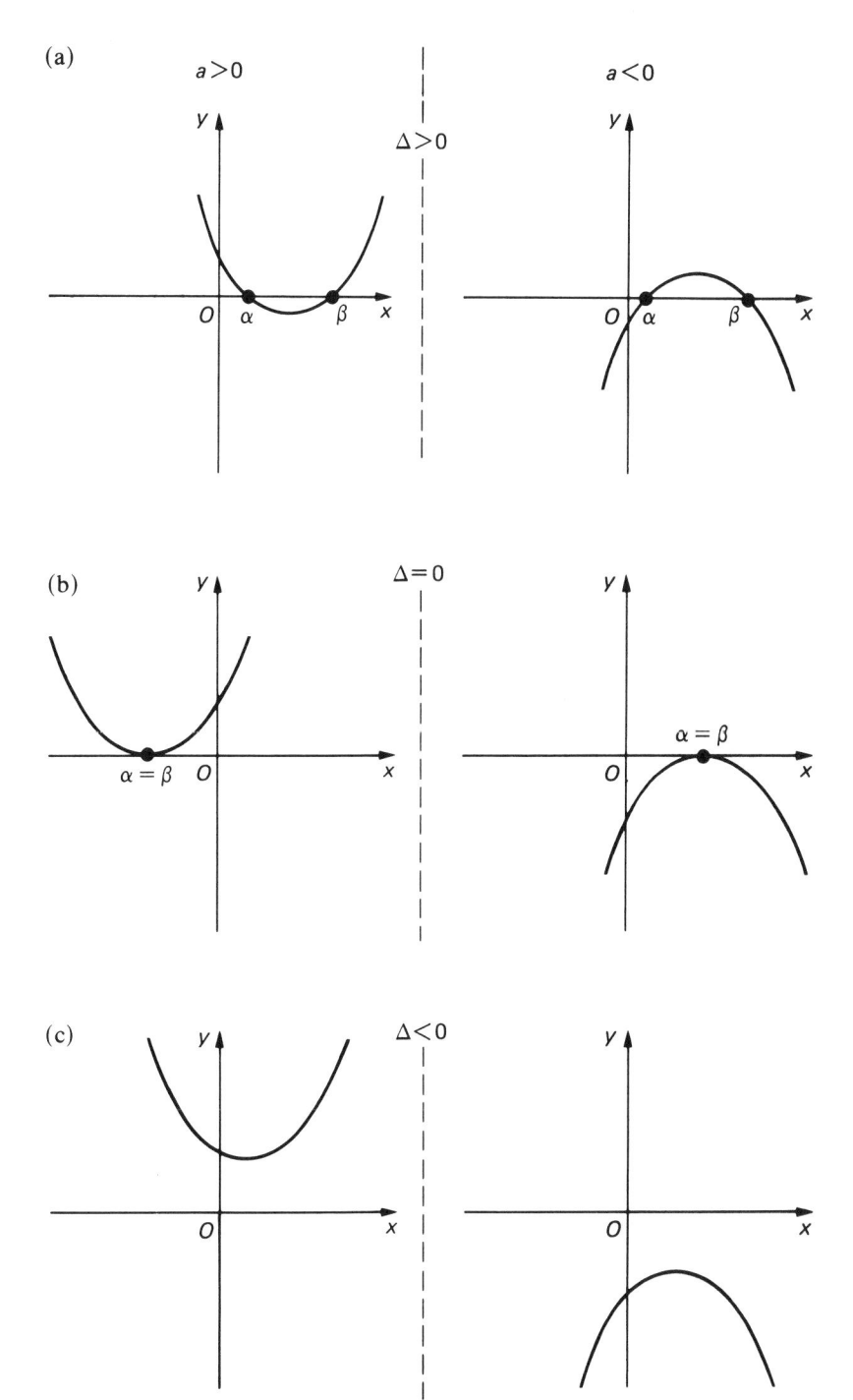

Fig. 2.1:1

78 Quadratic functions

(iii) At a turning point $dy/dx = 2ax+b = 0 \Rightarrow x = -b/2a$,
$y = (4ac-b^2)/4a = -\Delta/4a$. So

> the turning point is $(-b/2a, -\Delta/4a)$

$d^2y/dx^2 = 2a$ so if $a > 0$ it is a minimum, if $a < 0$ it is a maximum.
(iv) When $x \to \pm\infty$, $y \to +\infty$ if $a > 0$; $y \to -\infty$ if $a < 0$. (Why?)

> (v) The graph is symmetrical about the ordinate $x = -b/2a$,

i.e. the ordinate through the turning point (see question 8).
Thus there are six possible types of graph of $y = ax^2 + bx + c$, depending on the values of a and Δ. These are shown in fig. 2.1:1(a)-(f).

Exercise 2.1:2

Questions 1–3: Find the stationary value of $f(x)$, decide whether it is a maximum or a minimum value, find the coordinates of the point(s) (if any) where the curve $y = f(x)$ cuts the axes and sketch the curve, showing its axis of symmetry.

1. (i) $f(x) = (x+1)(x-5)$
   (ii) $f(x) = (2-x)(x+3)$
   (iii) $f(x) = x^2 - 5x + 6$
   (iv) $f(x) = 4 - 7x - 2x^2$
   (v) $f(x) = 1 - x^2$
   (vi) $f(x) = 4x^2 - 9$
   (vii) $f(x) = x(x+2)$
   (viii) $f(x) = 9x^2 - 12x$
   (ix) $f(x) = x^2 - 3x + 1$
   (x) $f(x) = 2x^2 + 2x - 3$

2. (i) $f(x) = x^2 - 4x + 4$
   (ii) $f(x) = (x+1)^2$
   (iii) $f(x) = 2x - x^2 - 1$

3. (i) $f(x) = x^2 + 9$
   (ii) $f(x) = -x^2 - 1$
   (iii) $f(x) = 2x^2 + x + 1$
   (iv) $f(x) = 6x - 10 - x^2$
   (v) $f(x) = 3x^2 + 7x + 4$

4. Sketch on the same axes the graphs of:
   (i) $y = (x+1)(x-2)$ and $y = (x+1)(2-x)$
   (ii) $y = (x-3)^2$ and $y = -(x-3)^2$
   (iii) $y = x^2 - 2x + 2$ and $y = 2x - 2 - x^2$

5. Sketch on the same axes the graphs of:
   (i) $y = x^2 - 4$, $y = x^2$, $y = x^2 + 4$
   (ii) $y = x^2 - 4x + 3$, $y = x^2 - 4x + 4$, $y = x^2 - 4x + 5$
   (iii) $y = 3\{(x+2)^2 - 1\}$, $y = 3(x+2)^2$, $y = 3\{(x+2)^2 + 1\}$

6. Sketch on the same axes the graphs of:
   (i) $y = x(x+3)$ and $y = x(x-3)$
   (ii) $y = x^2 - 1$, $y = x(x-2)$, $y = (x-1)(x-3)$

7. When $f(x) = x^2 - 4x + 3$, show that the minimum value of $f(x)$ occurs when $x = 2$.
   Show that $f(2-x) = f(2+x)$, i.e. that the graph of $f(x)$ is symmetrical about $x = 2$, the x-ordinate through the turning point.

8. Show that if $f(x) = ax^2 + bx + c$, then
$$f\left(-\frac{b}{2a} - x\right) \equiv f\left(-\frac{b}{2a} + x\right)$$
What is the significance of this result in terms of the graph of $f(x)$?

9. The curve $y = ax^2 + bx + c$ passes through the points $(-2, 15)$, $(1, -3)$ and $(4, 15)$. Find the values of a, b and c and show that $(1, -3)$ is the turning point of the curve.
   Sketch the curve, showing the three points.

10. Find the quadratic function which takes the value -6 when $x = -1$, the value 4 when $x = 1$ and the value -10 when $x = 3$. Sketch a graph of the function, marking in these three points.

11. Show that the stationary value of the function $f(x) = ax^2 + bx + c$ is $(4ac - b^2)/4a$. $f(x)$ takes the value zero when $x = 1$, the value 10 when $x = 0$ and has maximum value 18. Find two possible functions satisfying these conditions. Sketch on the same axes the graphs of the two functions.

12. The function $ax^2 + bx + c$ takes the value 10 when $x = -2$ and when $x = 4$ and its minimum value is -8. Sketch the graph of the function.

13. The supporting cable of a suspension bridge is in the form of a parabola. Its height above the bridge is 10 m at either end, 2 m in the middle. What is its height one quarter of the way across?

14 When a certain car factory produces n cars per day its profit £p per day is given by $p = 5n^2 - 100n$. How many cars per day must the factory produce

(i) to make a profit?
(ii) to make a profit of £24 000 per day?
(iii) to make the worst possible loss?

15 A batsman hits a cricket ball from a point O just above level ground. Referred to axes Ox and Oy horizontally and vertically upwards, the cartesian equation of its path is $y = \frac{1}{2}x - \frac{1}{80}x^2$ (1 unit = 1 m).

(i) Sketch the path.

(ii) Find the greatest height of the ball above the ground during its flight.

(iii) A fielder who can reach to a height of 2.5 m is standing at the point $(25, 0)$. Show that he cannot catch the ball without moving. What is the least distance he can run to give himself a chance of catching it?

16 When a ball is thrown vertically upwards from a point O with speed 30 m s^{-1}, its height h m above O after t s is given by $h = 30t - 5t^2$.

(i) What is its greatest height above O?
(ii) For how long is it more than 25 m above O?
(iii) How long does it take to return to O?

2.1:3 Quadratic inequalities

Figure 2.1:1 shows that if $\Delta > 0$, i.e. the roots of $ax^2 + bx + c = 0$ are real, $y = ax^2 + bx + c$ is sometimes negative, sometimes positive.

Worked example 3 Sketch the graph of $y = 3x^2 - 5x - 2$. Find the set of values of x for which y is (i) negative; (ii) positive.

When $x = 0$, $y = -2$ and when $y = 0$, $3x^2 - 5x - 2 = 0$, so $x = -\frac{1}{3}$ or 2 (see fig. 2.1:2).

The roots of $y = 0$ are $-\frac{1}{3}$ and 2, looking at the diagram, we see that

(i) $y < 0$, i.e. the graph is below the axis, when x is between the roots, i.e. $-\frac{1}{3} < x < 2$
(ii) $y > 0$, i.e. the graph is above the axis, when x is outside the roots, i.e. $x < -\frac{1}{3}$ or $x > 2$ (See 1.2:7 WE14.)

In general, suppose that the equation $ax^2 + bx + c = 0$ has distinct real roots, α and β, $\alpha < \beta$ say, then,

> if $a > 0$,
> (i) the set of values for which $ax^2 + bx + c > 0$ is $x < \alpha$ or $x > \beta$
> (ii) the set of values for which $ax^2 + bx + c < 0$ is $\alpha < x < \beta$ **(5)**

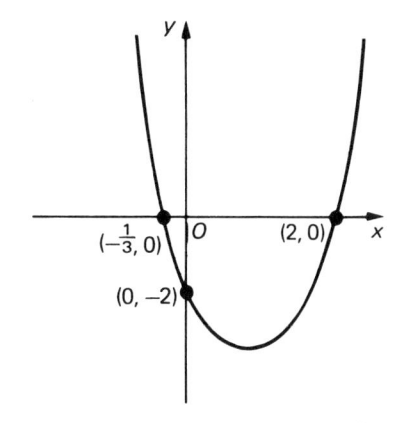

Fig. 2.1:2

Worked example 4 Find the set of values of x for which $2x + 1 > x^2$.

Since we would like to make use of statement (5), we first write the inequality in one of the forms $ax^2 + bx + c > 0$, $ax^2 + bx + c < 0$, with $a > 0$.

$$2x - x^2 > -1 \Rightarrow x^2 - 2x - 1 < 0$$

Now $x^2 - 2x - 1 = 0 \Rightarrow x = 1 - \sqrt{2}$ or $x = 1 + \sqrt{2}$. Then, by statement (5),

$$x^2 - 2x - 1 < 0 \Rightarrow 1 - \sqrt{2} < x < 1 + \sqrt{2}$$

Worked example 5 Find the set of values of x for which $0 < x^2 + x < 2x + 6$.

$0 < x^2 + x < 2x + 6$ means $0 < x^2 + x$ and $x^2 + x < 2x + 6$. By the method of worked example 4,

$$x^2 + x > 0 \Rightarrow x < -1 \text{ or } x > 0$$
$$x^2 + x < 2x + 6 \Rightarrow x^2 - x - 6 < 0 \Rightarrow -2 < x < 3$$

80 Quadratic inequalities

So x must satisfy $x<-1$ or $x>0$ *and* $-2<x<3$. We plot these sets of values on the real line and look for their overlap, as in fig. 2.1:3. So the solution is $\{-2<x<-1\}\cup\{0<x<3\}$.

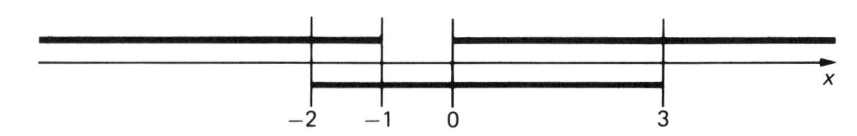

Fig. 2.1:3

Figure 2.1:1 shows that if $\Delta<0$, i.e. the roots of $ax^2+bx+c=0$ are non-real, the graph of $y=ax^2+bx+c$ lies entirely on one side of the x-axis:

$a>0, \Delta<0 \Rightarrow ax^2+bx+c>0$ for all real values of x	(6)
$a<0, \Delta<0 \Rightarrow ax^2+bx+c<0$ for all real values of x	(7)

Worked example 6 Find the minimum value of the function $f(x)=2x^2-4x+5$:

(i) by solving the equation $f'(x)=0$;
(ii) by expressing $f(x)$ in the form $2(x+p)^2+q$.

(i) $f'(x)=4x-4$. $f'(x)=0 \Rightarrow x=1$. When $x=1$, $f(x)=2\cdot 1^2-4\cdot 1+5=3$, so the stationary value of $f(x)$ is 3. The value is a minimum since $f''(1)=4>0$.

(ii)
$$f(x) \equiv 2x^2-4x+5 \equiv 2(x^2-2x)+5$$
$$\equiv 2\{(x-1)^2-1\}+5$$
$$\equiv 2(x-1)^2+3$$

(We have completed the square.)
The minimum value of $2(x-1)^2+3$ occurs when $(x-1)^2=0$, i.e. when $x=1$; the minimum value is 3.

The graph of $f(x)=2x^2-4x+5$ is shown in fig. 2.1:4.

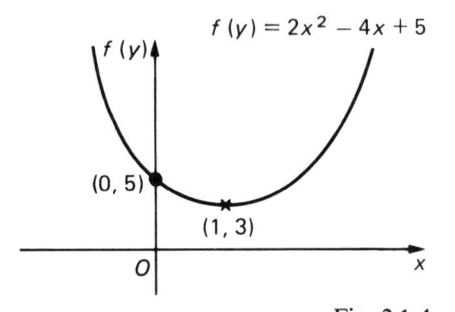

Fig. 2.1:4

Worked example 7 Find the set of values of k for which the equation
$$(k-1)x^2-(2k+1)x+(4k+2)=0$$
has non-real roots.
$$\Delta = (2k+1)^2-4(k-1)(4k+2) = -12k^2+12k+9$$
When the equation has non-real roots, $\Delta<0$, i.e. $-12k^2+12k+9<0$. This occurs when $k<-\tfrac{1}{2}$ or $k>\tfrac{3}{2}$ (by the method of worked example 4).

Worked example 8 Show that, for all real values of k, the equation
$$kx^2-(4k+1)x+(k+1)=0$$
has real roots.
$$\Delta = (4k+1)^2-4k(k+1)$$
$$= 12k^2+4k+1$$
$$= 12(k+\tfrac{1}{6})^2+\tfrac{2}{3} \qquad \text{(see worked example 6(ii))}$$
$$\geq 0 \text{ for all real values of } k$$

So the equation has real roots for all real values of k.

Algebraic, exponential and logarithmic functions 81

Exercise 2.1:3

③ **1** Sketch the graph of $y = f(x)$ and find the set of values of x for which y is (a) negative; (b) positive.

    (i) $f(x) = (x-1)(x+4)$
    (ii) $f(x) = (x-2)(5-x)$
→ (iii) $f(x) = 2x^2 + 5x - 3$
    (iv) $f(x) = 3 - 13x - 10x^2$
    (v) $f(x) = 25 - 4x^2$
    (vi) $f(x) = 7x - 3x^2$

④ **2** Find the set of values of x for which:

    (i) $(x+1)(x-3) > 0$
    (ii) $x^2 - 3x + 2 > 0$
    (iii) $2x^2 + 9x - 5 \leq 0$
    (iv) $8 + 2x - 3x^2 > 0$
    (v) $3x - 4x^2 \geq 0$
→ (vi) $6x^2 + 7x < 5$
    (vii) $x^2 + 6x - 5 > 2$
R (viii) $5x^2 + 4x < 6x + 3$

3 Find the set of values of x for which:

④ • (i) $x^2 > 9$
    (ii) $x^2 < 3$
→ (iii) $5x^2 > 4$

④ **4** Find the set of values of x for which:

    (i) $x^2 - 2x - 2 > 0$
    (ii) $x^2 + 1 > 3x$
    (iii) $x^2 < 4x + 3$
    (iv) $5 - x - 2x^2 < 0$

④ **R 5** Find the set of values of x for which $2x^2 + 3 < x + 4$. Illustrate your answer by sketching on the same axes the graphs of $y = 2x^2 + 3$ and $y = x + 4$, marking in the coordinates of their points of intersection. (C.f. 1.2:7 WE15.)

⑤ **6** Find the set of values of x for which:

    (i) $0 < x^2 - x < 6$
    (ii) $-2 < x^2 - 3x < 4$
    (iii) $-3 < x^2 + 4x < 5$
    (iv) $4 < x^2 < 9$
    (v) $3 < x^2 - 2x < 2x$
R (vi) $3 - x < 2x^2 < 14 - 3x$
→ (vii) $4x + 2 < x^2 + 3x < 6x$

⑤ **7** Find the set of values of x for which $2 < 3x^2 + 5x < 12$. Illustrate your result by sketching the curve $y = 3x^2 + 5x$ and the lines $y = 2$ and $y = 12$, marking in the coordinates of their points of intersection.

⑤ **8** Find the set of values of x for which $2x + 1 < 2x^2 + x < 4x + 5$. Illustrate your result by sketching on the same axes the curve $y = 2x^2 + x$ and the lines $y = 2x + 1$ and $y = 4x + 5$, marking in the coordinates of their points of intersection.

⑥ **9** If $f(x) = x^2 + 4x + 5$, sketch a graph of $y = f(x)$, marking in the coordinates of the turning point. Express $f(x)$ in the form $(x+p)^2 + q$ and hence verify that the least value of $f(x)$ is 1.

⑥ **10** If $f(x) = 2x^2 - 8x + 14$, express $f(x)$ in the form $2(x+p)^2 + q$. Hence show that $f(x)$ is always positive, and show that the least value of $f(x)$ is 6. Sketch a graph of $y = f(x)$, marking in the coordinates of the turning point.

⑥ → **11** Express $f(x) = 5x - 3x^2 - 3$ in the form $-3\{(x+p)^2 + q\}$. Hence show that $f(x)$ is always negative and find its greatest value. Sketch a graph of $y = f(x)$.

⑥ **R 12** Show that $1 \leq x^2 + 6x + 10 < \infty$. Deduce that $0 < 1/(x^2 + 6x + 10) \leq 1$.

⑥ **13** Sketch the graph of $y = x^2 - 6x + 9$. For what values of x is:

    (i) $x^2 - 6x + 9 > 0$
    (ii) $x^2 - 6x + 9 \geq 0$

⑥ + **14** Show that:

$$ax^2 + bx + c \equiv a\left(x + \frac{b}{2a}\right)^2 + \frac{-\Delta}{4a}, \text{ where } \Delta = b^2 - 4ac.$$

    (i) Hence show that

        if $a > 0$, $\Delta < 0$, $ax^2 + bx + c > 0$ for all x
        if $a < 0$, $\Delta < 0$, $ax^2 + bx + c < 0$ for all x

    (ii) Show also that, for any value of a except zero, the stationary value of $ax^2 + bx + c$ is $-\Delta/4a$.

⑦ **15** Find the sets of values of k for which the equation $f(x) = 0$ has (a) equal roots; (b) distinct real roots; (c) non-real roots:

    (i) $f(x) = 3x^2 - kx + 3$
    (ii) $f(x) = x^2 + (k-1)x + (k+2)$
    (iii) $f(x) = x^2 + (1-3k)x + (4-7k)$
    (iv) $f(x) = kx^2 + (k-2)x + k$
→ (v) $f(x) = (k-1)x^2 + 4x + (k+2)$

16 (i) If $f(x) = x^2 + (k-3)x - (k-6)$, find the sets of values of k for which the equation $f(x) = 0$ has (a) equal roots; (b) distinct real roots; (c) non-real roots.

    (ii) Sketch the graph of $y = f(x)$ in the cases $k = 6$, $k = 5$ and $k = 4$.
    Explain how these graphs illustrate your answers to (i).

⑦ **R 17** Find the set of values of k for which $f(x) = 4x^2 + 4kx - (3k^2 + 4k - 3)$ is positive for all real values of x. Sketch the graph of $y = f(x)$ in the cases $k = 0$ and $k = 1$. Explain how these graphs illustrate your result.

⑦ **18** If $f(x) = kx^2 - 2x + (3k - 2)$ find the set of values of k for which $f(x)$ is: (i) always positive; (ii) always negative.

⑦ **19** If $f(x) = (k-2)x^2 + (k+1)x - 1$, find the set of values of k for which $f(x)$ is always negative.

⑦→ **20** If $f(x) = px^2 - 2x + (3p + 2)$, find the set of value of p for which $f(x)$ is:

(i) negative for all real values of x;
(ii) positive for all real values of x.

Sketch the graph of $y = f(x)$ for each of the cases $p = -2$, $p = 1$.

⑧ **21** Show that the following equations have real roots for all real values of k:

(i) $(k^2 + 1)x^2 + (2k - 1)x - 1 = 0$
(ii) $(k - 3)x^2 + 2kx + 1 = 0$
(iii) $kx^2 + (3k - 2)x + (k - 1) = 0$
→ (iv) $kx^2 - (k + 1)x + (1 - 2k) = 0$

⑧ R **22** Show that if $f(x) = kx^2 + 2x - (k - 2)$ then the equation $f(x) = 0$ has real roots for all real values of k. Sketch the curve $y = f(x)$ in the cases $k = -1$, $k = 2$.

→ **23** Show that, at the points of intersection of the line $y = mx + c$ and the curve $y = x^2 + x + 4$,

$$x^2 + (1 - m)x + (4 - x) = 0$$

Hence show that if the line is tangent to the curve, then

$$m^2 - 2m + (4c - 15) = 0$$

Hence find the equations of the tangents to the curve which pass through the point $(0, 3)$.
Sketch the curve and these tangents.

24 Show that if the line $y = mx + c$ is tangent to the curve $y = x^2$, then $m^2 + 4c = 0$. Hence find the equations of the tangents to the curve which pass through the point $(-1, -3)$.

25 Find the set of values of m for which the straight line $y = mx$ meets the curve $y = 2x^2 - 3x + 2$ in real points. Deduce the equations of the tangents to the curve which pass through the origin.

26 Sketch on the same axes the curve $y = 4x - x^2$ and the lines $y = 8 - 2x$, $y = 9 - 2x$, $y = 10 - 2x$. Find the number of real roots of each of the equations:

(i) $4x - x^2 = 8 - 2x$
(ii) $4x - x^2 = 9 - 2x$
(iii) $4x - x^2 = 10 - 2x$

Write down the maximum value of k for which the line $k - 2x$ cuts the curve $y = 4x - x^2$.
Find also the equation of the trangent of gradient -4 to the curve.

27 Find the equations of the tangents to the circle $x^2 + y^2 = 25$ which pass through the point $(1, 7)$.
Sketch on the same axes the circle (see section 1.2:8) and these tangents.

2.1:4 Roots of a quadratic equation

Consider the quadratic equation

$$ax^2 + bx + c = 0 \quad (a \neq 0)$$

i.e.
$$x^2 + \frac{b}{a}x + \frac{c}{a} = 0 \quad (8)$$

It has two roots, α and β say. Now, we can write down a quadratic equation whose roots are α and β:

$$(x - \alpha)(x - \beta) = 0$$

i.e.
$$x^2 - (\alpha + \beta)x + \alpha\beta = 0 \quad (9)$$

Equations (8) and (9) have the same roots and, since the expressions of the LHS have the same coefficient of x^2, they must also have the same coefficient of x and the same constant term.

Equating coefficients of x: $\quad \alpha + \beta = -b/a$
Equating constant terms: $\quad \alpha\beta = c/a \quad$ (10)

Algebraic, exponential and logarithmic functions 83

Worked example 9 The roots of the equation $3x^2+4x-5=0$ are α and β. Find:

(i) $1/\alpha + 1/\beta$
(ii) $\alpha^2+\beta^2$
(iii) $\alpha^3+\beta^3$

We express the given function of α and β in terms of $\alpha+\beta$ and $\alpha\beta$, both of which we know from equation (10): $\alpha+\beta = -\frac{4}{3}$ and $\alpha\beta = -\frac{5}{3}$

(i) $1/\alpha + 1/\beta = (\alpha+\beta)/(\alpha\beta) = -\frac{4}{3}/-\frac{5}{3} = \frac{4}{5}$
(ii) $\alpha^2+\beta^2 = (\alpha+\beta)^2 - 2\alpha\beta = \frac{16}{9} + \frac{10}{3} = \frac{46}{9}$
(iii) $\alpha^3+\beta^3 = (\alpha+\beta)^3 - 3\alpha\beta(\alpha+\beta) = -\frac{64}{27} - \frac{20}{3} = -\frac{244}{27}$

Worked example 10 The roots of the equation $2x^2-7x+4=0$ are α and β. Without finding α and β, find an equation whose roots are α/β and β/α.

From equation (10), $\alpha+\beta = \frac{7}{2}$ and $\alpha\beta = 2$.

The second equation has two roots and so it must be quadratic. Let it be $x^2+px+q=0$. (How is it that we can assume that the coefficient of x^2 is 1?) Then from equation (10) (applied to $x^2+px+q=0$):

$$-p = \frac{\alpha}{\beta} + \frac{\beta}{\alpha} = \frac{\alpha^2+\beta^2}{\alpha\beta} = \frac{(\alpha+\beta)^2 - 2\alpha\beta}{\alpha\beta} = \frac{33}{8}$$

$$q = \frac{\alpha}{\beta} \cdot \frac{\beta}{\alpha} = 1$$

Thus the second equation is $x^2 - \frac{33}{8}x + 1 = 0$, or $8x^2 - 33x + 8 = 0$.

Exercise 2.1:4

1 (i) Explain why, if the equation $ax^2+bx+c=0$ has roots α and β, then $ax^2+bx+c \equiv a(x-\alpha)(x-\beta)$.

(ii) Using the result of Theorem 1 (section 2.1:1):
(a) solve the equation $x^2+2x-1=0$ and hence factorize x^2+2x-1;
(b) solve the equation $2x^2+4x-3=0$ and hence factorize $3x^2+5x+1$.

(iii) Factorize:
(a) x^2+4x-2; (b) x^2+x-4; (c) $5x^2-4x-2$.

2 Using equation (10), find the sum and product of the roots of:

(i) $x^2-4x+7=0$
(ii) $5x^2-4x+3=0$
(iii) $-3x^2+2x=6$

3 Solve the following equations and in each case find the sum and product of the roots. Do your answers agree with equation (10)?

(i) $x^2-4x+3=0$
(ii) $2x^2+x-15=0$
(iii) $3x^2+6x+1=0$
(iv) $x^2-2px-q=0$

4 If $\alpha = \{-b+\sqrt{(b^2-4ac)}\}/2a$
$\beta = \{-b-\sqrt{(b^2-4ac)}\}/2a$
show that $\alpha+\beta = -b/a$, $\alpha\beta = c/a$.

Questions 5–7: Find the values of:

(i) $\dfrac{1}{\alpha^2} + \dfrac{1}{\beta^2}$ (ii) $(\alpha-\beta)^2$

(iii) $\alpha^3\beta + \alpha\beta^3$ (iv) $\dfrac{1}{\alpha^2\beta} + \dfrac{1}{\alpha\beta^2}$

(v) $\dfrac{\alpha^2}{\beta} + \dfrac{\beta^2}{\alpha}$ (vi) $\dfrac{1}{\alpha+2} + \dfrac{1}{\beta+2}$

when α and β are the roots of the given equation.

⑨ **5** $x^2-3x-6=0$

⑨→ **6** $4x^2+8x-1=0$

⑨ **7** $x^2-px+q=0$

⑨ **8** The equation $ax^2+bx+c=0$ has roots α and β. Find the values of:

(i) $\dfrac{1}{\alpha} + \dfrac{1}{\beta}$ (ii) $\dfrac{1}{\alpha^2} + \dfrac{1}{\beta^2}$

(iii) $\alpha^2\beta + \alpha\beta^2$ (iv) $\alpha^3+\beta^3$

84   *Roots of a quadratic equation*

(v) $\dfrac{\alpha}{\beta^2}+\dfrac{\beta}{\alpha^2}$ (vi) $\dfrac{\alpha}{\beta+1}+\dfrac{\beta}{\alpha+1}$

Note: all these functions of α and β remain unchanged when α and β are reversed. They are called **symmetrical functions** of α and β. Explain why only symmetrical functions of α and β can be expressed in terms of $\alpha+\beta$ and $\alpha\beta$.

Questions 9–11: Find (quadratic) equations whose roots are:

(i) α^2, β^2 (ii) $\alpha-2, \beta-2$

(iii) $\dfrac{1}{\alpha}, \dfrac{1}{\beta}$ (iv) $\dfrac{\alpha}{\beta}, \dfrac{\beta}{\alpha}$

(v) $\alpha^2\beta, \alpha\beta^2$ (vi) $(\alpha+\beta)^2, (\alpha-\beta)^2$

when α and β are the roots of the given equation.

9 $x^2+3x+1=0$

10 $2x^2-4x+1=0$

11 The roots of the equation $ax^2+bx+c=0$ are α and β. Find equations with roots:

(i) α^2, β^2 (ii) $3\alpha, 3\beta$

(iii) $\dfrac{1}{\alpha}, \dfrac{1}{\beta}$ (iv) $\dfrac{1}{\alpha^2}, \dfrac{1}{\beta^2}$

R (v) $\dfrac{\alpha}{\beta}, \dfrac{\beta}{\alpha}$ (vi) $\dfrac{\alpha^2}{\beta}, \dfrac{\beta^2}{\alpha}$

12 The roots of the equation $3x^2+5x-2$ are α and β.

(i) Find an equation with integer coefficients whose roots are $1/\alpha^2$ and $1/\beta^2$.
(ii) Find the values of α and β and check by factorization that the new equation really does have roots $1/\alpha^2$ and $1/\beta^2$.

• **13** The roots of the equation $x^2+5x+k=0$ are α and $\alpha-1$. Find k.

14 The roots of the equation $3x^2+5x-k=0$ differ by 2. Find k.

→ **15** (i) One root of the equation $ax^2+bx+c=0$ is twice the other. Show that $2b^2=9ac$.

(ii) Give an example of a quadratic equation for which one root is twice the other and show that the result of (i) holds in this case.

R **16** If one of the roots of the equation $27x^2+kx-8=0$ is the square of the other, find k.

17 One root of the equation $6x^2+7x+k=0$ is $-\tfrac{1}{2}$. Find the other root and the value of k.

18 One root of the equation $3x^2+kx+5=0$ is 1. Find the other root and the value of k.

19 Find the set of values of k for which the roots of the equation $x^2+2kx+(2-k)=0$ are:

(i) real (see worked example 7);
(ii) real and positive.

(Hint: the roots α and β are positive $\Leftrightarrow \alpha+\beta$ positive and $\alpha\beta$ positive.)

20 Find the set of values of k for which the equation $x^2+(k-3)x+k=0$ has:

(i) real distinct roots;
(ii) real distinct roots of the same sign.

21 The equation $(x-k)(x-2k)=1$, $k\in\mathbb{R}$, has roots α and β.

(i) Show that, whatever the value of k, α and β are real.
(ii) Show that $-\alpha$ and $-\beta$ are the roots of the equation $(x+k)(x+2k)=1$.
(iii) Find an equation whose roots are $1/\alpha$ and $1/\beta$.

22 (i) Find an equation whose roots are the reciprocals of the roots of the equation

$$ax^2+bx+c=0.$$

Show that the roots of the equation

$$ax^2+bx+a=0$$

are the reciprocals of each other.

23 The roots of the equation $x^2+ax+b=0$ are a^2 and b. Find three possible sets of values of a and b.

24 (i) The roots of the equation

$$(x-a)(x-b)=c$$

are α and β. Show that the roots of the equation

$$(x+a)(x+b)=c$$

are $-\alpha$ and $-\beta$.

(ii) Find the roots of the equations
(a) $(x-1)(x-2)=3$
(b) $(x+1)(x+2)=3$
Sketch on the same axes the curves $y=(x-1)(x-2)$, and $y=(x+1)(x+2)$ and the line $y=3$, and label their points of intersection.

2.2 Polynomial functions

2.2:1 Multiplication and division

The function $P(x) \equiv a_n x^n + a_{n-1} x^{n-1} + a_{n-2} x^{n-2} + \ldots + a_2 x^2 + a_1 x + a_0$, where the a_i are constants, $a_n \neq 0$, is called a **polynomial of degree n**, e.g.

$x^3 + 3x$ is a polynomial of degree 3 (cubic)
$3x^4 - 7x^3 + 5$ is a polynomial of degree 4 (quartic)

We can divide one polynomial (of degree n) by another (of degree n or less) by using a method similar to that of long division.

Worked example 1 Find $(2x^4 - 5x^3 - 6) \div (x^2 - x + 2)$.

$$\begin{array}{r} 2x^2 - 3x - 7 \\ x^2 - x + 2 \overline{\smash{\big)}\, 2x^4 - 5x^3 - 6} \\ \underline{2x^4 - 2x^3 + 4x^2 } \\ -3x^3 - 4x^2 \\ \underline{-3x^3 + 3x^2 - 6x } \\ -7x^2 + 6x - 6 \\ \underline{-7x^2 + 7x - 14} \\ -x + 8 \end{array}$$

So

$$\frac{2x^4 - 5x^3 - 6}{x^2 - x + 2} \equiv 2x^2 - 3x - 7 + \frac{8 - x}{x^2 - x + 2}$$

$2x^2 - 3x - 7$ is called the **quotient**, $x^2 - x + 2$ the **divisor**, and $8 - x$ the **remainder**.

Exercise 2.2:1

1 What is the degree of each of the following polynomials? What is the coefficient of x^2?

(i) $5x^3 + x^2 + x + 1$
(ii) $4x^5 - x^4 + 6x^3 + x - 1$
(iii) x (iv) 6
(v) $3x^2 - x^4$ (vi) $(1 + x)(1 - x^2)$
(vii) $(2x^2 + x + 2)(x^2 + x - 1)$.

2 Find the product of:

(i) $(3x^2 - 2x - 1)$ and $(x^3 - x)$
(ii) $(2x^3 + x^2 + 5x - 4)$ and $(2x^2 - 3x + 1)$
(iii) $(x^3 + x^2 + x + 1)$ and $(x - 1)$
(iv) $(x^3 - x^2 + x - 1)$ and $(x + 1)$

① **3** Find the quotient and remainder when

→ (i) $2x^3 + 3x^2 - 4x + 5$ is divided by $x + 2$
(ii) $4x^3 - 6x^2 + 5$ is divided by $2x - 1$
(iii) x^5 is divided by $x^2 + 1$

(iv) $x^4 + 2x^3 + 3x^2 + 7$ is divided by $x^2 + x + 1$
→ (v) $x^3 - 5x^2 + 11x + 6$ is divided by $2x^2 - 3$
(vi) $x^4 + x^3 - x^2 + 1$ is divided by $(x - 1)(x^2 + 1)$

① **4** If $P(x) = x^4 - 1$, show that when $P(x)$ is divided by $x - 1$ the remainder is zero. Show that when $P(x)$ is divided by $x + 1$ the remainder is zero. Hence find the factors of $P(x)$.

① **5** Show that $P(x) = 2x^4 + x^3 - 11x^2 - 4x + 12$ is divisible by $x^2 + x - 2$. Hence find the four linear factors of $P(x)$. Hence solve the equation $P(x) = 0$.

① → **6** Find the remainder when $x^3 + 2x^2 + x$ is divided by:

(i) $x^2 - 1$ (ii) $x^2 + 1$ (iii) $x^2 + x$

What can you say about the degree of the remainder when a polynomial is divided by another polynomial of (a) degree 2; (b) degree n?

2.2:2 The remainder theorem

Theorem 1 (the remainder theorem): When a polynomial $P(x)$ is divided by a linear function $k(x-\alpha)$, the remainder is $P(\alpha)$.

Proof
Suppose that when $P(x)$ is divided by $k(x-\alpha)$, the quotient is $Q(x)$ and the remainder is R; then

$$P(x) \equiv k(x-\alpha)Q(x) + R \qquad (1)$$

Putting $x = \alpha$ in equation (1) gives

$$P(\alpha) = R$$

For example, suppose $P(x) = x^3 + 3x - 5$. If $P(x)$ is divided by $x+2$, the remainder is

$$P(-2) = (-2)^3 + (-2) - 5 = -19$$

If $P(x)$ is divided by $3x-5$, i.e. $3(x-\tfrac{5}{3})$, the remainder is

$$P(\tfrac{5}{3}) = (\tfrac{5}{3})^3 + 3(\tfrac{5}{3}) - 5 = \tfrac{125}{27}$$

Note: (i) the theorem applies only when the divisor is a *linear* function; (ii) if we want to find the quotient, we must actually do the division, as in worked example 1.

Worked example 2 When a polynomial $P(x)$ is divided by $(x-2)$ the remainder is 2, and when it is divided by $(x-3)$ the remainder is 5. Find the remainder when it is divided by $(x-2)(x-3)$.

Let the remainder be $ax + b$ (see 2.2:1 Q6), i.e.

$$P(x) \equiv (x-2)(x-3)Q(x) + (ax+b) \qquad (2)$$

By Theorem 1, $P(2) = 2$ and $P(3) = 5$.
Putting $x = 2$ in equation (2) gives $P(2) = 2a + b \Rightarrow 2a + b = 2$
Putting $x = 3$ in equation (2) gives $P(3) = 3a + b \Rightarrow 3a + b = 5$
From this pair of equations we find that $a = 3$ and $b = -4$, and so the remainder is $3x - 4$.

Exercise 2.2:2

1 Using the remainder theorem, find the remainder when

(i) $2x + 3x^2 - 4x + 5$ is divided by $x + 2$

(ii) $4x^3 - 6x^2 + 5$ is divided by $2x - 1$.
(Do your answers agree with 2.2:1 Q3?)

(iii) $25x^5 + 7x$ is divided by $5x + 4$

R 2 When $x^5 + 4x^4 - 6x^2 + ax + 2$ is divided by $x + 2$ the remainder is 6. Find a.

② → **3** When a certain polynomial is divided by $x - 3$ the remainder is 3. When it is divided by $x + 2$ the remainder is 13. Find the remainder when it is divided by $(x-3)(x+2)$.

② **R 4** When a certain polynomial is divided by $2x - 1$ the remainder is -1. When it is divided by $3x + 1$ the remainder is $+1$. Find the remainder when it is divided by $6x^2 - x - 1$.

② → **5** When a certain polynomial is divided by $x - 1$, x, $x + 1$, the remainders are 1, 2 and 3 respectively. Find the remainder when it is divided by $x(x-1)(x+1)$. (Note: what will the degree of the remainder be?)

Algebraic, exponential and logarithmic functions 87

② • **6** When the quadratic expression ax^2+bx+c is divided by x, the remainder is 1. When it is divided by $x-1$ the remainder is 2. When it is divided by $x-2$ the remainder is 5. Find a, b and c.

② **7** When x^3+ax^2+bx+c is divided by $x+3$ the remainder is -26, and when it is divided by x^2-x-2 the remainder is 14. Find a, b and c.

② **8** Find a cubic polynomial $P(x)$ which has a remainder -1 when divided by $2x-1$, and a remainder $+1$ when divided by $3x+1$. Divide $P(x)$ by $6x^2-x-1$. Does your answer agree with the result of question 4?

② **9** Find the values of a and b if $ax^4+bx^3-8x^2+6$ has remainder $2x+1$ when divided by x^2-1.

2.2:3 The factor theorem; solutions of polynomial equations

> **Theorem 2** (the factor theorem): If α is a root of the equation $P(x)=0$, i.e. if $P(\alpha)=0$ then $(x-\alpha)$ is a factor of $P(x)$.

Proof
Suppose that the remainder when $(x-\alpha)$ divides $P(x)$ is R. Then, by Theorem 1, $R = P(\alpha) = 0$ i.e. $(x-\alpha)$ is a factor of $P(x)$.

Worked example 3 If $P(x) \equiv 2x^3+x^2-13x+6$, solve the equation $P(x)=0$.

First we try out values of x in the equation, hoping to find a simple integer solution, i.e. $+1$, ± 2, ± 3, etc.
$P(1) = 2+1-13+6 = -4 \neq 0$, so $x=1$ is not a solution.
$P(-1) = -2+1+13+6 = 18 \neq 0$, so $x=-1$ is not a solution.
$P(2) = 16+4-26+6 = 0$; so $x=2$ is a solution, and by the factor theorem $(x-2)$ is a factor of $P(x)$, i.e. $P(x) \equiv (x-2)Q(x)$, where $Q(x)$ is a polynomial of degree 2 (a quadratic).
(Note: in fact we can see that the only values of x worth trying are factors of the constant term of $P(x)$, i.e. 6.)
We find $Q(x)$ either by division or by inspection:

$$P(x) \equiv 2x^3+x^2-13x+6 \equiv (x-2)(2x^2+5x-3) \equiv (x-2)(2x-1)(x+3)$$

So $P(x) = 0 \Rightarrow x = 2$ or $x = \frac{1}{2}$ or $x = -3$.

Worked example 4 Factorize the polynomial $P(x) \equiv x^3+x-2$.

$P(1) = 0$; so $(x-1)$ is a factor of $P(x)$. Thus $P(x) \equiv x^3+x-2 \equiv (x-1)(x^2+x+2)$.
Now $x^2+x+2=0$ has no real roots; so x^2+x+2 has no real linear factors. So we cannot factorize $(x-1)(x^2+x+2)$ any further.

> **Definition** An **irreducible polynomial** is a polynomial with no real factors other than itself and 1, e.g. x^2+4, $x^2+6x+13$.

In fact, the only irreducible polynomials turn out to be quadratic (see section 6.5:3).

Exercise 2.2:3

1 Using the factor theorem, show that:
 (i) $x-1$ is a factor of x^4-1
 (ii) $x+1$ is a factor of x^4-1
 (iii) $2x+1$ is a factor of $4x^5+4x^4-7x^3-2x^2+3x+1$
 (iv) $(x-2)(3x+1)$ is a factor of $6x^4-10x^3-x^2-5x-2$

2 Show that $(x+1)$, $(x-2)$, $(2x-1)$ and x are factors of $2x^5-11x^4+9x^3+14x^2-8x$. What is the other factor?

3 Find the value of a for which the polynomial $x^4+x^3+ax^2+5x-10$ has $x+2$ as a factor.

④ 4 Using the factor theorem, factorize completely the polynomial $P(x)$ and hence solve the equation $P(x)=0$ when:
 (i) $P(x)=x^3-6x^2+11x-6$
 (ii) $P(x)=x^3+4x^2+5x+2$
 (iii) $P(x)=2x^3-5x^2-14x+8$
 (iv) $P(x)=x^3-3x^2-2x+6$
 (v) $P(x)=x^3-3x^2-3x+1$
 → (vi) $P(x)=x^3+4x^2+3x-2$
 (vii) $P(x)=12x^3+47x^2-62x+15$

③ 5 Show that the polynomial $2x^4-5x^3-23x^2+38x+24$ has factors $(x-2)$ and $(x+3)$. Hence find its other factors.

③ 6 Factorize completely the polynomial $P(x)$ and hence solve the equation $P(x)=0$ when:
 (i) $P(x)=x^4-2x^2+1$
 (ii) $P(x)=2x^4-5x^3-14x^2+23x+30$
 (iii) $P(x)=3x^5+11x^4+5x^3-15x^2-8x+4$

③ 7 If $P(x)=x^4+hx^3+gx^2-16x-12$ has factors $(x+1)$ and $(x-2)$, find the constants h and g and the remaining factors.

④ 8 Find a real root of each of the following polynomial equations and show that they have no other real roots:
 (i) $x^3+x=0$
 → (ii) $x^3-5x^2+12x-18=0$
 (iii) $x^3-x^2-5x+2=0$
 (iv) $x^3+3x^2+16=0$

④ 9 (i) Show that x^3-1 has a factor $x-1$. Hence factorize x^3-1, i.e. express it as the product of a linear factor and an irreducible quadratic factor.
 (ii) Factorize x^3+1.
 (iii) Factorize a^3-b^3 and a^3+b^3. Hence factorize $27x^3+8$.

④ 10 Factorize the following polynomials as far as possible:
 (i) x^4-1
 (ii) x^4+2x^3-x-2
 (iii) $x^4+x^3+3x^2+2x+2$

• 11 Let $P(x)=2x^4+ax^2+bx-60$. The remainder when $P(x)$ is divided by $(x-1)$ is -94. One factor of $P(x)$ is $(x-3)$. Find a and b.

12 When $P(x)=2x^3+ax^2+bx+c$ is divided by x^2-1, the remainder is 6. Given that $(x-2)$ is a factor of $P(x)$, find a, b and c, and solve the equation $P(x)=0$.

13 Using the substitution $y=x^2$, find the real roots of the equations:
 (i) $x^4-5x^2+4=0$
 (ii) $x^4-3x^2+2=0$
 (iii) $x^4-x+2=0$
 (iv) $4x^4-21x^2+5=0$

14 Solve the equation $\{f(x)\}^2=16$, where $f(x)=x^2+3x-14$. (Hint: $\{f(x)\}^2=16 \Rightarrow f(x)=\pm 4$; solve two quadratic equations.)

15 Solve the equations:
 (i) $(6x^2-x-1)^2=1$
 (ii) $(3x^2+x-6)^2=16$

16 Give two examples of an irreducible polynomial. Explain why a cubic polynomial cannot be irreducible.
 Show that ax^2+bx+c is irreducible $\Leftrightarrow b^2-4ac<0$. What kind of *numbers* have similar properties to irreducible polynomials?

2.2:4 Graphs of polynomial functions

Worked example 5 Sketch the curve $y=2x^3+x^2-13x+6$. For what values of x is y (i) negative, (ii) positive?

When $x=0$, $y=6$. When $y=0$, $2x^3+x^2-13x+6=0$, so $x=2$ or $x=\frac{1}{2}$ or $x=-3$ (see worked example 3).
As $x\to\infty$, $y\to\infty$ and as $x\to-\infty$, $y\to-\infty$.

Algebraic, exponential and logarithmic functions 89

We can see that:

(i) $y < 0$ when $x < -3$ or $\frac{1}{2} < x < 2$
(ii) $y > 0$ when $-3 < x < \frac{1}{2}$ or $x > 2$

The graph of $y = 2x^3 + x^2 - 13x + 6$ is shown in fig. 2.2:1.

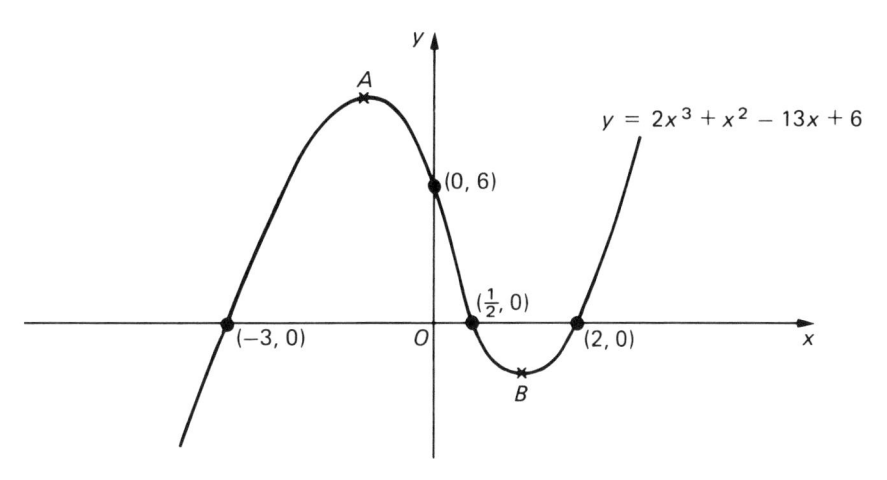

Fig. 2.2:1

Worked example 6 Sketch the curve $y = x^3 + x - 2$. For what values of x is y (i) negative; (ii) positive?

When $x = 0$, $y = -2$ and when $y = 0$, $x^3 + x - 2 = 0$ so the only real root is $x = 1$ (see worked example 4).

$dy/dx = 3x^2 + 1$. There is no real value of x for which $dy/dx = 0$, hence there are no stationary points.

As $x \to \infty$, $y \to \infty$, and as $x \to -\infty$, $y \to -\infty$.

The graph of $y = x^2 + x - 2$ is shown in fig. 2.2:2. Note that there is a point of inflexion at $(0, -2)$, i.e. where $d^2y/dx^2 = 0$.

We can see that:

(i) $y < 0$ when $x < 1$;
(ii) $y > 0$ when $x > 1$.

(This is obvious if we consider the factorization $y = (x - 1)(x^2 + x + 2)$ and note that $x^2 + x + 2 > 0 \; \forall x$—by the method of 2.1:3 WE6(ii).)

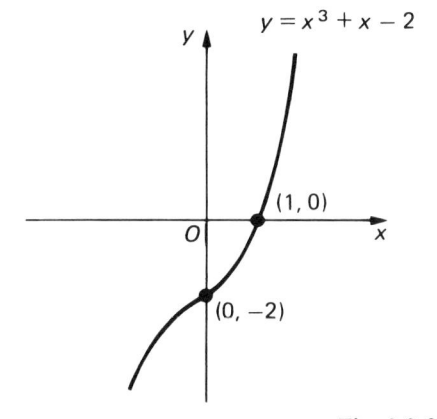

Fig. 2.2:2

> **Theorem 3:** If $P(x)$ has a factor $(x - \alpha)^2$ then $P'(x)$ has a factor $(x - \alpha)$.

Proof
If $P(x)$ has a factor $(x - \alpha)^2$, we can write

$$P(x) \equiv (x - \alpha)^2 Q(x)$$

Thus

$$P'(x) \equiv 2(x - \alpha)Q(x) + (x - \alpha)^2 Q'(x) \quad \text{(by product rule)}$$
$$\equiv (x - \alpha)H(x)$$

where $H(x) = 2Q(x) + (x - \alpha)Q'(x)$. So $P'(x)$ has a factor $(x - \alpha)$.

> **Corollary:** If $P(x)$ has a factor $(x - \alpha)^2$ then $(\alpha, 0)$ is a stationary point of $y = P(x)$. (3)

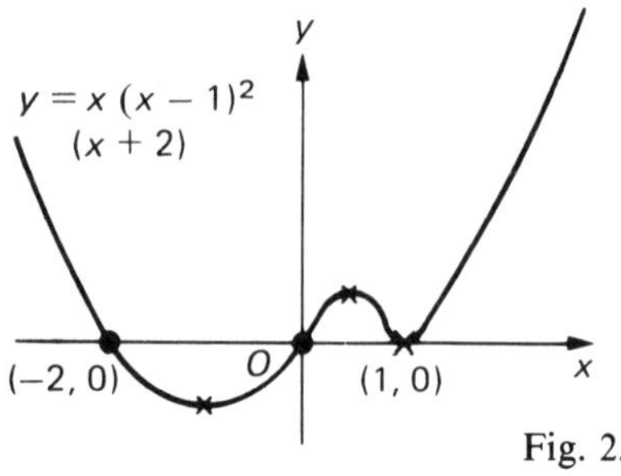

90 *Graphs of polynomial functions*

Worked example 7 Sketch the curve $y = x(x-1)^2(x+2)$.

When $x = 0$, $y = 0$, and when $y = 0$, $x(x-1)^2(x+2) = 0 \Rightarrow x = 0$, $x = 1$ or $x = -2$.

By corollary (3), the curve has a turning point at $(1, 0)$—we need not bother to find the other turning points.

As $x \to \pm\infty$, $y \to +\infty$.

The curve $y = x(x-1)^2(x+2)$ is shown in fig. 2.2:3.

Fig. 2.2:3

Exercise 2.2:4

⑤ **1** Sketch a graph of each of the functions $f(x)$. For what values of x is $f(x)$ (a) negative? (b) positive?

  (i) $f(x) = x(x^2 - 1)$
  (ii) $f(x) = (x-2)(3x-2)(5x+4)$
→ (iii) $f(x) = (1-x)(3-x)(5-x)$
  (iv) $f(x) = (x-1)(x+2)(x+1)(4-x)$

⑤ **2** Sketch the curve $y = (x+4)(x+1)(x-2)$. Show that the turning points of the curve are *not* given by $x = -\frac{5}{2}$, $x = \frac{1}{2}$ (i.e. symmetrically between the roots).

⑤ **3** In each of the following cases, solve the equation $f(x) = 0$ (see 2.2:3 WE3). Sketch a graph of the function $f(x)$. For what values of x is $f(x)$ (a) negative; (b) positive?

  (i) $f(x) = x^3 - x^2 - 4x + 4$
→ (ii) $f(x) = 2x^3 + 5x^2 - 4x - 3$
  (iii) $f(x) = x^4 - 5x^2 + 4$

⑤ **4** Find the set of values of x for which

  (i) $(x-1)(x-2)(x-3) > 0$
  (ii) $(2-3x)(x+4)(x-2) < 0$
  (iii) $(x+3)(x^2-1) \leq 0$
  (iv) $x^3 - 4x^2 - 5x < 0$
→ (v) $x^3 - 7x - 6 \geq 0$
  (vi) $x^3 - 3x^2 - 3x + 1 > 0$
  (vii) $x^4 - 4x^2 + 3 \leq 0$
→ (viii) $x(2x+3)(x^2-4) > 0$

Questions 5 and 6: Sketch on the same axes the graphs of the functions f and g. Show that the graphs intersect in three points, and find the coordinates of these points.
Find the set of values of x for which $f(x) > g(x)$.

⑤ **5** $f(x) = 2x^3 + x^2$, $g(x) = 13x - 6$

⑤ **6** $f(x) = 4x - x^3$, $g(x) = 2x^2 - x - 6$

⑥ **7** Find the stationary values (if any) of $f(x)$, find the real roots of the equation $f(x) = 0$ and hence sketch a graph of $f(x)$. For what values of x is $f(x)$ (a) negative; (b) positive?

  (i) $f(x) = x^3 - 1$
  (ii) $f(x) = x^3 + 3x$
→ (iii) $f(x) = x^3 + x + 2$
  (iv) $f(x) = 18 - 12x + 5x^2 - x^3$
  (v) $f(x) = x^3 + 3x^2 - 9x + 20$
  (vi) $f(x) = x^4 - 1$

⑥ **8** Find the set of values of x for which:

  (i) $(x-3)(x^2+1) > 0$
  (ii) $(2x-5)(x^2+2x+5) < 0$
→ (iii) $x^3 - x^2 + x - 6 \geq 0$
  (iv) $3x^3 - 4x^2 + 7x - 22 > 0$
  (v) $x^3 + 1 < 0$
  (vi) $x^4 - 3x^2 - 4 \geq 0$

⑥ **9** Sketch on the same axes the graphs of the functions f and g, where $f(x) = 2x + x^2 - x^3$ and $g(x) = x + 10$.
Show that the graphs intersect at only one point, and find its coordinates.
Find the set of values of x for which $f(x) \leq g(x)$.

10 Sketch the curve $y = x^3 - 3x + 2$. On the same axes, sketch also the curves $y = x^3 - 3x + 4$ and $y = x^3 - 3x$. How many real roots has each of the equations $x^3 - 3x + 2 = 0$, $x^3 - 3x + 4 = 0$ and $x^3 - 3x = 0$?

: **11** (i) Prove the corollary to Theorem 3.

  (ii) Show that, if $P(x)$ has a factor $(x - \alpha)^3$, then $(\alpha, 0)$ is, in general, a point of inflexion of $y = P(x)$. Why 'in general'?

⑦ **12** (i) Find the stationary points of the curve $y = x(x-1)^2(x+2)$. Sketch the curve.

  (ii) Find the stationary points of the curve $y = x(x-1)^3$. Verify that the point $(1, 0)$ is a point of inflexion. Sketch the curve.

⑦ **13** With the help of Theorem 3, sketch a graph of the function $f(x)$. For what values of x is $f(x)$ (a) negative; (b) positive; (c) non-negative?

  (i) $f(x) = (x-1)^2(x+2)$
  (ii) $f(x) = x(2-x)^3$
→ (iii) $f(x) = (x+1)^2(x-2)(x-3)$
  (iv) $f(x) = (x+1)^2(x+2)^2$
  (v) $f(x) = x^3 - 6x^2 + 9x - 4$

Algebraic, exponential and logarithmic functions 91

⑦ 14 If $f(x) = x(2-x)$, sketch on the same axes the graphs of $f(x)$ and $\{f(x)\}^2$.

⑦ 15 Find the set of values of x for which
(i) $(2x-1)(3x-1)^2 > 0$
(ii) $x-1)(x-2)^3 \leq 0$
(iii) $(x-1)^2(x-2)^2 \geq 0$
(iv) $2x^3 - 9x^2 + 12x - 4 \leq 0$

Questions 16 and 17: Find the equation of the tangent to the given curve at the given point. Where does the tangent meet the curve again?

→ 16 $y = x^3$; (1, 1)

17 $y = x^3 - 3x$; $(-2, -2)$

18 Explain why, if the line $y = mx + c$ is a tangent to the curve $y = P(x)$, the equation $P(x) - mx - c = 0$ has a repeated root.
Hence show that the line $y = 4x - 2$ is a tangent to the curve $y = x^3 + x$, and find the coordinates of the point of contact.

19 Show that the line $y = 3x + 9$ is a tangent to the curve $y = 5x^2 - x^3$. Sketch the line and the curve on the same axes and find the coordinates of the point at which the line meets the curve again.
Find the sets of values of x for which
(i) $3x + 9 > 5x^2 - x^3$
(ii) $3x + 9 \leq 5x^2 - x^3$

20 The equation $x^3 - 6x^2 + 9x + k = 0$ has a repeated root. Using Theorem 3, show that this root must be either $x = 1$ or $x = 3$. Hence find the two possible values of k.
In each case find the other root.

R 21 Find the real roots of the equation $f(x) = 0$, sketch the curve $y = f(x)$ and hence find the set of values of x for which $f(x) > 0$ when:
(i) $f(x) = 2x^3 - 7x^2 + x + 10$
(ii) $f(x) = 6 + 4x + x^2 - x^3$
(iii) $f(x) = 2x^3 - 3x^2 + 1$

22 Sketch, where you think it possible, the graph of a cubic function with:
(i) 3 real roots and 2 turning points
(ii) 3 real roots and no turning point
(iii) 2 real roots
(iv) 1 real root and 2 turning points
(v) 1 real root and no turning point
(vi) no real roots

Where you think it impossible, explain why.

Questions 23-25: Sketch on the same axes the curves $y = f(x)$ and $y^2 = f(x)$ when:

23 $f(x) = 4x$

24 $f(x) = x(x-1)(x-2)$

→ 25 $f(x) = x(x-1)^2(x-2)$

26 Show that the distance from the point (1, 0) to the curve $y^2 = 2x - x^2$ is constant. Sketch on the same axes the curves $y = 2x - x^2$ and $y^2 = 2x - x^2$.

2.3 Rational functions

2.3:1 Graphs of rational functions

A **rational function** is a function of the form

$$f(x) = \frac{P(x)}{Q(x)}$$

where $P(x)$ and $Q(x)$ are polynomials. (Note: to avoid ambiguity, we shall insist that $P(x)$ and $Q(x)$ have no real factors in common, e.g. $(x^2 + x - 2)/(x^2 - 1)$ is not a rational function, though $(x+2)/(x+1)$ is.)

Worked example 1 Sketch the graph of $y = (2x-1)/(3x+2)$.

When $x = 0$, $y = -\frac{1}{2}$, and when $y = 0$, $2x - 1 = 0 \Rightarrow x = \frac{1}{2}$ (fig. 2.3:1).
$3x + 2 = 0$ when $x = -\frac{2}{3}$, so the ordinate $x = -\frac{2}{3}$ is an asymptote. Using the notation of section 1.2:5:
as $x \to -\frac{2}{3}^+$, $y \to -\infty$, and as $x \to -\frac{2}{3}^-$, $y \to +\infty$.
As $x \to \pm\infty$, $y \to \frac{2}{3}$, so $y = \frac{2}{3}$ is an asymptote. (C.f. 1.2:6 WE13.)

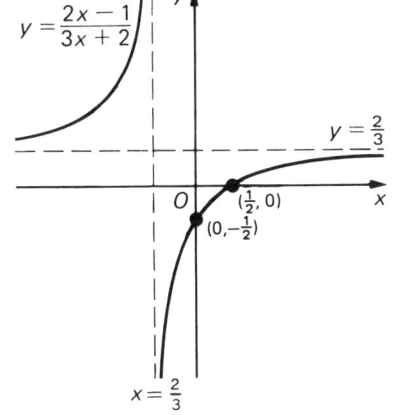

Fig. 2.3:1

Worked example 2 Sketch the graph of

$$y = \frac{1}{(x+1)(x-5)}$$

When $x = 0$, $y = -\frac{1}{5}$. There is no value of x for which $y = 0$, so the graph does not cut the x-axis (fig. 2.3:**2**).

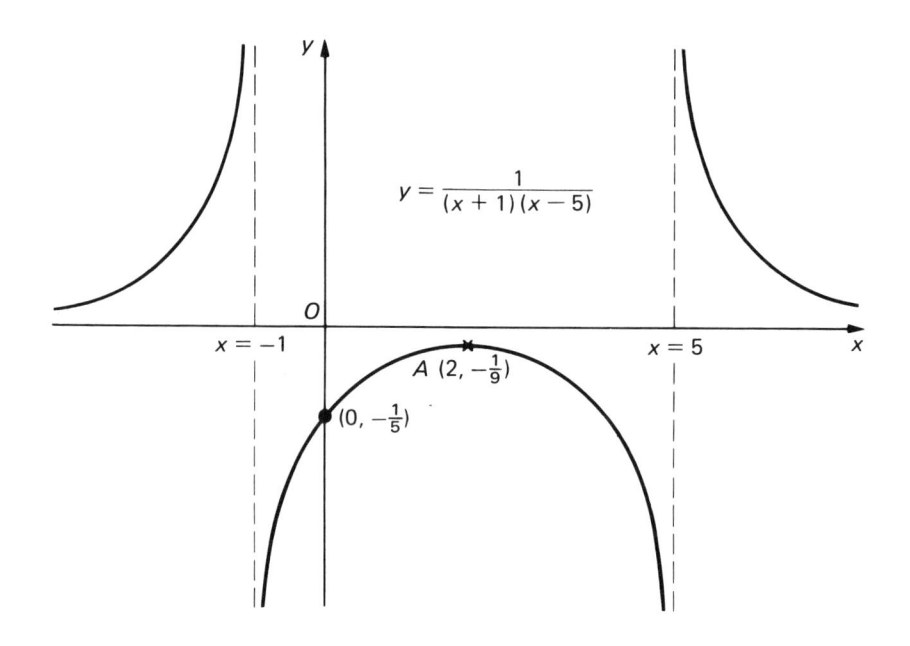

Fig. 2.3:**2**

$(x+1)(x-5) = 0$ when $x = -1$ or $x = 5$, so the ordinates $x = -1$ and $x = 5$ are asymptotes. Using the notation of section 1.2:5:
as $x \to 5^+$, $y \to +\infty$; as $x \to 5^-$, $y \to -\infty$;
as $x \to -1^+$, $y \to -\infty$; as $x \to -1^-$, $y \to +\infty$.
    As $x \to \pm\infty$, $y \to 0$; so $y = 0$ is an asymptote.

Note: We can find the turning point A $(2, -\frac{1}{9})$ of the curve by differentiation—but see question 8.

A slick method of finding the range of a rational function without drawing its graph is given in worked example 3.

Worked example 3 Find the range of the function $f: \mathbb{R} \to \mathbb{R}$ defined by $f(x) = (2x+1)/(x^2+2)$.
    Let $y = (2x+1)/(x^2+2)$. First, rearrange this equation as a quadratic in x:

$$x^2 y + 2y = 2x + 1$$
$$\Rightarrow \qquad yx^2 - 2x + (2y - 1) = 0$$

For each value of y in the range, x must be *real*, i.e. $\Delta \geq 0$.

$$4 - 4y(2y-1) \geq 0 \Rightarrow 2y^2 - y - 1 \leq 0 \Rightarrow -\tfrac{1}{2} \leq y \leq 1$$

So the range of the function is $[-\tfrac{1}{2}, 1]$. (See question 22.)

Exercise 2.3:1

① **1** Sketch the following graphs:

(i) $y = \dfrac{1}{x}$

(ii) $y = \dfrac{1}{x+2}$

→ (iii) $y = \dfrac{4}{3x-2}$

① **2** Sketch the following graphs:

(i) $y = \dfrac{x}{x+1}$

→ (ii) $y = \dfrac{2x-1}{x+3}$

R (iii) $y = \dfrac{3x-4}{4x+3}$

(iv) $y = \dfrac{3x+1}{x}$

(v) $y = \dfrac{1-3x}{x-2}$

① **3** Sketch on the same axes the graphs $y = 1/(x+1)$ and $y = (x+2)/(x+1)$. What do you notice? Explain it.

① **4** Sketch on the same axes the graphs $y = 1/(x-2)$ and $y = (3-2x)/(x-2)$.

① R **5** The function $f: \{x: x \in \mathbb{R}, x \neq -1\} \to \mathbb{R}$ is defined by $f(x) = (3x-2)/(x+1)$. What is the range of f? Sketch the graph of f.
Find $f^{-1}(x)$. What are the domain and range of the function f^{-1}? Sketch its graph.

① **6** Show that the function $f(x) = (x+3)/(2x-1)$ is its own inverse. Sketch on the same axes the curve $y = f(x)$ and the line $y = x$. What is the relationship between the curve and the line?
Under what circumstances is the function $g(x) = (ax+b)/(px+q)$ its own inverse?

② **7** Sketch the following graphs:

(i) $y = \dfrac{4}{(x+2)(x-6)}$

(ii) $y = \dfrac{1}{x^2-1}$

→ (iii) $y = \dfrac{1}{(2x+1)(3x-5)}$

(iv) $y = \dfrac{4}{x^2-5x+4}$

(v) $y = \dfrac{1}{x(x-2)(x+3)}$

② → **8** If $f(x) = (x-1)(x-5)$, find the turning point of the curve $y = f(x)$. Hence (or by differentiation) show that the curve $y = 1/f(x)$ has a maximum at the point $(3, -\tfrac{1}{4})$.
Sketch on the same axes graphs of $y = f(x)$ and $y = 1/f(x)$, marking in the line which is an axis of symmetry of both graphs.

② **9** Sketch on the same axes the graphs of

(i) $y = x^2 - 9$ and $y = \dfrac{1}{x^2-9}$

R (ii) $y = (x+1)(2x-5)$ and $y = \dfrac{1}{(x+1)(2x-5)}$

(iii) $y = 3x^2 - 8x + 4$ and $y = \dfrac{1}{3x^2-8x+4}$

In each case mark in the line which is the axis of symmetry of both graphs.

② **10** Sketch on the same axes the graphs $y = 6x^2 - x - 1$ and $y = 1/(6x^2 - x - 1)$. Mark in the coordinates of their points of intersection (see 2.2:3 Q15(i)).

11 Find the turning point of the curve $y = f(x)$, and sketch it.

→ (i) $f(x) = \dfrac{1}{x^2+1}$

(ii) $f(x) = \dfrac{1}{x^2+2x+5}$

12 Find the turning point of the curve $y = f(x)$, and sketch it.

(i) $f(x) = \dfrac{x}{x^2+1}$

→ (ii) $f(x) = \dfrac{x-1}{x^2+3}$

(iii) $f(x) = \dfrac{8-3x}{x^2+4}$

13 Find the coordinates of the turning point of the curve $y = x^2 - 2x + 2$.
Sketch on the same axes this curve and the curve $y = 1/(x^2 - 2x + 2)$.

14 Sketch on the same axes the graphs of

(i) $y = x^2 + 4$ and $y = \dfrac{1}{x^2+4}$

(ii) $y = 2x^2 + x + 3$ and $y = \dfrac{1}{2x^2+x+3}$

15 Find the turning points of the curve $y = (2x+1)/(x^2+2)$, and sketch it. Show that, as long as x takes real values, y always lies between $-\tfrac{1}{2}$ and 1.

94 *Inequalities involving rational functions*

- **16** Find $\lim_{x \to 2^+} \dfrac{1}{(x-2)^2}$ and $\lim_{x \to 2^-} \dfrac{1}{(x-2)^2}$

  Sketch the graph $y = 1/(x-2)^2$.

17 Sketch on the same axes the graphs of:

→ (i) $y = 4x^2 - 4x + 1$ and $y = \dfrac{1}{4x^2 - 4x + 1}$

  (ii) $y = 6x - x^2 - 9$ and $y = \dfrac{1}{6x - x^2 - 9}$

18 Find the coordinates of the turning point of the curve $y = f(x)$, and sketch the curve.

  (i) $f(x) = \dfrac{x}{(x-1)^2}$ (ii) $f(x) = \dfrac{2x-1}{(3x+2)^2}$

19 Sketch the graphs:

  (i) $y = \dfrac{3}{x-3}$ (ii) $y = \dfrac{x+2}{x-3}$

  (iii) $y = \dfrac{5}{(x+2)(x-3)}$ (iv) $y = \dfrac{5}{(x-3)^2}$

  (v) $y = \dfrac{3x-1}{x^2+2}$

20 Sketch the graphs:

  (i) $y = \dfrac{1}{x^2 - 4x + 3}$ (ii) $y = \dfrac{1}{x^2 - 4x + 4}$

  (iii) $y = \dfrac{1}{x^2 - 4x + 5}$ (C.f. 2.1:2 Q5(ii).)

21 Sketch diagrams showing on the same axes the form of the graphs $y = ax^2 + bx + c$ and $y = 1/(ax^2 + bx + c)$ in the cases:

  (i) $\Delta > 0$ (ii) $\Delta = 0$ (iii) $\Delta < 0$

  Assume that $a > 0$.

③ **22** Find the range of the function:

  (i) $\dfrac{3x}{2x^2 + 1}$ → (ii) $\dfrac{x^2 + 1}{x^2 + 4}$

  (iii) $\dfrac{4x - 8}{4x^2 + 9}$ (iv) $\dfrac{3x}{(x-1)^2}$

③ **23** If $f(x) = 2x/(4x^2 + 1)$,

  (i) find the stationary values of $f(x)$;
  (ii) sketch the curve $y = f(x)$;
  (iii) find the range of $f(x)$, using the method of worked example 3.

  Check that your answer to (iii) agrees with your answer to (i).

③ **R 24** If $f(x) = x/(x^2 + 4)$,

  (i) prove that $-\tfrac{1}{4} \leq f(x) \leq \tfrac{1}{4}$ for all real values of x;
  (ii) find the stationary values of $f(x)$;
  (iii) sketch on the same axes the graphs of $y = f(x)$ and $y = x + 1$ and hence show that the equation $x = (x^2 + 4)(x + 1)$ has only one real root.

③ **25** If $f(x) = x/(x-2)^2$,

  (i) prove that $f(x) \geq -\tfrac{1}{8}$ for all real values of x;
  (ii) find the stationary values of $f(x)$;
  (iii) sketch the curve $y = f(x)$.

26 If $f(x) = 2x/(1 + x^2)$, $x \in \mathbb{R}$, sketch the curve $y = f(x)$. Hence, or otherwise, find the range of f.

  Explain why f has no inverse.

  Show that if f is restricted to a suitable subdomain of $\mathbb{R}$, then f has an inverse f^{-1} and find an expression for $f^{-1}(x)$. What is the domain of f^{-1}?

R 27 If $f(x) = 4x^2 - 12x + 9$, show that $f(x) \geq 0 \; \forall x \in \mathbb{R}$. Sketch on the same axes the graphs of $y = f(x)$ and $y = 4/f(x)$. Deduce from your diagram that the equation $\{f(x)\}^2 = 4$ has two real roots and find them, correct to 2 dec. pls. (See 2.2:3 Q14.)

28 Using the method of worked example 3, find the range of $f(x)$ when

  (i) $f(x) = x^2 - 4x - 5$ (ii) $f(x) = 4x - x^2 - 5$

  In each case check your answer by finding the stationary value of $f(x)$.

2.3:2 Inequalities involving rational functions

We can solve inequalities involving rational functions either analytically or graphically.

Worked example 4 (analytical method) Find the set of values of x for which $(2x + 3)/(2x - 1) > 5$.

It is tempting to multiply each side of the inequality by $(2x - 1)$; but this is a dangerous step, since the sign of $(2x - 1)$ depends on the value of x. So instead we multiply each side by $(2x - 1)^2$, which is certainly a

Algebraic, exponential and logarithmic functions 95

positive number:
$$(2x+3)(2x-1) > 5(2x-1)^2$$
$$(2x-1)\{(2x+3)-5(2x-1)\} > 0$$
$$(2x-1)(-8x+8) > 0$$
So
$$\tfrac{1}{2} < x < 1$$

Note: We can simplify the solution by first rearranging the inequality into the form $f(x) > 0$:

$$\frac{2x+3}{2x-1} > 5 \Rightarrow \frac{2x+3}{2x-1} - 5 > 0 \Rightarrow \frac{-8x+8}{2x-1} > 0$$

and then multiplying both sides by $(2x-1)^2$.

Worked example 5 (graphical method) Sketch on the same axes the graphs of $y = 2/(x-1)$ and $y = x/(x+2)$. Find their points of intersection and hence find the set of values of x for which $2/x-1 > x/x+2$.

The graphs, shown in fig. 2.3:3, intersect at A and B, where

$$\frac{2}{x-1} = \frac{x}{x+2} \Rightarrow 2x+4 = x^2 - x$$
$$x^2 - 3x - 4 = 0$$

So
$$x = -1 \text{ or } x = 4$$

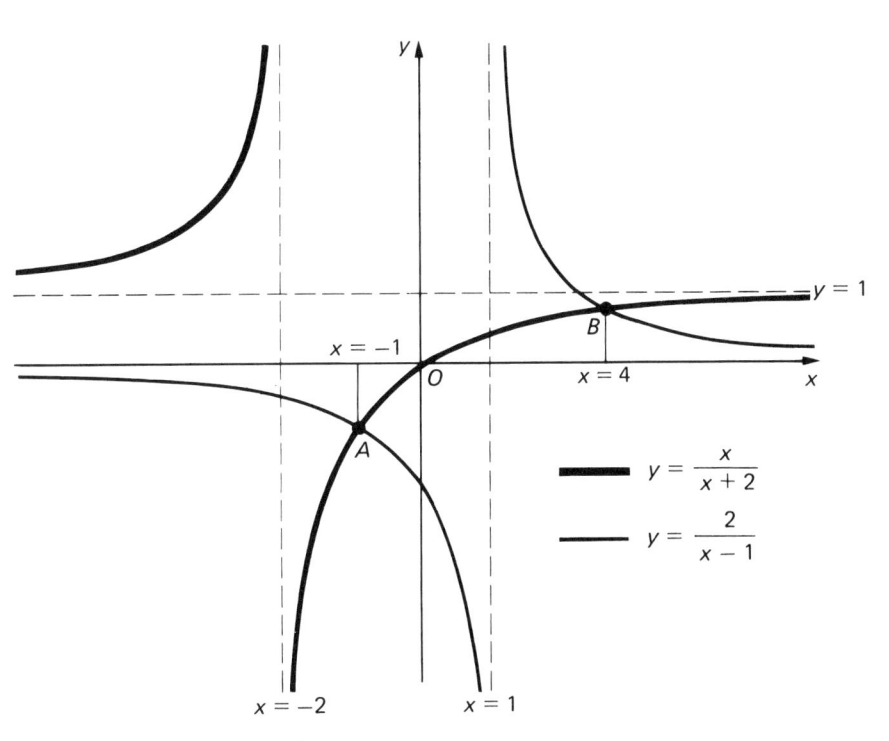

Fig. 2.3:3

The points of intersection are $A(-1, 1)$ and $B(4, \tfrac{2}{3})$. The inequality holds only when the thin line in fig. 2.3:3 is above the thick line, i.e. when $-2 < x < -1$ or $1 < x < 4$.
(C.f. 1.2:7 WE15.)

Exercise 2.3:2

④ **1** Arrange these inequalities into the form $f(x) > 0$ (or $f(x) < 0$) and hence solve them:

(i) $\dfrac{3x}{x+1} < 2$ → (ii) $\dfrac{2x-4}{x-1} > 1$

(iii) $\dfrac{4x+3}{3x-4} < 2$ (iv) $\dfrac{(2x-1)(x+4)}{x-5}$

(v) $\dfrac{4}{x+2} > x-1$ (vi) $6x+5 < \dfrac{5}{2x+1}$

(vii) $\dfrac{10-2x}{x-2} > x+1$ → (viii) $\dfrac{x+7}{2x-1} < x+1$

④ → **2** Find the set of values of x for which $\dfrac{2}{x-1} > \dfrac{x}{x+2}$

(Hint: arrange the inequality into the form $f(x) > 0$; then multiply both sides by $(x-1)^2(x+2)^2$.)

④ **3** Find the set of values of x for which:

(i) $\dfrac{1}{x+1} < \dfrac{x}{x-2}$ → (ii) $\dfrac{2x}{x-1} + \dfrac{x-5}{x-2} > 3$

R (iii) $\dfrac{x-2}{2x-3} < \dfrac{x}{3x-2}$ (iv) $\dfrac{x+3}{x-2} < \dfrac{4}{3x-3}$

⑤ → **4** Sketch on the same axes the curve $y = (2x+3)/(2x-1)$ and the line $y = 5$. Find their points of intersection and hence find the set of values of x for which

$$\dfrac{2x+3}{2x-1} > 5$$

(c.f. worked example 4).

⑤ **5** Sketch on the same axes the curve $y = (2x-4)/(x-1)$ and the line $y = 1$. Find their points of intersection and hence find the set of values of x for which

$$\dfrac{2x-4}{x-1} < 1.$$

⑤ R **6** Sketch on the same axes the curve $y = (x+5)/(x-1)$ and the line $y = x+1$. Find their points of intersection and hence find the set of values of x for which

$$\dfrac{x+5}{x-1} < x+1.$$

⑤ → **7** Sketch on the same axes the curves $y = 1/(x+1)$ and $y = x/(x-2)$. Hence find the set of values of x for which

$$\dfrac{1}{x+1} < \dfrac{x}{x-2}$$

8 Find the ranges of the functions f and g, where

$$f(x) = \dfrac{3}{x^2+2x+2} \quad \text{and} \quad g(x) = 4 - f(x)$$

Sketch on the same axes the graphs of $f(x)$ and $g(x)$ and find the set of values of x for which $f(x) > g(x)$.

9 Sketch on the same axes the graphs of f and g, where $f(x) = \dfrac{3}{(x-1)(x+3)}$ and $g(x) = \tfrac{4}{5}x - 1$

Find the set of values of x for which $f(x) \geq g(x)$.

R **10** The functions f and g are defined by

$$f(x) = \dfrac{4(x+1)}{(x-1)^2} \quad g(x) = x+1$$

(i) Show that, for all real values of x, $f(x) \geq -\tfrac{1}{2}$.
(ii) Show that the line $y = g(x)$ is a tangent to the curve $y = f(x)$. Find the coordinates of the point of contact and of the point at which the line meets the curve again.
(iii) Sketch the curve and the line on the same axes.
(iv) For what values of x is $f(x) > g(x)$?

2.3:3 Partial fractions

We add rational functions in the same way that we add fractions, i.e. by first looking for a common denominator, e.g.

$$\dfrac{x-1}{x^2+1} + \dfrac{3}{x-3} \equiv \dfrac{(x-1)(x-3)+3(x^2+1)}{(x^2+1)(x-3)} \equiv \dfrac{4x^2-4x+6}{(x^2+1)(x-3)}$$

Often, we want to reverse this process, i.e. to express the function $3x/[(x+1)(x+2)]$ in the form $A/(x+1) + B/(x+2)$, where A and B are constants that can be found. $A/(x+1)$ and $B/(x+2)$ are called **partial fractions** for $3x/[(x+1)(x+2)]$.

Algebraic, exponential and logarithmic functions 97

When finding partial fractions for a rational function, we choose which type to use by looking at the denominator of the function.

Type 1: real linear factors in denominator

Worked example 6 Find partial fractions for $3x/[(x+1)(x+2)]$.

Let
$$\frac{3x}{(x+1)(x+2)} \equiv \frac{A}{x+1} + \frac{B}{x+2}$$

i.e.
$$\frac{3x}{(x+1)(x+2)} \equiv \frac{A(x+2)+B(x+1)}{(x+1)(x+2)}$$

$$3x \equiv A(x+2) + B(x+1)$$

Since this is an identity, we can put into it any value of x we choose; in particular,

putting $x = -2$: $\quad -6 = B(-1) \Rightarrow B = 6$

putting $x = -1$: $\quad -3 = A(1) \Rightarrow A = -3$

(Why did we choose these particular values of x?)
So
$$\frac{3x}{(x+1)(x+2)} \equiv \frac{6}{x+2} - \frac{3}{x+1}$$

Type 2: denominator containing irreducible quadratic factor

Worked example 7 Find partial fractions for $(x-3)/[(x-1)(x^2+1)]$.

Let
$$\frac{x-3}{(x-1)(x^2+1)} \equiv \frac{A}{x-1} + \frac{Bx+C}{x^2+1}$$

i.e. $\quad x - 3 \equiv A(x^2+1) + (Bx+C)(x-1)$

Since this is an identity we can substitute any chosen value for x.

Putting $x = 1$: $\quad -2 = 2A \Rightarrow A = -1$

Since there is no real value of x for which $x^2 + 1 = 0$, we cannot find B and C immediately in this way; however we can find them by either of two methods:

(i) by comparing coefficients of x^2, x and constants in the identity.

x^2: $\quad 0 = A + B = -1 + B \Rightarrow B = 1$

x^0: $\quad -3 = A - C = -1 - C \Rightarrow C = 2$

(ii) by substituting two values of x (other than $x = 1$) for x in the second identity to obtain two simultaneous equations for B and C.

Putting $x = 0$: $\quad -3 = A - C = -1 - C \Rightarrow C = 2$

Putting $x = 2$: $-1 = 5A + 2B + C = -5 + 2B + 2 \Rightarrow B = 1$

So
$$\frac{x-3}{(x-1)(x^2+1)} \equiv -\frac{1}{x-1} + \frac{x+2}{x^2+1}$$

Type 3: degree of numerator ⩾ degree of denominator

Worked example 8 Find partial fractions for:

$$\frac{x^4 + x^3 - x^2 + 1}{(x-1)(x^2+1)}$$

We must divide first:

$$\frac{x^4 + x^3 - x^2 + 1}{(x-1)(x^2+1)} \equiv x + 2 - \frac{x-3}{(x-1)(x^2+1)} \quad \text{(see 2.2:1 Q3(vi))}$$

$$\equiv x + 2 + \frac{1}{x-1} - \frac{x+2}{x^2+1} \quad \text{(see worked example 7)}$$

Exercise 2.3:3

When finding partial fractions for a rational function, a useful rule of thumb is: if the degree of the denominator is n, we need n constants.

⑥ **1** Find partial fractions for:

(i) $\dfrac{1}{(x-1)(x-2)}$

(ii) $\dfrac{1}{x(x+4)}$

(iii) $\dfrac{2}{x^2-1}$

→ (iv) $\dfrac{x+3}{(x-2)(x+1)}$

(v) $\dfrac{3x+2}{x^2-4}$

R (vi) $\dfrac{x+1}{(2x+3)(3x-5)}$

⑥ **2** (i) Find the constants A, B and C in the identity

$$\frac{x^2}{(x-1)(x-2)(x-3)} \equiv \frac{A}{x-1} + \frac{B}{x-2} + \frac{C}{x-3}$$

(ii) Find partial fractions for:

→ (a) $\dfrac{1}{(x+2)(2x-1)(x-5)}$

(b) $\dfrac{x-1}{x(x^2-4)}$

⑦ **3** Try to find the constants A and B when

$$\frac{x-3}{(x-1)(x^2+1)} \equiv \frac{A}{x-1} + \frac{B}{x^2+1}$$

What goes wrong? (C.f. worked example 7.)

⑦ **4** Find partial fractions for:

→ (i) $\dfrac{5}{(x+1)(x^2+4)}$

R (ii) $\dfrac{1}{(x+2)(2x^2+3)}$

(iii) $\dfrac{10x+1}{(x-2)(x^2+x+1)}$

(iv) $\dfrac{3x}{(x^2+1)(x^2+4)}$

(v) $\dfrac{3x^2+3x+9}{(x^2+1)(x^2+4)}$

⑧ **5** Try to find the constants A, B and C in the identity

$$\frac{x^3}{(x-1)(x^2+1)} \equiv \frac{A}{x-1} + \frac{Bx+C}{x^2+1}$$

What goes wrong? Why?

⑧ **6** Find partial fractions for:

(i) $\dfrac{x^2}{(x-1)(x-2)}$

→ (ii) $\dfrac{x^3}{(x+1)(x-2)}$

(iii) $\dfrac{x^2+1}{x^2-1}$

R (iv) $\dfrac{x^3+1}{x(x^2+1)}$

7 Find partial fractions for:

$$\frac{x-1}{(x+1)(x-3)}$$

Sketch on the same axes the graphs

$$y = \frac{x-1}{(x+1)(x-3)}$$

$$y = \frac{1}{x+1} \quad \text{and} \quad y = \frac{1}{x-3}$$

What do you conclude?

2.4 Modulus functions

2.4:1 Graphs of modulus functions

Suppose that on the real line the point O represents the number zero and the point A the number a. The **modulus** $|a|$ of a is the distance OA, e.g. $|20| = 20$, $|-12| = 12$. Note that

$$|a| = \begin{cases} a & \text{if } a \geq 0 \\ -a & \text{if } a < 0 \end{cases} \qquad (1)$$

$$= +\sqrt{a^2} \qquad (2)$$

A **modulus function** is a function of the form $y = |f(x)|$.

Given the graph of $y = f(x)$, how do we find the graph of $y = |f(x)|$? From equation (1),

(i) if $f(x) \geq 0$ then $|f(x)| = f(x)$, i.e. the part of $y = f(x)$ above the x-axis remains unchanged;

(ii) if $f(x) < 0$, then $|f(x)| = -f(x)$, i.e. the part of $y = f(x)$ below the x-axis is reflected in the x-axis.

Worked example 1 Sketch on the same axes the curves $y = (x+1)(x-2)$ and $y = |(x+1)(x-2)|$.

The solution is given in fig. 2.4:1. The two curves coincide for $x \leq -1$ and $x \geq 2$.

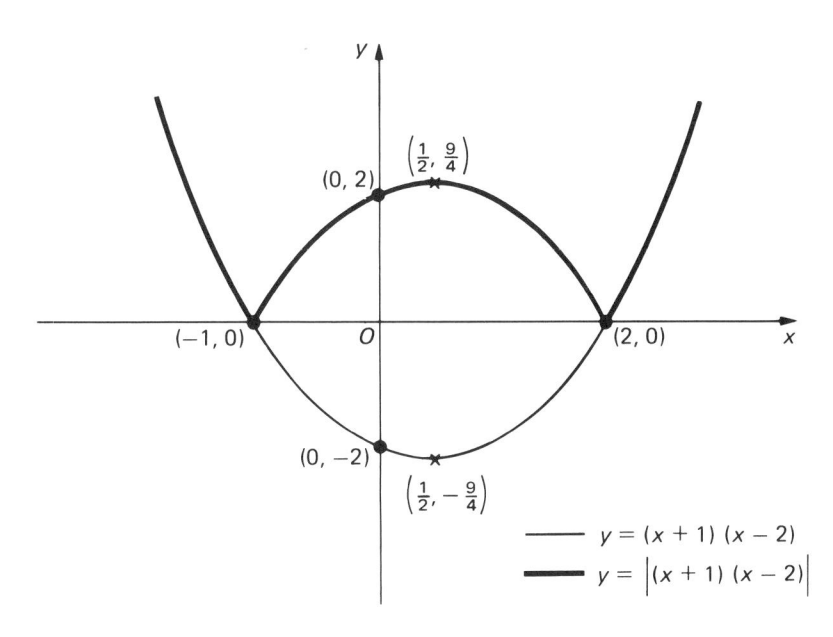

Fig. 2.4:1

Exercise 2.4:1

① 1 Sketch the following graphs:
(i) $y = |x|$
(ii) $y = |x - 4|$
(iii) $y = |4x + 6|$
→ (iv) $y = |4 - 5x|$
(v) $y = |\tfrac{1}{3}x + 1|$
(vi) $y = -2|x - 1|$

① 2 Sketch the following graphs:
(i) $y = |(x-1)(x-4)|$
(ii) $y = |x^2 + x|$
→ (iii) $y = |2x^2 - 5x - 3|$

① **3** Sketch the following graphs:
  (i) $y = |x^3|$
→ (ii) $y = |(x+3)(x-1)(x-5)|$
  (iii) $y = |x^3 + x - 2|$
  (iv) $y = |(x+1)^2(x-1)(x+3)|$

① **4** Sketch the following graphs:
→ (i) $y = \left|\dfrac{x}{x+4}\right|$
  (ii) $y = \left|\dfrac{2x+1}{x-2}\right|$
  (iii) $y = \left|\dfrac{1}{(x+1)(x-3)}\right|$

① R **5** Sketch the following graphs:
→ (i) $y = |5+3x|$
  (ii) $y = |5-3x|$
  (iii) $y = |(5+3x)(5-3x)|$
  (iv) $y = \left|\dfrac{5+3x}{5-3x}\right|$

① **6** Sketch on the same axes the curves $y = |1/(x+1)|$ and $y = |x/(x+2)|$. Hence find the number of real roots of the equation

$$\left|\dfrac{1}{x+1}\right| = \left|\dfrac{x}{x+2}\right|$$

① **7** Using the method of question 6, find the number of real roots of the equation

$$\left|\dfrac{x+2}{x-2}\right| = \left|\dfrac{1}{x-3}\right|$$

① **8** Find the turning points of the function $y = (x+3)(x-4)(x-5)$. Find the number of real roots of the equation $|(x+3)(x-4)(x-5)| = 1$.

9 Sketch the curves:
  (i) $|y| = x - 1$
  (ii) $|y| = (x-1)(x-2)$
  (iii) $|y| = (x-1)(x-2)(x-3)$

2.4:2 Equations and inequalities

Worked example 2 Find the values of x for which $|x-4| = 3$.

Remembering that $|x-4| \equiv \sqrt{(x-4)^2}$ (equation (2)), we square both sides of the equation:

$$|x-4| = 3$$
$$(x-4)^2 = 9$$
$$x^2 - 8x + 7 = 0$$
$$(x-7)(x-1) = 0$$

So $x = 1$ or $x = 7$ (see fig. 2.4:2).

Note: a possible interpretation of the equation is: 'the distance from x to 4 is 3' (see 1.1:4 QB).

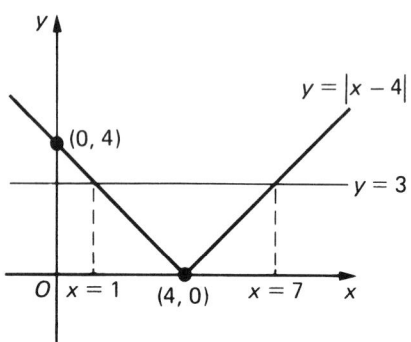

Fig. 2.4:2

Worked example 3 Find the set of values of x for which $|3-2x| < |x+4|$.

Following the method of worked example 2, we square both sides of the inequality:

$$|3-2x| < |x+4|$$
$$(3-2x)^2 < (x+4)^2$$
$$9 - 12x + 4x^2 < x^2 + 8x + 16$$
$$3x^2 - 20x - 7 < 0$$

i.e. $-\tfrac{1}{3} < x < 7$

Exercise 2.4:2

② **1** Solve the equations:
  (i) $|x-5|=4$
→ (ii) $|4-5x|=6$
  (iii) $|2x-1|=1$
  (iv) $|2x-1|=-1$

②→ **2** Sketch on the same axes the graphs of $y=|2x-7|$ and $y=3$.
  Find the values of x for which $|2x-7|=3$.

② **3** Solve the equations:
  (i) $|x-3|=|x+5|$
  (ii) $|2x-1|=|x+2|$
  (iii) $|5x-3|=|1-3x|$
  (iv) $|3-2x|=|x+4|$
  (v) $|2-x|=\frac{1}{2}|x|$
  (vi) $|x+2|=3|x-4|$

②→ **4** Sketch on the same axes the graphs of $y=|2x-6|$ and $y=|3x+1|$.
  Find the values of x for which $|2x-6|=|3x+1|$.

② **5** Show that x^4+2x^3-4x-4 has a factor x^2-2. Hence solve the equation
$$\left|\frac{1}{x+1}\right| = \left|\frac{x}{x+2}\right|$$

③ **6** Find the set of values of x for which:
  (i) $|x-4|>3$
  (ii) $|3-2x|<5$
→ (iii) $3|x-1|>1$
R (iv) $2|2x-1|\leq 7$

③→ **7** Sketch on the same axes the graphs of $y=|3x-5|$ and $y=4$.
  Find the set of values of x for which $|3x-5|>4$.

③ **8** Find the set of values of x for which:
  (i) $|x+3|>|x+2|$
→ (ii) $|3-2x|>|x-4|$
  (iii) $|1-5x|<|3x-4|$
  (iv) $|x+2|\leq 2|x+1|$
R (v) $|3x-4|>|2x-1|$

③→ **9** Sketch on the same axes the graphs of $y=|3x-2|$ and $y=|x+2|$.
  Find the set of values of x for which $|3x-2|<|x+2|$.

③ **10** Find the set of values of x for which:
  (i) $|x^2-4|\geq|x^2-1|$
  (ii) $|2x^2-3|>|x^2+2|$

③→ **11** Sketch on the same axes the graphs of $y=\sqrt{(8-x^2)}$ and $y=|x|$. Using the method of worked example 2, find the set of values of x for which $\sqrt{(8-x^2)}<|x|$. (Note: $\sqrt{(8-x^2)}$ is real only for certain values of x—which ones?)

③ **12** Find the set of values of x for which:
R (i) $|9-x^2|<8x^2$
  (ii) $\sqrt{(9-x^2)}<2\sqrt{2}|x|$

③ **13** Find the set of values of x for which:
  (i) $\sqrt{(4x)}>|x-3|$
  (ii) $\sqrt{(2x-1)}<\sqrt{(x+3)}$

③ **14** Sketch on the same axes the graphs of
$$y=\left|\frac{1}{x-2}\right| \quad \text{and} \quad y=\left|\frac{1}{x+3}\right|$$
Find the set of values of x for which
$$\left|\frac{1}{x-2}\right| < \left|\frac{1}{x+3}\right|$$

15 If $f(x)=|x-1|+|x-4|$, show, using equation (1), that
$$f(x)=\begin{cases} 2x-5 & \text{when } x\geq 4 \\ 3 & \text{when } 1<x<4 \\ -2x+5 & \text{when } x\leq 1 \end{cases}$$
Hence sketch a graph of $f(x)$.

16 Sketch the graphs of the following functions:
  (i) $y=|x+2|+|x-3|$
  (ii) $y=|2x-1|+|x-5|$
  (iii) $y=|x-1|-|x-4|$

17 If $f(x)=|2x-1|-|x+2|$, show that
$$f(x)=\begin{cases} x-3 & \text{when } x\geq \frac{1}{2} \\ -3x-1 & \text{when } -2<x<\frac{1}{2} \\ 3-x & \text{when } x\leq -2 \end{cases}$$
Find the set of values of x for which $f(x)\leq 2$.
Sketch on the same axes the graphs of $y=f(x)$ and $y=2$, and label their points of intersection. What is the range of function $f(x)$?
What is the value of $f'(4)$? Explain why $f'(-2)$ is undefined. For which other value of x is $f'(x)$ undefined?

2.5 Exponential and logarithmic functions

2.5:1 Power functions

A power function is a drinks party at which a lot of ambitious and ruthless people attempt to make each other feel small... but back to Mathematics...

A **power function** is a function in which the independent variable occurs as a power (i.e. an index), e.g. 2^x, 3^{-4x}.

The most important power functions are functions of the form a^x, where a is a positive constant.

Worked example 1 Sketch the curve $y = f(x)$, where $f(x) = 2^x$.

x	-3	-2	-1	0	1	2	3
$f(x)$	$\frac{1}{8}$	$\frac{1}{4}$	$\frac{1}{2}$	1	2	4	8

As $x \to +\infty$, $2^x \to +\infty$ very quickly.
As $x \to -\infty$, $2^x \to 0$; so $y = 0$ is an asymptote of the curve, which is shown in fig. 2.5:1.

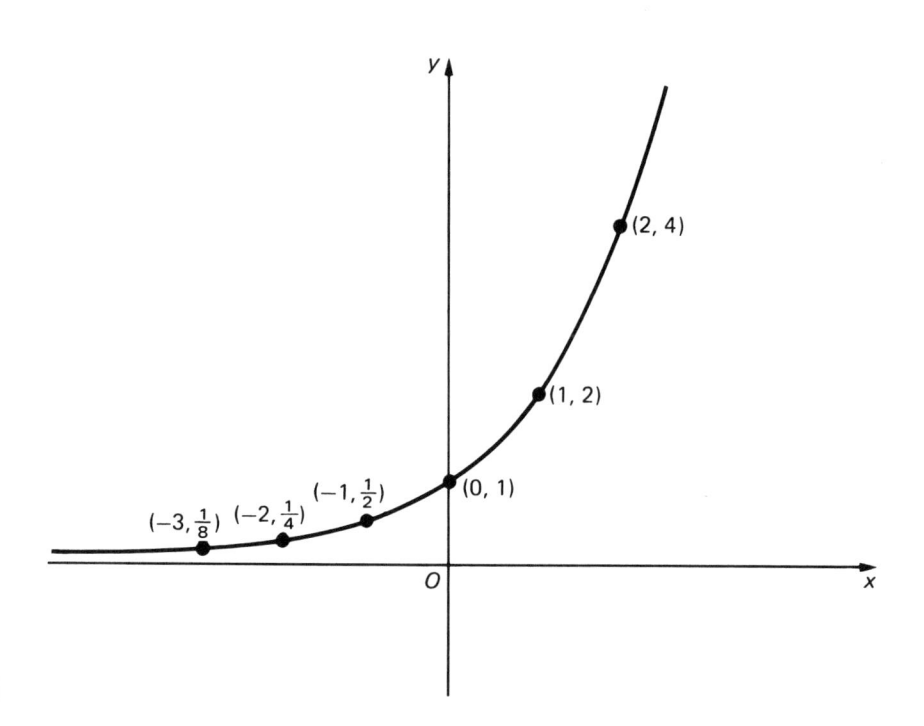

Fig. 2.5:1

Notice that 2^x is always positive; so the graph lies entirely above the x-axis (i.e. the range of 2^x is $\mathbb{R}^+$).

Worked example 2 The total number of nuclear missiles available to the two superpowers increases by 4% every month. Find how long it takes for the number to double.

At the end of each month, the number of available missiles is 1.04 times what it was at the beginning of the month.

Suppose that initially the number of missiles is N.
After 1 month, the number is $1.04N$.
After 2 months, the number is $(1.04)(1.04N)$, i.e. $(1.04)^2 N$.
After t months, the number, x say, is $(1.04)^t N$.
How long does it take for the number to double, i.e. to reach $2N$?

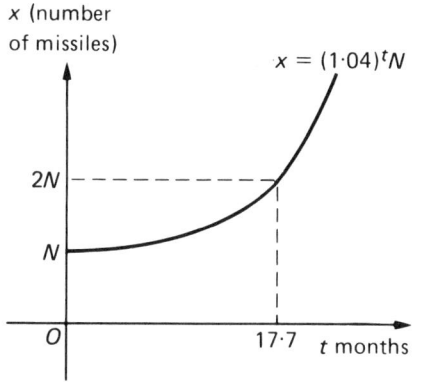

$$x = 2N \Rightarrow (1.04)^t N = 2N$$
$$\Rightarrow (1.04)^t = 2$$
$$\Rightarrow t = \log_{1.04} 2$$
$$= 17.7$$

so the number of missiles doubles in 17.7 months, i.e. just less than a year and a half (see fig. 2.5:2).

Notice that it takes this amount of time for the number N to double, whatever its initial value.

We have assumed that the number of missiles increases *continuously*. In fact, the number increases in jumps (of one) and at irregular intervals, but the jumps are so small compared with the actual number that we can sensibly think of the increase as continuous.

Fig. 2.5:2

Exercise 2.5:1

1 The function $f : \mathbb{R} \to \mathbb{R}$ is defined by $f(x) = 3^x$. Find $f(x)$ when $x = -3, -2, -1, 0, 1, 2, 3$. Find $\lim_{x \to -\infty} f(x)$ and $\lim_{x \to \infty} f(x)$. Sketch the graph of f. What is the range of f?

2 Sketch on the same axes the curves $y = 2^x$ and $y = (\frac{1}{2})^x$. At what point do they intersect?
Explain why $y = (\frac{1}{2})^x$ is the mirror image of $y = 2^x$ in the y-axis.

3 Show that all functions of the type $f : \mathbb{R} \to \mathbb{R}$ defined by $f(x) = a^x$, $a > 0$, have the same range. Sketch a general graph of a^x in the cases:

(i) $a > 1$ (ii) $a = 1$ (iii) $a < 1$

Explain the problems of defining a function of the type $f(x) = a^x$, $a < 0$.

4 Sketch on the same axes the graphs of the functions:

(i) $2^x, 3^x, 4^x$
(ii) $2^x, 2^x + 1, 2^x - 2$
(iii) $3^x, 3^{x-2}, 3^{x+1}$
(iv) $4^x, 4^{-x}, -4^x, -4^{-x}$
(v) $2^x, 2^{2x}, 2^{3x}$

showing clearly their asymptotes.

5 Sketch the curves:

(i) $y = 3^x - 2^x$ (ii) $y = 2^{1/x}$

6 The population x of a colony of insects increases by 50% every day. If there are initially 100 insects in the colony, how many insects are there after t days?

(i) Sketch a graph of x against t.

(ii) Find the population after:
(a) 5 days; (b) 3.1 days; (c) 1.7 days.
(iii) How long does the population take to reach:
(a) 500; (b) 1000; (c) 1500?

7 A man invests £1000 in a deposit account. The (compound) interest rate is 10% per year. After t years he has £x in his account.

(i) Sketch a graph of x against t.
(ii) Find the amount in his account after:
(a) 18 months; (b) 3 years; (c) $4\frac{1}{2}$ years.
(iii) How long does it take before the amount is:
(a) £1300; (b) £1500; (c) £2000?

8 A man buys a car for £5000. The car depreciates at 20% per year. After t years it is worth £x.

(i) Sketch a graph of x against t.
(ii) How long does it take for the car to depreciate to:
(a) 75%; (b) 60%; (c) 50%
of its original value?

9 The cost of living goes up by 12% per year. By how much does it go up in:

(i) 1 month (ii) 6 months (iii) 5 years?

How long does it take for the cost of living to double?

10 The population of a colony of gorillas increases by 6% per year. How long does it take for the population to:

(i) double (ii) treble?

104 *The exponential function* e^x

② → **11** A sample of radium loses mass at a rate of 4% per century. Find its half-life, i.e. the time taken for its mass to be halved.

② R **12** The half-life of carbon-14 is 5600 years.

(i) How long would it take for a sample of carbon-14 to lose 20% of its mass?

(ii) What is the percentage decrease in the mass per century?

② **13** 3% of light is absorbed in passing through a window with glass 0.6 cm thick. How much would be absorbed by a window where the glass is 2 cm thick?

② **14** The frequencies of the notes middle C and c (the octave above middle C) are 256 Hz and 512 Hz. The octave between them is divided into twelve intervals by the notes C♯, D, D♯, E, F, F♯, G, G♯, A, A♯ and B, and the frequencies of successive notes are in constant ratio. Find this ratio.

What are the frequencies of the notes E, F and G?

② **15** A businessman on a transatlantic aeroplane flight, depressed by the low standard of the food, jumps out of the plane. His velocity, v m s^{-1}, t seconds after he has jumped from the plane is given by

$$v = 200(1 - a^{-t})$$

where a is a positive constant. If after 5 seconds his velocity is 30 m s^{-1}, find a, correct to 3 decimal places. Find his speed after:

(i) 10 seconds (ii) 40 seconds

Show, with the aid of a graph, that his velocity approaches a certain value. What is this value? What are his chances of survival?

2.5:2 The exponential function e^x

Consider the graph of the power function $f(x) = a^x$ (where a is a positive constant) shown in fig. 2.5:3. Whatever the value of a, $a^0 = 1$, so the graph passes through the point $A(0, 1)$. Using the definition of the derivative (1.3:1 equation (1a)), with $x = 0$, the gradient of the graph at $A(0, 1)$ is given by

$$f'(0) = \lim_{\delta x \to 0} \frac{f(\delta x) - f(0)}{\delta x} = \lim_{\delta x \to 0} \frac{a^{\delta x} - a^0}{\delta x} = \lim_{\delta x \to 0} \frac{a^{\delta x} - 1}{\delta x} \quad (1)$$

and the gradient of the graph at $P(x, a^x)$ by

$$f'(x) = \lim_{\delta x \to 0} \frac{a^{x+\delta x} - a^x}{\delta x}$$

$$= \lim_{\delta x \to 0} a^x \left\{ \frac{a^{\delta x} - 1}{\delta x} \right\}$$

$$= a^x \lim_{\delta x \to 0} \frac{a^{\delta x} - 1}{\delta x}$$

$$= a^x f'(0) \quad \text{from equation (1)}$$

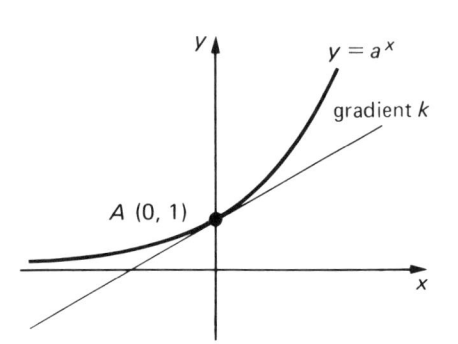

Fig. 2.5:3

Thus

$$f(x) = a^x \Rightarrow f'(x) = a^x f'(0) \quad (2)$$

(Note that in deriving $f'(x)$ we have assumed that

$$\lim_{\delta x \to 0} f(x) g(\delta x) = f(x) \lim_{\delta x \to 0} g(\delta x),$$

which is true since $f(x)$ does not depend on δx.)

Now there must be a value of a for which $f'(0) = 1$, i.e. the gradient of the graph at $(0, 1)$ is 1. We call this value e. Thus

$$\frac{d}{dx}(e^x) = e^x \quad (3)$$

Algebraic, exponential and logarithmic functions 105

In fact e is an irrational number. We shall see later that $e \approx 2.718$.

e^x is called the **exponential function**. It is sometimes written as exp x.

Worked example 3 Differentiate:
(i) e^{5x} (ii) $e^{(x^3+x)}$

We use the chain rule:

(i) $\dfrac{d}{dx}(e^{5x}) = e^{5x} \cdot 5 = 5e^{5x}$

(ii) $\dfrac{d}{dx}(e^{x^3+x}) = e^{x^3+x} \cdot (3x^2+1) = (3x^2+1)\,e^{x^3+x}$

Worked example 4 Sketch the graph $y = xe^{-x}$.

When $x = 0$, $y = 0$, and when $y = 0$, $xe^{-x} = 0$, so $x = 0$ since $e^{-x} > 0\ \forall x$.
By the product rule, $dy/dx = (1-x)\,e^{-x}$. At a turning point $dy/dx = 0$, so $x = 1$, $y = 1/e$.

$$\dfrac{d^2y}{dx^2} = (x-2)\,e^{-x}$$

When $x = 1$, $d^2y/dx^2 < 0$, so $(1, 1/e)$ is a maximum point.
When $x \to -\infty$, $e^{-x} \to +\infty$, so $xe^{-x} \to -\infty$.
When $x \to +\infty$, $e^{-x} \to 0$ very rapidly compared to x; so $xe^{-x} \to 0$.
The graph of $y = xe^{-x}$ is shown in fig. 2.5:4. Note: $d^2y/dx^2 = (x-2)\,e^{-x} = 0$ when $x = 2$, $y = 2/e^2$; so $(2, 2/e^2)$ is a point of inflexion.

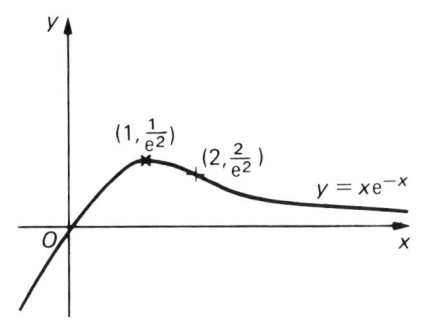

Fig. 2.5:4

Exercise 2.5:2

1 Using graph paper, draw on the same axes the graphs of $y = 2^x$ and $y = 3^x$ for $-2 < x < 2$ (see 2.5:1 Q4(i).

Draw the tangent to each graph at the point $(0, 1)$ and measure the gradient of each tangent.

• → **2** Using tables or a calculator, make a table of values of e^x for $-3 < x < 3$. Using graph paper, draw the graph of $y = e^x$ for $-3 < x < 3$. Draw the tangent to the graph at the point $A(0, 1)$ and measure its gradient.

Find the gradients of the lines AB, AC, AD and AE where B, C, D and E are the points on the curve with x-coordinates 0.5, 0.1, 0.05 and 0.01.

3 Sketch on the same axes the curves:
 (i) $y = e^x$, $y = e^{2x}$
 (ii) $y = e^x$, $y = e^{-x}$
→ (iii) $y = e^x$, $y = -e^{-x}$, $y = \tfrac{1}{2}(e^x - e^{-x})$
 (iv) $y = e^x$, $y = 3e^x$

4 On the same axes, sketch the graphs of $y = e^{2x}$ and $y = e^{x^2}$. Find their points of intersection.

③ **5** Differentiate:
 (i) e^{3x}
 (ii) e^{-x}
 (iii) $e^{-3x/2}$
→ (iv) $x^2 e^x$
R (v) xe^{-5x}

③ **6** Differentiate:
 (i) e^{x^3}
 (ii) $e^{\sqrt{x}}$
→ (iii) $e^{-2/x}$
R (iv) $\dfrac{e^x}{e^x + 1}$

7 Differentiate:
 (i) e^{ax+b}, a and b constant (ii) $e^{f(x)}$

Questions 8–12: Sketch the curve $y = f(x)$, showing any points of intersection with the axes, turning points and asymptotes. Find any points of inflexion.

④ **8** $y = xe^x$

④ **9** $y = (x+1)\,e^{-x}$

④ → **10** $y = xe^{1-x^2}$

④ **11** $y = \dfrac{e^x}{x}$

④ **12** $y = xe^{-x^2/2}$

④ R **13** Find the coordinates of the turning points of the curve $y = x^2 e^{-x}$. Sketch the curve.

Find the equation of the tangent to the curve at the point $(1, 1/e)$.

④ **14** The tangents to the curve $y = e^x(ax^2 + bx + c)$, a, b and c constant, are parallel to the x-axis when $x = 1$ and $x = 3$, and the curve cuts the y-axis at $y = 9$. Find a, b and c, and sketch the curve.

15 It is a common experience, when learning a new skill, such as the saxophone, or French, or bank robbery, that improvement happens quickly at first and more slowly later. The amount of expertise E is approximately related to the learning time t by an equation of the form $E = 1 - e^{-kt}$ where k is a constant. Sketch the graph of E against t.

We could call this curve the **learning curve**. What is the significance of the value $E = 1$? show that $dE/dt = k(1 - E)$. Interpret this equation.

Suppose that you are learning a new skill and that after 6 months you are 50% expert ($E = 0.5$). Find how long, approximately, you will take to become (i) 95% expert; (ii) 99% expert.

16 A certain obsessively vain young man makes a record every morning of the number of hairs on his chest. After years of careful observation, he discovers that the number of hairs on his chest n is approximately related to the time t years since his tenth birthday by the formula

$$n = \frac{1000 e^{t/4}}{999 + e^{t/4}}$$

Sketch a graph of n against t.

How many hairs does he have on his chest on (i) his tenth birthday; (ii) his twentieth birthday?

Show that he can expect to finish his life with 1000 hairs on his chest. Show that $dn/dt \propto n(1000 - n)$, and interpret this equation.

2.5:3 Logarithmic functions

Consider the **logarithmic function** $\log_a x$ (a is a positive constant)

$$y = \log_a x \Leftrightarrow x = a^y \qquad (4)$$

so

$\log_a x$ is the inverse function of a^x

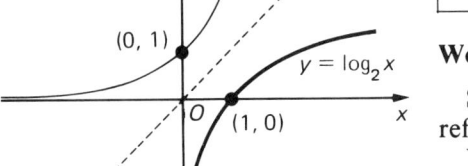

Worked example 5 Sketch the curve $y = f(x)$, where $f(x) = \log_2 x$.

Since $\log_2 x$ is the inverse function of 2^x, the graph of $y = \log_2 x$ is the reflection of $y = 2^x$ in the line $y = x$.

When $x \leq 0$, $\log_2 x$ is undefined (i.e. the domain of $\log_2 x$ is $\mathbb{R}^+$).
$\log_2 1 = 0$, so the curve passes through the point $(1, 0)$.
As $x \to 0^+$, $\log_2 x \to -\infty$; so $x = 0$ is an asymptote of the curve.
As $x \to +\infty$, $\log_2 x \to +\infty$, but more slowly than x.
The graphs of $y = \log_2 x$ and $y = 2^x$ are shown in fig. 2.5:5.

Fig. 2.5:5

Exercise 2.5:3

⑤ **1** Sketch on the same axes the graphs of the functions:

(i) 3^x and $\log_3 x$
(ii) $(\frac{1}{2})^x$ and $\log_{1/2} x$

⑤ **2** Sketch on the same axes the graphs of the functions:

(i) $\log_2 x$, $\log_2 (x - 3)$, $\log_2 (x + 1)$
(ii) $\log_3 (x)$, $\log_3 (-x)$, $\log_3 (1 - x)$
(iii) $\log_2 x$, $\log_3 x$, $\log_4 x$

2.5:4 The logarithmic function ln x

We write $\log_e x$ as $\ln x$ (l for log, n for natural. John Napier (1550–1617) invented natural logarithms. You may think that natural logarithms are about as natural as a MacDonald's milkshake, but to Napier they were as natural as natural yoghurt). $y = \ln x = \log_e x \Leftrightarrow x = e^y$

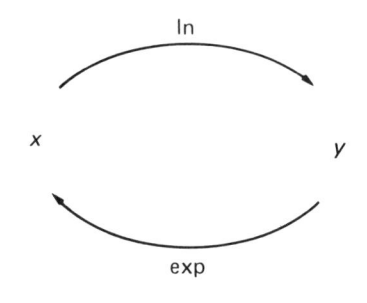

so

$\ln x$ is the inverse function of e^x

In particular,

$$\ln(e^x) = x \quad (5)$$
$$e^{\ln y} = y \quad (6)$$

Derivative of ln x

$$y = \ln x \Rightarrow x = e^y$$

Differentiating both sides with respect to x gives

$$1 = e^y \frac{dy}{dx} \quad \text{(by implicit function rule)}$$

$$\frac{dy}{dx} = \frac{1}{e^y} = \frac{1}{x}$$

$$\frac{d}{dx}(\ln x) = \frac{1}{x} \quad (7)$$

Worked example 6 Differentiate:
(i) (a) $\ln 3x$; (b) $\ln(1+x^2)$ (ii) $\ln x\sqrt{(x-1)}$

(i) Using the function of a function rule (1.3:6 Theorem 2):

(a) $\dfrac{d}{dx}(\ln 3x) = \dfrac{1}{3x} \cdot 3 = \dfrac{1}{x}$

(b) $\dfrac{d}{dx}\{\ln(1+x^2)\} = \dfrac{1}{1+x^2} \cdot 2x = \dfrac{2x}{1+x^2}$

(ii) It is often helpful to simplify the function first, using the rules of logarithms (see 1.1:7 Theorem 5).

$$y = \ln x\sqrt{(x-1)} = \ln x + \ln \sqrt{(x-1)} = \ln x + \tfrac{1}{2}\ln(x-1)$$

Thus $\quad \dfrac{dy}{dx} = \dfrac{1}{x} + \dfrac{1}{2(x-1)}$

Worked example 7 Sketch the graph $y = (\ln x)/x$.

The largest possible domain of the function $(\ln x)/x$ is $\mathbb{R}^+$, since $\ln x$ is not real when x is negative.
As $x \to 0$, $\ln x \to -\infty$; so $(\ln x)/x \to -\infty$; and $x = 0$ is an asymptote.
When $y = 0$, $(\ln x)/x = 0 \Rightarrow \ln x = 0 \Rightarrow x = 1$. So $dy/dx = 1 - \ln x/x^2$.
At the turning points $dy/dx = 0$, i.e. $\ln x = 1 \Rightarrow x = e$, $y = 1/e$.

$$\frac{d^2y}{dx^2} = \frac{2\ln x - 3}{x^3}$$

When $x = e$, $d^2y/dx^2 < 0$, so $(e, 1/e)$ is a maximum point.
As $x \to +\infty$, $\ln x \to +\infty$, but very slowly compared with x; so $(\ln x)/x \to 0$.
The graph of $y = (\ln x)/x$ is shown in fig. 2.5:6.

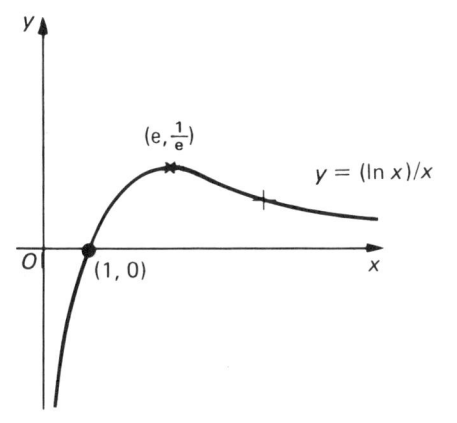

Fig. 2.5:6

Exercise 2.5:4

1 Find the values of:
 (i) $\ln e$ (ii) $\ln e^2$
 (iii) $\ln 1/e$ (iv) $\ln \sqrt{e}$

2 Find the values of:
 (i) $e^{\ln 2}$ (ii) $e^{2\ln 3}$
 (iii) $e^{\frac{1}{3}\ln 8}$ (iv) $e^{\ln 12 - \ln 3}$
 (v) $e^{-\ln 4}$ (vi) $e^{\ln 2} + e^{-\ln 2}$
 (vii) $e^{-\frac{1}{4}\ln 4}$

3 Simplify:
 (i) $e^{2\ln x}$ (ii) $e^{\ln x + \ln y}$
 (iii) $e^{-\ln x}$ (iv) $e^{-\frac{1}{2}\ln x}$

4 Sketch on the same axes the graphs of $y = e^x$, $y = x$ and $y = \ln x$. What is:
 (i) $\ln 1$ (ii) $\lim_{x \to 0^+} (\ln x)$
 (iii) $\lim_{x \to \infty} (\ln x)$

Notice that $\ln x$ is undefined when $x \leq 0$.
For what values of x is $\ln x$: (a) positive; (b) negative?

5 Sketch on the same axes the graphs of:
 (i) $y = \ln x$ and $y = \ln 2x$
 (ii) $y = \ln x$ and $y = 2\ln x$
 (iii) $y = \ln x$ and $y = \ln(-x)$
 (iv) $y = \ln x$, $y = \ln(1+x)$ and $y = \ln(2+x)$

What are the domains of the functions $\ln 2x$, $2\ln x$, $\ln(-x)$, $\ln(1+x)$, $\ln(2+x)$?

6 Sketch the graphs of:
 (i) $\ln(1-x)$ (ii) $\ln(3+4x)$
 R (iii) $\ln(3-4x)$

7 What are the domain and the range of the function $\ln(\sin x)$? Is it a periodic function? If so, what is its period?
Sketch the graph of the function, showing its asymptotes.

8 Differentiate:
 (i) $\ln 2x$ (ii) $\ln \frac{1}{3}x$
 (iii) $\ln(2+3x)$ (iv) $\ln(1+x^3)$
 (v) $\ln(1+e^{2x})$ (vi) $1/\ln x$
 (vii) $(\ln x)^{1/2}$ (viii) $x \ln x$
 (ix) $e^x \ln x$ (x) $(\ln x)/(\ln 3x)$

9 Differentiate:
 (i) $\ln(ax+b)$, a, b constant
 (ii) $\ln f(x)$

10 Simplify these functions first; then differentiate them:
 (i) $\ln x^2$ (ii) $\ln \sqrt{x}$
 (iii) $\ln \dfrac{1}{x}$ (iv) $\ln \dfrac{1+x}{1-x}$
 (v) $\ln x^2 \sqrt{(1-x^2)}$
 R (vi) $\ln \sqrt{\{(1+x^2)/(1-x^2)\}}$

11 Differentiate $\ln kx$, k constant. Show that, if $x < 0$,
$$\frac{d}{dx}(\ln(-x)) = \frac{1}{x}$$

12 (i) Using the change of base rule for logarithms (1.1:7 Theorem 6), differentiate $\log_a x$ where a is a constant.
 (ii) Differentiate $\log_{10} x^2$.

13 (i) By writing $10 = e^{\ln 10}$, differentiate 10^x. (Remember that $\ln 10$ is a constant.)
 (ii) Similarly, differentiate x^x.

14 $f(x) = x \ln x$. Find $f(x)$ when $x = 1$, 0.5, 0.2, 0.1, 0.01, 0.001.
What is $\lim_{x \to 0} f(x)$?
Sketch the curve $y = f(x)$.

Questions 15-17: Write down the largest possible domain of $f(x)$, and sketch the graph of $y = f(x)$, showing any points of intersection with the axes, turning points and asymptotes.

R 15 $f(x) = x^2 \ln x$

16 $f(x) = \ln(x^2 + x + 1)$

17 $f(x) = (\ln x)/x^2$

18 Solve the equations:
 (i) $2e^x - 2e^{-x} - 3 = 0$ (ii) $e^{2x} - 6e^{-2x} - 1 = 0$
 (iii) $e^{3x} - 4e^x + 3e^{-x} = 0$
 (See 1.1:6 WE8.)

19 Find the stationary values of the functions:
 R (i) $e^{2x} - 5e^x + 2x$
 (ii) $3e^{3x} - 9e^{2x} + 8e^x$
and determine their nature.

20 Sketch on the same axes the graphs $y = e^x$ and $y = 2 + 3e^{-x}$ and find their point of intersection.

21 If $f(x) = (1 - e^{2x})/(1 + e^{2x})$, $x \in \mathbb{R}$, find $\lim_{x \to -\infty} f(x)$ and $\lim_{x \to +\infty} f(x)$. Show that $f'(x)$ is always negative. Hence find the range of $f(x)$. Sketch the curve $y = f(x)$.
Find $f^{-1}(x)$. What is the domain of $f^{-1}(x)$? Sketch the curve $y = f^{-1}(x)$.

22 The function $f: \mathbb{R} \to \mathbb{R}$ is defined by $f(x) = \frac{1}{2}(e^x + e^{-x})$. Sketch on the same axes the curves $y = e^x$, $y = e^{-x}$ and $y = f(x)$. What is the range of f?
Solve the equations:
 (i) $f(x) = 1$ (ii) $f(x) = 2$

Show that, if the domain of f is restricted to $\mathbb{R}^+$, then f has an inverse, and find it.

2.5:5 Logarithmic differentiation

In differentiating functions of the type $f(x)^{g(x)}$, we first put $y = f(x)^{g(x)}$ and take ln of both sides.

Worked example 8 Differentiate 10^x.

$y = 10^x \Rightarrow \ln y = x \ln 10$. Using the implicit function rule,

$$\frac{1}{y}\frac{dy}{dx} = \ln 10 \quad \text{(since ln 10 is constant)}$$

$$\frac{dy}{dx} = y \ln 10$$

$$= 10^x \ln 10$$

Thus $\quad \dfrac{d}{dx}(10^x) = 10^x \ln 10$

Worked example 9 Differentiate x^x.

$y = x^x \Rightarrow \ln y = x \ln x$. Using the implicit function rule,

$$\frac{1}{y}\frac{dy}{dx} = \ln x + 1$$

So $\quad \dfrac{dy}{dx} = y(\ln x + 1)$

$$= x^x(\ln x + 1)$$

Exercise 2.5:5

⑧ **1** Differentiate:
 (i) 3^x (ii) 8^x

⑧ → **2** In section 2.5:2 we found that $(d/dx)(a^x) = ka^x$, where k is a constant which depends on a. Use the method of worked example 8 to find k.

⑧ **3** Differentiate:
 (i) 10^{3x} (ii) 2^{5x+1} (iii) 3^{x^2}

⑨ **4** Differentiate:
 (i) x^{2x} (ii) $3x^x$ (iii) $(5x)^x$

⑨ **5** Differentiate:
 (i) $(x+1)^x$
 (ii) $(1+2x)^{3x-1}$
 (iii) $(3x-1)^{1+2x}$

⑨ + **6** Sketch the graph of the function x^x, $x > 0$.

7 The growth rate of a plant is defined as the rate of increase, dW/dt, of its dry weight W. Its relative growth rate is the rate of increase of dry weight per unit of dry weight, i.e. $(1/W) \, dW/dt$.

(i) Show that the relative growth rate is
$$(d/dt)(\ln W).$$

(ii) Find the growth rate and relative growth rate of a plant whose weight W is given by:
(a) $W = at + b$; (b) $\ln W = at + b$;
(c) $\ln W = at^2 + bt + c$;
where a, b and c are constants

8 Suppose that the quantity q of some product that the public would like to buy if the price were p is given by $q = f(p)$. The elasticity of demand e is defined by

$$e = \frac{dq/dp}{q/p}$$

(i) Show that $e = \dfrac{d\{\ln f(p)\}}{d(\ln p)}$.

(ii) Show that, if $f(p) = kp^\alpha$ ($k > 0$, $\alpha < 0$, k and α constant) then $e = \alpha$.
(iii) Sketch the graph of q against p when $f(p) = 10/p$ and find e in this case.

R 9 (i) If $y = kx^\alpha$ (α and k constant), show that $\ln y = \ln k + \alpha \ln x$.
Hence show that small increases δx in x and δy in y are related by

$$\frac{\delta y}{y} \approx \alpha \frac{\delta x}{x}$$

(ii) Hence find the percentage increase in the volume of a sphere caused by a 0.5% increase in its radius.

→ **10** If $y = \ln x$, show that small increases δx in x and δy in y are related by $\delta y \approx \delta x / x$.
Given that $\ln 2 \approx 0.693\ 15$, estimate (to 5 dec. pls.):
(i) $\ln 2.0001$ (ii) $\ln 1.9997$

2.6 Approximate solution of equations

2.6:1 Curve sketching

In sketching the graph of a function $f(x)$ we have, until now, always solved the equation $f(x) = 0$ to find where the graph cuts the x-axis.

Now let us suppose that we are trying to solve the equation

$$f(x) = x^3 - 3x + 1 = 0 \tag{1}$$

The equation has no obvious integer roots; so it would be difficult to solve it (...though not impossible—see miscellaneous exercise 6, Q49). However, we can sketch the graph of $f(x)$ (without, obviously, trying to solve equation (1)), and hence find approximately the real roots of the equation by seeing where the graph cuts the x-axis. We can then refine these approximations by using the Newton–Raphson method or the $x = \phi(x)$ method (section 2.6:2).

Worked example 1 Sketch the curve $y = f(x)$ where $f(x) = x^3 - 3x + 1$. Hence find the number of real roots of the equation $f(x) = 0$.

When $x = 0$, $y = 1$.
$dy/dx = 3x^2 - 3$. At a turning point $dy/dx = 0$ so $x = 1$ or $x = -1$. When $x = 1$, $y = -1$; when $x = -1$, $y = 3$. Thus the turning points are $(-1, 3)$ and $(1, -1)$.
As $x \to +\infty$, $y \to +\infty$, and as $x \to -\infty$, $y \to -\infty$.
The curve (fig. 2.6:1) cuts the x-axis three times (at A, B and C), so the equation $f(x) = 0$ has three real roots.

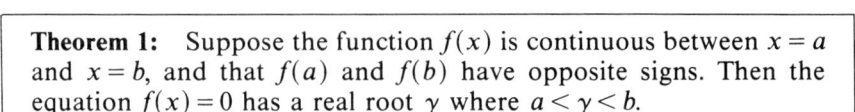

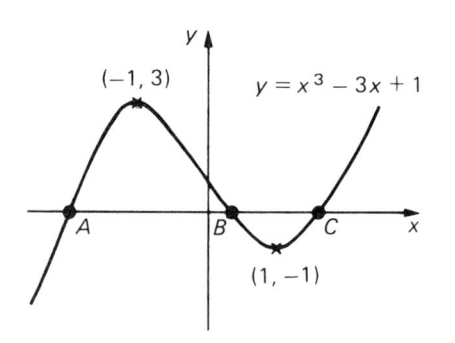

Fig. 2.6:1

> **Theorem 1:** Suppose the function $f(x)$ is continuous between $x = a$ and $x = b$, and that $f(a)$ and $f(b)$ have opposite signs. Then the equation $f(x) = 0$ has a real root γ where $a < \gamma < b$.

Proof
(i) Figure 2.6:2a shows the case $f(a) < 0$, $f(b) > 0$.
(ii) Figure 2.6:2b shows the case $f(a) > 0$, $f(b) < 0$.
The points $(a, f(a))$ and $(b, f(b))$ are on opposite sides of the x-axis, so any continuous line joining them, such as the graph of $f(x)$, must pass through the x-axis, at the point $(\gamma, 0)$ say; then the equation $f(x) = 0$ has a real root γ.

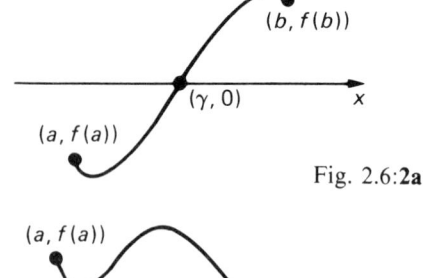

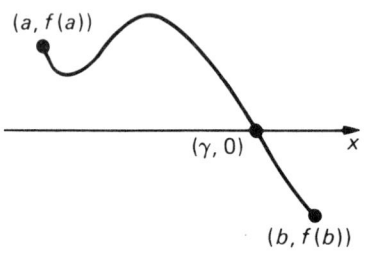

Fig. 2.6:2a

Fig. 2.6:2b

Algebraic, exponential and logarithmic functions 111

Worked example 2 Find integers a, b and c such that the roots of the equation $f(x) = x^3 - 3x + 1 = 0$ lie between a and $a+1$, b and $b+1$, c and $c+1$.

As indicated by Theorem 1, we look for changes of sign in $f(x)$:

x	-4	-3	-2	-1	0	1	2	3	4
$f(x)$	-51	-17	-1	3	1	-1	3	19	53

$f(-2) = -1$, $f(-1) = 3$, so there is a root between -2 and -1.
$f(0) = 1$, $f(1) = -1$, so there is a root between 0 and 1.
$f(1) = -1$, $f(2) = 3$, so there is a root between 1 and 2.
Thus $a = -2$, $b = 0$, $c = 1$. The graph of $y = x^3 - 3x + 1$ is shown in fig. 2.6:3.

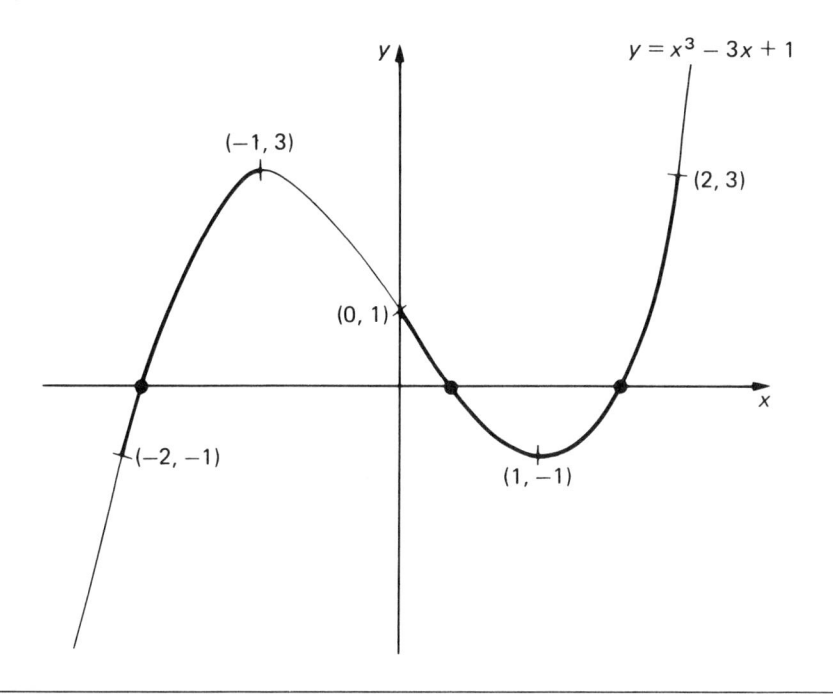

Fig. 2.6:3

Exercise 2.6:1

1 Find the stationary values of $f(x)$ (if any), sketch its graph, and hence find the number of real roots of the equation $f(x) = 0$.
  → (i) $f(x) = x^3 - 3x + 4$
  R (ii) $f(x) = x^3 + 3x - 6$
  (iii) $f(x) = x^3 - 3x^2 + 1$
  → (iv) $f(x) = x^3 - 6x + 2$
  (v) $f(x) = 2x^3 - 9x^2 + 12x - 1$
  (vi) $f(x) = 8x^3 - x^2 - 2x - 1$
  (vii) $f(x) = 4x^4 - 2x - 1$
  → (viii) $f(x) = 3x^4 - 4x^3 - 12x^2 + 1$
  (ix) $f(x) = x^5 - 5x + 3$
  (x) $f(x) = x^5 - 5x + 7$
  (xi) $f(x) = 16x^5 - 5x^2 + 1$

① → 2 Find the number of real roots of the equation $x^3 - 12x + k = 0$ when k is given by:
  (i) -20 (ii) -16 (iii) -4
  (iv) 4 (v) 16 (vi) 20

Show that $x^3 - 12x + k = 0$ has three real roots if $|k| \le 16$, one if $|k| > 16$.

① 3 (i) Find the number of real roots of the equation $3x^4 - 4x^3 - 12x^2 + k = 0$ when k is given by:
(a) -1; (b) 3; (c) 10; (d) 40.

(ii) Find the sets of values of k for which the equation has:
(a) four real roots; (b) two real roots; (c) no real roots.

① 4 Sketch the curve $y = x^3 - x^2 - 5x$ and find the coordinates of its turning points.

By considering the number of points of intersection of the curve and the line $y = k$ for different values of k, find the set of values of k for which the equation $x^3 - x^2 - 5x = k$ has three real roots.

① **5** Sketch the curve $y=(x+1)^2(3x-7)(x-5)$ and find the coordinates of its turning points. Hence find the set of values of k for which $(x+1)^2(3x-7)(x-5)=k$ has four real roots.

①: **6** Consider the curve $y=f(x)$ where $f(x)=ax^3+bx^2+cx+d$.

(i) Show that if $b^2>3ac$ the curve has two turning points. Suppose that the x-coordinates of these turning points are α and β. Show by sketching graphs that, if $f(\alpha)$ and $f(\beta)$ have opposite signs, then the equation $f(x)=0$ has three distinct real roots, and that, if $f(\alpha)$ and $f(\beta)$ have the same sign, the equation $f(x)=0$ has one real root. Illustrate this result by sketching the curve in the cases:
(a) $f(x)=x^3+3x^2-9x+8$;
(b) $f(x)=x^3+3x^2-9x-2$.

(ii) Show that if $b^2=3ac$ the curve has one turning point and that it is a point of inflexion. Discuss the number of real roots of the equation $f(x)=0$ in this case. Illustrate your answer by sketching the curve in the cases:
(a) $f(x)=(x-1)^3$; (b) $f(x)=(x-1)^3+1$.

(iii) Show that, if $b^2<3ac$, the equation $f(x)$ has exactly one real root.

② → **7** For each real root γ of $f(x)=0$ for each function $f(x)$ in question 1, find the integer k such that $k<\gamma<k+1$.

② **8** If $f(x)=ax^3+bx^2+cx+d$, $a\neq 0$, show, by considering

$$\lim_{x\to+\infty}f(x) \quad \text{and} \quad \lim_{x\to-\infty}f(x)$$

and using Theorem 1, that the equation $f(x)=0$ must have at least one real root.

Explain why it cannot have exactly two real roots.

② **9** Sketch the graph of the function $f(x)=1/x$.

Show that $f(-2)$ and $f(+2)$ are of opposite signs but that the equation $f(x)=0$ does not have a root between -2 and $+2$.

Explain the necessity for the condition '$f(x)$ is continuous' in Theorem 1.

② **R 10** (i) Show that the equation $x^3+3px+q=0$, $p>0$, $q>0$, has exactly one negative root and no positive root.

(ii) Show that if the equation $x^3-3px+q=0$, $p>0$, $q>0$ has three real roots then $4p^3>q^2$.

2.6:2 Iterative methods

Newton–Raphson method

> **Theorem 2:** If $x=a$ is an approximation to a root of the equation $f(x)=0$, then $x=a-f(a)/f'(a)$ is, in general, a better one.

Proof

Suppose that the equation $f(x)=0$ has a root $x=\gamma$, so the curve $y=f(x)$ cuts the x-axis at $Q(\gamma, 0)$ (fig. 2.6:4); and suppose that $x=a$ is an approximation to this root, so $N(a, 0)$ is near Q.

Let P be the point $(a, f(a))$ on the curve, so $NP=f(a)$; and let the tangent at P cut the x-axis at R. Then

$$\frac{NP}{NR}=\text{gradient of tangent at }P=f'(a)$$

$$NR=\frac{NP}{f'(a)}=\frac{f(a)}{f'(a)}$$

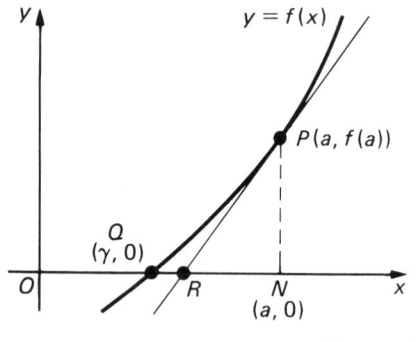

Fig. 2.6:4

So the x-coordinate of R is $a-f(a)/f'(a)$.

In general, R is nearer to Q than N; i.e. $a-f(a)/f'(a)$ is a better approximation to γ than a. (Why 'in general'? See question 7.)

We can apply the formula repeatedly to obtain better and better approximations:

$$a \qquad b=a-\frac{f(a)}{f'(a)} \qquad c=b-\frac{f(b)}{f'(b)} \qquad d=c-\frac{f(c)}{f'(c)} \qquad \text{etc.}$$

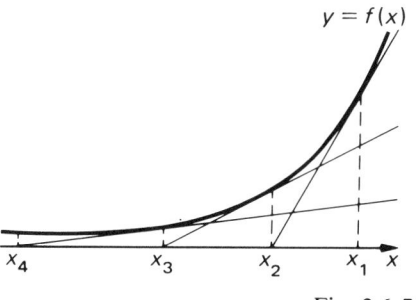

Algebraic, exponential and logarithmic functions 113

In general, if we call the *n*th approximation x_n, then

$$x_{n+1} = x_n - \frac{f(x_n)}{f'(x_n)}; \quad n = 1, 2, 3, 4, \ldots \quad (2)$$

(See fig. 2.6:**5**.)

Worked example 3 Consider the root of the equation $x^3 - 3x + 1 = 0$ lying between $x = 1$ and $x = 2$ (see worked example 2 (2.6:1)). Taking $x = 1.5$ as the first approximation to this root, find two successively better approximations, using the Newton–Raphson method.

$$f(x) = x^3 - 3x + 1 \Rightarrow f'(x) = 3x^2 - 3.$$

First approximation is $x_1 = 1.5$.

By equation (2), the second approximation is $x_2 = 1.5 - \dfrac{f(1.5)}{f'(1.5)} \approx 1.53$

and the third approximation is $x_3 = 1.53 - \dfrac{f(1.53)}{f'(1.53)} \approx 1.532$

How accurate is this third approximation? Well; $f(1.532) = -0.0036$ and $f(1.533) = +0.0036$, so the root certainly lies between 1.532 and 1.533.

$x = \phi(x)$ method

Theorem 3: If $x = a$ is an approximation to a root of the equation $x = \phi(x)$, then $x = \phi(a)$ is a better one, so long as $|\phi'(a)| < 1$.

Proof

We will not prove this theorem rigorously, but simply justify it by sketching graphs. Note that the exact root of the equation $x = \phi(x)$ corresponds to the point of intersection of the line $y = x$ and the curve $y = \phi(x)$.

(i) When $|\phi'(a)| < 1$ (fig. 2.6:**6a**) R is nearer to Q than N is.
(ii) When $|\phi'(a)| > 1$ (fig. 2.6:**6b**) R is further from Q than N is.

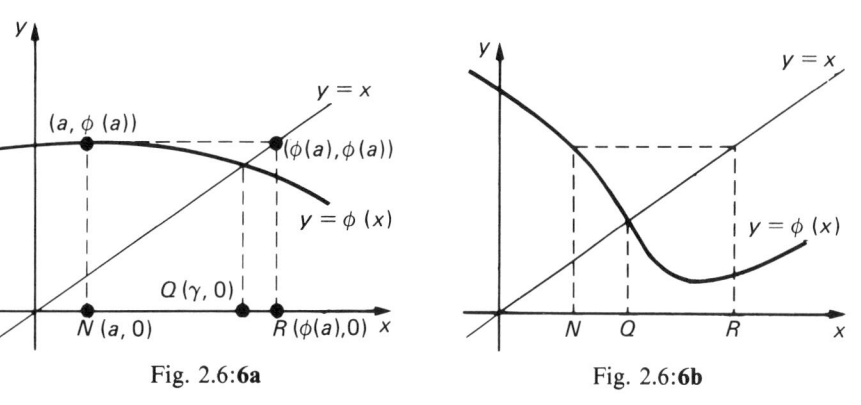

Fig. 2.6:**6a**   Fig. 2.6:**6b**

We can apply the formula repeatedly to obtain better and better approximations (see fig. 2.6:**7**):

$$x_{n+1} = \phi(x_n); \quad n = 1, 2, 3, 4, \ldots \quad (3)$$

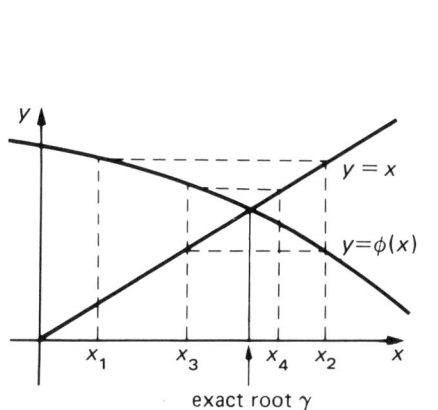

Fig. 2.6:**7**   Note: the smaller $|\phi'(x_1)|$ is, the more rapid the rate of convergence.

Iterative methods

Worked example 4 Consider the root of the equation $x^3 - 3x + 1 = 0$ lying between $x = 1$ and $x = 2$ (worked example 2). Taking $x = 1.5$ as the first approximation to this root, find two successive better approximations, using the $x = \phi(x)$ method.

We want to write the equation in the form $x = \phi(x)$. This can be done in several ways: we want to choose the form which makes $|\phi'(1.5)|$ the smallest.

	$\phi(x)$	$\phi'(x)$	$\phi'(1.5)$
(i) $x =$	$\frac{1}{3}(x^3 + 1)$	x^2	2.25
(ii) $x =$	$(3x - 1)^{1/3}$	$(3x - 1)^{-2/3}$	0.43
(iii) $x =$	$\dfrac{1}{3 - x^2}$	$\dfrac{2x}{(3 - x^2)^2}$	5.33

In fact, $x = (3x - 1)^{1/3}$ is the only form which is suitable. First approximation is $x_1 = 1.5$. By equation (3),

$$\text{the second approximation is} \quad x_2 = (3 \times 1.5 - 1)^{1/3} \approx 1.52$$
$$\text{the third approximation is} \quad x_3 = (3 \times 1.52 - 1)^{1/3} \approx 1.527$$

Exercise 2.6:2

③ **1** Taking the given number as the first approximation to a root of the equation, find two successive better approximations to the root using the Newton-Raphson method:

→ (i) $x^3 - x - 5 = 0$; 2
(ii) $x^3 - 28 = 0$; 3
(iii) $x^3 - 3x^2 - 2 = 0$; 3
(iv) $x^3 - 4x - 5 = 0$; 2.5
→ (v) $x^3 - 4x^2 - x - 12 = 0$; 5
R (vi) $x^4 - x - 15 = 0$; 2
(vii) $x^4 - 2x - 1 = 0$; 1.5
(viii) $x - \ln(2 - x) = 0$; 0.45
(ix) $xe^x = 1 - x^2$; 0.5
(x) $xe^{2-x} = 2$; 0.4

Comment on the accuracy of your answer in each case.

③ **2** Show that the equation $x^2 + x - 10 = 0$ has a root near $x = 3$. Using the Newton-Raphson method, find three successive better approximations to the root. Check the accuracy of your final approximation by solving the equation exactly.

③ → **3** (i) By considering the root of the equation $x^3 - 65 = 0$ near $x = 4$, use the Newton-Raphson method to find $\sqrt[3]{65}$ correct to 2 dec. pls.

(ii) Similarly find:
(a) $\sqrt{17}$ to 2 dec. pls.; (b) $\sqrt[5]{30}$ to 2 dec. pls.

③ **4** Show that if x_n is an approximation to the root of the equation

$$\frac{1}{x} - \lambda = 0 \quad (\lambda \in \mathbb{R})$$

then $x_n(2 - \lambda x_n)$ is a better one.

Taking 0.03 as the first approximation to the reciprocal of the number 31, use this procedure to find three successive better approximations. Check the accuracy of your answer.

③ **5** Show that the equation $x^4 + 20x - 4 = 0$ has a root near $x = 0$. By neglecting the term x^4, find an approximation to this root. Find the root correct to 4 dec. pls., using the Newton-Raphson method as many times as is necessary.

③ **6** Show that, when k is small, the equation $x^5 + x = k$ has a root near $x = 0$. Show that a better approximation is $x = k$.

③ → **7** (i) Show by sketching graphs that the Newton-Raphson method may break down if:
(a) $f'(a)$ is small; (b) $f''(a)$ is large.
Explain why the formula is useless when $f'(a)$ is small.

(ii) Explain with the aid of a graph why the method fails when $x = 1$ is taken as the first approximation to the root of $x^3 - 3x + 1 = 0$ lying between $x = 1$ and $x = 2$.

Algebraic, exponential and logarithmic functions 115

(iii) Show that the equation $x^3 - 15x + 20 = 0$ has a root between $x = 2$ and $x = 3$. Explain with the aid of a graph why the method fails when $x = 2$ is taken as the first approximation to the root.

Taking $x = 3$ as the first approximation, find two successive better approximations.

8 Taking the given number a as the first approximation to a root of the equation, rewrite the equation in the form $x = \phi(x)$ in such a way that $|\phi'(a)| < 1$; hence find three successive better approximations to the root:

(i) $x^3 - x - 5 = 0$; 2
(ii) $x^3 - 5x - 8 = 0$; 3
(iii) $4x^4 - 2x^2 - 3 = 0$; 1
R (iv) $x^4 - x - 15 = 0$; 2
(v) $x^5 + 5x + 7 = 0$; -1
(vi) $x - \ln(2 - x) = 0$; 0.45
(vii) $e^{-2x} = x + 8$; -1
(viii) $\ln(3 + x) = x^2$; 1.2

9 Given that $f(x) = x^3 + x^2 + 2x - 2$, show that the equation $f(x) = 0$ has only one real root and that this root lies between 0 and 1.

Taking $x = 1$ as the first approximation to this root, obtain two better approximations using:

(i) the Newton-Raphson method
(ii) the $x = \phi(x)$ method

10 By sketching on the same axes the curve $y = e^x$ and the line $y = x + 2$, show that the equation $e^x = x + 2$ has two real roots. Show that one of these roots lies between 1.1 and 1.2.

Taking $x = 1.1$ as the first approximation to this root, obtain a better approximation using:

(i) the Newton-Raphson method
(ii) the $x = \phi(x)$ method

11 Sketch on the same axes the curves $y = \ln x$ and $y = x/(x - 1)$. Show that the equation $(x - 1) \ln x = x$ has roots near 3.8 and 0.45.

Use the Newton-Raphson method to find better approximations to each of these roots.

12 By sketching on the same axes the curves $y = 4^x$ and $y = (x + 2)/x$, show that the equation $x4^x = x + 2$ has two real roots, one of which is negative and the other positive.

Show that the positive root lies between 0.8 and 0.9. Taking $x = 0.85$ as an approximation to this root, find a better approximation using:

(i) the Newton-Raphson method
(ii) the iterative formula $x_{n+1} = \log_4(1 + 2/x_n)$

2.7 Reduction of the relationship between two variables to linear form

Worked example 1 Suppose that in performing an experiment we let some quantity x vary and measure the corresponding values of some other quantity y.

Suppose that we obtain the readings:

x	0	1	2	3	4	5	6
y	0.33	0.29	0.20	0.13	0.09	0.08	0.05

and we suspect that x and y are related by an equation of the form:

$$y = \frac{1}{ax^2 + b} \quad (1)$$

where a and b are constant.

We would like to check whether this suspicion is justified and if so to find a and b.

If we use the readings to plot a graph of y against x, as in fig. 2.7:1, it is difficult to see whether or not this is of the form in equation (1). However, note that

$$y = \frac{1}{ax^2 + b} \Rightarrow \frac{1}{y} = ax^2 + b \quad (2)$$

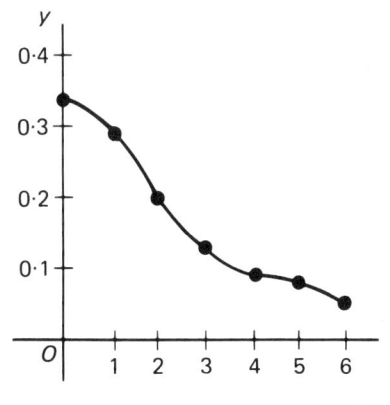

Fig. 2.7:1

116 *Reduction of the relationship between two variables to linear form*

so the graph of $1/y$ against x^2 should be a straight line of gradient a, with y-intercept b.

x	0	1	2	3	4	5	6
x^2	0	1	4	9	16	25	36
$1/y$	3.03	3.45	5.00	7.69	11.11	12.50	20.00
y	0.33	0.29	0.20	0.13	0.09	0.08	0.05

Let's use the readings to plot a graph of $1/y$ against x^2, as in fig. 2.7:**2**. The points all lie close to a straight line, except for the point (25, 12.50). This suggests that our suspicion is justified but that the reading of $y = 0.08$ when $x = 5$ is subject to experimental error, or blurred vision. We find a by measuring the gradient of the line (being careful to choose two points which are actually *on* the line rather than two points based on the original readings), and b by measuring the y-intercept:

$$a = \tfrac{10}{20} = \tfrac{1}{2}, \quad b = 3$$

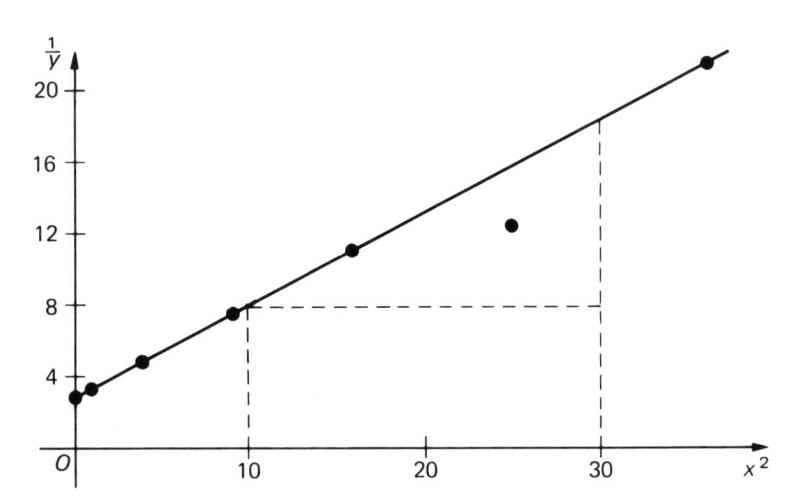

Fig. 2.7:**2**

In general, given a set of readings for the quantities x and y and a possible relation of the form

$$y = f(x, a, b) \tag{3}$$

we try to convert equation (3) into the form

$$Y = mX + c \tag{4}$$

where X and Y are both variables related to x and y and m and c are constants related to a and b. We then plot Y against X, hoping to obtain a straight line.

Worked example 2 Convert the equations:
(i) $y = ax^b$ (ii) $y = ax^2 + bx$
into the form $Y = mX + c$.

(i) $\qquad y = ax^b \Rightarrow \lg y = \lg(ax^b)$
$\qquad\qquad\qquad\qquad = \lg a + \lg x^b$
$\qquad\qquad\qquad\qquad = b \lg x + \lg a$

Algebraic, exponential and logarithmic functions 117

Thus $Y = \lg y$, $X = \lg x$ and $m = b$, $c = \lg a$.

(ii) $$y = ax^2 + bx \Rightarrow \frac{y}{x} = ax + b$$

Thus $Y = y/x$, $X = x$ and $m = a$, $c = b$.
Notice that both X and Y may be functions of x and y.

Exercise 2.7:1

Questions 1–7: By drawing a straight-line graph based on the tabulated data, find approximate values for a and b.

① **1** $y = ax + b$

x	0.1	0.2	0.3	0.4	0.5	0.6
y	0.15	0.26	0.40	0.51	0.65	0.77

① **2** $y = ax^2 + b$

x	1.1	2.5	3.2	6.5	7.0
y	7.85	13.92	18.70	57.18	64.89

① → **3** $y = 1/(ax + b)$

x	2	4	6	8	10
y	2.47	1.96	1.68	1.43	1.26

① R **4** $y = 1/(ax^2 + b)$

x	0.15	0.64	0.83	1.27	2.17	4.56
y	0.016 75	0.013 26	0.011 42	0.008 03	0.004 03	0.001 12

① **5** $ay^2 = 1/(x - b)$

x	1	2	3	4	5
y	1.612	0.960	0.749	0.634	0.559

① → **6** $y = ax^2 + b$

x	2.5	2.6	2.7	2.8	2.9
y	26.5	28.6	30.7	32.8	35.2

(Note: your graph should not include the y-axis; so it will be impossible to read off the y-intercept. However, once you have found a, choose a point on the graph and substitute its coordinates into the equation to find b.)

① **7** $y^2 = a/x + b$

x	0.156	0.161	0.178	0.181	0.189
y	13.632	13.456	12.916	12.830	12.607

Questions 8–15. Reduce each of the given relationships to the form $Y = mX + c$. Give the functions equivalent to X and Y and the constants equivalent to m and c.

② **8** (i) $y = ax^2 + b$
→   (ii) $y = ax^3 + bx$
  (iii) $y = a\sqrt{x} + b$
  (iv) $y = x^2 + ax + b$
→   (v) $y = (x - a)(x - b)$

② **9** (i) $y = a/x + b$
  (ii) $y = ax + b/x$
→   (iii) $y = ax^{1/2} + bx^{-1/2}$
  (iv) $y = 1/(ax + b)$
  (v) $y = a/(x + b)$
→   (vi) $y = 1/[(x - a)(x - b)]$

② → **10** $1/x + 1/y = 1/a$

② **11** (i) $y^2 = ax + b$
  (ii) $ay^2 = x^2 + bx$
→   (iii) $y(y - a) = x - b$

② **12** (i) $ay = x^b$
→   (ii) $y = a(1 + x)^b$

② → **13** (i) $y = ab^x$
  (ii) $y = ab^{-2x}$
  (iii) $y = a^{x+b}$

② **14** (i) $y = \lg(ax + b)$
  (ii) $y = \lg(ax^2 + bx)$
  (iii) $y = a \lg bx$
→   (iv) $ay^2 = (x + b) \lg x$

② **15** (i) $y(y - a) = b^x$
  (ii) $y(y - a) = b10^x$
  (iii) $\left(\dfrac{y}{x}\right)^a = b10^{-x}$
  (iv) $a2^y = x^2 - bx$

Questions 16–18: The variables x and y are related by the equation

$$\frac{1}{x} + \frac{1}{y} = \frac{1}{a}$$

Plot a graph of $1/y$ against $1/x$. Check that the gradient of your graph is -1 and estimate the value of a.

118 *Reduction of the relationship between two variables to linear form*

16

x	1	2	3	4	5
y	5.00	1.42	1.15	1.05	1.00

→ **17**

x	14	18	22	24	26
y	35.0	22.5	18.3	17.1	16.2

18

x	2.41	2.95	4.62	7.19	8.2
y	50.2	10.4	4.6	3.4	3.2

Questions 19–32: By drawing a straight line graph based on the tabulated data, find approximate values of a and b.

→ **19** $y = ax^2 + bx$

x	1	2	3	4	5
y	1.0	8.4	22.3	39.1	65.8

20 $a/x + b/y = 1$

x	6.0	6.5	8.0	10.0	14.0
y	30.0	21.7	13.4	10.0	7.8

R **21** $y = (x-a)(x-b)$

x	0.15	0.25	0.35	0.45	0.55
y	0.053	0.012	−0.008	−0.007	0.014

22 $y = 200/[(x-a)(x-b)]$

x	50	60	70	80	90
y	0.52	0.21	0.13	0.07	0.05

23 $y = ax^{\frac{1}{2}} + bx^{-\frac{1}{2}}$

x	0.2	0.4	0.6	0.8	1.0
y	16.4	10.8	8.0	6.2	5.0

→ **24** $y = ax^b$

x	2	4	6	8	10
y	1.70	9.54	26.43	54.37	94.87

25 $y = ax^b$

x	30	35	40	45	50
y	4.624	4.151	3.763	3.496	3.239

26 $y = ax^b$

x	0.1	0.2	0.3	0.4	0.5
y	1.9	4.4	7.1	10.0	13.0

27 $y = ab^x$

x	1	2	3	4	5
y	0.561	0.473	0.432	0.401	0.354

R **28** $y = ab^x$

x	1	2	3	4	5
y	10.2	26.4	66.1	163.2	414.7

→ **29** $y = ab^x$

x	1.46	3.63	7.29	12.32	14.53
y	0.4	0.6	1.1	2.8	4.3

30 $y = \lg(ax^2 + b)$

x	0.2	0.4	0.6	0.8	1.0
y	−0.75	−0.39	0.09	0.15	0.31

→ **31** $y = \lg(ax^2 + bx)$

x	1	2	3	4	5
y	1.278	1.521	1.674	1.741	1.789

32 $y(y-a) = b \lg x$

x	6.37	30.6	70.1	85.3	103.5
y	17.32	18.93	19.71	19.85	20.03

Questions 33–35: Two of the readings are incorrect. Identify them, ignore them, and find the values of a and b.

33 $y = a(1 + 2x)^b$

x	1	2	3	4	5	6
y	2.08	1.38	1.11	0.86	0.71	0.64

→ **34** $y - a = x(x - b)$

x	1	2	3	4	5	6
y	1.1	3.6	8.5	15.2	24.2	34.6

35 $y - a = 2^{x+b}$

x	1.00	1.23	1.62	2.24	2.51	3.43
y	0.45	0.49	0.61	0.84	0.91	1.55

36 An elastic string is fixed at one end and a variable weight is hung at the other. It is believed that the length l of the string and the weight W are related by a law of the form

$$l = aW + b$$

Using the following experimental data, confirm this belief and find a and b.

W (newtons)	1	2	3	4	5	6
l (metres)	0.46	0.51	0.57	0.64	0.70	0.75

37 A car moves along a straight road. The distance s moved by the car after it passes a radar point is measured at one-second intervals:

t (seconds)	1	2	3	4	5
s (metres)	18.7	43.6	75.1	112.4	156.2

It is believed that s and t are related by a law of the form $s = at + bt^2$. Confirm this belief and estimate the values of a and b, giving your answers to one dec. pl.

38 The population of a colony of insects is measured at daily intervals:

number of days t	0	1	2	3	4	5
population p	40	50	80	110	150	220

It is believed that p and t are related by an equation of the form $p = ab^t$.

(i) Estimate the values of a and b.
(ii) Estimate the population of the colony after 6 days.
(iii) What is the percentage growth rate per day?
(iv) How long does the population take to double?

→ **39** The table below shows the period T of a planet's orbit (measured in Earth years) against the average distance r of the planet from the Sun (taking the distance of the Earth from the Sun as one unit) for the five inner planets of the solar system:

	Mercury	Venus	Earth	Mars	Jupiter
r	0.39	0.72	1.00	1.52	5.20
T	0.24	0.62	1.00	1.89	11.86

T and r are related by an equation of the form

$$T = ar^b$$

(i) Find a and b.

(ii) Given that the average distance of Saturn from the Sun is 9.54 units, find the period of its orbit.

40 The paranoid husband of a famous, beautiful, neurotic, seven-times-married Hollywood film actress suspects that the length l years of each of her marriages is related to the number n of the marriage by the formula

$$l = ae^{-bn} \quad (a, b \text{ constant})$$

He rings up each of the six previous husbands and collects the data:

n	1	2	3	4	5	6
l	12.2	7.0	4.0	2.3	1.3	0.8

By drawing a suitable straight-line graph, verify that his suspicion appears to be justified and find approximate values of a and b.
How long can he expect to last?

R **41** The period T of oscillation of a pendulum and its length l are related by a law of the form

$$T = al^b$$

Using the following experimental data, estimate

(i) a and b.
(ii) the period of oscillation of a pendulum of length 0.9 m.
(iii) the length of a pendulum whose period is 1 second.

l (metres)	0.4	0.6	0.8	1.0	1.2
T (seconds)	1.25	1.55	1.76	2.01	2.19

Miscellaneous exercise 2

1 Given that α and β are the roots of the equation $x^2 - 4x + 1 = 0$, show that an equation whose roots are $(\alpha - 3)$ and $(\beta - 3)$ is $x^2 + 2x - 2 = 0$.
Find the coordinates of the point of intersection of the curves $y = x^2 - 4x + 1$ and $y = x^2 + 2x - 2$. Find the coordinates of the turning point of each curve. Sketch both curves on the same axes.

2 (i) The equation $x^2 + ax + b = 0$ has roots a and b. Find two possible pairs of values of a and b.

(ii) The roots of the equation $4x^2 + 8x + p = 0$ differ by 3. Find p.

3 Find the set of real values of x (if any) for which $x^2 - 2x + k \le 0$ when:

(i) $k = -3$ (ii) $k = -1$ (iii) $k = 1$ (iv) $k = 3$

4 (i) Find the set of values of k for which the equation $kx^2 - 4x + 2k = 7$ has real roots.

(ii) If the roots of the equation are α and β, find an equation whose roots are $1/\alpha$ and $1/\beta$.

5 Given that $f(x) = 4kx^2 - 6x + (k - 4)$, find:

(i) the values of k for which the equation $f(x) = 0$ has real roots.
(ii) the set of values of k for which $f(x)$ is negative for all real values of x.

Sketch the curve $y = f(x)$ for each of the cases $k = -1$, $k = -\frac{1}{2}$.

6 Given that $f(x) = 2x^2 - x - 6$,

(i) solve the equation $f(x) = 0$;

Miscellaneous exercise 2

(ii) solve the equation $f(x-2) = 0$;

(iii) find the set of real values of k for which the equation $f(x) + k = 0$ has real roots.

7 Given that $f(x) = kx^2 + (3k-7)x + k$,

(i) show that, for all real values of k, the equation $f(x) = 3$ has real roots;
(ii) find the set of values of k for which the equation $f(x) = 0$ has real roots.

8 Given that $f(x) = 4x^2 + 8x - 21$,
(i) find the minimum value of $f(x)$ and the values of x for which $f(x) = 0$;
(ii) sketch on the same axes the curves $y = f(x)$ and $y = 1/f(x)$;
(iii) find the set of possible values of y when $y = 1/f(x)$ and x is real;
(iv) write down the number of real roots of the equation $\{f(x)\}^2 = 1$.

9 Sketch on the same axes the graphs of $y = x^2 - 6x + k$ and $y = 1/(x^2 - 6x + k)$ when:

(i) $k = 8$ (ii) $k = 9$ (iii) $k = 10$

10 Given that $f(x) = 2 - 8/(x^2 - 2x + 5)$, and x is real, show that $0 \leq f(x) < 2$. Sketch the curve $y = f(x)$.

11 When a polynomial $f(x)$ is divided by $(2x-1)$ the remainder is 2; when it is divided by $x+1$, the remainder is 3. Find the remainder when it is divided by $2x^2 + x - 1$.

12 $f(x) = 4x^3 + ax^2 + bx - 6$. When $f(x)$ is divided by $(x-2)$ the remainder is a. When $f(x)$ is divided by $(x+1)$ the remainder is b. Find the remainder when $f(x)$ is divided by $(x+2)$.

13 The polynomial $f(x) = x^3 + px + q$ has remainder -12 when divided by $(x-1)$ and remainder 30 when divided by $(x-4)$. Find p and q. Express $f(x)$ as a product of linear factors. Solve the inequality $f(x) \geq 0$.

14 Sketch the curve $y = x^3 - 4x$.

(i) Find the coordinates of the points of intersection of the curve with the line $y = 3x + 6$. Hence solve the inequality $x^3 - 4x > 3x + 6$.
(ii) Find the equation of the tangent to the curve at the point $(2, 0)$ and the coordinates of the point at which this tangent cuts the curve again.
(iii) Show graphically that the equation
$$x^3 - 4x - 5 = 0$$
has only one real root. Taking $x = 2.5$ as a first approximation to this root, use the Newton-Raphson method to find a second approximation, giving your answer to 2 dec. pls.

15 (i) Given that $f(x) = x^3 - 19x + 30$, factorize $f(x)$ completely, sketch the curve $y = f(x)$ and hence solve the inequality $f(x) \geq 0$.

(ii) Given that $g(x) = x^3 + x^2 - 16x + 20$, factorize $g(x)$ completely, sketch the curve $y = g(x)$ and hence solve the inequality $g(x) \geq 0$.

16 Find the set of values of x for which
$$(x-1)(x+2)(kx-2) < 0$$
when:

(i) $k = 2$ (ii) $k = 1$ (iii) $k = 0$ (iv) $k = -1$

17 Solve the inequalities:

(i) $x^2 - 3x < 4$
(ii) $3/(x-1) < x + 1$
(iii) $(x-2)/(2x-1) > 2/(x+1)$
(iv) $|3x+1| < |1-2x|$

18 Given that $f(x) = (4x-3)/(x+2)$,

(i) sketch the graphs of $y = f(x)$ and $y = |f(x)|$;
(ii) solve the inequalities
(a) $f(x) < 1$; (b) $|f(x)| < 1$.

19 Sketch on the same axes the graphs $y = x/(x-2)$ and $y = 6/(x-1)$. Find the set of values of x for which

$$\frac{x}{x-2} < \frac{6}{x-1}$$

Sketch also the graph of $y = |x/(x-2)|$.

20 Sketch on the same axes the graphs $y = |3-x|$ and $y = 2|2x-1|$. Find the set of values of x for which $|3-x| > 2|2x-1|$.

21 Find the set of values of x for which:

(i) $x > |2x - 3|$
(ii) $2x > |2x - 3|$

22 Prove that if $f(x)$ is a polynomial and the equation $f(x) = 0$ has a repeated root α, then $f'(\alpha) = 0$. Hence:

(i) sketch the curve $y = x^3 - 3x^2$;
(ii) show that if the equation $x^3 - 3x^2 - 9x + k = 0$ has a repeated root then either $k = -5$ or $k = 27$.

23 (i) Find the set of values of x for which:

$$0 < x^2 + 5x < 7 - x$$

(ii) Sketch on the same axes the curve

$$y = \frac{3x - 5}{2x - 4}$$

and the line $y = x - 1$ and find their points of intersection. Hence find the set of values of x for which $(3x-5)/(2x-4) > x - 1$.

(iii) Sketch the curve $y = |(x-1)(x+3)|$.

24 (i) The function $f: \mathbb{R} \to \mathbb{R}$ is defined by
$$f(x) = \frac{x^2 - 1}{2x^2 + 1}$$
Find $f'(x)$. Hence find the coordinates of the turning point of the curve $y = f(x)$. Write down the equation of the asymptote to the curve. Sketch the curve. What is the range of f?

(ii) The function $f^*: [1, 2] \to \mathbb{R}$ is defined by $f^*(x) = (x^2 - 1)/(2x^2 + 1)$. What is the range of f^*?

25 Find the range of $f(x)$ and sketch the graph of $y = f(x)$ when:
(i) $f(x) = 2x^2 + 1$
(ii) $f(x) = x/(2x^2 + 1)$
(iii) $f(x) = (x^2 + 1)/(2x^2 + 1)$

+ 26 If $f(x) = 2|x - 1| - 3|x - 2|$,
(i) find the set of values of x for which $f(x) > 0$;
(ii) find separate linear expressions for $f(x)$ in terms of x when $x < 1$, $1 \le x \le 2$, $x > 2$, then sketch the graph of f;
(iii) find the set of values of x for which $f(x) > 1$.

27 (i) Solve for x the equation $e^{\ln x} - 3e^{-\ln x} = 2$.

(ii) Find the set of values of x for which
$$|x + 1| < 2|x - 1|$$

(iii) Sketch on the same axes the graphs of $y = 2/x$ and $y = \ln(x + 2)$. Hence find the number of real roots of the equation $x \ln(x + 2) = 2$.

28 The function $f: (0, \pi) \to \mathbb{R}$ is defined by
$$f(x) = \ln \sin x$$

Sketch on the same axes the curves $y = f(x)$ and $y = f'(x)$. Estimate the change in $f(x)$ caused by a change of 0.01 in x when:
(i) $x = \tfrac{1}{4}\pi$ (ii) $x = \tfrac{1}{2}\pi$

29 Show that the gradient of the curve $y = \ln(2 - x)$ is always negative. Sketch the curve. Find the equation of the tangent to the curve at the point $(1, 0)$.

30 The percentage y of the population which votes Conservative is a function of the number x of IRA bomb attacks during the previous month:
$$y = 100(1 - 2e^{-\frac{1}{4}x - 1}) \quad x \ge 0$$

Sketch a graph of y against x. How many attacks will ensure the Conservatives get 50% of the vote?

31 Differentiate:
(i) $\log_2 x / \sqrt{(1 - x^2)}$ (ii) $x 2^{1 - x^2}$

32 Find the stationary value(s) of the functions:
(i) $2 + 2e^{-2x} - e^{-3x}$ (ii) $(1 + x)^2 e^{-2x}$

33 The function $f: \mathbb{R} \to \mathbb{R}$ is defined by
$$f(x) = \frac{e^x - 1}{e^x + 1}$$

(i) Find $\lim_{x \to -\infty} f(x)$ and $\lim_{x \to +\infty} f(x)$. Show that $f'(x)$ is always positive. Hence find the range of $f(x)$. Sketch the curve $y = f(x)$.

(ii) Find $f^{-1}(x)$. What is the domain of $f^{-1}(x)$? Sketch the curve $y = f^{-1}(x)$.

+ 34 The function $f: \mathbb{R}^+ \to \mathbb{R}$ is defined by $f(x) = x^2 \ln x$.
(i) Show that the minimum value of $f(x)$ is $-1/2e$. Investigate the behaviour of $f(x)$ and $f'(x)$ as $x \to 0$ and as $x \to +\infty$. Sketch the curve $y = x^2 \ln x$.
(ii) Show graphically that the equation
$$x^2 \ln x + x - 5 = 0$$
has a root near $x = 2$. Taking $x = 2$ as an approximation to this root and using the Newton–Raphson method, find a better approximation. Give your answer to 2 dec. pls.

35 Find the coordinates of the turning point and the point of inflexion of the curve $y = (1 + x)e^{-2x}$. Sketch the curve.

36 Sketch on the same axes the graphs of $y = e^x$ and $y = 10 - x$. Hence show that the equation
$$e^x + x - 10 = 0$$
has just one real root.
Taking $x = 2$ as a first approximation to this root, find second and third approximations, using
(i) the Newton–Raphson method
(ii) the iterative procedure $x_{n+1} = \ln(10 - x_n)$
Give your answers to 3 dec. pls.

37 Sketch on the same axes the curve $y = \ln x$ and the line $y = 4 - x$, and hence show that the equation $\ln x - 4 + x = 0$ has only one real root. Prove that this root lies between 2.9 and 3.0.
Taking 2.9 as a first approximation to this root, find a second approximation, using
(i) the Newton–Raphson method
(ii) the iterative procedure $x_{n+1} = 4 - \ln x_n$
Give your answers to 3 dec. pls.

38 (i) Consider this method of finding successively better approximations to $\sqrt{2}$:
Starting with a positive number, find the quotient when the number is divided into 2; take the average of this quotient and the original number;

starting with the number you have just obtained, repeat the process, etc.

Write down an iterative formula p which is a symbolic representation of this method, and use it to find $\sqrt{2}$ to 3 dec. pls.

(ii) Consider the equation $x^2 - 2 = 0$. Write down an iterative formula q which is a symbolic representation of the Newton-Raphson method for finding the positive root of this equation. Compare it to your formula p.

Have you used a calculator to help find the approximation in (i)? If so, explain the conceptual difference between what you have done and simply pressing the two (three?) buttons necessary to find $\sqrt{2}$ on the calculator.

39 A certain publishing company decides to put a new monthly magazine on the market. It estimates that the number n of people who will buy each issue will be related to the price £p of the magazine by the formula

$$n = \begin{cases} 1000(16 + 3p^2 - p^3) & 0 \leq p \leq 4 \\ 0 & p > 4 \end{cases}$$

(i) What is the maximum price at which the magazine could be sold without killing the demand?
(ii) What price ensures maximum circulation? What would the circulation be in this case? Sketch a graph of n against p.
(iii) Show that if £p is the price which ensures maximum *turnover* per month (i.e. the maximum value of np), then $16 + 9p^2 - 4p^3 = 0$. Use the Newton-Raphson method to find p, correct to 2 dec. pls.

Is the maximum turnover per month more or less than £50,000?

40 If $f(x) = (x - 10) \ln x$, find $f'(x)$. By considering the roots of the equation $f'(x) = 0$, show that the curve $y = f(x)$ has exactly one turning point and that the value of the x-coordinate of this turning point lies between 4 and 5.

Find this value, correct to 2 sig. figs, and hence find the minimum value of $f(x)$, correct to 2 sig. figs.

41 Show graphically that the equation

$$x^3 + 3x - 15 = 0$$

has only one real root and that this root is near $x = 2$. Decide which of the forms:

(i) $x = \frac{1}{3}(15 - x^3)$

(ii) $x = 15/(x^2 + 3)$
(iii) $x = (15 - 3x)^{1/3}$

is suitable for applying the iterative formula

$$x_{n+1} = \phi(x_n)$$

Taking $x = 2$ as a first approximation, use the formula to find second and third approximations. Give your answers to 2 dec. pls.

42 In an experiment to determine the focal length f cm of a lens, the distance u cm of the object and the distance v cm of its image from the centre of the lens are measured:

u	10	15	20	25	30	35
v	66.9	20.7	15.4	13.3	12.3	11.6

u, v and f are related by the formula $1/u + 1/v = 1/f$. By plotting a suitable graph, estimate f to 2 sig. figs.

43 The weight W g of a certain plant is measured at weekly intervals:

t (weeks)	1	2	3	4	5
W (grams)	15.4	74.7	194.8	388.5	652.2

Show that these data approximately fit the model $W = at^b$ and find a and b. Give your answers to 2 sig. figs.

Using the method of small increments, estimate the increase in weight per hour after

(i) 2 weeks (ii) 5 weeks

What would you feel about the model if the weight of the plant after 6 weeks turned out to be 722.4 g?

44 The proportion p of the population which has a certain infectious disease is measured at monthly intervals:

p	1	2	3	4	5
t months	0.090	0.102	0.115	0.129	0.146

Show that these data approximately fit the model $p = ae^{bt}$, and find a and b. Give your answers to 2 dec. pls. Estimate the increase per day in p after (i) 3 weeks (ii) 3 months.

Discuss the deficiencies of the model, with particular reference to the behaviour of p when t is large. Discuss in detail the possible conclusions which could be drawn if the value of p after 6 months turned out to be (a) 0.203; (b) 0.143?

3 Trigonometry

3.1 Trigonometric functions

3.1:1 Definitions of sin θ, cos θ and tan θ

When $0 < \theta < 90°$ we can define the sine, cosine and tangent of θ by drawing any right-angled triangle ABC with $\widehat{B} = \theta$, $\widehat{C} = 90$ (fig. 3.1:1).

Definition

$$\sin \theta = \frac{b}{c} \qquad \cos \theta = \frac{a}{c} \qquad \tan \theta = \frac{b}{a} \qquad (1)$$

Fig. 3.1:1

We would like to define the sine, cosine and tangent of any angle θ. Taking perpendicular axes, Ox, Oy we draw a circle centre O, radius 1. Let P be the point on the circle for which $\widehat{POx} = \theta$ (fig. 3.1:2). (Conventionally, we consider anticlockwise to be positive, clockwise to be negative (fig. 3.1:3).)

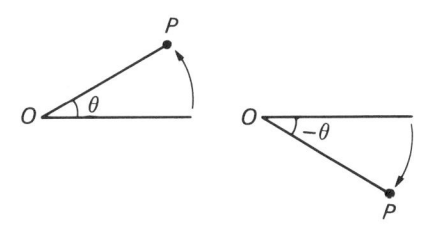

Fig. 3.1:2   Fig. 3.1:3   Anticlockwise is positive, clockwise is negative

Definition

$$\cos \theta = x\text{-coordinate of } P$$

$$\sin \theta = y\text{-coordinate of } P \qquad (2)$$

$$\tan \theta = \frac{\sin \theta}{\cos \theta} = \text{gradient of } OP$$

Exercise 3.1:1

1 We have defined $\sin \theta$, $\cos \theta$ and $\tan \theta$ in definitions (1) as ratios of the lengths of the sides of a right-angled triangle. Explain with the aid of a diagram why the definitions are independent of the size of the triangle.

2 Using definitions (1), show that, if $0° < \alpha < 90°$,

$$\sin \alpha \equiv \cos(90° - \alpha)$$

$$\cos \alpha \equiv \sin(90° - \alpha)$$

Hence, given that $\sin 20° \approx 0.3$, $\cos 20° \approx 0.9$, find $\tan 70°$, without using a calculator.
Show also that $\tan 45° = 1$.

124 *Definitions of* sin θ, cos θ *and* tan θ

3 If $\sin\alpha = \frac{3}{5}$ and α is acute, find $\cos\alpha$ and $\tan\alpha$.

4 Find $\sin\alpha$ and $\cos\alpha$ when α is acute and:
(i) $\tan\alpha = \frac{12}{5}$ (ii) $\tan\alpha = \frac{1}{2}$ (iii) $\tan\alpha = 3$

5 Find $\sin\alpha$ and $\tan\alpha$ when α is acute and:
(i) $\cos\alpha = \frac{3}{5}$ (ii) $\cos\alpha = \frac{8}{17}$ (iii) $\cos\alpha = \frac{1}{4}$

6 Using definitions (1), show that, if $0° < \alpha < 90°$, $\sin^2\alpha + \cos^2\alpha \equiv 1$ ($\sin^2\alpha$ means $(\sin\alpha)^2$). Hence, given that $\sin 60° = \frac{1}{2}\sqrt{3}$, find $\cos 60°$ and $\tan 60°$ without using a calculator. Hence, using the identities of question 2, find $\tan 30°$.

7 Sketch an isosceles $\triangle ABC$ with $A\hat{B}C = 90°$. If $AB = r$, find AC and hence use definitions (1) to write down $\sin 45°$, $\cos 45°$ and $\tan 45°$ in surd form. (Your answers should be independent of r.)

8 Sketch an equilateral $\triangle ABC$, and let AD be the perpendicular from A to BC. If $AB = 2r$, find BD and AD in terms of r. Hence, without using your calculator, write down $\sin 60°$, $\cos 60°$, $\tan 60°$, $\sin 30°$, $\cos 30°$ and $\tan 30°$ in surd form.

9 Complete this table, giving your answers in surd form:

α	0°	30°	45°	60°	90°
$\sin\alpha$	0				1
$\cos\alpha$	1				0
$\tan\alpha$					

(i) Using the table, sketch on the same axes the graphs of $\sin\alpha$ and $\cos\alpha$ for $0° \leq \alpha \leq 90°$.
Find the coordinates of their point of intersection. Explain how the graphs demonstrate the result
$$\sin\alpha \equiv \cos(90° - \alpha)$$
(ii) What happens to $\tan\alpha$ as $\alpha \to 90°$? What can you say about $\tan\alpha$ when $\alpha = 90°$? Sketch a graph of $\tan\alpha$ for $0° \leq \alpha < 90°$.

10 Using the table of question 9, sketch on the same axes the graphs of $\sin^2\alpha$ and $\cos^2\alpha$ for $0° \leq \alpha \leq 90°$. Explain how the graphs demonstrate the result
$$\sin^2\alpha + \cos^2\alpha \equiv 1$$

11 When $0° < \theta < 90°$, we define
$$\cot\theta = \frac{a}{b}, \quad \sec\theta = \frac{c}{a}, \quad \mathrm{cosec}\,\theta = \frac{c}{b}$$
(see fig. 3.1:1).

(i) What is the relationship between (a) $\cot\theta$ and $\tan\theta$; (b) $\sec\theta$ and $\cos\theta$; (c) $\mathrm{cosec}\,\theta$ and $\sin\theta$?

(ii) Write down the values of $\cot 30°$, $\sec 45°$, $\mathrm{cosec}\,60°$.

(iii) Find $\sin\alpha$ and $\sec\alpha$ when α is acute and
$$\cot\alpha = \frac{8}{15}$$

(iv) Show that, when α is acute,
(a) $\cot\alpha \equiv \tan(90° - \alpha)$
(b) $\sec^2\alpha \equiv 1 + \tan^2\alpha$

12 Using definitions (2), and with the aid of a diagram, write down the values of

(i) $\sin 0°$, $\cos 0°$, $\tan 0°$
(ii) $\sin 90°$, $\cos 90°$ (What can you say about $\tan 90°$?)
(iii) $\sin 180°$, $\cos 180°$, $\tan 180°$
(iv) $\sin 270°$, $\cos 270°$ (What can you say about $\tan 270°$?)

13 On graph paper, with origin O in the middle of the paper and perpendicular axes Ox, Oy, draw a large circle centre O, radius, say, 20 cm = 1 unit. Using only this diagram and definitions (2), write down approximate values (to 2 dec. pls.) of

(i) $\cos 40°$, $\cos(-40°)$, $\cos 140°$
(ii) $\sin 20°$, $\sin 160°$, $\sin 200°$, $\sin(-20°)$
(iii) $\tan 70°$, $\tan 250°$, $\tan 290°$, $\tan 430°$

14 Show with the aid of a diagram that, if $0° < \theta < 90°$ (θ acute), i.e. if P is in the first quadrant, then definitions (2) agree with the original definitions (1) of $\sin\theta$, $\cos\theta$ and $\tan\theta$.

15 (i) Show with the aid of a diagram that, if
$$90° < \theta < 180°$$
(θ obtuse), i.e. if P is in the second quadrant, then
$$\cos\theta = -\cos\beta,\ \sin\theta = \sin\beta,\ \tan\theta = -\tan\beta \quad (3)$$
where $\beta = 180° - \theta$ (an acute angle).

(ii) Using equations (3), write down, without using a calculator, the values of $\sin 150°$, $\cos 150°$, $\tan 150°$.

16 (i) Show with the aid of a diagram that, if
$$180° < \theta < 270°$$
(θ reflex), i.e. if P is in the third quadrant, then
$$\cos\theta = -\cos\gamma,\ \sin\theta = -\sin\gamma,\ \tan\theta = \tan\gamma \quad (4)$$
where $\gamma = \theta - 180°$ (an acute angle).

(ii) Using equations (4), write down, without using a calculator, the values of $\sin 240°$, $\cos 240°$, $\tan 240°$.

17 (i) Show with the aid of a diagram that, if

$$270° < \theta < 360°$$

(θ reflex), i.e. if P is in the fourth quadrant, then

$$\cos\theta = \cos\delta, \sin\theta = -\sin\delta, \tan\theta = -\tan\delta \quad (5)$$

where $\delta = 360° - \theta$ (an acute angle).

(ii) Using equations (5), write down, without using a calculator, the values of sin 315°, cos 315°, tan 315°.

18 Simplify:
   (i) $\sin(180° - \theta)$, $\cos(180° - \theta)$, $\tan(180° - \theta)$
   (ii) $\sin(180° + \theta)$, $\cos(180° + \theta)$, $\tan(180° + \theta)$
   (iii) $\sin(360° - \theta)$, $\cos(360° - \theta)$, $\tan(360° - \theta)$

19 (i) Explain why sin 390° = sin 30°.
   (ii) Without using a calculator write down sin 405°, tan 420°, cos 450°, tan 510°, sin 540°, cos 570°, sin 720°, tan 765°.

20 (i) Explain why cos (−50°) = cos 310°.
   (ii) Without using a calculator write down cos (−30°), sin (−90°), tan (−135°), sin (−240°).

21 Suppose θ is an angle which does not lie between 0° and 360° and that ϕ is the angle between 0° and 360° such that $\theta = \phi + 360n°$ (n integer). Explain why the trigonometrical ratios of θ are the same as those of ϕ. Hence, by finding the appropriate value of n, write down, without using a calculator
   (i) cos 2475°
   (ii) sin 1950°
   (iii) tan (−3360°)

22 Let θ be an acute angle.
   (i) Simplify $\sin(90° + \theta)$.
       (Hint: $90° + \theta = 180° - (90° - \theta)$ and see questions 15 and 2.)
   (ii) Simplify $\cos(90° + \theta)$. Hence find cos 110°, given that sin 20° ≈ 0.34.
   (iii) Show that $\tan(90° + \theta) = -1/\tan\theta$.

23 Let θ be an acute angle.
   (i) Simplify $\sin(270° - \theta)$.
       (Hint: $270° - \theta = 180° + (90° - \theta)$ and see questions 6 and 2.) Hence find sin 230°, given that cos 40° ≈ 0.77.
   (ii) Simplify $\cos(270° - \theta)$, $\tan(270° - \theta)$.

• → **24** Construct a table giving the values, in surd form, of sin θ, cos θ and tan θ for θ = −90°, −60°, −45°, −30°, 0°, 30°, 45°, 60°, 90°, 120°, 135°, 150°, 180°, 180° ... 450°.

3.1:2 Graphs of sin θ and cos θ

Sin θ and cos θ are functions of θ. We can usefully sketch their graphs (fig. 3.1:4).

(i) Sin θ and cos θ are both continuous functions.
(ii) The range of both sin θ and cos θ is the set [−1, 1], i.e.

$$-1 \leq \sin\theta \leq 1, \; -1 \leq \cos\theta \leq 1 \; \forall \theta$$

(iii) Sin θ and cos θ are both periodic with period 360°, i.e.

$$\sin\theta \equiv \sin(\theta + 360°), \cos\theta \equiv \cos(\theta + 360°)$$

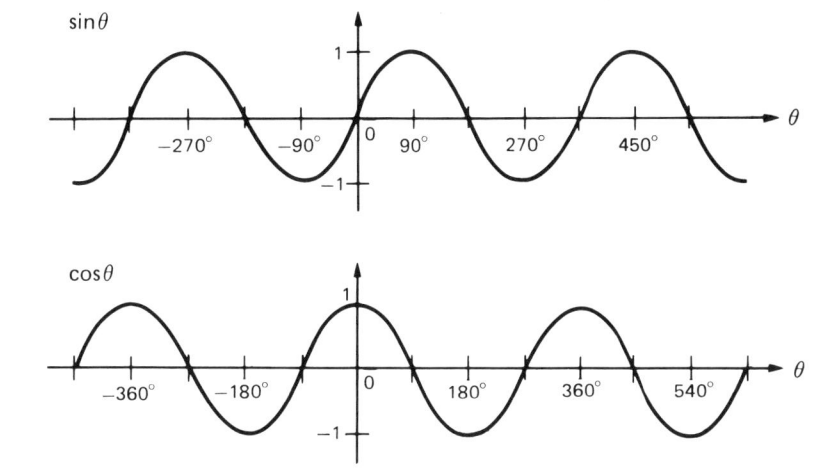

Fig. 3.1:4

126   *Graphs of* sin θ *and* cos θ

> **Theorem 1:** (i) $\sin(90° + \theta) \equiv \cos\theta$ (6)
> (ii) $\cos(90° + \theta) \equiv -\sin\theta$ (7)

The graph of sin θ is simply the graph of cos θ displaced 90° to the right. We say: 'There is a phase difference of +90° between sin θ and cos θ', e.g. sin 120° = cos 30°.

Proof
(i) (θ acute) By similar triangles (*ONQ*, *OMP* in fig. 3.1:5)

$$ON = OM$$

i.e.   *y*-coord of *Q* = *x*-coord of *P*

i.e. $\sin(90° + \theta) = \cos\theta$

We can extend the proof to all angles θ (see question 2).

Fig. 3.1:**5**   (ii) See question 3.

> **Theorem 2:**
> $$\sin(-\theta) \equiv -\sin\theta \quad (8)$$
> i.e. sin θ is an odd function of θ.
> $$\cos(-\theta) \equiv \cos\theta \quad (9)$$
> i.e. cos θ is an even function of θ. (See 1.2:4 Q21.)

Proof
See questions 4 and 5.

> **Theorem 3:**
> $$\sin(90° - \theta) \equiv \cos\theta \quad (10)$$
> $$\cos(90° - \theta) \equiv \sin\theta \quad (11)$$

Proof
See question 6.

Worked example 1   Sketch on the same axes the graphs of
(i) sin θ and sin 3θ   (ii) cos θ and 4 cos θ

(i) The graph of sin 3θ repeats itself 3 times as often as the graph of sin θ (fig. 3.1:6); its period is 360°/3, i.e. 120°.
(ii) The range of 4 cos θ is [−4, 4] (fig. 3.1:7); we say that 4 cos θ has **amplitude** 4. (Note that 4 cos θ has period 360°.)

In general,

> the functions $R\sin n\theta$, $R\cos n\theta$ have amplitude R and period $360°/n$.

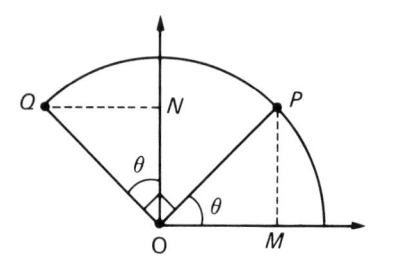

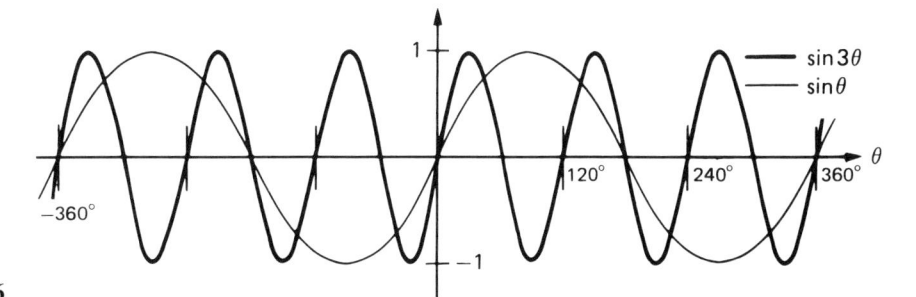

Fig. 3.1:**6**

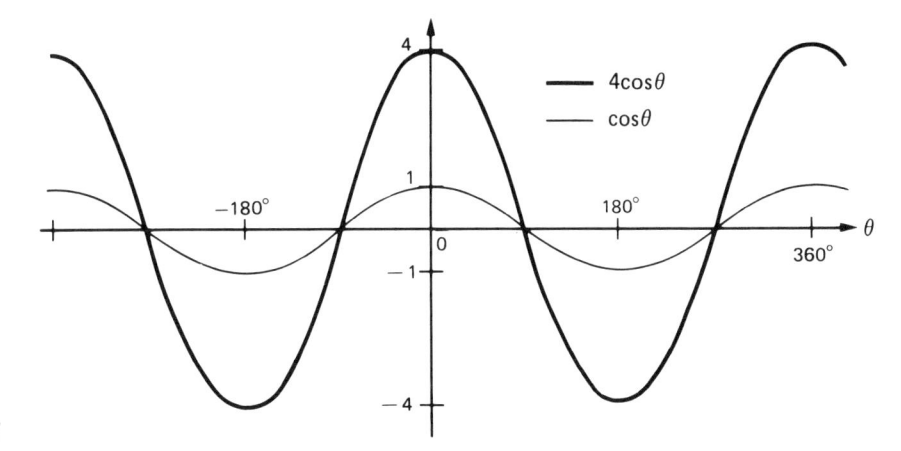

Fig. 3.1:**7**

Worked example 2 Sketch on the same axes the graphs of
(i) $\cos\theta$ and $\cos(\theta+60°)$ (ii) $\sin 2\theta$ and $\sin(2\theta-30°)$

(i) Note that $\cos\theta = 1$ when $\theta = 0°$, $360°$, etc., but that $\cos(\theta+60°) = 1$ when $(\theta+60°) = 0°$, $360°$, etc., i.e. when $\theta = -60°$, $300°$, etc.

The graph of $\cos(\theta+60°)$ is simply the graph of $\cos\theta$ displaced $60°$ to the left (fig. 3.1:**8**). We say that there is a **phase difference** of $60°$ between $\cos\theta$ and $\cos(\theta+60°)$.

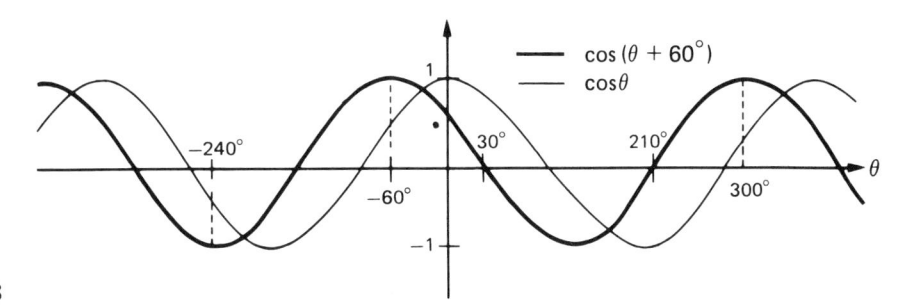

Fig. 3.1:**8**

(ii) Note that $\sin 2\theta = 0$ when $\theta = 0°$, $90°$, $180°$, etc., but that $\sin(2\theta-30°) = 0$ when $\theta = 15°$, $105°$, $195°$, etc.

The graph of $\sin(2\theta-30°)$ is simply the graph of $\sin 2\theta$ displaced $15°$ to the right (see fig. 3.1:**9**).

In general

$$\begin{array}{ccccc} R\sin(n\theta+\varepsilon) & \xleftarrow{\varepsilon/n \text{ to the}} & R\sin n\theta & \xrightarrow{\varepsilon/n \text{ to the}} & R\sin(n\theta-\varepsilon) \\ R\cos(n\theta+\varepsilon) & \text{left} & R\cos n\theta & \text{right} & R\cos(n\theta-\varepsilon) \end{array}$$

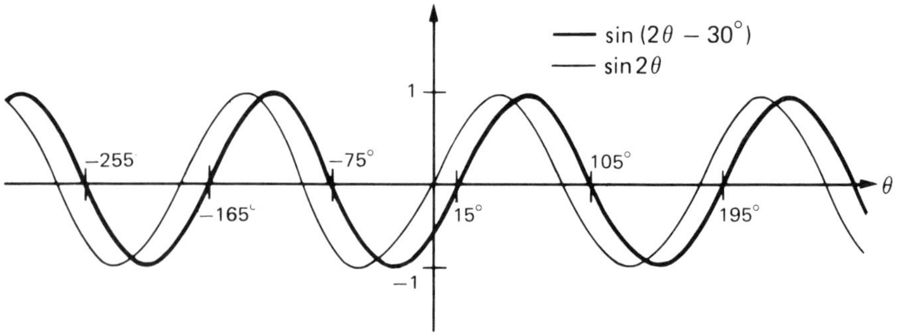

128 *Graphs of* $\sin \theta$ *and* $\cos \theta$

Fig. 3.1:9

Exercise 3.1:2

1 With the aid of your table (3.1:1 Q24), draw on graph paper graphs of $\sin \theta$ and $\cos \theta$ for

$$-90° \leqslant \theta \leqslant 450°$$

From your graphs read off approximate values (2 dec. pls.) of $\cos 40°$, $\cos(-40°)$, $\cos 140°$, $\sin 20°$, $\sin 160°$, $\sin 200°$, $\sin(-20°)$. Now look at your answers to 3.1:1 Q13.

• **2** (i) Show with the aid of a diagram that when θ is obtuse, $\sin(90° + \theta) \equiv \cos \theta$.

(ii) Using the results of 3.1:1 Q16 and Q17 show that the identity holds for all angles θ.

3 Show with the aid of a diagram that, when θ is acute, $\cos(90° + \theta) \equiv -\sin \theta$.

4 Show with the aid of a diagram that, when θ is acute, $\sin(-\theta) \equiv -\sin \theta$ and $\cos(-\theta) \equiv \cos \theta$.

5 Show with the aid of a diagram that, when θ is obtuse, $\sin(-\theta) \equiv -\sin \theta$ and $\cos(-\theta) \equiv \cos \theta$.

• **6** Prove Theorem 3:

(i) with the aid of a diagram (see proof of Theorem 1);

(ii) by replacing θ by $-\theta$ in Theorem 1 and using Theorem 2.

① **7** Sketch on the same axes the graphs of:

→ (i) $\cos \theta$ and $\cos 2\theta$
R (ii) $\sin \theta$ and $\sin 5\theta$

for $-360° \leqslant \theta \leqslant 360°$ marking the axes carefully to show the scales.

① **8** State the period of each of the following functions and sketch their graphs for $-360° \leqslant \theta \leqslant 360°$:

(i) $\sin \tfrac{1}{2}\theta$ (ii) $\cos \tfrac{3}{2}\theta$ (iii) $\cos \tfrac{1}{3}\theta$

① **9** Sketch on the same axes the graphs of

(i) $\sin \theta$ and $3 \sin \theta$
(ii) $\cos 3\theta$ and $2 \cos 3\theta$

for $-360° \leqslant \theta \leqslant 360°$.

① → **10** State the amplitude and period of each of the following functions:

(i) $5 \sin \tfrac{3}{2}\theta$
(ii) $\tfrac{3}{5} \cos \tfrac{2}{3}\theta$
(iii) $4 \sin \theta$

What are the maximum and minimum values of each function?

① **11** Sketch on the same axes the graphs of

(i) $2 \sin \theta$ and $1 + 2 \sin \theta$
(ii) $\sin \theta$, $-\sin \theta$ and $2 - \sin \theta$

for $-360° \leqslant \theta \leqslant 360°$.

① **12** Write down the maximum and minimum values of the following functions and sketch their graphs for $-360° \leqslant \theta \leqslant 360°$:

(i) $5 \sin 3\theta$
(ii) $1 + 3 \sin \theta$
(iii) $1 - 3 \sin \theta$
→ (iv) $4 + 2 \cos 2\theta$

① → **13** Using graph paper, draw on the same axes the graphs of $\sin \theta$ and $1 + \cos 2\theta$ for $0° \leqslant \theta \leqslant 180°$. Hence find an approximate solution of the equation

$$\sin \theta - \cos 2\theta = 1$$

① **14** By drawing approximate graphs, find approximate solutions, in the range $0° \leqslant \theta \leqslant 180°$, of the following equations:

(i) $\cos \theta = \tfrac{3}{5}$
(ii) $\sin \theta = \tfrac{3}{5}$
(iii) $\sin 2\theta = \tfrac{3}{5}$
(iv) $3 \sin \theta + 2 \cos \theta = 1$
(v) $2 \sin \theta - \sin 2\theta = 1$

② **15** Sketch on the same axes the graphs of:

(i) $\sin \theta$ and $\sin(\theta + 45°)$
(ii) $\cos \theta$ and $\cos(\theta - 120°)$

for $-360° \leqslant \theta \leqslant 360°$, marking carefully the points at which they cut the axes.

Trigonometry

② → **16** Sketch on the same axes the graphs of
   (i) $\cos 2\theta$ and $\cos (2\theta + 60°)$
   (ii) $5 \sin 3\theta$ and $5 \sin (3\theta - 75°)$
   (iii) $3 \cos \frac{1}{2}\theta$ and $3 \cos (\frac{1}{2}\theta - 20°)$
   for $-360° \leq \theta \leq 360°$.

② **17** Sketch on the same axes the graphs of $5 \sin 4\theta$, $5 \sin (4\theta + 120°)$, $5 \sin (4\theta - 120°)$ for $-180° \leq \theta \leq 180°$.

② **18** Write down the amplitude and period of each of the following functions and sketch their graphs:
   (i) $\sin (3\theta + 60°)$
   (ii) $3 \sin (\frac{1}{2}\theta - 30°)$
   R (iii) $5 \cos (\frac{2}{3}\theta + 45°)$

marking carefully the points at which they cut the axes.

② **19** Using graph paper, draw on the same axes the graphs of $1 - \cos \theta$ and $\sin (\theta + 20°)$ for $0° \leq \theta \leq 180°$. Hence find an approximate solution of the equation $\sin (\theta + 20°) + \cos \theta = 1$.

20 Sketch on the same axes the graphs of the following functions:
   (i) $\sin \theta$ (ii) $|\sin \theta|$ (iii) $\frac{1}{2}(|\sin \theta| + \sin \theta)$
for $-180° < \theta < 540°$. Which of the functions are even? Which are odd?

21 What functions are represented by the following graphs?

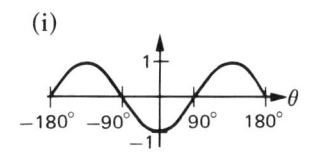

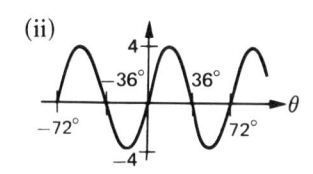

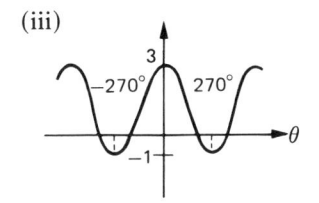

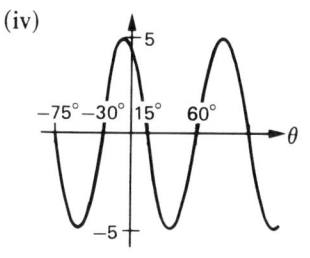

(i) [graph with values 1, -1 at $\pm 180°, \pm 90°, 90°, 180°$]

(ii) [graph with values 4, -4 at $\pm 72°, \pm 36°$]

(iii) [graph with values 3, -1 at $\pm 270°$]

(iv) [graph with values 5, -5 at $-75°, -30°, 15°, 60°$]

22 A function f is defined by
$$f(x) = \begin{cases} 1 - \cos x & 0° \leq x < 90° \\ \sin x & 90° \leq x < 180° \end{cases}$$
and f is periodic with period $180°$. Sketch the graph of f in the interval $-360° \leq x < 360°$.

R 23 A function f is defined by
$$f(x) = \begin{cases} \sin 3x & 0° \leq x < 60° \\ -1 - 2 \cos 2x & 60° \leq x < 90° \\ 1 + \cos x & 90° \leq x < 180° \end{cases}$$
and f is periodic with period $180°$. Sketch the graph of f in the interval $-180° \leq x < 360°$.

3.1:3 Graphs of tan θ, cot θ, sec θ and cosec θ

The graph of $\tan \theta$ is shown in Fig. 3.1:10.

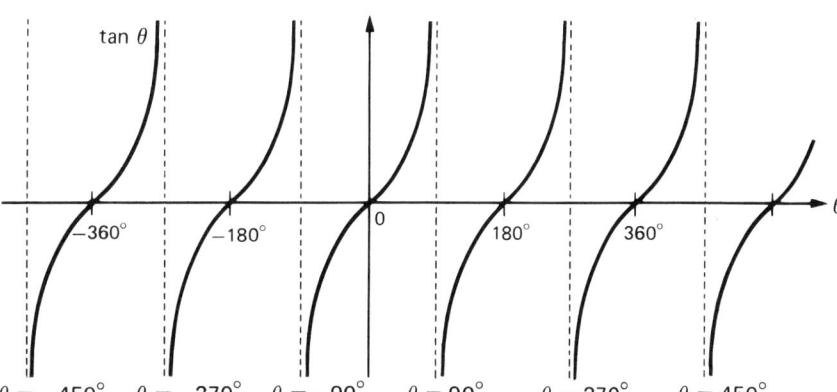

Fig. 3.1:10 $\theta = -450°$ $\theta = -270°$ $\theta = -90°$ $\theta = 90°$ $\theta = 270°$ $\theta = 450°$

(i) Tan θ is *not* a continuous function. Since $\tan \theta = \sin \theta / \cos \theta$, its graph has asymptotes wherever $\cos \theta = 0$, i.e. when $\theta = \pm 90°, \pm 270°$, etc.

130 *Graphs of* $\tan \theta$, $\cot \theta$, $\sec \theta$ *and* $\csc \theta$

(ii) The range of $\tan \theta$ is the whole of $\mathbb{R}$, i.e. $\tan \theta$ can take any real value.
(iii) Tan θ is a periodic function with period 180°, i.e.

$$\tan(\theta + 180°) \equiv \tan \theta$$

Definition

$$\cot \theta = \frac{\cos \theta}{\sin \theta} \qquad \sec \theta = \frac{1}{\cos \theta} \qquad \csc \theta = \frac{1}{\sin \theta} \qquad (12)$$

Worked example 3 Sketch on the same axes the graphs of $\cos \theta$ and $\sec \theta$.

The graphs are shown in fig. 3.1:11. Sec $\theta = 1/\cos \theta$ has asymptotes where $\cos \theta = 0$, i.e. where $\theta = \pm 90°$, $\pm 270°$, etc.
When $\theta = 0°$, $\pm 360°$, $\pm 720°$, etc., $\cos \theta = 1$, so $\sec \theta = 1$.
When $\theta = \pm 180°$, $\pm 540°$, etc., $\cos \theta = -1$, so $\sec \theta = -1$.
The range of $\sec \theta$ is $(-\infty, -1] \cup [1, \infty)$. Like $\cos \theta$, $\sec \theta$ is an even function, and has period 360°.

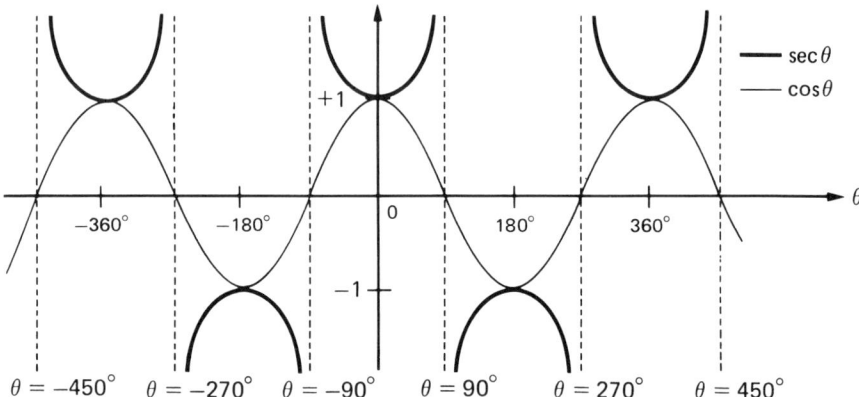

Fig. 3.1:11 $\theta = -450°$ $\theta = -270°$ $\theta = -90°$ $\theta = 90°$ $\theta = 270°$ $\theta = 450°$

Exercise 3.1:3

1 With the aid of the table from 3.1:1 Q24 draw on graph paper the graph of $\tan \theta$ for $-90° < \theta < 450°$, marking carefully the asymptotes to the graph. From your graph read off approximate values of $\tan 70°$, $\tan 250°$, $\tan 290°$, $\tan 430°$. Now look at your answers to 3.1:1 Q13(iii).

2 (i) Write down the domain of the function $\tan \theta$.
(ii) Draw a circular diagram (see proof of Theorem 1) to show that $\tan(-\theta) \equiv -\tan \theta$.
(iii) Draw a circular diagram to show that

$$\tan \theta \equiv \tan(\theta + 180°)$$

→ **3** Write down the period of the following functions and sketch their graphs for $-270° < \theta < 270°$:

(i) $\tan \frac{1}{2}\theta$ (ii) $\tan(\frac{5}{3}\theta - 45°)$

4 Sketch on the same axes the graphs of:

R (i) $\tan \theta$ and $\tan 3\theta$
(ii) $3 \tan 2\theta$ and $3 \tan(2\theta + 60°)$

for $-270° < \theta < 270°$.

5 Using graph paper, draw on the same axes the graphs of $\tan \theta$ and $\cos(\theta + 10°)$ for $-90° < \theta < 90°$. Hence find an approximate solution of the equation $\cos(\theta + 10°) = \tan \theta$.

6 By drawing appropriate graphs, find approximate solutions, in the range $-90° < \theta < 90°$, of the following equations:

(i) $\tan \theta = 2$
(ii) $\tan 2\theta = 1$
(iii) $\tan(\theta - 20°) = \sin \theta$

7 Do you have a calculator, a box of After Eights and several hours to spare? Try this one: find by trial and error the solution, to 6 dec. pls. of the equation $\tan x° = x$ which is near $x = 90$.

③ **8** Without using a calculator, write down the values of cot 60°, sec 30°, cosec 45°, cot 90°, sec 120°, cosec 240°, cot (−30°), sec (−45°).

③ **9** Simplify:

(i) $\cot \theta \tan \theta$
(ii) $\cot 2\theta \sec 2\theta$
(iii) $\sin \theta \sec \theta$
(iv) $\operatorname{cosec} \tfrac{1}{2}\theta \tan \tfrac{1}{2}\theta$
(v) $\tan \theta \operatorname{cosec} \theta / \sec \theta$

③ → **10** Using the same axes sketch the graphs of:

(i) $\sin \theta$ and $\operatorname{cosec} \theta$
(ii) $\tan \theta$ and $\cot \theta$

for −450° < θ < 450°, marking carefully the asymptotes to the curves.

Write down the domain, the range and the period of the functions cosec θ and cot θ. Are these functions even or odd or neither?

③ **11** Show that
$$\tan(90° - \theta) \equiv \cot \theta \text{ and } \cot(90° - \theta) \equiv \tan \theta$$

③ **12** Sketch graphs of:

→ (i) $\sec 3\theta$
R (ii) $4 \cot 2\theta$

for −180° < θ < 180°.

③ **13** Using graph paper, draw on the same axes graphs of tan 2θ and 2 cot θ for 0° < θ < 180°. Hence find an approximate solution of the equation

$$\tan \theta \tan 2\theta = 2$$

3.2 Trigonometric identities and equations

3.2:1 The equations sin θ = k, cos θ = k and tan θ = k

Principal solutions

Consider the equation

$$\sin \theta = k, \quad -1 \leq k \leq 1 \tag{1}$$

The equation has an infinite number of solutions, but whatever the value of k, exactly one of these solutions lies in the interval $-90° \leq \theta \leq 90°$ (fig. 3.2:1). This particular solution is called $\sin^{-1} k$ or **arcsin** k, the **principal solution** of equation (1).

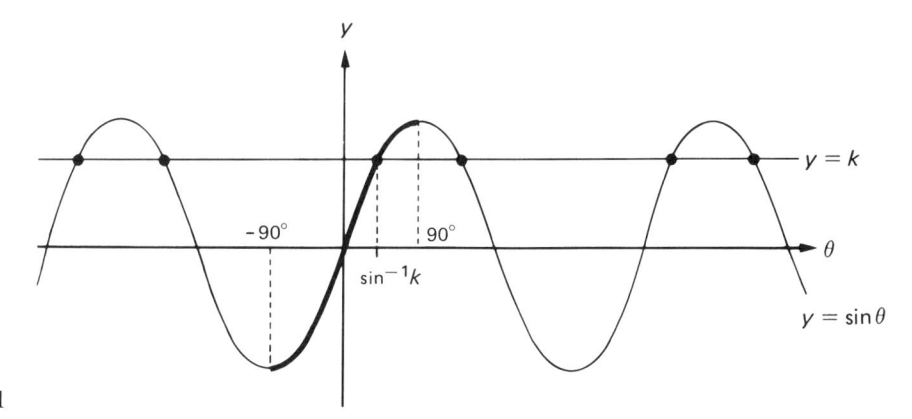

Fig. 3.2:1

We shall see later, in section 3.5:1, that $g(x) = \sin^{-1} x$ is a function, the inverse of the function $f(x) = \sin x$. Note that $\sin^{-1} x$ means 'inverse sine x' or 'the angle whose sine is x'; it does *not* mean $1/\sin x$.

132 *The equations* $\sin \theta = k$, $\cos \theta = k$ *and* $\tan \theta = k$

Definition $\text{Sin}^{-1} k$ is the angle between $-90°$ and $90°$ whose sine is k.

For example,
the principal solution of $\sin \theta = \frac{1}{2}$ is $\sin^{-1}(\frac{1}{2})$, i.e. $30°$;
the principal solution of $\sin \theta = -\frac{4}{5}$ is $\sin^{-1}(-\frac{4}{5})$, i.e. $-53.13°$.

Definition The principal solution, $\cos^{-1} k$, of the equation

$$\cos \theta = k, \quad -1 \leqslant k \leqslant 1 \tag{2}$$

is the solution lying in the interval $0° \leqslant \theta \leqslant 180°$ (see fig. 3.2:2).
For example,
the principal solution of $\cos \theta = -\frac{1}{3}$ is $\cos^{-1}(-\frac{1}{3})$, i.e. $109.47°$.

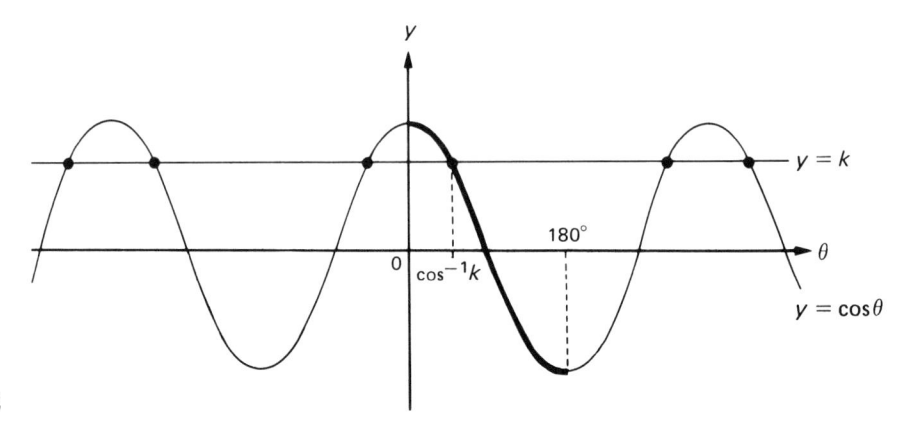

Fig. 3.2:**2**

Definition The principal solution, $\tan^{-1} k$, of the equation

$$\tan \theta = k, \quad k \in \mathbb{R} \tag{3}$$

is the solution lying in the interval $-90° < \theta < 90°$ (fig. 3.2:3).
For example,
the principal solution of $\tan \theta = -5$ is $\tan^{-1}(-5)$, i.e. $-78.69°$.

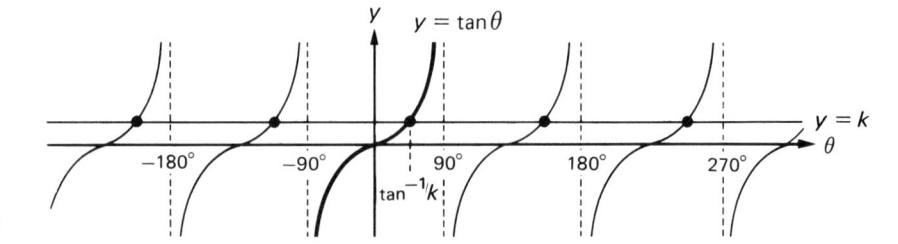

Fig. 3.2:**3**

General solutions

Worked example 1 Find the general solution of the equation $\sin \theta = \frac{1}{2}$.

First we look for the principal solution of the equation, $\theta = \sin^{-1}(\frac{1}{2})$, i.e. $\theta = 30°$, corresponding to the point P_1 (fig. 3.2:4). Another solution, corresponding to the point P_2, is $\theta = 180° - 30°$, i.e. $\theta = 150°$.

Trigonometry   133

The general solution is given by adding multiples of 360° to each of these solutions, i.e.

$$\left.\begin{array}{l}\theta = 30° + 360n° \\ \text{or } \theta = 150° + 360n°\end{array}\right\} n = 0, \pm 1, \pm 2, \text{ etc., i.e. } n \in \mathbb{Z}$$

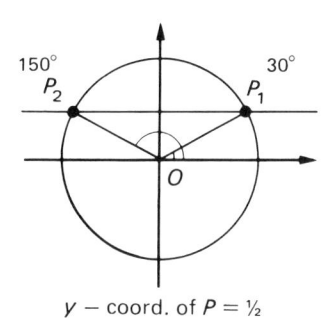

y − coord. of $P = \frac{1}{2}$

Fig. 3.2:4   In general,

$$\sin \theta = \sin \alpha \Rightarrow \theta = \alpha + 360n° \text{ or } (180° - \alpha) + 360n° \quad \text{(4)}$$

i.e. $\sin \theta = k \Rightarrow \theta = \sin^{-1} k + 360n° \text{ or } (180° - \sin^{-1} k) + 360n°$ **(4a)**

Worked example 2   Find the general solution of the equation $\cos \theta = -\frac{4}{5}$.

First we look for the principal solution of the equation, $\theta = \cos^{-1}(-\frac{4}{5})$, i.e. $\theta = 143.87°$, corresponding to the point P_1 (fig. 3.2:5). Another solution, corresponding to the point P_2, is $\theta = -143.13°$.

The general solution is given by adding multiples of 360° to each of these solutions, i.e.

$$\theta = \pm 143.87° + 360n°, \quad n \in \mathbb{Z}$$

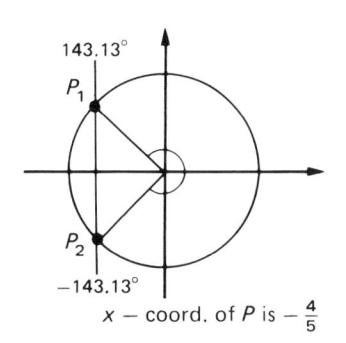

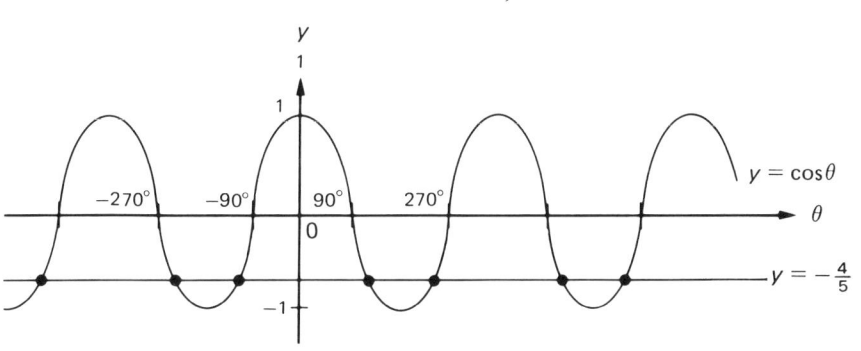

x − coord. of P is $-\frac{4}{5}$

Fig. 3.2:5   In general,

$$\cos \theta = \cos \alpha \Rightarrow \theta = \pm \alpha + 360n° \quad \text{(5)}$$

i.e. $\cos \theta = k \Rightarrow \theta = \pm \cos^{-1} k + 360n°$ **(5a)**

Worked example 3   Find the general solution of the equation $\tan \theta = -2$.

First we look for the principal solution of the equation, $\theta = \tan^{-1}(-2)$, i.e. $\theta = -63.43°$, corresponding to the point P_1 (fig. 3.2:6).

The general solution is given by adding multiples of 180° to this solution, i.e.

$$\theta = -63.43° + 180n°, \quad n \in \mathbb{Z}$$

134 *The equations* $\sin\theta = k$, $\cos\theta = k$ *and* $\tan\theta = k$

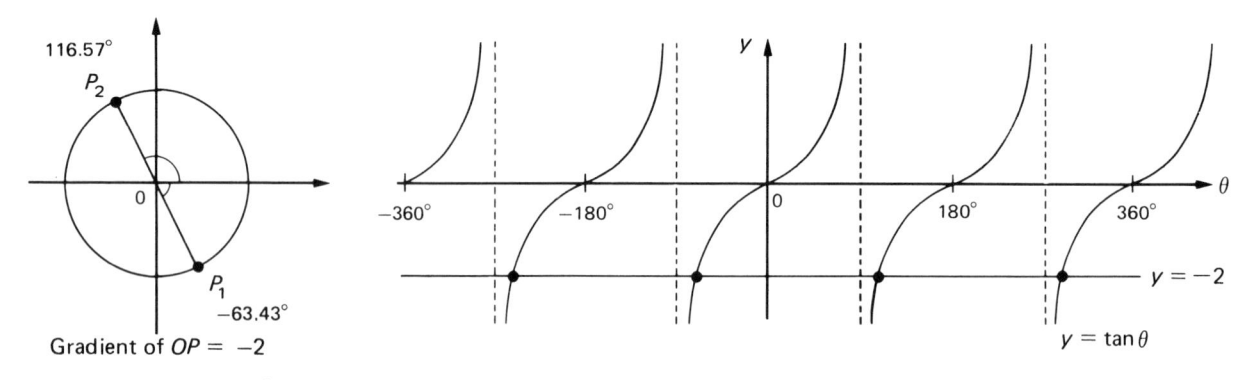

Fig. 3.2:6

In general

$$\tan\theta = \tan\alpha \Rightarrow \theta = \alpha + 180n° \quad (6)$$
$$\text{i.e. } \tan\theta = k \Rightarrow \theta = \tan^{-1} k + 180n° \quad (6a)$$

Worked example 4 Find the general solution of the equation $\cos 5\theta = -\frac{1}{2}$.

$$\cos 5\theta = -\tfrac{1}{2} \Rightarrow 5\theta = \pm 120° + 360n°$$
$$\Rightarrow \theta = \pm 24° + 72n°$$

Notice that, since the period of $\cos 5\theta$ is 72°, the pairs of solutions differ by multiples of 72° (fig. 3.2:7). Notice also that there are *ten* solutions between 0° and 360°.

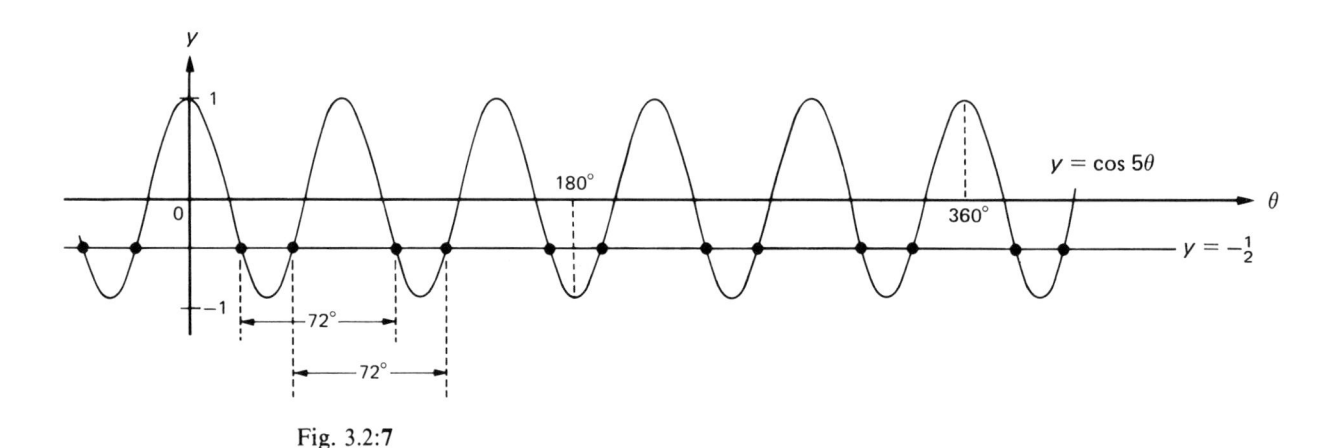

Fig. 3.2:7

Worked example 5 Find the general solution of the equation

$$\cos\theta = \cos 2\theta$$

$$\cos\theta = \cos 2\theta \Rightarrow \theta = \pm 2\theta + 360n° \quad \text{by (5)}$$

i.e. either $3\theta = 360n°$ or $-\theta = 360n°$
$$\Rightarrow \theta = 120n° \qquad \Rightarrow \theta = -360n°$$

Note that the second set of solutions is contained in the first (fig. 3.2:8). So we can write the general solution as $\theta = 120n°$.

Fig. 3.2:**8**

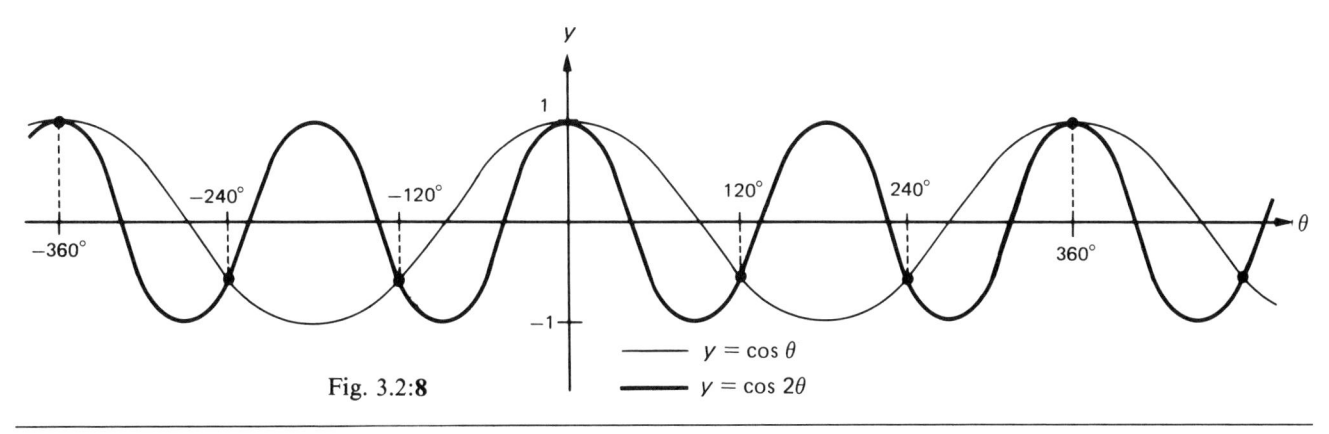

$y = \cos\theta$
$y = \cos 2\theta$

Exercise 3.2:1

→ **1** Without using a calculator but with the aid of the table from 3.1:1 Q24, write down the values of:

(i) $\sin^{-1}\frac{1}{2}\sqrt{3}$ (ii) $\sin^{-1}(-\frac{1}{2})$
(iii) $\sin^{-1}0$ (iv) $\cos^{-1}\frac{1}{2}\sqrt{3}$
(v) $\cos^{-1}(-\frac{1}{2})$ (vi) $\cos^{-1}0$
(vii) $\tan^{-1}1$ (viii) $\tan^{-1}(-\sqrt{3})$
(ix) $\tan^{-1}0$

2 Using tables, a calculator or whatever is to hand, find the values, to dec. pls., of:

(i) $\sin^{-1}(0.3)$ (ii) $\cos^{-1}\frac{3}{4}$
(iii) $\tan^{-1}\frac{1}{2}$ (iv) $\tan^{-1}5$
(v) $\sin^{-1}(-\frac{2}{3})$ (vi) $\cos^{-1}(0.9)$
(vii) $\tan^{-1}(-\frac{3}{2})$

3 Find the values of:

(i) $\sin^{-1}(0.4)$ and $\sin^{-1}(-0.4)$
(ii) $\cos^{-1}(0.2)$ and $\cos^{-1}(-0.2)$
(iii) $\cos^{-1}\frac{1}{2}\sqrt{2}$ and $\cos^{-1}(-\frac{1}{2}\sqrt{2})$
(iv) $\tan^{-1}3$ and $\tan^{-1}(-3)$

① *Questions 4–6:* Solve (i.e. find the general solution) of the following equations, showing in each case the solutions on a circular diagram and a graph as in worked example 1.

4 (i) $\sin\theta = \frac{1}{2}\sqrt{3}$
→ (ii) $\sin\theta = 0.2$
(iii) $\sin\theta = -0.41$
(iv) $\sin\theta = 1$

In what way is (iv) different from (i), (ii) and (iii)?

② **5** (i) $\cos\theta = \frac{1}{4}$
→ (ii) $\cos\theta = -\frac{1}{2}$
(iii) $\cos\theta = 0$
(iv) $\cos\theta = -1$

In what way is (iv) different from (i), (ii) and (iii)?

③ **6** (i) $\tan\theta = \dfrac{1}{\sqrt{3}}$
→ (ii) $\tan\theta = 3$
(iii) $\tan\theta = -\frac{1}{3}$
(iv) $\tan\theta = -1$

7 Solve the equation $\sin\theta = 0.65$:

(i) by the method of worked example 1;
(ii) by writing $\sin\theta \equiv \cos(90° - \theta)$ and using the method of worked example 2.

8 Solve the equations:

(i) $\cos\theta = \frac{1}{3}$ (ii) $\cos\theta = -\frac{1}{3}$

9 Solve the equations:

(i) $\tan\theta = 2$ (ii) $\cot\theta = 2$

10 Find the general solution of each of the following equations, and hence write down the solution(s) in the range $0° < \theta < 360°$:

• → (i) $\sin\theta = 0.7$
(ii) $\cos\theta = -\frac{3}{5}$
R (iii) $\tan\theta = -2$
(iv) $\sin\theta = -0.1$
(v) $\sin\theta = -1$

In what way is (v) different from (i), (ii), (iii) and (iv)?

④ **11** Find the general solution of the following equations and hence write down the solutions, if any, in the range $0° < \theta < 360°$:

(i) $\sin 2\theta = -\frac{1}{2}$
→ (ii) $\cos 3\theta = 1/\sqrt{2}$
(iii) $\tan\frac{1}{2}\theta = \sqrt{3}$
(iv) $3\tan 5\theta = 4$
(v) $2\sin\frac{2}{5}\theta = -1$
(vi) $\cos(\theta + 60°) = -\frac{4}{5}$
(vii) $5\sin(3\theta - 30°) = 2$
(viii) $\sec 4\theta = -2$

④ **12** Find the solutions in the range $-360° < \theta < 360°$ of the equations:

(i) $\sin(\theta - 45°) = \frac{1}{2}$
(ii) $\cos(\theta - 45°) = -\frac{1}{2}$

Illustrate your answers by sketching the appropriate graphs.

136 *The identity* $\sin^2 \theta + \cos^2 \theta \equiv 1$

④ **13** Find the solutions in the range $-180° < \theta < 180°$ of the equations:
→ (i) $\sin(2\theta + 80°) = \tfrac{1}{2}\sqrt{3}$
(ii) $\tan(2\theta - 80°) = 1/\sqrt{3}$
(iii) $\sec(3\theta + 45°) = \sqrt{2}$

Illustrate your answers by sketching the appropriate graphs.

④ **14** Find the values of θ between 0° and 360° for which:
(i) $\sin \theta \leq \tfrac{1}{2}$ (ii) $5 \cos 2\theta \geq 3$

④ : **15** What is wrong with the following 'solution'?
$$2 \sin \theta = \tan \theta \Rightarrow 2 = 1/\cos \theta$$
$$\Rightarrow \cos \theta = \tfrac{1}{2} \Rightarrow \theta = \pm 60° + 360n°$$
Correct it.

④ **16** Find the general solution of the equations:
(i) $\cos^2 \theta = \tfrac{1}{2}$
R (ii) $2 \sin \theta = \sqrt{3} \tan \theta$
(iii) $2 \sin^2 \theta = \sin \theta$
(iv) $\sin \theta + 4 \sin \theta \cos \theta = 0$
(v) $2 \cos^2 \theta + 3 \cos \theta + 1 = 0$
(vi) $\tan^3 \theta - \tan \theta - 2 = 0$
→ (vii) $\tan \theta + 3 \cot \theta = 4$
(viii) $2 \sin \theta \cos \theta + \sin \theta = 2 \cos \theta + 1$

⑤ **17** Solve the equations:
(i) $\cos \theta = \cos 3\theta$
(ii) $\cos 2\theta = \cos 5\theta$
(iii) $\sin \theta = \sin 3\theta$

(iv) $\sin \theta + \sin 2\theta = 0$
(v) $\tan \theta = \tan 5\theta$
(vi) $\tan \theta = -\tan 3\theta$
(vii) $\cos(\theta + 30°) = \cos(60° - 3\theta)$

⑤ • **18** Using the identity $\sin \theta \equiv \cos(90° - \theta)$, solve the equation $\sin \theta = \cos 3\theta$.

⑤ **19** Solve the equations:
(i) $\cos \theta = \sin 2\theta$
→ (ii) $\cos 2\theta = \sin 3\theta$
(iii) $\sin 3\theta = \cos 4\theta$
(iv) $\sin \theta + \cos 2\theta = 0$
(v) $\tan \theta \tan 2\theta = 1$
(vi) $\cos \theta + \cos 2\theta = 0$
(vii) $\sin(30° + \theta) = \cos(45° + \theta)$
(viii) $\sin(\theta + 20°) = \cos 3\theta$

⑤ **20** Sketch on the same axes the graphs of $\sin \theta$ and $\sin 2\theta$ for $-360° < \theta < 360°$, and find their points of intersection.

⑤ R **21** Solve the equations: (i) $\sin k\theta = \sin \theta$ (k constant); (ii) $\cos 3\theta = \cos 2\theta$.
Find the value of k for which the equations have exactly six common solutions in the range
$$0° \leq \theta \leq 360°$$

22 Solve the simultaneous equations:
• → (i) $\cos \theta + \cos \phi = 1$, $\sec \theta + \sec \phi = 4$
(Hint: let $\cos \theta = a$, $\cos \phi = b$)
(ii) $\tan \theta + \tan \phi = 1$, $\cot \theta + \cot \phi = -1$
(iii) $\cos \theta + \cos \phi = \tfrac{1}{2}$, $\cos \theta \cos \phi = -\tfrac{1}{2}$

3.2:2 The identity $\sin^2 \theta + \cos^2 \theta \equiv 1$

Theorem 1:
$$\sin^2 \theta + \cos^2 \theta \equiv 1 \tag{7}$$
$$\tan^2 \theta + 1 \equiv \sec^2 \theta \tag{8}$$
$$1 + \cot^2 \theta \equiv \operatorname{cosec}^2 \theta \tag{9}$$

Proof
See question 1.

Worked example 6 Find the values of θ between 0° and 360° for which $2 \sin^2 \theta + \cos \theta = 1$.
We use identity (7) to transform this equation into a quadratic equation in $\cos \theta$:
$$\sin^2 \theta \equiv 1 - \cos^2 \theta$$

So we can write
$$2(1 - \cos^2 \theta) + \cos \theta = 1$$
$$2 \cos^2 \theta - \cos \theta - 1 = 0$$
$$(2 \cos \theta + 1)(\cos \theta - 1) = 0$$

So either $\cos\theta = -\frac{1}{2}$ or $\cos\theta = 1$
$\Rightarrow \theta = 120°$ or $240°$ $\Rightarrow \theta = 0°$ or $360°$

Worked example 7 Eliminate θ from the equations
$$x = 3\cos\theta, \quad y = 1 + 4\sin\theta$$

We write $\cos\theta$ in terms of x and $\sin\theta$ in terms of y, and then use identity (7).
$$x = 3\cos\theta \Rightarrow \cos\theta = \tfrac{1}{3}x$$
$$y = 1 + 4\sin\theta \Rightarrow \sin\theta = \tfrac{1}{4}(y-1)$$

By identity (7), $\cos^2\theta + \sin^2\theta \equiv 1$.

So
$$\frac{x^2}{9} + \frac{(y-1)^2}{16} = 1$$
i.e.
$$16x^2 + 9(y-1)^2 = 144$$

Worked example 8 If $\sin\alpha = \tfrac{3}{5}$, find $\cos\alpha$ and $\tan\alpha$.

Identity (7) $\Rightarrow \cos^2\alpha = 1 - \sin^2\alpha = 1 - \tfrac{9}{25} = \tfrac{16}{25} \Rightarrow \cos\alpha = \pm\tfrac{4}{5}$.

$$\tan\alpha = \frac{\sin\alpha}{\cos\alpha} = \pm\tfrac{3}{4}$$

Note: if we know that α is acute (as is often the case in applied maths problems) then we can simply draw a triangle:

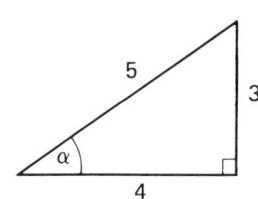

$\cos\alpha = \tfrac{4}{5}$, $\tan\alpha = \tfrac{3}{4}$

but notice that there is an angle $\alpha \approx 143.13°$ in the second quadrant for which $\sin\alpha = \tfrac{3}{5}$, $\cos\alpha = -\tfrac{4}{5}$, $\tan\alpha = -\tfrac{3}{4}$).

Exercise 3.2:2

• → **1** (i) Prove identity (7) when θ is (a) acute and (b) obtuse. (Hint: use (a) and 3.1:2 identity (10).)

(ii) Divide identity (7) through by $\cos^2\theta$. What do you get?

(iii) Prove identity (9).

⑥ **2** Solve the equations:

(i) $2\cos^2\theta - \sin\theta = 1$
(ii) $5\cos\theta - 4\sin^2\theta = 2$
R (iii) $6\cos^2\theta + 7\sin\theta = 8$
→ (iv) $6\sin^2\theta - \cos\theta = 5$
(v) $7\sin^2\theta - 5\sin\theta + \cos^2\theta = 0$
(vi) $\cos^2\theta = 3\sin\theta$

⑥ • **3** Using the identity $\sec^2\theta \equiv 1 + \tan^2\theta$, transform the equation $2\sec^2\theta = 2 + \tan\theta$ into a quadratic equation in $\tan\theta$ and hence solve it.

⑥ **4** Solve the equations:

(i) $3\tan^2\theta - 5\sec\theta + 1 = 0$
R (ii) $\sec^2\theta = \tan\theta - 3$
→ (iii) $4\sec^2\theta - 3\tan\theta = 5$
(iv) $\sec\theta = 1 - 2\tan^2\theta$

⑥ **5** Using the identity $\operatorname{cosec}^2\theta \equiv \cot^2\theta + 1$, solve the equations:

(i) $\cot^2\theta = \operatorname{cosec}\theta + 1$
(ii) $4\cot^2\theta + 39 = 24\operatorname{cosec}\theta$
(iii) $4\cot^2\theta + 4\operatorname{cosec}\theta + 1 = 0$

138 *The identity* $\sin^2\theta + \cos^2\theta \equiv 1$

⑥→ **6** By first writing each of the functions $\tan\theta$, $\cot\theta$ and $\sec\theta$ in terms of $\sin\theta$ and $\cos\theta$, transform the equation
$$\tan\theta + 4\cot\theta = 4\sec\theta$$
into a quadratic equation in $\sin\theta$ and hence solve it.

⑥ **7** Solve the equations:
R (i) $\tan\theta + 3\cot\theta = 5\sec\theta$
(ii) $2\sec\theta + 3\sin\theta = 4\cos\theta$
(iii) $5\cos\theta - \sec\theta = \tan\theta$
(iv) $3\sin^2\theta - \sin\theta\cos\theta - 4\cos^2\theta = 0$

⑥ **8** Find the values of θ between 0° and 360° satisfying the equation $\tan\theta = \cos\theta$. Illustrate your result by sketching on the same axes the graphs of $\tan\theta$ and $\cos\theta$.

Questions 9 *and* 10: Prove the given identities.

• **9** (i) $\cot\theta + \tan\theta \equiv \sec\theta\csc\theta$
→ (ii) $\tan^2\theta + \cos^2\theta \equiv (\sec\theta - \sin\theta)(\sec\theta + \sin\theta)$
(iii) $\cos^4\theta - \sin^4\theta \equiv \cos^2\theta - \sin^2\theta$
• (iv) $(\csc\theta - \sin\theta)(\sec\theta - \cos\theta) \equiv \dfrac{1}{\tan\theta + \cot\theta}$

(Hint: show that LHS $\equiv \cos\theta\sin\theta$; then show that RHS $\equiv \cos\theta\sin\theta$.)

10 (i) $\dfrac{\cos\theta}{1 - \tan\theta} + \dfrac{\sin\theta}{1 - \cot\theta} \equiv \sin\theta + \cos\theta$

→ (ii) $\dfrac{\tan\theta + \cot\theta}{\sec\theta + \csc\theta} \equiv \dfrac{1}{\sin\theta + \cos\theta}$

For which values of θ are the identities undefined?

11 Prove that
$$\sec\theta + \tan\theta \equiv 1/(\sec\theta - \tan\theta)$$
Hence, given that $\sec\theta - \tan\theta = \tfrac{1}{3}$, find the values of $\sec\theta$ and $\tan\theta$. (Don't try to find θ.)

12 Show that
$$(\cot\theta + \csc\theta)^2 \equiv (1 + \cos\theta)/(1 - \cos\theta)$$
Hence solve the equation
$$(\cot\theta + \csc\theta)^2 = \sec\theta$$

⑦ **13** Eliminate θ from the following pairs of equations:
(i) $x = 2\cos\theta$, $y = \sin\theta$
→ (ii) $x = 2 + \cos\theta$, $y = 1 + \sin\theta$
→ (iii) $x = 3\tan\theta$, $y = 4\sec\theta$
(iv) $x = a\sec\theta$, $y = b\sin\theta$

⑦ **14** If $x = 3\sin\theta$, $y = 4\tan\theta$, express $\csc\theta$ in terms of x and $\cot\theta$ in terms of y. Hence eliminate θ from the equations.

⑦ **15** Eliminate θ from the following pairs of equations:
(i) $x = 5\cos\theta$, $y = 2\tan\theta$
(ii) $x = 3\cos\theta$, $y = 4\sec\theta + 5\tan\theta$

⑦ **16** (i) If $x = \sin\theta\cos\theta$, $y = \sin\theta$, express x^2 in terms of y^2.
R (ii) If $x = \tan\theta - \sin\theta$, $y = \tan\theta + \sin\theta$, show that $(x^2 - y^2)^2 = 16xy$.
(iii) If $x = \sin\theta - \cos\theta$, $y = \tan\theta + \cot\theta$, express x^2 in terms of y.

⑦ **17** Eliminate θ from the following pairs of equations:
(i) $x = \sin\theta + \cos\theta$, $y = \sin\theta - \cos\theta$
(ii) $x = \csc\theta + \cot\theta$, $y = \csc\theta - \cot\theta$
(iii) $x = \sec\theta\tan\theta$, $y = 1 - \tan\theta$
→ (iv) $x = \sec\theta + \tan\theta$, $y = \sec\theta - \tan\theta$

⑧→ **18** If $\sin\alpha = \tfrac{5}{13}$, find without using a calculator or tables the possible values of $\cos\alpha$ and $\tan\alpha$.
If $\sin\alpha = \tfrac{5}{13}$ and α is obtuse, find $\cos\alpha$, $\tan\alpha$.

⑧ **19** (i) If $\tan\alpha = -\sqrt{3}$ and α is reflex, find $\cos\alpha$, $\csc\alpha$.
(ii) If $\cos\beta = -\tfrac{7}{25}$ and β is obtuse, find $\sin\beta$, $\cot\beta$.
(iii) If $\cot\gamma = 2$ and γ is reflex, find $\cos\gamma$, $\csc\gamma$.
(iv) If $\sin\delta = \tfrac{1}{3}$ and δ is obtuse, find $\cos\delta$, $\cot\delta$.

⑧ **20** If $\tan\alpha = 3$, find the possible values of $\sin\alpha$ and $\cos\alpha$.
If $\tan\alpha = 3$ and α is acute, draw an appropriate triangle to find the values of $\sin\alpha$ and $\cos\alpha$.

⑧ **R 21** If $8\sin^2\theta + 2\cos\theta - 5 = 0$ show that $\cos\theta = \tfrac{3}{4}$ or $-\tfrac{1}{2}$. Hence find the possible values of $\tan\theta$.

22 A swimming race consists of three lengths of a rectangular pool, $ABCD$, where $AB = 10$ m, $CD = 14$ m. One of the competitors, a wonderful swimmer but something of a dingbat, starts from a point E on AB ($AE = 6$ m) and swims at an angle θ (45° < θ < 90°) to AB. He arrives at a point F on BC and, thinking he has completed a length, swims straight across the pool to the point G on AD. Thinking he has now completed two lengths, he swims of at an angle θ to AD and arrives, by a wonderful coincidence, at the point H on CD exactly opposite his starting point (i.e. $DH = 6$ m). Find θ.
How far has he swum?
The other competitors have just begun their third length. He claims an astonishing victory and buys drinks for everyone at the bar. His friends have some difficulty in explaining why he has been disqualified.

23 A golf ball is struck from a point O. Referred to axes Ox, Oy (horizontal and vertically upwards respectively), the equation of its path is
$$y = x\tan\alpha - \tfrac{1}{32}x^2\sec^2\alpha$$
where α is the angle of projection ($\alpha < 90°$) (1 unit = 1 m). The ball just clears a tree 4 m high whose base is at a distance 8 m from O. Find two possible values of α.
Sketch a diagram illustrating your result.

3.2:3 Compound angle formulae

Theorem 2:

$$\sin(\theta+\phi) \equiv \sin\theta\cos\phi + \cos\theta\sin\phi \quad \textbf{(10a)}$$

$$\sin(\theta-\phi) \equiv \sin\theta\cos\phi - \cos\theta\sin\phi \quad \textbf{(10b)}$$

$$\cos(\theta+\phi) \equiv \cos\theta\cos\phi - \sin\theta\sin\phi \quad \textbf{(11a)}$$

$$\cos(\theta-\phi) \equiv \cos\theta\cos\phi + \sin\theta\sin\phi \quad \textbf{(11b)}$$

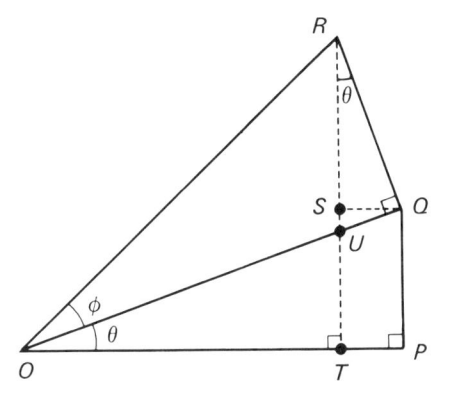

$$\tan(\theta+\phi) \equiv \frac{\tan\theta + \tan\phi}{1 - \tan\theta\tan\phi} \quad \textbf{(12a)}$$

$$\tan(\theta-\phi) \equiv \frac{\tan\theta - \tan\phi}{1 + \tan\theta\tan\phi} \quad \textbf{(12b)}$$

Proof

Identity (10a): Suppose that θ and ϕ are both acute (fig. 3.2:9).
Look at $\triangle OQR$: let $OR = 1$, then $RQ = \sin\phi$, $OQ = \cos\phi$.
Look at $\triangle RQU$: $RQ = \sin\phi$, so $RS = \sin\phi\cos\theta$. **(i)**
Look at $\triangle OQR$: $OQ = \cos\phi$, so $QP = \cos\phi\sin\theta$. **(ii)**
Look at $\triangle ORT$: $OR = 1$, so $RT = \sin(\theta+\phi)$. **(iii)**
But note that $RT = RS + ST = RS + QP$, i.e. from (i), (ii) and (iii),

$$\sin(\theta+\phi) = \sin\theta\cos\phi + \cos\theta\sin\phi$$

See question 2 for proofs of the identites (10b)–(12b).

Fig. 3.2:9

Worked example 9 Find $\sin(-15°)$ in surd form.

$$\begin{aligned}\sin(-15°) &= \sin(45° - 60°) \\ &= \sin 45° \cos 60° - \cos 45° \sin 60° \quad \text{by } \textbf{(10b)} \\ &= \tfrac{1}{2}\sqrt{2} \cdot \tfrac{1}{2} - \tfrac{1}{2}\sqrt{2} \cdot \tfrac{1}{2}\sqrt{3} \\ &= \tfrac{1}{4}\sqrt{2}(1 - \sqrt{3})\end{aligned}$$

(negative, as one would expect).

Worked example 10 If $\sin\alpha = \tfrac{3}{5}$ and $\cos\beta = \tfrac{12}{13}$, find the possible values of $\cos(\alpha+\beta)$.

$$\sin\alpha = \tfrac{3}{5} \Rightarrow \cos\alpha = \pm\tfrac{4}{5} \quad \text{(see worked example 8)}$$

similarly, $\cos\beta = \tfrac{12}{13} \Rightarrow \sin\beta = \pm\tfrac{5}{13}$.

So
$$\begin{aligned}\cos(\alpha+\beta) &= \cos\alpha\cos\beta - \sin\alpha\sin\beta \quad \text{by } \textbf{(11a)} \\ &= (\pm\tfrac{4}{5})(\tfrac{12}{13}) - (\tfrac{3}{5})(\pm\tfrac{5}{13}) \\ &= -\tfrac{63}{65} \text{ or } -\tfrac{33}{65} \text{ or } \tfrac{33}{65} \text{ or } \tfrac{63}{65}\end{aligned}$$

140 Compound angle formulae

Exercise 3.2:3

1 Is $\sin(\theta+\phi) \equiv \sin\theta + \sin\phi$? Prove it (if you dare) or find a counter example.

• 2 (i) Prove identity (10b). (Hint: replace ϕ by $-\phi$ in identity (10a).)

(ii) Prove identity (11a): (a) by replacing θ by $90°-\theta$ and ϕ by $-\phi$ in identity (10a); (b) by using the diagram and a proof similar to those of identity (10a).
Hence prove identity (11b).

(iii) Using the results for identities (10a) and (11a), prove identity (12a). Hence prove identity (12b).
For what values of θ and ϕ are identities (12a) and (12b) undefined?

• → 3 Simplify: (i) $\cos(90°+\theta)$; (ii) $\sin(90°+\theta)$. Hence simplify $\tan(90°+\theta)$. What is wrong with using identity (12a) to simplify $\tan(90°+\theta)$?

4 (i) Expand: (a) $\sin(45°+\theta)$; (b) $\cos(30°-\theta)$; (c) $\tan(60°-\theta)$.

→ (ii) Prove that $\tan(\theta+45°)\tan(\theta-45°) = -1$.

5 Simplify:

(i) $\sin 2\theta \cos\theta - \cos 2\theta \sin\theta$
(ii) $\cos 2\theta \cos 3\theta + \sin 2\theta \sin 3\theta$
(iii) $\sin^2 2\theta \sin^2\theta - \cos^2 2\theta \cos^2\theta$
(iv) $\left(\dfrac{\sin 3\theta}{\sin\theta}\right)^2 - \left(\dfrac{\cos 3\theta}{\cos\theta}\right)^2$

6 Simply $\dfrac{\tan\theta - 1}{1 + \tan\theta}$ (remember that $\tan 45° = 1$). Hence solve the equation

$$\tan\theta - 1 = \tan 2\theta + \tan\theta \tan 2\theta$$

⑨ 7 Find in surd form, without using a calculator:

→ (i) $\cos 75°$
(ii) $\tan 75°$
(iii) $\tan(-15°)$
R (iv) $\cos 105°$
(v) $\sin 165°$

• → 8 Find, without using a calculator, the value of $\tan\theta$ for which $\tan(\theta - 45°) = \tfrac{1}{3}$.

9 Solve the equations:

(i) $2\cos\theta = \sin(\theta + 30°)$
(ii) $3\sin\theta = \cos(\theta + 60°)$
(iii) $\cos(45° - \theta) = \sin(30° + \theta)$
(iv) $\cos(\theta - 30°) + \sin(\theta - 60°) = \cos\theta$

10 Solve the equation $\sin(\theta + 60°) = \cos\theta$:

(i) by expanding the LHS;
(ii) using the identity $\sin\theta \equiv \cos(90° - \theta)$.

R 11 If $\sin(\theta+\alpha) = \cos(\theta+\beta)$, find $\tan\theta$ in terms of α and β. Hence, if $\sin(\theta+\alpha) = \cos(\theta-\alpha)$, show that $\tan\theta = 1$.

⑩ → 12 If $\sin\alpha = \tfrac{4}{5}$ and $\sin\beta = -\tfrac{7}{25}$, find the possible values of $\sin(\alpha+\beta)$ and $\cos(\alpha+\beta)$.

⑩ 13 If $\sin\alpha = \tfrac{3}{5}$, α obtuse, and $\cos\beta = \tfrac{5}{13}$, β acute, find $\sin(\alpha+\beta)$ and $\cos(\alpha-\beta)$.

⑩ 14 If $\tan\alpha = -\tfrac{7}{24}$, α reflex, and $\sin\beta = \tfrac{24}{25}$, β acute, find $\cos(\alpha+\beta)$ and $\tan(\alpha+\beta)$.

⑩ R 15 If $\sin\alpha = \tfrac{2}{3}$ and $\cos\beta = -\tfrac{2}{7}$, find the possible values of $\tan(\alpha+\beta)$.

⑩ 16 If $\tan\alpha = \tfrac{1}{3}$ and $\tan\beta = \tfrac{1}{2}$, find the possible values of $\sin(\alpha-\beta)$ and $\cos(\alpha+\beta)$. If $(\alpha+\beta)$ is acute, find it.

⑩ 17 Given that $\sin\alpha = 0.6$ and α is obtuse, find $\cos(\alpha+270°)$ and $\cos(\alpha+540°)$.

• → 18 If $\tan\alpha$ and $\tan\beta$ are the roots of the equation $t^2 - 8t + 9 = 0$, show, without actually solving the equation, that $\tan(\alpha+\beta) = -1$.

• → 19 If θ is the angle between the lines $y = mx$ and $y = m'x$ ($m > m'$) show that

$$\tan\theta = \dfrac{m - m'}{1 + mm'}$$

(Hint: let α, β be the angles between the lines and the x-axis. We require $\tan(\alpha - \beta)$.)

20 At the world championship snooker final the surface of the table is a rectangle with vertices $A(0,0)$, $B(0, 2a)$, $C(a, 2a)$ and $D(a, 0)$. The score is 17 frames all, with the scores equal in the 35th frame, and only the black ball left in play. The black ball is at the point (ka, a), where $\tfrac{1}{2} < k < 1$; the white cue ball is on the lip of the pocket at B.

The popular underdog, Sean 'Typhoon' O'Nolan, prepares to pot the black and win several million pounds. He hits the white ball so that it collides with the black, which goes into the pocket at $E(a, a)$. Jubilation! The white ball trickles into the pocket at D....

Show that, if θ is the angle through which the path of the white ball is deflected by its collision with the black,

$$\tan\theta = \dfrac{2k - 1}{1 + k - k^2}$$

Sketch a graph of $\tan\theta$ against k for $\tfrac{1}{2} < k < 1$.

3.2:4 Double angle formulae

Theorem 3:

$$\sin 2\theta \equiv 2 \sin \theta \cos \theta \qquad (13)$$

$$\cos 2\theta \equiv \cos^2 \theta - \sin^2 \theta \qquad (14)$$

$$\tan 2\theta \equiv \frac{2 \tan \theta}{1 - \tan^2 \theta} \qquad (15)$$

Proof

Put $\phi = \theta$ in identities (10a), (11a) and (12a).

Using identity (7), we obtain the two alternative versions of identity (14):

$$\cos 2\theta = 2 \cos^2 \theta - 1 \qquad (14a)$$

$$\cos 2\theta \equiv 1 - 2 \sin^2 \theta \qquad (14b)$$

Worked example 11 If $\tan 2\alpha = 2$, find $\tan \alpha$ in surd form.

By identity (15),

$$2 = \frac{2 \tan \alpha}{1 - \tan^2 \alpha}$$

$\Rightarrow \qquad \tan^2 \alpha + \tan \alpha - 1 = 0$

$\Rightarrow \qquad \tan \alpha = \tfrac{1}{2}(-1 \pm \sqrt{5})$

Worked example 12 Find the general solution of the equation

$$\cos 2\theta + 5 \cos \theta = 2$$

Due to the presence of $\cos \theta$ we use identity (14a)—rather than identity (14) or (14b)—to reduce the equation to a quadratic equation in $\cos \theta$:

$$(2 \cos^2 \theta - 1) + 5 \cos \theta = 2$$

$\Rightarrow \qquad 2 \cos^2 \theta + 5 \cos \theta - 3 = 0$

$\Rightarrow \qquad (2 \cos \theta - 1)(\cos \theta + 3) = 0$

$\Rightarrow \qquad$ either $\cos \theta = \tfrac{1}{2}$ or $\cos \theta = -3$, which is impossible

$\Rightarrow \qquad \theta = \pm 60° + 360n°$

Exercise 3.2:4

1 Find:
- (i) $2 \sin 15° \cos 15°$
- (ii) $1 - 2 \sin^2 75°$
- (iii) $\dfrac{2 \tan 67.5°}{1 - \tan^2 67.5°}$

2 Find $\cos 2\theta$ when
- (i) $\cos \theta = -\tfrac{1}{2}$
- (ii) $\sin \theta = \tfrac{3}{5}$

→ • **3** If $\sin \alpha = \tfrac{4}{5}$, α obtuse, find $\cos \alpha$.
Hence find $\sin 2\alpha$ and $\cos 2\alpha$.
Hence find $\sin 4\alpha$.

R 4 If $\cos \alpha = \tfrac{5}{13}$, α acute, find $\sin 4\alpha$.

5 Find the possible values of $\sin 2\theta$ and $\cos 2\theta$, given that:
- (i) $\cos \theta = \tfrac{3}{5}$
- (ii) $\sin \theta = -\tfrac{7}{25}$
- (iii) $\tan \theta = 2$

6 If $\tan \alpha = \frac{3}{4}$, find $\tan 2\alpha$ and $\tan 4\alpha$.

7 Find $\tan 2\theta$ when:
   (i) $\tan \theta = -\frac{4}{3}$ (ii) $\cos \theta = -\frac{5}{13}$

⑪ **8** If $\tan 2\alpha = \frac{1}{3}$, find the two possible values of $\tan \alpha$.

⑪ → **9** If $t = \tan 22.5°$, use identity (15) to show that $t = -1 \pm \sqrt{2}$. Which of these solutions is correct, and why? Find $\tan 67.5°$ by a similar method.

⑪ R **10** Express $\cot 2\theta$ in terms of $\cot \theta$. Hence find $\cot 67.5°$ in surd form.

⑪ → **11** If $\tan 2\alpha = \frac{3}{4}$ and α is acute, find $\tan \alpha$. Hence find $\sin \alpha$ and $\cos \alpha$. Hence find $\sin 2\alpha$ and $\cos 2\alpha$ and check that these values agree with the given value of $\tan 2\alpha$.

⑪ **12** Express $\sin^2 \theta$ and $\cos^2 \theta$ in terms of $\cos 2\theta$. Hence, without using a calculator, find $\sin 15°$ and $\cos 15°$ in surd form.

⑫ **13** Solve the equations:
   (i) $\cos 2\theta + 3 \sin \theta = 2$
   R (ii) $3 \cos 2\theta - 7 \cos \theta + 5 = 0$
   → (iii) $5 \cos 2\theta + 11 \cos \theta + 8 = 0$
   (iv) $\sin 2\theta (2 \sin \theta + 3) = 4 \cos \theta$
   (v) $4 \cos^3 \theta + 2 \cos \theta = 5 \sin 2\theta$
   (vi) $5 \cos \theta \sin 2\theta + 4 \sin^2 \theta = 4$
   (vii) $3 \cos 2\theta + 5 \cos + 1 = 0$
   (viii) $\sin 2\theta - 1 = \cos 2\theta$
   (ix) $\tan 2\theta = 4 \cot \theta$
   → (x) $\cot 2\theta = 2 + \cot \theta$

⑫ + **14** Express $1 - \cos 2\theta$ in terms of $\sin^2 \theta$. Hence solve the equation
$$5(1 + \sin 2\theta - \cos 2\theta) - 6 \sin \theta (1 + \tan \theta) = 0$$

⑫ **15** Find the values of θ between $0°$ and $360°$ satisfying
$$2 \sin \theta = \cos 2\theta$$
Illustrate your answer by sketching on the same axes the groups of $2 \sin \theta$ and $\cos 2\theta$ in this interval.

⑫ **16** Solve the equation $\cos 2\theta = \sin \theta$:
   (i) using the appropriate double angle formula;
   (ii) using the identity $\sin \theta \equiv \cos(90° - \theta)$ (c.f. 3.2:1 Q19).

⑫ **17** By expressing $\tan 2\theta$ and $\sec^2 \theta$ each in terms of $t = \tan \theta$, solve the equation
$$4 \tan 2\theta + 3 \cot \theta \sec^2 \theta = 0$$

18 Prove the following identities. Find the values of θ for which each one is undefined.
   • (i) $\tan \theta + \cot \theta \equiv 2 \csc 2\theta$
   (ii) $\dfrac{1}{1 - \tan \theta} - \dfrac{1}{1 + \tan \theta} \equiv \tan 2\theta$
   → (iii) $\dfrac{\sin 2\theta - \cos 2\theta + 1}{\sin 2\theta + \cos 2\theta + 1} \equiv \tan \theta$
   (iv) $\sec 2\theta + \tan 2\theta \equiv \dfrac{\cos \theta + \sin \theta}{\cos \theta - \sin \theta}$

+ **19** Prove that
$$\sin 2\theta \equiv \frac{2t}{1+t^2} \qquad \cos 2\theta \equiv \frac{1-t^2}{1+t^2}$$
where $t = \tan \theta$. Hence show that
$$\sqrt{\left(\frac{1 - \sin 2\theta}{1 + \sin 2\theta}\right)} \equiv \frac{1-t}{1+t}$$

20 Simplify:
   (i) $\dfrac{1 - \cos 2\theta}{1 + \cos 2\theta}$
   (ii) $\dfrac{\sin \theta + \sin 2\theta}{1 + \cos \theta + \cos 2\theta}$
   (iii) $\tan 2\theta (\cot \theta - \tan \theta)$
   (iv) $\cot \theta - 2 \cot 2\theta$

21 Find without using tables or a calculator:
   (i) $\cos^2 15° - \sin^2 15°$
   (ii) $\cos^4 75° - \sin^4 75°$

• → **22** Express $\cos 2\theta$ in terms of $\cos \theta$, $\cos 4\theta$ in terms of $\cos 2\theta$ and $\cos 8\theta$ in terms of $\cos 4\theta$. Hence show that $1 + 2 \cos 4\theta + \cos 8\theta \equiv 4 \cos 4\theta \cos^2 2\theta$.

23 Eliminate θ from the following pairs of equations:
   (i) $x = \cos 2\theta$, $y = \cos \theta$
   (ii) $x = \tan 2\theta$, $y = 2 \tan \theta$
   (iii) $x = \cos 2\theta$, $y = \csc \theta$
   (iv) $x = \sin 2\theta$, $y = \sec 4\theta$
   (v) $x = \cos \theta + \sin \theta$, $y = \cos 2\theta$
   (vi) $x = \sin 2\theta$, $y = \cos \theta$

• → **24** By first writing $\sin 3\theta \equiv \sin(2\theta + \theta)$, show that
$$\sin 3\theta \equiv 3 \cos^2 \theta \sin \theta - \sin^3 \theta$$
Hence express $\sin 3\theta$ in terms of $\sin \theta$. Show that if $s = \sin 50°$, $8s^3 - 6s + 1 = 0$.

+ → **25** Show that $\cos 3\theta \equiv 4 \cos^3 \theta - 3 \cos \theta$. Hence show that if $\cos 3\theta = -\frac{1}{2}\sqrt{3}$, then $8c^3 - 6c + \sqrt{3} = 0$ where $c = \cos \theta$. Hence use the substitution $x = 2\sqrt{3} \cos \theta$ to find the roots of the equation
$$x^3 - 9x + 9 = 0$$

26 Show that $36°$ is a root of the equation
$$\cos 3\theta + \cos 2\theta = 0$$

Hence show that cos 36° is a root of the equation

$$4c^3 + 2c^2 - 3c - 1 = 0$$

Find an integer root of this equation and hence find cos 36° in surd form.

27 (i) Show that $\tan 3\theta \equiv \dfrac{3\tan\theta - \tan^3\theta}{1 - 3\tan^2\theta}$.

For what values of θ is this identity undefined?

(ii) Hence show that the three roots of the equation $t^3 - 3t^2 - 3t + 1 = 0$ are tan 15°, tan 75° and tan 135°. Find tan 15° in surd form.

28 Solve the equation $\tan 3\theta = 4\tan\theta$.

29 Solve the equations:
(i) $\cos 3\theta + 2\cos\theta = 0$
→ (ii) $\cos 3\theta + \cos\theta = \cos 2\theta + 1$
R (iii) $2\sin 3\theta - 7\cos 2\theta + \sin\theta + 1 = 0$
(iv) $2\cos 3\theta + \cos 2\theta + 1 = 0$

30 Verify that the equation $\sin 3\theta = 2\cos 2\theta$ is satisfied by $\theta = 30°$. Find the other angles between 0° and 360° which satisfy the equation.

31 Show that $\dfrac{\sin 3\theta}{1 + 2\cos 2\theta} \equiv \sin\theta$.

Hence show that $\sin 15° = (\sqrt{3} - 1)/2\sqrt{2}$.

3.2:5 *t* formulae

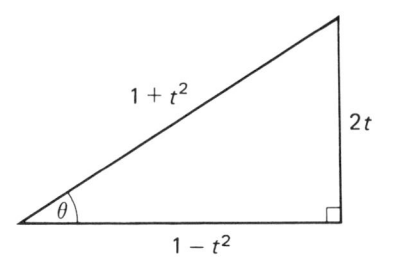

Fig. 3.2:10 A useful mnemonic

Theorem 4:

$$\sin\theta \equiv \frac{2t}{1+t^2} \quad (16)$$

$$\cos\theta \equiv \frac{1-t^2}{1+t^2} \quad (17)$$

$$\tan\theta \equiv \frac{2t}{1-t^2} \quad (18)$$

where $t = \tan\tfrac{1}{2}\theta$.

Proof
Identity (16):

$$\sin 2\theta \equiv 2\sin\theta\cos\theta$$
$$\equiv 2\frac{\sin\theta}{\cos\theta}\cdot\cos^2\theta$$
$$\equiv 2\tan\theta\cos^2\theta$$
$$\equiv \frac{2\tan\theta}{\sec^2\theta}$$
$$\equiv \frac{2\tan\theta}{1+\tan^2\theta}$$

Writing $\tfrac{1}{2}\theta$ instead of θ in this identity gives identity (16). See question 1 for proofs of identities (17) and (18).

Worked example 13 Find the general solution of the equation

$$3\sin\theta + 4\cos\theta = -2.5$$

We use identies (16) and (17) to transform this equation into an equation in one variable, $t = \tan \tfrac{1}{2}\theta$:

$$3 \cdot \frac{2t}{1+t^2} + 4 \cdot \frac{1-t^2}{1+t^2} = -\frac{5}{2}$$

$$\Rightarrow \quad 6t + 4(1-t^2) = -\frac{5}{2}(1+t^2)$$

$$\Rightarrow \quad 3t^2 - 12t - 13 = 0$$

$$\Rightarrow \quad t = \tan \tfrac{1}{2}\theta = 4.887 \quad \text{or} \quad -0.887$$

$$\Rightarrow \quad \tfrac{1}{2}\theta = 78.435° + 180n° \quad \text{or} \quad -41.565° + 180n°$$

$$\Rightarrow \quad \theta = 156.87° + 360n° \quad \text{or} \quad -83.13° + 360n°$$

(Note: there are two solutions in the range $0° < \theta < 360°$: $\theta = 156.87°$ and $\theta = 276.87°$.)

Exercise 3.2:5

- **1** (i) Using identity (14), prove identity (17).
  (ii) Show that identity (18) follows
  (a) from identity (15); (b) from identities (16) and (17).

- → **2** If $\tan \tfrac{1}{2}\theta = -\tfrac{2}{3}$, find $\sin \theta$ and $\cos \theta$. Hence find $\sin 2\theta$ and $\cos 2\theta$.

R 3 If $\tan \tfrac{1}{2}\theta = \tfrac{1}{3}$, find $\tan 2\theta$.

4 If $\tan \theta = \tfrac{12}{5}$, find the possible values of $\tan \tfrac{1}{2}\theta$.

→ **5** (i) By expressing each side of the identity in terms of $t = \tan \tfrac{1}{2}\theta$, prove that

$$\sec \theta + \tan \theta \equiv \tan(45° + \tfrac{1}{2}\theta)$$

Hence find $\tan 75°$.

(ii) By replacing θ by $-\theta$ in the identity, find a similar expression for $\sec \theta - \tan \theta$. Hence find $\tan 15°$.

6 Express in terms of $t = \tan \tfrac{1}{2}\theta$:

(i) $\dfrac{\sin \theta}{1 - \cos \theta}$ (ii) $\dfrac{1 - \cos \theta}{1 + \cos \theta}$

(iii) $\cot \theta \cot \tfrac{1}{2}\theta$ (iv) $\dfrac{1 - 2 \sin \theta}{2 \cos \theta + 1}$

(v) $\dfrac{1 - \sec \theta + \tan \theta}{1 + \sec \theta - \tan \theta}$

7 Express $(3 + \cos \theta)/\sin \theta$ in terms of t. Hence show that this function cannot take any value between $-2\sqrt{2}$ and $+2\sqrt{2}$.

⑬ **8** By expressing $\sin \theta$ and $\cos \theta$ in terms of t, solve these equations:

(i) $2 \sin \theta + \cos \theta = -2$
R (ii) $3 \sin \theta - 2 \cos \theta = 1$

→ (iii) $\sin \theta + 7 \cos \theta = 5$
(iv) $12 \cos \theta - 5 \sin \theta = 10$
(v) $1 + \cos \theta + \sin \theta = \tan(\tfrac{1}{2}\theta + 45°)$

⑬ **9** (i) Show that $\theta = 180°$ is a solution of the equation $2 \sin \theta - \cos \theta = 1$. Solve the equation using the method of worked example 13.
Why does this method not give the solution $\theta = 180°$?
(ii) Solve the equation $\sin \theta + 3 \cos \theta + 3 = 0$.

⑬ **10** Solve the equation $2 \sin \theta - \cos \theta = 2$. Illustrate your answer by sketching on the same axes the graphs of $2 \sin \theta$ and $2 + \cos \theta$.

⑬ • **11** On the same axes, sketch the circle $x^2 + y^2 = 1$ and the straight line $y - 2x = 1$.
(i) Show that the point $P(\cos \theta, \sin \theta)$ always lies on the circle. What is the significance of the angle θ?
(ii) Find the points of intersection of the line and the circle.
(iii) Hence solve the equation $\sin \theta - 2 \cos \theta = 1$.

⑬ → **12** Solve the equation $4 \sin \theta + 3 \cos \theta = 3$:
(i) by expressing $\sin \theta$ and $\cos \theta$ in terms of t;
(ii) by finding the points of intersection of the circle $x^2 + y^2 = 1$ and the line $4y + 3x = 3$.

⑬ **13** (i) Show that the point $P(\sec \theta, \tan \theta)$ lies on the curve $x^2 - y^2 = 1$.
(ii) Find the points of intersection of the line $y = (x - 1)\sqrt{3}$ and the curve.
(iii) Hence or otherwise solve the equation $3 \sec \theta - \tan \theta = \sqrt{3}$.
(Otherwise ... ?)

14 By expressing $\sec 2\theta$ and $\tan 2\theta$ in terms of $t = \tan \theta$, show that the equation

$$2 \tan \theta + \sec 2\theta = 2 \tan 2\theta$$

is satisfied by $\tan \theta = \frac{1}{2}$. Are there any other solutions?

15 Solve the equations:
  (i) $5 \cos 2\theta - 2 \sin 2\theta = 3$
  (ii) $3 \cot 2\theta + 7 \tan \theta = 5 \csc 2\theta$
  (iii) $(1 - \tan \theta)(1 + \sin 2\theta) = 1 + \tan \theta$

16 Prove that $\csc \theta - \cot \theta \equiv \tan \frac{1}{2}\theta$. Hence solve the equation $3 \csc \theta = 3 \cot \theta - 2$.

17 If α and β are two solutions of the equation $3 \sin \theta + 4 \cos \theta = 1$, show that $\tan \frac{1}{2}(\alpha + \beta) = \frac{4}{3}$.

18 If α and β are two solutions of the equation $a \sin \theta + b \cos \theta = c$, show that $\tan \frac{1}{2}(\alpha + \beta) = a/b$.

19 Solve the equations:
  (i) $\sin \theta - 3 \cos \theta = 1$ (ii) $\sin \theta - 3 \cos \theta = 0$

3.2:6 The form $a \sin \theta + b \cos \theta \equiv R \sin (\theta + \alpha)$

Worked example 14 Express $3 \sin \theta + 4 \cos \theta$ in the form $R \sin (\theta + \alpha)$.

Let $3 \sin \theta + 4 \cos \theta \equiv R \sin (\theta + \alpha) \equiv R \cos \alpha \sin \theta + R \sin \alpha \cos \theta$ **(i)**
Equating coefficients of $\sin \theta$ and $\cos \theta$ in (i),

$\sin \theta$: $\qquad R \cos \alpha = 3$ **(ii)**
$\cos \theta$: $\qquad R \sin \alpha = 4$ **(iii)**

(iii) ÷ (ii) ⇒ $\tan \alpha = \frac{4}{3} \Rightarrow \alpha = 53.13°$ or $223.13°$.
(ii)^2 + (iii)^2 ⇒ $R^2 = 25 \Rightarrow R = 5$ or -5.
From (ii) we can see that $R \cos \alpha$ is positive; so when $R = 5$, $\alpha = 53.13°$ ($\cos \alpha$ positive) and when $R = -5$, $\alpha = 223.13°$ ($\cos \alpha$ negative). Thus $3 \sin \theta + 4 \cos \theta \equiv 5 \sin (\theta + 53.13°) \equiv -5 \sin (\theta + 223.13°)$.
Usually we take R positive and α acute for convenience.

Worked example 15 Find the general solution of the equation

$$3 \sin \theta + 4 \cos \theta = -2.5$$

From worked example 14, $3 \sin \theta + 4 \cos \theta \equiv 5 \sin (\theta + 53.13°)$. Using this identity, the equation becomes

$$5 \sin (\theta + 53.13°) = -2.5$$
$\Rightarrow \qquad \sin (\theta + 53.13°) = -\frac{1}{2}$
$\Rightarrow \qquad \theta + 53.13° = -30° + 360n° \quad$ or $\quad 210° + 360n°$
$\qquad\qquad \theta = -83.13° + 360n° \quad$ or $\quad 156.87 + 360n°$

(See worked example 13.)

Worked example 16 Find the maximum and minimum values of the functions:

(i) $f(\theta) = 3 \sin \theta + 4 \cos \theta$

(ii) $g(\theta) = \dfrac{1}{3 \sin \theta + 4 \cos \theta + 7}$

(i) $f(\theta) = 3 \sin \theta + 4 \cos \theta \equiv 5 \sin (\theta + 53.13°)$ from worked example 14. Its maximum value, 5, occurs when $\sin (\theta + 53.13°) = 1$, i.e. when $\theta = 36.87° + 360n°$. Its minimum value, -5, occurs when $\sin (\theta + 53.13°) = -1$, i.e. when $\theta = 216.87° + 360n°$.

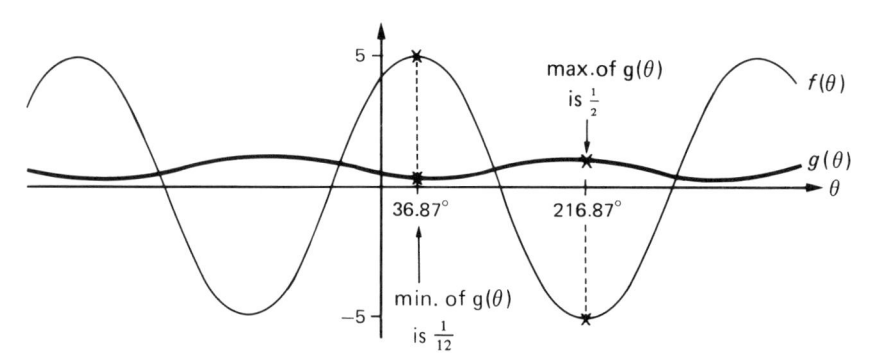

(ii) $g(\theta) = \dfrac{1}{f(\theta)+7}$; the maximum value of $g(\theta)$ occurs when $f(\theta)$ has its minimum value -5, i.e. the maximum value of $g(\theta)$ is $1/(-5+7) = \tfrac{1}{2}$.

Similarly, the minimum value of $g(\theta)$, which occurs when $f(\theta)$ has its maximum value 5, is $1/(5+7) = \tfrac{1}{12}$. These results are shown in fig. 3.2:11.

Fig. 3.2:11

Exercise 3.2:6

1 Find R (positive) and α (acute) when:
 (i) $\sqrt{3}\sin\theta + \cos\theta \equiv R\sin(\theta+\alpha)$
 (ii) $5\sin\theta + 12\cos\theta \equiv R\cos(\theta-\alpha)$
 (iii) $2\sin\theta - \cos\theta \equiv R\sin(\theta-\alpha)$
 (iv) $4\cos\theta - 3\sin\theta \equiv R\cos(\theta+\alpha)$

2 Express $3\sin\theta + 4\cos\theta$ in the form $R\cos(\theta-\beta)$, R positive and θ acute. Compare your working to that of worked example 14.

3 If $a\sin\theta + b\cos\theta \equiv R\sin(\theta+\alpha)$ (a, b positive), find R and $\tan\alpha$ in terms of a and b.
If $a\sin\theta + b\cos\theta \equiv R'\cos(\theta-\beta)$, find R' and $\tan\beta$ in terms of a and b.
Is $R = R'$? Is $\alpha = \beta$?

4 Express:
 (i) $3\cos 3\theta + \sin 3\theta$ in the form $R\cos(3\theta-\alpha)$
 (ii) $8\cos 2\theta - 15\sin 2\theta$ in the form $R\cos(2\theta+\alpha)$

5 Suggest the most appropriate form in which to express:
 (i) $\sin\theta + 4\cos\theta$
 (ii) $3\sin\theta - 2\cos\theta$
 (iii) $2\cos 5\theta - 5\sin 5\theta$

6 Express $4\cos\theta - 3\sin\theta$ in the form $R\cos(\theta+\alpha)$. Hence solve the equation $4\cos\theta - 3\sin\theta = 3$.

7 Solve the equations:
 (i) $\sin\theta + \cos\theta = \tfrac{1}{2}\sqrt{2}$
 (ii) $8\cos\theta - 15\sin\theta = 8.5$
 (iii) $8\sin\theta - 6\cos\theta = -5\sqrt{3}$
 (iv) $\sin\theta + 2\cos\theta = 1$
 (v) $\sqrt{3}\sin 2\theta - \cos 2\theta = -\sqrt{2}$
 (vi) $2\cos 5\theta + 3\sin 5\theta = -3$

8 Find the values of θ between $0°$ and $360°$ satisfying the equations:
 (i) $3\cos\theta + 4\sin\theta = 2.5$
 (ii) $\sin\theta + \sqrt{3}\cos\theta = 1$
 (iii) $7\cos\theta - 24\sin\theta = 15$
 (iv) $2\sin\theta - \cos\theta = -2$
 (v) $2\cos 2\theta + \sin 2\theta = \tfrac{1}{2}\sqrt{5}$

9 Find θ in terms of α if
$$\sqrt{3}\sin\theta + \cos\theta = \sin\alpha + \sqrt{3}\cos\alpha$$

10 Express the function $f(\theta) = 6\sin^2\theta + 8\sin\theta\cos\theta$ in terms of $\sin 2\theta$ and $\cos 2\theta$. Write $f(\theta)$ in the form
$$R\sin(2\theta - \alpha) + k$$
where R, α and k are constants.
Hence solve the equation $f(\theta) = 5$.

11 Solve the equations:
 (i) $2\cos\theta(3\cos\theta - \sin\theta) = 1$
 (ii) $\sin\theta(\sin\theta + \cos\theta) = 1$

12 Explain why the equation $5\cos\theta + 7\sin\theta = 9$ has no solution.

13 Express $f(\theta) \equiv \sqrt{3}\cos\theta + \sin\theta$ in the form $R\cos(\theta-\alpha)$. Hence:
 (i) find the maximum and minimum values of $f(\theta)$ and state the values of θ for which they occur;
 (ii) sketch the graph of $f(\theta)$ (See 3.1:2 WE2.)

14 Express $5\cos\theta - 12\sin\theta$ in the form
$$R\cos(\theta+\alpha)$$

Hence find the maximum and minimum values of:

(i) $5\cos\theta - 12\sin\theta + 20$

(ii) $\dfrac{1}{5\cos\theta - 12\sin\theta + 20}$

⑯ **15** Find the maximum and minimum values of these functions:

(i) $7\sin\theta + 24\cos\theta$
(ii) $3\cos 4\theta + 2\sin 4\theta$
(iii) $6\sin\theta - 8\cos\theta + 12$
(iv) $\dfrac{1}{6\sin\theta - 8\cos\theta + 12}$
(v) $\dfrac{1}{\sin\theta + \sqrt{3}\cos\theta + 3}$

⑯ **16** Show that, for all values of θ,

$$(\sin\theta + 2\cos\theta)^2 \le 5$$

For what values of θ does $(\sin\theta + 2\cos\theta)^2 = 5$?

⑯ **17** Find the maximum and minimum values of the function $a\sin\theta + b\cos\theta$, a and b constant.

By making the substitution $t = \tan\frac{1}{2}\theta$, show that the quadratic equation $(b+c)t^2 - 2at + (c-b) = 0$ has real roots $\Leftrightarrow a^2 + b^2 \geqslant c^2$

⑯ **18** Sketch graphs of:

(i) $f(\theta) = \cos\theta - \sqrt{3}\sin\theta$
(ii) $f(\theta) = 3\cos 2\theta + \sin 2\theta$

⑯ **19** Solve the equation $2\sin\theta + \cos\theta = 1$. Illustrate your answer by sketching the graph of $2\sin\theta + \cos\theta$.

⑯ **20** Find the values of θ between $0°$ and $360°$ for which:

(i) $4\cos\theta + 3\sin\theta \le 2$
(ii) $3\cos 2\theta - 5\sin 2\theta \le 4$

⑯ → **21** Express $f(\theta) = 3\sin\theta - 2\cos\theta - 2\cos\theta$ in the form $R\sin(\theta - \alpha)$. Hence sketch the graph of $f(\theta)$.

Find the asymptotes of the graph of $1/f(\theta)$ and sketch the graph.

⑯ **22** Sketch on the same axes, for $-360° < \theta < 360°$, the graphs of:

(i) $y = 2\cos 3\theta + \sin 3\theta$ and $y = \dfrac{1}{2\cos 3\theta + \sin 3\theta}$

(ii) $y = 4\sin\theta + 3\cos\theta + 6$ and

$$y = \dfrac{1}{4\sin\theta + 3\cos\theta + 6}$$

R 23 Express $f(\theta) = 4\cos\theta + \sin\theta$ in the form $R\cos(\theta - \alpha)$. Hence:

(i) sketch the graph of $f(\theta)$ for $-360° < \theta < 360°$;
(ii) show that $\{f(\theta)\}^2 \le 17$;
(iii) solve the equation $f(\theta) = 3$;
(iv) explain why the equation $f(\theta) = 5$ has no real roots.

R 24 Solve the equation $3\sin\theta - \cos\theta = 1$

(i) by using the t formulae;
(ii) by expressing the LHS in the form $R\sin(\theta - \alpha)$.

25 A dog moves along the x-axis in such a way that its coordinates at time t s are $(3\sin 3t + 2\cos 3t, 0)$. The extreme points of its path are A and B. Find the coordinates of A and B.

At what times does the dog first pass through A? When does it first pass through B?

Describe the motion of the dog. How long does the dog take to move directly from A to the point $(-1, 0)$?

26 Express $\cos^2\theta$ in terms of $\cos 2\theta$. Hence express $f(\theta) = 6\cos^2\theta - 4\sin 2\theta$ in the form $R\cos(2\theta + \alpha°) + p$, where R, α and p are constants. Find the maximum and minimum values of $f(\theta)$ and sketch its graph.

27 If $\sin 2\theta - 4\sin^2\theta \equiv R\sin(2\theta + \alpha°) + p$, where R, α and p are constants, $(R > 0, 0 < \alpha < 90)$, find R, α and p. Hence find the general solution of the equation

$$\sin 2\theta - 4\sin^2\theta + 1 = 0.$$

3.2:7 Factor formulae

Theorem 5:

$$2\sin X \cos Y \equiv \sin(X+Y) + \sin(X-Y) \quad \text{(19a)}$$

$$2\cos X \sin Y \equiv \sin(X+Y) - \sin(X-Y) \quad \text{(19b)}$$

$$2\cos X \cos Y \equiv \cos(X+Y) + \cos(X-Y) \quad \text{(19c)}$$

$$-2\sin X \sin Y \equiv \cos(X+Y) - \cos(X-Y) \quad \text{(19d)}$$

148 Factor formulae

Proof
See question 1.

> **Theorem 6** (factor formulae):
>
> $$\sin \theta + \sin \phi \equiv 2 \sin \tfrac{1}{2}(\theta + \phi) \cos \tfrac{1}{2}(\theta - \phi) \quad \text{(20a)}$$
> $$\sin \theta - \sin \phi \equiv 2 \cos \tfrac{1}{2}(\theta + \phi) \sin \tfrac{1}{2}(\theta - \phi) \quad \text{(20b)}$$
> $$\cos \theta + \cos \phi \equiv 2 \cos \tfrac{1}{2}(\theta + \phi) \cos \tfrac{1}{2}(\theta - \phi) \quad \text{(20c)}$$
> $$\cos \theta - \cos \phi \equiv -2 \sin \tfrac{1}{2}(\theta + \phi) \sin \tfrac{1}{2}(\theta - \phi) \quad \text{(20d)}$$

Proof
Put $X = \tfrac{1}{2}(\theta + \phi)$, $Y = \tfrac{1}{2}(\theta - \phi)$ in Theorem 5.

Worked example 17 Find the general solution of the equation

$$\cos \theta + \cos 2\theta + \cos 3\theta = 0$$

$\Rightarrow \qquad \cos 2\theta + (\cos \theta + \cos 3\theta) = 0 \qquad$ (Why this grouping?)

By identity (20c) $\cos 2\theta + 2 \cos 2\theta \cos \theta = 0$

$\Rightarrow \qquad \cos 2\theta (1 + 2 \cos \theta) = 0$

$\Rightarrow$ either $\qquad \cos 2\theta = 0 \qquad$ or $\quad \cos \theta = -\tfrac{1}{2}$

$\qquad \Rightarrow 2\theta = \pm 90° + 360n° \qquad \Rightarrow \theta = \pm 120° + 360n°$

$\qquad \Rightarrow \theta = \pm 45° + 180n°$

Exercise 3.2:7

• **1** Expand $\sin(X + Y)$ and $\sin(X - Y)$ using identities (10a) and (10b) (page 139). Hence prove identity (19a).
Similarly, prove identities (19b), (20a) and (20b).

2 Express as a sum or difference:

• (i) $2 \sin 2\theta \cos \theta$
(ii) $2 \cos 5\theta \sin \theta$
(iii) $\cos 3\theta \cos 4\theta$
→ (iv) $2 \sin 4\theta \sin 2\theta$
(v) $-2 \sin 4\theta \sin 2\theta$

3 Find without using tables or a calculator:

(i) $2 \cos 75° \cos 15°$
(ii) $2 \sin 37\tfrac{1}{2}° \sin 7\tfrac{1}{2}°$

4 Express $2 \cos 2\theta \sin \theta$ as the difference of two sines. Hence show that

$$\sin \theta (\cos 2\theta + \cos 4\theta + \cos 6\theta) \equiv \sin 3\theta \cos 4\theta$$

5 Express $2 \sin \theta \cos(\theta + 30°)$ as the difference of two sines. Hence show that the greatest value of the function is $\tfrac{1}{2}$. For what values of θ does this greatest value occur?

R 6 Express $2 \sin \theta \cos(\theta + \alpha)$ as the difference of two sines. Hence show that the greatest value, as θ varies, of $2 \sin \theta \cos(\theta + \alpha)/\cos^2 \alpha$ is $1/(1 + \sin \alpha)$. For what values of θ does this greatest value occur?

7 Express both $\cos X \cos Y$ and $\sin X \sin Y$ in terms of $\cos(X + Y)$ and $\cos(X - Y)$. Hence solve the equation $\cot \theta \cot(\theta - 30°) = 2 + \sqrt{3}$.

8 Solve the equation

$$\sin(\theta + 60°) \cos \theta = 2 \sin(\theta + 45°) \cos(\theta + 15°)$$

9 Factorize:

• (i) $\sin 3\theta + \sin \theta$
(ii) $\cos 4\theta + \cos 2\theta$
→ (iii) $\cos(\theta + 60°) + \cos(\theta - 60°)$
(iv) $\cos(\theta + 45°) - \cos(\theta - 45°)$

10 Find, without using tables or a calculator:

(i) $\sin 75° - \sin 15°$
(ii) $\dfrac{\cos 20° - \cos 100°}{\sin 100° - \sin 20°}$

11 Show that:

(i) $\dfrac{\cos\theta - \cos 3\theta}{\sin 3\theta - \sin\theta} \equiv \tan 2\theta$

→ (ii) $\dfrac{\sin 3\theta + \sin\theta}{\cos 3\theta + \cos\theta} \equiv \tan 2\theta$

R (iii) $\dfrac{\sin 2\theta + \sin 3\theta}{\cos 2\theta - \cos 3\theta} \equiv \cot\tfrac{1}{2}\theta$

(iv) $\dfrac{\cos 3\theta - \cos\theta}{\cos 4\theta - \cos 2\theta} \equiv \dfrac{\sin 2\theta}{\sin 3\theta}$

For what values of θ are the identities undefined?

12 Show that:

• → (i) $\cos 2\theta + \cos 3\theta + \cos 4\theta \equiv \cos 3\theta(1 + 2\cos\theta)$

(ii) $\sin\theta + \sin 2\theta + \sin 3\theta \equiv \sin 2\theta(1 + 2\cos\theta)$

(iii) $1 + 2\cos 2\theta + \cos 4\theta \equiv 4\cos^2\theta\cos 2\theta$

(iv) $\cos\theta - \cos 3\theta - \cos 5\theta + \cos 7\theta \equiv -4\sin\theta\sin 2\theta\cos 4\theta$

13 Show that:

(i) $\dfrac{\cos\theta + 2\cos 2\theta + \cos 3\theta}{\cos\theta - 2\cos 2\theta + \cos 3\theta} \equiv -\cot^2\tfrac{1}{2}\theta$

(ii) $\dfrac{\sin\theta + 2\sin 3\theta + \sin 5\theta}{\cos\theta - \cos 5\theta} \equiv \cot\theta$

14 Show that:

→ (i) $\cos\theta + \cos(\theta + 120°) + \cos(\theta + 240°) \equiv 0$

(ii) $\sin\theta + \sin(\theta + 120°) + \sin(\theta + 150°) \equiv 0$

(17) **15** Solve the equations:

(i) $\sin\theta + \sin 2\theta - \sin 3\theta = 0$

(ii) $\sin\theta + \sin 5\theta = \sin 3\theta$

→ (iii) $\cos\theta + \cos 3\theta + \cos 5\theta = 0$

(iv) $\cos\theta + \cos 2\theta = \sin\theta + \sin 2\theta$

(v) $\sin\theta + \sin 2\theta + \sin 3\theta + \sin 4\theta = 0$

(17) **16** Find the values of θ between 0° and 360° for which:

(i) $\sin 3\theta - \sin\theta = \cos 2\theta$

(ii) $\sin\theta - 2\sin 2\theta + \sin 3\theta = 0$

(iii) $\cos(\theta + 30°) - \sin\theta = \tfrac{1}{2}$

(iv) $\sin 60° + \sin(60° + \theta) + \sin(60° + 2\theta) = 0$

(17) R **17** Find the values of θ between 0° and 360° for which

$$\cos\theta + \cos(\theta + \alpha) + \cos(\theta + 2\alpha) = 1 + 2\cos\alpha$$

where $0° < \alpha < 90°$.

(17) → **18** Solve the equation $\sin\theta - \sin 2\theta = 0$:

(i) by expressing the LHS in factor form;
(ii) by writing it as $\sin\theta = \sin 2\theta$ and using the method of 3.2:1 WE5;
(iii) by using the identity $\sin 2\theta \equiv 2\sin\theta\cos\theta$.

19 Express $\cos n\theta + \cos(n-2)\theta$ as a product. Deduce that

$$\cos n\theta \equiv 2\cos\theta\cos(n-1)\theta - \cos(n-2)\theta$$

and express $\cos 3\theta$, $\cos 4\theta$ and $\cos 5\theta$ in terms of $\cos\theta$. (Hint: put $n = 2, 3, 4, 5$ successively in the identity.)

20 If A, B and C are the angles of a triangle, prove that:

(i) $\sin A + \sin B + \sin C = 4\cos\tfrac{1}{2}A\cos\tfrac{1}{2}B\cos\tfrac{1}{2}C$

(ii) $\cos A + \cos B + \cos C = 1 + 4\sin\tfrac{1}{2}A\sin\tfrac{1}{2}B\sin\tfrac{1}{2}C$

(iii) $\sin^2 A + \sin^2 B + \sin^2 C = 2 + 2\cos A\cos B\cos C$

3.3 Solution of triangles

3.3:1 Sine and cosine formulae

Given some of the sides and angles of a triangle, we would like to find the others. This process is called **solving the triangle**.

> **Theorem 1:** In any $\triangle ABC$, if $BC = a$, $AC = b$, $AB = c$,
>
> $$\dfrac{a}{\sin A} = \dfrac{b}{\sin B} = \dfrac{c}{\sin C} \qquad \text{(sine formula)} \qquad (1)$$
>
> and
>
> $$a^2 = b^2 + c^2 - 2bc\cos A \qquad \text{(cosine formula)} \qquad (2)$$
>
> (Similarly $b^2 = a^2 + c^2 - 2ac\cos B$ and $c^2 = a^2 + b^2 - 2ab\cos C$.)

150 Sine and cosine formulae

Fig. 3.3:1a A acute

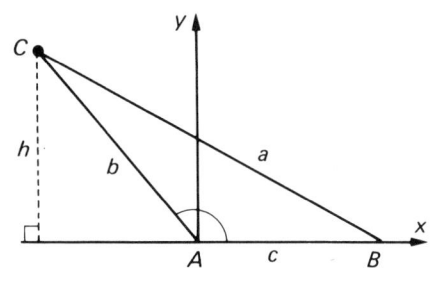

Fig. 3.3:1b A obtuse

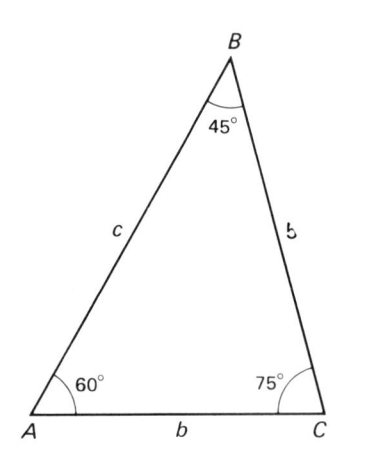

Fig. 3.3:2

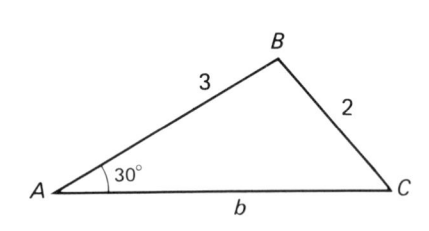

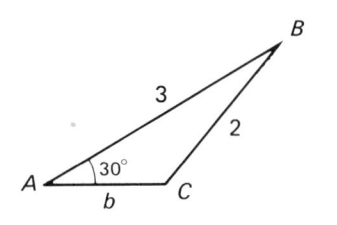

Fig. 3.3:3

Proof
Take A as the origin and Ax along AB. A may be acute (fig. 3.3:**1a**) or obtuse (fig. 3.3:**1b**), but in both cases the coordinates of B are $(c, 0)$ and those of C are $(b \cos A, b \sin A)$.

Equation (1): Let the height of C above AB be h. Then $h = b \sin A$ and similarly $h = a \sin B$. So

$$a \sin B = b \sin A$$

$$\frac{a}{\sin A} = \frac{b}{\sin B}$$

similarly

$$\frac{a}{\sin A} = \frac{c}{\sin C}$$

Equation (2):

$$a^2 = BC^2 = (c - b \cos A)^2 + (b \sin A)^2$$
$$= b^2(\cos^2 A + \sin^2 A) + c^2 - 2bc \cos A$$
$$= b^2 + c^2 - 2bc \cos A$$

Which formula we use depends on the data you have. The sine formula (the easier to use) is useful if you are told either:

(i) 2 angles and a side, or
(ii) 2 sides and an angle (as long as one of the sides is opposite the angle).

Worked example 1 Solve the $\triangle ABC$ in which $A = 60°$, $B = 45°$, $a = 5$.

$A = 60°$, $B = 45° \Rightarrow C = 75°$ (fig. 3.3:**2**).

By equation (1), $\quad \dfrac{c}{\sin 75°} = \dfrac{5}{\sin 60°} \Rightarrow c = 5.58$

and $\quad \dfrac{b}{\sin 45°} = \dfrac{5}{\sin 60°} \Rightarrow b = 4.08$

Worked example 2 Solve $\triangle ABC$ in which $A = 30°$, $a = 2$, $c = 3$.

There are *two* triangles satisfying the data (fig. 3.3:**3**). In one C is acute, in the other obtuse. By equation (1),

$$\frac{3}{\sin C} = \frac{2}{\sin 30°} \Rightarrow \sin C = \tfrac{3}{4} \Rightarrow C = 48.59° \text{ or } 131.41°$$

When $C = 48.59°$, $B = 101.41°$; so

$$\frac{b}{\sin 101.41°} = \frac{2}{\sin 30°} \Rightarrow b = 3.92$$

When $C = 131.41°$, $B = 18.59°$; so

$$\frac{b}{\sin 18.59°} = \frac{2}{\sin 30°} \Rightarrow b = 1.28$$

The cosine formula is useful if we are told either:

(i) 3 sides, or
(ii) 2 sides and an angle.

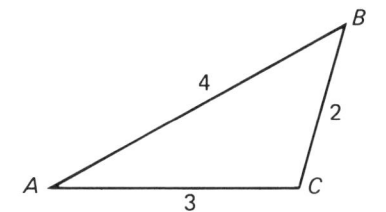

Worked example 3 Solve $\triangle ABC$ in which $a = 2$, $b = 3$, $c = 4$ (fig. 3.3:4).

By equation (2), $4^2 = 2^2 + 3^2 - 2 \cdot 2 \cdot 3 \cos C \Rightarrow \cos C = -\frac{1}{4} \Rightarrow C = 104.48°$.
It is then easier to use the sine formula to find A:

$$\frac{2}{\sin A} = \frac{4}{\sin 104.48°} \Rightarrow \sin A = 0.48 \Rightarrow A = 28.95°.$$

Fig. 3.3:4   So $B = 180° - 28.95° - 104.48° = 46.57°$.

Exercise 3.3:1

① **1** Solve $\triangle ABC$ in which:
→ (i) $A = 65°$, $B = 50°$, $a = 8$
   (ii) $B = 75°$, $C = 60°$, $c = 6$
   (iii) $A = 120°$, $C = 40°$, $c = 3$

② **2** Show by drawing diagrams that there are two possible triangles satisfying each of the following sets of data and solve them:
→ (i) $A = 45°$, $a = 4$, $c = 5$
   (ii) $A = 30°$, $a = 5$, $c = 6$
   (iii) $C = 50°$, $b = 5$, $c = 4.5$

② **3** Show that there is (only) one triangle satisfying each set of data and solve it:
→ (i) $A = 45°$, $a = 5$, $c = 3$
   (ii) $A = 120°$, $a = 8$, $c = 4$

② **4** If A, a and c are given, show that there are two possible triangles if $c \sin A < a < c$. What happens if $a > c$? What happens if $a < c \sin A$?

③ **5** Solve $\triangle ABC$ in which:
→ (i) $a = 2$, $b = 4$, $c = 5$
   (ii) $a = 9$, $b = 10$, $c = 12$
   (iii) $a = 8$, $b = 10$, $c = 15$

③ **6** Find the largest angle of $\triangle ABC$ in which $a = 5$, $b = 7$, $c = 8$.

③ **7** Find the smallest angle of $\triangle ABC$ in which $a = 4.5$, $b = 4$, $c = 5$.

③ **8** Solve $\triangle ABC$ in which:
•→ (i) $a = 2$, $b = 3$, $C = 60°$
   (ii) $a = 15$, $b = 10$, $C = 40°$
   (iii) $a = 5$, $b = 6$, $B = 48°$
   (iv) $b = 12$, $c = 17.5$, $A = 120°$

③→ **9** In $\triangle PQR$, $PR = QR = r$, $P\hat{R}Q = \alpha$.
• (i) Using equation (2), show that
$$PQ = r\sqrt{[2(1 - \cos \alpha)]}$$
(ii) If N is the midpoint of PQ, show that $P\hat{R}N = \frac{1}{2}\alpha$, and hence find PQ in terms of $\sin \frac{1}{2}\alpha$.

Show that the two expressions you have found for PQ are the same.

10 Explain why it is impossible to solve $\triangle ABC$ in which:
(i) $a = 7$, $b = 3$, $c = 2$
(ii) $A = 45°$, $a = 4$, $c = 6$
(iii) $A = 40°$, $B = 75°$, $C = 65°$

R **11** Solve, if possible, $\triangle ABC$ when:
(i) $a = 5$, $b = 3$, $B = 30°$
(ii) $a = 5$, $b = 2$, $B = 30°$
(iii) $a = 2$, $b = 3$, $B = 30°$
(iv) $a = 2$, $b = 3$, $B = 150°$

12 A, B and C are three triangulation points used in Ordnance Survey map-making. B is 3 km due south of A. C is 4 km on a bearing of 300° from B. Find the distance CA and the bearing of A from C. (Note that 'on a bearing of θ°' means 'in the direction N θ° E,' e.g. 'on a bearing of 210°' means 'in the direction N 210° E, i.e. S 30° W.')

13 A real-ale freak goes on a pub crawl. He leaves his home at O on his bicycle and travels 3 km due east to the Dog and Calculator for a couple of pints of their own Special Brew. Later he leaves the Dog and Calculator and travels 5 km on a bearing of 50° to the Lost Jockey for a jar or two of their Powerhouse. Much later he is asked to leave the Lost Jockey and travels 8 km on a bearing of 170° to the Ligger's Arms for a quick half of Stoat, a local beer which is an amazing 100% proof.

Later, his wife, alerted by a call from the owner of the Ligger's Arms, comes to collect him from the pavement outside the pub. How far and on what bearing does she have to travel?

•→ **14** Two ships, P and Q, leave a harbour O simultaneously. P travels at 25 km h^{-1} due N, Q at 30 km h^{-1} on a bearing of 120°. Find the distance PQ and the bearing of Q from P four hours later.

15 Two ships, P and Q, leave a harbour O simultaneously. P travels at 20 km h^{-1} on a bearing of 340°, Q at a constant speed on a bearing of 60°. Two and a half hours later the distance between the ships is 85 km. Find the speed at which Q travels and the bearing of Q from P at this time.

16 In the cyclic quadrilateral $ABCD$, $AB = 7$, $BC = 8$, $CD = 8$, $DA = 15$. Calculate $A\hat{D}C$ and the length of AC.

• **17** If $A = 2B$, show that:

(i) $\cos B = a/2b$ (ii) $a^2 - b^2 = bc$

18 In $\triangle ABC$, $a = x^2 - 1$, $b = x^2 - x + 1$, $c = x^2 - 2x$. Show that $B = 60°$.

→ **19** In $\triangle ABC$, $AB = x$, $BC = x + 1$, $CA = x + 2$. Show that
$$\cos B = \frac{x-3}{2x}$$

AD is an altitude of the triangle, i.e. the perpendicular from A to BC. If $AD = 2\sqrt{6}$, find x.

R 20 In $\triangle ABC$, $AB = x - y$, $BC = x$, $CA = x + y$. Show that:

(i) $\cos A = \dfrac{x - 4y}{2(x - y)}$

(ii) $\sin A - 2 \sin B + \sin C = 0$

• **21** A **median** of a triangle is the line joining a vertex to the midpoint of the opposite side.

(i) AD is a median of $\triangle ABC$. By using the cosine formula in $\triangle ABD$ and $\triangle ACD$, show that:
$$4AD^2 = 2b^2 + 2c^2 - a^2$$

(ii) A triangle has sides 4 cm, 5 cm and 7 cm. Find the length of its shortest median.

→ **22** The point X divides the side BC of $\triangle ABC$ internally in the ratio $m : n$ where $m + n = 1$. Show that
$$AX^2 = mb^2 + nc^2 - mna^2$$

23 The **circumcircle** of a $\triangle ABC$ is the circle passing through A, B and C. The centre of this circle is called the **circumcentre** of $\triangle ABC$.

O is the circumcentre of $\triangle ABC$ in which $a = 3$, $b = 4$, $c = 6$. Find $A\hat{O}B$, $B\hat{O}C$, $C\hat{O}A$ and the radius of the circumcircle.

24 O is the circumcentre (see question 23) of $\triangle ABC$, A acute. BO cuts the circumcircle at D. Let the radius of the circumcircle be R.

Show that $a = 2R \sin B\hat{D}C$. Hence show that $a = 2R \sin A$. Hence prove the sine formula.

Show that the proof still works when A is obtuse.

25 In $\triangle ABC$, $\hat{A} = 45°$, $\hat{B} = \beta$. M is the midpoint of AB, $B\hat{M}C = \gamma$. Show that:
$$\tan \gamma = \tan \beta (\tan \gamma - 2)$$

Find γ when:

(i) $\beta = 45°$ (ii) $\beta = 90°$

• → **26** (*Cotangent formula*) The point X divides the side BC of $\triangle ABC$ in the ratio $m : n$ where $m + n = 1$. Let $B\hat{A}X = \alpha$, $C\hat{A}X = \beta$, $A\hat{X}C = \theta$. Use the sine formula in $\triangle ABX$, $\triangle ACX$ to show that
$$\frac{m \sin (\theta - \alpha)}{\sin \alpha} = \frac{n \sin (\theta + \beta)}{\sin \beta}$$

Hence, using the compound angle formulae, show that
$$\cot \theta = m \cot \alpha - n \cot \beta$$

In particular show that if X is the midpoint of BC then $2 \cot \theta = \cot \alpha - \cot \beta$.

27 Show that in $\triangle ABC$ the bisector of the angle A divides BC in the ratio $c : b$.

28 Show with the aid of a diagram that in any $\triangle ABC$
$$a \cos B = c - b \cos A$$

3.3:2 Area of a triangle

Theorem 2: The area of a triangle ABC is given by
$$\text{area } \triangle ABC = \tfrac{1}{2} bc \sin A = \tfrac{1}{2} ac \sin B = \tfrac{1}{2} ab \sin C \qquad (3)$$

Proof

Look at fig. 3.3:1 (page 150) again:
$$h = b \sin A$$
$$\text{area of } \triangle ABC = \tfrac{1}{2}(\text{base})(\text{height})$$
$$= \tfrac{1}{2} ch$$
$$= \tfrac{1}{2} bc \sin A$$

Similarly $\quad$ area $\triangle ABC = \frac{1}{2}ac \sin B = \frac{1}{2}ab \sin C$

Worked example 4 Find the area of $\triangle ABC$ when $b = 6$, $c = 5$, $C = 30°$.

We first use the sine formula to find A and B.

$$\frac{6}{\sin B} = \frac{5}{\sin 30°} \Rightarrow \sin B = \frac{3}{5} \Rightarrow B = 36.87° \Rightarrow A = 113.13°$$

So by equation (3), area $= \frac{1}{2} \cdot 6 \cdot 5 \sin 113.13° = 13.79$.

Exercise 3.3:2

1 Find the area of $\triangle ABC$ in which:
→ (i) $a = 4$, $b = 5$, $C = 60°$
(ii) $b = 3$, $c = 4$, $C = 50°$
(iii) $a = 8$, $b = 3$, $A = 105°$
→ (iv) $a = 5$, $b = 7$, $c = 11$
(v) $a = 5$, $A = 40°$, $C = 65°$

R 2 The area of an acute-angled $\triangle ABC$ is 12. Given that $a = 5$ and $b = 8$, find c.

→ **3** The area of $\triangle ABC$ is 6. Given that $a = 3$, $b = 4.5$, find two possible values of c.

4 Find the area of $\triangle ABC$ when A, B and C are the points:

(i) $(-1, 1)$, $(5, 1)$, $(2, -3)$
(ii) $(-2, 3)$, $(6, 1)$, $(8, -4)$

5 In the $\triangle ABC$, D is the midpoint of BC, and E is the point on BC such that AE bisects the angle A. Show that:

(i) the areas of $\triangle ABD$ and $\triangle ADC$ are equal;
(ii) the ratio of the area of $\triangle ABE$ to the area of $\triangle AEC$ is c/b.

If G is the centroid of $\triangle ABC$, show that the areas of $\triangle AGB$, $\triangle AGC$ and $\triangle BGC$ are equal.

3.4 Derivatives of trigonometric functions

3.4:1 Radians, arcs and sectors

We are going to use a new unit for measuring angles: the **radian**. Instead of dividing one revolution into 360°, we divide it into 2π radians (2π rad), i.e.

$$\boxed{2\pi \text{ rad} = 360°} \tag{1}$$

So x rad $= (180/\pi)x°$ and $y° = (\pi/180)y$ rad; and, e.g.,

$$\sin(\tfrac{1}{6}\pi \text{ rad}) = \sin 30° = \tfrac{1}{2}$$
$$\tan(\tfrac{3}{4}\pi \text{ rad}) = \tan 135° = -1$$

From now on we will work almost exclusively in radians (and omit the 'rad' except occasionally for emphasis). The main reason for this is that the simplicity of important results such as

$$\frac{d}{dx}(\sin x) = \cos x$$

(see 3.4:3 Theorem 4) depends upon x being measured in radians.

154 *Radians, arcs and sectors*

Theorem 1: If the arc PQ of a circle centre O, radius r, subtends an angle α rad at O, then

$$\text{length of arc } PQ = \overset{\frown}{PQ} = r\alpha \qquad (2)$$

$$\text{area of sector } OPQ = \tfrac{1}{2}r^2\alpha \qquad (3)$$

Proof

$$P\hat{O}Q = \frac{\alpha}{2\pi} \times (\text{complete angle at } O) \text{ (fig. 3.4:1)}$$

$\Rightarrow$ equation (2) $\quad \overset{\frown}{PQ} = \dfrac{\alpha}{2\pi} \times (\text{circumference of circle})$

$$= \frac{\alpha}{2\pi} \times 2\pi r$$

$$= r\alpha$$

and $\Rightarrow$ equation (3) $\quad$ area of sector $OPQ = \dfrac{\alpha}{2\pi} \times (\text{area of circle})$

$$= \frac{\alpha}{2\pi} \times \pi r^2$$

$$= \tfrac{1}{2}\alpha r^2$$

Fig. 3.4:1

(Note: equation (2)$\Rightarrow$ when $\alpha = 1$, $\overset{\frown}{PQ} = r$; i.e. 1 radian is the angle subtended by an arc whose length is equal to the radius. By equation (1), 1 radian $\approx 57°$.)

Worked example 1 The arc PQ of a circle, centre O, radius 10, subtends an angle 150° at O (fig. 3.4:2). Find the area of the minor segment cut off by the chord PQ.

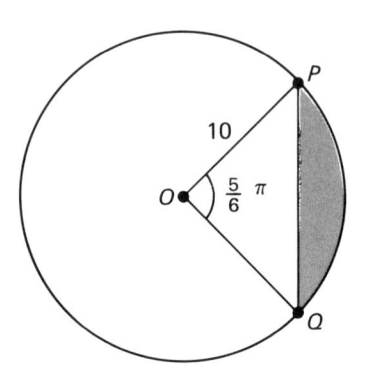

Fig. 3.4:2

$$P\hat{O}Q = 150° = \tfrac{5}{6}\pi \text{ rad}$$

By equation (2),

$$\text{area of sector } OPQ = \tfrac{1}{2} \cdot 10^2 \cdot \tfrac{5}{6}\pi = 130.90$$

$$\text{area of } \triangle OPQ = \tfrac{1}{2} \cdot 10^2 \sin \tfrac{5}{6}\pi = 25$$

$\Rightarrow \quad$ area of segment = area of sector – area of triangle

$$= 105.90$$

Trigonometry 155

Exercise 3.4:1

1 Express the following angles in radians:

(i) (a) 180° (b) 90° (c) 30° (d) 20°
(e) 18° (f) 135° (g) 150° (h) 270°
(i) 300° (j) 225° (k) 450° (l) 720°

(ii) (a) 1° (b) 41.2° (c) 214.73°

2 Express the following angles in degrees:

(i) (a) $\frac{1}{2}\pi$ rad (b) $\frac{3}{4}\pi$ rad
(c) $\frac{2}{3}\pi$ rad (d) $\frac{1}{5}\pi$ rad
(e) $\frac{7}{6}\pi$ rad (f) $\frac{11}{6}\pi$ rad
(g) $\frac{1}{12}\pi$ rad (h) $\frac{1}{15}\pi$ rad
(i) $\frac{3}{10}\pi$ rad (j) 5π rad

(ii) to 2 dec. pls.
(a) 1 rad (b) 0.14 rad (c) 4.38 rad

3 Without using a calculator write down the values of:

(i) $\sin \frac{1}{2}\pi$ (ii) $\cos \frac{1}{4}\pi$
(iii) $\tan \frac{1}{3}\pi$ (iv) $\sin(-\frac{1}{6}\pi)$
(v) $\sec \frac{2}{3}\pi$ (vi) $\cos \pi$
(vii) $\text{cosec} \frac{3}{4}\pi$ (viii) $\cot \frac{3}{2}\pi$
(ix) $\sin \frac{7}{3}\pi$

4 Write down the values of:

(i) $\sin(1.357 \text{ rad})$
(ii) $\cos(-2.04 \text{ rad})$
(iii) $\cot(3.7 \text{ rad})$

5 Simplify:

(i) $\cos(\frac{1}{2}\pi - x)$
(ii) $\sin(\frac{1}{2}\pi + x)$
(iii) $\tan(\frac{1}{2}\pi - x)$
(iv) $\sin(\pi - x)$
(v) $\cos(\pi + x)$
(vi) $\cot(\pi + x)$
(vii) $\sin(\frac{3}{2}\pi - x)$

6 Find the general solution in *radians* of the following equations. Hence write down the solutions for which $0 < x < 2\pi$.

(i) $\cos x = -\frac{1}{2}\sqrt{3}$
(ii) $\tan 3x = \frac{1}{4}$
(iii) $\cos 3x = \sin x$
→ (iv) $3\tan^2 x + 5 = 7\sec x$
R (v) $\sin 2x = 1 + \cos 2x$
→ (vi) $2\sin x - 3\cos x = 1$
(vii) $\sin 2x - \sin x = \cos x + \cos 2x$

• → **7** Sketch on the same axes the curve $y = \sin x$ and the line $y = \frac{3}{4}x$. Show that the equation has exactly 3 solutions.
On graph paper, draw the curve and the line accurately for $-\frac{1}{2}\pi \le x \le \frac{1}{2}\pi$, and hence find approximately (to 2 dec. pls.) the two non-zero roots of the equation.

8 Sketch on the same axes the curves $y = \tan x$ and $y = 1/x$. Hence show that the equation $x \tan x = 1$ has an infinite number of solutions. How many of these solutions lie between -2π and 2π?

9 On graph paper, draw accurate graphs of $y = \cos x$ and $y = x$ for $0 \le x \le \pi$ and hence find approximately the root of the equation $x = \cos x$.

10 By drawing the appropriate graphs find approximately the solution(s) of these equations in the range $0 \le x \le \pi$:

(i) $\tan x = 2x$ (ii) $\sin x = x - 2$
(iii) $2\sin x - \cos x = x$

+ **11** A function $f : \mathbb{R} \to \mathbb{R}$ is defined by $f(x) = x^2 \sin x$. Sketch a graph of this function.

+ **12** A function $g : \mathbb{R}^+ \to \mathbb{R}$ is defined by

$$f(x) = x \sin \frac{1}{x}$$

Sketch a graph of this function.

① **13** The arc PQ of a circle centre O, radius 20 subtends an angle of 120° at O. Find the area of the minor segment cut off by the chord PQ.

① • **14** The arc PQ of a circle centre O, radius r subtends an angle α rad ($\alpha < \pi$) at O. Show that the length of the chord PQ is $2r \sin \frac{1}{2}\alpha$.
Write down the lengths PQ and $\widehat{PQ}$ in the cases $\alpha = 0, \frac{1}{6}\pi, \frac{1}{4}\pi, \frac{1}{3}\pi, \frac{1}{2}\pi$.

① → **15** The arc PQ of a circle centre O, radius r subtends an angle α rad at O. Show that the area of $\triangle OPQ$ is $\frac{1}{2}r^2 \sin \alpha$. Hence find the area of:

(i) the minor segment cut off by PQ;
(ii) the major segment cut off by PQ.

Look at your answer to (i). What do you deduce about α and $\sin \alpha$ when α is small?

① **16** A circular cone with base radius r and slant height l is unrolled into a circular sector. Find the angle of this sector, in radians. Hence show that the curved surface area of the cone is $\pi r l$.

① → **17** A circle of radius r is drawn with its centre on
• the circumference of another circle of radius r. Show that the area common to both circles is

$$2r^2(\tfrac{1}{3}\pi - \tfrac{1}{4}\sqrt{3})$$

① **18** Two circles, centres A and B, with radii 10 cm and 6 cm respectively, intersect at P and Q. $P\widehat{A}B = \frac{1}{6}\pi$. Show that $\sin P\widehat{B}A = \frac{5}{6}$.
Find the area common to both circles.

① **19** Two circles, centres A and B, each of radius r, intersect at P and Q. If $P\hat{A}Q = P\hat{B}Q = 2\theta$, find the area Δ common to both circles. If $\Delta = \frac{1}{2}\pi r^2$, show that $\sin 2\theta = 2\theta - \frac{1}{2}\pi$.

By drawing appropriate graphs, find an approximate solution of this equation in the range $0 < \theta < \frac{1}{2}\pi$.

① **20** A circle is divided into two segments by a chord of length equal to the radius. Find the ratio of the areas of the segments.

① → **21** The chord PQ of a circle centre O divides the
• circle into two segments whose areas are in the ratio $2:1$. If $P\hat{O}Q = x$, show that $3 \sin x = 3x - 2\pi$.

① R **22** The chord PQ of a circle centre O divides the circle into two segments whose areas are in the ratio $5:1$. If $P\hat{O}Q = x$, show that $\sin x = x - \frac{1}{3}\pi$.

By sketching graphs of $y = \sin x$ and $y = x - \frac{1}{3}\pi$ in the range $0 \leq x \leq \pi$, find an approximate solution of this equation.

① → **23** At 3 p.m. every day in the park, Lionel the famous
• performing dog does his celebrated minimalist performance 'Pool Walk'. Accompanied by music from the park's resident brass band, Lionel starts from a point 10 m from the edge of the Round Pond (radius 10 m), walks to a point on the edge of the pond, around the pond and back to his starting point, taking the shortest possible route. Find the length of his route.

① **24** A post-modern neo-Norman housing complex consists of three identical circular towers, each of radius 6 m, surrounding a central enclosed space. Find the floor area A of the enclosed space.

A man in search of the meaning of life runs around the complex 500 times every day, taking the shortest possible route. Find the distance he runs.

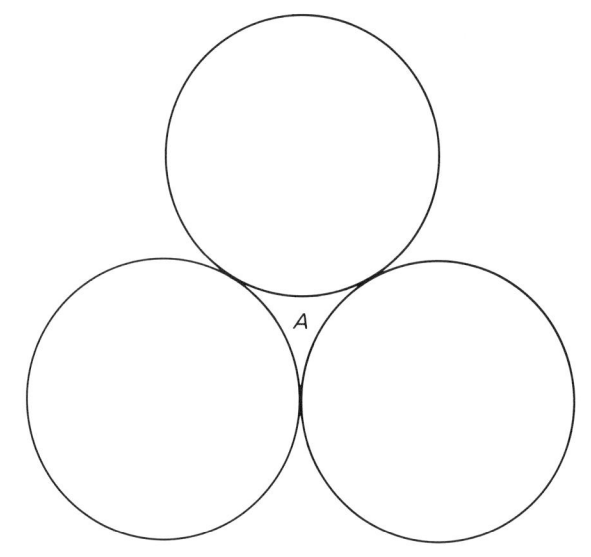

① **25** A circular lake has centre O and radius 30 m. A man, who swims at 1 m s^{-1} and walks at 2 m s^{-1}, starts at a point A on the edge of the lake. He swims to a point B on the edge and then walks along the edge to the point C exactly opposite A. He returns from C to A by swimming back to B and walking from B to A.

If $A\hat{O}B = \frac{1}{4}\pi$, find the time he takes for:

(i) the outward journey (from A to C)
(ii) the return journey (from C to A)

Find $A\hat{O}B$, given that the times for the outward journey and the return journey are the same.

① **26** The cross section of a tunnel is the major segment of a circle cut off by a chord of length 10 m. If the height of the tunnel is 15 m, find its cross-sectional area.

3.4:2 Small angles

Theorem 2: When θ is small

$$\sin \theta \approx \theta \qquad (4a)$$
$$\tan \theta \approx \theta \qquad (4b)$$
$$\cos \theta \approx 1 - \tfrac{1}{2}\theta^2 \qquad (4c)$$

Proof

Let P and Q be two points on a circle centre O, radius 1 (fig. 3.4:3), such that $POQ = \theta$.

$$\sin \theta = \frac{PN}{OP} = PN$$

$$\theta = \overset{\frown}{PQ}$$

$$\tan \theta = \frac{RQ}{OQ} = RQ$$

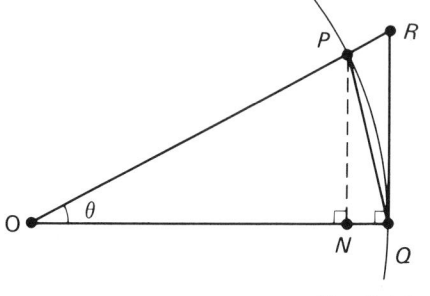

Fig. 3.4:3

When θ is small, $PN \approx \overset{\frown}{PQ} \approx RQ$, i.e.

$$\sin\theta \approx \theta \approx \tan\theta$$

and

$$\begin{aligned}\cos\theta &= 1 - 2\sin^2 \tfrac{1}{2}\theta \\ &\approx 1 - 2(\tfrac{1}{2}\theta)^2 = 1 - \tfrac{1}{2}\theta^2\end{aligned}$$

Worked example 2 Find an approximation, when θ is small, for

$$\frac{1-\cos 3\theta}{\theta\sin\theta}$$

When θ is small, $1-\cos 3\theta \approx 1 - \{1 - \tfrac{1}{2}(3\theta)^2\} = \tfrac{1}{2}\cdot 9\theta^2$

and $$\theta\sin\theta \approx \theta^2$$

So $$\frac{1-\cos 3\theta}{\theta\sin\theta} \approx \frac{9\theta^2}{2\theta^2} = \frac{9}{2}$$

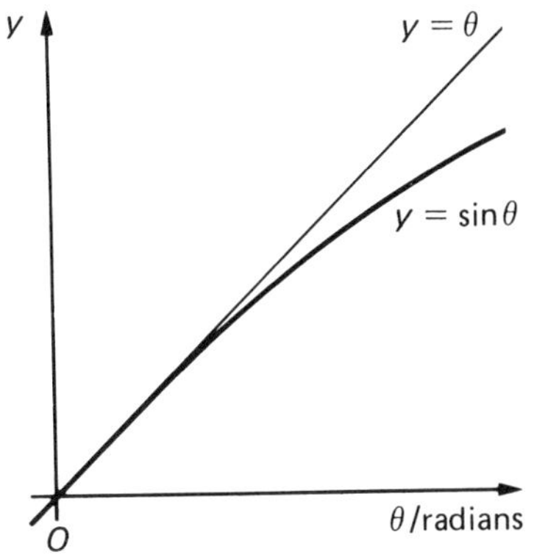

Fig. 3.4:4

Theorem 3:

$$\lim_{\theta\to 0} \frac{\sin\theta}{\theta} = 1 \qquad \textbf{(5)}$$

This is a more precise way of expressing approximation (4a). It says not only that $\sin\theta$ is near θ when θ is small, but also that $\sin\theta$ gets nearer and nearer to θ as θ gets smaller (fig. 3.4:**4**).

Proof

Look at fig. 3.4:**3** again. Area of $\triangle OPQ <$ area of sector $OPQ <$ area of $\triangle ORQ$, i.e.

$$\tfrac{1}{2}\sin\theta < \tfrac{1}{2}\theta < \tfrac{1}{2}\tan\theta$$

$$1 < \frac{\theta}{\sin\theta} < \sec\theta$$

As $\theta \to 0$, $\sec\theta \to 1$, so $\theta/\sin\theta \to 1$ and also $(\sin\theta)/\theta \to 1$. Thus

$$\lim_{\theta\to 0} \frac{\sin\theta}{\theta} = 1$$

Worked example 3 Find:

(i) $\displaystyle\lim_{\theta\to 0} \frac{\sin 3\theta}{\sin 5\theta}$ (ii) $\displaystyle\lim_{\theta\to 0} \frac{\cos 3\theta - \cos\theta}{\cos 4\theta - \cos 2\theta}$

(Note: if we actually let $\theta = 0$ in either case, the quotient is undefined.)

We want to avoid using approximations (4a), (4b) and (4c) as they are not sufficiently precise.

158 Small angles

$$\text{(i)} \lim_{\theta\to 0}\frac{\sin 3\theta}{\sin 5\theta} = \lim_{\theta\to 0}\frac{\sin 3\theta}{3\theta}\cdot\frac{5\theta}{\sin 5\theta}\cdot\frac{3}{5}$$

$$= \frac{3}{5}\left\{\lim_{\theta\to 0}\frac{\sin 3\theta}{3\theta}\right\}\left\{\lim_{\theta\to 0}\frac{5\theta}{\sin 5\theta}\right\}$$

$$= \frac{3}{5} \qquad \text{(by equation (5))}$$

(Note: we have used the result

$$\lim_{\theta\to 0} f(\theta)g(\theta) = \left\{\lim_{\theta\to 0} f(\theta)\right\}\left\{\lim_{\theta\to 0} g(\theta)\right\}$$

which holds so long as the limits exist and are finite.)

$$\text{(ii)} \lim_{\theta\to 0}\frac{\cos 3\theta - \cos\theta}{\cos 4\theta - \cos 2\theta} = \lim_{\theta\to 0}\frac{-2\sin 2\theta\sin\theta}{-2\sin 3\theta\sin\theta} \qquad \text{(by 3.2:7 identity (20d))}$$

$$= \lim_{\theta\to 0}\frac{\sin 2\theta}{\sin 3\theta}$$

$$= \frac{2}{3}$$

Exercise 3.4:2

② **1** Find approximations for the following functions when θ is small.

   (i) $\cot\theta(1-\cos\theta)$

   (ii) $\dfrac{2\sin\theta - \theta}{\sin 2\theta}$

→ (iii) $\dfrac{1-\cos 2\theta}{1-\cos\theta}$

   (iv) $\dfrac{\sin 4\theta}{\sin 2\theta}$

   (v) $\dfrac{\sin\theta\tan\theta}{1-\cos 4\theta}$

   (vi) $\cos(\tfrac{1}{3}\pi + \theta)$

   (vii) $\tan(\tfrac{1}{4}\pi - \theta)$

② **R 2** The radius OB of a circle centre O bisects the angle between the radii OA and OC. The points P and Q are the feet of the perpendiculars to OA from B and C respectively. Express the ratio $AP:PQ$ in terms of θ, the angle AOB, and find its value approximately when θ is small.

• → **3** Show that, when θ is small, $\cos 2\theta \approx 1 - 2\theta^2$. Hence find an approximate solution, near $\theta = 0$, of the equation $5\cos 2\theta = 49\theta$.

4 Find an approximate solution, near $\theta = 0$, of the equation $2\cos 3\theta = 17\theta$.

5 Find:
$$\lim_{\theta\to 0}\frac{\sin a\theta}{\sin b\theta}$$
where a and b are constant.

6 Find:

(i) $\displaystyle\lim_{\theta\to 0}\frac{1-\cos a\theta}{1-\cos b\theta}$ (ii) $\displaystyle\lim_{\theta\to 0}\frac{\cos a\theta - \cos b\theta}{\sin a\theta + \sin b\theta}$

where a and b are constant.

③ • **7** Prove that $\displaystyle\lim_{\theta\to 0}\frac{\tan\theta}{\theta} = 1$.

③ **8** Find:

(i) $\displaystyle\lim_{\theta\to 0}\frac{\sin 2\theta}{\sin 3\theta}$

(ii) $\displaystyle\lim_{\theta\to 0}\frac{\sin\tfrac{1}{2}\theta}{2\theta}$

(iii) $\displaystyle\lim_{\theta\to 0}\frac{(\sin\theta + \cos\theta)^2 - 1}{\theta}$

(iv) $\displaystyle\lim_{\theta\to 0}\frac{\sin 4\theta + \sin 2\theta}{5\sin 3\theta}$

→ (v) $\displaystyle\lim_{\theta\to 0}\frac{\cos\theta - \cos 2\theta}{\cos\theta - \cos 3\theta}$

R (vi) $\displaystyle\lim_{\theta\to 0}\frac{\cos 3\theta - \cos 2\theta}{\theta(\sin 3\theta + \sin 2\theta)}$

③ 9 Express $\cos 2\theta$ in terms of $\sin \theta$.
→ Hence find $\lim\limits_{\theta \to 0} \dfrac{1-\cos 2\theta}{\theta^2}$.

③ 10 Find:

R (i) $\lim\limits_{\theta \to 0} \dfrac{\sin^2 \theta}{1-\cos \theta}$ (ii) $\lim\limits_{\theta \to 0} \dfrac{1-\cos 2\theta}{1-\cos 4\theta}$

③ 11 By factorizing the numerators, find

• (i) $\lim\limits_{\theta \to 0} \dfrac{\sin(\frac{1}{3}\pi + \theta) - \sin\frac{1}{3}\pi}{\sin 2\theta}$

→ (ii) $\lim\limits_{\theta \to 0} \dfrac{\cos(\alpha + \theta) - \cos \alpha}{\theta}$, α constant

3.4:3 Derivatives of trigonometric functions

Theorem 4:

$$\frac{d}{dx}(\sin x) = \cos x \qquad (6)$$

$$\frac{d}{dx}(\cos x) = -\sin x \qquad (7)$$

Proof
Equation (6):

$$\frac{d}{dx}(\sin x) = \lim_{\delta x \to 0} \frac{\sin(x + \delta x) - \sin x}{\delta x} \qquad \text{by 1.3:1 equation (1)}$$

$$= \lim_{\delta x \to 0} \frac{2 \sin(\frac{1}{2}\delta x) \cos(x + \frac{1}{2}\delta x)}{\delta x} \qquad \text{by 3.2:7 identity (20b)}$$

$$= \lim_{\delta x \to 0} \cos(x + \tfrac{1}{2}\delta x) \qquad \text{by approximation (4a)}$$

$$= \cos x$$

Equation (7):

$$\frac{d}{dx}(\cos x) = \frac{d}{dx}(\sin(\tfrac{1}{2}\pi - x))$$

$$= -\cos(\tfrac{1}{2}\pi - x) \qquad \text{by chain rule}$$

$$= -\sin x$$

Note: the results are valid only when x is measured in radians (see question 3 (xiii).

Worked example 4 Differentiate: (i) $\sin 3x$ (ii) $\cos^3 x$

(i) $\qquad \dfrac{d}{dx}(\sin 3x) = 3 \cos 3x \qquad$ by the chain rule

(ii) $\qquad \dfrac{d}{dx}(\cos^3 x) = (3 \cos^2 x)(-\sin x) \qquad$ by the chain rule

$$= -3 \cos^2 x \sin x$$

Worked example 5 Differentiate:
(i) $\cot x$ (ii) $\cot(3x^2 + 2)$

160 *Derivatives of trigonometric functions*

(i) $y = \cot x = \dfrac{\cos x}{\sin x} \Rightarrow \dfrac{dy}{dx} = \dfrac{-\sin^2 x - \cos^2 x}{\sin^2 x}$ by the quotient rule

$$= \dfrac{-1}{\sin^2 x}$$

$$= -\operatorname{cosec}^2 x$$

(ii) Using the result of (i):

$$\dfrac{d}{dx} \cot(3x^2+2) = \{-\operatorname{cosec}^2(3x^2+2)\} \cdot \{6x\} \quad \text{by the chain rule}$$

$$= -6x \cos^2(3x^2+2)$$

Exercise 3.4:3

- **1** Sketch the graph of $f(x) = \sin x$ for $-2\pi < x < 2\pi$. By considering the gradient of the graph at intervals of $\tfrac{1}{6}\pi$, sketch a graph of $f'(x)$. Do you recognize this graph?

- **2** Starting from the definition (1.3:1 equation (1)) of the derivative, show that

$$\dfrac{d}{dx}(\cos x) = -\sin x$$

④ **3** Differentiate with respect to x:
 (i) $\sin 2x$
→ (ii) $\cos 3x$
 (iii) $\sin \tfrac{3}{5}x$
 (iv) $\sin^2 x$
 (v) $\cos^4 x$
 (vi) $\cos^2 5x$
 (vii) $\sin^5 \tfrac{1}{3}x$
→ (viii) $\sqrt{(\sin x)}$
 (ix) $1/\sqrt{(\cos x)}$
R (x) $(\sin 2x + \cos x)^2$
 (xi) $\sin \sqrt{x}$
 (xii) $\cos(1+x)$
 (xiii) $\sin x°$ (remember that $x° = \tfrac{1}{180}\pi x$ rad)

④ **4** Differentiate with respect to x:
 (i) $\sin(ax+b)$
 (ii) $\cos(ax+b)$
 (iii) $\sin^n x$
 (iv) $\cos^n x$
 (v) $\sin^n(ax+b)$
 (vi) $\cos^n(ax+b)$

where a, b and n are constant.

④ **5** Differentiate with respect to x:
 (i) $x \sin 3x$ (ii) $\sin 2x \cos 3x$
→ (iii) $\sin^2 x \cos^3 x$ (iv) $\dfrac{\cos 4x}{x}$
 (v) $\dfrac{x^2}{\sin x + \cos x}$

④ **6** Show that

$$\dfrac{d}{dx}(\sin^2 x) = -\dfrac{d}{dx}(\cos^2 x)$$

Illustrate the result by drawing graphs of $y = \sin^2 x$ and $y = \cos^2 x$ for $-\tfrac{1}{2}\pi \le x \le \tfrac{3}{2}\pi$.

④ **7** Given that $y = (A+x)\sin 2x$, where A is a constant, show that $d^2y/dx^2 + 4y$ is independent of A.

④ R **8** Given that $y = A \cos 3x + B \sin 3x$, where A and B are constant, show that

$$\dfrac{d^2y}{dx^2} + 9y = 0$$

⑤ → **9** (i) Prove that $d/dx(\tan x) = \sec^2 x$:
 (a) starting from the definition of the derivative (1.3:1 equation (1));
 (b) using Theorem 4 and the quotient rule.
 (ii) Differentiate:
 (a) $\operatorname{cosec} x$; (b) $\sec x$.

Hence complete this table:

y	dy/dx
$\sin x$	$\cos x$
$\cos x$	
$\tan x$	
$\cot x$	
$\sec x$	
$\operatorname{cosec} x$	

⑤ **10** With the help of your table (question 9), differentiate:
 (i) $\tan 3x$ (ii) $\tan \tfrac{1}{2}x$
 (iii) $\tan x°$ (iv) $\sec 4x$
 (v) $\cot \tfrac{2}{3}x$ (vi) $\operatorname{cosec} 5x$
 (vii) $\tan \sqrt{x}$ (viii) $\sec(x^2+1)$

(ix) $\cot\left(\dfrac{1+x}{1-x}\right)$ (x) $\tan^2 x$

(xi) $\sqrt{(\cot x)}$ → (xii) $\sec^3 x$

R (xiii) $\cot^3 3x$ (xiv) $\operatorname{cosec}^3 5x$ etc.

⑤ 11 Differentiate:

(i) $\cot x \sec x$ (ii) $\cos 2x \cot 4x$
(iii) $x \tan x$ (iv) $x^2 \tan^2 x$
(v) $\dfrac{\sec x}{1+\sin x}$

⑤ 12 Show that
$$\frac{d}{dx}(\sec^2 x) = \frac{d}{dx}(\tan^2 x)$$

Illustrate by drawing graphs of $y = \sec^2 x$ and $y = \tan^2 x$ for $-\frac{1}{2}\pi \leq x \leq \frac{3}{2}\pi$.

13 Differentiate with respect to x:

(i) (a) $e^x \cos x$ → (b) $e^{-\frac{1}{3}x} \sin 3x$
(c) $(\tan x)/e^x$

(ii) (a) $e^{\cos x}$ → (b) $e^{\tan x + \sec x}$
(c) $x e^{\sin x}$ (d) $e^{\sec^3 2x}$

14 If $y = e^{-\frac{1}{2}x} \sin 2x$, show that
$$4\frac{d^2 y}{dx^2} + 4\frac{dy}{dx} + 17y = 0$$

R 15 If $y = e^{3x} \sin 4x$, show that:

(i) $\dfrac{d^2 y}{dx^2} - 6\dfrac{dy}{dx} + 25 = 0$

(ii) dy/dx can be expressed in the form $Re^{3x} \sin(3x + \alpha)$. Find R and α.

16 Differentiate with respect to x:

(i) (a) $\ln \cos x$
→ (b) $\ln \sin 3x$
(c) $\ln \operatorname{cosec} 2x$
(d) $\ln \sqrt{(1+\sin x)}$
→ (e) $\ln\left(\dfrac{1+\sin x}{1+\sin 2x}\right)$
R (f) $\ln(\sec 2x + \tan 2x)$

(ii) (a) $x^{\sin x}$
(b) $(\sin x)^{\cos x}$
(c) $(\cos x)^{\tan x}$ (see section 2.5:5)

17 Sketch graphs of $f(x)$ and $f'(x)$ when:

(i) $f(x) = 1 - \cos 2x$

(ii) $f(x) = \begin{cases} 1-x^2 & 0 \leq x < 1 \\ \sin \pi x & 1 \leq x < 2 \end{cases}$
and f is periodic with period 2.

(iii) $f(x) = \begin{cases} \tan x & 0 \leq x < \frac{1}{4}\pi \\ 1 - \tan x & \frac{1}{4}\pi \leq x < \frac{1}{2}\pi \end{cases}$
and f is periodic with period $\frac{1}{2}\pi$.

•→ 18 Find the maximum and minimum values of the function $f(x) = 2\sin x - \cos 2x$ and distinguish between them.

Solve the equation $f(x) = 0$ for values of x between 0 and 2π. Hence sketch a graph of $f(x)$ in this range.

19 Find the maximum and minimum values of the function $2\sin x - x$.

Sketch a graph of the function for values of x between $-\pi$ and π, and hence find approximately the non-zero solutions of the equation $2\sin x = x$.

20 Find the stationary values of the function
$$f(x) = 3\cos x - \cos 3x \quad (0 \leq x \leq 2\pi)$$

Solve the equation $f(x) = 0$ and hence sketch a graph of $f(x)$. (See 3.2:4 Q25).

21 Find the turning points and zeroes of the function $e^x \sin x$. Sketch on the same axes the graphs $y = e^x$, $y = -e^x$, $y = e^x \sin x$.

22 Sketch on the same axes the graphs $y = e^{-x}$, $y = -e^{-x}$, $y = e^{-x} \cos 2x$.

23 A cylinder is cut from a solid sphere so that its total surface area is as large as possible. What is the ratio of its height to its diameter?

R 24 Show that the volume of a cone of slant height 3 m and semi-vertical angle θ is $9\pi \sin^2 \theta \cos \theta$ m^3.

Show that the maximum volume of the cone as θ varies is $2\pi\sqrt{3}$ m^3. Find its surface area in this case.

→ 25 An isosceles triangle is inscribed in a circle of radius r. Show that the area of the triangle is
$$r^2(1 + \cos\theta)\sin\theta$$
where θ is the angle between the equal sides. Find the maximum possible area of the triangle as θ varies.

26 A wire of length 1 m is shaped into the arc and two radii of a circular sector. Find the maximum area it can enclose.

27 A millionaire owns a huge estate. Weary of his monotonous lifestyle, he commissions four world-famous architects, each of whom is to design a lavish house at one corner of a square of side 5 km. He wants to make a system of roads joining them, on which he can potter from house to house in his custom-built Rolls Royce.

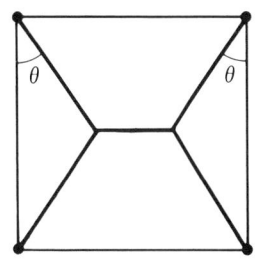

162 *Definitions and graphs*

Wanting to save on tarmac, he makes a system of the shortest possible total length. What is this length? (Hint: show that the length of the system shown in the diagram is $5(1 + 2\sec\theta - \tan\theta)$; then find the minimum value of this expression.)

28 A man can run at 8 m s^{-1} and swim at 2 m s^{-1}. He wants to cross a river of width 16 m to a point 80 m away on the opposite bank. Find the shortest possible time he can take. (Hint: suppose that he sets off at an angle θ to the perpendicular width; show that the time taken is $10 + 8\sec\theta - 2\tan\theta$.)

29 In the park, Lionel the performing dog is signing autographs. A huge circular crowd of adoring fans has formed around him. A man standing on the edge of the crowd wants to reach the point exactly opposite him on the other side of the crowd in the least possible time. If he can run round the outside of the crowd twice as fast as he can move through the crowd, describe what course he should take.

30 A millionaire builds a swimming pool in the shape of two intersecting circles, each of radius 10 m. If the area of the pool is 600 m^2, find its length, correct to 1 dec. pl.

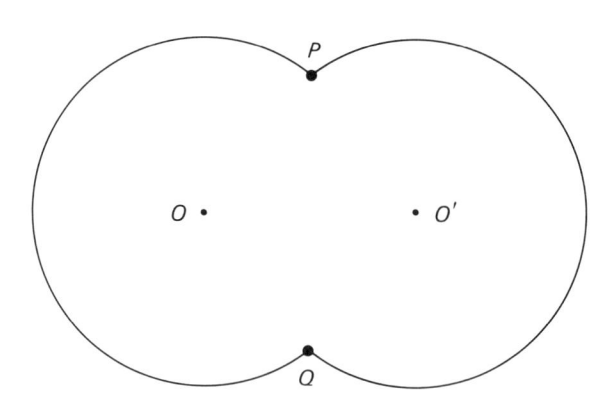

(Hint: let $P\hat{O}Q = 2\alpha$, where O and O' are the centres of the circles and show that α satisfies the equation

$$2\alpha - \sin 2\alpha = 2\pi - 6$$

Solve this equation approximately.)

3.5 Inverse trigonometric functions

3.5:1 Definitions and graphs

We saw in section 3.2:1 that, when $-1 \leq x \leq 1$, $\sin^{-1} x$ means 'the angle between $-\tfrac{1}{2}\pi$ and $\tfrac{1}{2}\pi$ whose sine is x', i.e.

$$\sin(\sin^{-1} x) = x \quad -1 \leq x < 1 \quad (1)$$
$$\sin^{-1}(\sin y) = y \quad -\tfrac{1}{2}\pi \leq y \leq \tfrac{1}{2}\pi \quad (2)$$

So if we restrict the domain of sine to $[-\tfrac{1}{2}\pi, \tfrac{1}{2}\pi]$, then sine has an inverse function:

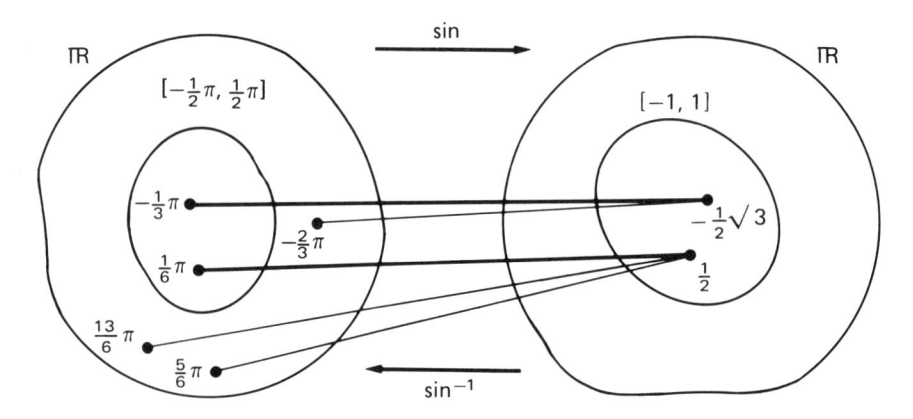

$$\sin^{-1}: [-1, 1] \to [-\tfrac{1}{2}\pi, \tfrac{1}{2}\pi] \quad \text{(fig. 3.5:1)}$$

Fig. 3.5:1

Note: we are restricting sine to a domain, i.e. $[-\tfrac{1}{2}\pi, \tfrac{1}{2}\pi]$, on which it is *into* (see 1.2:3 Q59).

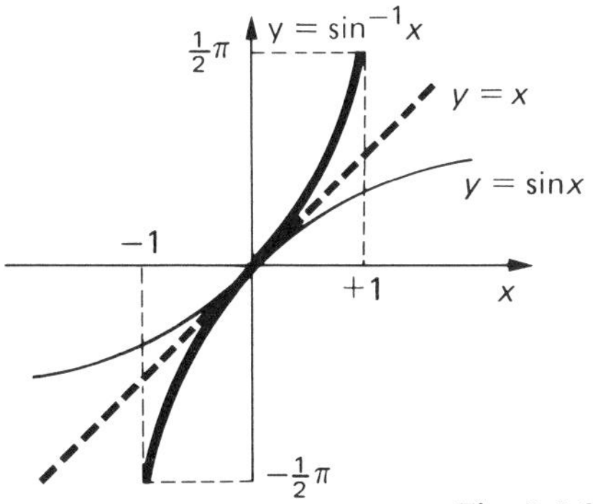

Graph of $\sin^{-1} x$

We reflect the graph of $y = \sin x$ ($-\frac{1}{2}\pi < x \leq \frac{1}{2}\pi$) in the line $y = x$ (1.2:4 Q27) to obtain the graph of $y = \sin^{-1} x$ (fig. 3.5:2).

Worked example 1 Show that $\sin(2 \tan^{-1} x) = 2x/(1+x^2)$.

Let $\tan^{-1} x = \phi$; so $x = \tan \phi$.

$$\sin(2 \tan^{-1} x) = \sin 2\phi = \frac{2 \tan \phi}{1 + \tan^2 \phi} \quad \text{from 3.2:5 identity (16)}$$

$$= \frac{2x}{1+x^2}$$

Fig. 3.5:2

Theorem 1:

$$\sin^{-1} x + \cos^{-1} x \equiv \tfrac{1}{2}\pi \qquad (3)$$

$$\tan^{-1} x + \tan^{-1} y \equiv \tan^{-1}\left(\frac{x+y}{1-xy}\right) \qquad (4)$$

Proof
See questions 4 and 9.

Exercise 3.5:1

• → **1** Suggest a suitable range for the function $\cos^{-1} x$ (see 3.2:1).
Sketch on the same axes the graphs of $y = \cos x$, $y = x$ and $y = \cos^{-1} x$. What is the domain of $\cos^{-1} x$? Sketch a diagram (like fig. 3.5:2) to show what is going on.

• → **2** Suggest a suitable range for the function $\tan^{-1} x$.
Sketch on the same axes the graphs of $y = \tan x$, $y = x$ and $y = \tan^{-1} x$. What are the asymptotes of $y = \tan^{-1} x$? What is the domain of $\tan^{-1} x$?

3 Without using a calculator or tables, write down the values of:

(i) $\sin^{-1} 0$ (ii) $\cos^{-1} \tfrac{1}{2}\sqrt{3}$
(iii) $\sin^{-1}(-\tfrac{1}{2})$ (iv) $\tan^{-1}(-1)$
(v) $\cot^{-1}\sqrt{3}$ (vi) $\sec^{-1} 2$
(vii) $\operatorname{cosec}^{-1} 1$

① → **4** (i) Sketch on the same axes graphs of $\sin^{-1} x$ and $\cos^{-1} x$.

(ii) Show that $\sin^{-1} x + \cos^{-1} x \equiv \tfrac{1}{2}\pi$ and explain how your diagram illustrates this.
(Hint: consider first the case $0 \leq x < 1$ and then the case $-1 \leq x < 0$.)

① **5** Show that $\cos(\sin^{-1} x) \equiv \sqrt{(1-x^2)}$. (Hint: let $\sin^{-1} x = \phi$ and consider $\cos \phi$.) Explain why $-\sqrt{(1-x^2)}$ is not a possible solution. (Consider the definition of the function $y = \sin^{-1} x$.)

① R **6** Simplify $\sin(\cos^{-1} x)$.

① **7** Show that, when $|x| \leq \pi$, $\cos^{-1}(\sin x) = \tfrac{1}{2}\pi - x$ and simplify $\sin^{-1}(\cos x)$.

① → **8** Simplify:

(i) $\cos(2 \tan^{-1} x)$ (ii) $\tan(2 \tan^{-1} x)$

① → **9** Show that

$$\tan^{-1} x + \tan^{-1} y \equiv \tan^{-1}\left(\frac{x+y}{1-xy}\right)$$

(Hint: let $\tan^{-1} x = \theta$, $\tan^{-1} y = \phi$ and consider $\tan(\theta + \phi)$.)

For what values of x and y is the RHS undefined?

10 Using the result of question 9, show that:

(i) $\tan^{-1} x - \tan^{-1} y \equiv \tan^{-1}\left(\dfrac{x+y}{1+xy}\right)$

(ii) $2 \tan^{-1} x \equiv \tan^{-1}\left(\dfrac{2x}{1-x^2}\right)$

Without using the result of question 9, but by a similar method, show that, if $x \neq 0$

(iii) $\tan^{-1} x + \tan^{-1}(1/x) \equiv \tfrac{1}{2}\pi$

Why is this not a valid deduction of the result of question 9, with $y = 1/x$?
Check the results of (ii) and (iii) when $x = 1/\sqrt{3}$, $x = \tfrac{1}{2}$.

11 Evaluate:
(i) $\tan^{-1}\frac{1}{2}+\tan^{-1}\frac{1}{3}$
R (ii) $\tan^{-1}\frac{1}{2}+\tan^{-1}\frac{2}{3}+\tan^{-1}\frac{4}{7}$
→ (iii) $\tan^{-1}\frac{1}{3}+\tan^{-1}\frac{1}{5}+\tan^{-1}\frac{1}{7}+\tan^{-1}\frac{1}{8}$
(iv) $2\tan^{-1}\frac{1}{3}+\tan^{-1}\frac{1}{7}$

12 Solve the equations:
(i) $\tan^{-1}2+\tan^{-1}x=\tan^{-1}4$
(ii) $\sin^{-1}x+\sin^{-1}2x=\frac{1}{3}\pi \quad (x>0)$
(iii) $\tan^{-1}(1+x)+\tan^{-1}(1-x)=8$
(iv) $\tan^{-1}\left(\frac{1-x}{1+x}\right)=\frac{1}{2}\tan^{-1}x$

13 Which of the following statements are true?
(i) $\tan^{-1}x \equiv \sin^{-1}x/\cos^{-1}x$
(ii) $\sin^{-1}x$ is an odd function
(iii) $\cos^{-1}x$ is a periodic function
(iv) $\cot^{-1}x \equiv \tan(1/x)$
(v) $\operatorname{cosec}^{-1}x+\sec^{-1}x \equiv \frac{1}{2}\pi \quad (x \geq 1)$

Prove the true statements and hit the false ones on the head with a counterexample.

3.5:2 Derivatives of inverse trigonometric functions

Theorem 2:

$$\frac{d}{dx}(\sin^{-1}x) = \frac{1}{\sqrt{(1-x^2)}} \qquad (5)$$

$$\frac{d}{dx}(\cos^{-1}x) = -\frac{1}{\sqrt{(1-x^2)}} \qquad (6)$$

$$\frac{d}{dx}(\tan^{-1}x) = \frac{1}{1+x^2} \qquad (7)$$

Proof
Equation (5): $y = \sin^{-1}x \Rightarrow x = \sin y$. Differentiating both sides with respect to x gives:

$$1 = (\cos y)\frac{dy}{dx} \quad \text{(using the implicit function rule)}$$

So $$\frac{dy}{dx} = \frac{1}{\cos y} = \frac{1}{\sqrt{(1-\sin^2 y)}} = \frac{1}{\sqrt{(1-x^2)}}$$

See question 1 for proof of equations (6) and (7).

Exercise 3.5:2

• → 1 (i) Show that

$$\frac{d}{dx}(\cos^{-1}x) = -\frac{1}{\sqrt{(1-x^2)}}$$

(a) by putting $y = \cos^{-1}x$ and differentiating $x = \cos y$;
(b) by differentiating the identity

$$\sin^{-1}x + \cos^{-1}x \equiv \frac{1}{2}\pi$$

and using equation (5).
(ii) Show that

$$\frac{d}{dx}(\tan^{-1}x) = \frac{1}{1+x^2}$$

→ 2 Show that

$$\frac{d}{dx}(\sec^{-1}x) = \frac{1}{x\sqrt{(x^2-1)}}$$

and differentiate $\operatorname{cosec}^{-1}x$ and $\cot^{-1}x$.

3 Differentiate:
• (i) $\sin^{-1}2x$
→ (ii) $\tan^{-1}(1-2x)$
R (iii) $\cos^{-1}(\tan x)$
(iv) $\cos(\cot^{-1}x)$

4 (i) Show that

$$\frac{d}{dx}(\sin^{-1}(\cos x)) = -1$$

Simplify the function $\sin^{-1}(\cos x)$.

(ii) Show that

$$\frac{d}{dx}\left(\tan^{-1}\left(\frac{2x}{1-x^2}\right)\right) = \frac{2}{1+x^2}$$

Simplify the function $\tan^{-1}\{2x/(1-x^2)\}$.

→ 5 (i) If $y = \sin(\lambda \sin^{-1} x)$, show that

$$(1-x^2)\frac{d^2y}{dx^2} - x\frac{dy}{dx} + \lambda^2 y = 0$$

(ii) If $y = \sin^{-1} x + (\sin^{-1} x)^2$, show that

$$(1-x^2)\frac{d^2y}{dx^2} - x\frac{dy}{dx} - 2 = 0$$

R (iii) If $y = (1+x^2)\tan^{-1} x$, show that

$$(1+x^2)\frac{d^2y}{dx^2} - 2y - 2x = 0$$

6 The ruthless owner of a vast raspberry plantation employs his grandmother to pick the raspberries. He pays her by the amount she picks, not by the hour, and so she picks only the fruit which is readily to hand, leaving any which is hidden or to which she needs to stoop.

One morning, the owner tethers her to a post situated 6 m from one end of the plantation by a rope of length l m ($l > 6$) and leaves here to do the picking. Show that she can cover an area A m^2, where

$$A = l^2(\pi - \cos^{-1}(6/l)) + 6\sqrt{(l^2-36)}$$

If $A = 180$, use the Newton-Raphson method to find l, correct to 1 dec. pl. (Take $l = 8$ as the first approximation.)

That night the plucky grandmother takes several sackfuls of raspberries to her grandson's dressing room. Gradually she introduces the raspberries into the drawers full of his clothes, starting with his enormous collection of shirts ...

Miscellaneous exercise 3

1 (i) Sketch on the same axes, for $-240° < x < 240°$, the curves $y = \sin\frac{3}{2}x$ and $y = 2\sin 3x$. Find the set of values of x between 0° and 240° for which $\sin\frac{3}{2}x > 2\sin 3x$.

(ii) Find a positive number R and an acute angle α such that $\sin 3x + \sqrt{3}\cos 3x = R\sin(3x+\alpha)$. Hence sketch, for $-240° < x < 240°$, the curve $y = \sin 3x + \sqrt{3}\cos 3x$.

2 (i) f, g and h are three functions such that $f(x) = g(x) + h(x)$. What is the period of $f(x)$ if the periods of $g(x)$ and $h(x)$ are
(a) π and π;   (b) 2π and 3π;
(c) 4π and 6π;   (d) a and b?

(ii) Sketch on the same axes, for $-360° < x < 360°$, the curves $y = 2 + 2\cos x$, $y = 2\sin 2x$ and $y = 1 + \cos x + \sin 2x$. What is the period of the function

$$f(x) = 1 + \cos x + \sin 2x$$

3 (i) The function $f: \mathbb{R} \to \mathbb{R}$ is defined by

$$f(x) = \begin{cases} 1 - \cos x & 0 \leq x < \frac{1}{2}\pi \\ 5 - \frac{8x}{\pi} & \frac{1}{2}\pi \leq x < \frac{3}{4}\pi \\ \sin 2x & \frac{3}{4}\pi \leq x < \pi \end{cases}$$

and f is periodic with period 2π. Sketch the graph of $f(x)$ for $-2\pi \leq x < 4\pi$, given that f is
(a) an even function;
(b) an odd function.

(ii) Which of the following functions are even, which odd and which neither?
(a) $3\sin 2x$;   (b) $1 - \cos 2x$;
(c) $1 - \cos 2x + 3\sin 2x$;
(d) $3\sin 2x (1 - \cos 2x)$.
What is the period of each function?

4 The function $f: \mathbb{R} \to \mathbb{R}$ is defined by

$$f(x) = \begin{cases} \sin x & x < 0 \\ x - x^2 & x \geq 0 \end{cases}$$

Sketch graphs of $f(x)$ and $f'(x)$ for $-2\pi < x < 2\pi$. Is f continuous? Is f' continuous? Is f'' continuous?

Questions 5-7: Find the general solutions, in degrees, of the equations given.

5 (i) $\sin 2x = 2\cos 4x$
(ii) $6\sin^2 x + 7\cos x = 3$
(iii) $\sin x + \sin 3x + \sin 5x = 0$

6 (i) $6\sin x + 8\cos x = 5$
(ii) $\sin 3x - \sin x = \cos 2x$
(iii) $3\sec^2 x + \tan x = 5$

7 (i) $\sin x - 7\cos x + 5 = 0$
(ii) $\sin(x + 30°) = \cos(x + 45°)$
(iii) $2\cot x + 2\csc x = \tan x$

8 (i) If $\cot 2x = \frac{3}{4}$, find, without using a calculator, two possible values of $\tan x$.

(ii) Given that $\sin x = \frac{1}{4}$, find, without using a calculator, $\cos 4x$.

(iii) Given that $\sin \alpha = \lambda$ and $\cos \beta = \mu$, find, in terms of λ and μ, the possible values of $\cos(\alpha + \beta)$.

9 Show that $\tan A \equiv \operatorname{cosec} 2A - \cot 2A$ $(A \neq n\pi/2)$. Why the restriction $A \neq n\pi/2$?
Hence:

(i) find $\tan 75°$ without using a calculator;
(ii) find the general solution of the equation $\operatorname{cosec} 4x - \cot 4x = 4 \tan x$.

10 (i) Show that $\cos 3A \equiv 4 \cos^3 A - 3 \cos A$.

(ii) Hence find the general solution, in degrees, of the equation $\cos 3x = 4 \cos^2 x$.

(iii) Sketch, on the same axes, the graphs $y = \cos 3x$ and $y = 4 \cos^2 x$ for $0° < x < 360°$. Hence find the set of values of x between $0°$ and $360°$ for which $\cos 3x > 4 \cos^2 x$.

11 Express $f(x) = 8 \cos x - 6 \sin x$ in the form $R \cos(x + \alpha)$, where R is positive and α is an acute angle. Hence:

(i) find the maximum and minimum values of $f(x)$
(ii) find the general solution of the equation $f(x) = 5$
(iii) sketch the graph of $y = f(x)$ for $-360° < x < 360°$.

12 Express $f(x) = 24 \cos x + 7 \sin x$ in the form $R \cos(x - \beta)$, where R is positive and β is an acute angle. Hence:

(i) show that $\{f(x)\}^2 \leq 625$ for all x

(ii) find the maximum and minimum values of $1/(f(x) + 30)$

(iii) find the general solution of the equation $f(x) = 10$.

13 (i) Given that $f(x) = 10/(5 + 2 \cos x - \sin x)$, show that the difference between the maximum and minimum values of $f(x)$ is $\sqrt{5}$.

(ii) Explain why the equation $4 \cos x - 2 \sin x = 5$ has no solutions.

(iii) Show that the maximum value of the function $g(x) = 4 \cos^2 x - \sin 2x$ is $\sqrt{5} + 2$.

14 Find the general solution of the equation $4 \cos x + \sin x = 1$ by expressing $4 \cos x + \sin x$:

(i) in terms of t, where $t = \tan \frac{1}{2}x$
(ii) in the form $R \cos(x - \beta)$, where R is positive and β is an acute angle.

15 Given that

$$f(x) = \sin x + \sin(x + \alpha) + \sin(x + 2\alpha)$$

(i) show that, when $\cos \alpha = -\frac{1}{2}$, $f(x) = 0$ for all x

(ii) when $\cos \alpha \neq \frac{1}{2}$ find, in terms of α, the general solution of the equation $f(x) = \frac{1}{2} + \cos \alpha$.

16 Show that

$$\cos(A+B)\cos(A-B) \equiv \cos^2 A - \sin^2 B$$

Hence:

(i) show that $\cos 15° \cos 105° = -\frac{1}{4}$
(ii) find the general solution of the equation $\cos^2 3x = \sin^2 x$.

17 Find the general solutions, in degrees, of the equations:

(i) $\sin 2x - \cos 2x = 0$
(ii) $\sin 2x - \cos 2x = 1$
(iii) $\sin x - \cos 2x = 1$
(iv) $\sin x - \cos 2x = \sin 3x$

18 Show that

$$(\cos A + \cos B)^2 + (\sin A + \sin B)^2 \equiv 4 \cos^2 \tfrac{1}{2}(A - B)$$

Hence:

(i) show that $\cos 15° = \frac{1}{4}(\sqrt{2} + \sqrt{6})$
(ii) find the general solution of the equation

$$(\cos x + \cos 3x)^2 + (\sin x + \sin 3x)^2 = 1$$

19 A runner in a cross-country race covers a triangular course between three checkpoints A, B and C. Starting from A he runs 4 km due north to B; from B he runs on a bearing of $300°$ (i.e. in the direction N60°W) to C; from C he runs on a bearing of $140°$ (i.e. in the direction S40°E) back to A. Find the length of his course.

Another runner, gambling on the fact that the race judge stationed at C will be asleep, cheats. He runs from A to B from B to D, the point halfway between B and C, from D to E, the point halfway between C and A, and from E back to A. Find the length of his course.

20 Every Friday at 5 p.m. a man is paid £200. Every Friday he goes out and spends it all. Every Saturday morning he comes home at 5 a.m. and apologizes to his wife, who forgives him and lends him £50, which he fritters gradually, with praiseworthy restraint, during the rest of the week.

t days after 5 p.m. on a certain Friday he has £x, where

$$x = \begin{cases} 200 \cos \pi t & 0 \leq t < \frac{1}{2} \\ 55 - 10t & \frac{1}{2} \leq t < \frac{11}{2} \\ 0 & \frac{11}{2} \leq t < 7 \end{cases}$$

Explain why x is a periodic function of t. Plot a graph of x against t, for $0 \leq t < 21$. At what times during the week does he have:

(i) £100 (ii) £20?

21 The chord PQ of a circle centre O divides the area of the circle in the ratio $1:3$. If $P\hat{O}Q = x$, show that $2x - 2\sin x = \pi$.

Show graphically that this equation has one root in the interval $0 < x < \pi$. Taking $x = 2$ as a first approximation to this root, use the Newton-Raphson method to find a second approximation, giving your answer to 2 dec. pls.

22 The line AB is a tangent to the circle centre O, touching the circle at B. The line OA cuts the circle at C, and $B\hat{O}C = x$. Given that the area of the sector OBC is half the area of $\triangle OAB$, show that $\tan x = 2x$. Show graphically that this equation has a root near $x = 1$. Taking $x = 1.2$ as a first approximation to this root, use

(i) the Newton-Raphson method
(ii) the iterative formula $x_{n+1} = \tan^{-1} 2x_n$

to obtain a further approximation. Give your answers to 2 dec. pls.

23 Show that, when x is small, $\sin x \approx x$ and $\cos 2x \approx 1 - 2x^2$. Hence find an approximate solution, near $x = 0$, of the equation

$$4\cos 2x - 4x \sin x = 47 \sin x$$

24 Find:

(i) $\lim\limits_{\theta \to 0} \dfrac{\sin 5\theta + \sin 7\theta}{\sin 5\theta + \sin \theta}$

(ii) $\lim\limits_{\theta \to 0} \dfrac{1 - \cos 2\theta + \theta \sin 3\theta}{\sin^2 \theta}$

(iii) $\lim\limits_{\theta \to 0} \dfrac{\sin(\frac{1}{3}\pi + 2\theta) - \sin \frac{1}{3}\pi}{\theta}$

(iv) $\lim\limits_{\theta \to 0} \dfrac{3\tan\theta + \sin\theta \sec 2\theta}{\theta}$

25 (i) If $y = e^{-\lambda x} \sin \mu x$, where λ and μ are constants, show that

$$\frac{d^2 y}{dx^2} + 2\lambda \frac{dy}{dx} + (\lambda^2 + \mu^2)y = 0$$

(ii) Given that $\sin x \cos y - \sin y \cos x = 1$, find dy/dx.

26 If $f(x) = 3\sin x / (2 - \cos x)$,

(i) find the greatest and least values of $f(x)$
(ii) find the values of x for which $f(x) = 1$

+ 27 The function $f:[0, 2\pi] \to \mathbb{R}$ is defined by

$$f(x) = \cos 2x - 2\sin x - 1$$

Find:

(i) the roots of the equation $f(x) = 0$
(ii) the coordinates of the turning points of the curve $y = f(x)$.

Hence find the number of real roots of the equation $\cos x \cos 2x - \sin 2x = \sin x + \cos x$ which lie in the interval $\frac{1}{2}\pi \leq x \leq \frac{3}{2}\pi$.

28 The depth x m of the water in a tidal river t hours after noon is given by $x = 8 + 4\sin \frac{1}{6}\pi t + 3\cos \frac{1}{6}\pi t$.

(i) Find the times of the high tides and the depth of the water at these times. Find the time which elapses between two high tides.

(ii) Find the rate of change in the depth of the water (a) at 3 p.m.; (b) when the depth of the water is 10 m. What is the maximum rate of change in the depth of the water?

29 A playboy sits by the edge of a circular lake of radius 2 km, sipping Diet Pepsi, trimming his chest hair and staring aimlessly through his binoculars. Suddenly he notices a stunningly beautiful girl apparently drowning on the other side of the lake. His amphibious motor boat can move at 50 m s^{-1} on land and 25 m s^{-1} on water. Find the least time he can take to reach the point exactly opposite him on the edge of the lake.

He arrives feeling seasick, to find that the girl has already swum to the shore and is in a meeting with the other editors of a respected feminist magazine ...

+ 30 Show graphically that

(i) the equation $\sin x = x$ has only one solution
(ii) the equation $\sin x = 1/x$ has an infinite number of solutions

Taking $x = 1.1$ as the first approximation to the smallest positive root of the equation $\sin x = 1/x$, find second and third approximations, using the Newton-Raphson method. Show that the use of the iterative formula $x_{n+1} = \sin^{-1} 1/x_n$ gives a less good second approximation. Explain this.

31 The function $f:(-\infty, \frac{1}{4}\pi) \to \mathbb{R}$ is defined by

$$f(x) = \begin{cases} 2x & x \leq 0 \\ \tan 2x & 0 < x < \frac{1}{4}\pi \end{cases}$$

(i) What is the range of f? Sketch the graph of f.

(ii) Find the inverse $g(x)$ of $f(x)$. What is the domain of $g(x)$? Sketch the graph of g.

(iii) Solve the equations
(a) $f'(x) = 4$; (b) $g'(x) = \frac{1}{4}$.

+ 32 The function $f:[0, \frac{1}{2}\pi] \to \mathbb{R}$ is defined by $f(x) = 1/(1 + \sin x)$.

(i) Find the range of $f(x)$.

(ii) Find $f^{-1}(x)$. What are the domain and range of $f^{-1}(x)$?

(iii) Sketch on the same axes the curves $y = f(x)$ and $y = f^{-1}(x)$.

(iv) Show that the equation $f(x) = x$ has a root near $x = 0.6$. Taking $x = 0.6$ as a first approximation to this root, find a better approximation.

4 Integration and differential equations

4.1 Integration

4.1:1 The reverse of differentiation

We know:

$$y = x^2 \Rightarrow \frac{dy}{dx} = 2x \qquad (1)$$

i.e. the gradient of the curve $y = x^2$ at the point (x, y) is $2x$.

Consider the question: 'Given that $dy/dx = 2x$, what is y?' i.e. given that the gradient of a curve at the point (x, y) is $2x$, what is the equation of the curve? Obviously $y = x^2$ is an answer, but so are $y = x^2 + 1$, $y = x^2 + 20$, $y = x^2 - 5$; in fact $y = x^2 + c$ where c can be any real number; i.e.

$$\frac{dy}{dx} = 2x \Rightarrow y = x^2 + c \qquad (2)$$

c is called an **arbitrary constant**.

The cases $c = -3, -1, +1$ are shown in fig. 4.1:**1**. The gradient of each of the curves at the point (x, y) is $2x$; e.g. the gradient of each of the curves at the point where $x = 1$ is 2. These curves are called **integral curves**.

We can find the exact value of c if we are given one extra piece of information, called an **initial condition**. For example, given that $x = 1$ when $y = -2$, we find from equation (2) that $-2 = (1)^2 + c$, i.e. $c = -3$. The solution is then $y = x^2 - 3$ (the lowest curve in fig. 4.1:**1**).

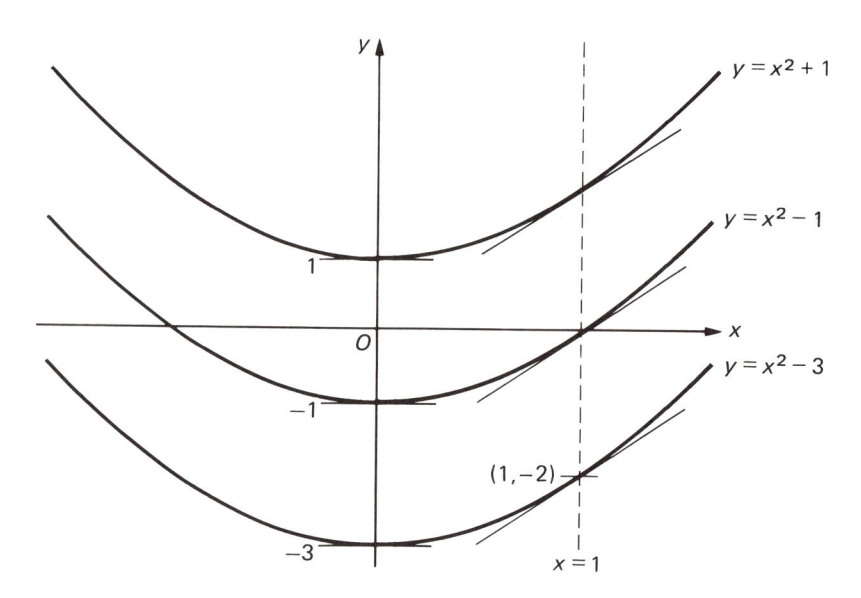

Fig. 4.1:1

Integration and differential equations 169

In general, given the gradient of a curve, can we find the equation of the curve?

Suppose $dy/dx = f(x)$; we write $y = \int f(x)\,dx$.

> **Definition** $\int f(x)\,dx$ is called the **integral of $f(x)$ with respect to x**.
> The process of finding $\int f(x)\,dx$ given $f(x)$ is called **integration with respect to x**; $f(x)$ is called the **integrand**.

(Note: the symbols $\int$ and dx are meaningless by themselves.)

The integral $\int f(x)\,dx$ is another function of x; call it $F(x)$. What do we know about $F(x)$? We know that $F'(x) = f(x)$, i.e.

$$F'(x) = f(x) \Leftrightarrow \int f(x)\,dx = F(x) + c \qquad (3)$$

e.g.
$$\frac{d}{dx}(\tfrac{1}{3}x^3) = x^2 \Leftrightarrow \int x^2\,dx = \tfrac{1}{3}x^3 + c$$

> So the problem of integrating a function $f(x)$ is one of finding a function $F(x)$ such that $F'(x) = f(x)$.

4.1:2 Basic integrals

Integral of x^α

Case 1: $\alpha \neq -1$

$$\frac{d}{dx}\left\{\frac{1}{\alpha+1}x^{\alpha+1}\right\} = \frac{1}{\alpha+1}\cdot(\alpha+1)x^\alpha = x^\alpha \qquad \text{(see 1.3:2)}$$

So by statement (3)

$$\int x^\alpha\,dx = \frac{1}{\alpha+1}x^{\alpha+1} + c \qquad (4)$$

e.g.
$$\int x^5\,dx = \tfrac{1}{6}x^6 + c$$

$$\int \frac{1}{\sqrt{x}}\,dx = \int x^{-1/2}\,dx = 2x^{1/2} + c$$

Note: when α is negative, the result (4) is true only when $x \neq 0$ (since the integrand is undefined when $x = 0$).

Case 2: $\alpha = -1$

If $x > 0$, $\qquad\qquad \dfrac{d}{dx}(\ln x) = \dfrac{1}{x} \qquad$ (from 2.5:4)

If $x < 0$, $\qquad\qquad \dfrac{d}{dx}(\ln(-x)) = \dfrac{1}{x} \qquad$ (from 2.5:4 Q11)

So by statement (3)

$$\int \frac{1}{x} dx = \begin{cases} \ln x + c & x > 0 \\ \ln(-x) + c & x < 0 \end{cases}$$

$$\int \frac{1}{x} dx = \ln |x| + c \quad (x \neq 0) \tag{5}$$

Integral of e^x

$$\frac{d}{dx}(e^x) = e^x \quad \text{(see 2.5:2)}$$

So by equation (3)

$$\int e^x \, dx = e^x + c \tag{6}$$

Integrals of $\sin x$ and $\cos x$

$$\frac{d}{dx}(\sin x) = \cos x \quad \text{(see 3.4:4)}$$

So by equation (3)

$$\int \cos x \, dx = \sin x + c \tag{7}$$

Moreover, $\quad \dfrac{d}{dx}(-\cos x) = \sin x \quad$ (see 3.4:4)

So by equation (3)

$$\int \sin x \, dx = -\cos x + c \tag{8}$$

Theorem 1:
(i) If a is a constant,

$$\int a f(x) \, dx = a \int f(x) \, dx \tag{9}$$

(ii) $\quad \displaystyle\int \{f(x) + g(x)\} \, dx = \int f(x) \, dx + \int g(x) \, dx \tag{10}$

Proof
This follows from 1.3:2 Theorem 1.

Worked example 1 Find $\int (8x^3 - 3x^2)\,dx$

$$\int (8x^3 - 3x^2)\,dx = \int 8x^3\,dx - \int 3x^2\,dx \qquad \text{by (10)}$$
$$= (2x^4 + c) - (x^3 + c') \qquad \text{by (9)}$$
$$= 2x^4 - x^3 + C$$

where $C = c - c'$; C is an arbitrary constant.

In section 4.2 we shall formulate some methods of integration; for the moment we shall content ourselves with integrating by inspection: to integrate a function $f(x)$ we shall look for a function $F(x)$ such that $F'(x) = f(x)$ (see statement (3)).

Worked example 2 Find: (i) $\int \sec^2 x\,dx$ (ii) $\int x\,e^{x^2}\,dx$.

(i) $\dfrac{d}{dx}(\tan x) = \sec^2 x \Rightarrow \int \sec^2 x\,dx = \tan x + c$

(ii) $\dfrac{d}{dx}(e^{x^2}) = 2x\,e^{x^2} \Rightarrow \int 2x\,e^{x^2}\,dx = e^{x^2}$

So $\int x\,e^{x^2}\,dx = \tfrac{1}{2}(e^{x^2} + c)$ by (9)

Exercise 4.1:2

1 (i) Find:

(a) $\int x^7\,dx$

→ (b) $\int \dfrac{1}{x^4}\,dx$

(c) $\int \sqrt{x}\,dx$

(d) $\int (x\sqrt{x})\,dx$

(e) $\int \dfrac{1}{x^2\sqrt{x}}\,dx$

(v) $\int \dfrac{5u + 4}{u^3}\,du$

(vi) $\int \dfrac{7 - x^2}{x^2}\,dx$

(vii) $\int (\sqrt{p} + 1)\,dp$

(viii) $\int \dfrac{x + 1}{\sqrt{x}}\,dx$

(ix) $\int \dfrac{2\sqrt{s} - 3}{s^2}\,ds$

→ (ii) Explain why equation (4) is invalid when $\alpha = -1$.

① **2** Find:

(i) $\int (4x^2 - 5x^3)\,dx$

(ii) $\int t^2(t - 1)\,dt$

→ (iii) $\int (2x - 1)(3x + 2)\,dx$

(iv) $\int (x^2 + 5)^2\,dx$

① → **3** Integrate with respect to x:

(i) $ax + b$ (ii) $ax^2 + bx + c$
(iii) $a\sqrt{x} + b$ (iv) $(ax + b)/\sqrt{x}$

where a and b are constants.

① **4** Find y in terms of x when:

(i) $dy/dx = 1/x^2$ and $y = 3$ when $x = 1$
(ii) $dy/dx = 5x\sqrt{x}$ and $y = 12$ when $x = 2$
(iii) $dy/dx = 3x^2 + 4x^3$ and $y = 0$ when $x = 1$

① → **5** If $dy/dx = 3x^2$, show that $y = x^3 + c$.
• Sketch the curves $y = x^3 + c$ in the cases $c = -2, -1, 0, 1, 2$. Draw the tangent to each curve at the point where $x = 1$.

172 *Basic integrals*

① → **6** If $dy/dx = -4x$, find y given that:
   (i) $y = 3$ when $x = 0$
   (ii) $y = -1$ when $x = 1$
   (iii) $y = -2$ when $x = 1$

   On the same axes sketch the graphs of y against x in these three cases and draw the tangents to the curves at the points where $x = 1$ and $x = 3$.

① **7** Find y in terms of x if $dy/dx = 2x + 3$ and $y = 8$ when $x = 2$.
   Sketch the graph of y against x and find the minimum value of y.

① → **8** The gradient of a curve at the point (x, y) is $4x + 1/x^2$ and the curve passes through the point $(1, 1)$. Find the equation of the curve.

① R **9** The gradient of a curve at the point (x, y) is $3x^2 - 4x + 1$ and the curve passes through the point $(2, 3)$.
   (i) Show that the curve passes through the point $A(-1, -3)$ and find the equation of the tangent at A.
   (ii) Find the maximum and minimum values of y.
   (iii) Sketch the curve.

① **10** $f'(x) = \begin{cases} 2 & x < 1 \\ 2x & x \geq 1 \end{cases}$
   Given that $f(-1) = -4$ and $f(3) = 8$, find f and show that f is continuous.

① **11** $f'(x) = \begin{cases} 6 & x \leq -2 \\ 2 - 2x & -2 < x < 4 \\ 6 & x \geq 4 \end{cases}$
   Given that $f(-3) = -6$, $f(2) = 8$, $f(5) = 6$, find f.
   Write down $f(-4)$, $f(0)$, $f(4)$, $f(8)$.
   Sketch a graph of f. Is f continuous?

② **12** (i) Differentiate $(3x+5)^4$. Hence integrate $(3x+5)^3$.
   → (ii) Differentiate $\sqrt{(1-x^2)}$. Hence integrate $x/\sqrt{(1-x^2)}$.
   (iii) Differentiate $(4+x^2)^{2/3}$. Hence integrate $x\sqrt[3]{(4+x^2)}$.

② **13** (i) Differentiate $1/(x-1)$. Hence integrate $1/(x-1)^2$.
   (ii) Show that $(x^2-2x)/(x-1)^2 \equiv 1 - 1/(x-1)^2$. Hence integrate $(x^2-2x)/(x-1)^2$.

② **14** Differentiate $1/(2x+1)$. Hence integrate $1/(2x+1)^2$. Hence integrate $(x^2+x)/(2x+1)^2$ (c.f. question 13).

② → **15** Differentiate e^{ax}, a constant. Hence find $\int e^{ax} dx$.
   • Write down the integrals of:
   (i) e^{5x} (ii) $e^{\frac{3}{2}x}$ (iii) e^{-x} (iv) $e^{-\frac{4}{5}x}$

② → **16** (i) Differentiate e^{x^3}. Hence integrate $x^2 e^{x^3}$.
   (ii) Differentiate $e^{\sin x}$. Hence integrate $\cos x \, e^{\sin x}$.

② **17** Differentiate $x e^x$. Hence find $\int x e^x dx$.

② **18** Find y in terms of x when:
   (i) $dy/dx = e^{4x}$ and $y = 1$ when $x = 0$
   → (ii) $dy/dx = e^{-x}$ and $y = 0$ when $x = \ln 2$
   (iii) $dy/dx = e^{\frac{1}{2}x}$ and $y = 6$ when $x = \ln 9$

② → **19** (i) Differentiate a^x, a constant (see section 2.5:5). Hence find $\int a^x dx$.
   (ii) Write down the integrals of
   (a) 10^x; (b) 3^{2x}.
   (iii) Differentiate 5^{x^2}. Hence find $\int x 5^{x^2} dx$.

② **20** (i) Differentiate $\cos 2x$; hence find $\int \sin 2x \, dx$.
   (ii) Differentiate $\sin 5x$; hence find $\int \cos 5x \, dx$.
   Find $\int \cos ax \, dx$ and $\int \sin ax \, dx$, a constant. Hence write down the integrals of:
   (a) $\cos \frac{1}{2}x$; (b) $\sin \frac{5}{3}x$.

② **21** Differentiate $\tan x$; hence find $\int \sec^2 x \, dx$. Hence, using the identity $1 + \tan^2 x \equiv \sec^2 x$, find $\int \tan^2 x \, dx$. Similarly find $\int \csc^2 x \, dx$ and $\int \cot^2 x \, dx$.

② **22** (i) Differentiate $\sin^4 x$; hence find $\int \sin^3 x \cos x \, dx$.
   (ii) Differentiate $\cos^3 x$; hence find $\int \cos^2 x \sin x \, dx$.
   → (iii) Differentiate $\tan^4 x$; hence find $\int \tan^3 x \sec^2 x \, dx$.

23 If $f(x) = x^4$, $g(x) = x^3$, find:
   (i) $\int f(x) \, dx$
   (ii) $\int g(x) \, dx$
   (iii) $\int f(x) g(x) \, dx$
   (iv) $\int \frac{f(x)}{g(x)} \, dx$

   Repeat the exercise when $f(x) = x + 1$, $g(x) = \sqrt{x}$. Investigate the following statements:
   (a) $\int f(x) g(x) \, dx = \left\{ \int f(x) \, dx \right\} \left\{ \int g(x) \, dx \right\}$
   (b) $\int \frac{f(x)}{g(x)} \, dx = \frac{\int f(x) \, dx}{\int g(x) \, dx}$
   (c) $\int \frac{f(x)}{g(x)} \, dx = \frac{1}{\int \frac{g(x)}{f(x)} \, dx}$

24 Consider the following:

(i) $\int x\, e^{2x}\, dx = \tfrac{1}{2}x^2 \cdot \tfrac{1}{2} e^{2x} + c$
$\phantom{(i) \int x\, e^{2x}\, dx} = \tfrac{1}{4} x^2 e^{2x} + c$

(ii) $\int \tan x\, dx = \int \dfrac{\sin x}{\cos x}\, dx = -\dfrac{\cos x}{\sin x} + c$
$ = -\cot x + c$

(iii) $\int x\, e^{x^2}\, dx = \tfrac{1}{2} e^{x^2} \Rightarrow \int e^{x^2}\, dx = \dfrac{\tfrac{1}{2} e^{x^2} + c}{2x}$

In each case, check by differentiation that the solution is incorrect and write down the 'non-rule' that has been used to obtain the solution. Make sure you never use these non-rules.

4.1:3 Area under a curve

The integral $\int f(x)\, dx$ is called the **indefinite integral** of $f(x)$ with respect to x.

Suppose $\int f(x)\, dx = F(x) + c$. Then the **definite integral** of $f(x)$ between the limits $x = a$ and $x = b$ is given by

$$\int_a^b f(x)\, dx = [F(x)]_a^b = F(b) - F(a) \qquad (11)$$

e.g. $\displaystyle\int_2^3 x^2\, dx = [\tfrac{1}{3} x^3]_2^3 = \tfrac{27}{3} - \tfrac{8}{3} = \tfrac{19}{3}$

Theorem 2: The area of the region bounded by the finite curve $y = f(x)$, the lines $x = a$ and $x = b$ and the x-axis is

$$\int_a^b f(x)\, dx \qquad (12)$$

(See questions 29 and 30 for some problem cases.)

Note that we often write more simply 'the area bounded by the curve $y = f(x)$, the lines $x = a$ and $x = b$ and the x-axis', or even 'the area under the curve $y = f(x)$ between $x = a$ and $x = b$'.

Proof

Let $A(x)$ be the area of the shaded region in fig. 4.1:2. Note that $A(a) = 0$. The required area is $A(b)$. As x increases, so does $A(x)$.

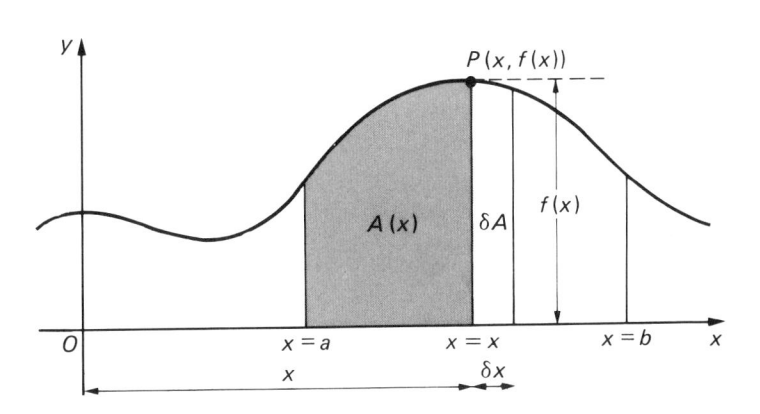

Fig. 4.1:2

174 *Area under a curve*

Consider a small increase δx in x and a corresponding increase δA in the area. Then, since the extra region is approximately a rectangle, height $f(x)$, width δx,

$$\delta A \approx f(x)\delta x \quad \text{i.e.} \quad \frac{\delta A}{\delta x} \approx f(x)$$

In the limit, as $\delta x \to 0$, the approximation gets better and better and we can write

$$\frac{dA}{dx} = f(x) \qquad \text{(see 1.3:1)}$$

So by 4.1:1 statement (3),

$$A(x) = \int f(x)\, dx = F(x) + c \qquad (13)$$

To find c, put $x = a$ in equation (13): $A(a) = F(a) + c$; $A(a) = 0$ so $c = -F(a)$.
So required area $A(b) = F(b) + c = F(b) - F(a) = \int_a^b f(x)\, dx$.

Worked example 3 Find the area of the region bounded by the curve $y = x^2$, the line $x = 2$ and the x-axis (fig. 4.1:3).

The region is bounded by $x = 2$ on the right and by $x = 0$ on the left; so the limits of integration are 0 and 2.

$$\text{Area} = \int_0^2 x^2\, dx = \left[\frac{1}{3}x^3\right]_0^2 = \frac{8}{3}$$

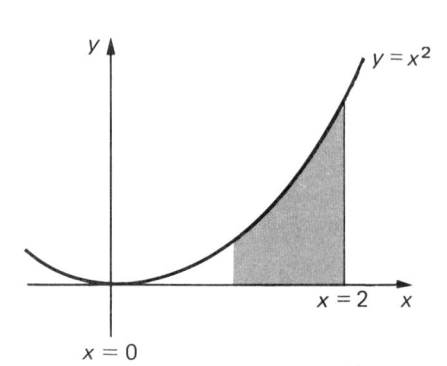

Worked example 4 Find the area of the region bounded by the curve $y = x^3$, the lines $x = -1$, $x = 2$ and the x-axis (fig. 4.1:4).

Area A_1 is bounded by the curve, the line $x = -1$ and the x-axis.

$$A_1 = \int_{-1}^0 x^3\, dx = \left[\frac{1}{4}x^4\right]_{-1}^0 = -\frac{1}{4}$$

(Note: areas below the x-axis are negative.)
Area A_2 is bounded by the curve, the line $x = 2$ and the x-axis.

$$A_2 = \int_0^2 x^3\, dx = \left[\frac{1}{4}x^4\right]_0^2 = 4$$

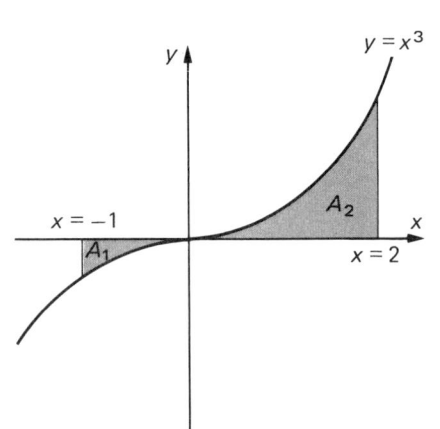

The **algebraic area** bounded by the curve, the lines $x = -1$, $x = 2$ and the x-axis is $A_1 + A_2$.

$$A_1 + A_2 = -\tfrac{1}{4} + 4 = \tfrac{15}{4}$$

Fig. 4.1:4   The **numerical area** bounded by the curve, the lines $x = -1$, $x = 2$ and the x-axis is $|A_1| + |A_2|$.

$$|A_1| + |A_2| = \tfrac{1}{4} + 4 = \tfrac{17}{4}$$

(Note: we can find the algebraic area simply by evaluating $\int_{-1}^2 x^3\, dx$; to find the numerical area we need to find A_1 and A_2 separately.)

Fig. 4.1:3

Area bounded by two curves

Suppose that we want to find the area of the region bounded by the curves $y = f(x)$ and $y = g(x)$ (fig. 4.1:5) and suppose that the curves intersect at the points A and B where $x = a$, $x = b$.

Area between $y = f(x)$ and $y = g(x)$

$= \{\text{area under } y = f(x) \text{ between } x = a \text{ and } x = b\}$

$\quad - \{\text{area under } y = g(x) \text{ between } x = a \text{ and } x = b\}$

$= \int_a^b f(x) \, dx - \int_a^b g(x) \, dx$

$= \int_a^b \{f(x) - g(x)\} \, dx \qquad (14)$

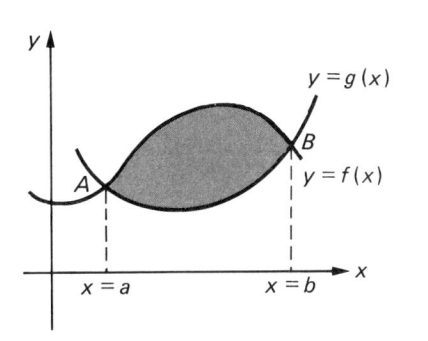

Fig. 4.1:5

by 2.2:2 equation (5).

Worked example 5 Find the area of the finite region enclosed by the curve $y = 4 - x^2$ and the line $y = 3x$ (fig. 4.1:6).

The two curves intersect at A and B where $y = 4 - x^2$ and $y = 3x$.

So $\quad 4 - x^2 = 3x \Rightarrow x^2 + 3x - 4 = 0 \Rightarrow (x-1)(x+4) = 0$

The curves intersect where $x = -4$ or $x = 1$. So the required area is

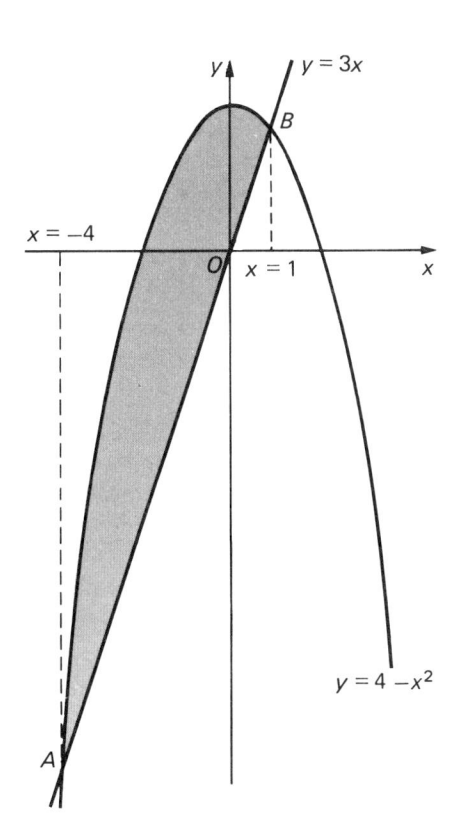

$$\int_{-4}^{1} \{4 - x^2 - 3x\} \, dx = [4x - \tfrac{1}{3}x^3 - \tfrac{3}{2}x^2]_{-4}^{1} = \tfrac{125}{6}$$

Fig. 4.1:6

Exercise 4.1:3

1 Evaluate:

(i) $\int_1^2 2x\,dx$
(ii) $\int_{-1}^3 x(x+1)\,dx$
(iii) $\int_2^4 \sqrt{x}\,dx$
(iv) $\int_{-2}^3 (x^3-x)\,dx$
(v) $\int_2^5 3x\sqrt{x}\,dx$

2 Evaluate:

(i) $\int_1^4 \frac{1}{x}\,dx$
(ii) $\int_{1/2}^2 \frac{1}{x}\,dx$
(iii) $\int_{-3}^{-2} \frac{1}{x}\,dx$
(iv) $\int_e^{e^2} \frac{1}{x}\,dx$
(v) $\int_1^2 \frac{x+1}{x}\,dx$

3 Evaluate:

(i) $\int_0^1 e^x\,dx$
(ii) $\int_{\ln 2}^{\ln 3} e^x\,dx$
(iii) $\int_{-\ln(1+\sqrt{2})}^{\ln(1+\sqrt{2})} e^x\,dx$

4 Evaluate:

(i) $\int_{\frac{1}{6}\pi}^{\frac{1}{2}\pi} \cos x\,dx$
(ii) $\int_0^{\pi} \sin x\,dx$
(iii) $\int_{\frac{1}{4}\pi}^{\frac{2}{3}\pi} \sin x\,dx$
(iv) $\int_{-\frac{1}{6}\pi}^{\frac{1}{6}\pi} (\cos x - \sin x)\,dx$

③ **5** Find the area of the region bounded by:

(i) $y = x^3$; $x = 1$, $x = 2$, x-axis
→ (ii) $y = 4/x^2$; $x = 2$, $x = 3$, x-axis
(iii) $y = x\sqrt{x}$; $x = 1$, $x = 4$, x-axis
(iv) $y = x^2 + 1$; y-axis, $x = 1$, x-axis
(v) $y = x^2$; $x = 3$, x-axis
(vi) $y = 1/x$; $x = 2$, $x = 6$, x-axis
(vii) $y = 4/x$; $x = 1$, $x = e$, x-axis
(viii) $y = e^x$; y-axis, $x = \ln 2$, x-axis
(ix) $y = \sin x$; $x = \frac{1}{3}\pi$, $x = \frac{2}{3}\pi$, x-axis
(x) $y = \cos x$; y-axis, $x = \frac{1}{4}\pi$, x-axis

③ → **6** Sketch the curve $y = (x+1)(3-x)$. Find the area of the region bounded by the curve and the x-axis.

③ **7** Find the area of the region bounded by each of the following curves and the x-axis:

(i) $y = 4 - x^2$
R (ii) $y = x^2(3-x)$

③ **8** Find the area of the region bounded by each of the following curves and the x-axis:

(i) $y = x(3-x)$
(ii) $y = (x+2)(1-x)$

Explain your answers.

③ **9** Find the area of the region bounded by the curve $x(2-x)$ and the x-axis.
Show that the line $x = 1$ divides the region into two parts with equal areas, and that the line $x = \frac{1}{2}$ divides the region into two parts whose areas are in the ratio 5:27.

③ **10** (i) Sketch the curve $y = f_1(x)$ where

$$f_1(x) = \begin{cases} x+2 & x \leq 1 \\ x(4-x) & x > 1 \end{cases}$$

Find the area of the region bounded by the curve and the x-axis.

(ii) Sketch the curve $y = f_2(x)$ where

$$f_2(x) = \begin{cases} 3-x & x < 0 \\ (x+1)(3-x) & 0 \leq x \leq 1 \\ 4x & x > 1 \end{cases}$$

Find the area of the region bounded by the curve, the lines $x = -1$, $x = 2$ and the x-axis.

11 Show that $|x-4| = \begin{cases} x-4 & x \geq 4 \\ 4-x & x < 4 \end{cases}$ (see 2.4:1).

Hence find $\int_2^7 |x-4|\,dx$.

12 Using the method of question 11, find:

(i) $\int_{-2}^1 |x|\,dx$
(ii) $\int_0^1 |2x-1|\,dx$
(iii) $\int_0^b |x-a|\,dx$ where a and b are constants with $0 < a < b$.

13 Find:

(i) $\int_0^3 |x^2 - 5x + 4|\,dx$
(ii) $\int_b^d |(x-a)(x-c)|\,dx$

where a, b, c and d are constants with $0 < a < b < c < d$.

③ **14** Find the area bounded by the line $y = -\frac{1}{2}x + 6$, the x-axis and the y-axis:

(i) by integration
(ii) without using integration

• **15** Sketch the curve $y = x(x-4)$. Show that area of the region bounded by the curve and the x-axis is $-\frac{32}{3}$. Explain this answer.

16 Explain, with reference to the proof of Theorem 2, why you would always expect the area of a region below the x-axis to be negative.

17 Find the areas of the regions bounded by:
  (i) $y = x^3$; $x = -2$, $x = -1$, x-axis
  (ii) $y = x^2 - 4$; $x = -1$, $x = 1$, x-axis
→ (iii) $y = x(1-x)$; $x = 3$, x-axis
  (iv) $y = -3\sqrt{x}$; $x = 1$, $x = 9$, x-axis
  (v) $y = \cos x$, $x = \frac{5}{6}\pi$, $x = \pi$, x-axis

18 Find the areas of the regions bounded by each of the following curves and the x-axis:
  (i) $y = 4x^2 - 1$ (ii) $y = 1 - 4x^2$

Explain your answers.

19 Find the areas of the regions bounded by:
  (i) $y = (x+3)(x-1)$; x-axis
R (ii) $y = x(2-x)$; x-axis, $x = 3$
→ (iii) $y = x(x-4)$; $x = 2$, x-axis between $x = 2$ and $x = 4$
  (iv) $y = x(x-4)$; $x = 6$, x-axis

20 A function f is defined by:
$$f(x) = 5 - 2x^2 \quad -1 \leq x < 1$$
and f is periodic with period 2.
Sketch a graph of f in the interval $-3 \leq x < 7$. Find:
  (i) $\int_{-1}^{1} f(x)\, dx$
  (ii) $\int_{-1}^{2} f(x)\, dx$
  (iii) $\int_{-2}^{7} f(x)\, dx$
  (iv) $\int_{-3/2}^{3/2} f(x)\, dx$

21 A function f is defined by:
$$f(x) = \begin{cases} x^2 & 0 \leq x < 2 \\ 8 - 2x & 2 \leq x < 4 \end{cases}$$
and f is periodic with period 4.
Sketch a graph of f in the interval $0 \leq x < 12$. Find:
  (i) $\int_{0}^{4} f(x)\, dx$
  (ii) $\int_{0}^{6} f(x)\, dx$
  (iii) $\int_{1}^{11} f(x)\, dx$

R **22** A function f is defined by:
$$f(x) = \begin{cases} 1 + 2x/\pi & 0 \leq x < \frac{1}{2}\pi \\ 2 + \cos x & \frac{1}{2}\pi \leq x < \pi \end{cases}$$

and f is periodic with period π. Find:
  (i) $\int_{\frac{1}{3}\pi}^{\frac{3}{2}\pi} f(x)\, dx$
  (ii) $\int_{\frac{5}{6}\pi}^{2\pi} f(x)\, dx$

④ **23** Find:
  (i) $\int_{0}^{3} x^3\, dx$
  (ii) $\int_{-2}^{0} x^3\, dx$
  (iii) $\int_{-2}^{3} x^3\, dx$

What is the (a) algebraic, (b) numerical area of the region bounded by $y = x^3$, $x = -2$, $x = 3$ and the x-axis?

④ → **24** Find:
  (i) $\int_{1}^{2} x(x-2)\, dx$
  (ii) $\int_{1}^{3} x(x-2)\, dx$
  (iii) $\int_{2}^{3} x(x-2)\, dx$

What is the (a) algebraic, (b) numerical area of the region bounded by $y = x(x-2)$, $x = 1$, $x = 3$ and the x-axis?

④ **25** Find the numerical area of the region bounded by:
  (i) $y = x^3$; $x = -1$, $x = 4$, x-axis
  (ii) $y = 9 - x^2$; $x = 1$, $x = 5$, x-axis
  (iii) $y = x(x-3)$; $x = -1$, $x = 4$, x-axis
R (iv) $y = x(x^2 - 4)$; x-axis
  (v) $y = x(x+1)(x-2)$; x-axis
  (vi) $y = x(x-1)(x-2)$; x-axis
  (vii) $y = \sin x$; $x = -\frac{1}{3}\pi$, $x = \frac{1}{2}\pi$, x-axis
  (viii) $y = \cos x$; y-axis, $x = \pi$, x-axis

④ **26** Show that the *numerical* area of the region under the curve $y = f(x)$ between $x = a$ and $x = b$ is equal to the area under the curve $y = |f(x)|$ between $x = a$ and $x = b$. (See 2.4:1 Q13–Q17.)

Under what circumstances is the numerical area of a region under a curve different from the algebraic area?

④ → **27** Find the area bounded by the curve $y = 2\sqrt{x}$, the lines $x = 1$, $x = 9$ and the x-axis.

Hence find the (numerical) area bounded by the curve $y^2 = 4x$ and the lines $x = 1$ and $x = 9$.

④ **28** Find the area bounded by the curve $y^2 = 6x$ and the line $x = 6$.

Area under a curve

29 (i) Try to evaluate $\int_{-1}^{2} 1/x^2 \, dx$.

(ii) Sketch the curve $y = 1/x^2$. Explain, with reference to your sketch, why the 'integral' $\int_{-1}^{2} 1/x^2 \, dx$ is meaningless.

Note: the fact that we can write the symbols $\int_{-1}^{2} 1/x^2 \, dx$ does not necessarily imply that they have a meaning, just as the fact that we can write the sentence 'Bananas hurt inspiration guitars kangaroo kangaroo' does not necessarily imply that it has a meaning.

(iii) Show that if $0 < \alpha < 2$,

$$f(\alpha) = \int_{\alpha}^{2} \frac{1}{x^2} \, dx = \frac{2 - \alpha}{2\alpha}$$

Show that as $\alpha \to 0$, $f(\alpha) \to \infty$.

30 (i) Sketch the curve $y = 1/x$. Discuss the integrals:

(a) $\int_{-1}^{2} \frac{1}{x} \, dx$ (b) $\int_{-2}^{2} \frac{1}{x} \, dx$.

(ii) Show that if $0 < \alpha < 2$,

$$f(\alpha) = \int_{\alpha}^{2} \frac{1}{x} \, dx = \ln \frac{2}{\alpha}$$

Show that as $\alpha \to 0$, $f(\alpha) \to \infty$.

31 Sketch the curve $y = 1/\sqrt{x}$ $(x > 0)$.

Show that if

$$f(\alpha) = \int_{\alpha}^{2} \frac{1}{\sqrt{x}} \, dx \quad (0 < \alpha < 2)$$

$f(\alpha) \to 2\sqrt{2}$ as $\alpha \to 0$.

Discuss the integral $\int_{-2}^{2} \frac{1}{\sqrt{x}} \, dx$.

32 Let $f(\alpha) = \int_{1}^{\alpha} \frac{1}{x^2} \, dx$, $\alpha \in \mathbb{R}$.

What happens to $f(\alpha)$ as $\alpha \to \infty$?

Let $g(\alpha) = \int_{1/\alpha}^{1} \frac{1}{x^2} \, dx$.

What happens to $g(\alpha)$ as $\alpha \to \infty$?

Illustrate your results on a sketch of the graph of $1/x^2$.

33 (i) Show that if $f(x)$ is an even function, then for any real number a,

$$\int_{-a}^{a} f(x) \, dx = 2 \int_{0}^{a} f(x) \, dx$$

What can you say about $\int_{-a}^{a} f(x) \, dx$ if $f(x)$ is an odd function? (See 1.2:4 Q21).

(ii) Find:

(a) $\int_{-1}^{1} x^3 \, dx$; (b) $\int_{-2}^{2} (7x^5 - 5x^3 + x) \, dx$

(iii) Find:

(a) $\int_{-a}^{a} \sin x \, dx$; (b) $\int_{0}^{a} \cos x \, dx$;

(c) $\int_{-a}^{a} \cos x \, dx$.

Explain your results.

34 (i) Show graphically that, if a, b and c are real numbers and $a < c < b$,

$$\int_{a}^{c} f(x) \, dx + \int_{c}^{b} f(x) \, dx = \int_{a}^{b} f(x) \, dx$$

(ii) Show that, if a and b are real numbers,

$$\int_{a}^{b} f(x) \, dx = -\int_{b}^{a} f(x) \, dx$$

(Hint: express both integrals in terms of $F(a)$ and $F(b)$.)

(iii) Express as single integrals:

(a) $\int_{-1}^{2} x^3 \, dx + \int_{2}^{3} x^3 \, dx$

(b) $\int_{0}^{1} x^3 \, dx - \int_{2}^{1} x^3 \, dx$

35 (i) Evaluate:

(a) $\int_{1}^{2} (x^3 + 3x) \, dx$

(b) $\int_{1}^{2} (t^3 + 3t) \, dt$

(c) $\int_{1}^{2} (\phi^3 + 3\phi) \, d\phi$

(ii) Find:

$$\int_{-2}^{0} (x^3 + 3x) \, dx + \int_{0}^{2} (t^3 + 3t) \, dt$$

(iii) Show that

$$\int_{a}^{b} f(x) \, dx = \int_{a}^{b} f(t) \, dt$$

(For this reason the variable x (or t or whatever) is often called the **dummy variable** in the definite integral.)

(iv) Is it true to say that

$$\int f(x) \, dx = \int f(t) \, dt?$$

36 Find the area of the region bounded by $y = 4 + x^2$, y-axis, x-axis and $x = 1$.

Hence find the area of the region bounded by $y = 4 + x^2$ and $y = 5$.

37 Find the points of intersection of the curve $y = (x+2)(4-x)$ and the line $y = 5$. Hence find the area of the (finite) region bounded by the curve and the line.

Find the area of the (finite) region bounded by the curve $y = 4 - x^2$ and the line $y = 3$.

38 Find the areas of the regions bounded by:

(i) $y = x(5-x)$ and $y = 6$
(ii) $y = (x-2)(3-x)$ and x-axis

Explain your answers.

39 Show that the area of the region bounded by the curve $y = (x-1)^2(4-x)$, the x-axis and the y-axis is equal to the area of the region bounded by the curve, the line $x = 4$ and the line $y = 4$.

40 Find the point at which the tangent to the curve $y = 3x^2$ at the point $(1, 3)$ cuts the x-axis. Hence find the area bounded by the curve, the tangent and the x-axis.

41 Find the area bounded by the curve $y = x(2-x)$ and the tangents to the curve at the points $(1, 1)$ and $(2, 0)$.

42 For each of the following curves find the area bounded by the curve, and the tangents at the points where the curve cuts the x-axis:

(i) $y = (x+2)(1-x)$
(ii) $y = 6x - x^2$

→ **43** The tangents at the points A and B, where the curve $y = (x-1)(x-3)$ cuts the x-axis, meet at T. Show that the curve divides $\triangle ABT$ into two parts whose areas are in the ratio $2:1$.

R 44 The tangents at the points A and B, where the curve $y = (x-\alpha)(x-\beta)$, $\alpha, \beta \in \mathbb{R}$, cuts the x-axis, meet at T. Show that the curve divides $\triangle ABT$ into two parts whose areas are in the ratio $2:1$.

45 Sketch the curve $y = \sin x$ and the tangents to the curve at the origin and the point $(\pi, 0)$. Find the area enclosed by the curve and these tangents.

46 The tangents to the curve $y = \cos x$ at $A(0, 1)$ and $B(\frac{1}{2}\pi, 0)$ meet at C. Find the coordinates of C. Hence find the area of the region enclosed by the lines AC, CB and the arc BA of the curve.

• **47** Suppose that we want to find the area of the region bounded by the curves $y = f(x)$ and $y = g(x)$ shown in the diagram. (Note: $g(x)$ is negative for some values of x between $x = a$ and $x = b$.)

Write down $\int_a^b f(x)\,dx$ and $\int_a^b g(x)\,dx$ in terms of A_1, A_2, A_3 and A_4. Hence show that

$$\int_a^b \{f(x) - g(x)\}\,dx = |A_2| + |A_3|$$

i.e. that the formula $\int_a^b \{f(x) - g(x)\}\,dx$ gives the *numerical area* of the region between the curves.

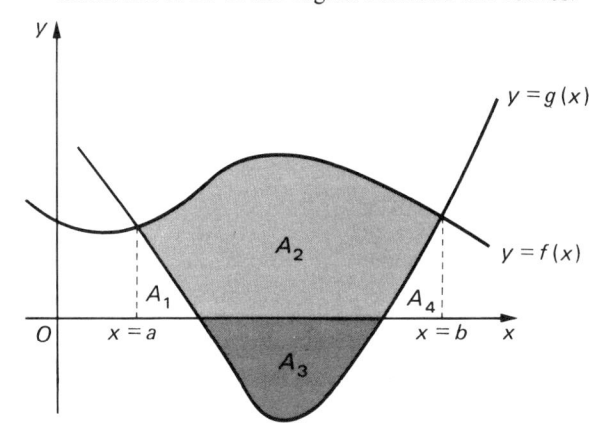

⑤ **48** Find the area of the finite region bounded by:

(i) $y = x(2-x)$ and $y = \frac{3}{2}x$
(ii) $y = x(4-3x)$ and $y = x$
→ (iii) $y = (3-x)(1+x)$ and $y = 1+x$
R (iv) $y = (x-1)(x-5)$ and $y = x-1$
(v) $y = (x+1)/x$ and $y = -2x+4$

⑤ **49** Find the area bounded by the lines $y = \frac{1}{2}x - 2$, $y = x+1$, $x = 6$:

(i) by integration
(ii) without using integration

⑤ **50** Find the area of the finite region bounded by the curves:

⑤ → (i) $y = x(3-x)$ and $y = x(2x+1)$
(ii) $y = x^2$ and $y = x^3 - x^2$
(iii) $y = 1/x^2$ and $y = 5 - 4x^2$
(iv) $y = x^2$ and $y^2 = 8x$

→ **51** Find the area of the finite region bounded by $y = x^2$ and:

(i) $y = 2$ (ii) $y = x$
(iii) $y = x+2$ (iv) $y^2 = x$
(v) $y = 8 - x^2$

⑤ **52** Find the area of the finite region bounded by:

(i) $y^2 = 4x$ and $y = \frac{1}{2}x + \frac{3}{2}$
(ii) $y^2 = x$ and $y = x - 2$
(iii) $y^2 = 6x$ and $y = -2x + 6$

⑤ **53** Farmer A owns a plot of land bounded by the x-axis, the line $x = 9$ and the line $y = \frac{4}{3}x$ (1 unit = 1000 m). Farmer B owns the neighbouring plot, bounded by the y-axis, the line $y = 12$ and the line $y = \frac{4}{3}x$. Relations between the two men are strained.

One night, farmer A moves the fence separating the farms from $y = \frac{4}{3}x$ to $y = 4\sqrt{x}$. Find the area of farmland he gains.

The next night, farmer B moves the fence to $y = \frac{4}{27}x^2$. What is the area of farmer B's farm now?

The night after, farmer A goes up in a crop-sprayer helicopter and sprays farmer B's crops with concentrated hydrochloric acid...

4.1:4 Integral as the limit of a sum; volume of a solid of revolution

We have seen in section 4.1:3 that the area A of the region bounded by the curve $y = f(x)$, the lines $x = a$ and $x = b$ and the x-axis is given by

$$A = \int_a^b f(x)\,dx$$

Consider an easier way of looking at the area under the curve. The region is made up of rectangles like the shaded ones in fig. 4.1:7.

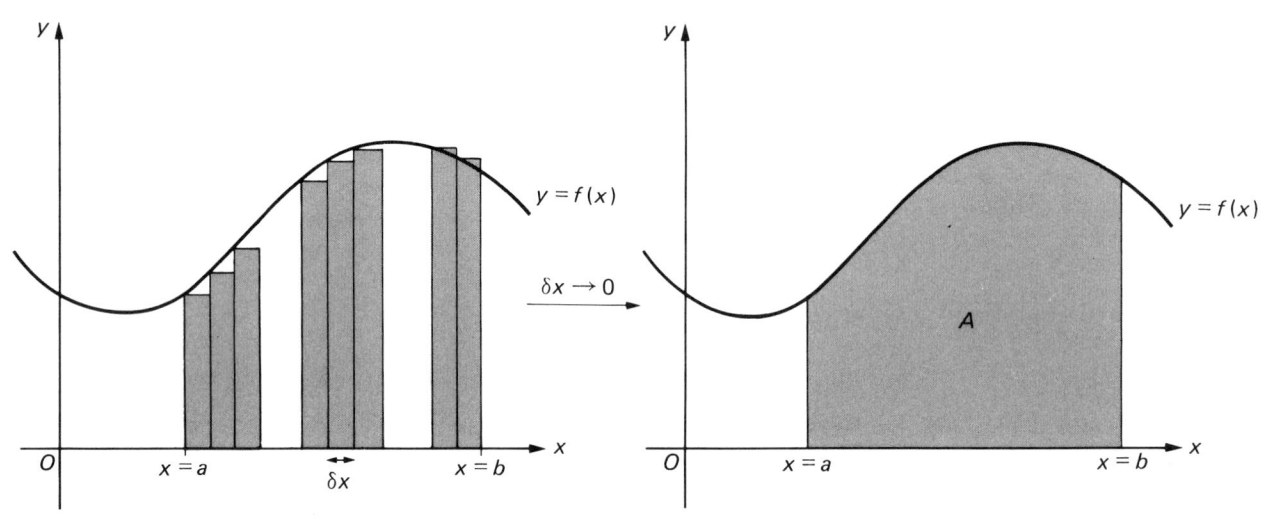

Fig. 4.1:7

Each rectangle has height $f(x)$ and width δx and thus has area $f(x)\delta x$. So

$$A \approx \sum_{x=a}^{x=b} f(x)\delta x$$

(Note: the Greek letter $\sum$ (sigma) stands for 'sum'.)

As $\delta x \to 0$ the approximation gets better and better, i.e.

$$A = \lim_{\delta x \to 0} \sum_{x=a}^{x=b} f(x)\delta x$$

Thus

$$\int_a^b f(x)\,dx = \lim_{\delta x \to 0} \sum_{x=a}^{x=b} f(x)\delta x \tag{15}$$

(Note: the integral symbol $\int$ is an elongated 'S' (for 'sum').)

Theorem 3: The volume V of the solid of revolution formed when the region bounded by the curve $y = f(x)$, the lines $x = a$ and $x = b$ and the x-axis is rotated completely about the x-axis is given by

$$V = \pi \int_a^b \{f(x)\}^2\,dx \tag{16}$$

Integration and differential equations 181

Proof
We divide the solid into elements (i.e. small sections) like the ones shown in fig. 4.1:**8**.

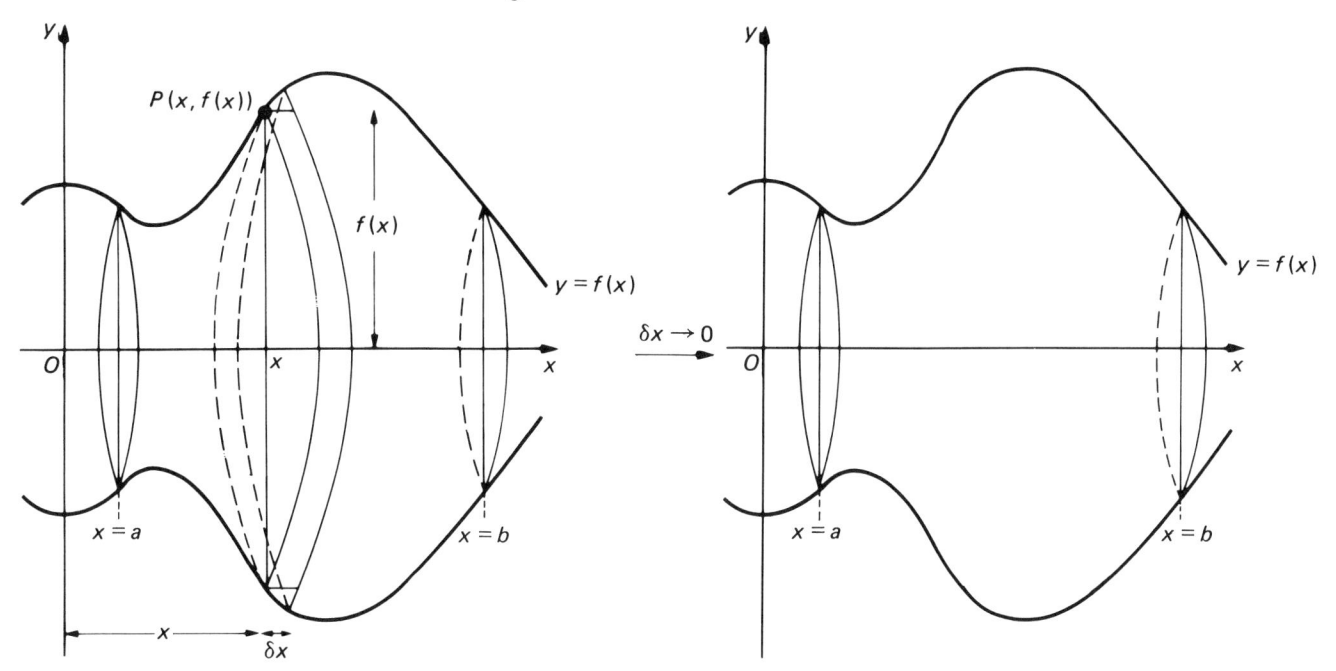

Fig. 4.1:**8**   Each element is approximately a cylinder, base radius $f(x)$ and thickness δx. So the volume of each element $\approx \pi\{f(x)\}^2 \, \delta x$. Thus

$$V = \lim_{\delta x \to 0} \sum_{x=a}^{x=b} \pi\{f(x)\}^2 \, \delta x = \pi \int_a^b \{f(x)\}^2 \, dx$$

(Note: we often write $A = \int_a^b y \, dx$, $V = \int_a^b \pi y^2 \, dx$ (bearing in mind that $y = f(x)$).)

Worked example 6   Find the volume of the solid of revolution formed when the region bounded by the curve $y = 2\sqrt{x}$ and the line $x = 1$ is rotated completely about the x-axis.

The solid of revolution is shown in fig. 4.1:**9**.

$$\text{Volume} = \int_0^1 \pi y^2 \, dx = \int_0^1 4\pi x \, dx$$
$$= \left[2\pi x^2 \right]_0^1$$
$$= 2\pi$$

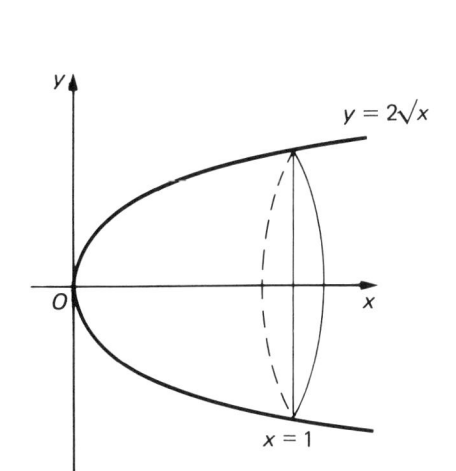

Fig. 4.1:**9**

Exercise 4.1:4

⑥ **1** Find the volume of the solid of revolution formed when the region bounded by the given lines is rotated completely about the x-axis:

(i) $y = x$, $x = 1$, $x = 3$, x-axis
(ii) $y = x^2$; $x = 1$, x-axis
→ (iii) $y = 1/x$, $x = 2$, $x = 3$, x-axis
→ (iv) $y = \sqrt{x}$; $x = 1$, $x = 9$, x-axis
(v) $y = \sqrt{(6x)}$; $x = 6$, x-axis
R (vi) $y = 2x - x^2$; x-axis
(vii) $y = x - 2/x$; $x = 2$, $x = 3$, x-axis

⑥ → **2** Show that the volume of a cone of base radius r and height h is $\frac{1}{3}\pi r^2 h$. (Hint: rotate the line $y = rx/h$ about the x-axis.)

182 *Integral as the limit of a sum; volume of a solid of revolution*

⑥ **3** The height of the frustrum of a cone of base radius 20 cm is 30 cm. The radius of its other end is 5 cm. Find its volume.

⑥ → **4** Show that the volume of a sphere of radius r is $\frac{4}{3}\pi r^3$. (Hint: rotate the circle $x^2+y^2=r^2$ about the x-axis.)

⑥ R **5** A cap of depth $\frac{1}{2}r$ is cut off a sphere of radius r. Find the volume of the cap.

⑥ **6** Find the volume of water needed to fill a hemispherical bowl of radius 20 cm to a depth of 12 cm.

⑥ **7** A wok (Chinese saucepan) is in the shape of the cap of a sphere of radius a cm. Hot fat is poured into the wok to a depth of x cm. Show that the volume of fat is $\frac{1}{3}\pi x^2(3a-x)$ cm^3.
The wok is full when $\frac{36}{125}\pi a^3$ cm^3 of fat has been poured into it. Find the depth of the wok.

⑥ **8** A cylindrical hole of radius 0.3 cm is bored symmetrically through a sphere of radius 0.5 m. Find the volume of the part which remains.

⑥ **9** A cylindrical hole of length 6 cm is bored symmetrically through a sphere. Find the volume of the part that remains.
(A virtuoso solution of this problem goes: 'The fact that the radius of the sphere has not been given suggests that the answer must be independent of the radius of the sphere. Assume then, that the sphere has radius 3 cm. The radius of the cylindrical hole must therefore be negligible, if it is still to be 6 cm long. The volume of the part that remains after boring this negligible hole is clearly the volume of the whole sphere, i.e. 36π cm^3.' Clever, huh? Be warned: this is the kind of cleverness that gets you strung upside down in the toilets by people of small intellect and great strength...)

⑥ **10** The regions bounded by the following lines are rotated completely about the x-axis. In each case find the volume of the solid formed.
   • (i) $y=x^2$ and $y=x$
   → (ii) $y=x(6-x)$ and $y=3x$
   (iii) $y=x(2a-x)$ and $y=ax$ (a constant)
   (iv) $y^2=4x$ and $y=2x$
   (v) $y^2=4x$ and $x^2=4y$

• → **11** Find the area of the region bounded by $y=1+x^2$, $x=0$, $x=1$, $y=1$.
Find the volume generated when this region is rotated completely about the line $y=1$.

12 Find the volume of the solid generated when the region bounded by the curve $y=x(5-x)$ and the line $y=6$ is rotated completely about the line $y=6$. (See 4.1:3 Q36).

13 Find the volume generated when the region bounded by the curve $y=4x-x^2$ and the line $y=3$ is rotated completely about the line $y=3$.

14 Find the area of the region bounded by the given lines, and also the volume generated when the region is rotated completely about the x-axis:
   (i) $y=\sqrt{x}$; $x=4$, x-axis
R (ii) $y=\sqrt{(8x)}$; $y=-x+6$, x-axis
   (iii) $y=\sqrt{(2x)}$; $y=\frac{1}{4}x^3$

• → **15** (i) Show that:

> (a) the area of the region bounded by the curve $y=f(x)$, the lines $y=c$ and $y=d$ and the y-axis is
> $$\int_c^d x\,dy \qquad (17)$$
>
> (b) the volume of the solid of revolution formed when this region is rotated completely about the y-axis is
> $$\int_c^d \pi x^2\,dy \qquad (18)$$

(Hint: divide the region into elements width δy parallel to the x-axis.)

Note that the variable of integration in each case is y. In order to evaluate these integrals you must first express x in terms of y.

(ii) If $y=x^3$, express x as a function of y. Hence:
(a) find the area of the region bounded by the curve $y=x^3$, the lines $y=1$ and $y=8$ and the y-axis;
(b) find the volume of the solid of revolution formed when this region is rotated completely about the y-axis.

16 Find the area of the region bounded by the given lines and the volume generated when the region is rotated completely about the y-axis:
→ (i) $y=\sqrt{x}$; $y=1$, $y=2$, y-axis
   (ii) $x=\sqrt{y}$; $y=9$
   (iii) $y^3=x^2$; $y=2$, $y=4$, y-axis
   (iv) $y=1/x^4$; $y=1$, $y=16$

17 Find the volume generated when the regions bounded by the following lines are rotated completely about the y-axis:
   (i) $x=\sqrt{(y-4)}$; $y=8$ (ii) $y=x^2-4$; x-axis
   (iii) $y^2=20x$; $y=2x$ (iv) $y=x^3$, $x=3$, $y=1$

18 Find the volume generated when the region bounded by the curves $y=x^2$, $y^2=x$ is rotated completely about:

(i) the x-axis   (ii) the y-axis.

Explain your answers.

19 Find the volume generated when the region bounded by the curve

$$\frac{x^2}{a^2}+\frac{y^2}{b^2}=1$$

(an ellipse of length $2a$ and height $2b$) is rotated through $180°$ about:

(i) the x-axis (ii) the y-axis.

Discuss these results in the case $a = b$.

20 Find the volume generated when the region bounded by $y^2 = 4x$ and $x = 1$ is rotated completely about $x = 1$.

21 (i) The region bounded by the curve $y = f(x)$, the lines $x = a$ and $x = b$ and the x-axis is rotated completely about the *y-axis*. By dividing the region into elements of width δx parallel to the y-axis, show that the volume V generated is

$$V = \int_a^b 2\pi xy\, dx$$

(ii) Find the volume generated when the region bounded by the curve $y = \sqrt{x}$, the lines $x = 1$, $x = 9$ and the x-axis is rotated completely about the y-axis.

Questions 22 and 23: We can use the methods of this section even when the solid does not necessarily have a circular cross-section.

22 Show that the volume of a pyramid of base area A and height h is $\frac{1}{3}Ah$. (Hint: show that the volume of a slice of thickness δx parallel to the base at a depth x below the vertex is $(x^2/h^2)A\delta x$.)

Hence write down the volume of

(i) a (right circular) cone of base radius r and height h;
(ii) a right square pyramid of height h whose base has side a;
(iii) a tetrahedron whose sides are all of length a.

23 (i) A solid of height h has a cross-sectional area $A(x)$ at a height x from its base. What is its volume?

(ii) A vase is 30 cm high. When the depth of liquid in it is x cm, the area of the surface of the liquid is $\frac{1}{5}x(40-x)$. Find the volume of the vase.

(iii) A man with a remarkable beer gut has a torso whose cross-sectional area at a distance x m below his shoulders is $\frac{1}{20}e^{3x/2}$, $0 \le x \le 0.5$. Find the volume of his torso in this region.

24 The coordinates $(\bar{x}, \bar{y})$ of the **centroid** of the region bounded by the curve $y = f(x)$, the lines $x = a$ and $x = b$ and the x-axis are given by

$$A\bar{x} = \int_a^b xy\, dx, \qquad A\bar{y} = \int_a^b \tfrac{1}{2}y^2\, dx$$

where A is the area of the region. (Note: This definition of a centroid gives a sensible precise meaning to the idea of the 'geometric centre' of the region (e.g. the centroid of a circle is at its centre, that of a rectangle is at the point of intersection of its diagonals).

Find the coordinates of the centroid of the region bounded by:

- (i) $y = x^2$; $x = 1$, x-axis
→ (ii) $y = \sqrt{x}$; $x = 4$, x-axis
R (iii) $y = x(1-x)$; x-axis

25 Show that the centroid of a triangle ABC in which $BC = a$, $AC = b$, $\hat{C} = 90°$ is at a distance $\tfrac{1}{3}a$ from BC and $\tfrac{1}{3}b$ from AC.

26 The coordinates $(\bar{x}, \bar{y})$ of the centroid of the solid of revolution formed when the region bounded by the curve $y = f(x)$, the lines $x = a$ and $x = b$ and the x-axis is rotated completely about the x-axis are given by:

$$V\bar{x} = \int \pi xy^2\, dx, \quad \bar{y} = 0$$

where V is the volume of the solid.

Find the coordinate of the centroid of the solid of revolution formed when the regions bounded by the following lines are related completely about the x-axis:

- (i) $y = x$; $x = 1$, x-axis
→ (ii) $y^2 = x$; $x = 4$
  (iii) $y = x(1-x)$; x-axis

27 Show that the centroid of:

→ (i) a cone of base radius r and height h is on its axis of symmetry at a distance $\tfrac{1}{4}h$ from its base;

R (ii) a hemisphere of radius r is on its axis of symmetry at a distance $\tfrac{3}{8}r$ from its plane face.

28 The region bounded by $y = x^2 + 1$ and $y = 10$ is rotated completely about the y-axis. Find the volume and the coordinates of the centroid of the solid formed.

4.2 Methods of integration

4.2:1 Integration by substitution

Worked example 1 Find

$$\int \frac{x}{\sqrt{(3x+1)}} \, dx$$

Let us put $u = 3x+1$ and make u the variable of integration: instead of $x/\sqrt{(3x+1)}$ we write $\frac{1}{3}(u-1)/u^{1/2}$; what do we write instead of dx?

Well, $x = \frac{1}{3}(u-1)$, so $dx/du = \frac{1}{3}$; let us write $dx = \frac{1}{3} du$ (as if dx/du meant $dx \div du$). Then

$$\int \frac{x}{\sqrt{(3x+1)}} \, dx = \int \frac{u-1}{3u^{1/2}} \cdot \frac{du}{3} \quad (1)$$

$$= \tfrac{1}{9} \int (u^{1/2} - u^{-1/2}) \, du$$

$$= \tfrac{1}{9}(\tfrac{2}{3} u^{3/2} - 2u^{1/2}) + c$$

$$= \tfrac{2}{27} u^{1/2}(u-3) + c$$

$$= \tfrac{2}{27}(3x+1)^{1/2}(3x-2) + c$$

We must justify this method; dx/du does *not* mean $dx \div du$ (see section 1.3:1), so the step

$$\frac{dx}{du} = \frac{1}{3} \Rightarrow dx = \frac{du}{3} \quad (2)$$

is a dangerous one.

Theorem 1:

$$\int f(x) \, dx = \int f(x) \frac{dx}{du} \, du \quad (3)$$

Proof

Let $y = \int f(x) \, dx$. Then $\dfrac{dy}{dx} = f(x)$

Now $\dfrac{dy}{du} = \dfrac{dy}{dx} \cdot \dfrac{dx}{du}$

by the chain rule (see section 1.3:6), i.e.

$$\frac{dy}{du} = f(x) \frac{dx}{du}$$

Integrating this equation with respect to u:

$$y = \int f(x) \frac{dx}{du} \, du$$

Integration and differential equations 185

This theorem says that dx and $(dx/du) \cdot du$ are equivalent operators; we can justify (2) by writing

$$\frac{dx}{du} = \frac{1}{3} \Rightarrow dx \equiv \frac{dx}{du} du \equiv \frac{1}{3} du$$

A rigorous solution of worked example 1 would begin:

$$\int \frac{x}{\sqrt{(3x+1)}} dx = \int \frac{x}{\sqrt{(3x+1)}} \frac{dx}{du} du$$

$$= \int \frac{u-1}{3u^{1/2}} \cdot \frac{1}{3} du$$

but this is the same as (1) in our original solution, so we will accept the validity of step (2) and use our original method of solution. The advantage of using the shorthand method in this example is minimal, but in later examples it will be much easier.

Worked example 2 Evaluate:

$$\int_2^6 x\sqrt{(2x-3)} \, dx$$

Note: the limits 2 and 6 are limits of x. When we change the variable to u, we need to change the limits to limits of u.

$u = 2x - 3 \Rightarrow$ when $x = 2$, $u = 1$ and when $x = 6$, $u = 9$; the corresponding limits of u are 1 and 9. Thus:

$$\int_2^6 x\sqrt{(2x-3)} \, dx = \int_1^9 \tfrac{1}{2}(u+3) \cdot u^{1/2} \cdot \tfrac{1}{2} \, du$$

$$= \tfrac{1}{4} \int_1^9 (u^{3/2} + 3u^{1/2}) \, du$$

$$= \tfrac{1}{4}[\tfrac{2}{5}u^{5/2} + 2u^{3/2}]_1^9$$

$$\approx 37.1$$

Theorem 2: If $\int f(x) \, dx = F(x) + c$ and a and b are constants, then

$$\int f(ax+b) \, dx = \frac{1}{a} F(ax+b) + C \tag{4}$$

Proof

Consider $\int f(ax+b) \, dx$; let $u = ax + b$, so $dx = (1/a) \, du$.

$$\int f(ax+b) \, dx = \int f(u) \frac{1}{a} \, du$$

$$= \frac{1}{a} \int f(u) \, du$$

$$= \frac{1}{a} \{F(u) + c\}$$

$$= \frac{1}{a} \{F(ax+b) + c\}$$

186 Integration by substitution

For example

$$\int e^{4x-5} \, dx = \tfrac{1}{4} e^{4x-5} + c$$

$$\int \sin 3x \, dx = -\tfrac{1}{3} \cos 3x + c$$

and in particular, when $f(u) = 1/u$,

$$\int \frac{1}{ax+b} \, dx = \frac{1}{a} \ln|ax+b| + c$$

(see question 3(ii)).

Exercise 4.2:1

① 1 Using the appropriate substitutions, integrate:
 (i) (a) $(3x+2)^6$ (b) $\sqrt{(2x+1)}$
 (c) $\dfrac{1}{(5x-4)^2}$

 (ii) (a) $x\sqrt{(3x-1)}$ R (b) $\dfrac{x}{\sqrt{(x-2)}}$
 (c) $\dfrac{x}{\sqrt{(3-2x)}}$ (d) $x^2\sqrt{(x+1)}$
 (e) $x(x-1)^7$ (f) $\dfrac{4x-5}{x+3}$
 (g) $\dfrac{x+3}{4x-5}$ (h) $\dfrac{3x+1}{2x+1}$
 (i) $\dfrac{2x^2+3x+5}{2x+1}$ (j) $\dfrac{x}{(x+2)^2}$

② 2 Evaluate:
 (i) $\displaystyle\int_1^2 \sqrt{(x-1)} \, dx$
 (ii) $\displaystyle\int_0^1 \frac{4}{(3-2x)^3} \, dx$
 (iii) $\displaystyle\int_1^2 x\sqrt{(x-1)} \, dx$
 (iv) $\displaystyle\int_1^2 x^2\sqrt{(x-1)} \, dx$
 → (v) $\displaystyle\int_1^5 \frac{1+x}{\sqrt{(2x-1)}} \, dx$
 R (vi) $\displaystyle\int_0^1 \frac{3+x}{1+2x} \, dx$
 (vii) $\displaystyle\int_0^1 \frac{3+x}{(1+2x)^2} \, dx$
 (viii) $\displaystyle\int_3^6 \frac{x^2-2x+3}{(x-2)\sqrt{(x-2)}} \, dx$

→ 3 (i) Using Theorem 2, integrate $(ax+b)^\alpha$, a, b, α constants, $\alpha \neq -1$.
 Hence write down the integrals of:
 (a) $(3x+2)^6$; (b) $(5x+1)^{12}$;
 (c) $(1-4x)^5$; (d) $\sqrt{(2x-1)}$;
 (e) $1/(x+1)^2$; (f) $1/(3x-1)^2$;
 (g) $1/(5-2x)^3$.

 (ii) Using Theorem 2, show that

$$\int \frac{1}{ax+b} \, dx = \frac{1}{a} \ln|ax+b| + c$$

 explaining the use of the modulus signs.
 Hence write down the integrals of:
 (a) $1/(x+2)$; (b) $4/(4x+5)$;
 (c) $2/(6-3x)$

 (iii) Integrate:
 (a) $1/(3x+1)^3$; (b) $1/(3x+1)^2$;
 (c) $1/3x+1$; (d) $1/\sqrt{(3x+1)}$.

4 Using Theorem 2, integrate e^{ax+b}, a, b constants. Hence write down the integrals of:
 (i) e^{5x} (ii) e^{3x-2} (iii) $e^{\frac{3}{2}x}$ (iv) e^{-4x} (v) e^{1-5x}

5 Using Theorem 2, integrate $\sin(ax+b)$ and $\cos(ax+b)$, a, b constants. Hence write down the integrals of:
 (i) $\sin 2x$ (ii) $\cos 5x$
 (iii) $\sin \tfrac{1}{3}x$ (iv) $5\cos \tfrac{1}{2}x$
 (v) $\sin \pi x$ (vi) $\cos(5-3x)$

6 Using Theorem 2, integrate:
 (i) $\sec^2 4x$
 (ii) $\operatorname{cosec} 5x \cot 5x$
 (iii) $\sec \tfrac{1}{2}x \tan \tfrac{1}{2}x$

7 Evaluate:

(i) $\int_0^1 e^{2x+1}\, dx$

→ (ii) $\int_\pi^{2\pi} 4 \sin \tfrac{1}{3}x\, dx$

(iii) $\int_0^1 \cos \tfrac{1}{2}\pi x\, dx$

(iv) $\int_1^3 (4x-3)^{3/2}\, dx$

→ (v) $\int_0^1 \dfrac{1}{(5-4x)^2}\, dx$

→ (vi) $\int_0^1 \dfrac{4}{5-4x}\, dx$

(vii) $\int_{-3}^{-2} \dfrac{1}{3+2x}\, dx$

(viii) $\int_1^2 \dfrac{4}{1-3x}\, dx$

•→ 8 (i) By writing $\cos x \equiv 2\cos^2 \tfrac{1}{2}x - 1$, find $\int \sqrt{(1+\cos x)}\, dx$.

(ii) By writing $\cos x \equiv 1 - 2\sin^2 \tfrac{1}{2}x$, find $\int \sqrt{(1-\cos x)}\, dx$.

9 Write $\cos 3x$ in terms of $\cos \tfrac{3}{2}x$. Hence, find

(i) $\int \sqrt{(1+\cos 3x)}\, dx$

(ii) $\int \dfrac{1}{1+\cos 3x}\, dx$

10 Evaluate:

(i) $\int_0^{\pi/3} \sqrt{(1+\cos 2x)}\, dx$

(ii) $\int_0^{\pi/3} \sqrt{(1-\cos 3x)}\, dx$

R (iii) $\int_0^{\pi/2} \dfrac{1}{(1+\cos x)}\, dx$

11 Using the substitution $u = \tfrac{1}{2}\pi - x$, find $\int \sqrt{(1+\sin x)}\, dx$.

•→ 12 (i) Express $\sin^2 x$ in terms of $\cos 2x$. Hence find $\int \sin^2 x\, dx$.

(ii) Find $\int \cos^2 x\, dx$:
(a) by expressing $\cos^2 x$ in terms of $\cos 2x$;
(b) by using the result of part (i).

R 13 Express $\cos^2 2x$ in terms of $\cos 4x$. Hence find:

(i) $\int \cos^2 2x\, dx$ (ii) $\int_0^{\frac{1}{4}\pi} \sin^4 x\, dx$

(iii) $\int_0^{\frac{1}{6}\pi} \cos^4 x\, dx$

14 Find:

(i) $\int \sin^2 2x\, dx$ (ii) $\int \sin^2 x \cos^2 x\, dx$

(iii) $\int \sin^2 3x\, dx$

•→ 15 Express $2\sin 5x \cos 3x$ as the sum of two sines. Hence find $\int \sin 5x \cos 3x\, dx$.

16 Using the appropriate factor formula, find:

(i) $\int \sin 3x \cos x\, dx$

(ii) $\int \cos x \cos 2x\, dx$

(iii) $\int \sin 4x \sin 2x\, dx$

(iv) $\int \sin 3x \cos^2 x\, dx$

17 Evaluate:

R (i) $\int_0^{\frac{1}{4}\pi} \sin 3x \cos 5x\, dx$

(ii) $\int_{\frac{1}{6}\pi}^{\frac{1}{4}\pi} 2\sin 3x \cos 2x\, dx$

(iii) $\int_0^{\frac{1}{2}\pi} \sin^2 x \cos 3x\, dx$

18 Find the area of the region bounded by the curve $y = 1/(3+2x)$, the y-axis, the ordinate $x = 1$ and the x-axis.

Find the volume generated when this region is rotated completely about the x-axis.

• 19 Find the volume generated when the region bounded by the arc of the curve $y = 2\sin 3x$ between $x = 0$, $x = \tfrac{1}{9}\pi$ and the x-axis is rotated completely about the x-axis.

R 20 Show that the function $\sin^2 x$ has stationary values when $x = \tfrac{1}{2}k\pi$, k an integer.

Sketch the curve $y = \sin^2 x$ for $0 \leqslant x \leqslant \pi$. Find the area of the region which lies beneath this part of the curve and above the line $y = \tfrac{1}{4}$.

→ 21 Sketch the curve $y = 2 + \sin x$. Calculate:

(i) the area of the region bounded by the curve, the x-axis and the lines $x = 0$ and $x = 2\pi$;

(ii) the volume generated when this region is rotated completely about the x-axis.

22 (i) Find the area of the finite region bounded by the curves $y = e^x$, $y = e^{2x}$ and the ordinate $x = \ln 2$.

(ii) Find the volume generated when this region is rotated completely about the x-axis.

4.2:2 Integrand containing a function and its derivative

Worked example 3 Find:

(i) $\int \dfrac{x}{\sqrt{(x^2+1)}} dx$ (ii) $\int \cos^4 x \sin x \, dx.$

(i) $u = x^2 + 1 \Rightarrow du = 2x \, dx \Rightarrow x \, dx = \tfrac{1}{2} du$; so we can substitute completely for $x \, dx$ in the integral:

$$\int \dfrac{x}{\sqrt{(x^2+1)}} dx = \int u^{-1/2} (\tfrac{1}{2} du)$$
$$= u^{1/2} + c$$
$$= \sqrt{(x^2+1)} + c$$

(ii) $u = \cos x \Rightarrow du = -\sin x \, dx \Rightarrow \sin x \, dx = -du$; so we can substitute completely for $\sin x \, dx$ in the integral:

$$\int \cos^4 x \sin x \, dx = -\int u^4 \, du$$
$$= -\tfrac{1}{5} u^5 + c$$
$$= -\tfrac{1}{5} \cos^5 x + c$$

Both the solutions in worked example 3 depended on the fact that the integrand contained a function $f(x)$ and its derivative $f'(x)$. In part (i) $f(x) = x^2 + 1$, in part (ii) $f(x) = \cos x$ and in both cases we substituted $u = f(x)$.

In general,

> suppose we are trying to find $I = \int g\{f(x)\} \cdot f'(x) \, dx$, we substitute $u = f(x)$, so that $du = f'(x) \, dx$; then
>
> $$I = \int g(u) \, du \qquad (5)$$

Worked example 4 Find

$$\int \tan^3 x \sec^2 x \, dx$$

The integrand contains the function $f(x) = \tan x$ and its derivative $f'(x) = \sec^2 x$. $g(u) = u^3$. We substitute $u = \tan x$, so that $du = \sec^2 x \, dx$. Then

$$\int \tan^3 x \sec^2 x \, dx = \int u^3 \, du$$
$$= \tfrac{1}{4} u^4 + c$$
$$= \tfrac{1}{4} \tan^4 x + c$$

Two particular cases
$g(u) = e^u$:

$$\int f'(x) e^{f(x)} dx = e^{f(x)} + c \qquad (6)$$

$g(u) = 1/u$:

$$\int \frac{f'(x)}{f(x)} dx = \ln |f(x)| + c \qquad (7)$$

Worked example 5 Find:

(i) $\int \frac{x+1}{x^2+2x+5} dx$ (ii) $\int \tan x \, dx$ (iii) $\int \sec x \, dx$

Using equation (7),

(i) $\int \frac{x+1}{x^2+2x+5} dx = \frac{1}{2} \ln |x^2+2x+5| + c$

(ii) $\int \tan x \, dx = \int \frac{\sin x}{\cos x} dx$

$\qquad = -\ln |\cos x| + c$

$\qquad = \ln \left| \frac{1}{\cos x} \right| + c$

$\qquad = \ln |\sec x| + c$

(iii) $\int \sec x \, dx = \int \frac{\sec^2 x + \sec x \tan x}{\sec x + \tan x} dx$

$\qquad = \ln |\sec x + \tan x| + c$

Exercise 4.2:2

④ **1** Using the appropriate substitutions, integrate with respect to x:

→ (i) $x\sqrt{(x^2-1)}$

(ii) $\frac{x^2}{\sqrt{(x^3+1)}}$

(iii) $\frac{x^2}{x^3+1}$

(iv) $x e^{x^2}$

→ (v) $\sec^2 x \, e^{\tan x}$

(vi) $\cos^3 x \sin x$

(vii) $\tan^4 x \sec^2 x$

(viii) $\frac{x^2+1}{\sqrt{(x^3+3x+4)}}$

(ix) $\sin x \sqrt{(\cos x)}$

(x) $\frac{\sin \sqrt{x}}{\sqrt{x}}$

④ **2** Integrate with respect to x:

(i) $\sin^n x \cos x$

(ii) $\cos^n x \sin x$

(iii) $\tan^n x \sec^2 x$

(iv) $\sec^n x \tan x$

190 *Integrand containing a function and its derivative*

④ **3** Evaluate:

(i) $\int_1^3 x|x^2-1|^{1/3}\,dx$

→ (ii) $\int_0^1 \dfrac{1+x}{\sqrt{(2+2x+x^2)}}\,dx$

(iii) $\int_0^{\frac{1}{6}\pi} \sin^4 x \cos x\,dx$

(iv) $\int_0^{\frac{1}{3}\pi} \sec^4 x \tan x\,dx$

(v) $\int_1^2 \dfrac{\ln x}{x}\,dx$

→ (vi) $\int_{-\frac{1}{4}\pi}^{\frac{1}{4}\pi} \sec^2 x \sqrt{(2+\tan x)}\,dx$

R (vii) $\int_0^{\pi} \sin x \sqrt{(1+\cos x)}\,dx$

(viii) $\int_1^4 \dfrac{\sqrt{(1+\sqrt{x})}}{\sqrt{x}}\,dx$

⑤ **4** Using equation (6) write down the integrals of:

(i) $\sec^2 x\,e^{\tan x}$

(ii) $\dfrac{e^{1/x}}{x^2}$

(iii) $\dfrac{e^{\sqrt{x}}}{\sqrt{x}}$

⑤ **5** Evaluate:

(i) $\int_{-1}^2 x\,e^{x^2}\,dx$

(ii) $\int_{\frac{1}{3}\pi}^{\frac{1}{2}\pi} \sin 2x\,e^{\cos 2x}\,dx$

⑤ **6** Using equation (7), write down the integrals of:

(i) $\dfrac{2x}{x^2+1}$

(ii) $\dfrac{x^2}{x^3+1}$

(iii) $\dfrac{x-2}{x^2-4x+7}$

(iv) $\dfrac{\sec^2 x}{1+\tan x}$

(v) $\dfrac{\cos x}{2-\sin x}$

(vi) $\dfrac{e^{3x}}{e^{3x}+1}$

(vii) $\dfrac{1}{x \ln x}$

⑤ **7** Evaluate:

(i) $\int_0^1 \dfrac{x}{x^2+2}\,dx$

(ii) $\int_2^3 \dfrac{x^2+1}{x^3+3x}\,dx$

→ (iii) $\int_0^{\frac{1}{6}\pi} \dfrac{\sin x}{2+\cos x}\,dx$

(iv) $\int_0^{\frac{1}{6}\pi} \tan 2x\,dx$

(v) $\int_0^{\ln 2} \dfrac{e^x}{e^x+1}\,dx$

R (vi) $\int_{\ln 2}^{\ln 3} \dfrac{e^x+e^{-x}}{e^x-e^{-x}}\,dx$

⑤ **8** Using the substitution $u=x^2+2$, integrate:

(i) $\dfrac{x}{(x^2+2)^3}$

(ii) $\dfrac{x}{(x^2+2)^2}$

(iii) $\dfrac{x}{x^2+2}$

(iv) $\dfrac{x}{\sqrt{(x^2+2)}}$

⑤ → **9** (i) Using equation (7), write down the integral of cot x. Evaluate:

(a) $\int_{\frac{1}{6}\pi}^{\frac{1}{2}\pi} \cot x\,dx$; (b) $\int_{-\frac{1}{4}\pi}^{-\frac{1}{6}\pi} \cot x\,dx$.

(ii) Integrate cosec x by the method of worked example 5(iii). Hence evaluate $\int_{\pi/4}^{\pi/3}$ cosec $x\,dx$.

(iii) Complete this table:

$f(x)$	sin x	cos x	tan x	cot x	sec x	cosec x
$\int f(x)\,dx$						

⑤ **10** Explain why the following integrals are meaningless:

(i) $\int_{-1}^3 \dfrac{x}{x^2-1}\,dx$

(ii) $\int_{\frac{1}{4}\pi}^{\frac{3}{4}\pi} \tan x\,dx$

⑤ **11** Prove that

$$\dfrac{d}{dx}\{\ln(x+\sqrt{(x^2+1)})\} = \dfrac{1}{\sqrt{(x^2+1)}}$$

Hence show that

$$\int_0^1 \dfrac{dx}{\sqrt{(x^2+1)}} = \ln(1+\sqrt{2})$$

⑤ **12** Write down $\int \dfrac{e^x}{1+e^x}\,dx$. Show that

$$\dfrac{1}{1+e^x} = 1 - \dfrac{e^x}{1+e^x}$$

Hence find $\int \dfrac{1}{1+e^x}\,dx$.

⑤ **13** Find $\int \dfrac{1}{2+e^x}\,dx$:

(i) by first multiplying the numerator and denominator of the integrand by e^{-x};

(ii) by the method of question 12.

14 Write down $\int \dfrac{2\cos x - \sin x}{\cos x + 2\sin x}\,dx$.

Find the constants a and b such that

$$\cos x \equiv a(\cos x + 2\sin x) + b(2\cos x - \sin x)$$

Hence show that

$$\dfrac{\cos x}{\cos x + 2\sin x} \equiv \dfrac{1}{5}\left\{1 + \dfrac{2(2\cos x - \sin x)}{\cos x + 2\sin x}\right\}$$

Hence find $\int \dfrac{\cos x}{\cos x + 2\sin x}\,dx$.

15 Using the methods of questions 12 and 14, find:

(i) $\int \dfrac{1}{1+e^{2x}}\,dx$

(ii) $\int \dfrac{\cos x}{\cos x + \sin x}\,dx$

(iii) $\int \dfrac{e^x + e^{3x}}{2e^x + 3e^{3x}}\,dx$

(iv) $\int \dfrac{3\cos x + 2\sin x}{2\cos x + 3\sin x}\,dx$

(v) $\int_0^{\frac{1}{2}\pi} \dfrac{\sin x}{3\sin x + 4\cos x}\,dx$

16 Evaluate:

(i) $\int_0^{\frac{1}{2}\pi} \cos x\, \sqrt{(1-\sin x)}\,dx$

(ii) $\int_{\frac{1}{6}\pi}^{\frac{1}{3}\pi} \sin^3 x \cos x\,dx$

(iii) $\int_{-\frac{1}{3}\pi}^{\frac{1}{3}\pi} \sec^3 x \tan x\,dx$

(iv) $\int_{\frac{1}{4}\pi}^{\frac{1}{2}\pi} \csc^2 x\, \sqrt{(3+\cot x)}\,dx$

(v) $\int_{\frac{1}{3}\pi}^{\frac{1}{2}\pi} (1 - 3\cos^2 x)^{1/2} \sin 2x\,dx$

R 17 Integrate $\cos^3 x \sin x$ and $\cos^5 x \sin x$. Use these results to help find

$$\int_0^{\frac{1}{6}\pi} \cos^3 x \sin^3 x\,dx$$

18 Integrate $\sin^2 x \cos x$ and $\sin^4 x \cos x$ (see question 2).

Hence, by writing $\cos^3 x \equiv (1 - \sin^2 x)\cos x$, find $\int \cos^3 x\,dx$ and by writing $\cos^5 x \equiv (1 - \sin^2 x)^2 \cos x$, find $\int \cos^5 x\,dx$.

19 Evaluate:

(i) $\int_0^{\frac{2}{3}\pi} \cos^3 x\,dx$

(ii) $\int_{\frac{1}{3}\pi}^{\frac{1}{2}\pi} \sin^5 x\,dx$

(iii) $\int_0^{\frac{1}{2}\pi} \sin^3 x \cos^2 x\,dx$

(iv) $\int_0^{\frac{1}{2}\pi} \sin^4 x \cos^3 x\,dx$

20 Integrate $\sec^2 x \tan^2 x$. Hence, by writing $\sec^4 x \equiv (1 + \tan^2 x)\sec^2 x$, find $\int \sec^4 x\,dx$.

Evaluate $\int_{-\pi/4}^{3\pi/4} \tan^4 x\,dx$.

R 21 Integrate $\tan x \sec^2 x$. Hence find:

(i) $\int \tan^3 x\,dx$

(ii) $\int \tan^5 x\,dx$

22 By first dividing the numerator and denominator of the integrand by $\cos^2 x$, find:

(i) $\int \dfrac{1}{2\cos^2 x - \sin x \cos x}\,dx$

(ii) $\int \dfrac{2\tan x}{\cos^2 x - \sin^2 x}\,dx$

23 Find the area of the region bounded by:

(i) $y = x/(x^2+1)$; y-axis, $x = \sqrt{3}$, x-axis

R (ii) $y = x\,e^{-x^2}$; x-axis, $x = 1$

24 Sketch on the same axes the graphs of $y = 2\sin x$ and $y = \tan x$ for $-\frac{1}{2}\pi < x < \frac{1}{2}\pi$.

Find the finite (numerical) area of the region bounded by the curves in the interval $-\frac{1}{3}\pi < x < \frac{1}{3}\pi$.

4.2:3 Substitutions of the type $x = a\sin\theta$

When the integrand contains the function $\sqrt{(a^2 - x^2)}$ (a is a constant), we substitute $x = a\sin\theta$, so that

$$\sqrt{(a^2 - x^2)} = \sqrt{(a^2 - a^2\sin^2\theta)} = \sqrt{(a^2\cos^2\theta)} = a\cos\theta$$

Note: $dx = a\cos\theta\,d\theta$.

Similarly, when the integrand contains $\sqrt{(a^2 + x^2)}$ we substitute $x = a\tan\theta$, and when the integrand contains $\sqrt{(x^2 - a^2)}$, we substitute $x = a\sec\theta$.

192 *Substitutions of the type $x = a \sin \theta$*

Worked example 6 Evaluate

$$\int_{-3/2}^{3/2} \frac{1}{(9-x^2)^{3/2}} \, dx$$

$x = 3 \sin \theta \Rightarrow dx = 3 \cos \theta \, d\theta$, and $(9-x^2)^{3/2} = (3 \cos \theta)^3 = 27 \cos^3 \theta$.
When $x = \frac{3}{2}$, $\sin \theta = \frac{1}{2} \Rightarrow \theta = \frac{1}{6}\pi$.
When $x = -\frac{3}{2}$, $\sin \theta = -\frac{1}{2} \Rightarrow \theta = -\frac{1}{6}\pi$.
(Note: we have a choice of limits for θ, so to avoid ambiguity we choose values of θ between $-\frac{1}{2}\pi$ and $\frac{1}{2}\pi$, i.e. the principal solutions of the equations $\sin \theta = \frac{1}{2}$ and $\sin \theta = -\frac{1}{2}$ (see section 4.2:1).)
So

$$\int_{-3/2}^{3/2} \frac{1}{(9-x^2)^{3/2}} \, dx = \int_{-\frac{1}{6}\pi}^{\frac{1}{6}\pi} \frac{3 \cos \theta}{27 \cos^3 \theta} \, d\theta$$

$$= \frac{1}{9} \int_{-\frac{1}{6}\pi}^{\frac{1}{6}\pi} \sec^2 \theta \, d\theta$$

$$= \frac{1}{9} \Big[\tan \theta \Big]_{-\frac{1}{6}\pi}^{\frac{1}{6}\pi}$$

$$= \frac{2\sqrt{3}}{27}$$

Worked example 7 Find

$$\int \frac{1}{(9-x^2)^{3/2}} \, dx$$

As in worked example 6, using the substitution $x = 3 \sin \theta$,

$$\int \frac{1}{(9-x^2)^{3/2}} \, dx = \frac{1}{9} \int \sec^2 \theta \, d\theta$$

$$= \frac{1}{9} \tan \theta + c$$

We want the answer in terms of x, i.e. we want to find $\tan \theta$ in terms of x:

$$\sin \theta = \tfrac{1}{3}x \Rightarrow \cos \theta = \sqrt{(1 - \tfrac{1}{9}x^2)} \Rightarrow \tan \theta = \frac{x}{3\sqrt{(1 - \tfrac{1}{9}x^2)}} = \frac{x}{\sqrt{(9-x^2)}}$$

So

$$\int \frac{1}{(9-x^2)^{3/2}} \, dx = \frac{x}{9\sqrt{(9-x^2)}} + c$$

$$\boxed{\begin{aligned} \int \frac{1}{\sqrt{(a^2 - x^2)}} \, dx &= \sin^{-1} \frac{x}{a} + c \qquad &(8) \\ \int \frac{1}{a^2 + x^2} \, dx &= \frac{1}{a} \tan^{-1} \frac{x}{a} + c \qquad &(9) \end{aligned}}$$

(See question 15.)

Exercise 4.2:3

⑥ → **1** (i) Using the substitution $x = \sin\theta$, evaluate:
 (a) $\int_0^{\frac{1}{2}\sqrt{2}} \frac{1}{(1-x^2)^{3/2}} \, dx$; (b) $\int_{-1/2}^{1/2} \frac{1}{\sqrt{(1-x^2)}} \, dx$.

 (ii) Integrate $\cos^2\theta$ with respect to θ. Using the substitution $x = \sin\theta$, evaluate $\int_0^1 \sqrt{(1-x^2)} \, dx$.

⑥ **2** Using the substitution $x = 2\sin\theta$, evaluate:
 (i) $\int_{-1}^{\sqrt{3}} \frac{1}{(4-x^2)^{3/2}} \, dx$ (ii) $\int_0^1 \frac{1}{\sqrt{(4-x^2)}} \, dx$

 R (iii) $\int_{-1}^1 \sqrt{(4-x^2)} \, dx$

⑥ **3** (i) Find $\int_0^1 x\sqrt{(1-x^2)} \, dx$ using the substitution
 (a) $x = \sin\theta$; (b) $u = 1 - x^2$.

 (ii) Find $\int_0^{3/2} x/(9-x^2)^{3/2} \, dx$ using the substitution
 (a) $x = 3\sin\theta$; (b) $u = 9 - x^2$.

⑥ **4** (i) Find $\int \sin^2\theta \cos^2\theta \, d\theta$ (c.f. 4.2:1 Q14).

 (ii) Hence, using the substitution $x = \sin\theta$, find $\int_0^{1/2} x^2\sqrt{(1-x^2)} \, dx$.

⑥ **5** Using the substitution $x = 3\tan\theta$, evaluate:
 (i) $\int_0^3 \frac{1}{\sqrt{(x^2+9)}} \, dx$

 (ii) $\int_0^3 \frac{1}{x^2+9} \, dx$

 → (iii) $\int_0^3 \frac{1}{(x^2+9)^{3/2}} \, dx$ (see worked example 5(iii))

⑥ **6** Evaluate:
 (i) $\int_1^{\sqrt{3}} \frac{1}{\sqrt{(x^2+1)}} \, dx$

 (ii) $\int_1^{\sqrt{3}} \frac{1}{x^2+1} \, dx$

 R (iii) $\int_1^{\sqrt{3}} \frac{1}{x\sqrt{(x^2+1)}} \, dx$ (see 4.2:2 Q9)

⑥ **7** Using the substitution $x = \sec\theta$, evaluate:
 → (i) $\int_{\sqrt{2}}^2 \frac{1}{x\sqrt{(x^2-1)}} \, dx$

 (ii) $\int_{\frac{2}{3}\sqrt{3}}^2 \frac{1}{\sqrt{(x^2-1)}} \, dx$ (see worked example 5(iii)).

⑥ **8** Using the appropriate substitutions, find
 (i) $\int_1^2 \frac{1}{\sqrt{(9-x^2)}} \, dx$ (ii) $\int_1^2 \frac{1}{(x^2+4)^{3/2}} \, dx$
 (iii) $\int_4^5 \frac{1}{x\sqrt{(x^2-1)}} \, dx$

⑥ R **9** Find $\int \sin^2\theta \, d\theta$ (see 4.2:1 Q12(i)). Hence, using the substitution $x = \sin^2\theta$, find
$$\int_0^{1/2} \sqrt{\left(\frac{x}{1-x}\right)} \, dx$$

⑥ **10** Using the substitution $x = 2(1 + \cos^2\theta)$, show that
$$\int_2^3 \sqrt{\left(\frac{x-2}{4-x}\right)} \, dx = \tfrac{1}{2}\pi - 1$$

Questions 11–13: Find the integrals using the appropriate substitutions.

⑦ **11** (i) $\int \frac{1}{(4-x^2)^{3/2}} \, dx$

 (ii) $\int \frac{1}{(x^2+9)^{3/2}} \, dx$

 • → (iii) $\int \frac{1}{x\sqrt{(x^2-9)}} \, dx$

⑦ **12** (i) $\int \frac{1}{\sqrt{(4-9x^2)}} \, dx$

 (ii) $\int \frac{1}{25+x^2} \, dx$

⑦ → **13** $\int \sqrt{(1-x^2)} \, dx$

⑦ **14** Show, by substituting $x = \tan\theta$, that
$$\int \frac{1}{\sqrt{(1+x^2)}} \, dx = \ln|x + \sqrt{(1+x^2)}| + c$$

(See 4.2:2 Q11.)

→ **15** Differentiate, with respect to x, $\sin^{-1}(x/a)$ and $\tan^{-1}(x/a)$, where a is a constant (see 3.5:2). Hence integrate with respect to x:

 (i) $\dfrac{1}{\sqrt{(a^2-x^2)}}$ (ii) $\dfrac{1}{a^2+x^2}$

 Check these results by integrating using the appropriate substitutions.

16 Using equations (8) and (9) write down the integrals of:
 (i) $\dfrac{1}{\sqrt{(9-x^2)}}$ (ii) $\dfrac{1}{16+x^2}$
 (iii) $\dfrac{1}{3+x^2}$

→ **17** Evaluate:
 (i) $\int_0^{\sqrt{3}} \frac{1}{\sqrt{(4-x^2)}} \, dx$ (ii) $\int_{-5}^5 \frac{1}{25+x^2} \, dx$

18 Write down the integral of $2x/(x^2+4)$. Hence integrate $(2x+3)/(x^2+4)$.
   Similarly, find:
   (i) $\displaystyle\int \frac{x+5}{x^2+5}\,dx$
   (ii) $\displaystyle\int \frac{3x+2}{3+x^2}\,dx$

19 By first taking out an appropriate factor, find:
• (i) $\displaystyle\int \frac{1}{\sqrt{(4-9x^2)}}\,dx$
   (ii) $\displaystyle\int \frac{1}{\sqrt{(1-25x^2)}}\,dx$
   (iii) $\displaystyle\int \frac{1}{9+16x^2}\,dx$
→ (iv) $\displaystyle\int \frac{1}{5+4x^2}\,dx$

20 Evaluate:
R (i) $\displaystyle\int_{-3/4}^{3/4} \frac{1}{\sqrt{(9-4x^2)}}\,dx$
   (ii) $\displaystyle\int_0^1 \frac{1}{16+9x^2}\,dx$
   (iii) $\displaystyle\int_0^1 \frac{1}{1+25x^2}\,dx$

21 Sketch on the same axes the curve $y=1/(1+x^2)$ and the line $x+2y=2$. Find the points of intersection of the curve and the line and hence find the finite area bounded by the curve and the line.

+ **22** Sketch the curve $y=1/\sqrt{(1-x^2)}$ for values of x between 0 and 1. If $I(\alpha)=\int_0^\alpha 1/\sqrt{(1-x^2)}\,dx$, find $I(\alpha)$ in terms of α. Show that as $\alpha \to 1$, $I(\alpha) \to \pi/2$.
   This suggests that we can meaningfully write the integral $\int_0^1 1/\sqrt{(1-x^2)}\,dx$ and that this integral is equal to $\pi/2$. Can we write meaningfully the integral $\int_0^\infty 1/(1+x^2)\,dx$?

4.2:4 The substitution $t = \tan \tfrac{1}{2}x$

When integrating functions of the type

$$f(x) = \frac{1}{a\sin x + b\cos x + c}$$

we make the substitution $t = \tan \tfrac{1}{2}x$ (see section 4.2:4); then

$$x = 2\tan^{-1} t \Rightarrow dx = \frac{2\,dt}{1+t^2} \quad \text{(see section 3.5:2)}$$

Remember that $\sin x = 2t/(1+t^2)$ and $\cos x = (1-t^2)/(1+t^2)$.

Worked example 8   Evaluate

$$\int_0^{\frac{1}{2}\pi} \frac{1}{1+\sin x + \cos x}\,dx$$

When $x = \tfrac{1}{2}\pi$, $t = \tan\tfrac{1}{4}\pi = 1$. When $x = 0$, $t = \tan 0 = 0$. So

$$\int_0^{\frac{1}{2}\pi} \frac{1}{1+\sin x + \cos x}\,dx = \int_0^1 \frac{1}{1 + \dfrac{2t}{1+t^2} + \dfrac{1-t^2}{1+t^2}} \cdot \frac{2\,dt}{1+t^2}$$

$$= \int_0^1 \frac{2}{2+2t}\,dt$$

$$= [\ln(1+t)]_0^1$$

$$= \ln 2$$

Exercise 4.2:4

⑧ **1** Use the substitution $t = \tan\tfrac{1}{2}x$ to find:

(i) $\displaystyle\int_0^{\frac{1}{2}\pi} \frac{1}{2+3\sin x+2\cos x}\,dx$

(ii) $\displaystyle\int_0^{\frac{1}{2}\pi} \frac{1}{1+\cos x}\,dx$

→ (iii) $\displaystyle\int_{\frac{1}{3}\pi}^{\frac{1}{2}\pi} \frac{1}{1-\cos x}\,dx$

(iv) $\displaystyle\int_{\frac{1}{2}\pi}^{\frac{2}{3}\pi} \frac{1}{1+\sin x}\,dx$

R (v) $\displaystyle\int_{\frac{1}{3}\pi}^{\frac{1}{2}\pi} \frac{3}{5-4\cos x-3\sin x}\,dx$

⑧ → **2** Use the substitution $t = \tan\tfrac{1}{2}x$ to find $\int \operatorname{cosec} x\,dx$. Check your answer against that for 4.2:2 Q9(ii).

⑧ **3** Using the substitution $t = \tan\tfrac{1}{2}x$, find

$$\int \frac{1}{1-\cos x}\,dx$$

Hence, using the method of 4.2:2 Q12 find

$$\int \frac{\cos x}{1-\cos x}\,dx$$

Hence find

$$\int \frac{\sin x - 2\cos x}{1-\cos x}\,dx$$

⑧ **4** Find

$$\int_0^{\frac{1}{3}\pi} \frac{1}{1+\cos x}\,dx$$

(i) using the substitution $t = \tan\tfrac{1}{2}x$;
(ii) by writing $\cos x$ in terms of $\cos\tfrac{1}{2}x$ (see 4.2:1 Q8).

⑧ R **5** Find $\int 1/(1+9t^2)\,dt$ (see 4.2:3 Q20(ii)). Hence find

$$\int_{\frac{1}{3}\pi}^{\frac{2}{3}\pi} \frac{1}{5-4\cos x}\,dx$$

4.2:5 Integration of rational functions

Worked example 9 Find

$$\int \frac{4x+3}{2x+1}\,dx$$

We could use the substitution $u = 2x+1$ (see section 4.2:1), but it is easier to begin by dividing out the integrand:

$$\int \frac{4x+3}{2x+1}\,dx = \int 2 + \frac{1}{2x+1}\,dx \qquad \text{(see section 2.2:1)}$$

$$= 2x + \tfrac{1}{2}\ln|2x+1| + c$$

Worked example 10 Find:

(i) $\displaystyle\int \frac{x+2}{(x+3)(x+4)}\,dx$

(ii) $\displaystyle\int \frac{1-x}{(x+1)(x^2+1)}\,dx$

(iii) $\displaystyle\int \frac{x^3}{(x+1)(x-2)}\,dx$

In each case we split the integrand into partial fractions (see section 2.3:3).

(i) $\displaystyle\int \frac{x+2}{(x+3)(x+4)}\,dx = \int \left\{\frac{-1}{x+3} + \frac{2}{x+4}\right\}dx$

$$= -\ln|x+3| + 2\ln|x+4| + c$$

$$= \ln\left|\frac{(x+4)^2}{x+3}\right| + c$$

196 *Integration of rational functions*

(ii) $\displaystyle\int \frac{1-x}{(x+1)(x^2+1)}\,dx = \int \left\{\frac{1}{x+1} - \frac{x}{x^2+1}\right\} dx$

$= \ln|x+1| - \tfrac{1}{2}\ln|x^2+1| + c$

$= \ln\left|\dfrac{x+1}{\sqrt{(x^2+1)}}\right| + c$

(iii) $\displaystyle\int \frac{x^3}{(x+1)(x-2)}\,dx = \int\left\{x+1+\frac{3x+2}{(x+1)(x-2)}\right\}dx$

$= \displaystyle\int\left\{x+1+\frac{\tfrac{1}{3}}{x+1}+\frac{\tfrac{8}{3}}{x-2}\right\}dx$

$= \tfrac{1}{2}x^2 + x + \tfrac{1}{3}\ln|x+1| + \tfrac{8}{3}\ln|x-2| + c$

Exercise 4.2:5

1 Integrate with respect to x:

(i) $\dfrac{x+1}{x-1}$ (ii) $\dfrac{3x}{x+2}$

(iii) $\dfrac{4x}{4-x}$ (iv) $\dfrac{3x-2}{x+1}$

(v) $\dfrac{x-4}{2x+1}$ (vi) $\dfrac{4x-3}{3x-4}$

(vii) $\dfrac{5x}{1-6x}$

2 Integrate the function $f(x) = \dfrac{5x+2}{3x+1}$:

(i) using the substitution $u = 3x+1$;
(ii) by the method of worked example 9.

3 Integrate with respect to x:

(i) $\dfrac{x^2}{x-1}$

(ii) $\dfrac{3x^2+5x-7}{3x-1}$

(iii) $\dfrac{x^2+x-1}{2x+1}$

(iv) $\dfrac{x^3-x+1}{x+2}$

4 Evaluate:

(i) $\displaystyle\int_{-1}^{0}\frac{2x}{3x+4}\,dx$

R (ii) $\displaystyle\int_{0}^{1}\frac{x-2}{x+2}\,dx$

(iii) $\displaystyle\int_{-1}^{1}\frac{1-x}{3+2x}\,dx$

(iv) $\displaystyle\int_{1}^{5}\frac{x^2+1}{x+3}\,dx$

5 Find:

(i) $\displaystyle\int\frac{x}{(x+1)^2}\,dx$ (Hint: use the substitution $u = x+1$.)

(ii) $\displaystyle\int\frac{x-2}{(x+3)^2}\,dx$

(iii) $\displaystyle\int\frac{x^3}{(x-1)^2}\,dx$

6 Evaluate:

(i) $\displaystyle\int_{0}^{1}\frac{1}{(x+2)^2}\,dx$

(ii) $\displaystyle\int_{0}^{1}\frac{x+1}{(x+2)^2}\,dx$

(iii) $\displaystyle\int_{0}^{1}\left(\frac{x+1}{x+2}\right)^2 dx$

Questions 7–9: Integrate the given expressions with respect to x.

7 (i) $\dfrac{1}{(x+1)(x+2)}$ (ii) $\dfrac{2}{x^2-1}$

(iii) $\dfrac{2x}{x^2-1}$ (iv) $\dfrac{2x-3}{x(x-3)}$

R (v) $\dfrac{x-2}{x^2-x-6}$ (vi) $\dfrac{x+1}{(x-1)(2x+3)}$

8 (i) $\dfrac{1}{x(x^2+1)}$

(ii) $\dfrac{x+1}{(x-2)(x^2+2)}$

(iii) $\dfrac{x^2-4x}{(x+1)(x^2+4)}$

(iv) $\dfrac{3x+7}{(x-1)(x^2+x+3)}$

Integration and differential equations 197

⑩ 9 (i) $\dfrac{x^3}{x^2+3x+2}$

→ (ii) $\dfrac{2x^3-3x^2-x+5}{2x^2+x-1}$

⑩ 10 Evaluate:

(i) $\displaystyle\int_1^2 \dfrac{2}{(2x+1)(2x+3)}\,dx$

(ii) $\displaystyle\int_0^1 \dfrac{1}{4-x^2}\,dx$

(iii) $\displaystyle\int_2^3 \dfrac{1}{x(x^2-1)}\,dx$

R (iv) $\displaystyle\int_3^4 \dfrac{2x+1}{(x-2)(x^2+1)}\,dx$

(v) $\displaystyle\int_3^4 \dfrac{x(2x+1)}{(x-2)(x^2+1)}\,dx$

→ (vi) $\displaystyle\int_0^1 \dfrac{x^4+x^3+4x^2+5x-4}{(x+1)(x^2+4)}\,dx$

⑩ 11 Use the substitution $t=\tan\tfrac{1}{2}x$ to find:

• (i) $\displaystyle\int_0^{\frac{1}{6}\pi} \dfrac{1}{5\cos x-4}\,dx$

(ii) $\displaystyle\int_0^{\frac{2}{3}\pi} \dfrac{1}{3+5\cos x}\,dx$

→ (iii) $\displaystyle\int_0^{\frac{1}{4}\pi} \dfrac{1}{4\cos x-3\sin x}\,dx$

R (iv) $\displaystyle\int_0^{\frac{1}{3}\pi} \dfrac{1}{1-\sin x+2\cos x}\,dx$

⑩→ 12 Use the substitution $t=\tan\tfrac{1}{2}x$ to find $\int\sec x\,dx$. Prove that your answer agrees with the result of worked example 5(iii).

⑩ 13 By first dividing the numerator and denominator of the integrand by $\cos^2 x$, and then using the substitution $u=\tan x$, find:

(i) $\displaystyle\int_0^{\frac{1}{6}\pi} \dfrac{1}{4\cos^2 x-9\sin^2 x}\,dx$

(ii) $\displaystyle\int_0^{\frac{1}{4}\pi} \dfrac{1}{9-8\sin^2 x}\,dx$

14 Find:

(i) $\displaystyle\int \dfrac{1}{x^2-a^2}\,dx$ (ii) $\displaystyle\int \dfrac{x}{a^2-x^2}\,dx$

(iii) $\displaystyle\int \dfrac{1}{x^2+a^2}\,dx$ (iv) $\displaystyle\int \dfrac{x}{a^2+x^2}\,dx$

15 Find the area of the region bounded by the lines $x=-1$, $x=1$, the x-axis and the curve:

(i) $y=\dfrac{1}{4-x^2}$

(ii) $y=\dfrac{1}{x^2-4}$

R 16 Find the area of the region bounded by the curve $y=x/(x+1)$, the x-axis and the line $x=1$.
Find the volume generated when the region is rotated completely about the x-axis.

4.2:6 Integration by parts

Theorem:

$$\int u\dfrac{dv}{dx}\,dx = uv - \int v\dfrac{du}{dx}\,dx \qquad (10)$$

Proof
Consider the product rule for differentiation:

$$\dfrac{d}{dx}(uv) = u\dfrac{dv}{dx} + v\dfrac{du}{dx}$$

Integrating this equation with respect to x:

$$uv = \int u\dfrac{dv}{dx}\,dx + \int v\dfrac{du}{dx}\,dx$$

Rearranging this equation gives equation (10).

When faced with integrating the product of two functions by means of this formula, we must choose which will be u and which dv/dx.
(a) We must be able to integrate dv/dx (to get v).
(b) If possible du/dx should be simpler than u. (Why?)

Worked example 11 Find $\int x \cos 2x \, dx$.

$u = x \Rightarrow du/dx = 1$; $dv/dx = \cos 2x \Rightarrow v = \tfrac{1}{2} \sin 2x$. So

$$\int x \cos 2x \, dx = \tfrac{1}{2} x \sin 2x - \tfrac{1}{2} \int \sin 2x \, dx$$
$$= \tfrac{1}{2} x \sin 2x + \tfrac{1}{4} \cos 2x + c$$

Worked example 12 Evaluate:

(i) $\int x^2 \ln x \, dx$ (ii) $\int \ln x \, dx$

(i) We must put $u = \ln x$ since we cannot integrate $\ln x$—yet.

$$u = \ln x \Rightarrow \frac{du}{dx} = \frac{1}{x}. \quad \frac{dv}{dx} = x^2 \Rightarrow v = \frac{1}{3} x^3.$$

So
$$\int x^2 \ln x \, dx = \tfrac{1}{3} x^3 \ln x - \tfrac{1}{3} \int x^3 \frac{1}{x} \, dx$$
$$= \tfrac{1}{3} x^3 \ln x - \tfrac{1}{9} x^3 + c$$

(ii) We use the method of part (i):

$$u = \ln x \Rightarrow \frac{du}{dx} = \frac{1}{x}. \quad \frac{dv}{dx} = 1 \Rightarrow v = x.$$

So
$$\int \ln x \, dx = x \ln x - \int x \frac{1}{x} \, dx$$
$$= x \ln x - x + c$$

Worked example 13 Find $\int e^{3x} \sin x \, dx$.

Let $I = \int e^{3x} \sin x \, dx$.

$$u = e^{3x} \Rightarrow \frac{du}{dx} = 3 e^{3x} \quad \frac{dv}{dx} = \sin x \Rightarrow v = -\cos x$$

$$I = -e^{3x} \cos x + 3 \int e^{3x} \cos x \, dx$$

$$u = e^{3x} \Rightarrow \frac{du}{dx} = 3 e^{3x} \quad \frac{dv}{dx} = \cos x \Rightarrow v = \sin x$$

$$I = -e^{3x} \cos x + 3(e^{3x} \sin x - 3 \int e^{3x} \sin x \, dx)$$
$$= -e^{3x} \cos x + 3 e^x \sin x - 9I$$
$$10I = -e^{3x} \cos x + 3 e^{3x} \sin x$$
$$I = \tfrac{1}{10}(-e^{3x} \cos x + 3 e^{3x} \sin x)$$

Exercise 4.2:6

⑪ : **1** (i) Find $\int x\,e^x\,dx$ using integration by parts with $u = x$, $dv/dx = e^x$.

(ii) Now try to find the same integral using $u = e^x$, $dv/dx = x$. What goes wrong?

⑪ **2** Integrate with respect to x:

(i) $x \sin x$
(ii) $x\,e^{3x}$
(iii) $x \cos 5x$
→ (iv) $x \sec^2 x$

⑪ **3** Find $\int 2^x\,dx$. Hence find $\int x\,2^x\,dx$.

⑪ **4** Using integration by parts twice in each case, integrate with respect to x:

• (i) $x^2\,e^x$
(ii) $x^2\,e^{-2x}$
(iii) $x^2 \sin x$
→ (iv) $x^2 \cos 3x$

⑪ **5** Evaluate:

• (i) $\int_0^{\frac{1}{2}\pi} x \cos x\,dx$

R (ii) $\int_0^{\frac{1}{4}\pi} x \sin 2x\,dx$

→ (iii) $\int_0^{\ln 2} x\,e^{-x}\,dx$

(iv) $\int_{\frac{1}{4}\pi}^{\frac{1}{2}\pi} x \operatorname{cosec}^2 x\,dx$

(v) $\int_0^{\frac{1}{2}\pi} x^2 \cos x\,dx$

R (vi) $\int_0^1 x^2\,e^{5x}\,dx$

⑪ **6** Evaluate $\int_0^{\frac{1}{4}\pi} x \sec^2 x\,dx$.
Hence evaluate $\int_0^{\frac{1}{4}\pi} x \tan^2 x\,dx$.

⑫ **7** (i) Integrate the following with respect to x:
→ (a) $x^3 \ln x$
  (b) $\ln 4x$
  (c) $\ln\{x/(x+1)\}$
  (d) $\ln 1/\sqrt{x}$

R (ii) Integrate $x^n \ln x$ $(n \neq -1)$ with respect to x.

(iii) Integrate $\ln x/x$ with respect to x using
 (a) integration by parts;
 (b) the substitution $u = \ln x$.

⑫ → **8** Evaluate:

(i) $\int_1^{e^2} \ln\sqrt{x}\,dx$

(ii) $\int_1^e (\ln x)^2\,dx$

⑫ **9** By using the change of base formula for logarithms, find:

(i) $\int_1^{10} \log_{10} x\,dx$

(ii) $\int_1^{10} x \log_{10} x\,dx$

⑫ **10** Using the method of worked example 12, find:

(i) $\int \tan^{-1} x\,dx$

(ii) $\int \sin^{-1} x\,dx$

⑬ : **11** Find $\int e^x \sin x\,dx$.
What is wrong with the following method?

$$I = \int e^x \sin x\,dx$$

$$u = e^x \Rightarrow \frac{du}{dx} = e^x;\ \frac{dv}{dx} = \sin x \Rightarrow v = -\cos x$$

$$I = -e^x \cos x + \int e^x \cos x\,dx$$

$$u = \cos x \Rightarrow \frac{du}{dx} = -\sin x;\ \frac{dv}{dx} = e^x \Rightarrow v = e^x$$

$$I = -e^x \cos x + e^x \cos x + \int e^x \sin x\,dx$$

$$= \int e^x \sin x\,dx$$

$$= I$$

⑬ **12** Integrate with respect to x:

(i) $e^x \cos x$
→ (ii) $e^{2x} \sin x$
(iii) $e^{\frac{1}{2}x} \sin x \cos x$
R (iv) $e^{-3x} \sin 4x$

⑬ **13** If $C = \int e^{ax} \cos bx\,dx$, and $S = \int e^{ax} \sin bx\,dx$, where a and b are constants, prove that

$$aC - bS = e^{ax} \cos bx \quad \text{and} \quad aS + bC = e^{ax} \sin bx$$

Hence find C and S.

⑬ + **14** (i) Using integration by parts once only, show that if

$$I = \int \sec^3 \theta\,d\theta$$

then

$$I = \sec \theta \tan \theta + \int \sec \theta\,d\theta - I$$

Hence find I (see worked example 5(iii)).

(ii) Hence, using the substitution $x = \tan\theta$, find $\int_0^1 \sqrt{(x^2+1)}\,dx$.

Questions 15–17: Use the formulae of 4.1:4 Q23 to find the coordinates of the centroid of the region bounded by the given lines.

15 The curve $y = e^x$, the y-axis, the line $x = \ln 3$ and the x-axis.

R 16 The arc of the curve $y = \sin 2x$ between $x = 0$ and $x = \tfrac{1}{3}\pi$, the x-axis and the line $x = \tfrac{1}{3}\pi$.

17 The curve $y = \ln x$, the line $x = e$ and the x-axis.

4.3 Approximate integration

Suppose we want to find an area bounded by a curve, but either (i) we don't know an equation $y = f(x)$ for the curve, or (ii) we know $f(x)$ but can't integrate it, e.g. $f(x) = (\sin x)^{1/2}$, $f(x) = e^{x^2}$.

We can find an approximate numerical value of the area: we approximate the curve by a succession of either (a) straight line segments **(trapezium rule)** or (b) parabola segments **(Simpson's rule)**.

Trapezium rule

Suppose that the curve passes through the points $B(x_k, y_k)$ and $C(x_k + d, y_{k+1})$ (see fig. 4.3:1). We approximate the curve between B and C by the straight line segment BC.

Area under curve between $x = x_k$ and $x = x_k + d \approx$ area of trapezium
$$= \tfrac{1}{2}d(y_k + y_{k+1}) \qquad (1)$$

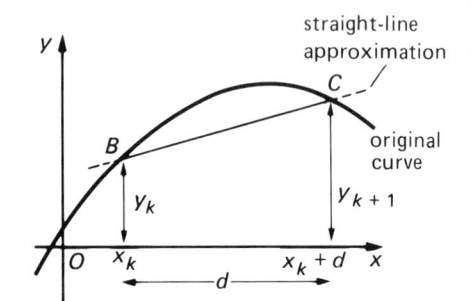

Fig. 4.3:1

We can divide the whole area under a curve into n strips each of width d as in fig. 4.3:2. (Note: the width of all the strips must be the same.) We approximate the curved boundary of each strip by a straight line segment.

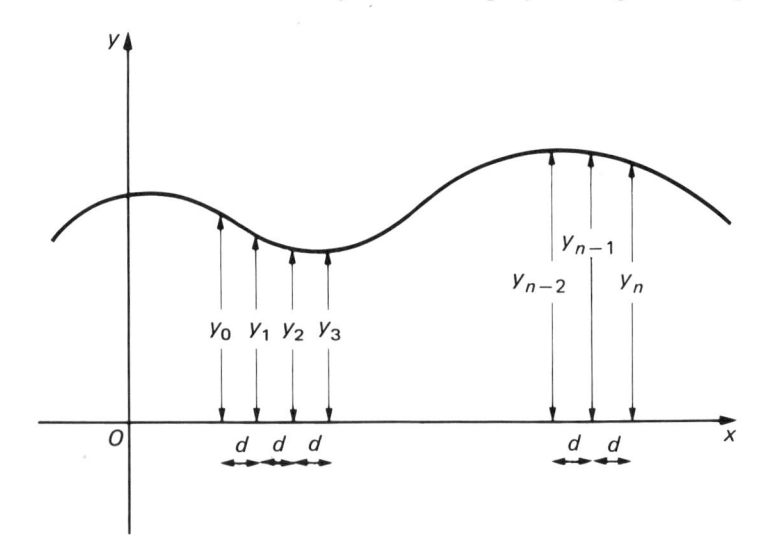

Fig. 4.3:2

Area under curve $\approx$ sum of areas of trapezia
$$= \tfrac{1}{2}d(y_0 + y_1) + \tfrac{1}{2}d(y_1 + y_2) + \tfrac{1}{2}d(y_2 + y_3) + \cdots$$
$$+ \tfrac{1}{2}d(y_{n-1} + y_n) \qquad \text{by equation (1)}$$

$$\boxed{\text{area} \approx \tfrac{1}{2}d\{y_0 + 2y_1 + 2y_2 + 2y_3 + \cdots + 2y_{n-1} + y_n\}} \qquad (2)$$

Integration and differential equations 201

Worked example 1 Find $\int_1^2 (1+x^2)^{1/2}\,dx$ using the trapezium rule with five ordinates.

We divide the area into four strips ($n=4$) each of width 0.25 ($d=0.25$), as in fig. 4.3:**3**.

x	1	1.25	1.5	1.75	2
$(1+x^2)^{1/2}$	1.4142	1.6008	1.8027	2.0156	2.2361
	y_0	y_1	y_2	y_3	y_4

$$\int_1^2 (1+x^2)^{1/2}\,dx \approx \tfrac{1}{2}d\{y_0 + 2y_1 + 2y_2 + 2y_2 + 2y_3 + y_4\}$$

$$= \frac{0.25}{2}\{1.4142 + 2(1.6008 + 1.8028 + 2.0156) + 2.2361\}$$

$$= 1.8111$$

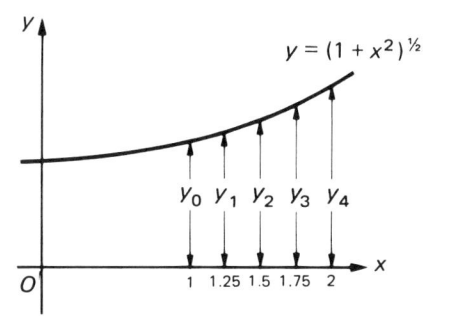

Fig. 4.3:**3**

Simpson's rule

Just as any two successive points on the curve can be joined by a straight line segment, any three points can be joined by a parabola segment (i.e. a short section of a parabola).

Suppose that the curve passes through successive points $A(x_k - d, y_{k-1})$, $B(x_k, y_k)$ and $C(x_k + d, y_{k+1})$, as in fig. 4.3:**4**. We approximate the curve between A and C by the parabola which passes through A, B and C (there is only one—why?).

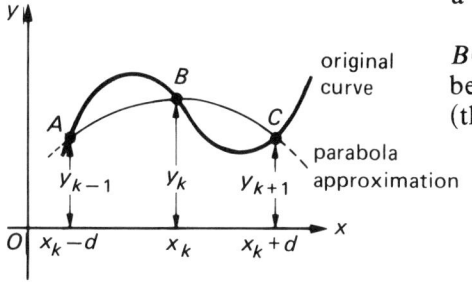

Area under curve between $x = x_k - d$ and $x = x_k + d$

$\approx$ area under parabola between $x = x_k - d$ and $x = x_k + d$

$$= \tfrac{1}{3}d(y_{k-1} + 4y_k + y_{k+1}) \qquad (3)$$

Fig. 4.3:**4** (See question 6.)

We can now divide the whole area under the curve into an *even* number n of strips, each of width d. We approximate the curve by a succession of $\tfrac{1}{2}n$ parabola segments, as in fig. 4.3:**5**.

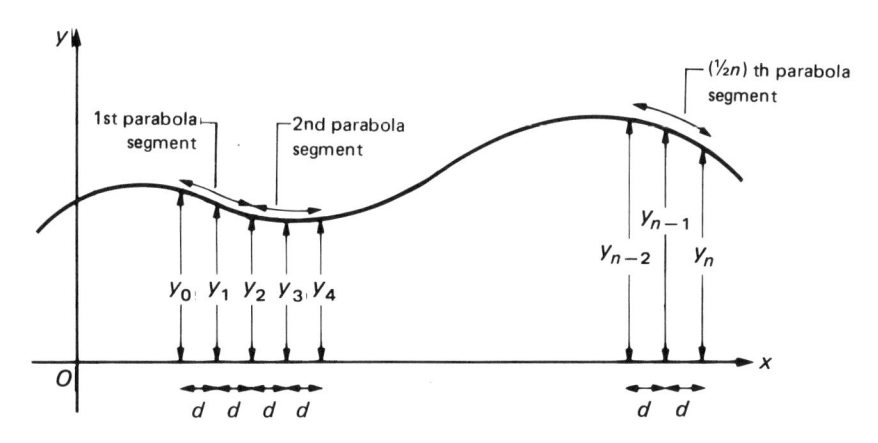

Fig. 4.3:**5**

Let the ordinates of the curve at successive points be $y_0, y_1, \ldots, y_n$. (Note: there is an *odd number* ($n+1$) of ordinates.)

area under curve $\approx$ sum of areas under parabola segments

$$= \tfrac{1}{3}d(y_0 + 4y_1 + y_2) + \tfrac{1}{3}d(y_2 + 4y_3 + y_4) + \ldots$$
$$+ \tfrac{1}{3}d(y_{n-2} + 4y_{n-1} + y_n) \qquad \text{by equation (3)}$$

202   Approximate integration

$$\text{area} = \tfrac{1}{3}d\{y_0 + 4y_1 + 2y_2 + 4y_3 + 2y_4 + 4y_5 + \ldots$$
$$+ 2y_{n-2} + 4y_{n-1} + y_n\} \qquad (4)$$

(Note: the sequence of coefficients is 1, 4, 2, 4, 2, 4, ..., 4, 2, 4, 2, 4, 1.)

Worked example 2   Find $\int_1^2 (1+x^2)^{1/2}\, dx$ using Simpson's rule with five ordinates.

From worked example 1, $d = 0.25$; $y_0 = 1.4142$, $y_1 = 1.6008$, $y_2 = 1.8028$, $y_3 = 2.0156$, $y_4 = 2.2361$.

$$\int_1^2 (1+x^2)^{1/2}\, dx \approx \tfrac{1}{3}d\{y_0 + 4y_1 + 2y_2 + 4y_3 + y_4\}$$

$$= \frac{0.25}{3}\{1.4142 + 4(1.6008 + 2.0156) + 2(1.8028) + 2.2361\}$$

$$= 1.8101$$

Actually, we can find this integral exactly, by using the substitution $x = \tan\theta$ (see 4.2:6 Q14). Its value (to four places of decimals) is 1.8101. We can see that here Simpson's rule gives a more accurate result than the trapezium rule; being a more sophisticated method, it usually does.

Exercise 4.3 :1

1 Using the trapezium rule with the given number of ordinates, find, to 4 sig. fig., approximations for:

(i) $\int_1^3 \dfrac{1}{x}\, dx$ (5 ordinates)

(ii) $\int_0^2 \dfrac{1}{1+x^2}\, dx$ (5 ordinates)

→ (iii) $\int_0^2 \dfrac{x}{1+x^3}\, dx$ (3 ordinates)

(iv) $\int_0^1 \dfrac{1}{\sqrt{(1+x^2)}}\, dx$ (3 ordinates)

(v) $\int_0^{0.4} \sqrt{(1-x^2)}\, dx$ (5 ordinates)

(vi) $\int_0^1 \sqrt{(x-x^3)}\, dx$ (5 ordinates)

→ (vii) $\int_0^\pi \sqrt{(\sin x)}\, dx$ (5 ordinates)

R (viii) $\int_0^{0.6} e^{x^2}\, dx$ (7 ordinates)

(ix) $\int_0^{\frac{1}{2}\pi} e^{\sin x}\, dx$ (5 ordinates)

(x) $\int_1^4 x \ln\, dx$ (7 ordinates)

2 Using the trapezium rule with 6 ordinates, find, to 5 sig. figs., approximations for:

(i) $\int_1^6 \sqrt{(x^2 - x)}\, dx$ (ii) $\int_0^2 2^x\, dx$

(iii) $\int_0^1 e^{x^3}\, dx$

3 Find approximations, to 4 sig. fig. for:

(i) $\int_0^4 \dfrac{1}{3+2x^2}\, dx$ (ii) $\int_0^1 e^{\sqrt{x}}\, dx$

(iii) $\int_0^\pi \sqrt{(1+\cos x)}\, dx$

using the trapezium rule with
(a) 3 ordinates; (b) 5 ordinates; (c) 9 ordinates.

4 The values of two related quantities x and y are shown in the table. Assuming that y is a continuous function of x, sketch a graph of y against x.

Use the trapezium rule to find approximations, to 3 sig. fig. to the given integrals.

• → (i)

x	0	1	2	3	4	5	6	7	8
y	1.31	1.38	1.47	1.57	1.70	1.88	2.13	2.46	3.00

(a) $\int_0^4 y\, dx$;   (b) $\int_0^8 y\, dx$.

R (ii)

x	0.3	0.45	0.6	0.75	0.9	1.05	1.2
y	8.2	10.7	12.6	13.3	11.9	8.5	1.3

(a) $\int_{0.3}^{0.9} y\,dx$; (b) $\int_{0.45}^{1.05} y\,dx$; (c) $\int_{0.6}^{1.2} y\,dx$.

(iii)

x	0	0.1	0.2	0.3	0.4	0.5	0.6
y	1.0000	1.414	1.792	2.139	2.461	2.761	2.041

(a) $\int_0^{0.6} \sqrt{y}\,dx$; (b) $\int_0^{0.6} \frac{1}{y}\,dx$; (c) $\int_{0.2}^{0.6} \frac{y^2}{x}\,dx$

5 Sketch the curve $y = x^3$.
Find $\int_0^1 x^3\,dx$:

(i) approximately, using the trapezium rule with 6 ordinates;
(ii) exactly, by direct integration.
Explain your answers.

② 6 (i) The curve $y = ax^2 + bx + c$ passes through the points (1, 3), (2, 10) and (3, 9). Find a, b and c (see 2.1:2 Q9), and hence find the area under the curve between $x = 1$ and $x = 3$.

(ii) The curve $y = ax^2 + bx + c$ passes through the points $(x_k - d, y_{k-1})$, (x_k, y_k) and $(x_k + d, y_{k+1})$. Show that $y_k = ax_k^2 + bx_k + c$ and find similar expressions for y_{k-1} and y_{k+1}. Hence show that the area under the curve between $x_k - d$ and $x_k + d$ is $\frac{1}{3}d(y_{k-1} + 4y_k + y_{k+1})$.

② → 7 Repeat question 1 using Simpson's rule.

② R 8 Repeat question 3 using Simpson's rule.

② R 9 Repeat question 4 using Simpson's rule.

10 The values of two related quantities x and y are shown in the table below.

x	0	1	2	4	7	9	10
y	4.0	4.2	4.4	4.9	5.7	6.3	6.6

Estimate $\int_0^{10} y\,dx$.

11 A straight canal of width 20 m and length 1000 m has a constant cross-section. The depth h m at a depth distance x m from one bank is shown in the table:

x	0	4	8	12	16	20
h	1.83	2.41	3.82	5.20	3.46	2.66

Estimate the volume of water in the canal.

Questions 12–18: Find the following integrals:

(i) approximately, to 4 sig. fig. using: (a) the trapezium rule; (b) Simpson's rule with the given number of ordinates;
(ii) exactly, correct to 4 dec. pls., by direct integration.

12 $\int_0^4 x^2\,dx$ (5 ordinates)

(The answer given by Simpson's rule should be exact—why?)

13 $\int_0^1 \frac{x}{x^2+1}\,dx$ (5 ordinates)

→ 14 $\int_0^2 x\,e^x\,dx$ (5 ordinates)

R 15 $\int_0^\pi \sin x\sqrt{(1+\cos x)}\,dx$ (7 ordinates)

16 $\int_2^5 \frac{1}{x^2-1}\,dx$ (7 ordinates)

17 $\int_0^1 10^x\,dx$ (5 ordinates)

→ 18 $\int_0^1 \frac{4}{1+x^2}\,dx$ (5 ordinates)

4.4 Differential equations

4.4:1 Formation of differential equations

A **differential equation** is an equation connecting x, y and the differential coefficients dy/dx, d^2y/dx^2, etc. For example,

$$y^2 \frac{dy}{dx} + 2xy = x \qquad (1)$$

$$\frac{d^2y}{dx^2} - 3\frac{dy}{dx} + 4y = \sin x \qquad (2)$$

204 Formation of differential equations

The **order** of a differential equation is the order of the differential coefficient of highest order in the equation (see section 1.3:4). So equation (1) is of first order; equation (2) of second order.

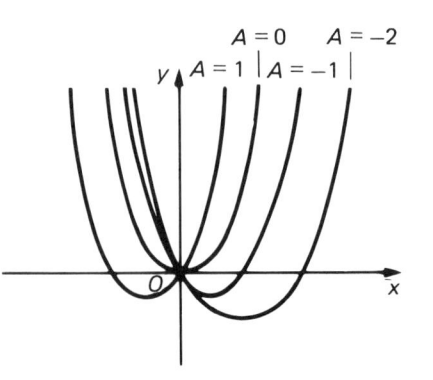

Worked example 1 Consider the family of curves
$$y = Ax + x^2 \qquad (3)$$
where A can take any value (see fig. 4.4:1).
$$\frac{dy}{dx} = A + 2x$$
Now $\frac{y}{x} = A + x$ from equation (3), so
$$\frac{dy}{dx} = \frac{y}{x} + x \qquad (4)$$

Fig. 4.4:1

Equation (4) is a **first order differential equation**. It gives a relationship between the coordinates (x, y) of any point P on any of the curves defined by equation (3), and the gradient of the curve at P. Equation (4) is satisfied by each member of the family of curves $y = Ax + x^2$.

The family of curves is called the **general solution** of the differential equation (4). The curves are called **integral curves** (see section 4.4:1). An extra piece of information, called an **initial condition**, specifies a particular curve, e.g. $y = 3$ when $x = 1$ specifies $A = 2$, i.e. the curve $y = 2x + x^2$.

The problem is, given a first order differential equation such as equation (4), to find its general solution. Generally this is very difficult, but as we shall see there are certain simple types which can be solved easily.

Exercise 4.4:1

① **1** (i) If $y = A\,e^{4x}$, show that
$$\frac{dy}{dx} = 4y$$

(ii) If $y = Ax\,e^{-x^2}$, show that
$$x\frac{dy}{dx} = (1 - 2x^2)y$$

→ (iii) If $y = \tfrac{1}{2}x + A/x$, show that
$$\frac{dy}{dx} + \frac{y}{x} = 1$$

① **2** In each of the following cases, find a first order differential equation for y which does not contain the constant A:

(i) $y = Ax^2$
(ii) $xy = A$
(iii) $y = A \sin x$
(iv) $y^2 = x + A$
(v) $x^2 - y^2 = A$
(vi) $y^2 = A \cos 2x$
(vii) $\dfrac{1+y}{1-y} = A\,e^{3x}$
R (viii) $y = \sin x + A \cos x$
→ (ix) $y = 1 + A\,e^{-x^2}$

① **3** Sketch the curves $y = Ax^2 - x$ in the cases $A = -1$, $A = -\tfrac{1}{2}$, $A = 0$, $A = \tfrac{1}{2}$, $A = 1$.
Show that the gradient dy/dx of any of the curves at the point $P(x, y)$ satisfies the equation
$$x\frac{dy}{dx} = 2y - x$$

① **4** Sketch the curves $y = A\,e^{x^2}$ in the cases $A = 2$, $A = 1$, $A = 0$, $A = -1$, $A = -2$.
Show that the gradient dy/dx of any of the curves at the point $P(x, y)$ satisfies the equation
$$\frac{dy}{dx} = 2xy$$

Which of the curves passes through the point:
(i) $(0, 3)$ (ii) $(1, 2)$?

4.4:2 First order differential equations with separable variables

Worked example 2 Find the general solution of the equation

$$\frac{dy}{dx} = 2xy \quad (y > 0) \tag{5}$$

$$\frac{dy}{dx} = 2xy \Rightarrow \frac{1}{y} \cdot \frac{dy}{dx} = 2x \Rightarrow \int \frac{1}{y} \cdot \frac{dy}{dx} dx = \int 2x \, dx$$

By section 4.2:1 Theorem 1,

$$\int \frac{1}{y} dy = \int 2x \, dx \tag{6}$$

$$\Rightarrow \qquad \ln y = x^2 + c$$

So
$$y = e^{x^2 + c}$$
$$= e^{x^2} \cdot e^c$$
$$= A\, e^{x^2} \text{ where } A = e^c$$

(Note: equation (6) only requires one arbitrary constant c. If we write $\ln y + k = x^2 + k'$, then we can put $c = k' - k$. Note also that we have written $\ln y$ rather than $\ln |y|$ since we are assuming that $y > 0$.)

In practice we can skip safely from equation (5) to equation (6), writing

$$\frac{dy}{dx} = 2xy \Rightarrow \int \frac{1}{y} dy = \int 2x \, dx, \text{ etc.}$$

We have effectively **separated the variables** of the original equation (5), putting all the terms containing x (including dx) on one side, and all the terms containing y (including dy) on the other.

In general, any differential equation which can be written in the form

$$f(y) \frac{dy}{dx} = g(x) \tag{7}$$

is called a **separable variable differential equation**. We solve it by separating the variables x and y, i.e. by writing

$$\int f(y) \, dy = \int g(x) \, dx$$

Worked example 3 Solve the equation

$$x \frac{dy}{dx} - xy = y \quad (x > 0, y > 0)$$

given that $y = 1$ when $x = 1$.

$$x \frac{dy}{dx} - xy = y$$

So
$$x \frac{dy}{dx} = (1 + x) y$$

$$\Rightarrow \quad \frac{1}{y}\frac{dy}{dx} = \frac{1+x}{x}$$

$$\Rightarrow \quad \int \frac{1}{y}\,dy = \int \frac{1+x}{x}\,dx \qquad (8)$$

$$= \int \left(\frac{1}{x}+1\right) dx$$

Integrating:

$$\ln y = \ln x + x + c \qquad (9)$$

(We have written $\ln x$, $\ln y$ instead of $\ln |x|$, $\ln |y|$ since $x > 0$, $y > 0$.)

$y = 1$ when $x = 1 \Rightarrow c = -1$. So from equation (9),

$$\ln y = \ln x + x - 1 \qquad (10)$$

$$\Rightarrow \quad y = e^{\ln x + x - 1}$$

$$= e^{\ln x} \cdot e^{x-1}$$

$$= x\, e^{x-1}$$

Exercise 4.4:2

1 Solve each of the following differential equations and sketch the integral curves in five or six simple cases (as in worked example 1):

(i) $\dfrac{dy}{dx} = y$

(ii) $\dfrac{dy}{dx} = -2y$

(iii) $\dfrac{dy}{dx} = -y^2$

(iv) $x\dfrac{dy}{dx} = y$

(v) $y\dfrac{dy}{dx} + x = 0$

For each part of questions 2, 3 and 4 solve each differential equation with the given initial conditions.

2 (i) $\dfrac{dy}{dx} = 4y$; $y = 3$ when $x = 0$

(ii) $\dfrac{dy}{dx} = 3y + 1$; $y = 5$ when $x = \ln 2$

(iii) $\dfrac{dy}{dx} = 2 - y$; $y = 1$ when $x = 0$

3 (i) $\dfrac{y}{x} \cdot \dfrac{dy}{dx} = -1$; $y = 3$ when $x = 4$

(ii) $x\dfrac{dy}{dx} = \dfrac{y^2 + 1}{y}$; $y = 0$ when $x = 1$

(iii) $x\dfrac{dy}{dx} = y(2x^2 + 1)$; $y = e$ when $x = 1$

(iv) $2y(x+1)\dfrac{dy}{dx} - y^2 = 4$; $y = 2$ when $x = 3$

(v) $\dfrac{dy}{dx} = y^2(1 + x)$; $y = 1$ when $x = 1$

(vi) $x^2\dfrac{dy}{dx} = e^y$; $y = 0$ when $x = 1$

(vii) $2\dfrac{dy}{dx} = 1 - y^2$; $y = 0$ when $x = 0$

(viii) $(1+x)^2\dfrac{dy}{dx} + y^2 = 1$; $y = 0$ when $x = 0$

4 (i) $x^2\dfrac{dy}{dx} = y(y-1)$; $y = 1/(1-e)$ when $x = -1$

(ii) $\sqrt{(x^2+1)}\dfrac{dy}{dx} - xy = 0$; $y = e$ when $x = 0$

(iii) $\dfrac{dy}{dx} - y \tan x = 0$; $y = 1$ when $x = \tfrac{1}{3}\pi$

(iv) $(\tan x)\dfrac{dy}{dx} = 1 - y^2$; $y = 0$ when $x = \tfrac{1}{6}\pi$

(v) $(1+x^2)\dfrac{dy}{dx} = x(1 - y^2)$; $y = 0$ when $x = 1$

(vi) $(1 + \cos 2x)\dfrac{dy}{dx} = y \sin 2x$; $y = 2$ when $x = \tfrac{1}{4}\pi$

(vii) $x^2\dfrac{dy}{dx} + 1 + y^2 = 0$; $y = 1$ when $x = 4/\pi$

(viii) $x\dfrac{dy}{dx} = \tan y$; $y = \tfrac{1}{6}\pi$ when $x = 1$

③ **5** In each of the following cases find y in terms of x, sketch the graph of y against x and find the equation of the tangent at the point $(1, 1)$:

(i) $\dfrac{dy}{dx} + 2xy = 0;\ y = 1$ when $x = 1$

→ (ii) $x\dfrac{dy}{dx} + (2x^2 - 1)y = 0;\ y = 1$ when $x = 1$

③ **6** If $dy/dx = 4 - x/y - 3$ and $y = 0$ when $x = 0$, show that (x, y) lies on a circle. Sketch the circle and find the equation of the tangent to the circle at the origin.

③ **R 7** Find y in terms of x when

$$x\frac{dy}{dx} + y = xy;\ y = 1/e \text{ when } x = 1$$

Show that the maximum value of y is $1/e$. Sketch the graph of y against x. Find the equation of the tangent at the point of inflexion.

③ **8** Solve the equation $dx/dt = kx(N - x)$, where k and N are positive constants, given that $x = 1$ when $t = 0$. Show that as $t \to \infty$, $x \to N$, and sketch a graph of x against t, for $t \geq 0$.

9 (i) Explain why it is impossible to separate the variables in the differential equation

$$\frac{dy}{dx} + y = 2x$$

If $u = 2x - y$, find dy/dx in terms of du/dx. Hence use the substitution $u = 2x - y$ to solve the equation, given that $y = 0$ when $x = 0$.

(ii) Use the substitution $u = x^2 + y$ to solve the differential equation

$$(1 - x)\frac{dy}{dx} + 2y + 2x = 0$$

given that $y = e$ when $x = 0$.

• → **10** If $dy/dx = p$, show that $d^2y/dx^2 = dp/dx$. Hence use the substitution $dy/dx = p$ to solve the (second order) differential equations:

(i) $(1 + x^2)\dfrac{d^2y}{dx^2} = 2x\dfrac{dy}{dx}$ (ii) $x^2\dfrac{d^2y}{dx^2} = \left(\dfrac{dy}{dx}\right)^2$

leaving two arbitrary constants in your answer in each case.

11 If $dy/dx = p$, show that

$$\frac{d^2y}{dx^2} = p\frac{dp}{dy}$$

Hence use the substitution $dy/dx = p$:

(i) to solve the equation

$$y\frac{d^2y}{dx^2} = \left(\frac{dy}{dx}\right)^2$$

leaving two arbitrary constants in your answer;

(ii) to show that, if $d^2y/dx^2 + 4y = 0$, then

$$\frac{dy}{dx} = \pm 2\sqrt{(a^2 - y^2)}$$

where a is constant.

12 The gradient of a curve at the point (x, y) is proportional to the product of x and y. The point $(0, 3)$ is a turning point of the curve.
Find the equation of the curve, and sketch it.

4.4:3 Natural occurrence of differential equations

Worked example 4 At time t, the rate of increase of the number of micro-organisms in a controlled environment is k times the number of micro-organisms present, where k is a positive constant. Initially the number of micro-organisms present is a. If after 3 hours the number present is $2a$, find the time that elapses before the number present is $3a$.

Let the number of micro-organisms present at time t be x.
Let's translate into mathematical notation the first two sentences of the question. The first gives:

$$\frac{dx}{dt} = kx \quad (k > 0) \tag{11}$$

The second gives:

$$\text{when } t = 0,\ x = a \tag{12}$$

Equation (11) is a differential equation governed by the initial condition

208 *Natural occurrence of differential equations*

(12); we solve it by the method of 4.4:2, and find

$$x = a\, e^{kt} \tag{13}$$

When $t = 3$, $x = 2a$, so $2a = a\, e^{3k} \Rightarrow e^{3k} = 2 \Rightarrow 3k = \ln 2 \Rightarrow k = \tfrac{1}{3}\ln 2$. Thus

$$x = a\, e^{(\frac{1}{3}\ln 2)t}$$

When $x = 3a$, $3a = a\, e^{(\frac{1}{3}\ln 2)t} \Rightarrow e^{(\frac{1}{3}\ln 2)t} = 3 \Rightarrow (\tfrac{1}{3}\ln 2)t = \ln 3$. Thus

$$t = 3\,\frac{\ln 3}{\ln 2}$$

$$\approx 4.755 \text{ hours}$$

The graph of x against t is shown in fig. 4.4:**2**. (C.f. 2.5:1 WE2.)

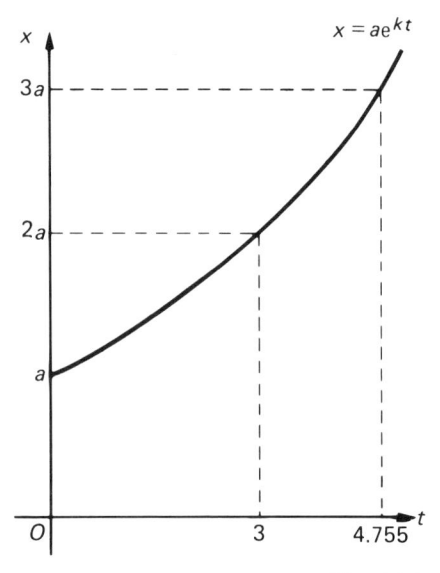

Fig. 4.4:**2**

Exercise 4.4:3

1 The rate of increase of the number of cells of a yeast is proportional to the number of cells present.

(i) If the number of cells at time t is n, show that $n = n_0\, e^{kt}$, where n_0 is the initial number of cells and k is a constant.

(ii) If the yeast takes 30 minutes to double in volume, show that $k \approx 0.023$ min^{-1}.

(iii) How long will it take to treble in volume? (See 2.5:1 WE2.)

2 A sample of radium loses mass at a rate which is proportional to the amount present.

(i) If its mass at time t is m, show that $m = m_0\, e^{-kt}$, where m_0 is its initial mass and k is a constant. Sketch a graph of m against t.

(ii) If its half-life (i.e. the time taken for its mass to be halved) is 1600 years, show that $k \approx 4.33 \times 10^{-4}$ years^{-1}.

(iii) Show that the percentage decrease in the mass per century is constant—about 4.3% (see 2.5:1 Q12).

3 If the half-life (see question 2) of a radioactive element is 1000 years, find the time taken for the mass of the element to reduce to one quarter of its original value.

R 4 Newton's Law of Cooling states that the rate of decrease of the temperature of a substance is proportional to its excess temperature over its surroundings; i.e. if the excess temperature is θ, then

$$\frac{d\theta}{dt} = -k\theta$$

where k is constant. Show that if the initial excess temperature is θ_0, then $\theta = \theta_0\, e^{kt}$. Draw a graph of θ against t.

5 A bathfull of water is at a temperature of 80 °C in a room where the temperature remains constant at 20 °C. Using Newton's Law of Cooling (see question 4), show that after time t its temperature θ is given by

$$\theta = 20 + 60\, e^{-kt}$$

where k is constant. If it takes 5 minutes to reach 70 °C, show that $k \approx 0.036$ min^{-1}. Hence find

(i) the temperature after 10 minutes;

(ii) the time taken for the temperature to reach 60 °C.

Draw a graph of θ against t.

6 A man reads in the newspaper that humans can be kept alive at very low temperatures for hundreds of years. Excited at the prospect of the year 2300, he climbs into his deep freeze and closes the lid.

In the first hour, the temperature of his body falls by 10 °C, in the second hour by 6 °C. Find the fall in temperature in the third hour (see question 4).

In the fourth hour he begins to experience some discomfort...

7 A direct e.m.f. E is applied to a coil of inductance L and resistance R. The current i in the circuit satisfies the equation

$$E - L\frac{di}{dt} = iR$$

(E, L and R are constants).

(i) If $i = 0$ when $t = 0$, show that
$$i = \frac{E}{R}(1 - e^{-Rt/L})$$

Show that the current settles to a steady value $i_0 = E/R$.

(ii) The circuit is then broken, and the current now satisfies the equation
$$-L\frac{di}{dt} = iR$$

Show that $i = i_0 e^{-Rt/L}$.

Sketch a rough graph of i against t. (L/R is called the **time constant** of the circuit.)

④ 8 The change in gene frequency in the mosquito may be described by the differential equation
$$\frac{dq}{dn} = \frac{q(1-q)}{3+q}$$
where q is the gene frequency after n generations. If the initial gene frequency is q_0, show that
$$n = 3\ln\frac{q}{q_0} + 4\ln\left\{\frac{1-q_0}{1-q}\right\}$$

④ 9 The greatest chef in the world is making his masterpiece, a completely spherical oyster and kiwi fruit soufflé, which he serves with three complementary sauces. No-one knows how he does it, but visiting gourmets and scientists have noticed that when the soufflé is in the oven it retains its spherical shape and its volume increases at a rate which is proportional to its radius.

Show that the radius r cm of the soufflé at time t minutes after it has gone into the oven satisfies the differential equation
$$\frac{dr}{dt} = \frac{k}{r}$$
where k is a constant.

Given that the radius of the soufflé is 8 cm when it goes in the oven and an unbelievable 12 cm when it is done, 30 minutes later, find its radius after 15 minutes in the oven.
(See 1.3:6 WE6.)

④ 10 Coal is falling from a chute, making a pile in the shape of a cone of constant semi-vertical angle, at a rate which is inversely proportional to the height of the pile. Show that the height h of the pile at time t satisfies the differential equation
$$\frac{dh}{dt} = \frac{k}{h^3}$$
where k is a constant.

Initially the pile has height H. After time T the pile has grown to a height $2H$. Find, in terms of T, the time after which it has grown to a height $3H$.

④ 11 At any instant a meteorite is gaining volume due to two effects: matter is condensing onto the meteorite at a rate proportional to its surface area, and matter is attracted onto the meteorite by its gravitational field at a rate proportional to its volume.

Assuming that the two effects can be added together and that the meteorite remains spherical, show that its radius r at time t satisfies the differential equation
$$\frac{1}{\lambda}\frac{dr}{dt} = k + r$$
where k and λ are constants.

If the radius of the meteorite is initially r_0, find its radius at time $t = \lambda/\ln 2$.

④ 12 A lock on a canal is filled by opening the sluices; water then pours into the lock at a rate which is inversely proportional to the depth of the water in the lock. The minimum depth of the water is 2 m and the maximum depth 8 m. Given that the lock can be half-filled (depth 5 m) in 30 s, find the time taken to fill it.

④ 13 A certain drug is being administered to a patient in a hospital at a constant rate. The rate at which the drug is lost from his body is proportional to the amount x of the drug present in this body. Show that
$$\frac{dx}{dt} = R_0 - kx$$
and explain the significance of the constants R_0 and k. If his body is initially free of the drug, find x in terms of t.

Show that as $t \to \infty$, $x \to R_0/k$, and explain the significance of this result. Sketch a graph of x against t.

The half-life T of the drug is defined to be the time taken for the amount of drug in the body to fall by one half when administration is stopped. Show that $T = (\ln 2)/k$.

④ → 14 A disease is spreading through a population. It is thought that the rate of increase of the number of infected individuals at any time is proportional to the product of the number of infected individuals and the number of uninfected individuals at that time. If initially only one person is infected, show that eventually the whole population will become infected. (See 4.4:2 Q24.)

④ 15 In a chemical reaction in which a compound X is formed from a compound Y and other substances, the masses of X and Y present at time t seconds are x gram and y gram respectively. At any time the sum of the two masses is 20 g, and the rate at which x is increasing is proportional to the product of the two masses present at that time.

Write down a differential equation governing the reaction.

Initially there are 2 g of X present; at time $t = \ln 3$ there are 5 g of X present. Find the time at which there are 5 g of Y present.

④ 16 The pressure P, density ρ and temperature T of the atmosphere at height x above the Earth's surface are related by the equations

$$\frac{dP}{dx} = -g\rho, \quad P = k\rho T$$

where g is the (constant) acceleration due to gravity and k is a constant. At the Earth's surface, $P = P_0$, $\rho = \rho_0$, $T = T_0$.

(i) Given that

$$\frac{P}{P_0} = \left(\frac{\rho}{\rho_0}\right)^\gamma$$

where γ is a constant greater than 1, show that

$$\frac{dP}{dx} = -g\rho_0 \left(\frac{P}{P_0}\right)^{1/\gamma}$$

(ii) Hence show that

$$P = P_0 \left\{ 1 - \left(\frac{\gamma - 1}{\gamma}\right) \frac{gx}{kT_0} \right\}^{\gamma/(\gamma - 1)}$$

and find T in terms of x, g, k T_0 and γ.

Show that these results are meaningless if x exceeds a certain value. Why is this?

Miscellaneous exercise 4

Questions 1–10: You are stranded on a desert island with a Swiss army knife, a boxed set of old Abba hits and a pair of Calvin Klein jeans. How would you integrate the following with respect to x?

1 (i) $x^2 e^{-x}$ (ii) $x e^{-x^2}$

2 (i) $x\sqrt{(4-x)}$ (ii) $x\sqrt{(4-x^2)}$
    (iii) $\sqrt{(4-x^2)}$ (iv) $x^2\sqrt{(4-x^2)}$

3 (i) $\sec x \tan x$ (ii) $\sec x \tan^2 x$
    (iii) $\sec^3 x \tan x$ (iv) $x \sec x \tan x$

4 (i) $\dfrac{x-2}{\sqrt{(x+1)}}$ (ii) $\dfrac{x-2}{x+1}$
    (iii) $\dfrac{x-2}{(x+1)^2}$ (iv) $\dfrac{x-2}{(x+1)(x-3)}$

5 (i) $\dfrac{x}{3x-1}$ (ii) $\dfrac{x}{\sqrt{(3x-1)}}$
    (iii) $\dfrac{x}{\sqrt{(3x^2-1)}}$

6 (i) $\sin^2 x \cos x$ (ii) $\sin x \cos^2 x$
    (iii) $\sin^3 x$ (iv) $\sin^2 x$
    (v) $\sin 5x \cos x$

7 (i) $\sin x\sqrt{(1-\cos x)}$ (ii) $\sqrt{(1-\cos x)}$
    (iii) $\dfrac{1}{1-\cos x}$

8 (i) $x \cos^2 x$ (ii) $\sin^2 x \cos^2 x$

9 (i) $\dfrac{1}{4+x^2}$ (ii) $\dfrac{1}{4-x^2}$
    (iii) $\dfrac{1}{\sqrt{(4-x^2)}}$ (iv) $\dfrac{x}{4+x^2}$

10 (i) $\dfrac{2}{(x+1)(x+3)}$ (ii) $\dfrac{2}{(x^2+1)(x+3)}$
    (iii) $\dfrac{x}{(x^2+1)(x+3)}$

Questions 11–14: Evaluate the following integrals:

11 (i) $\displaystyle\int_0^\pi x \sin x \, dx$

    (ii) $\displaystyle\int_0^{\frac{1}{2}\pi} \sin 2x \cos x \, dx$

    (iii) $\displaystyle\int_1^2 \dfrac{1}{x(x+2)} \, dx$

12 (i) $\displaystyle\int_0^1 x\sqrt{(1+x^2)} \, dx$

    (ii) $\displaystyle\int_0^{\frac{1}{4}\pi} \tan^2 x \, dx$

    (iii) $\displaystyle\int_3^4 \dfrac{x+1}{(x-1)(x-2)} \, dx$

13 (i) $\displaystyle\int_1^5 \dfrac{x}{\sqrt{(3x+1)}} \, dx$

    (ii) $\displaystyle\int_0^1 (x+1) e^{-2x} \, dx$

    (iii) $\displaystyle\int_0^{\frac{1}{4}\pi} \tan^3 x \sec^2 x \, dx$

14 (i) $\displaystyle\int_1^e x^2 \ln x \, dx$

    (ii) $\displaystyle\int_0^{\frac{1}{6}\pi} \cos^2 2x \, dx$

    (iii) $\displaystyle\int_0^1 \dfrac{4}{\sqrt{(4-x^2)}} \, dx$

15 Find:

(i) $\displaystyle\int_2^3 \frac{2x+4}{x^3-x}\,dx$ (ii) $\displaystyle\int_2^3 \frac{2x+4}{x^3+x}\,dx$

16 Show that:

(i) $\displaystyle\int_0^2 \frac{x(2-x)}{(x+2)(x^2+4)}\,dx = \tfrac{1}{4}\pi - \ln 2$

(ii) $\displaystyle\int_0^2 \frac{2(2-x)}{(x+2)(x^2+4)}\,dx = \tfrac{1}{2}\ln 2$

17 Find:

(i) $\displaystyle\int_0^{\frac{1}{2}\pi} \cos 6x \sin 4x\,dx$

(ii) $\displaystyle\int_0^{\frac{1}{2}\pi} \sin x \cos^4 x\,dx$

(iii) $\displaystyle\int_0^{\frac{1}{2}\pi} \frac{\cos x}{\sqrt{(1+\sin x)}}\,dx$

(iv) $\displaystyle\int_0^{\frac{1}{2}\pi} x \cos 3x\,dx$

18 Show that:

(i) $\displaystyle\int_0^{\frac{1}{2}} \frac{1}{\sqrt{(1-x^2)}}\,dx = \frac{\pi}{6}$

(ii) $\displaystyle\int_0^{\frac{1}{2}} \frac{x}{\sqrt{(1-x^2)}}\,dx = 1 - \tfrac{1}{2}\sqrt{3}$

(iii) $\displaystyle\int_0^{\frac{1}{2}} \frac{x^2}{\sqrt{(1-x^2)}}\,dx = \tfrac{1}{12}\pi - \tfrac{1}{8}\sqrt{3}$

(iv) $\displaystyle\int_0^{\frac{1}{2}} \frac{1}{1-x^2}\,dx = \tfrac{1}{2}\ln 3$

(v) $\displaystyle\int_0^{\frac{1}{2}} \frac{x^2}{1-x^2}\,dx = \tfrac{1}{2}\ln 3 - \tfrac{1}{2}$

19 (i) Using the substitution $x = \sin\theta$, evaluate

$$\int_0^{\frac{1}{2}} \sqrt{(1-x^2)}\,dx$$

(ii) Using the substitution $x = \sec^2\theta$, evaluate

$$\int_2^4 \frac{dx}{x^2\sqrt{(x-1)}}$$

(iii) Using the substitution $x = 3\tan\theta$, evaluate

$$\int_0^3 \frac{dx}{(9+x^2)^{3/2}}$$

20 (i) Using the substitution $x = 4\sin^2\theta$, show that

$$\int_2^3 \frac{dx}{\sqrt{(4x-x^2)}} = \frac{1}{3}\pi$$

(ii) Using the substitution $t = \tan\tfrac{1}{2}x$, show that

$$\int_0^{\frac{1}{3}} \frac{dx}{3 - 3\sin x + \cos x} = \ln\tfrac{1}{4}(5+\sqrt{3})$$

(iii) Using the substituion $u = 1/x$, show that

$$\int_{1/\sqrt{3}}^{1/\sqrt{2}} \frac{dx}{x^2\sqrt{(1-x^2)}} = \sqrt{2} - 1$$

21 (i) Using the substitution $u = \tfrac{1}{2}\pi - x$, evaluate

$$\int_0^{\frac{1}{2}\pi} \frac{dx}{1+\tan x}$$

(ii) Show that

$$\int_1^2 \lg x\,dx = \lg \frac{4}{e}$$

(iii) Find

$$\int_{-\frac{1}{4}\pi}^{\frac{1}{4}\pi} x \cos x\,dx$$

Explain your answer.

+ 22 Differentiate $\ln(\sec x + \tan x)$. Hence find $\int \sec(x-\alpha)\,dx$, where α is constant.

Express $3\cos x + 4\sin x$ in the form $R\cos(x-\beta)$, and hence find

$$\int_0^{\frac{1}{2}\pi} \frac{dx}{3\cos x + 4\sin x}$$

23 Integrate with respect to x: (i) $\ln x$; (ii) $(\ln x)^2$.

Find the area of the region bounded by the curve $y = e^x$, the line $y = 2$ and the y-axis.

Find the volume generated when this region is rotated completely about the y-axis.

24 Find the area of the finite region enclosed by the line $y = 4x$ and the curve $y^2 = 16x$.

Find the volume generated when this region is rotated completely about (i) the x-axis; (ii) the y-axis.

25 Find the area of the region bounded by:

(i) the curve $y = 2x - x^2$ and the x-axis;
(ii) the curve $y^2 = 4x$ and the line $x - 2y + 3 = 0$;
(iii) the curve $y = \sin x$ and the line $y = 2x/\pi$.

26 (i) Sketch the curve $y = 2x/(x^2+1)$. Find the area of the region bounded by the curve, the x-axis and the line $x = 1$.

(ii) Sketch the curve $y = 1/(1+x)(3-x)$. Find the area of the region bounded by the curve, the x-axis, the y-axis and the line $x = 2$.

212   Miscellaneous exercise 4

27 Given that $f(x) = (x-1)/(x+1)$, sketch the graphs of $y = f(x)$ and $y = |f(x)|$.
Find:
(i) $\int_0^2 f(x)\,dx$ (ii) $\int_0^2 |f(x)|\,dx$

28 The function $f : \mathbb{R} \to \mathbb{R}$ is periodic with period 2 and
$$f(x) = \begin{cases} x+1 & 0 \leq x < 1 \\ 2/x & 1 \leq x < 2 \end{cases}$$
Sketch the graph of $f(x)$ for $-4 < x < 6$.
Find:
(i) $\int_{-1}^{2} f(x)\,dx$ (ii) $\int_{-1/2}^{5/2} f(x)\,dx$

29 Sketch on the same axes the curve $y = 2x/(3-x)$ and the line $3x - 4y + 1 = 0$. Find the area of the finite region bounded by the curve and the line.

30 Sketch the curve $y = x\,e^{-x}$, marking in the coordinates of the turning point. Find the area of the finite region bounded by the curve, the y-axis and the line $y = 1/e$.

+ **31** (i) Using integration by parts, find $\int_0^1 x \tan^{-1} x\,dx$.
(ii) The function $f : \mathbb{R} \to \mathbb{R}$ is defined by $f(x) = \tan^{-1} x$. Sketch the curve $y = f(x)$. Find the area of the region bounded by the curve, the x-axis and the line $x = 1$.
(iii) Sketch also the curve $y = f'(x)$. Find $\int_0^1 f'(x)\,dx$.

32 (i) Sketch the graph of $y = |2 - 4x|$. Find
$$\int_0^2 |2 - 4x|\,dx$$
(ii) Find the volume generated when the region bounded by the curve $y = e^x$, $y = 1$ and $x = \ln 2$ is rotated completely about the line $y = 1$.
(iii) Find the area bounded by the curve $y = 3x/(x+2)$, the line $y = 2$ and the y-axis.

33 Show that $\int_0^1 4\,dx/(1+x^2) = \pi$.
Using the trapezium rule with 6 ordinates, estimate the value of $\int_0^1 4\,dx/(1+x^2)$. Hence find, to 2 dec. pls., an approximate value of π.

34 Find $\int_0^2 x\,e^{x^2}\,dx$:
(i) approximately, using the trapezium rule with 5 ordinates;
(ii) approximately, using Simpson's rule with 5 ordinates;
(iii) exactly, using the substitution $u = x^2$.
Give your answers to 3 dec. pls.

35 Differentiate $x\,e^{x^2}$ with respect to x.
If $I = \int_0^1 e^{x^2}\,dx$ and $J = \int_0^1 x^2 e^{x^2}\,dx$, show that $I + 2J = e$.
Using the trapezium rule with 5 ordinates, find an approximate value of I. Hence find an approximate value of J.

36 (i) Obtain estimates for $\int_0^{\pi/6} x \cos^{\frac{1}{2}} x\,dx$:
(a) by using the approximation $\cos x \approx 1 - \tfrac{1}{2}x^2$;
(b) by using Simpson's rule with 3 ordinates.
(ii) Using the trapezium rule with 5 ordinates, find approximately the area of the finite region bounded by the x-axis, the arc of the curve $y = 1 - \cos^{\frac{1}{2}} x$ between $x = 0$ and $x = \pi$, and the line $x = \pi$.

37 (i) Using the substitution $t = \tan \tfrac{1}{2}x$, show that
$$\int_{\frac{1}{2}\pi}^{\frac{2}{3}\pi} \frac{dx}{1 + \sin x} = 2 - \sqrt{3}$$
(ii) Using the trapezium rule with 5 ordinates, find an approximate value for
$$\int_0^2 \frac{1}{\sqrt{(1+x^2)}}\,dx$$
giving your answer to 2 dec. pls.

38 (i) Given that $y^2 = A \cos 3x$, find a first order differential equation for y which does not contain the constant A.
(ii) Find y in terms of x when
$$x\frac{dy}{dx} = (x+1)y$$
given that $y = 0$ when $x = 0$. Sketch a graph of y against x and find the minimum value of y.

39 (i) Using the substitution $u = \sin x$, show that
$$\int_0^{\frac{1}{2}\pi} \frac{\cos x}{3 + \cos^2 x}\,dx = \tfrac{1}{4}\ln 3$$
(ii) Solve the differential equation
$$\tan x \frac{dy}{dx} - y = 0$$
given that $y = 1$ when $x = \tfrac{1}{6}\pi$.

40 Find y in terms of x given that
$$x\frac{dy}{dx} = y(y+1)$$
and $y = 1$ when $x = 2$. Sketch a graph of y against x, showing its asymptotes.

41 Given that

$$2\frac{dy}{dx} + xy^2 = y^2$$

and that $y = -1$ when $x = 1$, show that

$$y = \frac{4}{(x-3)(x+1)}$$

Sketch the graph of y against x.

42 The gradient of a curve at the point (x, y) is proportional to the product of x and y^2. The curve passes through the points $(0, 2)$ and $(1, -\frac{2}{3})$. Find the equation of the curve. Write down the equations of its asymptotes, and sketch it.

+ 43 Show that $(d/dx)(\ln \tan x) = 2 \cos 2x$.
 Find the solution $y = f(x)$ of the differential equation

$$(\sin 2x)\frac{dy}{dx} = 2y(1-y)$$

for which $y = \frac{1}{3}$ when $x = \frac{1}{4}\pi$.
 Show that $f'(x)$ is always positive. Hence find the maximum and minimum values of $f(x)$ as x varies between $\frac{1}{4}\pi$ and $\frac{1}{2}\pi$.

44 Differentiate $\ln \sec x$ with respect to x. Hence

(i) find $\displaystyle\int_0^{\frac{1}{6}\pi} \frac{2 \sin x \, dx}{\cos^2 x - \sin^2 x}$

(ii) find y in terms of x, given that

$$\frac{dy}{dx} = y(1 + \tan x)$$

and that $y = 1$ when $x = 0$.

45 Find the solution $y = f(x)$ of the differential equation

$$(1+x^2)\frac{dy}{dx} + 2xy = 0$$

given that $y = 1$ when $x = 0$. Sketch the curve $y = f(x)$, marking in the coordinates of its turning point. Find the area of the region bounded by the curve, the x-axis and the ordinates $x = -1$ and $x = 1$.

46 Find y in terms of x, given that:

(i) $(\sin y \cos x)\dfrac{dy}{dx} = 1 + \cos 2x$ and $y = 0$ when $x = 0$

(ii) $x\dfrac{dy}{dx} = \cos^2 y$ and $y = 0$ when $x = 1$

47 Find the solution $y = f(x)$ of the differential equation

$$x\frac{dy}{dx} + (2x^2 - 1)y = 0 \quad (x > 0)$$

for which $y = 2/e$ when $x = 1$.
 Sketch the curve $y = f(x)$ for $x > 0$. Find $\int_1^{\sqrt{2}} f(x) \, dx$.

+ 48 Show that the volume of a cap, of depth h cm, cut from a sphere of radius a cm $(a > h)$ is $\frac{1}{3}\pi h^2(3a - h)$ cm^3.
 A bowl is in the shape of a cap of depth $\frac{1}{2}a$ cm cut from a spherical shell of radius a cm. Water is being poured into the bowl at a constant rate of $\frac{1}{24}\pi a^3$ cm^3 s^{-1}. Show that, when the depth of water in the bowl is x cm,

$$24\frac{dx}{dt} = \frac{a^3}{x(2a - x)}$$

Hence show that, if the bowl is initially empty, it will take 5 s to fill it up.

49 A harassed hostess places a hot cake, whose temperature is 35 °C, in the fridge. After t minutes the temperature x °C of the cake satisfies the differential equation

$$10\frac{d^2x}{dt^2} + \frac{dx}{dt} = 0$$

Using the substitution $y = dx/dt$, find x in terms of t, given that when $t = 10 \ln 2$, $x = 15$.
 Show that, when the temperature of the cake has reached an acceptable 5 °C, its rate of cooling has fallen to one degree per minute.
 The hostess forgets about the cake and leaves it in the fridge. How cool does the cake get eventually?

50 A new and wonderful computer comes on the market. The rate of increase in the number of people x who own one is proportional to the product of x and $(N - x)$, where N is constant. Given that initially $x = \frac{1}{10}N$, and that after one week of aggressive selling $x = \frac{1}{4}N$, show that after t weeks

$$\frac{x}{N - x} = 3^{t-2}$$

What happens to x as $t \to \infty$? Sketch the graph of x against t. Find the time at which $\frac{3}{4}N$ people own one of these computers.
 Several weeks later, a newer and more wonderful computer comes on the market, a computer that solves any problem you set it *and* makes delicious coffee . . . .

5 Series and probability

5.1 Sequences and series

5.1:1 Sequences

> **Definition** A **sequence** is a list of numbers: $u_1, u_2, u_3, u_4, \ldots$ which may be finite or infinite.

For example:

(i) 2, 5, 8, 11, 14, ...
(ii) 1, 4, 9, 16, 25, ...
(iii) 2, 4, 8, 16, 32, ...
(iv) $1, \frac{1}{2}, \frac{1}{3}, \frac{1}{4}, \frac{1}{5}, \ldots$
(v) $5, \pi, -7.3, \sqrt{2}, \ldots$

Sometimes we can find a formula for the rth term u_r of the sequence; the formulae for sequences (i)–(v) above are:

(i) $u_r = 3r - 1$ (ii) $u_r = r^2$
(iii) $u_r = 2^r$ (iv) $u_r = 1/r$

We can then write the sequence $u_1, u_2, u_3, \ldots$ as $\{u_r\}$, e.g. 1, 4, 9, 16, 25, ... can be written as $\{r^2\}$.

Sometimes we cannot find a sequence, e.g. look at example (v).

Exercise 5.1:1

1 Suggest the next three terms in each of the following sequences. Explain how you get your answers.

(i) 1, 3, 7, 9, ...
(ii) 3, −1, −5, ...
(iii) 2, 8, 18, 32, 50, ...
(iv) 0, 7, 26, 63, 124, ...
(v) $3, 1, \frac{1}{3}, \frac{1}{9}, \ldots$
(vi) $3, -1, \frac{1}{3}, -\frac{1}{9}, \ldots$
(vii) $\frac{1}{2}, \frac{1}{4}, \frac{1}{6}, \frac{1}{8}, \ldots$
(viii) $-\frac{1}{2}, \frac{1}{4}, -\frac{1}{6}, \frac{1}{8}, \ldots$
(ix) $\frac{1}{2 \cdot 5}, \frac{1}{3 \cdot 6}, \frac{1}{4 \cdot 7}, \ldots$
(x) $\frac{3}{1 \cdot 5}, \frac{5}{3 \cdot 7}, \frac{7}{5 \cdot 9}, \ldots$

2 Find the rth term of each of the sequences in question 1.

3 Suggest the next three terms in each of the following sequences. Explain how you get your answers.

(i) 1, 1, 2, 3, 5, 8, 13, 21, ... (Hint: look at sets of three consecutive terms.)

(ii) 1, 1, 3, 5, 11, 21, 43, ...
(iii) 1, 2, 3, 7, 22, 153, ...
(iv) 1, 3, 6, 10, 15, ...

4 Write down u_1, u_2, u_3, u_{50} (simplifying it if possible), $u_{n-1}, u_n, u_{n+1}, u_{2n}$ when u_r is given by:

(i) $3r + 1$

(ii) $\dfrac{1}{(2r+1)(2r-1)}$

(iii) $\dfrac{r+1}{(r+2)(r+3)}$

(iv) 2^r

(v) 5^{2-r}

(vi) $\dfrac{(-1)^r}{3^{2r-1}}$

(vii) $\cos r\pi$

5 For any sequence $\{u_r\}$, the relation $f: r \to u_r$ defines a function $f: \mathbb{Z}^+ \to \mathbb{R}$. Plot graphs of this function (i.e. plot u_r against r) when u_r is given by:

- (i) $2r - 1$
- (ii) $5 - 3r$
- (iii) r^2
→ (iv) $(r+1)/r$
- (v) $1/2^r$

In each case write down $\lim_{r \to \infty} u_r$.

6 Sketch on the same axes the graphs of:

(i) $1/x$, $x > 0$
(ii) $f: r \to u_r$, where $\{u_r\} = 1, \frac{1}{2}, \frac{1}{3}, \frac{1}{4}, \frac{1}{5}, \ldots$

(see 1.2:4 Q8).

7 Sketch on the same axes the graphs of:

(i) $1/x^2$, $x > 0$
(ii) $-1/x^2$, $x > 0$
(iii) $f: r \to u_r$, where $\{u_r\} = 1, -\frac{1}{4}, \frac{1}{9}, -\frac{1}{16}, \frac{1}{25}, \ldots$

8 Sketch on the same axes the graphs of:

(i) 3^x, $x > 0$
(ii) $f: r \to u_r$, where $\{u_r\} = 3, 9, 27, 81, \ldots$

• → **9** When r is a positive integer we define the number $r!$ (called 'r factorial') by

$$r! = r(r-1)(r-2)\ldots 5 \cdot 4 \cdot 3 \cdot 2 \cdot 1$$

Write down and evaluate the first seven terms in the sequence $\{r!\}$.

10 Simplify:

(i) $7 \cdot (6!)$
(ii) $8 \cdot 7 \cdot (6!)$
(iii) $(n+1)n!$
(iv) $\dfrac{7!}{7}$
(v) $\dfrac{7!}{7 \cdot 6}$
(vi) $\dfrac{7!}{5!}$
(vii) $\dfrac{10!}{7!3!}$
(viii) $\dfrac{(n+1)!}{n}$
(ix) $\dfrac{(n+1)!}{n!}$
(x) $9! - 4(8!)$
→ (xi) $(n+1)! - (n-1)!$

11 Write in factorial form:

(i) $15 \cdot 14 \cdot 13 \cdot 12 \cdot 11$
(ii) $\dfrac{10 \cdot 9 \cdot 8 \cdot 7}{4 \cdot 3 \cdot 2}$
(iii) $n(n-1)(n-2)(n-3)$
(iv) $\dfrac{n(n-1)(n-2)(n-3)(n-4)}{5 \cdot 4 \cdot 3 \cdot 2}$

12 Write down the first six terms of the sequence $\{u_r\}$ when:

- (i) $u_{r+1} \equiv u_r + 3$ $(r > 1)$ and $u_1 = 2$
→ (ii) $u_{r+1} \equiv 3u_r$ $(r > 1)$ and $u_1 = \frac{2}{9}$
- (iii) $u_{r+1} \equiv 1 - 2u_r$ $(r > 1)$ and $u_1 = -2$
- (iv) $u_{r+1} \equiv (r+1)u_r$ $(r > 1)$ and $u_1 = 1$
→ (v) $u_{r+2} \equiv u_{r+1} + u_r$ $(r > 2)$ and $u_1 = 1$, $u_2 = 1$
- (vi) $u_{r+2} \equiv u_{r+1} - u_r$ $(r > 2)$ and $u_1 = 3$, $u_2 = -1$
- (vii) $u_{r+2} \equiv 2u_{r+1} + 3u_r$ $(r > 2)$ and $u_1 = 1$, $u_2 = 1$

These equations are called **recurrence relations**. The sequence in part (v) is called the **Fibonacci sequence**.

13 (i) Write down the first five terms of the sequence defined by the recurrence relation

$$u_{r+1} \equiv 2u_r + 1 \quad (r > 1), \ u_1 = 1$$

and check that each term is given by the formula $u_r = 2^r - 1$.

(ii) Write down the first five terms of the sequence defined by the recurrence relation

$$u_{r+2} \equiv 5u_{r+1} - 6u_r \quad (r > 2), \ u_1 = 3, \ u_2 = 7$$

and check that each term is given by the formula $u_r = 2^r + 3^{r-1}$.

14 (i) Show that the solution $u_r = A\lambda^r$, where A is an arbitrary constant, satisfies the recurrence relation $u_{r+1} = \lambda u_r$.

(ii) Hence find a formula for u_r when
(a) $u_{r+1} \equiv 2u_r$ $(r > 1)$, $u_1 = 2$
(b) $u_{r+1} \equiv \frac{1}{2}u_r$ $(r > 1)$, $u_1 = 2$
(c) $u_{r+1} \equiv -3u_r$ $(r > 1)$, $u_1 = 1$

15 (i) Assuming that $u_r = \alpha^r$ satisfies the recurrence relation

$$u_{r+2} \equiv 5u_{r+1} - 6u_r$$

show that α satisfies the equation $\alpha^2 - 5\alpha + 6 = 0$. Hence find α_1, α_2 such that $u_r = \alpha_1^r$ and $u_r = \alpha_2^r$ satisfy the recurrence relation.

(ii) Show that $u_r = A\alpha_1^r + B\alpha_2^r$, where A and B are arbitrary constants, satisfies the recurrence relation in part (i).

(iii) Find A and B given that $u_1 = 5$, $u_2 = 13$.

16 Using the method of question 15, find a formula for u_r when

(i) $u_{r+2} \equiv 8u_{r+1} - 15u_r$ $(r > 2)$
$u_1 = 8$, $u_2 = 34$
(ii) $u_{r+2} \equiv 4u_{r+1} - 3u_r$ $(r > 2)$
$u_1 = 2$, $u_2 = 20$
+ (iii) $u_{r+2} \equiv u_{r+1} + u_r$ $(r > 2)$
$u_1 = 1$, $u_2 = 1$

216 *Arithmetic progressions*

5.1:2 Series

Consider the sum of the first n terms of the sequence $\{u_r\}$:

$$s_n = u_1 + u_2 + \cdots + u_n$$

s_n is called a **series**.

For example, consider the sequence 2, 5, 8, 11, 14, ... = $\{3r - 1\}$:

$$s_1 = 2, \; s_2 = 2 + 5, \; s_3 = 2 + 5 + 8 = 15, \; s_4 = 2 + 5 + 8 + 11 = 26, \ldots$$

The numbers s_n themselves form a sequence 2, 7, 15, 26, and we shall see (in section 5.1:3) that it is possible to find a formula for s_n:

$$s_n = \tfrac{1}{2} n(3n + 1)$$

$\sum$ notation

We often abbreviate $u_m + u_{m+1} + \cdots + u_n$ to $\sum_{r=m}^{n} u_r$, e.g.

$$\sum_{r=1}^{10} r^2 = 1^2 + 2^2 + \cdots + 10^2$$

$$\sum_{r=3}^{7} \tfrac{1}{r} = \tfrac{1}{3} + \tfrac{1}{4} + \tfrac{1}{5} + \tfrac{1}{6} + \tfrac{1}{7}$$

Exercise 5.1:2

1 Find $s_1, s_2, s_3, s_4, s_5, s_6$ when $\{u_r\}$ is given by:

- (i) $\{r\}$
- (ii) $\{5 - 2r\}$
- (iii) $\{(r+2)/(r+1)\}$
- → (iv) $\{2^{r-1}\}$
- (v) $\{(-1)^r\}$
- (vi) 1, 3, 6, 10, 15, ...

→ **2** Plot graphs of s_n against n ($n \leq 6$) for each of the sequences in question 1.

3 Write out in full and evaluate:

(i) $\sum_{r=1}^{7} r$

(ii) $\sum_{r=1}^{5} 2^r$

(iii) $\sum_{r=0}^{10} 2r + 1$

→ (iv) $\sum_{r=3}^{8} r^2$

4 Write the following series in $\sum$ notation:

(i) $1 + 2 + 3 + 4 + 5 + 6 + 7 + 8 + 9 + 10$

(ii) $\tfrac{1}{5} + \tfrac{1}{6} + \tfrac{1}{7} + \tfrac{1}{8}$

(iii) $1 + 3 + 5 + 7 + 9 + 11 + 13$

(iv) $4 + 2 + 1 + \tfrac{1}{2} + \tfrac{1}{4} + \tfrac{1}{8} + \tfrac{1}{16} + \tfrac{1}{32}$

(v) $m^2 + (m+1)^2 + \cdots + (2m)^2$

• → (vi) $1^3 - 2^3 + 3^3 - 4^3 + 5^3 + \cdots + (2m-1)^3 - (2m)^3$

(vii) $2^2 + 4^2 + 6^2 + 8^2 + \cdots + 100^2$

(viii) $\tfrac{1}{1 \cdot 3} + \tfrac{1}{2 \cdot 4} + \tfrac{1}{3 \cdot 5} + \cdots + \tfrac{1}{100 \cdot 102}$

5 Write out in longhand (using dots to represent many of the terms):

(i) $\sum_{r=1}^{n} 5 \cdot 2^r$

(ii) $\sum_{r=1}^{n} \dfrac{1}{(r+1)(r+2)}$

(iii) $\sum_{r=n}^{2n} r^2$

(iv) $\sum_{r=25}^{75} (-1)^r \dfrac{1}{r^3}$

5.1:3 Arithmetic progressions

> **Definition** An **arithmetic progression** (A.P.) is a sequence $\{u_r\}$ whose successive terms differ from each other by a fixed number d, called the **common difference** of the progression, i.e.
>
> $$u_r - u_{r-1} = d \quad \text{for all } r \qquad (1)$$

So the A.P. with first term a and common difference d is

$$a, a+d, a+2d, a+3d, \ldots$$

Its rth term u_r is given by

$$u_r = a + (r-1)d \qquad (2)$$

Worked example 1 The fifth term of an A.P. is 19, the ninth term is 35. Find its first term and common difference.

Let the first term be a and the common difference d. By equation (2),

$$a + 4d = 19$$
$$a + 8d = 35$$

So $a = 3$ and $d = 4$.

Worked example 2 The rth term of a sequence is $5r - 3$. Show that the sequence is an A.P. and find its first term and common difference.

$$u_r = 5r - 3 \Rightarrow u_{r-1} = 5(r-1) - 3 = 5r - 8$$
$$u_r - u_{r-1} = (5r - 3) - (5r - 8) = 5$$

So, by definition (1), the sequence is an A.P. with common difference 5. The first term $u_1 = 5 - 3 = 2$.

Theorem 1: The sum to n terms of the A.P., whose first term is a and whose nth term is l, is s_n where

$$s_n = \tfrac{1}{2}n(a + l) \qquad (3)$$

If the common difference of the A.P. is d, then $l = u_n = a + (n-1)d$; so

$$s_n = \tfrac{1}{2}n\{2a + (n-1)d\} \qquad (4)$$

Proof
$u_n = l$, so $u_{n-1} = l - d$, $u_{n-2} = l - 2d$, etc.

$$s_n = a + (a+d) + (a+2d) + \cdots + (l-2d) + (l-d) + l \qquad (5)$$
$$s_n = l + (l-d) + (l-2d) + \cdots + (a+2d) + (a+d) + a \qquad (6)$$

Adding series (5) and (6):

$$2s_n = (a+l) + (a+l) + (a+l) + \cdots + (a+l) + (a+l) + (a+l)$$
$$= n(a+l) \text{ since there are } n \text{ terms}$$

So $\quad s_n = \tfrac{1}{2}n(a+l)$

Worked example 3 How many terms of the series $1 + 4 + 7 + 10 + \cdots$ must be taken to make the sum exceed 10^4?

218 *Arithmetic progressions*

$a = 1$, $d = 3$, so by equation (4), $s_n = \frac{1}{2}n\{2 + 3(n-1)\} = \frac{1}{2}n(3n-1)$. We require:

$$\frac{1}{2}n(3n-1) > 10^4$$
$$3n^2 - n - 2 \cdot 10^4 > 0$$
$$\Rightarrow n < -81.5 \text{ or } n > 81.8 \qquad \text{(see 2.1:3 WE4)}$$

Since n is a positive integer, the sum of the series will exceed 10^4 when $n \geq 82$.

Finding the sequence given the sum

Consider the problem: given the sum to n terms of a sequence, s_n, can we find the underlying sequence $\{u_r\}$?

$$u_1 + u_2 + u_3 + \cdots + u_{n-1} + u_n = s_n \qquad (7)$$
$$u_1 + u_2 + u_3 + \cdots + u_{n-1} = s_{n-1} \qquad (8)$$

Subtracting equation (8) from (7) gives

$$\boxed{u_n = s_n - s_{n-1}} \qquad (9)$$

Worked example 4 The sum to n terms of a sequence is $n(2n+1)$. Show that the sequence is an A.P. and find its tenth term.

$s_n = n(2n+1) \Rightarrow s_{n-1} = (n-1)\{2(n-1)+1\}$. By equation (9),

$$u_n = s_n - s_{n-1} = n(2n+1) - (n-1)(2n-1)$$
$$= 4n - 1$$
$$u_n - u_{n-1} = (4n-1) - (4n-5)$$
$$= 4$$

So by definition (1) the sequence is an A.P. (with common difference 4).

$$u_n = 4n - 1, \text{ so } u_{10} = 4 \cdot 10 - 1 = 39$$

Exercise 5.1:3

1 Write down the next three terms of each of these sequences, given that they are A.P.s:

- (i) $-3, 1, 5, 9, \ldots$
- (ii) $5, 3, 1, \ldots$
- (iii) $2, \frac{4}{3}, \ldots$
- (iv) $x, 2x-1, 3x-2, \ldots$
- (v) $x, y, \ldots$
- (vi) $-\sin^2\theta, \cos^2\theta, \ldots$

2 Write down the next term and the rth term of each of the following sequences, given that they are A.P.s:

- (i) $1, 4, 7, 10, \ldots$
- → (ii) $-5, -\frac{10}{3}, -\frac{5}{3}, \ldots$
- (iii) $-1, -\frac{5}{2}, -4, \ldots$
- (iv) $x, 3x, 5x, \ldots$
- (v) $\ln 3, \ln 6, \ln 12, \ldots$

3 Write down the first six terms of the A.P. with first term a and common difference d when:

- (i) $a = 2$, $d = \frac{1}{3}$
- (ii) $a = \frac{1}{2}$, $d = -2$
- (iii) $a = x - 3$, $d = 1 - x$

(Notice that an A.P. is completely defined by its first term and common difference.)

4 Find the number of terms in each of the following A.P.s:

- (i) $2, 5, 8, 11, \ldots, 326$ (Hint: write down the rth term u_r of the sequence; put $u_r = 326$ and find r.)
- (ii) $7, 9, 11, \ldots, 193, 195$
- → (iii) $75, 69, 63, \ldots, -69, -75$
- (iv) $(n+1), (n+2), (n+3), \ldots, 2n-1, 2n$
- (v) $n, (n-3), (n-6), \ldots, (3-2n) - 2n$
- (vi) $(2m+1), (2m+3), \ldots, (4m-5)$

① **5** Find the first term and common difference of the A.P. whose:

→ (i) third term is 4 and sixth term 13
(ii) fifth term is 17 and ninth term 33
(iii) sixth term is 2 and eighteenth term −4
(iv) fourth term is $3x$ and eighth term $2x$

① R **6** In an A.P. the seventh term is three times the second term and the third term is 14. Find the fifth term.

① **7** The third and seventh terms of an A.P. are 1 and −5 respectively. Find the twenty-fifth term.

• → **8** If three numbers x, y and z are consecutive terms of an A.P., show that
$$y = \tfrac{1}{2}(x+z)$$
y is called the **arithmetic mean** of x and z (i.e. what is normally called the average of x and z).
Find the arithmetic mean of
(i) 12 and 30   (ii) −11 and 5
(iii) x and $7x$ (iv) $x-y$ and $-x+3y$

R **9** If 9, x, y, 1 are in A.P., find x and y.

10 If −2, x, y, 28 are in A.P., find x and y.

② **11** (i) Show that the following sequences are A.P.s. In each case, find the first term and common difference.
(a) $\{2r+3\}$
→ (b) $\{7-3r\}$
R (c) $\{pr+q\}$
(ii) Show that the sequence $\{2r^2+r\}$ is not an A.P.
(iii) Describe the graph of u_r against r when $\{u_r\}$ is an A.P.

• → **12** (i) Let $S = 1+2+3+4+\cdots+99+100$.   (10)
Write out this series backwards (to give equation (11)); add equations (10) and (11) and hence find S.
(ii) Find the number of terms in the series
$$S' = 3+7+11+\cdots+155+159$$
Hence, using the method of (i), find S'.

13 Using the method of question 12, find the sums of the following A.P.s to the given number of terms. Check your answers in formula (4).
(i) $7+10+13+\cdots$ to 20 terms
→ (ii) $30+28+26+\cdots$ to 50 terms
(iii) $2+(3-x)+(4-2x)+\cdots$ to 15 terms
→ (iv) $\lg 2+\lg 6+\lg 18+\cdots$ to 10 terms

14 Find the number of terms in each of the following A.P.s (see question 4). Hence find their sums.
→ (i) $4+9+14+\cdots+404$
(ii) $100+97+94+\cdots-200$

(iii) $-2-\tfrac{7}{2}-5-\cdots-62$
(iv) $1+\tfrac{3}{5}+\tfrac{1}{5}+\cdots-41$

15 Find:
(i) $\sum_{r=1}^{10}(4r-1)$ → (ii) $\sum_{r=3}^{8}(3-5r)$
R (iii) $\sum_{r=1}^{n}1-\tfrac{1}{3}r$ (iv) $\sum_{r=1}^{n}(p+qr)$

→ **16** Find $\sum_{r=1}^{n}r$, the sum of the first n integers.

17 Find:
(i) $1+2+3+\cdots+2m$
(ii) $(m+1)+(m+2)+\cdots+2m$
(iii) $1-2+3-4+\cdots+(2m-1)-2m$
(iv) $2\cdot 1+2+2\cdot 3+4+2\cdot 5+6+\cdots+2\cdot(2m-1)+2m$

18 Find the sum of all the integers:
(i) between 1 and 300 (inclusive)
(ii) between 1 and 300 which are divisible by 3
(iii) between 1 and 300 which are not divisible by 3

19 Find the sum of all the integers between 200 and 500 which are not divisible by 4.

20 Find the sum of all the integers between 1 and 400 inclusive:
(i) which are not divisible by 7
(ii) which are not divisible by 5
• (iii) which are not divisible by 7 or 5

→ **21** Find the sum of all the integers between 1 and 500 which are not divisible by 3 or 4.

R **22** Find the sum of all the integers between 1 and $15n$ (n integer) which are not divisible by 3 or 5.

① • **23** The fourth term of an A.P. is 3 and the eighth term is −7. Find the sum of the first twenty terms.

→ **24** The sum to 3 terms of an A.P. is 3 and the sum to 5 terms is −5. Find its first term and common difference.

25 The sum to 8 terms of an A.P. is twice the sum to 5 terms. If the first term is 3, find the second term.

R **26** The fifth term of an A.P. is 24 and the sum of the first five terms is 80. Find the sum to 15 terms.

③ **27** How many terms of the following series must be taken to make the sum exceed 10^5?
(i) $1+2+3+4+5+\cdots$
(ii) $-\tfrac{1}{2}+1+\tfrac{5}{2}+\cdots$
→ (iii) $-4+(-1)+2+5+\cdots$
(iv) $3+\tfrac{11}{2}+8+\cdots$

Series and probability 219

③ **28** For what values of n is:

R (i) $\sum_{r=1}^{n} (4r-1) > 10^7$?

(ii) $\sum_{r=1}^{n} \frac{1}{2}(1-3r) < -10^4$?

③ **29** How many terms of the series:

(i) $50 + 49 + 48 + \cdots$
(ii) $100 + 96 + 92 + \cdots$

must be taken to make the sum negative?

③ **30** Find the number of terms of the series with the given sum:

• (i) $1 + 2 + 3 + 4 + 5 + \cdots$; 210
(ii) $1 + 2 + 3 + 4 + 5 + \cdots$; 1295
→ (iii) $3 + 1 - 1 - 3 - \cdots$; -252
(iv) $-1 - \frac{7}{3} - \frac{11}{3} - \cdots$; -425

④ **31** The sum to n terms of a sequence $\{u_r\}$ is s_n.

(i) Show that $\{u_r\}$ is an A.P. when s_n is given by:
(a) $n(3n+1)$
→ (b) $\frac{1}{2}n(5-3n)$
R (c) $n(pn+q)$

(ii) Is $\{u_r\}$ an A.P. when s_n is given by:
(a) n^2; (b) n^3?

5.1:4 Geometric progressions

Definition A **geometric progression** (G.P.) is a sequence $\{u_r\}$ whose successive terms are in a **common ratio** ρ, i.e.

$$\frac{u_{r+1}}{u_r} = \rho \quad \text{for all } r \tag{12}$$

So the G.P. with first term a and common difference ρ is

$$a, a\rho, a\rho^2, a\rho^3, \ldots$$

Its rth term u_r is given by

$$u_r = a\rho^{r-1} \tag{13}$$

Worked example 5 The third term of a G.P. is 4, the fifth term is $\frac{16}{9}$. Find the first term and two possible values of the common ratio.

Let the first term be a and the common ratio ρ. By equation (13),

$$a\rho^2 = 4$$
$$a\rho^4 = \tfrac{16}{9}$$

So $\rho^2 = \frac{4}{9} \Rightarrow \rho = \pm\frac{2}{3}$, $a = 9$.

Thus there are two G.P.s satisfying the given conditions:

$$9, 6, 4, \tfrac{8}{3}, \tfrac{16}{9}, \ldots \text{ and } 9, -6, 4, -\tfrac{8}{3}, \tfrac{16}{9}, \ldots$$

Theorem 2: The sum to n terms of the G.P. whose first term is a and whose common ratio is ρ, is s_n where

$$s_n = \frac{a(1-\rho^n)}{1-\rho} = \frac{a(\rho^n - 1)}{\rho - 1} \tag{14}$$

(Note: the second form is more convenient when $\rho > 1$.)

Proof

$$s_n = a + a\rho + a\rho^2 + \cdots + a\rho^{n-1} \quad (15)$$

$$\rho s_n = \phantom{a + {}} a\rho + a\rho^2 + \cdots + a\rho^{n-1} + a\rho^n \quad (16)$$

Taking equation (16) from (15):

$$(1-\rho)s_n = a - a\rho^n = a(1-\rho^n)$$

So

$$s_n = \frac{a(1-\rho^n)}{1-\rho}$$

Worked example 6 How many terms of the series $3 + 6 + 12 + 24 + \cdots$ must be taken to make the sum exceed 10^{10}?

$a = 3$, $\rho = 2$, so by equation (14), $s_n = 3(2^n - 1)$. We require:

$$3(2^n - 1) > 10^{10}$$

$$2^n > \frac{10^{10}}{3} + 1 \approx 3.33 \times 10^9$$

$$n \lg 2 > \lg(3.33 \times 10^9) = 9 + \lg 3.33$$

$$n > \frac{9 + \lg 3.33}{\lg 2} \approx 31.6$$

Since n is an integer the sum of the series will exceed 10^{10} when $n \geqslant 32$.

Exercise 5.1:4

1 Write down the next three terms of each of these sequences, given that they are G.P.s:

- (i) $1, 3, 9, \ldots$
- (ii) $4, 2, 1, \ldots$
- (iii) $\frac{3}{4}, -1, \frac{4}{3}, \ldots$
- (iv) $1, \sqrt{2}, 2, \ldots$
- (v) $2, 1.2, \ldots$
- (vi) $3x, -6x^2, 12x^3$
- (vii) $x^2, 2x, 4$
- (viii) $-3, 3, \ldots$
- (ix) $\sec^2 \theta, \sin \theta, \ldots$

2 Write down the next term and the rth term of each of the following sequences, given that they are G.P.s:

- (i) $12, 6, 3, \ldots$
- (ii) $\frac{6}{5}, -2, \ldots$
- (iii) $x, -x^2, x^3, -x^4, \ldots$
- (iv) $e^x, e^{-x}, e^{-3x}, \ldots$
- (v) $-\csc \theta, -\sin \theta, \ldots$

3 Write down the given term of each of the following G.P.s:

- (i) $4, 1, \frac{1}{4}, \ldots$; twelfth term
- (ii) $4, -1, \frac{1}{4}, \ldots$; twelfth term
- (iii) $2, -14, 98, \ldots$; tenth term
- (iv) $\frac{1}{2}, x, 2x^2, \ldots$; seventh term
- (v) $e^x, e^{4x}, e^{7x}, \ldots$; fifteenth term

4 Which of the following sequences are A.P.s, which G.P.s, which both and which neither? Explain your answers.

- (i) $3, 6, 9, 12, \ldots$
- (ii) $1, 3, 6, 10, 15, \ldots$
- (iii) $-2, -1, -\frac{1}{2}, \ldots$
- (iv) $1, 1, 1, 1, \ldots$
- (v) $1, 2x, 3x^2, 4x^3, \ldots$
- (vi) $\{r^2\}$
- (vii) $\{1/r\}$
- (viii) $\{3^{r-1}/2^r\}$
- (ix) $\{r \cdot 2^r\}$

5 Write down the first six terms of the G.P. with first term a and common ratio ρ when:

- (i) $a = \frac{1}{9}$, $\rho = \frac{3}{2}$
- (ii) $a = 10$, $\rho = -0.2$
- (iii) $a = x^2$, $\rho = 2/x$

(Notice that a G.P. is completely defined by its first term and common ratio.)

6 Find the number of terms in each of the following G.P.s:

- (i) $27, 81, \ldots, 3^{15}$
- (ii) $1024, 512, \ldots, \frac{1}{1024}$
- (iii) $(\frac{1}{3})^2, (\frac{1}{3})^5, \ldots, (\frac{1}{3})^{98}$
- (iv) $2^{2m}, 2^{2m+2}, 2^{2m+4}, \ldots, 2^{4m-2}$

222 Geometric progressions

⑤ **7** Find the first term and common ratio of the G.P. whose:
→ (i) third term is 10 and sixth term 80
R (ii) fifth term is 2 and seventh term 18
(iii) third term is 162 and fifth term 18
(iv) fourth term is x^5 and eighth term $1/x^3$

8 A man drops a ball onto the ground from a height of 10 m. It rebounds to a height of 8 m. What height does it reach after the tenth bounce, assuming it rebounds to the same proportion of its original height each time?

9 The Moon is 38×10^4 km away. You have a piece of paper 0.05 mm thick. Assuming the piece of paper is large enough, how many times would you have to fold it to reach the Moon?

10 A man invests £1000 at 15% compound interest per annum. How long must he wait before he has

(i) £2000
(ii) £3000
(iii) £4000?

(c.f. 2.5:1 Q7).

11 A woman measures the height of her child at birth and at monthly intervals afterwards. The child's height increases by 5% per month. How many measurements has she made before the child's height becomes twice what it was at birth?

12 Find x if these numbers are consecutive terms of a G.P.:

• (i) $x-2, x, x+3$
(ii) $x, 2x+3, 4x-2$
(iii) $x-6, 2x, 8x+20$
R (iv) $2^{2x+1}, 4^x, 64$

In each case find the next term in the G.P.

13 The **geometric mean** of two positive numbers x and z is defined as

$$y = \sqrt{(xz)}$$

Show that in this case x, y and z are consecutive terms of a G.P.
Find the geometric mean of:

(i) 8 and 32 (ii) 18 and 50
(iii) $\frac{1}{27}$ and 12 (iv) x^2y and x^6y^3

→ **14** Find the arithmetic mean (A.M.) and geometric mean (G.M.) of:

(i) 6 and 24 (ii) 1 and 25
(iii) $\frac{1}{4}$ and 36

Verify that A.M. ≥ G.M. in each case.

• → **15** Show that for any two positive numbers a and b, the arithmetic mean of a and b ≥ the geometric mean of a and b. (Hint: let $a = x^2$, $b = y^2$. Warning: don't start with the answer.)

R **16** Using the result of question 15, show that

$$a^2 + b^2 + c^2 \geq ab + bc + ca$$

for any positive numbers a, b and c.

+ **17** Show that for any real numbers a and b

$$a^3b + ab^3 \leq a^4 + b^4$$

+ **18** Show that for any real numbers a, b, c and d
(i) $(ab+cd)^2 \leq (a^2+c^2)(b^2+d^2)$
(ii) $a^4+b^4+c^4+d^4 \geq 4abcd$

19 x, y and z are three positive numbers. Show that x, y and z are consecutive terms of a G.P. ⇔ $\log_a x$, $\log_a y$ and $\log_a z$ are consecutive terms of an A.P.

20 If the first, second and fifth terms of an A.P. are consecutive terms of a G.P., show that the common ratio of the G.P. is twice the first term of the A.P.

21 If 3, x, y and $\frac{64}{9}$ are in G.P., find x and y.

22 The arithmetic mean of two numbers is $\frac{13}{2}$ and their geometric mean is 6. Find the numbers.

23 Find the sum of all the integers between 1 and 100 which are not powers of 2.

24 (i) Show that the following sequences are G.P.s. In each case find the first term and common ratio.

• (a) $\{3^r\}$
R (b) $\{5 \cdot 2^{-3r}\}$
→ (c) $\left\{\dfrac{3^{2r}}{2^r}\right\}$
(d) $\{pq^r\}$

(ii) Show that the sequence $\{r \cdot 2^r\}$ is not a G.P.

(iii) Describe the graph of u_r against r when $\{u_r\}$ is a G.P.

• → **25** Let

$$S = 2 + 4 + 8 + \cdots + 1024 \quad (17)$$

Write out the series for $2S$ (equation (18)); subtract (17) from (18) and hence find S.

26 Using the method of question 25, find the sums of the following G.P.s:

(i) $81 + 27 + 9 + \cdots + \frac{1}{81}$
→ (ii) $\frac{8}{81} - \frac{4}{27} + \frac{2}{9} - \cdots - \frac{27}{16}$
(iii) $\dfrac{1}{x^{10}} + \dfrac{1}{x^7} + \cdots + x^{20}$

27 Using Theorem 2, find the sums of the following G.P.s to the given number of terms:

(i) $5 + 10 + 20 + \cdots$; 8 terms
→ (ii) $\frac{1}{2} + \frac{1}{6} + \frac{1}{18} + \cdots$; 6 terms
R (iii) $e^{-x} + e^x + e^{3x} + \cdots$; 10 terms

28 Find:

(i) $\sum_{r=1}^{10} 3 \cdot 2^{r-1}$

(ii) $\sum_{r=3}^{8} \frac{5}{3^r}$

→ (iii) $\sum_{r=10}^{20} (\tfrac{1}{2})^r$

R (iv) $\sum_{r=1}^{6} 3 \cdot 2^r - 2^{2r-1}$

29 The wedding of the year. A famous rock star marries a famous model. The unique twenty-tier wedding cake consists of 20 cylindrical cakes each of height 10 cm. The diameter of the lowest cake is 25 cm. The diameter of each cake is 4% less than the cake below. If each guest is given 50 cm^3 of cake, how many guests can be fed?

30 Each year on June 1st, beginning in 1987, a woman invests £5000 in a deposit account. What is the balance of her account just after her deposit of June 1st 1992, if the rate of interest is:

(i) 20% per year?
(ii) 10% per half-year?
(iii) 5% per quarter?

31 Each year a man invests £10 000 in a deposit account at 10% per annum compound interest. What is the balance of his account immediately after his fifteenth deposit:

(i) if he never makes any withdrawals?
(ii) if he makes a yearly alimony payment of £4000 to his ex-wife?

32 A woman's house is infested by rats. There are 500 rats in the house and their number is increasing by 5% daily. The woman decides to fight back. Every day she sets twenty rat-traps, each of which kills one rat without fail. How long is it before the number of rats has doubled?
What is the least number of rats the woman needs to kill every day if she is eventually to rid the house of them completely?

33 A competitor in a single-handed round-the-world yacht race gets becalmed in the Sargasso Sea. The only food on board is 20 kg of tinned shark meat. She decides that each day she will eat 5% of her remaining food. How long is it before only half her food is left?

When she is finally picked up by a pleasure cruiser she has just 0.01 kg of food left. After how long is this?

Another competitor is also becalmed in the Sargasso Sea. He has nothing but 20 kg of deep-frozen Boeuf Stroganoff on board and he too decides to eat 5% of his remaining food per day. However, his sponsors are airlifting in 0.5 kg of Boeuf Stroganoff by helicopter to him every day. (The helicopter pilots misinterpret his frantic signals as friendly waves.) How long is it before he has only 10 kg of food left?

34 Find the first term and common ratio of the G.P. for which:

• (i) the sum of the second and third terms is 6, the sum of the third and fourth is -12
(ii) the sum to 3 terms is 171, the sum to 6 terms is $221\tfrac{2}{3}$
R (iii) the sum to 6 terms is 1, the sum to 12 terms is 65

→ **35** The sum to six terms of a G.P. is 9 times the sum to 3 terms. Find the common ratio of the G.P.

36 (i) If $S(x) = 1 + 2x + 3x^2 + \cdots + nx^{n-1}$, consider $(1-x)S(x)$ and hence find $S(x)$ in terms of x and n.

(ii) Sum the series $1 + x + x^2 + \cdots + x^n$ ($x \neq 1$). By differentiation with respect to x, find $S(x)$ in terms of x and n again, and verify that your answer agrees with (i).

(iii) Deduce the value of $1 \cdot 2 + 2 \cdot 2^2 + 3 \cdot 2^3 + \cdots + n \cdot 2^n$.

37 How many terms of the following series must be taken to make the sum exceed 10^5?

(i) $2 + 4 + 8 + \cdots$
R (ii) $2 + 8 + 32 + \cdots$
→ (iii) $4 + 6 + 9 + \cdots$
(iv) $512 + 256 + 128 + \cdots$

+ **38** Show that

$$\sum_{r=1}^{n} (3^r + 3^{r-1} + \cdots + 1) = \tfrac{1}{4}(3^{n+2} - 2n - 9)$$

39 The sum to n terms of a sequence $\{u_r\}$ is s_n. Show that $\{u_r\}$ is a G.P. when s_n is given by:

(i) $2 \cdot 5^n$ (ii) $3\{1 - (\tfrac{2}{3})^n\}$

(See worked example 4.)

5.1:5 Convergence of series

Definition A sequence $\{u_r\}$ is said to **tend to a limit** l if u_r becomes closer and closer to l as $r \to \infty$. We write

$$u_r \to l \text{ as } r \to \infty \quad \text{or} \quad \lim_{r \to \infty} u_r = l \qquad (19)$$

224 *Convergence of series*

For example

$$\frac{1}{r} \to 0 \quad \text{as} \quad r \to \infty$$

$$\frac{5r}{3r+1} \to \frac{5}{3} \quad \text{as} \quad r \to \infty$$

Note: we could use a more formal definition of a limit: $u_r \to l$ if, given any small number δ, $\exists\, n$ such that $|u_n - l| < \delta$; but since we shall not be trying to prove any formal theorems about the limits of sequences, we may as well stick to the intuitive definition given above.

> **Definition** A series $s_n = u_1 + u_2 + \cdots + u_n$ is **convergent** if s_n tends to some finite limit S as $n \to \infty$.

S is called the **sum to infinity** s_∞ of the series. We write $s_\infty = \sum_{r=1}^{\infty} u_r$

Worked example 7 Consider the series $1 + \frac{1}{2} + \frac{1}{4} + \frac{1}{8} + \cdots$
$s_1 = 1$, $s_2 = 1 + \frac{1}{2} = 1.5$, $s_3 = 1 + \frac{1}{2} + \frac{1}{4} = 1.75$, $s_4 = 1.875$, $s_5 = 1.9375$,
$s_6 = 1.96875, \ldots$

As n gets larger, s_n gets closer and closer to 2 (without ever actually getting there—see fig. 5.1:1), i.e.

$$s_n \to 2 \quad \text{as} \quad n \to \infty$$

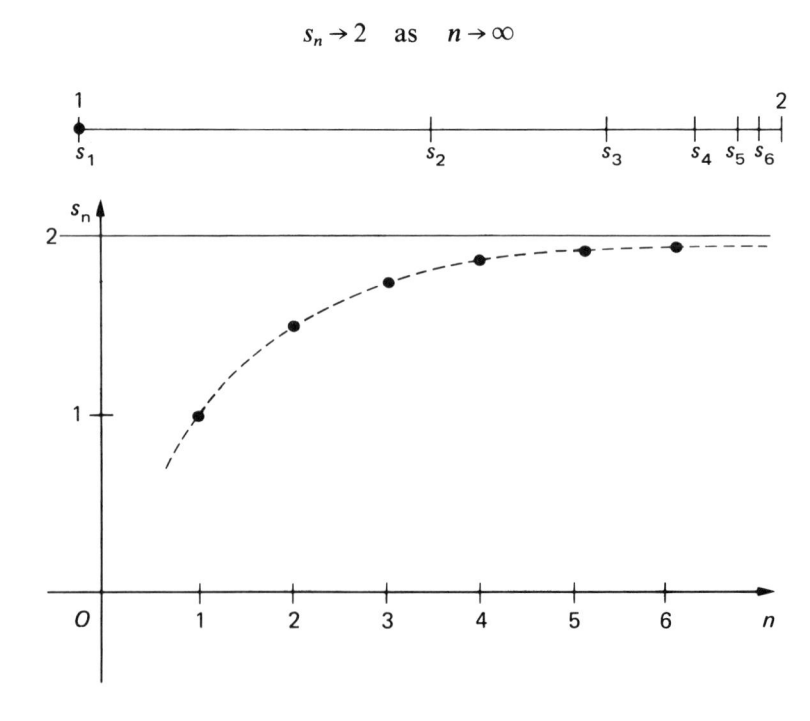

Fig. 5.1:1

So the sum to infinity of the series is 2.
We can write $1 + \frac{1}{2} + \frac{1}{4} + \frac{1}{8} + \cdots = 2$.

> **Definition** A series is **divergent** if it is not convergent. For example, the series $1 + 3 + 5 + 7 + \cdots$ is divergent.

Series and probability 225

It is extremely tempting to think that if $u_r \to 0$ as $r \to \infty$ then s_n must be convergent. However...

Worked example 8 Show that the series $1+\frac{1}{2}+\frac{1}{3}+\frac{1}{4}+\cdots$ is divergent.

$$1+\tfrac{1}{2}+\tfrac{1}{3}+\tfrac{1}{4}+\cdots = 1+\tfrac{1}{2}+(\tfrac{1}{3}+\tfrac{1}{4})+(\tfrac{1}{5}+\tfrac{1}{6}+\tfrac{1}{7}+\tfrac{1}{8})+\cdots$$
$$> 1+\tfrac{1}{2}+(\tfrac{1}{4}+\tfrac{1}{4})+(\tfrac{1}{8}+\tfrac{1}{8}+\tfrac{1}{8}+\tfrac{1}{8})+\cdots$$
$$> 1+\tfrac{1}{2}+\quad\tfrac{1}{2}\quad+\quad\tfrac{1}{2}\quad+\cdots$$

The series is always greater than the (divergent) series $1+\frac{1}{2}+\frac{1}{2}+\frac{1}{2}+\cdots$, so it must itself be divergent. Admittedly, the series diverges very slowly. The sum of the first million terms is about 15.

Exercise 5.1:5

1 (i) Find $\lim_{r\to\infty} u_r$ when:

(a) $u_r = \dfrac{1}{r^2}$; (b) $u_r = \dfrac{1}{2^r}$; (c) $u_r = \dfrac{r}{r+1}$;

(d) $u_r = \dfrac{1-2r}{4r+1}$; (e) $u_r = \dfrac{3r^2+2}{r^2-1}$.

(ii) Show that if $u_r = (r^2+1)/r$, $u_r \to \infty$ as $r \to \infty$.

2 The sequence $\{u_r\}$ is defined by the recurrence relation

$$u_{r+1} \equiv \frac{1}{2}\left(u_r+\frac{3}{u_r}\right) \quad (r>1),\ u_1=1$$

(i) Write down the first five terms of the sequence, correct to 4 dec. pls.
(ii) Show that as $r\to\infty$, $u_r \to \sqrt{3}$.

(Hint: suppose that as $r\to\infty$, $u_r \to l$; then the terms of the sequence become closer and closer to l and we can put $u_{r+1}=u_r=l$ in the recurrence relation.)

3 (i) The sequence $\{u_r\}$ is defined by the recurrence relation

$$u_{r+1} \equiv \frac{1}{2}\left(u_r+\frac{5}{u_r}\right) \quad (r>1),\ u_1=2$$

Show that as $r\to\infty$, $u_r \to \sqrt{5}$.

(ii) Taking 2 as the first approximation to the positive root of the equation $x^2-5=0$, use the Newton-Raphson method to obtain three successive better approximations.
If x_r is the rth approximation, write down the $(r+1)$th approximation x_{r+1} in terms of x_r.

4 The sequence $\{u_r\}$ is defined by the recurrence relation

$$u_{r+2} \equiv u_{r+1}+u_r \quad (r>2),\ u_1=1,\ u_2=1$$

The sequence $\{a_r\}$ is given by $a_r \equiv u_r/u_{r+1}$.

(i) Write down the first six terms of the sequence $\{a_r\}$, correct to 5 dec. pls.
(ii) Show that as $r\to\infty$, $a_r \to \frac{1}{2}(\sqrt{5}-1)$.

⑦ **5** If $s_n = \sum_{r=1}^{n} u_r$, find s_n for $n=1, 2, 3, 4, 5, 6$ and plot a graph of s_n against n for these values of n when $\{u_r\}$ is given by:

→ (i) $1, \tfrac{1}{3}, \tfrac{1}{9}, \tfrac{1}{27}, \ldots$
(ii) $1, -\tfrac{1}{3}, \tfrac{1}{9}, -\tfrac{1}{27}, \ldots$
(iii) $1, -\tfrac{1}{2}, \tfrac{1}{4}, -\tfrac{1}{8}, \ldots$
(iv) $1, 1.1, (1.1)^2, \ldots$
(v) $1, 4, 7, 10, \ldots$
(vi) $\dfrac{1}{1\cdot 2}, \dfrac{1}{2\cdot 3}, \dfrac{1}{3\cdot 4}, \ldots$
(vii) $\{1/r^2\}$
(viii) $1, -\tfrac{1}{2}, \tfrac{1}{3}, -\tfrac{1}{4}, \ldots$

Which of the series is convergent? Estimate the sum to infinity of the series if it exists.

⑧ + **6** Show that the series $1+\frac{1}{3}+\frac{1}{5}+\cdots$ is divergent.

⑧ + **7** By comparing it with the convergent series $1+\frac{1}{2}+\frac{1}{4}+\frac{1}{8}+\cdots$, show that the series

$$\frac{1}{1^2}+\frac{1}{2^2}+\frac{1}{3^2}+\frac{1}{4^2}+\cdots$$

is convergent. (In fact its sum to infinity is $\frac{1}{6}\pi^2$, but this is difficult to prove.)

5.1:6 Convergence of geometric series

Consider the G.P. $\{a\rho^{r-1}\}$. By equation (14),

$$s_n = \frac{a(1-\rho^n)}{1-\rho}$$

226 *Convergence of geometric series*

If $|\rho|<1$, $\rho^n \to 0$ as $n \to \infty$; so $s_n \to a/(1-\rho)$.
If $|\rho|>1$, $\rho^n \to \pm\infty$ as $n \to \infty$; so $s_n \to \pm\infty$.
So

> the series $a + a\rho + a\rho^2 + a\rho^3 + \cdots$ converges to the sum $a/(1-\rho)$
> when $|\rho|<1$. Otherwise the series diverges. (20)

Worked example 9 Find the sum to infinity of the following G.P.s:

(i) $3 + 1 + \frac{1}{3} + \frac{1}{9} + \cdots$
(ii) $3 - 1 + \frac{1}{3} - \frac{1}{9} + \cdots$
(iii) $1 + 3 + 9 + 27 + \cdots$

(i) $a = 3, \rho = \frac{1}{3} \Rightarrow s_n \to 3/(1-\frac{1}{3}) = \frac{9}{2}$ (fig. 5.1:**2a**)

(ii) $a = 3, \rho = -\frac{1}{3} \Rightarrow s_n \to 3/(1+\frac{1}{3}) = \frac{9}{4}$ (fig. 5.1:**2b**)

(iii) $a = 1, \rho = 3 \Rightarrow s_n \to \infty$ (fig. 5.1:**2c**)

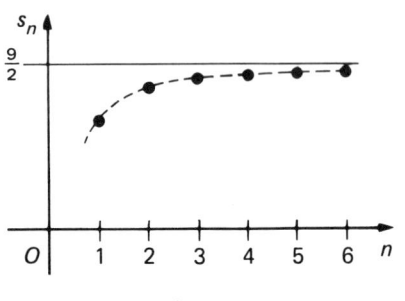

Fig. 5.1:**2a**

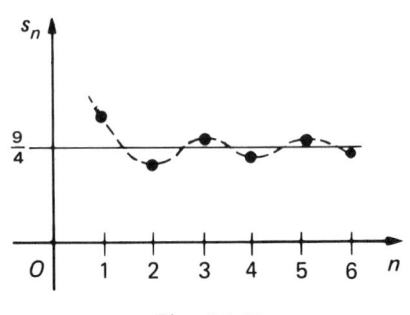

Fig. 5.1:**2b**

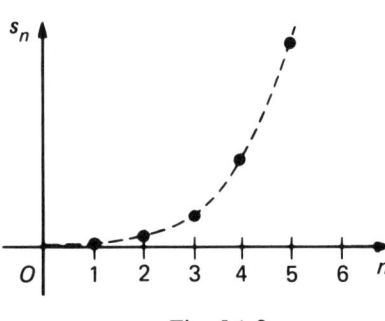

Fig. 5.1:**2c**

Worked example 10 Find the smallest integer n for which the sum to n terms of the series $2 + 1 + \frac{1}{2} + \frac{1}{4} + \frac{1}{8}$ differs from its sum to infinity by less than 10^{-5}.

$a = 2, \rho = \frac{1}{2}$, so by equation (14)

$$s_n = \frac{2(1-(\frac{1}{2})^n)}{1-\frac{1}{2}} = 4\{1 - (\frac{1}{2})^n\}$$

As $n \to \infty$, $s_n \to 4$, i.e. $s_\infty = 4$

$$s_\infty - s_n = 4(\tfrac{1}{2})^n = 2^{2-n}$$

We require:

$$2^{2-n} < 10^{-5}$$

So
$$(2-n)\lg 2 < -5$$

$$2 - n < -\frac{5}{\lg 2} = -16.61$$

$$n > 18.61$$

Thus the smallest integer n for which $s_\infty - s_n < 10^{-5}$ is 19.

Worked example 11 For what range of values of x is the series

$$\frac{1}{x} + \frac{1}{x-1} + \frac{x}{(x-1)^2} + \cdots$$

convergent?

The common ratio $\rho = x/(x-1)$. The series converges if $|\rho| < 1$, i.e.

$$\left|\frac{x}{x-1}\right| < 1$$

$\Rightarrow \qquad \dfrac{x^2}{(x-1)^2} < 1 \quad$ (see 2.4:2 WE3)

$\Rightarrow \qquad x^2 < (x-1)^2$

$\Rightarrow \qquad x^2 < x^2 - 2x + 1$

$\Rightarrow \qquad 2x < 1$

$\Rightarrow \qquad x < \tfrac{1}{2}$

Exercise 5.1:6

1 If $s_n = \sum_{r=1}^{n} u_r$, plot a graph of s_n against n when $\{u_r\}$ is given by:

- (i) $4, 2, 1, \tfrac{1}{2}, \ldots$
- (ii) $3, -2, \tfrac{4}{3}, -\tfrac{8}{9} \ldots$
- (iii) $10, -2, \tfrac{2}{5} \ldots$
- (iv) $1, 0.9, 0.81, \ldots$
- (v) $1, -0.9, 0.81, \ldots$

What are the sums to infinity of these series?

2 Which of the following G.P.s converge? Write down the sum to infinity of the G.P. when it exists.

(i) $\tfrac{4}{3} + 1 + \tfrac{3}{4} + \tfrac{9}{16} + \cdots$
(ii) $9 - 6 + 4 - \cdots$
(iii) $1 - 2 + 4 - 8 + \cdots$
(iv) $\tfrac{8}{81} - \tfrac{4}{27} + \tfrac{2}{9} - \cdots$
(v) $1 + 0.7 + 0.49 + \cdots$
(vi) $1 + 1.05 + (1.05)^2 + \cdots$

R 3 (i) A G.P. has first term a and sum to infinity S. Find the common ratio of the G.P. if S is:
(a) $3a$; (b) $5a$.

(ii) Can the sum to infinity of a G.P. be $\tfrac{2}{3}$ of the first term? Can it be $\tfrac{1}{3}$ of the first term?

4 Express $0.\dot{4}$ (i.e. $0.4444\ldots$) as a fraction. (Hint: $0.\dot{4} = \tfrac{4}{10} + \tfrac{4}{100} + \tfrac{4}{1000} + \cdots$).

5 Express as fractions:
(i) $0.\dot{1}$
(ii) $0.\dot{1}\dot{2}$
(iii) $0.0\dot{1}\dot{2}$
(iv) $0.1\dot{2}\dot{5}$

6 Express as fractions:
(i) $0.\dot{6}$
(ii) $0.6\dot{4}$
(iii) $0.\dot{1}2\dot{8}$

7 A tortoise sets out from London Zoo. He walks 2 m in the first hour. Every hour he walks 10% less than in the previous hour. How far does he walk?

R 8 A man drops a ball onto the ground. It takes 4 s to hit the ground, bounces, and takes a further 6 s to hit the ground for the second time, and so on. For how long is it in the air before it stops bouncing altogether?

9 Show that the sum to 10 terms of the series

$$3 + 2 + \tfrac{4}{3} + \cdots$$

differs from the sum to infinity by less than $\tfrac{1}{6}$.

10 Show that the sum to 12 terms of the series

$$8 - 2 + \tfrac{1}{2} - \tfrac{1}{8} + \cdots$$

differs from the sum to infinity by less than 4×10^{-7}.

R 11 The sum of the first four terms of a G.P. of positive terms is 80 and the sum to infinity is 81. Show that the sum of the first seven terms of the series differs from the sum to infinity by $\tfrac{1}{27}$.

12 Find the smallest integer n for which the sum to n terms of each of these geometric series differs from the sum to infinity by less than 10^{-10}:

(i) $4 + 2 + 1 + \tfrac{1}{2} + \cdots$
(ii) $3 + 1 + \tfrac{1}{3} + \tfrac{1}{9} + \cdots$
R (iii) $3 + \tfrac{1}{3} + \tfrac{1}{27} + \cdots$
(iv) $3 - \tfrac{1}{3} + \tfrac{1}{27} - \cdots$

13 Find the smallest integer n for which the sum to n terms of each of these geometric series differs from the sum to infinity by less than 2^{-20}:
(i) $3 + \frac{3}{2} + \frac{3}{4} + \cdots$
(ii) $3 + \frac{1}{4} + \frac{1}{48} + \cdots$

14 (i) (a) If $|x| < 1$, find $1 + x + x^2 + x^3 + \cdots$
  (b) By differentiating your result, find $1 + 2x + 3x^2 + 4x^3 + \cdots$.
  (c) By integrating your result, find $x + \frac{1}{2}x^2 + \frac{1}{3}x^3 + \cdots$.

(ii) Find the sums of the series:
(a) $1 + \frac{2}{2} + \frac{3}{2^2} + \frac{4}{2^3} + \cdots$
(b) $\frac{1}{2} + \frac{1}{8} + \frac{1}{24} + \frac{1}{64} + \cdots$
(c) $\frac{1}{2} - \frac{1}{8} + \frac{1}{24} - \frac{1}{64} + \cdots$

15 If $|x| \neq 1$, find the sum s_n of the series $x + x^3 + x^5 + \cdots + x^{2n-1}$.
Given that $x = \frac{1}{2}$, show that $s_n \to \frac{2}{3}$ as $n \to \infty$.

16 Find the set of values of x for which each of these G.P.s is convergent and state the sum to infinity (when this sum exists).
(i) $1 - x + x^2 - x^3 + \cdots$
(ii) $1 + 2x + 4x^2 + 8x^3 + \cdots$
(iii) $1 + (x+2) + (x+2)^2 + \cdots$
(iv) $\frac{x+1}{x^2} + \frac{1}{x} + \frac{1}{x+1} + \cdots$ ($x \neq 0$ or -1)
(v) $1 + \frac{x}{1-x} + \frac{x^2}{(1-x)^2} + \cdots$
(vi) $\sum_{r=0}^{\infty} (-1)^r \frac{x^{2r}}{(x+2)^r}$
(vii) $1 + e^{2x} + e^{4x} + \cdots$
(viii) $e^{-x} + e^{-2x} + e^{-3x} + \cdots$
(ix) $1 + 2\sin x + 4\sin^2 x + \cdots$
(x) $\cos x + \sin x + \cdots$

17 Find the set of values of x for which the G.P.
$$1 + \frac{x+1}{x} + \left(\frac{x+1}{x}\right)^2 + \cdots$$
is convergent. If the sum to infinity is $\frac{3}{4}$, find x.

18 Find the set of values of x for which the series
$$\sum_{r=0}^{\infty} \left(\frac{2x-1}{x+2}\right)^r$$
is convergent. If the sum to infinity is 4, find x.

19 Find the set of values of x for which the G.P.s
$$x + x^2 + x^3 + \cdots$$
and
$$3 + \frac{3(x-1)}{x} + \frac{3(x-1)^2}{x^2} + \cdots$$
converge simultaneously. Find the value of x for which both series converge to the same sum.

20 State the values of x for which each of these series is convergent:
(i) $1 + x + x^2 + \cdots$ (ii) $1 + 3x + 9x^2 + \cdots$

Find their sums to infinity and hence find the sum to infinity of the series:
$$5 + 7x + 13x^2 + 31x^3 + \cdots$$

21 For what values of θ is the series $1 + \sin\theta + \sin^2\theta + \cdots$ convergent? Find the sum of the series. Hence find $1 + 2\sin\theta + 3\sin^2\theta + \cdots$.

22 For what values of x is the series $1 + e^x + e^{2x} + \cdots$ convergent? Find the sum of the series. Hence find:
(i) $e^x + 2e^{2x} + 3e^{3x} + \cdots$
(ii) $e^x + \frac{1}{2}e^{2x} + \frac{1}{3}e^{3x} + \cdots$

23 The sequence $\{u_r\}$ is a G.P. with common ratio λ, and the series $\sum_{r=1}^{\infty} u_r$ is convergent. If $v_r = u_r u_{r+1}$ for all r, show that the sequence $\{v_r\}$ is a G.P. Show that the series $\sum_{r=1}^{\infty} v_r$ is convergent and that it converges to the sum $u_1^2/1 - \lambda$.

24 $\sum_{r=1}^{\infty} u_r$ is a convergent geometric series. Discuss the convergence of the series:
(i) $\sum_{r=1}^{\infty} (-1)^{r+1} u_r$
(ii) $\sum_{r=1}^{\infty} u_r^2$
(iii) $\sum_{r=1}^{\infty} \frac{1}{u_r}$
(iv) $\sum_{r=1}^{\infty} (3u_r + 2u_{r+1})$
(v) $\sum_{r=1}^{\infty} (u_r + 1)$

25 (i) Draw graphs to illustrate the divergence of the G.P. with first term 1 and common ratio
(a) 2; (b) 1; (c) -2.

(ii) Consider the G.P. with first term 1 and common ratio -1. If s_n is the sum to n terms of this G.P., find s_n when n is (a) odd; (b) even. Is this G.P. convergent?

5.1:7 Method of differences

> **Theorem 3:** Suppose we are trying to find the sum
> $$s_n = u_1 + u_2 + \cdots + u_n = \sum_{r=1}^{n} u_r$$
> and suppose we can find a sequence $\{f_r\}$ such that
> $$u_r = f_r - f_{r-1} \quad \forall r \tag{21}$$
> Then
> $$s_n = f_n - f_0 \tag{22}$$

Proof
Consider the equation

$$u_r = f_r - f_{r-1} \text{ for } r = 1, 2, 3, \ldots, (n-1), n$$

$r = 1$: $\quad u_1 = f_1 - f_0$
$r = 2$: $\quad u_2 = f_2 - f_1$
$r = 3$: $\quad u_3 = f_3 - f_2$
$\qquad\qquad\vdots$
$r = n-1$: $\quad u_{n-1} = f_{n-1} - f_{n-2}$
$r = n$: $\quad u_n = f_n - f_{n-1}$

Adding: $\quad u_1 + u_2 + \cdots + u_n = f_n - f_0$

So $\quad s_n = f_n - f_0$

Worked example 12 If $f_r = r(r+1)(2r+1)$, find $f_r - f_{r-1}$. Hence find $\sum_{r=1}^{n} r^2$.

$$\begin{aligned} f_r - f_{r-1} &= r(r+1)(2r+1) - (r-1)r(2r-1) \\ &= r\{2r^2 + 3r + 1 - (2r^2 - 3r + 1)\} \\ &= 6r^2 \end{aligned}$$

By Theorem 3:

$$\sum_{r=1}^{n} 6r^2 = f_n - f_0 = n(n+1)(2n+1)$$

so $\quad \sum_{r=1}^{n} r^2 = \tfrac{1}{6} n(n+1)(2n+1)$

It is instructive to write out the equation

$$6r^2 = r(r+1)(2r+1) - (r-1)r(2r-1)$$

for $r = 1, 2, 3, \ldots, (n-1), n$, as in the proof of the theorem, to show clearly how the cancelling takes place:

$r = 1$: $\quad 6 \cdot 1^2 = 1 \cdot 2 \cdot 3 - 0 \cdot 1 \cdot 1$

$r = 2$: $\quad 6 \cdot 2^2 = 2 \cdot 3 \cdot 5 - 1 \cdot 2 \cdot 3$

$r = 3$: $\quad 6 \cdot 3^2 = 3 \cdot 4 \cdot 7 - 2 \cdot 3 \cdot 5$

$\quad \vdots$

$r = n-1$: $\quad 6 \cdot (n-1)^2 = (n-1)n(2n-1) - (n-2)(n-1)(2n-3)$

$r = n$: $\quad 6 \cdot n^2 = n(n+1)(2n+1) - (n-1)n(2n-1)$

Adding: $\quad \sum 6r^2 = n(n+1)(2n+1)$

If the terms of the original sequence $\{u_r\}$ are *fractions* e.g. $u_r = 1/r(r+1)$, we often find that f_0 is undefined; however, we can use Theorem 3A, a more convenient version of Theorem 3.

Theorem 3A: Suppose we are trying to find the sum

$$s_n = u_1 + u_2 + \cdots + u_n = \sum_{r=1}^{n} u_r$$

and suppose we can find a sequence f_r such that

$$u_r = f_r - f_{r+1} \qquad (23)$$

Then $\qquad s_n = f_1 - f_{n+1} \qquad (24)$

Proof
See question 31.

Worked example 13 Express $3/[(3r-1)(3r+2)]$ in partial fractions. Hence find

$$\sum_{r=1}^{n} \frac{3}{(3r-1)(3r+2)}, \quad \text{i.e.} \quad \frac{3}{2 \cdot 5} + \frac{3}{5 \cdot 8} + \cdots + \frac{3}{(3n-1)(3n+2)}$$

$$\frac{3}{(3r-1)(3r+2)} = \frac{1}{3r-1} - \frac{1}{3r+2} \quad \text{(see 2.3:3 WE6)}$$

$r = 1$: $\qquad \dfrac{3}{2 \cdot 5} = \dfrac{1}{2} - \dfrac{1}{5}$

$r = 2$: $\qquad \dfrac{3}{5 \cdot 8} = \dfrac{1}{5} - \dfrac{1}{8}$

$r = 3$: $\qquad \dfrac{3}{8 \cdot 11} = \dfrac{1}{8} - \dfrac{1}{11}$

$\qquad \vdots$

$r = n-1$: $\qquad \dfrac{3}{(3n-4)(3n-1)} = \dfrac{1}{3n-4} - \dfrac{1}{3n-1}$

$r = n$: $\qquad \dfrac{3}{(3n-1)(3n+2)} = \dfrac{1}{3n-1} - \dfrac{1}{3n+2}$

Adding: $\qquad \displaystyle\sum_{r=1}^{n} \frac{3}{(3r-1)(3r+2)} = \frac{1}{2} - \frac{1}{3n+2}$

Note: if we let $\dfrac{3}{(3r-1)(3r+2)} = u_r$ and $\dfrac{1}{3r-1} = f_r$

then $\{f_r\}$ is the sequence required in Theorem 3A, and so

$$\sum_{r=1}^{n} u_r = f_1 - f_{n+1} = \frac{1}{2} - \frac{1}{3n+2}$$

For the first time we have actually *found* f_r.

Exercise 5.1:7

(12) R **1** If $f_r = r(r+1)$ find $f_r - f_{r-1}$. Hence verify that

$$\sum_{r=1}^{n} r = \tfrac{1}{2} n(n+1)$$

(12) → **2** If $f_r = r^2(r+1)^2$ find $f_r - f_{r-1}$. Hence find $\sum_{r=1}^{n} r^3$.

(12) **3** If $f_r = r^2$ find $f_r - f_{r-1}$. Hence find the sum of the first n odd numbers.

(12) **4** If $f_r = r^3$ find $f_r - f_{r-1}$. Hence find
$1 + 7 + 19 + \cdots + (3n^2 - 3n + 1)$.

(12) → **5** If $f_r = r(r+1)(r+2)$ show that $f_r - f_{r-1} = 3r(r+1)$.
Hence find $\sum_{r=1}^{n} r(r+1)$.

Write down $\sum_{r=1}^{n} r$ (see 5.1:3 Q16). Hence verify that

$$\sum_{r=1}^{n} r^2 = \tfrac{1}{6} n(n+1)(2n+1)$$

(12) **6** If $f_r = r(r+1)(r+2)(r+3)$, find $f_r - f_{r-1}$. Hence find $\sum_{r=1}^{n} r(r+1)(r+2)$.

Write down $\sum_{r=1}^{n} r$ and $\sum_{r=1}^{n} r^2$ (see question 5), and hence verify that

$$\sum_{r=1}^{n} r^3 = \tfrac{1}{4} n^2 (n+1)^2$$

(12) **7** (i) If $u_r = r(r+1)(r+2)(r+3)$, find f_r such that $f_r - f_{r-1} = u_r$. Find $\sum_{r=1}^{n} u_r$ and hence show that

$$\sum_{r=1}^{n} r^4 = \tfrac{1}{30} n(n+1)(2n+1)(3n^2 + 3n - 1)$$

(ii) Write down $\sum_{r=1}^{n} r(r+1)(r+2)\ldots(r+k)$.

(12) + **8** By equating the coefficients of r^3, r^2, r and the constants in the identity, find constants a, b and c such that:

$$24r^2 \equiv (ar+1)^3 - (br-1)^3 + c$$

Hence verify that

$$\sum_{r=1}^{n} r^2 = \tfrac{1}{6} n(n+1)(2n+1)$$

(12) + **9** Find constants a, b, c and d such that:

$$5r^4 + r^2 \equiv ar^2(r+1)^3 + br^3(r+1)^2 + cr^2(r-1)^3 + dr^3(r-1)^2$$

Hence verify that

$$\sum_{r=1}^{n} r^4 = \tfrac{1}{30} n(n+1)(2n+1)(3n^2 + 3n - 1)$$

(12) + **10** If $f_r = ar^3(r+1)^3 + br^2(r+1)^2 + cr(r+1)$, find a, b and c such that $f_r - f_{r-1} = r^5$.
Hence show that

$$\sum_{r=1}^{n} r^5 = \tfrac{1}{12} n^2 (n+1)^2 (2n^2 + 2n - 1)$$

(12) R **11** If $f_r = (r+1)!$, find $f_r - f_{r-1}$. Hence find $\sum_{r=1}^{n} r \cdot r!$

(12) → **12** Using the appropriate factor formula (see section 3.2:7) show that if $f_r = \sin(2r+1)\theta$, then $f_r - f_{r-1} = 2\sin 2r\theta \cos\theta$.
Hence find $\sum_{r=1}^{n} \sin 2r\theta$.

(12) R **13** If $f_r = \cos 2r\theta$, find $f_r - f_{r-1}$. Hence find

$$\sum_{r=1}^{n} \sin(2r-1)\theta$$

Method of differences

14 (i) Show that $\sum_{r=1}^{n} r^0 = n$. Write down the formulae for

$$\sum_{r=1}^{n} r, \quad \sum_{r=1}^{n} r^2, \quad \sum_{r=1}^{n} r^3$$

(ii) Use these formulae to find:

- (a) $\sum_{r=1}^{n} r(r+4)$
- (b) $\sum_{r=1}^{n} r(2r-1)$
- **R** (c) $\sum_{r=1}^{n} r^2(r+1)$
- (d) $\sum_{r=1}^{n} (r-1)(r+1)$
- → (e) $\sum_{r=1}^{n} (2r+1)(r-3)$

15 Find $n + 2(n-1) + 3(n-2) + \cdots + n$.

16 Show that

$$r(r+1)(r+3) \equiv r(r+1)(r+2) + r(r+1)$$

Hence, using the results of questions 1 and 5, find

$$\sum_{r=1}^{n} r(r+1)(r+3)$$

Similarly, find

$$\sum_{r=1}^{n} r(r+1)^2$$

• → **17** Write down $\sum_{r=1}^{n} r^2$ (see question 5). Hence write down $\sum_{r=1}^{2n} r^2$. Hence find $\sum_{r=n+1}^{2n} r^2$. Find $51^2 + 52^2 + \cdots + 100^2$.

18 Find $\sum_{r=n}^{2n} r^3$. Find $10^3 + 11^3 + \cdots + 20^3$.

19 Find

(i) $\sum_{r=1}^{n+2} r^2$ (ii) $\sum_{r=1}^{n} (r+2)^2$

(iii) $\sum_{r=1}^{3n} r^2$ (iv) $\sum_{r=1}^{n} (3r)^2$

R 20 (i) Using the formulae for $\sum_{r=1}^{n} r$ and $\sum_{r=1}^{n} r^2$, find

$$\sum_{r=1}^{n} (r-2)(r+3)$$

Hence find

$$\sum_{r=n+1}^{2n} (r-2)(r+3)$$

(ii) Find

$$\sum_{r=16}^{30} (2r-1)(2r+1)$$

21 Show that

$$\sum_{r=1}^{n} (n+r)^2 = \tfrac{1}{6} n(2n+1)(7n+1)$$

Find

$$1 + (1+2) + (1+2+3) + \cdots + (1+2+3+4+\cdots+n).$$

• → **22** (i) Find:
(a) $1^2 + 2^2 + 3^2 + \cdots + (2n)^2$
(b) $2^2 + 4^2 + 6^2 + \cdots + (2n)^2$
Hence find: (c) $1^2 + 3^2 + 5^2 + \cdots + (2n-1)^2$.

(ii) Check your answer to (c) by writing it as $\sum_{r=1}^{n} (2r-1)^2$ and using the method of question 14.

(iii) Find $1^2 + 3^2 + 5^2 + \cdots + 99^2$.

R 23 Find the sum of the cubes of the first n odd numbers. Find $1^3 + 3^3 + 5^3 + \cdots + 99^3$.

24 (i) Using the method of question 22, show that

$$-1^2 + 2^2 - 3^2 + 4^2 - \cdots + (-1)^n n^2$$
$$= (-1)^n \tfrac{1}{2} n(n+1)$$

(ii) Check this result by using the fact that $a^2 - b^2 = (a+b)(a-b)$.

(iii) Find $-2^2 + 3^2 - 4^2 + 5^2 - \cdots - 98^2 + 99^2$.

25 Show that:
(i) $1^2 + 2 + 3^2 + 4 + 5^2 + 6 + \cdots + (2n-1)^2 + 2n$
$= \tfrac{1}{3} n(4n^2 + 3n + 2)$
(ii) $1^2 + 2 \cdot 2^2 + 3^2 + 2 \cdot 4^2 + 5^2 + 2 \cdot 6^2 + \cdots$
$+ (2n-1)^2 + 2 \cdot (2n)^2 = n(2n+1)^2$

26 Find the sum of the squares of the numbers between 1 and 200 (inclusive) which are not divisible by 3.

27 Find the sum of the cubes of the numbers between 1 and 100 which are not divisible by 5 or 7.

28 Show that the sum of the cubes of the numbers between 100 and 200 which are divisible by 3 is a multiple of 4200.

29 How many terms of the series $1^3 + 2^3 + 3^3 + \cdots$ must be taken to make the sum exceed 10^6?

30 If $s_n = \sum_{r=1}^{n} u_r$, find u_r when s_n is given by:

(i) $n(n+1)(n+3)$
(ii) $n^2(n+1)$
(iii) n^3

(See 5.1:3 WE 4.)

• **31** If $f_r - f_{r+1} = u_r$ for all r, show that

$$\sum_{r=1}^{n} u_r = f_1 - f_{n+1}$$

(See proof of Theorem 3, page 229.)

• → **32** If $f_r = 1/r$, show that $f_r - f_{r+1} = 1/r(r+1)$. Hence find

$$\sum_{r=1}^{n} \frac{1}{r(r+1)}$$

33 If $f_r = 1/r(r+1)$, find $f_r - f_{r+1}$. Hence find

$$\sum_{r=1}^{n} \frac{1}{r(r+1)(r+2)}$$

34 If $f_r = 1/r!$, find $f_r - f_{r+1}$. Hence find

$$\sum_{r=1}^{n} \frac{r}{(r+1)!}$$

35 If $f_r = (r+2)/r(r+1)$, find $f_r - f_{r+1}$. Hence find

$$\sum_{r=1}^{n} \frac{r+4}{r(r+1)(r+2)}$$

36 (i) If $f_r = (2r+3)/r(r+1)$, find $f_r - f_{r+1}$.

(ii) Hence find

$$\sum_{r=1}^{n} \frac{2(r+3)}{r(r+1)(r+2)}$$

(iii) Deduce the sum to infinity of the series.

• → **37** (i) If $f_r = 1/[(2r-1)(2r+1)]$, find $f_r - f_{r+1}$.

(ii) Hence find

$$\sum_{r=1}^{n} \frac{1}{(2r-1)(2r+1)(2r+3)}$$

(iii) Deduce the sum to infinity s_∞ of the series.

(iv) Find the smallest integer n such that s_n differs from s_∞ by less than 10^{-4}.

R 38 (i) If $f_r = 1/[(3r-2)(3r+1)]$, find $f_r - f_{r+1}$.

(ii) Hence find

$$\sum_{r=1}^{n} \frac{1}{(3r-2)(3r+1)(3r+4)}$$

(iii) Deduce the sum to infinity s_∞ of the series.

(iv) Find the smallest integer n such that s_n differs from s_∞ by less than 2^{-10}.

(13) **39** By expressing u_r in partial fractions, find $\sum_{r=1}^{n} u_r$ when u_r is given by:

(i) $\dfrac{1}{r(r+1)}$

→ (ii) $\dfrac{1}{4r^2-1}$

(iii) $\dfrac{4}{(4r-3)(4r+1)}$

• → (iv) $\dfrac{1}{r(r+2)}$

(v) $\dfrac{1}{r(r+3)}$

(vi) $\dfrac{1}{r(r+1)(r+2)}$

(vii) $\dfrac{r+4}{r(r+1)(r+2)}$

• → (viii) $\dfrac{1}{(2r-1)(2r+1)(2r+3)}$

(ix) $\dfrac{r+1}{(2r-1)(2r+1)(2r+3)}$

(x) $\dfrac{1}{r(r+2)(r+4)}$

(xi) $\dfrac{2r-1}{r(r+1)(r+2)}$

(Hint: write out an array as in worked example 13 to make clear how the cancelling proceeds.)
In each case find the sum to infinity of the series.

(13) **R 40** (i) By splitting $u_r = r/[(r+2)(r+3)(r+4)]$ into partial fractions, find $\sum_{r=1}^{n} u_r$.

(ii) If $f_r = (r+1)/[(r+2)(r+3)]$, show that $u_r = f_r - f_{r+1}$. Hence check your answer to (i).

(iii) Write down $\sum_{r=1}^{\infty} u_r$.

5.2 Method of induction

Worked example 1 Prove that, for all positive integers n,

$$1^2+2^2+3^2+\cdots+n^2=\tfrac{1}{6}n(n+1)(2n+1) \qquad (1)$$

(i) First note that the statement is true when $n=1$, since L.H.S. $=1^2=1$, R.H.S. $=\tfrac{1}{6}\cdot 1\cdot 2\cdot 3=1$.
(ii) Let us now *assume* that the statement is true for some value of n, e.g. $n=k$; i.e. that

$$1^2+2^2+3^2+\cdots+k^2=\tfrac{1}{6}k(k+1)(2k+1) \qquad (2)$$

Now, adding $(k+1)^2$ to both sides of this equation:

$$\begin{aligned}1^2+2^2+3^2+\cdots+k^2+(k+1)^2 &= (1^2+2^2+3^2+\cdots+k^2)+(k+1)^2\\ &=\tfrac{1}{6}k(k+1)(2k+1)+(k+1)^2 \\ &\qquad\text{by assumption (2)}\\ &=\tfrac{1}{6}(k+1)\{k(2k+1)+6(k+1)\}\\ &=\tfrac{1}{6}(k+1)(2k^2+7k+6)\\ &=\tfrac{1}{6}(k+1)(k+2)(2k+3) \qquad (3)\end{aligned}$$

But equation (3) is exactly the same as statement (1) with $n=k+1$. We have proved that:

if the statement is true when $n=k$, it is true when $n=k+1$ **(4)**

But the statement is true when $n=1$.
So, by statement (4) with $k=1$, it is true when $n=2$.
So, by statement (4) with $k=2$, it is true when $n=3$.
So, by statement (4) with $k=3$, it is true when $n=4$...
Continuing this argument, the statement is true for all integer values of n.
In general,

> suppose we are trying to prove that the statement P_n is true for all positive integers n. Then
> (i) we check that P_1 is true;
> (ii) we show that if P_k is true then P_{k+1} is true.

This method of proof is called the **method of induction**.

Worked example 2 Prove that, for all positive integers n, $2^{n+1}+3^{2n-1}$ is divisible by 7.

The symbol $|$ is commonly used to mean 'is a factor of'. So we wish to prove that

$$7|(2^{n+1}+3^{2n-1}) \text{ for all } n\in\mathbb{Z}^+$$

Let this statement be P_n.

(i) When $n=1$, $2^{n+1}+3^{2n-1}=2^2+3=7$. Now $7|7$, so P_1 is true.

(ii) Assume P_k is true, i.e. that $7|(2^{k+1}+3^{2k-1})$, i.e. that for some positive integer p,

$$2^{k+1}+3^{2k-1} = 7p$$

so $\qquad 2^{k+1} = 7p - 3^{2k-1} \qquad$ (5)

Consider $\quad 2^{k+2}+3^{2k+1} = 2 \cdot 2^{k+1}+3^{2k+1}$

$$= 2(7p-3^{2k-1})+3^{2k+1}$$

by assumption (5)

$$= 14p - 2 \cdot 3^{2k-1}+3^{2k+1}$$
$$= 14p - 2 \cdot 3^{2k-1}+9 \cdot 3^{2k-1}$$
$$= 14p + 7 \cdot 3^{2k-1}$$
$$= 7(2p+3^{2k-1})$$

$2p+3^{2k-1}$ is an integer; so $7|(2^{k+2}+3^{2k+1})$, i.e. P_{k+1} is true.
So, by the principle of induction, P_n is true for all positive integers n.

Exercise 5.2

Questions 1–39: Prove by induction that, for all positive integers n, the following are true.

① **1** $1+2+3+4+\cdots+n = \frac{1}{2}n(n+1)$

① **2** The sum of the first n odd numbers is n^2.

①R **3** $1^3+2^3+3^3+\cdots+n^3 = \frac{1}{4}n^2(n+1)^2$

①→ **4** $1 \cdot 2 + 2 \cdot 3 + 3 \cdot 4 + \cdots + n(n+1)$
$= \frac{1}{3}n(n+1)(n+2)$

① **5** $\sum_{r=1}^{n} (2r-1)^2 = \frac{1}{3}n(4n^2-1)$

① **6** $\sum_{r=1}^{n} r(r+1)(r+2) = \frac{1}{4}n(n+1)(n+2)(n+3)$

① **7** $\sum_{r=1}^{n} r(r+2)(r+4) = \frac{1}{4}n(n+1)(n+4)(n+5)$

① **8** $\sum_{r=1}^{n} r(r+1)(2r+1) = \frac{1}{2}n(n+1)^2(n+2)$

① **9** $\sum_{r=1}^{n} (n+r)^2 = \frac{1}{6}n(2n+1)(7n+1)$

①+ **10** $2\sum_{r=1}^{n} r^5 + \sum_{r=1}^{n} r^3 = 3\left\{\sum_{r=1}^{n} r^2\right\}^2$

① **11** $\sum_{r=1}^{n} 3^r = \frac{3}{2}(3^n-1)$

① **12** $\sum_{r=1}^{n} r \cdot 2^r = 2+2^{n+1}(n-1)$

① **13** $\sum_{r=1}^{n} r \cdot r! = (n+1)!-1$

① **14** $\sum_{r=1}^{n} (r^2+1)r! = n(n+1)!$

①+ **15** $\sum_{r=1}^{n} (3^r+3^{r-1}+\cdots+3+1) = \frac{1}{4}(3^{n+2}-2n-9)$

①→ **16** $\sum_{r=1}^{n} \frac{1}{r(r+1)} = 1 - \frac{1}{n+1}$

①R **17** $\sum_{r=1}^{n} \frac{1}{(2r-1)(2r+1)} = \frac{1}{2}\left(1 - \frac{1}{2n+1}\right)$

① **18** $\sum_{r=1}^{n} \frac{1}{r(r+1)(r+2)} = \frac{n(n+3)}{4(n+1)(n+2)}$

①→ **19** $\sum_{r=1}^{n} \sin(2r+1)\theta$
• $= \sin(n+2)\theta \sin n\theta \operatorname{cosec} \theta$

①R **20** $\sum_{r=1}^{n} \cos(2r-1)\theta = \frac{1}{2}\sin 2n\theta \operatorname{cosec} \theta$

① **21** $\sum_{r=1}^{n} \sin^2 r\theta = \frac{1}{2}\left\{n - \frac{\cos(n+1)\theta \sin n\theta}{\sin \theta}\right\}$

②• **22** $3|(7^n+5)$

② **23** $8|(9^n-1)$

②R **24** $16|(17^n-1)$

② → 25 $14|(8^n+6)$

② 26 $3|(n^3+6n^2+8n)$

② 27 $6|(n^3-n)$

② 28 $9|[(n-1)^3+n^3+(n+1)^3]$

② → 29 $8|(5^{2n}-3^{2n})$

② R 30 $3|(7^n-2^{2n})$

② 31 $9|(5^{2n}+3n+8)$

② 32 $6|(7^{2n}-5\cdot 7^n-2)$

② 33 $7|(5^{6n}+2^{3n+1}-3)$

② + 34 $25|(7^{2n}+2^{3n-3}\cdot 3^{n-1})$

② 35 $(x-y)|(x^n-y^n)$ for all integers x and y.

36 $\dfrac{d}{dx}(x^n)=nx^{n-1}$

37 $\dfrac{d^n}{dx^n}(\sin ax)=a^n\sin(ax+\tfrac{1}{2}n\pi)$

+ 38 $\dfrac{d^n}{dx^n}(x^{n-1}\ln x)=\dfrac{(n-1)!}{x}$

39 An n-sided polygon has $\tfrac{1}{2}n(n-3)$ diagonals.

• 40 If $u_{r+1}=2u_r+1$ $(r>1)$, $u_1=1$, prove that for all positive integers n, $u_n=2^n-1$ (see 5.1:1 Q13(i)).

→ 41 If $u_{r+2}=5u_{r+1}-6u_r$ $(r>2)$, $u_1=3$, $u_2=7$, prove that for all positive integers n, $u_n=2^n+3^{n-1}$ (see 5.1:1 Q13(ii)).

42 If $u_{r+1}=(r+1)u_r$ $(r>1)$, $u_1=1$, prove that for all positive integers n, $u_n=n!$.

5.3 Permutations and combinations

5.3:1 Permutations of objects which are all different

Worked example 1 How many possible arrangements are there of the letters A, B, C and D?

We want to write down 4 letters in order, e.g.

1st 2nd 3rd 4th
B   C   A   D

The first letter may be any one of 4: A or B or C or D.
For each choice of the first letter, there are 3 choices for the second, e.g. if the first letter is B, the second may be A or C or D.
For each choice of the first two letters, there are 2 choices for the third, e.g. if the first two letters are B, C, the third may be A or D.
For each choice of the first three letters there remains just 1 possibility for the fourth letter, e.g. if the first three letters are B, C, A, the fourth must be D.

Number of arrangements $= 4\times 3\times 2\times 1=4!=24$

This is illustrated by a **tree diagram** (see fig. 5.3:1).

In general:

number of **arrangements**, or **permutations**, of n different objects
$$= n(n-1)(n-2)\cdots 3\cdot 2\cdot 1 = n! \qquad (1)$$

Series and probability 237

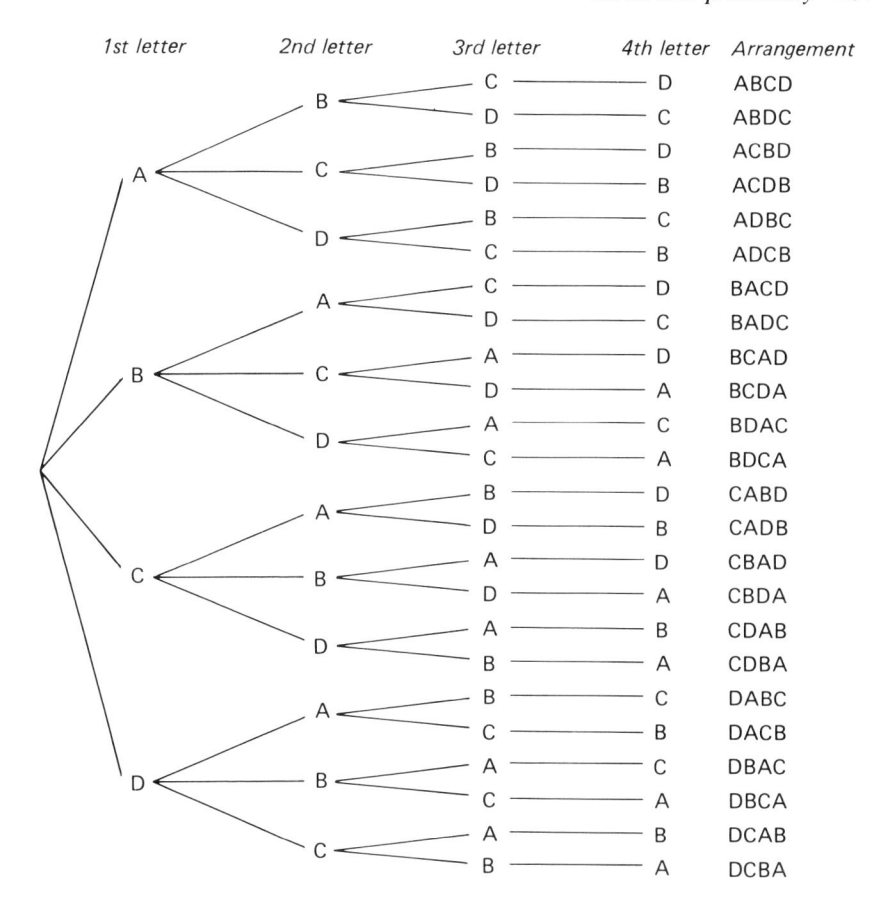

1st letter   2nd letter   3rd letter   4th letter   Arrangement

```
                    C ——————— D    ABCD
            B <
                    D ——————— C    ABDC
                    B ——————— D    ACBD
    A <       C <
                    D ——————— B    ACDB
                    B ——————— C    ADBC
            D <
                    C ——————— B    ADCB
                    C ——————— D    BACD
            A <
                    D ——————— C    BADC
                    A ——————— D    BCAD
    B <       C <
                    D ——————— A    BCDA
                    A ——————— C    BDAC
            D <
                    C ——————— A    BDCA
                    B ——————— D    CABD
            A <
                    D ——————— B    CADB
                    A ——————— D    CBAD
    C <       B <
                    D ——————— A    CBDA
                    A ——————— B    CDAB
            D <
                    B ——————— A    CDBA
                    B ——————— C    DABC
            A <
                    C ——————— B    DACB
                    A ——————— C    DBAC
    D <       B <
                    C ——————— A    DBCA
                    A ——————— B    DCAB
            C <
                    B ——————— A    DCBA
```

Fig. 5.3:**1**

Worked example 2 How many arrangements can be made using two different letters chosen from the letters A, B, C, D and E?

We want to write down 2 letters in order, e.g.

$$\begin{array}{cc} 1st & 2nd \\ E & B \end{array}$$

There are 5 choices for the first letter: A or B or C or D or E.

For each choice of the first letter, there are 4 choices for the second letter, e.g. if the first letter is E, the second may be A or B or C or D (see fig. 5.3:**2**).

Number of arrangements $= 5 \times 4 = 5!/3! = 20$

In general:

^nP_r = number of arrangements of r different objects chosen from n different objects

$ = n(n-1)(n-2) \cdots (n-r+1)$

$ = \dfrac{n!}{(n-r)!}$ **(2)**

238 *Permutations of objects which are all different*

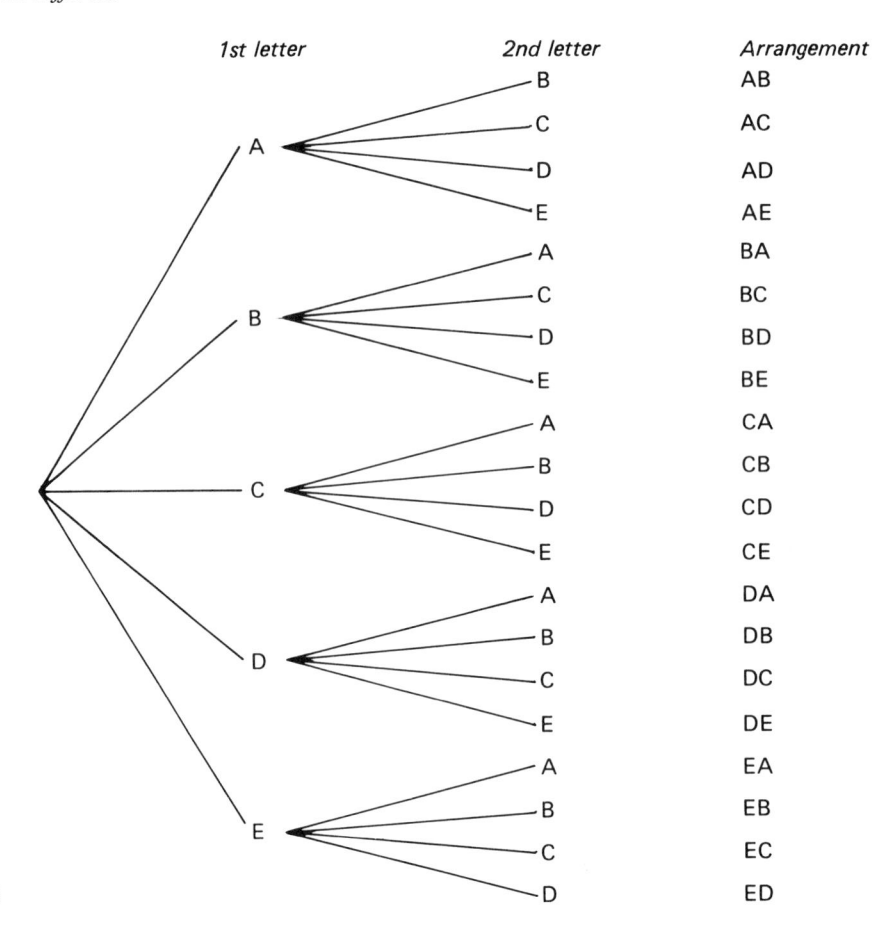

1st letter	2nd letter	Arrangement
A	B	AB
	C	AC
	D	AD
	E	AE
B	A	BA
	C	BC
	D	BD
	E	BE
C	A	CA
	B	CB
	D	CD
	E	CE
D	A	DA
	B	DB
	C	DC
	E	DE
E	A	EA
	B	EB
	C	EC
	D	ED

Fig. 5.3:**2**

Note: if we put $r = n$ in equation (2) we should get equation (1) for all n. So we define

$$0! = 1$$

Worked example 3 How many arrangements of the letters T, R, I, F, L and E begin with a vowel?

There are 2 choices for the first letter: I and E.
For each choice of the first letter, there are 5 choices for the second letter, e.g. if the first letter is E, the second may be T, R, I, F or L.
There are 4 choices for the third letter, 3 for the fourth, 2 for the fifth, and 1 for the sixth.

So   number of arrangements $= 2 \times 5 \times 4 \times 3 \times 2 \times 1 = 240$

Worked example 4 In how many arrangements of the letters T, R, I, F, L and E are the vowels together?

Consider the arrangements containing the ordered pair IE. There are effectively 5 letters to arrange T, R, F, L and 'IE'. So the number of arrangements is 5!
Similarly, there are 5! arrangements containing the ordered pair EI.
So the number of arrangements of the letters T, R, I, F, L and E in which the vowels are together is $5! \times 2 = 240$.

Worked example 5 In how many arrangements of the letters O, R, A, N, G, E and S are the three vowels separated?

Number of ways of arranging the consonants R, N, G, S is 4! For each of these arrangements, e.g. SGRN, the vowels must be arranged within the five positions shown below:

$$* S * G * R * N *$$

The first vowel may fill any of 5 positions, the second vowel any of 4, the third any of 3.
So the vowels may be arranged in $(5 \times 4 \times 3)$ or $^5P_3$ ways.
Thus the number of ways of arranging the letters O, R, A, N, G, E and S so that the vowels are separated is $4! \times {}^5P_3 = 1440$.

Worked example 6 How many four-digit numbers can be made from the digits 0, 1, 2, 3, 4 and 5, if repetition *is* allowed?

The first digit may be any one of 5: 1, 2, 3, 4 or 5 (since 0534, for example, is not a four-digit number), the second any one of 6, the third any one of 6, and the fourth any one of 6.
So the number of four-digit numbers is $5 \times 6 \times 6 \times 6 = 1080$.

In general:

the number of arrangements of r objects from n different objects if repetition *is* allowed is n^r **(3)**

Exercise 5.3:1

① → **1** How many ways are there of getting from A to C, passing through each point at most once?

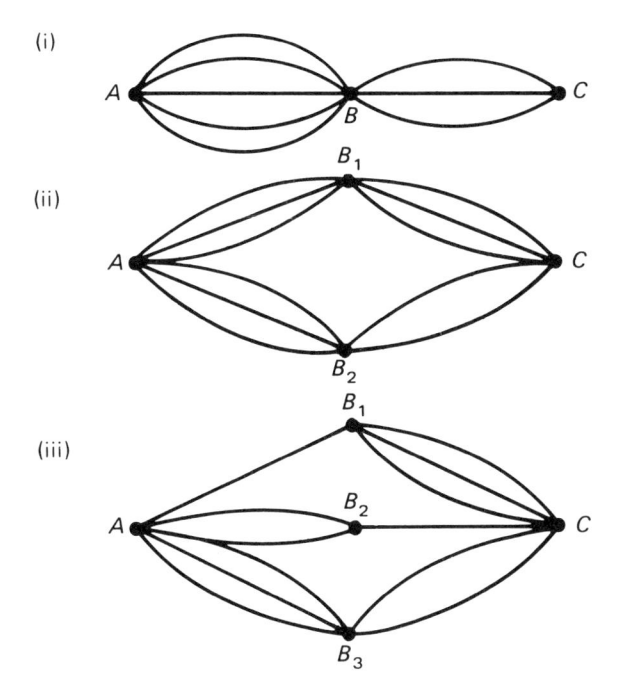

(i)

(ii)

(iii)

① **2** How many arrangements are there of the letters:
(i) DOG   (ii) SPAM?

Draw tree diagrams illustrating these results.

① → **3** How many arrangements are there of the letters:
(i) DUCKLING   (ii) SANDWICH?

① **4** How many five-digit numbers can be made from the digits 2, 3, 4, 7 and 9 if no digit may be repeated?

① **5** In a competition on the back of a Corn Flakes packet, seven desirable qualities for a kitchen (e.g. spaciousness, versatility) must be put in order of importance. How many different entries must be completed to ensure a winning order?

① **6** Ten men and ten women write to a computer dateline service in search of love and fulfillment. In how many ways can they be paired off?

② **7** How many arrangements are there of:
(i) two letters from the word OVEN?
(ii) three letters from the word FRIDGE?

Draw tree diagrams illustrating these results, if you can be bothered.

240 *Permutations of objects which are all different*

② → **8** How many arrangements are there of:

 (i) four letters from the word MICROWAVE?
 (ii) five letters from the word GARNISH?

② **9** How many four-digit numbers can be made from the digits 2, 3, 4, 5, 7 and 9 if no digit may be repeated?

② → **10** How many numbers can be made from the digits 2, 3, 4 and 5 if no digit may be repeated? (Hint: consider 4 cases: four-digit numbers, three-digit numbers, two-digit numbers, single-digit numbers.)

② **11** How many numbers greater than 100 can be made from the digits 1, 2, 3, 5, 7 and 9 if no digit may be repeated?

② **12** Seventeen countries enter the Eurovision Song Contest. How many ways are there of filling:

 (i) the first three places?
 (ii) the last four places?

Assume that Monaco always comes last.

② **13** In a competition on the back of a Weetabix packet, nine qualities desirable in a car (e.g. low fuel consumption) are given, of which five must be listed in order of importance. How many entries must be completed to ensure a winning order?

+ **14** Santa has 15 different presents which he is delivering to 6 children. How many different distributions of the presents are possible if:

 (i) each child gets one present and Santa keeps the rest?
 (ii) each child gets two presents and Santa keeps the rest?

• **15** 6 actors and 8 actresses are available for a play with 4 male roles and 3 female roles. How many different cast lists are possible?

16 Santa has 5 (different) dolls and 7 (different) plastic sub-machine guns which he is delivering to 3 girls and 4 boys. How many different distributions of the presents are possible if:

 (i) the girls get the dolls and the boys get the guns?
 (ii) the boys get the dolls and the girls get the guns?
 (iii) the boys get the girls and the dolls get the guns?

• → **17** How many different possibilities are there when eight people form:

 (i) one queue of three and one queue of five?
 (ii) two queues of four?

R **18** How many different possibilities are there when twelve people form:

 (i) a queue of seven and another queue of five?
 (ii) two queues of six?
 (iii) three queues of four?
 (iv) two queues of five and a queue of two?

③ → **19** How many four-digit numbers can be made with the digits 0, 1, 2 and 3 if no digit may be repeated?

③ **20** How many numbers between 2000 and 4000 can be made with the digits 1, 2, 3 and 4 if no digit may be repeated?

③ **21** How many odd five-digit numbers can be made with the following digits if no digit may be repeated?

 (i) 1, 2, 3, 4, 5, 6 and 7
 (ii) 2, 3, 4, 5, 6, 7 and 8

③ → **22** How many arrangements of the letters in the word DUCKLING:

 (i) start with a vowel?
 (ii) end with a consonant?
 (iii) start with a vowel and end with a consonant?

③ **23** How many arrangements of the letters in the word MUSTARD start with a vowel and end with an M or an S?

③ **24** How many arrangements of the letters in the word GARNISH start with a vowel and end with an A or an N? (Hint: consider two cases: (i) first letter A; (ii) first letter I. In each case start by considering the choice of seventh letter.)

③ → **25** How many arrangements of the letters in the word TRIFLE start with a vowel and end with an E or an F?

③ **26** How many even four-digit numbers can be made with the digits 0, 1, 2 and 3, if no digit may be repeated?

③ **27** How many four-digit numbers divisible by 5 can be made with the digits 0, 2, 3, 5 and 7, if no digit may be repeated?

③ R **28** How many numbers can be made with the digits 0, 1, 2, 3 and 4 if no digit may be repeated? How many of these numbers are between 9 and 2000?

④ **29** In how many arrangements of the letters:

 (i) ZEBRA
 (ii) TURBOT
→ (iii) STURGEON
 (iv) GUNFIRE

are the vowels together?

④ R **30** How many arrangements of the letters LAMB CHOP are there?

In how many of these arrangements are the vowels:

 (i) together (ii) separated?

④ **31** In how many arrangements of the letters BAKING are the vowels separated?

32 In how many arrangements of the letters:

(i) ROASTING
(ii) MICROWAVE

are the vowels separated?

33 In how many arrangements are all the girls separated when:

(i) 3 boys and 2 girls
(ii) 7 boys and 5 girls
(iii) r boys and s girls ($r+1 \geqslant s$)

are lined up in a row? Why the restriction in part (iii)?

34 12 different books are to be arranged on a bookshelf, 3 by Dickens, 4 by Martin Amis, 4 by Barbara Cartland, 1 by me. How many possible arrangements are there if:

(i) the books by Dickens must be next to each other?
(ii) the books by Martin Amis must be kept separate?

R 35 Alice, Brian, Carol, Donald, Eric and Frances go to the cinema. How many arrangements are possible when they sit in six adjacent seats if:

(i) Donald insists on sitting next to Alice?
(ii) Frances refuses to sit next to Brian?

36 There are 7 seats around a circular table, and 7 people to be seated at the table. How many different seating plans are possible, assuming that the orientation of the final arrangement is (i) important (ii) unimportant?

37 There are 5 seats around a circular table and 8 people hanging around waiting to be seated, all of them too lazy to fetch another chair. How many possible seating plans are there if all 5 seats are used, assuming that the orientation of the final arrangement is unimportant?

38 A necklace manufacturer has a stock of identical circular rings and stocks of beads—red, green, blue and black—which are identical except for colour. How many distinct necklaces can be manufactured if each is made by threading four different-coloured beads onto a ring?

R 39 A necklace manufacturer has a stock of identical circular rings and stocks of beads—red, green, blue, yellow, brown, black and white—which are identical except for colour. How many distinct necklaces can be manufactured if each is made by threading four different-coloured beads onto a ring?

40 The 6 friends of question 35 go to a restaurant. They sit at a circular table with 6 seats. How many possible seating plans are there, assuming that the orientation of the final arrangement is unimportant, if:

(i) Alice insists on sitting next to Eric?
(ii) Brian insists on sitting exactly opposite Carol to facilitate eye-to-eye contact?
(iii) Frances refuses to sit next to lecherous Brian?

41 How many seating plans are possible when m people are arranged around a circular table with n seats, assuming that the orientation of the final arrangement is unimportant, if:

(i) there is no restriction on where anyone sits?
(ii) two particular people have to be kept apart?

42 How many four-digit numbers can be made from the digits 1, 2, 3, 4 and 5:

(i) if no digit may be repeated?
(ii) if repetitions are allowed?

43 How many four-digit numbers can be made from the digits 0, 1, 2, 3, 4, 5, 6 and 7:

(i) if no digit may be repeated?
(ii) if repetitions are allowed?

44 In a class of 8 students, how many possibilities are there if we award:

(i) a first prize, a second prize and a third prize?
(ii) a Maths prize, a Physics prize and a Chemistry prize?

(Assume that all the students are capable of winning any of the prizes—an unusual class.)

R 45 Find how many numbers less than 2000 can be made from the digits 1, 2, 4 and 7:

(i) if no digit may be repeated?
(ii) if repetitions are allowed?

How many of the numbers are:

(iii) even?
(iv) odd?

46 Ten third-rate rock groups are available to play at the Dog and Aardvark. The publican wants to book one group every night for a week. How many possible schedules can he make:

(i) if he wants a different group each night?
(ii) if he doesn't make this restriction?

5.3:2 Combinations of objects which are all different

The number of possible arrangements of r different objects which are chosen from n different objects is ^nP_r. We can consider this in two steps. We first *choose* the r objects, and then *arrange* them.

242 *Combinations of objects which are all different*

Let the number of ways of choosing r objects from n be ^nC_r. (This is sometimes written as $\binom{n}{r}$.)

Since the number of ways of arranging r given objects is $r!$,

$$^nP_r = {^nC_r} \times r!$$

$$^nC_r = \frac{^nP_r}{r!}$$

$$= \frac{n!}{(n-r)!\,r!}$$

> ^nC_r, the number of possible **choices**, or **selections**, or **combinations** of r objects from n different objects is given by
>
> $$^nC_r = \frac{n(n-1)(n-2)\cdots(n-r+1)}{r!} = \frac{n!}{(n-r)!\,r!} \qquad (4)$$

Note that a choice, selection or combination of r objects does not depend on their order. An arrangement does depend on order.

Worked example 7 A group of four journalists is to be chosen to cover an important espionage trial. There are 5 male and 7 female journalists available. How many possible groups can be formed:

(i) consisting of 2 men and 2 women?
(ii) containing at least one woman?

(i) There are 5 men available; 2 must be chosen. Number of ways of choosing the men $= {^5C_2}$. For each choice of two men, there are $^7C_2$ ways of choosing the two women. So

$$\text{number of choices} = {^5C_2} \times {^7C_2} = \frac{5 \cdot 4}{2 \cdot 1} \times \frac{7 \cdot 6}{2 \cdot 1} = 210$$

(ii) The only possibility excluded is an all-male group. Number of ways of choosing all men $= {^5C_4} = 5$. Total number of ways of choosing the group, i.e. of choosing 4 people from 12, is

$$^{12}C_4 = \frac{12 \cdot 11 \cdot 10 \cdot 9}{4 \cdot 3 \cdot 2 \cdot 1} = 495$$

So the number of ways of choosing the group so that it contains at least one woman $= 495 - 5 = 490$.

Exercise 5.3:2

1 How many possibilities are there when a 5-card hand is dealt from a pack of 52?

• **2** How many possibilities are there when 4 cards are dealt from a pack of 52, given that one of them is the ace of hearts?

3 A theatre company consisting of 6 players is to be chosen from 15 actors. How many possibilities are there if the company must include the most pushy actor?

4 An England cricket touring party consists of 17 members. In how many ways can 11 players be selected for a match:

(i) if Ian Botham must be selected?
(ii) if Ian Botham or David Gower must be selected, but not both?

→ **5** There are 20 players on the Men's World Championship Tennis circuit. In how many ways can 8 players be selected for a tournament if both Boris Becker and John McEnroe must play, or neither?

6 How many possible results are there when 12 people are divided into two groups:

(i) one of five and one of seven?
(ii) each of six?

7 How many possible results are there when 16 people are divided into two groups:

(i) one of ten and one of six?
(ii) each of eight?

R 8 Eight people are divided into two groups and two particular people, Carol and Eric, have to be in different groups. How many possible results are there when the groups are:

(i) one of three and one of five;
(ii) two groups of four?

9 Find the number of possible results when 10 children are divided into two groups of 5, if the two youngest children:

(i) must be in the same group;
(ii) must not be in the same group.

10 How many possible results are there when 12 people are divided into three groups:

(i) of 3, 4 and 5?
(ii) each of 4?

11 A tennis team consists of six players. How many different teams are possible if the team consists of:

(i) three pairs?
(ii) first, second and third pairs?

12 How many possibilities are there when 12 people are divided into two groups?

13 A rich aristocrat has ten country houses which he wants to leave in his will to his two sons. How many different distributions are possible if each son receives:

(i) the same number of houses?
(ii) an odd number of houses?

14 A committee of three is to be chosen from 5 men and 4 women. How many possible committees are there?
How many of these committees will consist of:

(i) one man and two women?
(ii) two men and one woman?

15 Four fish are to be chosen to fill an aquarium. If the only fish available are 7 piranhas and 9 sharks, how many possibilities are there? For how many of these possibilities will the aquarium contain

(i) no sharks?
(ii) exactly one shark?
(iii) exactly two sharks?

(iv) exactly three sharks?
(v) four sharks?

What is the sum of your answers to (i), (ii), (iii), (iv) and (v)?

16 A committee of five is to be chosen from 6 men and 6 women. How many possibilities are there if the committee must contain:

(i) at least one man?
(ii) at least one woman?
(iii) at least one man and one woman?

17 A committee of four is to be chosen from 5 men and 7 women. How many possibilities are there if the committee must contain

(i) the most aggressive man?
(ii) the most power-crazed woman or the most aggressive man, but not both?

18 An all-night showing at a cinema is to consist of five films. There are fourteen different films available, ten disaster movies and four horror movies. How many possible schedules include:

(i) at least one horror movie?
(ii) at least three disaster movies?
(iii) both *Showering Inferno* and *Omen X*?
(iv) either *Showering Inferno* or *Omen X* but not both?

19 Five different letters are chosen from the letters A, B, C, D, E, F, G, H and I. Find the number of choices containing:

(i) no vowels
(ii) exactly one vowel
(iii) exactly two vowels
(iv) exactly three vowels

20 Five married couples want to go on holiday to the Hotel Ambre Solaire, Torremelinos, but there are only four places left in the hotel. How many possible groups of four are there if:

(i) two married couples
(ii) exactly one married couple
(iii) no married couple

must go?

21 Repeat question 20 with five places left in the hotel.

22 An EEC committee of seven people, containing at least one representative from each country, is to be chosen from five Frenchmen, three Italians, six Britons and one German.

(i) How many different committees are possible?

(ii) How many different committees are possible, given that the German will not serve if more than one Italian is chosen?

⑦ **23** How many different six-person sub-committees can be chosen from 9 Conservative, 6 Labour and 5 Alliance MPs if:

(i) there are no restrictions on membership?

(ii) the committee must consist of 3 Conservative, 2 Labour and 1 Alliance MPs?

(iii) there must be at least one representative from each party?

⑦ **24** A football touring party consists of 20 players: 3 goalkeepers, 7 backs, 6 forwards and the rest midfield players. How many possible teams are there if there must be 1 goalkeeper, 4 backs and 4 forwards?

⑦ R **25** There are 12 balls in a bag, 5 red, 4 green and 3 blue. How many selections of 4 balls contain:

(i) no red balls?
(ii) at least 1 green ball?
(iii) at least 1 green and 1 blue ball?
(iv) at least one ball of each colour?

Questions 26–31: Four cards are dealt from a pack of 52. How many possible hands are there if the following restrictions hold?

• → **26** (i) All four are spades.
(ii) Exactly three are spades.
(iii) Exactly two are spades.
(iv) Exactly one is a spade.
(v) None is a spade.

27 (i) Two are hearts and two diamonds.
(ii) One is a heart, one a spade and two are clubs.
(iii) All the cards are different suits.

28 (i) At least one is a spade.
(ii) At least two are spades.
(iii) At least three are spades.

29 (i) Exactly three are Queens.
(ii) All four are court cards (King, Queen, Jack).
(iii) None is a court card.

30 (i) Either a spade of a heart or both are dealt.
(ii) Either a spade or a heart, but not both, are dealt.

R **31** (i) Either an ace or a heart or both are dealt.
(ii) Neither an ace nor a heart are dealt.

• → **32** Find the number of different sets of 3 pairs that can be selected to form a tennis team of 6 players, if each pair is to consist of a boy and a girl, and there are 5 boys and 4 girls from whom to choose.

(What is wrong with this solution? 'There are 20 possible pairings of which 3 must be chosen. So the number of sets of pairs is $^{20}C_3$'.)

R **33** 7 men and 5 women are available to form a team of three mixed doubles pairs for a tennis match. How many different teams are possible?

34 Find how many four-digit numbers can be made from the digits 1, 2, 3, 4, 5, 6, 7 and 9 if no digit may be repeated.

Find how many of the numbers include

(i) two odd digits and two even digits
(ii) three odd digits and one even digit

In each case find how many such numbers are even.

• **35** How many different selections of any number of different letters can be made from the letters:

(i) ABCDE
(ii) ABCDEFGH?

36 Show that $\sum_{r=1}^{n} {}^nC_r = 2^n - 1$.

37 How many different integers can be made by multiplying together two, three or four different digits from 2, 3, 5 and 7?

→ **38** How many different integers can be made by multiplying together two or more different digits from the digits 2, 3, 5, 7, 11 and 13?

R **39** How many factors (other than 1 and itself) has the number 390? (Hint: find the prime factors first.)

40 How many possible results are there when a coin is tossed:

(i) 3 times (ii) 5 times (iii) n times?

41 How many possible results are there for 10 football matches if each one can finish in a home win, a draw or an away win?

42 How many possible results are there when a die is thrown:

(i) twice (ii) 3 times (iii) n times?

43 How many possible results are there when an experiment with k possible results is repeated n times?

• **44** A coin is tossed 3 times. How many different outcomes include:

(i) 3 heads?
(ii) exactly 2 heads?
(iii) exactly 1 head?
(iv) no heads?

45 A coin is tossed 5 times. How many different outcomes include:

(i) 5 heads?
(ii) exactly 4 heads?
(iii) exactly 3 heads?
(iv) exactly 2 heads?
(v) exactly 1 head?
(vi) no heads?

What is the sum of your answers?

→ **46** A coin is tossed n times. How many different outcomes include r heads?

47 A coin is tossed 8 times. How many different outcomes include:

(i) at least one head?
(ii) at least two heads?

• **48** In a multiple-choice paper of 10 questions, each question has one correct answer and four wrong answers.

How many possibilities are there if a candidate selects:

(i) (a) a wrong answer to every question?
   (b) exactly nine wrong answers?
   (c) exactly eight wrong answers?
(ii) (a) at least one wrong answer?
   (b) at least two wrong answers?

→ **49** A forecast is to be made of 11 First Division football matches, each of which may result in a home win, a draw or an away win. Find the number of forecasts containing:

(i) exactly nine correct results
(ii) at least nine correct results.

50 A forecast is to be made of five First Division football matches, each of which may result in a home win, a draw or an away win. Find the number of forecasts containing:

(i) 5   (ii) 4   (iii) 3
(iv) 2   (v) 1   (vi) 0

correct results.
What is the sum of your answers?

51 A test consists of ten questions. A candidate can score 0, 1 or 2 marks for each question. How many different distributions of marks result in a total of:

(i) 20   (ii) 19   (iii) 18   (iv) 17?

52 A die is thrown three times.

(i) How many possible outcomes are there?

(ii) How many result in:
(a) three sixes; (b) exactly two sixes; (c) exactly one six; (d) no sixes?

What is the relationship between the answer to (i) and the answers to (ii)?

• **53** Eight points, $A_1, A_2, \ldots, A_8$, are marked on a line. How many different directed line segments A_iA_j are there joining pairs of points on the line?

→ **54** Ten parallel lines are drawn on a page and twelve parallel lines are drawn perpendicular to them, forming a grid. Find the total number of rectangles (of all sizes) formed by these lines.

R **55** How many rectangles (of all sizes) are there on:

(i) an 8×8 chess board?
(ii) a 19×19 'Go' board?

How many of these rectangles are squares?

56 (i) $A = \{A_1, A_2, A_3\}$ and $B = \{B_1, B_2, B_3, B_4\}$ are two distinct sets of points. Find the number of directed line segments A_iB_j joining a point in A to a point in B.

(ii) Repeat (i) if A is the set $\{A_1, A_2, \ldots, A_r\}$ and B the set $\{B_1, B_2, \ldots, B_s\}$.

Discuss the case in which A and B are not distinct.

(iii) $A = \{A_1, A_2, A_3\}$, $B = \{B_1, B_2, B_3\}$ and $C = \{C_1, C_2\}$ are three distinct sets of points. Find the number of line segments joining a point in one set to a point in a different set.

(iv) Repeat (iii) if A, B and C are the sets $\{A_1, \ldots, A_r\}$, $\{B_1, \ldots, B_s\}$ and $\{C_1, \ldots, C_t\}$.

Discuss the case in which A, B and C are not distinct.

57 There are ten players available for a tennis tournament. How many possible:

(i) singles matches;
(ii) doubles matches

can be organised?

58 Each member of a family of twelve gives every other member a Christmas present. How many presents are given?

59 In a football league consisting of twelve sides, each team plays every other team once. How many matches are played?

60 How many diagonals has a polygon with:

(i) 5 vertices?
(ii) 12 vertices?
(iii) n vertices?

61 What is the greatest possible number of points of intersection of:

(i) 3   (ii) 4   (iii) 5   (iv) n

straight lines?
Find for each case the number of regions into which the lines divide the plane.

5.3:3 Permutations and combinations of objects which are not all different

Worked example 8 How many arrangements are there of the letters T, A, T, T and Y?

There are 5! arrangements of the objects T_1, A, T_2, T_3 and Y where the Ts are distinguished by subscripts. These arrangements can be divided into groups of arrangements which will be identical when the subscripts are removed, e.g. $T_1YT_2AT_3$, $T_1YT_3AT_2$, $T_2YT_1AT_3$, $T_2YT_3AT_1$, $T_3YT_1AT_2$, $T_3YT_2AT_1$.

Each group contains 3! = 6 arrangements, i.e. the number of arrangements of the objects T_1, T_2, T_3 in three fixed positions.

So the number of arrangements of T, A, T, T and $Y = \dfrac{5!}{3!} = \dfrac{120}{6} = 20$

In general:

number of arrangements of n objects of which p are identical = $n!/p!$ (5)

By extension:

number of arrangements of n objects of which p are identical and q are identical (but different from the p) = $n!/p!q!$ (6)

For example, number of arrangements of the word DREADED (7 letters, 2 Es, 3 Ds) is $\dfrac{7!}{3!2!}$.

Worked example 9 How many (i) selections, (ii) arrangements are there of 4 letters of the word ERGOMETER?

Consider the four cases in Table 5.3:1.

	Number of selections	Number of arrangements per selection	Total number of arrangements
(i) with all letters different (choose 4 from ERGOMT; e.g. RGOT)	$^6C_4$	4!	$^6C_4 \times 4!$
(ii) containing exactly one pair (2Es: choose 2 from RGOMT; or 2Rs: choose 2 from EGOMT; e.g. RREM)	$2 \times {}^5C_2$	$\dfrac{4!}{2!}$	$2 \times {}^5C_2 \times \dfrac{4!}{2!}$
(iii) containing two pairs (2Es and 2Rs, e.g. EERR)	1	$\dfrac{4!}{2!2!}$	$1 \times \dfrac{4!}{2!2!}$
(iv) containing a trio (3Es, e.g. EEEG, choose 1 from RGOMT)	$^5C_1$	$\dfrac{4!}{3!}$	$^5C_1 \times \dfrac{4!}{3!}$
Totals	41		626

Thus there are 41 selections and 626 arrangements of 4 letters from ERGOMETER.

Exercise 5.3:3

1 How many different arrangements are there of the letters:
  (i) COOKER?
  (ii) POPPY?
  (iii) COURGETTE?
  (iv) BANANA?
→ (v) TARAMASALATA?

2 In how many different orders can colours appear when five balls are drawn without replacement from a bag containing:
  (i) one red ball, one yellow, one blue and two white?
  (ii) one red ball, one yellow and three white?
  (iii) two red balls and three white?

• **3** Repeat question 2, supposing that each ball drawn is immediately replaced.

4 A race at the Olympics has 7 runners. In how many orders can their countries finish if:
  (i) there are 3 Americans, 2 East Germans, 1 Russian and 1 Pole?
  (ii) 4 are American and 3 East German?

→ **5** Naval signals are made by arranging coloured flags in a vertical line and the flags are then read from top to bottom.
   How many signals using six flags can be made if:
  (i) one green, 3 red and 2 blue flags
  (ii) 2 green, 2 red and 2 blue flags
  (iii) unlimited supplies of green, red and blue flags
   are available?

6 How many different six-digit numbers can be made from the numbers:
  (i) 3, 3, 3, 4, 4 and 7?
  (ii) 5, 5, 5, 5, 8 and 9?
  (iii) 0, 0, 1, 1, 1 and 1?

7 (i) Using expression (6), write down the number of ways of arranging 5 green bananas and 4 yellow bananas in a row. Explain why your answer is equal to $^9C_5$.
  (ii) In how many ways can r green bananas and $(n-r)$ yellow bananas be arranged in a row?
  (iii) How many ways are there of getting exactly r heads when n coins are tossed?
  (iv) An experiment which can result in either success or failure is repeated n times. How many ways are there of getting exactly:
     (a) r successes; (b) r failures?

8 A town has 10 streets running from north to south and 8 streets running from west to east. A man wishes to drive from the extreme south-west intersection to the extreme north-east intersection, moving either north or east along one of the streets. Find the number of different routes he can take. (Hint: every time he reaches an intersection he can choose to go either north or east.)

9 In how many different orders can colours appear when four balls are drawn from a bag, each ball drawn being immediately replaced, if the bag contains:
  (i) one red, one yellow, one blue, one green and one brown ball?
  (ii) one red, one yellow, one blue, one green and three brown balls?
  (iii) unlimited supplies of red, yellow, blue, green and brown balls?

10 (i) In how many different arrangements of the letters POODLE are the two Os adjacent?
  (ii) In how many different arrangements of the letters MUGGING are the three Gs adjacent?
  (iii) In how many different arrangements of the letters HERRING are the Rs separated?

11 In how many arrangements of the letters
• (i) KIPPER
  (ii) ROSEMARY
  (iii) AROMATIC
  (iv) POODLEFAKER
→ (v) PINEAPPLE
are all the vowels adjacent?

12 In how many arrangements of the letters LETTUCE are:
  (i) the vowels adjacent?
  (ii) a T and an E adjacent?

13 In how many arrangements of the letters CUCUMBER are:
  (i) the Us adjacent?
  (ii) the Cs adjacent?
  (iii) the Us and the Cs adjacent?

14 In how many different arrangements of the letters
• (i) MONOTONY
  (ii) WATERCRESS
  (iii) ARROWROOT
R (iv) DEMERARA
  (v) MACDONALDS
are all the vowels separated? (See worked example 5.)

→ **15** In how many arrangements of the letters TOPOLOGY are:
  (i) the Os separated?
  (ii) the Os together?
  (iii) exactly two Os together?

16 How many (i) selections; (ii) arrangements are there of:
(a) 3 letters from the word POODLE?
→ (b) 4 letters from the word SAUCEPAN?
(c) 3 letters from the word COFFEE?
(d) 4 letters from the word HONEYBEE?
(e) 4 letters from the word TATTOOING?
(f) 5 letters from the word MARMALADE?

17 Seven counters are identical except for their colour. Three are blue, two red, one yellow and one green. How many distinguishable sets of:
(i) three counters (ii) four counters
can be selected?

18 Nine counters are identical except for their colour. Three are blue, four red and two yellow. How many distinguishable sets of:
(i) five counters (ii) four counters
can be selected?

19 Nine counters are identical except for their colour. Two are black, two red, two yellow, two green and one pink. How many arrangements are there of:
(i) three (ii) four (iii) five
of these counters?

R 20 A bag contains ten counters, three red, two green and one each of blue, yellow, black, white and brown. How many:
(i) selections (ii) arrangements
are there of four of these counters?

21 How many five-digit numbers can be made from the digits:
(i) 3, 3, 3, 4, 4, 7, 7;
(ii) 3, 3, 3, 4, 4, 7, 7, 7?

→ **22** Naval signals (see question 5) are being made from one green, three red and two blue flags. How many signals can be made using:
(i) all six flags?
(ii) exactly five flags?
(iii) exactly four flags?
(iv) exactly three flags?

23 How many different arrangements are possible using three of each of the following sets of letters?
(i) AABBCC   (ii) AAABBCC
(iii) AABBCCDD   (iv) AAABBCCD

24 (i) You have at your disposal n different letters, and two of each letter. How many different arrangements are possible of:
(a) 2 letters; (b) 3 letters?

(ii) You have at your disposal n different letters, and three of each letter. How many different arrangements are possible of:
(a) 3 letters; (b) 4 letters; (c) 5 letters?

+ **25** You have at your disposal n different letters, and r of each letter. How many different arrangements are possible of:
(i) r letters?
(ii) $(r+1)$ letters?
(iii) $(r+2)$ letters?

• **26** How many different selections of any number of letters can be made from the letters:
(i) AAAABBBCC?
(ii) AAAABBBCDE?

27 How many different selections of any number of counters can be made from four red, three green and two white counters? How many of these contain:
(i) exactly one red counter?
(ii) exactly two red counters?
(iii) at least one red counter?

28 How many different sums of money can be made from three £1 coins, one 50p piece, three 10p pieces and (i) four 2p pieces; (ii) five 2p pieces?

29 How many different integers can be made by multiplying together any two to eight of the digits 2, 3, 3, 5, 5, 7, 7 and 7?

30 How many factors (other than 1 and itself) has the number:
(i) 144 (ii) 720 (iii) 2100?
(Hint: find the prime factors first.)

31 How many:
(i) four-digit;
(ii) five-digit;
(iii) n-digit
numbers can be made from the digits 1 and 2?

32 How many:
(i) three-digit
(ii) four-digit
(iii) n-digit
numbers can be made from the digits 2, 3 and 4?

5.4 Binomial expansions

5.4:1 Binomial theorem

Theorem 1 (binomial theorem):   If n is a positive integer,

$$(a+b)^n \equiv a^n + {}^nC_1 a^{n-1}b + {}^nC_2 a^{n-2}b^2 + \cdots$$
$$+ {}^nC_{n-2} a^2 b^{n-2} + {}^nC_{n-1} ab^{n-1} + b^n$$
$$\equiv \sum_{r=0}^{n} {}^nC_r a^{n-r} b^r \qquad (1)$$

Proof

$$(a+b)^n \equiv \underbrace{(a+b)(a+b)\cdots(a+b)}_{n \text{ brackets}}$$

Either an a or a b from each bracket is multiplied together to form the distinct terms in the expansion. Consider the term in $a^{n-r}b^r$. We must choose the bs from r brackets; then the as will come from the other $(n-r)$ brackets. Number of ways of choosing r bs from n is ${}^nC_r$. So the coefficient of $a^{n-r}b^r$ in the expansion is ${}^nC_r$.

Look at the coefficients in the expansion of $(a+b)^n$ for $n = 1, 2, 3, \ldots$

$$\begin{array}{ccccccccc}
& & & & 1 & 1 & & & \\
& & & 1 & {}^2C_1 & 1 & & & \\
& & 1 & {}^3C_1 & {}^3C_2 & 1 & & & \\
& 1 & {}^4C_1 & {}^4C_2 & {}^4C_3 & 1 & & & \\
1 & {}^5C_1 & {}^5C_2 & {}^5C_3 & {}^5C_4 & 1 & & &
\end{array}$$

$$\vdots$$

$$1 \quad {}^nC_1 \quad {}^nC_2 \quad \cdots \quad {}^nC_{r-1} \quad {}^nC_r \quad \cdots \quad {}^nC_{n-1} \quad 1$$
$$1 \quad {}^{n+1}C_1 \quad {}^{n+1}C_2 \quad \cdots \quad {}^{n+1}C_r \quad \cdots \quad {}^{n+1}C_n \quad 1$$

$$\vdots$$

i.e.

$$\begin{array}{ccccccc}
& & & 1 & 1 & & \\
& & 1 & 2 & 1 & & \\
& 1 & 3 & 3 & 1 & & \\
& 1 & 4 & 6 & 4 & 1 & \\
1 & 5 & 10 & 10 & 5 & 1 &
\end{array}$$

$$\vdots$$

This triangular array is called **Pascal's Triangle** (see 1.1:4 Q7).

Theorem 2:

(i) $\qquad\qquad {}^nC_{n-r} = {}^nC_r \qquad\qquad (2)$

so the numbers in each row of Pascal's Triangle are symmetrically arranged.

(ii) $\qquad\qquad {}^{n+1}C_r = {}^nC_{r+1} + {}^nC_r \qquad\qquad (3)$

so each number can be found by adding the two numbers immediately above it.

Proof
(i) When we choose r objects from n we automatically reject the other $(n-r)$; so we have automatically put these $(n-r)$ objects in a group. Thus when choosing r objects, we are also automatically choosing $(n-r)$ objects.
(ii) When we choose r objects from $(n+1)$ we can either:
(a) choose the first and then choose $(r-1)$ from the remaining n (number of ways: $^nC_{r-1}$), or
(b) reject the first and choose all the r from the remaining n (number of ways: ^nC_r).

Using equation (2), we can rewrite equation (1):

Corollary to Theorem 1:

$$(a+b)^n \equiv a^n + {^nC_1}a^{n-1}b + {^nC_2}a^{n-2}b^2 + \cdots$$
$$+ {^nC_2}a^2b^{n-2} + {^nC_1}ab^{n-1} + b^n \quad (4)$$

Putting $a = 1$, $b = x$:

$$(1+x)^n \equiv 1 + {^nC_1}x + {^nC_2}x^2 + {^nC_3}x^3 + \cdots + {^nC_2}x^{n-2} + {^nC_1}x^{n-1} + x^n \quad (5)$$

Worked example 1 Expand $(3x-2y)^3$.

Using equation (4) with $a = 3x$, $b = -2y$, $n = 4$:

$$(3x-2y)^4 \equiv (3x^4) + {^4C_1}(3x)^3(-2y) + {^4C_2}(3x)^2(-2y)^2$$
$$+ {^4C_1}(3x)(-2y)^3 + (-2y)^4$$
$$\equiv 81x^4 - 216x^3y + 216x^2y^2 - 96xy^3 + 16y^4$$

Worked example 2 Expand $(1-2x)^5$ in ascending powers of x. Hence find $(0.98)^5$ correct to 4 dec. pls.

Using equation (4) with $a = 1$, $b = -2x$, $n = 5$:

$$(1-2x)^5 \equiv 1 - 10x + 40x^2 - 80x^3 + 80x^4 - 32x^5 \quad (6)$$

Putting $x = 0.01$ in equation (6):

$$(0.98)^5 = 1 - 10(0.01) + 40(0.0001) - 80(0.000\,001)$$
$$+ 80(0.000\,000\,01) - 32(0.000\,000\,000\,1)$$
$$= 0.903\,920\,796\,8$$
$$= 0.9039 \text{ correct to 4 dec. pls.}$$

Notice that we need only have used the first four terms in the expansion, since the fifth and sixth are too small to affect the fourth decimal place.

In general:

---

when x is small, we can get a good approximation to the value of $(1+kx)^n$ by taking the first few terms of the expansion

$$1 + {^nC_1}(kx) + {^nC_2}(kx)^2 + {^nC_3}(kx)^3 + \cdots + (kx)^n$$

Notice that the approximation gets better as more terms are taken.

Exercise 5.4:1

→ 1 Do 1.1:4 Q7 if you haven't already done it.

① 2 Expand (in ascending powers of x):
→   (i) $(1+x)^3$
→   (ii) $(1+x)^4$
  (iii) $(1+\frac{1}{2}x)^5$
  (iv) $(1-\frac{2}{3}x)^4$
  (v) $(4-3x)^3$

① 3 Expand:
  (i) $(2x+y)^3$
  (ii) $(x-y)^3$
→   (iii) $(2x-3y)^3$
  (iv) $(4x+\frac{1}{2}y)^4$
  (v) $(3x-y)^6$

① 4 Expand:
  (i) $(x+1/x)^3$
→   (ii) $(x-1/x)^4$
  (iii) $(x^2+1/x)^5$

① 5 Write down the first three terms when each of the following expressions is expanded in ascending powers of x:
  (i) $(1+x)^8$
  (ii) $(1+2x)^6$
→   (iii) $(1-\frac{1}{3}x)^{20}$
  (iv) $(2-3x)^5$
  (v) $(\frac{3}{2}-2x)^{10}$

① 6 Find the coefficients of the given terms in the expansions of the following expressions
  (i) $(1-2x)^{12}$; x^3, x^7
  (ii) $(3+\frac{1}{2}x)^9$; x^5
  (iii) $(2-1/x)^8$; $1/x^5$
→   (iv) $(x^2+1/x)^{10}$; x^{11}, $1/x$
  (v) $(3x-y)^7$; $x^4 y^3$

① 7 Find the constant term in the expansion of:
  (i) $(x-2/x)^2$
  (ii) $(x+1/x)^4$
  (iii) $(2x+1/x^2)^3$
R (iv) $(x^2-1/x)^6$
  (v) $(x^2-1/x)^{3n}$

① 8 Find the ratio of the coefficient of the term in x^5 to the coefficient of the term in x^6 in the expansion of $(2x+3)^{20}$.

① 9 Find x if the middle term in the expansion of:
  (i) $(1+x)^6$ (ii) $(1+x)^{24}$ ($x \neq 0$)

in ascending powers of x is the arithmetic mean of the term immediately before and the term immediately after it.

① 10 Find n when:
  (i) the coefficients of x^2 and x^3 are equal in the expansion of $(1+x)^n$, $n>1$.
  (ii) the coefficient of x^3 in the expansion of $(1+x)^n$ is 5 times the coefficient of x.

① → 11 Given that n is a positive integer and that the coefficient of x^2 in the binomial expansion of $(5+2x)^n$ is twice the coefficient of x, find n.

① 12 Find a, b and n given that the first four terms in the expansion of $(1+ax)^n$ in ascending powers of x are
  (i) $1+6x+\frac{27}{2}x^2+bx^3$
R (ii) $1-15x+90x^2+bx^3$

• → 13 Expand $(1+x+x^2)^6$ in ascending powers of x as far as the term in x^3. (Hint: write $(1+x+x^2)^6 = \{1+(x+x^2)\}^6$; don't do too much work.)

14 Expand the following expressions in ascending powers of x as far as the given term:
  (i) $(1-x+x^2)^5$; x^3
  (ii) $(1+x+3x^2)^4$; x^4
  (iii) $(1+\frac{1}{2}x+\frac{1}{3}x^2)^6$; x^2
  (iv) $(2+x+x^2)^5$; x^3
  (v) $(1+x+x^2)^n$; x^2

15 Expand:
• (i) $(1+x)(1-x)^3$
  (ii) $(3-2x)(1+x)^4$
→   (iii) $(1+x-x^2)(2+x)^3$

16 Expand the following expressions in ascending powers of x as far as the given term:
  (i) $(1+x)^4(1-3x)^5$; x^2
  (ii) $(1+x)(1+2x)^{10}$; x^3
  (iii) $(1-x)(1+x)^n$; x^3

17 Find the constant term in the expansion of $(x^2+1/x)^4(x+2/x)^6$.

18 Find the coefficient of:
  (i) x^{10} in the expansion of $(1-2x)(1+x)^{18}$
  (ii) x^8 in the expansion of $(1+3x)(1-\frac{1}{2}x)^{12}$

R 19 In the expansion of $(1+ax)(1+bx)^5$, where a and b are real constants, $b \neq 0$, the coefficient of x^2 is zero and the coefficient of x^3 is -80. Find the values of a and b.

• → 20 Show that the coefficient of x^r in the expansion of $(1+2x)(1+x)^n$ is

$$\frac{n!(n+r+1)}{(n-x+1)!r!}$$

21 Find the coefficient of x^r in the expansion of:
  (i) $(1+x)^n + (1+2x)^n$
  (ii) $(2+x)(1+x^n)$
R (iii) $(1-x)(1+2x)^n$
  (iv) $(1+x+x^2)(1+x)^n$

22 Find the coefficient of x^r in the expansion of
  (i) $(1-3x)^n$
  (ii) $(2+3x)(1-x)^n$
  (iii) $(1+x-x^2)(1-2x)^n$

23 (i) Put $a = b = 1$ in equation (1). Hence find

$$\sum_{r=0}^{n} {}^nC_r. \text{ (C.f. 5.3:2 Q36.)}$$

(ii) Put $a = 1$ in equation (1) and differentiate it with respect to b. Then put $b = 1$, and hence find

$$\sum_{r=1}^{r} r {}^nC_r.$$

24 Using equation (1), find:

$$\sum_{r=0}^{n} (-1)^r {}^nC_r$$

25 Without using tables or a calculator, find the values of:
  (i) $(\sqrt{3}+1)^4 + (\sqrt{3}-1)^4$
  (ii) $(2+1/\sqrt{2})^5 + (2-1/\sqrt{2})^5$

+ 26 Show that, if n is a non-negative integer,

$$\sin^{2n+1} x \equiv (1 - \cos^2 x)^n \sin x$$

Hence, using the binomial theorem, find $\int \sin^{2n+1} x \, dx$. Use your result to evaluate $\int_0^{\frac{1}{2}\pi} \sin^9 x \, dx$ (c.f. 4.2:2 Q19.)

Using the binomial theorem, find $\int \sec^{2n+2} x \, dx$ where n is a non-negative integer. Use your result to evaluate

(i) $\int_0^{\frac{1}{4}\pi} \sec^6 x \, dx$

(ii) $\int_0^{\frac{1}{4}\pi} \sec^{12} x \, dx$

27 (i) Expand $(x + \delta x)^4$. Hence, using 1.3:1 definition (1a) of the derivative, show that

$$\frac{d}{dx}(x^3) = 3x^2$$

(ii) Expand $(x + \delta x)^n$ in ascending powers of δx as far as the term in $(\delta x)^3$.
Hence show that

$$\frac{d}{dx}(x^n) = nx^{n-1}$$

② **28** Write down the expansion of $(1+x)^6$ in ascending powers of x. Hence find the value of $(1.001)^6$ correct to 5 dec. pls.

② **29** Write down the first four terms in the expansion of $(1 + \frac{1}{2}x)^{10}$ in ascending powers of x. Hence find the value of $(1.005)^{10}$, correct to 4 dec. pls.

② **R 30** Write down the first four terms in the expansion of $(1 + 3x)^9$ in ascending powers of x. Hence find the values of $(1.003)^9$, correct to 5 dec. pls.

② → **31** Expand $(1-x)^6$ in ascending powers of x up to the term in x^4. Using only the first four terms of your expansion, find an approximate value of $(0.99)^6$.
Explain why the answer you have found is correct to 6 dec. pls.

② **32** Use the binomial theorem to find the exact value of:
  (i) $(9.9)^5$ (ii) $(102)^4$

② **33** Write down the first five terms in the expansion of $(1+x)^7$ in ascending powers of x.
Hence find the value, to the nearest £1, at the end of 7 years of a car costing £3000 if:
  (i) it increases in value by 10% per year
  (ii) it decreases in value by 20% per year

5.4:2 Binomial series

We want to extend the binomial theorem by looking at the expansion of $(1+x)^\alpha$ when α is not a positive integer, e.g. $(1+x)^{-1/2}$.
Consider these expansions:

(i) $(1+x)^4 \equiv 1 + 4x + 6x^2 + 4x^3 + x^4$ (for all x)
(ii) $(1+x)^{-1} \equiv 1 - x + x^2 - x^3 + \cdots$ (when $|x| < 1$)   (see section 5.1:6)

The first is a finite series, valid for all values of x. The second is an infinite series which converges only when $|x| < 1$.

Series and probability 253

> **Theorem 3:** If α is any rational number then
>
> $$(1+x)^\alpha \equiv 1 + \alpha x + \frac{\alpha(\alpha-1)}{2!}x^2 + \frac{\alpha(\alpha-1)(\alpha-2)}{3!}x^3 + \cdots \quad (7)$$
>
> (i) If $\alpha = n$, a positive integer, the series terminates at the term in x^n and the expansion given by identity (7) agrees with the expansion given by identity (5).
> (ii) If α is not a positive integer, the series is infinite, and converges only when $|x| < 1$.

Proof
We will not prove this theorem.

Worked example 3 Find the first four terms in the series expansion of:

(i) $\sqrt{(1+x)}$ (ii) $\dfrac{1}{1-2x}$

(i) $\sqrt{(1+x)} \equiv (1+x)^{1/2}$

$$\equiv 1 + \tfrac{1}{2}x + \frac{(\tfrac{1}{2})(-\tfrac{1}{2})}{2!}x^2 + \frac{(\tfrac{1}{2})(-\tfrac{1}{2})(-\tfrac{3}{2})}{3!}x^3 + \cdots$$

$$\equiv 1 + \tfrac{1}{2}x - \tfrac{1}{8}x^2 + \tfrac{1}{16}x^3 + \cdots$$

The series converges only when $|x| < 1$.

(ii) $\dfrac{1}{1-2x} \equiv (1-2x)^{-1}$

We use Theorem 3 with $\alpha = -1$, replacing x by $-2x$:

$$(1-2x) \equiv 1 + (-1)(-2x) + \frac{(-1)(-2)}{2!}(-2x)^2 + \frac{(-1)(-2)(-3)}{3!}(-2x)^3 + \cdots$$

$$\equiv 1 + 2x + 4x^2 + 8x^3 + \cdots$$

The series converges only when $|-2x| < 1$, i.e. when $|x| < \tfrac{1}{2}$.

Notice that the series is a G.P. with common ratio $2x$. We have already seen (section 5.1:6) that this series converges when $|2x| < 1$, i.e. when $|x| < \tfrac{1}{2}$.

Worked example 4 Find the first four terms in the series expansion of $1/(2+x)^2$.

$$\frac{1}{(2+x)^2} \equiv \frac{1}{\{2(1+\tfrac{1}{2}x)\}^2}$$

$$\equiv \tfrac{1}{4}(1+\tfrac{1}{2}x)^{-2}$$

$$\equiv \tfrac{1}{4}(1 - x + \tfrac{3}{4}x^2 - \tfrac{1}{2}x^3 + \cdots)$$

The series converges only when $|\tfrac{1}{2}x| < 1$, i.e. when $|x| < 2$.

Worked example 5 Express

$$f(x) = \frac{1}{(1+x)(1-2x)}$$

in partial fractions. Hence find the first four terms in the series expansion of $f(x)$.

$$\frac{1}{(1+x)(1-2x)} \equiv \frac{\frac{1}{3}}{1+x} + \frac{\frac{2}{3}}{1-2x}$$

$$\equiv \tfrac{1}{3}(1+x)^{-1} + \tfrac{2}{3}(1-2x)^{-1}$$

$$\equiv \tfrac{1}{3}\{1-x+x^2-x^3+\cdots\} + \tfrac{2}{3}\{1+2x+4x^2+8x^3+\cdots\}$$

$$\equiv 1+x+3x^2+5x^3+\cdots$$

The series converges only when $|x|<1$ and $|x|<\tfrac{1}{2}$, i.e. when $|x|<\tfrac{1}{2}$.

Worked example 6 Find the first four terms in the series expansion of $(1-x)^{1/2}$. Hence find:

(i) (0.99) to 5 dec. pls.   (ii) $\sqrt{11}$ to 4 dec. pls.

From Theorem 3:

$$(1-x)^{1/2} \equiv 1 - \tfrac{1}{2}x - \tfrac{1}{8}x^2 - \tfrac{1}{16}x^3 - \cdots$$

(i) Putting $x = 0.01$:

$$\sqrt{(0.99)} = (0.99)^{1/2} = 1 - \tfrac{1}{2}(0.01) - \tfrac{1}{8}(0.0001) - \cdots$$

$$= 0.994\,99 \text{ to 5 dec. pls.}$$

The next term, $\tfrac{1}{16}(0.01)^3$, does not affect the fifth decimal place.

(ii) $\sqrt{11} = \sqrt{\left(0.99 \times \frac{11}{0.99}\right)} = \sqrt{\left(0.99 \times \frac{100}{9}\right)}$

$$= \tfrac{10}{3}\sqrt{(0.99)}$$

$$= \tfrac{10}{3} \cdot 0.994\,99$$

$$= 3.3167 \text{ to 4 dec. pls.}$$

Exercise 5.4:2

1 Find the sums of these infinite series, stating in each case for what values of x the series converges (see section 5.1:6):

(i) $1+x^2+x^3+\cdots$
(ii) $1+3x+9x^2+27x^3+\cdots$
(iii) $1-x+x^2-x^3+\cdots$
(iv) $1+x^2+x^4+x^6+\cdots$
(v) $1+2x+3x^2+4x^3+\cdots$
(vi) $2+x+2x^2+x^3+2x^4+\cdots$
(vii) $S = 1+x+3x^2+5x^3+7x^4+\cdots$
(Hint: consider $(1+x)S$.)

2 (i) Using identity (7), write down the series expansions of $(1+x)^4$ and $(1+x)^5$.

(ii) Using identity (7) write down the terms in x^0, x^1, x^2, x^3, x^{n-1}, x^n, x^{n+1} and x^{n+2} in the expansion of $(1+x)^n$ when n is a positive integer.

What can you say about the term in x^r when $r > n$? Show, using identity (7), that the series expansion of $(1+x)^n$ agrees with the binomial expansion (identity (5)).

3 (i) Find the sum of the G.P. $1-x+x^2-x^3+\cdots$. For what values of x does the series converge?

(ii) By putting $\alpha = -1$ in identity (7), find the series expansion of $1/(1+x)$. For what values of x, according to Theorem 3, does the series converge?

Questions 4–8: Find the first four terms in the series expansions of the following functions. (Note: this is not necessarily the same as finding the series up to the term in x^3, e.g. $(1+x^3)^{1/2} \equiv 1+\tfrac{1}{2}x^3$ (up to the term in x^3), but its first four terms are: $1+\tfrac{1}{2}x^3-\tfrac{1}{8}x^6+\tfrac{1}{16}x^9$.) In each case state the values of x for which the series converges.

④ **4**
  (i) $(1+x)^{2/3}$
  (ii) $(1+x)^{-4}$
  (iii) $\dfrac{1}{(1+x)^2}$
  (iv) $(1-x)^{1/3}$
  (v) $(1-2x)^{3/2}$
→ (vi) $1/\sqrt{(1+3x)}$
R (vii) $\sqrt{(1-\tfrac{2}{3}x)}$

④ **5**
  (i) $1/(2-x)$
  (ii) $1/\sqrt{(4+x)}$
→ (iii) $\sqrt{(9+4x)}$
  (iv) $1/(3+x)^2$
  (v) $1/(x-1)$

⑤ **6** (First split the function into partial fractions.)
→ (i) $\dfrac{5x}{(1-2x)(1+3x)}$
  (ii) $\dfrac{3}{(1-x)(1+2x)}$
  (iii) $\dfrac{x+2}{x^2-1}$
R (iv) $\dfrac{x+5}{(1+3x)(2-x)}$
  (v) $\dfrac{1}{x^2+5x+4}$
  (vi) $\dfrac{x^2-x}{(1+x)(1+x^2)}$
  (vii) $\dfrac{1}{(1-2x)(1-x)^2}$

• **7**
  (i) $(1+2x)\sqrt{(1+x)}$
  (ii) $\dfrac{1+x}{1-x}$
  (iii) $\dfrac{3-x}{3+x}$
  (iv) $\dfrac{1-x}{\sqrt{(1+x)}}$
→ (v) $\dfrac{1+2x}{(1-x)^2}$

8
  (i) $\dfrac{\sqrt{(1-x)}}{1+x}$
  (ii) $\dfrac{1}{(1-x)\sqrt{(1-4x)}}$
R (iii) $\sqrt{\left(\dfrac{1+2x}{1-2x}\right)}$

•→ **9** Find a, b and c if:
$$\dfrac{1+ax+bx^2}{(1-x)^2} \equiv 1+x^2+cx^3+\cdots$$

10 Find a and b if:
(i) $\dfrac{(1+x)^{1/2}}{1-ax} \equiv 1+bx^2+\cdots$

(ii) the expansions of $\sqrt{(1-4x)}$ and $1-2x(1-ax)^b$ are identical up to and including the term in x^3.

11 Write down the first five terms in the series expansions of:
(i) $\dfrac{1}{(1-x)^2}$ (ii) $\dfrac{1}{(1-x)^3}$

Check your answers by differentiating the series for $1/(1-x)$.

12 The first four terms in the expansion of $(1+ax)^b$ are 1, c, $\tfrac{3}{4}x^2$ and cx^3. Find a, b and c. State the set of values of x for which the expansion is valid.

⑥→ **13** Write down the series expansion of $(1+x)^{1/3}$ up to the term in x^3.
Hence find $(\tfrac{9}{8})^{1/3}$ to 5 dec. pls.
Hence find $\sqrt[3]{9}$ to 3 dec. pls.

⑥ **14** Write down the series expansion of $(1-x)^{1/3}$ up to the term in x^3.
Hence find $\sqrt[3]{(0.999)}$ to 10 dec. pls.
Hence find $\sqrt[3]{(37)}$ to 6 dec. pls. (Hint: $37 = 0.999/0.027$.)

⑥ **15** Write down the series expansion of $(1-x^2)^{1/2}$ up to the term in x^6.
Hence find $\sqrt{(0.99)}$ to 7 dec. pls.
Hence find $\sqrt{11}$ to 5 dec. pls.

⑥ **16** Write down the series expansion of $(1-x)^{-1/2}$ up to the term in x^3.
Hence find $\sqrt{2}$ to 5 dec. pls.

⑥ R **17** Write down the series expansion of $(1-2x)^{1/2}$ up to the term in x^2.
Hence find $\sqrt{10}$ to 3 dec. pls.

⑥ **18** Write down the first four terms in the series expansion of $(1+x)^\alpha$ when $\alpha=-2$ and when $\alpha=\tfrac{1}{2}$.
Hence find
(i) $\dfrac{1}{(1.003)^2}$ (ii) $\sqrt{(4.02)}$
to 6 dec. pls.

⑥ **19** By multiplying the numerator and denominator by $\sqrt{(1+x)}$, find the expansion of $\sqrt{\left(\dfrac{1+x}{1-x}\right)}$ as far as the term in x^2. Putting $x=\tfrac{1}{9}$, prove that $\sqrt{5} \approx 181/81$.

⑥ **20** Expand $\sqrt{\left(\dfrac{1+2x}{1-2x}\right)}$ up to the term in x^2. Putting $x=\tfrac{1}{100}$, find $\sqrt{51}$ to 3 dec. pls.

21 Expand the function $f(x) = \sqrt{\left(\dfrac{1+2x}{1-x}\right)}$ up to the term in x^2.

In using the formula $T = 2\pi\sqrt{(l/g)}$ to find the period of a pendulum a student makes an error of $+2\%$ in measuring l, the length of the pendulum, and an error of -1% in the estimate of g, the acceleration due to gravity. Show that the resulting error in T is slightly more than $1\tfrac{1}{2}\%$.

22 Expand the function $f(x) = \sqrt{[(1-4x)(1+3x)^3]}$ up to the term in x^2.

The number I of people injured in riots per year in a certain problematic urban borough is given by $I = k\sqrt{(np^3)}$, where n is the number of people living in the borough, p is the number of arrests per year made by the police, and k is a constant. The number of people living in the borough decreases by 4% and the number of arrests per year made by the police increases by 3%. Find the approximate percentage increase in the number of people injured in riots per year.

•→ **23** Write down the first four terms in the series expansion of $1/\sqrt{(1+x^2)}$.

Hence find approximately to 5 dec. pls. the value of
$$\int_0^{0.4} \dfrac{1}{\sqrt{(1+x^2)}}\, dx$$

24 Find $\displaystyle\int_0^1 \dfrac{1}{4+x^2}\, dx$:

(i) approximately, to 5 dec. pls., by expanding the integrand up to the term in x^{10};

(ii) by direct integration.

Hence find an approximate value of $\tan^{-1}\tfrac{1}{2}$.

R 25 Find $\displaystyle\int_0^{0.5} \sqrt{(1-x^2)}\, dx$:

(i) approximately, to 5 dec. pls., by expanding the integrand up to the term in x^6;

(ii) by direct integration, using the substitution $x = \sin\theta$.

5.5 Probability

5.5:1 Sample space, probability

Suppose that an unbiased die is thrown and the result noted. This process is called a **trial**.

The trial will result in one of six possible **outcomes**, i.e. one of the numbers 1, 2, 3, 4, 5 or 6 appearing on the upper face.

The set $\mathscr{E}$ of all possible outcomes is called the **possibility space** or **sample space**. Here $\mathscr{E} = \{1, 2, 3, 4, 5, 6\}$. $\mathscr{E}$ is the universal set whose elements are the outcomes of the trial. $n(\mathscr{E})$ is the number of elements in $\mathscr{E}$. So here $n(\mathscr{E}) = 6$.

The word 'unbiased' suggests that the outcomes are **equally likely**. We say that the **probability** of each equally likely outcome is the same:

$$p(1) = p(2) = p(3) = p(4) = p(5) = p(6) = \tfrac{1}{6}$$

In other words, if the die is thrown many times, the proportion of times each number appears on the upper face will gradually approach $\tfrac{1}{6}$.

An **event** A is a set of outcomes, i.e. a subset of $\mathscr{E}$. For example, {the die shows an even number} is an event.

Definition If the outcomes in a sample space are equally likely, the **probability** $p(A)$ that the event A occurs is given by

$$p(A) = \dfrac{n(A)}{n(\mathscr{E})} \tag{1}$$

For example,

$$p\text{ (die shows an even number)} = p(\{2, 4, 6\})$$
$$= \frac{n(\{2, 4, 6\})}{n(\{1, 2, 3, 4, 5, 6\})}$$
$$= \frac{3}{6} = \frac{1}{2}$$

Note: by equation (1),

$$p(\mathscr{E}) = \frac{n(\mathscr{E})}{n(\mathscr{E})} = 1, \quad p(\varnothing) = \frac{n(\varnothing)}{n(\mathscr{E})} = \frac{0}{n(\mathscr{E})} = 0$$

For example, if a ball is chosen from a bag of marbles, $p(\text{marble}) = 1$ and $p(\text{football}) = 0$.

Worked example 1 An integer is chosen at random from the integers 1 to 9 inclusive. What is the probability that it is even?

$$\mathscr{E} = \{1, 2, 3, 4, 5, 6, 7, 8, 9\}; \; n(\mathscr{E}) = 9$$

The words 'at random' suggest that each outcome is equally likely.
Let A be the event: the number chosen is even. Then

$$A = \{2, 4, 6, 8\}; \; n(A) = 4$$

So by equation (1),

$$p(A) = \frac{n(A)}{n(\mathscr{E})} = \frac{4}{9}$$

Worked example 2 Three coins are tossed. What is the probability of getting at least two heads?
There are 8 equally likely outcomes:

$$\mathscr{E} = \{\text{HHH, HHT, HTH, THH, HTT, THT, TTH, TTT}\}$$

(Note: if you don't like this idea, think of the coins as looking different e.g. a 2p, a 5p, a 10p; 2pH, 5pT, 10pH is clearly a different outcome from 2pT, 5pH, 10pH. But then notice that the outcomes are not altered if the coins look identical.)
Let A be the event: at least two heads appear face up.

$$A = \{\text{HHH, HHT, HTH, THH}\}; \; n(A) = 4.$$

So by equation (1),

$$p(A) = \frac{n(A)}{n(\mathscr{E})} = \frac{4}{8} = \frac{1}{2}$$

Note: since this question is only concerned with the number of heads, it is tempting to consider the 4 cases:

3 heads, 2 heads and 1 tail, 2 tails and one head, 3 tails.

However, since these four *choices* are not equally likely, we must consider the possible *arrangements*, and see how many of these consist of at least two heads.

Worked example 3 Five different letters are chosen in turn, at random, from the letters A, B, C, D, E, F, G, H and I. What is the probability that exactly two are vowels?

$$\mathscr{E} = \{\text{choices of 5 letters}\}; \quad n(\mathscr{E}) = {}^9C_5$$

The outcomes in $\mathscr{E}$ are all equally likely.

There are 3 vowels and 6 consonants to choose from. Number of choices containing 2 vowels (and 3 consonants) $= {}^3C_2 \times {}^6C_3$.

So by equation (1) $\quad p(\text{exactly 2 vowels}) = \dfrac{{}^3C_2 \times {}^6C_3}{{}^9C_5}$

$$\approx 0.476$$

(c.f. 5.3:2 Q19).

Look back at worked example 2. Think of the 8 possible outcomes as being *arrangements* of 3 letters using the letters H and T only (repetition allowed). Now look at worked example 3. Here the outcomes are *choices* of letters. Why the difference?

In worked example 2, in which repetition is allowed, there are different choices of letters—3H, 2H and 1T, 1H and 2T, 3T—but they are *not* equally likely. These choices consist of, respectively, 1, 3, 3, 1 equally likely arrangements.

In worked example 3 there are no repetitions and the different choices *are* equally likely, since each choice consists of 5! equally likely arrangements. (We could also have used the sample space $\mathscr{E}'$ of arrangements, $n(\mathscr{E}') = {}^9P_5$, but the working would have been slightly more laborious.)

Exercise 5.5:1

Assume, unless you are told otherwise, that all dice and coins are fair, that all packs of cards contain 52 cards and are carefully shuffled, that a child is just as likely to be a boy as a girl, and that no policeman is crooked.

① **1** An integer is chosen at random from the integers 1 to 7 (inclusive). Find the probability that it is:

(i) even   (ii) odd

What is the sum of these probabilities?

① **2** A letter is chosen at random from the letters MICROWAVE.

(i) Find the probability that it is:
(a) a vowel; (b) a consonant.
What is the sum of these probabilities?

(ii) Find the probability that the chosen letter is an O or an E.

① **3** A bag contains 4 red balls and 5 blue balls. A ball is drawn at random. What is the probability that it is blue?

① **4** A bag contains 4 red, 3 blue and 3 green balls. A ball is drawn at random. What is the probability that it is:

(i) blue   (ii) blue or green   (iii) yellow?

① **5** An integer is chosen at random from the integers 1 to 12 (inclusive). Find the probability that it is:

(i) even
(ii) divisible by 3
(iii) even and divisible by 3
(iv) even or divisible by 3, or both
(v) even or divisible by 3, but not both.

① **6** A card is drawn at random from a pack. What is the probability that it is:

(i) an ace?
(ii) a heart?
(iii) a red Queen?
(iv) a court card (i.e. King, Queen or Jack)?
(v) a black card?

① **7** A card is drawn at random from a pack. What is the probability that it is:

(i) the ace of hearts?
(ii) an ace or a heart (or both)?
(iii) an ace but not a heart?
(iv) a heart but not an ace?
(v) neither an ace nor a heart?

(Hint: draw a Venn diagram.)

① •→ **8** A die is marked 1, 1, 2, 3, 3, 3. What is the probability of scoring:

(i) 1   (ii) 2   (iii) 3?

What is the sum of these probabilities?

① **9** An integer is chosen at random from the set {1, 2, 2, 3, 4, 4, 4, 5, 5, 6}. Find the probability that it is:

(i) a 4   (ii) a 5
(iii) greater than 3   (iv) odd

① **10** There are 500 tickets sold for a raffle; there are 7 third prizes, 3 second prizes and one first prize. You buy a ticket. What is the probability that:

(i) you win first prize?
(ii) you win a prize?
(iii) you don't win a prize?

② **11** A coin is tossed three times. Find the probability of getting:

(i) more heads than tails
(ii) exactly two consecutive heads
(iii) at least one head

② → **12** Four coins are tossed. Write down the 16 equally likely outcomes. Find the probability of getting exactly:

(i) 4 heads   (ii) 3 heads
(iii) 2 heads   (iv) one head
(v) no heads

② **13** A coin is tossed four times. Find the probability of getting:

(i) exactly two consecutive heads
(ii) at least two consecutive heads
(iii) more heads than tails

② **14** A man and his wife have two children. What is the probability that:

(i) one is a boy and one is girl?
(ii) at least one is a girl?

② **15** A woman has three children. What is the probability that:

(i) exactly one is a boy?
(ii) the eldest is a boy?
(iii) there are more boys than girls?
(iv) all the boys are older than all the girls?

② •→ **16** Two dice are thrown. Show in a 6 × 6 array the 36 equally likely outcomes.

(i) Find the probabilities that the total of the two numbers on the dice is 0, 1, 2, 3, 4, ..., 11, 12 respectively.
(ii) What is the sum of these probabilities?
(iii) Which is the most likely total?
(iv) Let p_r be the probability that the total is r. Draw a graph of p_r against r ($r = 0, 1, 2, \ldots, 12$).

② **17** Two dice are thrown.

(i) Find the probability of scoring a total of 10 or more. (What is wrong with the following argument? 'The total score could be 2, 3, ..., 12. Of these eleven possible totals, three—10, 11, 12—are equal to or more than 10. So the probability is $\frac{3}{11}$.')
(ii) Find the probability of scoring a total of 5 or less.

② → **18** Two dice, one red and one blue, are thrown. Find the probability that:

(i) the numbers on the dice are the same.
(ii) the red die shows a larger number than the blue die.
(iii) the number on the red die is more than twice the number on the blue die.

② **19** Two dice are thrown. What is the probability that the number on one die is greater than the number on the other by:

(i) 0   (ii) 1   (iii) 2   (iv) 3   (v) 4   (vi) 5?

② **20** Two dice are thrown. Find the probability of getting:

(i) two sixes
(ii) exactly one six
(iii) at least one six
(iv) no sixes

21 Four dice A, B, C and D are marked as follows:

A: 1 1 4 4 4 4
B: 2 2 2 2 2 2
C: 3 3 3 3 6 6
D: 2 2 2 5 5 5

Find the probability that in a single throw:

(i) $(A) > (B)$ (i.e. A has a higher number showing than B)
(ii) $(B) > (C)$
(iii) $(C) > (D)$
(iv) $(D) > (A)$

In a game between two players, one player chooses one of the four dice, the other player then chooses another die and each player throws his die once. The player whose die shows a higher number wins. If you were playing this game would you rather have first choice of die or second choice? Explain why.

② **R 22** Two numbers (not necessarily different) are chosen at random from the integers 1 to 5.

(i) The integers are combined to make a two-digit number. What is the probability that this number is: (a) even; (b) odd; (c) divisible by 5; (d) greater than 24?

(ii) What is the probability that the sum of the integers is: (a) 3; (b) 8; (c) 8 or more?

(iii) What is the probability that the product of the integers is: (a) prime; (b) even; (c) divisible by 5; (d) greater than 14?

Question 23 et seq.: We are getting to the stage where it is too laborious to write down all the possible outcomes; instead they must be kept in the mind.

• → **23** Two numbers, x and y, are selected at random from the integers 1 to 10. If the same number may be selected twice, what is the probability that $|x - y|$ is:

(i) 0   (ii) 1   (iii) 7 or more?

What is the probability that:

(iv) $x > y$ (v) $x \geq y$?

→ **24** Three dice are thrown. How many possible outcomes are there? Find the probability that the total score on the dice is:

(i) either 9 or 10
(ii) from 7 to 14 inclusive
(iii) even

25 Three tetrahedral dice, each numbered 1 to 4, are thrown and the number face down on each die noted. How many possible outcomes are there?
What is the probability that:

(i) the total score is 3 or 4?
(ii) the total score is less than 10?
(iii) the score on each die is the same?
(iv) the total score for two of the dice is the same as the score on the third?

R 26 A die marked 1, 2, 2, 3, 3, 3 is thrown three times. What is the probability of getting a total score of:

(i) 3   (ii) 4   (iii) 5?

Question 27 et seq.: We are getting to the stage where it is impossible to keep in mind a list of possible outcomes. We need to work out $n(A)$ and $n(\mathscr{E})$ using some of the methods of section 5.3:2.

• **27** Two cards are drawn from a pack, the first one drawn being immediately replaced and the pack reshuffled. How many possible outcomes are there?
In how many of these:

(i) are both cards aces?
(ii) is one card an ace and the other not?
(iii) is neither card an ace?

Hence write down the probabilities of these events.

→ **28** Three cards are drawn from a pack, each one drawn being immediately replaced and the pack reshuffled. Find the probability of getting:

(i) three clubs
(ii) two hearts and one diamond
(iii) three cards of different suits
(iv) three court cards

R 29 Four cards are drawn from a pack, each one drawn being immediately replaced and the pack reshuffled. Find the probability of getting:

(i) no spades
(ii) four spades
(iii) two clubs and two spades
(iv) one card from each suit

30 A trial has t equally likely outcomes of which s result in the event A taking place. The trial is repeated n times. What is the probability that A takes place exactly:

(i) once   (ii) twice
(iii) n times   (iv) exactly r times?

③ **31** Three different letters are chosen at random from the letters MICROWAVE. What is the probability that exactly:

(i) 3   (ii) 2   (iii) 1   (iv) none

are vowels? What is the sum of these probabilities?

③ → **32** Four different letters are chosen at random from the letters BEHAVIOUR. What is the probability that:

(i) all four are vowels?
(ii) exactly three are vowels?
(iii) at least two are vowels?

③ **33** Two different integers are chosen at random from the integers 1 to 10. Find the probability that the larger number is:

(i) 5   (ii) 8 or more

What is the probability that the second number chosen is larger than the first?

③ **34** A man goes to the dentist. Five of his teeth are rotten and the other 27 are in good condition. The dentist removes two teeth at random. What is the probability that they are both rotten?

③ **35** At a supermarket, two of the fifteen cans of tuna fish on display have mercury contamination. If a customer chooses three cans of tuna, what is the probability that she takes a contaminated one?

36 A team of five is chosen at random from five men and six women. Find the probability that it contains:

(i) all men
(ii) exactly two men
(iii) exactly two women

37 Three different integers are chosen at random from the integers 1 to 8. Find the probability that:

(i) all three are even
(ii) exactly two are even
(iii) exactly one is even
(iv) all three are odd

38 Three cards are drawn, without replacement, from a pack. What is the probability of getting:

(i) three spades?
(ii) exactly two spades?
(iii) two spades and one heart?
(iv) one spade and two red cards?

39 In the game of bridge, each player is dealt a hand of 13 cards. What is the probability that, in one hand:

(i) all the cards of the same suit?
(ii) none of the cards is a court card?
(iii) one of the cards is an ace?

40 Three letters are chosen at random from the letters A, B C, D and E. What is the probability of choosing exactly two consonants if the letters:

(i) must be different?
(ii) need not be different?

41 Two cards are selected at random from ten cards numbered 1 to 10. Find the probability that the sum is odd if the two cards are drawn one after the other:

(i) without replacement (ii) with replacement

42 Five cards are drawn from an ordinary pack. What is the probability of getting exactly three aces if the cards are drawn:

(i) without replacement?
(ii) with replacement?

Explain the problem involved in using the sample space $\mathscr{E} = \{\text{choices of 5 cards}\}$ in (ii).

43 Five cards are drawn from an ordinary pack, without replacement. What is the probability that an ace appears for the first time on:

(i) the fifth draw
(ii) the fourth draw?

(Hint: think carefully about what sample space to use.)

44 The letters of the word BEHAVIOUR are written down in random order. What is the probability that the vowels are:

(i) adjacent (ii) separated?

45 The letters of the word PARANOIA are written on eight separate cards and the cards are put in a hat. Three cards are drawn at random, without replacement. Find the probability of getting:

(i) three As
(ii) exactly two As
(iii) exactly one A
(iv) no A

(Hint: Treat the As as 3 different letters (A_1, A_2, A_3), find the total number of selections of three letters, and then the number of these selections containing three As, two As etc.)

46 The letters of the word BANANA are written on six separate cards and the cards are put in a hat. Three cards are drawn at random, without replacement. Find the probability of getting:

(i) three As
(ii) exactly two As
(iii) no A

How are the probabilities affected if each card drawn is immediately replaced? (C.f. question 45).

47 From a bag containing three white, one red, one green and one blue ball, three balls are drawn at random, without replacement. Find the probability that the three balls drawn are:

(i) all white (ii) of different colours

48 A bag contains 4 red balls and 5 blue balls. 4 balls are drawn at random without replacement. Find the probability of getting:

(i) at least 2 red balls
(ii) at least 2 blue balls

49 Twelve chairs are placed round a circular table. Find the probability that three men who are allocated seats at random occupy adjacent chairs.

R 50 Eight trees are planted in a circle in random order. If two of the trees are diseased and later die, what is the probability that the two dead trees are next to each other?

If four of them are diseased find:

(i) the probability that at least two of them are next to each other;
(ii) the probability that all four are next to each other.
(C)

51 Seven chairs are placed round a circular table. Four men, two women and a child are allocated seats at random. Find the probability that the child is seated:

(i) between the two women
(ii) between two men
(iii) between a man and a women

5.5:2 $p(A')$, $p(A \cup B)$

Theorem 1:
$$p(A') = 1 - p(A) \qquad (2)$$

Proof
From fig. 5.5:1,
$$p(A') = \frac{n(A')}{n(\mathscr{E})} = \frac{n(\mathscr{E}) - n(A)}{n(\mathscr{E})}$$
$$= 1 - \frac{n(A)}{n(\mathscr{E})} = 1 - p(A)$$

Fig. 5.5:1

Worked example 4 Four cards are drawn at random from a pack, without replacement. What is the probability that at least one is an ace?

Number of choices of cards $= {}^{52}C_4$.
Number of choices containing no aces $= {}^{48}C_4$.
So, by equation (1),
$$p(\text{no ace}) = \frac{{}^{48}C_4}{{}^{52}C_4} \approx 0.719$$

And by equation (2),
$$p(\text{at least one ace}) \approx 1 - 0.719$$
$$= 0.281$$

Theorem 2:
$$p(A \cup B) = p(A) + p(B) - p(A \cap B) \qquad (3)$$

Proof
A and B are shown in fig. 5.5:2.
$$n(A \cup B) = n(A) + n(B) - n(A \cap B)$$
$$\frac{n(A \cup B)}{n(\mathscr{E})} = \frac{n(A)}{n(\mathscr{E})} + \frac{n(B)}{n(\mathscr{E})} - \frac{n(A \cap B)}{n(\mathscr{E})}$$

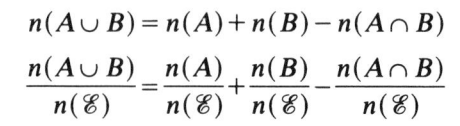

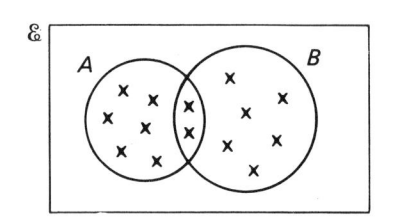

Fig. 5.5:2

So by definition (1),
$$p(A \cup B) = p(A) + p(B) - p(A \cap B)$$

Worked example 5 At a second-hand car showroom, 20% of the cars have no engine, 40% have bald tyres, and 15% have no engine and bald tyres. What is the probability that a car chosen at random has good tyres and an engine?

Let A be the event: the car has no engine; then A' is the event: the car has an engine.

Let B be the event: the car has bald tyres; then B' is the event: the car has good tyres.

$$p(A) = \tfrac{20}{100} = \tfrac{1}{5} \quad p(B) = \tfrac{40}{40} = \tfrac{2}{5} \quad p(A \cap B) = \tfrac{15}{200} = \tfrac{3}{20}$$

The event {the car has good tyres and an engine} is $A' \cap B'$ (see fig. 5.5:3).

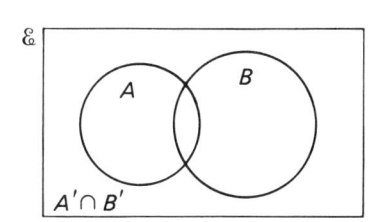

$$\begin{aligned} p(A' \cap B') &= p[(A \cup B)'] & \text{by de Morgan's laws (see 1.1:2 Q12)} \\ &= 1 - p(A \cup B) & \text{by equation (2)} \\ &= 1 - \{p(A) + p(B) - p(A \cap B)\} & \text{by equation (3)} \\ &= 1 - (\tfrac{1}{5} + \tfrac{2}{5} - \tfrac{3}{20}) \\ &= \tfrac{11}{20} \end{aligned}$$

Fig. 5.5:3

Exercise 5.5:2

1 Two dice are thrown. Find the probability of scoring a total of:

(i) 3 or less (ii) 4 or more

2 Three cards are drawn, without replacement, from a pack. Find the probability of getting:

(i) no spades
(ii) at least one spade

3 Four cards are drawn, without replacement, from a pack. Find the probability of getting:

(i) at least one King
(ii) at most three Kings

4 Four different letters are chosen at random from the letters BEHAVIOUR. What is the probability that:

(i) at least one is a vowel?
(ii) at least two are vowels?

5 Four balls are taken at random, without replacement, from a box containing five white and four red balls. Find the probability that at least two of the balls are red.

6 The headmaster of a primary school is going to show three films to his pupils. He chooses them from a catalogue of 15 films of which 5 have X-certificates. He chooses at random. What is the probability that of the films he chooses:

(i) none (ii) exactly one (iii) at least one

has an X-certificate?

7 (i) Three dice are thrown. What is the probability of getting:
(a) at least one six;
(b) at most two sixes?

(ii) Twelve coins are tossed. What is the probability of getting:
(a) at least two heads;
(b) at most ten heads?

8 Two people are chosen at random from a crowd. What is the probability that their birthdays are on:

(i) the same day of the week?
(ii) different days of the week?
(iii) consecutive days of the week?

9 Repeat question 8 assuming that three people are chosen.

10 There are n people in a room. What is the probability that at least two people have their birthdays on the same day of the year?

How many people must be in the room for this probability to be more than $\tfrac{1}{2}$? (Ignore leap years.)

11 Show that the probability that a six-digit number includes at least two consecutive digits the same is about $\tfrac{2}{5}$.

12 A man who knows nothing about changing a plug is changing a plug. He attaches the three strands of wire at random to the three connectors. What is the probability that:

(i) he wires the plug correctly?
(ii) he wires at least one connection correctly?

13 A secretary types 4 letters and addresses 4 envelopes for them, but then, due to an unconscious desire to get the sack, stuffs the letters at random into the envelopes. What is the probability that at least one letter goes to the right person?

What if there were 5 letters?

14 A card is drawn from a pack. What is the probability of getting:

(i) an ace or a heart?
(ii) neither an ace nor a heart?

15 A card is drawn from a pack. What is the probability of getting:

(i) a court card or a red card?
(ii) neither a court card nor a red card?

16 Let A and B be events with
$$p(A) = \tfrac{3}{8}, \quad p(B) = \tfrac{1}{2}, \quad p(A \cap B) = \tfrac{1}{4}$$
Find:
(i) $p(A \cup B)$
(ii) $p(A')$ and $p(B')$
(iii) $p(A' \cap B')$
(iv) $p(A' \cup B')$
(v) $p(A \cap B')$
(vi) $p(B \cap A')$

17 Let A and B be events with
$$p(A \cup B) = \tfrac{3}{4}, \quad p(A') = \tfrac{2}{3}, \quad p(A \cap B) = \tfrac{1}{4}$$
Find:
(i) $p(A)$ (ii) $p(B)$
(iii) $p(A \cap B')$ (iv) $p(A' \cap B)$

18 In a class of 30 boys, 15 have bicycles, 10 have motorbikes, and 4 have both. If a student is picked at random, what is the probability that:

(i) he has neither a bicycle nor a motorbike
(ii) he has a bicycle but no motorbike

19 A certain town has a corrupt council. The mayor and council members have accepted bribes from local businessmen. Of the 10 000 people living in the town, 2500 support the imprisonment of the mayor and 6000 support the dismissal of the entire council. 3000 people are happy with the situation and support neither the mayor's imprisonment nor the council's dismissal.

If a person is chosen at random from the population of the town, what is the probability that he supports

(i) the mayor's imprisonment and the council's dismissal?

(ii) the mayor's imprisonment but not the council's dismissal?

20 An opinion poll shows that 30% of people in England vote Conservative, 40% want capital punishment reintroduced, and 25% vote Conservative and want capital punishment reintroduced. If a person is chosen at random from the population, what is the probability that he votes Conservative but doesn't want capital punishment reintroduced?

21 An integer x is chosen at random. Let A be the event $\{x$ is divisible by 4$\}$, B the event $\{x$ is divisible by 7$\}$.

(i) Write down $p(A)$, $p(B)$.
(ii) Describe the event $A \cap B$. Write down $p(A \cap B)$.
(iii) Hence find the probability that x is divisible by 4 or 7.

22 An integer is chosen at random. Find the probability that:

(i) it is not divisible by 3 or 5
(ii) it is not divisible by 4 or 6

5.5:3 Mutually exclusive and independent events

> **Definition** Two events which can never happen together are called **mutually exclusive**.

If two events A and B are mutually exclusive,
$$A \cap B = \varnothing, \quad \text{so} \quad p(A \cap B) = 0$$
thus by equation (3)
$$p(A \cup B) = p(A) + p(B) \tag{4}$$

For example, when a card is drawn from a pack, the events {the card is a spade} and {the card is red} are mutually exclusive; so

$p\{$the card is a spade or a red card$\}$
$= p\{$the card is a spade$\} + p\{$the card is red$\}$
$= \tfrac{1}{4} + \tfrac{1}{2}$
$= \tfrac{3}{4}$

Note: we used this idea implicitly in formulating definition (1). In a sample space $\mathscr{E}$ of equally likely outcomes, the probability of each outcome

Series and probability 265

is $1/n(\mathscr{E})$. An event A consists of $n(A)$ mutually exclusive outcomes; so

$$p(A) = \frac{1}{n(\mathscr{E})} + \frac{1}{n(\mathscr{E})} + \cdots + \frac{1}{n(\mathscr{E})} \quad (n(A) \text{ times})$$
$$= \frac{n(A)}{n(\mathscr{E})}$$

Definition Two events which can happen together but have no bearing on each other are called **independent events**.

For example, if a red die and a blue die are thrown, the events

$A\{\text{red die shows a 6}\}$ and $B\{\text{blue die shows an even number}\}$

are independent. Note that

$$p(A) = \tfrac{1}{6},\ p(B) = \tfrac{1}{2},\ p(A \cap B) = p\{(6,2),(6,4),(6,6)\} = \tfrac{3}{36} = \tfrac{1}{12};$$

so

$$p(A)p(B) = \tfrac{1}{12} = p(A \cap B)$$

In general:

---
events A and B are independent $\Leftrightarrow p(A \cap B) = p(A)p(B)$ **(5)**

This will become clearer when we have considered conditional probability; see 5.5:4 Q4.

Worked example 6 A and B are independent events. $p(A) = \tfrac{1}{3}$ and $p(A \cup B) = \tfrac{3}{5}$. Find $p(B)$.

A and B independent $\Rightarrow p(A \cap B) = p(A)p(B) = \tfrac{1}{3}p(B)$.
From equation (3),

$$p(A \cup B) = p(A) + p(B) - p(A \cap B)$$

i.e. $$\tfrac{3}{5} = \tfrac{1}{3} + p(B) - \tfrac{1}{3}p(B)$$

$$\tfrac{2}{3}p(B) = \tfrac{4}{15}$$

$$p(B) = \tfrac{2}{5}$$

Independent trials
Suppose we carry out a sequence of trials in which the outcome of any trial is independent of the outcome of any other trial, e.g. tossing a coin repeatedly (or, equivalently, tossing several coins). We call these trials **independent trials**.

Worked example 7 A coin is biassed in such a way that $p(\text{H}) = \tfrac{2}{3}, p(\text{T}) = \tfrac{1}{3}$. The coin is tossed three times. What is the probability of getting at least two heads?

We can display the outcomes and their probabilities on a **probability tree** (fig. 5.5:4). The eight outcomes are not equally likely. But since the trials are independent, we can find the probability of each outcome by multiplying the probabilities along the branches leading to that outcome.

266 *Mutually exclusive and independent events*

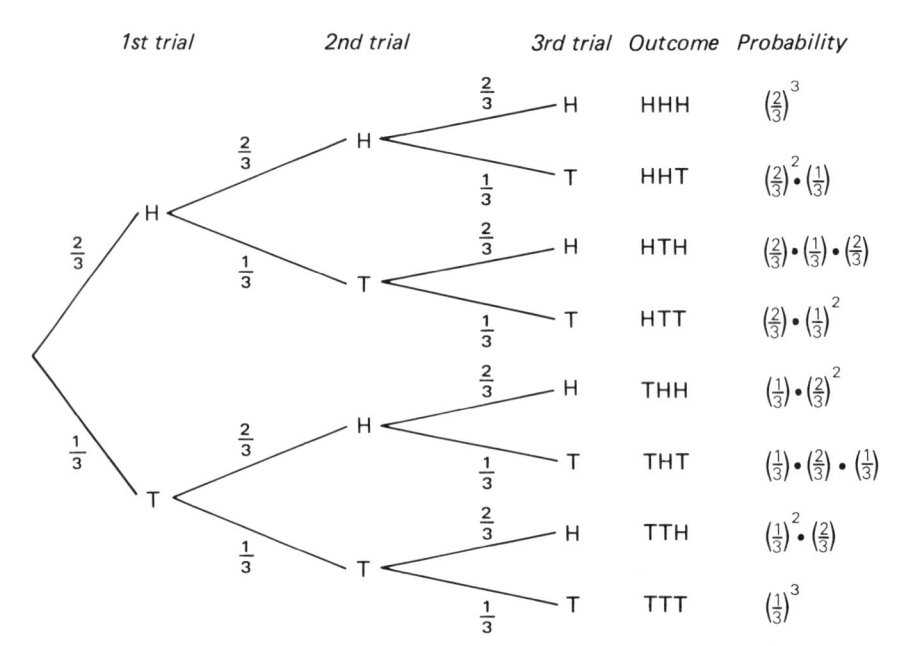

Fig. 5.5:4

Thus, for example,

$$p(\text{THT}) = p(T_1 \cap H_2 \cap T_3)$$
$$= p(T_1)p(H_2)p(T_3)$$
$$= (\tfrac{1}{3})(\tfrac{2}{3})(\tfrac{1}{3})$$
$$= \tfrac{2}{27}$$

where H_2 means a head is obtained on the second toss, etc.

$$p(\text{at least two heads}) = p(\text{HHH, HHT, HTH, THH})$$
$$= (\tfrac{2}{3})^3 + 3(\tfrac{2}{3})^2(\tfrac{1}{3})$$
$$= \tfrac{20}{27}$$

Worked example 8 A coin is biassed in such a way that $p(H) = \tfrac{2}{3}$, $p(T) = \tfrac{1}{3}$. The coin is tossed 12 times. What is the probability of getting exactly 5 heads?

Number of ways of getting 5 heads = $^{12}C_5$ (see 5.3:3 Q7(iii)). Each of these ways consists of 5 heads and 7 tails, and so has probability $(\tfrac{2}{3})^5(\tfrac{1}{3})^7$. So

$$p(5 \text{ heads}) = {}^{12}C_5(\tfrac{2}{3})^5(\tfrac{1}{3})^7$$

Exercise 5.5:3

Questions 1–3: Which of the pairs of events A and B are mutually exclusive? In each case find $p(A)$, $p(B)$, $p(A \cup B)$.

1 An integer x is chosen at random.
 (i) $A = \{x \text{ odd}\}$, $B = \{x \text{ even}\}$
 (ii) $A = \{x \text{ divisible by 6}\}$
    $B = \{x \text{ divisible by 9}\}$

2 A card is drawn from a pack.
 (i) $A = \{\text{card is a heart}\}$
    $B = \{\text{card is a diamond}\}$
 (ii) $A = \{\text{card is a spade}\}$
    $B = \{\text{card is a King}\}$

3 A red die and a blue die are thrown. Let T = total score on dice, S = difference between scores on dice.

(i) $A = \{T = 10\}$
$B = \{$red die shows 4$\}$
(ii) $A = \{T = 10\}$
$B = \{T = 5\}$
(iii) $A = \{T = 9\}$
$B = \{S = 4\}$

4 Are A and B mutually exclusive when:

(i) $p(A) = \frac{1}{2}$, $p(B) = \frac{1}{3}$, $p(A \cup B) = \frac{3}{4}$?
(ii) $p(A' \cap B') = \frac{1}{2}$, $p(B) = \frac{1}{2}$?
(iii) $p(A' \cup B') = 0$?

5 A card is drawn from a pack. Let A, B and C be the events {card is a spade}, {card is a court card} and {card is a black card}. Which of the pairs A and B, B and C, A and C are independent?

6 Two cards are drawn from a pack. Let A and B be the events {first card is an ace}, {second card is an ace} respectively. Are A and B independent:

(i) if the first card drawn is replaced before the second is drawn?

(ii) if the first card drawn is not replaced?

7 An integer x is chosen at random. Let A, B and C be the events $\{x$ is divisible by $4\}$, $\{x$ is divisible by $7\}$ and $\{x$ is divisible by $6\}$ respectively. Which of the pairs A and B, B and C, A and C are independent?
If D is the event $\{x$ is a perfect square$\}$, are A and D independent?

8 A red die and a blue die are thrown. Let A, B and C be the events {1 or 2 on red die}, {total of 10 or more} and {even number on blue die} respectively. Which of the pairs A and B, B and C, A and C are:

(i) mutually exclusive (ii) independent?

9 If $p(A) = \frac{1}{3}$, $p(B) = \frac{2}{5}$ and $p(A \cup B) = \frac{3}{5}$, show that A and B are independent.

10 If $p(A) = \frac{1}{3}$, $p(B) = \frac{1}{2}$ and $p(A \cup B) = \frac{3}{5}$, show that A and B are neither mutually exclusive nor independent.

11 If $p(A') = \frac{3}{8}$, $p(B) = \frac{1}{3}$ and $p(A' \cap B') = \frac{1}{4}$, show that:

(i) A and B (ii) A' and B'

are independent.

12 Under what circumstances, if any, can two events A and B be mutually exclusive and independent?

13 A and B are independent events. Find $p(B)$ if:

(i) $p(A) = \frac{1}{2}$, $p(A \cup B) = \frac{3}{4}$
(ii) $p(A) = \frac{2}{3}$, $p(A \cup B) = \frac{2}{3}$
(iii) $p(A) = \frac{1}{2}$, $p(A \cap B') = \frac{1}{6}$
(iv) $p(A \cap B') = \frac{4}{25}$, $p(A' \cap B) = \frac{9}{25}$

14 If $p(A) = \frac{1}{3}$, $p(B) = \frac{2}{5}$ and A and B are independent, find:

(i) $p(A \cap B)$ (ii) $p(A \cup B)$
(iii) $p(A' \cap B')$ (iv) $p(A' \cap B)$

15 (i) If $p(A) = p_1$, $p(B) = p_2$ and A and B are independent, find:
(a) $p(A' \cap B')$; (b) $p(A' \cap B)$.

(ii) Hence show that if two events A and B are independent, then:
(a) A' and B are independent;
(b) A' and B' are independent.
Give an example to illustrate these results.

16 Two women set out independently to seduce the same man, their probabilities of success being $\frac{2}{3}$ and $\frac{1}{4}$ respectively. Find the probability that at least one of them will succeed.

17 The probability that A hits a target is $\frac{1}{4}$ and the independent probability that B hits it is $\frac{2}{5}$. What is the probability that the target will be hit if A and B each shoot at it?

18 The probability that a man will live 10 more years is $\frac{1}{4}$ and the probability that his wife will live 10 more years is $\frac{1}{3}$. Find the probability that:

(i) both will be alive in 10 years
(ii) at least one will be alive in 10 years
(iii) neither will be alive in 10 years
(iv) only the wife will be alive in 10 years

Comment on any assumption you have made in solving this problem.

19 Urn A contains 5 red marbles and 3 white marbles; urn B contains 2 red marbles and 6 white marbles.

(i) If a marble is drawn from each urn, what is the probability that both marbles are of the same colour?

(ii) If two marbles are drawn from each urn, what is the probability that all four marbles are of the same colour?

20 A, B and C are events. A and B are independent, B and C are independent, A and C are mutually exclusive, and $p(A \cap B) = \frac{1}{4}$; $p(B) = \frac{2}{3}$; $p(C) = \frac{1}{4}$. Find:

(i) $p(A)$ (ii) $p(B' \cap C)$ (iii) $p(A \cup B \cup C)$

21 A biassed coin, for which $p(H) = \frac{3}{5}$, $p(T) = \frac{2}{5}$, is tossed three times. Draw a tree diagram showing the possible outcomes and their probabilities.
What is the probability of getting:

(i) exactly one head?   (ii) exactly two heads?

22 A rifleman hits his target with probability $\frac{3}{4}$. He fires four times. Draw a tree diagram showing the possible outcomes and their probabilities.
What is the probability that the man hits the target:

(i) exactly twice   (ii) at least once?

⑦ → **23** A die is thrown three times. Thinking of the outcomes of each trial as being $A = \{\text{a six}\}$ and $A' = \{\text{not a six}\}$, draw a tree diagram showing the possible outcomes and their probabilities.

Find the probability of throwing exactly two sixes.

⑦ → **24** A football team has probabilities $\frac{1}{4}$ of winning each match it plays, $\frac{1}{8}$ of drawing, $\frac{5}{8}$ of losing. The team plays three matches. Draw a tree diagram showing the possible results and their probabilities.

What is the probability of the team:

(i) losing two of the matches and drawing the other?
(ii) winning exactly one match?

⑦ **R** **25** A die is biassed so that $p(r) = kr$, $r = 1, 2, 3, 4, 5, 6$. Find k.

(i) If the die is thrown twice, calculate the probability that the total score is:
(a) 4; (b) 7; (c) 10 or more.

(ii) If the die is thrown three times, what is the probability of getting fewer than two sixes?

We are now able to do worked example 2 and its associated questions by thinking of the experiments as being **sequences of independent trials**. Before we relied on the fact that the outcomes were equally likely, now this is no longer important.

→ **26** Four coins are tossed. Find the probability of getting exactly:

(i) 4 heads   (ii) 3 heads
(iii) 2 heads (iv) 1 head
(v) no heads

assuming that:
(a) the coin is fair; (b) $p(\text{H}) = 0.7$, $p(\text{T}) = 0.3$.

(c.f. 5.5:1 Q12.)

27 A bag contains 4 red balls and 5 blue balls. A ball is drawn and replaced. A ball is drawn again. What is the probability that:

(i) the ball is red each time?
(ii) the second ball is red?
(iii) the same ball is drawn each time?

28 Repeat question 27, assuming that the bag contains 4 red, 4 blue and 2 green balls.

⑧ **29** A coin is tossed 9 times. What is the probability of getting exactly 5 heads if:

(i) the coin is fair?
(ii) $p(\text{H}) = \frac{2}{5}$, $p(\text{T}) = \frac{3}{5}$?

⑧ **30** A die is thrown 12 times. What is the probability of getting:

(i) exactly six sixes?
(ii) at least three sixes?

⑧ **31** In a multiple-choice paper there are ten questions. Each question has five suggested answers of which one is right. A student has wandered into the wrong exam but doesn't realise it. He chooses answers entirely at random. What is the probability that he gets:

(i) all ten questions right?
(ii) at least three questions right?

⑧ → **32** 15% of a consignment of 'beef' joints is actually kangaroo meat. If ten joints of meat are selected at random, what is the probability that:

(i) at least one is kangaroo?
(ii) exactly one is kangaroo?

Assume the consignment is so big that the selections can be regarded as independent.

⑧ **33** In a raffle there are 2000 tickets. A man buys 300 of them. There are three prizes. What is the probability that he wins:

(i) a prize   (ii) all three prizes?

What assumption have you made in doing this question?

⑧ **R** **34** A certain type of car includes three components A, B and C. It is found that 3% of components of type A are defective, 5% of type B and 1% of type C.

(i) What is the probability that in a car chosen at random at least one component is defective?

What is the probability that if five cars are chosen at random:
(a) none has a defective component; (b) each has at least one defective component? Assume that n, the number of cars, is so large that the selections may be regarded as independent.

(ii) Repeat part (i) without the final assumption, giving your answers to (a) and (b) in terms of n. Show that as $n \to \infty$ these answers approach those obtained in (i).

⑧ → **35** An experiment consists of m independent trials, each of which may result in just two outcomes, success or failure. (This kind of trial is called a **Bernoulli trial**.)

(i) Let $p(\text{success}) = p$, $p(\text{failure}) = q$. What is $p + q$?

(ii) Let p_r be the probability of exactly r successes in the n trials. Find p_0, p_1, p_2, p_n.

(iii) Find p_r. Show that p_r is the coefficient of $q^r p^{n-r}$ in the expansion of $(q+p)^n$.

⑧ **36** 15% of a large consignment of apples are bad. What is the probability that

(i) if 4 apples are chosen at random, exactly 2 are bad?
(ii) if 12 apples are chosen at random, exactly 6 are bad?
(iii) if 20 apples are chosen at random, exactly 10 are bad?

37 Each time a man fires at a target he has a probability of $\frac{1}{5}$ of hitting it. How many times must he fire in order to ensure that the probability that he hits the target at least once is more than $\frac{9}{10}$?

→ 38 How many times must a coin be tossed in order to ensure that the probability of getting at least one head is more than $\frac{95}{100}$?

R 39 In a bag of hydrangea seeds the probability that a seed produces white flowers is $\frac{1}{4}$. How many seeds must be sown to ensure that the probability of getting at least one white flower is more than 99%?

• → 40 A football team has probabilities $\frac{3}{5}$ of winning, $\frac{3}{10}$ of drawing, $\frac{1}{10}$ of losing each match it plays. If the team plays six matches in a tournament what is the probability that:

(i) it wins exactly four matches?
(ii) it loses exactly four matches?
(iii) it wins two matches, loses two and draws two?
(iv) it wins three matches, loses one and draws two?
(v) it draws every match?

41 In a general election, 50% of the country vote Conservative, 30% vote Labour and 20% vote SDP/Liberal Alliance.

(i) If seven people are chosen at random, what is the probability that:
(a) exactly three voted Conservative, exactly four Labour?
(b) exactly three voted Conservative, exactly two Labour?

(ii) If three people are chosen at random what is the probability they voted for:
(a) the same party?
(b) different parties?

R 42 On any particular day, the probability that a certain local government official accepts no bribes is $\frac{2}{5}$, that he accepts exactly one bribe is $\frac{1}{2}$ and that he accepts exactly two bribes is $\frac{1}{10}$. What is the probability that he accepts exactly five bribes in:

(i) a three-day working week?
(ii) a five-day working week?

43 At a horse show, a horse is attempting to clear five fences, each one more difficult than the last. The probability that the horse clears the rth fence is $1 - \frac{1}{8}r$, independently of its performance over the other fences. Find the probability that the horse clears:

(i) all five fences   (ii) no fences
(iii) exactly two fences

44 Alice and Brenda are playing tennis. Alice has a probability of $\frac{3}{4}$ of winning each set. The match is won by the player who first wins three sets. Draw a tree diagram showing the possible outcomes and their probabilities. Find the probability that Alice wins the match.

• 45 Alice and Brenda toss a coin in turn. Alice starts. The first person to toss a head wins. Draw a tree diagram to show the possibilities.
Find the probability that Alice wins:

(i) at first attempt
(ii) at second attempt
(iii) at third attempt
(iv) at fourth attempt.

Hence, by finding the sum to infinity of the appropriate G.P., find the probability that Alice wins.

→ 46 Alice and Brenda roll a die in turn. Alice starts. The first person to throw a six wins. Find the probability that:

(i) Alice wins   (ii) Brenda wins

47 In their continuing struggle for all-round supremacy... Alice and Brenda fight a duel. They fire at each other alternately until one hits the other—and wins. With each shot Alice has probability p_1 of hitting Brenda, and with each shot Brenda has probability p_2 of hitting Alice.

If Alice starts,

(i) find the probability that she wins
(ii) show that if $p(\text{Alice wins}) = p(\text{Brenda wins})$ then $p_2 = p_1/(1 - p_1)$.

→ 48 Alice, Brenda and Carol toss a coin in turn. The first person to toss a head wins. If Alice starts, followed (if necessary) by Brenda, find the probability that:

(i) Alice wins
(ii) Brenda wins
(iii) Carol wins

R 49 Repeat question 48, assuming that the coin is biassed, with:

(a) $p(H) = \frac{3}{5}$ (b) $p(H) = p$

Is it possible to choose p so that they each have an equal chance of winning?

• 50 Each time a man marries, the marriage has a probability of $\frac{5}{6}$ of ending on the rocks. Find the probability that his fourth marriage is his first successful one.

51 In a game of darts, the probability that a girl scores a double with any particular dart is $\frac{1}{4}$. Find the probability that she scores her first double with:

(i) her fifth dart
(ii) her eighth dart

52 Each time a man takes his driving test, he has a probability of $\frac{1}{3}$ of passing. Find the probability that he has to take the test:

(i) three times
(ii) five times
(iii) ten times or more

270 *Conditional probability*

→ **53** A coin is tossed repeatedly. Find the probability that the first head appears at the *r*th toss if:

(i) the coin is fair (ii) $p(H) = \frac{2}{3}$, $p(T) = \frac{1}{3}$

54 An experiment in which the probability of success is p is performed many times. What is the probability that the experiment is first successful on the *r*th time?

R 55 Each time that a doctor performs a certain heart operation she has a probability of $\frac{2}{5}$ of succeeding. What is the probability that she takes exactly six operations to achieve:

(i) her first success?
(ii) her second success?
(iii) two successes running for the first time?

+→ **56** A die is thrown until it shows a six twice running. Find the probability that this requires r tosses.

57 A man smuggles commodities between two countries A and B. He takes gold from A to B and brings back arms from B to A. The probability of his being caught when he goes through customs is $\frac{1}{20}$ when he is travelling from A to B and $\frac{1}{10}$ when he is travelling from B to A.

If he starts with a gold run (thereafter making alternate arms and gold runs), find the probability that:

(i) he is caught on the ninth run
(ii) he is caught on the tenth run
(iii) he is still free after twenty runs

5.5:4 Conditional probability

Let $p(A|B)$ be the probability that the event A takes place given that the event B has taken place.

Theorem 3:

$$p(A|B) = \frac{p(A \cap B)}{p(B)} \qquad (6)$$

Proof

$$p(A|B) = \frac{n(A \cap B)}{n(B)}$$

$$= \frac{n(A \cap B)}{n(\mathscr{E})} \Big/ \frac{n(B)}{n(\mathscr{E})}$$

$$= \frac{p(A \cap B)}{p(B)}$$

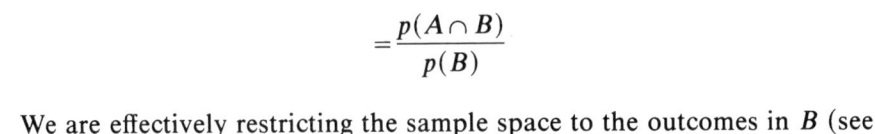

We are effectively restricting the sample space to the outcomes in B (see fig. 5.5:5).

Fig. 5.5:5

Worked example 9 Three fair coins are tossed. What is the probability of getting at least two heads, given that the first coin comes up heads?

Let A be the event {at least two heads}, i.e. {HHH, HHT, HTH, THH}.
Let B be the event {first coin is a head}, i.e. {HHH, HHT, HTH, HTT}; $p(B) = \frac{1}{2}$.
$A \cap B = $ {at least two heads and first coin head}, i.e. {HHH, HHT, HTH}; $p(A \cap B) = \frac{3}{8}$.

$$p(A|B) = \frac{p(A \cap B)}{p(B)} = \frac{\frac{3}{8}}{\frac{1}{2}} = \frac{3}{4}$$

i.e. of the 4 outcomes in B, 3 are also in A (see fig. 5.5:6).

Fig. 5.5:6

Dependent trials

Suppose that in a sequence of trials the outcome of some trial depends on the outcome of another trial. We call these trials **dependent**.

Worked example 10 A coin, weighted so that $p(H) = \frac{2}{3}$ and $p(T) = \frac{1}{3}$, is tossed. If the head appears, then a number is selected at random from the numbers 1 to 9; if the tail appears, then a number is selected at random from the numbers 1 to 5.

(i) What is the probability that an even number is selected?
(ii) Given that an even number is selected, what is the probability that the coin came up heads?

The trials are not independent, but we can still find the probability of each outcome by multiplying the probabilities along the branches of the tree in fig. 5.5:7 using equation (6), e.g.

$$p(H \cap O) = p(H)p(O|H) = \frac{2}{3} \cdot \frac{5}{9} = \frac{10}{27}$$

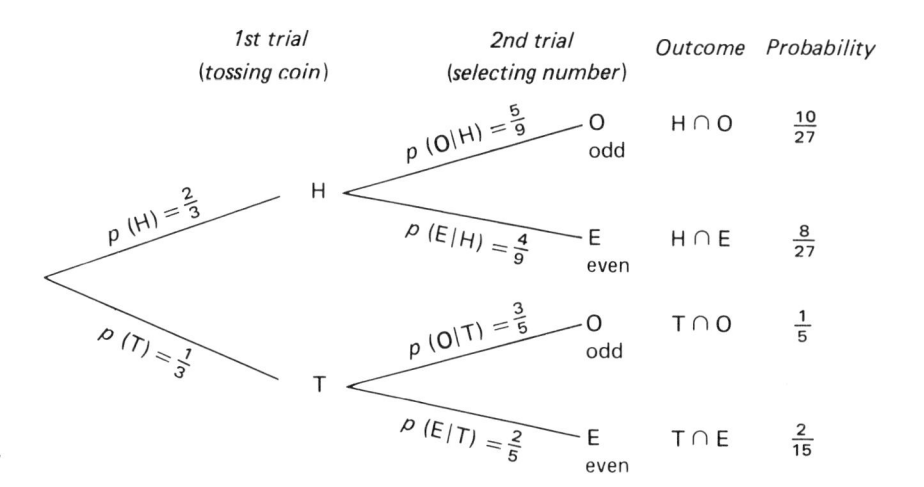

Fig. 5.5:7

where O denotes an odd number selected and E denotes an even number selected. Note: the second trial may have either 9 or 5 possible outcomes; but we are interested only in the events {odd} and {even}.

(i) $p(E) = p(H \cap E) + p(T \cap E) = \frac{8}{27} + \frac{2}{15} = \frac{58}{135}$

(ii) $p(H|E) = \dfrac{p(H \cap E)}{p(E)} = \dfrac{\frac{8}{27}}{\frac{58}{135}} = \dfrac{20}{29}$

Exercise 5.5:4

1 A red die and a blue die are thrown. Let A, B and C be the events {total score is 6}, {red die shows 1 or 2} and {one of the dice shows a 2} respectively. Find:

(i) $p(A|B)$, $p(B|A)$
(ii) $p(A|C)$, $p(C|A)$
(iii) $p(B|C)$, $p(C|B)$

2 Two dice are thrown. Let A, B and C be the events {sum of the scores is 7}, {difference between the scores is 4} and {product of the scores is 12}. Find:

(i) $p(A|B)$, $p(B|A)$
(ii) $p(A|C)$, $p(C|A)$
(iii) $p(B|C)$, $p(C|B)$

3 (i) Under what circumstances is $p(B|A) = p(A|B)$?

(ii) Show that $p(A \cap B) = p(B)p(A|B)$. Write this equation in words.

(iii) What can you say about $p(A|B)$ and $p(B|A)$ when A and B are mutually exclusive?

4 (i) Explain how the original definition of independent events, i.e. as events which have no bearing on each other, implies that if two events A and B are independent $p(A|B) = p(A)$.

(ii) Hence show that if A and B are independent $p(A \cap B) = p(A)p(B)$.

(iii) Show that if $p(A \cap B) = p(A)p(B)$ then A and B are independent.

5 Let A and B be events with $p(A) = \frac{1}{2}$, $p(B) = \frac{1}{3}$, $p(A \cap B) = \frac{1}{4}$. Find:

(i) $p(A|B)$
(ii) $p(B|A)$
(iii) $p(A \cup B)$
(iv) $p(A' \cap B')$
(v) $p(A'|B')$
(vi) $p(B'|A')$

6 Let A and B be events with $p(A) = \frac{2}{5}$, $p(B) = \frac{1}{2}$, $p(A \cup B) = \frac{4}{5}$. Find:

(i) $p(A \cap B)$
(ii) $p(A|B)$
(iii) $p(B|A)$
(iv) $p(A' \cap B')$
(v) $p(A'|B')$
(vi) $p(B'|A')$

7 Let A and B be events with $p(A) = \frac{3}{5}$, $p(B) = \frac{1}{4}$, $p(A \cup B) = \frac{3}{4}$. Show that A and B are neither mutually exclusive nor independent. Find $p(A|B)$ and $p(A'|B')$.

8 Let A and B be events with $p(A) = \frac{2}{5}$, $p(A|B) = \frac{3}{8}$, $p(B|A) = \frac{1}{2}$. Find:

(i) $p(A \cap B)$ (ii) $p(A' \cap B')$

9 Let A and B be events with $p(A) = \frac{8}{15}$, $p(A \cap B) = \frac{1}{3}$, $p(A|B) = \frac{4}{7}$. Find:

(i) $p(B|A)$ (ii) $p(B|A')$

Are A and B:

(iii) mutually exclusive (iv) independent?

10 In a certain school, all the students are forced to do Maths O level. 15% of the students have parents who are divorced. 25% of the students failed the exam. 10% of the students failed the exam and have parents who are divorced. A student is chosen at random.

(i) If her parents are divorced, what is the probability that she failed the exam?

(ii) If she failed the exam, what is the probability that her parents are still married?

(iii) If her parents are still married, what is the probability that she passed the exam?

11 When the Masked Raider performs one of his terrifying and daring bank robberies, he sometimes rings the local police and leaves a challenging message. M is the event that he leaves a message with the police, and C is the event that the police catch him before he robs another bank. Explain in words the meaning of $p(C|M)$.

Given that $p(M) = 0.6$ and $p(C \cap M) = 0.1$, find $p(C|M)$. Given also that $p(C) = 0.15$ find:

(i) $p(C|M')$ (ii) $p(M|C)$

explaining in words the meaning of each. Find also the probability that:

(iii) he robs exactly ten banks before being caught
(iv) he is still free after robbing fifteen banks

12 For two diseases d_1 and d_2 there are two possible treatments t_1 and t_2. If a patient has d_1 and is given t_1, the probability of improvement is $\frac{3}{8}$. If she has d_1 and is given t_2, the probability of improvement is $\frac{3}{4}$. If, however, the patient has d_2 and is given t_1, she is certain to improve. If she has d_2 and is given t_2, the probability of improvement is only $\frac{1}{8}$.

A patient goes to a doctor with symptoms which would indicate d_1 in seven cases out of ten and d_2 in three cases out of ten. What treatment should the doctor prescribe to give the patient the best chance of recovery?

13 (i) A coin is tossed three times. Find the probability of getting exactly two heads given that:
(a) at least one of the tosses results in a head;
(b) the first toss results in a head.

(ii) There are three children in a family. Find the probability that two are girls and the other is a boy, given that:
(a) at least one of the children is a girl;
(b) the eldest child is a girl.

⑨ 14 (i) A man says 'I have two children. At least one of them is a boy.' What is the probability that his other child is also a boy?

(ii) A man says 'I have two children. The younger one is a boy.' What is the probability that his other child is also a boy?

⑨ + 15 At the beginning of a game of bridge, each player is dealt a hand of 13 cards. Mr East and Mr West each look at their own hands, and each notices that he has no ace. Mr North carelessly drops one of his cards and hurries to pick it up. However...

(i) Mr West has noticed that it is an ace. What, according to Mr West, is the probability that Mr North has another ace?

(ii) Mr East has noticed that it is the ace of spades. What, according to Mr East, is the probability that Mr North has another ace?

Assume that Mr West and Mr East, apart from being ruthless card players, are equipped with staggering powers of mental arithmetic.

⑩ 16 Of two bags A and B, A contains 5 red, 3 white and 8 blue marbles, and B contains 3 red and 4 white marbles. A die is thrown: if a three or a six appears, a marble is chosen from A, otherwise a marble is chosen from B. Find the probability that:
(i) the chosen marble is red
(ii) the chosen marble is blue
(iii) the marble was chosen from bag A, given that it is red
(iv) the marble was chosen from bag A, given that it is white

⑨ → 17 The probability that a man has an argument with his wife over breakfast is $\frac{1}{4}$. If he has an argument with his wife, the probability that he then takes his secretary out to lunch is $\frac{2}{3}$, whereas if he has no argument the probability is $\frac{1}{6}$. On any particular day, what is the probability that he takes his secretary out to lunch?
Given that he is having lunch with his secretary, what is the probability that he had a peaceful breakfast with his wife?

⑨ 18 The probability that a group have their first single played regularly on the radio is $\frac{2}{5}$. If the single is played on the radio, the probability that it is a hit is $\frac{5}{8}$, if not the probability is $\frac{1}{8}$.
Given that the single is a hit, what is the probability that it was not played regularly on the radio?

⑨ 19 At a street market somewhere in the East End of London, you are buying a watch from a charming Cockney stall-holder. He has three boxes of watches. Box A contains 10 watches of which 4 don't work, Box B has 6 watches of which one doesn't work, Box C has 8 watches of which 3 don't work. He tells you that they are all in perfect condition; so you select a box at random and then choose a watch at random. What is the probability that:
(i) it doesn't work?
(ii) you chose Box A, given that it doesn't work?

⑨ 20 Of three coins, two are fair and the other is biassed so that the probability of obtaining a head is three times as great as that of obtaining a tail. One of the coins is chosen at random and tossed three times. Find the probability that the chosen coin was the biassed one given that:
(i) a head was obtained on each throw
(ii) a tail was obtained on each throw

⑨ R 21 On a particular day the probabilities that a man travels to work by car, by bus or by bike are $\frac{1}{2}, \frac{1}{6}$ and $\frac{1}{3}$ respectively. The probabilities that he is late for work when he uses these means of transport are $\frac{3}{4}, \frac{5}{8}$ and $\frac{1}{8}$ respectively.
(i) Find the probability that he is late for work. Would you employ this man?
(ii) Given that he is late for work, find the probability that he came to work by:
(a) car; (b) bus; (c) bike.
(iii) Given that he arrives on time, find the probability that he came by:
(a) car; (b) bus; (c) bike.

⑨ 22 An experiment consists of two trials. The possible outcomes of the first trial are A and A', of the second B and B'.

$$p(A) = \tfrac{3}{5} \quad p(B|A) = \tfrac{2}{3} \quad p(B|A') = \tfrac{1}{2}$$

Draw a tree diagram showing the possible outcomes and their probabilities. Find:
(i) $p(A \cap B), p(A \cap B'), p(A' \cap B), p(A' \cap B')$
(ii) $p(B), p(B')$
(iii) $p(A|B), p(A|B')$

• 23 Alice and Brenda play tennis regularly against each other. The probability that Alice wins a match is $\frac{3}{5}$ if she won the previous match, $\frac{1}{2}$ if she lost it. During one week they play four times. Alice wins the first match. What is the probability that:
(i) she wins the other three?
(ii) she loses the other three?

→ 24 The probability that a particular day in summer will have dry weather is $\frac{3}{4}$ if the previous day was dry, $\frac{2}{3}$ if it was wet. What is the probability that 16, 17 and 18 July are all dry, given that 15 July was:
(i) wet   (ii) dry?

25 Alice and Brenda are now getting heavily into chess. The probability that Alice wins a match is $\frac{3}{4}$ if she won the previous match, and $\frac{5}{8}$ if she lost it. During a series of matches, what is the probability that Alice wins:

(i) two consecutive matches?
(ii) three consecutive matches?

We are now able to do worked example 3 and its associated questions by thinking of the experiments as being *sequences of dependent trials*.

• → **26** Three cards are drawn, without replacement, from a pack. Thinking of the outcomes of each draw as being $A = \{$a spade$\}$ and $A' = \{$not a spade$\}$, draw a tree diagram showing the possible outcomes and their probabilities. What is the probability of getting:

(i) three spades (ii) exactly two spades?

27 A bag contains 5 red, 4 white and 3 green balls. 3 balls are chosen at random, without replacement. Find the probability that the balls are of:

(i) the same colour
(ii) different colours

28 Five different letters are chosen in turn, at random, from the letters A, B, C, D, E, F, G, H and I. Thinking of the outcome of each choice as being $V = \{$vowel$\}$ or $V' = \{$consonant$\}$, find, using a tree diagram if necessary, the probability that two of the chosen letters are vowels. (See worked example 3.)

• → **29** A bag contains 5 red and 3 blue balls. Balls are drawn at random, without replacement. Find the probability that the first blue ball appears at the fifth draw.

30 A bag contains 20 balls, of which 3 are white. Balls are drawn at random, without replacement. Find the probability that the first white ball appears at:

(i) the third draw (ii) the rth draw

31 A TV producer is looking for a secretary. His method is to go to the office typing pool and choose the first girl he sees with brown eyes. There are 50 girls in the pool of whom 20 have brown eyes. What is the probability that he must look into the eyes of:

(i) 3 girls (ii) 4 girls (iii) 6 girls

before finding a secretary?

32 A bag contains 7 white and 3 black counters. Counters are taken out at random, without replacement. Find the probability that:

(i) the first taken is black
(ii) the second is black
(iii) the first two are black
(iv) at least one of the first two is black
(v) the first was black, given that the second is black

33 A bag contains s red balls and t blue balls ($s \geq 3$, $b \geq 2$). Three balls are chosen at random, without replacement. Find the probability that:

(i) the first draw is a red ball
(ii) the first two draws are both red balls
(iii) the second draw is a blue ball, given that the first was a red ball
(iv) the first draw was a red ball, given that the third is also a red ball

R **34** A man draws 5 cards from a pack, without replacement. What is the probability that he draws:

(i) no Queens?
(ii) at least one Queen?
(iii) exactly two Queens?
(iv) exactly two Queens, given that the first card he draws is a Queen?
(v) no Queen given that the first card he draws is a club?

What is the probability that he gets his first Queen on his fifth draw?

Miscellaneous exercise 5

1 (i) A sequence of numbers $\{u_r\}$ is defined by the iterative procedure $u_{r+1} \equiv u_r + 2r + 1$, $u_1 = 1$. Find u_2, u_3 and u_4 and show by induction that $u_n = n^2$ for all integers n.

(ii) A sequence of numbers $\{v_r\}$ is defined by the iterative procedure $3v_{r+1} \equiv 2v_r$, $v_1 = 9$. Find v_2, v_3, v_4. Find a formula for v_n, and prove it. Find $\sum_{n=1}^{\infty} v_n$.

2 The sum to n terms of a sequence is $n^2 + 4n$. Find the nth term of the sequence. Show that the sequence is an A.P. How many terms of the sequence must be taken to make a sum greater than 10^4?

3 For what values of x does the geometric series $x + (x-4) + (x-4)^2/x + \cdots$ converge? When $x = 6$, find the smallest value of n for which the sum to n terms differs from the sum to infinity by less than 10^{-5}.

4 Every year a man gives his son a goldfish tank containing two goldfish for his birthday. On average, the number of goldfish in each tank increases by 10% every three months. Find how many goldfish the boy has just after his twelfth birthday:

(i) if he keeps all the goldfish

(ii) if, in a desperate attempt to keep the numbers down, he sells 100 goldfish on his seventh birthday.

5 (i) The numbers x, $x+2$, $2x+1$ are consecutive terms of a G.P. Find two possible values of x and the common ratio of the G.P. in each case.

(ii) The arithmetic mean of two positive numbers x and y ($x > y$) is twice their geometric mean. Find the ratio $x:y$.

6 Find the set of values of x for which the geometric series $e^{3x} + 4e^x + 16e^{-x} + \cdots$ is convergent. Find the sum to infinity, $s(x)$, of the series.
Evaluate:

(i) $s(\ln 3)$ (ii) $\lim_{x \to \infty} \dfrac{s(x)}{e^{3x}}$

Sketch on the same axes the curves $y = s(x)$ and $y = e^{3x}$.

7 $\{u_r\}$ is a G.P.

(i) If $v_r = (-1)^{r+1} u_r$ for all r and $w_r = u_r^2$ for all r, show that $\{v_r\}$ and $\{w_r\}$ are also G.P.s.

(ii) Given that
$$\sum_{r=1}^{\infty} u_r = \lambda \quad \text{and} \quad \sum_{r=1}^{\infty} v_r = \mu$$
find $\sum_{r=1}^{\infty} w_r$.

(iii) If $\mu = 2\lambda$, find the common ratios of the three sequences.

8 (i) Find the set of values of x for which $\sum u_r$ is convergent when u_r is given by:

(a) $(1+x)^{r-1}$ (b) $\dfrac{1}{(1+x)^{r-1}}$

(c) $\left(\dfrac{2x}{1+x}\right)^{r-1}$

and in each case find $\sum_{r=1}^{\infty} u_r$ when it exists.

(ii) The sum to n terms of a sequence is
$$\dfrac{16}{3^{n-2}}(3^n - 2^n)$$
Find the nth term of the sequence. Show that the sequence is a G.P. Find its common ratio and its sum to infinity.

9 (i) The sum to infinity of a G.P. is four times its second term. Find its common ratio.

(ii) The sequence $\{u_n\}$ is such that $u_1 = 2$ and $u_n = \lambda \sum_{r=1}^{n-1} u_r$, λ constant. Find u_2, u_3 and u_4 in terms of λ. Show that $\{u_2, u_3, u_4, \ldots\}$ is a G.P. and find its common ratio. Find its sum to n terms.
For what values of λ does the series $u_2 + u_3 + u_4 + \cdots$ converge? Find the sum to infinity when it exists.

+ **10** A man deposits £x in a building society. The money is invested at $p\%$ compound interest per year. At the end of each year he takes out £y and sends it to his ex-wife as an alimony payment.
If $px < y$, find the number of years for which he can continue paying in this way.
If $px \geq y$, show that he can continue the payments for ever from this source.

11 (i) If $f_r = r(r+1)(r+2)$, find $f_r - f_{r-1}$. Hence find $\sum_{r=1}^{n} r(r+1)$. Hence show that
$$\sum_{r=1}^{n} r^2 = \tfrac{1}{6} n(n+1)(2n+1)$$

(ii) If $f_r = 1/r^2$, find $f_r - f_{r+1}$. Hence find
$$\sum_{r=1}^{n} \dfrac{2r+1}{r^2(r+1)^2}$$

12 If $f_r = 1/[(2r-1)(2r+1)]$, find $f_r - f_{r+1}$. Hence show that
$$\sum_{r=1}^{n} \dfrac{1}{(2r-1)(2r+1)(2r+3)} = \dfrac{1}{12} - \dfrac{1}{4(2n+1)(2n+3)}$$

Verify this result using the method of induction.

13 If $u_r = 1/[(2r+1)(2r+3)]$, express u_r in partial fractions. Hence show that
$$\sum_{r=1}^{n} u_r = \dfrac{n}{3(2n+3)}$$

Find $\sum_{r=1}^{\infty} u_r$.

14 If $u_r = 1/r(r+2)$, express u_r in partial fractions. Hence show that
$$\sum_{r=1}^{n} u_r = \dfrac{3n^2 + 5n}{4(n+1)(n+2)}$$

Verify this result using the method of induction.

15 If $u_r = (3r+1)/[(r-1)r(r+1)]$, show that
$$\sum_{r=2}^{n} u_r = \dfrac{5n^2 - n - 4}{2n(n+1)}$$

Verify this result using the method of induction.

16 If $u_r = 1/[r(r+1)(r+2)]$, find:

(i) $\sum_{r=1}^{n} u_r$ (ii) $\sum_{r=1}^{\infty} u_r$

Find the smallest positive integer n such that the sum to n terms differs from the sum to infinity by less than 10^{-4}.

17 If $S_\lambda = \sum_{r=1}^n r^\lambda$, write down S_0, S_1, S_2 and S_3. Hence find:

(i) $\sum_{r=1}^n (r+1)(r+3)$

(ii) $\sum_{r=1}^n r(r+1)(r+3)$

(iii) $\sum_{r=n}^{2n} r(r+1)$

(iv) $1^2 - 2^2 + 3^2 - 4^2 + \cdots + (2n-1)^2 - (2n)^2$

(v) $\sum_{r=1}^{2n} (-1)^{r+1}(2r-1)^2$

18 (i) Show that
$$\sum_{r=1}^n r(r+1) = \tfrac{1}{3}n(n+1)(n+2)$$
Find n such that
$$\sum_{r=1}^{25} r(r-1) = \sum_{r=1}^n r(r+1)$$
Hence evaluate $\sum_{r=1}^{25} r(r-1)$.

(ii) Show that
$$\sum_{r=1}^n \frac{1}{r(r+1)} = \frac{n}{n+1}$$
Evaluate:

(a) $\sum_{r=10}^{19} \frac{1}{r(r+1)}$; (b) $\sum_{r=1}^{\infty} \frac{1}{r(r+1)}$.

19 Find:

(i) $\sum_{r=5}^{20} (4-3r)$ (ii) $\sum_{r=5}^{20} (r-1)(r+2)$

(iii) $\sum_{r=5}^{20} \frac{2^{r+1}}{3^{r-1}}$ (iv) $\sum_{r=5}^{20} \frac{1}{r(r+1)}$

20 Prove by induction that for $n \in \mathbb{Z}^+$:

(i) $5^{2n} + 3n - 1$ is divisible by 9
(ii) $5^n - 2^n$ is divisible by 3
(iii) $7^n - 5^n - 2n$ is divisible by 4

21 (i) The nth term of a G.P. is equal to one third of the sum of all the terms after (but not including) the nth term. Show that the sum to infinity of the G.P. is four times its first term.

(ii) Prove by induction that, for $n \in \mathbb{Z}^+$,
$$\sum_{r=1}^n (r^2+1)r! = n(n+1)!$$

22 Prove by induction that:

(i) $\sum_{r=1}^n r(3r+1) = n(n+1)^2$

(ii) $\sum_{r=1}^n r^2(r+1) = \tfrac{1}{12}n(n+1)(3n^2+7n+2)$

23 (i) Car registration numbers consist of a letter followed by three digits followed by three more letters, e.g. A 212 XBC. Any letter except I and O and any digit may be used. Letters and digits may be repeated. How many such numbers are there? How many of these numbers contain no repeated letters or digits?

(ii) London phone numbers consist of seven digits. The only numbers not allowed are those beginning with the digit 1 or those beginning with the digits 999. How many numbers are possible? How many contain no repeated digits?

+ **24** Find the number of different permutations using all of the letters of the word DIFFERENT. Find the number of different:

(i) selections (ii) arrangements

which can be made using only 5 of the letters in the word DIFFERENT.

25 Twelve people are at a party. The conversation is so stilted that the host suggests playing party games. How many possibilities are there if the people are divided into:

(i) two groups, each of six people?
(ii) two groups, one of five, one of seven people?
(iii) two groups?
(iv) three groups, each of four people?
(v) three groups, of five, five and two people?

+ **26** (i) How many distinguishable ways are there of numbering the faces of a die?

(ii) The street layout of a new town consists of a grid of 10 streets running north–south which intersect with 12 streets running east–west, as shown in the diagram.

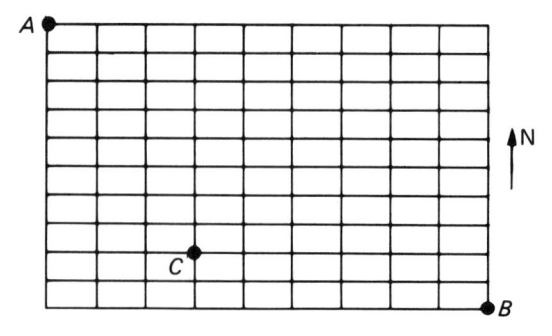

How many ways are there of driving from A to B without doubling back, i.e. by travelling always towards the east or towards the south? How many of these routes pass through C?

27 (i) Find the coefficients of x^4 and x^6 in the expansion of $(1+x)^{10}$.

(ii) Show that the error in using the first four terms of this expansion to find the value of $(1.001)^{10}$ is about 2.1×10^{-10}.

(iii) A coin is tossed 10 times. Find the number of ways of getting:
(a) exactly 4 heads; (b) exactly 4 tails.

(iv) There are 10 people in a room. From these 10 a smaller group is selected. How many possibilities are there, if the group is to contain:
(a) 4 people; (b) 6 people; (c) n people?

28 (i) Find the term independent of x in the expansion of $(x^2-1/x)^6(x+1/x)^4$. Find also the coefficient of x in the expansion.

(ii) Show that
$$\sum_{r=0}^{n} \binom{n}{r} 2^r = 3^n$$

Find
$$\sum_{r=0}^{n} \binom{n}{r} r 2^r$$

29 The first four terms in the expansion of $(1+ax)^b$ are $1, -2x, -2x^2$ and cx^3. Find a, b and c. State the set of values of x for which the expansion is valid.

30 The first three terms in the expansion of $(1+ax)/\sqrt{(1+bx)}$ are $1, -2x$ and $\frac{5}{2}x^2$. Find two possible pairs of values for a and b. In each case state the set of values of x for which the expansion is valid.

31 Given that $f(x) = (1+2x)/[(1+x)(1-2x^2)]$, express $f(x)$ in partial fractions. Hence find the first four terms in the expansion of $f(x)$ in ascending powers of x. State the set of values of x for which the expansion is valid.

32 Given that $f(x) = 3/(2-x-x^2)$, express $f(x)$ in partial fractions. Hence find the first four terms in the expansion of $f(x)$ in ascending powers of x. State the set of values of x for which the expansion is valid.

33 Find the first four terms in the expansions of
(i) $\dfrac{1}{1-\lambda x}$ (ii) $\dfrac{1}{(1-\lambda x)^2}$, λ constant.

Hence, or otherwise, find
(iii) $1 + \dfrac{2x}{3} + \dfrac{4x^2}{9} + \dfrac{8x^3}{27} + \cdots$ ($|x| < \frac{3}{2}$)
(iv) $1 - 2x + 3x^2 - 4x^3 + \cdots$ ($|x| < 1$)
(v) $2 + 3x + 4x^2 + 5x^3 + \cdots$ ($|x| < 1$)

34 Given that $f(x) = x/\sqrt{(1-2x)}$, find $\int_0^{0.2} f(x)\,dx$:

(i) approximately, by expanding $f(x)$ in ascending powers of x up to the term in x^3;

(ii) exactly, using the substitution $u = 1 - 2x$.

Give your answer to 4 dec. pls.

35 Find the first three terms in the expansion in ascending powers of x of $\sqrt{(1+4x^2)}$. Hence, without using a calculator or tables, find, correct to 3 dec. pls., approximations for:

(i) $\sqrt{26}$ (ii) $\displaystyle\int_0^{0.3} \sqrt{(1+4x^2)}\,dx$

36 Find the first three terms in the expansion of $1/\sqrt{(1+y)}$ in ascending powers of y. For what values of y is the expansion valid?

Find approximations for:
(i) $\displaystyle\int_0^{0.01} \dfrac{1}{\sqrt{(1+\sqrt{x})}}\,dx$
(ii) $\displaystyle\int_0^{0.01} \dfrac{1}{\sqrt{(1+\sin x)}}\,dx$

Give your answers to 6 dec. pls.

+ 37 (i) Given that $f(x) = \sqrt{(4+x^2)}$, find the first three non-zero terms in the binomial expansion of $f(x)$ in ascending powers of x. For what values of x is the expansion valid?

(ii) By writing $f(x) = x\sqrt{(1+4/x^2)}$, find the first three non-zero terms in the binomial expansion of $f(x)$ in ascending powers of $1/x$. For what values of x is the expansion valid? Use this expansion:
(a) to find an approximate value of $\sqrt{104}$;
(b) to show that
$$\int_{10}^{20} f(x)\,dx \approx 150 + \ln 4$$

38 A and B are events. $p(A) = 0.4$, $p(B) = 0.45$, $p(A \cup B) = 0.67$. Show that A and B are independent. Find:

(i) $p(A \cap B')$ (ii) $p(A \cup B')$

39 A, B and C are events. $p(A) = \frac{1}{5}$, $p(A \cup B) = \frac{2}{5}$, $p(B \cup C) = \frac{3}{8}$, $p(C) = \frac{1}{6}$ and events A and B are independent. Find $p(B)$. Show that the events B and C are independent.

40 A, B and C are events. A and B are mutually exclusive. A and C are independent. $p(A) = \frac{1}{4}$, $p(B) = \frac{1}{4}$, $p(A \cup C) = \frac{5}{9}$, $p(B \cup C) = \frac{7}{12}$. Find:

(i) $p(A \cup B)$ (ii) $p(A \cap C)$ (iii) $p(B \cap C)$

Are B and C independent?

41 Two dice are thrown, a red one and a blue one. Find the probability that:

(i) the total score is at least 7
(ii) the total score is divisible by 4
(iii) the product of the scores is less than 7
(iv) the score on the red die is bigger than the score on the blue die
(v) the score on the red die is more than twice the score on the blue die

42 A player throws a die repeatedly until he *fails* to obtain a six. His total score T is the sum of the scores on all his throws. Find the probability that:

(i) $T < 6$ (ii) $T = 9$
(iii) $T = 12$ (iv) $T = 25$
(v) $T > 9$ (vi) $12 < T < 18$
(vii) $T > 25$

43 Two cards are drawn, without replacement, from ten cards which are numbered from 1 to 10. Find the probability that:

(i) the number on both cards is even
(ii) the number on one card is odd and the number on the other card is even
(iii) the sum of the numbers on the two cards is greater than 4
(iv) the difference between the numbers on the two cards is greater than 4
(v) neither of the two numbers is divisible by 5

44 A bag contains 3 red, 4 blue and 5 yellow balls. Three balls are taken at random from the bag, without replacement. Find the probability that:

(i) the three balls are all of the same colour
(ii) the three balls are all of different colours
(iii) the third ball chosen is yellow

45 A bag contains 5 red and 4 blue balls. Five balls are taken at random from the bag, without replacement. Find the probability that:

(i) no red balls are left in the bag
(ii) no blue balls are left
(iii) two red and two blue balls are left

46 (i) Twelve relatives meet for Christmas dinner. Every person embraces every other person once. Find the total number of embraces which take place.

(ii) Two numbers x and y are chosen at random from the numbers 1 to 5. A number z is chosen at random from the numbers 1 to 10. Find:
(a) $p(x+y > z)$; (b) $p(x+y \geq z)$.

47 A husband and wife decide to have three children and do so. A is the event {at least two of the children are boys}. B is the event {the eldest child is a girl and the middle one a boy}. C is the event {the eldest child is a boy}. Assuming that the wife is just as likely to give birth to a boy as to a girl, show that A and B are independent, B and C are mutually exclusive and A and C are neither independent nor mutually exclusive.

Are these results affected if, due to some genetic characteristic, the probability that the wife gives birth to a boy is $\frac{3}{5}$?

+ **48** A hand of 4 cards is dealt at random from a pack of playing cards. Find the probability that the hand contains:

(i) 4 cards of the same suit
(ii) 2 cards of one suit and 2 of another
(iii) 4 cards of different suits
(iv) 4 cards of the same denomination (e.g. 4 aces)
(v) 2 cards of one denomination and 2 of another
(vi) 4 cards of different denominations

49 A and B are events. $p(A) = \frac{1}{2}$, $p(B) = \frac{1}{4}$, $p((A \cap B') \cup (A' \cap B)) = \frac{1}{3}$. Find $p(A \cap B)$.

Are A and B mutually exclusive, independent, or neither?

Find also $p(A' \cap B)$. Show that $p(A|B) = \frac{5}{6}$ and find $p(B|A')$.

50 A and B are events. $p(A) = 0.3$, $p(B) = 0.5$, $p(A \cup B) = 0.6$. Show that A and B are not independent. Find:

(i) $p(A|B)$ (ii) $p(A|B')$ (iii) $p(A'|B')$

51 Two cards are drawn, without replacement, from a pack of playing cards.

(i) Find the probability that
(a) both cards are aces;
(b) exactly one card is an ace.

(ii) Given that at least one ace is drawn, find the probability that the two cards are of different suits.

(iii) Given that the two cards are of different suits, find the probability that at least one ace is drawn.

52 (i) The probability that the Russians will start a nuclear war in the next twenty years is $\frac{1}{6}$. The probability that the Americans will start a nuclear war in the next twenty years is $\frac{1}{4}$. The events are mutually exclusive. The probability that the human race is wiped out, given that there is a nuclear war, is $\frac{9}{10}$. Find the probability that the human race still exists in twenty years time.

(ii) The method the President of the United States uses to decide whether or not to press the button is as follows: he has a bag containing 10 sweets, of which 9 are red and the other is green; he draws the sweets in turn at random from the bag, without replacement. If the last sweet chosen is green, he presses the button. Find the probability that he presses the button if he performs the process:
(a) once; (b) ten times.

53 A die is biased in such a way that the probability of obtaining a score of k is proportional to k. What is the probability of obtaining a six?

This die is thrown six times. What is the probability of obtaining:

(i) exactly four sixes?
(ii) at least four sixes?
(iii) the first six on the final throw?
(iv) the fourth six on the final throw?
(v) a different score on each throw?

54 A wife promises that she will not leave her husband unless he has an affair with another woman, but that if he does have an affair, she will walk out and take him to the cleaners over the alimony. There are three women with whom he might have an affair: his secretary A, his wife's sister B and his wife's best friend C. The probabilities of him having an affair with A, B and C are $\frac{3}{8}, \frac{1}{8}$ and $\frac{1}{8}$ respectively. If he has an affair with A, B or C, the probability of his wife finding out is $\frac{1}{2}, \frac{2}{3}$ and 1 respectively. What is the probability that his wife leaves him? Find the probability that he has had an affair with his secretary, given that he is on the phone to his best friend and he is saying:

(i) 'The wife has just left me. I never thought she'd do it'.

(ii) 'Incredible! The missus still hasn't found out'.

(iii) 'I've decided to walk out on the wife'.

55 Three cards are drawn, without replacement, from a pack of playing cards. Find the probability of the events $A = \{$all three cards are aces$\}$; $B = \{$none of the cards is a heart$\}$; $C = \{$at least one of the cards is a spade$\}$. Find:

(i) $p(A|B)$
(ii) $p(B|C)$
(iii) $p(C|A)$

56 Saturday night. Three self-confident teenaged boys A, B and C are in a nightclub. They all fancy the same girl and decide to take turns in trying to ask her to dance. A is spotty and shy, while B and C are handsome and charismatic; so A is given the opportunity to start. If A fails, B tries; if B fails, C tries; if C fails, A tries again and so on. The probabilities of the girl saying yes to any one attempt are $\frac{1}{3}, \frac{2}{3}, \frac{2}{3}$ for A, B, C respectively.

(i) Find the probability for each boy that he is the successful one.

(ii) The outrageously handsome and irresistible D, whose probability of success when asking any girl to dance is 1, decides he will wait until after A, B and C have each had two attempts before asking her to dance. What is the probability that the girl dances with D?

+ **57** If $|x|<1$, find $\sum_{r=0}^{\infty} x^r$. Hence find:

(i) $\sum_{r=1}^{\infty} rx^{r-1}$

(ii) $\sum_{r=2}^{\infty} r(r-1)x^{r-2}$

In a chess match between two grandmasters P and Q, the winner is the person who first wins a total of two games. The probabilities that P wins, draws or loses any particular game are $\frac{1}{3}, \frac{1}{2}, \frac{1}{6}$. The games are independent. Show that the probability that P wins the match after r games ($r \geq 1$) is

$$\tfrac{1}{54}\{6r(\tfrac{1}{2})^{r-1} + r(r-1)(\tfrac{1}{2})^{r-2}\}$$

Hence show that the probability that P wins the match is $\frac{20}{27}$. What is the probability that Q wins the match? Explain your answer.

6 Vectors, coordinate geometry and complex numbers

6.1 Vectors

6.1:1 Displacements

Consider the *xy* plane; think of the *x*-axis as being due east, the *y*-axis due north.

Worked example 1 Suppose that every point in the plane is shifted 60 m NE (see fig. 6.1:1). This shift is called a **displacement**.

Let us denote this particular displacement by **u**; so **u** is the displacement that shifts every point on the plane 60 m NE.

Suppose that **u** shifts A to B, A' to B'. We write the displacement that shifts A to B as $\overrightarrow{AB}$, the displacement that shifts A' to B' as $\overrightarrow{A'B'}$.

$\overrightarrow{AB}$, $\overrightarrow{A'B'}$ and **u** have the same magnitude, 60 m, and the same direction, NE. We write

$$\overrightarrow{AB} = \overrightarrow{A'B'} = \mathbf{u} \qquad (1)$$

A displacement is specified completely by its magnitude and direction.

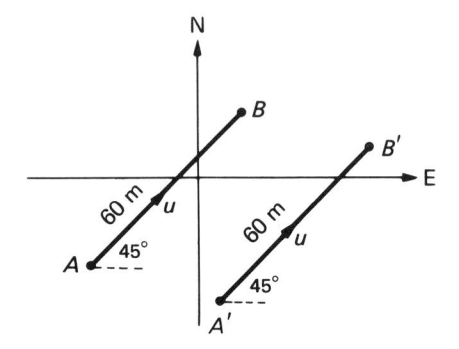

Fig. 6.1:1

Worked example 2 Suppose that every point on the plane is shifted by

                    60 m NE: displacement **u**

and then   40 m due E: displacement **v**

and suppose that **u** shifts A to B, and **v** shifts B to C (see fig. 6.1:2). Then **u** together with **v** shifts A to C. We write

$$\overrightarrow{AB} + \overrightarrow{BC} = \overrightarrow{AC} \qquad (2)$$

This equation is called **the triangle law of addition**. So displacements obey the triangle law of addition.

Note: the symbol '+' used in equation (2) means 'together with'. Why do we write it as '+'? Well, as we shall see later, it has many characteristics in common with + as applied to real numbers; in fact, addition of displacements is a generalization of addition of real numbers.

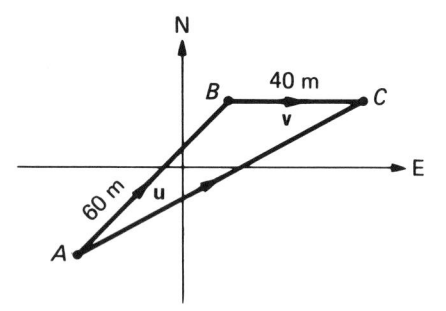

Fig. 6.1:2

6.1:2 Vectors

> **Definition** A **vector** is a quantity which has magnitude and direction, and obeys the triangle law of addition.

Vectors, coordinate geometry and complex numbers 281

A displacement is a vector. We will eventually meet other kinds of vectors, e.g. velocity, acceleration and force, but displacement vectors are the only kind we will consider now.

Note: a vector with a specific location, such as the displacement vector $\overrightarrow{AB}$ in worked example 1, is called a **tied** or **localized vector**.

A vector with no specific location, such as the displacement vector **u** in worked example 1, is called a **free vector**.

Representation of vectors

We represent all vectors by displacement vectors as in fig. 6.1:3. The length of the line segment represents the magnitude of the vector. The direction of the line segment (taking into account the arrow) represents the direction of the vector.

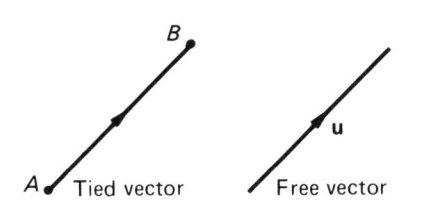

Fig. 6.1:3

Magnitude of a vector

The magnitude of the vector **u** is written as |**u**| or u. Note that the magnitude of the displacement vector $\overrightarrow{AB}$ is the *distance* between A and B, i.e.

$$|\overrightarrow{AB}| = AB$$

A vector of magnitude 1 is called a **unit vector**.

Scalars

> **Definition** A quantity which is specified completely by its magnitude only is called a **scalar**, e.g. time, mass, distance, (real) number.

Equal vectors

Fig. 6.1:4

> **Definition** Two vectors are **equal** if they have the same magnitude and the same direction (in fig. 6.1:4, $\overrightarrow{AB} = \overrightarrow{CD} = \mathbf{u}$, but $\mathbf{u} \neq \mathbf{v}$ although $|\mathbf{u}| = |\mathbf{v}|$).

Multiplication of a vector by a scalar

> **Definition** If λ is a positive number, then
> (i) $\lambda\mathbf{u}$ is the vector of magnitude $\lambda|\mathbf{u}|$ in the same direction as **u**;
> (ii) $-\lambda\mathbf{u}$ is the vector of magnitude $\lambda|\mathbf{u}|$ in the opposite direction to **u**.
> (See fig. 6.1:5.)

Fig. 6.1:5

Addition of vectors

> **Definition** If **u** is represented by $\overrightarrow{AB}$, **v** by $\overrightarrow{BC}$ (fig. 6.1:6), then $\mathbf{w} = \mathbf{u} + \mathbf{v}$ is represented by $\overrightarrow{AC}$. (3)

w is called the **resultant** of **u** and **v**. This rule for adding vectors is called the **triangle law of addition**.

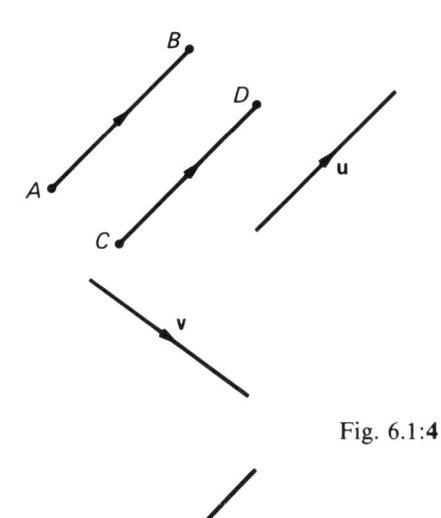

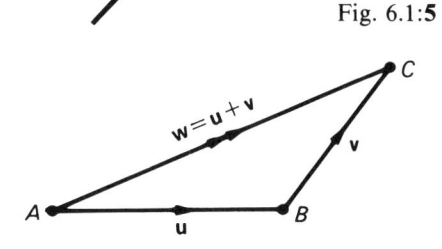

Fig. 6.1:6

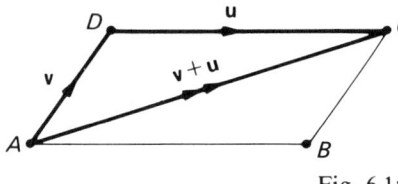

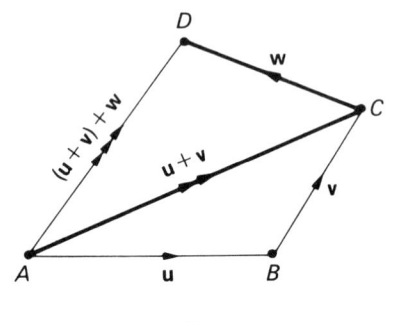

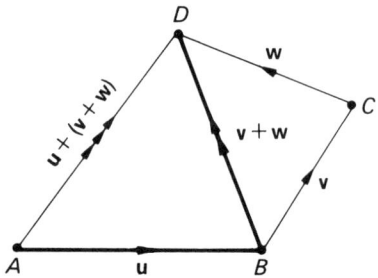

(i) Notice that, if $ABCD$ is a parallelogram, **u** is also represented by $\overrightarrow{DC}$, **v** by $\overrightarrow{AD}$ (fig. 6.1:7).

$$\mathbf{u}+\mathbf{v} = \overrightarrow{AB}+\overrightarrow{BC} = \overrightarrow{AC}$$
$$= \overrightarrow{AD}+\overrightarrow{DC} = \mathbf{v}+\mathbf{u}$$

i.e. $\mathbf{u}+\mathbf{v}=\mathbf{v}+\mathbf{u}$, so vector addition is **commutative**.

(ii) If **u**, **v** and **w** are three vectors, represented by $\overrightarrow{AB}$, $\overrightarrow{BC}$ and $\overrightarrow{CD}$ (fig. 6.1:**8**), then

$$(\mathbf{u}+\mathbf{v})+\mathbf{w} = \overrightarrow{AC}+\overrightarrow{CD} = \overrightarrow{AD} = \overrightarrow{AB}+\overrightarrow{BD} = \mathbf{u}+(\mathbf{v}+\mathbf{w})$$

i.e. $$(\mathbf{u}+\mathbf{v})+\mathbf{w} = \mathbf{u}+(\mathbf{v}+\mathbf{w})$$

so vector addition is **associative**, and we can write $\mathbf{u}+\mathbf{v}+\mathbf{w}$ unambiguously.

The expression $\mathbf{u}+\mathbf{v}+\mathbf{w}$ is meaningful even if **u**, **v** and **w** are not coplanar (i.e. even if **u**, **v** and **w** do not lie in the same plane—see questions 15 and 31).

In fact all our definitions and results about vectors are as meaningful in 3D (and for that matter 4D and 5D...) as in 2D. Thus we shall apply the results of this section when we consider planar coordinate geometry in sections 6.2 and 6.3 *and* when we consider 3D coordinate geometry in section 6.4.

(iii) Note also that if **u** and **v** are vectors, λ and μ scalars, then

$$(\lambda+\mu)\mathbf{u} = \lambda\mathbf{u}+\mu\mathbf{u}$$

and

$$\lambda(\mathbf{u}+\mathbf{v}) = \lambda\mathbf{u}+\lambda\mathbf{v}$$

(Note: the operation '+' for vectors is not the same as the operation '+' for real numbers (see note on p. 280); so some people prefer to write the vector addition symbol as $\oplus$; however, the operation has, as we shall see, so many characteristics in common with real number addition that we shall continue to write it as '+'.)

Subtraction of vectors

Definition $\mathbf{u}-\mathbf{v}$ means $\mathbf{u}+(-\mathbf{v})$

If **u** is represented by $\overrightarrow{AB}$, **v** by $\overrightarrow{BC}$ and $ABCD$ is a parallelogram, as in fig. 6.1:9, then **u** can be represented by $\overrightarrow{DC}$, $-\mathbf{v}$ by $\overrightarrow{CB}$, and so, by definition (3)

$$\mathbf{u}-\mathbf{v} \text{ is represented by } \overrightarrow{DB}$$

Zero vector The zero vector **0** is the vector with zero magnitude. Note that for any vector **u**,

(i) $0\mathbf{u}=\mathbf{0}$ (ii) $\mathbf{u}-\mathbf{u}=\mathbf{0}$

See question 16.

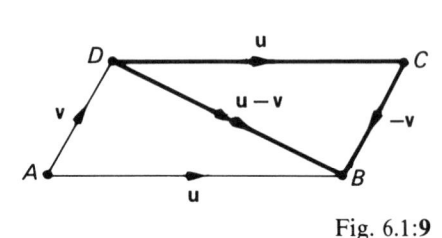

Vectors, coordinate geometry and complex numbers 283

Worked example 3 *ABCDEF* is a regular hexagon, centre *O* (fig. 6.1:10). Find single displacement vectors equivalent to
(i) $\overrightarrow{CB} + \overrightarrow{CD}$
(ii) $\overrightarrow{AC} + \overrightarrow{CD} + \overrightarrow{DE}$
(iii) $2\overrightarrow{AO} + \overrightarrow{OB}$
(iv) $\overrightarrow{AB} - \overrightarrow{AF}$
(v) $\overrightarrow{AB} - \overrightarrow{BC}$

(i) $\overrightarrow{CD} = \overrightarrow{BO}$ so $\overrightarrow{CB} + \overrightarrow{CD} = \overrightarrow{CB} + \overrightarrow{BO} = \overrightarrow{CO}$
(ii) $\overrightarrow{AC} + \overrightarrow{CD} + \overrightarrow{DE} = \overrightarrow{AD} + \overrightarrow{DE} = \overrightarrow{AE}$
(iii) $2\overrightarrow{AO} = \overrightarrow{AD}$ so $2\overrightarrow{AO} + \overrightarrow{OB} = \overrightarrow{AD} + \overrightarrow{OB} = \overrightarrow{AD} + \overrightarrow{DC} = \overrightarrow{AC}$
(iv) $-\overrightarrow{AF} = \overrightarrow{FA}$ so $\overrightarrow{AB} - \overrightarrow{AF} = \overrightarrow{AB} + \overrightarrow{FA} = \overrightarrow{FA} + \overrightarrow{AB} = \overrightarrow{FB}$
(v) $\overrightarrow{AB} - \overrightarrow{BC} = \overrightarrow{OC} - \overrightarrow{BC} = \overrightarrow{OC} + \overrightarrow{CB} = \overrightarrow{OB}$

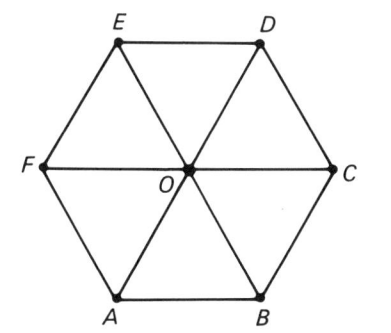

Fig. 6.1:10

Worked example 4 **u** and **v** are vectors of magnitude 3 and 4. Find the magnitude and direction of **u** + **v** if the angle between **u** and **v** is:
(i) 90° (ii) 60°

(i) If **u** is represented by $\overrightarrow{AB}$, **v** by $\overrightarrow{BC}$, **u** + **v** is represented by $\overrightarrow{AC}$ (fig. 6.1:11).

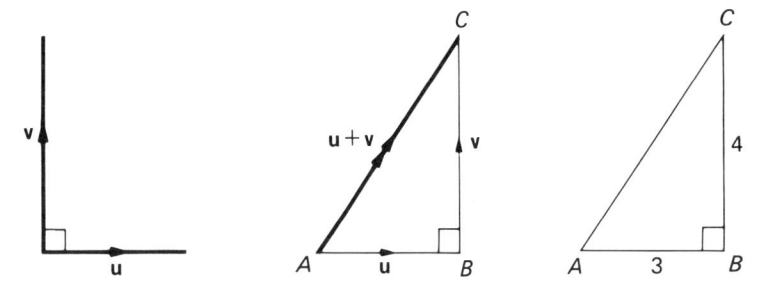

Fig. 6.1:11

$AB = 3$, $BC = 4$, $A\hat{B}C = 90°$, so $AC = \sqrt{(3^2 + 4^2)} = 5$, and $\tan A = \frac{4}{3}$ so $A \approx 53.13°$.
Magnitude of **u** + **v**: $|\mathbf{u} + \mathbf{v}| = AC = 5$.
Direction of **u** + **v**: **u** + **v** makes an angle of 53.13° with **u**.

(ii) $AB = 3$, $BC = 4$, $A\hat{B}C = 120°$ (fig. 6.1:12).

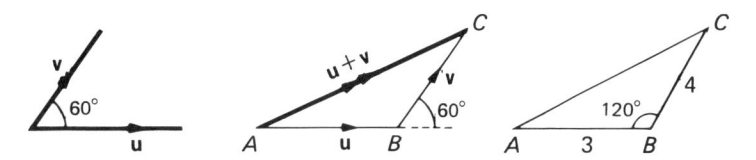

Fig. 6.1:12

Using the cosine formula (3.3:1 equation (2))

$$AC^2 = 3^2 + 4^2 - 2 \cdot 3 \cdot 4 \cos 120 = 37$$
$$AC = \sqrt{37}$$

Using the sine formula (3.3:1 equation (1))

$$\frac{4}{\sin A} = \frac{\sqrt{37}}{\sin 120°}$$
$$\sin A = 0.569 \Rightarrow A = 34.72°$$

Magnitude of **u** + **v**: $|\mathbf{u} + \mathbf{v}| = AC = \sqrt{37}$.
Direction of **u** + **v**: **u** + **v** makes an angle of 34.72° with **u**.

284   Vectors

Scalar product of two vectors

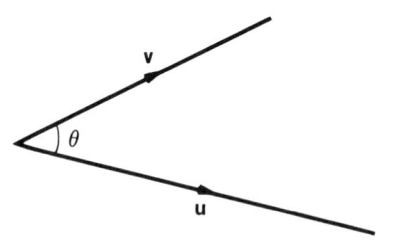

Definition The **scalar** or **dot** product **u . v** of two vectors **u** and **v** is defined as

$$\mathbf{u} \cdot \mathbf{v} = uv \cos \theta \qquad (4)$$

where θ is the angle between **u** and **v** (fig. 6.1:**13**).

Fig. 6.1:**13**

Note that **u . v** is a *scalar*.

The scalar product is:

(i) *commutative*:

$$\mathbf{v} \cdot \mathbf{u} = vu \cos \theta = uv \cos \theta = \mathbf{u} \cdot \mathbf{v}$$

(ii) *associative*:

$$(\lambda \mathbf{u}) \cdot \mathbf{v} = (\lambda u)(v) \cos \theta \qquad \text{since } \lambda \mathbf{u} \text{ is parallel to } \mathbf{u}$$
$$= \lambda uv \cos \theta$$
$$= \lambda (\mathbf{u} \cdot \mathbf{v})$$

(iii) *distributive over addition*:

from fig. 6.1:**14**,

$$\mathbf{u} \cdot (\mathbf{v} + \mathbf{w}) = u|\mathbf{v} + \mathbf{w}| \cos \psi$$
$$= u \cdot ON$$
$$= u(OM + MN)$$
$$= u(v \cos \theta + w \cos \phi)$$
$$= uv \cos \theta + uw \cos \phi$$
$$= \mathbf{u} \cdot \mathbf{v} + \mathbf{u} \cdot \mathbf{w}$$

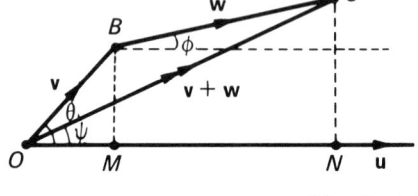

Fig. 6.1:**14**

Theorem 1: If **u** and **v** are non-zero vectors, then

$$\mathbf{u} \text{ perpendicular to } \mathbf{v} \Leftrightarrow \mathbf{u} \cdot \mathbf{v} = 0 \qquad (5)$$

Proof
Suppose that the angle between **u** and **v** is θ.

$$\mathbf{u} \text{ is perpendicular to } \mathbf{v} \Leftrightarrow \theta = 90° \text{ (or 270°)}$$
$$\Leftrightarrow \cos \theta = 0$$
$$\Leftrightarrow \mathbf{u} \cdot \mathbf{v} = 0$$

Exercise 6.1:2

- **1** *ABCD* is a rhombus. Which of the following statements are true?

  (i) $\overrightarrow{AB} = \overrightarrow{DC}$
  (ii) $\overrightarrow{AB} = \overrightarrow{CD}$
  (iii) $\overrightarrow{AB} = \overrightarrow{AD}$
  (iv) $\overrightarrow{AC} = \overrightarrow{BD}$
  (v) $|\overrightarrow{AB}| = |\overrightarrow{DC}|$
  (vi) $|\overrightarrow{AB}| = |\overrightarrow{BC}|$

- → **2** If **a** is a vector, what can you say about the magnitude and direction of the vector **â** defined by

  $$\hat{\mathbf{a}} = \frac{1}{a} \mathbf{a}?$$

3 ABC is a straight line such that $BC = 2AB$. If the vector **a** is represented by $\overrightarrow{AB}$, find the vectors represented by:

(i) $\overrightarrow{BA}$ (ii) $\overrightarrow{AC}$
(iii) $\overrightarrow{CB}$ (iv) $\overrightarrow{CA}$

4 **u** is a vector of magnitude 4 in the direction NE. Find the magnitude and direction of:

(i) $3\mathbf{u}$ (ii) $\tfrac{1}{2}\mathbf{u}$
(iii) $-\mathbf{u}$ (iv) $-\tfrac{3}{2}\mathbf{u}$

5 (i) P, Q and R are three points. The vectors **u** and **v** are represented by $\overrightarrow{PQ}$ and $\overrightarrow{PR}$ respectively. Describe with the aid of a diagram how to represent the vector $\mathbf{u} + \mathbf{v}$ by a single displacement vector.

(ii) J, K, L and M are four points. The vectors **u** and **v** are represented by $\overrightarrow{JK}$ and $\overrightarrow{LM}$ respectively. Show with the aid of a diagram how to represent the vector $\mathbf{u} + \mathbf{v}$ by a single displacement vector.

6 A, B, C and D are four points. Simplify:

(i) $\overrightarrow{AB} + \overrightarrow{BC}$
(ii) $\overrightarrow{AD} + \overrightarrow{DB}$
(iii) $\overrightarrow{AB} + \overrightarrow{BC} + \overrightarrow{CD}$
(iv) $\overrightarrow{BD} - \overrightarrow{CD}$
(v) $\overrightarrow{AB} - \overrightarrow{DB} + \overrightarrow{DC}$

7 A, B, C and D are four points. Simplify:

(i) $\overrightarrow{AC} + \overrightarrow{CB}$
(ii) $\overrightarrow{AC} + \overrightarrow{CB} + \overrightarrow{BD}$
(iii) $\overrightarrow{AC} - \overrightarrow{DC}$
(iv) $\overrightarrow{CB} + \overrightarrow{BD} - \overrightarrow{AD}$

8 A, B, C, D and E are five points. Simplify:

(i) $\overrightarrow{AB} + \overrightarrow{BC} + \overrightarrow{CE} + \overrightarrow{ED}$
(ii) $\overrightarrow{AB} + \overrightarrow{BC} - \overrightarrow{DC} - \overrightarrow{ED}$
(iii) $\overrightarrow{AB} + \overrightarrow{AE} - \overrightarrow{DB} - \overrightarrow{DE}$
(iv) $2\overrightarrow{AB} - 2\overrightarrow{CB} - \overrightarrow{CD} - \overrightarrow{DE} + \overrightarrow{AE}$

Questions 9 and 10: $ABCD$ is a parallelogram. O is the point of intersection of its diagonals.

9 Simplify:

(i) $2\overrightarrow{AB} + \overrightarrow{BD}$
(ii) $\tfrac{1}{2}\overrightarrow{AC} + \tfrac{1}{2}\overrightarrow{DB}$
(iii) $\overrightarrow{AC} + \overrightarrow{AO} - \overrightarrow{CO}$
(iv) $\overrightarrow{CO} - \overrightarrow{OB}$
(v) $2\overrightarrow{AC} - \overrightarrow{BC} - 2\overrightarrow{AB}$
(vi) $\overrightarrow{AB} + \overrightarrow{AD} + \overrightarrow{BC} + \overrightarrow{DC}$

R 10 If $\overrightarrow{AB} = \mathbf{u}$ and $\overrightarrow{AD} = \mathbf{v}$, find in terms of **u** and **v**:

(i) $\overrightarrow{AC}$ (ii) $\overrightarrow{OC}$ (iii) $\overrightarrow{OD}$ (iv) $\overrightarrow{OB}$

11 Would the results of questions 9 and 10 be the same if $ABCD$ were a trapezium whose diagonals met at O?

Questions 12-14: $ABCDEF$ is a regular hexagon, centre O.

12 Simplify:

(i) $\overrightarrow{FA} + \overrightarrow{FE}$
(ii) $\overrightarrow{AO} + \overrightarrow{FO}$
(iii) $\overrightarrow{FA} + 2\overrightarrow{BC} + \overrightarrow{DE}$
(iv) $2\overrightarrow{BO} - \overrightarrow{CO}$
(v) $\overrightarrow{CD} - \overrightarrow{AB}$
(vi) $\overrightarrow{ED} - \overrightarrow{DC}$
(vii) $\tfrac{1}{2}\overrightarrow{AD} - \overrightarrow{OF}$

13 Let $\overrightarrow{AB} = \mathbf{u}$, $\overrightarrow{BC} = \mathbf{v}$. Write in terms of **u** and **v**:

(i) $\overrightarrow{FD}$ (ii) $\overrightarrow{CD}$ (iii) $\overrightarrow{OF}$ (iv) $\overrightarrow{CE}$

R 14 Let $\overrightarrow{OA} = \mathbf{a}$, $\overrightarrow{OB} = \mathbf{b}$. Find in terms of **a** and **b**:

(i) $\overrightarrow{AB}$ (ii) $\overrightarrow{CD}$ (iii) $\overrightarrow{CE}$ (iv) $\overrightarrow{FD}$ (v) $\overrightarrow{CF}$

15 $ABCDEFGH$ is a rectangular box. $\overrightarrow{AB} = \mathbf{u}$, $\overrightarrow{AD} = \mathbf{v}$, $\overrightarrow{AE} = \mathbf{w}$. Find in terms of **u**, **v** and **w** the vectors:

(i) $\overrightarrow{AC}$ (ii) $\overrightarrow{AG}$ (iii) $\overrightarrow{FH}$ (iv) $\overrightarrow{HB}$ (v) $\overrightarrow{GE}$

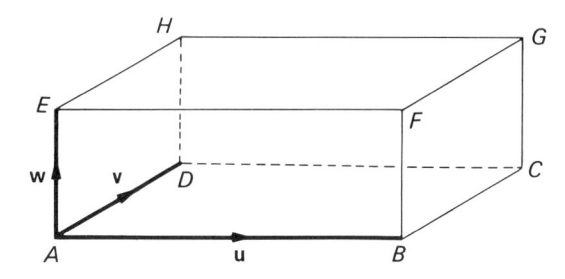

16 $ABCD$ is a quadrilateral. Simplify:

(i) $\overrightarrow{AB} + \overrightarrow{BA}$
(ii) $\overrightarrow{AB} + \overrightarrow{BC} + \overrightarrow{CA}$
(iii) $\overrightarrow{AB} + \overrightarrow{BC} + \overrightarrow{CD} + \overrightarrow{DA}$

17 The vectors **u** and **v** are perpendicular. Find the magnitude of $\mathbf{u} + \mathbf{v}$ and the angle between $\mathbf{u} + \mathbf{v}$ and **u** when:

(i) $|\mathbf{u}| = 1$, $|\mathbf{v}| = 1$
(ii) $|\mathbf{u}| = 2\sqrt{3}$, $|\mathbf{v}| = 2$
(iii) $|\mathbf{u}| = 5$, $|\mathbf{v}| = 12$

18 Find the magnitude and direction of the resultant of the displacement vectors $\mathbf{u} = 2$ m due north and $\mathbf{v} = 4$ m due west.

19 The vectors $\mathbf{u} + \mathbf{v}$ and **v** are perpendicular; $|\mathbf{u}| = 25$, $|\mathbf{u} + \mathbf{v}| = 7$. Find $|\mathbf{v}|$.

20 The vectors **u** and **v** are inclined at $60°$ to one another. Find the magnitude of $\mathbf{u} + \mathbf{v}$ and the angle between $\mathbf{u} + \mathbf{v}$ and **u** when:

(i) $|\mathbf{u}| = 1$, $|\mathbf{v}| = 1$
(ii) $|\mathbf{u}| = 5$, $|\mathbf{v}| = 8$

21 Find the magnitude and direction of the resultant of the displacement vectors **u** and **v** if:

(i) $\mathbf{u} = 2$ m due east, $\mathbf{v} = 6$ m N30°E
(ii) $\mathbf{u} = 2$ m due east, $\mathbf{v} = 6$ m N30°W

R 22 The vectors **u** and **v** have magnitudes 2 and 3 respectively. Find the magnitude and direction of **u**+**v** if **u** and **v** are inclined at:

(i) 60° to each other
(ii) 120° to each other

23 The vectors **u** and **v** have magnitudes 3 and 5 respectively. Find the magnitude and direction of **u**+**v** if **u** and **v** are inclined at:

(i) 0° (ii) 40° (iii) 150° (iv) 180°

to each other

24 **u** and **v** are vectors; $|\mathbf{u}|=2$, $|\mathbf{v}|=3$, $|\mathbf{u}+\mathbf{v}|=4$. Find the angle between **u** and **v**.

25 **u** and **v** are vectors; $|\mathbf{u}|=5$, $|\mathbf{v}|=7$, $|\mathbf{u}+\mathbf{v}|=3$. Find the angle between **u** and **v**.

26 Show geometrically that, for any vectors **u** and **v**,

$$|\mathbf{u}+\mathbf{v}| \leq |\mathbf{u}|+|\mathbf{v}|$$

When does the equality hold?

27 The vectors **u** and **v** are perpendicular. If $|\mathbf{u}|=7$ and $|\mathbf{v}|=24$, find $|\mathbf{u}+\mathbf{v}|$ and $|\mathbf{u}-\mathbf{v}|$.

R 28 The vectors **u** and **v** are inclined at 60° to one another. If $|\mathbf{u}|=4$ and $|\mathbf{v}|=5$, find $|\mathbf{u}+\mathbf{v}|$ and $|\mathbf{u}-\mathbf{v}|$.

29 Find the angle between **u** and **v** if $|\mathbf{u}|=4$, $|\mathbf{v}|=5$ and $|\mathbf{u}-\mathbf{v}|=8$.

30 Find the magnitude and direction of the resultant of **u**, **v** and **w** when:

(i) **u** = 3 m due east, **v** = 2 m due north
    w = 5 m due west
(ii) **u** = 4 m due east, **v** = 4 m due north
    w = 4√2 m north-west
(iii) **u** = 4 m due east, **v** = 6 m N30°E
    w = 5 m south-east

31 Find the magnitude of the resultant of **u**, **v** and **w** when:

(i) **u** = 3 m due east, **v** = 2 m due north
    w = 5 m vertically upwards
(ii) **u** = 2 m due north, **v** = 5 m N60°E
    w = 2 m vertically upwards

32 **u** and **v** are unit vectors in directions due east and due north respectively. Find the magnitude and direction of:

(i) 3**u**+2**v**
→ (ii) 3**u**−2**v**
(iii) 2**u**+2**v**
(iv) 2(**u**+**v**)

33 If **p** and **q** are unit vectors in directions due south and S30°W, find the magnitude and direction of:

(i) 3**p**+**q**
(ii) 6**p**+2**q**
(iii) −3**p**−**q**

34 Show by drawing diagrams that, if **u** and **v** are any vectors and λ and μ are any scalars, then:

(i) $(\lambda+\mu)\mathbf{u} = \lambda\mathbf{u}+\mu\mathbf{u}$
(ii) $\lambda(\mathbf{u}+\mathbf{v}) = \lambda\mathbf{u}+\lambda\mathbf{v}$

35 **u** and **v** are vectors; $|\mathbf{u}|=2$, $|\mathbf{v}|=3$. Find **u** . **v** if the angle between **u** and **v** is:

(i) 0°   (ii) 30°   (iii) 60°   (iv) 90°
(v) 120°   (vi) 150°   (vii) 180°

36 **u**, **v** and **w** are vectors of magnitudes 2, 5 and 4 respectively. The angles between **u** and **v** and between **v** and **w** are both 30°.

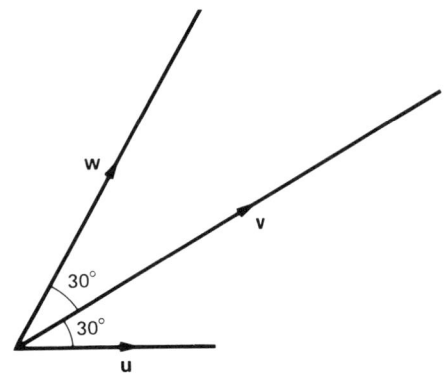

Find:

(i) **u** . **v**
(ii) **u** . **w**
(iii) **v** . **w**
(iv) (**u**+**v**) . **w**
(v) (3**u**−**v**) . **w**
(vi) (**u**+**w**) . **v**
(vii) (2**u**−**w**) . **v**

37 $ABCD$ is a square, centre O, of side a. Find, in terms of a:

(i) $\overrightarrow{AB} . \overrightarrow{AD}$
(ii) $\overrightarrow{AB} . \overrightarrow{AC}$
(iii) $\overrightarrow{OA} . \overrightarrow{OB}$
(iv) $\overrightarrow{OA} . \overrightarrow{AB}$

38 $ABCDEF$ is a regular hexagon, centre O, of side a. Find, in terms of a:

(i) $\overrightarrow{OA} . \overrightarrow{OB}$
(ii) $\overrightarrow{OA} . \overrightarrow{OC}$
(iii) $\overrightarrow{BC} . \overrightarrow{BE}$
(iv) $\overrightarrow{BC} . \overrightarrow{BF}$
(v) $\overrightarrow{AB} . \overrightarrow{AC}$

39 In $\triangle ABC$, $AB=AC=a$ and $\hat{A}=120°$. Find, in terms of a, $\overrightarrow{AB} . \overrightarrow{AC}$ and $\overrightarrow{AB} . \overrightarrow{CB}$. Verify that $\overrightarrow{AB} . \overrightarrow{AC} + \overrightarrow{AB} . \overrightarrow{CB} = a^2$ and explain this result.

6.1:3 Use of vectors for geometry

We have based our definition of a vector and our development of vector arithmetic on geometrical ideas (e.g. triangle law of addition). Now we are going to reverse the process and use vectors as a tool for simplifying geometrical proofs. One of the advantages of using vectors is that problems in three (or for that matter four) dimensions are no more difficult to solve than problems in two dimensions.

When we are considering the location of points in space, it is useful to refer to their positions relative to a fixed origin O.

> **Definition** If P is a point in space and O is the origin, the displacement vector $\overrightarrow{OP}$ is called the **position vector** of P.

Note: the position vector of a point is a tied vector.

The position vector $\overrightarrow{OA}$ of the fixed point A is often written as $\mathbf{a}$, $\overrightarrow{OB}$ as $\mathbf{b}$, etc. The position vector of the variable point P is written as $\mathbf{r}$.

Line segments

If A and B are two points with position vectors $\mathbf{a}$ and $\mathbf{b}$ (fig. 6.1:**15**) then the displacement vector $\overrightarrow{AB} = \overrightarrow{AO} + \overrightarrow{OB} = -\mathbf{a} + \mathbf{b}$, i.e.

$$\overrightarrow{AB} = \mathbf{b} - \mathbf{a} \qquad (6)$$

and

$$\overrightarrow{BA} = -\overrightarrow{AB} = \mathbf{a} - \mathbf{b}$$

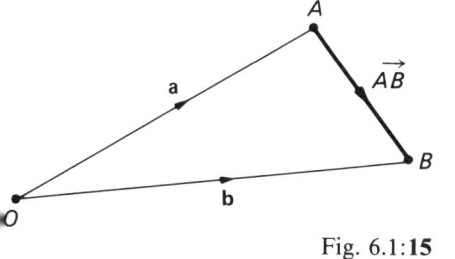

Fig. 6.1:**15**

Division of line segments

> **Theorem 2:** If the point P divides the line segment AB in the ratio $m:n$ (i.e. $AP:PB = m:n$), then the position vector $\mathbf{r}$ of P is given by
>
> $$\mathbf{r} = \frac{n\mathbf{a} + m\mathbf{b}}{m+n} \qquad (7)$$

Proof
From fig. 6.1:**16**,

$$\mathbf{r} = \overrightarrow{OP} = \overrightarrow{OA} + \overrightarrow{AP}$$

$$= \mathbf{a} + \frac{mk}{(m+n)k}\overrightarrow{AB}$$

$$= \mathbf{a} + \frac{m}{m+n}(\mathbf{b} - \mathbf{a})$$

$$= \frac{n\mathbf{a} + m\mathbf{b}}{m+n}$$

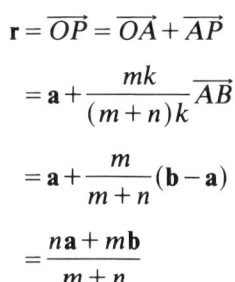

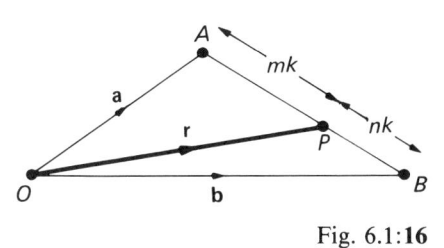

Fig. 6.1:**16**

> **Corollary:** The position vector $\mathbf{m}$ of the midpoint M of the line segment AB is given by
>
> $$\mathbf{m} = \tfrac{1}{2}(\mathbf{a} + \mathbf{b}) \qquad (8)$$

288   *Use of vectors for geometry*

Proof
Put $m = n = 1$ in equation (7).

Worked example 5 *A, B, C* and *D* are four points. If *P, Q, R* and *S* are the midpoints of *AB, BC, CD* and *DA*, show that *PQRS* is a parallelogram.

By equation (8) and fig. 6.1:17, $\mathbf{p} = \frac{1}{2}(\mathbf{a}+\mathbf{b})$, $\mathbf{q} = \frac{1}{2}(\mathbf{b}+\mathbf{c})$; so

$$\overrightarrow{PQ} = \mathbf{q} - \mathbf{p} = \frac{1}{2}(\mathbf{b}+\mathbf{c}) - \frac{1}{2}(\mathbf{a}+\mathbf{b}) = \frac{1}{2}(\mathbf{c}-\mathbf{a})$$

Similarly, $\mathbf{r} = \frac{1}{2}(\mathbf{c}+\mathbf{d})$, $\mathbf{s} = \frac{1}{2}(\mathbf{d}+\mathbf{a})$; so

$$\overrightarrow{SR} = \mathbf{r} - \mathbf{s} = \frac{1}{2}(\mathbf{c}+\mathbf{d}) - \frac{1}{2}(\mathbf{d}+\mathbf{a}) = \frac{1}{2}(\mathbf{c}-\mathbf{a})$$

i.e. $\overrightarrow{PQ} = \overrightarrow{SR}\ (= \frac{1}{2}\overrightarrow{AC})$

So the line segments *PQ* and *SR* are parallel and of the same length; thus *PQRS* is a parallelogram.

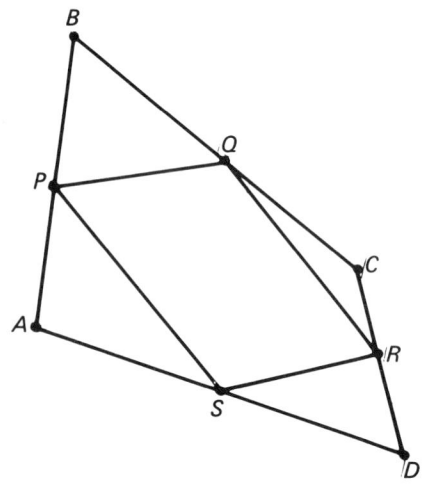

Fig. 6.1:17

A **median** of a triangle is a line joining a vertex to the midpoint of the side opposite the vertex.

Theorem 3: The medians of a triangle meet at a point called the centroid of the triangle. If the position vectors of the vertices *A, B* and *C* of the triangle are $\mathbf{a}$, $\mathbf{b}$ and $\mathbf{c}$, the position vector $\mathbf{g}$ of the centroid is

$$\mathbf{g} = \tfrac{1}{3}(\mathbf{a}+\mathbf{b}+\mathbf{c}) \qquad (9)$$

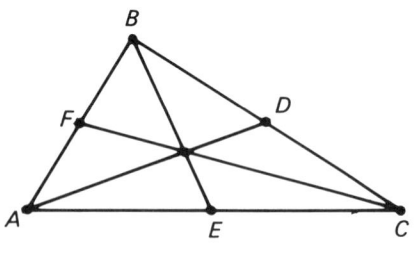

Fig. 6.1:18a

Proof
Suppose that the medians of $\triangle ABC$ are *AD*, *BE* and *CF* (fig. 6.1:18a). *D* is the midpoint of *BC*, so $\mathbf{d} = \frac{1}{2}(\mathbf{b}+\mathbf{c})$.

Consider the point *G* which divides *AD* in the ratio 2:1 (fig. 6.1:18b). By Theorem 2, its position vector $\mathbf{g} = \frac{1}{3}(2\mathbf{d}+\mathbf{a}) = \frac{1}{3}(\mathbf{a}+\mathbf{b}+\mathbf{c})$.

We can show similarly that *G* lies on *BE* and *CF* (and, in fact, divides them both in the ratio 2:1). So the medians intersect at *G*.

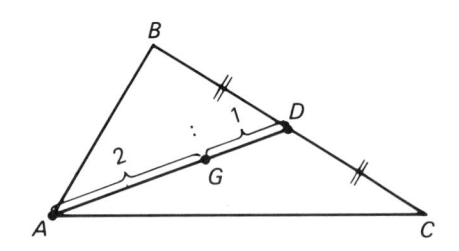

Fig. 6.1:18b

Vector equation of a line
Suppose that Λ is the line passing through the point *A* and parallel to the vector $\mathbf{d}$ (see fig. 6.1:19). If *P* is any point on Λ with position vector $\mathbf{r}$, then $\overrightarrow{AP}$ is parallel to $\mathbf{d}$, so $\overrightarrow{AP} = \lambda\mathbf{d}$ (λ scalar). Now $\mathbf{r} = \overrightarrow{OA} + \overrightarrow{AP}$, i.e.

$$\mathbf{r} = \mathbf{a} + \lambda\mathbf{d} \qquad (10)$$

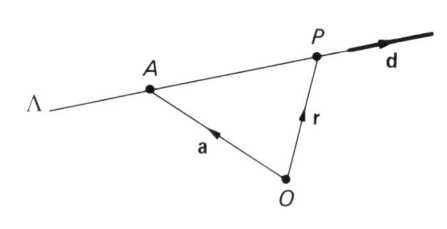

Fig. 6.1:19

This gives a condition that a general point *P* lies on Λ, so it is a **vector equation** of Λ. Each value of the **parameter** λ corresponds to a particular point on Λ (in particular $\lambda = 0$ corresponds to *A*).

Note: any equation of the form $\mathbf{r} = \mathbf{a}' + \mu\mathbf{d}'$, where $\mathbf{a}'$ is a point on Λ and $\mathbf{d}'$ is a vector in the direction of Λ, is also a vector equation of Λ.

Exercise 6.1:3

1 A, B, C and D are four points. Write down, in terms of **a**, **b**, **c** and **d**, the conditions for the quadrilateral $ABCD$ to be:

(i) a parallelogram (ii) a rhombus (iii) a square

2 A, B, C and D are four points. What can you say about the quadrilateral $ABCD$ if:

(i) $\mathbf{a}+\mathbf{c} = \mathbf{b}+\mathbf{d}$
(ii) $|\mathbf{a}-\mathbf{c}| = |\mathbf{b}-\mathbf{d}|$
(iii) $\mathbf{a}+\mathbf{c} = \mathbf{b}+\mathbf{d}$ and $|\mathbf{a}-\mathbf{c}| = |\mathbf{b}-\mathbf{d}|$?

3 A, B and C are three points; $\mathbf{c} = 2\mathbf{a}+\mathbf{b}$. Find $\overrightarrow{AB}$, $\overrightarrow{BC}$ and $\overrightarrow{CA}$ in terms of **a** and **b**. Check that $\overrightarrow{AB}+\overrightarrow{BC}+\overrightarrow{CA} = \mathbf{0}$.

4 A, B, C and D are four points; $\mathbf{c} = \mathbf{a}+3\mathbf{b}$, $\mathbf{d} = 2\mathbf{a}+2\mathbf{b}$. Find $\overrightarrow{AB}$ and $\overrightarrow{DC}$. What do you deduce?

5 Show that the points A, B and C are collinear $\Leftrightarrow \overrightarrow{AB} = \lambda \overrightarrow{AC}$.

6 Show that the points with position vectors $\mathbf{a}+\mathbf{b}$, $3\mathbf{a}+4\mathbf{b}$, $\frac{3}{5}\mathbf{a}+\frac{2}{5}\mathbf{b}$ are collinear.

7 Are the following sets of points collinear? (Assume that O, **a** and **b** are not collinear.)

(i) **a**, $\mathbf{a}+\mathbf{b}$, **b**
(ii) $\mathbf{a}+\mathbf{b}$, $3\mathbf{a}$, $-3\mathbf{a}+3\mathbf{b}$, $-5\mathbf{a}+4\mathbf{b}$

8 Show that the points **a**, **b** and $\alpha\mathbf{a}+\beta\mathbf{b}$ are collinear $\Leftrightarrow \alpha+\beta = 1$.

9 $ABCD$ is a parallelogram; X and Y are the midpoints of AB and CD respectively. If $\overrightarrow{AB} = \mathbf{u}$, $\overrightarrow{AD} = \mathbf{v}$, find in terms of **u** and **v** the vectors:

(i) $\overrightarrow{XY}$ (ii) $\overrightarrow{CX}$ (iii) $\overrightarrow{YA}$
(iv) $\overrightarrow{BY}$ (v) $\overrightarrow{XD}$

What do you deduce?

10 $OABC$ is a parallelogram; L and M are the midpoints of AB and BC. If $\overrightarrow{OA} = \mathbf{u}$, $\overrightarrow{OC} = \mathbf{v}$ show that $\overrightarrow{OL}+\overrightarrow{OM} = \frac{3}{2}(\mathbf{u}+\mathbf{v})$.

11 $OABC$ is a parallelogram; J and K are the points of trisection of AC. Show that $OJBK$ is a parallelogram.

12 (i) In $\triangle ABC$, L is the midpoint of AB and M is the midpoint of AC. Show that LM is parallel to BC and that $LM = \frac{1}{2}BC$.

(ii) In $\triangle ABC$, X divides AB internally in the ratio $m:n$, and Y divides AC internally in the ratio $m:n$. Show that XY is parallel to BC and that

$$XY = \frac{m}{m+n}BC$$

13 Show that the diagonals of a rhombus bisect one another at right angles.

14 AOB is the diameter of a circle and C is a point on the circumference. The position vectors of A, B and C relative to O, the centre of the circle, are **a**, **b** and **c** respectively. Show that $\mathbf{b} = -\mathbf{a}$ and $|\mathbf{c}| = |\mathbf{a}|$. Hence show that $\overrightarrow{BC}$ is perpendicular to $\overrightarrow{CA}$, i.e. that the angle in a semicircle is a right angle.

15 If the sides $\overrightarrow{AB}$, $\overrightarrow{AC}$ and $\overrightarrow{CB}$ of $\triangle ABC$ are represented by the vectors **p**, **q** and **r**, show that $\mathbf{r} = \mathbf{p}-\mathbf{q}$. Hence show that $r^2 = p^2+q^2-2\mathbf{p}\cdot\mathbf{q}$, and so prove the cosine formula for the $\triangle ABC$. (Remember that $r^2 = \mathbf{r}\cdot\mathbf{r}$, etc.)

16 If $OABC$ is a parallelogram, show that
$$AC^2+OB^2 = 2(OA^2+OC^2)$$

17 If $ABCD$ is a cyclic quadrilateral (i.e. if A, B, C and D are points on a circle), show that
$$AC\cdot BD = AB\cdot CD+BC\cdot AD$$

18 An **altitude** of a triangle is a perpendicular from a vertex to the side opposite the vertex. In $\triangle ABC$ the altitudes through A and B meet in a point O. Let **a**, **b** and **c** be the position vectors of A, B and C relative to O as origin. Show that $\mathbf{a}\cdot\mathbf{b} = \mathbf{a}\cdot\mathbf{c}$ and $\mathbf{b}\cdot\mathbf{a} = \mathbf{b}\cdot\mathbf{c}$. Deduce that $\mathbf{c}\cdot\mathbf{a} = \mathbf{c}\cdot\mathbf{b}$ and hence show that the altitudes of the triangle intersect at O.

19 $OABC$ is a tetrahedron (a triangular-based pyramid, of which all the faces are triangles). If L, M and N are the midpoints of OB, AB and AC, find $\overrightarrow{LM}$, $\overrightarrow{MN}$ and $\overrightarrow{NC}$ in terms of **a**, **b** and **c**.

20 $OABCDEFG$ is a parallelepiped (i.e. it has six faces and opposite faces are parallel). $\overrightarrow{OA} = \mathbf{u}$, $\overrightarrow{OC} = \mathbf{v}$, $\overrightarrow{OD} = \mathbf{w}$.

(i) Find the position vectors of the midpoints L and M of AC and EG.

(ii) Find the position vectors of the midpoints of AG, CE, OF and BD. What do you deduce?

(iii) Find the position vector of the midpoint of LM.

21 If two pairs of opposite edges of a tetrahedron $OABC$ are perpendicular, show that the third pair is also perpendicular.

22 The points L, M and N divide the sides BC, CA and AB of $\triangle ABC$ in the ratio $1:2$. Show that $\triangle LMN$ and $\triangle ABC$ have the same centroid.
Can you generalize this result?

23 $\triangle ABC$ and $\triangle DEF$ have centroids at G and H respectively. Show that $\overrightarrow{AD}+\overrightarrow{BE}+\overrightarrow{CF} = 3\overrightarrow{GH}$.

24 If AD, BE and CF are the medians of $\triangle ABC$, show that $\overrightarrow{AD} + \overrightarrow{BE} + \overrightarrow{CF} = \mathbf{0}$.

→ **25** Show that the lines joining the midpoints of the opposite edges of a tetrahedron $ABCD$ are concurrent at a point G with position vector $\frac{1}{4}(\mathbf{a} + \mathbf{b} + \mathbf{c} + \mathbf{d})$.

Show that the lines joining the vertices of the tetrahedron to the centroids of the opposite faces are concurrent at G.

26 Prove that the perpendicular bisectors of the sides of a triangle are concurrent.

+ **27** The points A, B and C have position vectors $\mathbf{a}$, $\mathbf{b}$ and $\mathbf{c}$ relative to O, and OC bisects the angle AOB internally. Show that $\mathbf{c} = \lambda(\hat{\mathbf{a}} + \hat{\mathbf{b}})$, where $\hat{\mathbf{a}}$ and $\hat{\mathbf{b}}$ are unit vectors in the directions $\mathbf{a}$ and $\mathbf{b}$.

Hence prove that the internal bisectors of the angles of a triangle are concurrent.

→ **28** Let O be the circumcentre of $\triangle ABC$, i.e. the point such that $OA = OB = OC$, and let the position vectors of A, B and C relative to O be $\mathbf{a}$, $\mathbf{b}$ and $\mathbf{c}$.

(i) Show that the point H with position vector $\mathbf{a} + \mathbf{b} + \mathbf{c}$ is the orthocentre of $\triangle ABC$ (i.e. the point of intersection of the altitudes of $\triangle ABC$).

(ii) Hence show that O, the centroid G and orthocentre H are collinear, and that G divides OH in the ratio $1:2$.

→ • **29** (i) Show that $\mathbf{r} = \mathbf{a} + \lambda(\mathbf{b} - \mathbf{a})$ is a vector equation of the line AB where A and B have position vectors $\mathbf{a}$ and $\mathbf{b}$.

(ii) Draw a diagram to show the points on the line at which:
(a) $\lambda = -2$; (b) $\lambda = -1$; (c) $\lambda = 0$;
(d) $\lambda = 1$; (e) $\lambda = 2$.

(iii) Find the values of λ at the two points P on the line for which $AP = 2PB$ and hence write down the position vectors of these points.

→ **30** The points A and B have position vectors $2\mathbf{p} + \mathbf{q}$ and $4\mathbf{p} + 3\mathbf{q}$ respectively. M and N are the midpoints of OA and AB. Write down a vector equation of:
(i) the line AB (ii) the line MN

• → **31** Show that if $\mathbf{u}$ and $\mathbf{v}$ are non-parallel vectors and $\alpha \mathbf{u} + \beta \mathbf{v} = \alpha' \mathbf{u} + \beta' \mathbf{v}$, where $\alpha, \beta, \alpha', \beta'$ are scalars, then $\alpha = \alpha'$, $\beta = \beta'$. (Hint: assume that $\alpha \neq \alpha'$ and hence show that $\mathbf{u}$ is parallel to $\mathbf{v}$—a contradiction.)

32 If $\mathbf{u}$ and $\mathbf{v}$ are non-parallel vectors, and
$$(1 - \lambda)\mathbf{u} + 2\lambda \mathbf{v} = (2 - 3\mu)\mathbf{u} + (2 + 2\mu)\mathbf{v}$$
find λ and μ.

• → **33** A and B are two points, and O, A and B are non-collinear. Find in terms of $\mathbf{a}$ and $\mathbf{b}$ the position vector of the point of intersection of the lines
$$\mathbf{r} = (4 - \lambda)\mathbf{a} + \lambda \mathbf{b} \text{ and } \mathbf{r} = (1 + 3\mu)\mathbf{a} + (-5 + \mu)\mathbf{b}$$

Questions 34–36: $OACB$ is a parallelogram; $\overrightarrow{OA} = \mathbf{a}$, $\overrightarrow{OB} = \mathbf{b}$.

34 L and M are the midpoints of BC and AC. Find in terms of $\mathbf{a}$ and $\mathbf{b}$ the position vectors of the points of intersection of:

(i) the lines LM and OC
(ii) the lines AL and OM

35 L and M are the midpoints of BC and AC. Find in terms of $\mathbf{a}$ and $\mathbf{b}$ the position vectors of the points J and K of intersection of:

(i) the lines OL and AB
(ii) the lines OM and AB

Are J and K the points of trisection of AB?

36 X and Y divide CB and CA respectively in the ratio $2:1$. Find in terms of $\mathbf{a}$ and $\mathbf{b}$ the position vectors of the points of intersection of:

(i) the lines OX and AB
(ii) the lines OY and AB

→ **37** The points A and B have position vectors $\mathbf{a}$ and $\mathbf{b}$ respectively. Write down a vector equation of the line AB.

C and D are the points with position vectors $\frac{1}{2}\mathbf{a}$ and $\frac{3}{2}\mathbf{b}$ respectively. Find the position vector of the point of intersection of the lines AB and CD.

What happens if $\mathbf{a}$ is parallel to $\mathbf{b}$?

R **38** The points A, B and C have position vectors $\mathbf{a}$, $\mathbf{b}$ and $\mathbf{c}$. Find:

(i) the equation of the line joining L, the midpoint of OA, to M, the midpoint of BC;

(ii) the position vector of the point in which LM meets the line joining the midpoint of OB to the midpoint of AC.

• → **39** AD, BE and CF are the medians of $\triangle ABC$. Show that a vector equation of AD is
$$\mathbf{r} = (1 - \lambda)\mathbf{a} + \tfrac{1}{2}\lambda(\mathbf{b} + \mathbf{c})$$
and find similar vector equations, with parameters μ and ν, for BE and CF.

By choosing appropriate values of λ, μ and ν, find the position vector of the point of intersection G of the medians, i.e. the centroid, of $\triangle ABC$. Show that G divides AD, BE and CF in the ratio $2:1$.

R **40** A, B, C and D are four coplanar points, with position vectors $\mathbf{a}$, $\mathbf{b}$, $\mathbf{c}$ and $\mathbf{d}$. P is the midpoint of AB, Q divides BC internally in the ratio $2:1$ and R divides CD externally in the ratio $2:1$. Find the position vectors of P, Q and R, and write down vector equations for the lines through:

(i) A and B (ii) P and Q

Find the position vector of the point of intersection of these lines.

6.2 Coordinate geometry in two dimensions

6.2:1 Components of a vector; vectors in a plane

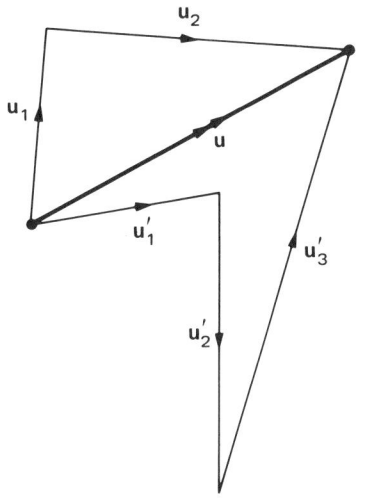

If $\mathbf{u} = \mathbf{u}_1 + \mathbf{u}_2 + \mathbf{u}_3 + \cdots + \mathbf{u}_n$, then $\mathbf{u}_1, \ldots, \mathbf{u}_n$ are called the **components** of $\mathbf{u}$; i.e. a vector is the sum of its components.

Given a vector $\mathbf{u}$, can we find its components? This is not a sensible question, for a vector does not have a unique set of components, as shown in fig. 6.2:1; $\mathbf{u} = \mathbf{u}_1 + \mathbf{u}_2$ and $\mathbf{u} = \mathbf{u}'_1 + \mathbf{u}'_2 + \mathbf{u}'_3$.

A more sensible question is: given a vector $\mathbf{u}$, can we find its components *in specified directions*? We first consider the problem for vectors in a plane.

Vectors in a plane

> **Theorem 1:** If $\mathbf{u}$ and $\mathbf{v}$ are any two non-parallel vectors in a plane then, for any other vector $\mathbf{a}$ in the plane, there exist unique scalars α and β such that
>
> $$\mathbf{a} = \alpha\mathbf{u} + \beta\mathbf{v} \qquad (1)$$

Fig. 6.2:1

$\alpha\mathbf{u}$ and $\beta\mathbf{v}$ are the components of $\mathbf{a}$ in the $\mathbf{u}$ and $\mathbf{v}$ directions; so the theorem says: given any two non-parallel directions $\mathbf{u}$ and $\mathbf{v}$ in the plane, we can find uniquely the components of any vector $\mathbf{a}$ in the $\mathbf{u}$ and $\mathbf{v}$ directions.

Proof

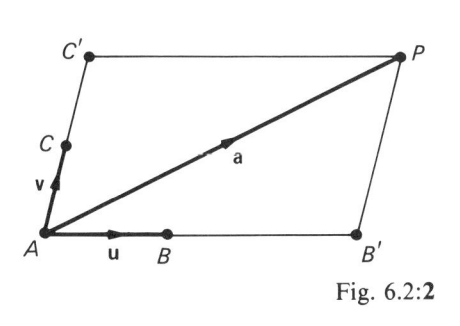

Suppose that $\mathbf{u}$ and $\mathbf{v}$ are represented by $\overrightarrow{AB}$ and $\overrightarrow{AC}$, and $\mathbf{a}$ by $\overrightarrow{AP}$ (see fig. 6.2:2). Produce AB to B' so that $\overrightarrow{B'P}$ is parallel to $\overrightarrow{AC}$. Then

$\overrightarrow{AB'}$ is parallel to $\overrightarrow{AB}$ so $\overrightarrow{AB'} = \alpha\mathbf{u}$ (α a scalar)

$\overrightarrow{B'P}$ is parallel to $\overrightarrow{AC}$ so $\overrightarrow{B'P} = \beta\mathbf{v}$ (β a scalar)

and $\quad \mathbf{a} = \overrightarrow{AP} = \overrightarrow{AB'} + \overrightarrow{B'P} = \alpha\mathbf{u} + \beta\mathbf{v}$

Fig. 6.2:2

To prove uniqueness we suppose that

$$\mathbf{a} = \alpha\mathbf{u} + \beta\mathbf{v} \qquad \text{(i)}$$
$$\mathbf{a} = \alpha'\mathbf{u} + \beta'\mathbf{v} \qquad \text{(ii)}$$

and $\alpha \neq \alpha'$, say.

Taking equation (ii) from equation (i) gives:

$$\mathbf{0} = (\alpha - \alpha')\mathbf{u} + (\beta - \beta')\mathbf{v}$$

$$\mathbf{u} = \frac{\beta - \beta'}{\alpha - \alpha'}\mathbf{v}$$

i.e. $\mathbf{u}$ is a scalar multiple of $\mathbf{v}$ and is therefore parallel to $\mathbf{v}$. This is a contradiction, so we can conclude that the scalars α and β are unique.

The set $\{\mathbf{u}, \mathbf{v}\}$ is called a **basis** for the plane. Theorem 1 says that any pair of non-parallel vectors forms a basis for the plane.

When considering vectors in a plane, we usually choose an origin and two mutually perpendicular axes Ox and Oy. Let $\mathbf{i}$ be the unit vector

292 *Components of a vector; vectors in a plane*

(vector of magnitude 1) in the direction Ox, $\mathbf{j}$ the unit vector in the direction Oy. Then any vector $\mathbf{a}$ in the plane can be expressed in the form

$$\mathbf{a} = a_1\mathbf{i} + a_2\mathbf{j} \qquad (2)$$

where a_1 and a_2 are scalars, since the set $\{\mathbf{i}, \mathbf{j}\}$ forms a basis for the plane (fig. 6.2:**3a**).

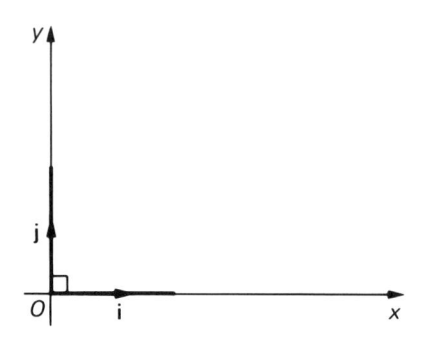

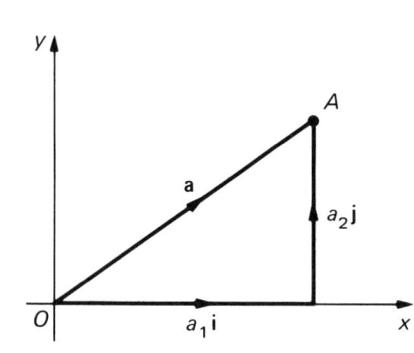

Fig. 6.2:**3a**

Fig. 6.2:**3b**

Look at fig. 6.2:**3b**. The vector $\mathbf{a}$ is the resultant of $a_1\mathbf{i}$ and $a_2\mathbf{j}$. The vectors $a_1\mathbf{i}$ and $a_2\mathbf{j}$ are the components of $\mathbf{a}$ in the $\mathbf{i}$ and $\mathbf{j}$ directions.

We have represented $\mathbf{a}$ by the position vector $\overrightarrow{OA}$. If $\mathbf{a}$ is a free vector, then it can be represented by any displacement vector equal to $\overrightarrow{OA}$ as shown in fig. 6.2:**4**.

Magnitude and direction of a vector

If $\mathbf{a} = a_1\mathbf{i} + a_2\mathbf{j}$ (fig.6.2:**5**), the magnitude $|\mathbf{a}|$ (or a) of $\mathbf{a}$ is given by

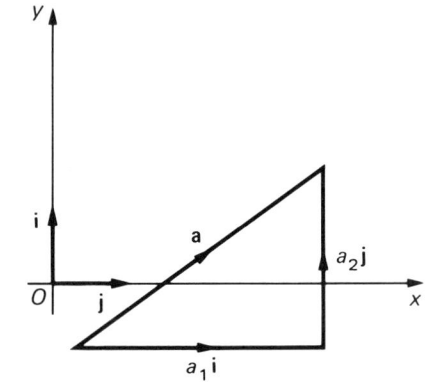

$$|\mathbf{a}| = \sqrt{(a_1^2 + a_2^2)} \qquad (3)$$

e.g.
$$|3\mathbf{i} + 4\mathbf{j}| = \sqrt{(3^2 + 4^2)} = \sqrt{25} = 5$$
$$|-\sqrt{3}\mathbf{i} + \mathbf{j}| = \sqrt{((-\sqrt{3})^2 + 1^2)} = \sqrt{4} = 2$$

Fig. 6.2:**4**

The unit vector $\hat{\mathbf{a}}$ in the direction of $\mathbf{a}$ is given by

$$\hat{\mathbf{a}} = \frac{1}{a}\mathbf{a} = \frac{a_1}{a}\mathbf{i} + \frac{a_2}{a}\mathbf{j} \qquad (4)$$

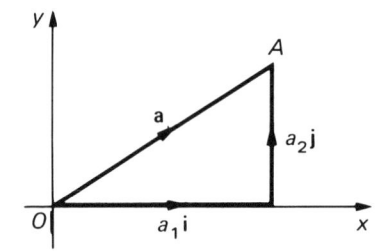

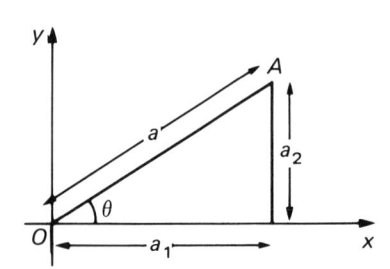

Fig. 6.2:**5**

Note: (i) $\hat{\mathbf{a}}$ is in the direction of $\mathbf{a}$ since $\hat{\mathbf{a}}$ is a scalar multiple of $\mathbf{a}$;

(ii) $\hat{\mathbf{a}}$ is a unit vector since $|\hat{\mathbf{a}}| = \frac{1}{a}|\mathbf{a}| = 1$.

For example, if $\mathbf{a} = 3\mathbf{i} + 4\mathbf{j}$, $\hat{\mathbf{a}} = \frac{1}{5}\mathbf{a} = \frac{3}{5}\mathbf{i} + \frac{4}{5}\mathbf{j}$.

Vectors, coordinate geometry and complex numbers 293

Worked example 1 Find the vector **u** of magnitude 20 in the direction of the vector **a** = 3**i** + 4**j**.

$$\mathbf{u} = 20\hat{\mathbf{a}} = 20(\tfrac{3}{5}\mathbf{i} + \tfrac{4}{5}\mathbf{j}) = 12\mathbf{i} + 16\mathbf{j}$$

We can describe the direction of the vector $\mathbf{a} = a_1\mathbf{i} + a_2\mathbf{j}$ by specifying the angle θ it makes with Ox:

$$\tan\theta = \frac{a_2}{a_1} \tag{5}$$

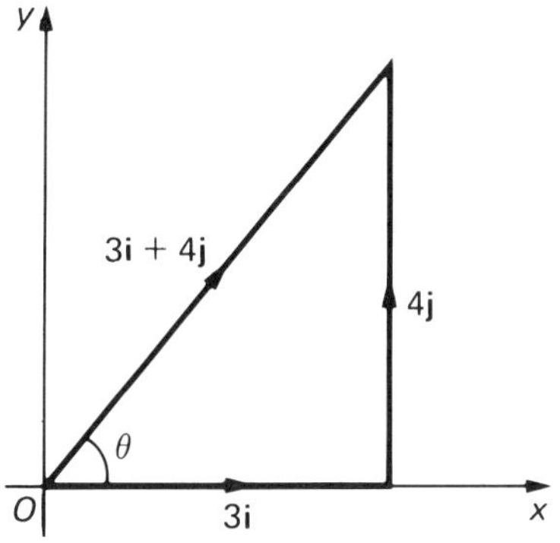

Usually we choose the value of θ between $-180°$ and $180°$.
(Note: although $\tan\theta = a_2/a_1$, it is not necessarily true that $\theta = \tan^{-1}(a_2/a_1)$. The equation $\tan\theta = a_2/a_1$ has an infinite number of solutions $\theta = \tan^{-1}[(a_2/a_1) + 180k°]$; we need to choose the appropriate value of k (-1, 0 or 1) to ensure that $-180° < \theta \leq 180°$.)

Fig. 6.2:**6a**

Worked example 2 Find the angle between Ox and the vector:

(i) $3\mathbf{i} + 4\mathbf{j}$ (ii) $-\sqrt{3}\mathbf{i} + \mathbf{j}$

(i) Fig. 6.2:**6a**: $\tan\theta = \tfrac{4}{3}$; so $\theta = 53.13°$.
(ii) Fig. 6.2:**6b**: $\tan\theta = -1/\sqrt{3}$; so $\theta = 150°$ since the vector makes an obtuse angle with the (positive) Ox direction.

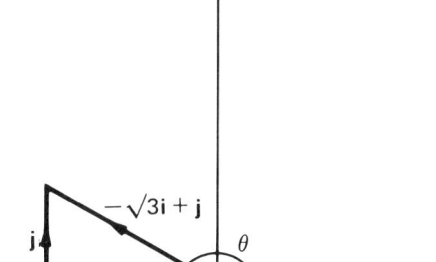

Resultant of a set of vectors
If $\mathbf{a} = a_1\mathbf{i} + a_2\mathbf{j}$, $\mathbf{b} = b_1\mathbf{i} + b_2\mathbf{j}$ (fig. 6.2:7), then

$$\begin{aligned}\mathbf{a} + \mathbf{b} &= (a_1\mathbf{i} + a_2\mathbf{j}) + (b_1\mathbf{i} + b_2\mathbf{j}) \\ &= (a_1 + b_1)\mathbf{i} + (a_2 + b_2)\mathbf{j}\end{aligned} \tag{6}$$

For example, $(3\mathbf{i} + 4\mathbf{j}) + (5\mathbf{i} - \mathbf{j}) = (3+5)\mathbf{i} + (4-1)\mathbf{j} = 8\mathbf{i} + 3\mathbf{j}$; and we can extend the result to three or more vectors:

$$(3\mathbf{i}) + (5\mathbf{i} + 2\mathbf{j}) - (2\mathbf{i} + \mathbf{j}) + 4\mathbf{j} = (3+5-2)\mathbf{i} + (2-1+4)\mathbf{j} = 6\mathbf{i} + 5\mathbf{j}$$

Fig. 6.2:**6b**

In general this is a much easier way of adding vectors than using a vector polygon.

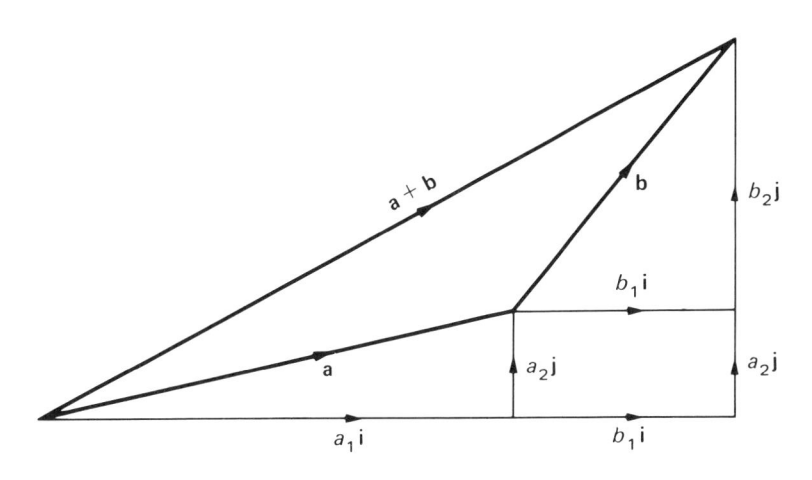

Fig. 6.2:**7**

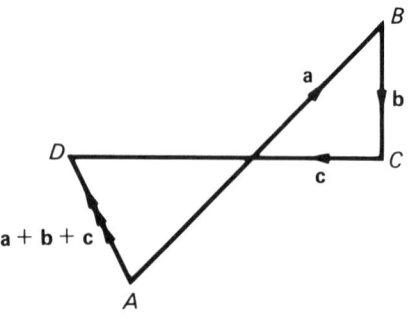

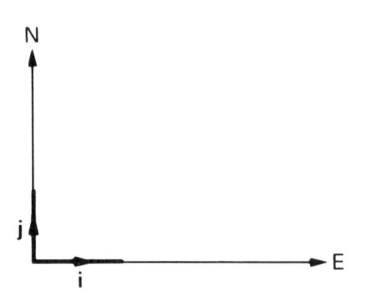

Worked example 3 Find the resultant of vectors $\mathbf{a} = 4\sqrt{2}$ m north-east, $\mathbf{b} = 2$ m due south, $\mathbf{c} = 5$ m due west.

If $\mathbf{a}$ is represented by $\overrightarrow{AB}$, $\mathbf{b}$ by $\overrightarrow{BC}$, $\mathbf{c}$ by $\overrightarrow{CD}$, $\mathbf{a} + \mathbf{b} + \mathbf{c}$ is represented by $\overrightarrow{AD}$ (fig. 6.2:8a). But it is difficult to find the magnitude and direction of $\overrightarrow{AD}$ using only geometry.

Instead, let $\mathbf{i}$ and $\mathbf{j}$ be unit vectors in the directions east and north (fig. 6.2:8b). Then $\mathbf{a} = 4\mathbf{i} + 4\mathbf{j}$ (see question 6), $\mathbf{b} = -2\mathbf{j}$, $\mathbf{c} = -5\mathbf{i}$, so

$$\mathbf{a} + \mathbf{b} + \mathbf{c} = (4\mathbf{i} + 4\mathbf{j}) + (-2\mathbf{j}) + (-5\mathbf{i}) = -\mathbf{i} + 2\mathbf{j}$$

So the resultant of $\mathbf{a}$, $\mathbf{b}$ and $\mathbf{c}$ is $-\mathbf{i} + 2\mathbf{j}$.

Note: we can find the magnitude $\sqrt{5}$ and direction N26.57°W of this vector using the method of worked example 2.

Fig. 6.2:8a

Fig. 6.2:8b

Calculation of scalar product
Since $\mathbf{i}$ is perpendicular to $\mathbf{j}$,

$$\mathbf{i} \cdot \mathbf{j} = \mathbf{j} \cdot \mathbf{i} = 0 \qquad \text{by 6.1:2 Theorem 1}$$

and

$$\mathbf{i} \cdot \mathbf{i} = \mathbf{j} \cdot \mathbf{j} = 1 \qquad \text{by 6.1:2 equation (4).}$$

So if $\mathbf{a} = a_1\mathbf{i} + a_2\mathbf{j}$ and $\mathbf{b} = b_1\mathbf{i} + b_2\mathbf{j}$,

$$\boxed{\mathbf{a} \cdot \mathbf{b} = (a_1\mathbf{i} + a_2\mathbf{j}) \cdot (b_1\mathbf{i} + b_2\mathbf{j}) = a_1 b_1 + a_2 b_2} \qquad (7)$$

For example,

$$(3\mathbf{i} + 2\mathbf{j}) \cdot (5\mathbf{i} - 4\mathbf{j}) = (3)(5) + (2)(-4) = 7$$

Worked example 4 If $\mathbf{a} = \mathbf{i} + 3\mathbf{j}$, $\mathbf{b} = 2\mathbf{i} - 2\mathbf{j}$, find the angle θ between $\mathbf{a}$ and $\mathbf{b}$.

By 6.1:2 equation (4):

$$\cos \theta = \frac{\mathbf{a} \cdot \mathbf{b}}{ab} = \frac{(\mathbf{i} + 3\mathbf{j}) \cdot (2\mathbf{i} - 2\mathbf{j})}{\sqrt{10}\sqrt{8}} = -\frac{4}{\sqrt{80}} = -\frac{1}{\sqrt{5}}$$

So

$$\theta = 116.57°$$

Exercise 6.2:1

- **1** If $\mathbf{a} = 3\mathbf{u} + 2\mathbf{v}$, $\mathbf{b} = \mathbf{u} - 3\mathbf{v}$, find:

  (i) $\mathbf{a} + \mathbf{b}$ (ii) $\mathbf{a} - \mathbf{b}$ (ii) $2\mathbf{a} + 3\mathbf{b}$

  in terms of $\mathbf{u}$ and $\mathbf{v}$.

2 If $\mathbf{a} = 2\mathbf{u} + \mathbf{v}$, $\mathbf{b} = \mathbf{u} - 3\mathbf{v}$, find $\mathbf{u}$ and $\mathbf{v}$ in terms of $\mathbf{a}$ and $\mathbf{b}$.

3 If $\mathbf{i}$ and $\mathbf{j}$ are unit vectors in directions due east and due north, find the magnitude and direction of the following vectors:

(i) $3\mathbf{i}$ (ii) $-2\mathbf{i}$ (iii) $-\mathbf{j}$
(iv) $\mathbf{i} + \mathbf{j}$ (v) $\mathbf{i} - \mathbf{j}$
(vi) $3\mathbf{i} + 4\mathbf{j}$ (vii) $-3\mathbf{i} + 4\mathbf{j}$

4 If $\mathbf{p}$ and $\mathbf{q}$ are unit vectors in directions south-west and south-east, find the magnitude and direction of the following vectors:

(i) $3\mathbf{p}$ (ii) $3(\mathbf{p} + \mathbf{q})$
(iii) $-3\mathbf{p} + 3\mathbf{q}$ (iv) $\mathbf{p} + 2\mathbf{q}$

5 The diagram shows a grid of equal parallelograms; $\overrightarrow{OA} = \mathbf{a}$, $\overrightarrow{OB} = \mathbf{b}$.

Express in terms of $\mathbf{a}$ and $\mathbf{b}$ the position vectors of C, D, E, F, G and H. Hence express in terms of $\mathbf{a}$ and $\mathbf{b}$ the vectors $\overrightarrow{CD}$, $\overrightarrow{GA}$ and $\overrightarrow{HF}$.

If $\overrightarrow{DC} = \mathbf{u}$, $\overrightarrow{AJ} = \mathbf{v}$, explain why it is not possible to express $\overrightarrow{CE}$ in the form $\alpha \mathbf{u} + \beta \mathbf{v}$.

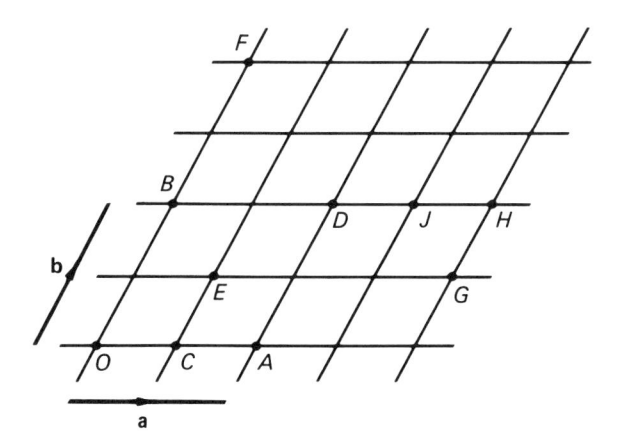

Vectors, coordinate geometry and complex numbers 295

6 (i) If **i** and **j** are unit vectors in the directions east and north respectively, write the following vectors in the form $a_1\mathbf{i} + a_2\mathbf{j}$:
(a) 5 m due east; (b) 4 m north-east;
(c) 2 m N30°W; (d) 10 m S40°W.

(ii) A vector of magnitude v makes an angle α with Ox. Write the vector in the form $a_1\mathbf{i} + a_2\mathbf{j}$.

(iii) Write each of the four (displacement) vectors in the diagram in the form $a_1\mathbf{i} + a_2\mathbf{j}$.

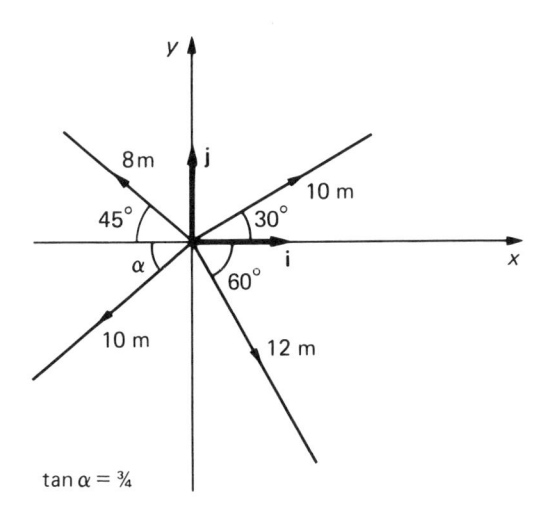

7 If $\mathbf{a} = 8\mathbf{i} + p\mathbf{j}$ and $|\mathbf{a}| = 10$, find two possible values of p.

8 Find the unit vector in the direction of each of the following vectors:
(i) $3\mathbf{i} - 4\mathbf{j}$
(ii) $-\mathbf{i} + 2\mathbf{j}$
(iii) $-3\mathbf{i} - 3\mathbf{j}$

9 If the vector $p\mathbf{i} + \tfrac{1}{2}\mathbf{j}$ is a unit vector, find two possible values of p.

10 If the vector $\tfrac{4}{5}\mathbf{i} + p\mathbf{j}$ is a unit vector, find two possible values of p.

11 Find the vector of the given magnitude in the given direction:
(i) 10; $3\mathbf{i} + 4\mathbf{j}$
(ii) 39; $12\mathbf{i} - 5\mathbf{j}$
(iii) $4\sqrt{5}$; $-2\mathbf{i} + \mathbf{j}$
(iv) 4; $-\sqrt{3}\mathbf{i} - \mathbf{j}$

12 Which of the following pairs of vectors are parallel?
(i) $\mathbf{i} + 3\mathbf{j}$ and $12\mathbf{i} + 4\mathbf{j}$
(ii) $3\mathbf{i} - 2\mathbf{j}$ and $9\mathbf{i} - 6\mathbf{j}$
(iii) $3\mathbf{i} - 2\mathbf{j}$ and $9\mathbf{i} + 6\mathbf{j}$
(iv) $\mathbf{i} + \mathbf{j}$ and $-5\mathbf{i} - 5\mathbf{j}$

13 Find the magnitude and direction of the following vectors:
(i) $3\mathbf{i}$
(ii) $-6\mathbf{j}$
(iii) $\mathbf{i} + 2\mathbf{j}$
(iv) $12\mathbf{i} - 5\mathbf{j}$
(v) $-\mathbf{i} + \sqrt{3}\mathbf{j}$
(vi) $-4\mathbf{i} + \mathbf{j}$
(vii) $-\mathbf{i} - \mathbf{j}$
(viii) $-2\mathbf{i} - 3\mathbf{j}$

14 Find $\mathbf{u} + \mathbf{v}$ when:
(i) $\mathbf{u} = 4\mathbf{i} - 3\mathbf{j}$, $\mathbf{v} = 5\mathbf{i} + 4\mathbf{j}$
(ii) $\mathbf{u} = -\mathbf{i} - 7\mathbf{j}$, $\mathbf{v} = \mathbf{i} + 2\mathbf{j}$

Verify in each case that $|\mathbf{u} + \mathbf{v}| < |\mathbf{u}| + |\mathbf{v}|$.

15 If $\mathbf{u} = 2\mathbf{i} - \mathbf{j}$, $\mathbf{v} = 6\mathbf{i} - 3\mathbf{j}$, find $\mathbf{u} + \mathbf{v}$ and show that $|\mathbf{u} + \mathbf{v}| = |\mathbf{u}| + |\mathbf{v}|$. Illustrate your result by drawing a diagram.

16 If $\mathbf{u} = 3\mathbf{i} + \mathbf{j}$, $\mathbf{v} = -2\mathbf{i} + 5\mathbf{j}$, find:
(i) $\mathbf{u} + \mathbf{v}$
(ii) $3\mathbf{u} + \mathbf{v}$
(iii) $\mathbf{u} - \mathbf{v}$
(iv) $\tfrac{1}{2}\mathbf{u} - \tfrac{3}{2}\mathbf{v}$

17 Find $\mathbf{u} + \mathbf{v} + \mathbf{w}$ when:
(i) $\mathbf{u} = 4\mathbf{i} - 7\mathbf{j}$, $\mathbf{v} = \mathbf{i} + 6\mathbf{j}$, $\mathbf{w} = -2\mathbf{i} + 4\mathbf{j}$
(ii) $\mathbf{u} = \mathbf{i} + \mathbf{j}$, $\mathbf{v} = 3\mathbf{i} - 3\mathbf{j}$, $\mathbf{w} = -4\mathbf{i} + 2\mathbf{j}$

18 Find in the form $a_1\mathbf{i} + a_2\mathbf{j}$, where **i** and **j** are unit vectors in directions due east and due north, the resultant of the vectors **u** and **v** when:
(i) $\mathbf{u} = 4\sqrt{2}$ m north-west, $\mathbf{v} = 6$ m due south
(ii) $\mathbf{u} = 8$ m S60°W, $\mathbf{v} = 4\sqrt{3}$ m S30°E

19 Find the magnitude and direction of the resultant of the vectors **u** and **v** when:
(i) $\mathbf{u} = \mathbf{i} + 2\mathbf{j}$, $\mathbf{v} = 2\mathbf{i} - 5\mathbf{j}$
(ii) **u** is a vector of magnitude 15 in the direction $-4\mathbf{i} + 3\mathbf{j}$, **v** is a vector of magnitude $10\sqrt{2}$ in the direction $\mathbf{i} + 7\mathbf{j}$.

③ **20** Find the magnitude and direction of the resultant of the vectors **u**, **v** and **w** when:

(i) $\mathbf{u} = 3\mathbf{i} + 4\mathbf{j}$, $\mathbf{v} = -\mathbf{i} - 2\mathbf{j}$, $\mathbf{w} = 2\mathbf{i} + \mathbf{j}$

R (ii) **u** is a vector of magnitude $4\sqrt{5}$ in the direction $\mathbf{i} + 2\mathbf{j}$, **v** is a vector of magnitude $2\sqrt{5}$ in the direction $\mathbf{i} - 2\mathbf{j}$ and **w** is a vector of magnitude 26 in the direction $-5\mathbf{i} + 12\mathbf{j}$.

• → **21** (i) If $a_1\mathbf{i} + a_2\mathbf{j} = b_1\mathbf{i} + b_2\mathbf{j}$, what can you say about a_1, a_2, b_1 and b_2?
Explain your reasoning.

(ii) If $x\mathbf{i} + y\mathbf{j} = (1+\lambda)\mathbf{i} + (3-2\lambda)\mathbf{j}$ show that $y = -2x + 5$.

(iii) The resultant of the vectors $p\mathbf{i} + 2\mathbf{j}$, $q\mathbf{i} + 2p\mathbf{j}$, $4\mathbf{i} - q\mathbf{j}$ is $2\mathbf{i} + 9\mathbf{j}$. Find p and q.

• **22** If $\mathbf{u} = 2\mathbf{i} + \mathbf{j}$, $\mathbf{v} = \mathbf{i} - 2\mathbf{j}$, express the vector $\mathbf{a} = 4\mathbf{i} + 7\mathbf{j}$ in the form $\alpha\mathbf{u} + \beta\mathbf{v}$.

23 If $\mathbf{u} = \mathbf{i} + \mathbf{j}$, $\mathbf{v} = -\mathbf{i} + 3\mathbf{j}$, express the vector $\mathbf{a} = 3\mathbf{i} - 5\mathbf{j}$ in the form $\alpha\mathbf{u} + \beta\mathbf{v}$.

→ **24** A vector of magnitude 10 parallel to the vector $4\mathbf{i} + 3\mathbf{j}$ is the resultant of two vectors **u** and **v** which are parallel to the vectors $2\mathbf{i} + \mathbf{j}$ and $\mathbf{i} + \mathbf{j}$ respectively. Find the magnitudes of **u** and **v**.

→ **25** Find $\mathbf{a} \cdot \mathbf{b}$ when:

(i) $\mathbf{a} = 3\mathbf{i} + 2\mathbf{j}$, $\mathbf{b} = 4\mathbf{i} + 5\mathbf{j}$
(ii) $\mathbf{a} = -3\mathbf{i} - \mathbf{j}$, $\mathbf{b} = 2\mathbf{i} - \mathbf{j}$
(iii) $\mathbf{a} = 4\mathbf{i}$, $\mathbf{b} = \mathbf{i} + 5\mathbf{j}$
(iv) $\mathbf{a} = 3\mathbf{i} + 2\mathbf{j}$, $\mathbf{b} = -2\mathbf{i} + 3\mathbf{j}$

What do you deduce in (iv)?

26 If $\mathbf{a} = 2\mathbf{i} - \mathbf{j}$, $\mathbf{b} = 3\mathbf{i} + \mathbf{j}$, $\mathbf{c} = 4\mathbf{i} + 5\mathbf{j}$, find:

(i) $\mathbf{a} \cdot \mathbf{b}$ (ii) $\mathbf{a} \cdot \mathbf{c}$ (iii) $\mathbf{a} \cdot (\mathbf{b} + \mathbf{c})$

Verify that $\mathbf{a} \cdot (\mathbf{b} + \mathbf{c}) = \mathbf{a} \cdot \mathbf{b} + \mathbf{a} \cdot \mathbf{c}$.

27 (i) Show that if $\mathbf{a} = \lambda^2 \mathbf{b}$ (λ scalar), then $\mathbf{a} \cdot \mathbf{b} = ab$. In particular show that $\mathbf{a} \cdot \mathbf{a} = a^2$. Verify this result when $\mathbf{a} = 3\mathbf{i} - 2\mathbf{j}$.

(ii) Show that if $\mathbf{a} = -\lambda^2 \mathbf{b}$, $\mathbf{a} \cdot \mathbf{b} = -ab$.

28 Which of the following pairs of vectors are perpendicular?

(i) $\mathbf{i} + \mathbf{j}$ and $\mathbf{i} - \mathbf{j}$
(ii) $2\mathbf{i} + \mathbf{j}$ and $2\mathbf{i} - \mathbf{j}$
(iii) $3\mathbf{i} - 4\mathbf{j}$ and $4\mathbf{i} + 3\mathbf{j}$

R **29** If the vectors $\mathbf{i} + 3\mathbf{j}$ and $6\mathbf{i} + \lambda\mathbf{j}$ are perpendicular, find λ.

→ **30** Show that the vectors $a\mathbf{i} + b\mathbf{j}$ and $b\mathbf{i} - a\mathbf{j}$ are perpendicular.

31 Write in the most general form the vector perpendicular to the vector $\mathbf{u} = \mathbf{i} + 2\mathbf{j}$. Hence find the two unit vectors perpendicular to **u**.

④ **32** Find the cosine of the angle between **a** and **b** when:

→ (i) $\mathbf{a} = 3\mathbf{i} + \mathbf{j}$, $\mathbf{b} = \mathbf{i} + 2\mathbf{j}$
(ii) $\mathbf{a} = \mathbf{i} + \mathbf{j}$, $\mathbf{b} = -2\mathbf{i} + \mathbf{j}$
(iii) $\mathbf{a} = -3\mathbf{i} + \mathbf{j}$, $\mathbf{b} = 6\mathbf{i} - 2\mathbf{j}$
R (iv) $\mathbf{a} = 8\mathbf{i} - 4\mathbf{j}$, $\mathbf{b} = 2\mathbf{i} + 4\mathbf{j}$

What do you deduce in (iii) and in (iv)?

④ **33** Find two unit vectors inclined at 60° to the vector:

(i) **i** (ii) $2\mathbf{i} + \mathbf{j}$

6.2:2 Position vector of a point in a plane

We saw in section 1.2:1 that, having chosen an origin O and mutually perpendicular axes Ox and Oy, we can specify the position of any point P in the plane by its cartesian coordinates (x, y).

If we now let the unit vectors in the directions Ox and Oy be **i** and **j**, then the position vector $\overrightarrow{OP}$ of P has components $x\mathbf{i}$ and $y\mathbf{j}$ (fig. 6.2:9). So

$$P \text{ is the point } (x, y) \Leftrightarrow \overrightarrow{OP} = x\mathbf{i} + y\mathbf{j} \qquad (8)$$

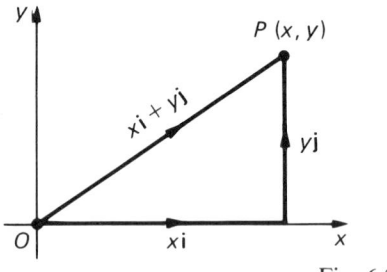

Fig. 6.2:9

(Note: we may refer loosely to, e.g., 'the point $3\mathbf{i} + 4\mathbf{j}$', meaning the point with position vector $3\mathbf{i} + 4\mathbf{j}$, i.e. the point $(3, 4)$.)

Worked example 5 A and B are the points $(-3, 4)$ and $(3, 1)$ shown in fig. 6.2:10. Find:

(i) $\overrightarrow{AB}$ (ii) $\overrightarrow{BA}$ (iii) the distance AB

(i) $\overrightarrow{AB} = \mathbf{b} - \mathbf{a} = (3\mathbf{i} + \mathbf{j}) - (-3\mathbf{i} + 4\mathbf{j}) = 6\mathbf{i} - 3\mathbf{j}$
(ii) $\overrightarrow{BA} = -\overrightarrow{AB} = -6\mathbf{i} + 3\mathbf{j}$
(iii) $AB = |\overrightarrow{AB}| = \sqrt{(6^2 + (-3)^2)} = \sqrt{45} = 3\sqrt{5}$

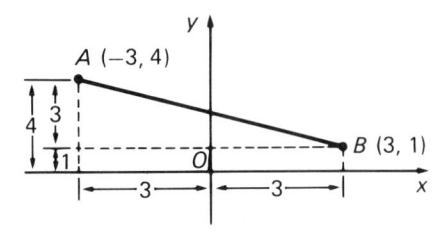

Fig. 6.2:10

Vectors, coordinate geometry and complex numbers 297

Worked example 6 *A*, *B* and *C* are the points $(-3, 4)$, $(3, 1)$ and $(-1, 5)$ shown in fig. 6.2:**11**. Find the cosine of the angle θ between *AB* and *AC*.

$\overrightarrow{AB} = 6\mathbf{i} - 3\mathbf{j}; \; \overrightarrow{AC} = 2\mathbf{i} + \mathbf{j}$, so

$$\cos\theta = \frac{\overrightarrow{AB} \cdot \overrightarrow{AC}}{(AB)(AC)}$$

$$= \frac{(6\mathbf{i} - 3\mathbf{j}) \cdot (2\mathbf{i} + \mathbf{j})}{(3\sqrt{5})(\sqrt{5})}$$

$$= \frac{3}{5}$$

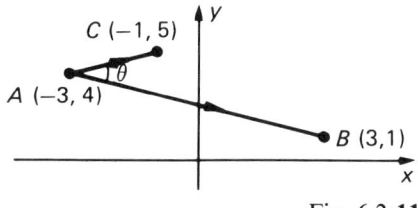

So $\theta = 53.13°$

Fig. 6.2:**11**

Worked example 7 *A* and *B* are the points $(-3, 4)$ and $(3, 1)$. Find the position vector of:

(i) the point *S* which divides *AB* internally (i.e. *S* between *A* and *B*) in the ratio $2:1$ (fig. 6.2:**12a**);
(ii) the point *T* which divides *AB* externally (i.e. *T* on *AB* produced) in the ratio $5:2$ (fig. 6.2:**12b**).

(i) $AS:SB = 2:1$ so $AS:AB = 2:3$. Thus

$$\mathbf{s} = \mathbf{a} + \tfrac{2}{3}\overrightarrow{AB}$$
$$= \mathbf{a} + \tfrac{2}{3}(\mathbf{b} - \mathbf{a})$$
$$= \tfrac{1}{3}(\mathbf{a} + 2\mathbf{b}) = \mathbf{i} + 2\mathbf{j}$$

(ii) $AT:TB = 5:2$ so $AT:AB = 5:3$. Thus

$$\mathbf{t} = \mathbf{a} + \tfrac{5}{3}\overrightarrow{AB}$$
$$= \mathbf{a} + \tfrac{5}{3}(\mathbf{b} - \mathbf{a})$$
$$= \tfrac{1}{3}(-2\mathbf{a} + 5\mathbf{b}) = 7\mathbf{i} - \mathbf{j}$$

Alternatively, we can use 6.1:3 Theorem 2 with:
(i) $m = 2, n = 1$ (ii) $m = 5, n = -2$

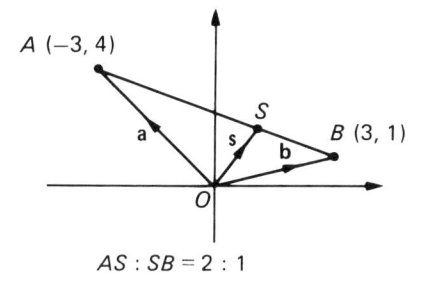

Fig. 6.2:**12a**

AS : SB = 2 : 1

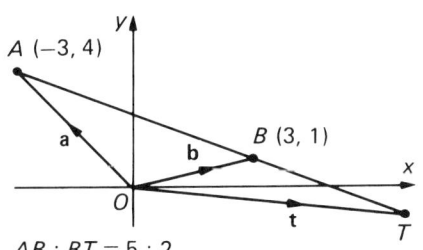

AB : BT = 5 : 2

Fig. 6.2:**12b**

Exercise 6.2:2

1 Find the positions vectors of the points:

(i) $(1, 2)$ (ii) $(3, -1)$ (iii) $(-\tfrac{1}{2}, 4)$
(iv) $(4, 0)$ (v) $(0, -1)$

2 Find the cartesian coordinates of the points with position vectors:

(i) $3\mathbf{i} + 7\mathbf{j}$ (ii) $-3\mathbf{i} - 2\mathbf{j}$
(iii) $5\mathbf{j}$ (iv) $-2\mathbf{i}$

⑤ **3** Sketch a diagram showing *A* and *B*, find $\overrightarrow{AB}$, $\overrightarrow{BA}$ and the distance *AB* when *A* and *B* are the points:

(i) $A(1, 3), B(4, 7)$
(ii) $A(-2, 5), B(2, 1)$

(iii) $A(0, 0), B(-3, -1)$
→ (iv) $A(-4, -2), B(-3, 5)$
(v) $A(3, 5), B(3, -4)$

(See 1.2:1 Q2.)

⑤ **4** Find $\overrightarrow{AB}$ and the distance *AB* when *A* and *B* are the points (x_1, y_1) and (x_2, y_2).

⑤ **5** Write down a condition, in terms of displacement vectors, for the quadrilateral *ABCD* to be a parallelogram.

If *A*, *B*, *C* and *D* are the points $(1, 2)$, $(5, 3)$, $(2, 6)$ and $(-2, 5)$, show that *ABCD* is a parallelogram.

298 *Position vectors of a point in a plane*

⑤ → **6** A, B and C are the points with position vectors $\mathbf{i}$, $3\mathbf{i}-2\mathbf{j}$ and $-\mathbf{i}-4\mathbf{j}$. Find the position vectors of the points D and D' if $ABCD$ and $ABD'C$ are parallelograms.

⑤ **7** If A, B, C and D are the points $(-2, 2)$, $(5, 3)$, $(10, 8)$ and $(3, 7)$, show that $ABCD$ is a rhombus.

⑤ **8** If A, B and C are the points $(-5, -1)$, $(2, -5)$, $(3, 0)$, show that ABC is an isosceles triangle.

⑤ R **9** A, B and M are the points $(2, 0)$, $(7, -1)$ and $(6, 2)$. If $ABCD$ is a parallelogram and M is the point of intersection of its diagonals, find the coordinates of C and D.

⑤ • **10** A and B are the points $(9, 7)$ and $(1, 9)$. Find the distances from A and B of the point $(x, 0)$. Hence find the point on the x-axis which is equidistant from A and B. (See 1.2:1 Q40).

⑤ → **11** A and B are the points $(-1, 6)$ and $(-9, -2)$. Find

   (i) the point on the x-axis
   (ii) the point on the y-axis

equidistant from A and B.

⑤ → **12** If the point $P(x, y)$ is equidistant from the points $A(4, 1)$ and $B(-2, 5)$, show that $2x - 3y + 4 = 0$. Describe the set of points P.

⑤ R **13** If the point $P(x, y)$ is equidistant from the points $A(3, 4)$ and $B(-1, 2)$, find a relation between x and y. Illustrate your answer by drawing a diagram.

⑥ **14** Find the angle between AB and AC when A, B and C are the points:

   (i) $(0, 0)$, $(1, 1)$, $(3, 4)$
→ (ii) $(2, 3)$, $(4, 7)$, $(5, 4)$
   (iii) $(-1, 3)$, $(5, 1)$, $(-3, -1)$

⑥ **15** A, B, C and D are the points $(-1, -2)$, $(3, 1)$, $(5, 6)$ and $(4, 3)$. Show that:

   (i) $AC = 2AB$ (ii) $D\hat{A}B = D\hat{A}C$

⑥ **16** A, B and C are the points $(0, 2)$, $(2, 1)$ and $(4, 5)$. Show that AB is perpendicular to BC. Hence find the area of $\triangle ABC$.

⑥ → **17** A, B and C are the points $(-1, 2)$, $(4, 1)$ and $(-2, 5)$. Find AB, BC and $A\hat{B}C$. Hence find the area of $\triangle ABC$.

⑥ **18** A, B, C and D are the points $(-4, -5)$, $(5, -3)$, $(12, 3)$ and $(3, 1)$. Show that $ABCD$ is a rhombus. Verify that its diagonals are perpendicular.

⑥ **19** A, B, C and D are the points $(3, -1)$, $(7, 0)$, $(6, 4)$ and $(2, 3)$. Show that $ABCD$ is a square.

⑥ → **20** If $\mathbf{u} = a\mathbf{i} + b\mathbf{j}$, find two vectors $\mathbf{v}$ such that $\mathbf{u}$ is perpendicular to $\mathbf{v}$ and $|\mathbf{u}| = |\mathbf{v}|$.
   A and B are the points $(-1, 2)$ and $(4, -1)$. If $ABCD$ is a square, find two pairs of positions for C and D.

⑥ R **21** A and B are the points $(-5, 2)$ and $(1, 3)$. If $ABCD$ is a square, find two pairs of positions for C and D.

⑥ • **22** A and C are the points $(-1, 1)$ and $(5, 3)$. Find a vector which is perpendicular to $\overrightarrow{AC}$ and of the same magnitude as $\overrightarrow{AC}$.
   Find the midpoint of AC.
   If $ABCD$ is a square, find the coordinates of B and D.

⑥ → **23** A and C are the points $(2, 1)$ and $(-4, 7)$ and $ABCD$ is a square. Find the coordinates of B and D.

⑥ R **24** The centre of a square (i.e. the point of intersection of its diagonals) is the point $(3, 2)$ and one of its vertices is the point $(1, -4)$. Find the coordinates of its other vertices.

⑥ **25** A, B and P are the points $(3, -1)$, $(-2, 2)$ and (x, y). If AP is perpendicular to BP, find an equation connecting x and y. Describe the set of points P.

⑦ → **26** A and B are the points with position vectors $-5\mathbf{i} + 2\mathbf{j}$ and $7\mathbf{i} - 4\mathbf{j}$.

   (i) Find the position vectors of the points which divide AB *internally* in the ratio:
   (a) $1:1$; (b) $2:1$; (c) $3:1$; (d) $1:3$.

   (ii) Find the position vectors of the points which divide AB *externally* in the ratio:
   (a) $4:3$; (b) $3:1$; (c) $5:1$.

Mark these points on a diagram.

⑦ **27** A and B are the points with position vectors $2\mathbf{i} + 3\mathbf{j}$ and $-4\mathbf{i} + 6\mathbf{j}$. Find the position vectors of the points which divide:

   (i) AB (ii) BA

externally in the ratio:
(a) $3:2$; (b) $5:2$; (c) $4:1$.
Mark these points on a diagram.

⑦ R **28** A and B are the points with position vectors $-\mathbf{i} + \mathbf{j}$ and $2\mathbf{i} + 2\mathbf{j}$. Find the position vectors of the points which divide AB:

   (i) internally   (ii) externally

in the ratio $3:2$.

⑦ **29** A and B are the points (x_1, y_1) and (x_2, y_2). Find the coordinates of the point which divides AB in the ratio $m:n$.
   Hence write down the coordinates of the midpoint of AB.

Vectors, coordinate geometry and complex numbers 299

⑦ 30 A and B are the points with position vectors $4\mathbf{i} - \mathbf{j}$, $\mathbf{i} + 3\mathbf{j}$. Show that $AB = 5$.

(i) C is the point on AB produced such that $AC = 15$. Show that C divides AB externally in the ratio $3:2$. Hence find the position vector of C.

(ii) D is the point on BA produced such that $AD = 15$. Show that D divides BA externally in the ratio $4:3$. Hence find the position vector of D.

⑦ 31 A and B are the points with position vectors $-3\mathbf{i}$, $\mathbf{i} + 2\mathbf{j}$. Find the position vectors of the two points on the line AB distant $3\sqrt{5}$ units from A.

⑦ 32 A and B are the points $(3, 4)$ and $(12, 5)$. C is the point on AB such that OC bisects $A\hat{O}B$ internally. Find the position vector of C. (Hint: use the angle bisector theorem; see 6.1:3 Q27).

⑦ 33 A, B and C are the points $(-1, 1)$, $(1, 2)$ and $(2, 7)$. D is the point on BC such that AD bisects $B\hat{A}C$ internally. Find the position vector of D.

34 Show that the points A, B and C are collinear if $\overrightarrow{AB} = \lambda \overrightarrow{AC}$, and vice-versa. Hence investigate whether or not the following sets of points are collinear:

• (i) $\mathbf{i} + \mathbf{j}$, $2\mathbf{i} + 3\mathbf{j}$, $5\mathbf{i} + 9\mathbf{j}$
   (ii) the origin, $2\mathbf{i} + \mathbf{j}$, $-6\mathbf{i} - 3\mathbf{j}$
   (iii) $(2, 1)$, $(3, 4)$, $(-1, 0)$

Sketch diagrams to illustrate your results. (C.f. 6.1:3 Q5-7.)

R 35 If the points $2\mathbf{i} + 3\mathbf{j}$, $\mathbf{i} + \mathbf{j}$ and $-2\mathbf{i} + p\mathbf{j}$ are collinear, find p.

→ 36 If the points $-3\mathbf{i} + \mathbf{j}$, $-2\mathbf{i} + q\mathbf{j}$ and $2\mathbf{i} + 2\mathbf{j}$ are collinear, find q.

37 A, B and C are the points with position vectors $\mathbf{a} = 4\mathbf{i} - \mathbf{j}$, $\mathbf{b} = 2\mathbf{i} + \mathbf{j}$ and $\mathbf{c} = -2\mathbf{i} + 5\mathbf{j}$. Express $\mathbf{c}$ in the form $\mathbf{a} + \lambda(\mathbf{b} - \mathbf{a})$. Find the position vector of the point H on the line ABC for which $AH = AC$.

38 If the points $-\mathbf{i} + \mathbf{j}$, $x\mathbf{i} + y\mathbf{j}$ and $3\mathbf{i} + 2\mathbf{j}$ are collinear, find an equation connecting x and y. What is the significance of this equation?

39 Write down the position vectors of A, B and C and hence, using 6.1:2 Theorem 1, find the coordinates of the centroid of $\triangle ABC$ when A, B and C are the points.

• → (i) $(-1, 4)$, $(3, 5)$, $(4, 0)$
    (ii) $(3, 2)$, $(2, -1)$, $(7, 5)$
    (iii) $(0, 0)$, $(4, 1)$, $(1, 4)$
    (iv) $(-3, 4)$, $(5, 0)$, $(-2, -1)$

Illustrate your answers by sketching diagrams.
Write down the coordinates of the centroid of the triangle whose vertices are (x_1, y_1), (x_2, y_2) and (x_3, y_3).

→ 40 Find the position vector of the centroid G of the $\triangle ABC$ where A, B and C are the points with position vectors $-2\mathbf{i} + 3\mathbf{j}$, $6\mathbf{i} - \mathbf{j}$ and $2\mathbf{i} - 5\mathbf{j}$.
If AD, BE and CF are the medians of the triangle, verify that G divides each of AD, BE and CF internally in the ratio $2:1$.

R 41 A, B and C are the points $(1, 5)$, $(-3, -7)$ and $(9, 9)$. Find the centroid of $\triangle ABC$.
The points L, M and N divide internally AB, BC and CA respectively in the ratio $1:3$. Find the centroid of $\triangle LMN$. What do you notice?

6.2:3 Equation of a line: parametric form

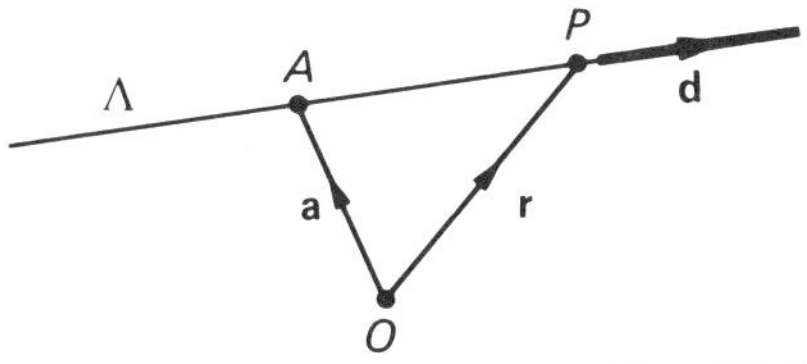

We have seen (section 6.1:3) that a vector equation of the line Λ passing through the point A and parallel to the vector $\mathbf{d}$ (fig. 6.2:13) is

$$\mathbf{r} = \mathbf{a} + \lambda \mathbf{d}$$

Fig. 6.2:13

Worked example 8

(i) Find a vector equation of the line passing through the points $A(3, -2)$ and $B(-1, 4)$.
(ii) Hence find the cartesian equation of the line.

(i) Direction of line $\mathbf{d} = \overrightarrow{AB} = \mathbf{b} - \mathbf{a} = -4\mathbf{i} + 6\mathbf{j}$.

So a vector equation of the line is:

$$\mathbf{r} = \mathbf{a} + \lambda(\mathbf{b} - \mathbf{a}) = 3\mathbf{i} - 2\mathbf{j} + \lambda(-4\mathbf{i} + 6\mathbf{j}) \tag{9}$$

(Note: $\lambda = 0$ corresponds to the point A, $\lambda = 1$ to the point B.)

300 *Equation of a line: parametric form*

(ii) Let the general point P on the line have coordinates (x, y); so $\mathbf{r} = x\mathbf{i} + y\mathbf{j}$. From equation (9):

$$x\mathbf{i} + y\mathbf{j} = 3\mathbf{i} - 2\mathbf{j} + \lambda(-4\mathbf{i} + 6\mathbf{j})$$

So
$$\left.\begin{array}{l} x = 3 - 4\lambda \\ y = -2 + 6\lambda \end{array}\right\} \qquad (10)$$

These are the **parametric equations** of the line. Eliminating λ we have:

$$\frac{x-3}{-4} = \frac{y+2}{6}$$

$$3x + 2y = 5 \qquad (11)$$

$$y = -\tfrac{3}{2}x + \tfrac{5}{2} \qquad (12)$$

Note: we can write the cartesian equation of a line in either of the forms

$$Ax + By = C \quad \text{or} \quad y = mx + c$$

(See 1.2:2, equations (8) and (9).)

Worked example 9 Find a vector equation of the line with cartesian equation $3x - 4y = -8$.

$$3x - 4y = -8 \Rightarrow y = \tfrac{3}{4}x + 2$$

The gradient of the line is $\tfrac{3}{4}$; so the line is parallel to the vector $4\mathbf{i} + 3\mathbf{j}$ (see fig. 6.2:**14**).

The line passes through the point $A(0, 2)$ with position vector $2\mathbf{j}$. So a vector equation of the line is

$$\mathbf{r} = 2\mathbf{j} + \lambda(4\mathbf{i} + 3\mathbf{j})$$

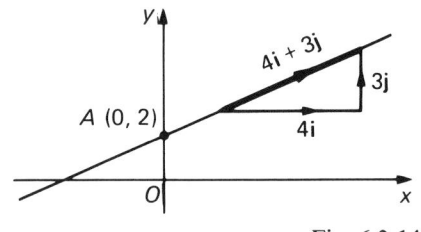

Fig. 6.2:14

Exercise 6.2:3

⑧ **1** Find a vector equation and the cartesian equation of the line:

- (i) parallel to the vector $\mathbf{i} + 2\mathbf{j}$ and passing through the point $\mathbf{i} + \mathbf{j}$
- (ii) passing through the origin and the point $3\mathbf{i} - \mathbf{j}$
- → (iii) passing through the points $4\mathbf{i} - \mathbf{j}$ and $\mathbf{i} + 2\mathbf{j}$
- (iv) passing through the points $-3\mathbf{i} - 2\mathbf{j}$, $3\mathbf{j}$

⑧ **2** (i) Show that the line $\mathbf{r} = 3\mathbf{i} + \mathbf{j} + \lambda(4\mathbf{i} - 6\mathbf{j})$ is parallel to the line $\mathbf{r} = \mathbf{i} - \mathbf{j} + \mu(-2\mathbf{i} + 3\mathbf{j})$.
Find the cartesian equations of these lines.

(ii) Find a vector equation of the line passing through the point $2\mathbf{i} + \mathbf{j}$, parallel to the line $\mathbf{r} = 4\mathbf{i} + \lambda(-3\mathbf{i} + 4\mathbf{j})$.

⑧ **R 3** Find vector and cartesian equations of the lines through the point $\mathbf{i} + 3\mathbf{j}$ which are:

(i) parallel   (ii) perpendicular

to the line $\mathbf{r} = 2\mathbf{i} - \mathbf{j} + \lambda(\mathbf{i} - \mathbf{j})$.

⑧ **4** Find vector and cartesian equations of the line perpendicular to the vector $A\mathbf{i} + B\mathbf{j}$ passing through the point $(C/B)\mathbf{j}$.

- **5** A and B are the points with position vectors $-\mathbf{i} + \mathbf{j}$ and $3\mathbf{i} + 3\mathbf{j}$. Find:

  (i) the position vector of the midpoint M of AB
  (ii) the vector $\overrightarrow{AB}$
  (iii) a vector perpendicular to $\overrightarrow{AB}$

  Hence find a vector equation and the cartesian equation of the line through M perpendicular to AB, i.e. the perpendicular bisector of AB (see 1.2:2 Q19).

6 Find a vector equation and the cartesian equation of the perpendicular bisector of AB, when A and B are the points with position vectors

→ (i) $-3\mathbf{i} - \mathbf{j}$ and $7\mathbf{i} + \mathbf{j}$ (ii) $a\mathbf{i} + b\mathbf{j}$ and $2a\mathbf{i} + 3b\mathbf{j}$

⑧ : 7 Which of the following pairs of equations represent the same line? Explain your answers.

(i) $\mathbf{r}=2\mathbf{i}+\mathbf{j}+\lambda(\mathbf{i}-\mathbf{j})$; $\mathbf{r}=2\mathbf{i}+\mathbf{j}+\mu(-2\mathbf{i}+2\mathbf{j})$
(ii) $\mathbf{r}=\mathbf{i}-4\mathbf{j}+\lambda(-2\mathbf{i}+3\mathbf{j})$; $\mathbf{r}=-3\mathbf{i}+2\mathbf{j}+\mu(2\mathbf{i}-3\mathbf{j})$
(iii) $\mathbf{r}=\lambda(4\mathbf{i}-3\mathbf{j})$; $\mathbf{r}=\mu(3\mathbf{i}-4\mathbf{j})$

⑧ → 8 (i) Find a vector equation of the line through the points (x_1, y_1) and (x_2, y_2). Show that the cartesian equation of this line may be written

$$\frac{y-y_1}{x-x_1}=\frac{y_2-y_1}{x_2-x_1}$$

(ii) Find a vector equation of the line whose intercepts on the x- and y-axes are a and b. Show that the cartesian equation of this line may be written

$$\frac{x}{a}+\frac{y}{b}=1$$

⑨ 9 Find vector equations of the lines with cartesian equations:

(i) $y=x+1$ (ii) $y=-2x-3$
→(iii) $3x+4y=12$ (iv) $x-3y+2=0$
(v) $x=5$ (vi) $y=-1$

⑨ 10 Find vector equations of the lines with cartesian equations:

(i) $y=mx+c$ (ii) $Ax+By=C$

⑨ 11 Find, in the form $a\mathbf{i}+b\mathbf{j}$, the direction of the line:

(i) $y=4x+3$ (ii) with gradient $-\frac{1}{2}$
(iii) $3x-5y+7=0$

⑨ 12 Find the gradient of the line whose direction is:

→ (i) $3\mathbf{i}+2\mathbf{j}$ (ii) $-\mathbf{i}+\mathbf{j}$
(iii) $\mathbf{i}$ (iv) $a\mathbf{i}+b\mathbf{j}$

• → 13 Find values of λ and μ satisfying the equation

$$\mathbf{i}-\mathbf{j}+\lambda(3\mathbf{i}+2\mathbf{j})=\mathbf{j}+\mu(\mathbf{i}+2\mathbf{j})$$

Hence find the position vector of the point of intersection of the lines $\mathbf{r}=\mathbf{i}-\mathbf{j}+\lambda(3\mathbf{i}+2\mathbf{j})$ and $\mathbf{r}=\mathbf{j}+\mu(\mathbf{i}+2\mathbf{j})$.
Sketch a diagram illustrating your result.

14 (i) Using the method of question 13, find the position vector of the point of intersection of the lines $\mathbf{r}=(\mathbf{i}+2\mathbf{j})+\lambda(\mathbf{i}-\mathbf{j})$ and $\mathbf{r}=-3\mathbf{j}+\mu(-\mathbf{i}+4\mathbf{j})$.

(ii) Write down the cartesian equations of the lines and by solving these equations simultaneously, check your answer to (i).

• → 15 (i) Attempt, using the method of question 13, to find the point of intersection of the lines $\mathbf{r}=3\mathbf{i}+2\mathbf{j}+\lambda(-\mathbf{i}+3\mathbf{j})$ and $\mathbf{r}=\mathbf{i}-5\mathbf{j}+\mu(2\mathbf{i}-6\mathbf{j})$. What goes wrong? Why?

(ii) Comment on the possible solutions of the following pairs of simultaneous equations, with reference to their geometric representation:
(a) $2x+y=1$; $4x+2y=3$
(b) $3x-2y=2$; $6x-4y=4$

16 A, B and C are the points $(6,0)$, $(-2,-4)$ and $(2,10)$. AD, BE and CF are the medians of $\triangle ABC$. Find the cartesian equations of AD and BE and hence find their point G of intersection.
Show that G lies on CF.
Find the ratio $CG:GF$.

R 17 A and B are the points $4\mathbf{i}-2\mathbf{j}$ and $6\mathbf{i}+2\mathbf{j}$. $OABC$ is a square. L and M are the midpoints of AB and BC. Find vector equations of the lines OL and AM. Hence find the position vector of their point of intersection K. If BK meets OA at N, show that $ON=\frac{2}{3}OA$.

6.2:4 Equation of a line: scalar product form

Suppose the line Λ passes through the point A and is perpendicular to the vector $\mathbf{n}$, as shown in fig. 6.2:15. Then, if P is any point on the line with position vector $\mathbf{r}$, $\overrightarrow{AP}$ is perpendicular to $\mathbf{n}$; so $\overrightarrow{AP}\cdot\mathbf{n}=0$, i.e. $(\mathbf{r}-\mathbf{a})\cdot\mathbf{n}=0$, so

$$\mathbf{r}\cdot\mathbf{n}=\mathbf{a}\cdot\mathbf{n} \tag{13}$$

This gives a condition that a general point P lies on Λ, so it is a *vector equation* of Λ.
If $\mathbf{a}=a_1\mathbf{i}+a_2\mathbf{j}$, $\mathbf{n}=A\mathbf{i}+B\mathbf{j}$, the vector equation of Λ is

$$\mathbf{r}\cdot(A\mathbf{i}+B\mathbf{j})=C \quad \text{where } C=a_1A+a_2B \tag{14}$$

Let P be the point (x, y), so that $\mathbf{r}=x\mathbf{i}+y\mathbf{j}$; then $(x\mathbf{i}+y\mathbf{j})\cdot(A\mathbf{i}+B\mathbf{j})=C$

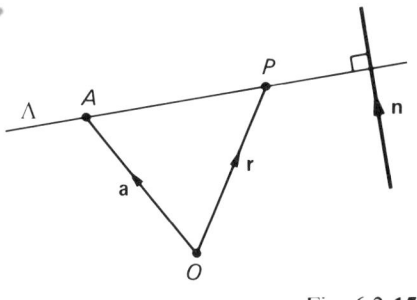

i.e. $$Ax+By=C \tag{15}$$

Fig. 6.2:15

This is the cartesian equation of Λ in a familiar form; we can now see the significance of the numbers A and B in the equation $Ax + By = C$ of a straight line: $A\mathbf{i} + B\mathbf{j}$ is a vector perpendicular to the line. For example the vector $3\mathbf{i} - 4\mathbf{j}$ is perpendicular to the line $3x - 4y = 8$.

Worked example 10 Find the vector equation (scalar product form) and the cartesian equation of the line passing through the points $A(3, -2)$ and $B(-1, 4)$.

Direction of line $\mathbf{d} = \overrightarrow{AB} = \mathbf{b} - \mathbf{a} = -4\mathbf{i} + 6\mathbf{j}$.
A vector perpendicular to $\mathbf{d}$ is $\mathbf{n} = 6\mathbf{i} + 4\mathbf{j}$ (see 6.2:1 Q30). So the vector equation of L is

$$\mathbf{r} \cdot (6\mathbf{i} + 4\mathbf{j}) = (3\mathbf{i} - 2\mathbf{j}) \cdot (6\mathbf{i} + 4\mathbf{j}) \Rightarrow \mathbf{r} \cdot (6\mathbf{i} + 4\mathbf{j}) = 10$$

and so its cartesian equation is $6x + 4y = 10$, i.e. $3x + 2y = 5$.

Theorem 2: The shortest distance from the point L with position vector $\mathbf{l}$ to the line Λ: $\mathbf{r} \cdot \mathbf{n} = d$ is

$$\left| \frac{d - \mathbf{l} \cdot \mathbf{n}}{n} \right| \quad (16)$$

Proof
Let M be the point on the line closest to L (fig. 6.2:16). Then LM is perpendicular to Λ, so $\overrightarrow{LM}$ is in the direction of $\mathbf{n}$.

$$p = LA \cos \theta = \frac{\overrightarrow{LA} \cdot \mathbf{n}}{n} = \frac{(\mathbf{a} - \mathbf{l}) \cdot \mathbf{n}}{n}$$

$$= \frac{\mathbf{a} \cdot \mathbf{n} - \mathbf{l} \cdot \mathbf{n}}{n}$$

$$= \frac{d - \mathbf{l} \cdot \mathbf{n}}{n}$$

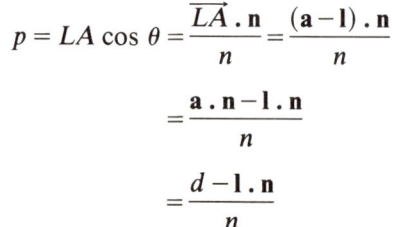

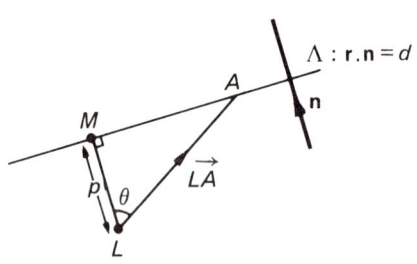

Fig. 6.2:16

Note: p positive $\Leftrightarrow \overrightarrow{LM}$ is in the same direction as $\mathbf{n}$;
p negative $\Leftrightarrow \overrightarrow{LM}$ is in the opposite direction to $\mathbf{n}$.

Corollary: The shortest distance from the point (h, k) to the line $Ax + By = C$ is

$$\left| \frac{Ah + Bk - C}{\sqrt{(A^2 + B^2)}} \right| \quad (17)$$

Proof
See question 11.

Worked example 11 Find the distance of the line Λ with equation $3x - 4y + 8 = 0$ from the points: (i) $L(5, 2)$ (ii) $L'(-2, 3)$

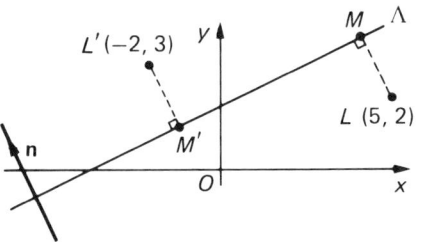

The line Λ and points L and L' are shown in fig. 6.2:17.

(i) Distance from L to $\Lambda = \left|\dfrac{3 \cdot 5 - 4 \cdot 2 + 8}{\sqrt{(3^2 + 4^2)}}\right|$ from expression (17)

$$= \left|\dfrac{15}{5}\right| = |3| = 3$$

Fig. 6.2:17

(ii) Distance from L' to $\Lambda = \left|\dfrac{3(-2) - 4 \cdot 3 + 8}{\sqrt{3^2 + 4^2}}\right| = \left|\dfrac{-10}{5}\right| = |-2| = 2$

The opposite signs within the moduli occur because L and L' are on opposite sides of Λ; in particular,
$\overrightarrow{LM}$ is in the same direction as $\mathbf{n} = -3\mathbf{i} + 4\mathbf{j}$
$\overrightarrow{L'M'}$ is in the opposite direction to $\mathbf{n}$.

Angle between two lines

Consider the lines $\Lambda_1: \mathbf{r} \cdot \mathbf{n}_1 = d_1$ and $\Lambda_2: \mathbf{r} \cdot \mathbf{n}_2 = d_2$ (fig. 6.2:18).

angle θ between Λ_1 and Λ_2 = angle between $\mathbf{n}_1$ and $\mathbf{n}_2$

So

$$\cos\theta = \dfrac{\mathbf{n}_1 \cdot \mathbf{n}_2}{n_1 n_2} \quad (= \hat{\mathbf{n}}_1 \cdot \hat{\mathbf{n}}_2) \tag{18}$$

Fig. 6.2:18

Worked example 12 Find the angle θ between the lines $3x - 4y + 8 = 0$ and $x + y - 3 = 0$.

A vector perpendicular to $3x - 4y + 8 = 0$ is $3\mathbf{i} - 4\mathbf{j}$. A vector perpendicular to $x + y - 3 = 0$ is $\mathbf{i} + \mathbf{j}$. So

$$\cos\theta = \dfrac{(3\mathbf{i} - 4\mathbf{j}) \cdot (\mathbf{i} + \mathbf{j})}{|3\mathbf{i} - 4\mathbf{j}||\mathbf{i} + \mathbf{j}|} = -\dfrac{1}{5\sqrt{2}}$$

So $\theta \approx 98.13°$.
(Note: we have found the obtuse angle between the lines. The acute angle between them is $180° - 98.13° \approx 81.87°$.)

Exercise 6.2:4

(10) **1** Find the vector equation, in scalar product form, and hence the cartesian equation of the line:

(i) perpendicular to the vector $3\mathbf{i} - 2\mathbf{j}$, passing through the point $\mathbf{i} + \mathbf{j}$
(ii) perpendicular to the vector $\mathbf{i} + 4\mathbf{j}$, passing through the point $3\mathbf{i}$
→ (iii) passing through the points $-\mathbf{i} + 2\mathbf{j}$ and $3\mathbf{i} - \mathbf{j}$
(iv) passing through the points $5\mathbf{i}$ and $-3\mathbf{j}$

(10) **2** Write down a vector perpendicular to the line:

→ (i) $3x + 2y - 7 = 0$
(ii) $x - 3y + 2 = 0$
(iii) $y = 4x + 1$
(iv) $y = -\tfrac{3}{7}x - 1$
(v) $y = 5$

(10) **3** Write down, in both parametric and scalar product form, vector equations of the line passing through the points:

(i) $\mathbf{j}$ and $4\mathbf{i} - 5\mathbf{j}$
(ii) $\mathbf{i} + \mathbf{j}$ and $-3\mathbf{i} - 3\mathbf{j}$

(10) **4** Write down the cartesian equations of each of these lines and hence find the position vector of their point of intersection:

$$\mathbf{r} \cdot (4\mathbf{i} - 3\mathbf{j}) = 5; \quad \mathbf{r} \cdot (\mathbf{i} + \mathbf{j}) = 3$$

Check that the vector satisfies both equations.
Illustrate your answer on a diagram.

304 *Equation of a line: scalar product form*

(10) → **5** Find vector and cartesian equations of the line through the point with position vector $\mathbf{a} = p\cos\alpha\,\mathbf{i} + p\sin\alpha\,\mathbf{j}$, perpendicular to the vector $\mathbf{a}$ (p and α constants). Sketch the line on a diagram showing the significance of p and α.

(10) R **6** Sketch the line $\Lambda: \tfrac{1}{2}\sqrt{3}x + \tfrac{1}{2}y = 2$.
If N is the point on Λ closest to O find:
 (i) the distance ON
 (ii) the angle between ON and Ox
Write down a unit vector perpendicular to Λ.

(10) → **7** Write down the cartesian equation of the straight line:
 (i) with gradient m and y-intercept c
 (ii) passing through (x_1, y_1), with gradient m
 (iii) passing through (x_1, y_1) and (x_2, y_2)
 (iv) whose shortest distance ON from O is p, where $N\hat{O}x = \alpha$
 (v) with x-intercept a and y-intercept b

Illustrate your answers with sketches.

8 In each case, find the distance from the point to the line:
 • → (i) $-4\mathbf{i}+\mathbf{j}$; $\mathbf{r}\cdot(3\mathbf{i}+4\mathbf{j}) = 2$
 (ii) $3\mathbf{i}-3\mathbf{j}$; $\mathbf{r}\cdot(\mathbf{i}-\mathbf{j}) = 4$
 (iii) the origin; $\mathbf{r}\cdot(\mathbf{i}+7\mathbf{j}) = 10$
 (iv) $\mathbf{i}+2\mathbf{j}$; $\mathbf{r}\cdot(2\mathbf{i}+3\mathbf{j}) = 8$

What do you deduce in (iv)? Check it.

9 Write the equation of the line $\mathbf{r} = 3\mathbf{i}+\mathbf{j}+\lambda(\mathbf{i}+2\mathbf{j})$ in scalar product form. Hence find the distance from the line to the point $\mathbf{i}-\mathbf{j}$.

R **10** Find the distance from the line $\Lambda: \mathbf{r}\cdot(\mathbf{i}+3\mathbf{j}) = 2$ to the point:
 (i) L_1 with position vector $\mathbf{i}+2\mathbf{j}$
 (ii) L_2 with position vector $-\mathbf{j}$

Are L_1 and L_2 on opposite sides of Λ or the same side? Is L_1 the mirror image of L_2 in Λ?
Illustrate your answers on a diagram.

11 Find the equation of the line $Ax + By = C$ in scalar product form. Hence show that the distance from the line to the point (h, k) is

$$\left| \frac{Ah + Bk - C}{\sqrt{(A^2 + B^2)}} \right|$$

Deduce the distance from the line $y = mx + c$ to the point (h, k).

(11) **12** In each case find the distance from the point to the line:
 → (i) $(2, 2)$; $5x + 12y + 5 = 0$
 (ii) $(0, -1)$; $4x - 3y = 7$
 (iii) $(3, 7)$; $y = -4x + 2$

(11) **13** Find the distance from the origin to the following lines:
 (i) $5x + 2y = 3$
 (ii) $5x - 2y = 3$
 (iii) $2x - 5y = 3$
 (iv) $2x + 5y = 3$

What do you deduce?
Illustrate your answers by sketching a diagram.

(11) **14** Determine, without using a diagram, whether L_1 and L_2 are on the same or opposite sides of Λ in each case:
 → (i) $L_1(1, 1)$; $L_2(3, -4)$; $\Lambda: 3x + y - 1 = 0$
 (ii) $L_1(-1, 0)$; $L_2(2, 3)$; $\Lambda: 7x - y + 5 = 0$

Check your answers by sketching diagrams.

(11) → **15** The vertices of a triangle are the point $A(1, 2)$, $B(3, 1)$, $C(-1, -2)$. Find the length BC and the shortest distance from A to BC. Hence find the area of $\triangle ABC$.

(11) **16** Find the area of $\triangle ABC$ when A, B and C are the points:
 (i) $(-1, 2)$, $(3, -1)$, $(2, 3)$
 (ii) $(0, 0)$, $(2, 6)$, $(5, 5)$
 (iii) (x_1, y_1), (x_2, y_2), (x_3, y_3)

(11) **17** Show that $C(7, -2)$ and $D(1, 6)$ are equidistant from the line $3x - 4y - 4 = 0$. Is C the mirror image of D in this line?

(11) **18** A point $P(X, Y)$ is equidistant from the line $4x + 3y - 3 = 0$ and the point $(1, 2)$. Find the equation connecting X and Y.

• **19** A point $P(X, Y)$ is equidistant from the lines $\Lambda_1: x + 2y - 1 = 0$ and $\Lambda_2: 2x - y + 2 = 0$. Show that $X + 2Y - 1 = \pm(2X - Y + 2)$.
Hence show that $X - 3Y + 3 = 0$ or $3X + Y + 1 = 0$. (Note: these are the equations of the two lines which bisect the angles between Λ_1 and Λ_2, called the **angle bisectors** of Λ_1 and Λ_2.)

(11) **20** Find the equations of the angle bisectors of the lines:
 (i) $3x + 2y + 1 = 0$ and $3x - 2y + 17 = 0$
 → (ii) $3x - 4y - 11 = 0$ and $12x + 5y - 2 = 0$
 (iii) $x - y - 1 = 0$ and $7x - y + 3 = 0$

(11) R **21** A, B and C are the points $(1, 0)$, $(4, -1)$ and $(3, -6)$. Find the equation of the line which bisects $B\hat{A}C$. If this line meets BC in D, show that D divides BC in the ratio $AB : AC$.

(11) • **22** Find the distances from the origin to the (parallel) lines:
 (i) $3x - y + 3 = 0$
 (ii) $3x - y + 8 = 0$

Hence find the distance between the lines.

23 Find the distance between the lines:
   (i) $5x+12y+7=0$ and $5x+12y-19=0$
   (ii) $x-2y-4=0$ and $2x-4y+7=0$

24 Find the equations of the lines parallel to and distant one unit from $3x-4y+2=0$.

25 Find the angles between the following pairs of lines:
   (i) $\mathbf{r}\cdot(\mathbf{i}-\mathbf{j})=5;\ \mathbf{r}\cdot(7\mathbf{i}-\mathbf{j})=-6$
   (ii) $\mathbf{r}\cdot(2\mathbf{i}+\mathbf{j})=7;\ \mathbf{r}\cdot(3\mathbf{i}-2\mathbf{j})=0$
   (iii) $\mathbf{r}\cdot(3\mathbf{i}-5\mathbf{j})=4;\ \mathbf{r}\cdot(5\mathbf{i}+3\mathbf{j})=8$
   (iv) $\mathbf{r}\cdot(\mathbf{i}-2\mathbf{j})=1;\ \mathbf{r}\cdot(-4\mathbf{i}+8\mathbf{j})=1$

What do you deduce in (iii); what in (iv)?

26 Find the angles between the following pairs of lines:
   (i) $x-y+9=0;\ 3x+y+7=0$
   (ii) $4x+3y=1;\ x+2y=6$
   (iii) $y=3x+5;\ y=-2x$

27 Suppose that the lines $y=mx+c$, $y=m'x+c'$ ($m>m'$) make angles α and β with the x-axis. Write down the relations between m and α, m' and β. If θ is the angle between the lines, find θ in terms of α and β. Hence show that

$$\tan\theta = \frac{m-m'}{1+mm'} \quad \text{(see 3.2:3 Q19)}$$

Hence write down the conditions for the lines to be:
   (i) parallel (ii) perpendicular.

28 Write down the vector equations of the lines $y=mx+c$ and $y=m'x+c'$ in scalar product form. Find the cosine of the angle between them; hence check your answer to question 27.

29 Using the formula you found in question 27, find the angles between the lines:
   (i) $y=3x+5,\ y=-2x$ (c.f. question 26(iii))
   (ii) $y=x+2,\ y=\tfrac{3}{4}x+5$

30 Using the formula you found in question 27, find the tangent of the angle between the lines $Ax+By=C$ and $A'x+B'y=C'$.

Hence find the angle between the lines $4x+3y=1$ and $x+2y=6$ (c.f. question 26(ii)).

6.2:5 Loci

Definition The **locus** of a variable point $P(x,y)$ is the set of points which P can occupy.

In the xy plane there are two kinds of loci:
(i) curves and straight lines: the cartesian equation of a curve or a line is the equation connecting x and y which is satisfied by all points $P(x,y)$ on it;
(ii) regions of the plane: these loci are defined by inequalities connecting x and y. (We will not be studying these in this book.)

Worked example 13 A and B are the points $(3,4)$ and $(-1,2)$. Find the cartesian equation of the locus of the point $P(x,y)$ which moves so that:
(i) $AP=BP$ (ii) $AP=2BP$

$$AP^2=(x-3)^2+(y-4)^2 \quad BP^2=(x+1)^2+(y-2)^2$$

(i) $AP=BP \Rightarrow AP^2=BP^2 \Rightarrow (x-3)^2+(y-4)^2=(x+1)^2+(y-2)^2$
$$\Rightarrow 2x+y-5=0$$

The locus is the perpendicular bisector of AB (fig. 6.2:**19a**).

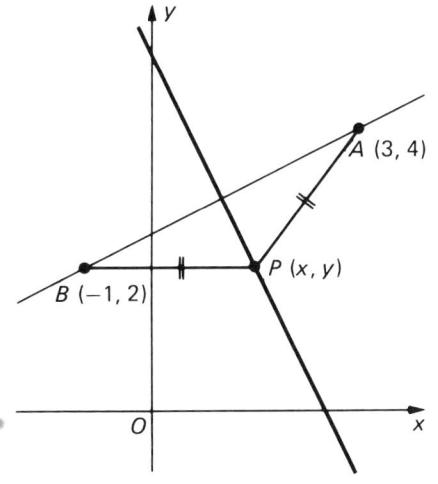

(ii) $AP=2BP \Rightarrow AP^2=4BP^2 \Rightarrow (x-3)^2+(y-4)^2=4\{(x+1)^2+(y-2)^2\}$
$$\Rightarrow 3x^2+3y^2+14x-8y-5=0$$

This, as we shall see in section 6.2:6, is the cartesian equation of a circle (fig. 6.2:**19b**). Note that there are two points C and D on AB which satisfy the equation, and that CD is the diameter of the circle.

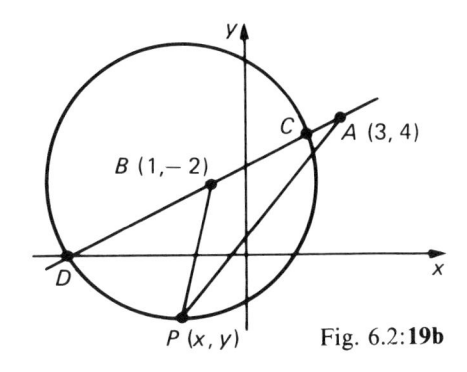

Exercise 6.2:5

1 C is the point $(4, 1)$. A point $P(x, y)$ moves so that the distance $CP = 2$. Describe the locus of P, sketch it and find its cartesian equation.

2 C is a fixed point. A point $P(x, y)$ moves so that the distance $CP = k$. Sketch the locus of P and find its cartesian equation when:

(i) C is the point $(3, 2)$, $k = 1$
(ii) C is the origin, $k = 4$
(iii) C is the point $(-3, 4)$, $k = 5$

In (iii) show that the origin lies on the locus.

3 A and B are fixed points. A point $P(x, y)$ moves so that $AP = BP$. Describe the locus of P, sketch it and find its cartesian equation when A and B are the points:

(i) $A(2, 1)$, $B(5, 4)$
(ii) $A(0, 0)$, $B(1, -2)$
(iii) $A(-3, 2)$, $B(5, 2)$

4 Repeat question 3 when $AP = 2BP$.

5 A and B are the points $(-1, -2)$ and $(3, 1)$. A point $P(x, y)$ moves so that the distance $AP = kBP$. Sketch the locus of P and find its cartesian equation when:

(i) $k = 1$ (ii) $k = 2$ (iii) $k = 3$

6 A and B are two points. A point $P(x, y)$ moves so that the distance $AP = kBP$. Sketch the locus of P when:

(i) $k = 4$ (ii) $k = 3$ (iii) $k = 2$
(iv) $k = 1$ (v) $k = \frac{1}{2}$ (vi) $k = \frac{1}{3}$

7 A and B are fixed points. A point $P(x, y)$ moves so that AP is perpendicular to BP. Describe the locus of P, sketch it and find its cartesian equation when A and B are the points:

(i) $(-2, 1)$ and $(4, 5)$
(ii) $(1, 0)$ and $(5, 0)$

(See 6.2:2 Q25.)

8 A and B are the points $(2, 0)$ and $(0, 2)$. Describe the locus of a point $P(x, y)$ which moves so that the angle $A\widehat{P}B$ is

(i) $0°$ (ii) $30°$
(iii) $60°$ (iv) $90°$

Sketch these loci on a single diagram, and find their cartesian equations.

9 A and B are the points $(-1, 0)$ and $(1, 0)$. A point $P(x, y)$ moves so that the sum of the distances AP and BP is always equal to 4. Find the cartesian equation of the locus of P, and sketch it.

10 S is the point $(3, 4)$ and Λ is the line $x + y - 2 = 0$. A point $P(x, y)$ moves so that $SP = PN$ where PN is the shortest distance from P to Λ (i.e. N is the point on Λ nearest to P). Find the cartesian equation of the locus of P. (Hint: write down SP; using 6.2:4 equation (17), write down PN. Put $SP^2 = PN^2$.)

Using the equation $SP = PN$, sketch the locus (it's a parabola—see section 6.2:7).

11 S is a fixed point and Λ is a fixed line. A point $P(x, y)$ moves so that $SP = PN$ where PN is the shortest distance from P to Λ. Sketch the locus of P and find its cartesian equation when:

(i) S is the point $(3, 3)$ and Λ is the line $x + y + 1 = 0$
(ii) S is the point $(1, 0)$ and Λ is the line $2x - y + 3 = 0$
(iii) S is the point $(1, 0)$ and Λ is the line $x = -1$

12 Λ_1 and Λ_2 are the lines $x + y - 1 = 0$ and $x - 2y + 2 = 0$. A point $P(x, y)$ moves so that its shortest distance from Λ_1 is equal to its shortest distance from Λ_2. Describe the locus of P and sketch it. Find the cartesian equations of the two lines on which P may lie (c.f. 6.2:4 Q19).

13 For each of the following pairs of lines, find the cartesian equations of the two angle bisectors:

(i) $3x + 4y - 1 = 0$; $4x - 3y + 1 = 0$
(ii) $3x + 4y - 1 = 0$; $3x - 4y + 1 = 0$
(iii) $x = 1$; $y = 2$
R (iv) $x + 3y + 1 = 0$; $2x - y - 1 = 0$

6.2:6 The circle

Definition A circle is the locus of a point P which moves so that its distance from a fixed point C is constant, i.e. $CP = r$, say.

If C is the point (h, k) and $P(x, y)$ (see fig. 6.2:**20**),

$$CP^2 = (x - h)^2 + (y - k)^2$$

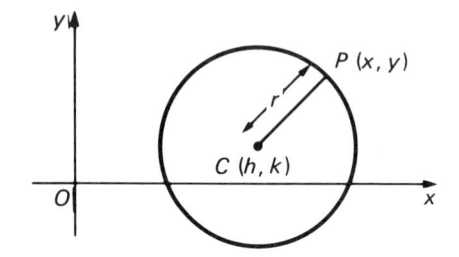

Fig. 6.2:**20**

Vectors, coordinate geometry and complex numbers 307

$CP = r \Rightarrow CP^2 = r^2$, so

$$(x-h)^2 + (y-k)^2 = r^2 \tag{19a}$$
i.e. $$x^2 + y^2 - 2hx - 2ky + (k^2 + k^2 - r^2) = 0 \tag{19b}$$

This is the equation of a circle centre $C(h, k)$, radius r. In particular, the equation of a circle centre O, radius r, is

$$x^2 + y^2 = r^2 \tag{20}$$

Worked example 14 Find the centre and radius of the circle with equation

$$x^2 + y^2 - 6x - 8y + 21 = 0 \tag{21}$$

Let the centre be (h, k) and the radius r. Then the equation is, from equation (19b)

$$x^2 + y^2 - 2hx - 2ky + (h^2 + k^2 - r^2) = 0 \tag{22}$$

Comparing equations (21) and (22):

term in x: $\quad -2h = -6 \Rightarrow h = 3$

term in y: $\quad -2k = -8 \Rightarrow k = 4$

constant term: $\quad h^2 + k^2 - r^2 = 21 \Rightarrow (3)^2 + (4)^2 - r^2 = 21$
$$\Rightarrow r = 2$$

So the centre of the circle is $(3, 4)$ and the radius is 2.

Worked example 15 Find the shortest distance from the point $L(-1, 2)$ to the circle $x^2 + y^2 - 6x - 8y + 21 = 0$.

The circle has centre $C(3, 4)$ and radius 2 (see worked example 14). The nearest point on the circle to L is M, where LMC is a straight line (see fig. 6.2:**21**).

$$LC = \sqrt{(4^2 + 2^2)} = 2\sqrt{5}$$

MC = radius of circle = 2, so

$$LM = LC - MC = 2\sqrt{5} - 2 = 2(\sqrt{5} - 1)$$

So the shortest distance from L to the circle is $2(\sqrt{5} - 1)$.

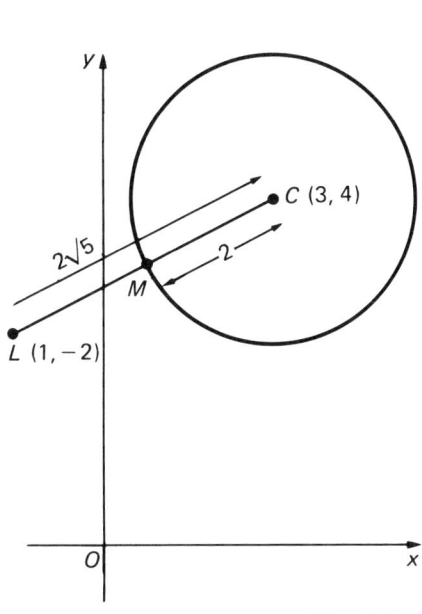

Fig. 6.2:**21**

308 *The circle*

Worked example 16   Find the equation of the circle which passes through the points $L(4, 8)$, $M(-2, 6)$ and $N(-4, 2)$.

The equations of the perpendicular bisectors of ML and MN (fig. 6.2:**22**) are $y = -3x + 10$ and $y = -\frac{1}{2}x + \frac{5}{2}$ (c.f. 6.2:3 Q6). These lines meet at $C(3, 1)$ which is therefore the centre of the circle.

Radius of circle $= CL = \sqrt{50}$ (c.f. 6.2:2 Q3), so equation of circle is

$$(x-3)^2 + (y-1)^2 = 50$$
$$x^2 + y^2 - 6x - 2y - 40 = 0$$

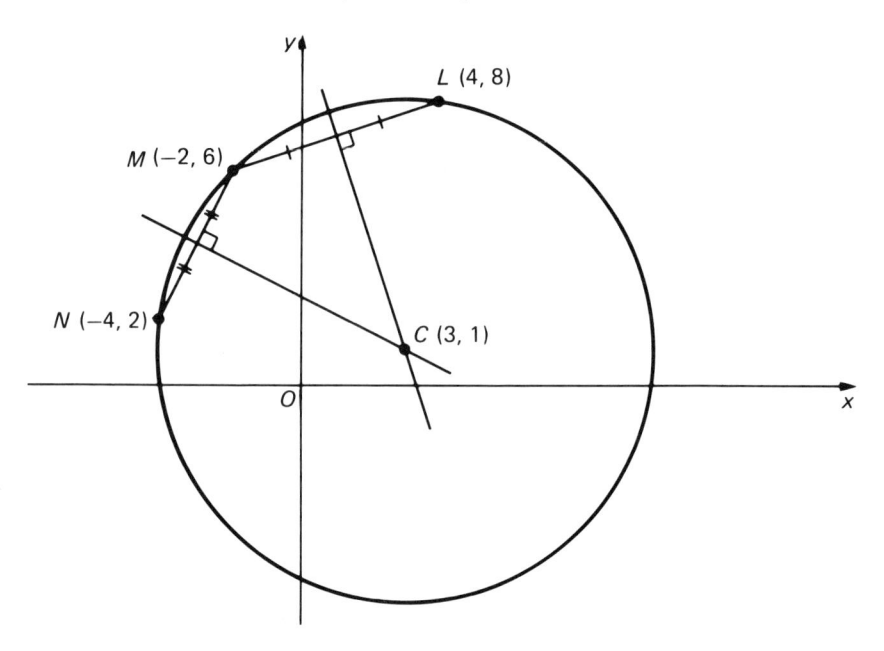

Fig. 6.2:**22**

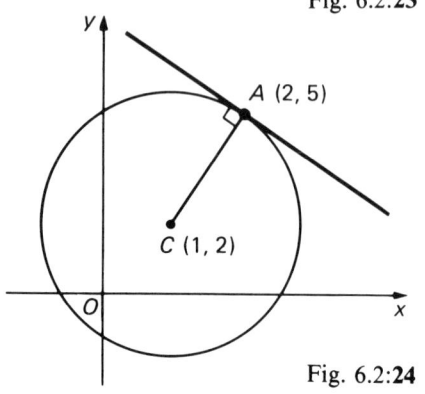

Worked example 17 A and B are the points $(-3, 2)$ and $(2, 1)$. Find the equation of the circle with diameter AB.

When $P(x, y)$ lies on the circle, AP is perpendicular to BP, since the angle in a semicircle is a right angle (see fig. 6.2:**23**).

Gradient of $AP = (y-2)/(x+3)$; gradient of $BP = (y-1)/(x-2)$; so

$$\frac{y-2}{x+3} \cdot \frac{y-1}{x-2} = -1$$

Fig. 6.2:**23**   So

$$(x+3)(x-2) + (y-2)(y-1) = 0$$
$$x^2 + y^2 + x - 3y - 4 = 0$$

Worked example 18   Find the gradient of the tangent at the point $A(2, 5)$ to the circle

$$x^2 + y^2 - 2x - 4y - 5 = 0 \tag{23}$$

The centre of the circle (fig. 6.2:**24**) is the point $C(1, 2)$ (see method of worked example 14). So the gradient of $CA = (5-2)/(2-1) = 3$.

The tangent at A is perpendicular to CA; so the gradient of the tangent is $-\frac{1}{3}$. (Note: CA is the normal to the circle at A.)

Fig. 6.2:**24**

Vectors, coordinate geometry and complex numbers 309

Alternatively: Gradient of tangent to circle at general point (x, y) is dy/dx. We find dy/dx by differentiating equation (23) using the implicit function rule:

$$2x + 2y\frac{dy}{dx} - 2 - 4\frac{dy}{dx} = 0$$

$$\Rightarrow \quad \frac{dy}{dx} = -\frac{2x-2}{2y-4} = \frac{1-x}{y-2}$$

At A, $x = 2$ and $y = 5$ so

$$\text{gradient of tangent} = \frac{1-2}{5-2} = -\frac{1}{3}$$

Exercise 6.2:6

1 Write down the coordinates of the centre and radius of each of the following circles and sketch each one. (Be alert for these possibilities: (a) the circle touches one or both of the axes; (b) the circle passes through O.)

(i) $(x-3)^2 + (y-2)^2 = 1$
(ii) $(x+2)^2 + (y-4)^2 = 3$
(iii) $(x-4)^2 + (y+4)^2 = 4$
(iv) $(x+3)^2 + (y-5)^2 = 3$
(v) $(x+3)^2 + (y+4)^2 = 5$
(vi) $(x+3)^2 + y^2 = 5$
(vii) $x^2 + (y+2)^2 = 2$
(viii) $x^2 + y^2 = 6$

2 Write down, in the form of both equation (19a) and equation (19b), the equation of the circle with the given centre and radius:

(i) $(-1, 4)$; 2
(ii) $(7, -6)$; 6
(iii) $(4, 2)$; $2\sqrt{5}$
(iv) $(0, -1)$; 4
(v) $(5, 0)$; 5

3 (i) Find the condition that the circle $(x-h)^2 + (y-k)^2 = r^2$:
(a) touches the x-axis;
(b) touches the y-axis;
(c) passes through the origin.

(ii) A circle of radius r touches both the coordinate axes. Find the possible positions of its centre.

(iii) The circle $(x+1)^2 + (y-k)^2 = 10$ passes through the origin. Find the possible values of k.

4 Find the centre and radius of the given circle:

(i) $x^2 + y^2 - 8x - 2y + 8 = 0$
(ii) $x^2 + y^2 + 6x - 14y + 33 = 0$
(iii) $x^2 + y^2 + 4x + 4y - 1 = 0$
(iv) $x^2 + y^2 - 2x + 10y + 25 = 0$
(v) $x^2 + y^2 - x = 6$

(vi) $x^2 + y^2 + 3x + 5y - 4 = 0$
(vii) $x^2 + y^2 - 8x - 6y = 0$
(viii) $2x^2 + 2y^2 + x + 3y - 1 = 0$ (Hint: divide the whole equation by 2 first.)
(ix) $5x^2 + 5y^2 + 4x - 10y - 4 = 0$

R 5 A and B are the points $(1, -1)$ and $(4, 2)$. A point P moves so that $AP = 2BP$. Show that the locus of P is a circle and find its centre and radius.
Illustrate your answer by sketching a diagram.

6 A and B are the points $(4, 5)$ and $(-4, 1)$. A point P moves so that $AP = 3BP$. Show that the locus of P is a circle and find its centre C and radius r. Find the coordinates of the two points D_1 and D_2 on AB which satisfy the equation. Verify that C is the midpoint of $D_1 D_2$.
Illustrate your answer by sketching a diagram.

7 Find the condition that the circle

$$x^2 + y^2 - 2hx - 2ky + c = 0$$

(i) touches the x-axis
(ii) touches the y-axis
(iii) passes through the origin

8 Find the centre and radius of the circle

$$x^2 + y^2 - 10x - 12y + 45 = 0$$

Hence find the shortest distance to the circle from:

(i) the point $(1, 3)$
(ii) the point $(13, 0)$

9 Find the shortest distance from:

(i) the point $(2, 3)$
(ii) the point $(0, 2)$

to the circle $x^2 + y^2 - 2x - 2y - 7 = 0$.

10 Find the shortest distance from:

(i) the origin
(ii) the point $(-1, 6)$
(iii) the point $(-3, 3)$

to the circle $x^2 + y^2 + 8x - 4y + 16 = 0$.

R 11 Find the shortest distance from:

(i) the origin
(ii) the point $(4, 0)$
(iii) the point $(-8, 2)$

to the circle $x^2 + y^2 - 8x + 6y - 39 = 0$.

12 (i) Find the distance of the point $L(-1, 1)$ to the point on the circle $x^2 + y^2 - 6x - 10y + 30 = 0$ that is: (a) closest to L; (b) furthest from L.

(ii) Find the distance of the point $L'(4, 6)$ to the point on the circle that is: (a) closest to L'; (b) furthest from L'.

13 Find the points of intersection of the given line with the given circle:

(i) $y = 2x + 4$; $x^2 + y^2 + 6x - 6y + 8 = 0$
(ii) $x + y = 5$; $x^2 + y^2 + 2x + 4y - 35 = 0$
(iii) $x - 3y + 5 = 0$; $x^2 + y^2 - 2x + 16y - 25 = 0$

Illustrate your answers by sketching diagrams.

14 Find the length of the chord of the circle $x^2 + y^2 = 25$ formed by the line $x + y = 4$.

15 Find the length of the chord of the circle

$$x^2 + y^2 - 12x - 6y + 20 = 0$$

formed by the line $2x + y - 5 = 0$.

16 The circle $x^2 + y^2 - 2hx - 2ky + c = 0$ passes through the points $L(4, 8)$, $M(-2, 6)$ and $N(-4, 2)$. Write down three equations which must be satisfied by h, k and c. Solve these equations and hence write down the equation of the circle LMN.

Check that your answer agrees with the answer to worked example 16. Do you prefer this method to the one given?

17 Find the equation of the circle passing through the points:

(i) $(4, 4)$, $(-4, 0)$, $(6, 0)$
(ii) $(1, 1)$, $(2, 4)$, $(3, 2)$
(iii) $(0, 4)$, $(0, 10)$, $(7, 3)$
(iv) $(0, 0)$, $(-6, 8)$, $(2, 14)$
(v) $(-5, 2)$, $(-3, -4)$, $(1, 8)$
R (vi) $(1, -1)$, $(2, -2)$, $(5, 3)$
(vii) $(0, 0)$, $(4a, 0)$, $(0, 2a)$; a constant

18 A and B are the points $(-3, 2)$ and $(2, 1)$. Find the midpoint M of AB and the distance AM. Hence find the equation of the circle with diameter AB. Do you prefer this method to the one given in worked example 17?

19 Find the equation of the circle with diameter AB when A and B are the points:

(i) $A(3, -1)$, $B(5, 3)$
(ii) $A(0, 4)$, $B(6, 0)$
(iii) $A(3, 4)$, $B(-2, 6)$
R (iv) $A(-5, 2)$, $B(1, 3)$
(v) $A(x_1, y_1)$, $B(x_2, y_2)$

20 A and B are two points in the xy plane. Show that the vector equation of the circle diameter AB is $(\mathbf{r} - \mathbf{a}) \cdot (\mathbf{r} - \mathbf{b}) = 0$.

By putting $\mathbf{r} = x\mathbf{i} + y\mathbf{j}$, $\mathbf{a} = 3\mathbf{i} - \mathbf{j}$, $\mathbf{b} = 5\mathbf{i} + 3\mathbf{j}$ in this equation, check your answer to question 19(i).

21 Find the equation of the circle passing through A and B with radius r when:

(i) A and B are the points $(2, 4)$ and $(4, 6)$; $r = \sqrt{10}$
(ii) A and B are the points $(-3, -2)$ and $(4, 5)$; $r = 5$

22 Find the equation of the circle passing through A and B, with its centre on the line Λ when:

(i) A and B are the points $(0, 0)$ and $(-2, 6)$ and Λ is the line $3x - y = 2$.
(ii) A and B are the points $(7, 3)$ and $(1, 5)$ and Λ is the line $2x + y - 7 = 0$.
(iii) A and B are the points $(1, 0)$ and $(4, 5)$ and Λ is the line $3x - 5y + 20 = 0$.

23 Find the equation of the tangent to the given circle at the given point:

(i) $x^2 + y^2 - 6x - 4y - 7 = 0$; $(1, 6)$
(ii) $x^2 + y^2 + 8x - 2y - 9 = 0$; $(-5, -4)$
R (iii) $x^2 + y^2 - 12x + 26 = 0$; $(3, 1)$
(iv) $x^2 + y^2 = 20$; $(-4, 2)$
(v) $x^2 + y^2 - 2hx - 2ky + c = 0$; (x_1, y_1)
(vi) $x^2 + y^2 + 8x - 4y = 0$; $(0, 0)$
(vii) $(x - 1)^2 + (y - 7)^2 = 50$; $(0, 0)$
(viii) $x^2 + y^2 - 2hx - 2gy = 0$; $(0, 0)$

24 Show that the tangents to the circle

$$x^2 + y^2 - 2x - 2y - 3 = 0$$

at the points $(3, 2)$ and $(2, -1)$ are perpendicular. Illustrate this result on a diagram.

25 Find the equations of the tangents to:

(i) the circle $x^2 + y^2 = 10$ at the points where $x = 1$
(ii) the circle $x^2 + y^2 - 10x - 6y + 26 = 0$ at the points where $x = 7$
(iii) the circle $x^2 + y^2 - 8x + 3 = 0$ at the points where $y = 2$

26 Find the equations of the tangents to the given circle which are parallel to either the x-axis or the y-axis:

(i) $x^2 + y^2 - 10x - 2y + 10 = 0$
R (ii) $x^2 + y^2 + 4x - 6y - 36 = 0$
(iii) $x^2 + y^2 - 8x = 0$

27 Find the centre and radius of the circle
$$x^2+y^2+2x-6y-6=0$$
Show that the distance from the centre to the line $3x+4y+11=0$ (see 6.2:4 WE11) is equal to the radius. What do you deduce?

28 Determine which of the given lines is a tangent to the circle $x^2+y^2-4x+2y-8=0$:
 (i) $2x+3y-14=0$
 (ii) $2x-3y+2=0$
 (iii) $3x-2y+5=0$

29 Determine whether or not the given line is a tangent to the given circle:
 (i) $x^2+y^2-4x-6y-37=0$; $x+y+5=0$
 (ii) $x^2+y^2-8x+11=0$; $x-2y+9=0$
 (iii) $x^2+y^2+2x+2y-7=0$; $3x+4y+6=0$

R 30 Show that the line $y=x+16$ is a tangent to the circle $x^2+y^2-6x-2y-22=0$ and find the coordinates of the point of contact.

31 Show that the line $2x+y+9=0$ is tangent to the circle $x^2+y^2+4x-10y+9=0$ and find the coordinates of the point of contact.

32 Find the values of c for which the line $y=2x+c$ is a tangent to the circle $x^2+y^2-4x-6y-7=0$.

33 Find the equations of the tangents to the given circle which are parallel to the given line.
 (i) $x^2+y^2+4x-6y+5=0$; $x=y$
 (ii) $x^2+y^2-12x-14y+75=0$; $x-3y+10=0$
 (iii) $x^2+y^2-8x-y+5=0$; $2x-y+3=0$
 (iv) $x^2+y^2-2x+y-5=0$; $4x-3y=0$

In each case find the coordinates of the points of contact.

34 Find the distance from the centre of the circle $x^2+y^2+6x+2y+9=0$ to the line $y=mx$. If this distance is equal to the radius of the circle, find the values of m. Hence find the equations of the tangents from the origin to the circle.
Illustrate your answer on a diagram.

35 Find the equations of the tangents from the origin to the circle:
 (i) $x^2+y^2-10x+16=0$
 (ii) $x^2+y^2-5x-5y+10=0$
 R (iii) $x^2+y^2+10x-6y+25=0$

36 Find the condition that the line $y=mx+c$ is a tangent to the circle $x^2+y^2-6x-4y=0$. Hence find the equations of the tangents to the circle from the point $(-2,+1)$.

37 Find the condition that the line $y=mx+c$ is a tangent to the circle $x^2+y^2-4x=0$. Hence find the equation of one of the tangents to the circle from the point $(0,1)$. Find the other one by drawing a sketch.

38 Find the equations of the tangents to the circle $x^2+y^2-8x-2y+9=0$ from the point $(0,1)$.

39 Show that the condition that the line $y=mx+c$ is a tangent to the circle $x^2+y^2=r^2$ is $c^2=r^2(1+m^2)$. Hence find:
 (i) the equations of the tangents of gradient 3 to the circle $9x^2+9y^2=10$;
 (ii) the equations of the tangents to the circle $x^2+y^2=20$ from the point $(0,5)$.

40 (i) The tangents to a circle centre C from a point L touch the circle at S and T. Show that:
  (a) LS is perpendicular to CS and LT is perpendicular to CT;
  (b) $LS=LT$.

 (ii) Find the centre and radius of the circle $x^2+y^2-4x-2y-4=0$.
 Find the distance from the point $(7,4)$ to the centre of the circle. Hence find the length of the tangents from the point to the circle.

41 Find the length of the tangents from the given point to the given circle:
 (i) $(4,-3)$; $x^2+y^2-2x-2y-14=0$
 (ii) $(6,7)$; $x^2+y^2+2x+6y-19=0$
 R (iii) $(0,-2)$; $x^2+y^2-10x-6y+33=0$

42 (i) Show that the length of the tangents from the point (x_1, y_1) to the circle $x^2+y^2-2hx-2ky+c=0$ is $\sqrt{(x_1^2+y_1^2-2hx_1-2ky_1+c)}$.

 (ii) The distance of a point $P(X,Y)$ from the origin is equal to the length of the tangents from P to the circle $x^2+y^2-8x+2y+8=0$. Find an equation connecting X and Y.

43 Find the equation of the circle which passes through the point A and touches the line Λ at the point B where:
 (i) $A(4,-3)$, $B(3,2)$, $\Lambda: x+2y-7=0$. (Hint: find the equation of the line through B perpendicular to Λ and the equation of the perpendicular bisector of AB. Find the point C of intersection of these lines; C is the centre of the circle—why? Note that the radius is the distance BC.)
 (ii) $A(0,0)$, $B(2,1)$, $\Lambda: 4x-3y-5=0$
 (iii) $A(1,9)$, $B(4,0)$, Λ: x-axis
 (iv) $A(-4,0)$, $B(6,0)$, $\Lambda: 2x+y-12=0$

44 Find the equation of the circle:
 (i) whose centre is the point $(5,4)$ and which touches the line $x+y=5$;
 R (ii) whose centre lies on the line $x-2y+2=0$ and which touches the y-axis at the point $(0,3)$.

45 Find the equations of the circles which:

- (i) pass through the points $(0, 2)$ and $(0, 8)$ and touch the x-axis;
- (ii) pass through the points $(-1, 0)$ and $(1, 0)$ and touch the line $y = x - 3$.

46 Find the equation of the circle:

(i) with its centre in the first quadrant which touches the x-axis, the y-axis and the line $3x - 4y - 12 = 0$;

(ii) with its centre in the third quadrant which touches the x-axis, the y-axis and the line $5x + 12y + 5 = 0$.

47 Find the equations of the circles which touch the x-axis, the line $4x - 3y = 0$ and the line $4x + 3y - 48 = 0$ (there are four). Sketch them.

48 The square $ABCD$ is bounded by the lines $3x - 4y + 8 = 0$, $4x + 3y - 26 = 0$, $3x - 4y - 2 = 0$ and $4x + 3y - 16 = 0$. Find the equation of the circle which touches all four sides of the square.

49 Find the equations of the circles which

(i) have their centres on the line $x - 2y + 2 = 0$ and touch the x-axis and the y-axis

(ii) have their centres on the line $x + y - 3 = 0$ and touch the x-axis and the line $3x - 4y + 3 = 0$

50 Show that at the points A and B of intersection of the circles $x^2 + y^2 - 2x - 2y + 1 = 0$ and $x^2 + y^2 + 2x + y - 2 = 0$, the equation $4x + 3y - 3 = 0$ **(24)** is satisfied.

Equation (24) is the equation of a straight line. Which straight line? Why?

51 Find the equation of the common chord of each of the following pairs of circles. Illustrate your answer on a diagram.

(i) $x^2 + y^2 - 2x - 4y - 4 = 0$; $x^2 + y^2 + 3x - y = 0$
(ii) $x^2 + y^2 - 10x + 9 = 0$; $x^2 + y^2 - 10y + 9 = 0$
(iii) $3x^2 + 3y^2 + x + 2y - 10 = 0$; $x^2 + y^2 - x + y - 4 = 0$

52 Find the equation of the common chord of the circles $x^2 + y^2 - 12x - 10y + 36 = 0$ and $x^2 + y^2 + 4x - 2y - 20 = 0$. Hence find the points of intersection of the circles. (Hint: find the points of intersection of the line and one of the circles.)

53 Find the points of intersection of the circles:

(i) $x^2 + y^2 = 20$; $x^2 + y^2 - 10x - 10y + 40 = 0$
(ii) $x^2 + y^2 - 10x - 14y + 24 = 0$;
$x^2 + y^2 - 4x + 4y - 12 = 0$
R (iii) $x^2 + y^2 - 2x + 8y - 9 = 0$;
$x^2 + y^2 + 4x - y - 12 = 0$

54 Find the equation of the common chord of the circles $x^2 + y^2 - 2h_1 x - 2k_1 y + c_1 = 0$ and $x^2 + y^2 - 2h_2 x - 2k_2 y + c_2 = 0$. Hence show that the common chord of two circles is always perpendicular to the line joining their centres.

55 (i) Two circles have centres C_1 and C_2 and radii r_1 and r_2 respectively. Show by drawing diagrams that:
  (a) the circles touch externally $\Leftrightarrow C_1 C_2 = r_1 + r_2$;
  (b) the circles touch internally $\Leftrightarrow C_1 C_2 = r_1 - r_2$.

(ii) Show that the circles $x^2 + y^2 + 8x + 4y - 5 = 0$ and $x^2 + y^2 - 7x - 16y + 20 = 0$ touch externally.

(iii) Show that the circles $x^2 + y^2 - 4x + 2y - 27 = 0$ and $x^2 + y^2 - 10x - 4y + 27 = 0$ touch internally.

56 (i) Two circles with centres C_1 and C_2, radii r_1 and r_2, touch externally at the point K. Show that K divides $C_1 C_2$ internally in the ratio $r_1 : r_2$.

(ii) Show that the circles $x^2 + y^2 + 2x - 2y - 15 = 0$ and $x^2 + y^2 - 10x - 5y + 27 = 0$ touch externally. Find the coordinates of their point of contact.

57 Show that the circles

$$x^2 + y^2 - 6x + 7 = 0 \quad \textbf{(25)}$$
$$x^2 + y^2 + 2x - 8y - 1 = 0 \quad \textbf{(26)}$$

touch externally. Find the coordinates of their point of contact and hence find the equation of their common tangent.

Show that the equation of this common tangent can be found by subtracting equation (25) from equation (26). Explain why this method works.

58 Find the common tangent of the circles $x^2 + y^2 + 8x + 4y - 80 = 0$ and $x^2 + y^2 - 16x - 14y + 88 = 0$. Hence find the coordinates of their point of contact, K.

Show that K divides the line joining their centres in the ratio of their radii.

59 Show that the following pairs of circles touch each other and find the equation of their common tangent and the coordinates of their point of contact:

(i) $x^2 + y^2 = 4$; $x^2 + y^2 - 6x - 8y + 16 = 0$
(ii) $x^2 + y^2 = 49$, $x^2 + y^2 - 6x - 8y + 21 = 0$
(iii) $x^2 + y^2 + 8x + 2y = 0$; $x^2 + y^2 + 4x + y = 0$
(iv) $x^2 + y^2 - 8x - 6y = 0$
$x^2 + y^2 - 24x - 18y + 200 = 0$
(v) $x^2 + y^2 - 2x - 2y - 14 = 0$
$x^2 + y^2 - 14x - 18y + 94 = 0$
R (vi) $x^2 + y^2 + 8x + 4y - 5 = 0$
$x^2 + y^2 - 7x - 6y + 20 = 0$

60 Find the distances from the origin to the points on the circle $x^2 + y^2 - 10x - 24y + 120 = 0$ closest to and furthest from the origin. Hence find the (positive) values of k for which the circle $x^2 + y^2 = k^2$ touches this circle.

61 Find the values of K for which the circle $x^2+y^2=K^2$ touches the circle:

(i) $x^2+y^2-8x+7=0$

R (ii) $x^2+y^2-8x-6y+24=0$

→ **62** Find the distances from the point $L(-3,-1)$ to the points on the circle $x^2+y^2-6y-7=0$ closest to and furthest from L. Hence find the values of c for which the circle $x^2+y^2+6x+2y+c=0$ touches the given circle.

• → **63** (i) Show that the circles $x^2+y^2+4x-6y-3=0$ and $x^2+y^2-10x-8y+32=0$ lie entirely outside each other.

(ii) Show that the circle $x^2+y^2+2x+2y-2=0$ lies entirely inside the circle $x^2+y^2-6y-38=0$.

6.2:7 The parabola

Definition A **parabola** is the locus of a point P which moves so that its distance from a fixed point S is equal to its distance from a fixed line Λ.

S is called the **focus** of the parabola, Λ its **directrix**.

The general parabola is shown in fig. 6.2:**25**.

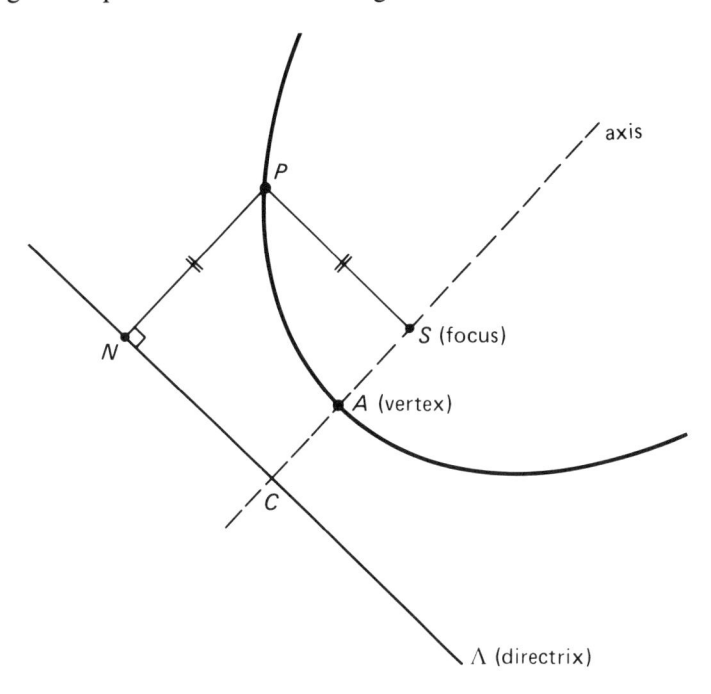

Fig. 6.2:**25**

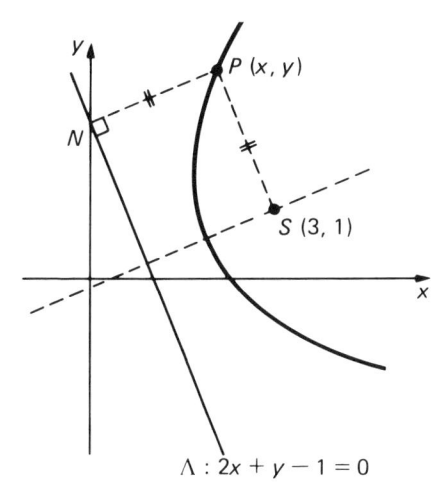

(i) The line through S perpendicular to Λ is the axis of symmetry of the parabola. It is called the **axis** of the parabola.

(ii) There is one point A of the parabola that is on the axis. It is called the **vertex** of the parabola.

(iii) The distance SA is called the **focal length** of the parabola.

Worked example 19 Find the cartesian equation of the parabola with focus $S(3,1)$ and directrix $\Lambda: 2x+y-1=0$.

Let $P(x,y)$ be the general point on the parabola (fig. 6.2:**26**). Then

$$SP^2 = (x-3)^2 + (y-1)^2, \qquad PN = \left|\frac{2x+y-1}{\sqrt{5}}\right|$$

Fig. 6.2:**26** by 6.2:4 equation (17).

$$SP = PN \Rightarrow SP^2 = PN^2 \Rightarrow (x-3)^2 + (y-1)^2 = \tfrac{1}{5}(2x+y-1)^2$$

So the equation of the parabola is

$$x^2 + 4y^2 - 4xy - 36x - 8y + 49 = 0$$

(c.f. 6.2:5 Q8.)

Standard parabola

We want to investigate the geometric properties of the parabola using coordinates; so if we are considering just one parabola, it is sensible to place the coordinate axes in such a way that the equation of the parabola is as simple as possible.

We let the vertex be the origin and the axis of the parabola be the x-axis; so that the y-axis is the tangent at the vertex (fig. 6.2:27).

Let S be the point $(a, 0)$ where a is a positive constant. Then the directrix Λ is the line $x = -a$.

$$SP^2 = (x-a)^2 + y^2, \qquad PN = PK + KN = x + a$$

$$SP = PN \Rightarrow SP^2 = PN^2 \Rightarrow (x-a)^2 + y^2 = (x+a)^2$$

$$\Rightarrow \qquad y^2 = 4ax \qquad (27)$$

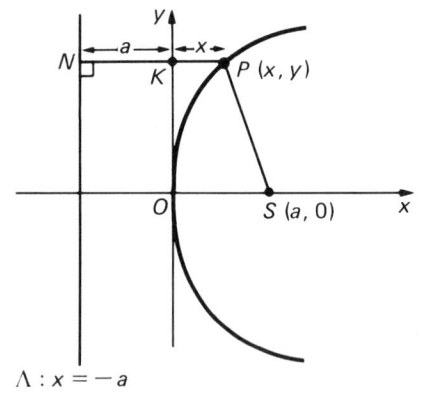

Fig. 6.2:27

Worked example 20 Find the equations of the tangent and normal to the parabola $y^2 = 4ax$ at the point $(a, 2a)$ (fig. 6.2:28).

$$y^2 = 4ax \Rightarrow 2y\frac{dy}{dx} = 4a \Rightarrow \frac{dy}{dx} = \frac{4a}{2y} = \frac{2a}{y}$$

So the gradient of the tangent at $(a, 2a)$ is $2a/2a$, i.e. 1. The equation of the tangent is

$$\frac{y - 2a}{x - a} = 1, \qquad \text{i.e. } y = x + a$$

The gradient of the normal is -1, so the equation of the normal is

$$\frac{y - 2a}{x - a} = -1, \qquad \text{i.e. } y = -x + 3a$$

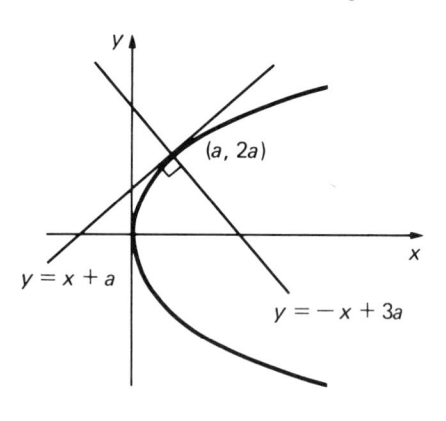

Fig. 6.2:28

Exercise 6.2:7

1 Find the cartesian equation of the parabola with focus S and directrix Λ:

- (i) $S(5, 0)$; Λ: $x - y - 3 = 0$
- **R** (ii) $S(-1, 2)$; Λ: $3x - y = 0$
- (iii) $S(0, 0)$; Λ: $x + 2y + 5 = 0$
- (iv) $S(0, \tfrac{1}{4})$; Λ: $y = -\tfrac{1}{4}$
- (v) $S(3, 0)$; Λ: $x = -3$

• → **2** Sketch on the same axes the given parabolae:

(i) $y^2 = x$, $y^2 = 4x$, $y^2 = 8x$
(ii) $y^2 = x$, $y^2 = -x$
(iii) $y^2 = x$, $x^2 = y$

showing clearly any points of intersection. (See 1.2:8 Q2.)

3 Find the focal lengths of the parabolae:

(i) $y^2 = 4x$ (ii) $y^2 = 6x$ (iii) $y^2 = -x$

• → **4** (i) Show that the equation of the parabola with vertex (h, k), focal length a, with its axis parallel to the x-axis is $(y - k)^2 = 4a(x - h)$.

(ii) Find the coordinates of the vertex, the focal length and the equation of the directrix of the following parabolae:

(a) $y^2 = 4x - 4$
(b) $(y - 1)^2 = 4(x - 2)$
(c) $(y + 2)^2 = 2x + 3$
(d) $y^2 + 6y - 6x + 13 = 0$

Sketch them.

(iii) Sketch on the same diagram the parabolae $y^2 = 4x$, $y^2 = 4x - 8$ and $y^2 = 8 - 4x$.

5 (i) Find the equation of the parabola with vertex (h, k), focal length a, with its axis parallel to the y-axis.

(ii) Find the coordinates of the vertex, the focal length and the equation of the directrix of the following parabolae:
(a) $(x+2)^2 = 2(y+1)$; (b) $y = 2x^2 + 3x + 5$
Sketch them.

(20) **6** Find the equations of the tangent and normal to the parabola $y^2 = 4ax$ at the point

(i) $(a, -2a)$ (ii) $(9a, 6a)$

(20) **7** Find the equations of the tangents to the parabola $y^2 = 8x$ at the points where $x = 2$.
Find the point of intersection of these tangents.

(20) → **8** Show that the tangents to the parabola $y^2 = 4ax$ at the points $(4a, 4a)$ and $(a, -2a)$ are perpendicular. Show that their point of intersection lies on the directrix.

(20) **R 9** The tangent and normal to the parabola $y^2 = 12x$ at the point $A(3, 6)$ meet the x-axis at L and M. Find:

(i) the coordinates of the centroid of $\triangle ALM$
(ii) the area of $\triangle ALM$

(20) **10** The normal to the parabola $y^2 = 4x$ at the point $A(9, 6)$ meets the parabola again at B. Find the coordinates of B.
Hence find the area of $\triangle OAB$.

6.2:8 Tangents

Consider the intersection of the straight line

$$y = mx + c \tag{28}$$

and the curve

$$f(x, y) = 0 \tag{29}$$

Solving equations (28) and (29) simultaneously leads to an equation in x:

$$f(x, mx + c) = 0 \tag{30}$$

If the roots of equation (30) are distinct, then the line $y = mx + c$ cuts the curve in distinct points (e.g. lines $\Lambda_1, \Lambda_2, \Lambda_3$ in fig. 6.2:29).

However, if equation (30) has a pair of *equal* roots, then the line is a tangent to the curve (Λ_4 in fig. 6.2:29).

Fig. 6.2:29

Worked example 21 (i) Find the condition for the line $y = mx + c$ to be a tangent to the parabola $y^2 = 4ax$.
(ii) Hence find the equation of the tangent of gradient 2 to the parabola.
(iii) Find the point of contact of this tangent with the parabola.

(i) The points of intersection of the line and the parabola are given by solving simultaneously the equations $y = mx + c$ and $y^2 = 4ax$. This gives

$$(mx + c)^2 = 4ax$$
$$m^2 x^2 + (2mc - 4a)x + c^2 = 0 \tag{31}$$

If the line is a tangent to the parabola, equation (31) has two equal roots; so

$$(2mc - 4a)^2 - 4m^2 c^2 = 0$$
$$16mca - 16a^2 = 0$$
$$mc = a$$

(ii) When $m = 2$, $c = a/m = \tfrac{1}{2}a$ (fig. 6.2:30). So the equation is

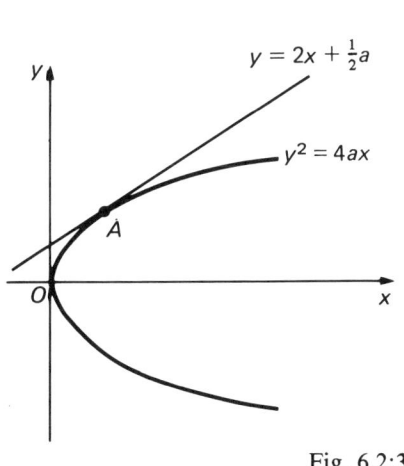

$$y = 2x + \tfrac{1}{2}a \tag{32}$$

Fig. 6.2:30

(iii) We can find the point of contact A by solving simultaneously the equations

$$y = 2x + \tfrac{1}{2}a$$
$$y^2 = 4ax$$

but notice that we have started to do this in (i) (in greater generality), obtaining equation (31). We know that equation (31) has equal roots and it is these roots that we would like to find. In the general case $ax^2 + bx + c = 0$,

$$x = \frac{-b \pm \sqrt{(b^2 - 4ac)}}{2a}$$

so if the roots are equal ($b^2 - 4ac = 0$), then $x = -b/2a$.
So, because the roots of equation (31) are equal,

$$x = \frac{-(2mc - 4a)}{2m^2} = \frac{a}{4} \quad (\text{since } m = 2, c = \tfrac{1}{2}a)$$

So $y = a$ from equation (32) and the point of contact is $A(\tfrac{1}{4}a, a)$.

Exercise 6.2:8

(21) **1** (i) Find the condition for the line $y = mx + c$ to be a tangent to the parabola $y^2 = 2x$.

(ii) Hence find the equation of the tangent of gradient $\tfrac{1}{4}$ to the parabola.

(iii) Find the point of contact of this tangent with the parabola.

(21) **2** Show that $x - 2y + 8 = 0$ is a tangent to the parabola $y^2 = 8x$. Find the coordinates of the point of contact.

(21) **3** Find the equations of the tangents to the parabola $y^2 = 16x$:

(i) parallel to the line $2x + y + 3 = 0$
(ii) perpendicular to the line $x + y - 1 = 0$
(iii) from the point $(2, 6)$

(21) **R 4** Find the equations of the tangents to the parabola $y^2 = 8x$ from the point $(-2, 3)$. Show that they are perpendicular.

(21) → **5** Show that the equation of the tangent of gradient m to the parabola $y^2 = 4ax$ is $y = mx + a/m$.

(i) Write down the tangent of gradient $-1/m$. Show that if two tangents to the parabola are perpendicular their point of intersection lies on the directrix.

(ii) Find the equations of the tangents to the parabola from the point $(6a, 5a)$.

(21) **6** The product of the gradients of the tangents from the point P to the parabola $y^2 = 4x$ is k. Find the cartesian equation of the locus of P when:

(i) $k = -1$ (ii) $k = 2$ (iii) $k = 3$

(21) → **7** Show that if the line $y = mx + c$ is tangent to the circle $x^2 + y^2 = r^2$; then $c^2 = r^2(1 + m^2)$. Hence find:

(i) the equations of the tangents of gradient 3 to the circle $2x^2 + 2y^2 = 5$ and the coordinates of their points of contact;

(ii) the equations of the tangents to the circle $x^2 + y^2 = 10$ from the point $(-10, 0)$, and the co-ordinates of their points of contact.

(21) **8** Find the condition that the line $y = mx$ is tangent to the circle $x^2 + y^2 + 6x + 2y + 9 = 0$. Hence find the equations of the tangents from the origin to the circle.

(21) **9** Prove that the line $y = x + 2$ is a common tangent of the parabola $y^2 = 8x$ and the circle $x^2 + y^2 = 2$. Draw a sketch to illustrate this.

(21) **R 10** Prove that the line $y = mx + 4/m$ is a tangent to the parabola $y^2 = 16x$ for all non-zero values of m. Hence find the equations of the common tangents to this parabola and the circle $x^2 + y^2 = 8$. Sketch a diagram showing the circle, the parabola and the tangents.

• **11** The line $y = mx + c$ meets the parabola $y^2 = 4x$ in the points $P(x_1, y_1)$ and $Q(x_2, y_2)$. Find an expression for $x_1 + x_2$. If the midpoint of PQ is $(2, 1)$, find m and c.
Illustrate your answer on a diagram.

12 The line $y = mx + c$ meets the parabola $y = 5x^2 + 3x - 2$ at the points $P(x_1, y_1)$ and $Q(x_2, y_2)$. Find an expression for $x_1 + x_2$. If the midpoint of PQ is $(-\tfrac{1}{2}, 2)$, find m and c.

Vectors, coordinate geometry and complex numbers 317

6.3 Parameters

6.3:1 Graphs

The equation of a curve (or line) may be given by expressing the coordinates of a general point $P(x, y)$ on the curve as functions of a third variable t called a **parameter**:

$$x = f(t) \quad y = g(t)$$

We have used the parametric form before. (See 6.2:3 WE8, for example, where we expressed the equation of a line in parametric form $x = 3 - 4\lambda$, $y = -2 + 6\lambda$ (parameter λ). Look also at 3.2:5 Q11 where we effectively expressed the equation of the circle $x^2 + y^2 = 1$ in parametric form $x = \cos\theta$, $y = \sin\theta$ (parameter θ).)

Worked example 1 Sketch the curve $x = 1 + 2t^2$, $y = 4t$.

t	−3	−2	−1	0	1	2	3
x	19	9	3	1	3	9	19
y	−12	−8	−4	0	4	8	12

The curve is shown in fig. 6.3:**1**.

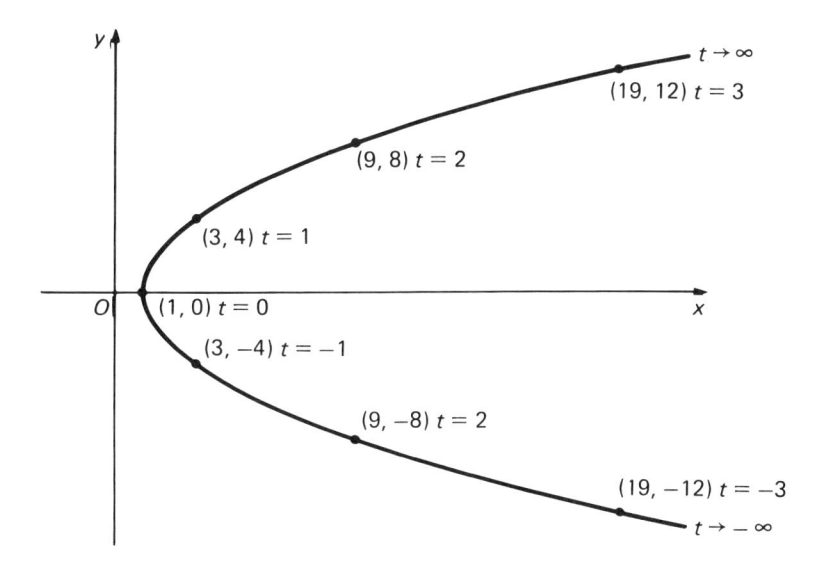

Fig. 6.3:**1**

Worked example 2 Sketch the curve $x = 3\cos\theta$, $y = 2\sin\theta$.

θ	0	$\pm\tfrac{1}{6}\pi$	$\pm\tfrac{1}{3}\pi$	$\pm\tfrac{1}{2}\pi$	$\pm\tfrac{2}{3}\pi$	$\pm\tfrac{5}{6}\pi$	$\pm\pi$
x	3	2.6	1.5	0	−1.5	−2.6	−3
y	0	±1	±1.7	±2	±1.7	±1	0

The curve is shown in fig. 6.3:**2**.

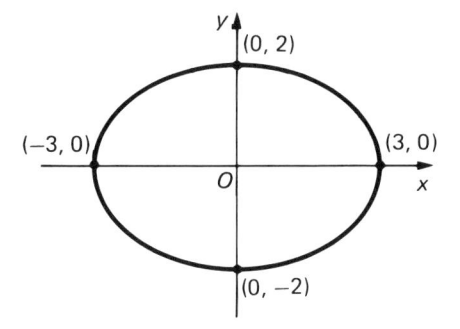

Fig. 6.3:**2**

Vector equation of a curve
If the parametric equations of a curve are $x = f(t)$, $y = g(t)$, then the position vector **r** of the general point P on the curve is given by

$$\mathbf{r} = x\mathbf{i} + y\mathbf{j} = f(t)\mathbf{i} + g(t)\mathbf{j}$$

i.e. the equation $\mathbf{r} = f(t)\mathbf{i} + g(t)\mathbf{j}$ is a **vector equation of the curve**.

Look at 6.2:3 WE8 again. The parametric equations of the line are $x = 3 - 4\lambda$, $y = -2 + 6\lambda$ and its vector equation is $\mathbf{r} = (3 - 4\lambda)\mathbf{i} + (-2 + 6\lambda)\mathbf{j}$.

Cartesian equation of a curve given in parametric form
Given the equation of a curve in parametric form

$$x = f(t) \qquad (1)$$
$$y = g(t) \qquad (2)$$

we find its cartesian equation by eliminating t from equations (1) and (2).

Worked example 3 Find the cartesian equations of the curves
(i) $x = 1 + 2t^2$, $y = 4t$ (ii) $x = 3\cos\theta$, $y = 2\sin\theta$
(i) $t = \frac{1}{4}y \Rightarrow x = 1 + 2(\frac{1}{4}y)^2$, i.e. $y^2 = 8(x - 1)$
(ii) $\cos\theta = \frac{1}{3}x$, $\sin\theta = \frac{1}{2}y \Rightarrow \frac{1}{9}x^2 + \frac{1}{4}y^2 = 1$ (see section 3.2:2)

(Note: these are the equations of worked examples 1 and 2.)

The point $(f(t), g(t))$ on the curve $x = f(t)$, $y = g(t)$ is called **the point with parameter** t; e.g. on the curve $x = t^2$, $y = t^3$,
the point with parameter t is (t^2, t^3);
the point with parameter 3 is $(3^2, 3^3)$, i.e. $(9, 27)$.

Worked example 4 Find the points of intersection of the curve $x = t^2$, $y = 2t$ and the line $2x - y - 4 = 0$.

At the points A and B of intersection (see fig. 6.3:3), all three equations are satisfied:

$$x = t^2 \qquad (3)$$
$$y = 2t \qquad (4)$$
$$2x - y - 4 = 0 \qquad (5)$$

Substituting into equation (5) the values of x and y from equations (3) and (4):

$$2t^2 - 2t - 4 = 0$$
$$(t + 1)(t - 2) = 0$$
$$t = -1 \text{ or } 2$$

So A and B are the points with parameters -1 and 2, i.e. $(1, -2)$ and $(4, 4)$.

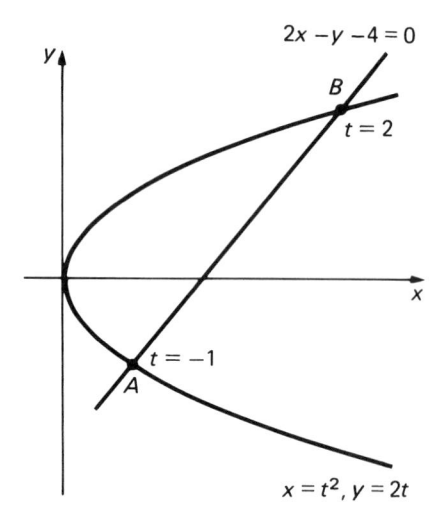

Fig. 6.3:3

Exercise 6.3:1

① **1** Find the values of x and y corresponding to $t = 0$, $\pm 1, \pm 2, \pm 3$ and hence sketch the curve (or line) given by the following pairs of parametric equations:

(i) $x = 5t, y = -2t$
(ii) $x = 3 - t, y = -4 + 3t$
(iii) $x = 3t^2, y = 6t$
→ (iv) $x = 1 + t^2, y = 2 + t$

① → **2** Show that the x-axis is the tangent to the curve $x = t^2, y = t^3$ at the origin. (Hint: show that as $t \to 0$, $y/x \to 0$.)

Sketch the curve (a **semi-cubical parabola**) (see 1.2:8 Q4(iv)).

① **3** If $x = 1 - t^2, y = t(1 - t^2)$, find the values of t for which:

(i) $x = 0$ (ii) $y = 0$.

Investigate the behaviour of x and y as $t \to \pm\infty$.
Find the values of x and y corresponding to $t = 0$, $\pm\frac{1}{2}, \pm 1, \pm 2$. Sketch the curve given by this pair of equations.

① **4** Sketch the curves:

(i) $x = t^2 - 3, y = t(t^2 - 3)$
(ii) $x = t(2 - t), y = t^2(2 - t)$

① **5** If $x = 4t, y = 4/t$, investigate the behaviour of x and y as $t \to 0$ and as $t \to \pm\infty$.

Find the values x and y corresponding to $t = \pm\frac{1}{2}$, $\pm 1, \pm 2$. Sketch the curve given by this pair of equations.

① **6** Sketch the curves:

(i) $x = t^2, y = 2/t$
(ii) $x = t + 1/t, y = t - 1/t$

① **7** If $x = t/(t - 1), y = t/(t - 2)$, investigate the behaviour of x and y as $t \to 1^+, t \to 1^-, t \to 2^+, t \to 2^-$, and as $t \to \pm\infty$. Hence sketch the curve given by this pair of equations.

② **8** For the following pairs of parameteric equations, find the values of x and y corresponding to $\theta = 0$, $\pm\frac{1}{6}\pi, \pm\frac{1}{3}\pi, \pm\frac{1}{2}\pi, \pm\frac{2}{3}\pi, \pm\frac{5}{6}\pi, \pm\pi$, and hence sketch the curves they represent:

(i) $x = 4 \cos\theta, y = 2 \sin\theta$
→ (ii) $x = 3 \cos\theta, y = 5 \sin\theta$
(iii) $x = \sin\theta, y = \sin 2\theta$
(iv) $x = 3 \cos 2\theta, y = 2 \cos\theta$

③ **9** Find the cartesian equations of the following curves:

(i) $x = 2t^2, y = 4t$
→ (ii) $x = 1 + 2t^2, y = -3 + 4t$
→ (iii) $x = 3t, y = 3/t$
(iv) $x = 1/t, y = 1 - t$
(v) $x = t^2 - 1, y = t(t^2 - 1)$

③ **10** Find the cartesian equations of the following curves:

(i) $x = 2 \cos\theta, y = 2 \sin\theta$
→ (ii) $x = 1 + 2 \cos\theta, y = 3 + 2 \sin\theta$
→ (iii) $x = 3 \cos\theta, y = 5 \sin\theta$
(iv) $x = \sin\theta, y = 2 \cos\theta$
(v) $x = \sin\theta, y = \sin 2\theta$

③ **11** Write down the parametric equations of the following curves and find their cartesian equations:

(i) $\mathbf{r} = 3\mathbf{i} + 2\mathbf{j} + t(\mathbf{i} - \mathbf{j})$
→ (ii) $\mathbf{r} = t^2\mathbf{i} + t^3\mathbf{j}$
→ (iii) $\mathbf{r} = 3\mathbf{i} - 4\mathbf{j} + (2 \cos\theta\mathbf{i} + 2 \sin\theta\mathbf{j})$
(iv) $\mathbf{r} = -\mathbf{j} + (3 \cos\theta\mathbf{i} + \sin\theta\mathbf{j})$

• → **12** Find the cartesian equation of the circle $x = r \cos\theta, y = r \sin\theta$ (r constant) and sketch it. Explain the significance of the parameter θ.

→ **13** (i) Find the cartesian equation of the circle $x = h + r \cos\theta, y = k + r \sin\theta$. Write down its radius and the coordinates of its centre. Explain the significance of θ.

(ii) Sketch the circles:
(a) $x = 5 + 3 \cos\theta, y = 4 + 3 \sin\theta$
(b) $x = -3 + 2 \cos\theta, y = -2 + 2 \sin\theta$

14 Sketch the circles with vector equations:

(i) $\mathbf{r} = (2 + 2 \cos\theta)\mathbf{i} + 2 \sin\theta\mathbf{j}$
R (ii) $\mathbf{r} = (3\mathbf{i} - 4\mathbf{j}) + (\cos\theta\mathbf{i} + \sin\theta\mathbf{j})$

In each case find the maximum and minimum values of $|\mathbf{r}|$:
(a) by the method of 6.2:5 WE15;
(b) by finding $|\mathbf{r}|$ in terms of θ and differentiating.

• → **15** Find the cartesian equation of the parabola $x = at^2, y = 2at$ (a positive constant). Sketch the curve (c.f. question 1(iii)).

16 Sketch the curves:

(i) $x = at^2 + a, y = 2at$
→ (ii) $x = at^2 - 2a, y = 2at + a$
(iii) $x = 2at, y = at^2$
R (iv) $\mathbf{r} = (3t^2 + 1)\mathbf{i} + (-2 + 6t)\mathbf{j}$

17 Find the cartesian equations of the curves:

(i) $x = 3 \cos\theta, y = 2 \sin\theta$
(ii) $x = 2 \cos\theta, y = 3 \sin\theta$

Sketch the curves (they are **ellipses**).

→ **18** Find the cartesian equation of the curve $x = a \cos\theta, y = b \sin\theta$. Sketch the curve when:

(i) $a > b$ (ii) $a = b$ (iii) $a < b$

•→ **19** Find the cartesian equation of the curve $x = ct$, $y = c/t$ (c positive constant). Investigate the behaviour of x and y as $t \to 0$, $t \to \pm\infty$. Hence find the asymptotes of the curve and sketch it (c.f. question 5).
The curve is called a **rectangular hyperbola**.

20 Sketch on the same axes the curves:

(i) $x = t$, $y = 1/t$ and $x = 3t$, $y = 3/t$
(ii) $x = 4t$, $y = 4/t$ and $x = -4t$, $y = 4/t$
(iii) $xy = 25$ and $xy = -25$

21 Find the cartesian equation of each of the following curves and sketch them:

(i) $x = 5 + 3\cos\theta$, $y = 4 + 2\cos\theta$
(ii) $x = 3 + t^2$, $y = -1 + t$
(iii) $x = -2 + t$, $y = 1 + 1/t$

•→ **22** Sketch the curve $x = a\cos^3\theta$, $y = a\sin^3\theta$ (an **astroid**). If P is the general point $(a\cos^3\theta, a\sin^3\theta)$ on the curve and $P\hat{O}x = \phi$, what is the relationship between θ and ϕ? When is θ equal to ϕ?

R 23 Sketch the curve $x = a(\theta - \sin\theta)$, $y = a(1 - \cos\theta)$ (a **cycloid**) for values of θ between -2π and 4π.

24 By putting $y = tx$, find parametric equations for the curve

• (i) $y^2 = x^2 + x$
→ (ii) $y^2 = x^3 - 4x^2$

25 Find the values of the parameter t at the given points on the curve:

• (i) $x = t^2$, $y = 3t$; (4, 6), (4, −6)
(ii) $x = at^2$, $y = 2at$; (9a, 6a), (9a, −6a)
→ (iii) $x = at^3$, $y = at^2$; $(\frac{1}{8}a, \frac{1}{4}a)$, $(-\frac{1}{8}a, \frac{1}{4}a)$
(iv) $x = t/(1+t)$, $y = t^2/(1+t)$; (2, −4), $(-1, \frac{1}{2})$
(v) $x = 3t$, $y = 3/t$; (1, 9)
(vi) $x = 2\cos t$, $y = 4\sin t$;
(2, 0), (−2, 0), (0, −4), $(1, -2\sqrt{3})$

26 Find the equation of the chord AB of each of the following curves, where A and B are the points with the given parameters:

→ (i) $x = 2t$, $y = 2/t$; 1, 2
(ii) $x = 2\cos\theta$, $y = \sin\theta$; $\frac{1}{6}\pi$, $-\frac{1}{2}\pi$
(iii) $x = at^2$, $y = 2at$; 1, −2
(iv) $x = t^2$, $y = t^3$; p, q

R 27 $P(ap^2, 2ap)$ and $Q(aq^2, 2aq)$ are two points on the parabola $y^2 = 4ax$.

(i) Find the equation of the chord PQ.

(ii) Find the point of intersection of PQ and the x-axis. Hence show that the area of $\triangle OPQ$ is $|a^2 pq(p+q)|$.
(iii) Find the coordinates of the centroid of $\triangle OPQ$.
(iv) S is the focus of the parabola and M is the midpoint of PQ. The line through S perpendicular to PQ meets the directrix at R. Prove that

$$2RM = SP + SQ$$

→ **28** P and Q are the points with parameters p and q on the rectangular hyperbola $x = ct$, $y = c/t$, and M is the midpoint of PQ. PQ meets the axes at R and S. Show that:

(i) $MQ = OM$
(ii) M is the midpoint of RS
(iii) $\triangle OMR$ is isosceles

29 P and Q are the points with parameters θ and ϕ on the ellipse $x = a\cos t$, $y = b\sin t$. If the gradient of PQ is fixed, show that $\theta + \phi$ is constant.

④ **30** Find the points of intersection of the curve and the line:

(i) $x = t^2$, $y = t^3$; $x = 4$
(ii) $x = 3\cos t$, $y = 2\sin t$; $2x + 3y + 2 = 0$
(iii) $x = t^2$, $y = 2t$; $x + y = 0$
→ (iv) $x = at^2$, $y = 2at$; $2x + 3y - 8a = 0$
R (v) $x = t + 1/t$, $y = t - 1/t$; $x - 2y + 2 = 0$
(vi) $x = 5\cos\theta$, $y = 5\sin\theta$; $3x - y + 5 = 0$

④ **31** Find the points of intersection of the curves:

(i) $x = 2t^2$, $y = 4t$; $x^2 + y^2 - 2x - 16 = 0$
(ii) $x = t^2 + 1$, $y = 2t - 1$; $x^2 + y^2 - 7x + 2y + 7 = 0$
(iii) $x = 2t$, $y = 2/t$; $x^2 + y^2 = 17$

④ **32** Find the point of intersection of the curves:

•→ (i) $x = 4t^2$, $y = 8t$; $x = 2s$, $y = 2/s$
(ii) $x = t^2 + 2$, $y = 2t + 1$; $x = 3s$, $y = 3/s$
(iii) $x = t^2 + 1$, $y = 2t$; $x = 2s$, $y = 2/s$

④ **33** Find the coordinates of the midpoint of the chord on the given line cut off by the given curve:

(i) $x + y + 3 = 0$; $x = t^2$, $y = 2t$
(ii) $x - y + 6 = 0$; $x = 4t$, $y = 4/t$
(iii) $2x - y - 16 = 0$; $x = 3t^2 + 2$, $y = 6t$

④ **R 34** P, Q, R and S are the points with parameters p, q, r and s on the rectangular hyperbola $x = ct$, $y = c/t$. If P, Q, R and S lie on a circle, show that $pqrs = 1$.

6.3:2 Tangent and normal to a curve given in parametric form

We found, as long ago as section 1.3:1, that the gradient of the tangent to a curve at a general point (x, y) on the curve is dy/dx. Now with the equation of the curve in the form $x = f(t)$, $y = g(t)$, we cannot differentiate

Vectors, coordinate geometry and complex numbers 321

y directly with respect to x since y is a function of t; but, using the chain rule,

$$\frac{dy}{dx} = \frac{dy}{dt} \cdot \frac{dt}{dx} = \frac{dy/dt}{dx/dt} \quad \left(\text{since } \frac{dt}{dx} = \frac{1}{dx/dt}\right)$$

Note also that

$$\frac{d^2y}{dx^2} = \frac{d}{dx}\left(\frac{dy}{dx}\right) = \frac{d}{dt}\left(\frac{dy}{dx}\right) \cdot \frac{dt}{dx} = \frac{\frac{d}{dt}(dy/dx)}{dx/dt}$$

Thus

$$\frac{dy}{dx} = \frac{dy/dt}{dx/dt} \tag{6}$$

$$\frac{d^2y}{dx^2} = \frac{\frac{d}{dt}(dy/dx)}{dx/dt} \tag{7}$$

Worked example 5 If $x = 3\cos\theta$, $y = 2\sin\theta$, find dy/dx and d^2y/dx^2 in terms of θ.

$$\frac{dx}{d\theta} = -3\sin\theta \qquad \frac{dy}{d\theta} = 2\cos\theta$$

By equation (6)

$$\frac{dy}{dx} = \frac{dy/d\theta}{dx/d\theta} = \frac{2\cos\theta}{-3\sin\theta} = -\tfrac{2}{3}\cot\theta$$

By equation (7)

$$\frac{d^2y}{dx^2} = \frac{(d/d\theta)(dy/dx)}{dx/d\theta} = \frac{\tfrac{2}{3}\operatorname{cosec}^2\theta}{-3\sin\theta} = -\tfrac{2}{9}\operatorname{cosec}^3\theta$$

Worked example 6 Find the equation of the tangent to the curve $x = 3t^2$, $y = 2t^3$ at: (i) the general point $P(3t^2, 2t^3)$; (ii) the point $A(12, -16)$.

(i) Gradient of tangent at P:

$$\frac{dy}{dx} = \frac{dy/dt}{dx/dt} = \frac{6t^2}{6t} = t$$

So equation of tangent at P is

$$\frac{y - 2t^3}{x - 3t^2} = t$$

$$tx - y = t^3 \tag{8}$$

(ii) At $A(12, -16)$, $t = -2$ so the equation of the tangent at A is equation (8) with $t = -2$:

$$-2x - y = (-2)^3$$

i.e. $$2x + y = 8$$

Worked example 7 Find the equations of the tangents to the curve $x = 3t^2$, $y = 2t^3$ from the point $(7, 6)$.

The equation of the tangent to the curve at the general point $P(3t^2, 2t^3)$ is $tx - y = t^3$ (see equation 8). When this tangent passes through the point

(7, 6), the values $x = 7$, $y = 6$ satisfy equation (8), i.e.

$$7t - 6 = t^3$$
$$t^3 - 7t + 6 = 0$$
$$(t-1)(t^2 + t - 6) = 0 \qquad \text{(see 2.2:3 WE3)}$$
$$(t-1)(t-2)(t+3) = 0$$
$$t = 1, 2 \text{ or } -3$$

So the tangents from (7, 6) touch the curve at the points with parameters 1, 2 and 3.

The equations of the tangents are simply equation (8) with $t = 1$, 2 and -3, i.e. $x - y = 1$, $2x - y = 8$, $3x + y = 27$.

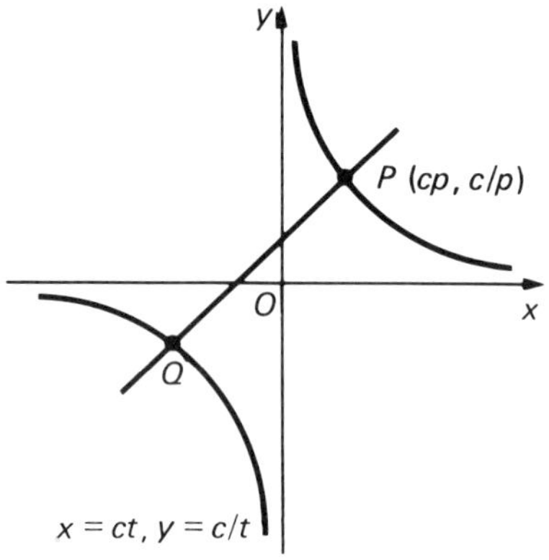

Worked example 8 Find the coordinates of the point Q where the normal to the curve $x = cp$, $y = c/p$ at the point $P(cp, c/p)$ meets the curve again.

Equation of normal at the point P is $y = p^2 x + (c/p - cp^3)$. We now use a new parameter q to represent the point $Q(cq, c/q)$ of intersection of the curve and the normal. We want to find q in terms of t. At the points of intersection of the normal and the curve (fig. 6.3:4)

$$x = cq \qquad (9)$$
$$y = c/q \qquad (10)$$
$$y = p^2 x + \left(\frac{c}{p} - cp^3\right) \qquad (11)$$

From equations (9), (10) and (11)

$$\frac{c}{q} = cp^2 q + \left(\frac{c}{p} - cp^3\right)$$
$$cp^2 q^2 + \left(\frac{c}{p} - cp^3\right)q - c = 0$$
$$c(q - p)\left(p^2 q + \frac{1}{p}\right) = 0$$

$$\Rightarrow \qquad q = p \text{ or } -\frac{1}{p^3}$$

Note: we could have predicted the solution $q = p$, since we knew that the normal intersects the curve at P, where $q = p$.

At Q, $q = -1/p^3$, so Q is the point $(-c/p^3, -cp^3)$.
(See 1.3:3 Q20(ii).)

Fig. 6.3:4

Exercise 6.3:2

⑤ **1** Find dy/dx and d^2y/dx^2 when:

→ (i) $x = at^2$, $y = 2at$
(ii) $x = ct$, $y = c/t$
(iii) $x = 3t^2$, $y = 2t^3$
(iv) $x = t + 1/t$, $y = t - 1/t$
(v) $x = t/(1 + t^2)$, $y = t^2/(1 + t^2)$
→ (vi) $x = 3 \cos \theta$, $y = \sin \theta$
(vii) $x = \sec \theta$, $y = \tan \theta$
R (viii) $x = a \cos^3 \theta$, $y = a \sin^3 \theta$

⑥ **2** Find the equations of the tangent and normal to the curve at the point with parameter p. Hence find the equations of the tangent and normal to the curve at the given points:

(i) $x = ct$, $y = c/t$; (c, c), $(3c, \frac{1}{3}c)$
(ii) $x = at^2$, $y = 2at$; $(a, 2a)$, $(9a, -6a)$
(iii) $x = 2a + at^2$, $y = 2at$; $(6a, 4a)$
(iv) $x = 3t$, $y = 3/t$; $(9, 1)$, $(-1, -9)$
→ (v) $x = at^2$, $y = at^3$; $(\frac{1}{4}a, -\frac{1}{8}a)$
(vi) $x = 4 \cos \theta$, $y = 2 \sin \theta$; $(2\sqrt{3}, 1)$, $(0, -2)$
(vii) $x = a \cos^3 \theta$, $y = a \sin^3 \theta$; $(-\frac{1}{4}a\sqrt{2}, -\frac{1}{4}a\sqrt{2})$

⑥ **3** Find the point of intersection of:

(i) the tangents to the curve $x = ct$, $y = c/t$ at the points (c, c), $(\frac{1}{2}c, 2c)$;

(ii) the normals to the curve $x = t^2$, $y = t^3$ at the points with parameters -1 and 2.

⑥ → **4** The tangent at the point P with parameter p to the curve $x = 4t$, $y = 4/t$ cuts the axes at L and M. Show that:

(i) P is the midpoint of LM
(ii) the area of $\triangle OLM$ is independent of p

⑥ **5** The tangent at the point with parameter ϕ to the astroid $x = a\cos^3\theta$, $y = a\sin^3\theta$ meets the axes at L and M. Show that $LM = a$.

⑥ → **6** Find the point of intersection of the tangents to the curve $x = at^2$, $y = 2at$ at the points P and Q with parameters p and q respectively.

The line through T parallel to the x-axis meets the parabola at R and the chord PQ at M.

(i) Find the coordinates of R. Show that the tangent at R is parallel to PQ.

(ii) Show that M is the midpoint of PQ.

⑥ **7** The normal at the point P with parameter p to the curve $x = at^2 + 2at$, $y = 2at + 2a$ meets the x-axis at L. M is the foot of the perpendicular from P to the x-axis. Show that $LM = 2a$.

⑥ **8** P is the point $(at^2, 2at)$ on the parabola $y^2 = 4ax$.

(i) K is the foot of the perpendicular from $S(a, 0)$ to the tangent at P. Show that K lies on the y-axis.

(ii) Find the equation of the chord CSD through S which is parallel to the tangent at P. Prove that $CD = 4PS$.

⑥ → **9** (i) The tangent to the parabola $y^2 = 4ax$ at $P(at^2, 2at)$ meets the x-axis at T. Show that $SP = ST$, where S is the point $(a, 0)$. Hence show that the tangent at P bisects the angle between the x-axis and PS.

(ii) If P is any point on a parabolic mirror, prove that a ray of light parallel to the axis of the mirror is reflected through the focus.

⑥ R **10** The normal to the parabola $y^2 = 4ax$ at $P(at^2, 2at)$ meets the x-axis at N.

(i) The foot of the perpendicular from P to the x-axis is M. Show that MN is independent of the position of P.

(ii) Show that the perpendicular from $S(a, 0)$ to NP bisects NP.

⑥ **11** Find the equation of the tangent to the curve $x = 1 + \cos\theta$, $y = \sin\theta$ at the point with parameter ϕ. The tangent cuts the axes at L and M. Show that the area of $\triangle OLM$ is $\cot^2\frac{1}{2}\phi \tan\phi$.

⑥ **12** The tangent at the point with parameter ϕ on the ellipse $x = a\cos\theta$, $y = b\sin\theta$ meets the axes at G and H. Show that the area of $\triangle OGH$ is $ab\operatorname{cosec} 2\phi$. Find the least possible area of the triangle as ϕ varies.

The tangent at the point with parameter ϕ on the hyperbola $x = a\sec\theta$, $y = b\tan\theta$ meets the axes at L and M. Show that the area of $\triangle OLM$ is a^2.

⑥ **13** P and Q are the points with parameters ϕ and $(\phi + \frac{1}{2}\pi)$ on the ellipse $x = a\cos\theta$, $y = b\sin\theta$.

(i) Show that $OP^2 + OQ^2 = a^2 + b^2$.
(ii) Find the point of intersection of the tangents at P and Q.
(iii) Show that the tangent at Q is parallel to OP.
(iv) Find the length of PQ. Hence find the greatest and least possible lengths of PQ as ϕ varies.

⑥ **14** The tangent and normal at the point P with parameter ϕ to the ellipse $x = a\cos\theta$, $y = b\sin\theta$ meet the x-axis at T and N. Show that $OT \cdot ON = a^2 - b^2$.

⑥ R **15** P is the point with parameter p on the rectangular hyperbola $x = ct$, $y = c/t$.

(i) The tangent at P meets the line $y = x$ at M, and N is the foot of the perpendicular from P to this line. Show that $OM \cdot ON = 2c^2$.

(ii) The tangent at P meets the axes at R and S and the normal at P meets the lines $y = x$, $y = -x$ at T and U. Show that, if $p^2 \ne 1$, then $RUST$ is a rhombus. What happens if $p^2 = 1$?

⑥ **16** P and Q are the points $(12, 3)$, $(18, 2)$ on the rectangular hyperbola $xy = 36$. Find the coordinates of the point T of intersection of the tangents at P and Q. Show that O, M (the midpoint of PQ) and T are collinear.

⑥ **17** P, Q, R and S are the points with parameters p, q, r and s on the rectangular hyperbola $x = ct$, $y = c/t$. If PQ is perpendicular to RS, show that $pqrs = -1$. Hence show that PS is perpendicular to RQ and QS is perpendicular to PR.

Show that the normals to the curve at the four points intersect at a point.

⑦ → **18** Find the equation of the tangent to the parabola $x = at^2$, $y = 2at$ at the point with parameter p. Hence:

(i) find the equation of the tangent of gradient $\frac{1}{3}$ and find the point of contact A;

(ii) find the equations of the tangents from the point $(8a, 6a)$ and find the points B and C of contact;

(iii) find the equations of the tangents from the point $(-a, -\frac{3}{2}a)$ and show that they are perpendicular.

⑦ **19** Find the equations of:

(i) the tangents to the parabola $y^2 = 4ax$ from the point $(6a, 5a)$

(ii) the normals to the parabola from the point $(5a, 2a)$

(See 6.2:8 Q5.)

⑦ **20** Find the equations of the tangents to the rectangular hyperbola $xy = c^2$ from the point:
  (i) $(8c, c)$ (ii) $(\tfrac{3}{2}c, \tfrac{1}{2}c)$

⑦ **R 21** Show that there is just one tangent to the curve $x = t^2$, $y = t^3$ from the point $(-1, 2)$. Draw a diagram to illustrate this result.

⑦ **22** Find the gradient of the tangent to the parabola $y^2 = 4x - 4$ at the point $(t^2 + 1, 2t)$.
  Hence find the coordinates of the point of contact of the tangent of gradient 2 with the parabola.
  Hence find the shortest distance from the line $2x - y + 3 = 0$ to the parabola.

⑦ **23** Find the shortest distance from the straight line $3x - 4y + 5 = 0$ to the parabola $y^2 = 4x$.

⑦ → **24** If the tangent to the parabola $y^2 = 4ax$ at the point $(at^2, 2at)$ passes through the point (h, k) show that
$$at^2 - kt + h = 0$$
Hence show that there are two tangents to the parabola from a point (h, k) so long as $k^2 > 4ah$. Sketch a diagram showing the region in which the point (h, k) may lie.
  Show that if the tangents meet the parabola at the points with parameters t_1 and t_2 then $h = at_1 t_2$, $k = a(t_1 + t_2)$.

⑦ **25** If the normal to the parabola $y^2 = 4ax$ at the point $(at^2, 2at)$ passes through the point (h, k), show that
$$at^3 + (2a - h)t - k = 0$$
Hence show that, in general, there are three normals to the parabola from a point (h, k). Show that if these normals meet the parabola at the points with parameters t_1, t_2 and t_3 then:
  (i) $t_1 + t_2 + t_3 = 0$
  (ii) $h = a(2 - t_1 t_2 - t_2 t_3 - t_3 t_1)$, $k = a t_1 t_2 t_3$

⑦ **26** Show that, in general, four normals can be drawn from any point (h, k) to the rectangular hyperbola $xy = c^2$ and that, if they cut the curve at the points P, Q, R and S, then the sum of the x-coordinates of P, Q, R and S is h and the sum of their y-coordinates is k.

• → **27** Show that, at the points where the line $y = mx + c$ meets the parabola $x = at^2$, $y = 2at$, $amt^2 - 2at + c = 0$.
  Hence show that if the line is tangent to the parabola then $a = mc$ (see section 6.2:8).
  Hence find the equations of the tangents to the parabola:
  (i) with gradient 2
  (ii) through the point $(2a, 3a)$
  (iii) through the point $(16a, 17a)$

28 Find the condition that the line $y = mx + c$ is a tangent to the rectangular hyperbola $x = 3t$, $y = 3/t$. Find the equations of the tangents to the hyperbola:
  (i) with gradient -1
  (ii) through the point $(5, -8)$

• → **29** Find the equation of the chord joining the points with parameters p and q on the curve $x = at^2$, $y = 2at$. By letting $q \to p$ find the equation of the tangent to the curve at the point with parameter p.

30 Find the equation of the chord joining the points with parameters p and q on the curve $x = ct$, $y = c/t$.
  Hence find the equation of the tangent to the curve at the point with parameter p.
  Hence write down the equations of the tangents to the curve at the points $(2c, \tfrac{1}{2}c)$ and $(-\tfrac{1}{2}c, -2c)$.

31 Find an equation for the parameters of the points where the circle $x^2 + y^2 = r^2$ cuts the rectangular hyperbola $x = 2t$, $y = 2/t$.
  Deduce the radius of the circle centre O which touches the hyperbola. Find the points of contact. Sketch this circle and the hyperbola.

⑧ **32** Find the coordinates of the point Q where the tangent/normal to the curve at the point P with parameter p meets the curve again:
→ (i) $x = at^2$, $y = 2at$, normal
  (ii) $x = t$, $y = t^3$, tangent
→ (iii) $x = at^2$, $y = at^3$, tangent at $(4a, 8a)$
  (iv) $x = t^3 - t^2 + 2$, $y = t^2 - 1$, tangent
  (v) $x = 1/t$, $y = t^2$, tangent

⑧ **33** Find the equation of the normal to the curve $x = t + 1$, $y = 4 - t^2$ at the point $P(2, 3)$. If this normal meets the curve again at Q and the x-axis at R, show that $PQ : QR = 5 : 7$.

⑧ **34** Find the coordinates of the point Q where the tangent to the curve $x = t^2$, $y = t^3$ at the point P with parameter p meets the curve again. If N is the foot of the perpendicular from P to the x-axis, and R is the point where the tangent at P cuts the y-axis, show that OQ and RH are equally inclined to the x-axis.

⑧ **R 35** Find the equation of the tangent to the curve $x = 3t^2$, $y = 2t^3$ at the point with parameter p. Find, in terms of p, the coordinates of the point Q at which this tangent meets the curve again. If this tangent is the normal to the curve at Q, show that $p = \pm\sqrt{2}$.

⑧ + **36** The tangent to the curve $x = t^2$, $y = t^3$ at the point P in the first quadrant is the normal to the curve at the point Q. Find the coordinates of P and Q.

6.3:3 Secondary loci

Worked example 9 (i) Find the coordinates of the point T of intersection of the tangents to the curve $x = 3t^2$, $y = 2t^3$ at the points $P(3p^2, 2p^3)$, $Q(3q^2, 2q^3)$.

Vectors, coordinate geometry and complex numbers 325

(ii) If the tangents are perpendicular, find the cartesian equation of the locus of *T*.

(i) Equation of tangent at general point $(3t^2, 2t^3)$ is

$$tx - y = t^3 \quad \text{(from equation 8 worked example 6)}$$

At *P*, $t = p$, so equation of tangent at *P* is $px - y = p^3$.
At *Q*, $t = q$, so equation of tangent at *Q* is $qx - y = q^3$.
At their point *T* of intersection

$$px - y = p^3 \quad \textbf{(12)}$$

$$qx - y = q^3 \quad \textbf{(13)}$$

(12)-(13):
$$px - qx = p^3 - q^3$$
$$x = (p^3 - q^3)/(p - q)$$
$$= p^2 + pq + q^2$$

From equation (12):
$$y = p(p^2 + pq + q^2) - p^3$$
$$= p^2 q + pq^2$$
$$= pq(p + q)$$

So *T* is the point $(p^2 + pq + q^2, pq(p + q))$.

(ii) If the tangents at *P* and *Q* are perpendicular, the product of their gradients must be -1, i.e. $pq = -1$. So *T* is $(p^2 + q^2 - 1, -(p + q))$.

As *p* and *q* vary, subject to the condition $pq = -1$, *T* moves along the curve:

$$x = p^2 + q^2 - 1$$
$$y = -(p + q)$$

In order to find the cartesian equation of this curve, we want to eliminate *p* and *q*, remembering that $pq = -1$. There seems to be a

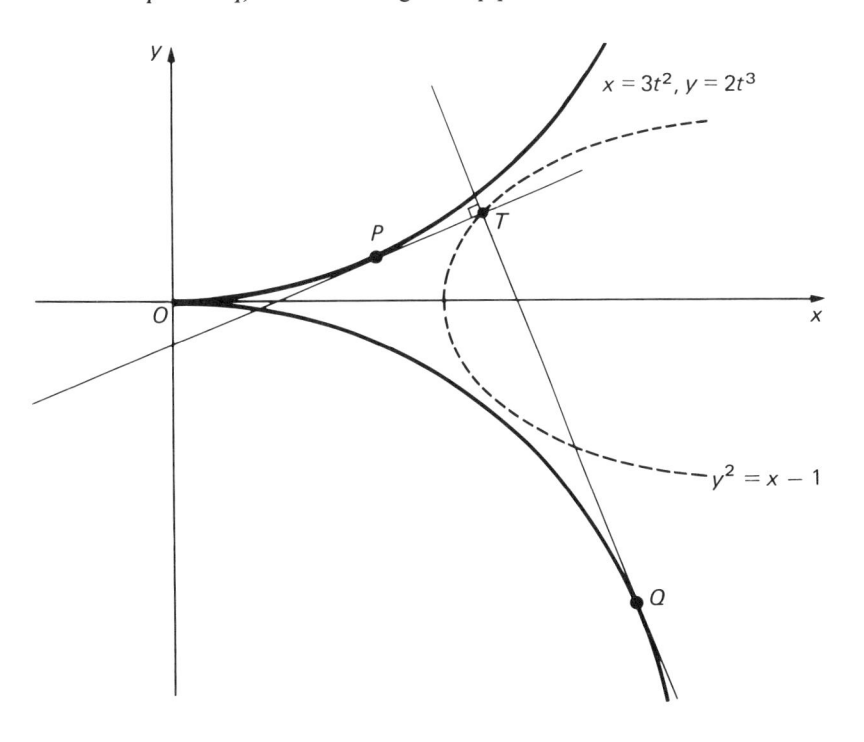

Fig. 6.3:5

relationship between y^2 and x, so we look at

$$y^2 = p^2 + 2pq + q^2$$
$$= p^2 + q^2 - 2$$
$$= x - 1$$

So the cartesian equation of the locus of T is

$$y^2 = x - 1$$

This locus is shown in fig. 6.3:**5**.

Exercise 6.3:3

• → **1** P is the general point with parameter t on the rectangular hyperbola $x = 3t$, $y = 3/t$.

(i) Find the coordinates of the point Q which divides OP internally in the ratio $1:2$. Find the cartesian equation of the locus of Q as t varies.

(ii) Find the cartesian equation of the locus of R, the point which divides AP in the ratio $1:2$, where A is the point $(6, 6)$.

2 P is any point on the ellipse $x = 4 \cos \theta$, $y = 3 \sin \theta$. Find the cartesian equation of the locus of:

(i) Q, where P is the midpoint of OQ
(ii) R, where P is the point on OR such that $OP : PR = 3 : 1$.

3 P is any point on the parabola $y^2 = 4ax$, and A is the point $(a, 0)$. Find the locus of the midpoint of AP.

4 P is any point on the rectangular hyperbola $xy = 25$, and A is the point $(5, 0)$. Find the locus of the point which divides AP internally in the ratio $3:2$.

5 The normal to the parabola $y^2 = 4ax$ at $P(at^2, 2at)$ cuts the x-axis at L. M is the midpoint of PL. Find the cartesian equation of the locus of M.

6 The perpendicular from the origin to the tangent to the parabola $y^2 = 4ax$ at $P(at^2, 2at)$ meets the tangent at R. Find the locus of R.

→ **7** Find the locus of the foot of the perpendicular from the origin to the tangent to the rectangular hyperbola $xy = c^2$.

→ **8** The tangent to the curve $x = 3t^2$, $y = 2t^3$ at the general point P with parameter t cuts the x-axis at Q. The normal at P cuts the y-axis at R. Find the locus of the midpoint of QR.

9 The tangent to the curve $x = ct$, $y = c/t$ at the general point P with parameter t cuts the x-axis at Q. The normal at P cuts the y-axis at R. Find the locus of the midpoint of QR.

10 The normal to the ellipse $x = 5 \cos \theta$, $y = 3 \sin \theta$ at the general point P with parameter θ meets the axes at L and M. Find the cartesian equation of the locus of the midpoint of LM and sketch it.

11 P is the general point with parameter θ on the ellipse $x = 5 \cos \theta$, $y = 4 \sin \theta$. Find the cartesian equation of the locus of:

(i) the foot of the perpendicular from $S(3, 0)$ to the tangent at P;

(ii) the foot of the perpendicular from the origin O to the normal at P.

R **12** (i) The tangent at the general point $P(ct, c/t)$ of the rectangular hyperbola $xy = c^2$ meets the axes at L and M. Show that $\triangle OLM$ has constant area $2c^2$ and that the locus of its centroid is the rectangular hyperbola $xy = 4c^2/9$.

(ii) The foot of the perpendicular from the origin O to the tangent at P is N. Show that $OP \cdot ON$ is constant.

13 The tangent to the parabola $y^2 = 4ax$ at the general point $(at^2, 2at)$ meets the circle $x^2 + y^2 = a^2$ at the points P and Q. Find the coordinates of the midpoint M of PQ. Find the cartesian equation of the locus of M as t varies.

14 Find the cartesian equation of the locus of $P(x, y)$ if:

• (i) $x = p + q$, $y = p^2 + q^2$; $pq = -1$
(ii) $x = p - q$, $y = p + q$; $pq = 4$
(iii) $x = p^2 + q^2$, $y = p^2q + pq^2$; $pq = -2$

⑨ → **15** Find the coordinates of the point of intersection T of the tangents to the parabola $y^2 = 4ax$ at the points $P(ap^2, 2ap)$ and $Q(aq^2, 2aq)$.

(i) If T lies on the line $x = -1$, show that $pq = -1$, and hence find the locus of M, the midpoint of PQ. Show also that:
(a) the tangents at P and Q are perpendicular;
(b) PQ passes through the focus $S(a, 0)$ of the parabola.

(ii) If OP is perpendicular to OQ, show that $pq = -4$. Find the locus of M and show that PQ always passes through a fixed point on the x-axis.

(iii) Find the area of $\triangle PTQ$. If P and Q move so that the area of $\triangle PTQ$ is $4a^2$, find the locus of T.

Vectors, coordinate geometry and complex numbers 327

⑨ **16** The normal at the point P of the rectangular hyperbola $xy = c^2$ meets the hyperbola again at Q, and the tangents at P and Q meet at T. Find the cartesian equation of the locus of T.

R 17 The normal at the point P of the parabola $y^2 = 4ax$ meets the parabola again at Q and the tangents at P and Q meet at T. Find the cartesian equation of the locus of T.

⑨ **18** $P(ap^2, 2ap)$ and $Q(aq^2, 2aq)$ are points on the parabola $y^2 = 4ax$, and A is the point $(-a, 0)$. OP cuts AQ at R. If PQ passes through $S(a, 0)$, find the cartesian equation of the locus of R.

⑨ **19** The tangents to the parabola $y^2 = 4ax$ at the points $P(ap^2, 2ap)$ and $Q(aq^2, 2aq)$ intersect at an angle $\tan^{-1} k$. Find the locus of their point of intersection.

⑨ **20** Find the coordinates of the point T of intersection of the tangents to the astroid $x = a\cos^3\theta$, $y = a\sin^3\theta$ at the points with parameters α and β. If $\alpha + \beta = \frac{1}{2}\pi$, show that T lies on the line $y = x$. Draw a diagram illustrating this result.

⑨→ **21** If the chord PQ of the rectangular hyperbola $x = ct$, $y = c/t$ passes through the point $(c, 2c)$, show that the tangents to the hyperbola at P and Q meet on the line $x + 2y = 2c$.

6.3:4 Area under a curve given in parametric form

Theorem 1: The area A of the region under the arc of the curve $x = f(t)$, $y = g(t)$ between the points with parameters t_1 and t_2 is given by

$$A = \int_{t_1}^{t_2} y \frac{dx}{dt} dt \qquad (14)$$

The volume V generated when this region is rotated completely about the x-axis is given by

$$V = \int_{t_1}^{t_2} \pi y^2 \frac{dx}{dt} dt \qquad (15)$$

Proof
See question 1.

Worked example 10 (i) Find the area of the region in the first quadrant bounded by the curve $x = 3t^2$, $y = 6t$, the lines $x = 3$, $x = 12$ and the x-axis.
(ii) Find the volume generated when this region is rotated completely about the x-axis.

The region is shown in fig. 6.3:6. Let C and D be the points of intersection of the lines $x = 3$ and $x = 12$ with the curve.
C is the point $(3, 6)$ with parameter 1.
D is the point $(12, 12)$ with parameter 2

$$\frac{dx}{dt} = 6t$$

(i) By equation (14):

$$A = \int_1^2 6t \cdot 6t \, dt$$
$$= \int_1^2 36t^2 \, dt = \left[12t^3\right]_1^2 = 84$$

(ii) By equation (15):

$$V = \int_1^2 \pi(6t)^2 \cdot 6t \, dt$$
$$= \int_1^2 216\pi t^3 \, dt = \left[54\pi t^4\right]_1^2 = 810\pi$$

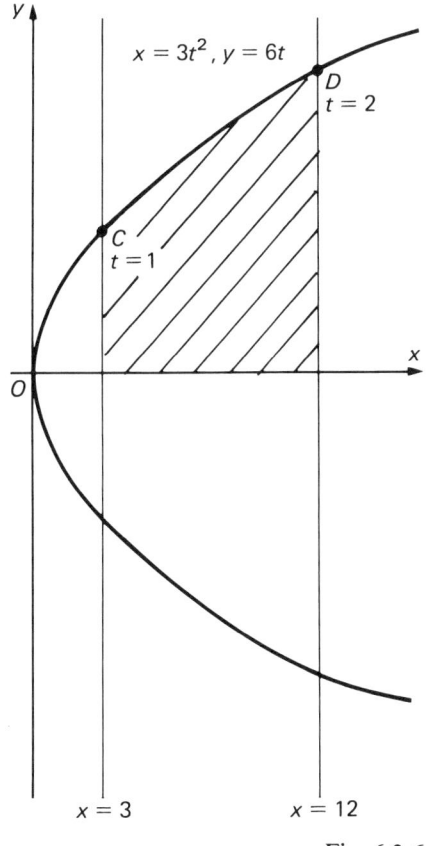

Fig. 6.3:6

Exercise 6.3:4

- → **1** Using the formulae of 4.1 and 4.2:1 Theorem 1 prove formulae (14) and (15), being careful to justify the limits of the integrals. Explain why it is necessary to adapt the formulae of 4.1 when the curve is given in parametric form.

⑩ → **2** (i) Find the area of the region in the first quadrant bounded by the parabola $x = at^2$, $y = 2at$, the line $x = a$ and the x-axis.

  (ii) Find the volume generated when this region is rotated completely about the x-axis.

⑩ → **3** Find $\int \sin^2 \theta \, d\theta$. Hence find the area bounded by the ellipse $x = a \cos \theta$, $y = b \sin \theta$.

→ **4** Find $\int \sin^3 \theta \, d\theta$. Hence find the volume generated when the interior of the ellipse $x = a \cos \theta$, $y = b \sin \theta$ is rotated through 180° about:

  (i) the x-axis (ii) the y-axis.

  Comment on the case $a = b$.

5 Show that the area bounded by the astroid $x = a \cos^3 \theta$, $y = a \sin^3 \theta$ is $\frac{3}{4} \pi a^2$.

⑩ **6** Show that the area bounded by the arc of the cycloid $x = a(\theta - \sin \theta)$, $y = a(1 - \cos \theta)$ from $\theta = 0$ to $\theta = 2\pi$ and the x-axis is $3\pi a^2$.

⑩ **7** Find the area of the finite region bounded by the curve $x = 4t$, $y = 4/t$ and the line $x + y = 10$.

⑩ **8** Find the area of the finite region bounded by the curve $x = 3t^2$, $y = 2t^3$ and the tangent to the curve at the point with parameter 2.

9 Sketch the curve with parametric equation $x = t(t^2 - 1)$, $y = t^2 - 1$. Find the area bounded by the loop of the curve.

10 Find the values of the parameters s and t at the points of intersection of the curves $x = as^2$, $y = as^3$ and $x = at^2$, $y = 2at$ (a constant). Show that the total area of the two finite regions bounded by the curves is $32a^2\sqrt{2}/15$.

6.4 Coordinate geometry in three dimensions

6.4:1 Vectors in three-dimensional space

> **Theorem 1:** If **u**, **v** and **w** are any three non-coplanar vectors then for any other vector **a** there exist unique scalars α, β and γ such that:
> $$\mathbf{a} = \alpha \mathbf{u} + \beta \mathbf{v} + \gamma \mathbf{w} \qquad (1)$$
> (See 6.2:1 Theorem 1.)

The theorem says: given any three non-coplanar directions **u**, **v** and **w**, we can find uniquely the components of any vector **a** in the **u**, **v** and **w** directions.

Fig. 6.4:1

Fig. 6.4:2

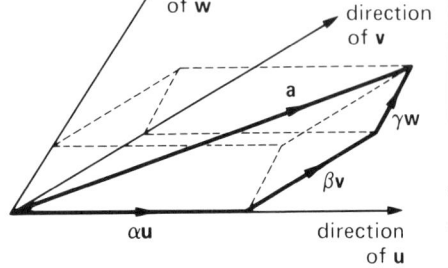

Proof
We will not attempt a formal proof. Figure 6.4:1 suggests how it can be done.

What goes wrong when **u**, **v** and **w** are coplanar? By 6.2:1 Theorem 1, any vector $\alpha \mathbf{u} + \beta \mathbf{v} + \gamma \mathbf{w}$ is in the same plane Π as **u**, **v** and **w**; so if **a** is not also in Π (fig. 6.4:2) we cannot express **a** in this form.

The set $\{\mathbf{u}, \mathbf{v}, \mathbf{w}\}$ is called a **basis** for 3D-space. The theorem says that any set of three non-coplanar vectors forms a basis for 3D-space.

Vectors, coordinate geometry and complex numbers 329

When considering vectors in 3D-space we choose an origin O and three mutually perpendicular axes Ox, Oy and Oz. We call the unit vectors in the directions Ox, Oy and Oz **i**, **j** and **k** respectively (fig. 6.4:**3a**). Then any vector **a** can be expressed in the form

$$a = a_1\mathbf{i} + a_2\mathbf{j} + a_3\mathbf{k} \tag{2}$$

since the set $\{\mathbf{i}, \mathbf{j}, \mathbf{k}\}$ forms a basis for 3D-space. (Note the relative orientation of the axes: Oz is in the sense of a right-hand screw from Ox to Oy—usually we think of Ox and Oy horizontal and Oz vertically upwards.)

In fig. 6.4:**3b**, **a** is the resultant of $a_1\mathbf{i}$ and $a_2\mathbf{j}$ and $a_3\mathbf{k}$. $a_1\mathbf{i}$, $a_2\mathbf{j}$ and $a_3\mathbf{k}$ are the components of **a** in the **i**, **j** and **k** directions.

The magnitude $|\mathbf{a}|$ (or a) of **a** is given by

$$|\mathbf{a}| = \sqrt{(a_1^2 + a_2^2 + a_3^2)} \tag{3}$$

e.g. $|\mathbf{i} - 2\mathbf{j} + 2\mathbf{k}| = \sqrt{(1^2 + (-2)^2 + 2^2)} = 3$.

From 6.2:1 equation (4), the unit vector $\hat{\mathbf{a}}$ in the direction of **a** is

$$\hat{\mathbf{a}} = \frac{1}{a}\mathbf{a} = \frac{a_1}{a}\mathbf{i} + \frac{a_2}{a}\mathbf{j} + \frac{a_3}{a}\mathbf{k} \tag{4}$$

Note: (i) $\hat{\mathbf{a}}$ is in the direction of **a** since $\hat{\mathbf{a}}$ is a scalar multiple of **a**;
(ii) $\hat{\mathbf{a}}$ is a unit vector since

$$|\hat{\mathbf{a}}| = \frac{1}{a}|\mathbf{a}| = 1$$

e.g. if $\mathbf{a} = \mathbf{i} - 2\mathbf{j} + 2\mathbf{k}$, $\hat{\mathbf{a}} = \tfrac{1}{3}\mathbf{i} - \tfrac{2}{3}\mathbf{j} + \tfrac{2}{3}\mathbf{k}$.

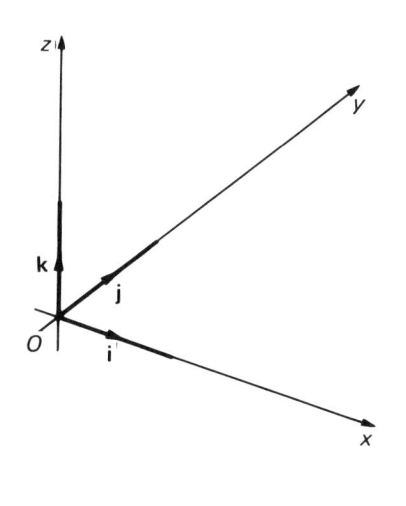

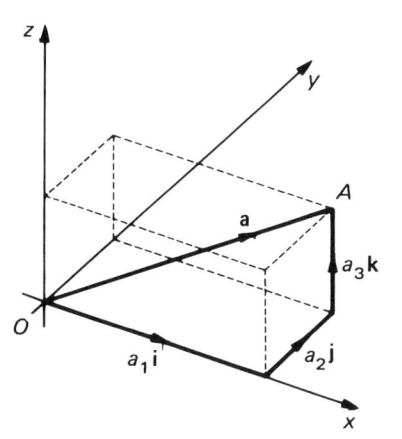

Fig. 6.4:**3**

Resultant of a set of vectors

If $\mathbf{a} = a_1\mathbf{i} + a_2\mathbf{j} + a_3\mathbf{k}$ and $\mathbf{b} = b_1\mathbf{i} + b_2\mathbf{j} + b_3\mathbf{k}$, then

$$\mathbf{a} + \mathbf{b} = (a_1 + b_1)\mathbf{i} + (a_2 + b_2)\mathbf{j} + (a_3 + b_3)\mathbf{k} \tag{5}$$

We can extend this result in an obvious way to three or more vectors, e.g.

$$(\mathbf{i} + 2\mathbf{j} - 2\mathbf{k}) + (3\mathbf{i} - \mathbf{j} + \mathbf{k}) + (-5\mathbf{i} + 4\mathbf{j} + \mathbf{k})$$
$$= (1 + 3 - 5)\mathbf{i} + (2 - 1 + 4)\mathbf{j} + (-2 + 1 + 1)\mathbf{k}$$
$$= -\mathbf{i} + 5\mathbf{j}$$

Calculation of the scalar product

$$\mathbf{i}.\mathbf{j} = \mathbf{j}.\mathbf{k} = \mathbf{k}.\mathbf{i} = 0 \qquad \text{by 6.1:2 Theorem 1}$$

$$\mathbf{i}.\mathbf{i} = \mathbf{j}.\mathbf{j} = \mathbf{k}.\mathbf{k} = 1 \qquad \text{by 6.1:2 equation (4)}$$

So if $\mathbf{a} = a_1\mathbf{i} + a_2\mathbf{j} + a_3\mathbf{k}$, $\mathbf{b} = b_1\mathbf{i} + b_2\mathbf{j} + b_3\mathbf{k}$,

$$\mathbf{a}.\mathbf{b} = (a_1\mathbf{i} + a_2\mathbf{j} + a_3\mathbf{k}).(b_1\mathbf{i} + b_2\mathbf{j} + b_3\mathbf{k})$$
$$= a_1b_1 + a_2b_2 + a_3b_3 \tag{6}$$

e.g. $(4\mathbf{i} + 5\mathbf{j} - 3\mathbf{k}).(2\mathbf{i} - 7\mathbf{j} - 7\mathbf{k}) = (4)(2) + (5)(-7) + (-3)(-7) = 6$

330 *Vectors in three-dimensional space*

Worked example 1   If $\mathbf{a} = \mathbf{i} + 2\mathbf{j} - 2\mathbf{k}$, $\mathbf{b} = 3\mathbf{i} - \mathbf{j} + \mathbf{k}$, find the angle θ between $\mathbf{a}$ and $\mathbf{b}$.

$$\cos\theta = \frac{\mathbf{a}.\mathbf{b}}{ab} = \frac{(\mathbf{i}+2\mathbf{j}-2\mathbf{k}).(3\mathbf{i}-\mathbf{j}+\mathbf{k})}{|\mathbf{i}+2\mathbf{j}-2\mathbf{k}||3\mathbf{i}-\mathbf{j}+\mathbf{k}|}$$

$$= -\frac{1}{3\sqrt{11}}$$

So $\theta \approx 95.76°$

(See 6.2:1 WE4.)

Exercise 6.4:1

1 Find $|\mathbf{a}|$ and $\hat{\mathbf{a}}$ when:

(i) $\mathbf{a} = 2\mathbf{i} + 6\mathbf{j} + 3\mathbf{k}$
(ii) $\mathbf{a} = -\mathbf{i} + 8\mathbf{j} - 4\mathbf{k}$
(iii) $\mathbf{a} = -\mathbf{i} - 2\mathbf{j} - 3\mathbf{k}$
(iv) $\mathbf{a} = -4\mathbf{j} + 3\mathbf{k}$

2 (i) If $|4\mathbf{i} + p\mathbf{j} + 2\mathbf{k}| = 6$ find two possible values of p.

(ii) If $q\mathbf{i} + \frac{1}{3}\mathbf{j} + \frac{2}{3}\mathbf{k}$ is a unit vector, find two possible values of q.

→ **3** If $\mathbf{a} = 6\mathbf{i} - 6\mathbf{j} + 3\mathbf{k}$, find:

(i) $|\mathbf{a}|$ (ii) $\hat{\mathbf{a}}$
(iii) the vector of magnitude 24 in the direction of $\mathbf{a}$

4 The vector $\mathbf{a} = a_1\mathbf{i} + a_2\mathbf{j} + a_3\mathbf{k}$ makes angles α, β and γ with Ox, Oy and Oz respectively.

(i) Show that $\cos\alpha = a_1/a$, $\cos\beta = a_2/a$, $\cos\gamma = a_3/a$ ($\cos\alpha$, $\cos\beta$, $\cos\gamma$ are called the **direction cosines** of $\mathbf{a}$ and are often written as l, m and n respectively).

(ii) Show that $l^2 + m^2 + n^2 = 1$

(iii) Find the direction cosines of the vector $2\mathbf{i} + 6\mathbf{j} + 3\mathbf{k}$.

(iv) A vector has direction cosines $\frac{1}{2}$, $-\frac{1}{2}$ and n. Find two possible values of n and hence find two possible angles (one acute, one obtuse) between the vector and Oz.

(v) Find the unit vector which makes equal acute angles with all three axes.

5 Which of the following pairs of vectors are parallel?

(i) $\mathbf{i} - \mathbf{j} - \mathbf{k}$ and $-\mathbf{i} + \mathbf{j} + \mathbf{k}$
(ii) $3\mathbf{i} - 2\mathbf{j} + \mathbf{k}$ and $3\mathbf{i} + 2\mathbf{j} - \mathbf{k}$
(iii) $4\mathbf{i} + 8\mathbf{j} - 2\mathbf{k}$ and $-2\mathbf{i} - 4\mathbf{j} + \mathbf{k}$

6 If $\mathbf{a} = \mathbf{i} + 2\mathbf{j} + \mathbf{k}$, $\mathbf{b} = 3\mathbf{i} + \mathbf{j} - \mathbf{k}$:

→ (i) find $\mathbf{a} + \mathbf{b}$, $\mathbf{a} - \mathbf{b}$, $3\mathbf{a} - 2\mathbf{b}$
(ii) find unit vectors in the directions of $\mathbf{a} + \mathbf{b}$ and $\mathbf{a} - \mathbf{b}$
(iii) verify that $|\mathbf{a} + \mathbf{b}| < |\mathbf{a}| + |\mathbf{b}|$

7 (i) If $x\mathbf{i} + y\mathbf{j} + z\mathbf{k} = \mathbf{0}$, what can you say about x, y and z? Why?

(ii) If $a_1\mathbf{i} + a_2\mathbf{j} + a_3\mathbf{k} = b_1\mathbf{i} + b_2\mathbf{j} + b_3\mathbf{k}$, show, using the result of (i), that $a_1 = b_1$, $a_2 = b_2$, $a_3 = b_3$. Explain how this result agrees with Theorem 1.

(iii) If $x\mathbf{i} + y\mathbf{j} + z\mathbf{k} = (1+\lambda)\mathbf{i} + 2\lambda\mathbf{j} + (4+3\lambda)\mathbf{k}$, show that

$$x - 1 = \tfrac{1}{2}y = \tfrac{1}{3}(z - 4)$$

8 The resultant of the vectors $\mathbf{i} + p\mathbf{j} + q\mathbf{k}$, $2\mathbf{i} - 3q\mathbf{j} + p\mathbf{k}$, $p\mathbf{i} + 4p\mathbf{j} - r\mathbf{k}$ and $r\mathbf{i} + \mathbf{k}$ is $4\mathbf{i} + \mathbf{j} + 7\mathbf{k}$. Find p, q and r.

9 Find $\mathbf{a}.\mathbf{b}$ when $\mathbf{a}$ and $\mathbf{b}$ are the vectors:

→ (i) $2\mathbf{i} - \mathbf{j} + 3\mathbf{k}$, $-\mathbf{i} + 4\mathbf{j} + 5\mathbf{k}$
(ii) $\mathbf{i} + \mathbf{j}$, $\mathbf{j} + \mathbf{k}$
(iii) $3\mathbf{i} + 5\mathbf{j} - \mathbf{k}$, $2\mathbf{i} - \mathbf{j} + \mathbf{k}$

What do you deduce in (iii)?

10 If $\mathbf{a} = \mathbf{i} + 2\mathbf{j}$, $\mathbf{b} = -3\mathbf{i} + 2\mathbf{j} - \mathbf{k}$, $\mathbf{c} = 5\mathbf{i} - \mathbf{j} + 4\mathbf{k}$, find:

(i) $\mathbf{a}.\mathbf{b}$ (ii) $\mathbf{a}.\mathbf{c}$ (iii) $\mathbf{a}.(\mathbf{b}+\mathbf{c})$

and verify that in this case $\mathbf{a}.(\mathbf{b}+\mathbf{c}) = \mathbf{a}.\mathbf{b} + \mathbf{a}.\mathbf{c}$.

11 If $a_1\mathbf{i} + a_2\mathbf{j} + a_3\mathbf{k} = b_1\mathbf{i} + b_2\mathbf{j} + b_3\mathbf{k}$, show by forming the scalar product of this equation successively with $\mathbf{i}$, $\mathbf{j}$ and $\mathbf{k}$, that $a_1 = b_1$, $a_2 = b_2$ and $a_3 = b_3$. (c.f. question 7(ii).)

→ **12** If $\mathbf{a} = 3\mathbf{i} - 2\mathbf{j} - \mathbf{k}$, $\mathbf{b} = 3\mathbf{i} - 5\mathbf{j} + 2\mathbf{k}$ and $\mathbf{c} = \mathbf{i} + p\mathbf{j} + q\mathbf{k}$, find p and q, given that $\mathbf{c}$ is perpendicular to $\mathbf{a}$ and $\mathbf{c}$ is perpendicular to $\mathbf{b}$. Hence find a unit vector perpendicular to both $\mathbf{a}$ and $\mathbf{b}$.

13 If $\mathbf{a} = \mathbf{i} + 2\mathbf{j} + \mathbf{k}$ and $\mathbf{b} = 3\mathbf{i} + 2\mathbf{j}$, find a unit vector perpendicular to both $\mathbf{a}$ and $\mathbf{b}$.

① **14** Find the cosine of the angle between $\mathbf{a}$ and $\mathbf{b}$ when

→ (i) $\mathbf{a} = \mathbf{i} + 2\mathbf{j} + \mathbf{k}$, $\mathbf{b} = 3\mathbf{i} + \mathbf{j} - \mathbf{k}$
R (ii) $\mathbf{a} = \mathbf{i} + \mathbf{k}$, $\mathbf{b} = \mathbf{j} - \mathbf{k}$
(iii) $\mathbf{a} = 4\mathbf{i} - \mathbf{j} + \mathbf{k}$, $\mathbf{b} = 2\mathbf{i} + 5\mathbf{j} - 3\mathbf{k}$
(iv) $\mathbf{a} = -2\mathbf{i} + 4\mathbf{j} + 2\mathbf{k}$, $\mathbf{b} = 3\mathbf{i} - 6\mathbf{j} - 3\mathbf{k}$

What do you deduce in (iii) and in (iv)?

6.4:2 Position vector of a point in space

Having chosen an origin O and axes Ox, Oy and Oz, we can specify the position of any point P in 3D-space by giving its cartesian coordinates (x, y, z).

> P is the point $(x, y, z) \Leftrightarrow$ the position vector of P is
> $$\overrightarrow{OP} = x\mathbf{i} + y\mathbf{j} + z\mathbf{k} \qquad (7)$$

Worked example 2 A, B and C are the points $(-1, 3, -1)$, $(3, 5, -5)$ and $(2, -2, 1)$. Find:

(i) the distance AB
(ii) the cosine of the angle θ between AB and AC
(iii) the position vector of the point S which divides AB internally in the ratio $1:3$

$\overrightarrow{AB} = \mathbf{b} - \mathbf{a} = (3\mathbf{i} + 5\mathbf{j} - 5\mathbf{k}) - (-\mathbf{i} + 3\mathbf{j} - \mathbf{k}) = 4\mathbf{i} + 2\mathbf{j} - 4\mathbf{k}$

$\overrightarrow{AC} = \mathbf{c} - \mathbf{a} = (2\mathbf{i} - 2\mathbf{j} + \mathbf{k}) - (-\mathbf{i} + 3\mathbf{j} - \mathbf{k}) = 3\mathbf{i} - 5\mathbf{j} + 2\mathbf{k}$

(i) $AB = |\overrightarrow{AB}| = |4\mathbf{i} + 2\mathbf{j} - 4\mathbf{k}| = 6$

(ii) $\cos \theta = \dfrac{\overrightarrow{AB} \cdot \overrightarrow{AC}}{AB \cdot AC} = \dfrac{(4\mathbf{i} + 2\mathbf{j} - 4\mathbf{k}) \cdot (3\mathbf{i} - 5\mathbf{j} + 2\mathbf{k})}{6\sqrt{38}} = \dfrac{-1}{\sqrt{38}}$

(iii) $\mathbf{s} = \mathbf{a} + \tfrac{1}{4}\overrightarrow{AB} = \mathbf{a} + \tfrac{1}{4}(\mathbf{b} - \mathbf{a}) = \tfrac{1}{4}(3\mathbf{a} + \mathbf{b}) = \tfrac{7}{2}\mathbf{j} - 2\mathbf{k}$

(See 6.2:2 WE5-7.)

Exercise 6.4:2

1 A and B are the points $(3, 5, -2)$ and $(1, -1, 1)$. Find:

(i) $\overrightarrow{AB}$
(ii) $\overrightarrow{BA}$
(iii) the distance AB

R 2 A, B and C are the points $(2, 0, 3)$, $(4, 1, 1)$ and $(-4, 2, 0)$. Find the cosine of the angle between AB and AC.

3 A, B and C are the points $(1, 1, 0)$, $(1, 3, -1)$ and $(7, 2, 2)$. Find $\overrightarrow{AB}$ and $\overrightarrow{AC}$ and show that $\overrightarrow{AB}$ is perpendicular to $\overrightarrow{AC}$. Hence find the area of $\triangle ABC$.

→ 4 A, B and C are the points $(4, -2, 3)$, $(4, 5, -1)$ and $(-2, 3, 5)$. Show that $AB = AC$ and find $B\hat{A}C$. Hence find the area of $\triangle ABC$.

5 The vertices of a tetrahedron are $A(1, 2, -1)$, $B(2, 3, 4)$, $C(3, 4, 3)$ and $D(2, 1, -1)$. Show that AC, BC and AD are mutually perpendicular. Hence find the volume of the tetrahedron.

6 Investigate whether or not the following sets of points are collinear:

→ (i) $\mathbf{i} + \mathbf{j} + \mathbf{k}$, $5\mathbf{i} + 2\mathbf{k}$, $-3\mathbf{i} + 2\mathbf{j}$
(ii) $(3, 2, 0)$, $(4, 3, -1)$, $(5, -4, 1)$
(iii) $(2, 1, 1)$, $(5, -2, 4)$, $(3, 0, 2)$

(See 6.2:2 Q34.)

7 A, B and C are the points $(4, -1, -1)$, $(2, 1, 0)$ and $(-2, 5, 2)$. Express $\mathbf{c}$ in the form $\mathbf{a} + \lambda(\mathbf{b} - \mathbf{a})$. Find the position vector of the point H on the line ABC for which $AH = AC$.

→ 8 A and B are the points $(-2, -3, 1)$ and $(1, 1, -2)$. Find the position vector of:

(i) the point S which divides AB internally in the ratio $2:1$;
(ii) the point T which divides AB externally in the ratio $5:2$.

② R 9 A and B are two points with position vectors $\mathbf{i}-\mathbf{j}+\mathbf{k}$ and $3\mathbf{i}+\mathbf{j}+2\mathbf{k}$. Find the position vectors of:

(i) the midpoint of AB;
(ii) the point which divides AB internally in the ratio $5:1$;
(iii) the point which divides AB externally in the ratio $4:3$.

② 10 A and B are the points $(1, 2, -4)$ and $(4, 2, 0)$. Find the position vector of the point which divides AB:

(i) internally (ii) externally in

the ratio $3:2$.

② 11 A and B are two points with position vectors $3\mathbf{i}+\mathbf{j}-\mathbf{k}$ and $\mathbf{i}-5\mathbf{j}+4\mathbf{k}$. Find the position vectors of the two points P on AB for which $AP=2AB$.

② 12 A, B, C and D are the points $(2, -3, 4)$, $(1, 2, -1)$, $(3, 1, 2)$ and $(4, 1, -2)$.

(i) Find the position vectors of the midpoints P, Q, R and S of AB, BC, CD and DA.

(ii) Hence show that $PQRS$ is a parallelogram.

(iii) Find the vectors $\frac{1}{4}(\mathbf{a}+\mathbf{b}+\mathbf{c}+\mathbf{d})$ and $\frac{1}{4}(\mathbf{p}+\mathbf{q}+\mathbf{r}+\mathbf{s})$ and the position vector of the point of intersection of the diagonals of $PQRS$.

② 13 Find the position vector of the centroid of $\triangle ABC$ when A, B and C are the points:

→ • (i) $(2, 3, -5)$, $(4, 1, 1)$, $(0, 5, 1)$
(ii) $(-1, 1, 1)$, $(-3, 3, 3)$, $(-4, 1, -4)$
(iii) $(3, -4, 5)$, $(1, 3, 1)$, $(-4, 1, -6)$

(See 6.1:3 Theorem 3.)

② 14 A, B and C are the points $(-3, 2, 1)$, $(9, -6, 9)$ and $(1, 10, 5)$. The points D, E and F divide AB, BC and CA internally in the ratio $1:3$. The points G, H and J divide AB, BC and CA externally in the ratio $5:1$. Find the position vectors of the centroids of $\triangle ABC$, $\triangle DEF$ and $\triangle GHJ$. (See 6.1:3 Q22.)

② R 15 $OABCDEFG$ is a cube of side $2a$ with OD, AE, BF, CG perpendicular to the base $OABC$. M and N are the centres of the faces $DEFG$ and $ABFE$ respectively. $\mathbf{i}$, $\mathbf{j}$, $\mathbf{k}$ are unit vectors along $\overrightarrow{OA}$, $\overrightarrow{OC}$ and $\overrightarrow{OD}$ respectively. Express in terms of a, $\mathbf{i}$, $\mathbf{j}$ and $\mathbf{k}$ the position vectors, relative to O, of the points F, M and N. Find:

(i) $F\hat{O}M$ (ii) $A\hat{O}N$

16 Prove that the angle between the longest diagonals of a cube is $\cos^{-1}\frac{1}{3}$.

6.4:3 Equations of a line

We have seen (section 6.1:3) that a vector equation of the line Λ passing through the point A and parallel to the vector $\mathbf{d}$ (fig. 6.4:4) is

$$\mathbf{r}=\mathbf{a}+\lambda\mathbf{d}$$

If $\mathbf{a}=a_1\mathbf{i}+a_2\mathbf{j}+a_3\mathbf{k}$, $\mathbf{d}=d_1\mathbf{i}+d_2\mathbf{j}+d_3\mathbf{k}$, a vector equation of Λ is

$$\mathbf{r}=a_1\mathbf{i}+a_2\mathbf{j}+a_3\mathbf{k}+\lambda(d_1\mathbf{i}+d_2\mathbf{j}+d_3\mathbf{k}) \qquad (8)$$

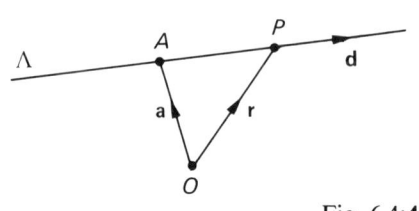

Fig. 6.4:4

Let P be the point (x, y, z), so that $\mathbf{r}=x\mathbf{i}+y\mathbf{j}+z\mathbf{k}$. From equation (8), $x\mathbf{i}+y\mathbf{j}+z\mathbf{k}=a_1\mathbf{i}+a_2\mathbf{j}+a_3\mathbf{k}+\lambda(d_1\mathbf{i}+d_2\mathbf{j}+d_3\mathbf{k})$, so

$$\left.\begin{array}{l}x=a_1+\lambda d_1\\ y=a_2+\lambda d_2\\ z=a_3+\lambda d_3\end{array}\right\} \qquad (9)$$

the **parametric equations** of Λ. Eliminating λ:

$$\frac{x-a_1}{d_1}=\frac{y-a_2}{d_2}=\frac{z-a_3}{d_3} \qquad (10)$$

the **cartesian equations** of Λ.

Notice that a line in 3D is represented by a *pair* of cartesian equations. Why?

Vectors, coordinate geometry and complex numbers 333

A general point in space has three degrees of freedom, a general point on a surface two degrees, a general point on a line one degree. An equation such as $x+2y+z=0$ reduces the freedom of a general point (x, y, z) by one degree, to two degrees, so it represents a surface (which in this case is a plane).

A pair of equations such as $\frac{1}{2}(x-1) = \frac{1}{5}(y-3) = \frac{1}{6}z$ reduces the freedom of a general point (x, y, z) by two degrees, to one degree, so it represents a line (in this case a straight line).

Worked example 3 (i) Find a vector equation of the line passing through the points $A(2, -2, -1)$ and $B(4, -3, 1)$.
(ii) Hence find the cartesian equations of the line.

(i) Direction of line $\mathbf{d} = \overrightarrow{AB} = \mathbf{b} - \mathbf{a} = 2\mathbf{i} - \mathbf{j} + 2\mathbf{k}$. So a vector equation of the line is

$$\mathbf{r} = \mathbf{a} + \lambda(\mathbf{b} - \mathbf{a}) = 2\mathbf{i} - 2\mathbf{j} - \mathbf{k} + \lambda(2\mathbf{i} - \mathbf{j} + 2\mathbf{k})$$

(Note that $\lambda = 0$ corresponds to the point A, $\lambda = 1$ to the point B.)

(ii) The parametric equations of the line are $x = 2 + 2\lambda$, $y = -2 - \lambda$, $z = -1 + 2\lambda$. Eliminating λ gives

$$\frac{x-2}{2} = \frac{y+2}{-1} = \frac{z+1}{2}$$

Worked example 4 Find the shortest distance p from the point L with position vector $-\mathbf{i} - \mathbf{j} + \mathbf{k}$ to the line $\Lambda: \mathbf{r} = 2\mathbf{j} - 3\mathbf{k} + \lambda(2\mathbf{i} - \mathbf{j} + 2\mathbf{k})$.

The position vector $\mathbf{r}$ of any point P on Λ is given by $\mathbf{r} = 2\mathbf{j} - 3\mathbf{k} + \lambda(2\mathbf{i} - \mathbf{j} + 2\mathbf{k})$, as shown in fig. 6.4:5, so

$$\overrightarrow{LP} = \mathbf{r} - \mathbf{l} = \mathbf{i} + 3\mathbf{j} - 4\mathbf{k} + \lambda(2\mathbf{i} - \mathbf{j} + 2\mathbf{k})$$

If P is the point on the line *closest* to L, then LP is perpendicular to Λ, so

$$\overrightarrow{LP} \cdot (2\mathbf{i} - \mathbf{j} + 2\mathbf{k}) = 0$$
$$\{(\mathbf{i} + 3\mathbf{j} - 4\mathbf{k}) + \lambda(2\mathbf{i} - \mathbf{j} + 2\mathbf{k})\} \cdot (2\mathbf{i} - \mathbf{j} + 2\mathbf{k}) = 0$$

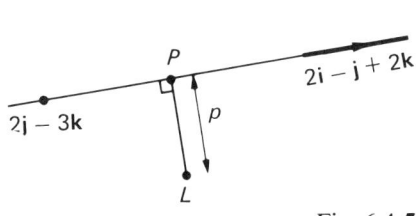

Fig. 6.4:**5**

i.e. $-9 + 9\lambda = 0$; so $\lambda = 1$.
Thus $\overrightarrow{LP} = 3\mathbf{i} + 2\mathbf{j} - 2\mathbf{k}$, so

$$p = LP = |3\mathbf{i} + 2\mathbf{j} - 2\mathbf{k}| = \sqrt{17}$$

Intersection of two lines
In 3D-space two lines may be:

(i) **parallel**
(ii) **intersecting**
(iii) **skew**, i.e. not parallel and not intersecting—by far the most likely case.

Worked example 5 Show that the lines $\mathbf{r} = \mathbf{i} + \lambda(6\mathbf{i} + 2\mathbf{j} - 3\mathbf{k})$ and $\mathbf{r} = \mathbf{i} + \mathbf{j} + \mathbf{k} + \mu(-2\mathbf{i} + \mathbf{j} - 2\mathbf{k})$ are skew.

Suppose that the lines intersect; then, at their point of intersection:

$$\mathbf{i} + \lambda(6\mathbf{i} + 2\mathbf{j} - 3\mathbf{k}) = \mathbf{i} + \mathbf{j} + \mathbf{k} + \mu(-2\mathbf{i} + \mathbf{j} - 2\mathbf{k})$$

334 *Equations of a line*

Thus

$$1 + 6\lambda = 1 - 2\mu \tag{11}$$
$$2\lambda = 1 + \mu \tag{12}$$
$$-3\lambda = 1 - 2\mu \tag{13}$$

Equations (11) and (12) give $\lambda = \frac{1}{5}$, $\mu = -\frac{3}{5}$. These values of λ and μ do *not* satisfy equation (13) and so the lines do not intersect. They are not parallel either. (How can you tell?) So they must be skew.

Angle between two lines

Consider the lines Λ_1: $\mathbf{r} = \mathbf{a}_1 + \lambda \mathbf{d}_1$ and Λ_2: $\mathbf{r} = \mathbf{a}_2 + \lambda \mathbf{d}_2$ shown in fig. 6.4:6. Even if Λ_1 and Λ_2 do not intersect, we can still talk about the angle θ between Λ_1 and Λ_2:

angle θ between Λ_1 and Λ_2 = angle between $\mathbf{d}_1$ and $\mathbf{d}_2$

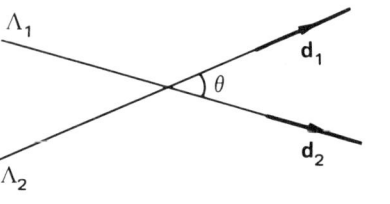

Fig. 6.4:6   So

$$\cos \theta = \frac{\mathbf{d}_1 \cdot \mathbf{d}_2}{d_1 d_2} \quad (= \hat{\mathbf{d}}_1 \cdot \hat{\mathbf{d}}_2) \tag{14}$$

Note that θ is independent of $\mathbf{a}_1$ and $\mathbf{a}_2$, i.e. of the positions of Λ_1 and Λ_2—it depends only on their directions. For example, the angle θ between the lines $(\mathbf{r} = 19\mathbf{i} + \mathbf{j} + \lambda(\mathbf{i} + 2\mathbf{j} - 2\mathbf{k}))$ and $\mathbf{r} = 3\mathbf{i} - 32\mathbf{k} + \mu(3\mathbf{i} - \mathbf{j} + \mathbf{k})$ is given by

$$\cos \theta = \frac{(\mathbf{i} + 2\mathbf{j} - 2\mathbf{k}) \cdot (3\mathbf{i} - \mathbf{j} + \mathbf{k})}{|\mathbf{i} + 2\mathbf{j} - 2\mathbf{k}| \cdot |3\mathbf{i} - \mathbf{j} + \mathbf{k}|} = -\frac{1}{3\sqrt{11}}$$

Exercise 6.4:3

③ 1 Find vector and cartesian equations of the line:

(i) parallel to the vector $2\mathbf{i} + 3\mathbf{j} + 2\mathbf{k}$ and passing through the point $-\mathbf{i} + 4\mathbf{j} + 5\mathbf{k}$
(ii) parallel to the line $\mathbf{r} = 3\mathbf{i} + \mathbf{j} + (\mathbf{i} + 2\mathbf{j} - 5\mathbf{k})$ and passing through the point $\mathbf{i} + 2\mathbf{j} + 3\mathbf{k}$
(iii) passing through the point $4\mathbf{i} + 5\mathbf{j} - \mathbf{k}$ and the origin
→ (iv) passing through points $3\mathbf{i} + 2\mathbf{j}$ and $-2\mathbf{i} + 4\mathbf{j} + 3\mathbf{k}$
R (v) passing through points $\mathbf{i} + 3\mathbf{j} + 4\mathbf{k}$ and $6\mathbf{i} - 2\mathbf{j} - \mathbf{k}$

③ 2 (i) Show that the line $\mathbf{r} = \mathbf{i} + 5\mathbf{j} + \lambda(3\mathbf{i} - 6\mathbf{j} + 15\mathbf{k})$ is parallel to the line $\mathbf{r} = 5\mathbf{j} + 3\mathbf{k} + \mu(-2\mathbf{i} + 4\mathbf{j} - 10\mathbf{k})$.

(ii) Show that the line through the points $(2, 0, 1)$ and $(-1, 4, -2)$ is parallel to the line through the points $(3, -2, -1)$ and $(6, -6, 2)$.

③ 3 Λ_1, Λ_2, Λ_3 and Λ_4 are the lines
$\mathbf{r} = \mathbf{i} + \alpha(2\mathbf{i} + \mathbf{j} - 3\mathbf{k})$, $\mathbf{r} = \mathbf{j} - 3\mathbf{k} + \beta(2\mathbf{i} + 5\mathbf{j} + 3\mathbf{k})$,
$\mathbf{r} = 2\mathbf{i} + \mathbf{j} + \mathbf{k} + \gamma(\mathbf{i} + \mathbf{j} + \mathbf{k})$ and $\mathbf{r} = \mathbf{i} - \mathbf{k} + \delta(3\mathbf{i} - 6\mathbf{j})$.
Show that Λ_2, Λ_3 and Λ_4 are each perpendicular to Λ_1.

③ → 4 The vector equation of a line is not unique.

• (i) If the equations $\mathbf{r} = \mathbf{a} + \lambda \mathbf{d}$ and $\mathbf{r} = \mathbf{a}' + \mu \mathbf{d}'$ represent the same line, what can you say about $\mathbf{a}'$ and $\mathbf{d}'$?

(ii) Show that the equations:
(a) $\mathbf{r} = 2\mathbf{i} + \mathbf{k} + \lambda(\mathbf{i} + \mathbf{j} - \mathbf{k})$ and
$\mathbf{r} = -\mathbf{i} - 3\mathbf{j} + 4\mathbf{k} + \mu(-3\mathbf{i} - 3\mathbf{j} + 3\mathbf{k})$
(b) $\mathbf{r} = \mathbf{i} + \mathbf{j} + \lambda(2\mathbf{i} - \mathbf{j} + 3\mathbf{k})$ and
$\mathbf{r} = 3\mathbf{i} + 3\mathbf{k} + \mu(4\mathbf{i} - 2\mathbf{j} + 6\mathbf{k})$
represent the same line.

③ : 5 Which of the following pairs of equations represent the same line?

(i) $\mathbf{r} = \mathbf{i} + 2\mathbf{j} + \mathbf{k} + \lambda(\mathbf{i} - \mathbf{j} + 3\mathbf{k})$ and
$\mathbf{r} = 3\mathbf{i} + 7\mathbf{k} + \mu(2\mathbf{i} - 2\mathbf{j} + 6\mathbf{k})$
(ii) $\mathbf{r} = \mathbf{i} + 2\mathbf{j} + \mathbf{k} + \lambda(\mathbf{i} - \mathbf{j} + 3\mathbf{k})$ and
$\mathbf{r} = -\mathbf{i} + 4\mathbf{j} - \mathbf{k} + \mu(2\mathbf{i} - 2\mathbf{j} + 6\mathbf{k})$
(iii) $\mathbf{r} = \mathbf{i} + \mathbf{k} + \lambda(\mathbf{i} - \mathbf{j} + \mathbf{k})$ and
$\mathbf{r} = 5\mathbf{i} - 4\mathbf{j} + 5\mathbf{k} + \mu(\mathbf{i} + \mathbf{j} - \mathbf{k})$

③ → **6** A line has cartesian equations

$$\frac{x-3}{-2} = \frac{y+5}{-1} = \frac{z+1}{4}$$

Write down a vector in the direction of the line and the position vector of a point on the line. (Hint: compare the equation with equation (10).) Hence write down a vector equation of the line.

③ **7** Find vector equations of the lines with cartesian equations:

(i) $\dfrac{x-3}{4} = \dfrac{y+2}{-5} = \dfrac{z-6}{3}$

(ii) $\dfrac{x}{1} = \dfrac{y}{2} = \dfrac{z}{3}$

(iii) $\dfrac{3x+1}{2} = \dfrac{1-2y}{1} = \dfrac{z}{2}$

• **8** A line has cartesian equations $2x - y = 1$, $3x + z = 2$. Rewrite these equations in the form

$$\frac{x - a_1}{d_1} = \frac{y - a_2}{d_2} = \frac{z - a_3}{d_3}$$

and hence find a vector equation of the line.

9 Find vector equations of the lines with cartesian equations:

(i) $5x - z = 20$, $4x - y = 14$
(ii) $2x + y = 3$, $y - z = 3$
(iii) $x - y - z = 1$, $3x + y - z = 3$
(iv) $x + y - z = 6$, $2x - 3y + 2z = 2$

③ **10** Find the cartesian equations of the line through the point $(1, -2, 4)$ parallel to the line

$$\frac{x+3}{2} = \frac{y+4}{-3} = \frac{z+5}{4}$$

③ → **11** Find the position vectors of the points where the line $\mathbf{r} = \mathbf{i} + 3\mathbf{j} + 2\mathbf{k} + \lambda(\mathbf{i} - \mathbf{j} + 4\mathbf{k})$ cuts:

• (i) the yz plane
(ii) the xz plane
(iii) the xy plane

③ • **12** A and B are the points $(2, -1, 1)$ and $(4, -3, 0)$.

(i) Show that a vector equation of the line AB is $\mathbf{r} = 2\mathbf{i} - \mathbf{j} + \mathbf{k} + \lambda(2\mathbf{i} - 2\mathbf{j} - \mathbf{k})$. What are the values of λ corresponding to A and B?

(ii) Show that $|\overrightarrow{AB}| = 3$.

(iii) Hence show that at the two points on AB distant 12 units from A, $\lambda = +4$ or -4. Hence find the position vectors of these two points.

(See 6.1:3 Q29.)

③ → **13** If A and B are the points $(-1, 3, 2)$ and $(5, 1, -1)$, find the two points on AB distant 21 units from A.

③ → **14** Write down a vector equation of the line joining
• the points $(1, 2, 1)$ and $(2, 2, 4)$.
Show that the cartesian equations of this line can be written

$$\frac{x-2}{1} = \frac{z-4}{3} \quad y = 2$$

This line lies entirely in the plane $y = 2$ (parallel to the xz plane).

③ **15** Find vector and cartesian equations of:

(i) the line joining $(3, 2, 1)$ and $(-1, -1, 1)$
(ii) the line parallel to the vector $\mathbf{j} + \mathbf{k}$ and passing through the point $2\mathbf{i} - \mathbf{j} + 5\mathbf{k}$

③ **16** Find vector equations of the lines with cartesian equations:

(i) $\dfrac{x}{3} = \dfrac{y-5}{-1}$, $z = 2$

(ii) $x = 1$, $z = -3$

④ **17** Find the shortest distance from the point to the line:

→ (i) $-\mathbf{i} + 3\mathbf{j} + 5\mathbf{k}$; $\mathbf{r} = 6\mathbf{i} + 3\mathbf{j} + 3\mathbf{k} + \lambda(3\mathbf{i} - 2\mathbf{j} + 2\mathbf{k})$
(ii) the origin; $\mathbf{r} = 7\mathbf{i} - 4\mathbf{j} - 12\mathbf{k} + \lambda(4\mathbf{i} + \mathbf{j} - 5\mathbf{k})$
(iii) $-4\mathbf{i} + 2\mathbf{j} + 5\mathbf{k}$; $\mathbf{r} = 9\mathbf{i} - 4\mathbf{k} + \lambda(4\mathbf{i} - 2\mathbf{j} + 3\mathbf{k})$
(iv) $-\mathbf{i} + 3\mathbf{j}$; $\mathbf{r} = \mathbf{i} - 8\mathbf{j} + 2\mathbf{k} + \lambda(\mathbf{i} + 2\mathbf{j} + \mathbf{k})$

④ R **18** Show that the point $3\mathbf{i} + 5\mathbf{j} - 2\mathbf{k}$ is the mirror image of the point $-\mathbf{i} + 3\mathbf{j} + 4\mathbf{k}$ in the line $\mathbf{r} = -4\mathbf{i} + 8\mathbf{j} - \mathbf{k} + \lambda(5\mathbf{i} - 4\mathbf{j} + 2\mathbf{k})$.

④ • **19** Show that the point $A(4, -1, 0)$ is the mirror image of the point $B(0, 5, 8)$ in the line $\mathbf{r} = 2\mathbf{i} - 3\mathbf{k} + \lambda(2\mathbf{j} - \mathbf{k})$.
Show that $\mathbf{r} = 4\mathbf{i} - \mathbf{j} + \mu(-2\mathbf{i} + 3\mathbf{j} + 4\mathbf{k})$ is a vector equation of the line passing through A and B.
What are the values of μ corresponding to the points A, M (the midpoint of AB) and B?

④ **20** Find the position vector of the point on the line $\Lambda: \mathbf{r} = 3\mathbf{i} - 4\mathbf{j} + 7\mathbf{k} + \lambda(-\mathbf{i} + 3\mathbf{j} - 2\mathbf{k})$ closest to the point $L(0, 1, 2)$. Hence find a vector equation of the perpendicular from L to Λ.
Show that the shortest distance from L to Λ is $\sqrt{3}$.
Find the mirror image of L in Λ.

④ **21** Find the mirror image of the point in the line:

→ (i) $\mathbf{i} + \mathbf{j} + \mathbf{k}$; $\mathbf{r} = \mathbf{i} + \mathbf{j} - 5\mathbf{k} + \lambda(\mathbf{i} + 2\mathbf{j} + 5\mathbf{k})$
(ii) $3\mathbf{i} - \mathbf{j} - 5\mathbf{k}$; $\mathbf{r} = -\mathbf{j} + \mathbf{k} + \lambda(2\mathbf{i} + \mathbf{j} - \mathbf{k})$

22 Investigate whether the following pairs of lines are parallel, intersecting or skew. If they are intersecting, find their point of intersection:

(i) $\mathbf{r} = \mathbf{i} + \mathbf{j} + \mathbf{k} + \lambda(2\mathbf{i} - \mathbf{j} + \mathbf{k})$ and
$\mathbf{r} = -5\mathbf{j} - 4\mathbf{k} + \mu(\mathbf{i} + \mathbf{j} + 2\mathbf{k})$
(ii) $\mathbf{r} = 2\mathbf{i} - \mathbf{j} + \lambda(-3\mathbf{i} + \mathbf{j} - \mathbf{k})$ and
$(x-4)/6 = y/-2 = (z+6)/2$
(iii) $\mathbf{r} = 2\mathbf{i} + \mathbf{j} + \lambda(\mathbf{i} + \mathbf{k})$ and
$\mathbf{r} = \mathbf{i} + \mathbf{j} + 3\mathbf{k} + \mu(3\mathbf{i} - \mathbf{j} - 2\mathbf{k})$
(iv) $\mathbf{r} = \mathbf{k} + \lambda(\mathbf{i} - \mathbf{j} + 2\mathbf{k})$ and
$\mathbf{r} = \mathbf{i} + \mathbf{j} + \mu(2\mathbf{i} + \mathbf{k})$
(v) $\mathbf{r} = 3\mathbf{i} + \mathbf{j} - 5\mathbf{k} + \lambda(-2\mathbf{i} + \mathbf{k})$ and
$(x-5)/2 = (z-1)/-1; y = 4$
(vi) $\mathbf{r} = -\mathbf{i} + 3\mathbf{j} + 5\mathbf{k} + \lambda(\mathbf{i} - 2\mathbf{j} - 4\mathbf{k})$ and
$\mathbf{r} = 3\mathbf{i} - 3\mathbf{k} + \mu(2\mathbf{i} + \mathbf{j})$
(vii) $\mathbf{r} = -\mathbf{k} + \lambda(2\mathbf{i} - \mathbf{j} + 5\mathbf{k})$ and
$\mathbf{r} = 6\mathbf{i} - \mathbf{j} + \mu(-4\mathbf{i} + 2\mathbf{j} - 10\mathbf{k})$
(viii) $\mathbf{r} = -2\mathbf{i} - \mathbf{j} + 2\mathbf{k} + \lambda(\mathbf{i} - \mathbf{j} + \mathbf{k})$ and
$\mathbf{r} = -2\mathbf{j} + \mathbf{k} + \mu(-2\mathbf{i} + 3\mathbf{j} - 5\mathbf{k})$

23 Explain how an attempt to find the point of intersection of the lines $\mathbf{r} = \mathbf{a} + \lambda \mathbf{d}$ and $\mathbf{r} = \mathbf{a}' + \mu \mathbf{d}'$ leads to three (scalar) equations for λ and μ.

Discuss the solutions of these three equations when the lines are:

(i) parallel (ii) intersecting (iii) skew

24 Two lines have equations

$$\mathbf{r} = \mathbf{i} + 2\mathbf{j} + \mathbf{k} + \lambda(\mathbf{i} - 2\mathbf{j} + 3\mathbf{k})$$

and

$$\mathbf{r} = -2\mathbf{i} + p\mathbf{j} + 7\mathbf{k} + \mu(-\mathbf{i} + \mathbf{j} + 2\mathbf{k}).$$

If the lines intersect, find the value of p and the position vector of their point of intersection.

R 25 Two lines have equations

$$\mathbf{r} = -2\mathbf{i} - 5\mathbf{j} + \lambda(3\mathbf{i} + \mathbf{j} - \mathbf{k})$$

and

$$\mathbf{r} = 5\mathbf{i} + p\mathbf{j} - 5\mathbf{k} + \mu(\mathbf{i} - \mathbf{j} + \mathbf{k}).$$

If the lines intersect, find the value of p and the position vector of their point of intersection.

26 The vertices of a $\triangle ABC$ are $A(1, 0, 1)$, $B(3, 2, 4)$ and $C(2, -1, 2)$. Find the vector equations of the three medians (see 6.1:3 Theorem 3) of $\triangle ABC$. Show that they intersect and find the coordinates of their point of intersection, i.e. the centroid of $\triangle ABC$.

27 Find the angles between the following pairs of lines:

(i) $\mathbf{r} = \mathbf{i} + \mathbf{j} + \lambda(2\mathbf{i} + 2\mathbf{j} - \mathbf{k})$ and $\mathbf{r} = 2\mathbf{j} - \mathbf{k} + \mu(4\mathbf{i} + 3\mathbf{j})$
R (ii) $\mathbf{r} = \mathbf{i} + \mathbf{j} + \lambda(2\mathbf{i} + 2\mathbf{j} - \mathbf{k})$ and
$\mathbf{r} = 3\mathbf{i} + 3\mathbf{j} - \mathbf{k} + \mu(5\mathbf{i} - 7\mathbf{j} - 4\mathbf{k})$
(iii) $\dfrac{x-3}{2} = \dfrac{y+1}{-1} = \dfrac{z-2}{3}$ and $\dfrac{x+5}{6} = \dfrac{y-1}{-3} = \dfrac{z}{2}$
(iv) $\mathbf{r} = \mathbf{i} + \mathbf{j} + \lambda(2\mathbf{i} + 2\mathbf{j} - \mathbf{k})$ and $\dfrac{x-7}{-8} = \dfrac{y+1}{-8} = \dfrac{z}{4}$

28 Describe and sketch the curve whose vector equation is $\mathbf{r} = (a \cos \theta)\mathbf{i} + (a \sin \theta)\mathbf{j} + b\theta\mathbf{k}$ where a and b are constants.

6.4:4 Equation of a plane: parametric form

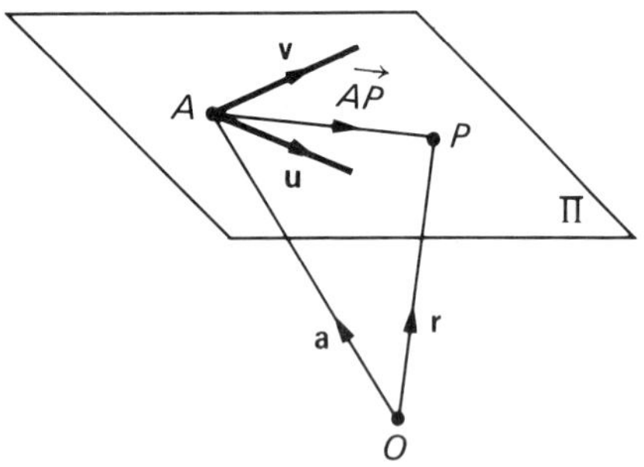

Suppose Π is the plane containing the point A and parallel to the vectors $\mathbf{u}$ and $\mathbf{v}$ (fig. 6.4:7). Let P be any point on Π, with position vector $\mathbf{r}$. Then the vector $\overrightarrow{AP}$ is in Π. So we can find scalars λ and μ such that $\overrightarrow{AP} = \lambda \mathbf{u} + \mu \mathbf{v}$ (see 6.2:1 Theorem 1), i.e.

$$\mathbf{r} - \mathbf{a} = \lambda \mathbf{u} + \mu \mathbf{v}$$

$$\mathbf{r} = \mathbf{a} + \lambda \mathbf{u} + \mu \mathbf{v} \qquad (15)$$

Fig. 6.4:7

This is a vector equation of the plane in parametric form. Notice that it contains two parameters λ and μ; the point P has two degrees of freedom.

We can find the cartesian equation of the plane by putting $\mathbf{r} = x\mathbf{i} + y\mathbf{j} + z\mathbf{k}$ in equation (15) and eliminating λ and μ.

Worked example 6 Find the vector (parametric form) and cartesian equations of the plane through the points $A(2, 0, -2)$, $B(-1, 1, 3)$ and $C(2, 1, -1)$.

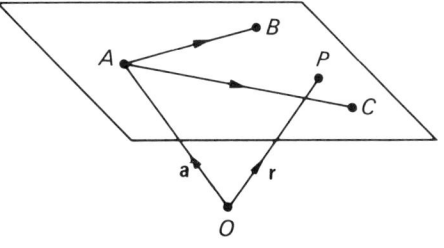

Vectors, coordinate geometry and complex numbers 337

The vectors $\overrightarrow{AB}$ and $\overrightarrow{AC}$ are in the plane (fig. 6.4:8); so a vector equation of the plane is

$$\mathbf{r} = \mathbf{a} + \lambda \overrightarrow{AB} + \mu \overrightarrow{AC}$$

i.e. $\quad \mathbf{r} = \mathbf{a} + \lambda(\mathbf{b} - \mathbf{a}) + \mu(\mathbf{c} - \mathbf{a})$

i.e. $\quad \mathbf{r} = 2\mathbf{i} - 2\mathbf{k} + \lambda(-3\mathbf{i} + \mathbf{j} + 5\mathbf{k}) + \mu(\mathbf{j} + \mathbf{k})$

So

$$x\mathbf{i} + y\mathbf{j} + z\mathbf{k} = (2 - 3\lambda)\mathbf{i} + (\lambda + \mu)\mathbf{j} + (-2 + 5\lambda + \mu)\mathbf{k}$$

and

$$x = 2 - 3\lambda \quad (16)$$
$$y = \lambda + \mu \quad (17)$$
$$z = -2 + 5\lambda + \mu \quad (18)$$

(17)−(18): $\quad y - z = 2 - 4\lambda \quad (19)$

$4 \times (16) - 3 \times (19)$: $\quad 4x - 3y + 3z = 2 \quad (20)$

This is the cartesian equation of the plane.

Fig. 6.4:8

Exercise 6.4:4

⑥ **1** Find the vector (parametric form) and cartesian equations of the plane:

(i) containing the point $2\mathbf{i} + \mathbf{j} + 3\mathbf{k}$, parallel to the directions $\mathbf{i} + \mathbf{j} + 3\mathbf{k}$ and $\mathbf{j} + 5\mathbf{k}$.

(ii) through the origin, parallel to the directions $\mathbf{i} + \mathbf{j}$ and $\mathbf{j} + \mathbf{k}$.

→ (iii) through the points $(3, 4, -1)$, $(2, 0, 1)$ and $(-1, 2, 0)$.

(iv) through the points $(-2, 4, 5)$, $(3, 4, 5)$ and $(-2, 4, 1)$.

(v) through the origin and the points $(1, 2, 3)$ and $(3, -2, 4)$.

⑥ R **2** Show that the lines $\mathbf{r} = \mathbf{i} - \mathbf{j} + \mathbf{k} + \lambda(2\mathbf{i} - \mathbf{j} + \mathbf{k})$ and $\mathbf{r} = -5\mathbf{j} - 4\mathbf{k} + \mu(\mathbf{i} + \mathbf{j} + 2\mathbf{k})$ are intersecting. Find a vector equation of the plane containing them.

⑥ **3** Find a vector equation of the plane containing the line $\mathbf{r} = 4\mathbf{i} + \mathbf{j} + \lambda(5\mathbf{i} - \mathbf{j} + 2\mathbf{k})$ and the point $3\mathbf{i} + 4\mathbf{j} + 5\mathbf{k}$.

⑥ **4** Show that the lines $\mathbf{r} = 2\mathbf{i} - \mathbf{j} + \lambda(-3\mathbf{i} + \mathbf{j} - \mathbf{k})$ and $\mathbf{r} = 4\mathbf{i} + \mathbf{j} - 6\mathbf{k} + \mu(6\mathbf{i} - 2\mathbf{j} + 2\mathbf{k})$ are parallel. Find a vector equation of the plane containing them.

⑥ **5** The vector equation of a plane, in parametric form, is not unique.

Show that $\mathbf{r} = 3\mathbf{i} - 2\mathbf{j} + 3\mathbf{k} + \lambda(\mathbf{i} + \mathbf{j} + \mathbf{k}) + \mu(\mathbf{i} + 2\mathbf{j})$ and $\mathbf{r} = 2\mathbf{i} - \mathbf{k} + \alpha(\mathbf{j} - \mathbf{k}) + \beta(3\mathbf{i} + 2\mathbf{j} + 4\mathbf{k})$ are equations of the same plane and explain why.

Show that the vectors $\mathbf{i} + \mathbf{j} + \mathbf{k}$, $\mathbf{i} + 2\mathbf{j}$, $\mathbf{j} - \mathbf{k}$ and $3\mathbf{i} + 2\mathbf{j} + 4\mathbf{k}$ are all perpendicular to the vector $2\mathbf{i} - \mathbf{j} - \mathbf{k}$ and explain the significance of this.

6 Find the position vector of the point of intersection of the plane $\mathbf{r} = 2\mathbf{i} - 4\mathbf{j} + \mathbf{k} + \lambda(2\mathbf{i} + \mathbf{k}) + \mu(-2\mathbf{i} + 3\mathbf{j})$ and the line $\mathbf{r} = \mathbf{i} + 2\mathbf{j} + s(\mathbf{i} - \mathbf{j} + \mathbf{k})$.

• **7** Find the vector equation of the line of intersection of the planes

$$\mathbf{r} = \mathbf{i} - \mathbf{j} + \mathbf{k} + \lambda(2\mathbf{i} + \mathbf{k}) + \mu(\mathbf{i} + \mathbf{j} - \mathbf{k}) \quad \text{(i)}$$
$$\mathbf{r} = -\mathbf{k} + s(3\mathbf{i} + \mathbf{j} - \mathbf{k}) + t(\mathbf{i} + \mathbf{j}) \quad \text{(ii)}$$

(Hint: using equations (i) and (ii), obtain three (scalar) equations for λ, μ, s and t. Show that $\mu = 1 + 2\lambda$; substitute this expression for μ into equation (i).)

8 Find the vector equation of the line of intersection of the planes

$$\mathbf{r} = 3\mathbf{i} - \mathbf{j} - \mathbf{k} + \lambda(\mathbf{i} + 2\mathbf{j}) + \mu(\mathbf{i} - \mathbf{k}) \text{ and}$$
$$\mathbf{r} = 2\mathbf{i} + \mathbf{k} + s(3\mathbf{i} - 5\mathbf{k}) + t(\mathbf{j} + \mathbf{k})$$

9 Show that the planes

$$\Pi_1: \mathbf{r} = 2\mathbf{i} - \mathbf{j} + \lambda(4\mathbf{i} - \mathbf{j} + 3\mathbf{k}) + \mu(\mathbf{j} + \mathbf{k})$$

and $\quad \Pi_2: \mathbf{r} = -3\mathbf{j} + 3\mathbf{k} + s(\mathbf{i} - \mathbf{j}) + t(3\mathbf{i} + \mathbf{j} + 4\mathbf{k})$

are parallel, by showing that the vector $\mathbf{n} = \mathbf{i} + \mathbf{j} - \mathbf{k}$ is perpendicular to both planes.

Write down the equation of the line Λ through the point $2\mathbf{i} - \mathbf{j}$ in the direction of $\mathbf{n}$. Find the position vector of the point of intersection of Λ and Π_2. Hence find

(i) the distance between the planes
(ii) a vector equation of the mirror image of Π_1 in Π_2.

6.4:5 Equation of a plane: scalar product form

Suppose Π is the plane containing the point A and perpendicular to the direction **n**, i.e. **n** is the **normal** to Π. (Note: we say here loosely: **n** is *the* normal to Π. In fact, any vector $\alpha\mathbf{n}$, where α is a scalar, is a normal to Π.)

Let P be any point on Π, with position vector **r** (fig. 6.4:9). Then the vector $\overrightarrow{AP}$ is in Π; so $\overrightarrow{AP}$ is perpendicular to **n**, i.e.

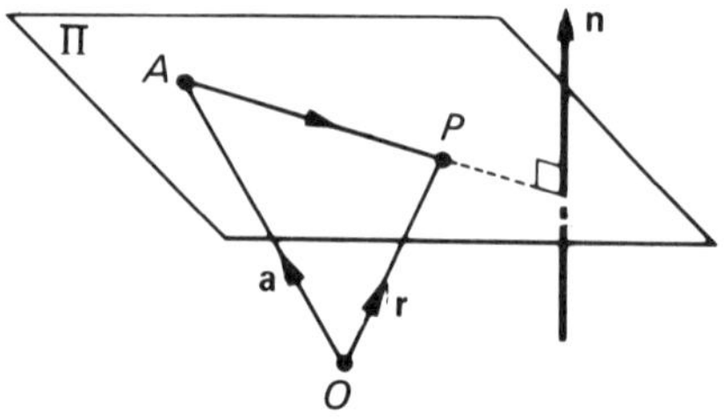

$$\overrightarrow{AP} \cdot \mathbf{n} = 0$$

$$(\mathbf{r} - \mathbf{a}) \cdot \mathbf{n} = 0$$

$$\mathbf{r} \cdot \mathbf{n} = \mathbf{a} \cdot \mathbf{n} \tag{21}$$

Fig. 6.4:9

This gives a condition that a general point P lies on Π, so it is a **vector equation** of Π.

If $\mathbf{a} = a_1\mathbf{i} + a_2\mathbf{j} + a_3\mathbf{k}$, $\mathbf{n} = A\mathbf{i} + B\mathbf{j} + C\mathbf{k}$, the vector equation of Π is

$$\mathbf{r} \cdot (A\mathbf{i} + B\mathbf{j} + C\mathbf{k}) = d \quad \text{where } d = a_1 A + a_2 B + a_3 C \tag{22}$$

Let P be the point (x, y, z), so that $\mathbf{r} = x\mathbf{i} + y\mathbf{j} + z\mathbf{k}$; then

$$(x\mathbf{i} + y\mathbf{j} + z\mathbf{k}) \cdot (A\mathbf{i} + B\mathbf{j} + C\mathbf{k}) = d$$

i.e. $$Ax + By + Cz = d \tag{23}$$

This is the *cartesian equation* of Π.

Look at worked example 6 again. We can now see the significance of the coefficients 4, -3 and 3 on the LHS of equation (20): $4\mathbf{i} - 3\mathbf{j} + 3\mathbf{k}$ is the normal to the plane.

(Note: the equation $\mathbf{r} \cdot \mathbf{n} = \mathbf{a} \cdot \mathbf{n}$ represents a line in 2D-space (see section 6.2:4) and a plane in 3D-space. Why?

An equation such as

$$\mathbf{r} \cdot (\mathbf{i} + 2\mathbf{j}) = 3, \quad \text{i.e. } x + 2y = 3 \tag{24}$$

reduces the number of degrees of freedom of a general point P by one. If P has two degrees of freedom (2D), equation (24) reduces the number to one and represents a line. If P has three degrees of freedom (3D), equation (24) reduces the number to two and represents a plane.)

Worked example 7 Find the vector (scalar product) and cartesian equations of the plane through the point $A(-2, 3, 5)$, perpendicular to the vector $3\mathbf{i} - \mathbf{j} + 2\mathbf{k}$.

$\mathbf{a} = -2\mathbf{i} + 3\mathbf{j} + 5\mathbf{k}$, $\mathbf{n} = 3\mathbf{i} - \mathbf{j} + 2\mathbf{k}$. The vector equation of the plane is

$$\mathbf{r} \cdot \mathbf{n} = \mathbf{a} \cdot \mathbf{n}$$

i.e. $\mathbf{r} \cdot (3\mathbf{i} - \mathbf{j} + 2\mathbf{k}) = (-2\mathbf{i} + 3\mathbf{j} + 5\mathbf{k}) \cdot (3\mathbf{i} - \mathbf{j} + 2\mathbf{k}) = 1$

So the cartesian equation of the plane is $3x - y + 2z = 1$.

Vectors, coordinate geometry and complex numbers 339

> **Theorem 2:** The shortest distance from the point L to the plane $\Pi: \mathbf{r} \cdot \mathbf{n} = d$ is
> $$\left| \frac{d - \mathbf{l} \cdot \mathbf{n}}{n} \right| \qquad (25)$$

Proof
See question 12. The proof is essentially identical to that of 6.2:4 Theorem 2.

> **Corollary:** The shortest distance from the origin to the plane $\Pi: \mathbf{r} \cdot \mathbf{n} = d$ is
> $$\left| \frac{d}{n} \right| \qquad (26)$$

Proof
Put $\mathbf{l} = \mathbf{0}$ in equation (25).

Look at worked example 6 yet again. We can now see the significance of the number 2 on the R.H.S. of equation (20): $2/|4\mathbf{i} - 3\mathbf{j} + 3\mathbf{k}|$, i.e. $2/\sqrt{34}$, is the shortest distance from the origin to the plane.

Worked example 8 Find the distance of the plane $\mathbf{r} \cdot (3\mathbf{i} - 6\mathbf{j} - 2\mathbf{k}) = -6$ from: (i) the point $L(-3, 1, -1)$; (ii) the origin O.

Show that L and O are on opposite sides of the plane.
The normal to the plane Π is $\mathbf{n} = 3\mathbf{i} - 6\mathbf{j} - 2\mathbf{k}$, so $n = 7$.

(i) Distance from L to $\Pi = \left| \dfrac{d - \mathbf{l} \cdot \mathbf{n}}{n} \right| = \left| \dfrac{-6 + 13}{7} \right| = |1| = 1.$

(ii) Distance from O to $\Pi = \left| \dfrac{d}{n} \right| = \left| \dfrac{-6}{7} \right| = \dfrac{6}{7}$

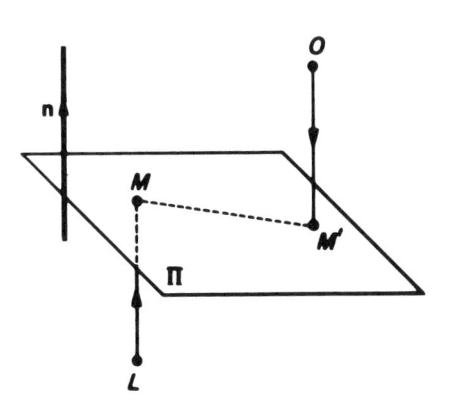

Fig. 6.4:**10**

The fact that L and O are on opposite sides of Π is shown by the opposite signs within the moduli: if M and M' are the points on Π closest to L and O respectively (see fig. 6.4:**10**), then

$$\overrightarrow{LM} \text{ is in the same direction as } \mathbf{n}$$
$$\overrightarrow{OM'} \text{ is in the opposite direction to } \mathbf{n}.$$

Intersection of a line and a plane

Worked example 9 Find the point of intersection of the plane $\mathbf{r} \cdot (3\mathbf{i} - \mathbf{j}) = 5$ and the line $\mathbf{r} = 3\mathbf{j} + \mathbf{k} + \lambda(2\mathbf{i} + 2\mathbf{j} + 5\mathbf{k})$.

At the point H of intersection of the plane and the line,

$$\mathbf{r} \cdot (3\mathbf{i} - \mathbf{j}) = 5 \qquad (27)$$
and $$\mathbf{r} = 3\mathbf{j} + \mathbf{k} + \lambda(2\mathbf{i} + 2\mathbf{j} + 5\mathbf{k}) \qquad (28)$$
i.e. $$\{3\mathbf{j} + \mathbf{k} + \lambda(2\mathbf{i} + 2\mathbf{j} + 5\mathbf{k})\} \cdot (3\mathbf{i} - \mathbf{j}) = 5$$
$\Rightarrow$ $$3 + 4\lambda = 5$$
$\Rightarrow$ $$\lambda = 2$$

To find the position vector **h** of H, we substitute $\lambda = 2$ into equation (28):

$$\mathbf{h} = 3\mathbf{j} + \mathbf{k} + 2(2\mathbf{i} + 2\mathbf{j} + 5\mathbf{k}) = 4\mathbf{i} + 7\mathbf{j} + 11\mathbf{k}$$

Intersection of two planes

Worked example 10 Find a vector equation of the line Λ of intersection of the planes $\Pi_1: \mathbf{r}.(\mathbf{i}+\mathbf{j}-\mathbf{k}) = 6$ and $\Pi_2: \mathbf{r}.(2\mathbf{i}-3\mathbf{j}+2\mathbf{k}) = 2$

Cartesian equation of Π_1: $\quad x + y - z = 6 \quad$ (29)

Cartesian equation of Π_2: $\quad 2x - 3y + 2z = 2 \quad$ (30)

$3 \times (29) + (30)$: $\quad 5x - z = 20$

$2 \times (29) + (30)$: $\quad 4x - y = 14$

The cartesian equations of Λ are

$$\frac{x}{1} = \frac{y+14}{4} = \frac{z+20}{5}$$

So a vector equation of Λ is

$$\mathbf{r} = -14\mathbf{j} - 20\mathbf{k} + \lambda(\mathbf{i} + 4\mathbf{j} + 5\mathbf{k})$$

Angle between a line and a plane

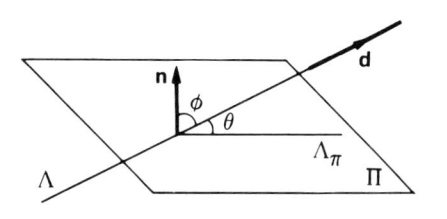

Fig. 6.4:**11**

Definition: The angle θ between a line Λ and a plane Π is the angle between Λ and the projection Λ_Π of Λ onto Π (i.e. θ is the *smallest* possible angle between Λ and any line on Π).

Now, if ϕ is the angle between Λ and **n**, $\cos \phi = \mathbf{d}.\mathbf{n}/dn$; but $\theta = \frac{1}{2}\pi - \phi$, so $\sin \theta = \cos \phi = \mathbf{d}.\mathbf{n}/dn$, i.e.

$$\sin \theta = \frac{\mathbf{d}.\mathbf{n}}{dn} \quad (= \hat{\mathbf{d}}.\hat{\mathbf{n}}) \quad (31)$$

For example, the angle θ between the plane $\mathbf{r}.(3\mathbf{i} - \mathbf{j}) = 5$ and the line $\mathbf{r} = 2\mathbf{i} + 13\mathbf{j} - \mathbf{k} + \lambda(2\mathbf{i} + 2\mathbf{j} + \mathbf{k})$ is given by

$$\sin \theta = \frac{(2\mathbf{i}+2\mathbf{j}+\mathbf{k}).(3\mathbf{i}-\mathbf{j})}{|2\mathbf{i}+2\mathbf{j}+\mathbf{k}||3\mathbf{i}-\mathbf{j}|} = \frac{4}{3\sqrt{10}}$$

Angle between two planes

Definition The angle θ between two planes is the angle between their respective normals.

We can see from fig. 6.4:**12** that this agrees with our intuitive idea of the angle between two planes. The angle θ between $\mathbf{r}.\mathbf{n} = d_1$ and $\mathbf{r}.\mathbf{n}_2 = d_2$

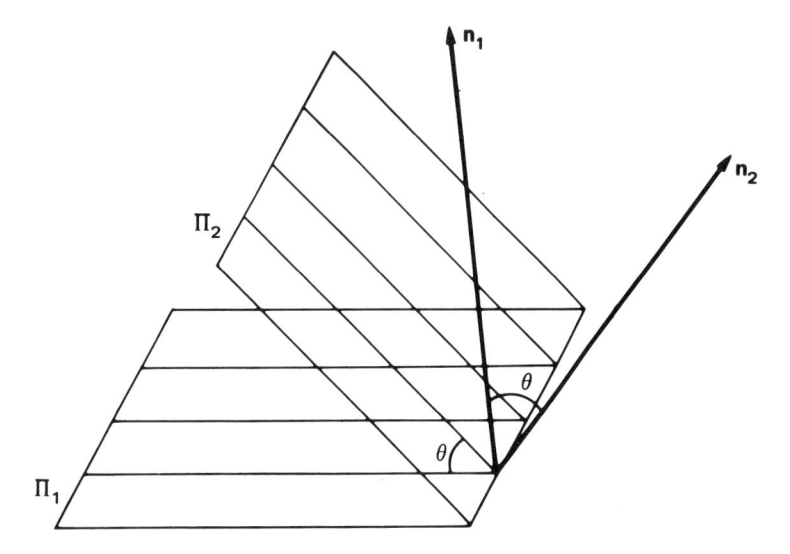

Vectors, coordinate geometry and complex numbers 341

Fig. 6.4:**12**

is given by

$$\cos\theta = \frac{\mathbf{n}_1 \cdot \mathbf{n}_2}{n_1 n_2} \quad (= \hat{\mathbf{n}}_1 \cdot \hat{\mathbf{n}}_2) \tag{32}$$

(see 6.2:3 equation 18).

For example. The angle θ between the planes $\mathbf{r} \cdot (\mathbf{i}+\mathbf{j}+\mathbf{k}) = 23$ and $\mathbf{r} \cdot (2\mathbf{i}-3\mathbf{j}+2\mathbf{k}) = 2$ is given by

$$\cos\theta = \frac{(\mathbf{i}+\mathbf{j}+\mathbf{k}) \cdot (2\mathbf{i}-3\mathbf{j}+2\mathbf{k})}{|\mathbf{i}+\mathbf{j}+\mathbf{k}||2\mathbf{i}-3\mathbf{j}+2\mathbf{k}|} = \frac{1}{\sqrt{51}}$$

Exercise 6.4:5

⑦ **1** Find the vector (scalar product) and cartesian equations of the plane:

→ (i) through the point $\mathbf{i}+\mathbf{j}-\mathbf{k}$ and perpendicular to the vector $4\mathbf{i}+\mathbf{j}+3\mathbf{k}$
(ii) through the origin and perpendicular to the vector $4\mathbf{i}+4\mathbf{j}+4\mathbf{k}$
(iii) through the point $3\mathbf{i}-\mathbf{j}+5\mathbf{k}$ and perpendicular to the vector $2\mathbf{i}-\mathbf{k}$
(iv) through the point $(1, -2, 3)$ and perpendicular to the line $(x-3)/4 = y/1 = z/-4$

R (v) which bisects at right angles the line joining the points $(-3, 2, 5)$ and $(1, 0, 3)$
(vi) which bisects at right angles the line joining the origin to the point $(4, -2, 6)$

⑦ **2** Write down a normal to each of the following planes:

(i) $3x + 5y - 2z = 7$
(ii) $x + y = z$
(iii) $x = 2y$
(iv) $z = -5$

⑦ **3** Find the vector (scalar product form) equation of the plane with cartesian equation:

→ (i) $3x - 4y - 5z = -2$ (ii) $x + 2z = 3$
(iii) $x = 2$

Write down the position vector of a point on each plane.

⑦ **4** Find a vector equation of the line passing through the point $(2, -4, -3)$ and perpendicular to the plane $2x + 5y - 2z + 8 = 0$.

⑦ **5** Find the cartesian equation of the plane

$$\mathbf{r} = \mathbf{i} + \mathbf{j} + \lambda(\mathbf{i}+\mathbf{j}-\mathbf{k}) + \mu(2\mathbf{i}+\mathbf{k})$$

and hence find its vector equation in scalar product form.

⑦ **6** Show that the vector $2\mathbf{i}-5\mathbf{j}+\mathbf{k}$ is perpendicular to both the vectors $\mathbf{i}+\mathbf{j}+3\mathbf{k}$ and $\mathbf{j}+5\mathbf{k}$.

Hence find the vector (scalar product form) and cartesian equations of the plane through the point $2\mathbf{i}+\mathbf{j}+3\mathbf{k}$ and parallel to the directions $\mathbf{i}+\mathbf{j}+3\mathbf{k}$ and $\mathbf{j}+5\mathbf{k}$ (c.f. 6.4:4 Q1(i)).

342 *Equation of a plane: scalar product form*

⑦ **7** A, B and C are the points $(3, 4, -1)$, $(2, 0, 1)$ and $(-1, 2, 0)$. Show that the vector $\mathbf{j} + 2\mathbf{k}$ is perpendicular to both the vectors $\overrightarrow{AB}$ and $\overrightarrow{AC}$. Hence find the vector (scalar product form) and cartesian equations of the plane through A, B and C (6.4:4 Q1(iii)).

⑦ → **8** Using the method of 6.4:1 Q12, find a vector perpendicular to both the vectors $\mathbf{i} + 2\mathbf{j} + 2\mathbf{k}$ and $2\mathbf{i} + 3\mathbf{k}$.
• Hence find the vector (scalar product form) and cartesian equations of the plane through the point $\mathbf{i} + \mathbf{j} - \mathbf{k}$ and parallel to these vectors.
Write down a vector equation of the plane in parametric form and check that this leads to the same cartesian equation.

⑦ **9** Using the method of question 8, find the vector (scalar product form) and cartesian equations of the plane:

(i) containing the origin and parallel to the directions $\mathbf{i} + \mathbf{j}$ and $\mathbf{j} + \mathbf{k}$
(ii) containing the point $-2\mathbf{i} + \mathbf{k}$ and parallel to the directions $\mathbf{i} - \mathbf{j} + \mathbf{k}$ and $2\mathbf{i} + \mathbf{j} + 2\mathbf{k}$
(iii) containing the origin and the points $3\mathbf{i} + 2\mathbf{k}$ and $-\mathbf{j} - \mathbf{k}$
R (iv) containing the points $-2\mathbf{i} - \mathbf{j}$, $\mathbf{i} + \mathbf{k}$ and $2\mathbf{i} - 4\mathbf{j} + 2\mathbf{k}$
(v) containing the points $(2, 0, -2)$, $(-1, 1, 3)$ and $(2, 1, -1)$

⑦ **10** Write down the position vectors of three points lying in the plane $\mathbf{r} \cdot (\mathbf{i} + 2\mathbf{j} + 3\mathbf{k}) = 4$.
Hence find a vector equation of the plane in parametric form.

⑧ **11** Find the distance from the origin to the plane:

→ (i) $\mathbf{r} \cdot (\mathbf{i} + 2\mathbf{j} + 2\mathbf{k}) = 4$
(ii) $\mathbf{r} \cdot (4\mathbf{i} - \mathbf{j} - 8\mathbf{k}) = -9$
(iii) $\mathbf{r} \cdot (2\mathbf{i} + \mathbf{j}) = -5$
(iv) $\mathbf{r} \cdot (4\mathbf{i} - \mathbf{j} - \mathbf{k}) = 0$
(v) $x + 2y - z = 12$
(vi) $3x - y = 5z$

Explain your answers to (iv) and (vi).

⑧ • **12** Prove Theorem 2. (Hint: look at the proof of 6.2:4 Theorem 2.)

⑧ → **13** Find the distance from each of the following points to the plane $\mathbf{r} \cdot (2\mathbf{i} - 2\mathbf{j} + \mathbf{k}) = -5$:

(i) $C(1, 0, 1)$ (ii) the origin
(iii) $D(5, 4, -2)$ (iv) $E(1, 6, -2)$

Are C and the origin on opposite sides of the plane or the same side? What about C and E?

⑧ **14** Show that $P(3, 0, 1)$ and $Q(2, 1, -2)$ lie on opposite sides of the plane $\mathbf{r} \cdot (2\mathbf{i} - \mathbf{j} + \mathbf{k})$.

⑧ R **15** A, B, C and D are the points $(1, 0, 1)$, $(3, 2, 4)$, $(2, -1, 2)$ and $(5, 3, 0)$. Show that the vector $5\mathbf{i} + \mathbf{j} - 4\mathbf{k}$ is perpendicular to both $\overrightarrow{AB}$ and $\overrightarrow{AC}$.
Hence find the distance from D to the plane ABC.

⑧ **16** Find the distance from the point (x, y, z) to each of the planes $\mathbf{r} \cdot (2\mathbf{i} - 2\mathbf{j} + \mathbf{k}) = -5$ and $\mathbf{r} \cdot (\mathbf{i} + 2\mathbf{j} - 2\mathbf{k}) = 4$. Hence find the cartesian equations of the locus of a point (x, y, z) which is equidistant from these two planes. Describe this locus.

⑧ • **17** (i) Show that the planes $\Pi_1: \mathbf{r} \cdot (2\mathbf{i} - \mathbf{j} + 2\mathbf{k}) = 8$ and $\Pi_2: \mathbf{r} \cdot (4\mathbf{i} - 2\mathbf{j} + 4\mathbf{k}) = 5$ are parallel.
Show that the point A with position vector $\mathbf{i} + 3\mathbf{k}$ lies on Π_1 and find the distance from A to Π_2. Hence find the distance between the planes.
Find the distance of each plane from the origin.

(ii) Show that the planes $\Pi_1: \mathbf{r} \cdot (2\mathbf{i} - \mathbf{j} + 2\mathbf{k}) = 8$ and $\Pi_3: \mathbf{r} \cdot (-4\mathbf{i} + 2\mathbf{j} - 4\mathbf{k}) = 1$ are parallel.
Find the distance from A to Π_3. Hence find the distance between the planes.
Find the distance of Π_3 from the origin. What is the essential difference between (i) and (ii)?

⑧ **18** Find the distances between the following pairs of (parallel) planes:

→ (i) $\mathbf{r} \cdot (\mathbf{i} + 3\mathbf{j} + \mathbf{k}) = 1$; $\mathbf{r} \cdot (\mathbf{i} + 3\mathbf{j} - \mathbf{k}) = 4$
(ii) $\mathbf{r} \cdot (6\mathbf{i} - 2\mathbf{j} - 3\mathbf{k}) = -2$; $\mathbf{r} \cdot (12\mathbf{i} - 4\mathbf{j} - 6\mathbf{k}) = 3$
(iii) $3x + 3y = 4$; $5x + 5y = 2$

⑧ R **19** Find the vector equations of the two planes which are parallel to the plane $\mathbf{r} \cdot (3\mathbf{i} + 6\mathbf{j} - 6\mathbf{k}) = 10$ and at a distance 2 units from it.

⑧ **20** Find the cartesian equations of the two planes which are parallel to the plane $2x + 3y - 6z = 5$ and at a distance 1 unit from it.

⑨ **21** Find the point of intersection of the plane and the line:

→ (i) $\mathbf{r} \cdot (\mathbf{i} + \mathbf{j} + \mathbf{k}) = 4$; $\mathbf{r} = \mathbf{i} + 2\mathbf{k} + \lambda(3\mathbf{i} - \mathbf{j} - \mathbf{k})$
(ii) $\mathbf{r} \cdot (2\mathbf{i} + 2\mathbf{j} + 5\mathbf{k}) = 11$; $\mathbf{r} = \mathbf{j} + \lambda(\mathbf{j} - \mathbf{k})$
R (iii) $\mathbf{r} \cdot (5\mathbf{i} - 4\mathbf{j}) = 7$; $\mathbf{r} = -\mathbf{i} + 6\mathbf{j} - 4\mathbf{k} + \lambda(\mathbf{i} - \mathbf{j} + 2\mathbf{k})$

⑨ **22** Show that the plane $\mathbf{r} \cdot (3\mathbf{i} + \mathbf{j} - \mathbf{k}) = -7$ is parallel to the line $\mathbf{r} = 5\mathbf{i} + 2\mathbf{j} + 3\mathbf{k} + \lambda(\mathbf{i} + 2\mathbf{j} + 5\mathbf{k})$ and contains the line $\mathbf{r} = \mathbf{i} - 4\mathbf{j} + 6\mathbf{k} + \mu(\mathbf{i} + 2\mathbf{j} + 5\mathbf{k})$. If the plane contains the line $\mathbf{r} = -2\mathbf{i} + p\mathbf{j} + 4\mathbf{k} + \nu(q\mathbf{i} + 4\mathbf{j} + 3\mathbf{k})$ find p and q.

⑨ **23** Show that the line $\mathbf{r} = 2\mathbf{i} + \mathbf{j} + \lambda(3\mathbf{i} - \mathbf{j} + 4\mathbf{k})$ is parallel to the plane $\mathbf{r} \cdot (2\mathbf{i} + 2\mathbf{j} - \mathbf{k}) = 2$ and find the distance between them.

⑨ **24** Explain how an attempt to find the point of intersection of the plane $\Pi: \mathbf{r} \cdot \mathbf{n} = p$ and the line $\Lambda: \mathbf{r} = \mathbf{a} + \lambda \mathbf{d}$ leads to a (scalar) equation of λ. Discuss the solution of this equation when:

(i) Λ is parallel to Π
(ii) Λ is in Π
(iii) Λ meets Π in one point

Show that in case (iii) the position vector of the point of intersection of Λ and Π is

$$\mathbf{a} + \frac{p - \mathbf{a} \cdot \mathbf{n}}{\mathbf{d} \cdot \mathbf{n}} \mathbf{b}$$

Vectors, coordinate geometry and complex numbers 343

⑨ • **25** Find a vector equation of the line which passes through the point $L(-7, -1, 12)$ and is perpendicular to the plane $\Pi: \mathbf{r} \cdot (4\mathbf{i} - \mathbf{j} - 5\mathbf{k}) = -3$.
Find the point M of intersection of this line with Π. Hence find:

(i) the distance from L to Π
(ii) the mirror image of L in Π

⑨ → **26** Find the mirror image of the point $(-3, 3, 2)$ in the plane $\mathbf{r} \cdot (\mathbf{i} - 2\mathbf{j} - 2\mathbf{k}) = 5$.

⑨ → **27** To find the projection Λ_Π of the line $\Lambda: \mathbf{r} = \mathbf{i} + \mathbf{j} + \lambda(2\mathbf{i} + 3\mathbf{j} - \mathbf{k})$ onto the plane $\Pi: \mathbf{r} \cdot (4\mathbf{i} - \mathbf{j} + \mathbf{k}) = -1$:

(i) find the point H of intersection of Λ and Π;

(ii) using the method of question 25, find the point G on Π closest to the point A, with position vector $\mathbf{i} + \mathbf{j}$, on Λ;

(iii) Hence find a vector equation of the line GH, i.e. the line Λ_Π.

⑨ **28** Find a vector equation of the projection of the line $\mathbf{r} = 6\mathbf{i} + \mathbf{j} + 11\mathbf{k} + \lambda(3\mathbf{i} - 3\mathbf{j} + 10\mathbf{k})$ on the plane $\mathbf{r} \cdot (\mathbf{i} + \mathbf{j} + 5\mathbf{k}) = 8$.

29 A, B and C are the points with position vectors $-\mathbf{i} + 2\mathbf{j} + \mathbf{k}, 3\mathbf{j} + 3\mathbf{k}, -2\mathbf{i} + 2\mathbf{j} + 2\mathbf{k}$. Find a vector perpendicular to both $\overrightarrow{AB}$ and $\overrightarrow{AC}$. Hence find the vector equation of the plane ABC. Find the position vector of the point of intersection of the plane with the line $\mathbf{r} = \mathbf{k} + \lambda(\mathbf{i} + \mathbf{j} - 5\mathbf{k})$. Find the angle between the line and the plane.

⑩ **30** Find a vector equation of the line of intersection of the planes:

(i) $\mathbf{r} \cdot (\mathbf{i} - \mathbf{j} - \mathbf{k}) = 1$ and $\mathbf{r} \cdot (3\mathbf{i} + \mathbf{j} - \mathbf{k}) = 3$

→ (ii) $\mathbf{r} \cdot (2\mathbf{i} + 3\mathbf{j}) = 4$ and $\mathbf{r} \cdot (\mathbf{i} + \mathbf{k}) = 2$
(iii) $\mathbf{r} \cdot (\mathbf{i} - \mathbf{j} + \mathbf{k}) = 0$ and $\mathbf{r} \cdot (\mathbf{i} + 3\mathbf{j} + 5\mathbf{k}) = 4$
(iv) $\mathbf{r} \cdot (\mathbf{i} + 3\mathbf{j} + \mathbf{k}) = 1$ and $\mathbf{r} \cdot (6\mathbf{i} - 2\mathbf{j} - 3\mathbf{k}) = 3$

⑩ R **31** Show that the line $L: \mathbf{r} = \mathbf{i} + \mathbf{j} + 4\mathbf{k} + \lambda(2\mathbf{i} - 3\mathbf{k})$ lies in the plane $\Pi_1: \mathbf{r} \cdot (3\mathbf{i} + 4\mathbf{j} + 2\mathbf{k}) = 15$. Find the point of intersection of the line L and the line L' of intersection of Π_1 and $\Pi_2: \mathbf{r} \cdot (\mathbf{i} + 3\mathbf{j} - \mathbf{k}) = 0$. Find the angle between L and L'.

32 Find the angle between the plane and the line:

→ (i) $\mathbf{r} \cdot (3\mathbf{i} + 6\mathbf{j} + 2\mathbf{k}) = 2$;
$\mathbf{r} = \mathbf{i} + \mathbf{j} - 5\mathbf{k} + \lambda(3\mathbf{i} - 2\mathbf{j} + 3\mathbf{k})$
(ii) $x + 2y + z = 7$; $\frac{1}{4}(x-1) = \frac{1}{3}(y+1)$, $z = 3$
R (iii) $\mathbf{r} \cdot (5\mathbf{i} + \mathbf{j} - \mathbf{k}) = -7$; $\mathbf{r} = 3\mathbf{k} + \lambda(5\mathbf{i} + \mathbf{j} - \mathbf{k})$

33 Find the angles between the following pairs of planes:

→ (i) $\mathbf{r} \cdot (\mathbf{i} + 5\mathbf{j} + \mathbf{k}) = 6$ and $\mathbf{r} \cdot (\mathbf{i} - \mathbf{j} + \mathbf{k}) = 7$
R (ii) $\mathbf{r} \cdot (3\mathbf{i} + 4\mathbf{j} + 5\mathbf{k}) = 4$ and $\mathbf{r} \cdot (2\mathbf{i} + \mathbf{j} - 2\mathbf{k}) = 3$
(iii) $3x + 2y + 2z = 3$ and $x + 2y + z = 7$
(iv) $x + y = 5$ and $y + z = 6$

• → **34** Find the vector and cartesian equations of a sphere centre O radius a.

35 Describe the surface represented by the equation:

(i) $x^2 + y^2 = a^2$ (ii) $\dfrac{x^2}{a^2} + \dfrac{y^2}{b^2} = 1$

36 A and B are two points. Show that the vector equation of the sphere diameter AB is

$$(\mathbf{r} - \mathbf{a}) \cdot (\mathbf{r} - \mathbf{b}) = 0$$

(See 6.2:6 Q20.)

6.5 Complex numbers

6.5:1 Definition of a complex number

The quadratic equation $x^2 + 1 = 0$, i.e. $x^2 = -1$ has no real solution since there is no real number whose square is -1. So that the equation $x^2 = -1$ does have a solution, we invent a new number $\sqrt{-1}$ which we call i for convenience.

$$i = \sqrt{-1}; \text{ so } i^2 = -1 \tag{1}$$

Then we can write

$$x^2 + 1 = 0 \Rightarrow x^2 = -1 \Rightarrow x = \pm\sqrt{-1}, \text{ i.e. } x = \pm i$$

The number i, and any real-number multiple of i, are called **imaginary numbers**. For example, i, $3i$, $-\sqrt{2}i$ and πi are all imaginary numbers.

Worked example 1 Solve the equations:
(i) $x^2 + 16 = 0$ (ii) $x^2 - 2x + 10 = 0$.

(i) $x^2 + 16 = 0 \Rightarrow x^2 = -16 \Rightarrow x = \pm\sqrt{-16}$
$\Rightarrow \qquad x = \pm\sqrt{16}\sqrt{-1}$
$\Rightarrow \qquad x = \pm 4i$

(ii) $x^2 - 2x + 10 \Rightarrow x = \tfrac{1}{2}(2 \pm \sqrt{(4-40)})$ by section 2.1 Theorem 1
$\Rightarrow x = \tfrac{1}{2}(2 \pm \sqrt{-36})$
$x \Rightarrow = \tfrac{1}{2}(2 \pm 6i)$
$\Rightarrow x = 1 \pm 3i$

Definition A **complex number** is a number of the form $z = a + bi$, $a, b \in \mathbb{R}$, e.g.

$$2 + 3i, \; -\sqrt{2} + i, \; 5.73 - 6.2i, \; -\sqrt{5}i, \; 6$$

The set of all complex numbers is denoted by $\mathbb{C}$:

$$\mathbb{C} = \{a + bi : a, b \in \mathbb{R}\}$$

a is called the **real part** of z: $a = \operatorname{Re} z$.
bi is called the **imaginary part** of z: $bi = \operatorname{Im} z$

Exercise 6.5:1

① **1** Solve the equations:
(i) $x^2 + 4 = 0$
(ii) $x^2 + 81 = 0$
(iii) $x^2 + 3 = 0$
→ (iv) $x^2 + 20 = 0$
(v) $9x^2 + 25 = 0$
(vi) $4x^2 + 7 = 0$

① **2** Solve the equation $x^2 + b^2 = 0$, $b \in \mathbb{R}$.
Hence write down the quadratic equations with roots:
(i) $\pm 3i$ (ii) $\pm 2\sqrt{3}i$

① **3** Solve the equations:
(i) $x^2 + 2x + 2 = 0$
(ii) $x^2 - 6x + 13 = 0$
(iii) $x^2 + x + 1 = 0$
(iv) $2x^2 + 3x + 2 = 0$
→ (v) $4x^2 - 12x + 25 = 0$

① **4** Solve the equation $x^2 - 2ax + (a^2 + b^2) = 0$, $a, b \in \mathbb{R}$.
Hence write down the quadratic equations with roots:
(i) $3 \pm i$ (ii) $2 \pm \sqrt{3}i$ (iii) $-3 \pm 4i$

5 Write down Re z and Im z when:
(i) $z = \tfrac{3}{2} + 2i$ (ii) $z = -5 + \sqrt{7}i$
(iii) $z = 5$ (iv) $z = -9i$
(v) $z = -4i + 5$

6 Draw a Venn diagram showing the following sets:

$\mathbb{C} = \{\text{complex numbers}\}$, $\mathbb{R} = \{\text{real numbers}\}$
$\mathbb{I} = \{\text{imaginary numbers}\}$, $\mathbb{Q} = \{\text{rational real numbers}\}$

Are the following statements true? Explain your answers.

(i) Any real number is a complex number, i.e. $\mathbb{R} \subset \mathbb{C}$.
(ii) Any complex number is a real number, i.e. $\mathbb{C} \subset \mathbb{R}$.
(iii) There is no number which is both real and imaginary, i.e. $\mathbb{R} \cap \mathbb{I} = \varnothing$.

6.5:2 Simple operations on complex numbers

We need to define how to add, subtract, multiply and divide complex numbers.

Addition and subtraction
We treat the real and imaginary parts separately.

Worked example 2

(i) $(2+i)+(3-2i) = (2+3)+(1-2)i = 5-i$

(ii) $(2+i)-(3-2i) = (2-3)+(1+2)i = -1+3i$

Multiplication
We expand the brackets, as for real numbers, bearing in mind that $i^2 = -1$.

Worked example 3

$$(2+i)(3-2i) = 6+3i-4i-2i^2$$
$$= 6-i+2$$
$$= 8-i$$

> **Definition** The **complex conjugate** of a complex number $z = a+bi$ is the complex number $\bar{z} = a-bi$, e.g. if $z = 5+2i$, $\bar{z} = 5-2i$.

Note: the complex conjugate of z is also written as z^*.

Division
We multiply the numerator and denominator of the fraction by the conjugate of the denominator.

Worked example 4

$$\frac{2+i}{3-2i} = \frac{(2+i)(3+2i)}{(3-2i)(3+2i)}$$
$$= \frac{4+7i}{13}$$
$$= \frac{4}{13}+\frac{7}{13}i$$

Note: the resulting numbers in all four operations are themselves complex numbers.

The complex number zero
There is just one complex number zero, i.e. $0+0i$, so

> $$x+iy = 0 \Rightarrow x=0, y=0 \qquad (2)$$

Note: zero is both a real number and an imaginary number.

Theorem 1:

$$a + bi = c + di \Rightarrow a = c \text{ and } b = d \qquad (3)$$

i.e. if two complex numbers are equal, they must have equal real parts and equal imaginary parts.

Proof

$$a + bi = c + di$$

$$\Rightarrow \qquad (a - c) + (b - d)i = 0$$

Squaring both sides of this equation:

$$(a - c)^2 = -(d - b)^2$$

Since $(a - c)^2$ is positive or zero, and $-(d - b)^2$ is negative or zero, it follows that both must be zero. So

$$a = c \text{ and } b = d$$

Worked example 5 Find the (two) square roots of $3 + 4i$.

Suppose that $x + iy$, $x, y \in \mathbb{R}$, is a square root of $3 + 4i$. Then

$$(x + iy)^2 = 3 + 4i$$
$$x^2 - y^2 + 2xyi = 3 + 4i$$

From identity (3)

$$x^2 - y^2 = 3 \qquad (4)$$
$$2xy = 4 \qquad (5)$$

Equation (5):
$$y = \frac{2}{x} \qquad (6)$$

Substituting for y in equation (4):

$$x^2 - \frac{4}{x^2} = 3$$
$$x^4 - 3x^2 - 4 = 0$$
$$(x^2 - 4)(x^2 + 1) = 0$$

$x^2 + 1 = 0$ is impossible since $x \in \mathbb{R}$.
$x^2 - 4 = 0 \Rightarrow \qquad\qquad x = 2 \quad \text{or } x = -2$
so by equation (6) $\qquad\qquad y = 1 \quad \text{or } y = -1$
   So the square roots of $3 + 4i$ are $2 + i$ and $-2 - i$.
(Note: do not write $x = \pm 2$, $y = \pm 1$, as this suggests four solutions: $2 + i$, $2 - i$, $-2 + i$, $-2 - i$.)

Exercise 6.5:2

② **1** Write the following complex numbers in the form $a + bi$:

(i) $(3+7i) + (2+5i)$
(ii) $(4+3i) + (7-2i)$
(iii) $(3+7i) - (2+5i)$
(iv) $(4+3i) - (7-2i)$
(v) $(2\sqrt{3} - \sqrt{5}i) + (-\sqrt{3} - 3\sqrt{5}i)$
(vi) $(2\sqrt{3} - \sqrt{5}i) - (-\sqrt{3} - 3\sqrt{5}i)$
(vii) $(4a + 7ai) + (5a - ai)$
(viii) $(4a + 7ai) - (5a - ai)$
(ix) $(p+qi) + (p-qi)$
(x) $(p+qi) - (p-qi)$
→ (xi) $(p+qi) + (r+si)$
→ (xii) $(p+qi) - (r+si)$

③ **2** Write the following complex numbers in the form $a + bi$:

(i) $(2+3i)(3+i)$
(ii) $(-4+5i)(-2-4i)$
(iii) $(5+2i)i$
(iv) $(1+\sqrt{3}i)(1-\sqrt{3}i)$
(v) $(p+qi)(p-qi)$
(vi) $(3+i)^2$
(vii) $(3+i)^3$
(viii) $(1+i)(1+2i)(1+3i)$
(ix) $(p+qi)(q+pi)$
→ (x) $(p+qi)(r+si)$

③ • **3** Simplify:

(i) i^3 (ii) i^4 (iii) i^5 (iv) i^6 (v) i^{29}

③ **4** Simplify:

(i) i^{4n} (ii) i^{4n+1} (iii) i^{4n+2} (iv) i^{4n+3}

(n an integer).

③ **5** Find z^2, z^3 and z^4 when:

(i) $z = 1 + 2i$ (ii) $z = -\sqrt{3} + i$

③ **6** Find $(2+i)^2$. Hence find the two square roots of $3 + 4i$.

③ **7** Simplify:

(i) $(\cos 60° + i \sin 60°)(\cos 30° + i \sin 30°)$
(ii) $(\cos 60° + i \sin 60°)^3$

③ **8** Simplify:

(i) $(\cos \theta + i \sin \theta)(\cos \theta - i \sin \theta)$
(ii) $(\cos \theta + i \sin \theta)^2$
(iii) $(\cos \theta + i \sin \theta)(\cos \phi + i \sin \phi)$

③ **9** Find integers x and y such that $(3+2i)(x+iy)$ is:

(i) real (ii) imaginary

10 z is a complex number.

(i) What is $\bar{\bar{z}}$?
(ii) Show that $z + \bar{z}$ and $z\bar{z}$ are both real numbers. What about $z - \bar{z}$?

④ **11** Write in the form $a + bi$:

(i) $\dfrac{2+3i}{3+i}$ (ii) $\dfrac{-4+5i}{-2-4i}$

(iii) $\dfrac{3i}{5+i}$ (iv) $\dfrac{\sqrt{3}+i}{\sqrt{3}-i}$

(v) $\dfrac{5+2i}{i}$ (vi) $\dfrac{x-iy}{x+iy}$

(vii) $\dfrac{1}{3+4i}$ (viii) $\dfrac{2+3i}{-2i}$

→ (ix) $\dfrac{p+qi}{r+si}$

④ **12** Simplify:

(i) $1/i$ (ii) $1/i^2$
(iii) $1/i^3$ (iv) $1/i^{15}$

④ **13** If $z = 2 - i$, find $1/z$, $1/(z+1)$, $1/(2z-i)$.

④ **14** If $z = x + iy$, find:

(i) z^2 (ii) $1/z$
(iii) $(z-1)/z$ (iv) $z/(z-1)$

④ **15** If $z = 1 + 3i$, find z^2, $1/z$, $1/z^2$, $(1/z)^2$.

④ **16** If $z_1 = 3 + i$, $z_2 = 2 - 3i$, find $z_1 + z_2$, $z_2 - z_1$, $z_1 z_2$, z_1/z_2, z_1^2, z_2^2, z_1^2/z_2^2.

④ **17** If $z_1 = 5 + 4i$, $z_2 = 4 - 3i$, find z_1/z_2, z_2/z_1.

④ **18** Simplify:

(i) $\dfrac{1}{\cos \theta - i \sin \theta}$

(ii) $\dfrac{1}{1 - \cos \theta - i \sin \theta}$

(iii) $\dfrac{1}{1 + \cos \theta - i \sin \theta}$

(iv) $\dfrac{\cos \theta + i \sin \theta}{\cos \phi + i \sin \phi}$

④ → **19** If $z = \cos \theta + i \sin \theta$, find:

(i) $z + \dfrac{1}{z}$ (ii) $z - \dfrac{1}{z}$ (iii) $z^2 - \dfrac{1}{z^2}$

348 *Simple operations on complex numbers*

④ **20** Simplify:
   (i) $\dfrac{(3+4i)(2-i)}{3-i}$
   (ii) $\dfrac{1+5i}{(1+i)(3-2i)}$
   (iii) $\dfrac{(2+i)^3}{(1-i)^2}$

④ **21** Simplify:
   (i) $\dfrac{1}{2-i} - \dfrac{1}{2+i}$
   (ii) $\dfrac{1}{x+3i} + \dfrac{1}{x-3i}$
   (iii) $\dfrac{1}{x+iy} + \dfrac{1}{x-iy}$

④ R **22** Simplify:
   (i) $\dfrac{1}{2-i} - \dfrac{1}{3+i}$
   (ii) $\dfrac{1}{(1+i)^2} - \dfrac{1}{(1-i)^2}$

④ **23** Find z when:
   (i) $z(2+3i) = 4+i$
   (ii) $(z+1)(2-i) = 3-4i$
   → (iii) $\dfrac{1}{z} = \dfrac{1}{3+4i} + \dfrac{2}{5+5i}$
   R (iv) $\dfrac{z}{1+2i} + \dfrac{z-1}{5i} = \dfrac{-2+i}{1-2i}$

④ **24** If the two branches of an electrical circuit wired in parallel have impedances z_1 and z_2 ohms, the total impedance of the circuit, z ohms, is given by
$$\dfrac{1}{z} = \dfrac{1}{z_1} + \dfrac{1}{z_2}$$
Find z when:
   (i) $z_1 = 1+i$, $z_2 = 2+i$
   (ii) $z_1 = 2-3i$, $z_2 = \tfrac{4}{3}+i$

④ **25** The complex numbers z_1 and z_2 are defined by $z_1 = R_1 + iwL$ and $z_2 = R_2 - i/wC$, where R_1, R_2, L and C are real numbers, and the number z is given by
$$\dfrac{1}{z} = \dfrac{1}{z_1} + \dfrac{1}{z_2}$$
For what values of w is z a real number?

④ **26** If $z = x+iy$, find the real and imaginary parts of $z + 1/z$.
If $z + 1/z$ is a real number, show that either z is a real number or $x^2 + y^2 = 1$.

④ **27** Find $z\bar{z}$ and $z/\bar{z}$ when:
   (i) $z = 3+4i$
   (ii) $z = 1/(2+3i)$
   (iii) $z = \cos\theta + i\sin\theta$

④ **28** Explain carefully why the method of simplifying the quotient z_1/z_2 by multiplying the numerator and denominator by $\bar{z}_2$ always produces a complex number of the form $a+bi$.

29 Find real values of x and y when:
   • (i) $x+iy = -3+5i$
   (ii) $(x-y) + iy = 1+2i$
   R (iii) $(1+3i)x + (3-4i)y - 5 + 2i = 0$
   → (iv) $(-2+3i)x + (1+2i)y - 7i = 0$

30 If $(2-i)/(3+2i) = x+iy$, $x, y \in \mathbb{R}$, find x and y by multiplying across and using Theorem 1.

31 Using the method of question 29, find:
   (i) $\dfrac{3+4i}{5-2i}$ (ii) $\dfrac{1+i}{i}$

32 Find real values of x and y when:
   (i) $\dfrac{x}{1-i} + \dfrac{y}{1+3i} = 2$
   R (ii) $\dfrac{x}{3-2i} + \dfrac{iy}{2+i} = \dfrac{2}{1+8i}$

33 Find z when:
   (i) $(2-i)z - 3\bar{z} = 1+7i$
   (ii) $z\bar{z} + 2iz = 12+6i$
   (Hint: put $z = x+iy$.)

⑤ **34** Find the two square roots of the number:
   → (i) $5+12i$
   (ii) $3-4i$
   (iii) $2i$
   (iv) $1+\sqrt{3}i$
   R (v) $7-24i$
   (vi) $-8+6i$

Check that your solutions are correct by squaring them again.

⑤ **35** Find the two square roots of the number $a+bi$. Hence write down the square roots of:
   (i) $2+2\sqrt{3}i$ (ii) $2-2\sqrt{3}i$

⑤ **36** Find the four (complex) roots of the equation:
   → (i) $x^4 - 6x^2 + 25 = 0$
   (ii) $x^4 + 16 = 0$
   (iii) $x^4 - x^2 + 1 = 0$
   (Hint: put $x^2 = y$.)

37 Solve the equation $(2+i)x^2 - x + (2+i) = 0$.

6.5:3 Complex roots of polynomial equations

Quadratic equations
In section 2.1:1 we found that, if $b^2 < 4ac$, the quadratic equation

$$ax^2 + bx + c = 0$$

has two non-real roots. Now we can be more precise: these roots are complex numbers. Thus:

> the quadratic equation $ax^2 + bx + c = 0$, $a, b, c \in \mathbb{R}$ has:
> (i) two real roots when $b^2 \geq 4ac$;
> (ii) two complex roots when $b^2 < 4ac$.

> **Theorem 2A:** If a quadratic equation with real coefficients has complex roots, these roots are conjugates.

Proof
Suppose the equation $ax^2 + bx + c = 0$ has complex roots; then $b^2 < 4ac$, so

$$\sqrt{(b^2 - 4ac)} = i\sqrt{(4ac - b^2)}$$

Then

$$ax^2 + bx + c = 0 \Rightarrow x = \frac{-b \pm \sqrt{(b^2 - 4ac)}}{2a} = -\frac{b}{2a} \pm \frac{i\sqrt{(4ac - b^2)}}{2a}$$

Let $\gamma = -b/2a$, $\delta = \sqrt{(4ac - b^2)}/2a$; then

$$x = \gamma \pm \delta i$$

Worked example 6 If one of the roots α of the equation $z^2 + pz + q = 0$ is $3 - 2i$, find p and q.

Note: if we expect a polynomial equation to have complex roots we often write the variable as z rather than x.

By Theorem 2A, the other root β is $3 + 2i$. So

$$-p = \alpha + \beta = (3 - 2i) + (3 + 2i) = 6$$
$$q = \alpha\beta = (3 - 2i)(3 + 2i) = 13$$

So $p = -6$, $q = 13$.

Cubic equations

> **Theorem 2B:** If a cubic equation with real coefficients has complex roots these roots are conjugates.

Proof
We found (in section 2.2:2) that every cubic equation has at least one real root.

Suppose that the equation $ax^2 + bx^2 + cx + d = 0$ has a real root α. Then, by the Factor Theorem, $ax^3 + bx^2 + cx + d$ has a factor $(x - \alpha)$. Suppose its other factor is $ax^2 + lx + m$. Then

$$(x - \alpha)(ax^2 + lx + m) = 0$$

i.e. $x = \alpha$ or $ax^2 + lx + m = 0$.

This quadratic equation has either two real roots or two complex roots, which, by Theorem 2A, are conjugates.

> Thus, a cubic equation has either:
> (i) three real roots, or
> (ii) one real root and two conjugate complex roots.

Worked example 7 Given that $1 - i$ is a root of the equation

$$z^3 - 5z^3 + 8z + p = 0$$

find the other two roots and the value of p.

By Theorem 2B, the other complex root is $1 + i$. Suppose the real root is α. Then, by the Factor Theorem,

$$z^3 - 5z^2 + 8z + p \equiv (z - 1 + i)(z - 1 - i)(z - \alpha)$$

Equating terms in z^2:

$$-5 = (-1 + i) + (-1 - i) - \alpha \Rightarrow \alpha = 3$$

Equating the constant terms:

$$p = (-1 + i)(-1 - i)(-\alpha) = -2\alpha = -6$$

General polynomial equations

Taking complex numbers into account, we know that:

(i) a quadratic equation (degree 2) always has 2 roots;
(ii) a cubic equation (degree 3) always has 3 roots.

> **Theorem 3** (Fundamental Theorem of Algebra): A polynomial equation of degree n has n roots (real or complex).

Proof
See question 29.

We can generalize Theorems 2A and 2B:

> **Theorem 4:** The complex roots of any polynomial equation with real coefficients occur in conjugate pairs.

Proof
See 6.5:8 Q49.

Exercise 6.5:3

⑥ → **1** If one of the roots of the equation $z^2 + pz + q = 0$ is $2 + 5i$, find p and q.

⑥ R **2** If one of the roots of the equation $z^2 + pz + q = 0$ is $1 - \sqrt{3}i$, find p and q.

⑥ **3** Find quadratic equations with roots:
  (i) $2i, -2i$
  (ii) $3 + i, 3 - i$
  (iii) $-2 + \sqrt{3}i, -2 - \sqrt{3}i$
→ (iv) $a + bi, a - bi, \ a, b \in \mathbb{R}$
(See 6.5:1 Q4.)

⑥ **4** Find the roots α and β of the equation
$$2z^2 + z + 5 = 0$$
Verify that $\alpha + \beta = -b/a$; $\alpha\beta = c/a$.

⑥ •→ **5** Solve the equation $z^2 + 2z + 10 = 0$. Hence express $z^2 + 2z + 10$ as the product of two complex linear factors.

⑥ **6** Solve the equation $z^2 - z + 2 = 0$. Hence factorize $z^2 - z + 2$.

⑥ **7** Factorize:
  (i) $z^2 + 16$
  (ii) $z^2 + 4z + 5$
  (iii) $z^2 + (-1 + 4i)z + (-5 + i)$
  (iv) $4z^2 - 4z + 10$

⑦ **8** For the following cubic equations, each with a given complex root, find the other two roots and the value of p:
  (i) $z^3 - 4z^2 + 6z + p = 0$; $1 - i$
→ (ii) $z^3 - 3z^2 + 9z + p = 0$; $2 + 3i$
  (iii) $2z^3 - 7z^2 + 10z + p = 0$; $1 + i$
  (iv) $z^3 + z + p = 0$; $1 - 2i$

⑦ **9** For the following cubic equations, each with a given complex root, find the value of p simply by substituting the given root into the equation:
•→ (i) $z^3 - z^2 + p = 0$; $1 + i$
  (ii) $z^3 - z^2 - 7z + p = 0$; $2 - i$
  (iii) $2z^3 + z^2 + 4z + p = 0$; $-1 + 2i$

⑦ •→ **10** Show that $1 - i$ is a root of the equation
$$z^3 - 3z^2 + 4z - 2 = 0$$
and find the other two roots.

⑦ R **11** Show that $2 + i$ is a root of the equation
$$2z^3 - 9z^2 + 14z - 5 = 0$$
and find the other two roots.

⑦ •→ **12** Given that $1 + i$ is a root of the equation
$$z^3 + pz^2 + qz + 6 = 0$$
find the other two roots and the values of p and q.

⑦ **13** Given that $3 - 2i$ is a root of the equation
$$z^3 + pz^2 + 7z + q = 0$$
find the other two roots and the values of p and q.

⑦ •→ **14** Let us clarify the method of worked example 7.
  (i) Suppose that the cubic equation
  $$ax^3 + bx^2 + cx + d = 0$$
  has roots α, β, γ. Using the factor theorem,
  $$ax^3 + bx^2 + cx + d \equiv a(x - \alpha)(x - \beta)(x - \gamma)$$
  By equating coefficients, show that:
  (a) $\alpha + \beta + \gamma = -b/a$;
  (b) $\alpha\beta + \beta\gamma + \gamma\alpha = c/a$;
  (c) $\alpha\beta\gamma = -d/a$.
  (ii) Given that the equation $z^3 - z^2 + 3z + p = 0$ has a root $1 + 2i$, use (a) and (c) to find the other roots and the value of p. Check that (b) holds for this equation.

  (Effectively this is the same method as we used in worked example 7, but the working is simplified by using the general formulae (a), (b) and (c).)

⑦ **15** For the following cubic equations, each with a given root, find the other two roots and the values of p and q.
  (i) $z^3 + 3z^2 + pz + q = 0$; $-2 - i$
  (ii) $z^3 - 4z^2 + pz + q = 0$; $1 + 2i$
  (iii) $z^3 - 5z^2 + pz + q = 0$; $3 - i$
R (iv) $2z^3 - 5z^2 + pz + q = 0$; $1 + 3i$
  (v) $12z^3 - 28z^2 + pz + q = 0$; $\frac{3}{2} + i$
  (vi) $z^3 + pz^2 + qz - 25 = 0$; $4 - 3i$
  (vii) $z^3 + pz^2 - 7z + q = 0$; $2 - i$

16 Find an integer real root of each of the following equations, and hence find all its roots:
• (i) $z^3 + z^2 + z + 1 = 0$
→ (ii) $z^3 - 4z^2 + 6z - 4 = 0$
  (iii) $z^3 + z^2 - z + 2 = 0$
  (iv) $2z^3 - 5z^2 - 2z - 3 = 0$

→ **17** (i) Find the real root of the equation $z^3 - 1 = 0$. Hence find the two complex roots.
  Notice that the roots are the three cube roots of 1.
  (ii) If either of the complex roots is denoted by w, show that the other complex root is w^2.
  (iii) Show that w and w^2 are conjugates.
  (iv) Show that $1 + w + w^2 = 0$. (Hint: see question 14(i)(a).)

R 18 (i) Find the three cube roots of -1 (one real and two complex).

(ii) If either of the complex roots is denoted by λ, show that the other complex root is $-\lambda^2$.

(iii) Show that λ and $-\lambda^2$ are conjugates.

(iv) Show that $1 + \lambda^2 = \lambda$.

19 Find a cubic equation with roots:

• → (i) $1+i, 1-i, 3$
(ii) $i, -i, 2$
(iii) $3+4i, 3-4i, -1$
(iv) $-1+\sqrt{3}i, -1-\sqrt{3}i, 2$

20 Find the three linear factors of:

(i) $x^3 - 8$ (ii) $x^3 - x^2 + x - 1$

21 Suppose the cubic equation $ax^3 + bx^2 + cx + d = 0$, $a, b, c, d \in \mathbb{R}$, has roots α, β, γ. Using question 14(i), and also the fact that one of the roots, α say, must be real, prove that β and γ are either real or complex conjugates. Explain why your proof doesn't work if a, b, c or d is complex.

22 Prove that a cubic equation cannot have more than three roots. (Hint: suppose it has more than three roots and, using the factor theorem, obtain a contradiction.)

• → **23** Find the roots of $z^4 - 1 = 0$.

24 Find the other three roots of the quartic equation with the given root

• → (i) $z^4 - 4z^3 + 3z^2 + 2z - 6 = 0; 1-i$
(ii) $z^4 - 6z^3 - 23z^2 - 34z + 26 = 0; 2+3i$
(iii) $z^4 + 3z^2 - 6z + 10 = 0; -1+2i$

R 25 Show that $3-i$ is a root of the equation
$$4z^4 - 24z^3 + 39z^2 + 6z - 10 = 0$$
Hence find the other three roots.

26 Show that $1+2i$ is a root of the equation
$$z^4 - 6z^3 + 18z^2 - 30z + 25 = 0$$
Hence find the other three roots.

27 Find an integer root of the equation
$$z^5 - z^4 + 4z^3 - 2z^2 + 8 = 0$$
Given that $1+i$ is also a root of this equation, find the other three roots.

28 Why do Theorems 2A, 2B and 4 specify *real* coefficients?

Solve the equations:

(i) $z^2 - (2+i)z + 2i = 0$
(ii) $z^2 + (-1+i)z + (2+i) = 0$
(iii) $z^3 + iz^2 + 2i = 0$

29 Show that a polynomial equation of degree n cannot have more than n roots (see question 22).

(The proof that a polynomial equation of degree n has at least n roots is more difficult.)

6.5:4 Argand diagram

Just as we can represent real numbers by points on a line, we can represent complex numbers by points on a plane (see fig. 6.5:**1**). The complex number $z = x + iy$ is represented by the point $P(x, y)$.

Note: all points $(x, 0)$ on the x-axis represent real numbers, and all points $(0, y)$ on the y-axis represent pure imaginary numbers. So the x-axis is called the **real axis** and the y-axis the **imaginary axis**.

This type of diagram is called an **Argand diagram**.

If P is the point (x, y), then the position vector of P is given by $\overrightarrow{OP} = x\mathbf{i} + y\mathbf{j}$; so we find it convenient to think of $z = x + iy$ as also being represented by the position vector $\overrightarrow{OP} = x\mathbf{i} + y\mathbf{j}$; e.g. $3 - 2i$ is represented by the point $P(3, -2)$ and the position vector $\overrightarrow{OP} = 3\mathbf{i} - 2\mathbf{j}$.

In fact we can more usefully think of $3 - 2i$ as being represented by the *free* vector $3\mathbf{i} - 2\mathbf{j}$ (see fig. 6.5:**2**).

Thus the complex number $z = x + iy$ is represented by the vector $x\mathbf{i} + y\mathbf{j}$:

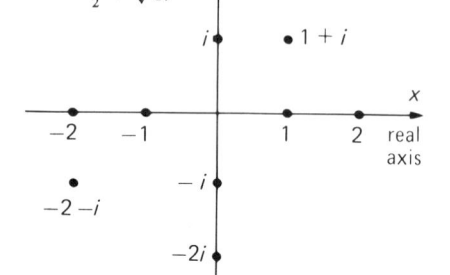

$$x + iy \equiv (x, y) \equiv x\mathbf{i} + y\mathbf{j} \qquad (7)$$

Fig. 6.5:**1**

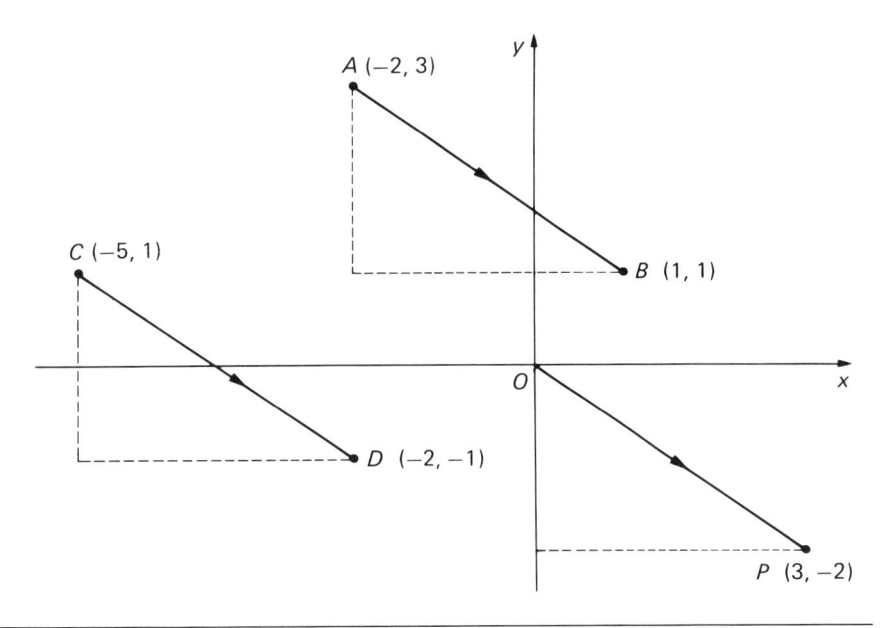

Vectors, coordinate geometry and complex numbers 353

Fig. 6.5:**2**

Exercise 6.5:4

→ **1** Mark on an Argand diagram the points representing the following numbers:

3; $2i$; $1-i$; $4+i$; $-2+3i$; $3-2i$; $-2i$.

2 Mark on an Argand diagram the points representing:

(i) $\frac{1}{2}+i, 1+2i, \frac{5}{2}+5i, -1-2i$
(ii) $-3+2i, 3-2i, \frac{3}{2}-i$

3 Mark a a point in the first quadrant on an Argand diagram. Suppose it represents the number z. Mark points representing the numbers $2z, 3z, \frac{1}{2}z, -z, -\frac{3}{2}z$.

4 If the point P represents a complex number z and Q the number $1/z$, show that O, P and Q are collinear.

5 Mark on an Argand diagram the points representing:

(i) $4, 4+i, 4+2i, 4-2i$
(ii) $2i, 1+2i, -1+2i, -3+2i$

6 Mark a point in the first quadrant of an Argand diagram. Suppose it represents the number $x+yi$. Mark the points representing the numbers:
$x-yi, -x+yi, -x-yi, x, 2x+2yi, -x+2yi$.

7 Suppose the number z is represented by the point (x, y). What points represent:

(i) iz (ii) $-z$ (iii) $-iz$?

Illustrate this on an Argand diagram.

8 Write down the point and the free vector representing the number:

(i) $3+4i$ (ii) $-2-5i$ (iii) -3 (iv) $2i$
(v) $3i-7$

9 Find the complex number represented by the vector $\overrightarrow{AB}$ when A and B are the points:

→ (i) $(1, 0)$ and $(5, 6)$
(ii) $(4, -1)$ and $(-2, 3)$
(iii) $(3, -5)$ and $(-4, -5)$

6.5:5 Addition and subtraction in the Argand diagram

Suppose that z_1 and z_2 are two complex numbers, and that z_1 is represented by $\overrightarrow{OA}$, z_2 by $\overrightarrow{OB}$. Then (see section 6.1:2) z_1+z_2 is represented by $\overrightarrow{OA}+\overrightarrow{OB}$, i.e. $\overrightarrow{OC}$, where $OACB$ is a parallelogram (fig. 6.5:**3a**), and z_1-z_2 is represented by $\overrightarrow{OA}-\overrightarrow{OB}$ i.e. $\overrightarrow{BA}$ (fig. 6.5:**3b**).

$$z_1+z_2 \equiv \overrightarrow{OC} \qquad (8)$$
$$z_1-z_2 \equiv \overrightarrow{BA} \qquad (9)$$

354 *Addition and subtraction in the Argand diagram*

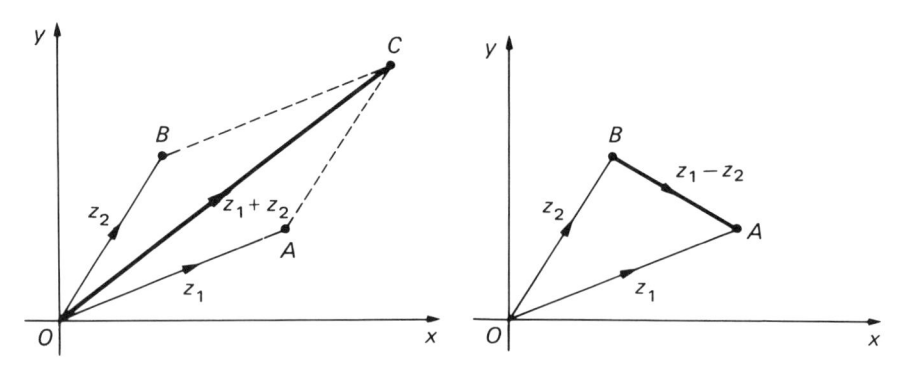

Fig. 6.5:3a
Fig. 6.5:3b

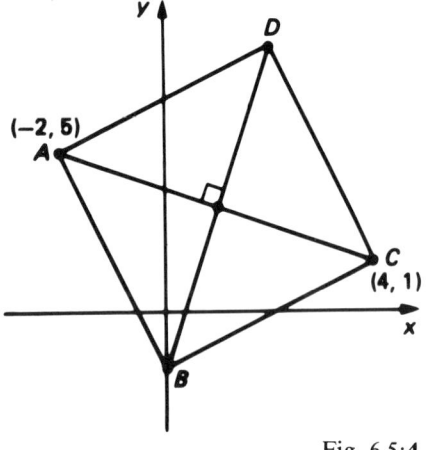

Fig. 6.5:4

Worked example 8 The points A and C represent the complex numbers $z_1 = -2 + 5i$ and $z_3 = 4 + i$, and $ABCD$ is a square (fig. 6.5:4). Find the complex numbers z_2 and z_4 represented by B and D.

$\overrightarrow{AC} \equiv 6 - 4i$; so $\overrightarrow{BD} \equiv 4 + 6i$ (see 6.2:1 Q30). The centre of the square is $\frac{1}{2}(z_1 + z_3)$, i.e. $1 + 3i$. So

$$z_4 \equiv 1 + 3i + \tfrac{1}{2}\overrightarrow{BD} \equiv 3 + 6i$$
$$z_2 \equiv 1 + 3i - \tfrac{1}{2}\overrightarrow{BD} \equiv -1$$

Exercise 6.5:5

1 If $z_1 = 3 + i$, $z_2 = 1 + 2i$, show on an Argand diagram vectors corresponding to

$$z_1, \quad z_2, \quad z_1 + z_2, \quad z_1 - z_2$$

2 Repeat question 1 when $z_1 = -5 + i$, $z_2 = -3i$.

→ **3** If z_1 and z_2 are represented by the points A and B in the Argand diagram, show how to construct the points representing:

$$z_1 + z_2, \quad z_1 - z_2, \quad 2z_1, \quad 2z_1 + z_2, \quad z_1 - 3z_2$$

4 Mark two points on an Argand diagram. Suppose these points represent the numbers z_1 and z_2. Mark in the points representing:

$$2z_1 + z_2, \quad 2z_1 - z_2, \quad \tfrac{2}{3}z_1 + \tfrac{1}{3}z_2, \quad -z_1 + \tfrac{1}{2}z_2$$

5 If z_1 and z_2 are represented by the points A and B in the Argand diagram, describe the points representing:

$$\tfrac{1}{2}z_1 + \tfrac{1}{2}z_2, \quad \tfrac{3}{4}z_1 + \tfrac{1}{4}z_2$$
$$(\mu z_1 + \lambda z_2)/(\lambda + \mu)(\lambda, \mu \in \mathbb{R})$$

6 If z_1, z_2 and z_3 are represented by the points A, B and C in the Argand diagram, show how to construct the points representing:

$$z_1 + z_2 + z_3, \quad z_1 + z_2 - z_3, \quad z_1 - z_2 - z_3$$

7 If z_1, z_2 and z_3 are represented by the points A, B and C in the Argand diagram, find the point representing $\tfrac{1}{3}(z_1 + z_2 + z_3)$.

Mark on an Argand diagram the points representing $1 + i$, $-2 + 3i$, $-5 - i$, $-2 + i$, showing clearly their relationship.

8 Mark a point in the first quadrant of an Argand diagram. Suppose it represents the number z. Mark in points representing:

$$z + 1, \quad z + 1 + i, \quad z - 2, \quad z + 3i,$$
$$2z + 1, \quad -z + 2i, \quad -\tfrac{1}{2}z + 2 - i$$

+ **9** If z lies on a circle centre O radius 1, what are the loci of:

$$z - 1, \quad z + 3 + 4i, \quad 2z, \quad \tfrac{1}{2}z, \quad 3z + 1?$$

Questions 10–13: The points A and B represent the numbers z_1 and z_2.

10 Find the distance AB when:

• (i) $z_1 = 3 + 4i$, $z_2 = 3 - 6i$
 (ii) $z_1 = 1 - 7i$, $z_2 = -4 + 5i$

11 If $z_1 = 1 - i$, $z_2 = 4 + 3i$ find:

(i) AB (ii) the angle between AB and Ox

12 If $z_1 = 1 + i$ and $z_2 = -1 + \sqrt{3}i$, find:

(i) $A\hat{O}B$ (ii) $O\hat{B}A$ (iii) $O\hat{A}B$

13 (i) Show that $A\hat{O}B = 90°$ when:

(a) $z_1 = 1 + 4i$, $z_2 = 4 - i$
(b) $z_1 = -2 - 3i$, $z_2 = 6 - 4i$
(c) $z_1 = p + qi$, $z_2 = q - pi$

(See 6.2:1 Q28.)

(ii) The complex number z is represented by the point P in the Argand diagram. What is the effect on $\overrightarrow{OP}$ when z is multiplied (a) by i; (b) by $-i$?

Questions 14–25: The points A, B, C and D represent the numbers z_1, z_2, z_3 and z_4.

14 $OABC$ is a parallelogram. $z_1 = 1 + i$, $z_2 = 4 + 5i$. Find z_3.

15 $ABCD$ is a parallelogram $z_1 = -3i$, $z_2 = 5 - i$, $z_3 = -2 + 2i$. Find z_4, and the number represented by the point of intersection of the diagonals.

16 Find z_2 and z_4 when $ABCD$ is a square and

(i) $z_1 = 2 + i$, $z_3 = 6 + 7i$
(ii) $z_1 = 6 - 2i$, $z_3 = 6i$
(iii) $z_1 = 1 + 4i$, $z_3 = -3 - 4i$

17 $ABCD$ is a square. If $z_1 = 6 - 3i$, $z_3 = -4 + 5i$, find the numbers represented by $\overrightarrow{CA}$, $\overrightarrow{AD}$, $\overrightarrow{BD}$. (Assume D is in the first quadrant.)

18 $OABC$ is a square. If $z_1 = 3 + i$, $z_2 = 4 + pi$, find z_3 and the value of p.

19 The centre of a square is the point $3 + 2i$ and one vertex is $-1 + 5i$. Find the others.

20 The centre of a square is the point $-2 + i$ and one vertex is $1 + 3i$. Find the others.

21 $ABCD$ is a square. Find z_2 and z_4 in terms of z_1 and z_3.

22 $ABCD$ is a square. Find the possible positions of C and D when:

(i) $z_1 = 1$, $z_2 = i$
(ii) $z_1 = 2 + i$, $z_2 = -1 + 7i$

23 If $ABCD$ is a square, show that:

(i) $z_2 - z_1 = i(z_3 - z_2)$
(ii) $z_3 - z_1 = i(z_2 - z_4)$

24 $ABCD$ is a rhombus; $AC = 2BD$. If $z_2 = 2 + i$, $z_4 = 1 - 2i$, find z_1 and z_3.

25 ABC is an equilateral triangle. Find the two possible positions of C when:

(i) $z_1 = 3 + i$, $z_2 = -5 + i$
(ii) $z_1 = 3 - 2i$, $z_2 = -3 + 2i$
(iii) $z_1 = 3 + i$, $z_2 = -1 + 4i$
(iv) $z_1 = -2$, $z_2 = 2\sqrt{2}i$

26 Find the area of $\triangle ABC$ when:

(i) $z_1 = -2 + i$, $z_2 = 3 + i$, $z_3 = 2 + 5i$
(ii) $z_1 = -1 - 2i$, $z_2 = 5 + i$, $z_3 = -4i$
(iii) $z_1 = 2 - i$, $z_2 = 8$, $z_3 = 7 + 4i$

6.5:6 Modulus and argument of a complex number

Suppose that the complex number z is represented by the point P in the Argand diagram (Fig. 6.5:**5**).

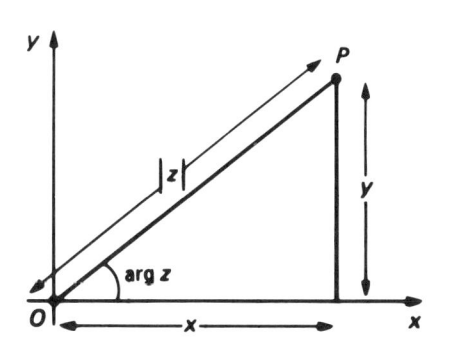

> **Definition** The **modulus** of z, $|z| = |\overrightarrow{OP}|$.
> The **argument** of z, arg z, is the angle between Ox and OP.

In order to avoid ambiguity, we choose arg z so that

$$-180° < \arg z \leq 180° \quad (10)$$

Fig. 6.5:**5**   If $z = x + iy$,

$$|z| = \sqrt{(x^2 + y^2)} \quad (11)$$

$$\tan(\arg z) = \frac{y}{x} \quad (12)$$

356 *Modulus and argument of a complex number*

Note that although $\tan(\arg z) = y/x$, it is not necessarily true that $\arg z = \tan^{-1} y/x$. The equation $\tan(\arg z) = y/x$ has an infinite number of solutions $\arg z = \tan^{-1}(y/x) + 180k°$; we need to choose the appropriate value of k ($-1, 0,$ or 1) to ensure that $-180° < \arg z \leq 180°$. An Argand diagram is helpful, not to say essential (see section 6.2:1).

Worked example 9 Find the modulus and argument of:
(i) $z = 3 + 4i$ (ii) $z' = -\sqrt{3} + i$.

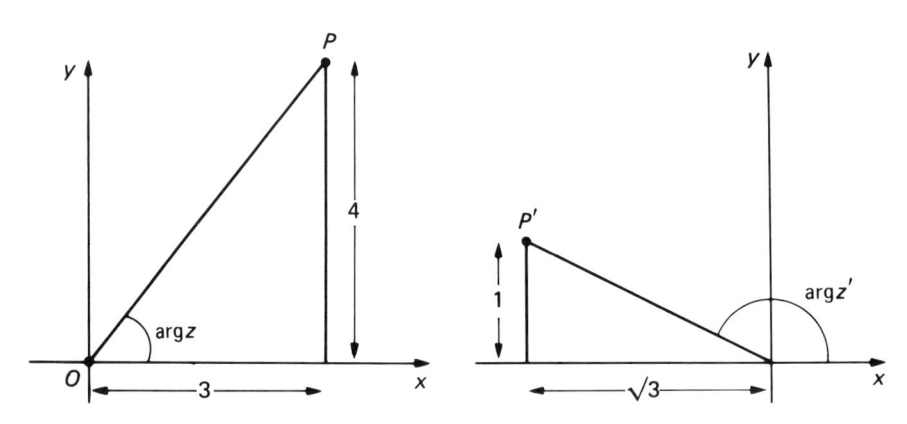

Fig. 6.5:6a

Fig. 6.5:6b

First draw the Argand diagrams for the numbers (fig. 6.5:6).

(i) $|z| = \sqrt{(3^2 + 4^2)} = \sqrt{25} = 5$
$\tan(\arg z) = \frac{4}{3} \Rightarrow \arg z = 54.13°$, since P is in the first quadrant.

(ii) $|z'| = \sqrt{((-\sqrt{3})^2 + 1^2)} = \sqrt{4} = 2$.
$\tan(\arg z') = -\frac{1}{\sqrt{3}} \Rightarrow \arg z' = 150°$, since P' is in the second quadrant.

(See 6.2:1 WE2.)

Exercise 6.5:6

⑨ • **1** (i) Find the modulus and argument of $-4 + 3i$. What is wrong with the following?
(a) $|-4 + 3i| = \sqrt{(-16 + 9)} = \sqrt{-7} = i\sqrt{7}$
(b) $\tan \arg(-4 + 3i) = -\frac{3}{4}$, so $\arg(-4 + 3i) = \tan^{-1}(-\frac{3}{4}) = -36.87°$

→ (ii) Find the modulus and argument of:
(a) $2 - i$; (b) $-2 + i$.

⑨ **2** Find the modulus and argument of:

(i) $1 + i$
(ii) $1 + \sqrt{3}i$
(iii) $4\sqrt{3} - 4i$
(iv) $-3 + 3i$
(v) $-4 - 3i$
(vi) $3 - 2i$

⑨ **3** Find the modulus and argument of:

(i) (a) $2i$; (b) -1; (c) -5; (d) $-\sqrt{3}i$; (e) 6.
(ii) (a) λ^2; (b) $\lambda^2 i$; (c) $-\lambda^2$; (d) $-\lambda^2 i$ (λ real).

⑨ **4** Sketch on an Argand diagram circles centre O, radii 1, 2, 4 and 5, and mark in the points representing the numbers:

(i) $5i$
(ii) $2\sqrt{2} + 2\sqrt{2}i$
(iii) $-3 + 4i$
(iv) $1 - \sqrt{3}i$
(v) $\frac{4}{5} + \frac{3}{5}i$
(vi) $-3 - \sqrt{7}i$
(vii) -4

⑨ **5** Find the modulus and argument of $z_1, z_2, z_1 z_2$ and z_1/z_2 when:

(i) $z_1 = 1 + i$, $z_2 = \sqrt{3} + i$
(ii) $z_1 = 2 - 2\sqrt{3}i$, $z_2 = -3 - 4i$

Do you notice any connections?

⑨ **6** Show that:

(i) $|\bar{z}| = |z|$
(ii) $\arg \bar{z} = -\arg z$

Illustrate these results on an Argand diagram.

⑨ 7 Find the modulus and argument of:

(i) $(2+i)(3+i)$ (ii) $\dfrac{-1-7i}{3-4i}$

(iii) $\dfrac{5+12i}{3+4i}$

⑨ 8 Given that $z_1 = 3-2i$ and $z_2 = -2+4i$, represent the numbers z_1, z_2, $z_1 + z_2$ and $z_1 - z_2$ on an Argand diagram. Find the modulus and argument of z_1, z_2, $z_1 + z_2$ and $z_1 - z_2$.

⑨ 9 If $z_1 = 5-4i$ and $z_2 = -3+8i$, show that

$$|z_1 + z_2| < |z_1 - z_2| < |z_1| + |z_2|$$

Do the inequalities hold when $z_1 = 1+2i$ and $z_2 = 3+i$?

6.5:7 The polar form of a complex number

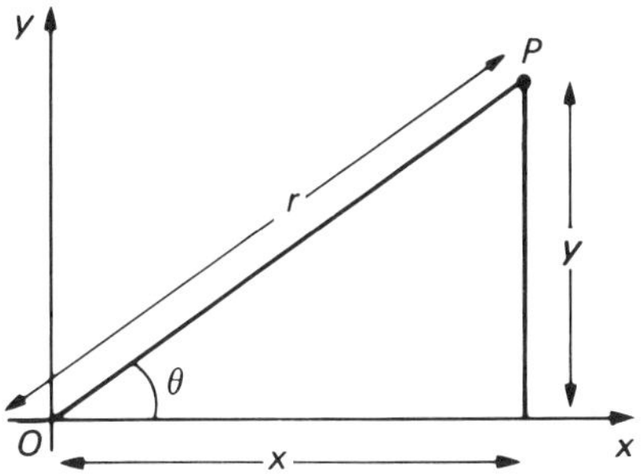

Suppose that $z = x + iy$ and that $|z| = r$, arg $z = \theta$ (fig. 6.5:7). Then

$$x = r \cos \theta$$
$$y = r \sin \theta \qquad (13)$$

So $z = x + iy = r \cos \theta + i(r \sin \theta) = r(\cos \theta + i \sin \theta)$.

The complex number $\cos \theta + i \sin \theta$ is abbreviated by cis θ; so we can write

$$z = r \text{ cis } \theta \qquad (14)$$

Fig. 6.5:7

This is called the **modulus-argument** or **polar** form for z.

Worked example 10 Express the number $-\sqrt{3} + i$ in the form r cis θ.

From worked example 9,

$$r = |-\sqrt{3} + i| = 2$$

and

$$\theta = \text{arg}(-\sqrt{3} + i) = \tfrac{5}{6}\pi$$

So

$$-\sqrt{3} + i \equiv 2 \text{ cis } \tfrac{5}{6}\pi$$

Worked example 11 Express the number $4 \text{ cis }(-\tfrac{3}{4}\pi)$ in the form $x + iy$.

$$4 \text{ cis }(-\tfrac{3}{4}\pi) = 4\{\cos(-\tfrac{3}{4}\pi) + i \sin(-\tfrac{3}{4}\pi)\}$$
$$= 4\left\{-\dfrac{1}{\sqrt{2}} + i\left(-\dfrac{1}{\sqrt{2}}\right)\right\}$$
$$= -\dfrac{4}{\sqrt{2}} - \dfrac{4}{\sqrt{2}}i$$
$$= -2\sqrt{2} - 2\sqrt{2}i$$

Exercise 6.5:7

⑩ 1 Write in the form r cis θ:

(i) $2\sqrt{3} + 2i$
(ii) $1 - i$
(iii) $-2 + i$
(iv) $-\tfrac{1}{2} + \tfrac{1}{2}\sqrt{3}i$
(v) -5
(vi) $6i$
(vii) $-3 - 4i$
(viii) $-i$
(ix) $1 - 3i$
(x) $-\tfrac{4}{5} + \tfrac{3}{5}i$

⑩ → 2 Write in the form r cis θ:

(i) $3 + 2i$
(ii) $3 - 2i$
(iii) $-3 + 2i$
(iv) $-3 - 2i$

⑩ 3 Write in the form r cis θ:

(i) $\cos \alpha - i \sin \alpha$
(ii) $\sin \alpha + i \cos \alpha$
(iii) $\sin \alpha - i \cos \alpha$

⑪ 4 Write in the form $a + bi$:
  (i) $4 \operatorname{cis} \tfrac{1}{3}\pi$
  (ii) $2 \operatorname{cis} 20°$
  (iii) $5 \operatorname{cis} \tfrac{1}{2}\pi$
  (iv) $3 \operatorname{cis} \pi$
  (v) $3\sqrt{2} \operatorname{cis}(-\tfrac{3}{4}\pi)$
  (vi) $3 \operatorname{cis}(-140°)$
  (vii) $2 \operatorname{cis} 100°$
  (viii) $\tfrac{1}{2} \operatorname{cis} \tfrac{2}{3}\pi$

⑪ 5 Write in the form $a + bi$:
  (i) $8 \operatorname{cis} 2\pi$
  (ii) $4 \operatorname{cis}(-\tfrac{11}{6}\pi)$
  (iii) $2 \operatorname{cis} \tfrac{19}{4}\pi$
  (iv) $\operatorname{cis} \tfrac{40}{3}\pi$
  (v) $3 \operatorname{cis} 15\pi$

⑪ 6 (i) Show that the complex number $\operatorname{cis}\theta$ is represented by a point P lying on a circle centre O, radius 1. What is the significance of the angle θ? (See 6.3:1 Q12.)

(ii) Write in the form $a + bi$:
$\operatorname{cis} 0$, $\operatorname{cis} \tfrac{1}{2}\pi$, $\operatorname{cis} \pi$, $\operatorname{cis} -\tfrac{1}{2}\pi$, $\operatorname{cis} \tfrac{1}{3}\pi$, $\operatorname{cis} \tfrac{3}{4}\pi$, $\operatorname{cis}(-\tfrac{2}{5}\pi)$, $\operatorname{cis}(-\tfrac{5}{6}\pi)$.

(iii) Sketch on an Argand diagram the circle centre O, radius 1, and mark in the points representing the numbers in (ii).

⑪ → 7 Sketch on an Argand diagram circles centre O, radii $\tfrac{1}{2}$, 1, 2 and 4, and mark in the points representing the numbers:
$2 \operatorname{cis} \tfrac{1}{4}\pi$, $\operatorname{cis} \tfrac{1}{10}\pi$, $\tfrac{1}{2} \operatorname{cis}(-\tfrac{1}{3}\pi)$, $4 \operatorname{cis}(-\tfrac{1}{2}\pi)$, $2 \operatorname{cis}(-\tfrac{3}{5}\pi)$, $\tfrac{1}{2} \operatorname{cis} \tfrac{5}{6}\pi$.

⑪ 8 Solve the following equations. Give the roots in the form $r \operatorname{cis} \theta$ and mark the points representing the roots on an Argand diagram.
  (i) (a) $z^2 + z + 1 = 0$;
      (b) $z^2 + 4z + 8 = 0$
      (c) $z^2 - i = 0$
  (ii) (a) $z^2 + 1 = 0$
   → (b) $z^3 - 1 = 0$
  (iii) $z^4 - 1 = 0$

⑪ 9 Solve the equation $z^2 - (2\cos\alpha)z + 1 = 0$. Hence find
  (i) a quadratic equation with roots $\operatorname{cis} \tfrac{1}{6}\pi$, $\operatorname{cis}(-\tfrac{1}{6}\pi)$;
  (ii) a cubic equation with roots 2, $\operatorname{cis} \tfrac{3}{4}\pi$, $\operatorname{cis}(-\tfrac{3}{4}\pi)$;
  (iii) a quintic (order 5) equation with roots 1, $\operatorname{cis} \tfrac{2}{5}\pi$, $\operatorname{cis} \tfrac{4}{5}\pi$, $\operatorname{cis}(-\tfrac{2}{5}\pi)$, $\operatorname{cis}(-\tfrac{4}{5}\pi)$.

6.5:8 Multiplication and division of complex numbers in polar form; De Moivre's theorem

Theorem 4: If $z_1 = r_1 \operatorname{cis} \theta_1$ and $z_2 = r_2 \operatorname{cis} \theta_2$, then

$$z_1 z_2 = r_1 r_2 \operatorname{cis}(\theta_1 + \theta_2) \tag{15}$$

$$\frac{z_1}{z_2} = \frac{r_1}{r_2} \operatorname{cis}(\theta_1 - \theta_2) \tag{16}$$

Proof
Equation (15):

$$\begin{aligned} z_1 z_2 &= (r_1 \operatorname{cis} \theta_1)(r_2 \operatorname{cis} \theta_2) \\ &= \{r_1(\cos\theta_1 + i\sin\theta_1)\}\{r_2(\cos\theta_2 + i\sin\theta_2)\} \\ &= r_1 r_2 \{(\cos\theta_1 \cos\theta_2 - \sin\theta_1 \sin\theta_2) \\ &\quad + i(\cos\theta_1 \sin\theta_2 + \sin\theta_1 \cos\theta_2)\} \\ &= r_1 r_2 \{\cos(\theta_1 + \theta_2) + i\sin(\theta_1 + \theta_2)\} \\ &= r_1 r_2 \operatorname{cis}(\theta_1 + \theta_2) \end{aligned}$$

Equation (16): see question 1.

Worked example 12 Find in the form r cis θ the numbers:

$$2+2i, \quad 1+\sqrt{3}i, \quad (2+2i)(1+\sqrt{3}i), \quad \frac{2+2i}{1+\sqrt{3}i}$$

$2+2i = 2\sqrt{2}$ cis $45°$; $1+\sqrt{3}i = 2$ cis $60°$, so, by equation (15),

$$(2+2i)(1+\sqrt{3}i) = (2\sqrt{2} \text{ cis } 45°)(2 \text{ cis } 60°) = 4\sqrt{2} \text{ cis } 105°$$

and by equation (16):

$$\frac{2+2i}{1+\sqrt{3}i} = \frac{2\sqrt{2} \text{ cis } 45°}{2 \text{ cis } 60°} = \sqrt{2} \text{ cis } (-15°)$$

Theorem 5 (De Moivre's Theorem):   If n is a positive integer,

$$(r \text{ cis } \theta)^n = r^n \text{ cis } n\theta \qquad (17)$$

Proof
See question 28.
(Note that
if $\qquad z = r$ cis θ,
then $\qquad z^2 = (r \text{ cis } \theta)(r \text{ cis } \theta) = r^2 \text{ cis } 2\theta \qquad$ by equation (15)
$\qquad z^3 = (r^2 \text{ cis } 2\theta)(r \text{ cis } \theta) = r^3 \text{ cis } 3\theta \qquad$ by equation (15)
etc.

so it seems like a plausible result.)
  De Moivre's Theorem also holds when n is any negative integer (see question 28) and indeed for any real number.

Worked example 13   Find $(\sqrt{3}-i)^8$ in the form $a+bi$.

$$(\sqrt{3}-i)^8 = (2 \text{ cis } (-\tfrac{1}{6}\pi))^8$$
$$= 2^8 \text{ cis } (-\tfrac{4}{3}\pi) \qquad \text{by Theorem 5}$$
$$= 256\left(-\frac{1}{2}+\frac{\sqrt{3}}{2}i\right) \qquad \text{(see worked example 11)}$$
$$= 128 + 128\sqrt{3}\,i$$

Exercise 6.5:8

- **1** If $z_1 = r_1$ cis θ_1, $z_2 = r_2$ cis θ_2, show, using the method of worked example 4, that

$$\frac{z_1}{z_2} = \frac{r_1}{r_2} \text{ cis } (\theta_1 - \theta_2)$$

(12) **2** Find in the form r cis θ the numbers:
  (i) $1+i$, $\sqrt{3}+i$, $(1+i)(\sqrt{3}+i)$, $(1+i)/(\sqrt{3}+i)$
  (ii) $\tfrac{1}{2}+\tfrac{1}{2}\sqrt{3}i$, $-\tfrac{1}{2}\sqrt{}+\tfrac{1}{2}i$, $(\tfrac{1}{2}+\tfrac{1}{2}\sqrt{3}i)(-\tfrac{1}{2}\sqrt{3}+\tfrac{1}{2}i)$, $(\tfrac{1}{2}+\tfrac{1}{2}\sqrt{3}i)/(-\tfrac{1}{2}\sqrt{3}+\tfrac{1}{2}i)$
→ (iii) $4+3i$, $3-4i$, $(4+3i)(3-4i)$, $(4+3i)/(3-4i)$
R (iv) $3-i$, $1+2i$, $(3-i)(1+2i)$, $(3-i)/(1+2i)$

(12) **3** Find the modulus and argument of the numbers $z_1 = 1+i$, $z_2 = 1-\sqrt{3}i$.
  Hence find in the form r cis θ the numbers:
  (i) $z_1 z_2$ (ii) z_1/z_2
  (iii) z_2/z_1 (iv) z_1^2
  (v) z_2^3 (vi) z_1^2/z_2^4

(12) **4** Find the modulus and argument of the numbers $z_1 = 2\sqrt{2} - 2\sqrt{2}i$, $z_2 = -1+\sqrt{3}i$.
  Hence find in the form r cis θ the numbers:
  (i) $z_1 z_2$ (ii) z_1/z_2
  (iii) z_1^3 (iv) z_2^5
  (v) $(z_1/z_2)^4$ (vi) z_2^5/z_1^3

5 Express the numbers $1, 3i, -4, z = 2+\sqrt{5}i$ in $r \operatorname{cis} \theta$ form. Hence, using Theorem 4, express in $r \operatorname{cis} \theta$ form:

(i) $1/z$ (ii) $3iz$ (iii) $z/3i$
(iv) $-4z$ (v) $-4/z$

6 If $z = r \operatorname{cis} \theta$, find the modulus and argument of:

(i) $\bar{z}$ (ii) z^2 (iii) $1/z$ (iv) $-z$

7 Find in $r \operatorname{cis} \theta$ form:

(i) $\dfrac{2i}{1-i}$ (ii) $\dfrac{-8}{-\sqrt{3}+i}$

(iii) $\dfrac{(1-i)^2(\sqrt{3}+i)}{1-\sqrt{3}i}$

8 (i) Use Theorem 4 to write down equations connecting $|z_1|, |z_2|, |z_1 z_2|$; $|z_1|, |z_2|, |(z_1/z_2)|$; $\arg z_1$, $\arg z_2$, $\arg z_1 z_2$; $\arg z_1$, $\arg z_2$, $\arg(z_1/z_2)$.

(ii) Find:

(a) $\left|\dfrac{3+4i}{1+i}\right|$;   (b) $\left|\dfrac{\sqrt{3}+i}{\sqrt{3}-i}\right|$;

(c) $\arg\left(\dfrac{\sqrt{3}+i}{\sqrt{3}-i}\right)$;   (d) $\arg\left(\dfrac{1-i}{-1+\sqrt{3}i}\right)$.

9 If $z_1 = 1-i$, $z_2 = 7+i$, find the modulus of:

(i) $z_1 - z_2$
(ii) $z_1 z_2$
(iii) $(z_1 - z_2)/z_1 z_2$

10 Simplify:

(i) $\operatorname{cis} \tfrac{1}{4}\pi \operatorname{cis} \tfrac{1}{3}\pi$
(ii) $\operatorname{cis} \tfrac{2}{5}\pi \operatorname{cis} \tfrac{3}{5}\pi$
(iii) $(\operatorname{cis} \tfrac{2}{3}\pi)^2$
(iv) $\{\operatorname{cis}(-\tfrac{2}{3}\pi)\}^3$

11 If $z = \operatorname{cis} \theta$, find $1/z$. Show that:

(i) $z + 1/z$ is a real number
(ii) $\bar{z} = 1/z$

12 If $z = \operatorname{cis} \theta$, find in polar form:

(i) $z+1$
(ii) $z-1$
(iii) $(z-1)/(z+1)$
R (iv) $2z/1-z^2$

13 If $z = \operatorname{cis} \theta$, find in the form $a + bi$:

(i) $1/(z+1)$
(ii) $2z/(1+z^2)$
(iii) $(1-z^2)/(1+z^2)$

14 Sketch a circle centre O, radius 1, on an Argand diagram. If $z_1 = \operatorname{cis} \tfrac{2}{5}\pi$, $z_2 = \operatorname{cis} \tfrac{1}{10}\pi$, mark on the diagram the vectors representing $z_1, z_2, z_1 z_2, z_1^2, 1/z_2, 1/z_2^3$.

15 Sketch on an Argand diagram circles centre O, radii $\tfrac{1}{2}, 1, 2, 4$ and 8. If $z_1 = 4 \operatorname{cis} 40°$, $z_2 = 2 \operatorname{cis}(-50°)$, mark on the diagram the vectors representing $z_1, z_2, z_1/z_2, z_1/z_2^2, z_1/z_2^3, z_2^3$.

16 Mark on an Argand diagram the vectors representing the numbers:

(i) $1+2i, 1-2i, (1+2i)(1-2i), (1+2i)/(1-2i)$
(ii) $1+7i, 4+3i, (1+7i)(4+3i), (1+7i)/(4+3i)$

Show any equal angles.

17 If $z_1 = -1 + i\sqrt{3}$ and $z_2 = \sqrt{3} + i$, show on an Argand diagram vectors representing the numbers: z_1, z_2, $z_1 + z_2, z_1 - z_2, z_1 z_2$ and z_1/z_2.

18 Mark a point P in the first quadrant of an Argand diagram. Sketch the circle centre O through P.

(i) Suppose P represents the number $z = r \operatorname{cis} \theta$. Mark on the diagram the points representing:
(a) $z \operatorname{cis} \tfrac{1}{3}\pi$;   (b) $z \operatorname{cis} \tfrac{1}{2}\pi$;
(c) $z \operatorname{cis} \tfrac{3}{4}\pi$;   (d) $z \operatorname{cis}(-\tfrac{1}{3}\pi)$.

(ii) If Q represents the number $z \operatorname{cis} \alpha$, show that $P\widehat{O}Q = \alpha$. Describe the effect on the vector $\overrightarrow{OP}$ when z is multiplied by: (a) $\operatorname{cis} \alpha$; (b) $k \operatorname{cis} \alpha$.

(iii) Hence describe the effect on $\overrightarrow{OP}$ when z is multiplied by $i, -1, -i$.

R 19 Given that the number z is represented by the point P on an Argand diagram, show how to construct the positions of:

(i) $2z$ (ii) $-3z$ (iii) $|z|$ (iv) $\bar{z}$
(v) iz (vi) $-iz$ (vii) z^2 (viii) iz^2
(ix) $(z-i)^2$ (x) $z^2 + 3i$

20 Using Theorem 5, find in polar form:

(i) $(\sqrt{3} \operatorname{cis} \tfrac{1}{3}\pi)^6$
(ii) $\{\sqrt{2} \operatorname{cis}(-\tfrac{1}{6}\pi)\}^{10}$

21 Find $\sqrt{3}+i$ in the form $r \operatorname{cis} \theta$. Hence find:

(i) $(\sqrt{3}+i)^3$
(ii) $(\sqrt{3}+i)^8$

in the form $a + bi$.

22 Find in the form $a + bi$:

(i) $(1-i)^4$
(ii) $(1-i)^7$
(iii) $(1-i)^{10}$

23 Find in the form $a + bi$:

(i) $(1+\sqrt{3}i)^5$
(ii) $(\sqrt{3}-i)^{10}$
(iii) $(-\sqrt{3}+i)^8$

24 Find $(1+i)^5$:

(i) by the method of worked example 13
(ii) by using Pascal's Triangle

(See section 5.4:1.)

25 Find, in the form $a+bi$, z^2, z^4 and z^6 when:

(i) $z = 1+(\sqrt{2}-1)i$
(ii) $z = (\sqrt{2}+1)+i$

⑬ **R 26** Simplify $(1+i)^{10} - (1-i)^{10}$.
If n is a positive integer, show that
$$(1+i)^{4n} - (1-i)^{4n} = 0$$

⑬ **27** If n is a positive integer, prove that:

(i) $(\sqrt{3}+i)^n + (\sqrt{3}-i)^n$ is a real number
(ii) $(-1+i\sqrt{3})^n + (-1-i\sqrt{3})^n = 2^{n+1} \cos(\tfrac{2}{3}n\pi)$

28 Prove De Moivre's Theorem, for all positive integers n, by induction.

• **29** (i) If $z = r \operatorname{cis} \theta$ show, by expressing the number 1 in polar form and using Theorem 5, that
$$z^{-n} = r^{-n} \operatorname{cis}(-n\theta)$$
i.e. that we can extend the theorem to negative integers.

(ii) Find in the form $a+bi$:

(a) $\dfrac{1}{(1+i)^6}$; (b) $\dfrac{1}{(2 \operatorname{cis} 15°)^8}$.

30 Simplify:

(i) $\dfrac{(\operatorname{cis} \tfrac{1}{9}\pi)^4}{(\operatorname{cis}(-\tfrac{1}{9}\pi))^5}$ (ii) $\dfrac{(1+\sqrt{3}i)^6}{(1-\sqrt{3}i)^4}$

• → **31** Sketch the circle centre O, radius 1, on an Argand diagram. Mark on the diagram the vectors representing z, z^2, z^3, z^4, z^5, $1/z$ and $1/z^2$, where $z = \operatorname{cis} \tfrac{1}{5}\pi$.

32 Sketch on an Argand diagram the circles centre O, radii $\tfrac{1}{4}$, $\tfrac{1}{2}$, 1, 2, 4 and 8. Mark on the diagram the vectors representing z, z^2, z^3, $1/z$, $1/z^2$, where $z = 2 \operatorname{cis}(-\tfrac{1}{3}\pi)$.

R 33 If $z = \tfrac{1}{2}\sqrt{3} + \tfrac{1}{2}i$, simplify z^2, z^3 and z^4 and mark these points on an Argand diagram.

34 When:

(i) $z = \sqrt{3}+i$ (ii) $z = 1-i$

find z, z^2, z^3, $1/z$ and $1/z^2$ and mark the vectors representing these numbers on an Argand diagram.

35 If $z = 2 \operatorname{cis} \tfrac{1}{6}\pi$, illustrate on an Argand diagram the points representing z, $\bar{z}$, $(\bar{z})^2$, $(\bar{z})^2/|z|$.

36 If $z = \operatorname{cis} \tfrac{1}{3}\pi$, show that $1+z^2 = z$. Draw a diagram to illustrate this.

37 Show that if r is any integer then $(\operatorname{cis} \tfrac{2}{3}\pi r)^3 = 1$. Hence find the three cube roots of 1, i.e. solve the equation $z^3 - 1 = 0$.

(i) Show that the roots may be written 1, w, w^2.

(ii) Show that $1 + w + w^2 = 0$.
(iii) Mark the points A, B and C representing the roots on an Argand diagram. Show that $\triangle ABC$ is equilateral. (c.f. 6.5:3 Q17.)

Questions 38–48: The points A, B, C and D represent the numbers z_1, z_2, z_3 and z_4.

• → **38** Show that $\arg(z_1/z_2) = A\hat{O}B$. Hence show that if $z_1 = iz_2$, then OA is perpendicular to OB.
What happens if $z_1 = -iz_2$? Draw a diagram to illustrate this.

39 If $z_1 = 4+5i$, $z_2 = 1+i$, $z_3 = -2-i$, find:

(i) the angle between BA and Ox
(ii) the angle between CA and Ox
(iii) $B\hat{A}C$
(iv) $\arg\left(\dfrac{z_1-z_2}{z_1-z_3}\right)$

• → **40** Show that:
$$\arg\left(\dfrac{z_1-z_2}{z_1-z_3}\right) = B\hat{A}C$$

(i) What can you say about A, B and C if:

(a) $\arg\left(\dfrac{z_1-z_2}{z_1-z_3}\right) = \tfrac{1}{2}\pi$;

(b) $\arg\left(\dfrac{z_1-z_2}{z_1-z_3}\right) = 0$?

(ii) If $(z_1-z_2)/(z_1-z_3) = \operatorname{cis} \tfrac{1}{3}\pi$, show that $\triangle ABC$ is equilateral. Check that this equation holds when $z_1 = 1$, $z_2 = w$, $z_3 = w^2$ and that z_1, z_2 and z_3 are the three roots of the equation $z^3 - 1 = 0$.

(iii) What can you say about $\triangle ABC$ if
$$z_1 - z_2 = i(z_1 - z_3)?$$

R 41 If $z_1 = -2+2i$, $z_2 = 2$, $z_3 = -3$:

(i) show that
$$\arg\left(\dfrac{z_1-z_2}{z_1-z_3}\right) = \tfrac{1}{2}\pi$$
and hence show that A lies on the circle with diameter BC;

(ii) find the numbers represented by $\overrightarrow{BA}$ and $\overrightarrow{CA}$ and deduce that BA is perpendicular to CA.

42 Show that
$$\arg\left(\dfrac{z_1-z_2}{z_1+z_2}\right) = \tfrac{1}{2}\pi \Leftrightarrow |z_1| = |z_2|$$

43 If $|z_1 - z_2| = |z_1 + z_2|$ show that
$$\arg \dfrac{z_1-z_2}{z_1+z_2} = \pm\tfrac{1}{2}\pi$$

44 If $z_1 = 1$, $z_2 = 4$, $z_3 = 2i$, find $A\hat{B}C$ and $A\hat{C}O$. Hence show that the circle through ABC touches the y-axis at C.

45 In the $\triangle ABC$, $BC = p$, $AC = q$, $AB = r$, $A\hat{C}B = \gamma$. Draw $\triangle ABC$ on an Argand diagram with C at the origin and A on the positive real axis. Write down the numbers represented by A and B.
  Hence prove the cosine rule, i.e.
$$r^2 = p^2 + q^2 - 2pq \cos \gamma$$

46 Find the angles of $\triangle ABC$ and the area of $\triangle ABC$ when:

(i) $z_1 = 1 - i$, $z_2 = 5 + i$, $z_3 = 2 - 3i$
(ii) $z_1 = 1 + i$, $z = 2 + 5i$, $z_3 = -2 + 4i$

47 What can you say about A, B, C and D when:

(i) $z_1 - z_2 = z_3 - z_4$ (ii) $z_1 - z_2 = i(z_3 - z_4)$

48 If P, Q, R and S are the midpoints of the sides AB, BC, CD and DA of the quadrilateral $ABCD$, write down the numbers represented by P, Q, R and S. Show that $PQRS$ is a parallelogram.

49 If $z = r \operatorname{cis} \theta$, find z in modulus-argument form. Hence show that for all real values of n, $\bar{z}^n = \overline{z^n}$.
  Hence show that if
$$f(z) = a_n z^n + a_{n-1} z^{n-1} + \cdots + a_1 z + a_0$$
where a_i is real for all i, then $f(\bar{z}) = \overline{f(z)}$. Hence show that if the equation $f(z) = 0$ has a complex root z, it also has a root $\bar{z}$, i.e. that the complex roots of polynomial equations occur in conjugate pairs.

Miscellaneous exercise 6

1 The points P and Q have coordinates $(3a, 4a)$ and $(-8a, 6a)$, where a is a non-zero constant.

(i) Show that $P\hat{O}Q$ is a right angle. Find the area of $\triangle POQ$.

(ii) Show that, if X is a point on the line PQ, the position vector of X can be written in the form $a\{(1-\lambda)(3\mathbf{i} - 4\mathbf{j}) + \lambda(-8\mathbf{i} + 6\mathbf{j})\}$ and explain the significance of the parameter λ. By choosing suitable values of λ, find the position vectors of:
(a) M, the midpoint of PQ;
(b) L and N, the points of trisection of PQ.

(iii) Find the point of intersection of the line PQ and the line with vector equation
$$\mathbf{r} = a\{(1-\mu)(2\mathbf{i} - \mathbf{j}) + \mu(5\mathbf{i})\}$$

2 The line Λ passes through the point $A(3, 8)$ and is perpendicular to the vector $-2\mathbf{i} + \mathbf{j}$. Find a vector equation of Λ in the form $\mathbf{r} \cdot \mathbf{n} = d$. Find:

(i) the distance from the origin O to Λ;
(ii) the distance from the point $B(7, 6)$ to Λ.

Find the coordinates of the point C on Λ nearest to O and show that the point on Λ nearest to B is A.

3 Show that the lines $\Lambda_1: \mathbf{r} = -\mathbf{i} - 2\mathbf{j} + \lambda(5\mathbf{i} + \mathbf{j})$ and $\Lambda_2: \mathbf{r} = -2\mathbf{i} + 3\mathbf{j} + \mu(3\mathbf{i} - 2\mathbf{j})$ intersect and find the position vector of their point of intersection.
  Find the angle between the lines.
  Find a vector equation of the line which passes through the point A with position vector $2\mathbf{i} - 4\mathbf{j}$ and is perpendicular to Λ_2. Hence find:

(i) the distance from A to Λ_2;
(ii) the position vector of the mirror image of A in Λ_2.

4 O is the origin, and A and B are the points $(1, 7)$ and $(10, 10)$ respectively. If C is the point on AB which is equidistant from the lines OA and OB, show that C divides AB in the ratio $1:2$. Hence find the coordinates of C.

5 The distance of a variable point P from two fixed points $A(-a, 0)$ and $B(b, 0)$ is such that $PA : PB = 1 : k$ $(k > 0)$. Show that:

(i) if $k = 1$, the locus of P is a straight line;
(ii) if $k \neq 1$, the locus of P is a circle. Find its centre and radius when $k = 2$.

Sketch the locus of P in the cases $k = \frac{1}{2}$, $k = 1$, $k = 2$.

6 A and B are the points $(0, 8)$ and $(4, 0)$ respectively. Find the equation of:

(i) the circle Γ_1 with AB as diameter;
(ii) the circle Γ_2 which passes through A and touches the x-axis at B.

Does any part of Γ_1 lie outside Γ_2?

7 Show that the circles $x^2 + y^2 - 4y - 1 = 0$ and $x^2 + y^2 - 8x - 29 = 0$ touch internally. Find the equation of their common tangent T_1 and the coordinates of their point of contact.
  Find the equation of the tangent T_2 to the smaller circle which is parallel to T_1. Find the area of the minor segment of the larger circle cut off by T_2.

8 Sketch on the same axes the circle
$$\mathbf{r} = (5 + 5\cos\theta)\mathbf{i} + (8 + 5\sin\theta)\mathbf{j}$$
and the line
$$\mathbf{r} = 2t\mathbf{i} + (8 - t)\mathbf{j}$$

(i) Find the position vectors of the points A and B of intersection of the circle and the line.
(ii) Find the area of the minor segment cut off by AB.
(iii) Find $A\hat{P}B$, where P is a point on the major arc AB of the circle.

9 Find the equations of the tangents of gradient 4 and the tangents of gradient -4 to the circle

$$x^2 + y^2 - 4x - 6y - 4 = 0$$

Find the areas of:

(i) the rhombus contained by these tangents;
(ii) the rectangle whose vertices are the points of contact of the tangents with the circle.

10 A is the point $(5, 4)$ and Γ is the circle whose equation is $x^2 + y^2 + 2x - 4y - 4 = 0$. Find:

(i) the ratio of the greatest to the least distance from A to Γ;
(ii) the length of the tangents from A to Γ.

These tangents touch the circle at X and Y. Find:

(iii) the length of the minor arc XY;
(iv) the area of the finite region bounded by the tangents and this arc.

11 If $x = a(t - \sin t)$, $y = a(1 - \cos t)$, a constant, show that:

(i) $dy/dx = \cot \frac{1}{2} t$
(ii) $d^2y/dx^2 = -(1/4a) \operatorname{cosec}^4 \frac{1}{2} t$

12 If $x = 3t^2$, $y = 6t$, find:

(i) dy/dx
(ii) d^2y/dx^2 at the point $(12, -12)$

13 Sketch the curve $x = t(t-1)$, $y = t - 1$. Find the area of the finite region bounded by the curve and the y-axis.

14 Find the equation of the tangent to the curve $x = 4t + 5 \cos t$, $y = 5 + 5 \sin t$, $0 \le t \le \pi$, at the point with parameter t. Show that there are two points on the curve for which $y = 8$, and find the gradients of the tangents at these points.

15 Sketch the curve $x = t^2 - 4$, $y = t(t^2 - 4)$. Find:

(i) the equations of the tangents to the curve which are parallel to the x-axis,
(ii) the area bounded by the loop of the curve.

16 The tangent and normal at the point $P(at^2, 2at)$ on the curve $y^2 = 4ax$ meet the x-axis at T and N respectively. Find:

(i) the ratio $PT:PN$
(ii) the area of $\triangle PTN$

17 Find the coordinates of the points P and Q of intersection of the parabola $x = at^2$, $y = 2at$ and the circle $x = a + 5a \cos \theta$, $y = 5a \sin \theta$ (a constant). The tangents to the parabola at P and Q meet at T and the normals to the parabola at P and Q meet at N. Show that both T and N lie on the circle.

18 Find the coordinates of the point at which the normal to the parabola $y^2 = 4ax$ at the point $(at^2, 2at)$ cuts the x-axis. Hence show that the centre of the circle which touches the parabola at the points $A(4a, 4a)$ and $B(4a, -4a)$ is the point $C(6a, 0)$.

Find the equation of the circle. If the mutual tangents to the circle and parabola at A and B meet at D, find the area of $\triangle CDA$.

19 Given that the line $x - 3y + \lambda a = 0$ is a tangent to the parabola $x = at^2$, $y = 2at$, find λ and the coordinates of the point of contact. Find the area of the finite region bounded by the tangent, the curve and the x-axis.

20 The points $P(p^2, 2p)$ and $Q(q^2, 2q)$ lie on the parabola $y^2 = 4x$.

(i) The tangents to the parabola at P and Q meet at T. Show that T is the point $(pq, p+q)$.

(ii) The normals to the parabola at P and Q meet at N. Show that N is the point $(2 + p^2 + pq + q^2, pq(p+q))$.

Show that if T lies on the line $x = -2$, N lies on the parabola $y^2 = 4x - 16$. Find the cartesian equation of the locus of N if T lies on the line $x = 1$.

21 The tangent to the parabola $y^2 = 4ax$ at the point $P(at^2, 2at)$ meets the y-axis at T. Find the coordinates of the centre C of the circle through O, P and T. Show that, as t varies, the locus of C is the parabola $y^2 = \frac{1}{2} a(x-a)$. Sketch this parabola.

22 Show that the equation of the tangent to the circle $x^2 + y^2 = a^2$ at the point $P(a \cos \theta, a \sin \theta)$ is $x \cos \theta + y \sin \theta = a$.

Q is the foot of the perpendicular from the point $A(a, 0)$ to this tangent. Find the length AQ.

T is the point at which the tangent cuts the x-axis. Find the area of $\triangle APT$.

23 Show that the equation of the chord joining the points $P(cp, c/p)$ and $Q(cq, c/q)$ on the rectangular hyperbola $xy = c^2$ is $x + pqy = c(p+q)$.

The point $R(cr, c/r)$ also lies on the hyperbola and PQ is perpendicular to PR. Show that $p^2 qr = -1$. Hence show that QR is perpendicular to the tangent at P.

24 (i) The normal to the curve $xy = 4$ at the point $P(2t, 2/t)$ meets the curve again at Q. Find the coordinates of Q.

(ii) The tangent to the curve at P meets the axes at R and S. Find the area of $\triangle ORS$.

25 Show that the equation of the tangent to the curve $xy = c^2$ at the point $(ct, c/t)$ is $x + t^2 y = 2ct$. Find the coordinates of the foot P of the perpendicular from the origin to this tangent. Show that P lies on the curve $(x^2 + y^2)^2 = 4c^2 xy$.

26 (i) The tangent to the curve $y^2 = x^3$ at the point $P(t^2, t^3)$ meets the curve again at Q. Find the coordinates of Q.

(ii) The normal to the curve at P meets the x-axis at N. Find the area of $\triangle NPQ$.

27 Sketch the curve $x = at^2$, $y = at^3$, a constant. Find the coordinates of the points of intersection of the curve with the line $3x - y = 4a$. Find the area of the finite region bounded by the curve and the line.

28 Sketch the curve $x = 2a + at^2$, $y = 2at$, a constant. Find the equations of the tangents to the curve at the points $(6a, 4a)$ and $(6a, -4a)$. Find the area of the finite region bounded by the curve and these tangents.

29 Three non-collinear points A, B and C have position vectors $\mathbf{a}$, $\mathbf{b}$ and $\mathbf{c}$ respectively; the point D divides AB internally in the ratio $2:1$.

(i) The point E divides BC internally in the ratio $1:2$. Show that DE is parallel to AC.

(ii) The point F divides BC internally in the ratio $2:1$. Show that the lines DF and AC meet at the point with position vector $\frac{1}{3}(4\mathbf{c} - \mathbf{a})$.

(iii) The point G divides CA internally in the ratio $2:1$. Show that the centroids of $\triangle ABC$ and $\triangle DFG$ coincide.

+ **30** The vertices A, B, C and D of a tetrahedron have position vectors $\mathbf{a}$, $\mathbf{b}$, $\mathbf{c}$ and $\mathbf{d}$ respectively.

(i) Show that the lines joining the midpoints of opposite edges meet at the point with position vector $\frac{1}{4}(\mathbf{a} + \mathbf{b} + \mathbf{c} + \mathbf{d})$.

(ii) Given that two pairs of opposite edges are perpendicular, show that the third pair is also perpendicular and that $AB^2 + CD^2 = AC^2 + BD^2$.

31 $\mathbf{u} = 3\mathbf{i} + 4\mathbf{j}$, $\mathbf{v} = 4\mathbf{i} + \lambda\mathbf{j}$. Find λ if:

(i) $\mathbf{u}$ is perpendicular to $\mathbf{v}$
(ii) $\mathbf{u}$ is parallel to $\mathbf{v}$
(iii) the angle between $\mathbf{u}$ and $\mathbf{v}$ is $\frac{1}{4}\pi$

32 The position vectors of the points A, B, C and D are $2\mathbf{i} + \mathbf{k}$, $\mathbf{i} - \mathbf{j} + 2\mathbf{k}$, $-3\mathbf{j} + \mathbf{k}$ and $7\mathbf{k}$ respectively. Find:

(i) the position vector of the point of intersection E of the lines AB and CD
(ii) the cosine of the angle between the lines
(iii) the area of $\triangle ACE$

33 The lines $\mathbf{r} = 4\mathbf{i} + a\mathbf{j} + 2\mathbf{k} + \mu(\mathbf{i} + 3\mathbf{j} - 3\mathbf{k})$ and $\mathbf{r} = \mathbf{i} + \mathbf{j} + \mathbf{k} + \lambda(-2\mathbf{i} - \mathbf{j} + \mathbf{k})$ intersect. Find the value of a and the position vector of their point of intersection.

Show that the vector $\mathbf{j} + \mathbf{k}$ is perpendicular to both of the lines. Hence find the vector equation of the plane containing the lines.

Show that the plane is parallel to the line

$$\mathbf{r} = \mathbf{i} + \mathbf{k} + \lambda(\mathbf{i} - 2\mathbf{j} + 2\mathbf{k})$$

Find the distance from this line to the plane.

34 Given that the lines $\mathbf{r} = 5\mathbf{i} - 4\mathbf{k} + \lambda(-\mathbf{i} + \mathbf{j} + \mathbf{k})$ and $\mathbf{r} = 2\mathbf{i} + a\mathbf{j} + \mathbf{k} + \mu(\mathbf{j} + \mathbf{k})$ intersect, find the value of a and the position vector of their point of intersection. Find:

(i) in the form $\mathbf{r} = \mathbf{a} + s\mathbf{d} + t\mathbf{e}$
(ii) in the form $\mathbf{r} \cdot \mathbf{n} = d$, a vector equation of the plane containing the lines.

35 The plane Π has vector equation $\mathbf{r} \cdot (\mathbf{i} - 2\mathbf{j} + \mathbf{k}) = 4$. Show that Π is parallel to the line

$$\Lambda_1: \mathbf{r} = 3\mathbf{i} - \mathbf{k} + \lambda(4\mathbf{i} + \mathbf{j} - 2\mathbf{k})$$

and contains the line $\Lambda_2: \mathbf{r} = 3\mathbf{i} + \mathbf{k} + \lambda(4\mathbf{i} + \mathbf{j} - 2\mathbf{k})$. Find:

(i) the distance from Λ_1 to Π
(ii) the position vector of the point of intersection of Π with the line $\Lambda_3: \mathbf{r} = \mathbf{k} + \lambda(-\mathbf{j} + \mathbf{k})$
(iii) the angle between Λ_3 and Π

36 Find a vector equation of the line passing through the points $A(0, -2, 2)$ and $B(6, 0, -2)$. Show that this line lies in the plane $\mathbf{r} \cdot (\mathbf{i} + \mathbf{j} + 2\mathbf{k}) = 2$. Find a vector equation of the perpendicular bisector of AB which lies in this plane. Show that the point $C(1, 3, -1)$ lies on this bisector and verify that $AC = BC$. Find the volume of the tetrahedron $OABC$.

37 The plane Π contains the point A with position vector $3\mathbf{i} - \mathbf{j} + 2\mathbf{k}$ and the line Λ with vector equation $\mathbf{r} = 2\mathbf{i} + \lambda(\mathbf{j} + \mathbf{k})$. Find:

(i) a vector which is normal to Π
(ii) a vector equation of Π
(iii) the shortest distance from the origin to Π
(iv) the position vector of the point of intersection of Π and the line $\mathbf{r} = \mathbf{i} + \lambda(\mathbf{i} - \mathbf{j} + \mathbf{k})$
(v) the angle between Π and the plane $\mathbf{r} \cdot (2\mathbf{i} + 5\mathbf{j}) = 7$

38 A is the point with position vector $2\mathbf{i} + 4\mathbf{j} - 4\mathbf{k}$ and Π is the plane $\mathbf{r} \cdot (\mathbf{i} - 3\mathbf{j} + 2\mathbf{k}) = 10$. Write down a vector equation of the line Λ through A which is perpendicular to Π. Find the position vector of the point of intersection of Λ and Π. Hence find:

(i) the shortest distance from A to Π
(ii) the position vector of the mirror image B of A in Π

Show that the point C with position vector $\mathbf{i} - \mathbf{j} + 3\mathbf{k}$ lies in Π. Verify that $AC = BC$. Find the area of $\triangle ABC$.

39 A, B, C, D are the points $(1, 0, 1)$, $(2, 3, -1)$, $(1, 2, 4)$, $(-3, 5, 2)$.

(i) P, Q, R, S are the midpoints of the sides AB, BC, CD, DA respectively. Find the position vectors of P, Q, R, S. Show that $PQRS$ is a parallelogram.

(ii) Find the vector equations of the lines AB and CD. Find whether these lines are parallel, intersecting or skew. Find the angle between them.

(iii) Find, in the form $\mathbf{r} = \mathbf{a} + \lambda\mathbf{u} + \mu\mathbf{v}$, the vector equation of the plane ABC. Find its cartesian equation. Hence find the normal to the plane. Hence find the vector equation of the plane in the form $\mathbf{r}\cdot\mathbf{n} = d$. What is the distance from the origin to the plane?

(iv) Find the position vector of the point of intersection of the plane ABC and the line
$$\mathbf{r} = 3\mathbf{i} + \mathbf{k} + \lambda(\mathbf{i} + 2\mathbf{j} + 3\mathbf{k})$$

+ 40 In four-dimensional space, O is the origin and $\mathbf{i}$, $\mathbf{j}$, $\mathbf{k}$ and $\mathbf{l}$ are mutually perpendicular unit vectors in the directions Ox, Oy, Oz and Ow respectively.

Assume that in a four-dimensional space, words such as 'angle' and skew have meanings analogous to their meanings in three dimensions.

(i) Find a unit vector in the direction of the vector $\mathbf{i} + \mathbf{j} + \mathbf{k} + \mathbf{l}$. What is the angle between this vector and the x-axis?

(ii) Show that the lines $\Lambda_1: \mathbf{r} = \mathbf{i} + \mathbf{j} + \mathbf{k} + \lambda(\mathbf{j} + \mathbf{k} + \mathbf{l})$ and $\Lambda_2: \mathbf{r} = \mathbf{i} + 2\mathbf{k} + 3\mathbf{l} + \mu(\mathbf{k} + \mathbf{l})$ intersect and find the position vector of their point of intersection. Find the angle between the lines.

(iii) Show that the plane $\mathbf{r} = s(\mathbf{i} + \mathbf{j}) + t(\mathbf{k} + \mathbf{l})$ and the line Λ_1 are skew. Explain why in 4-D a plane and a line can be skew. Can two planes be skew?

(iv) How many degrees of freedom has a point whose position vector $\mathbf{r}$ satisfies the equation $\mathbf{r}\cdot(\mathbf{i} + \mathbf{j} + \mathbf{k} + \mathbf{l}) = 0$? Describe the space represented by this equation. Find the distance from the point $\mathbf{i} + \mathbf{j}$ to this space. Comment on this question (be polite).

(v) Describe the advantages and disadvantages of living in a 4-D world when everyone else is living in a 3-D region of that world.

(vi) 'In four-dimensional space, no-one can hear you scream...' Discuss.

41 The complex numbers z_1, z_2 and z_3 are represented in the Argand diagram by the points A, B and C.

(i) If $ABCD$ is a parallelogram, find the complex number represented by D.

(ii) If A, B and C are collinear, show that
$$(z_3 - z_2)/(z_3 - z_1)$$
is a real number.

(iii) If $z_3 = \bar{z}_1$ and $z_3 = z_1^3$, show that $A\hat{O}B = A\hat{O}C$, where O is the origin.

42 Given that $z = 1 + i$, plot the points A, B and C representing respectively -1, z and z^3 on an Argand diagram. Show that:

(i) $\arg\{(z^3 + 1)/(z + 1)\} = \frac{1}{2}\pi$
(ii) the circle with diameter BC passes through A.

Does this circle also pass through the origin?

43 Find z if:
(i) $2/z + 1/(1 + 2i) = 1$
(ii) $z\bar{z} = 5$ and $z - \bar{z} = 4i$
(iii) $|z| = 2$ and $\arg z = -\frac{3}{4}\pi$

44 (i) Find the square roots of the complex number $3 + 4i$.

(ii) Solve the equation $z^4 - 6z^2 + 25 = 0$. Show that the points on the Argand diagram representing the roots of the equation lie at the vertices of a rectangle.

45 (i) Solve the equation $z - 2\bar{z} = 2 + 3i$.

(ii) Find the two square roots of the number $15 + 8i$. Represent these two square roots by vectors in the Argand diagram.

(iii) One root of the equation $x^3 - 3x^2 + px + q = 0$ is $1 + i$. Find the other two roots and the values of p and q.

(iv) The points A, B, C, D represent the numbers $-2 + i$, $5 + 2i$, $10 + 7i$, $3 + 6i$. Show that $ABCD$ is a rhombus. Find the numbers represented by:
(a) the midpoint of AB,
(b) the point of intersection of the diagonals of the rhombus.

46 (i) Given that $(2 + 3i)$ is a root of the equation $z^3 - 6z^2 + 21z - p = 0$, find the other two roots and the value of p.

(ii) The cube roots of the number 8 are 2, z_1 and z_2. Find z_1 and z_2 in modulus-argument form. Show that the points in the Argand diagram representing the roots lie at the vertices of an equilateral triangle and find its area.

47 Mark on an Argand diagram the points representing the numbers z_1, z_2, $z_1 + z_2$ and $z_1 - z_2$, where $z_1 = \sqrt{3} + i$ and $z_2 = 1 - \sqrt{3}i$. Find the modulus and argument of:

(i) z_1 (ii) $\bar{z}_2$ (iii) $z_1\bar{z}_2$ (iv) $z_1^3/\bar{z}_2$.

By considering the position of the point on the diagram representing $(z_1 + z_2)$, show, without using a calculator or tables, that $\tan\frac{1}{12}\pi = 2 - \sqrt{3}$.

+ 48 If z_r is the complex number with modulus 1 and argument $\frac{1}{3}r\pi$ ($r \in \mathbb{N}$), show that $(z_r)^6 = 1$.

(i) Mark on an Argand diagram the points representing the numbers z_r ($r = 0, 1, 2, 3, 4, 5$) and show that these points are the vertices of a regular hexagon.

(ii) Show that $\arg\{(z_1 - z_3)/(z_5 - z_3)\} = \frac{1}{3}\pi$.

(iii) Show that the points representing the numbers iz_r ($r = 0, 1, 2, 3, 4, 5$) are also the vertices of a regular hexagon.

49 Given that $z = 2 + 2(\cos\theta + i\sin\theta)$, show that the locus of the points representing z, as θ varies, is a circle.

Show that there are two points on this circle representing complex numbers with real part 1. Find these numbers.

Find the greatest and least values of $|z|$. If $w = z + i$, find the greatest and least values of $|w|$.

50 (i) If z and w are complex numbers, show geometrically that $|z| + |w| \geqslant |z + w|$. Hence, by choosing suitable values of z and w, show that, if z_1, z_2 and z_3 are complex numbers,
$$|z_1| + |z_2| + |z_3| \geqslant |z_1 + z_2 + z_3|$$

(ii) The complex numbers z_1, z_2, z_3 and z_4 are represented on the Argand diagram by the points A, B, C and D, and $ABCD$ is a square. Given that $z_1 = 4 + i$ and $z_3 = +4 + 3i$, find z_2 and z_4 and the area of the square. Find
$$\arg\{(z_3 - z_1)/(z_4 - z_2)\}$$
Explain your answer.

51 The complex numbers z_1, z_2, z_3 and z_4 are represented by the points A, B, C and D and $ABCD$ is a square. Given that $z_1 = 2 - 5i$, $z_3 = 3 + 2i$, find z_2 and z_4.

(i) If w is any complex number, show that the points representing the numbers wz_1, wz_2, wz_3 and wz_4 also form a square.

(ii) Show that each of z_1, z_2, z_3 and z_4 satisfies the equation $|2z - 5 + 3i| = 5\sqrt{2}$. Explain this.

+ **52** Consider the cubic equation $x^3 + 3px - 2q = 0$. Suppose that λ and μ are two numbers such that $\lambda - \mu = 2q$ and $\lambda\mu = p^3$, and let $x = \lambda^{1/3} - \mu^{1/3}$. Find x^3. Solve the simultaneous equations for λ and μ and hence show that the original cubic equation is satisfied by the solution
$$x = \{q + \sqrt{(q^2 + p^3)}\}^{1/3} - \{-q + \sqrt{(q^2 + p^3)}\}^{1/3}$$

This formula was found by the Italian mathematician Tartaglia in the sixteenth century (and published by his 'friend' Cardano who specialised in taking the credit for other people's achievements).

Use the formula to find a solution of the equation $x^3 + x - 2 = 0$. What about the other two solutions?

Consider also the equation $x^3 - 7x + 6 = 0$. What are the problems of actually using the formula on it?

+ **53** Given that $f(x) = \cos x + i \sin x$ and $g(x) = e^{ix}$ (whatever that may mean), show that $f'(x) = if(x)$ and that, assuming imaginary powers of e to behave like real ones, $g'(x) = ig(x)$. Verify that $f(0) = g(0)$ and explain why it is sensible to deduce that $f(x) = g(x)$.

Taking $f(x)$ and $g(x)$ to be equal,

(i) show that $e^{i\pi} + 1 = 0$ (Euler's equation);

(ii) show that any complex number can be expressed in the form $re^{i\theta}$; hence construct a simple proof of De Moivre's theorem;

(iii) express the complex number $\sqrt{3} + i$ in the form $re^{i\theta}$ and hence find $\ln(\sqrt{3} + i)$. Similarly find $\ln(-1)$. Is it necessary, in the light of this result, to reconsider the graph of $\ln x$ ($x \in \mathbb{R}$)?

7 Kinematics of a particle

Introduction

Mechanics is concerned with the motion of objects. There are three different kinds of motion:

(i) **Translation** or **linear motion**, i.e. motion from one place to another, while keeping the same shape and orientation. (Note that linear motion does not necessarily mean motion in a *straight* line.)

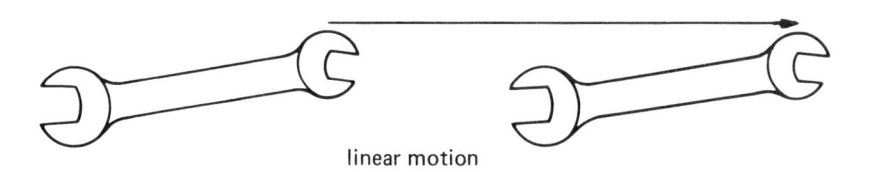

linear motion

(ii) **Rotation** or **angular motion**.

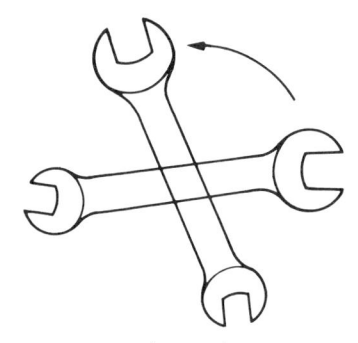

angular motion

(iii) **Change of shape** or **elastic motion**.

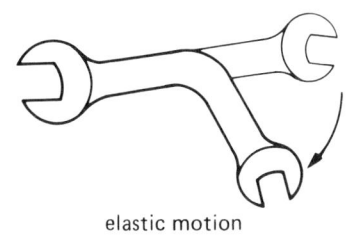

elastic motion

Most objects in the real world move in a complicated way which is a combination of all these three types. We want to be able to analyse this motion mathematically. Since it is so complicated, we need to make some simplifications before we can start. We first simplify the objects we are going to consider.

Rigid body: If an object is not very elastic, we can think of it as being a **rigid body**. A rigid body is an object which never changes its shape.

There is no such thing as a rigid body in the real world; every object in the real world is to some extent elastic, even a block of concrete. A rigid body is a mathematical simplification, or **mathematical model**, of a fairly inelastic object.

A rigid body is capable of only two different kinds of motion: translation and rotation.

Particle: If an object is not very elastic and not very big, compared to the distances being considered, we can think of it as being a **particle**. For example, if we consider a football in flight across a pitch we can sensibly think of the football as a particle, but if we consider the motion of a spider crawling over the football, then we need to think of the football as an object of finite size. A particle is an infinitely small object, a point body.

A particle is capable of only one kind of motion: translation.

7.1 Motion of a particle in a straight line

7.1:1 Displacement, velocity and acceleration

Consider a particle P moving in a straight line (fig. 7.1:1). We can describe the position of P as follows:

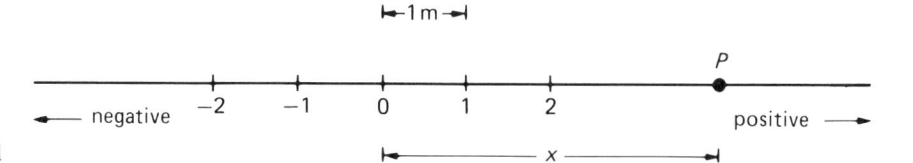

Fig. 7.1:1

Choose a fixed point O on the line, number the line as we did the real line (so that one direction is positive, the other negative). Then if P is at the point representing the number x, x is called the **displacement of P from O**. The unit of displacement is the metre (m).

The displacement of P can be positive or negative. Conventionally, we take left to right as positive. Note that the line may be vertical—or indeed in any other direction. The **distance** OP from O to P is the magnitude of the displacement:

$$OP = |x|$$

Suppose that t s after the particle P starts to move, its displacement is x m. Since P is moving, x is a function of t.

Definition The **velocity** v of P after t s is the rate of change of its displacement, i.e.

$$v = \frac{dx}{dt} \qquad (1)$$

Units of velocity: m s^{-1}.

Definition The **acceleration** a of P after t s is the rate of change of its velocity, i.e.

$$a = \frac{dv}{dt} = \frac{d^2x}{dt^2} \qquad (2)$$

Units of acceleration: m s^{-2}.

The quantity x describes where P is, v describes how it is moving, and a describes how its motion is varying. x, v and a are all functions of t.

$$\text{If } x = f(t), \text{ then } v = f'(t) \text{ and } a = f''(t) \qquad (3)$$

v is the gradient function of the x-t graph.
a is the gradient function of the v-t graph.

Definition The **speed** of P is the magnitude $|v|$ of its velocity.

Worked example 1 P moves so that $x = 6t^2 - t^3$. Find v and a. Sketch graphs of x, v and a against t ($t \geq 0$).

$$x = 6t^2 - t^3$$

$$\Rightarrow v = \frac{dx}{dt} = 12t - 3t^2$$

$$\Rightarrow a = \frac{dv}{dt} = 12 - 6t$$

The graphs are shown in fig. 7.1:**2**.

Worked example 2 P moves so that $x = 6t^2 - t^3$ (see worked example 1).
 (i) Show that P is initially at O and returns to O after 6 seconds.
 (ii) Find the maximum displacement of P from O in the time interval $0 \leq t \leq 6$.
 (iii) Sketch a graph of the speed of P against t. Find the maximum speed of P in the time interval $0 \leq t \leq 6$.

 (i) P is at $O \Rightarrow x = 0 \Rightarrow 6t^2 - t^3 = 0 \Rightarrow t = 0$ or 6.
 Thus P is initially at O; it returns to O after 6 s.
 (ii) x maximum $\Rightarrow dx/dt = 0$ (i.e. $v = 0$) $\Rightarrow 12t - 3t^2 = 0 \Rightarrow t = 0$ or 4.
 When $t = 4$, $x = 32$, i.e. the maximum displacement of P from O in the time interval $0 \leq t \leq 6$ is 32 m.

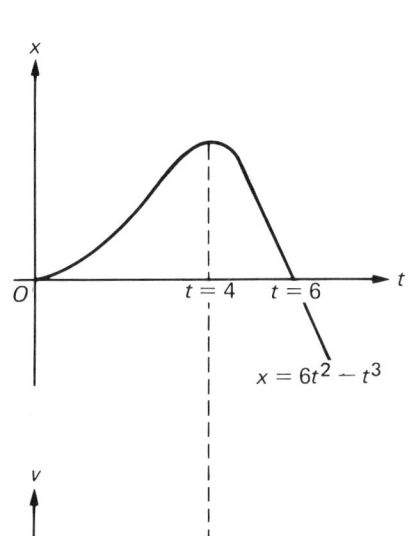

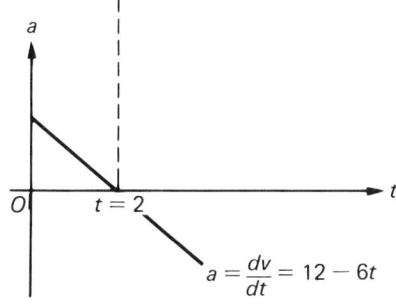

Fig. 7.1:**2**

370 *Displacement, velocity and acceleration*

(iii) The speed of $P = |v| = |12t - 3t^2|$. The graph of $|v|$ against t is shown in fig. 7.1:3. When $t = 6$,

$$|v| = |12 \cdot 6 - 3 \cdot 6^2| = 36$$

We can see from the graph that 36 m s^{-1} is in fact the maximum speed in the time interval $0 \leqslant t \leqslant 6$.

Note that when $t = 2$, i.e. when $dv/dt = 0$, the speed is only 12 m s^{-1}. The stationary point $(2, 12)$ is only a *local* maximum of the graph.

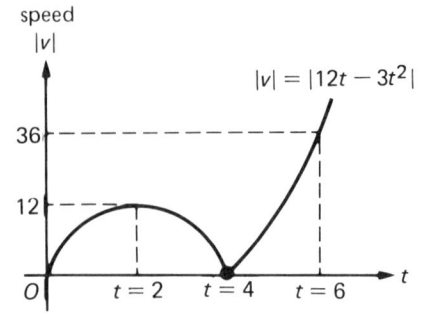

Fig. 7.1:3

Exercise 7.1:1

When in doubt, assume that the question is about a particle P moving in a straight line, whose displacement, velocity and acceleration at time t are x, v and a respectively.

Questions 1-5: Find v and a, and sketch graphs of x, v and a against t ($t \geqslant 0$) when P moves according to the equation:

① **1** $x = 4t - t^2$

① → **2** $x = t^3 - 12t^2$

① **3** $x = 5 \sin 2t$

① **4** $x = e^{-t} - 4e^{-2t}$

① **5** $x = 20 \ln (1 + t)$

② • **6** $x = 5t + 3$
  (i) Find x when $t = 0, 1, 2, 3, 4$.
  (ii) Sketch a graph of x against t.
  (iii) Show that the velocity of P is constant. What is its acceleration?
  (iv) Find x when $t = T$, $t = T + 1$, $t = T + \alpha$. How far does P move in the time intervals:
  (a) $T \leqslant t \leqslant T + 1$; (b) $T \leqslant t \leqslant T + \alpha$?

② → **7** A car moves along a straight road. At time t s, its displacement x m from a set O of traffic lights is given by $x = 60 - 20t$.
  (i) Sketch a graph of x against t.
  (ii) What is the velocity of the car? What is the speed of the car?
  (iii) How far from O is the car initially (i.e. when $t = 0$)?
  (iv) How long does the car take to reach O?

② **8** $x = pt + q$ (p, q constants).
  (i) Show that the velocity of P is constant. What is its acceleration?
  (ii) How far from O is P initially?
  (iii) How could you tell whether P moved initially towards or away from O?

② **9** Lionel the performing dog is attempting to become the first dog to swim across the English Channel. He swims with constant speed in a straight line from Dover to Calais, a distance of 35 km. At noon he is 10 km from Calais; at 6 p.m. he is 7 km from Calais. Find his speed:
  (i) in km h^{-1}
  (ii) in m s^{-1}

Find how long he takes for the whole crossing.

② **10** $x = 8t - t^2$.
  (i) Sketch a graph of x against t. Find x when $t = 0, 1, 2, 3, 4, 5, 6$. Find the times at which P is at O. For what values of t is P:
  (a) to the right of O; (b) to the left of O?
  (ii) Find v. Sketch a graph of v against t. Find the time at which P is (instantaneously) at rest, i.e. its velocity is zero. What can you say about the displacement of P at this time? Find the velocity when $t = 5$. What is the significance of the fact that it is negative? For what values of t is P moving:
  (a) from left to right; (b) from right to left?
  Find the speed of P when $t = 1, 3, 5, 7$.
  (iii) Show that a is constant. What is the significance of the fact that a is negative?

② **11** $x = 3t^2 - 24t + 36$.
  (i) At what times is P at O?
  (ii) What is the maximum distance of P from O:
  (a) in the time interval $0 \leqslant t \leqslant 6$; (b) in the time interval $2 \leqslant t \leqslant 6$?
  (iii) Sketch a graph of the speed of P against t. What is its initial speed?
  (iv) Show that the acceleration of P is constant.

Kinematics of a particle

② 12 A stone is dropped from a point O at the top of a cliff of height 80 m. The depth x m of the stone below O after t s is given by $x = 5t^2$.

(i) How long does the stone take to reach the bottom of the cliff?
(ii) Find the distance travelled by the stone in the 1st, 2nd, 3rd and 4th seconds of motion.
(iii) Find the velocity of the stone just before it reaches the ground.
(iv) Show that the acceleration of the stone is constant.

② 13 A stone is thrown vertically upwards from a point O. The height x m of the stone above O after t s is given by $x = 20t - 5t^2$.

(i) What is the greatest height above O reached by the stone? How long does it take to reach this height?
(ii) What is the initial speed of the stone? What is its speed on returning to O?
(iii) For how long is the stone more than 15 m above O?

② 14 $x = pt^2 + qt$, p, q constant.

(i) What is the initial displacement of P?
(ii) Find v. What is the initial velocity of P?
(iii) Show that a is constant.
(iv) By eliminating t from your equations for x and v, show that $v^2 = q^2 + 4px$.

② 15 A man runs from his house O along a straight road to a post office A, realises he has forgotten to bring the letter he meant to post, so turns round and runs back to O again. During the journey his distance x m from O is given by

$$x = \tfrac{1}{80}(30t^2 - t^3)$$

when he has been moving for t s.

(i) Find the distance from O to A.
(ii) Find the total time taken for the journey.
(iii) Find his maximum speed during the journey.

② 16 $x = t^3 - 6t^2 + 9t + 4$.

(i) Find v and a.
(ii) Find the times at which P is at rest. Find the distance of P from O at these times.
(iii) Find the time at which the velocity of P is a minimum.
(iv) Find the time at which P returns to its starting point.
(v) Sketch a graph of the speed of P against t.

② 17 A piston moves back and forth in a cylinder. At time t s its displacement x m from a fixed position O is given by $x = 12 \cos 3t$. Find:

(i) the initial position of the piston
(ii) the maximum displacement of the piston from O
(iii) the velocity of the piston at time t
(iv) its initial velocity
(v) its maximum velocity

② 18 $x = 5 \sin 2t$. Find:

(i) the acceleration of P at time t
(ii) an expression for the acceleration in terms of x
(iii) the maximum acceleration of P

② 19 $x = R \sin(nt + \varepsilon)$, R, n and ε constant. Find

(i) the initial position of P
(ii) the velocity of P in terms of x
(iii) the maximum velocity of P
(iv) the acceleration of P in terms of x

② 20 $x = 6 + 3 \sin t + 4 \cos t$. Find:

(i) the velocity and acceleration of P at time t
(ii) the time at which P first comes to rest
(iii) its acceleration and its distance from O at this instant

② → 21 $x = \sin 2t + 2 \cos t$. Find:

(i) the velocity and acceleration of P at time t
(ii) the time at which P first comes to rest
(iii) its acceleration and its distance from O at this instant

② R 22 $x = \lambda(1 + \cos^2 t)$, λ constant. Find:

(i) an expression for the acceleration of P in terms of x
(ii) the distance from O of P when its velocity is:
(a) zero; (b) maximum

② 23 A ball-bearing is dropped into a vat of treacle. At t s after it hits the surface of the treacle, its displacement x m from the surface is given by

$$x = 0.02(4t + \tfrac{1}{4} e^{-2t})$$

(i) Find v in terms of t. Describe the behaviour of v as $t \to \infty$. Sketch the graph of v against t.

(ii) Find a in terms of t. Hence show that $a = 0.16 - 2v$. Describe the behaviour of a as $t \to \infty$. Sketch the graph of a against t.

② + 24 A failed actor, weary of life, hangs himself by an elastic string from a point O on the ceiling of his dark and dismal bedsit. As he hangs motionless, a chunk of plaster falls from the ceiling, lands on his head, and sets him moving in a vertical plane. t s later, his displacement x m from O is given by

$$x = 2 + e^{-2t} \sin 4t$$

(i) Sketch a graph of x against t. Find the time at which the man first returns to his initial position.
(ii) Show that $a = 40 - 20x - 4v$.
(iii) Describe the motion of the man.

② : **25** Describe what is happening to P when
  (i) $x = 0$
  (ii) $x > 0$
  (iii) $x < 0$
  (iv) x is increasing—what can you say about v?
  (v) x is decreasing—what can you say about v?
  (vi) $v = 0$—what can you say about x?
  (vii) v is increasing—what can you say about a?
  (viii) v is decreasing—what can you say about a?
  (ix) $a = 0$—what can you say about v?

② : **26** Discuss each of the statements (prove it or give a counter-example).
  (i) $\dfrac{d}{dt}$ (distance) = speed.
  (ii) P is decelerating if v is decreasing.
  (iii) The velocity of P is maximum $\Rightarrow$ its acceleration is zero.

7.1:2 Use of integration

$$a = \frac{dv}{dt} \Rightarrow$$
$$v = \int a\,dt \quad (4)$$

$$v = \frac{dx}{dt} \Rightarrow$$
$$x = \int v\,dt \quad (5)$$

Worked example 3 P moves so that $a = 6t$. When $t = 1$, $v = 0$ and $x = 4$. Find v and x at time t.

By equation (4), $\quad v = \int 6t\,dt = 3t^2 + c$

$v = 0$ when $t = 1 \Rightarrow c = -3$; so $v = 3t^2 - 3$.

By equation (5), $\quad x = \int (3t^2 - 3)\,dt = t^3 - 3t + d$

$x = 4$ when $t = 1 \Rightarrow 4 = 1 - 3 + d \Rightarrow d = 6$; so $x = t^3 - 3t + 6$.

Worked example 4 A particle moves in a straight line with velocity $3t^2 - 6t - 9$ after t s. Find:

(i) its change in displacement between $t = 1$ and $t = 4$;
(ii) the distance it moves in that time.

(i) change in displacement $= \int_1^4 v\,dt$

$\qquad = \int_1^4 (3t^2 - 6t - 9)\,dt$

$\qquad = [t^3 - 3t^2 - 9t]_1^4$

$\qquad = -9$ (See fig. 7.1:4.)

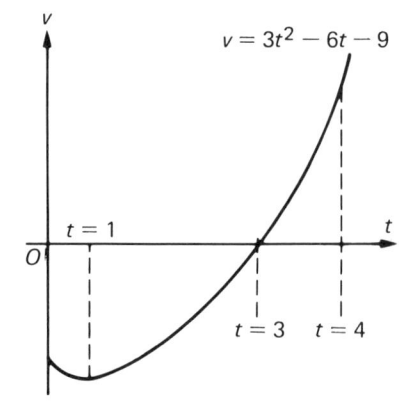

Fig. 7.1:4

(ii) $v = 3t^2 - 6t - 9 = 3(t^2 - 2t - 3) = 3(t+1)(t-3)$. When $t = 3$, $v = 0$, i.e. the particle is (instantaneously) at rest.
From $t = 1$ to $t = 3$, $v < 0$, i.e. the particle is moving from right to left.
From $t = 3$ to $t = 4$, $v > 0$, i.e. the particle is moving from left to right.

$$\text{distance moved} = \left| \int_1^3 v\, dt \right| + \left| \int_3^4 v\, dt \right|$$
$$= |-16| + |7|$$
$$= 23$$

Note:

change in displacement = *algebraic* area under v-t curve (6)

distance moved = *numerical* area under v-t curve (7)

Exercise 7.1:2

Assume where necessary that the question is about a particle P moving in a straight line, whose displacement, velocity and acceleration at time t are x, v and a respectively.

③ **1** Initially (i.e. when $t = 0$), $x = 3$, $v = 4$. Find v and x if:

(i) $a = 12$ (ii) $a = 12t$ (iii) $a = 12t^2$ (iv) $a = 12\sqrt{t}$

Questions 2–10: Find v and x at time t for the given conditions.

③ **2** $a = \lambda t$, λ constant. When $t = 0$, $v = 0$ and $x = 0$.

③ → **3** $a = 4/t^3$. When $t = 1$, $v = 3$ and $x = 4$.

③ **4** $a = 4 - 12t$. When $t = 0$, $v = 3$ and $x = 0$.

③ → **5** $a = -36 \cos 3t$. When $t = 0$, $v = 0$ and $x = 4$.

③ **6** $a = -2/t^2$. When $t = 1$, $v = 2$ and $x = 0$.

③ **7** $a = 12\, e^{-3t}$. When $t = 0$, $v = 0$ and $x = 0$.

③ **8** $a = -\lambda\omega^2 \sin \omega t$, λ, ω constant. When $t = 0$, $v = \lambda\omega$ and $x = \lambda$.

③ **9** $a = \lambda/(t+2)^2$, λ constant. When $t = 0$, $v = \lambda$ and $x = 0$.

③ R **10** $a = \lambda\omega^2\, e^{-\omega t}$, λ, ω constant. When $t = 0$, $v = -\lambda\omega$ and $x = 2\lambda$.

④ **11** $v = 3t^2 - 12$. Find the time at which $v = 0$. Sketch a graph of v against t and a graph of the speed of P against t.
Find the change in displacement of P and the distance moved by P between $t = 0$ and $t = 4$.

④ **12** $v = 3t^2 - 18t + 15$. Find the times at which $v = 0$. Sketch a graph of v against t and a graph of the speed of P against t.
Find the change in displacement of P and the distance moved by P between:

(i) $t = 0$ and $t = 1$
(ii) $t = 0$ and $t = 2$
(iii) $t = 2$ and $t = 5$
(iv) $t = 2$ and $t = 6$

④ **13** $a = 8$. Initially $x = -40$, $v = -12$.

(i) Find v.
(ii) Sketch graphs of v and of the speed of P against t. Find the time at which $v = 0$.
(iii) Find the distance covered by P between $t = 0$ and $t = 4$.
(iv) Find the time at which P is at the origin.

④ **14** $a = 6t$. Initially $x = 0$, $v = -12$.

(i) Find v.
(ii) At what time does P come (instantaneously) to rest for the first time? Find its distance from O at this time.
(iii) Find the time at which P is moving with velocity $15\, \text{m s}^{-1}$. Find its distance from O at this time.
(iv) Find the time at which P is moving through O again. What is its velocity at this time?

④ **15** $a = 2t - 6$. Initially $v = 8$. Find the times at which P is at rest and its distances from its initial position at these times.

16 $v = t(t-2)^2$. Find:
  (i) the times at which $a = 0$
  (ii) the distance covered by P between $t = 1$ and $t = 2$

17 $v = 1/t^2$ ($t > 0$). Find the distance covered by P between $t = 2$ and $t = 5$.

18 $v = 1/t^2$ ($t > 0$). When $t = \frac{1}{2}$, $x = 0$. Show that as $t \to \infty$, P approaches a certain point. Sketch a graph of x against t ($t > 0$). Why is the restriction $t > 0$ so important?

19 $v \propto t^2$. When $t = 3$, $a = 3.6$. Find the distance moved by P in the first three seconds of its motion.

20 $v \propto t^{-1/2}$. When $t = 4$, $a = -\frac{1}{2}$. Find the distance moved by P between $t = 1$ and $t = 4$.

21 $a \propto \sqrt{t}$. P starts from rest; when $t = 4$ its velocity is 32.
  (i) Find its velocity when $t = 2$.
  (ii) Find the distance P moves in the first four seconds of motion.

R 22 $a \propto 1/t^3$ ($t > 0$). When $t = 1$, $v = 2$; as $t \to \infty$, $v \to 7$.
  (i) Find v in terms of t.
  (ii) Find the distance travelled by P between $t = 1$ and $t = 2$.

23 A rocket is launched. After t minutes its acceleration a km min^{-2} is given by
$$a = \begin{cases} t^2 - \frac{1}{12}t^3 & 0 \leq t \leq 12 \\ 0 & t > 12 \end{cases}$$
Find its speed after:
  (i) 6 min
  (ii) 12 min
  (iii) 18 min
Find the distance the rocket travels before it ceases to accelerate.

24 $a = 36 \cos 3t$. When $t = 0$, $x = 0$ and $v = 12$.
  (i) Find x in terms of t.
  (ii) How far from the origin is P when it first comes to rest?

25 $a = \cos 2t - \cos t$. P starts from rest. Find the distance it covers between:
  (i) $t = 0$ and $t = 2\pi$
  (ii) $t = 0$ and $t = 3\pi$

26 A particle is moving in a straight line. At a certain instant it is 2 m from a point O on the line, and is moving away from O with speed 2 m s^{-1}. t seconds later its acceleration is $(8 \cos 2t + 4 \sin 2t)$ m s^{-2} towards O.

  (i) What is the relationship between its acceleration and its displacement from O after time t?
  (ii) What is the maximum distance of the particle from O?
  (iii) How long is it before the particle is at O for:
    (a) the first time; (b) the second time; (c) the nth time?
  (iv) What is the maximum speed of the particle?

27 $a = 1 - (2/t^2)$. When $t = 1$, $v = 3$.
  (i) Find the velocity of P at time t. Sketch a velocity-time graph. Find the minimum velocity of P.
  (ii) Find the distance travelled by P between $t = 2$ and $t = 3$.

28 $a = 1/(t+1)^2$. When $t = 0$, $v = 2$. Find the distance travelled by P in the first three seconds of motion.

29 A man is desperately pushing his way through a crowd at a rock concert. He starts from rest. t seconds later his acceleration $a = 6 e^{-2t}$ m s^{-2}.
  (i) Find his velocity v m s^{-1} after t seconds. Show that as $t \to \infty$, $v \to 3$. Sketch a graph of v against t.
  (ii) Show that $a = 6 - 2v$. How do you interpret this equation in terms of the difficulty of pushing your way through a crowd?

30 A car starts from rest at a petrol station O on a straight motorway and moves with an acceleration which is proportional to $e^{-0.1t}$ m s^{-1}, where t is the time for which the car has been travelling. If its limiting speed for large values of t is 50 m s^{-1}, find the time at which its speed is 25 m s^{-1}, and the distance it has then travelled. What is its average speed as it accelerates from 0 m s^{-1} to 25 m s^{-1}?

R 31 $v = e^{2t} - 5e^t + 2t$.
  (i) Find the acceleration of P at time t. When is its acceleration zero? What can you say about its velocity at this instant? What is its velocity at this instant?
  (ii) Find the distance covered by P between $t = 0$ and $t = \ln 2$.

32 The velocity of a moving car t seconds after it starts is v m s^{-1}. The values of v and t are shown in the table.

(i)
t	0	1	2	3	4	5	6	7	8
v	0	1.25	2.50	3.73	4.95	6.15	7.33	8.47	9.58

Using (a) the trapezium rule, (b) Simpson's rule (see section 4.3), estimate the distance covered by the car in the first 8 seconds of motion.

(ii)
t	0	5	10	15	20	25	30	35	40
v	0	1.20	1.44	1.73	2.07	2.49	2.99	3.58	4.30

Using Simpson's rule, estimate the distance covered by the car: (a) in the first 20 seconds of motion; (b) in the first 40 seconds of motion.

33 A particle moves from rest in a straight line. Its acceleration t seconds after it starts is a m s^{-2}. The values of a and t are shown in the table.

t	0	1	2	3	4	5	6	7	8
a	0	0.120	0.364	0.696	1.103	1.576	2.110	2.700	3.343

Using the trapezium rule, estimate:

(i) the velocity of the particle after 2 s, 4 s, 6 s, 8 s
(ii) the distance moved by the particle in 8 seconds

* **34** A particle P starts from rest at O and moves in a straight line with acceleration a m s^{-2} where

$$a = \begin{cases} 6t & 0 \leq t \leq 2 \\ -4 & t \geq 2 \end{cases}$$

(i) Find the velocity and displacement of P at time t ($t < 2$).

(ii) Show that at time $t = 2$ its velocity is 12 m s^{-1} and its displacement is 8 m.

(iii) Using these results as initial conditions for the second part of the motion, find the velocity and displacement of P at time t ($t \geq 2$). Hence find:
(a) the time at which P first comes to rest;
(b) the time at which P returns to O.

(iv) Sketch a graph of v against t in the interval $0 \leq t \leq 6$.

→ **35** A particle starts from O with velocity 2 m s^{-1} and moves in a straight line with acceleration a m s^{-2}, where

$$a = \begin{cases} 24t^2 & 0 \leq t < 1 \\ 12t & t \geq 1 \end{cases}$$

Find the velocity and displacement of the particle when $t = 3$.

7.1:3 Constant acceleration

Suppose that a particle starts from O with initial velocity u and moves with *constant* acceleration a.

By equation (4), $v = \int a \, dt = at + c$

$v = u$ when $t = 0 \Rightarrow c = u$; so

$$v = u + at \qquad (8)$$

By equation (5), $x = \int (u + at) \, dt = ut + \tfrac{1}{2}at^2 + d$

$x = 0$ when $t = 0 \Rightarrow d = 0$; so

$$x = ut + \tfrac{1}{2}at^2 \qquad (9)$$

From equations (8) and (9),

$$v^2 = u^2 + 2ax \qquad (10)$$
$$x = \tfrac{1}{2}(u + v)t \qquad (11)$$
$$x = vt - \tfrac{1}{2}at^2 \qquad (12)$$

Worked example 5 A particle P starts from O with velocity 12 m s^{-1} and moves with acceleration -3 m s^{-2}. Find
(i) the times at which it passes through the point A whose displacement from O is 18 m;
(ii) its speed as it passes through A.

$u = 12 \text{ m s}^{-1}, a = -3 \text{ m s}^{-2}, x = 18 \text{ m}$.

(i) $x = ut + \frac{1}{2}at^2 \Rightarrow 18 = 12t - \frac{3}{2}t^2 \Rightarrow t^2 - 8t + 12 = 0 \Rightarrow t = 2 \text{ or } 6$.
So P passes through A after 2 s and after 6 s.

(ii) $v^2 = u^2 + 2ax \Rightarrow v^2 = (12)^2 - 2 \cdot 3 \cdot 18 \Rightarrow v^2 = 36 \Rightarrow v = \pm 6$.
So the speed of P as it passes through A is 6 m s^{-1}.

Note that P passes through A for the first time after 2 s with velocity $+6 \text{ m s}^{-1}$ and for the second time after 6 s with velocity -6 m s^{-1}.

Worked example 6 A train stops at two stations A and B which are 2000 m apart. It accelerates uniformly from A at 1 m s^{-2} for 15 s and maintains a constant speed for a time T s before decelerating uniformly at $\frac{1}{2} \text{ m s}^{-2}$ to rest at B. Find T.

The acceleration of the train is not constant throughout the motion; however, we can divide the motion into three parts in each of which the acceleration *is* constant: 1 m s^{-2}, 0 m s^{-2}, $-\frac{1}{2} \text{ m s}^{-2}$ (fig. 7.1:5).

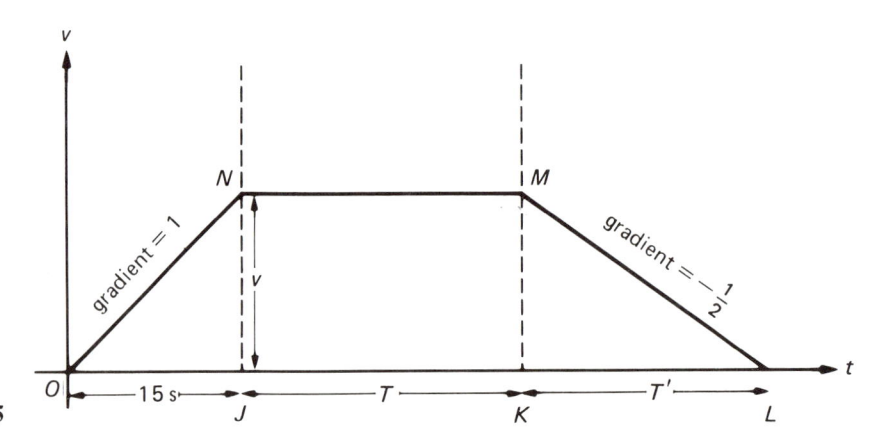

Fig. 7.1:5

Let the constant speed be $v \text{ m s}^{-1}$. Let the time for which the train is decelerating be T' s, as shown in fig. 7.1:5 (not to scale).
In $\triangle OJN$, $v/15 = 1 \Rightarrow v = 15$.
In $\triangle LKM$, $-v/T' = -\frac{1}{2} \Rightarrow T' = 30$.

Total distance travelled = (numerical) area under v-t graph
= area of trapezium $OLMN$ (by equation (8))

i.e. $2000 = \frac{1}{2}\{(15 + T + T') + T\}v$
$= \frac{1}{2}\{(15 + T + 30) + T\}15$
$= \frac{15}{2}(2T + 45)$

So $T \approx 111$

Exercise 7.1:3

Assume where necessary that the question is about a particle P moving in a straight line, whose displacement, velocity and acceleration at time t are x, v, and a respectively.

→ • **1** A particle starts from O with initial velocity u and constant acceleration a.

(i) By eliminating t from the equations $v = u + at$ (equation (8)) and $x = ut + \frac{1}{2}at^2$ (equation (9)), show that $v^2 = u^2 + 2ax$ (equation (10)). Similarly, show that $x = \frac{1}{2}(u+v)t$ (equation (11)) and $x = vt - \frac{1}{2}at^2$ (equation (12)).

(ii) Show that

$$a = \frac{\text{increase in velocity}}{\text{increase in time}}$$

$$\left(\text{c.f. constant velocity} = \frac{\text{increase in displacement}}{\text{increase in time}}\right)$$

(iii) Sketch rough graphs of v against t and x against t (assume u positive, a positive). Explain equation (11) in terms of the area under the v-t graph.

(iv) Explain the significance of equations (8)–(12) in the case $a = 0$.

Questions 2-4: A particle P starts with velocity u and moves in a straight line with constant acceleration a.

⑤ **2** $u = -10$, $a = 2$.

(i) At what time is P at rest?
(ii) At what time does P return to its starting point? At what speed is it then moving?

⑤ **3** $u = 6$, $a = -4$. Find the distance of P from its starting point when

(i) it is moving with speed 2
(ii) it first comes to instantaneous rest

⑤ → **4** $u = V$, $a = f$ ($V > 0$).

(i) If f is negative, show that P comes to instantaneous rest after time $V/|f|$ and that it is then at a distance $\frac{1}{2}V^2/f$ from its starting point.
Show that P returns through its starting point with velocity $-V$.

(ii) If f is positive, show that the velocity of P is always positive. Find the speed of P when its displacement from its starting point is:
(a) V^2/f, (b) $3V^2/2f$, (c) $2V^2/f$.

⑥ • **5** A car moves along a straight road with constant acceleration. At a certain instant it is moving with speed $4\,\text{m s}^{-1}$. Five seconds later it is moving in the same direction with speed $24\,\text{m s}^{-1}$. Find the distance it moves during this time.

⑤ **6** A particle, projected up the line of greatest slope of an inclined plane, moves with constant acceleration. At a certain instant it is moving up the plane with speed $5\,\text{m s}^{-1}$. Four seconds later it is moving down the plane with speed $7\,\text{m s}^{-1}$. Find the distance it moves during this time.

⑤ → **7** A particle moves in a straight line with constant acceleration. It moves a distance $8\,\text{m}$ in the third second of motion and $10\,\text{m}$ in the fourth second of motion. Find its acceleration. Hence find the initial velocity and the distance moved in the first four seconds of motion.

⑤ **8** A particle moves along Ox with constant deceleration. It passes through the points O, $A(1, 0)$ and $B(2, 0)$ at times $t = T$, $2T$ and $4T$ respectively. Find the coordinates of:

(i) the initial position of the particle;
(ii) the point at which it comes to rest.

⑤ **9** A particle moves in a straight line with constant deceleration. It travels $20\,\text{m}$ in the first $15\,\text{s}$ of motion, and another $20\,\text{m}$ in the next $30\,\text{s}$. Find its initial velocity.

⑤ **10** A particle moves in a straight line with constant acceleration. Find its acceleration:

(i) if it covers distances of $2\,\text{m}$ and $3\,\text{m}$ in successive seconds;
(ii) if it takes $2\,\text{s}$ and $3\,\text{s}$ to cover two successive distances of $1\,\text{m}$.

⑥ • **11** Look at worked example 6 again.

(i) Consider the first part of the motion (acceleration). Use the appropriate constant acceleration formulae to find the final velocity and the distance travelled.

(ii) Consider the second part of the motion (constant speed). Find the distance travelled, in terms of T.

(iii) Consider the third part of the motion (deceleration). Find the distance travelled (it should not be necessary to find the time taken).

Hence find T.
Do you prefer this analytical method or the graphical method given in the text?

⑥ • **12** A particle moves in a straight line ABC from rest at A to rest at C. It accelerates uniformly at $2\,\text{m s}^{-2}$ from A to B, where $AB = 9\,\text{m}$, and then decelerates uniformly for $2\,\text{s}$ from B to C. Find:

(i) its speed as it passes through B;
(ii) the distance AC.

⑥ 13 A particle moves in a straight line *ABC* from rest at *A* to rest at *C*. It accelerates uniformly to a speed of 5 m s^{-1} at *B* and then decelerates uniformly. The total time taken for the journey is 16 s. Find the distance *AC*.

⑥ → 14 A particle moves in a straight line *ABC* from rest at *A* to rest at *C*. It accelerates uniformly to a speed of 24 m s^{-1} at *B* and then decelerates uniformly at 3 m s^{-2}. If *AC* = 168 m, find the time for which the particle is:

(i) accelerating;
(ii) decelerating.

⑥ → 15 A particle starts from rest and moves in a straight line. It accelerates uniformly at 2 m s^{-2} for 10 s, moves with constant speed for 12 s and then decelerates uniformly to rest in 4 s. Find:

(i) the total distance travelled by the particle;
(ii) its deceleration.

⑥ 16 A train starts from rest at a station *A* and travels along a section of straight track to rest at a station *B*. It accelerates uniformly at 1 m s^{-2} for 30 s, next moves with constant speed, and then decelerates uniformly at 0.6 m s^{-2}. If *AB* = 6 km, find the time for which the train is travelling with constant speed.

⑥ 17 A scenic elevator travels from rest at the ground floor of a building to rest at the top floor, 50 m above, taking 18 s. It accelerates uniformly for 10 m, travels with constant speed for 32 m and then decelerates. Find:

(i) the constant speed;
(ii) the acceleration and the deceleration.

⑥ 18 A particle starts from rest at *A* and moves in a straight line to rest at *B*. *AB* = 800 m.

(i) If it accelerates uniformly at 2 m s^{-2} for $\frac{1}{5}$ of the time, moves with constant speed for $\frac{3}{5}$ of the time and then decelerates uniformly, find the constant speed and the total time taken.

(ii) If it accelerates uniformly at 2 m s^{-2} for $\frac{1}{5}$ of the distance, moves with constant speed for $\frac{3}{5}$ of the distance and then decelerates uniformly, find the constant speed and the total time taken.

⑥ R 19 A particle moves in a straight line from rest at *A* to rest at *B*. It accelerates uniformly at *f* m s^{-2}, moves with constant speed *V* m s^{-1} and then decelerates uniformly at $\frac{1}{2}f$ m s^{-2}. If the total time for the journey is *T* s and the distance *AB* is $\frac{5}{32}fT^2$ m, show that $V = \frac{5}{32}fT$ m s^{-1}.

+⑥ 20 A particle travels a distance *s* m in a straight line from rest at *A* to rest at *B*. Its maximum acceleration is *f* m s^{-2}, its maximum deceleration is *f'* m s^{-2} and its maximum velocity *V* m s^{-1}. Find the shortest time for the journey if:

(i) *s* is not large enough for the maximum velocity to be reached;
(ii) *s* is large enough for the maximum velocity to be reached.

21 A particle starts from rest and moves in a straight line. Its acceleration *a* m s^{-2} is given by

$$a = \begin{cases} 4 & 0 \leq t \leq 2 \\ 1 & 2 < t < 6 \\ -3 & 6 \leq t \leq 10 \end{cases}$$

(i) Find the speed of the particle when $t = 2$, $t = 6$, $t = 10$.
(ii) Find the distance travelled by the particle in the interval $0 \leq t \leq 10$.

22 A particle starts from rest and moves in a straight line. Its acceleration *a* m s^{-2} is given by

$$a = \begin{cases} f & 0 \leq t \leq T \\ \frac{1}{3}f & T < t < 4T \\ -\frac{1}{2}f & 4T \leq t \leq 6T \end{cases}$$

Find the distance travelled by the particle in the intervals:

(i) $0 \leq t \leq 6T$
(ii) $0 \leq t \leq 3T$

7.1:4 Free vertical motion under gravity

Galileo found experimentally that, if air resistance is ignored, any freely falling object, whatever its mass, has the same constant acceleration towards the centre of the Earth. This acceleration is only approximately constant—it increases slightly as the object gets nearer the centre of the Earth. (Thus it also varies from one part of the Earth's surface to another since the Earth is not precisely spherical. See exercise 8.1:3, questions 3-6.

We often refer to this acceleration as *g*. $g \approx 9.81$ m s^{-2}; in numerical examples we usually take $g = 10$ m s^{-2} to make the calculations easier.

Kinematics of a particle 379

Worked example 7 A ball is projected vertically upwards from ground level with a speed of 40 m s^{-1}. Find:

(i) the greatest height reached by the ball;
(ii) the time for which it is more than 60 m above the ground.

Take upwards as positive; then the acceleration of the ball is -10 m s^{-2}. $u = 40$ m s^{-1}, $a = -10$ m s^{-2}.

(i) When the ball reaches its greatest height its velocity is instantaneously zero, i.e. $v = 0$.
$v^2 = u^2 + 2ax \Rightarrow 0 = (40)^2 + 2(-10)x \Rightarrow x = 80$.
So the greatest height of the ball is 80 m.

(ii) When the height of the ball above the ground is 60 m, i.e. when $x = 60$,
$x = ut + \frac{1}{2}at^2 \Rightarrow 60 = 40t - 5t^2 \Rightarrow t^2 - 8t + 12 = 0 \Rightarrow (t-2)(t-6) = 0 \Rightarrow t = 2$ or 6.
The ball is more than 60 m above the ground between $t = 2$ and $t = 6$, i.e. for 4 s.

Exercise 7.1:4

1 A stone is dropped from the top of a cliff of height 50 m. Find:

(i) its speed after t s
(ii) the distance it has travelled after t s
(iii) its speed when it hits the ground
(iv) the time taken for it to hit the ground

(Take downwards as positive; then the acceleration of the stone is $+10$ m s^{-2}.)

2 A ball is thrown vertically upwards from ground level with a speed of 20 m s^{-1}. Find:

(i) the ball's greatest height
(ii) the time it takes to reach its greatest height
(iii) the ball's speed just before it hits the ground again
(iv) the time it takes to reach the ground

(Take upwards as positive; then the acceleration of the ball is -10 m s^{-2}.)

3 A ball is thrown vertically upwards from a point O with speed 6 m s^{-1}. Find the times at which the ball is 1 m above O. For how long is the ball more than 1 m above O?

4 An ecstatic man, who has just won £10 000 at the races, throws his hat vertically upwards with speed 10 m s^{-1}. For how long is the hat more than 2 m above the point of projection?

5 A ball is thrown vertically upwards from a point O at ground level with speed 15 m s^{-1}. Find the time for which the ball is:

(i) in the air
(ii) more than 7.2 m above O
(iii) more than 10 m above O

6 A ball is thrown vertically upwards from a point A 1 m above ground level with speed 10 m s^{-1}. Find:

(i) its greatest height above ground level
(ii) its speed when it passes through A again
(iii) its speed when it hits the ground
(iv) the time taken for it to hit the ground
(v) the time for which it is more than 1.8 m above A

7 A ball is thrown vertically upwards with speed u. Show that:

(i) its greatest height is $u^2/2g$
(ii) its velocity when it is at a height h above O is $\pm\sqrt{(u^2 - 2gh)}$
(iii) its velocity when it passes through O again is $-u$
(iv) it is above O for a time $2u/g$
(v) it is more than $3u^2/8g$ above O for a time u/g

8 A secretary, realising that shorthand and typing are simply the tools of repression of a ruling class of males, goes to the window of her office on the 119th floor of an office block and drops her typewriter over the sill. It takes 8 s to reach the ground.

Find the height of the window above the ground. How fast is the typewriter moving when it hits the ground?

⑦ **9** A Japanese tourist, deeply dissatisfied with the quality of the sweet and sour pork in a Chinese take-away meal, furiously throws the whole meal vertically downwards out of the window of his luxury hotel room. 4 s later it hits the ground with velocity 45 m s^{-1}.

Find the height of the window above the ground. Find the initial velocity of the meal.

⑦ → **10** A stone is dropped from the top of a cliff. In the last two seconds of its motion it falls through a distance of 80 m. Find the height of the cliff.

⑦ **11** A stone is dropped from the top of a cliff. In the last second of its motion if falls through a distance which is $\frac{15}{64}$ times the height of the cliff. Find the height of the cliff.

⑦ **R 12** A stone is dropped from the top of a cliff. In the last second of its motion it falls through a distance which is k times the height of the cliff ($k \leq 1$). Show that the time taken for the stone to reach the bottom of the cliff is $1/[1-\sqrt{(1-k)}]$ and find the height of the cliff.

⑦ **13** The acceleration due to gravity on the Moon is approximately 1.6 m s^{-2}.

(i) If you dropped a stone from 5 m above the surface of: (a) the Earth; (b) the Moon; with what velocity would it hit the ground?

(ii) If the Olympic Games were held on the Moon, estimate the high jump record.

7.1:5 Use of differential equations

When a particle moves in a straight line its acceleration a is given by:

$$a = \frac{dv}{dt} \quad (13)$$

$$a = \frac{d^2x}{dt^2} \quad (14)$$

$$a = v\frac{dv}{dx} \quad (15)$$

(See question 1.)

Worked example 8 A particle moves in a straight line with acceleration $6 - 2v$, where v is its velocity. If the particle is initially at O moving with velocity 1,

(i) find its velocity and displacement from O at time t;
(ii) show that its velocity tends to a limiting value.

(i) $$a = 6 - 2v \quad (16)$$

so from equation (13),

$$\frac{dv}{dt} = 6 - 2v$$

This is a separable-variable differential equation which can be solved by the method of section 4.4:2:

$$\int \frac{dv}{3-v} = \int 2 \, dt$$

$$\Rightarrow \qquad -\ln(3-v) = 2t + c$$

Kinematics of a particle 381

(We have written the integral of $1/(3-v)$ as $-\ln(3-v)$ rather than $-\ln|3-v|$ since we assume that $3-v$ is positive throughout—note that initially $3-v=2$, which is positive.)

When $t=0$, $v=1$; so $c=-\ln 2$. Thus

$$-\ln(3-v) = 2t - \ln 2$$
$$\ln[2/(3-v)] = 2t$$
$$2/(3-v) = e^{2t}$$
$$v = 3 - 2e^{-2t} \qquad (17)$$

We can integrate equation (17) directly to find x:

$$x = 3t + e^{-2t} + d$$

When $t=0$, $x=0$; so $d=-1$. Thus

$$x = 3t + e^{-2t} - 1 \qquad (18)$$

(ii) Consider equation (17). As $t \to \infty$, $e^{-2t} \to 0$; so $v \to 3$, i.e. v tends to a limiting value, 3, which is called the **terminal velocity** of the particle (fig. 7.1:6).

Note that as the velocity of the particle tends towards the terminal velocity, its acceleration tends to zero; and in fact we can find the terminal velocity simply by putting $a = 0$ in the original equation (16):

$$a = 0 \Rightarrow 0 = 6 - 2v \Rightarrow v = 3$$

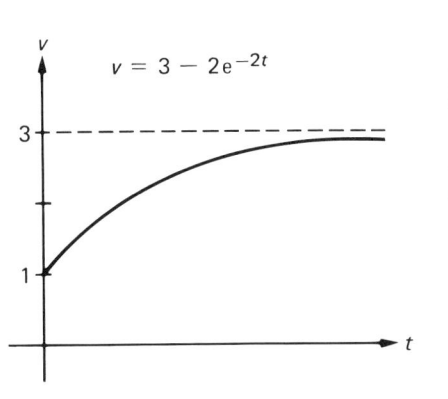

Fig. 7.1:6

Worked example 9 A particle moves in a straight line with acceleration $-(4+v^2)$ where v is its velocity. If the particle is initially at O moving with velocity 2, find its displacement from O when it first comes to instantaneous rest.

$$a = -(4+v^2)$$

Since the problem is concerned with the relationship between v and x, we put $a = v\,dv/dx$ (equation (15)).

$$v\frac{dv}{dx} = -(4+v^2)$$
$$\int \frac{v\,dv}{4+v^2} = -\int dx \qquad \text{(see section 4.4:2)}$$
$$\tfrac{1}{2}\ln(4+v^2) = -x + c$$

When $x=0$, $v=2$; so $c = \tfrac{1}{2}\ln 8$. Thus

$$\tfrac{1}{2}\ln(4+v^2) = -x + \tfrac{1}{2}\ln 8$$
$$x = \tfrac{1}{2}\ln[8/(4+v^2)]$$

When the particle first comes to instantaneous rest, $v=0$, so $x = \tfrac{1}{2}\ln\tfrac{8}{4} = \tfrac{1}{2}\ln 2$.

382 Use of differential equations

Worked example 10 A particle moves in a straight line so that its acceleration is $-3x^2$, where x is its displacement from a fixed point O on the line. Initially it is at O, moving with velocity 4. Find its displacement from O when it first comes to instantaneous rest.

$$a = -3x^2$$

i.e.
$$v \frac{dv}{dx} = -3x^2 \qquad \text{by equation (15)}$$

So
$$\int v \, dv = \int -3x^2 \, dx \qquad \text{(see section 4.4:2)}$$

$$\tfrac{1}{2}v^2 = -x^3 + c$$

When $x = 0$, $v = 4$; so $c = 8$. Thus

$$\tfrac{1}{2}v^2 = -x^3 + 8$$

So when $v = 0$, $-8 = -x^3$; i.e. $x = 2$.

Exercise 7.1:5

1 Using the definition $a = dv/dt$ and the function of a function rule (see section 1.3:6), show that $a = v \, dv/dx$.

(i) If $v = 2\sqrt{(9 - x^2)}$, show that $a = -4x$.
(ii) If $v^2 = 3x + 4$, show that a is constant.

2 If $v = k/(1 + x)$, k constant, show that $a = -v^3/k$.

3 If $v = e^{-kx}$, k constant, show that $a = -kv^2$.

4 A particle starts from the origin with velocity u and moves with constant acceleration a. Using equation (15), show directly that $v^2 = u^2 + 2ax$.

5 $a = -3v$. Initially $v = 6$, $x = 0$.

(i) Find v and x at time t.
(ii) Sketch a graph of v against t and a graph of x against t.
(iii) Show that as $t \to \infty$, P gradually approaches a certain point on the line. What is this point?

6 $a = 3 - 2v$. P starts from rest at O.

(i) Find v and x at time t.
(ii) Sketch graphs of v against t and x against t.
(iii) Find the terminal velocity of P.
(iv) Show that for large values of t, $x \approx \tfrac{3}{4}(2t - 1)$.

7 $a = -5 - 5v$. P starts from O with velocity 3. Find the time at which P first comes to instantaneous rest, and hence show that the maximum displacement of P from O is $\tfrac{1}{5}(3 - \ln 4)$.

8 $a = 4 - v^2$. P starts from rest.

(i) Find the terminal velocity of P.
(ii) Find v in terms of t. Verify your answer to (i).

9 $a = 4 - v^2$. P starts from rest at O.

(i) Show that $v^2 = 4 - 3e^{-2x}$.
(ii) Find v when $x = \tfrac{1}{2} \ln 2$, $\ln 2$, $\tfrac{3}{2} \ln 2$, $2 \ln 2$.
(iii) What happens to v as $x \to \infty$?

(C.f. question 8.)

10 $a = -4 - 9v^2$. Initially P is at O, moving with velocity $\tfrac{2}{3}$. Show that, when P first comes to instantaneous rest, $x = \tfrac{1}{18} \ln 2$.

11 $a = -kb^2 - kv^2$ (k, b constant). If the initial velocity of P is u, find the distance it travels before it:

(i) has velocity $\tfrac{1}{4}u$;
(ii) comes to instantaneous rest.

12 $a = 6 - 2v$. Initially P is at O, moving with velocity 1.

(i) Show that $x = \tfrac{1}{2}\{1 - v + 3 \ln \{2/(3 - v)\}\}$.
(ii) Hence find the displacement of P from O when it first comes to instantaneous rest.
(iii) Using the results of worked example 9, show that P comes to rest after time $\tfrac{1}{2} \ln \tfrac{2}{3}$, and hence verify your answer to (ii).

13 $a = -5 - 5v$. P starts from O with velocity 3.

(i) Find x in terms of v.
(ii) Hence show that the maximum displacement of P from O is $\frac{1}{5}(3 - \ln 4)$.

(C.f. question 7.)

14 $a = -k/(1+v)$ (k constant). If the initial velocity of P is u, find the distance P travels before coming to instantaneous rest.

R 15 $a = -k\, \mathrm{e}^{v/u}$ (k, u constant). If the initial velocity of P is u, find the time taken:

(i) for its velocity to decrease to $\frac{1}{2}u$;
(ii) before P comes to rest.

Find the distance it travels before it comes to rest.

16 When a particle falls vertically in a certain resisting medium its acceleration is $g - kv$, where v is its velocity, g is the acceleration due to gravity and k is a constant.

(i) Show that the terminal velocity of the particle is g/k.

(ii) If the particle starts from rest, show that its velocity after t s is given by

$$v = \frac{g}{k}(1 - \mathrm{e}^{-kt})$$

Verify that its terminal velocity is g/k.

(iii) Show that the distance x through which the particle falls in t s is given by

$$x = \frac{gt}{k} + \frac{g}{k^2}(\mathrm{e}^{-kt} - 1)$$

(iv) Show that

$$x = \frac{g}{k^2} \ln \frac{g}{g - kv} - \frac{v}{k}$$

Hence find the distance through which the particle has fallen when its velocity is: (a) $g/3k$; (b) $2g/3k$; (c) g/k.

17 When a particle is projected vertically upwards in a certain resisting medium, its acceleration is $-g - kv$ (upwards positive). If its initial speed is U, find the time taken:

(i) for its speed to be halved;
(ii) for it to reach its greatest height above the point of projection.

Hence find its greatest height.

18 A particle is projected vertically upwards with speed U in a resisting medium. Its acceleration is $-g - kv^2$ (upwards positive). Find its greatest height.
When it falls through the medium, its acceleration is $g - kv^2$ (downwards positive). Show that its speed V as it passes through the point of projection is given by

$$\frac{1}{V^2} - \frac{1}{U^2} = \frac{k}{g}$$

19 Consider question 18 again. Find the time the particle takes:

(i) to reach the highest point;
(ii) to return to the point of projection.

20 A parachutist jumps from a plane and falls freely under gravity for 2 seconds until she opens her parachute. When the parachute is open her downward acceleration is $(10 - \frac{1}{8}v)\,\mathrm{m\,s^{-2}}$, where $v\,\mathrm{m\,s^{-1}}$ is her speed. Find the speed and the distance she has travelled 10 seconds after she has jumped from the plane. (Assume that her initial velocity is zero.)

21 The school bully pushes the school intellectual gently off the top diving board of the swimming pool. The poor genius, still clutching a copy of Einstein's General Theory of Relativity, falls vertically from rest under gravity until he hits the water. As he moves through the water his downward acceleration is $g(1 - v/\lambda)$, where v is his speed and λ is a positive constant. The height of the diving board above the water is $\lambda^2/4g$ and the depth of the water is only $\lambda^2/16g$. Show that the speed v with which he hits the bottom of the pool is given by $v + \lambda \ln(\lambda - v) = \lambda(\frac{7}{16} + \ln\frac{1}{2}\lambda)$. Ouch! What are the units of the constant λ?
The school anarchist promptly saws through the diving board with a chain saw, and the bully plummets into the pool....

22 $a = 3x^2$. P starts from O with velocity 1.

(i) Find v in terms of x.
(ii) Show that when $x = 2$, $v = 3\sqrt{2}$.

23 $a = -9x^2$. P starts from O with velocity 12. Find its displacement from O when it first comes to instantaneous rest.

24 $a = -kx^2$ (k constant). P starts from rest when $x = A$ (A constant). Find the speed of P when it first reaches O.

R 25 $a = -u^2 A/(A + x)^2$ (u, A constant). P starts from O with velocity u. Find its velocity when $x = A$.

384 Collisions

(10) → 26 A particle, attached by an elastic string to a fixed point, hangs at rest at a point O. At $t=0$ it is projected downwards from O with speed An, where A and n are constants. Subsequently its acceleration is proportional to its distance from O, and is directed towards O. The constant of proportionality is n^2.

If at time t the velocity of the particle is v and its displacement from O is x,

(i) show that $v^2 = n^2(A^2 - x^2)$, and hence
(ii) show, by putting $v = dx/dt$, that $x = A \sin nt$.

Show that the maximum distance of the particle from O is A. What is (a) its maximum velocity; (b) its maximum acceleration?
Describe the motion.

(10) 27 A, L, M and B are the points on Ox at which $x = -4$, $x = 2$, $x = 2$ and $x - 4$ respectively. A particle starts from rest at B and moves on Ox with acceleration $-9x$. Find the velocity of the particle in terms of x, and show that $x = 4 \cos 3t$.

Find the speed of the particle as it passes through, and the time it takes to reach:
(i) M (ii) O (iii) L (iv) A

(10) 28 $$a = u^2\left(\frac{A}{x^2} - \frac{x}{A^2}\right)$$

(u and A constant). When $x = A$, the velocity of P is u. Show that, when P first comes to instantaneous rest,

$$x^3 - 4A^2x + 2A^3 = 0$$

Use the Newton–Raphson method (see section 2.6:2) to find an approximate solution of this equation.

(10) 29 A contestant in the London Marathon, keen but not very fit, dresses in a gorilla suit and stiletto heels and carries a tray with a bottle of champagne on it. When he has run a distance x km his speed v is $50/(5+x)$ km h^{-1}. Find his deceleration (in terms of x).

The length of the course is 42 km. Find the time he takes to complete it.

(10) 30 (i) A particle moves in a straight line. Its velocity when it is x m from a fixed point O on the line is v m s^{-1}. Show that the time taken to cover a displacement c from O is given by

$$\int_0^c \frac{1}{v} dx$$

(See 4.2:1 Theorem 1.)

(ii) A car moves along a straight road. Its velocity v m s^{-1} and its displacement x m from a fixed point A are related by the equation $v = 4x^{\frac{1}{2}}$. Find the time taken to travel:
(a) 50 m from A; (b) 100 m from A.

(iii) A car moves along a straight road. Its velocity at different distances from a fixed point A is shown in this table:

s (m)	0	10	20	30	40	50	60
v (m s^{-1})	1.5	1.9	2.6	3.1	3.5	4.0	4.7

Using Simpson's rule (see section 4.3), find approximately the time taken by the car to travel: (a) 40 m from A; (b) 60 m from A.

7.1:6 Collisions

Worked example 11 A particle P starts from O with velocity 3 m s^{-1} and moves with acceleration 2 m s^{-2}. Two seconds later another particle Q starts from O with velocity 4 m s^{-1} and moves with acceleration 4 m s^{-2}. Find their distance from O when they collide.

Suppose that they collide T s after P has started to move; then Q has been moving for $(T-2)$ s. Also suppose that their displacement from O is then X m (fig. 7.1:7).

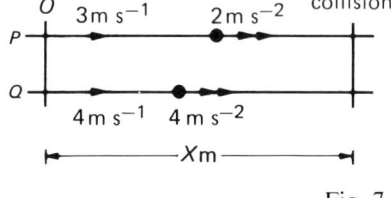

Fig. 7.1:7

P: $u = 3$, $a = 2$, $t = T$, $x = X$
$x = ut + \frac{1}{2}at^2 \Rightarrow X = 3T + \frac{1}{2} \cdot 2T^2$, i.e. $X = 3T + T^2$ (19)

Q: $u = 4$, $a = 4$, $t = T - 2$, $x = X$
$x = ut + \frac{1}{2}at^2 \Rightarrow X = 4(T-2) + \frac{1}{2} \cdot 4(T-2)^2$, i.e. $X = -4T + 2T^2$ (20)

From equations (19) and (20):

$$3T + T^2 = -4T + 2T^2$$
$$T^2 - 7T = 0$$
$$T(T-7) = 0$$
$$\Rightarrow T = 0 \text{ or } 7$$

$T = 0$: P and Q are both at O.
$T = 7$: P and Q collide. From equation (19), $X = 3 \cdot 7 + 7^2 = 70$. So they collide at a distance 70 m from O.

Exercise 7.1:6

Questions 1-6: Two particles P and Q move along Ox. Find the time (measured from the moment at which P starts to move) at which they collide and their displacement from O at this time.

(11) **1** P starts from O with velocity 5 m s^{-1} and moves with acceleration 4 m s^{-2}. Two seconds later Q starts from O with velocity 16 m s^{-1} and moves with acceleration 6 m s^{-2}.

(11) → **2** P starts from O with velocity 2 m s^{-1} and moves with acceleration 2 m s^{-2}. Two seconds later Q starts from rest at O and moves with acceleration 6 m s^{-2}. (Explain why the solution $T = 1$ is meaningless.)

(11) **3** P starts from O and moves with constant velocity 10 m s^{-1}. One second later Q starts from O with velocity 2 m s^{-1} and moves with acceleration 4 m s^{-2}.

(11) **4** P starts from O and moves with constant velocity 6 m s^{-1}. Simultaneously Q starts from O with velocity -6 m s^{-1} and moves with acceleration 8 m s^{-2}.

(11) **5** P starts from O and moves with constant velocity 4 m s^{-1}. Three seconds later Q starts from O and moves with constant velocity 6 m s^{-1}.

(11) → **6** P starts from O and moves with constant velocity V. Simultaneously Q starts from the point $(-6V, 0)$ and moves with constant velocity $3V$.

(11) **7** A car is travelling along a straight motorway at a constant speed of 40 m s^{-1}. Naughty! As the car passes a radar checkpoint O, a police car sets out from rest at O and travels with constant acceleration $f \text{ m s}^{-2}$ to catch the offender.

(i) If the interception occurs after one minute, find f.

(ii) If $f = 1$, find the distance from O at which the interception occurs.

(11) **8** A bus moves away from rest at a bus stop with acceleration 1 m s^{-2}. As the bus starts to move, a man who is 16 m behind the stop starts to run with constant speed after the bus.

(i) If he runs with speed 6 m s^{-1}, find the time he takes to catch the bus.

(ii) Find the least speed at which he can run if he is to catch the bus.

• → **9** A ball is thrown vertically upwards from a point O with speed 15 m s^{-1}. Two seconds later another ball is dropped from O. Find the distance below O at which the balls collide.

10 A ball is thrown vertically upwards from a point O with speed 10 m s^{-1}. One second later another ball is thrown vertically upwards from O with the same speed. Find the distance above O at which the balls collide.

11 At time $t = 0$, a particle is projected vertically upwards from a point O with speed u. At time $t = u/2g$ another particle is projected vertically upwards from O with speed v. Find the time at which the particles collide:

(i) when $v = u$
(ii) when $v = 2u$

Show that in both cases they collide at a height $15u^2/32g$ above O. Explain this.

R 12 At time $t = 0$ a particle is projected vertically upwards with speed U, and at time $t = \tau$ a second particle is projected vertically upwards from the same point with the same speed. Find the time at which the particles collide and their height above the point of projection when they collide.
Discuss the cases $\tau = 0$, $\tau = u/g$, $\tau > 2u/g$.

13 A man standing on a hydraulic platform throws a ball vertically upwards with velocity u. Immediately after the ball leaves his hand, the man and the platform descend vertically with constant velocity ku. Show that the time that elapses before the ball returns to his hand is $2(k+1)u/g$.

14 A stone is dropped from the top of a building. At the same time another stone is thrown vertically upwards from the bottom of the building with a speed of 20 m s^{-1}. They collide after 3 s. Find:

(i) the height of the building
(ii) the speed of each stone just before the collision

15 The headmaster of a public school pursues the most decadent boy in the school, a boy known for his involvement with sex, drugs and rock and roll, down a long corridor. At time $t = 0$, the boy sets out from the door of the assistant matron's room, and runs with constant speed V down the corridor while rolling a cigarette with one hand and adjusting the volume on his personal stereo with the other. At time $t = T$ the headmaster emerges from the assistant matron's room. Starting from rest, he runs, fuelled by indignation, with constant acceleration V/T.

Show that he catches up with the boy after he has been running for time $T(1+\sqrt{3})$, and that he and the evil-doer are then at a distance of $VT(2+\sqrt{3})$ from the door of the assistant matron's room.

• **16** Two particles P and Q move in a straight line. Their displacements s_P and s_Q from O at time t ($t \geqslant 0$) are given by $s_P = t^2 + 5t$, $s_Q = 3t + 8$.
On the same axes sketch graphs of s_P and s_Q against t. Find the time at which the particles collide.

17 Two particles P and Q move in a straight line. Their displacements s_P and s_Q from O at time t are given by $s_P = 4t - t^3$, $s_Q = 5t - t^2 - 6$.
On the same axes sketch graphs of s_P and s_Q against t. Show that the particles collide when $t = 2$. Find their velocities at this instant.

• → **18** Two particles P and Q move along Ox. At $t = 0$, P starts from the point $(15, 0)$ with velocity -6 m s^{-1} and moves with acceleration 2 m s^{-2}. Simultaneously Q starts from O with velocity 6 m s^{-1} and moves with acceleration -3 m s^{-2}.

(i) Show that at time t the distance between P and Q is $\frac{5}{2}t^2 - 12t + 15$.

(ii) Hence show (by the methods of section 1.3:5) that they are closest together after $\frac{12}{5}$ s and that their distance apart is then $\frac{3}{5}$ m.

19 Two particles P and Q move along Ox. At $t = 0$, P starts from O with velocity 10 m s^{-1} and moves with acceleration -4 m s^{-2}. At $t = 1$, Q starts from O with velocity 15 m s^{-1} and moves with acceleration -10 m s^{-2}. Show that they are closest together after $\frac{5}{2}$ s and that their distance apart is then $\frac{5}{4}$ m.

20 Two particles P and Q move along Ox. P starts from rest at O and moves with acceleration 1 m s^{-2}. Two seconds later Q starts from O and moves with constant velocity 3 m s^{-1}. Find their shortest distance apart in subsequent motion.

7.2 Motion of a particle in a plane

7.2:1 Use of differentiation

Consider the motion of a particle P along a curve (called the **path** of P) in the xy plane. Suppose that, at time t, P is at the point (x, y) with position vector $\mathbf{r}$. So

$$\mathbf{r} = x\mathbf{i} + y\mathbf{j}$$

x and y are (scalar) functions of t. $\mathbf{r}$ is a vector function of t.

Definition		
The **velocity** of P,	$\mathbf{v} = \dfrac{d\mathbf{r}}{dt}$	(1)
The **acceleration** of P,	$\mathbf{a} = \dfrac{d\mathbf{v}}{dt}$	(2)

Kinematics of a particle 387

P, **r**, **v** and **a** are shown in fig. 7.2:1.

What is meant by $d\mathbf{u}/dt$ when **u** is a vector? We could define $d\mathbf{u}/dt$ rigorously (as we defined dy/dx rigorously in section 1.3:1); but instead we shall work with a simple common-sense definition, i.e.

$$\text{if } \mathbf{u} = f(t)\mathbf{i} + g(t)\mathbf{j}, \text{ then } \frac{d\mathbf{u}}{dt} = f'(t)\mathbf{i} + g'(t)\mathbf{j} \qquad (3)$$

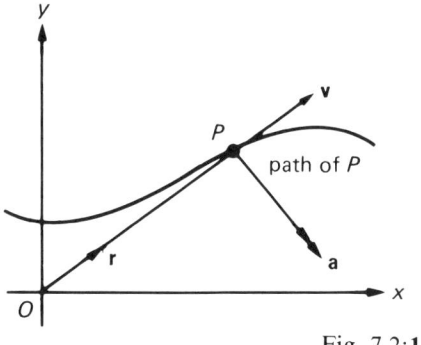

Fig. 7.2:1

e.g. if $\mathbf{u} = t^2\mathbf{i} + t^3\mathbf{j}$, then $d\mathbf{u}/dt = 2t\mathbf{i} + 3t^2\mathbf{j}$.

Thus, since $\mathbf{r} = x\mathbf{i} + y\mathbf{j}$, by definition (1)

$$\mathbf{v} = \frac{dx}{dt}\mathbf{i} + \frac{dy}{dt}\mathbf{j}$$

and by definition (2)

$$\mathbf{a} = \frac{d^2x}{dt^2}\mathbf{i} + \frac{d^2y}{dt^2}\mathbf{j}$$

We write

$$\mathbf{v} = \dot{x}\mathbf{i} + \dot{y}\mathbf{j} \qquad (4)$$

and

$$\mathbf{a} = \ddot{x}\mathbf{i} + \ddot{y}\mathbf{j} \qquad (5)$$

Worked example 1 At time t, the position vector of a particle is given by $\mathbf{r} = 3\cos 2t\,\mathbf{i} + 4\sin 2t\,\mathbf{j}$. Find its velocity and acceleration when $t = \frac{1}{6}\pi$.

$$\mathbf{r} = 3\cos 2t\,\mathbf{i} + 4\sin 2t\,\mathbf{j}$$

So by equation (4) $\quad \mathbf{v} = -6\sin 2t\,\mathbf{i} + 8\cos 2t\,\mathbf{j}$

and by equation (5) $\quad \mathbf{a} = -12\cos 2t\,\mathbf{i} - 16\sin 2t\,\mathbf{j}$

When $t = \frac{1}{6}\pi$, $\quad \mathbf{v} = -6\sin\frac{1}{3}\pi\,\mathbf{i} + 8\cos\frac{1}{3}\pi\,\mathbf{j}$

$$= -3\sqrt{3}\mathbf{i} + 4\mathbf{j}$$

and $\quad \mathbf{a} = -12\cos\frac{1}{3}\pi\,\mathbf{i} - 16\sin\frac{1}{3}\pi\,\mathbf{j}$

$$= -6\mathbf{i} - 8\sqrt{3}\mathbf{j}$$

Cartesian equation of the path
If $\mathbf{r} = f(t)\mathbf{i} + g(t)\mathbf{j}$, then the parametric equations of the path are:

$$x = f(t), \qquad y = g(t)$$

(C.f. section 6.3:1.)
To find the cartesian equation of the path, eliminate t.

Worked example 2 At time t the position vector of a particle is given by $\mathbf{r} = 3\cos 2t\mathbf{i} + 4\sin 2t\mathbf{j}$. Find the cartesian equation of its path.

$$\mathbf{r} = 3\cos 2t\mathbf{i} + 4\sin 2t\mathbf{j}$$

$x = 3\cos 2t \Rightarrow \frac{1}{3}x = \cos 2t$
$y = 4\sin 2t \Rightarrow \frac{1}{4}y = \sin 2t$

Thus

$$\tfrac{1}{9}x^2 + \tfrac{1}{16}y^2 = \cos^2 2t + \sin^2 2t = 1$$

So the cartesian equation of the path is $\frac{1}{9}x^2 + \frac{1}{16}y^2 = 1$ (an ellipse).

Velocity

Speed The speed v is a scalar function of t given by

$$v = |\mathbf{v}| = \sqrt{(\dot{x}^2 + \dot{y}^2)} \tag{6}$$

If s is the distance, measured along the path, of P from a fixed point A on the path (e.g. the initial position of the particle) as shown in fig. 7.2:**2**, then

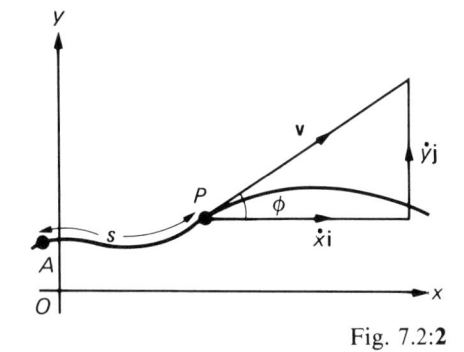

$$v = \frac{ds}{dt} \tag{7}$$

Fig. 7.2:**2**

(see question 16).

Direction of motion Let ϕ be the angle between the direction of motion of the particle and Ox (so ϕ is a scalar function of t). Then

$$\tan\phi = \frac{\dot{y}}{\dot{x}} = \frac{dy/dt}{dx/dt} = \frac{dy}{dx} \tag{8}$$

(see section 6.3:2). Note: this shows that *the direction of motion of the particle is always along the tangent to the path.*

Kinematics of a particle

Worked example 3 At time t, the position vector of a particle is given by $\mathbf{r} = 2t\mathbf{i} + (t - t^2)\mathbf{j}$. Find:

(i) its speed at time $t = 2$
(ii) its direction of motion (a) at time $t = \frac{1}{4}$; (b) when it is at the point $(4, -2)$.

The path of P is shown in fig. 7.2:3.
$\mathbf{r} = 2t\mathbf{i} + (t - t^2)\mathbf{j} \Rightarrow \mathbf{v} = 2\mathbf{i} + (1 - 2t)\mathbf{j}$

(i) At time $t = 2$, $\mathbf{v} = 2\mathbf{i} - 3\mathbf{j}$, so speed $v = |\mathbf{v}| = |2\mathbf{i} - 3\mathbf{j}| = \sqrt{13}$ (see equation (6)).

(ii) By equation (8), $\tan\phi = \frac{1}{2}(1 - 2t)$.
 (a) When $t = \frac{1}{4}$, $\tan\phi = \frac{1}{2}(1 - 2 \cdot \frac{1}{4}) = \frac{1}{4} \Rightarrow \phi = 14.04°$.
 (b) When the particle is at the point $(4, -2)$, $t = 2$ (why?), so $\tan\phi = \frac{1}{2}(1 - 2 \cdot 2) = -\frac{3}{2} \Rightarrow \phi = -56.31°$.
 Alternatively: $x = 2t$, $y = t - t^2 \Rightarrow y = \frac{1}{2}x - \frac{1}{4}x^2$ (the cartesian equation of the path (see worked example 2)); so $\tan\phi = dy/dx = \frac{1}{2} - \frac{1}{2}x = -\frac{3}{2}$ at the point $(4, -2)$.

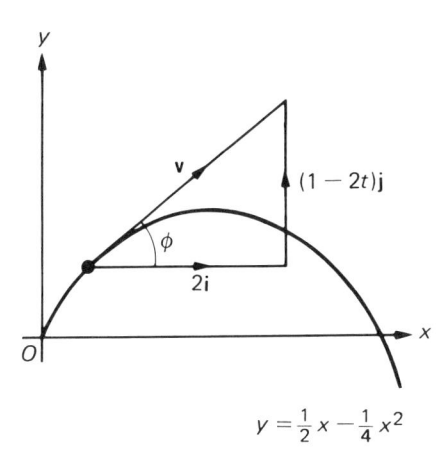

Fig. 7.2:3

Exercise 7.2:1

Questions 1–8: The position vector of a particle at time t is $\mathbf{r}$. Find its velocity and acceleration at time t. Hence write down its velocity and acceleration at the given times.

① → **1** $\mathbf{r} = 3t^2\mathbf{i} + t^3\mathbf{j}$; $t = 0$, $t = 1$, $t = 2$

① **2** $\mathbf{r} = 40t\mathbf{i} + (30t - 5t^2)\mathbf{j}$; $t = 1$, $t = 3$

① **3** $\mathbf{r} = \frac{2}{t}\mathbf{i} + \left(4 - \frac{1}{t^2}\right)\mathbf{j}$; $t = 1$

① **4** $\mathbf{r} = 5\cos t\mathbf{i} - 3\sin t\mathbf{j}$; $t = 0$, $t = \frac{1}{2}\pi$, $t = \pi$

① → **5** $\mathbf{r} = 2\cos 3t\mathbf{i} + 2\sin 3t\mathbf{j}$; $t = 0$, $t = \frac{1}{9}\pi$

① **6** $\mathbf{r} = \sin^2 t\mathbf{i} + (\sin t - \cos^2 t)\mathbf{j}$; $t = \frac{1}{3}\pi$

① **7** $\mathbf{r} = e^{-t}\mathbf{i} + e^{2t}\mathbf{j}$; $t = 0$, $t = 1$, $t = \ln 3$

① **8** $\mathbf{r} = \ln t\mathbf{i} + t\mathbf{j}$; $t = 3$

① R **9** The position vector $\mathbf{r}$ of a particle at time t is given by $\mathbf{r} = (t + \cos t)\mathbf{i} + (1 + \sin t)\mathbf{j}$. Find the velocity and acceleration of the particle at time t. At what times is the particle at rest? Show that the magnitude of the acceleration of the particle is constant.

② **10** Find the cartesian equation of the path, and sketch the path, of a particle whose position vector at time t is $\mathbf{r}$, when:

→ (i) $\mathbf{r} = t\mathbf{i} + (t - t^2)\mathbf{j}$
 (ii) $\mathbf{r} = t^2\mathbf{i} + (1 + 2t)\mathbf{j}$
 (iii) $\mathbf{r} = e^t\mathbf{i} + e^{-t}\mathbf{j}$
 (iv) $\mathbf{r} = \sin t\mathbf{i} + \cos t\mathbf{j}$
R (v) $\mathbf{r} = 5\sin 2t\mathbf{i} + 5\cos 2t\mathbf{j}$
→ (vi) $\mathbf{r} = (4 + \sin t)\mathbf{i} + (-3 + \cos t)\mathbf{j}$

② **11** Find the cartesian equation of the path of a particle whose position vector at time t is $\mathbf{r}$, when:

(i) $\mathbf{r} = 5\cos 3t\mathbf{i} + 4\sin 3t\mathbf{j}$
(ii) $\mathbf{r} = \sin t\mathbf{i} + \sin 2t\mathbf{j}$
(iii) $\mathbf{r} = \sec t\mathbf{i} + \tan t\mathbf{j}$
(iv) $\mathbf{r} = \sin t\mathbf{i} + \cos 2t\mathbf{j}$
(v) $\mathbf{r} = t^2\mathbf{i} + t^3\mathbf{j}$
(vi) $\mathbf{r} = e^{2t}\mathbf{i} + e^{3t}\mathbf{j}$

② **12** The position vector of a particle at time t is $3\cos\omega t\mathbf{i} + 3\sin\omega t\mathbf{j}$, where ω is a constant. Show that the cartesian equation of the path is independent of ω. Find the speed of the particle.

② **13** The position vectors of two particles A and B at time t are given by $\mathbf{r}_A = t\mathbf{i} + (t^2 - 2t)\mathbf{j}$ and $\mathbf{r}_B = (t - 2)\mathbf{i} + (t^2 - 6t)\mathbf{j}$ respectively. Show that A and B travel along the same path. What is the difference between the motion of the two particles?

② **14** The position vectors of two particles A and B at time t are given by $\mathbf{r}_A = t\mathbf{i} + (t^2 - 2t)\mathbf{j}$ and $\mathbf{r}_B = 2t\mathbf{i} + (4t^2 - 4t)\mathbf{j}$ respectively.

(i) Find the cartesian equation of the path of A, and sketch it. Mark the position of A at times $t = 0, \frac{1}{2}, 1, 2$.
(ii) Show that the cartesian equation of the path of B is the same as that of A. On another sketch of this path, mark the position of B at times $t = 0, \frac{1}{2}, 1, 2$.

390   *Use of differentiation*

② **15** The position vectors of two particles A and B at time t are given by $\mathbf{r}_A = \cos 2t\mathbf{i} + \sin 2t\mathbf{j}$ and $\mathbf{r}_B = (2 + \cos 2t)\mathbf{i} + (3 + \sin 2t)\mathbf{j}$. On one diagram sketch the paths of the particles. Show that their velocities are the same and that their accelerations are the same. Mark the position, velocity and acceleration of each particle at time $t = \tfrac{1}{4}\pi$ on the diagram.

16 Let $P(x, y)$ and $P'(x + \delta x, y + \delta y)$ be two adjacent points on the path of a moving particle; and let the arc length PP' be δs. Show that

$$\delta s^2 \approx \delta x^2 + \delta y^2$$

and hence show, using equation (6), that the speed v of the particle is given by

$$v = \frac{ds}{dt}$$

where s is the distance, measured along the path, of P from some fixed point on the path.
(Note: this provides a natural extension, for motion along a curve, to our definition of velocity in a straight line (7.1:1 definition (1)). We shall make use of it in section 7.4.)

③ **17** The position vector of a particle at time t is $\mathbf{r}$. Find its speed at time t. Hence write down its speed at the given times:

(i) $\mathbf{r} = 10t\mathbf{i} + (20 - 5t^2)\mathbf{j}$; $t = 0$, $t = 2$, $t = 4$
(ii) $\mathbf{r} = 3\cos 2t\mathbf{i} + 3\sin 2t\mathbf{j}$; $t = 0$, $t = \tfrac{1}{6}\pi$
→ (iii) $\mathbf{r} = \sec t\mathbf{i} + \tan t\mathbf{j}$; $t = 0$, $t = \tfrac{1}{3}\pi$

③ → **18** The position vector $\mathbf{r}$ of a particle at time t is given by $\mathbf{r} = 3t\mathbf{i} + (4t - 5t^2)\mathbf{j}$. Find the direction of its velocity

(i) at times $t = 0, \tfrac{1}{5}, \tfrac{2}{5}, \tfrac{3}{5}, \tfrac{4}{5}, 1$
(ii) when it is at the point $(6, -12)$.

③ **19** The position vector $\mathbf{r}$ of a particle at time t is given by $\mathbf{r} = (t - 1)\mathbf{i} + (2t^2 - t - 3)\mathbf{j}$. Find the direction of its velocity

(i) at time $t = 1$
(ii) when it is at the point $(1, 3)$.

Illustrate these results on a sketch of the path of the particle.

③ → **20** The position vector $\mathbf{r}$ of a particle at time t is given by $\mathbf{r} = 2\cos 3t\mathbf{i} + 2\sin 3t\mathbf{j}$. Show that the speed of the particle is constant. Show that its velocity $\mathbf{v}$ and its acceleration $\mathbf{a}$ are always perpendicular. (Hint: consider $\mathbf{v} \cdot \mathbf{a}$.)

③ **21** The position vector $\mathbf{r}$ of a particle at time t is given by $\mathbf{r} = 10t\mathbf{i} + (20t - 5t^2)\mathbf{j}$. Find the time at which the velocity of the particle is perpendicular to its initial velocity, and the position vector of the particle at this time.

③ **22** The position vector $\mathbf{r}$ of a particle at time t is given by $\mathbf{r} = (t^2 + 1)\mathbf{i} + (t^3 - 2t)\mathbf{j}$. Find the position vector of the particle when its velocity and acceleration are perpendicular.

③ **23** The position vectors of two particles A and B at time t are given by $\mathbf{r}_A = \cos t\mathbf{i} + \sin t\mathbf{j}$ and $\mathbf{r}_B = t\mathbf{i} + 2t\mathbf{j}$. Find the values of t for which the paths of the particles are perpendicular.

③ **R 24** A particle moves so that at time t its position vector is given by $\mathbf{r} = e^{2t}\cos 2t\mathbf{i} + e^{2t}\sin 2t\mathbf{j}$. Find the magnitudes of the velocity and acceleration vectors, and show that the acceleration vector is always perpendicular to the radius vector.

25 The position vector $\mathbf{r}$ m of a passenger riding on a particular section of a Big Dipper at a fun-fair is given by

$$\mathbf{r} = 2t\mathbf{i} + \tfrac{1}{4}(32 - 6t^2 + t^3)\mathbf{j} \quad 0 \leq t \leq 5$$

where $\mathbf{i}$ and $\mathbf{j}$ are unit vectors acting horizontally and vertically upwards. Sketch the path of the passenger $(0 \leq t \leq 5)$.

(i) Find the time he takes to travel from the highest point A to the lowest point B on the path. Find his speed at A and at B.

(ii) Find his greatest speed and the magnitude of his acceleration at this instant.

Questions 26–28: Suppose that $\mathbf{l}$ and $\mathbf{m}$ are two constant vectors in space—not necessarily $\mathbf{i}$ and $\mathbf{j}$—and that the position vector $\mathbf{r}$ of a particle P is given by

$$\mathbf{r} = f(t)\mathbf{l} + g(t)\mathbf{m}$$

By a slight extension of the idea of vector differentiation we can write

$$\mathbf{v} = f'(t)\mathbf{l} + g'(t)\mathbf{m}$$
$$\mathbf{a} = f''(t)\mathbf{l} + g''(t)\mathbf{m}$$

Note that (so long as $\mathbf{l}$ is not parallel to $\mathbf{m}$) P is moving in a plane—the plane of which $\mathbf{l}$ and $\mathbf{m}$ form a basis (see 6.2:1 Theorem 1).

26 The position vector of a particle P at time t is $(2t^2 - t)\mathbf{l} + t^2\mathbf{m}$, where $\mathbf{l}$ and $\mathbf{m}$ are constant vectors. Show that the acceleration of P is constant.

27 The position vector **r** of a particle P at time t is given by

$$\mathbf{r} = (5\cos 3t)\mathbf{l} + (5\sin 3t)\mathbf{m}$$

where **l** and **m** are constant vectors. Find the acceleration **a** of P, and show that $\mathbf{a} = -25\mathbf{r}$.

(i) Describe the path of P and find its cartesian equation when $\mathbf{l} = 2\mathbf{i}$, $\mathbf{m} = \mathbf{j}$.

(ii) Show that the velocity and acceleration of P are perpendicular at all times so long as $|\mathbf{l}| = |\mathbf{m}|$ and **l** is perpendicular to **m**.

28 When a particle is connected by a light elastic spring to the origin and moves in a smooth horizontal plane, its position vector **r** satisfies the differential equation

$$\frac{d^2\mathbf{r}}{dt^2} + n^2\mathbf{r} = 0 \quad n \text{ constant}$$

Verify that a possible solution of this equation is

$$\mathbf{r} = \mathbf{l}\sin nt + \mathbf{m}\cos nt$$

where **l** and **m** are constant vectors.
  Find **l** and **m** if the particle is initially at the point $2a\mathbf{i}$ (a constant), moving with velocity $an\mathbf{j}$. Find the cartesian equation of the path of the particle.

29 By extending the definition of the derivative of a vector to vectors in three dimensions, find the velocity and acceleration at time t of a particle whose position vector is:

- (i) $2t\mathbf{i} - 3t\mathbf{j} - 5t^2\mathbf{k}$
- (ii) $t^3\mathbf{i} + \frac{1}{2}t^2\mathbf{j} + (t^3 - 2)\mathbf{k}$
- (iii) $4\sin t\mathbf{i} + 4\cos t\mathbf{j} + 8t\mathbf{k}$

30 The position vector of a particle P at time t is

$$\mathbf{r} = (4 - 3t)\mathbf{i} + 2t\mathbf{j} + (3 + 2t)\mathbf{k}$$

Find the velocity and acceleration of P. Describe the path of P and find its cartesian equations.

R 31 The position vector of a particle at time t is $3e^t\mathbf{i} + 4e^{-2t}\mathbf{j} + 3t\mathbf{k}$. Find its speed when $t = \ln 2$.

32 The position vector of a particle at time t is $(a\cos nt)\mathbf{i} + (a\sin nt)\mathbf{j} + 2at\mathbf{k}$, where a and n are constant.

(i) Find the velocity and acceleration of the particle at time t. Show that the magnitude of its acceleration is constant. Is its speed constant?

(ii) Describe the path of the particle.

7.2:2 Use of integration

Bearing in mind our definition of $d\mathbf{u}/dt$, it seems sensible to make a corresponding definition of $\int \mathbf{u}\, dt$:

If $\mathbf{u} = f(t)\mathbf{i} + g(t)\mathbf{j}$, then

$$\int \mathbf{u}\, dt = \left\{\int f(t)\, dt\right\}\mathbf{i} + \left\{\int g(t)\, dt\right\}\mathbf{j} \qquad (9)$$

For example, if $\mathbf{u} = 3t^2\mathbf{i} + 2t\mathbf{j}$, then

$$\int_1^2 \mathbf{u}\, dt = [t^3\mathbf{i} + t^2\mathbf{j}]_1^2 = (8\mathbf{i} + 4\mathbf{j}) - (\mathbf{i} + \mathbf{j}) = 7\mathbf{i} + 3\mathbf{j}$$

Thus, since $\mathbf{a} = d\mathbf{v}/dt$,

$$\mathbf{v} = \int \mathbf{a}\, dt \qquad (10)$$

Use of integration

and since $\mathbf{v} = d\mathbf{r}/dt$,

$$\mathbf{r} = \int \mathbf{v}\, dt \qquad (11)$$

Worked example 4 The acceleration of a particle at time t is $6t\mathbf{i} - 4\mathbf{j}$. Initially the particle is moving with velocity $2\mathbf{i}$; when $t = 2$ it is at the point $-8\mathbf{j}$. Find its velocity and position vector at time t.

$$\mathbf{a} = 6t\mathbf{i} - 4\mathbf{j}$$

By equation (10),

$$\mathbf{v} = \int (6t\mathbf{i} - 4\mathbf{j})\, dt = 3t^2\mathbf{i} - 4t\mathbf{j} + \mathbf{c}$$

where $\mathbf{c}$ is an arbitrary *vector* constant.
$\mathbf{v} = 2\mathbf{i}$ when $t = 0 \Rightarrow \mathbf{c} = 2\mathbf{i}$; so

$$\mathbf{v} = (3t^2 + 2)\mathbf{i} - 4t\mathbf{j}$$

By equation (11),

$$\mathbf{r} = \int \{(3t^2 + 2)\mathbf{i} - 4t\mathbf{j}\}\, dt = (t^3 + 2t)\mathbf{i} - 2t^2\mathbf{j} + \mathbf{d}$$

where $\mathbf{d}$ is an arbitrary vector constant.
$\mathbf{r} = -8\mathbf{j}$ when $t = 2 \Rightarrow -8\mathbf{j} = 12\mathbf{i} - 8\mathbf{j} + \mathbf{d} \Rightarrow \mathbf{d} = -12\mathbf{i}$; so

$$\mathbf{r} = (t^3 + 2t - 12)\mathbf{i} - 2t^2\mathbf{j}$$

Exercise 7.2:2

Questions 1-7: A particle moves with acceleration $\mathbf{a}$. Find its velocity and position vector at time t.

1 $\mathbf{a} = 6t\mathbf{i}$. Initially the particle is at the origin, moving with velocity $\mathbf{j}$.

2 $\mathbf{a} = 2\mathbf{i} + \mathbf{j}$. Initially the particle is at rest at the point $3\mathbf{i} + \mathbf{j}$.

3 $\mathbf{a} = -4\mathbf{j}$. The initial velocity of the particle is $\mathbf{i} + \mathbf{j}$ and at time $t = 1$ it is at the point $2\mathbf{i} - \mathbf{j}$.

4 $\mathbf{a} = 12t^2\mathbf{i} + 6t\mathbf{j}$. Initially the particle is at the origin, moving with velocity $\mathbf{i} + 2\mathbf{j}$.

5 $\mathbf{a} = (15\sqrt{t})\mathbf{i} + (12/\sqrt{t})\mathbf{j}$. At time $t = 1$ the particle is at the point $4\mathbf{i} + 4\mathbf{j}$, moving with velocity $10\mathbf{i} + 24\mathbf{j}$.

6 $\mathbf{a} = (4 - 12t)\mathbf{i} - 36t^2\mathbf{j}$. At time $t = 1$ the particle is at the point $-3\mathbf{i} - 5\mathbf{j}$, moving with velocity $3\mathbf{i} + 2\mathbf{j}$.

R 7 $\mathbf{a} = 3\sin t\mathbf{i} + 4\cos t\mathbf{j}$. At time $t = \tfrac{1}{2}\pi$ the particle is at the point $5\mathbf{i} + \mathbf{j}$ moving with velocity $4\mathbf{j}$.

8 A particle moves in a plane with acceleration proportional to time t (i.e. $\mathbf{a} = \mathbf{k}t$, $\mathbf{k}$ constant). The velocity vector of the particle is zero at time $t = 0$ and is equal to $2\mathbf{i} - 6\mathbf{j}$ when $t = 1$. Find the velocity of the particle when $t = 3$.

9 A particle moves in a plane with acceleration proportional to t^2. Initially it is at rest at the point $\mathbf{i} - \mathbf{j}$. If its velocity when $t = 2$ is $16\mathbf{i} + 16\mathbf{j}$, find its position vector at this instant.

10 The acceleration at time t ($0 \leq t \leq 1$) of a particle P is $2\mathbf{i} + 6t\mathbf{j}$. Initially, P is at the origin moving with velocity $-\mathbf{j}$.

(i) Find the velocity $\mathbf{v}$ and position vector $\mathbf{r}$ of P at time t ($0 \leq t \leq 1$).

(ii) Show that, when $t = 1$, $\mathbf{v} = 2\mathbf{i} + 2\mathbf{j}$ and $\mathbf{r} = \mathbf{i}$.

At time $t = 1$, the acceleration of P becomes $-2\mathbf{j}$. Using the results of (ii) as initial conditions for the second part of the motion,

(iii) find $\mathbf{v}$ and $\mathbf{r}$ at time t ($t > 1$)

(iv) show that, when $t = 4$, $\mathbf{v} = 2\mathbf{i} - 4\mathbf{j}$ and $\mathbf{r} = 7\mathbf{i} - 3\mathbf{j}$

11 A particle is initially at rest at the point $2\mathbf{j}$. Subsequently its acceleration $\mathbf{a}$ is given by

$$\mathbf{a} = \begin{cases} 6t^2\mathbf{i} + 4\mathbf{j} & 0 \leq t \leq 2 \\ 6t\mathbf{j} & t > 2 \end{cases}$$

Find the velocity and position vector of the particle:

(i) at time t ($0 \leq t \leq 2$)
(ii) at time $t = 2$
(iii) at time t ($t > 2$)
(iv) at time $t = 3$

12 A particle starts from the point $3\mathbf{i} + \mathbf{j}$ and moves with constant velocity $-\mathbf{i} + 2\mathbf{j}$. Find its position vector at time t. Show that the path of the particle is a straight line and find its cartesian equation.

13 A particle moves with constant velocity from the point A with position vector $-2\mathbf{i} - 3\mathbf{j}$ to the point B with position vector $4\mathbf{i} + 5\mathbf{j}$. If the speed of the particle is 10 m s^{-1}, find its velocity and the time taken to travel from A to B.

14 At time $t = t_0$, a particle, moving with constant velocity $\mathbf{v}$, is at the point with position vector $\mathbf{r}_0$. Show that at time t its position vector $\mathbf{r}$ is given by

$$\mathbf{r} = \mathbf{r}_0 + \mathbf{v}(t - t_0) \qquad (12)$$

Note that this is *the vector equation of a straight line*. Find its cartesian equation(s) when

(i) $\mathbf{r}_0 = \mathbf{0}$, $\mathbf{v} = 5\mathbf{i}$, $t_0 = 0$
(ii) $\mathbf{r}_0 = 2\mathbf{j}$, $\mathbf{v} = -2\mathbf{i} + 4\mathbf{j}$, $t_0 = 3$
(iii) $\mathbf{r}_0 = -\mathbf{i} + 2\mathbf{j} - 3\mathbf{k}$, $\mathbf{v} = \mathbf{i} + \mathbf{j} + 4\mathbf{k}$, $t_0 = 0$

15 Find the position vector at time t of a particle which

(i) is at the point $-\mathbf{i} + 3\mathbf{j}$ at $t = 2$ and moves with constant velocity $-2\mathbf{i} + \mathbf{j}$
(ii) is at the point $6\mathbf{j}$ at $t = \tfrac{1}{2}$ and moves with constant velocity $4\mathbf{i} - 10\mathbf{j}$

16 A particle moves with constant acceleration $\mathbf{a}$. If its initial velocity is $\mathbf{u}$, show that at time t its velocity $\mathbf{v}$, and its displacement $\mathbf{r}$ from its initial position are given by

$$\mathbf{v} = \mathbf{u} + \mathbf{a}t \qquad (13)$$

and

$$\mathbf{r} = \mathbf{u}t + \tfrac{1}{2}\mathbf{a}t^2 \qquad (14)$$

Hence show that

$$\mathbf{r} = \tfrac{1}{2}(\mathbf{u} + \mathbf{v})t \qquad (15)$$

17 A particle moves with constant acceleration $-10\mathbf{j}$. Initially it is at the origin, moving with velocity $12\mathbf{i} + 16\mathbf{j}$. Find the cartesian equation of its path.

18 A particle moves with constant acceleration $6\mathbf{i}$. Initially it is at the point $3\mathbf{i} - \mathbf{j}$; when $t = 2$ its velocity is $8\mathbf{i} - 3\mathbf{j}$. Find the cartesian equation of its path.

19 A particle moves with constant acceleration $2\mathbf{i}$. Initially the particle is at the point $\mathbf{i} + \mathbf{j}$ moving with velocity $3\mathbf{i} + 4\mathbf{j}$. Find the position vector of the particle at time t.

20 A particle moves with constant acceleration $\mathbf{i} + 2\mathbf{j}$. Initially it is at the point $\mathbf{i} - \mathbf{j}$ moving with velocity $-2\mathbf{i} - 4\mathbf{j}$. Show that the particle travels in a straight line and find its cartesian equation.

21 A particle moves with constant acceleration. Initially the particle is at the origin moving with velocity $\mathbf{i} + \mathbf{j}$. If the velocity of the particle when $t = 2$ is $3\mathbf{i} - 2\mathbf{j}$, find its position vector at this time.

22 A particle moves with constant acceleration $\mathbf{a}$. Its initial velocity is $\mathbf{u}$.

(i) If $\mathbf{u} = \lambda \mathbf{a}$, where λ is constant, show that the particle travels in a straight line.

(ii) Show that if $\mathbf{u} \neq \lambda \mathbf{a}$, the path of the particle is a parabola.

(Hint: assume without loss of generality that $\mathbf{a} = -a\mathbf{j}$, $\mathbf{u} = b\mathbf{i} + c\mathbf{j}$ (a, b, c constant, $b \neq 0$), and that the particle is initially at the origin.)

7.2:3 Relative motion

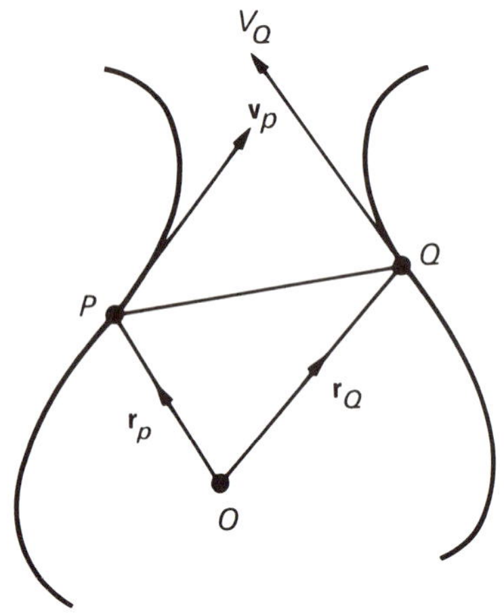

Consider two particles P and Q moving along curves in the xy plane. Suppose that their position vectors at time t are $\mathbf{r}_P$ and $\mathbf{r}_Q$ (fig. 7.2:4). Next consider the motion of Q relative to P, i.e. consider how the motion of Q would appear if you were moving with P:

The displacement of Q relative to P,

$$\overrightarrow{PQ} = \mathbf{r}_Q - \mathbf{r}_P \qquad (16)$$

Fig. 7.2:4

The velocity of Q relative to P,

$$_Q\mathbf{v}_P = \frac{d}{dt}\overrightarrow{PQ} = \frac{d}{dt}(\mathbf{r}_Q - \mathbf{r}_P)$$

$$= \frac{d\mathbf{r}_Q}{dt} - \frac{d\mathbf{r}_P}{dt} = \mathbf{v}_Q - \mathbf{v}_P$$

Thus

$$_Q\mathbf{v}_P = \mathbf{v}_Q - \mathbf{v}_P \qquad (17)$$

Similarly, the acceleration of Q relative to P,

$$_Q\mathbf{a}_P = \mathbf{a}_Q - \mathbf{a}_P \qquad (18)$$

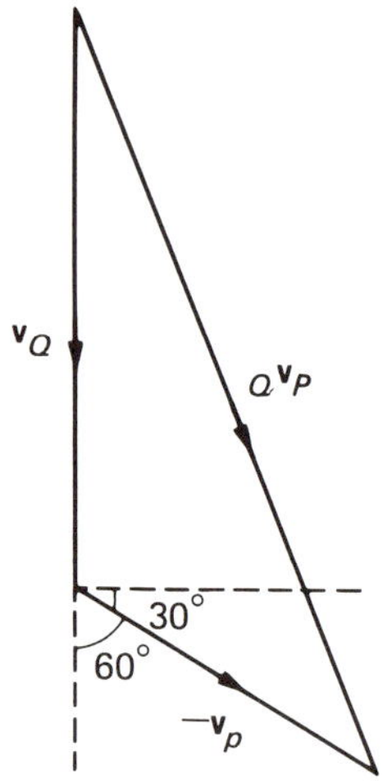

Fig. 7.2:5a

Worked example 5 The position vectors of particles P and Q at time t are $\cos t\mathbf{i} + \sin t\mathbf{j}$ and $t^2\mathbf{i} + t\mathbf{j}$ respectively. Find the velocity and acceleration of Q relative to P at time t.

$\mathbf{r}_P = \cos t\mathbf{i} + \sin t\mathbf{j} \Rightarrow \mathbf{v}_P = -\sin t\mathbf{i} + \cos t\mathbf{j} \Rightarrow \mathbf{a}_P = -\cos t\mathbf{i} - \sin t\mathbf{j}$.
$\mathbf{r}_Q = t^2\mathbf{i} + t\mathbf{j} \Rightarrow \mathbf{v}_Q = 2t\mathbf{i} + \mathbf{j} \Rightarrow \mathbf{a}_Q = 2\mathbf{i}$.

So $\qquad _Q\mathbf{v}_P = \mathbf{v}_Q - \mathbf{v}_P = (2t + \sin t)\mathbf{i} + (1 - \cos t)\mathbf{j}$

and $\qquad _Q\mathbf{a}_P = \mathbf{a}_Q - \mathbf{a}_P = (2 + \cos t)\mathbf{i} + \sin t\mathbf{j}$

Worked example 6 A ship P is steaming N60°W at 15 km h^{-1}. A ship Q is steaming due S at 25 km h^{-1}. Find the velocity of Q relative to P.

First draw the velocity triangle (fig. 7.2:5a) representing the equation

$$_Q\mathbf{v}_P = \mathbf{v}_Q - \mathbf{v}_P$$

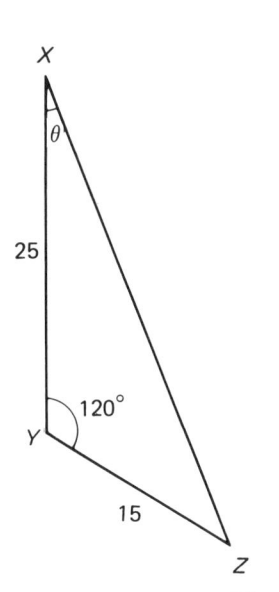

Kinematics of a particle 395

If $\mathbf{v}_Q$ is represented by $\overrightarrow{XY}$, $-\mathbf{v}_P$ by $\overrightarrow{YZ}$, then $\overrightarrow{XZ}$ represents $_Q\mathbf{v}_P$ (see fig. 7.2:**5b**). We can use the sine and cosine rules (see section 3.3:1) to find the magnitude and direction of $\overrightarrow{XZ}$.

Using the cosine rule,

$$(XZ)^2 = 25^2 + 15^2 - 2 \cdot 25 \cdot 15 \cos 120° \Rightarrow XZ = 35$$

Using the sine rule,

$$\frac{\sin \theta}{15} = \frac{\sin 120°}{35} \Rightarrow \theta = 21.8°$$

so the velocity of Q relative to P is 35 km h^{-1} in the direction S21.8°E.

Note that this method depends on the fact that the velocities are constant. For an alternative method see question 11.

Fig. 7.2:**5b**

Worked example 7 Relative to a ship P which is travelling due N at a speed of 10 km h^{-1}, the velocity of a speedboat Q is in the direction N45°E. Relative to a second ship R which is travelling due S at a speed of 10 km h^{-1}, the velocity of Q is in the direction N30°E. Find the actual velocity of Q.

Take unit vectors $\mathbf{i}$ and $\mathbf{j}$ in directions E and N respectively. Let $\mathbf{v}_Q = u\mathbf{i} + v\mathbf{j}$. We must find u and v.

Consider the motion relative to P:

$$\mathbf{v}_P = 10\mathbf{j}, \mathbf{v}_Q = u\mathbf{i} + v\mathbf{j} \Rightarrow {}_Q\mathbf{v}_P = \mathbf{v}_Q - \mathbf{v}_P = u\mathbf{i} + (v-10)\mathbf{j}$$

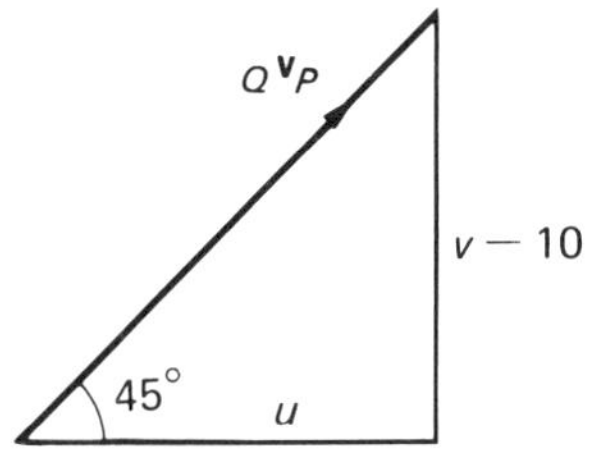

But we know that $_Q\mathbf{v}_P$ is in the direction N45°E (fig. 7.2:**6**), so

$$\frac{v-10}{u} = \tan 45° = 1$$

Fig. 7.2:**6**   i.e. $\hspace{4em} u = v - 10 \hspace{4em}$ **(19)**

Consider the motion relative to R:

$$\mathbf{v}_R = -10\mathbf{j}, \mathbf{v}_Q = u\mathbf{i} + v\mathbf{j} \Rightarrow {}_Q\mathbf{v}_R = \mathbf{v}_Q - \mathbf{v}_R = u\mathbf{i} + (v+10)\mathbf{j}$$

But we know that $_Q\mathbf{v}_R$ is in the direction N30°E (fig. 7.2:**7**), so

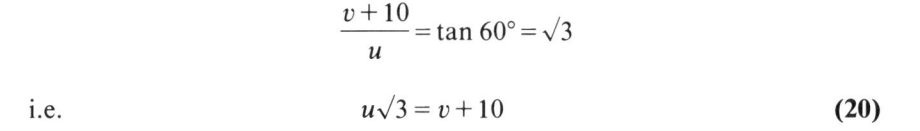

$$\frac{v+10}{u} = \tan 60° = \sqrt{3}$$

i.e. $\hspace{4em} u\sqrt{3} = v + 10 \hspace{4em}$ **(20)**

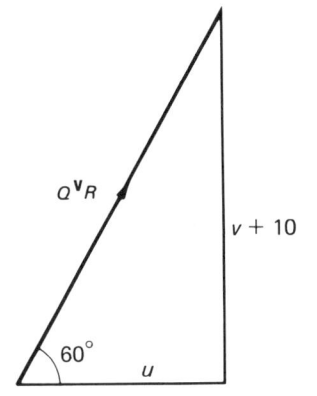

From equations (19) and (20), $u = 20/(\sqrt{3} - 1) \approx 27.32$, $v = 10(\sqrt{3} + 1)/(\sqrt{3} - 1) \approx 37.32$, so

Fig. 7.2:**7** $\hspace{8em} \mathbf{v}_Q \approx 27.32\mathbf{i} + 37.32\mathbf{j}$

Worked example 8 A plane whose speed in still air is 800 km h^{-1} flies at a constant height from A to B, where B is 1000 km S30°E of A. A wind of speed 100 km h^{-1} blows from the west. Find the course to be set and the time the plane takes to reach B.

We are asked to find the time taken to reach B, so we should find the actual speed of the plane, i.e. $|\mathbf{v}_P|$. We are also asked for the direction of $_P\mathbf{v}_W$, i.e. the direction in which the plane must steer in order to fly along AB.

We draw the velocity triangle (fig. 7.2:**8**) representing the equation

$$_P\mathbf{v}_W = \mathbf{v}_P - \mathbf{v}_W$$

i.e.
$$\mathbf{v}_P = \mathbf{v}_W + {}_P\mathbf{v}_W$$

Using the cosine rule (see section 3.3:1),

$$800^2 = 100^2 + (XY)^2 - 2 \cdot 100 \cdot XY \cos 60°$$
$$\Rightarrow XY \approx 850 \quad \text{(the other solution is negative)}$$

Thus $|\mathbf{v}_P|$, the actual speed of the plane, is 850 km h^{-1}, and

$$\text{time taken} = \frac{\text{distance}}{\text{speed}} = \frac{1000}{850} \approx 1.15 \text{ h}$$

Using the sine rule,

$$\frac{\sin \theta}{850} = \frac{\sin 60°}{800} \Rightarrow \theta = 113.12°$$

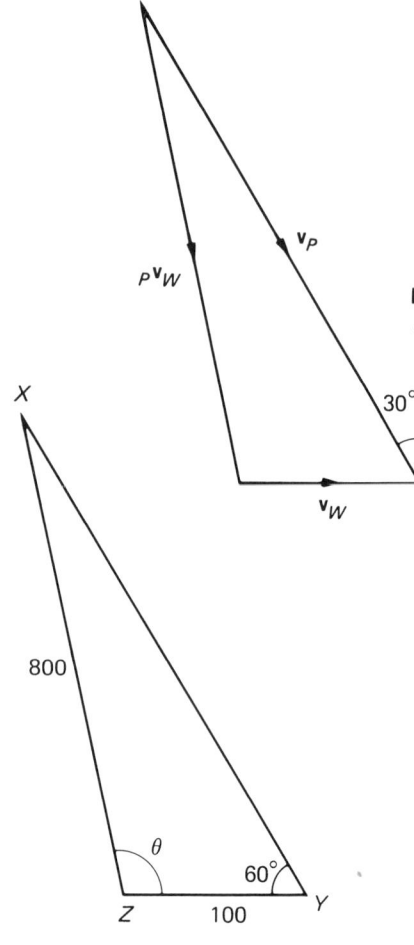

Fig. 7.2:**8**

so the course to be steered is S23.12°E.

Worked example 9 At $t = 0$ a particle P starts from the point $\mathbf{i} - 4\mathbf{j}$ and moves with constant velocity $\frac{1}{3}\mathbf{i} + \frac{1}{2}\mathbf{j}$. Simultaneously, a particle Q starts from the point $-3\mathbf{i} + 5\mathbf{j}$ and moves with constant velocity $\mathbf{i} + a\mathbf{j}$. If P and Q collide, find the time of collision, the value of a and the position vector of the point of collision.

At time t (by equation (12)),

$$\mathbf{r}_P = (\tfrac{1}{3}t + 1)\mathbf{i} + (\tfrac{1}{2}t - 4)\mathbf{j} \quad (21)$$
$$\mathbf{r}_Q = (t - 3)\mathbf{i} + (at + 5)\mathbf{j} \quad (22)$$

So $\overrightarrow{PQ} = \mathbf{r}_Q - \mathbf{r}_P = (\tfrac{2}{3}t - 4)\mathbf{i} + (at - \tfrac{1}{2}t + 9)\mathbf{j}$

First method: When P and Q collide, $\overrightarrow{PQ} = \mathbf{0}$, i.e.

$$(\tfrac{2}{3}t - 4)\mathbf{i} + (at - \tfrac{1}{2}t + 9)\mathbf{j} = \mathbf{0}$$
$$\Rightarrow \qquad t = 6, \quad a = -1$$

Substituting $t = 6$ into equation (21) (or (22)), the position vector of the point of collision is $3\mathbf{i} - \mathbf{j}$.

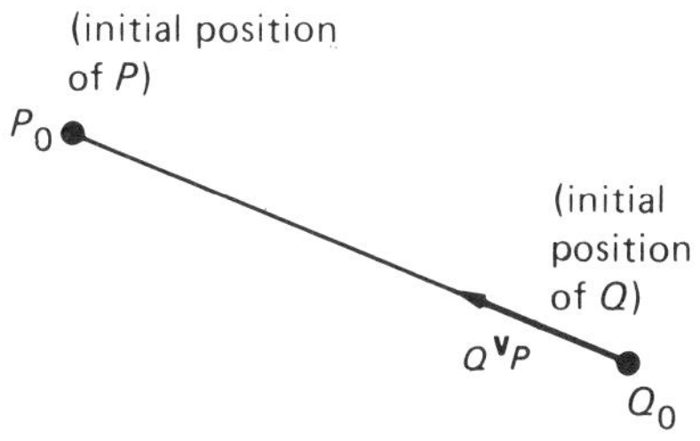

Kinematics of a particle 397

(initial position of P)

P_0

(initial position of Q)

$_Q\mathbf{v}_P$

Q_0

Fig. 7.2:9

Second method: Consider the motion of the particles relative to P.

$$_P\mathbf{v}_P = \mathbf{0} \quad \text{(i.e. } P \text{ is at rest)}$$

$$_Q\mathbf{v}_P = \mathbf{v}_Q - \mathbf{v}_P = \tfrac{2}{3}\mathbf{i} + (a - \tfrac{1}{2})\mathbf{j}$$

If P and Q are to collide, Q must appear to come straight towards P (fig. 7.2:9), i.e.

if P and Q are to collide, $_Q\mathbf{v}_P$ must be parallel to $\overrightarrow{Q_0P_0}$

Furthermore, if the collision occurs at time t,

$$t(_Q\mathbf{v}_P) = \overrightarrow{Q_0P_0}$$

i.e. $$t\{\tfrac{2}{3}\mathbf{i} + (a - \tfrac{1}{2})\mathbf{j}\} = 4\mathbf{i} - 9\mathbf{j}$$

$\Rightarrow \quad t = 6, \quad a = -1$

Worked example 10 A particle P starts from the point $3\mathbf{i} + 3\mathbf{j}$ and moves with constant velocity $5\mathbf{i} + 2\mathbf{j}$. One second later a particle Q stands from the point $9\mathbf{i} - 2\mathbf{j}$ and moves with constant velocity $2\mathbf{i} + 3\mathbf{j}$. Find the time at which the particles are closest together, their shortest distance apart and their position vectors at this time.

At time t (by equation (12)):

$$\mathbf{r}_P = (3 + 5t)\mathbf{i} + (3 + 2t)\mathbf{j} \quad (23)$$

$$\mathbf{r}_Q = (7 + 2t)\mathbf{i} + (-5 + 3t)\mathbf{j} \quad (24)$$

So $$\overrightarrow{PQ} = \mathbf{r}_Q - \mathbf{r}_P = (4 - 3t)\mathbf{i} + (-8 + t)\mathbf{j}$$

First method: Let the distance $PQ = l$; then

$$l^2 = |\mathbf{r}_Q - \mathbf{r}_P|^2 = (4 - 3t)^2 + (-8 + t)^2 = 80 - 40t + 10t^2$$

l is least when l^2 is least, i.e. when

$$\frac{d(l^2)}{dt} = -40 + 20t = 0 \Rightarrow t = 2$$

and then $l = \sqrt{(80 - 80 + 40)} = \sqrt{40}$.

Substituting $t = 2$ into equations (23) and (24), the position vectors of the particles are $13\mathbf{i} + 7\mathbf{j}$ and $11\mathbf{i} + \mathbf{j}$.

Second method: Consider the motion of the particles relative to P.

$$_P\mathbf{v}_P = \mathbf{0} \quad \text{(i.e. } P \text{ is at rest)}$$

$$_Q\mathbf{v}_P = \mathbf{v}_Q - \mathbf{v}_P = -3\mathbf{i} + \mathbf{j}$$

398 *Relative motion*

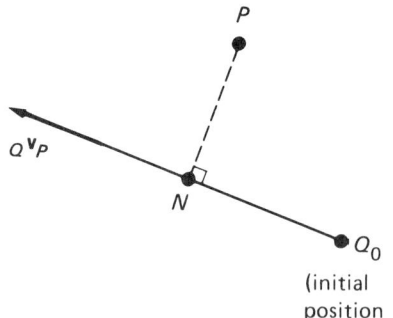

Q moves closer and closer to P until (relative to P) it reaches N, where PN is perpendicular to ${}_Q\mathbf{v}_P$, and then moves further and further away (fig. 7.2:**10**), i.e.

> P and Q are closest together when $\overrightarrow{PQ}$ is perpendicular to ${}_Q\mathbf{v}_P$

i.e. when

$$\overrightarrow{PQ} \cdot {}_Q\mathbf{v}_P = 0 \qquad \text{(see 6.1:1 Theorem 1)}$$
$$\Rightarrow \{(4-3t)\mathbf{i} + (-8+t)\mathbf{j}\} \cdot (-3\mathbf{i}+\mathbf{j}) = 0$$
$$\Rightarrow -12 + 9t - 8 + t = 0$$
$$\Rightarrow t = 2$$

Fig. 7.2:**10**

Exercise 7.2:3

1 The position vectors of two particles P and Q at time t are $3\cos 2t\,\mathbf{i} + 3\sin 2t\,\mathbf{j}$ and $4\sin t\,\mathbf{i} - 4\cos t\,\mathbf{j}$. Find the velocity and acceleration of Q relative to P when $t = \tfrac{1}{3}\pi$.

2 The position vectors of particles P and Q at time t are $t^3\mathbf{i} + t^2\mathbf{j}$ and $3t^3\mathbf{i} - t\mathbf{j}$ respectively. Find the velocity and acceleration of Q relative to P

(i) at time t
(ii) at time $t = 1$

Find the speed of Q relative to P at time $t = 1$.

3 The position vectors of particles P and Q at time t are $t\mathbf{i} + (\sin t)\mathbf{j}$ and $(\cos t)\mathbf{i} - t\mathbf{j}$ respectively. Find the velocity and acceleration of Q relative to P when $t = \tfrac{1}{2}\pi$.

4 The particle P starts from the origin with initial velocity $\mathbf{i} - \mathbf{j}$ and moves with acceleration $2t\mathbf{j}$. The particle Q starts from the point $\mathbf{i}$ with initial velocity $\mathbf{i} + \mathbf{j}$ and moves with acceleration $-2t\mathbf{i}$.

(i) Find the velocity of Q relative to P at time t. Hence find the speed of Q relative to P when $t = 2$.

(ii) Find $\overrightarrow{PQ}$ at time t. Hence find the distance between the particles when $t = 2$.

5 Two particles start simultaneously from O. Particle P moves with constant velocity $2\mathbf{i} + 6\mathbf{j}$. Particle Q has initial velocity $4\mathbf{i} - 8\mathbf{j}$ and moves with acceleration $\mathbf{i} + \mathbf{j}$. Find:

(i) the time at which their velocities are perpendicular;
(ii) the velocity of Q relative to P at time t;
(iii) the distance between them at time t.

6 The velocity of a particle P at time t is $6t - 2\mathbf{j}$. The velocity of Q relative to P is $-4t\mathbf{i} + 5\mathbf{j}$; find the velocity of Q.

7 The velocity of a particle P is $2\mathbf{i} - 3t^2\mathbf{j}$. The velocity of another particle Q relative to P is $2t\mathbf{i} + 2\mathbf{j}$. If Q is initially at the point $\mathbf{i} - \mathbf{j}$, find the position vector of Q at time t.

8 The position vector of a particle P at time t is $t\mathbf{i} + t^2\mathbf{j}$. The velocity of another particle Q relative to P is $\mathbf{i} + 4t\mathbf{j}$, and when $t = 0$, $\overrightarrow{PQ} = \mathbf{j}$. Find the cartesian equation of the path of Q.

9 P and Q are two moving particles. Find:

(i) ${}_P\mathbf{v}_P$ (ii) ${}_P\mathbf{v}_Q + {}_Q\mathbf{v}_P$

Is the speed of P relative to Q the same as the speed of Q relative to P? If the accelerations of P and Q are the same, what can you say about ${}_Q\mathbf{v}_P$?

10 Two particles P and Q start simultaneously from O with initial velocities $20\mathbf{i} + 30\mathbf{j}$ and $25\mathbf{i} + 15\mathbf{j}$ respectively and move with acceleration $-10\mathbf{j}$. Find:

(i) ${}_Q\mathbf{a}_P$ (ii) ${}_Q\mathbf{v}_P$ (iii) $\overrightarrow{PQ}$

at time t.

• **11** A ship P is steaming N60°W at 15 km h^{-1}. A ship Q is steaming due S at 25 km h^{-1}. Taking unit vectors $\mathbf{i}$ and $\mathbf{j}$ in directions E and N respectively, write $\mathbf{v}_P$ and $\mathbf{v}_Q$ in the form $\alpha\mathbf{i} + \beta\mathbf{j}$. Hence find ${}_Q\mathbf{v}_P$ in the form $a\mathbf{i} + b\mathbf{j}$. Hence find the magnitude and direction of ${}_Q\mathbf{v}_P$.

Compare this method with that of worked example 6. Which do you prefer?

Questions 12–19: Find the magnitude and direction of the velocity of Q relative to P when the speeds of P and Q are 4 m s^{-1} and 3 m s^{-1}, and their directions of motion are those given in the question.

⑥ **12** P: due N; Q: due S.

⑥ **13** P: due N; Q: due E.

⑥ **14** P: NE; Q: NW.

⑥ **15** P: N30°E; Q: N60°W.

⑥ **16** P: N30°E; Q: due E.

⑥ → **17** P: N30°E; Q: SE.

⑥ **18** P: N30°E; Q: S40°W.

⑥ **19** P: S60°W; Q: S40°W.

⑥ **20** A ship steams due S at 20 km h^{-1} while a power boat moves S60°E at 70 km h^{-1}. Find the apparent velocity of the power boat to an observer on the ship.

⑥ **21** A boat is moving N30°W at 15 km h^{-1} in a wind which comes from the SW at 10 km h^{-1}. Find the apparent speed and direction of the wind to an observer on the boat.

⑥ **22** A man is cycling along a level road at 6 m s^{-1}. Rain, inevitably, is falling. It falls vertically at 8 m s^{-1}. Find the apparent speed and direction of the rain to the cyclist.

⑥ → **23** A train is moving at 25 m s^{-1}. Rain is falling vertically at 5 m s^{-1}. Find the angle which the streaks on the carriage windows make with the vertical.

⑥ • **24** To an observer in a boat moving NE at 20 km h^{-1}, a plane appears to be flying due W at 100 km h^{-1}. What is the actual velocity of the plane?

⑥ → **25** A passenger in a train travelling NE at 100 km h^{-1} watches a car moving on a straight road. The car appears to be travelling S30°W at 75 km h^{-1}. What is the actual velocity of the car?

⑥ → **26** To a cyclist riding along a level road at 6 m s^{-1} the inevitable rain appears to be falling at 12 m s^{-1}. If the rain is actually falling vertically, find the actual speed of the rain.

⑥ **27** To a man walking due W at 5 km h^{-1}, the wind appears to blow from S30°E. If the wind actually blows from S60°E, find the actual speed of the wind.

⑦ **28** Three particles P, Q and R are moving with constant velocities. The velocities of P and Q are $\mathbf{i}+3\mathbf{j}$ and $-4\mathbf{i}+3\mathbf{j}$ respectively. The velocity of R relative to P is in the direction $3\mathbf{i}+2\mathbf{j}$ and the velocity of R relative to Q is in the direction $7\mathbf{i}+3\mathbf{j}$. Find the actual velocity of R.

⑦ → **29** Relative to a ship P which is travelling due E at a speed of 20 km h^{-1}, another ship R appears to be travelling due N. Relative to a ship Q which is travelling due S at a speed of 15 km h^{-1}, R appears to be travelling in the direction N30°E. Find the actual velocity of R.

⑦ **30** To a cyclist riding due S at 20 km h^{-1} the wind appears to be blowing from N60°E. When she reduces her speed to 15 km h^{-1}, without changing direction, the wind appears to blow from N30°W. Find the actual velocity of the wind.

⑦ **31** A boat P is moving due E at 9 knots and a boat Q is moving N30°E at 6 knots. A boat R appears to an observer on P to be sailing due S, and to an observer on Q to be sailing S30°E. Find the actual velocity of R.

⑦ **32** Two aircraft, A and B, are flying at the same height. Both have speed 400 km h^{-1}; A is flying in a direction N30°W and B is flying due E. A third aircraft, also flying at the same height, appears to the pilot of A to be on a course due S while to the pilot of B its course appears to be S60°W. In what direction is the third aircraft actually flying?

⑦ **33** A man walking due E finds that the wind appears to blow from due N. When he doubles his velocity he finds that the wind appears to blow from NE. Find the direction from which the wind is actually blowing.

⑧ **34** A ship which steams at 18 km h^{-1} has to travel NE through water in which there is a current flowing from due W at 4 km h^{-1}. Find the direction in which the ship should be headed and the actual speed of the ship.

⑧ → **35** A plane which can move in still air at 1000 km h^{-1} flies from A to B at constant height. B is 400 km N30°W of A. There is a wind blowing from due N at a speed of 100 km h^{-1}. Find in what direction the pilot must steer and the time taken by the plane to reach B.

⑧ **36** A man in a rowing boat is at a point A on the northern bank of a river 100 m wide which is flowing at 0.8 m s^{-1} from W to E between parallel banks. He wants to reach a point B on the south bank 75 m downstream from A. If he can row at 1 m s^{-1} in still water, find the course he must steer in order to reach B as fast as possible and the time he takes to reach B.

⑧ **37** A boat whose speed in still water is 10 km h^{-1} has to travel 20 km N60°E and back in a current of 3 km h^{-1} from due E. Find the time it takes for the complete journey.

⑧ → **38** A helicopter flies with constant airspeed 100 km h^{-1} from position A to position B which is 50 km northeast of A, and then flies back to A. Throughout the flight the wind velocity is 30 km h^{-1} from the west. Find the total time of the flight.

⑧ → **39** A boat can travel in still water at 5 m s^{-1}. It travels across a river flowing at 3 m s^{-1} which is 500 m wide. Find the time it takes if:

(i) it travels the shortest possible distance
(ii) it takes the shortest possible time

⑧ **40** A boy can swim in still water at $1\,\mathrm{m\,s^{-1}}$. He swims across a river flowing at $0.6\,\mathrm{m\,s^{-1}}$ which is 336 m wide. Find the time he takes if:

(i) he travels the shortest possible distance
(ii) he takes the shortest possible time

⑧+ **41** A river flows at a constant speed of $5\,\mathrm{m\,s^{-1}}$ between straight parallel banks which are 300 m apart. A boat, which has a maximum speed of $3.25\,\mathrm{m\,s^{-1}}$ in still water, leaves a point A on one bank and sails in a straight line to the opposite bank. Find the least time the boat can take to reach a point B on the opposite bank where $AB = 500$ m and B is downstream of A. Find the least time the boat can take to cross the river. Find the time taken to sail from A to B by the slowest boat capable of sailing directly from A to B.

Questions 42–46: Two particles P and Q move in the xy plane. Given that P and Q collide, find the time of collision, the value of the constant a, and the position vector of the point of collision. Illustrate the motion of the particles on a sketch. Try both the methods of worked example 9. Decide which method you prefer. Feed the other method to your pet boa constrictor.

⑨ **42** At $t = 0$, P starts from the point $4\mathbf{i} + 3\mathbf{j}$ and moves with velocity $-3\mathbf{i} + \mathbf{j}$. Simultaneously, Q starts from the point $-2\mathbf{i} - \mathbf{j}$ and moves with velocity $3\mathbf{i} + a\mathbf{j}$.

⑨ **43** At $t = 0$, P starts from the point $4\mathbf{i} - 4\mathbf{j}$ and moves with velocity $-5\mathbf{i} + 2\mathbf{j}$. Simultaneously, Q starts from the point $-2\mathbf{i} + 5\mathbf{j}$ and moves with velocity $a\mathbf{i} - \mathbf{j}$.

⑨ **44** At $t = 0$, P starts from the point $-3\mathbf{i} + 6\mathbf{j}$ and moves with velocity $5\mathbf{i} + \mathbf{j}$. At $t = 1$, Q starts from the point $a\mathbf{i} + 3\mathbf{j}$ and moves with velocity $-\mathbf{i} + 5\mathbf{j}$. (Hint: if you are using the second method here, you will need first to find the position of Q at $t = 0$.)

⑨→ **45** At $t = 0$, P starts from the point $-2\mathbf{i} + 5\mathbf{j}$ and moves with velocity $a\mathbf{i} + 3\mathbf{j}$. One second later, Q starts from the point $7\mathbf{i} - \mathbf{j}$ and moves with velocity $\mathbf{i} + 6\mathbf{j}$.

⑨ R **46** At $t = 0$, P starts from the point $\mathbf{i} + \mathbf{j}$ and moves with velocity $-\tfrac{1}{3}\mathbf{i} + \tfrac{2}{3}\mathbf{j}$. At $t = \tfrac{1}{2}$, Q starts from the point $-2\mathbf{i} - 2\mathbf{j}$ and moves with velocity $a\mathbf{i} + 2\mathbf{j}$.

⑨ **47** A particle P starts at rest at the point $-4\mathbf{i} - 3\mathbf{j}$ and moves with acceleration $2\mathbf{i} + 6t\mathbf{j}$. At the same time, a particle Q starts from the point $-6\mathbf{i} + \mathbf{j}$ and moves with constant velocity $3\mathbf{i} + 2\mathbf{j}$. Prove that P and Q collide, and find the time of collision and the position vector of the point of collision. (Note: the second method of worked example 9 is dangerous here—why?)

⑨ **48** At $t = 0$ a particle P starts from the point $-8\mathbf{i} + 4\mathbf{j}$ and moves with constant velocity $3\mathbf{i} - 2\mathbf{j}$. At $t = 1$ a particle Q starts from the point $-5\mathbf{i} - 10\mathbf{j}$ and moves with constant speed in the direction of the vector $3\mathbf{i} + 4\mathbf{j}$. If P and Q collide, find the time of collision, the speed of Q and the position vector of the point of collision. (Hint: use the second method of worked example 9.)

⑨→ **49** A particle P starts from the point $-5\mathbf{i} + 4\mathbf{j}$ and moves with constant velocity $7\mathbf{i} + 3\mathbf{j}$. At the same time a particle Q starts from the point $\mathbf{i} + 4\mathbf{j}$ and moves with constant speed 5 so as to intercept P. Find the velocity of Q, the time taken before interception and the position vector of the point of interception.

⑨ **50** A rock star, escaping from his adoring fans after a triumphant concert at which vast amounts of money are raised for various charities, and for the star's manager, emerges from the stage door of the theatre at the point $(10\mathbf{i} + 10\mathbf{j})$ m and saunters with constant velocity $(-\tfrac{3}{2}\mathbf{i} + \mathbf{j})\,\mathrm{m\,s^{-1}}$ towards his private helicopter. At the same time an 11-year-old girl breaks through the police cordon at the point $(-30\mathbf{i} - 40\mathbf{j})$ m and runs with constant speed $6.5\,\mathrm{m\,s^{-1}}$ so as to intercept the star.

Find the velocity of the girl, the time taken before the interception and the position vector of the point of interception.

He turns round, surprised, and kisses her tenderly on the cheek ... (It is his daughter.)

⑨ **51** A DC10, P, is flying at $600\,\mathrm{km\,h^{-1}}$ in the direction S30°W. Another DC10, Q, initially 10 km due S of P, is flying NW at the same height. Given that P and Q collide, find the speed of Q and the time that elapses before the disaster. (Full photos—p. 11; Widows tell tale of anguish—p. 17.)

⑨ **52** A horse is galloping across a field at $30\,\mathrm{km\,h^{-1}}$ in a direction N30°W. It is already 200 m due E of a rider who sets off at $40\,\mathrm{km\,h^{-1}}$. In what direction should she ride to catch the horse? How long will it take her?

Questions 53–56: Two particles P and Q move in the xy plane. Find the time at which they are closest together, their shortest distance apart and their position vectors at this time.

Try both the methods of worked example 10.

⑩ **53** At $t = 0$ P starts from the point $-3\mathbf{i} + 3\mathbf{j}$ and moves with velocity $5\mathbf{i} + 4\mathbf{j}$. Simultaneously, Q starts from the point $4\mathbf{i} + 7\mathbf{j}$ and moves with velocity $3\mathbf{i} + 5\mathbf{j}$.

⑩ **54** At $t = 0$ P starts from the point $2\mathbf{i}$ and moves with velocity $\mathbf{j}$. Simultaneously, Q starts from the point $-\mathbf{i} + \mathbf{j}$ and moves with velocity $\mathbf{i} + 2\mathbf{j}$.

⑩→ **55** At $t = 1$ P starts from the point $9\mathbf{i} + 6\mathbf{j}$ and moves with velocity $-3\mathbf{i} + \mathbf{j}$. At $t = 1$ Q starts from the point $3\mathbf{j}$ and moves with velocity $-2\mathbf{i} + 2\mathbf{j}$.

⑩ **56** At $t = 0$ P starts from the point $8\mathbf{j}$ and moves with velocity $5\mathbf{i} + \mathbf{j}$. At $t = \tfrac{1}{2}$ Q starts from the point $17\mathbf{i} + 2\mathbf{j}$ and moves with velocity $2\mathbf{i} + 4\mathbf{j}$.

Kinematics of a particle 401

(10) **57** A particle P starts from the point $9\mathbf{i} - 12\mathbf{j}$ and moves with constant velocity $\mathbf{i} + 5\mathbf{j}$. At the same time a particle Q starts from the point $-7\mathbf{i} + a\mathbf{j}$ and moves with constant velocity $4\mathbf{i} + 3\mathbf{j}$. If P and Q are closest together after 6 s, find the value of a and their shortest distance apart.

(10) **58** A particle P starts from the point $3\mathbf{i} + \frac{1}{2}\mathbf{j}$ with initial velocity $5\mathbf{i} + \mathbf{j}$ and moves with acceleration $4\mathbf{i} + \mathbf{j}$. At the same time a particle Q starts from the point $4\mathbf{i} + 3\mathbf{j}$ with initial velocity $2\mathbf{i} + \mathbf{j}$ and moves with acceleration $6\mathbf{i} - 2\mathbf{j}$. Show that P and Q are closest together after 1 s and find their shortest distance apart.

(10) • **59** At $t = 0$ a particle P starts from the point $-12\mathbf{i} + 8\mathbf{j}$ and moves with velocity $3\mathbf{i} + 3\mathbf{j}$. Simultaneously a particle Q starts from the point $10\mathbf{i} - 14\mathbf{j}$ and moves with velocity $-\mathbf{i} + 5\mathbf{j}$. Find $\overline{PQ}$ at time t. Hence find the time for which the distance between P and Q is less than 10.

(10) → **60** At $t = 0$ a particle P starts from the point $4\mathbf{i} - 2\mathbf{j}$ and moves with velocity $4\mathbf{i} + 5\mathbf{j}$. At $t = 1$ a particle Q starts from the point $\mathbf{i} - 8\mathbf{j}$ and moves with velocity $5\mathbf{i} + 8\mathbf{j}$. Find:

(i) $_Q\mathbf{v}_P$
(ii) $\overline{PQ}$
(iii) the shortest distance between P and Q
(iv) the time for which the distance between P and Q is less than 10.

(10) R **61** A particle P starts from the point $3\mathbf{i} - 5\mathbf{j}$ and moves with velocity $\mathbf{i} + 2\mathbf{j}$. Simultaneously a particle Q starts from the point $3\mathbf{i} + 5\mathbf{j}$ and moves with velocity $2\mathbf{i} - \mathbf{j}$. Find:

(i) the shortest distance between P and Q
(ii) the time for which the distance between P and Q is less than 5.

(10) → **62** A ship A is moving at a speed of 15 km h^{-1} in the direction $-3\mathbf{i} + 4\mathbf{j}$. A second ship B, which is initially $20\sqrt{2}$ km from A in the direction $-7\mathbf{i} + \mathbf{j}$, is moving in the direction $\mathbf{i} + 2\mathbf{j}$. Show that, if the speed of B is $5\sqrt{5}$ km h^{-1}, the ships will collide, and find the position vector of their point of collision. Show that if the speed of B is $3\sqrt{5}$ km h^{-1} the ships will not collide, and find their shortest distance apart.

(10) **63** A plane P, flying with velocity 200 km h^{-1} due W, is 40 km due N of another plane Q flying at the same height with velocity 300 km h^{-1} N30°W. After what time will P and Q be closest together? How far apart will they be at this time? Find the time for which they are within 30 km of each other.

(10) **64** A ship P steaming at 20 km h^{-1} in the direction N50°E is 120 km due W of a ship Q steaming at 12 km h^{-1} in the direction N30°W. Find:

(i) the shortest distance between them in subsequent motion and the time taken to reach this position
(ii) the time for which the ships are within a range of 50 km of each other.

(10) → **65** The position vectors of particles P and Q at time t are $2\mathbf{i} + \cos 2t\mathbf{j} + \sin 2t\mathbf{k}$ and $\sin 2t\mathbf{i} - \cos 2t\mathbf{j} + 3\mathbf{k}$ respectively. Find $\overline{PQ}$ and $_Q\mathbf{v}_P$ at time t. Hence find the smallest and greatest distances between P and Q. (See 7.2:1 Q28.)

+ **66** $ABCD$ is a rectangular field; $AD = 45$ m, $AB = 60$ m. A horse starts from A and gallops along AB at 6 m s^{-1}. At the same moment a man starts out from D to try to catch the horse. If the man can run at 4 m s^{-1}, show that he cannot get nearer to the horse than $5\sqrt{5}$ m. Show that the man would have to run at at least 7.5 m s^{-1} in order to catch the horse before it reached B.

7.3 Projectiles

7.3:1 Motion of a projectile

Consider a particle projected with initial velocity u at an angle α to the horizontal. We take the point of projection as the origin O, and axes Ox, Oy horizontally and vertically upwards. Throughout the flight, the acceleration $\mathbf{a}$ of the particle is g vertically downwards (see section 7.1:4), i.e.

$$\mathbf{a} = -g\mathbf{j} \quad (1)$$

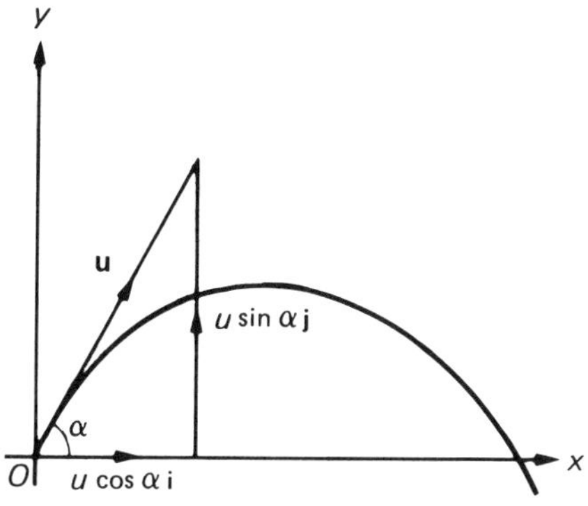

Fig. 7.3:1a Initially

Let the initial velocity of the particle P be $\mathbf{u}$; $\mathbf{u} = u \cos \alpha \mathbf{i} + u \sin \alpha \mathbf{j}$ (fig. 7.3:1a). After time t, since the acceleration $\mathbf{a}$ is constant (see 7.2:2 Q16),

402   *Motion of a projectile*

$$\left. \begin{array}{l} \mathbf{v} = \mathbf{u} + \mathbf{a}t = u\cos\alpha\,\mathbf{i} + (u\sin\alpha - gt)\mathbf{j} \\ \mathbf{r} = \mathbf{u}t + \tfrac{1}{2}\mathbf{a}t^2 = ut\cos\alpha\,\mathbf{i} + (ut\sin\alpha - \tfrac{1}{2}gt^2)\mathbf{j} \end{array} \right\} \quad (2)$$

i.e.

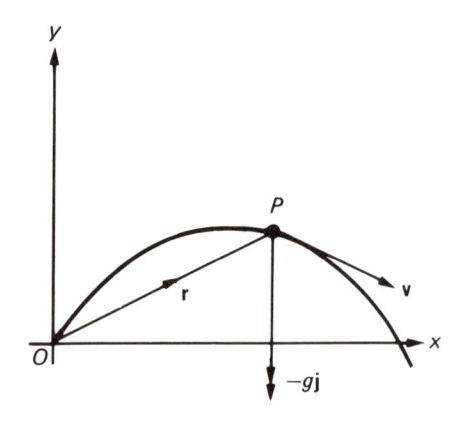

$$\left. \begin{array}{ll} \dot{x} = u\cos\alpha & \dot{y} = u\sin\alpha - gt \\ x = ut\cos\alpha & y = ut\sin\alpha - \tfrac{1}{2}gt^2 \end{array} \right\} \quad (3)$$

Fig. 7.3:**1b** At time t

See fig. 7.3:**1b**.

Worked example 1 A particle is projected from a point O on level ground with a velocity of 25 m s^{-1} at an angle $\tan^{-1}\tfrac{3}{4}$ to the horizontal. Find:

(i) the greatest height reached by the particle;
(ii) the distance from O at which the particle first strikes the ground (called the **range** of the particle).

The path of the particle is shown in fig. 7.3:**2**. At time t,

$$\dot{x} = 20 \qquad \dot{y} = 15 - 10t$$
$$x = 20t \qquad y = 15t - 5t^2$$

by equations (3).

(i) At B, the highest point on the path, the particle is moving horizontally, i.e. $\dot{y} = 0 \Rightarrow 15 - 10t = 0 \Rightarrow t = \tfrac{3}{2}$ s.

Fig. 7.3:2

Greatest height $BC = y(\tfrac{3}{2}) = 15(\tfrac{3}{2}) - 5(\tfrac{3}{2})^2 = \tfrac{45}{4}$ m

(ii) At A, where the particle strikes the ground,
$y = 0 \Rightarrow 15t - 5t^2 = 0 \Rightarrow t = 3$ s or 0 s.
(What is the significance of the solution $t = 0$?)

Range $OA = x(3) = 20 \cdot 3$ m $= 60$ m

Worked example 2 A ball is projected from the top of a cliff 40 m high with initial velocity 20 m s^{-1} at an angle 30° to the horizontal. Find the time of flight and the range of the ball (i.e. the distance from the bottom of the cliff to the point at which the ball falls into the sea).

The path of the ball is shown in fig. 7.3:**3**. At time t,

$$\dot{x} = 10\sqrt{3} \qquad \dot{y} = 10 - 10t$$
$$x = 10t\sqrt{3} \qquad y = 10t - 5t^2$$

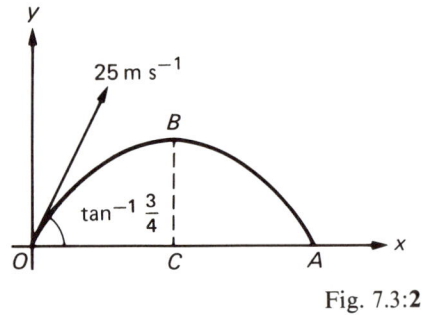

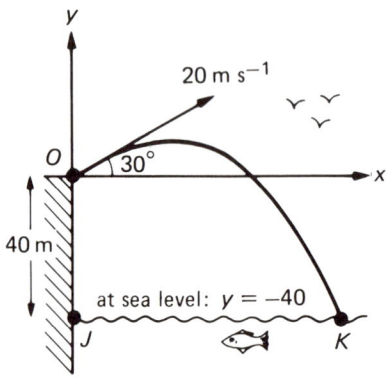

Fig. 7.3:3   by equations (3).

At K, where the ball falls into the sea, $y = -40$. (Why minus?) Let the time of flight be T. Then

$$-40 = 10T - 5T^2 \Rightarrow T^2 - 2T - 8 = 0 \Rightarrow T = 4 \text{ s or } -2 \text{ s}$$

$T = -2$ s is impossible, so $T = 4$ s. Thus

$$\text{range } JK = x(4) = 40\sqrt{3} \text{ m}$$

Exercise 7.3:1

1 A particle is projected from a point O with initial velocity **u**. Find its velocity and position vector at time t. (**i** and **j** are unit vectors horizontally and vertically upwards.)

(i) $\mathbf{u} = 3\mathbf{i} + 4\mathbf{j}$
(ii) $\mathbf{u} = 40\mathbf{i} + 35\mathbf{j}$
→ (iii) $\mathbf{u} = 40$ m s^{-1}, at an angle $30°$ to the horizontal
(iv) $\mathbf{u} = 15$ m s^{-1}, at an angle $\tan^{-1} \frac{3}{4}$ to the horizontal
(v) $\mathbf{u} = 6\sqrt{10}$ m s^{-1}, at an angle $\tan^{-1} 3$ to the horizontal
(vi) $\mathbf{u} = 4U\mathbf{i} + U\mathbf{j}$ (U constant)
(vii) $\mathbf{u} = 16$ m s^{-1}, at an angle $60°$ *below* the horizontal
→ (viii) $\mathbf{u} = 8\sqrt{5}$ m s^{-1}, at an angle $\tan^{-1} \frac{1}{2}$ *below* the horizontal
(ix) $\mathbf{u} = 20$ m s^{-1}, horizontally

2 A particle is projected from O with initial speed 24 m s^{-1}. Find its velocity and position vector after 2 s if the angle of projection is:

(i) $30°$ above the horizontal
(ii) $30°$ below the horizontal

3 A particle is projected from O with initial velocity 20 m s^{-1} at an angle $\tan^{-1} \frac{4}{3}$ to the horizontal. Find its velocity and position vector at time t. Hence find the cartesian equation of its path.

4 A particle is projected from O with an initial velocity whose horizontal and vertical components are U and $2U$ respectively. Find its velocity and position vector at time t. Find the cartesian equation of its path.

5 A particle is projected from O with initial velocity 10 m s^{-1} at an angle $\tan^{-1} \frac{3}{4}$ to the horizontal. Find the cartesian equation of its path. Hence find the coordinates of:

(i) the highest point on the path
(ii) the point where the path crosses the x-axis
(iii) the point where the path crosses the line $y = -27$

Questions 6–12: A particle is projected from a point O on level ground with initial velocity **u**. Find:

(i) the time τ that it takes to reach its greatest height
(ii) its greatest height
(iii) its time of flight T
(iv) its range

and verify that $T = 2\tau$

① **6** $\mathbf{u} = 15$ m s^{-1}, at an angle $\tan^{-1} \frac{3}{4}$ to the horizontal.

① → **7** $\mathbf{u} = 30$ m s^{-1}, at an angle $30°$ to the horizontal.

① **8** $\mathbf{u} = 8\sqrt{2}$ m s^{-1}, at an angle $45°$ to the horizontal.

① → **9** $\mathbf{u} = 20\mathbf{i} + 30\mathbf{j}$.

① **10** $\mathbf{u} = 3U\mathbf{i} + U\mathbf{j}$ (U constant).

① **11** $\mathbf{u} = U\mathbf{i} + V\mathbf{j}$ (U, V constant).

① **12** u m s^{-1} at an angle α to the horizontal.

① **13** A particle is projected from a point O on level ground with speed $\sqrt{(2gh)}$. Find its greatest height, its time of flight and its range if the angle of projection is:
(i) $90°$ (ii) $60°$ (iii) $45°$ (iv) $30°$

① **14** A particle is projected from a point Q on level ground with speed 20 m s^{-1}. If its time of flight is $\frac{2}{3}$ s, find its angle of projection and its range.

①R **15** A particle is projected from a point O on level ground with an initial velocity whose horizontal and vertical components are U and $3U$ respectively. Find its greatest height H and range R. Find the cartesian equation of its path. Check that:

(i) when $y = 0$, $x = R$
(ii) when $y = H$, $x = \frac{1}{2}R$

Illustrate these results on a sketch.

16 A particle is projected from a point O on level ground with initial velocity $26\,\mathrm{m\,s^{-1}}$ at an angle $\tan^{-1}\frac{12}{5}$ to the horizontal. Find its greatest height H and range R. Find the cartesian equation of its path. Check that

(i) when $y = 0$, $x = R$
(ii) when $y = H$, $x = \frac{1}{2}R$

Illustrate these results on a sketch.

17 A particle is projected with initial velocity $10\,\mathrm{m\,s^{-1}}$ at an angle $\tan^{-1}\frac{3}{4}$ to the horizontal from the top of a wall 8 m high. Find the time of flight and the range (i.e. the distance from the bottom of the wall to the point at which the particle hits the ground).

18 A ball is projected with initial velocity $10\sqrt{5}\,\mathrm{m\,s^{-1}}$ at an angle $\tan^{-1} 2$ to the horizontal from the top of a cliff 60 m high. Find the time of flight and the range of the ball.

19 A ball is projected with initial velocity $40\,\mathrm{m\,s^{-1}}$ at an angle $45°$ to the horizontal from the top of a cliff 50 m high. Find the time of flight and the range.

20 A popular pastime for bored teenagers in Paris is to go on to one of the bridges over the Seine and throw plastic bags full of detergent at the tourists on the sightseeing boats. A particularly vicious boy named Jean-Claude throws one of these missiles with speed $10\sqrt{2}\,\mathrm{m\,s^{-1}}$ at an angle of $45°$ to the horizontal from a bridge. If the height of the bridge above the deck of the oncoming boat is 15 m, show that the missile will strike a tourist 3 seconds later if the tourist is 30 m from the bridge.

21 A lunatic is on the roof of a tower block 70 m high, firing randomly at passers-by with a sub-machine gun. Bullets leave the gun with speed $500\,\mathrm{m\,s^{-1}}$. Find the range of the bullets if the man angles the gun at:

(i) $30°$ above the horizontal
(ii) $30°$ below the horizontal

22 A ball is projected with initial speed $8\sqrt{10}\,\mathrm{m\,s^{-1}}$ from the top of a cliff 36 m high. Find the time of flight and the range if the angle of projection is:

(i) $\tan^{-1}\frac{1}{3}$ above the horizontal
(ii) $\tan^{-1}\frac{1}{3}$ below the horizontal

23 A ball is projected with initial velocity $u\,\mathrm{m\,s^{-1}}$ at an angle α to the horizontal from the top of a cliff of height h m. Show that the range of the ball is

$$\frac{u}{2g}\{\sin 2\alpha + ug\cos\alpha\sqrt{(u^2\sin^2\alpha + 2gh)}\}$$

24 A ball is thrown horizontally at $40\,\mathrm{m\,s^{-1}}$ from the top of a cliff and falls into the sea at a distance 60 m from the bottom of the cliff. Find the height of the cliff.

25 A ball is thrown at an angle $\tan^{-1}\frac{3}{4}$ to the horizontal with initial velocity $30\,\mathrm{m\,s^{-1}}$ from the top of a cliff and falls into the sea at a distance 120 m from the bottom of the cliff. Find the height of the cliff.

26 A stone is thrown with initial velocity u at an angle α to the horizontal from the top of a cliff, and falls into the sea at a distance d from the bottom of the cliff. Find the height of the cliff.

27 A ball is projected at an angle $30°$ to the horizontal from the top of a wall 5 m high, and hits the ground 2 s later. Find the initial speed of the ball, and its distance from the wall when it hits the ground.

28 A ball is projected from the top of a tower of height 32 m. If it hits the ground 4 seconds later at a distance 64 m from the foot of the tower, find the speed and direction of projection.

29 A gang of creative 17-year-olds is meticulously dismantling the contents of a tower block. One of them throws a telephone from a twelfth-floor window, 96 m above the ground. If it hits the ground $4\sqrt{2}\,\mathrm{s}$ later at a distance 64 m from the foot of the tower, find the speed and direction of its projection.

30 A stone is thrown from the top of a cliff of height 75 m. If it falls into the sea 3 s later at a distance 60 m from the foot of the cliff, find the speed and direction of its projection.

31 A particle is projected from a point O on level ground with initial velocity $10\sqrt{5}\,\mathrm{m\,s^{-1}}$ at an angle $\tan^{-1} 2$ to the horizontal. Find its height above the ground when its horizontal distance from O is:

(i) 5 m (ii) 10 m (iii) 20 m (iv) 30 m (v) 35 m

32 A particle is projected from a point O on level ground with initial velocity $5\sqrt{(gh)}$ at an angle $\tan^{-1}\frac{3}{4}$ to the horizontal. Find its height above the ground when its horizontal distance from O is:

(i) $8h$ (ii) $12h$ (iii) $16h$

33 A golf ball is struck with initial velocity $20\mathbf{i} + 25\mathbf{j}$ from a point O on level ground. Show that it will clear a tree of height 28 m if the tree is 40 m from O, but not if it is 70 m from O.

34 A golf ball is struck from a point O on level ground with initial velocity $40\,\mathrm{m\,s^{-1}}$ at $45°$ to the horizontal. Show that it will just clear a tree of height 30 m if the tree is 40 m from O but not if it is 140 m from O. Draw a sketch illustrating this.

35 A particle is projected from a point O on level ground with initial velocity $40\,\mathrm{m\,s^{-1}}$ at an angle $\tan^{-1}\frac{3}{4}$ to the horizontal. Find the times at which the particle is 28 m above the ground. For how long is the particle more than 28 m above the ground?

36 A particle is projected from a point O on level ground with initial velocity 44 m s^{-1} inclined at 30° to the horizontal. For how long is the particle more than 8 m above the ground?

→ **37** A particle is projected from a point O on level ground with initial velocity 20 m s^{-1} at an angle α to the horizontal. It passes through two points A and B both at height 10 m above the ground. Find the time taken to travel from A to B. If $AB = 20$ m, find α.

38 A particle is projected from a point O on level ground with initial velocity 20 m s^{-1} at an angle $\tan^{-1} \frac{3}{4}$ to the horizontal. It hits a wall at a horizontal distance 32 m from O. Find the time that elapses before it hits the wall. Find the height above the ground at which it hits the wall.

• **39** A particle, projected from a point O with speed u, passes through the point (h, k) after time t. Show that:
$$h^2 = (u^2 - gk)t^2 - \tfrac{1}{4}g^2 t^4 - k^2$$

→ **40** A particle, projected with speed 30 m s^{-1} from a point O on ground level, hits the ground again after t s at a distance R from O. Show that:
$$R^2 = 25(36 t^2 - t^4)$$

Hence show that the maximum value of R is 90 m and that the corresponding angle of projection is 45°.

R **41** A particle, projected with speed u from a point O on ground level, hits the ground again after time t at a distance R from O. Show that:
$$R^2 = u^2 t^2 - \tfrac{1}{4}g^2 t^4$$

Hence show that the maximum value of R is u^2/g and that the corresponding angle of projection is 45°.

→ **42** A particle, projected with speed 40 m s^{-1} from the top of a cliff of height 20 m, falls into the sea after t s at a distance s from the bottom of the cliff. Show that:
$$s^2 = 25(72 t^2 - t^4 - 16)$$

Hence show that the maximum possible value of s is $80\sqrt{5}$ m and the corresponding angle of projection is $\tan^{-1} \frac{2}{3}$.

+ **43** A particle, projected with speed u from the top of a cliff of height h, falls into the sea after time t at a distance s from the bottom of the cliff. Find s^2 in terms of u, g, h and t. Show that the maximum possible value of s is
$$\frac{u\sqrt{(u^2 + 2gh)}}{g}$$
and that the corresponding angle of projection is $\tan^{-1}(u^2/gs)$.
By putting $h = 0$, check the answer to question 41.

44 Two particles are projected simultaneously with speed u from a point O at angles of elevation α and β. The particles travel in the same vertical plane through O. Show that $_Q\mathbf{v}_P$ is a constant vector and find it. Hence show that $\overrightarrow{PQ}$ is in a fixed direction.

45 Two particles P and Q are projected simultaneously from a point O. P is projected vertically upwards with speed $V\sqrt{3}$ and Q is projected with speed V at an angle of elevation 60°. Find the distance between the particles when P is at its highest point.

→ **46** Two particles are projected simultaneously from a point O with the same initial speed and with angles of elevation α and $90° - \alpha$. Show that the particles have the same range, and show that at any time during their flight the line joining them is inclined at 45° to the horizontal.

• **47** A particle is projected from a point O with initial velocity $20\mathbf{i} + 30\mathbf{j}$. Two seconds later another particle is projected from O with initial velocity $60\mathbf{i} + 50\mathbf{j}$. Show that the particles collide one second later and find the position vector of their point of collision.

→ **48** A particle is projected from a point O with initial velocity 20 m s^{-1} at an angle $\tan^{-1} \frac{4}{3}$ to the horizontal. Two seconds later another particle is projected from O. Given that it collides with the first particle one second later, find its initial velocity.

49 A particle P is projected from a point O with initial velocity 60 m s^{-1} at an angle 30° to the horizontal. At the same time a particle Q is projected in the opposite direction with initial speed 50 m s^{-1} from a point level with O and 100 m from O. Given that P and Q collide, find the angle of projection of Q, the time of collision and the position vector of the point of collision.

R **50** A particle P is projected from a point O with initial velocity 30 m s^{-1} at an angle of 45° to the horizontal. At the same time a particle Q is projected in the opposite direction with initial velocity 30 m s^{-1} at an angle of 45° to the horizontal from a point 120 m from O and level with O. Show that the particles collide. Find the time of collision and the height above O at which they collide.

51 The corners of a football pitch are at the points $(\pm 20, \pm 40)$ (distances measured in metres). At a certain instant, a sweeper kicks the ball with speed $31/\sqrt{2}$ m s^{-1} at an angle 45° to the horizontal from the point $(0, -10)$ straight up the ground (i.e. in the direction of the y-axis). At the same instant, the centre forward starts from the point $(-18, 36.5)$ and runs with speed 6 m s^{-1} straight across the ground (i.e. in the direction of the x-axis). Show that if the height of the centre forward is 2 m, he can expect to be able to intercept the ball. Find the height above the ground at which the ball will strike his body.

7.3:2 Properties of the flight of a projectile

A particle is projected from a point O with initial velocity u at an angle α to the horizontal. The equations for $x(t)$, $y(t)$, $\dot{x}(t)$ and $\dot{y}(t)$ were obtained on p. 402, equations (3).

Greatest height

Let the time taken to reach B, the highest point on the flight (fig. 7.3:4), be t_B.

At B, $\dot{y}=0 \Rightarrow u \sin \alpha - g t_B = 0$, so

$$t_B = (u \sin \alpha)/g \qquad (4)$$

Fig. 7.3:4

$$\text{greatest height } H = y(t_B) = u t_B \sin \alpha - \tfrac{1}{2} g t_B^2$$
$$= \frac{u^2 \sin^2 \alpha}{g} - \frac{u^2 \sin^2 \alpha}{2g}$$

So

$$H = \frac{u^2 \sin^2 \alpha}{2g} \qquad (5)$$

Horizontal range

Let the time taken to reach A, the point on the flight level with O (fig. 7.3:5), be T.

At A, $y = 0 \Rightarrow uT \sin \alpha - \tfrac{1}{2} g T^2 = 0$, so

$$T = (2u \sin \alpha)/g \qquad (6)$$

or $T = 0$. (Why also 0?)

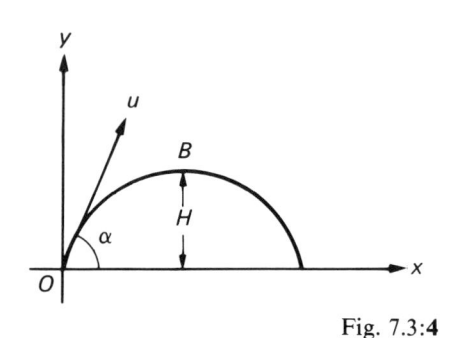

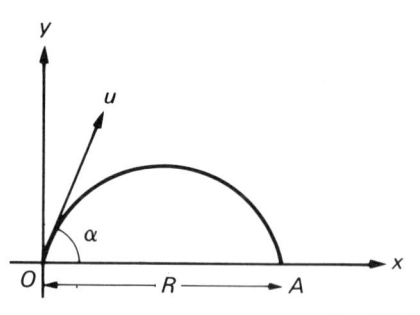

Fig. 7.3:5

$$\text{range } OA = R = x(T) = uT \cos \alpha$$
$$= \frac{2u^2 \sin \alpha \cos \alpha}{g}$$

So

$$R = \frac{u^2 \sin 2\alpha}{g} \qquad (7)$$

Maximum range

For a given value of u, the range $R = (u^2 \sin 2\alpha)/g$ is maximum when $\sin 2\alpha = 1$, i.e. when $\alpha = 45°$ (fig. 7.3:6). In this case $R = u^2/g$. Thus

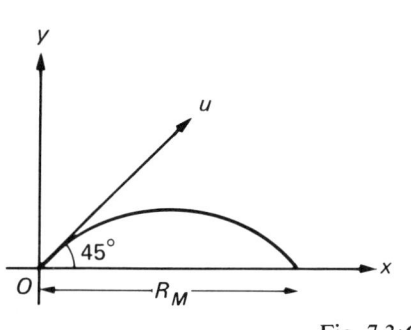

$$\text{maximum range } R_M = u^2/g \text{ and occurs when } \alpha = 45° \qquad (8)$$

Fig. 7.3:6

Kinematics of a particle 407

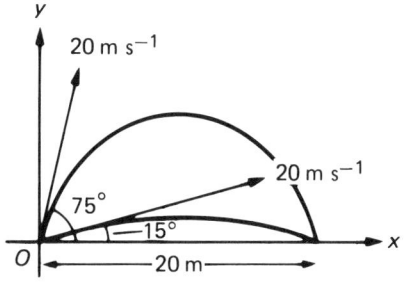

Worked example 3 A particle is projected with initial speed 20 m s^{-1}. Find the two possible angles of projection needed to achieve a range of 20 m.

With angle of projection α, range $R = (u^2 \sin 2\alpha)/g = 40 \sin 2\alpha$. When $R = 20$, $20 = 40 \sin 2\alpha \Rightarrow \sin 2\alpha = \tfrac{1}{2} \Rightarrow 2\alpha = 30°$ or $150°$. So

$$\alpha = 15° \text{ or } 75°$$

Fig. 7.3:7 The two possible paths giving a range of 20 m are shown in fig. 7.3:7.

Cartesian equation of the path

Consider again a particle projected from a point O with initial velocity u at an angle α to the horizontal. The path of the particle is shown in fig. 7.3:8.

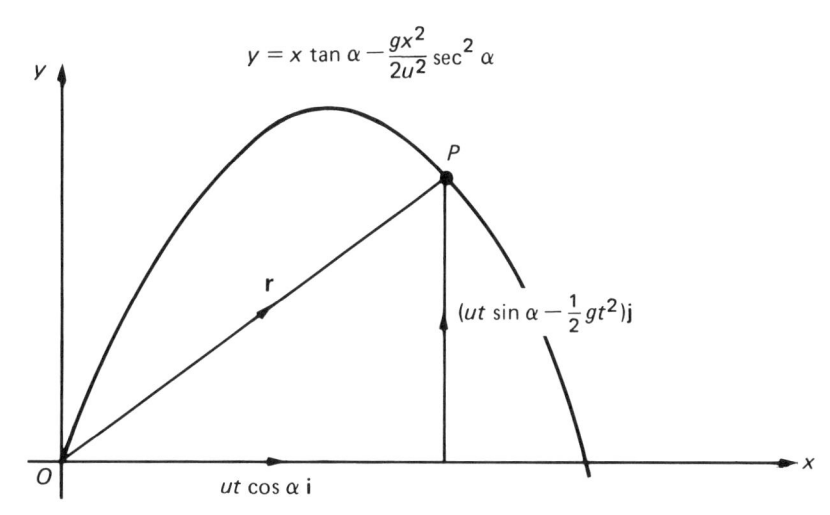

Fig. 7.3:8

$$\mathbf{r} = ut \cos \alpha \, \mathbf{i} + (ut \sin \alpha - \tfrac{1}{2}gt^2)\mathbf{j}$$

So $\qquad x = ut \cos \alpha, \qquad y = ut \sin \alpha - \tfrac{1}{2}gt^2$

are the *parametric equations* of the path of the particle. To find the cartesian equation, we eliminate t: $t = x/u \cos \alpha \Rightarrow y = x \tan \alpha - gx^2/(2u^2 \cos^2 \alpha)$, i.e.

$$y = x \tan \alpha - \frac{gx^2}{2u^2} \sec^2 \alpha \qquad (9)$$

Note that equation (9) is of the form $y = ax^2 + bx$. This shows that *the path of a projectile is a parabola* (see 6.2:7 Q5).

Worked example 4 A particle is projected from a point O with initial speed 20 m s^{-1} to pass through a point at 10 m from O horizontally and 10 m above O. Show that there are two possible angles of projection.

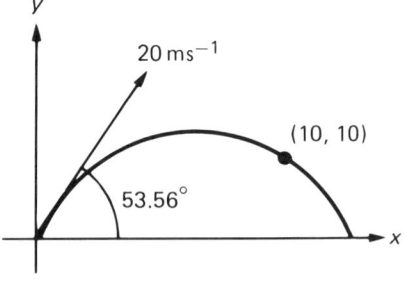

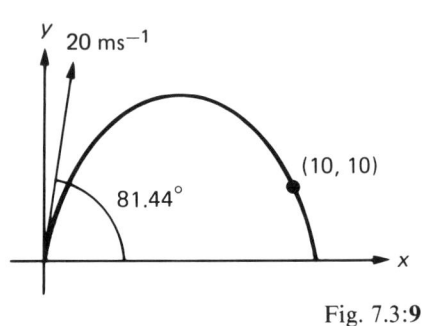

Let the angle of projection be α. The cartesian equation of the path is $y = x \tan \alpha - x^2 \sec^2 \alpha / 80$ (by equation (9)). The path of the particle passes through the point (10, 10). Thus

$$10 = 10 \tan \alpha - \tfrac{5}{4} \sec^2 \alpha$$
$$10 = 10 \tan \alpha - \tfrac{5}{4}(1 + \tan^2 \alpha)$$
$$\tan^2 \alpha - 8 \tan \alpha + 9 = 0$$
$$\Rightarrow \quad \tan \alpha = 1.354 \quad \text{or} \quad 6.646$$
$$\alpha = 53.56° \quad \text{or} \quad 81.44°$$

The two possible paths passing through the point (10, 10) are shown in fig. 7.3:9.

Fig. 7.3:9

Exercise 7.3:2

• **1** A particle is projected with speed 40 m s^{-1} from a point on level ground. Show that, when the angle of projection of the particle is θ, its range R is given by:

$$R = 160 \sin 2\theta$$

  (i) Sketch a graph of R against θ (for values of θ between 0° and 90°).
  (ii) Find the maximum value of R as θ varies. For what value of θ does this occur?
  (iii) Find R when $\theta = 15°, 75°, 30°, 60°$.
  (iv) Show the angles of projection α and $90° - \alpha$ give the same range.

2 A particle is projected with speed 40 m s^{-1} from a point on level ground. Show that, when the angle of projection of the particle is θ, its greatest height H is given by:

$$H = 80 \sin^2 \theta$$

  (i) Sketch a graph of H against θ (for values of θ between 0° and 90°).
  (ii) Find the maximum value of H as θ varies. For what value of θ does this occur?
  (iii) Find the values of θ for which $H = 20, 40, 60$.

3 A particle projected with initial speed 20 m s^{-1} just clears a wall 5 m high when travelling horizontally. Find the angle of projection of the particle.

4 The range of a projectile is six times its greatest height. Find the angle of projection of the projectile.

5 A particle is projected at an angle α to the horizontal. If its greatest height is 4 m and its range is 8 m, find α.

6 The maximum range of a projectile is 200 m. Find its initial speed and the greatest height reached when it achieves its maximum range.

→ **7** Show that when a particle achieves its maximum range this range is four times the greatest height.

8 A particle is projected with speed 40 m s^{-1} inside a tunnel of height 5 m. Find the greatest possible range of the particle.

9 Lionel the performing dog is preparing for his unique performance piece 'Crossing the river'. He is to be fired from a cannon on one bank of the River Thames at Wapping, where the river is 150 m wide, and hopes to reach the other side of the river, where 1000 mattresses and many more fans will be waiting for him. Find the minimum possible speed u of his projection.

The performance itself is not without its problems. The TV personality hired by Lionel's overenthusiastic agent to fire the cannon believes that Lionel will have a better chance of success if the angle of projection is greater. Consequently Lionel is projected with speed u at an angle of 60° to the horizontal. Find the distance from the opposite bank at which he enters the water.

The resourceful dog swims to the opposite bank and goes to sleep on the mattresses. 'Crossing the river' has been another memorable performance.

Questions 10–13: A particle is projected with initial speed u m s^{-1}. If its horizontal range is R m, find the two possible angles of projection. In each case find also the corresponding times of flight and the greatest heights.

③ 10 $u = 20$, $R = 20\sqrt{3}$

③ 11 $u = 30$, $R = 60$

③ 12 $u = 40$, $R = 80$

③ 13 $u = 50$, $R = 200$

③ 14 A particle is projected with initial speed 40 m s^{-1}. Show that it is not possible that it achieves a range of 200 m. If the horizontal range is 160 m, show that there is just one possible angle of projection and find it. Why is there a single solution?

③ R 15 If a particle projected with speed u has a horizontal range $3u^2/5g$, calculate the two possible angles of projection. Show that the difference in the maximum heights attained with these angles of projection is $2u^2/5g$.

③ → 16 A particle is projected with speed u to reach a
• given range R. Show that:

(i) if $R < u^2/g$, there are two possible angles of projection α_1 and α_2 and that $\alpha_1 + \alpha_2 = 90°$;
(ii) if $R = u^2/g$ there is one angle of projection. What is it?
(iii) What if $R > u^2/g$?

③ 17 A particle is projected with speed u to reach a given range R. Show that the greatest height H is given by the quadratic equation

$$16gH^2 - 8u^2H + gR^2 = 0$$

(i) Why are there in general two solutions to this equation?
(ii) In what circumstances is there just one solution? Check that in this case $R = 4H$.
(iii) In what circumstances is there no solution?

③ + 18 If a particle projected with speed u has a horizontal range ku^2/g ($k < 1$), find the two possible angles of projection. Show that the difference in the maximum heights attained with these angles of projection is $(u^2/2g)\sqrt{(1-k^2)}$.

Questions 19–21: A particle is projected at an angle α to the horizontal and passes through the point A. Find the initial speed of the particle.

→ • 19 $\alpha = \tan^{-1}\frac{3}{4}$; $A(32, 4)$ (Hint: use equation (9).)

20 $\alpha = 30°$; $A(40\sqrt{3}, 20)$

R 21 $\alpha = \tan^{-1} 2$; $A(60, -5)$

→ 22 A particle, projected from a point O on level ground at an angle of 45° to the horizontal, just clears a wall 5 m high at a horizontal distance 15 m from O. Find the initial speed of the particle and the greatest height it reaches during its flight.

+ 23 A particle, projected from a point O at an angle α to the horizontal, passes through the point (h, k). Show that the greatest height reached by the particle is $h^2 \tan^2 \alpha / 4(h \tan \alpha - k)$.

Questions 24–32: A particle is projected with speed u m s^{-1} and passes through the point A. Find the two possible angles of projection.

④ 24 $u = 40$; $A(80, 0)$

④ → 25 $u = 70$; $A(280, 40)$

④ 26 $u = 20$; $A(4, 7)$

④ 27 $u = 50$; $A(100, 50)$

④ 28 $u = 5\sqrt{2}$; $A(20, -40)$

④ 29 $u = 10\sqrt{6}$; $A(60, -60)$

④ 30 $u = \sqrt{(5gh)}$; $A(2h, -\frac{2}{5}h)$

④ 31 $u = \sqrt{(2gh)}$; $A(h, \frac{1}{2}h)$

④ 32 $u = \sqrt{(2gh)}$; $A(2\sqrt{2}h, -h)$

④ 33 A particle is projected with initial velocity 30 m s^{-1} at an angle α to the horizontal and passes through the point (40, 10). Show that:

$$8\tan^2 \alpha - 36 \tan \alpha + 17 = 0$$

Do not solve this equation, but, by considering the sum and product of its roots, show that if α_1 and α_2 are the two possible angles of projection, then:

$$\tan(\alpha_1 + \alpha_2) = -4$$

(See 3.2:3 Q18.)

④ 34 A particle is projected from a point O with initial speed 15 m s^{-1} to pass through a point at 5 m from O horizontally and 10 m above O. Show that there are two possible angles of projection α_1 and α_2 and that $\tan(\alpha_1 + \alpha_2) = -\frac{1}{2}$.

④ 35 A particle is projected from a point O with initial speed u m s^{-1} to pass through a point at h m from O horizontally and k m above O. Show that, if α_1 and α_2 are the two possible angles of projection, then $\tan(\alpha_1 + \alpha_2) = h/k$.

④ 36 A ball is thrown from ground level with speed 40 m s^{-1}. It just clears a wall 30 m high, 40 m from the point of projection. Show that there are two possible angles of projection, one of which is 45°. Find in each case:

(i) the greatest height
(ii) the range

reached on each trajectory.

④ R **37** A particle is projected with initial speed 70 m s^{-1} from the top of a cliff of height 40 m, and falls into the sea at a distance 200 m from the bottom of the cliff. Find the two possible angles of projection and the difference between the corresponding times of flight. (Take $g = 9.8 \text{ m s}^{-2}$.)

- **38** A ball is thrown from a point O on horizontal ground. Two walls of height $\tfrac{15}{4}$ m are at horizontal distances 15 m and 45 m from O. If the particle just clears them, find its initial speed and angle of projection.

39 A particle is projected from a point O on horizontal ground. Two walls of height h are of horizontal distances $2h$ and $4h$ from O. If the particle just clears them, show that:
(i) the angle of projection is $\tan^{-1}\tfrac{3}{4}$
(ii) the speed of projection is $\tfrac{5}{2}\sqrt{(gh)}$
What is the range of the particle?

→ **40** A particle is projected from a point O on horizontal ground. Two trees of heights 20 m and 30 m are at horizontal distances 50 m and 100 m from O. If the particle just clears them, find its initial speed, its angle of projection and its range.

41 Three parallel walls of heights h, $2h$ and h are spaced $h\sqrt{2}$ apart on horizontal ground. A particle, projected from ground level and moving in a vertical plane perpendicular to the walls, just clears the tops of the walls. Show that:
(i) the angle of projection is $\tan^{-1} 2$
(ii) the speed of projection is $\sqrt{(5gh)}$
What is the range of the particle?

+ **42** (i) A particle is projected with initial speed u to pass through the point (x, y). Show that, if α is the angle of projection,

$$\frac{gx^2}{2u^2}\tan^2\alpha - x\tan\alpha + \left(\frac{gx^2}{2u^2} + y\right) = 0$$

Hence show that if the point (x, y) can *just* be reached with initial speed u, then

$$y = -\frac{g}{2u^2}x^2 + \frac{u^2}{2g}$$

(Hint: think of the first equation as a quadratic in $\tan\alpha$; what can you say about its roots if (x, y) can *just* be reached?)
Sketch this curve. It is called the **parabola of safety**.
(ii) A golfer can hit the ball at 20 m s^{-1}. There is a tree 20 m in front of him. Show that he can clear the tree if it is less than 15 m high.

(iii) A particle is projected with speed u from the top of a cliff of height h, and falls into the sea at a distance s from the bottom of the cliff. Show that, for s to have its maximum value, the angle of projection should be $\tan^{-1}(u^2/gs)$, and that the maximum value of s is

$$(u/g)\sqrt{(u^2+2gh)}$$

(C.f. 7.3:1 Q43.)

43 A particle is projected from a point O on level ground with initial velocity u at an angle α to the horizontal.
(i) Find its speed at time t.
(ii) When is its speed at a minimum?
Show that the ratio of the least speed to the greatest speed during the motion is $\cos\alpha$.

- **44** A particle is projected from a point O with initial velocity 15 m s^{-1} at an angle $\tan^{-1}\tfrac{4}{3}$ to the horizontal. Find its speed when it is at a vertical height $\tfrac{20}{7}$ m above O.

→ **45** A particle is projected from a point O with initial velocity 30 m s^{-1} at an angle $\tan^{-1} 2$ to the horizontal. Find its height above O when its speed is 20 m s^{-1}.

46 A particle is projected from a point O with an initial velocity whose horizontal and vertical components are U and $3U$ respectively. Find the times at which the speed of the particle is $2U$.

R **47** A particle is projected from a point O. Its speed when at its greatest height h above O is $\sqrt{(\tfrac{2}{5})}$ times its speed when at a height $\tfrac{1}{2}h$ above O. Show that the angle of projection is $60°$.

- → **48** A particle is projected with initial velocity 50 m s^{-1} at an angle $\tan^{-1}\tfrac{3}{4}$ to the horizontal. Find the direction of its velocity when $t = 1$ s, 3 s, 5 s and 6 s. (See 7.2:1 WE3.)

49 A particle is projected from O with initial velocity $25\mathbf{i} + 20\mathbf{j}$. Show that it passes through the points $(25, 15)$, $(50, 20)$, $(75, 15)$, $(100, 0)$ and $(125, -25)$. Find the directions of its velocity as it passes through these points.

- → **50** A particle is projected with speed u at an angle α to the horizontal. If ϕ is the angle between the velocity of the particle and the horizontal after t s, find $\tan\phi$ in terms of:
(i) t (ii) x
(See 7.2:1 equation (7).)

51 A particle is projected with initial velocity 15 m s^{-1} at $45°$ to the horizontal from the top of a wall of height 2 m. Find the direction of its velocity when it hits the ground.

52 A particle is projected with initial velocity $10\mathbf{i}+30\mathbf{j}$. Find the angle through which the velocity of the particle turns in the first three seconds of its motion.

→ **53** A particle is projected with initial velocity 40 m s^{-1} at an angle $45°$ to the horizontal. Find the direction of its velocity when it is at a height 40 m above the point of projection.

54 A particle is projected at an angle of $45°$ to the horizontal; $3\sqrt{2}$ s later it is moving in the direction $\tan^{-1}\frac{1}{4}$ to the horizontal. Find its initial speed. After what time is it moving:

(i) at an angle $\tan^{-1}\frac{1}{2}$ to the horizontal?
(ii) horizontally?
(iii) at an angle $\tan^{-1}(-\frac{1}{2})$ to the horizontal?

55 A particle is projected with initial velocity 40 m s^{-1} at an angle of $60°$ to the horizontal. Find its distance from the point of projection when its direction of motion makes an angle of $30°$ with the horizontal.

Questions 56-61: A particle is projected with initial velocity $\mathbf{u}$. Find the time at which its velocity is at right angles to its initial velocity for the given values of $\mathbf{u}$.

• **56** $\mathbf{u}=15\mathbf{i}+30\mathbf{j}$ (Hint: write down its velocity at time t. Use 6.1:1 Theorem 1.)

57 $\mathbf{u}=40\mathbf{i}+30\mathbf{j}$

→ **58** $\mathbf{u}=2\sqrt{10}$ m s^{-1}, the particle is projected at an angle of $\tan^{-1}3$ to the horizontal.

59 $\mathbf{u}=40$, the particle is projected at an angle of $60°$ to the horizontal.

60 $\mathbf{u}=U\mathbf{i}+V\mathbf{j}$ (U,V constant)

61 $\mathbf{u}=u$ m s^{-1}, the particle is projected at an angle α to the horizontal.

R **62** A particle P is projected from a point O with an initial velocity whose horizontal and vertical components are U and $2U$ respectively. If θ is the angle between OP and the horizontal at time t, find θ when the velocity of P is at right-angles to its initial velocity.

→ **63** A particle is projected with velocity 20 m s^{-1} at an angle of elevation α. The particle passes through the point $A(30, 3.75)$.

(i) Find the two possible values of $\tan\alpha$.
(ii) Find the direction of the velocity of the particle at A in each of the two trajectories.

R **64** A particle is projected with velocity $7\sqrt{5}$ m s^{-1} at an angle of elevation α. The particle passes through the point $A(7,9)$.

(i) Find the two possible values of $\tan\alpha$.
(ii) Find the direction of the velocity of the particle at A in each of the two trajectories.

• **65** A ball is to be thrown over a building which is 12 m high and 10 m wide. If the ball is thrown from ground level, find the minimum velocity necessary and the distance of the point of projection from the nearest wall of the building. (Hint: at what angle to the horizontal should the ball be travelling when it passes over the nearest wall of the building?)

→ **66** A ball is to be thrown over a building which is 20 m high and 15 m wide. If the ball is thrown from ground level, find the minimum velocity necessary, the angle of projection, and the distance of the point of projection from the nearest wall of the building.

7.4 Angular velocity and circular motion

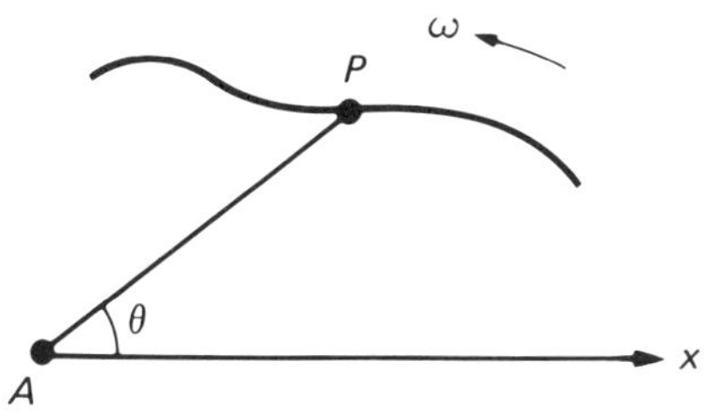

7.4:1 Angular velocity

Let A be a fixed point in the xy plane. Suppose that a particle P moves in the plane so that at time t the angle between AP and some fixed direction, Ax say, is θ (see fig. 7.4:1). θ is a (scalar) function of t.

Fig. 7.4:1

Definition The **angular velocity** ω of P about A is $d\theta/dt$:

$$\omega = \frac{d\theta}{dt} \quad (1)$$

ω is measured in radians per second (rad s^{-1}).

412   *Angular velocity*

The angular velocity of P about A may be positive or negative, depending on the sense of rotation—usually we take anticlockwise as positive.

The **angular speed** of P about A is the magnitude of its angular velocity about A.

Definition   The angular acceleration α of P about A is

$$\alpha = \frac{d\omega}{dt} = \frac{d^2\theta}{dt^2} \qquad (2)$$

α is measured in rad s^{-2}.

Worked example 1   A particle moves in a plane. At time t the radius vector OP has turned through an angle θ rad, where $\theta = t^3 - t$. Find the angular velocity and angular acceleration of the particle about θ when $t = 2$.

$\theta = t^3 - t \Rightarrow \omega = d\theta/dt = 3t^2 - 1 \Rightarrow \alpha = d\omega/dt = 6t.$
At time $t = 2$, $\omega = 3 \cdot 4 - 1 = 11$ rad s^{-1}; $\alpha = 6 \cdot 2 = 12$ rad s^{-2}.

Worked example 2   At time t the position vector of a particle is $t\mathbf{i} + (t - 2t^2)\mathbf{j}$. Find the angular velocity of the particle about O at time t.

The path of the particle is shown in fig. 7.4:2.

$$\mathbf{r} = t\mathbf{i} + (t - 2t^2)\mathbf{j} \Rightarrow x = t, \ y = t - 2t^2$$

Now

$$\tan \theta = y/x = (t - 2t^2)/t = 1 - 2t$$

$$\Rightarrow \qquad \theta = \tan^{-1}(1 - 2t)$$

Thus

$$\frac{d\theta}{dt} = \frac{-2}{1 + (1 - 2t)^2} = \frac{-1}{1 - 2t + 2t^2} \quad \text{(see 3.5:2 Q3(ii))}$$

Fig. 7.4:2

Relationship between speed and angular velocity for a particle travelling in a circle   Let A be a fixed point on the circle. Suppose that at time t the particle is at the point P, where $P\hat{O}A = \theta$ (fig. 7.4:3). If its angular velocity at this time is ω, then $\omega = d\theta/dt$.

(When a particle is moving in a circle and we talk simply about its angular velocity, we mean its angular velocity about the centre of the circle.)

Speed $v = ds/dt$, where s is the arc length AP (see 7.2:1 equation (7)). But $s = r\theta$ (see section 3.4:1); so

$$\frac{ds}{dt} = r\frac{d\theta}{dt} = r\omega$$

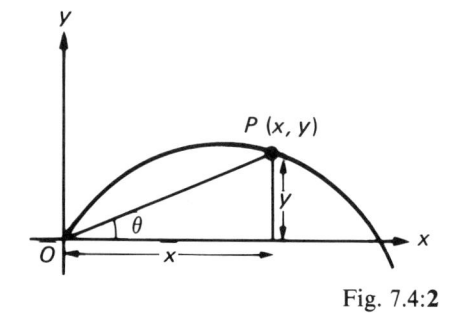

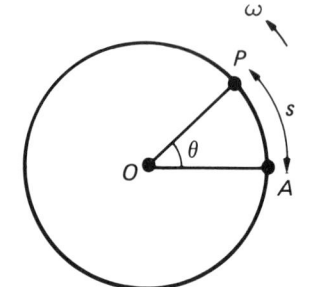

Fig. 7.4:3

---

i.e. $$v = r\omega \qquad (3)$$

Note: this equation is true whether or not ω is constant. If ω is constant, then so is v.

Exercise 7.4:1

① → **1** A particle P moves in a plane. At time t the radius vector OP has turned through θ rad, where $\theta = 5t - t^2$. Find the angular velocity of P about O at time t. When is the angular velocity zero? Show that the angular acceleration is constant.

① **2** A particle P moves in a plane. At time t the radius vector OP has turned through θ rad, where $\theta = 3e^t - e^{2t}$. When is the angular velocity zero? At this moment, is the velocity necessarily zero? Explain. Find the angular velocity and angular acceleration when $t = \ln 2$.

① **3** A particle P moves in a plane. The angular velocity of P about O after t s is $2t - 3/t^2$. Find the angle through which OP turns in the first 4 s of motion.

4 A particle P describes a circle centre O with constant angular velocity ω rad s^{-1}. Show that the time taken for OP to turn through θ rad is θ/ω. Find the time taken for the particle to make one complete revolution.

5 Two particles A and B are moving anticlockwise in two circles centre O. Their angular velocities are 1 rad s^{-1} and 3 rad s^{-1} respectively. At $t = 0$, O, A and B are collinear. After how long will they again be collinear?

② → **6** At time t a particle has position vector $t\mathbf{i} - t^2\mathbf{j}$. Show that at time $t = 1$ its angular velocity about O is $-\frac{1}{2}$ rad s^{-1} and its angular acceleration is $\frac{1}{2}$ rad s^{-2}.

② R **7** At time t a particle is at the point (e^t, e^{-t}). Show that at time $t = \ln 2$ its angular speed is $\frac{8}{17}$ rad s^{-1}.

② • **8** A particle moves so that at time t its position vector is $t^2\mathbf{i} - t^3\mathbf{j}$. Find the angular velocity of the particle at time t:

(i) about O (ii) about the point $-\mathbf{j}$

② **9** A particle moves so that its position vector at time t is $3\cos 2t\mathbf{i} + 3\sin 2t\mathbf{j}$. Sketch the path of the particle. Find the angular velocity of the particle at time t:

(i) about O (ii) about the point $(0, 3)$

② **10** A particle moves along the line $y = 4x$. Show that its angular velocity about O is always zero.

② **11** A particle starts from the point $-3\mathbf{i} + 4\mathbf{j}$ and moves with constant velocity $2\mathbf{i} + \mathbf{j}$. Find its angular velocity about O at time t. Find the position vector of the particle when its angular speed about O is at a maximum.

② → **12** (i) The position vector of a particle P moving in the xy plane is $x\mathbf{i} + y\mathbf{j}$. Show that its angular velocity ω about O is $(x\dot{y} - y\dot{x})/(x^2 + y^2)$. Hence show that

$$\omega = \frac{\text{component of velocity perpendicular to } \overrightarrow{OP}}{OP} \quad (4)$$

(ii) At a certain instant a particle is at the point $P(20, 10)$, moving with speed 8 m s^{-1} along the line $x = 20$ away from Ox. Find the component of its velocity perpendicular to $\overrightarrow{OP}$ and hence, using equation (4), find its angular velocity.

② R **13** A particle moving in a vertical plane xy (Ox horizontal, Oy vertically upwards) falls from rest at the point $(20, 20)$. Find its angular speed about O when it reaches the point $(20, 15)$.

14 A particle moves in a circle of radius 2 m with constant speed 5 m s^{-1}. Find the time it takes to complete one revolution.

• → **15** A particle P moves in a plane. Its initial angular velocity about O is ω_0, and it has *constant* angular acceleration α about O. At time t its angular velocity about O is ω, and the radius vector OP has turned through θ rad. Show by integration that

$$\omega = \omega_0 + \alpha t \quad \text{and} \quad \theta = \omega_0 t + \tfrac{1}{2}\alpha t^2$$

Hence show that

$$\omega^2 = \omega_0^2 + 2\alpha\theta, \quad \theta = \tfrac{1}{2}(\omega + \omega_0)t, \quad \theta = \omega t - \tfrac{1}{2}\alpha t^2$$

Compare these with the constant acceleration formulae of section 7.1:3.

16 A particle describes a circle with constant angular acceleration 3 rad s^{-2}. If it starts from rest, find its angular velocity after 2 s.

17 A particle describes a circle of radius 0.5 m with constant angular acceleration 5 rad s^{-2}. Its initial speed is 2 m s^{-1}. Find its speed after 3 s.

→ **18** A particle describes a circle with constant angular acceleration α. Starting with angular velocity 2π rad s^{-1}, it makes 3 complete revolutions in 5 s. Find α.

19 A particle describes a circle of radius 10 m with constant angular acceleration. Its initial speed is π m s^{-1}, and it makes one complete revolution in 5 s. Find its speed at this time.

20 A particle P describes a circle centre O, starting from rest, with constant angular acceleration $\tfrac{1}{4}\pi$ rad s^{-2}. Find the angle through which the radius vector OP turns:

(i) in the first second of motion
(ii) in the second second of motion
(iii) in the third second of motion

R **21** A particle describes a circle, starting from rest, with constant angular acceleration $\tfrac{1}{4}\pi$ rad s^{-2}. Find the time it takes to complete:

(i) one revolution
(ii) two revolutions
(iii) three revolutions

7.4:2 Circular motion

Consider a particle moving in a circle of radius r, centre O, with constant angular velocity ω. Its speed v is given by $v = r\omega$; so its speed is also constant. Suppose that at time t the particle is at the point P, and that the angle between OP and some fixed direction, OA say, is θ. Then, since ω is constant,

$$\omega = \theta/t$$

i.e. $$\theta = \omega t \tag{5}$$

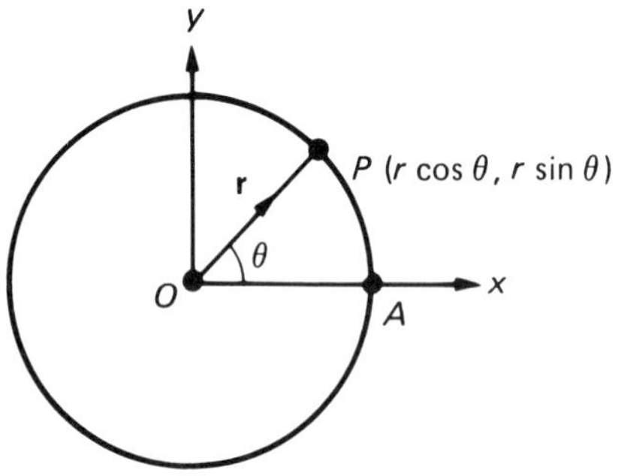

Taking axes Ox and Oy as shown in fig. 7.4:4,

$$\mathbf{r} = (r\cos\theta)\mathbf{i} + (r\sin\theta)\mathbf{j}$$

i.e. $$\mathbf{r} = (r\cos\omega t)\mathbf{i} + (r\sin\omega t)\mathbf{j} \quad \text{from equation (5)}$$

So $$\mathbf{v} = (-r\omega\sin\omega t)\mathbf{i} + (r\omega\cos\omega t)\mathbf{j}$$

and $$\mathbf{a} = (-r\omega^2\cos\omega t)\mathbf{i} + (-r\omega^2\sin\omega t)\mathbf{j}$$

Fig. 7.4:4

Velocity

Speed $\quad v = |\mathbf{v}| = \sqrt{(r^2\omega^2\sin^2\omega t + r^2\omega^2\cos^2\omega t)} = r\omega$

as we found before.

Direction of motion Note that $\mathbf{v} \cdot \mathbf{r} = 0$ (check this), i.e. $\mathbf{v}$ is perpendicular to $\mathbf{r}$; this confirms that the direction of motion is along the tangent to the circle (see section 7.2:1).

Acceleration
Note that $\mathbf{a} = -\omega^2 \mathbf{r}$.

Magnitude $\quad |\mathbf{a}| = |-\omega^2 \mathbf{r}| = \omega^2 r = v^2/r$

Direction $\mathbf{a}$ is a negative scalar multiple of $\mathbf{r}$; therefore $\mathbf{a}$ is in the opposite direction to $\mathbf{r}$, i.e. towards O. It has no component along the tangent.

When a particle P moves in a circle with constant angular velocity ω,

(i) its velocity is $r\omega$ $(= v)$ in the direction of the tangent to the circle; (6)
(ii) its acceleration is $r\omega^2$ $(= v^2/r)$ towards the centre of the circle. (7)

Exercise 7.4:2

- **1** The position vector **r** of a particle P at time t is given by $\mathbf{r} = 5\cos 2t\mathbf{i} + 5\sin 2t\mathbf{j}$.
  - (i) Find the cartesian equation of the path of P.
  - (ii) Find the velocity **v** of P at time t. Show that **v** is perpendicular to **r**. Show that the speed of P is constant. What is the angular velocity of P?
  - (iii) Find the acceleration **a** of P. Show that $\mathbf{a} = -4\mathbf{r}$.

→ **2** A particle P moves on the circle

$$\mathbf{r} = 3\cos\theta\mathbf{i} + 3\sin\theta\mathbf{j}$$

with constant angular velocity 2 rad s^{-1}. Write down the position vector of P at time t, given that P is initially at the point $3\mathbf{i}$. Find the velocity and acceleration of P at time $t = \tfrac{1}{6}\pi$. Show that the speed of P is constant, and find it. Find the time taken to make one complete revolution (called the **period** of the motion).

3 A particle moves on the circle $x^2 + y^2 = 4$ with constant speed 10 m s^{-1}. Write down the position vector of P at time t, given that:

(i) P is initially at the point $(2, 0)$
(ii) P is initially at the point $(0, 2)$

4 A particle P moves on the circle

$$r = (3 + 2\cos\theta)\mathbf{i} + (4 + 2\sin\theta)\mathbf{j}$$

with constant angular velocity $\tfrac{1}{4}\pi$ rad s^{-1}. Write down the position vector of P at time t, given that P is initially at the point $5\mathbf{i} + 4\mathbf{j}$. Show that the speed of P and the magnitude of its acceleration are both constant, and find them. Find the direction of the velocity and the direction of the acceleration at time t, and show that they are perpendicular.

5 A particle P starts from the point $2\mathbf{i}$ and moves in a circle of radius 2, centre O, with constant angular velocity 3 rad s^{-1}. A particle Q starts at the same time as P from the point $3\mathbf{j}$ and moves in a circle of radius 3, centre O, with the same angular velocity.

  - (i) Find the position vectors of the particles at time t. Hence find $\overrightarrow{PQ}$ at time t. Show that the distance between the particles is constant, and explain why this is so.
  - (ii) Find the velocity of Q relative to P and the speed of Q relative to P.

R 6 Two particles P and Q are moving on the circle $x^2 + y^2 = 4$. P is initially at the point $(2, 0)$ and moves with constant angular velocity ω. Q is initially at the point $(0, 2)$ and moves with constant angular velocity -2ω. Find:

  - (i) the velocity of Q relative to P at time t
  - (ii) the time at which the particles are first moving in the same direction
  - (iii) the acceleration of Q relative to P at time t

7 The Earth's orbit around the Sun is approximately a circle of radius 1.5×10^8 km. Find:

(i) its speed
(ii) the magnitude of its acceleration

8 A man swings a small bucket of water round and round at constant angular speed ω in a vertical circle of radius 1 m. Find the least possible value of ω, in revolutions per second (rev s^{-1}), which ensures that the water stays in the bucket.

9 A greyhound runs with speed 25 m s^{-1} around a circular track of radius 100 m. Find:

  - (i) his angular velocity
  - (ii) the magnitude of his acceleration
  - (iii) the time he takes to complete one circuit

Miscellaneous exercise 7

+ **1** The banks of a river run along the lines $x - 2y - 34 = 0$ and $x - 2y + 6 = 0$. Find the width of the river (distances measured in metres).

  In a TV contest of talent between sports personalities, there is a race in which each competitor must run from the point $A(-16, 15)$ to the river, catch a fish, swim directly across the river, drink a gallon of beer and then run to the point $B(52, -31)$. A competitor X can run at $3\sqrt{5}$ m s^{-1} swim at $\sqrt{5}$ m s^{-1}. He knows that it will take him at least 10 s to catch a fish and 15 s to drink the beer. What is the least amount of time in which he can expect to finish the race?

2 A particle moving in a straight line starts at time $t = 0$ s with velocity 8 m s^{-1}. After t s its acceleration is $(6t - 10)$ m s^{-2}. Find:

(i) the distance the particle moves before first coming to instantaneous rest;

(ii) the times t_1 and t_2 seconds ($t_1 < t_2$) at which it is 4 m from its starting point. Show that at time t_2 it is instantaneously at rest.

If v is the velocity of the particle and x its distance from its starting point after t seconds, sketch graphs of v against t and x against t ($t > 0$).

What is the maximum speed of the particle in the time interval $t_1 < t < t_2$?

3 At time $t=0$ a particle is projected vertically upwards with speed $2\sqrt{(ga)}$ from a point O. At time $t=\sqrt{(a/g)}$ another particle is projected vertically upwards from O with speed $6\sqrt{(ga)}$. Find the height above O at which the particles collide.

4 A particle P moves in a straight line Ox. When $OP = x$, the speed of P is v.
(i) If $a^2v^2 = u^2(a^2 - x^2)$, a and u constant, show that the acceleration of P is given by $-u^2x/a^2$.
(ii) If $e^{x/a} = 1 + v/u$, show that the acceleration of P is given by $v(v+u)/a$.

5 A particle P, moving in a straight line, experiences a retardation which is inversely proportional to the square of its speed. When P passes through a point O on the line, its speed is u and its retardation is $\frac{1}{4}g$. Find the time taken for its speed to be halved and the distance it travels in this interval.

6 A particle moving in a straight line with speed v experiences a retardation of magnitude $e^{(v-u)/u}$ (u constant). Initially the particle is moving with speed u. Show that:
(i) the time taken for the speed to be halved is $u(\sqrt{e} - 1)$
(ii) the time taken for the particle to come to rest is $u(e-1)$
(iii) the distance moved by the particle in decelerating to rest is $u^2(e-2)$

7 A particle moves on the x-axis with retardation of magnitude kv^2, where v is its speed and k is a positive constant. Initially it is at the origin and moving with speed u.
(i) Find an expression for v in terms of x. Sketch the graph of v against x for $x \geq 0$. Find the distance moved by the particle in decelerating to half its original speed.
(ii) Show that, at time t,
$$x = \frac{1}{k}\ln(1 + kut)$$
Sketch the graph of x against t for $t \geq 0$.
(iii) Describe the motion of the particle a long time after it starts.

8 A mad tourist on a coach tour of the Sahara Desert goes for a stroll and gets lost. Several days later, while sitting disconsolately on a tuft of scrub at a point O of the desert, he sees an oasis (or is it a mirage?) and starts to run with speed 4 m s^{-1} towards it. But the sun is hot and his pace gets slower and slower. He moves with a retardation $0.0025\sqrt{v} \text{ m s}^{-2}$, where $v \text{ m s}^{-1}$ is his speed. If the oasis is 2.1 km from O, find his speed when he arrives and the time it takes him to reach the oasis.

9 A particle moving in a straight line experiences a retardation of $(5 + \frac{1}{2}x)$, where x is its distance from a fixed point O of the line. If the particle starts from O with speed 20 m s^{-1}, find the distance moved by the particle:
(i) before its speed is halved
(ii) before it comes to rest

10 A particle moves on the x-axis with acceleration k/x^2, where k is a positive constant. The particle is at rest when $x = a$. Find its speed when $x = 2a$. What is its terminal speed?

11 Measurements are made of the velocity $v \text{ m s}^{-1}$ of a particle moving in a straight line and its corresponding distance x m from its starting point O:

v	27.5	23.7	20.4	17.6	15.1	13.0
x	10	20	30	40	50	60

Show that these values of v and x satisfy approximately a relationship of the form $v = ae^{bx}$, and find a and b, giving your answers to 2 sig. figs.

Sketch a graph of v against x ($x \geq 0$). What is the initial velocity of the particle? Find the deceleration of the particle when it is at a distance 30 m from O.

12 At time t a particle moving in the xy plane has position vector $t\mathbf{i} - t^3\mathbf{j}$. Find:
(i) the cartesian equation of its path
(ii) its speed at time $t = 1$
(iii) its angular speed about the origin at time $t = 1$
(iv) its acceleration at time $t = 1$

13 The position vector $\mathbf{r}$ of a particle at time t is $\mathbf{r} = a(t - \sin t)\mathbf{i} + a(1 - \cos t)\mathbf{j}$.
(i) Show that the speed of the particle at time t is $2a|\sin \frac{1}{2}t|$.
(ii) Show that the magnitude of the acceleration of the particle is constant.
(iii) Find the times at which the velocity and acceleration of the particle are perpendicular.

14 The position vector $\mathbf{r}$ of a particle at time t is
$$\mathbf{r} = (a \cos \omega t)\mathbf{i} + (a \sin \omega t)\mathbf{j} + bt\mathbf{k}$$
where a, b and ω are constant. Show that the speed of the particle is constant and find its acceleration at time $t = \pi/2\omega$.

$+$ **15** At time t the acceleration $\mathbf{f}$ of a particle moving in the xy plane is given by the equation $\mathbf{f} = n^2\mathbf{r}$ where n is constant and $\mathbf{r}$ is the position vector of the particle. Verify that the solution
$$\mathbf{r} = \mathbf{a}e^{nt} + \mathbf{b}e^{-nt}$$
where $\mathbf{a}$ and $\mathbf{b}$ are constant vectors, satisfies this equation.

Find **a** and **b**, given that at time $t=0$ the particle is at the point with position vector $2\mathbf{i}$ and moving with velocity $2n\mathbf{j}$.

Find the cartesian equation of the path of the particle, and sketch the path. Show that after a long time the particle is travelling close to the line $y=x$.

16 At time $t=0$ a particle is at rest at the point with position vector $-2\mathbf{i}+3\mathbf{j}$. Subsequently its acceleration **a** is given by

$$\mathbf{a} = \begin{cases} 24t^3\mathbf{i}+6\mathbf{j} & 0<t\leq2 \\ 6\mathbf{j} & t>2 \end{cases}$$

Find its velocity and position vector when

(i) $t=2$ (ii) $t=3$.

Show that, for $t>2$, the path of the particle is a parabola.

17 Two particles move in the xy plane in such a way that, after time t, their position vectors are $3(1-\cos 2t)\mathbf{i}+4\sin t\mathbf{j}$ and $(3\sin t)\mathbf{i}+4(1-\cos 2t)\mathbf{j}$.

(i) Find the position vector of the point at which the particles collide.
(ii) Find the relative speed of the particles at the instant of collision.

18 At time $t=0$ a particle P starts from the point with position vector $-2\mathbf{i}-5\mathbf{j}$ and moves with velocity $4\mathbf{i}+2\mathbf{j}$. Simultaneously another particle Q starts from the point with position vector $\mathbf{i}+4\mathbf{j}$ and moves with constant velocity. Subsequently P and Q collide. Find:

(i) the speed of Q if the collision occurs after one second
(ii) the least speed at which Q can travel

19 An officer of the law sees a criminal person, wearing dark glasses and carrying a bag marked Swag, at a distance d due north of him. The man is running with speed u in a direction Nα°E. The officer, who has only been in the Force for a short time, runs enthusiastically to intercept the felon, and moves with speed v in a direction Nθ°E. Taking unit vectors **i** and **j** in directions due E and due N respectively, show that after time t the position vector of the officer relative to the villain is

$(v\sin\theta - u\sin\alpha)t\mathbf{i}+\{(v\cos\theta-u\cos\alpha)t-d\}\mathbf{j}$

If $v=\sqrt{2}u$, $\alpha=45°$, show that successful interception (followed by conviction and a punitive, but just, prison sentence) occurs after time $d(1+\sqrt{3})/v$. How large must v/u be for it always to be possible to choose θ so that interception occurs?

+ **20** Two particles A and B are moving with constant velocities $\mathbf{v}_A$ and $\mathbf{v}_B$. At time $t=0$, $\overrightarrow{AB}=\mathbf{p}$.

(i) Show that, if the particles collide, the time taken to reach the point of collision is $t=|\mathbf{p}|/|\mathbf{v}_B-\mathbf{v}_A|$.

(ii) Show that, if the particles do not collide, they are closest together when $t=(\mathbf{v}_A-\mathbf{v}_B)\cdot\mathbf{p}/|\mathbf{v}_A-\mathbf{v}_B|^2$.

21 A particle A, moving with constant velocity $3\mathbf{i}-\mathbf{j}$, passes through the point with position vector $-3\mathbf{j}$ at the same instant as a particle B, moving with constant velocity $a\mathbf{i}-3\mathbf{j}$, passes through the point with position vector $4\mathbf{i}+5\mathbf{j}$. Find:

(i) the velocity of B relative to A
(ii) the displacement vector $\overrightarrow{AB}$ t seconds later

Hence find the value of a which ensures that A and B collide, and the position vector of the point of collision.

If $a=1$, find the minimum distance between A and B in the subsequent motion.

22 In an exciting new shopping complex, which wins several awards and then has to be demolished three years later, **i**, **j** and **k** are mutually perpendicular vectors, each of length 1 m, with **k** vertically upwards, and O is the exit from Foodos supermarket. Relative to O, the position vectors of the points A, B and C are $11\mathbf{j}+5\mathbf{k}$, $15\mathbf{i}+11\mathbf{j}$ and $15\mathbf{i}+30\mathbf{j}+10\mathbf{k}$ respectively. Pedestrian walkways W_1 and W_2 run from A directly to B and from O directly to C, respectively. At time $t=0$ a woman X starts from A and walks down W_1 with speed $\frac{1}{2}\sqrt{10}$ m s^{-1} towards B. Simultaneously her son Y, whom she has not seen for 20 years and believes to be dead, starts from O and walks up W_2 with speed $1\frac{3}{4}$ m s^{-1} towards C. Find the velocity of Y relative to X and the displacement vector $\overrightarrow{XY}$ at time t. Hence show that mother and son never meet, and find the shortest distance between them in subsequent motion. At time $t=7$ a friend of the family Z, who is standing at the point D with position vector $-9\mathbf{j}$, notices that X and Y are going to miss each other and starts running with speed 5 m s^{-1} directly towards B to intercept X. Show that Z arrives at B two seconds after X. Y subsequently disappears and is never seen again.

23 A cyclist is cycling due south along a long straight road. When he cycles with speed u, the wind appears to blow from S60°W. When he cycles with speed $2u$, the wind appears to blow from S30°W. Find the actual speed and direction of the wind. If he increases his speed to $3u$, find the apparent speed and direction of the wind.

+ **24** A river flows between parallel banks which are at a distance $2a$ apart. A motor boat, whose speed in still water is u, crosses the river. The speed of the current at any point is $ux/2a$, where x is the distance of the point from the nearer bank.

(i) If the boat is steered perpendicular to the banks, find how far it is carried downstream while making the crossing.

(ii) If the boat is steered so as to move directly across the river perpendicular to the banks, show that, when its distance from the nearer bank is x, its speed is

$u\sqrt{(4a^2 - x^2)}/4a$, and hence find the time it takes to cross the river.

25 At time $t = 0$ a particle is projected from a point O with speed u at an angle of elevation α. Taking unit vectors $\mathbf{i}$ and $\mathbf{j}$ horizontally and vertically upwards, write down the velocity and the position vector of the particle at time t. Find the time τ at which the particle is moving at right angles to its original direction of motion. Find the time T at which the particle strikes the horizontal plane through O.
If $\tau = \tfrac{2}{3} T$, find α and show that $u = gT/\sqrt{3}$.

26 A particle is projected from a point O with speed u.

(i) Find its maximum range X on the horizontal plane through O.

(ii) The particle is to be projected to strike a target A on the same horizontal level as O, where $OA = \tfrac{3}{4} X$. Show that there are two possible angles of projection, and show that the difference between the greatest heights reached on the two trajectories is $\sqrt{7} u^2/8g$.

27 A shell bursts on level ground, throwing fragments with speed u in all directions. After a time T a fragment hits the ground at a distance R from the shell. Show that

$$g^2 T^4 - 4u^2 T^2 + 4R^2 = 0$$

Given that $u = 30 \text{ m s}^{-1}$, find the period of time during which a man lying 30 m from the place where the shell bursts is in danger.

28 At time $t = 0$ a particle is projected from a point O on horizontal ground with speed u at an angle of elevation 45°. Write down expressions for the horizontal and vertical components of:

(i) the velocity
(ii) the displacement of the particle at time t

When travelling upwards at an angle β to the horizontal the particle passes through a point P. The line OP makes an angle θ with the horizontal. Show that:

$$2 \tan \theta - \tan \beta = 1$$

The point P is at a horizontal distance 20 m from O and is at a height of 15 m above the ground. Find:

(i) β (ii) u
(iii) the time at which the particle is travelling downwards at an angle β to the horizontal.

29 A particle is projected from a point O on horizontal ground with speed $2\sqrt{(2gh)}$ at an angle of elevation α. The particle passes through two points A and B at the same height h above O. Find:

(i) the speed of the particle at A and at B
(ii) the time it takes to travel from A to B

If $AB = 2h$, show that

$$64 \sin^4 \alpha - 80 \sin^2 \alpha + 17 = 0$$

Show that this equation has a root near $\alpha = \tfrac{1}{6}\pi$. Taking $\alpha = \tfrac{1}{6}\pi$ as a first approximation to this root, find a second approximation. Give your answer to 2 dec. pls.

30 A particle is projected with speed u from a point O on horizontal ground. If R is the range of the particle and H the greatest height it attains, show that:

$$gR^2 = 8H(u^2 - 2gH)$$

Sketch a graph of R^2 against H for $H < u^2/2g$. Show that R^2 is an increasing function of H so long as $H < u^2/4g$.

Hence find the greatest range which can be attained by a particle projected with speed 20 m s^{-1} from the floor of a tunnel if the height of the tunnel is:

(i) 8 m   (ii) 12 m

31 A particle is projected from a point O with speed u at an angle of elevation α. Show that the equation of its path, referred to horizontal and vertical axes Ox and Oy in the plane of the path, is:

$$y = x \tan \alpha - \frac{gx^2}{2u^2} \sec^2 \alpha$$

(i) A particle is projected at an angle of elevation $\tan^{-1} 3$ from a point on a horizontal plane distant 80 m from the foot of a vertical tower of height 20 m. The particle just clears the tower. Find the initial speed of the particle.

(ii) If, instead, the particle is projected from the same point with speed 40 m s^{-1}, find the set of possible angles of projection which will ensure that the particle clears the tower.

32 A hawk is flying with speed $4\sqrt{(ga)}$ at an angle $\tan^{-1} \tfrac{1}{2}$ above the horizontal. When passing through a point O at a height $2a$ above a horizontal plain, it drops a dove from its beak. The dove lands on the ground at A. Assuming that air resistance is negligible, find:

(i) the time of fall of the dove
(ii) the distance OA
(iii) the speed of the dove and the tangent of the angle between its direction of motion and the horizontal when it reaches A.

33 A ball is projected from a point O on level ground with speed u at an angle α to the horizontal. When moving upwards at an angle of 45° to the horizontal, it just clears a wall of height $2a$ which is at a horizontal distance a from O. The path of the ball is in a vertical plane which is perpendicular to the wall. Show that $u = \sqrt{(5ga)}$ and find α.

Subsequently, when moving downwards at an angle 45° to the horizontal, the ball just clears another wall, parallel to the first, of height $2a$. Show that the distance between the walls is $3a$, and find the distance from O at which the ball hits the ground for the first time.

34 A particle is projected with speed 20 m s^{-1} from a point O on horizontal ground. The foot of a vertical tower of height h m is at a distance 10 m from O.

(i) Show that the particle can clear the tower so long as $h \leq 18.75$.

(ii) Find the two possible angles of projection if $h = 10$ and the particle just clears the tower. Find the range in each case.

+ **35** A rugby player is attempting to convert a try from directly in front of the goal. The horizontal bar is at a height b from the ground. He can kick the ball with speed u. Show that if his distance from the goal is a, and he kicks the ball at an angle of elevation α, his attempt to convert the try will succeed so long as

$$\tan^2 \alpha - \frac{2u^2}{ga} \tan \alpha + \left(\frac{2u^2 b}{ga^2} + 1\right) \leq 0$$

Deduce that, for fixed b and u, he can expect to succeed only if

$$a \leq \frac{u^2}{g} \sqrt{(1 - 2gb/u^2)}$$

36 A particle A is projected from a point O with initial speed 30 m s^{-1} at an angle of elevation $\tan^{-1} \frac{4}{3}$. Two seconds later another particle B is projected from O. After a further two seconds the particles collide.

(i) Find the initial speed and angle of elevation of B.
(ii) Find the speed at which each particle is moving when they collide.

+ **37** Two boys stand on level ground at distance h apart. One of them throws a stone vertically upwards with speed $\sqrt{(\lambda gh)}$; the other waits till this stone is at its highest point and then throws a stone directly at it with speed v. Show that the stones will collide provided that $v \geq v_0$, where

$$v_0^2 = gh(\lambda + 4)/4\lambda$$

Find the value of λ for which v_0 is least.

38 On the Norfolk Broads at 4 a.m. It is still dark. A duck takes off from rest at a point A at water-level, rising vertically with acceleration f. Simultaneously an aristocrat in a punt, who has been waiting patiently all night for a bit of action, fires a bullet from a point B, also at water level, where $AB = h$. The bullet has initial speed u and angle of elevation α. Show that, if the bullet hits the duck,

$$u^2 \sin 2\alpha = h(f + g)$$

Meanwhile, several miles away, an ardent conservationist is covering the walls of the aristocrat's country mansion with graffiti...

39 (i) A particle P starts from the point A with position vector $8\mathbf{i}$ and moves on the circle

$$\mathbf{r} = 4(1 + \cos \theta)\mathbf{i} + 4 \sin \theta \mathbf{j}$$

with constant angular velocity $\frac{1}{2}\pi$ rad s^{-1} about the centre of the circle. Find the speed of P and the magnitude of its acceleration.
  Another particle Q starts from A and moves with constant velocity $8\mathbf{j}$. Show that, when $t = 1$, O, P and Q are collinear, and find the speed of Q relative to P at this instant.

(ii) Two points O and B are on the same horizontal level; $OB = 10$ m. A particle falls vertically from rest at A. Show that, t seconds later, the angular speed ω of P about O is $4t/(1 + t^4)$ rad s^{-1}. Show that ω is greatest when $A\hat{O}P = \frac{1}{6}\pi$.

+ **40** (i) At time $t = 0$ a particle P, moving with constant velocity $\mathbf{v}$, is at the point with position vector $\mathbf{r}_0$. Write down a vector equation of the path of P. Show that P is (or was) closest to the origin at time

$$t = -\frac{\mathbf{v} \cdot \mathbf{r}_0}{|\mathbf{v}|^2}$$

(ii) At time $t = 0$ a particle Q, moving with velocity $\mathbf{u}$, is at the point with position vector $\mathbf{r}_0$. Subsequently it moves with constant acceleration $\mathbf{f}$. Write down a vector equation of the path of Q. Show that the path of Q lies either in straight line or a plane.

8 Forces on a particle

8.1 Newton's laws

8.1:1 Newton's first law

Velocity and acceleration of an object
If an object is moving **linearly** then each point of the object moves with the same velocity **v** and the same acceleration **a** (see p. 367). So we can sensibly talk about the velocity **v** and acceleration **a** of the object, meaning the velocity and acceleration of each point.

An object is in **equilibrium** if it is at rest, or moving with constant velocity.

Why do objects move?
A pencil sharpener rests on a table. How do you move it?

You push it—you are exerting a force on it; or you pull it—you are exerting a force on it; or you put a magnet near it—the magnet exerts a force on it; or you take the table away—the Earth exerts a force on it; or you ring up Mercury Pencil Sharpener Removal Services....

Force is necessary to change the velocity of an object; i.e.

N1: Newton's First Law of Motion
An object in equilibrium stays in equilibrium until a force acts on it. **(1)**

We have said 'force is necessary to move an object.' Is the converse true, i.e. does an object necessarily move when a force is exerted on it? The answer is obviously 'no', e.g. when you lean on a wall it doesn't necessarily fall over. The reason is that there are other forces on the wall other than the force you are exerting; and overall the forces on the wall hold it in equilibrium.

We are interested not in the individual effect of one particular force on an object but in the *overall effect* of all the forces on the object.

Resultant of a system of forces
A force has a magnitude and a direction. So force is a vector quantity. Thus, if a system of forces acts on a body, we can find the resultant **R** of the forces using the methods of Chapter 6.

The single force **R** has the same overall linear effect as the system of forces.

Note: the unit of force is the newton (N); this is explained in section 8.1:2.

Forces on a particle 421

Worked example 1 Forces $\mathbf{F}_1$ and $\mathbf{F}_2$ of magnitude 4 N and 6 N respectively are inclined at 60° to one another (fig. 8.1:**1a**). Find the magnitude and direction of their resultant $\mathbf{R}$.

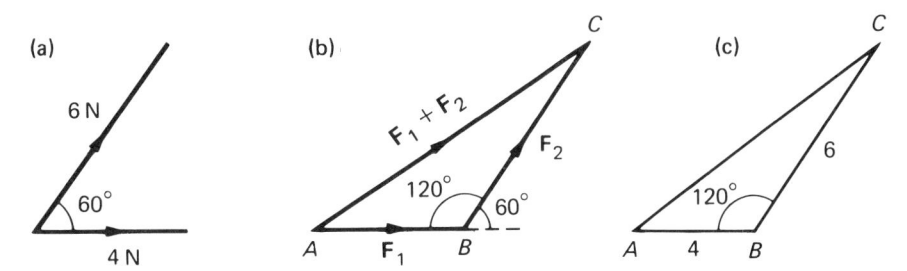

Fig. 8.1:1

Method 1: Draw a *vector triangle* (fig. 8.1:**1b**). If $\mathbf{F}_1$ is represented by $\overrightarrow{AB}$, $\mathbf{F}_2$ by $\overrightarrow{BC}$, then $\mathbf{F}_1 + \mathbf{F}_2$ is represented, in magnitude and direction, by $\overrightarrow{AC}$.

Using the cosine rule (section 3.3:1)

$$AC^2 = 4^2 + 6^2 - 2 \cdot 4 \cdot 6 \cos 120° = 76$$
$$\Rightarrow \quad AC = \sqrt{76} \approx 8.72$$

Using the sine rule (section 3.3:1)

$$\frac{\sin A}{6} = \frac{\sin 120°}{AC} \quad \Rightarrow \quad A \approx 36.6°$$

So $\mathbf{R}$ has magnitude 8.72 N and makes an angle of 36.6° with $\mathbf{F}_1$.

Note: In fig. 8.1:**1** we have denoted each force, as usual, by its *magnitude* only, and indicated its direction by the direction of the arrow.

Method 2: Take unit vectors $\mathbf{i}$ and $\mathbf{j}$ in convenient directions (fig. 8.1:**2a**) and resolve $\mathbf{F}_1$ and $\mathbf{F}_2$ into components in these directions (section 6.2:1).

$\mathbf{F}_1 = 4\mathbf{i}$, $\mathbf{F}_2 = 6 \cos 60° \mathbf{i} + 6 \sin 60° \mathbf{j} = 3\mathbf{i} + 3\sqrt{3}\mathbf{j}$. Thus

$$\mathbf{R} = \mathbf{F}_1 + \mathbf{F}_2 = 7\mathbf{i} + 3\sqrt{3}\mathbf{j} \quad \text{by 6.2:1 equation (6)}$$

So $\mathbf{R}$ has magnitude $\sqrt{(7^2 + (3\sqrt{3})^2)} = \sqrt{76} \approx 8.72$ N and makes an angle $\tan^{-1}\{3\sqrt{3}/7\}$, i.e. 36.6° with $\mathbf{F}_1$ (fig. 8.1:**2**).

We can extend either of these methods to systems of more than two forces: if $\mathbf{R}$ is the resultant of a system of forces $\{\mathbf{F}\}$, we write

$$\mathbf{R} = \sum \mathbf{F} \qquad (2)$$

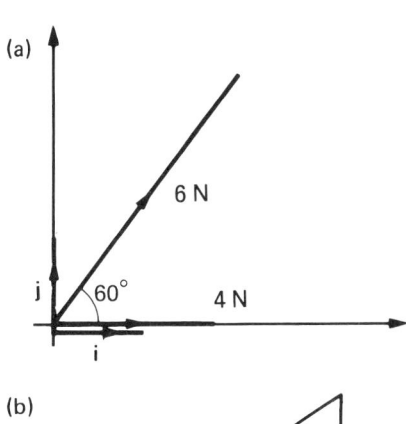

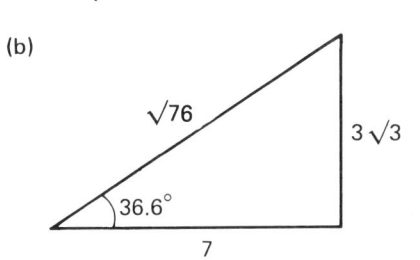

Fig. 8.1:2

Using Method 1, we would draw a *vector polygon*, as in fig. 8.1:**3**.

Method 2 is usually much easier, particularly when more than two forces are involved. If each force $\mathbf{F} = F_x \mathbf{i} + F_y \mathbf{j}$, and $\mathbf{R} = R_x \mathbf{i} + R_y \mathbf{j}$, then equation (2) becomes

$$R_x \mathbf{i} + R_y \mathbf{j} = \sum (F_x \mathbf{i} + F_y \mathbf{j}) = (\sum F_x)\mathbf{i} + (\sum F_y)\mathbf{j} \qquad (3)$$

by 6.2:1 equation (6). So

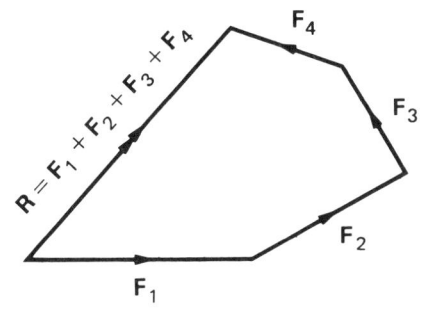

$$R_x = \sum F_x \quad \text{and} \quad R_y = \sum F_y$$

Fig. 8.1:3

422 Newton's first law

For example, the resultant of forces $\mathbf{F}_1 = \mathbf{i} + 3\mathbf{j}$, $\mathbf{F}_2 = 4\mathbf{i} - 2\mathbf{j}$, $\mathbf{F}_3 = -3\mathbf{i}$, $\mathbf{F}_4 = \mathbf{i} + 5\mathbf{j}$ is

$$\mathbf{R} = (\mathbf{i} + 3\mathbf{j}) + (4\mathbf{i} - 2\mathbf{j}) + (-3\mathbf{i}) + (\mathbf{i} + 5\mathbf{j})$$
$$= (1 + 4 - 3 + 1)\mathbf{i} + (3 - 2 + 5)\mathbf{j} = 3\mathbf{i} + 6\mathbf{j}$$

Exercise 8.1:1

• → 1 (i) A force of magnitude F makes an angle α with Ox. What are the components of the force in the $\mathbf{i}$ and $\mathbf{j}$ directions? (See 6.2:1 Q6.)

(ii) Write down each of the forces in the form $p\mathbf{i} + q\mathbf{j}$:

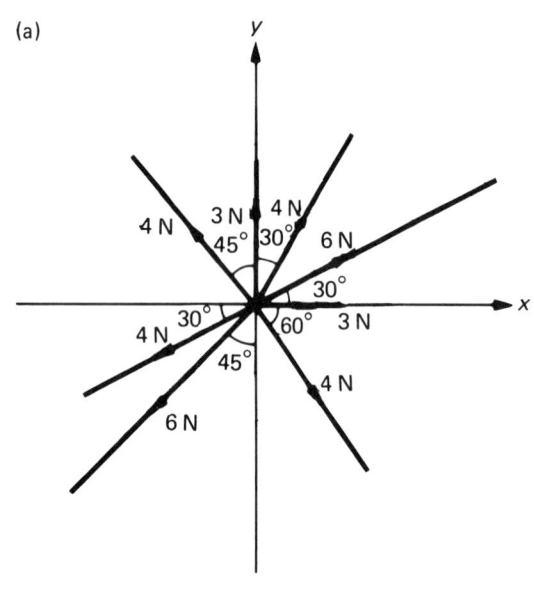

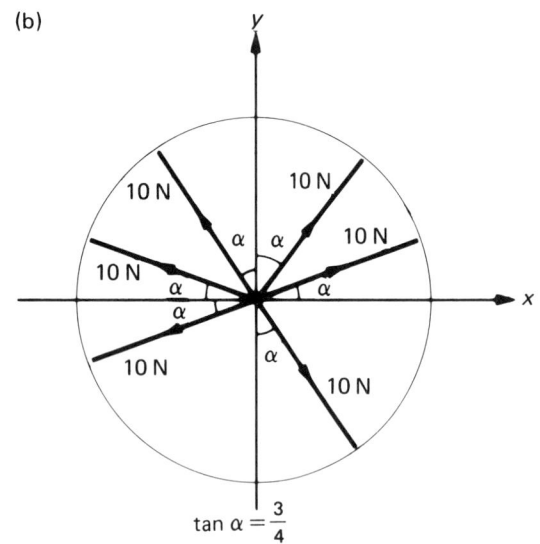

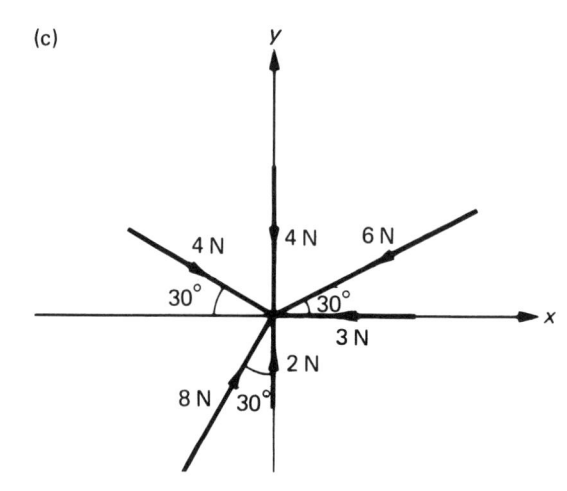

(a)

(b) $\tan \alpha = \dfrac{3}{4}$

(c)

2 A particle moving on a horizontal plane is acted on by the forces shown in the diagram. Find the components of each force in the directions $\mathbf{i}$ (horizontal) and $\mathbf{j}$ (vertically upwards).

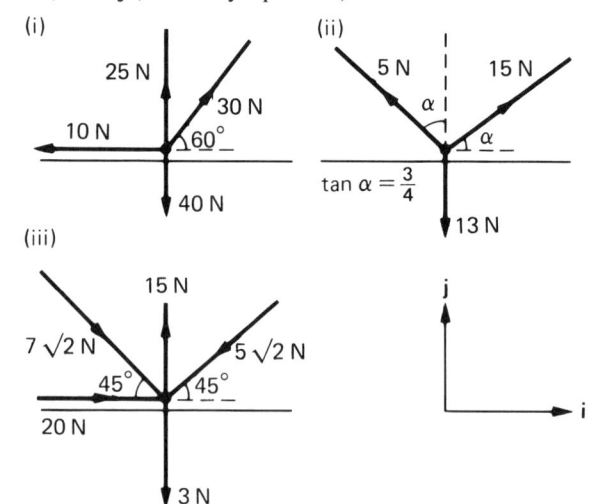

(i)

(ii) $\tan \alpha = \dfrac{3}{4}$

(iii)

3 A particle moving on an inclined plane is acted on by the forces shown in the diagram. Find the components of each force in the directions $\mathbf{i}$ (up the plane) and $\mathbf{j}$ (perpendicular to the plane).

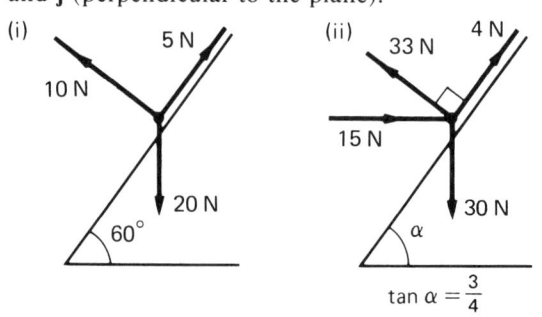

(i)

(ii) $\tan \alpha = \dfrac{3}{4}$

(iii) [diagram: 18 N at 30° from dashed line, 3√3 N at 60°, 12 N downward, 30° incline]

4 Find the magnitude and direction of each of the following forces:

- (i) $5\mathbf{i}+5\mathbf{j}$
- (ii) $-4\mathbf{i}+3\mathbf{j}$
- (iii) $4\mathbf{i}-3\mathbf{j}$
- (iv) $-2\mathbf{i}-\mathbf{j}$

(C.f. 6.2:1 Q13.)

5 The magnitude of a force $p\mathbf{i}+5\mathbf{j}$ is 13. Find the two possible values of p and the corresponding inclinations of the force to Ox.

6 The magnitude of a force $-4\mathbf{i}+q\mathbf{j}$ is 8. Find the two possible values of q and the corresponding inclinations of the force to Ox.

7 Find the unit vector in the direction of **F** when:

- (i) $\mathbf{F}=\mathbf{i}+\mathbf{j}$
- (ii) $\mathbf{F}=-12\mathbf{i}+5\mathbf{j}$
- (iii) $\mathbf{F}=-\mathbf{i}-3\mathbf{j}$
- (iv) $\mathbf{F}=-7\mathbf{i}$

8 If $\mathbf{u}=4\mathbf{i}+3\mathbf{j}$, write down the unit vector $\hat{\mathbf{u}}$ in the direction of $\mathbf{u}$.

A force **F** of magnitude 15 N acts in the direction of **u**. Express **F** in the form $p\mathbf{i}+q\mathbf{j}$.

9 Express in the form $p\mathbf{i}+q\mathbf{j}$:

(i) a force of magnitude $5\sqrt{2}$ N in the direction of the vector $\mathbf{i}-\mathbf{j}$
(ii) a force of magnitude 8 N in the direction of the vector $-\sqrt{3}\mathbf{i}+\mathbf{j}$
(iii) a force of magnitude 26 N in the direction $\overrightarrow{OA}$, where A is the point $(5, 12)$
(iv) a force of magnitude $\sqrt{40}$ N in the direction $\overrightarrow{OB}$, where B is the point $(3, -1)$
(v) a force of magnitude 5 N in the direction $\overrightarrow{CD}$, where C and D are the points $(-5, -6)$ and $(3, 2)$
(vi) a force of magnitude 34 N in the direction $\overrightarrow{EF}$, where E and F are the points $(8, 8)$ and $(-7, 0)$.

Questions 10–20: Find the magnitude and direction of the resultant of each of the following systems of forces, using either of the methods of worked example 1:

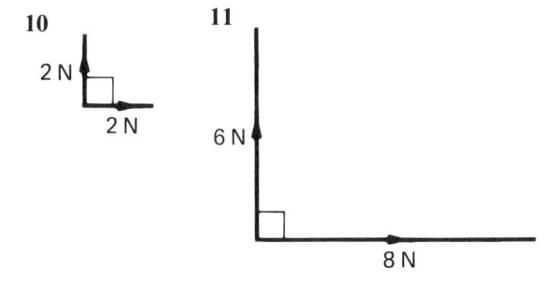

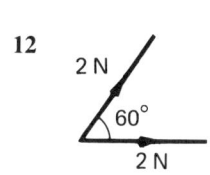

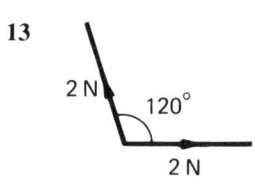

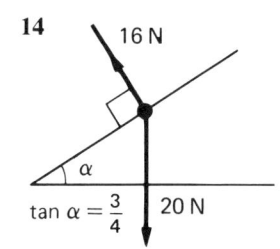

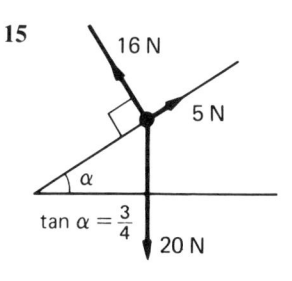

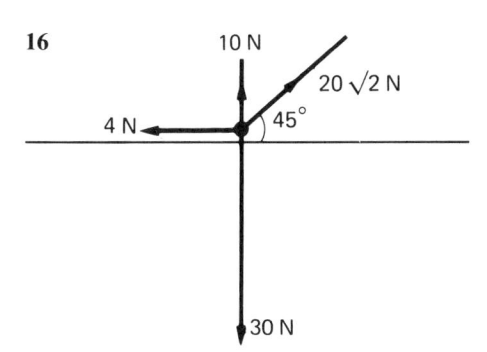

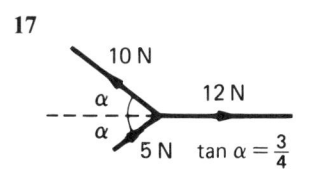

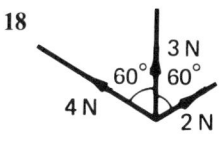

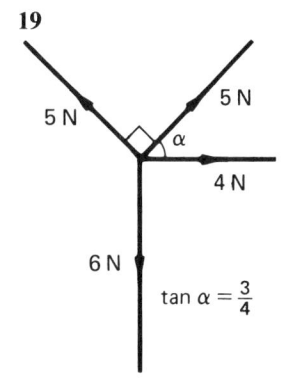

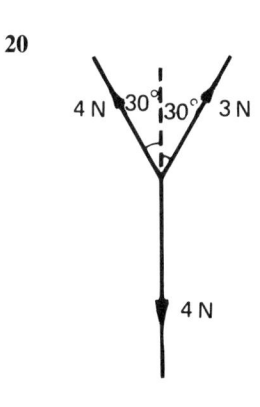

10 [diagram: 2 N up, 2 N right]

11 [diagram: 6 N up, 8 N right]

12 [diagram: 2 N at 60°, 2 N right]

13 [diagram: 2 N, 120°, 2 N right]

14 [diagram: 16 N along incline, 20 N down, $\tan\alpha = \frac{3}{4}$]

15 [diagram: 16 N up incline, 5 N perpendicular, 20 N down, $\tan\alpha = \frac{3}{4}$]

16 [diagram: 10 N up, 20√2 N at 45°, 4 N left, 30 N down]

17 [diagram: 10 N, 12 N, 5 N, angles α, $\tan\alpha = \frac{3}{4}$]

18 [diagram: 3 N, 4 N, 2 N, 60°, 60°]

19 [diagram: 5 N, 5 N, 4 N, 6 N, angle α, $\tan\alpha = \frac{3}{4}$]

20 [diagram: 4 N, 3 N at 30°, 30°, 4 N down]

21 Find the magnitude and direction of the resultant of two forces of magnitudes 3 N and 2 N if the angle between them is:
(i) 0° (ii) 60° (iii) 90° (iv) 120° (v) 180°

22 Two forces of magnitudes 2 N and 4 N are inclined at an angle θ. Find θ if the magnitude of their resultant is:
(i) 6 N (ii) 5 N (iii) 4 N (iv) 3 N (v) 2 N

424 *Newton's first law*

Questions 23–26: Set up unit vectors **i** and **j** in sensible directions and hence find the magnitude and direction of the resultant of each of the systems of forces.

23 *ABCD* is a square.

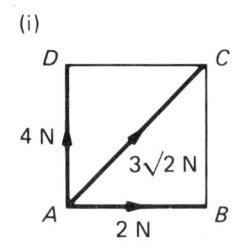

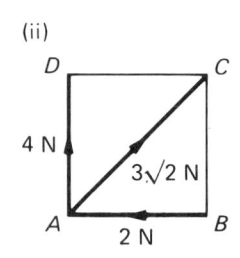

(i) [Square ABCD with 4 N along DA, 3√2 N along AC, 2 N along AB]

(ii) [Square ABCD with 4 N along AD, 3√2 N along AC, 2 N along AB]

24 *ABC* is an equilateral triangle. *D* is the midpoint of *BC*. *G* is the centroid of △*ABC*.

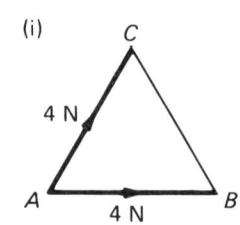

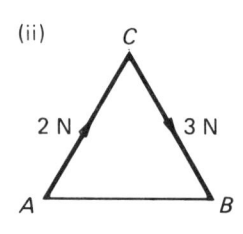

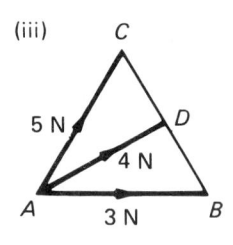

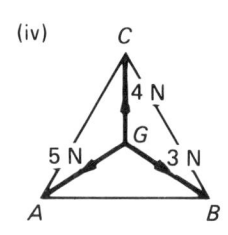

(i) [Triangle with 4 N along AC, 4 N along AB]

(ii) [Triangle with 2 N along AC, 3 N along BC]

(iii) [Triangle with 5 N along AC, 4 N along AD, 3 N along AB]

(iv) [Triangle with forces from G: 4 N towards C, 5 N towards A, 3 N towards B]

25 *ABCD* is a rectangle. *AB* = 4 m, *BC* = 3 m.

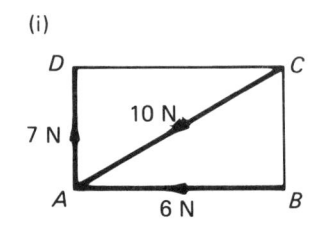

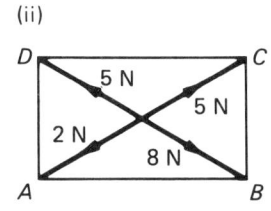

(i) [Rectangle with 7 N along AD, 10 N along AC, 6 N along AB]

(ii) [Rectangle with 5 N along DC, 5 N along BC, 2 N along CA, 8 N along AC]

26 *ABCDEF* is a regular hexagon.

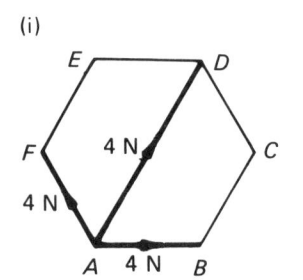

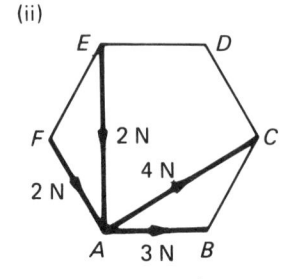

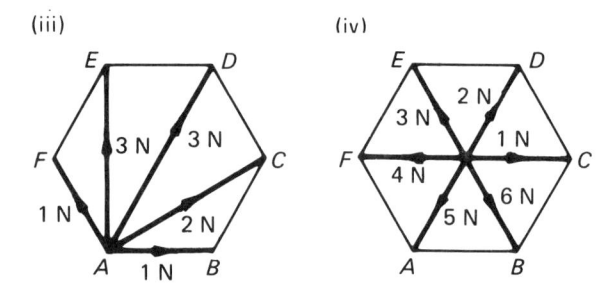

(i) [Hexagon with 4 N along FA, 4 N along AD]

(ii) [Hexagon with 2 N along FA, 2 N along AE, 4 N along AC, 3 N along AB]

(iii) [Hexagon with 3 N along AF, 3 N along AE, 3 N along AD, 2 N along AC, 1 N along AB]

(iv) [Hexagon with forces from centre: 2 N towards D, 1 N towards C, 6 N towards B, 5 N towards A reversed, 4 N towards F, 3 N towards E]

27 Find in the form $p\mathbf{i} + q\mathbf{j}$ the resultant of the forces:
(i) $\mathbf{i} + 7\mathbf{j}$, $4\mathbf{i} - 3\mathbf{j}$
(ii) $4\mathbf{i} + 4\mathbf{j}$, $-5\mathbf{j}$, $-\mathbf{i} + 6\mathbf{j}$, $3\mathbf{i} - 2\mathbf{j}$

→ **28** The resultant of forces $5\mathbf{i} + 3\mathbf{j}$, $p\mathbf{i} + 4\mathbf{j}$ and $\mathbf{i} + q\mathbf{j}$ is $3\mathbf{i} + 5\mathbf{j}$. Find p and q.

29 The resultant of forces $p\mathbf{i} + \mathbf{j}$, $2q\mathbf{i} + 3p\mathbf{j}$ and $\mathbf{i} + q\mathbf{j}$ is $-6\mathbf{i}$. Find p and q.

30 The resultant of forces $p\mathbf{i} + \mathbf{j}$ and $7\mathbf{i} - 9\mathbf{j}$ is parallel to the vector $\mathbf{i} + 2\mathbf{j}$. Find p.

31 Find the resultant of the following force:
• (i) $\mathbf{F}_1$ of magnitude 20 N in the direction $3\mathbf{i} + 4\mathbf{j}$
  $\mathbf{F}_2$ of magnitude $2\sqrt{5}$ in the direction $-\mathbf{i} + 2\mathbf{j}$
→ (ii) $\mathbf{F}_1$ of magnitude $3\sqrt{10}$ N in the direction $3\mathbf{i} + \mathbf{j}$
  $\mathbf{F}_2$ of magnitude $2\sqrt{10}$ N in the direction $\mathbf{i} - 3\mathbf{j}$
  $\mathbf{F}_3$ of magnitude 4 N in the direction $\mathbf{j}$
(iii) $\mathbf{F}_1$, $\mathbf{F}_2$ and $\mathbf{F}_3$, each of magnitude $10\sqrt{2}$ N, in the directions $\mathbf{i} + \mathbf{j}$, $7\mathbf{i} - \mathbf{j}$ and $-\mathbf{i} - 7\mathbf{j}$.

R 32 A force of 10 N acting parallel to the vector **i** is the resultant of two forces **P** and **Q** which act parallel to the vectors $4\mathbf{i} + 3\mathbf{j}$ and $-2\mathbf{i} - 4\mathbf{j}$ respectively. Calculate the magnitudes of **P** and **Q**.

33 *A* and *B* are the points $(1, 1)$ and $(4, 5)$. Find the resultant of forces $\mathbf{F}_1$ of magnitude $3\sqrt{2}$ N acting along $\overrightarrow{OA}$ and $\mathbf{F}_2$ of magnitude 15 N acting along $\overrightarrow{AB}$.

R 34 *A* and *B* are the points $(3, 4)$ and $(-1, 1)$. Find the resultant of the forces:
$\mathbf{F}_1$ of magnitude 5 N in the direction Ox,
$\mathbf{F}_2$ of magnitude 7 N in the direction Oy,
$\mathbf{F}_3$ of magnitude 8 N in the direction $\overrightarrow{OA}$,
$\mathbf{F}_4$ of magnitude 6 N in the direction $\overrightarrow{AB}$.

→ **35** *A* and *B* are the points $(2, 3)$ and $(-1, -1)$. Find the resultant of the forces:
$\mathbf{F}_1$ of magnitude $2\sqrt{13}$ N in the direction $\overrightarrow{OA}$,
$\mathbf{F}_2$ of magnitude $8\sqrt{}$ N in the direction $\overrightarrow{OB}$,
$\mathbf{F}_3$ of magnitude 10 N in the direction $\overrightarrow{BA}$.

8.1:2 Newton's second law

Definition
The **mass** of an object is the quantity of matter contained in it.

(This seems a vague definition, since we haven't explained what we mean by 'matter'. But then we mustn't expect too much of definitions, since we can only define a term in terms of other terms. Look at the definitions of some simple words—'the', 'take', 'tree'—in the dictionary. The important word in the definition of mass is the word 'quantity'; this implies that mass can be measured.)

Mass is a scalar quantity. The unit of mass is the kilogram (kg).

Definition
A **light object** is an object with no mass.

A light object is a mathematical model for an object with a very small mass compared with other objects that are being considered; e.g. a brussel sprout in an oil tanker.

Definition
The **linear momentum** of an object of mass m moving with velocity $\mathbf{v}$ is $m\mathbf{v}$. (4)

Linear momentum is a vector quantity. Its units are those of mass × velocity, i.e. kg m s^{-1}.

Newton found experimentally that the resultant $\mathbf{R}$ of a system of forces acting on an object is proportional to the rate of change of the linear momentum of the object.

$$\mathbf{R} \propto \frac{d}{dt}(m\mathbf{v}) \tag{5}$$

$$\mathbf{R} = k\frac{d}{dt}(m\mathbf{v}) \quad k \text{ constant}$$

If the mass of the object is constant,

$$\frac{d}{dt}(m\mathbf{v}) = m\frac{d\mathbf{v}}{dt} = m\mathbf{a};$$

so $$\mathbf{R} = km\mathbf{a} \tag{6}$$

In particular

$$|\mathbf{R}| = km|\mathbf{a}| \tag{7}$$

Let us define the unit of force to be the newton (N) where 1 N is the (resultant) force needed to give an object of mass 1 kg an acceleration of 1 m s^{-2}. From equation (7),

$$1 = k \cdot 1 \cdot 1 \quad \Rightarrow \quad k = 1$$

So with this choice of unit of force, $\mathbf{R} = m\mathbf{a}$, giving Newton's Second Law of Motion.

> **N2: Newton's Second Law of Motion**
> If a system of forces with resultant $\mathbf{R}$ acts on an object of constant mass m moving with acceleration $\mathbf{a}$, then
> $$\mathbf{R} = m\mathbf{a} \qquad (8)$$
> so long as $\mathbf{R}$ is measured in N, m in kg and $\mathbf{a}$ in m s^{-2}.

Worked example 2 A particle of mass 5 kg moves under the action of three forces, $\mathbf{F}_1 = 9\mathbf{i} - 2\mathbf{j}$, $\mathbf{F}_2 = 4\mathbf{i} + 3\mathbf{j}$ and $\mathbf{F}_3 = 2\mathbf{i} - 6\mathbf{j}$. Find its acceleration.

The resultant of the forces,
$$\mathbf{R} = \mathbf{F}_1 + \mathbf{F}_2 + \mathbf{F}_3$$
$$= (9\mathbf{i} - 2\mathbf{j}) + (4\mathbf{i} + 3\mathbf{j}) + (2\mathbf{i} - 6\mathbf{j})$$
$$= 15\mathbf{i} - 5\mathbf{j}$$

N2 $\Rightarrow 15\mathbf{i} - 5\mathbf{j} = 5\mathbf{a}$, so $\mathbf{a} = 3\mathbf{i} - \mathbf{j}$.

Exercise 8.1:2

- **1** A resultant force of magnitude 10 N acts on a particle of mass 25 kg. Find the magnitude of the acceleration of the particle.

2 A resultant force of magnitude 2 N acts on a particle of mass 0.16 kg. Find the magnitude of the acceleration of the particle.

3 A resultant force of magnitude $3mg$ acts on a particle of mass $4m$. Find the magnitude of the acceleration of the particle.

② • **4** A particle of mass 3 kg moves under the action of three forces $4\mathbf{i}$, $-2\mathbf{j}$ and $2\mathbf{i} + 5\mathbf{j}$. Find the resultant of the three forces. Hence show that the acceleration of the particle is $2\mathbf{i} + \mathbf{j}$.

② **5** Find the acceleration of a particle of mass 4 kg moving under the action of the following systems of forces:
  (i) $4\mathbf{i} - 3\mathbf{j}$, $8\mathbf{i} - 5\mathbf{j}$
  (ii) $\mathbf{i} + \mathbf{j}$, $-2\mathbf{i} + 3\mathbf{j}$, $-\mathbf{i} - 4\mathbf{j}$
  (iii) $3\mathbf{i} - \mathbf{j}$, $-7\mathbf{i} + 5\mathbf{j}$, $6\mathbf{i} + 9\mathbf{j}$, $-4\mathbf{i} - 3\mathbf{j}$

② **6** A particle of mass 2 kg moves under the action of two forces: $\mathbf{F}_1$, of magnitude 20 N in the direction $4\mathbf{i} - 3\mathbf{j}$, and $\mathbf{F}_2$, of magnitude 30 N in the direction $-3\mathbf{i} + 4\mathbf{j}$. Find its acceleration.

② → **7** A particle of mass 6 kg moves under the action of three forces:
  $\mathbf{F}_1$ of magnitude $6\sqrt{5}$ N in the direction $\mathbf{i} + 2\mathbf{j}$,
  $\mathbf{F}_2$ of magnitude $3\sqrt{5}$ N in the direction $\mathbf{i} - 2\mathbf{j}$,
  $\mathbf{F}_3$ of magnitude 39 N in the direction $-5\mathbf{i} - 12\mathbf{j}$.
  Find its acceleration.

② **8** A particle of mass 5 kg moves with acceleration $-3\mathbf{i} + \mathbf{j}$ under the action of two forces $\mathbf{F}_1 = -6\mathbf{i} - \mathbf{j}$ and $\mathbf{F}_2 = p\mathbf{i} + q\mathbf{j}$. Find p and q.

② **9** A particle moves with acceleration $6\mathbf{i} + a\mathbf{j}$ under the action of two forces:
  $\mathbf{F}_1$ of magnitude 26 N in the direction $12\mathbf{i} - 5\mathbf{j}$,
  $\mathbf{F}_2$ of magnitude $12\sqrt{5}$ N in the direction $-2\mathbf{i} + \mathbf{j}$.
  Find a and the mass of the particle.

- **10** A particle of mass 2 kg moves under the action of a force $4t\mathbf{i} - 6\mathbf{j}$. Initially the particle is at the point $\mathbf{i} + \mathbf{j}$ moving with velocity $\mathbf{i} - \mathbf{j}$. Find the position vector of the particle at time $t = 3$.

→ **11** A particle of mass 3 kg moves under the action of a force $(24/t^3)\mathbf{j}$. At time $t = 1$ the particle is at the point $4\mathbf{i} + 4\mathbf{j}$ moving with velocity $4\mathbf{i} - 4\mathbf{j}$. Find the velocity and position vector of the particle at time t. Find the cartesian equation of its path.

R **12** A particle of mass 2 kg moves under the action of a force $-4 \cos t\mathbf{i} - 4 \sin t\mathbf{j}$.

  (i) If the particle is initially at the point $-2\mathbf{i}$ moving with velocity $-2\mathbf{j}$, show that the path of the particle is a circle, centre the origin, and find its radius.

  (ii) If the particle is at the point $5\mathbf{i} + 4\mathbf{j}$ moving with velocity $2\mathbf{j}$, show that the path of the particle is again a circle, and find its centre and radius.

13 At time t the position vector of a particle of mass 3 kg is $5t^2\mathbf{i} - 10t\mathbf{j}$. Find:

(ii) the resultant force acting on the particle at time t

14 At time t the position vector of a particle of mass $\frac{1}{4}$ kg is $3\cos 4t\mathbf{i} + 5\sin 4t\mathbf{j}$. Find the momentum of the particle and the resultant force acting on it at time $t = \frac{1}{6}\pi$.

15 The momentum of a particle at time t is $40t^2\mathbf{i} - 20t^3\mathbf{j}$. Find the resultant force acting on the particle at time $t = \frac{1}{2}$.

16 At time t the position vector of a particle of mass 4 kg is $20t\mathbf{i} + (10t^2 - 12)\mathbf{j} - 10t^3\mathbf{k}$. Find the momentum of the particle and the resultant force acting on it at time t.

17 Forces $\mathbf{F}_1 = 4\mathbf{i} - 5\mathbf{j}$, $\mathbf{F}_2 = -3\mathbf{i} - \mathbf{j}$ and $\mathbf{F}_3 = -\mathbf{i} + 6\mathbf{j}$ act on a particle of mass 4 kg. Find its acceleration. What do you deduce about its velocity?

18 Forces $\mathbf{F}_1 = -2\mathbf{i} + 10\mathbf{j}$, $\mathbf{F}_2 = -6\mathbf{j}$ and $\mathbf{F}_3 = 2\mathbf{i} - 4\mathbf{j}$ act on a particle of mass 2 kg. Find the position vector of the particle at time $t = 3$ if it is initially:

(i) at rest at the point $3\mathbf{i} + 2\mathbf{j}$
(ii) at the origin moving with velocity $2\mathbf{i} + \mathbf{j}$

19 A system of forces for which $\sum \mathbf{F} = \mathbf{0}$ acts on an object. What can you say about the motion of the object if the object is:

(i) a particle?
(ii) a rigid body (of finite size)?

20 Forces $\mathbf{F}_1 = 3\mathbf{i} + p\mathbf{j}$, $\mathbf{F}_2 = q\mathbf{i} - 6\mathbf{j}$ and $\mathbf{F}_3 = -2\mathbf{i} + 4\mathbf{j}$ act on a particle. If the particle is in equilibrium, find p and q.

21 Forces $\mathbf{F}_1 = p\mathbf{i} + q\mathbf{j}$, $\mathbf{F}_2 = 2q\mathbf{i} - p\mathbf{j}$ and $\mathbf{F}_3 = -3\mathbf{i} + 2\mathbf{j}$ act on a particle of mass 2 kg. Find p and q if the particle

(i) is in equilibrium
(ii) moves with acceleration $2\mathbf{i} + 4\mathbf{j}$

8.1:3 Types of force

We can conveniently divide the forces on an object into two categories: **forces at a distance** and **contact forces**.

Forces at a distance
There are three main kinds of forces at a distance: magnetic, electromagnetic and gravitational. We are concerned here only with gravitational forces.

Weight: Suppose that a teapot falls off a mantelpiece. Its velocity increases as it falls, so there must be a force acting on it. This force is its **weight, W**.

The weight of a body is the force exerted by the Earth on the body (see question 3).

The acceleration of the teapot is (ignoring air resistance) $-g\mathbf{j}$, where $\mathbf{j}$ is a unit vector vertically upwards (fig. 8.1:4). By N2,

$$\mathbf{W} = -mg\mathbf{j}$$

Magnitude of **W** is $W = mg$.

Direction of **W** is $-\mathbf{j}$, i.e. vertically downwards.

Note: in numerical problems we take $g = 10$ m s^{-2}; so that, for example, the weight of an object of mass 3 kg is 30 N.

Contact forces
When two objects are touching they exert a force on one another.

(a) **Bodies attached** Consider the two parts of a crane, the fixed vertical section A and the movable section B. Section A exerts a force, **P** say, on B (fig. 8.1:5a)

We often split the force into components in two convenient directions, for example, horizontal and vertical (fig. 8.1:5b). Then

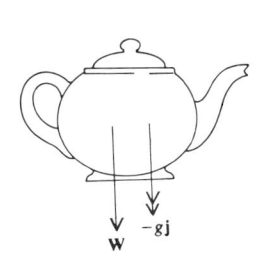

$$\mathbf{P} = X\mathbf{i} + Y\mathbf{j}$$

Fig. 8.1:4

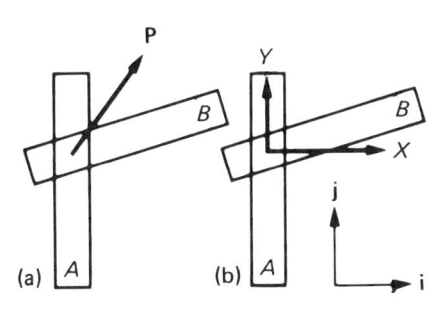

Fig. 8.1:5a Force exerted by A on B

Fig. 8.1:5b Force exerted by A on B (horizontal and vertical components)

428 *Types of force*

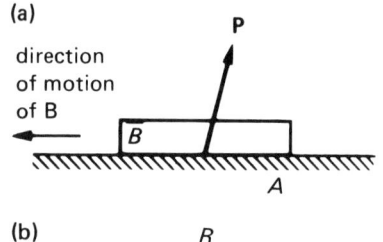

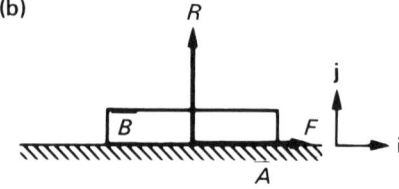

(a) *direction of motion of B*

Fig. 8.1:6a Force exerted by A on B

Fig. 8.1:6b Force exerted by A on B (horizontal and vertical components)

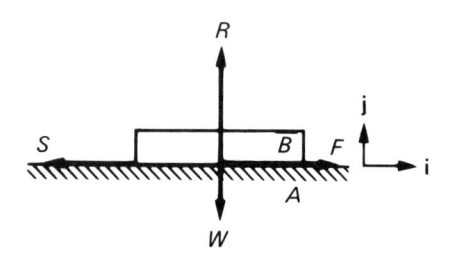

Fig. 8.1:7

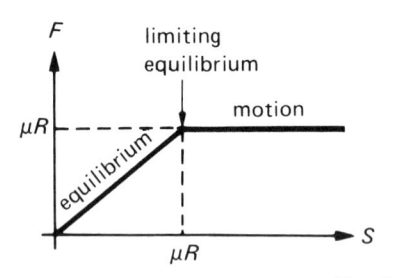

Fig. 8.1:8

Fig. 8.1:9

(b) **Bodies not attached** Consider a box B being pushed across a floor A. The floor A exerts a force **P** on B, which prevents B from falling through the floor, and resists the tendency for B to move across the floor (fig. 8.1:**6a**).

We emphasize the dual effect of **P** by splitting it into two components (fig. 8.1:**6b**):

(i) along the common tangent of A and B: F, the **frictional reaction**
(ii) along the common normal of A and B: R, the **normal reaction**

Then

$$\mathbf{P} = F\mathbf{i} + R\mathbf{j}$$

Dynamic friction We find experimentally that, when B is moving relative to A, the ratio F/R is a constant, which we call μ, **the coefficient of friction between A and B**:

$$\frac{F}{R} = \mu \quad \text{i.e.} \quad F = \mu R \qquad (9)$$

The magnitude of μ depends only on the materials of which A and B are made. The rougher A and B are, the larger μ is. Theoretically, μ can be any (positive) number; but in practice μ usually lies between 0 and $\frac{1}{2}$.

Static friction Suppose now that B is simply resting on A, and that we apply a gradually increasing horizontal force of magnitude S to B (fig. 8.1:7). At first B does not move. This implies that the resultant force on B is zero, i.e. $S = F$.

As S increases, F increases, and $S = F$. When $S = \mu R$, $F = \mu R$. But we find experimentally that F can get no bigger than μR. μR is called the limiting value of F. When $F = \mu R$, B is in **limiting equilibrium**, i.e. it is just about to move (fig. 8.1:**8**).

If S increases further, $S > \mu R$ while $F = \mu R$, so $S > F$ and B moves.

$S \leq \mu R$	B is in equilibrium $(S = F)$	$F \leq \mu R$
$S = \mu R$	B is in limiting equilibrium $(S = F)$	$F = \mu R$ (10)
$S > \mu R$	B is moving $(S > F)$	$F = \mu R$

See fig. 8.1:**8**.

Angle of friction When B is in limiting equilibrium or in motion, **P** makes a constant angle with the common normal of A and B (fig. 8.1:**9**). Note that $\tan \lambda = F/R = \mu$. The angle λ is called the **angle of friction between A and B**.

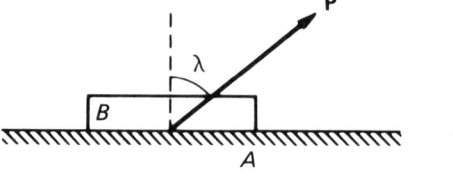

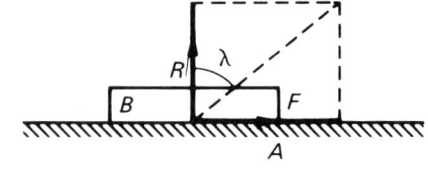

Forces on a particle 429

Smooth contact If $\mu = 0$ (so $F = 0$) the contact between A and B is called **smooth**; in this case the force of A on B is a normal reaction (fig. 8.1:**10**)

$$\mathbf{P} = R\mathbf{j}$$

(We often speak of a body A as being 'smooth'. This means that the contact between A and any other body is smooth.)

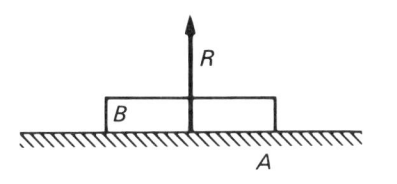

Fig. 8.1:**10**

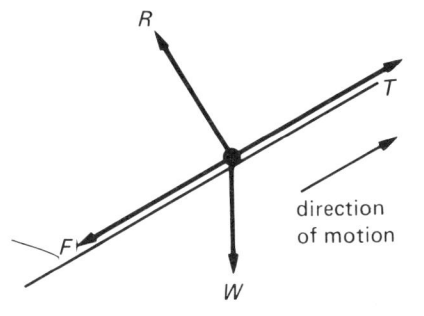

Smooth contact is a mathematical model for contact with a very small coefficient of friction.

Worked example 3 A particle is pulled up a rough inclined plane by a string parallel to a line of greatest slope of the plane. Draw a diagram showing the forces acting on the particle.

The forces (fig. 8.1:**11**) are:

the weight W of the particle;
the tension T in the string;
the reaction of the plane on the particle, which we have split into two components, R (normal) and F (frictional).

Fig. 8.1:**11**   Note that, since the particle is moving, $F = \mu R$.

Exercise 8.1:3

1 Find the weight (in N) of a body of mass:

(i) 0.001 kg   (ii) 0.3 kg   (iii) 5000 kg

2 Estimate the weight of:

(i) an apple   (ii) a pin   (iii) the Eiffel Tower

Questions 3–6: Newton's Universal Law of Gravitation states that between any two objects in the universe there is a force whose magnitude F is given by the equation

$$F = \frac{Gm_1 m_2}{r^2} \qquad (11)$$

where G is constant, m_1 and m_2 are the masses of the objects and r their distance apart (i.e. the distance between their centres of mass).

$$G \approx 6.7 \times 10^{-11} \text{ m}^3 \text{ kg}^{-1} \text{ s}^{-2}$$

3 Given that the mass of the Earth $\approx 6 \times 10^{24}$ kg and the radius of the Earth $\approx 6.4 \times 10^6$ m, show that the force exerted by the Earth on an object of mass m at the Earth's surface is mg where $g \approx 9.8$ m s^{-2}.

4 Given that the mass of the Sun $\approx 2 \times 10^{30}$ kg and the distance of the Earth from the Sun $\approx 1.5 \times 10^{11}$ m, find the force exerted by the Sun on a body of mass 80 kg at the Earth's surface.

5 Explain why the acceleration due to gravity on the Moon is approximately one-sixth of that on Earth.

6 Show that the force exerted by the Earth on a body of mass m at a height x above the Earth's surface is $mgR^2/(R+x)^2$, where R is the radius of the Earth. Find an approximation for this when x is small.

Questions 7–39: Draw a diagram showing the forces acting on the following systems.

7 A particle resting on a table.

8 A particle hanging by a string from a fixed point.

9 A particle being pulled along a smooth table by a horizontal string.

10 A particle being pulled along a rough table by a horizontal string.

11 A particle being pulled along a rough table by a string inclined at 30° to the horizontal.

12 A particle sliding down a smooth plane inclined at 30° to the horizontal.

13 A ring sliding down a smooth fixed straight wire inclined at 30° to the horizontal.

14 A particle sliding down a rough plane inclined at 30° to the horizontal.

15 A particle being pulled up a rough plane of inclination 30° by a string parallel to the plane.

16 A particle being pushed up a rough plane of inclination 30° by a horizontal force.

17 A particle being pulled down a rough plane of inclination 30° by a strong parallel to the plane.

18 A particle being pulled up a rough plane of inclination 30° by a string inclined at 20° to the plane.

19 A particle sliding down the smooth inside surface of a fixed hemisphere.

③ → **20** A ring sliding down a smooth circular hoop fixed in a vertical plane.

③ → **21** A particle sliding down the rough inside surface of a fixed sphere.

③ **22** A particle attached to a fixed point by a string, moving in a vertical circle.

③ → **23** A particle attached by a string to a fixed point, moving in a horizontal circle.

③ **24** The body of a yo-yo moving down the string.

③ **25** A ring *P*, threaded onto a smooth fixed vertical circular hoop centre *O*, held in equilibrium with *OP* making an angle 30° with the downward vertical by a string attached to the highest point of the hoop.

• **26** A ladder resting on rough (horizontal) ground and against a smooth wall.

27 A ladder resting on rough ground and against a smooth horizontal rail. (The rail is perpendicular to the vertical plane containing the ladder.)

→ **28** A ladder resting on rough ground and against a rough horizontal rail.

29 A rod resting on rough ground and against a smooth fixed sphere.

30 A ladder resting on rough ground and against a rough wall, with a man standing on one of the rungs.

31 A plank resting on rough supports:

(i) the plank is horizontal

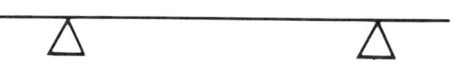

(ii) the plank is not horizontal

• **32** A car moving along a level road.

→ **33** A car accelerating up a hill.

34 A car decelerating along a level road with the engine off and the hand-brake on.

35 A car accelerating down a hill.

36 A car moving down a hill with the engine off and the hand-brake on.

37 A man standing on a (rough) slope.

38 A man walking along a level road.

39 A man running up a flight of steps.

8.1:4 Newton's third law

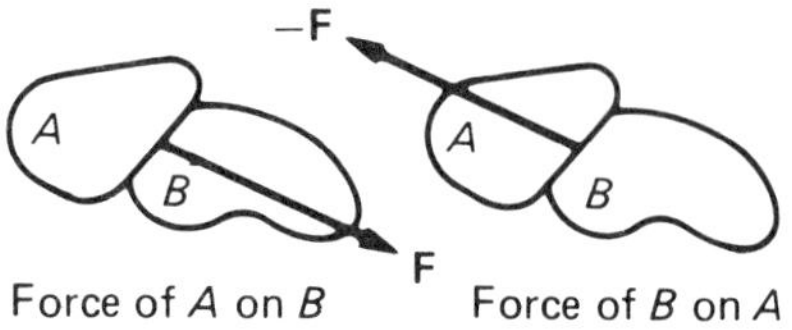

N3: Newton's Third Law of Motion
When an object *A* exerts a force **F** on an object *B*, *B* exerts a force −**F** on *A*, i.e. action and reaction are equal and opposite (fig. 8.1:**12**). **(12)**

Fig. 8.1:**12** Force of *A* on *B*   Force of *B* on *A*

N3 applies not only to contact forces but also to forces at a distance, e.g. the Earth exerts a force mg vertically downwards on a falling body of mass m; so, by N3, the body exerts a force mg vertically upwards on the Earth.

Worked example 4 A particle slides down a smooth wedge which is fixed to a table. Draw diagrams showing the forces on: (a) the particle (b) the wedge

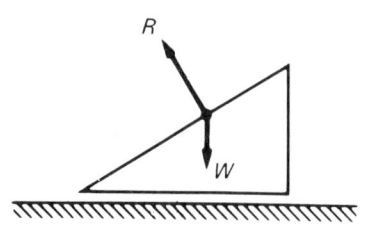

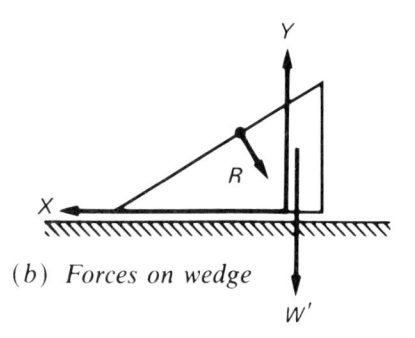

Fig. 8.1:**13** (*a*) *Forces on particle*   (*b*) *Forces on wedge*

Forces on a particle 431

Since the wedge exerts a force R on the particle, by N3 the force of the particle on the wedge is R (not W). We have split the reaction of the table on the wedge into components X and Y.

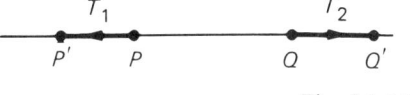

Tension in a string

When a string is taut each point P in the string exerts a force on the points next to it. The magnitude of this force is called the **tension of the string at** P.

Consider the tensions at two points P and Q of a *light* string (fig. 8.1:**14**). Let the tensions at P and Q be T_1 and T_2 respectively. By N3, the point P' to the left of P exerts a force T_1 on P, and the point Q' to the right of Q exerts a force T_2 on Q.

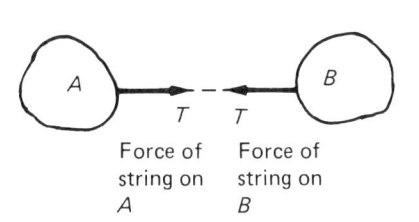

Consider the section PQ of the string. The external forces on PQ are T_1 and T_2, so by N2,

$$T_1 - T_2 = 0 \quad \text{(since the mass of } PQ \text{ is zero)}$$

i.e. $$T_1 = T_2$$

the tension at every point of a light string is the same.

If two objects, A and B, are attached by a taut light string and the tension at every point in the string is T (fig. 8.1:**15**), then in particular the tension at each endpoint is T; so the force of the string on A and B is T.

A string which stays the same length whatever the tension in it is called an **inextensible** or **inelastic** string. Inextensible strings do not exist in the real world; every string is to some extent elastic. An inextensible string is a mathematical model for a string which is only slightly elastic.

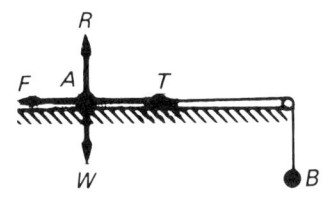

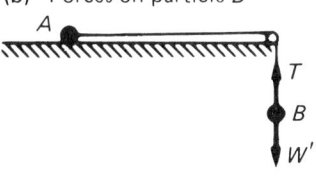

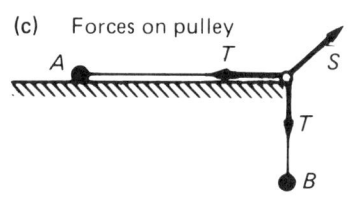

Worked example 5 A particle A rests on a rough table. A light inextensible string passing over a smooth pulley at the edge of the table connects the particle to another particle B which hangs freely. Draw diagrams showing the forces on (a) A, (b) B and (c) the pulley.

Figure 8.1:**16** shows the forces acting on the particles and the pulley. In fig. 8.1:**16**c, S is the reaction of the table on the pulley. (Think of the pulley as a smooth *fixed* object. The fact that the tension is the same throughout the string is due to the pulley being *smooth*; if the pulley were rough, the tensions in the two halves of the string would be different.)

Tension and thrust in a rod

A rod may be in:

(i) *tension*—each point on the rod exerts a pulling force on its neighbouring points;
(ii) *compression* (*thrust*)—each point on the rod exerts a pushing force on its neighbouring points.

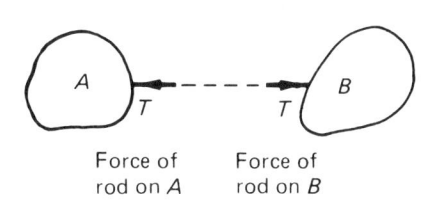

In particular, when two objects A and B are connected by a light rod in compression, the rod exerts equal and opposite pushing forces on A and B, as shown in fig. 8.1:**17**.

Exercise 8.1:4

④ **1** A wedge rests on a rough table. A particle slides down the rough inclined face of the wedge. Draw diagrams showing the forces on:

(i) the particle (ii) the wedge

④ **2** Two books X and Y rest, with X on top of Y, on a table. Draw diagrams showing the forces on:

(i) X (ii) Y

Newton's third law

3 Three books X, Y and Z rest in a pile, with X on top of Y on top of Z, on a table. Draw diagrams showing the forces on:

(i) X (ii) Y (iii) Z

4 A ladder rests on rough (horizontal) ground and against a rough wall. A man is climbing the ladder. Draw diagrams showing the forces on:

(i) the ladder
(ii) the man

5 A rough hemisphere rests with its plane face on rough ground. A rod rests on the ground and against the hemisphere. Draw diagrams showing the forces on:

(i) the hemisphere
(ii) the rod

6 Three identical smooth cylindrical logs A, B and C sit in a rectangular box, which rests on the ground.

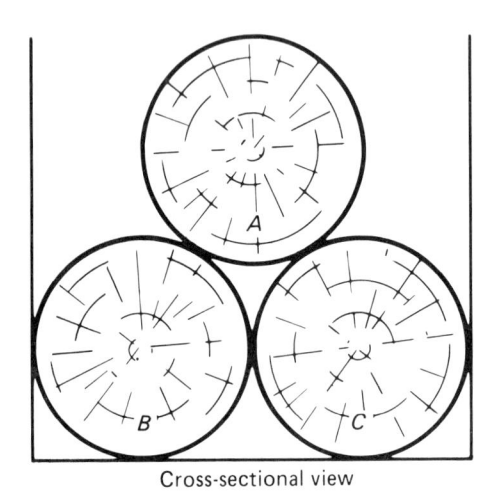

Cross-sectional view

Draw diagrams showing the forces on:

(i) A (ii) B (iii) C (iv) the box

7 A man sits in a chair. Draw diagrams showing the forces on:

(i) the man
(ii) the chair

when the man is (a) sitting upright; (b) slumped in the chair.

8 A tightrope walker, holding a rod horizontally to steady herself, walks the high wire. Draw diagrams showing the forces on:

(i) the rod
(ii) the tightrope walker
(iii) the wire

9 A dead tree consists of a trunk leading to a single branch leading to a single leaf. On the leaf sits a beetle. Draw diagrams showing the forces on:

(i) the trunk   (ii) the branch
(iii) the leaf   (iv) the beetle
(v) the tree

10 A waiter is carrying a tray with a cocktail glass containing a single wilting red rose. Draw diagrams showing the forces on:

(i) the rose
(ii) the glass
(iii) the tray

11 A bird sits on a perch. Draw diagrams showing the forces on:

(i) the bird   (ii) the perch

12 A bird sits on a perch in a birdcage. Draw diagrams showing the forces on:

(i) the bird
(ii) the perch
(iii) the birdcage

13 A bird sits on a perch in a birdcage which hangs in a lift....

14 Using N3, explain why the internal forces inside a rigid body are always in equilibrium.

Questions 15–19: Two particles A and B, of masses M and m, are attached to the ends of a light inextensible string passing over a smooth pulley. Draw diagrams showing the forces on:

(i) A (ii) B (iii) the pulley

15 The pulley is attached to the ceiling. The hanging parts of the string are vertical.

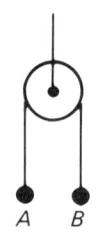

16 A moves on a smooth horizontal table. The pulley is attached to the edge of the table.

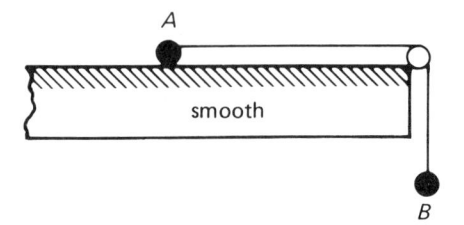

⑤ → **17** *A* moves on a smooth inclined plane. The pulley is attached to the upper edge of the plane.

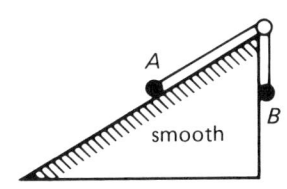

⑤ **18** *A* moves on a rough inclined plane. The pulley is attached to the upper edge of the plane.

(i) *A* moves down the plane.
(ii) *A* moves up the plane.

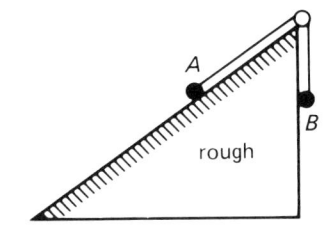

⑤ **19** *A* and *B* move on rough inclined planes. The pulley is attached to the common upper edge of the planes.

(i) *A* moves down the plane.
(ii) *A* moves up the plane.

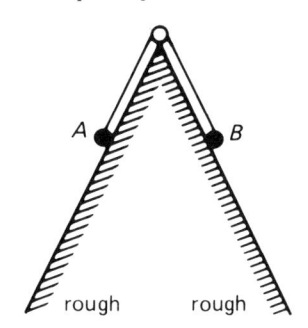

⑤ **20** A wedge rests on a rough table. Particles *A* and *B*, attached by a light inextensible string passing over a smooth pulley fixed to the upper edge of the wedge, move on its smooth inclined faces.

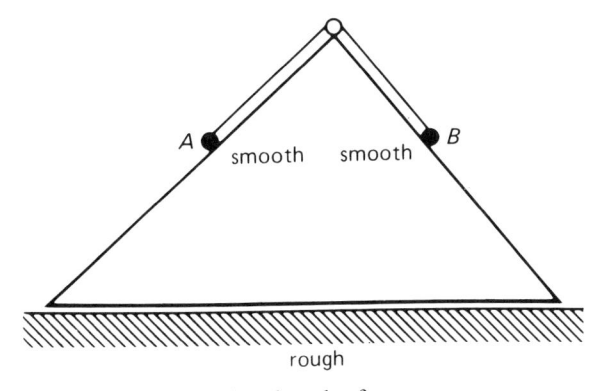

Draw diagrams showing the forces on:

(i) *A* (ii) *B* (iii) the wedge

⑤ → **21** A wedge rests on a rough table. A particle *A* which can move on the rough inclined face of the wedge, is attached by a light inextensible string passing over a smooth pulley (fixed to the upper edge of the wedge) to a particle *B* which hangs freely.

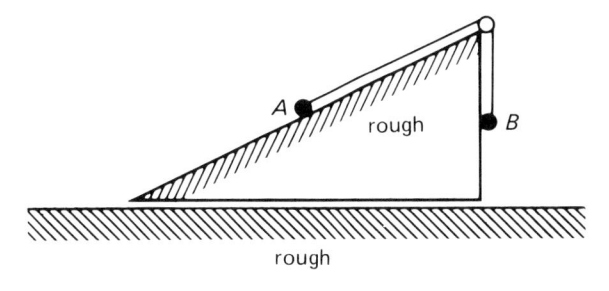

Draw diagrams showing the forces on:

(i) *A* (ii) *B* (iii) the wedge

when *B* moves (a) upwards; (b) downwards.

⑤ **22** A child is fishing, using a piece of stick with a (light) string tied to one end. He catches a fish. Sketch diagrams showing the forces on:

(i) the fish
(ii) the stick
(iii) the string

as he pulls the fish out of the water.

23 An object hangs from a fixed point *A* by a string whose mass is not negligible. Explain why the tension in the string is not the same at every point. At which point of the string is its tension greatest? Sketch an approximate graph of the tension in the string at the point *P* against the distance of *P* from *A*.

8.1:5 Hooke's law

The **natural length** l_0 of an elastic string or spring is its length when it is slack.

When the string is taut its length l is greater than l_0; the difference $l - l_0$ is the string's **extension** x (fig. 8.1:**18**). As x increases so does the tension T.

Robert Hooke found experimentally in 1676 that T/x is a constant, k:

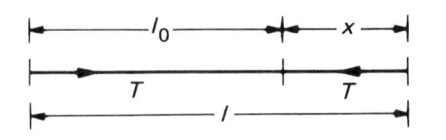

$$\frac{T}{x} = k \quad \text{i.e.} \quad T = kx$$

Fig. 8.1:**18**

For convenience we define a new constant $\lambda = k l_0$ called the **modulus of elasticity of the string**.

Hooke's Law

$$T = \frac{\lambda x}{l_0} \qquad (13)$$

In fact, when x reaches a certain value, called the **elastic limit**, a further increase in T results in a sudden increase in x, followed soon by the string breaking (i.e. Hooke's Law applies only up to the elastic limit).

The constant λ depends on the material of which the string is made. Note that when $x = l_0$, i.e. the string is twice its natural length,

$$T = \lambda l_0 / l_0 = \lambda$$

i.e. λ *is the tension needed to double the length of the string.*

Notice that a *spring* can be compressed as well as extended (fig. 8.1:**19**). The force in the spring is a *thrust*. Hooke's Law still applies:

$$T = \frac{\lambda x}{l_0}$$

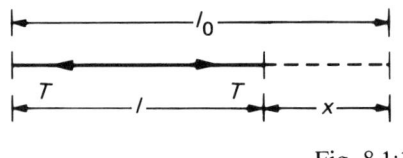

Fig. 8.1:**19**

where x is the compression of the spring and $x = l_0 - l$.

Worked example 6 A light elastic spring has natural length 0.4 m and modulus of elasticity 12 N. Find:

(i) the tension in the spring when its length is 0.6 m
(ii) the length of the spring when the thrust in it is 3 N

(i) Extension $= 0.6 - 0.4 = 0.2 \Rightarrow T = 12(0.2)/0.4 = 6$ N.
(ii) Let x be the compression of the spring.

$$3 = \frac{12x}{0.4} \quad \Rightarrow \quad x = 0.1 \text{ m}$$

So the length of the spring is 0.4 m $-$ 0.1 m $=$ 0.3 m.

Forces on a particle

Exercise 8.1:5

⑥ → **1** A light elastic string has natural length 2 m and modulus of elasticity 10 N.

(i) Find the length to which it is stretched when the tension in it is (a) 5 N; (b) 10 N; (c) 20 N.

(ii) Find the tension in it when it is stretched to a length of (a) 2.4 m; (b) 2.8 m; (c) 3.2 m.

(iii) The string snaps when the tension in it exceeds 25 N. Find the maximum possible length of the string.

⑥ **2** A light elastic spring has natural length 0.15 m and modulus of elasticity 60 N.

(i) Find the length of spring when
(a) the tension in it is 30 N;
(b) the thrust in it is 12 N.

(ii) Find the force in the spring when its length is
(a) 0.2 m; (b) 0.15 m; (c) 0.12 m; (d) 0.10 m;
and state whether the force is a thrust or a tension.

⑥ **3** A light elastic spring has natural length l and modulus of elasticity λ. Find the length of the spring when:

(i) the tension
(ii) the thrust

in it is (a) $\frac{1}{4}\lambda$; (b) $\frac{1}{2}\lambda$; (c) $\frac{3}{4}\lambda$.

⑥ **4** A force of 15 N stretches an elastic string of natural length 0.6 m. Find the stretched length of the string if its modulus of elasticity is:

(i) 5 N   (ii) 15 N   (iii) 25 N

⑥ R **5** A force F stretches an elastic string of natural length l. Find the stretched length of the string if its modulus of elasticity is:

(i) $\frac{1}{2}P$ (ii) P (iii) $2P$ (iv) $3P$

⑥ **6** An elastic string of natural length 1 m is stretched to a length of 1.4 m by a force of 2 N. Find its modulus of elasticity.

⑥ → **7** An elastic spring is stretched to a length of 0.48 m by a force of 30 N and is compressed to a length 0.20 m by a force of 40 N. Find its natural length and modulus of elasticity.

⑥ • **8** The end of an elastic string AB of natural length 0.8 m and modulus of elasticity 20 N is fastened to the end of another elastic string BC of natural length 0.8 m and modulus of elasticity 40 N.

(i) Find the stretched length of the complete string ABC when the tension in it is
(a) 10 N; (b) 20 N; (c) 30 N.

(ii) The ends A and C are attached to two points 2.55 m apart in a (horizontal) line. Find
(a) the extension of AB; (b) the tension in ABC.

⑥ → **9** The end of an elastic string AB of natural length 0.6 m and modulus of elasticity 8 N is fastened to the end of another elastic string BC of natural length 0.5 m and modulus of elasticity 12 N. The ends A and C are attached to two points 1.8 m apart in a line. Find the extension of AB.

⑥ **10** The end of an elastic string AB of natural length $2a$ and modulus of elasticity λ is fastened to the end of another elastic string BC of natural length $3a$ and modulus of elasticity 2λ. The ends A and C are attached to two points distant $6a$ apart. Find the extension of AB.

⑥ **11** A light elastic string has natural length 0.4 m and modulus of elasticity 24 N. When the extension of the string is x, the tension in it is T. The breaking tension of the string is 40 N. Draw a graph of x against T:

(i) assuming that Hooke's Law applies up to the moment the string breaks (What is the gradient of the graph?)

(ii) if the elastic limit of the string is 0.5 m (approximate graph only).

⑥ **12** A light elastic string has natural length l and modulus of elasticity λ. When the extension of the string is x the tension in it is T. Sketch a graph of x against T if the breaking tension of the string is 3λ, assuming that Hooke's Law applies up to the moment the string breaks. What is the gradient of the graph?

8.2 Particles in equilibrium

8.2:1 Equilibrium problems

Remember: a particle is in **equilibrium** if it is at rest or moving with constant velocity, i.e. if its acceleration is zero.

A particle is in equilibrium ⇔ the resultant force acting on it is zero, i.e.

436 Equilibrium problems

> a particle is in equilibrium under a system of forces $\{F\} \Leftrightarrow \sum \mathbf{F} = \mathbf{0}$ (1)

If each force $\mathbf{F} = F_x\mathbf{i} + F_y\mathbf{j}$,

$$\sum \mathbf{F} = \mathbf{0} \Leftrightarrow \sum (F_x\mathbf{i} + F_y\mathbf{j}) = \mathbf{0}$$
$$\Leftrightarrow (\sum F_x)\mathbf{i} + (\sum F_y)\mathbf{j} = 0$$
$$\Leftrightarrow \sum F_x = 0 \quad \text{and} \quad \sum F_y = 0 \qquad (2)$$

Writing the two scalar equations (2) instead of the vector equation (1) is called **resolving** equation (1) **into components**.

Worked example 1 A particle P of mass 5 kg is suspended from a ceiling by two strings AP and PB. AP makes an angle $\alpha = \tan^{-1}\frac{3}{4}$ with the ceiling and $A\widehat{P}B = 90°$. Find the tensions in the strings.

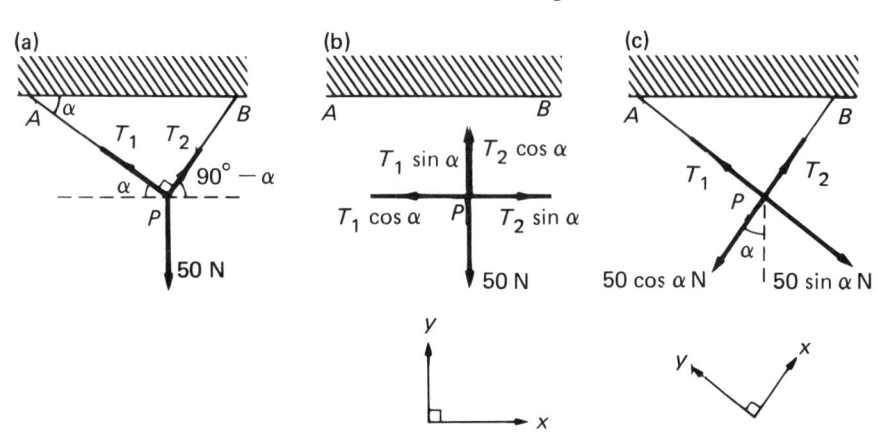

Fig. 8.2:1

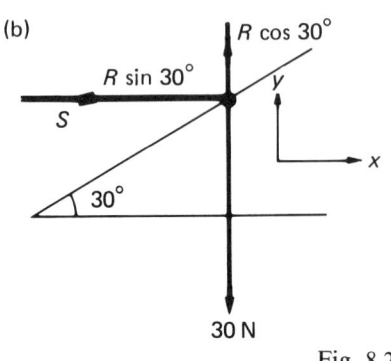

The forces on P are shown in fig. 8.2:1(a).
Take $g = 10$ m s^{-2}, then the weight of the particle $= 5 \times 10$ N $= 50$ N; $\tan \alpha = \frac{3}{4} \Rightarrow \sin \alpha = \frac{3}{5}$, $\cos \alpha = \frac{4}{5}$.

(i) Choosing Ox horizontal, Oy vertical (fig. 8.2:1(b)), and using equations (2):
↗) $\sum F_x = 0 \Rightarrow T_2 \sin \alpha - T_1 \cos \alpha = 0 \Rightarrow \frac{3}{5}T_2 - \frac{4}{5}T_1 = 0$
↖) $\sum F_y = 0 \Rightarrow T_1 \sin \alpha + T_2 \cos \alpha - 50 = 0 \Rightarrow \frac{4}{5}T_2 + \frac{3}{5}T_1 = 50$

Solving gives $T_1 = 30$ N and $T_2 = 40$ N

(ii) Alternatively, choosing Ox along PB, Oy along PA (fig. 8.2:1(c))
↗) $\sum F_x = 0 \Rightarrow T_2 - 50 \cos \alpha = 0 \Rightarrow T_2 = 40$ N
↖) $\sum F_y = 0 \Rightarrow T_1 - 50 \sin \alpha = 0 \Rightarrow T_1 = 30$ N

The second method is slightly simpler—but note that its simplicity depends on the fact that the strings are perpendicular.

Worked example 2 A particle of mass 3 kg rests on a smooth slope of inclination 30°, held in position by a horizontal force of magnitude S. Find S.

Since the slope is smooth, the reaction R is normal to the slope (fig. 8.2:2a). Choosing Ox horizontal, Oy vertical, and using equations (2): (fig. 8.2:2b)

→) $\sum F_x = 0 \Rightarrow S - R \sin 30° = 0 \Rightarrow R = 20\sqrt{3}$ N
↑) $\sum F_y = 0 \Rightarrow R \cos 30° - 30 = 0 \Rightarrow S = 10\sqrt{3}$ N

Fig. 8.2:2

Exercise 8.2:1

1 A particle rests on a horizontal table. Find the force exerted by the table on the particle if the mass of the particle is:

- (i) 40 kg (ii) 50 kg (iii) m kg

2 A particle hangs in equilibrium from the end of an inelastic string. If the tension in the string is 20 N find the mass of the particle.

→ **3** A man of mass 60 kg is in a lift which is moving upwards with constant velocity 2 m s^{-1}. Find the force exerted on the man by the floor of the lift.

4 A man of mass 80 kg is in a lift. Find the force exerted on the man by the floor of the lift when the lift is:

(i) at rest
(ii) moving upwards with constant velocity 3 m s^{-1}
(iii) moving downwards with constant velocity 5 m s^{-1}

Comment on your answers.

Questions 5–12: One end of a light, inextensible string is attached to a fixed point, the other to a particle. The particle is held in equilibrium by a force S.

- **5** Find S and the tension in the string.

6 $S = 90$ N. Find θ and the tension in the string.

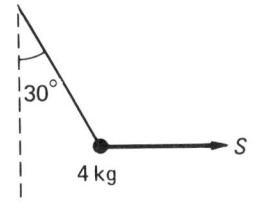

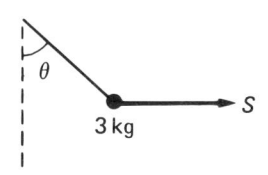

7 Tan $\alpha = 2$. Find S and the tension in the string, in terms of m (and g).

8 $S = 40\sqrt{3}$ N. Find x and the tension in the string.

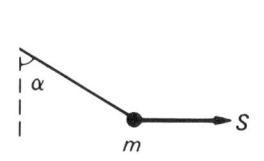

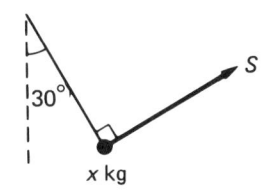

9 Find S and the tension in the string.

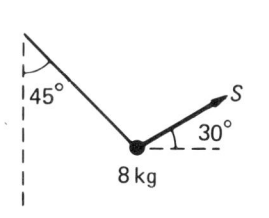

10 Find S in terms of m, θ, ϕ (and g).

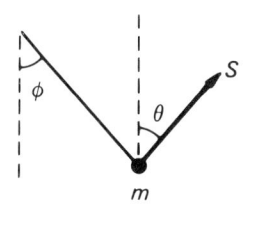

11 Tan $\gamma = \frac{3}{4}$. Find, in terms of m (and g), the least force S needed to keep the particle in equilibrium. (Hint: decide intuitively the direction in which S must act, then explain your decision.)

R 12 Tan $\gamma = \frac{3}{4}$. Find S in terms of m, θ (and g). Find the value of θ for which S is least.

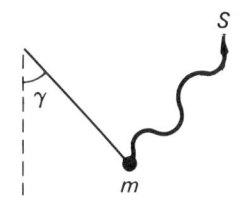

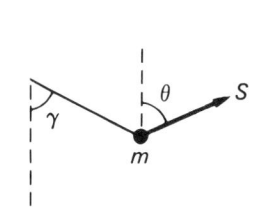

Questions 13–20: A particle is held in equilibrium by two light inextensible strings attached to two points on a ceiling.

① **13** Find the tensions in the strings.

① **14** Find the tensions in the strings (tan $\alpha = \frac{3}{4}$).

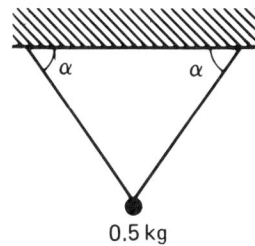

Equilibrium problems

① → **15** The tension in AP is 48 N. Find the tension in PB and the mass of the particle.

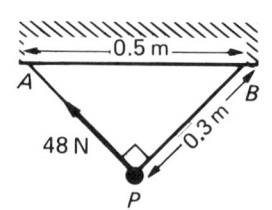

① **16** The tensions in AP and PB are 15 N and 36 N. Find θ and the mass of the particle.

23 Tan $\alpha = \frac{1}{2}$

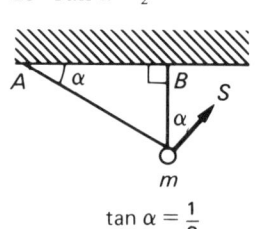

tan $\alpha = \frac{1}{2}$

→ **24** Tan $\beta = \frac{7}{24}$

tan $\beta = \frac{7}{24}$

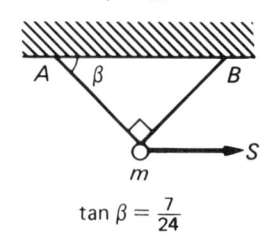

② • **25** Repeat worked example 2, this time choosing Ox up the slope and Oy perpendicular to the slope (in the direction of R). Is this method better than the one in the text?

Questions 26–29: A particle rests on a smooth slope, held in position by a force S.

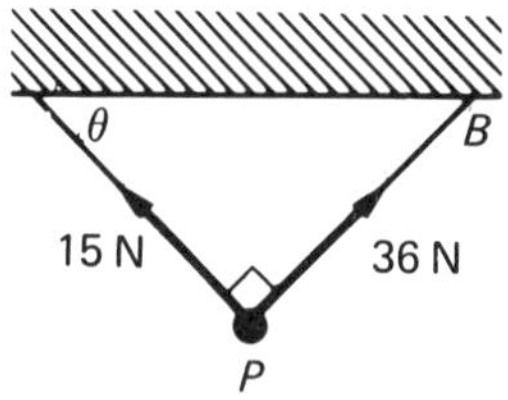

① **17** Find the tensions in the strings.

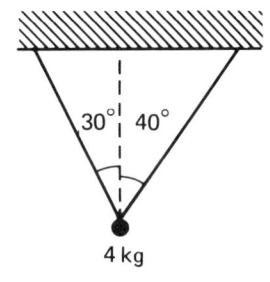

① **18** Find the tensions in the strings in terms of m, α, β (and g). Simplify your answers when $\alpha + \beta = 90°$.

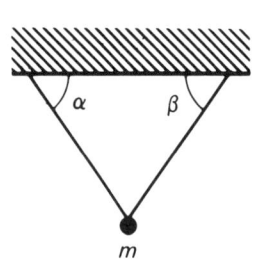

② **26** Find S.

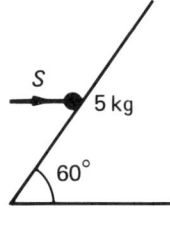

② → **27** Find S, in terms of n, α (and g).

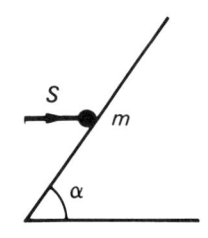

① **19** Find the tensions in the strings.

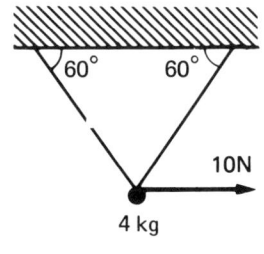

① **20** Find the tensions in the strings when $\tan \alpha = \frac{3}{4}$ and
 (i) $S = 50$ N
 (ii) $S = 100$ N
 (iii) $S = 150$ N

What is the biggest possible value of S?

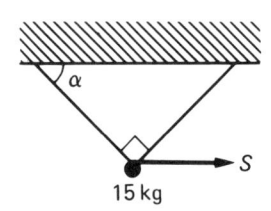

② **28** Tan $\alpha = \frac{5}{12}$. Find S. 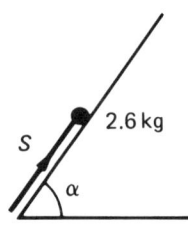 **R 29** Find S.

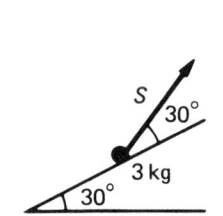

Questions 30–33: A smooth ring is threaded onto a fixed, smooth, vertical hoop. It is held in equilibrium by a light string attached to a point on the hoop. Find the tension in the string and the reaction of the hoop on the ring. (A is the highest point of the wire, B one end of a horizontal diameter.)

Questions 21–24: A smooth ring is threaded onto a light, inextensible string. The ends of the string are attached to fixed points A and B on a ceiling. The ring is held in equilibrium by a force S. Find S and the tension in the string.

30

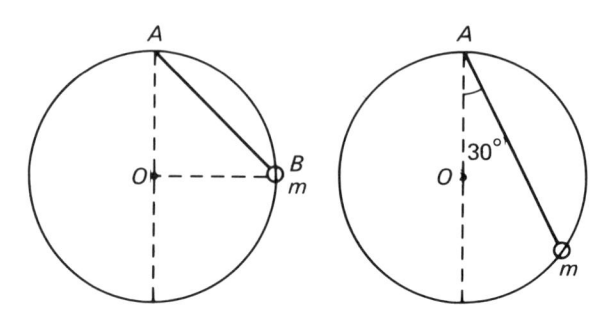

→ **31**

• **21**

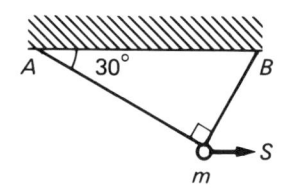

22 The string is of length $2a$

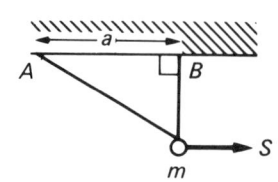

32 **33**

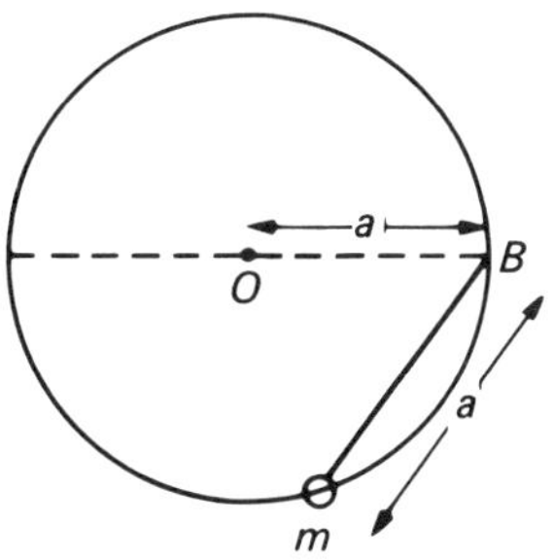

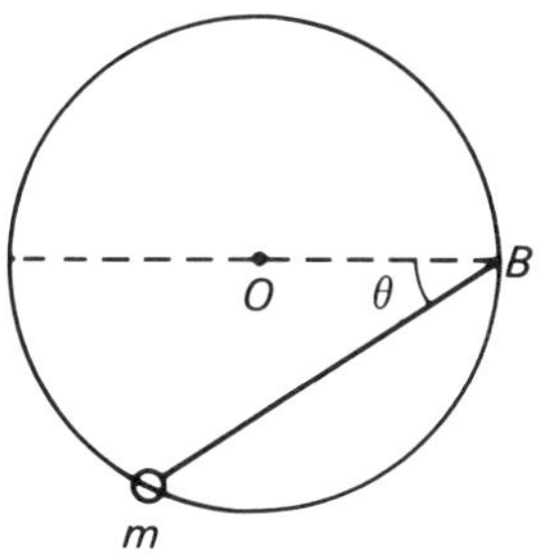

34 A smooth bead of mass m can slide on a fixed, smooth, straight, vertical wire. It is held in equilibrium by a light inextensible string which makes an angle θ with the vertical. Find the tension in the string, in terms of m, θ and g. Explain why it is not possible to hold the bead in equilibrium with the string horizontal.

8.2:2 Equilibrium problems (friction)

Worked example 3 A particle of mass 2 kg is being pulled with constant velocity up a rough slope of inclination $\tan^{-1}\frac{3}{4}$ by a string parallel to the slope (fig. 8.2:**3**). The coefficient of friction between the particle and the slope is $\frac{1}{2}$. Find the tension in the string.

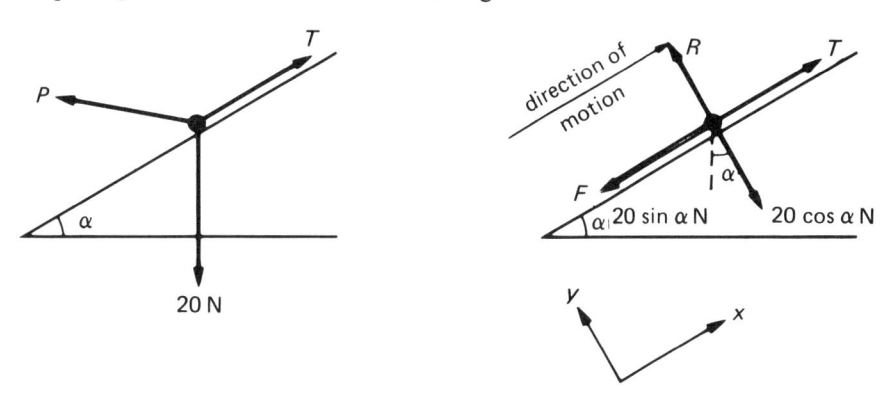

Fig. 8.2:**3**

Here it is definitely best to choose Ox and Oy parallel to and perpendicular to the slope. We split the reaction P of the slope on the particle into components F and R in these directions; notice that F is in the opposite direction to the motion.

$$\tan \alpha = \tfrac{3}{4} \Rightarrow \sin \alpha = \tfrac{3}{5}, \cos \alpha = \tfrac{4}{5}.$$

↗) $\sum F_x = 0 \Rightarrow T - F - 20 \sin \alpha = 0$

↖) $\sum F_y = 0 \Rightarrow R - 20 \cos \alpha = 0$

$$\Rightarrow \qquad R = 16 \text{ N}$$

Since the particle is moving, $F = \mu R$, i.e. $F = \tfrac{1}{2}R = 8$ N; so

$$T = F + 20 \sin \alpha = 20 \text{ N}$$

Exercise 8.2:2

Questions 1-4: A particle is pushed across a rough (horizontal) table at constant speed by a horizontal force S. The coefficient of friction between the particle and the table is μ. In each case find the force exerted by the particle on the table (use N3).

• **1** $\mu = \tfrac{1}{2}$. Find S.

  S 4 kg

→ **2** Find S in terms of m, μ (and g).

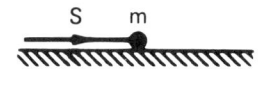

  S m

Equilibrium problems (friction)

3 $S = 30$ N. Find μ

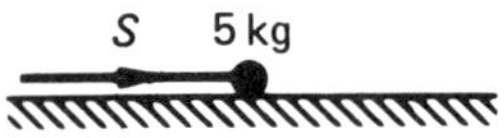

S 5 kg

4 $S = 0.1$ N. Find μ

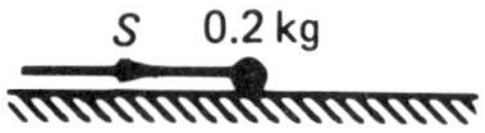

S 0.2 kg

5 A particle of mass 5 kg rests on a rough table. The coefficient of friction between the particle and the table is $\frac{1}{2}$. A gradually increasing horizontal force S is applied to the particle. Find the frictional reaction F when:

(i) $S = 10$ N   (ii) $S = 25$ N   (iii) $S = 60$ N

Describe what happens to the particle in each case.

→ **6** A particle of mass 3 kg rests on a rough table. The coefficient of friction between the particle and the table is $\frac{2}{3}$. A gradually increasing horizontal force S is applied to the particle. What is the largest value of S for which the particle remains in equilibrium? Find the reaction of the particle on the table when:

(i) $S = 5$ N   (ii) $S = 10$ N
(iii) $S = 20$ N   (iv) $S = 40$ N

• **7** A car is moving with constant velocity along a horizontal road. If the resistance to the motion of the car is 1500 N, find the driving force of the engine.

→ **8** A car is moving with constant velocity along a horizontal road. If the resistance to the motion of the car is R, find the driving force of the engine.

Questions 9–12: A particle is pulled across a rough table at constant speed by a light string (inclined to the horizontal). The coefficient of friction between the particle and the table is μ.

• **9** $\mu = 1/\sqrt{3}$. Find the tension in the string.

3 kg, 30°

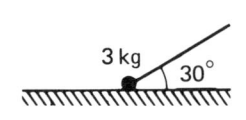

10 The tension in the string is 40 N. Find μ.

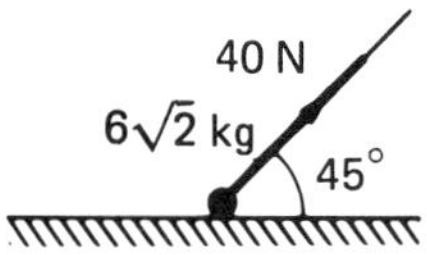

$6\sqrt{2}$ kg, 40 N, 45°

11 The mass of the particle is $3m$. The tension in the string is mg. Find μ.

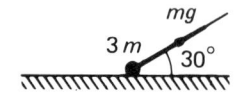

$3m$, mg, 30°

→ **12** The tension in the string is $2mg$. Find μ in terms of θ. Find the least possible value of μ.

m, $2mg$, θ

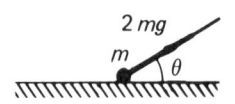

R 13 A particle of mass m is pulled across a rough table at constant speed by a light string inclined at an angle α to the horizontal. The coefficient of friction between the particle and the table is μ. Show that the tension in the string is $\mu mg / (\cos\alpha + \mu \sin\alpha)$.

+ **14** A particle of mass m rests on a rough horizontal table. A force S, inclined upwards at an angle θ to the horizontal, is exerted on the particle. The angle of friction between the particle and the table is λ.

(i) If the particle is in limiting equilibrium, show that $S = mg \sin\lambda / \cos(\theta - \lambda)$.

(ii) Hence find the minimum value of S for which the particle stays in equilibrium. Find the value of θ in this case.

+ **15** A particle of mass m rests on a rough horizontal table. A force S, inclined downwards at an angle θ to the vertical, is exerted on the particle. The angle of friction between the particle and the table is λ.

(i) Prove that the particle will not move unless $\theta > \lambda$.
(ii) If $\theta > \lambda$, show that the least force needed to move the particle is $mg \sin\lambda / \sin(\theta - \lambda)$.

Questions 16–19: A particle resting on a horizontal table is acted on by a system of horizontal forces. If the particle is in limiting equilibrium find μ. (Hint: find the resultant of the forces. The frictional reaction will be in the opposite direction to the resultant.)

• **16**

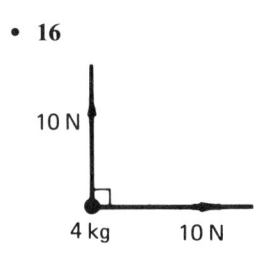

10 N, 4 kg, 10 N

17

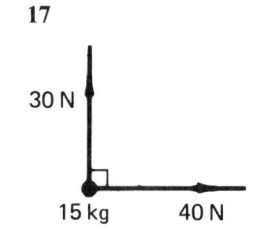

30 N, 15 kg, 40 N

→ **18**

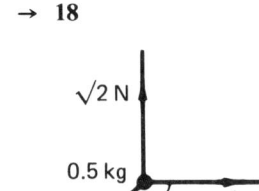

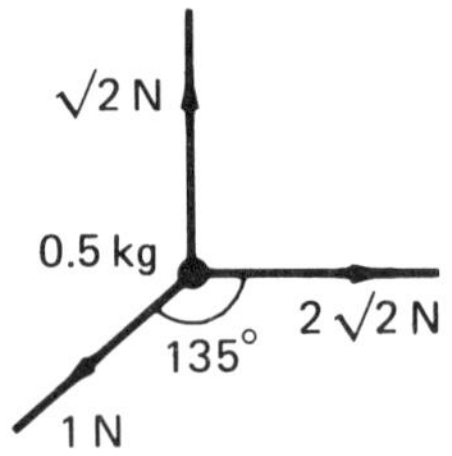

$\sqrt{2}$ N, 0.5 kg, $2\sqrt{2}$ N, 135°, 1 N

19

5 N, 10 N, 60°, 60°, 10 kg, 15 N

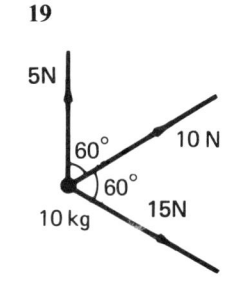

Questions 20 and 21: A particle of mass 5 kg resting on a rough horizontal table is acted on by two horizontal forces: a given force (of magnitude 20 N) and a force of variable magnitude S. The coefficient of friction between the particle and the table is 0.3. Find the greatest and least possible values of S.

20

S, 20 N

21 (Hint: draw triangles of forces.)

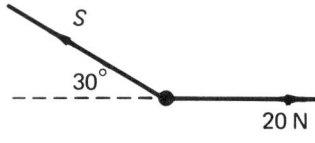

S, 30°, 20 N

Questions 22-25: A particle rests in limiting equilibrium on a rough inclined plane. The coefficient of friction between the particle and the plane is μ. In each case find the magnitude and direction of the reaction of the plane on the particle.

• **22** Find μ. → **23** Tan $\alpha = \tfrac{3}{4}$. Find μ.

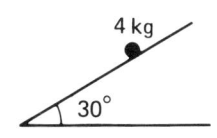

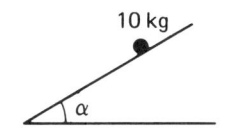

24 $\mu = \tfrac{1}{2}$. Find α. **25** Find μ in terms of α.

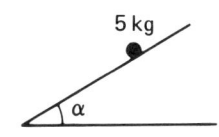

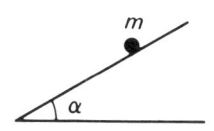

→ **26** A particle of mass m is placed on a rough plane of inclination α. The coefficient of friction between the particle and the plane is μ. Find the least possible value of μ for equilibrium to be possible.

27 A particle of mass m is placed on a rough plane. The coefficient of friction between the particle and the plane is $\tfrac{1}{3}$. The plane is gradually tilted. Find the inclination of the plane to the horizontal when the particle is about to slide.

• **28** A car of mass 1000 kg moves on a road of inclination 1 in 10 (i.e. $\sin^{-1}\tfrac{1}{10}$). The resistance to the motion of the car is 3000 N. Find the driving force of the engine when the car is moving:

(i) down the road with constant velocity
(ii) up the road with constant velocity

→ **29** A car of mass M moves on a road of inclination α. The resistance to the motion of the car is R. Find the driving force of the engine when the car is moving:

(i) down the road with constant velocity
(ii) up the road with constant velocity

Questions 30 and 31: A bead, attached to one end of a light inextensible string, is pulled with constant velocity up a fixed rough vertical straight wire. The coefficient of friction between the bead and the wire is μ.

30 Tan $\alpha = \tfrac{3}{4}$. The tension in the string is 40 N. Find μ.

31 $\mu = \tfrac{1}{2}$. Find the tension in the string.

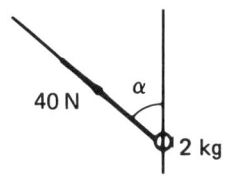

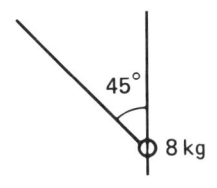

Questions 32-35: A particle is just about to slide *up* a rough inclined plane under the action of a force of magnitude S. The coefficient of friction between the plane and the particle is μ.

③ **32** $\mu = 1/\sqrt{3}$. Find S. ③ **33** $S = 40$ N. $\mu = \tfrac{1}{4}$. Find θ.

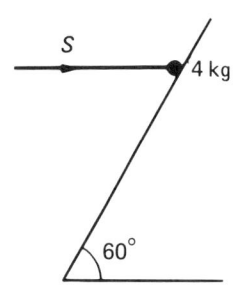

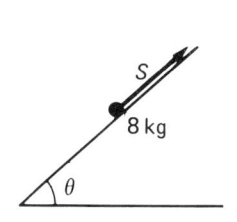

③ **34** $\mu = 1/2\sqrt{3}$. Find S. ③ **35** Find S in terms of m, α, β, μ and g.

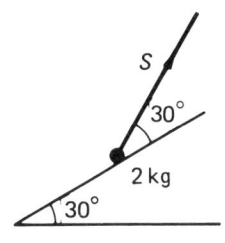

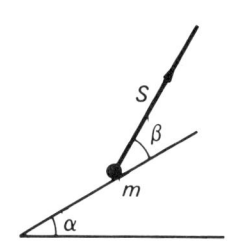

③ **36** A particle of mass 0.5 kg is placed on a rough plane of inclination $\tan^{-1}\tfrac{3}{4}$. Show that if the coefficient of friction between the particle and the plane is $\tfrac{1}{2}$, the particle cannot rest in equilibrium. Find the magnitude of the least possible force needed to keep the particle in equilibrium if the force is:

(i) horizontal
(ii) parallel to the slope

Questions 37-39: A particle is just about to slide *down* a rough inclined plane under the action of a force of magnitude S. The coefficient of friction between the particle and the plane is μ.

③ **37** $\mu = \tfrac{1}{8}$. Find S. ③ **38** $S = 20$ N. $\mu = \tfrac{1}{2}$. Find the mass of the particle.

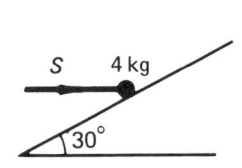

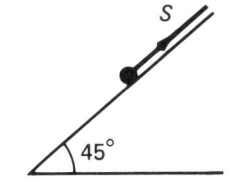

Equilibrium problems (friction)

③→ **39** $\tan\alpha = \frac{1}{2}$. $\mu = \frac{3}{4}$. Find S.

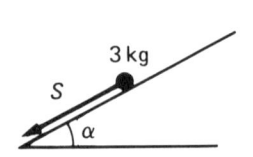

③ **40** Find S in terms of m, α, β, μ and g.

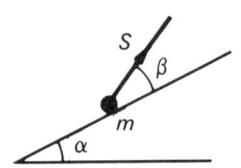

Questions 41 and 42: A particle rests in equilibrium on a rough inclined plane under the action of a horizontal force of magnitude S. Find the range of values of S. (Hint: consider two cases: (i) particle about to slide up plane; (ii) particle about to slide down plane.)

③ **41** $\tan\alpha = \frac{4}{3}$. $\mu = \frac{2}{3}$.

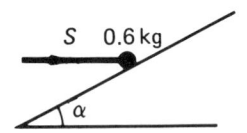

③→ **42** Find S in terms of m, α, μ and g. Assume that $\mu < \tan\alpha$. What is the significance of this assumption?

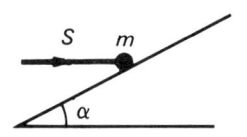

③ R **43** A particle of mass m is just about to slide up a rough plane of inclination α when acted on by a force of magnitude $mg\sin\lambda$ parallel to the plane, where λ is the angle of friction between the particle and the plane. Show that $\sin(\alpha+\lambda) = \sin\lambda\cos\lambda$.

③ **44** A particle rests on a rough plane of inclination α, held in equilibrium by a force parallel to the slope. If the greatest possible value of the force is twice its least possible value, show that the coefficient of friction between the particle and the plane is $\frac{1}{3}\tan\alpha$.

③ **45** A particle of mass m rests on a rough inclined plane. The angle of friction between the particle and the plane is λ. The particle is about to slide up the plane when acted on either by a horizontal force pmg or by a force qmg parallel to the slope. Show that $\cos\lambda = p/q\sqrt{(p^2+1)}$.

③+ **46** A particle of mass m rests on a rough slope of inclination α. A force S, inclined upwards at an angle θ to the line of greatest slope of the plane, is exerted on the particle. The angle of friction between the particle and the plane is λ.

(i) If the particle is about to move up the plane, show that $S = mg\sin(\alpha+\lambda)/\cos(\theta-\lambda)$. Hence find the minimum value of S required to move the particle up the plane.

(ii) Show that the particle will not slide down the slope unless $\alpha > \lambda$. If $\alpha > \lambda$, and the particle is about to slide down the plane, show that $S = mg\sin(\alpha-\lambda)/\cos(\theta+\lambda)$. Hence find the minimum value of S required to support the particle.

Questions 47–48: A ring, threaded onto a fixed, rough, circular, vertical hoop of radius a, rests in limiting equilibrium. The coefficient of friction between the ring and the hoop is μ. In each case find the magnitude and direction of the force exerted by the hoop on the ring.

→ **47** Find μ.

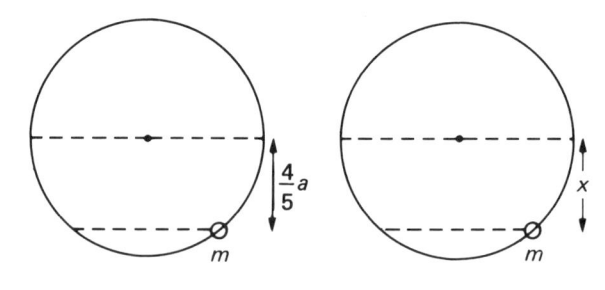

48 $\mu = \frac{1}{3}$. Find x.

49 A particle is placed on the inner surface of a fixed rough hollow sphere of internal radius a. Given that the coefficient of friction between the particle and the sphere is μ show that the particle rests in limiting at a depth $a/\sqrt{(1+\mu^2)}$ below the centre of the sphere.

Questions 50 and 51: A wedge rests on a rough table. A particle rests in limiting equilibrium on the (rough) sloping face of the wedge.

→ **50** (i) Find θ, and the magnitude and direction of the reaction of the particle on the wedge.

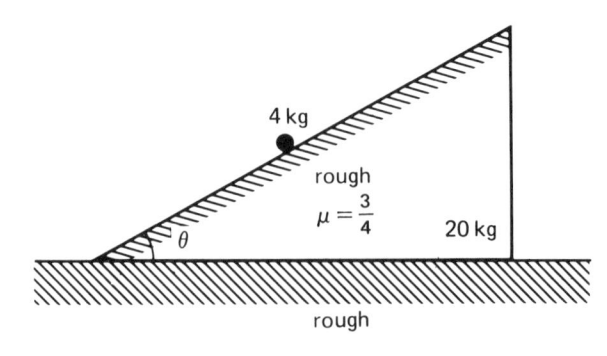

(ii) Draw a diagram showing the forces on the wedge. Find the magnitude and direction of the reaction of the table on the wedge.

+ **51** A horizontal force S is applied to the particle so that it is just about to move up the sloping face. Find S, and the magnitude and direction of the reaction of the table on the wedge.

8.2:3 Equilibrium problems (elastic strings)

Worked example 4 A particle P of mass 5 kg is suspended by two strings AP and BP attached to two fixed points A and B on a ceiling, where $AB = 1.3$ m. In equilibrium, $AP = 0.5$ m and $BP = 1.2$ m. AP is an inelastic string, BP is an elastic string of natural length 1 m (fig. 8.2:**4a**). Find the modulus of elasticity of BP.

The forces on P are shown in fig. 8.2:**4b**. Since $AB^2 = AP^2 + PB^2$, $A\widehat{P}B = 90°$ (by Pythagoras' Theorem). Let $A\widehat{B}P = \theta$. Then $\sin\theta = \frac{5}{13}$, $\cos\theta = \frac{12}{13}$.

The natural length of BP is 1 m, its stretched length is 1.2 m, so its extension is 0.2 m. Let the modulus of elasticity of BP be λ.

$\nearrow$) $\sum F_x = 0 \Rightarrow \qquad T_2 - 50\sin\theta = 0$

i.e. $\qquad T_2 = 250/13$ N

Hooke's Law $\Rightarrow \qquad T_2 = \dfrac{\lambda(0.2)}{1}$

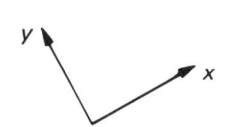

i.e. $\qquad \lambda = \dfrac{250}{13\cdot(0.2)} \approx \dfrac{1250}{13}$ N

Fig. 8.2:4

Note: since we are not interested in T_1, there is no need to resolve along Oy.

Exercise 8.2:3

Questions 1–4: One end of a light elastic string of natural length l_0 and modulus of elasticity λ is attached to a fixed point A, the other to a particle of mass m. When the particle hangs in equilibrium, the extension of the string is x_0.

- **1** $m = 3$ kg, $l_0 = 0.5$ m, $\lambda = 60$ N; find x_0.

- **2** $m = 4$ kg, $l_0 = 3$ m, $\lambda = 120$ N; find x_0.

- **3** $m = 2$ kg, $l_0 = 2$ m, $x_0 = 0.5$ m; find λ.

→ **4** Find x_0 in terms of n, λ, l_0 and g.

5 One end of a light elastic string is attached to a fixed point, the other to a scale pan of mass 2 kg. When the scale pan rests in equilibrium the stretched length of the string is 0.8 m. A particle of mass 1 kg is placed in the scale pan. When the system rests in equilibrium the stretched length of the string is 0.9 m. Find the natural length and modulus of elasticity of the string.

•→ **6** Two light elastic strings AB and BC of natural length 0.6 m and moduli of elasticity 30 N and 50 N are joined together at B. The end A is attached to a fixed point, and a particle of mass 4 kg is attached to C. Find the stretched length of AC.

7 Two light elastic strings, AB of natural length a and modulus of elasticity $2mg$, and BC of natural length $2a$ and modulus of elasticity $3mg$, are joined together at B. The end A is attached to a fixed point and a particle of mass km is attached to C. The stretched length of AC is $4a$. Find k.

Questions 8–11: One end of a light elastic string of natural length l_0 and modulus of elasticity λ is attached to a fixed point on a smooth plane of inclination α, the other to a particle of mass m. When the particle rests in equilibrium on the plane, the extension of the string is x_0.

- **8** $m = 4$ kg, $l_0 = 2$ m, $\lambda = 50$ N, $\alpha = 30°$; find x_0.

9 $l_0 = a$, $\alpha = 30°$, $x_0 = \frac{1}{4}a$; find λ in terms of m and g.

444   *Equilibrium problems (elastic strings)*

10 $m = 0.5$ kg, $l_0 = 1$ m, $\lambda = 50$ N, $x_0 = 0.08$ m; find α.

→ **11** Find x_0 in terms of m, l_0, λ, α and g.

Questions 12 and 13: Two strings AB and BC of natural lengths l_1 and l_2, moduli of elasticity λ_1 and λ_2, are joined together at B. The ends A and C are fixed to two points, with A vertically above C. A particle is attached to B.

• **12** $l_1 = l_2 = a$, $\lambda_1 = \lambda_2 = 2mg$. Find x.

→ **13** $l_1 = a$, $l_2 = 2a$, $\lambda_1 = 2mg$, $\lambda_2 = 3mg$. Find x.

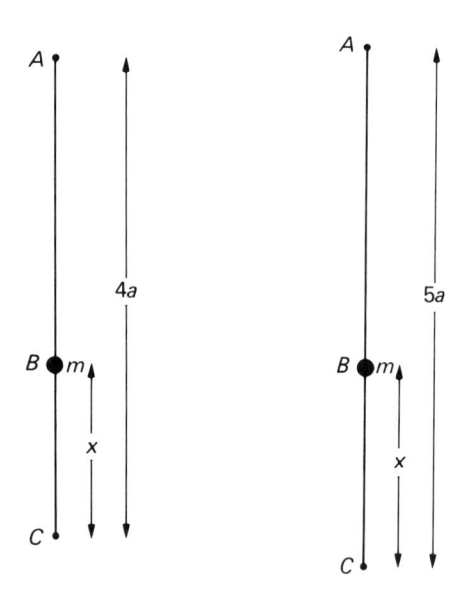

Questions 14–21: A particle P is suspended by two strings AP and BP attached to two fixed points A and B on a ceiling.

④ **14** BP is an inelastic string. AP is an elastic string of natural length $3l$. Find its modulus of elasticity.

④ **15** BP is an inelastic string. AP is an elastic string of natural length 2 m and modulus of elasticity 100 N. Find the tensions in AP and BP and the lengths of AP and BP.

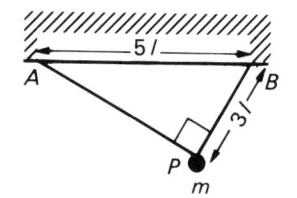

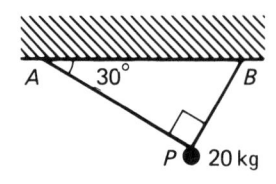

④ → **16** BP is an inelastic string. AP is an elastic string of natural length 1 m. Find its modulus of elasticity.

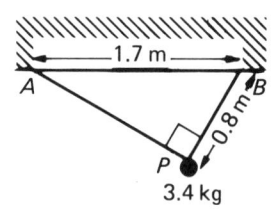

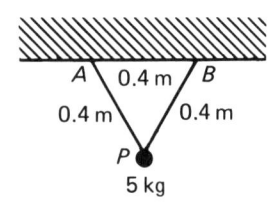

④ **17** AP and BP are elastic strings of natural length 0.3 m. Find their modulus of elasticity.

④ **18** AP and BP are elastic strings of natural length l, modulus of elasticity $4mg$. Find k.

④ **19** AP and BP are elastic strings of natural length l. Find their modulus of elasticity.

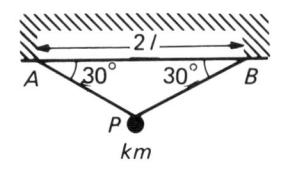

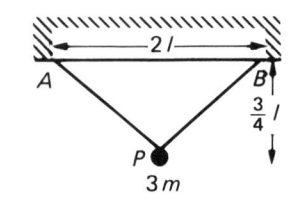

④ **20** AP and BP are elastic strings of natural length l and modulus of elasticity $2mg$; $\tan \alpha = 2$. Find k.

④ **R 21** AP and BP are elastic strings of natural lengths 0.9 m and 0.4 m. Find their moduli of elasticity.

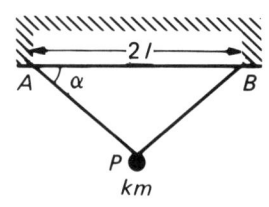

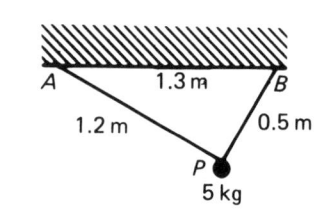

Questions 22 and 23: One end of a light elastic string of natural length l_0 and modulus of elasticity λ is attached to a fixed point, the other to a particle. The particle is held in equilibrium by a horizontal force S.

22 $l_0 = a$, $\lambda = 2mg$. Find S, the tension in the string and the extension of the string.

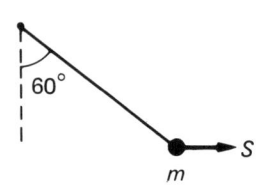

Forces on a particle 445

23 $l_0 = a$, $S = 2mg$. Find θ. If the extension of the string is $\frac{1}{2}a$, find λ.

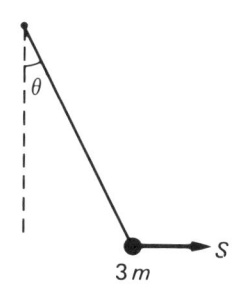

24 A bead P of mass m is threaded on to a smooth, fixed, vertical, circular hoop of radius a. The bead is held in equilibrium by a light elastic string of natural length $2a$ which is threaded through the bead and attached to the points A and B, where AB is the horizontal diameter of the wire and $P\hat{A}B = 30°$. Find the modulus of elasticity of the string and the reaction of the ring on the wire.

8.2:4 Connected particles

Worked example 5 A particle P of mass $2m$, resting on a rough horizontal table, is attached to a particle Q of mass m by a light inelastic string passing over a smooth pulley at the edge of the table. If the system is just about to move, find:

(a) the coefficient of friction μ between P and the table
(b) the force exerted by the string on the pulley

We consider P and Q separately, as shown in fig. 8.2:**5**.

We split the reaction of the table on P into components R and F (fig. 8.2:**5a**). Notice that F opposes the tendency of P to move.

$$\rightarrow) \qquad T - F = 0$$
$$\uparrow) \qquad R - 2mg = 0$$
$$\Rightarrow \qquad F = T \quad \text{and} \quad R = 2mg$$

Notice that the tension in the string is the same at P and at Q, since the pulley is smooth (see section 8.1: WE 5 and fig. 8.2:**5b**).

$$\uparrow) \qquad T - mg = 0 \Rightarrow T = mg$$

Thus $F = mg$

(a) Since P is in limiting equilibrium, $F = \mu R$, i.e.

$$\mu = \frac{F}{R} = \frac{mg}{2mg} = \frac{1}{2}$$

(b) Since the tensions in the string on either side of the pulley are equal, their resultant must bisect the angle between them (fig. 8.2:**6**). So

$$\text{resultant force} = T \cos 45° + T \cos 45° = 2T \cos 45°$$
$$= T\sqrt{2}$$
$$= mg\sqrt{2}$$

(a) Forces on P

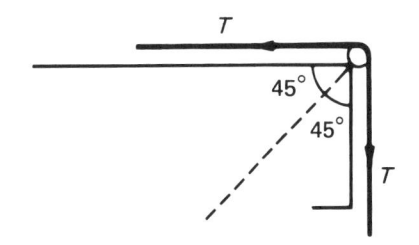

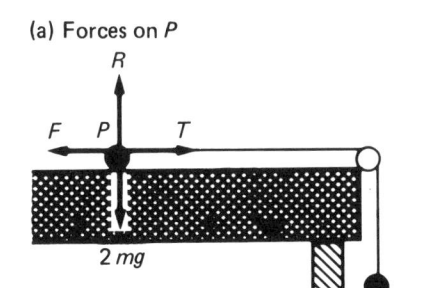

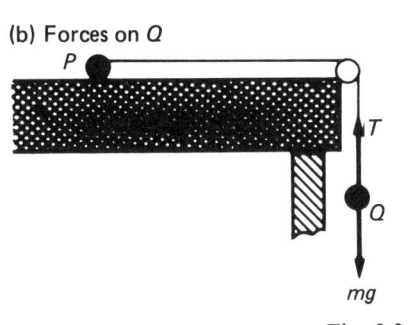

(b) Forces on Q

Fig. 8.2:**5**

Fig. 8.2:**6**

Exercise 8.2:4

Questions 1–14: Particles P and Q are connected by a light inelastic string passing over a smooth fixed pulley. In each case find also the magnitude and direction of the force of the string on the pulley.

⑤ **1** Find k if the coefficient of friction between P and the plane is $\tfrac{2}{3}$ and the system is in limiting equilibrium.

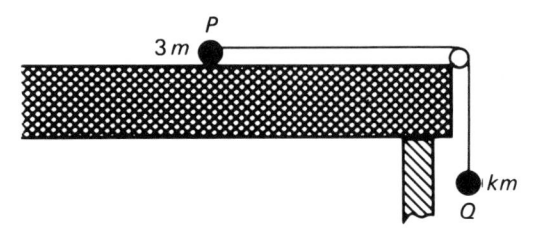

⑤ **2** The system is in equilibrium. Find θ.

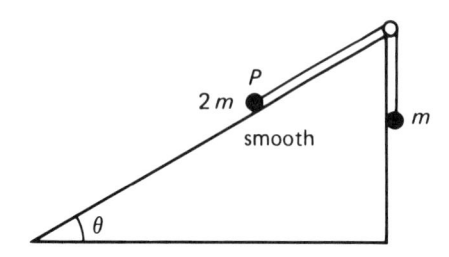

⑤ **3** The system is in equilibrium. Find k.

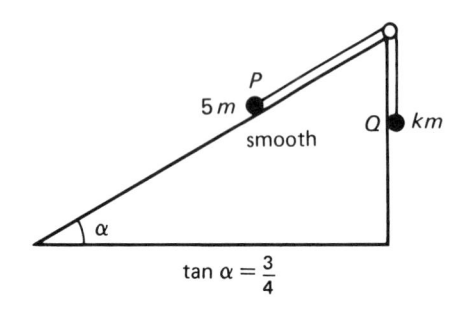

$\tan \alpha = \tfrac{3}{4}$

⑤ **4** The system is in limiting equilibrium. Find the coefficient of friction between P and the plane.

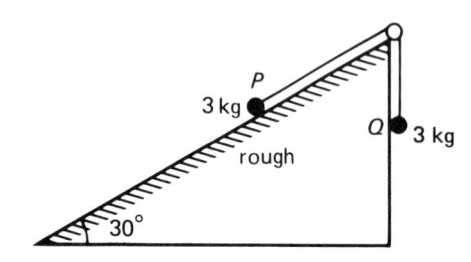

Questions 5–10: Particles P and Q are connected by a light elastic string passing over a smooth fixed pulley.

⑤ → **5** $\tan \alpha = \tfrac{3}{4}$ and $\mu = \tfrac{1}{2}$. Show that the system is in equilibrium if $1 \leq k \leq 5$. (Hint: consider two cases: (i) Q about to move upwards; (ii) Q about to move downwards.)

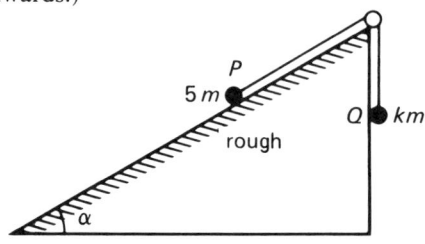

⑤ **6** $\tan \alpha = \tfrac{1}{2}$; $\mu = \tfrac{1}{4}$. Show that the system is in equilibrium if $1 \leq k \leq 3$.

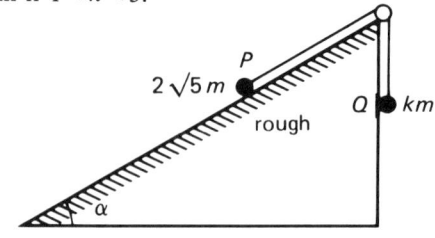

⑤ **7** The system is in equilibrium. Find θ.

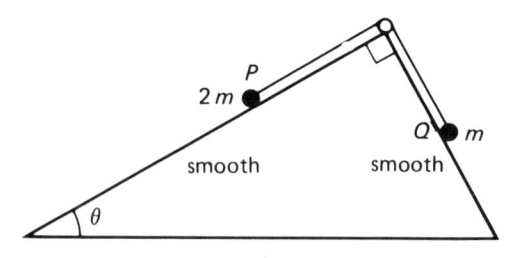

⑤ **8** $\tan \alpha = \tfrac{3}{4}$; the coefficient of friction at both contacts is $\tfrac{2}{5}$. Find the range of values of k for which the system is in equilibrium.

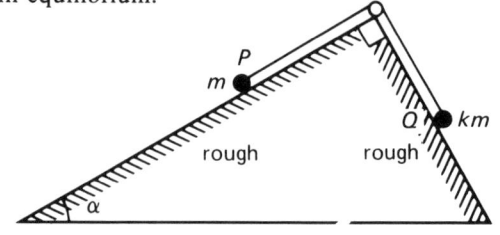

⑤ R **9** The coefficient of friction at both contacts is $\tfrac{1}{2}$. Find the range of values of k for which the system is in equilibrium.

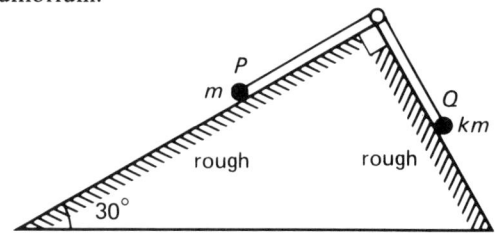

⑤ **10** The coefficient of friction μ is the same for both contacts; $\beta > \alpha$; Q is just about to slide down the plane. Show that

$$\mu = \frac{M \sin \beta - m \sin \alpha}{M \cos \beta + m \cos \alpha}$$

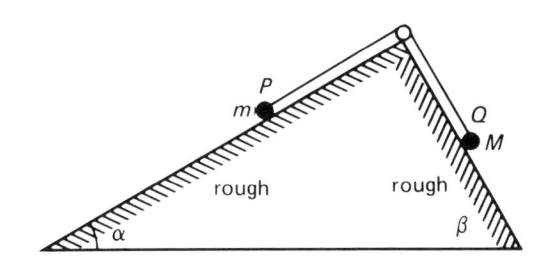

Questions 11-21: Each of the following systems of connected particles is in equilibrium.
× = fixed point
○ = smooth pulley

• **11** Find θ and the tension in AP.

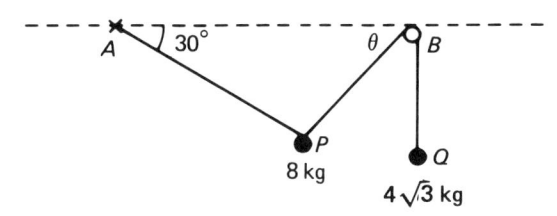

12 The string is threaded through the smooth ring P. Find θ.

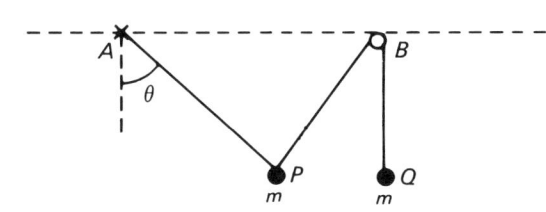

→ **13** Find k, the tensions in the strings, and the mass of the particle at Q.

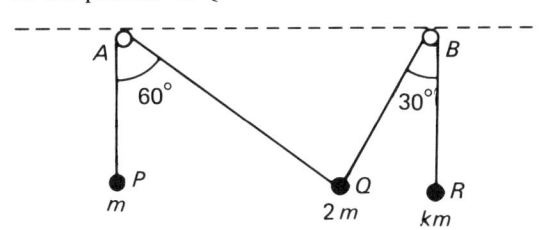

14 Find $\cos \theta$ and $\cos \phi$.

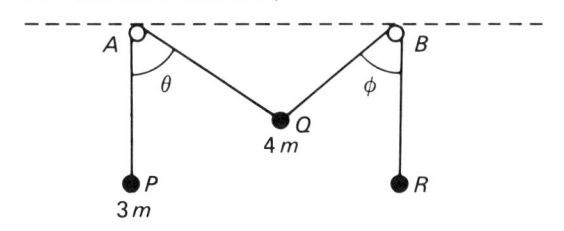

15 Show that $\sin \theta = (p^2 + q^2 - r^2)/2pq$.

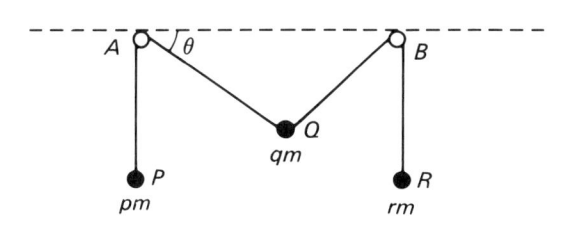

• **16** Find θ and the tensions in the strings.

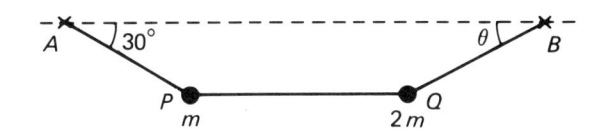

17 Find k.

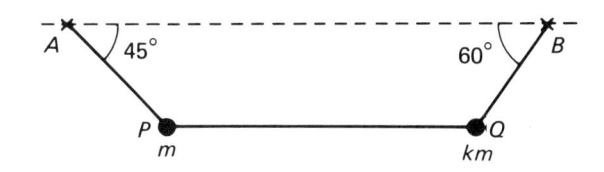

18 Find θ and ϕ.

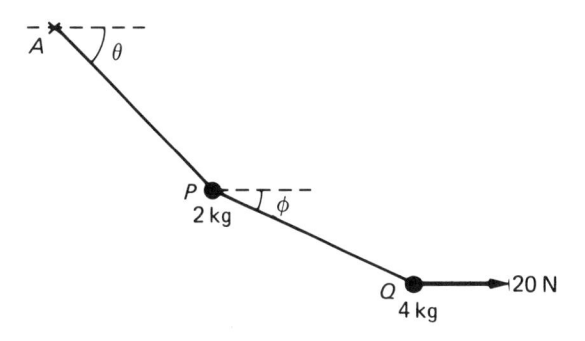

→ **19** Find θ and the tensions in the strings.

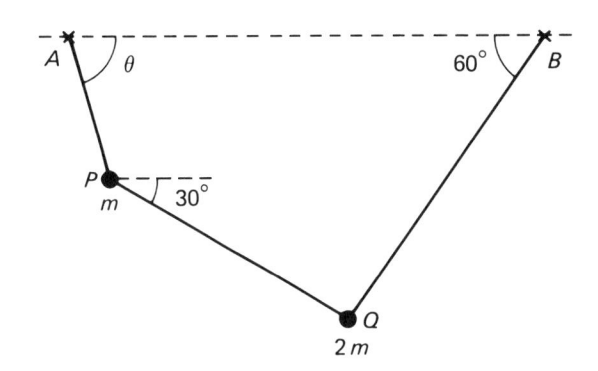

448 *Dynamics problems*

+ 20 Show that
$$k = \frac{\sin \alpha \sin (\beta + \gamma)}{\sin (\alpha - \beta) \sin \gamma}$$

R 21 Show that $5 \tan \phi = 3 \tan \theta$.

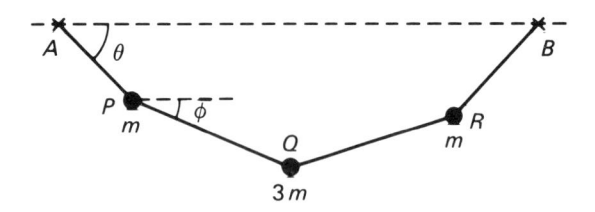

22 Two beads P and Q of equal mass m are threaded onto a rough fixed horizontal wire. The coefficient of friction between each bead and the wire is $\frac{2}{5}$. The beads are connected by a light inelastic string which passes through a smooth ring R of mass $3m$. If θ is the greatest possible angle between the two sections of the string, PR and PQ show that $\tan \frac{1}{2}\theta = \frac{2}{3}$.

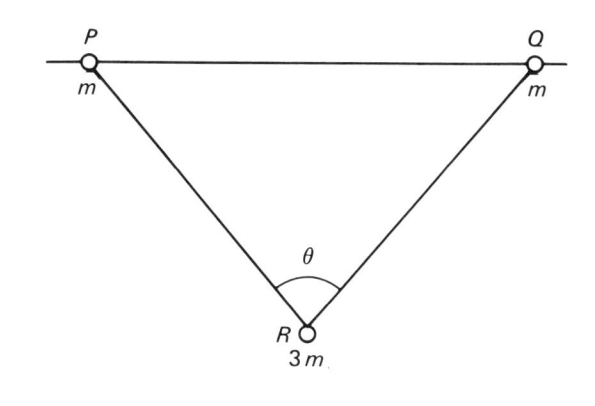

8.3 Motion of a particle

8.3:1 Dynamics problems

Newton's Second Law states that if a system of forces $\{\mathbf{F}\}$ acts on a particle of mass m, the particle moves with acceleration $\mathbf{a}$ where

$$\sum \mathbf{F} = m\mathbf{a} \qquad (1)$$

In practice we resolve the equation in two convenient perpendicular directions Ox and Oy:

$$\left. \begin{array}{l} \sum F_x = ma_x \\ \sum F_y = ma_y \end{array} \right\} \qquad (2)$$

where each force $\mathbf{F} = F_x \mathbf{i} + F_y \mathbf{j}$ and $\mathbf{a} = a_x \mathbf{i} + a_y \mathbf{j}$.

Worked example 1   A particle of mass 5 kg is pulled along a smooth table by a force of 20 N inclined at 30° to the table. Find the acceleration of the particle and the magnitude and direction of the force exerted by the particle on the table.

The force diagram is shown in fig. 8.3:**1a**. Resolving into components (fig. 8.3:**1b**):
$\rightarrow$) $\sum F_x = ma_x \Rightarrow 20 \cos 30° = 5a$, i.e. $a = 2\sqrt{3}$ m s^{-2}.
$\uparrow$) $\sum F_y = ma_y \Rightarrow R + 20 \sin 30° - 50 = 0$, i.e. $R = 40$ N.
The force of the table on the particle is 40 N vertically upwards; so, by

Fig. 8.3:1

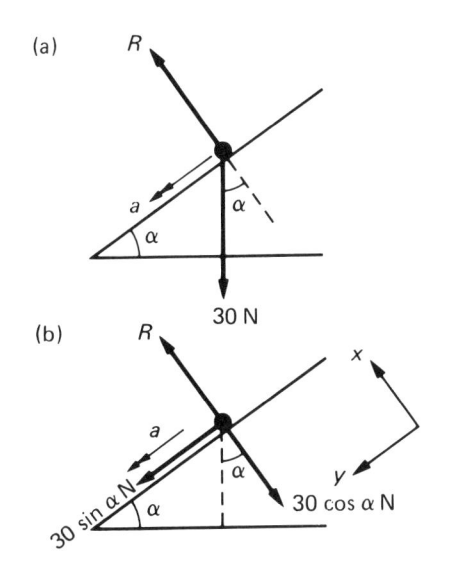

Forces on a particle 449

N3 (p. 430), the force of the particle on the table is 40 N vertically downwards. Notice that it is *not* equal to the weight of the particle.

Worked example 2 A particle of mass 3 kg slides down a smooth plane of inclination $\tan^{-1}\frac{3}{4}$. Find its acceleration.

Since the direction of the acceleration is down the plane (fig. 8.3:**2a**), we resolve the forces parallel to and perpendicular to the plane (fig. 8.3:**2b**) to avoid having to split the acceleration into components. (Tan $\alpha = \frac{3}{4} \Rightarrow \sin \alpha = \frac{3}{5}$ and $\cos \alpha = \frac{4}{5}$.)

↖) $\sum F_x = ma_x \Rightarrow R - 30 \cos \alpha = 0$.
↙) $\sum F_y = ma_y \Rightarrow 30 \sin \alpha = 3a$, i.e. $a = 10 \sin \alpha = 6$ m s^{-2}.

Fig. 8.3:**2**

Exercise 8.3:1

1 A particle is moved vertically by a string. Find its acceleration:

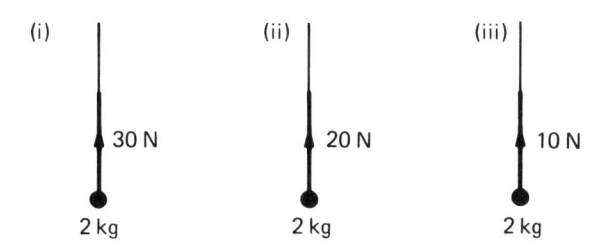

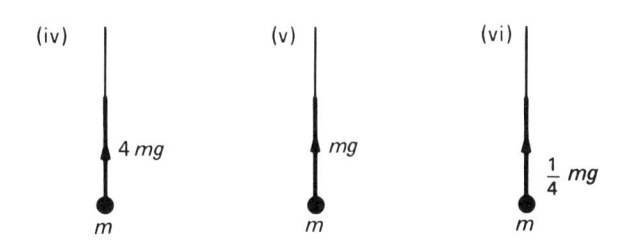

(i) 30 N, 2 kg
(ii) 20 N, 2 kg
(iii) 10 N, 2 kg
(iv) 4 mg, m
(v) mg, m
(vi) $\frac{1}{4}$ mg, m

2 A stone of mass 3 kg is falling vertically through the air.

(i) If the air resistance is 6 N, find its acceleration.
(ii) If its acceleration is only 5 m s^{-2}, find the air resistance.

→ **3** A man of mass 75 kg is travelling in a lift. Find the force exerted on the man by the floor of the lift when the lift accelerates:

- (i) upwards at 4 m s^{-2}
  (ii) upwards at 2 m s^{-2}
  (iii) downwards at 2 m s^{-2}
  (iv) downwards at 4 m s^{-2}

4 A dog of mass 12 kg is travelling in a lift. Find the force exerted on the dog by the floor of the lift when the lift accelerates downwards at:

(i) 1 m s^{-2} (ii) 5 m s^{-2}
(iii) 10 m s^{-2} (iv) 15 m s^{-2}

Explain your answers to parts (iii) and (iv). Where would you expect to find the dog in part (iv)? Draw a diagram showing the forces acting on the dog.

Questions 5-10: A particle is pulled across a smooth horizontal table by a string. Find the quantity indicated. In each case find also the magnitude and direction of the reaction of the particle on the table.

① **5** Find a. (4 kg, 6N)

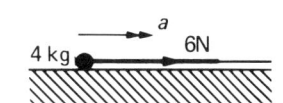

① **6** Find T. (5 kg, 3 m s^{-2}, T)

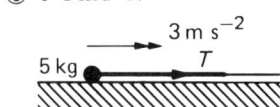

① **7** Find a (in terms of g). (3m, mg)

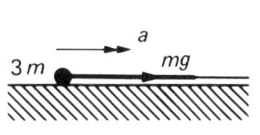

① **8** Find a. (8 kg, 10 N, 30°)

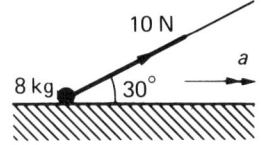

①→ **9** Tan $\alpha = \frac{3}{4}$. Find T. (0.5 kg, T, 8 m s^{-2}, α)

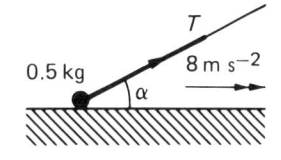

① **10** Find a in terms of T, m, θ. (m, T, θ)

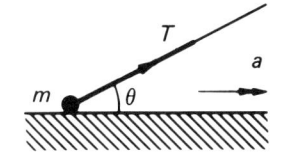

Dynamics problems

① **11** A particle is pulled across a smooth horizontal table by a string. What is the maximum possible horizontal acceleration of the particle (i.e. so that the particle does not leave the table) when the inclination of the string to the horizontal is:

(i) 30° (ii) 45° (iii) 60°?

① **12** A particle is attached by a light inextensible string to the ceiling of a coach of a train.

(i) If the string is inclined at 15° to the vertical, find the acceleration of the train.

(ii) If the train accelerates at 1 m s^{-2}, find the inclination of the string to the vertical.

Questions 13 and 14: A particle of mass 3 kg resting on a smooth table is acted on by the systems of forces shown. The forces act in a *vertical* plane through the particle. Find the acceleration of the particle and the magnitude and direction of the force it exerts on the table.

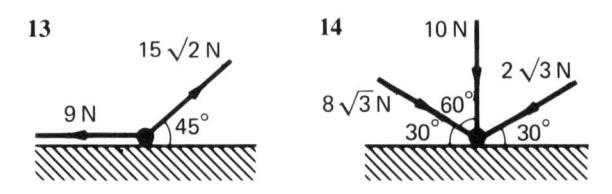

13 [Diagram: 15√2 N at 45°, 9 N horizontal]

14 [Diagram: 10 N vertical, 8√3 N at 30°, 2√3 N at 30°, 60°]

Questions 15–18: A particle of mass 4 kg resting on a smooth table is acted on by the following systems of *horizontal* forces. Find the magnitude and direction of the acceleration of the particle in each case.

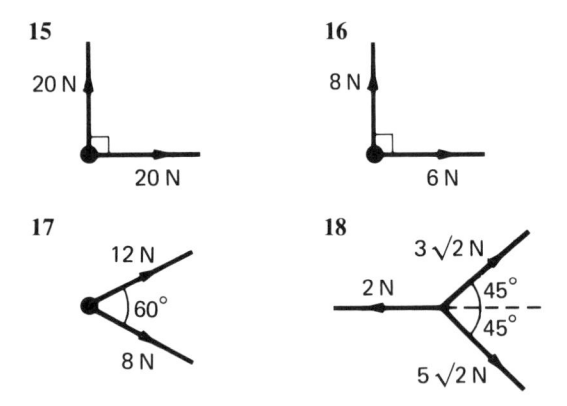

15 [20 N up, 20 N right]

16 [8 N up, 6 N right]

17 [12 N and 8 N at 60°]

18 [3√2 N at 45°, 2 N, 5√2 N at 45°]

Questions 19–22: A particle slides down a smooth inclined plane. Find the quantity indicated. Find also the magnitude and direction of the force exerted by the particle on the plane.

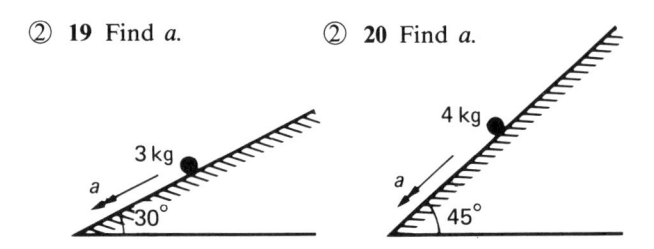

② **19** Find a. [3 kg, 30°]

② **20** Find a. [4 kg, 45°]

② **21** Find θ. [2 kg, 8 m s^{-2}]

② **22** Find a (in terms of g, α). [mass m, angle α]

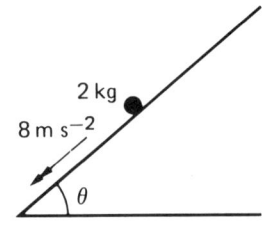

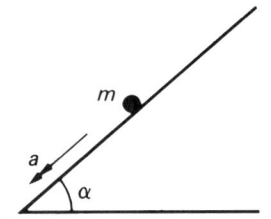

② **23** Comment on this problem: A particle slides down a smooth plane of inclination 30° with acceleration 2 m s^{-2}. Find the mass of the particle.

Questions 24–27: A particle of mass 5 kg moves on a smooth inclined plane under the action of a single force. Find its acceleration (which may be up or down the plane).

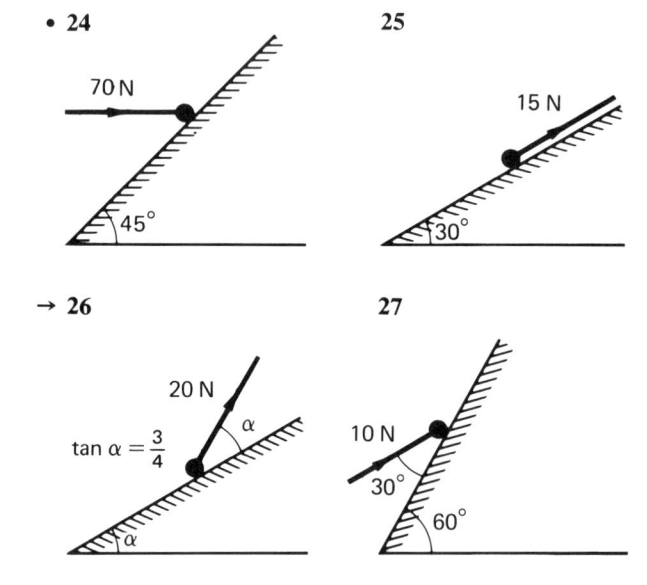

• **24** [70 N horizontal, 45° plane]

25 [15 N along plane, 30°]

→ **26** [20 N, $\tan\alpha = \frac{3}{4}$]

27 [10 N at 30°, 60° plane]

28 A particle of mass 2 kg is pulled up a smooth plane of inclination $\tan^{-1}\frac{4}{3}$ by a string parallel to the plane. Find the tension in the string and the force exerted by the particle on the plane when its acceleration is:

(i) 0 m s^{-2} (ii) 4 m s^{-2} (iii) 8 m s^{-2}

• **29** A scale pan of mass 2 kg, containing a particle of mass 4 kg, is pulled vertically upwards by a string. If the tension in the string is 90 N,

(i) find the acceleration of the system;
(ii) by considering the forces on the particle only find the reaction between the particle and the scale pan.

Check your answer to part (ii) by considering the forces on the scale pan only.

Forces on a particle 451

→ **30** A scale pan of mass $3m$, containing a particle of mass $2m$, is moved vertically by a string. Find the acceleration of the system and the reaction between the particle and the scale pan when the tension in the string is:

(i) $10mg$ (ii) $6mg$
(iii) $5mg$ (iv) $4mg$

31 A man of mass 80 kg stands in the basket of an air balloon of mass 400 kg. The balloon moves in a vertical line under the action of an upward force of magnitude S. Find the acceleration of the balloon, and the force exerted by the basket on the man, when:

(i) $S = 7200$ N
(ii) $S = 4800$ N
(iii) $S = 2400$ N

32 A scale pan of mass 2 kg, containing a block A of mass 2 kg on top of a mass B of 4 kg, is pulled vertically upwards by a string. If the tension in the string is 120 N, find:

(i) the acceleration of the system
(ii) the reaction between A and B
(iii) the reaction between B and the scale pan

8.3.2 Dynamics problems (friction)

Worked example 3 A particle of mass 5 kg is pulled along a rough table by a force of 20 N inclined at 30° to the table. The coefficient of friction between the particle and the table is $\frac{1}{4}$. Find the acceleration of the particle and the magnitude and direction of the force exerted by the particle on the table (c.f. worked example 1).

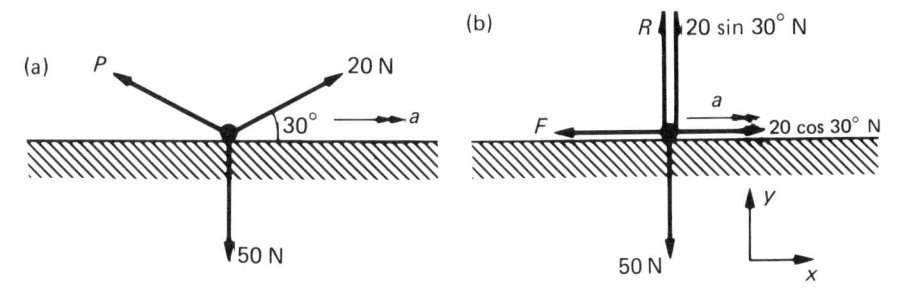

Fig. 8.3:3

The forces on the particle are shown in fig. 8.3:**3a**. We split the reaction P of the table on the particle into components F (frictional reaction) and R (normal reaction) as in fig. 8.3:**3b**.

→) $\sum F_x = ma_x \Rightarrow 20\cos 30° - F = 5a$.
↑) $\sum F_y = ma_y \Rightarrow R + 20\sin 30° - 50 = 0$, i.e. $R = 40$ N.

Since the particle is moving,

$$F = \mu R = \tfrac{1}{4} \cdot 40 = 10 \text{ N}$$

So $\qquad 5a = 10\sqrt{3} - 10$

Thus $\qquad a = 2(\sqrt{3} - 1) \text{ m s}^{-2}$

Fig. 8.3:4   (a) Reaction of table on particle   (b) Reaction of particle on table (by N3)

From fig. 8.3:**4b**, the magnitude of the reaction exerted by the particle on the table is $\sqrt{(10^2 + 40^2)} = 10\sqrt{17}$ N. This force makes an angle α with a downward vertical where

$$\tan \alpha = \tfrac{10}{40} = \tfrac{1}{4}$$

452 Dynamics problems (friction)

(a)

(b)

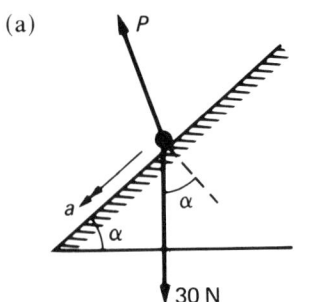

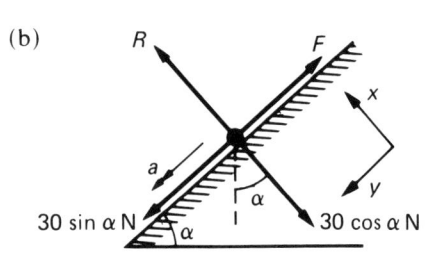

Fig. 8.3:**5**

Worked example 4 A particle of mass 3 kg slides down a rough plane of inclination $\alpha = \tan^{-1}\frac{3}{4}$ (fig. 8.3:**5a**). The coefficient of friction between the particle and the plane is $\frac{1}{3}$. Find the acceleration of the particle (c.f. worked example 2).

As in worked example 2, we resolve the forces parallel to and perpendicular to the plane. Notice that the frictional reaction F opposes the direction of motion (fig. 8.3:**5b**).

↖) $\sum F_x = ma_x \Rightarrow R - 30 \cos \alpha = 0$, i.e. $R = 24$ N.
↗) $\sum F_y = ma_y \Rightarrow 30 \sin \alpha - F = 3a$.

Since the particle is moving,

$$F = \mu R = \tfrac{1}{3} \cdot 24 = 8 \text{ N}$$

Thus
$$3a = 30 \cdot \tfrac{3}{5} - 8 = 10$$
$$a = \tfrac{10}{3} \text{ m s}^{-2}$$

Exercise 8.3:2

Questions 1–6: A particle is pulled across a rough table by a string. The coefficient of friction between the particle and the table is μ. Find the quantity indicated, and in each case find the magnitude and direction of the reaction of the particle on the table.

③ **1** $\mu = \tfrac{2}{5}$. Find a.

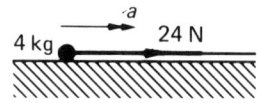

③ **2** Find μ.

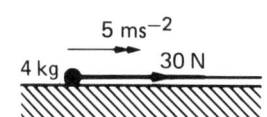

③ **3** $\mu = \tfrac{1}{2}$. Find a (in terms of g).

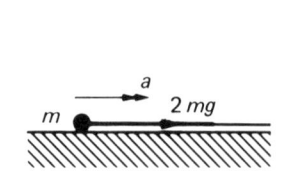

③ → **4** $\mu = \tfrac{1}{4}$, $\tan \alpha = \tfrac{3}{4}$. Find a.

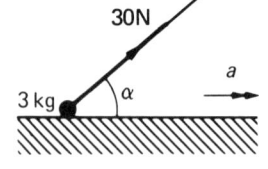

③ **5** $\mu = \tfrac{1}{2}$. Find T.

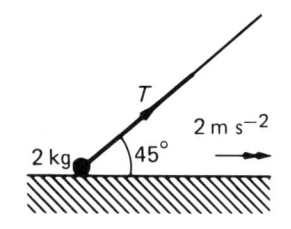

③ **R 6** Find a in terms of T, m, θ, μ and g.

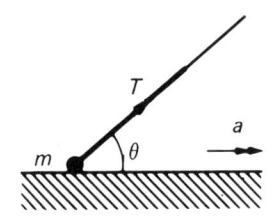

③ → **7** A particle of mass 6 kg resting on a rough table is acted on by a horizontal force of magnitude S. The coefficient of friction between the particle and the table is $\tfrac{1}{5}$. Describe what happens if:

(i) $S = 6$ N   (ii) $S = 12$ N
(iii) $S = 18$ N   (iv) $S = 24$ N

③ **8** A particle of mass 8 kg resting on a rough table is acted on by an upward force of magnitude S inclined at $\tan^{-1}\tfrac{4}{3}$ to the vertical. The coefficient of friction between the particle and the table is $\tfrac{1}{4}$. Describe what happens if:

(i) $S = 10$ N   (ii) $S = 25$ N
(iii) $S = 50$ N   (iv) $S = 200$ N

③ **9** A particle of mass 4 kg is pulled along a rough table by a horizontal force of magnitude 20 N. Find its acceleration if the coefficient of friction between it and the table is:

(i) 0 (ii) $\tfrac{1}{5}$ (iii) $\tfrac{2}{5}$ (iv) $\tfrac{1}{2}$ (v) $\tfrac{3}{5}$

③ **10** A particle of mass 2 kg resting on a rough table is acted on by a downward force of magnitude 40 N inclined at an angle θ to the horizontal. The coefficient of friction between the particle and the table is $\tfrac{1}{2}$. Find its acceleration when:

(i) $\theta = 0°$ (ii) $\theta = 30°$ (iii) $\theta = 60°$ (iv) $\theta = 90°$

Find the greatest value of θ for which motion occurs.

Forces on a particle

11 A particle of mass m resting on a rough table is acted on by a downward force of magnitude S inclined at an angle θ to the horizontal. The coefficient of friction between the particle and the table is $\frac{1}{2}$. Show that if the greatest value of θ for which motion occurs is $45°$, then $S = \sqrt{2}mg$. Find the acceleration of the particle when $\theta = 30°$.

Questions 12-15: A particle of mass 4 kg resting on a rough table is acted on by the following systems of *horizontal* forces. The coefficient of friction between the particle and the table is $\frac{1}{4}$. Find the magnitude and direction of the acceleration of the particle.

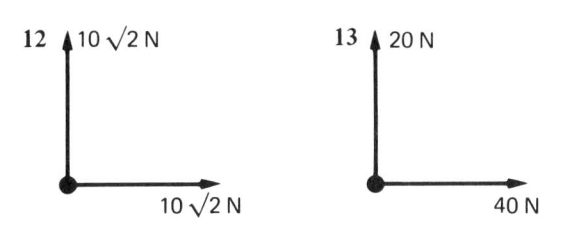

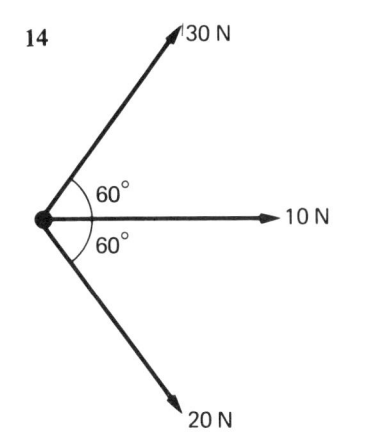

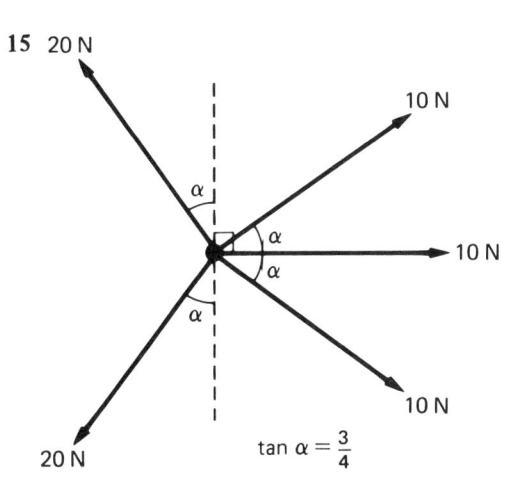

12 $10\sqrt{2}$ N (up), $10\sqrt{2}$ N (right)

13 20 N (up), 40 N (right)

14 30 N at 60° above horizontal (upper left), 10 N horizontal (right), 20 N at 60° below horizontal (lower left)

15 20 N (upper left), 10 N (upper right), 10 N (right), 10 N (lower right), 20 N (lower left); $\tan\alpha = \frac{3}{4}$

• **16** A car of mass 400 kg moves along a horizontal road. The resistance to motion of the car is 1000 N. Find the acceleration of the car when the driving force of the engine is:

(i) 1000 N (ii) 1500 N (iii) 2000 N

17 A car of mass 1800 kg moves with constant velocity up a slope of 1 in 6. The resistance to the motion of the car is 1200 N. Find the driving force exerted by the engine.

With the same driving force, what is the acceleration of the car when it moves:

(i) along a horizontal road?
(ii) down a slope of 1 in 6?

(Assume that the resistance to the motion of the car is always the same.)

→ **18** A car of mass M moves against a constant resistance of $\frac{1}{4}Mg$. Find the driving force necessary to give the car an acceleration $\frac{1}{10}g$:

(i) along a horizontal road
(ii) up a slope of inclination 1 in 6
(iii) up a slope of inclination 1 in 8

R 19 A car of mass M moves against a constant resistance R. Find the driving force necessary to give the car an acceleration a:

(i) along a horizontal road
(ii) up a slope of inclination α
(iii) down a slope of inclination α

Questions 20-23: A particle slides down a rough inclined plane. The coefficient of friction between the particle and the plane is μ. In each case find the quantity indicated, and also the magnitude and direction of the force exerted by the particle on the plane.

20 $\mu = 1/2\sqrt{3}$. Find a. (4 kg on 30° incline)

21 $\tan\alpha = \frac{4}{3}$. Find μ. (8 kg, 8 m s^{-2})

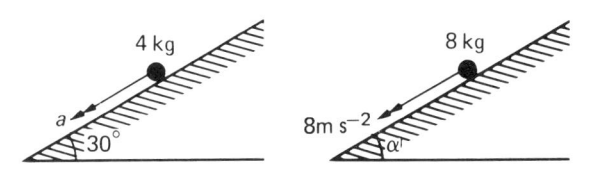

22 $\mu = \frac{1}{8}$. Show that $\tan\beta = \frac{7}{24}$. (2 kg, 1.6 m s^{-2})

→ **23** $\tan\beta = \frac{3}{4}$; $\mu = \frac{1}{2}$. Find a. (5 kg)

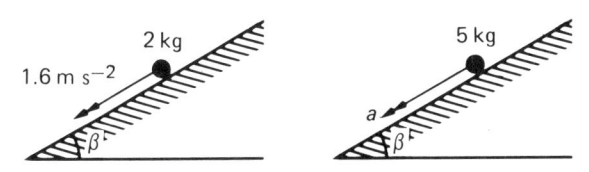

24 A particle rests on a rough plane. The coefficient of friction between the particle and the plane is $1/\sqrt{3}$. The plane is gradually tilted. Find the acceleration of the particle when the inclination of the plane is:

(i) 15° (ii) 30° (iii) 45° (iv) 60°

R 25 A particle slides down a rough plane of inclination α. The coefficient of friction between the particle and the plane is μ. Find its acceleration (in terms of g, α and μ).

Dynamics problems (friction)

(i) Deduce the maximum possible value of μ for motion to take place.

(ii) Deduce the acceleration of a particle sliding down a smooth plane of inclination α.

26 A particle is projected up a rough plane of inclination 30°. The coefficient of friction between the particle and the plane is μ. Find the retardation of the particle if:

- (i) $\mu = 1/3\sqrt{3}$
- (ii) $\mu = 1/2\sqrt{3}$
- (iii) $\mu = 1/\sqrt{3}$

Questions 27-29: A particle is projected up a plane of inclination α. The coefficient of friction between the particle and the plane is μ. Find its acceleration when it is:

(i) moving up the plane
(ii) moving back down the plane

under the given conditions.

27 $\alpha = 60°$, $\mu = \frac{1}{5}$.

28 $\alpha = \tan^{-1}\frac{4}{3}$, $\mu = \frac{1}{3}$.

29 $\alpha = 45°$, $\mu = \frac{1}{2}\sqrt{2}$.

Questions 30-33: A particle moves on a rough inclined plane under the action of a single force. The coefficient of friction between the particle and the plane is μ.

30 $\mu = \frac{1}{4}$. Find the acceleration of the particle.

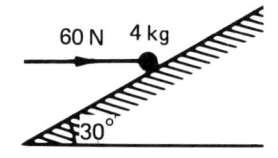

31 $\mu = \frac{1}{4}$. Find the acceleration of the particle.

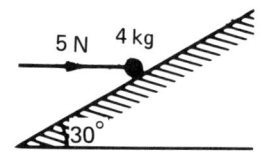

R 32 $\tan \alpha = \frac{7}{24}$; $\mu = \frac{3}{8}$. Find the acceleration of the particle.

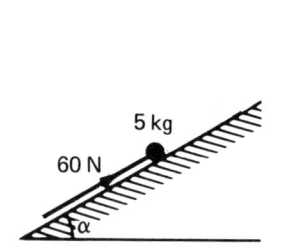

33 Find μ.

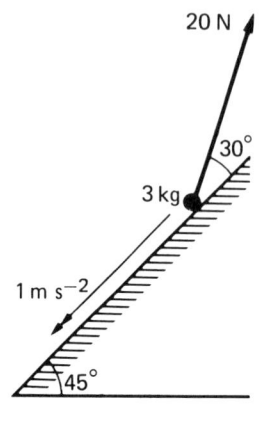

Questions 34-37: A wedge is fixed to a horizontal plane. A particle moves on the sloping face of the wedge. Find:

(i) the acceleration of the particle
(ii) the reaction of the particle on the wedge
(iii) the magnitude and direction of the force of the floor on the wedge

34

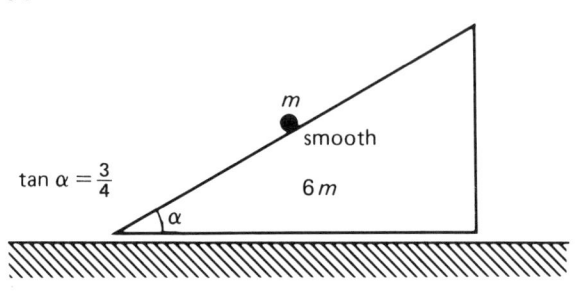

35

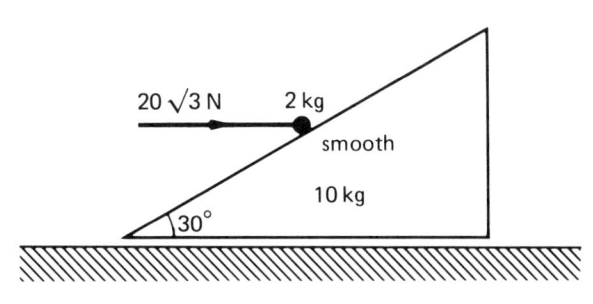

36

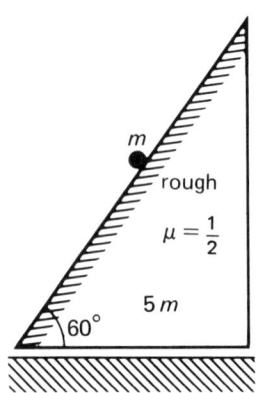

37 $\tan \beta = \frac{4}{3}$. Consider two cases:

(a) The sloping face of the wedge is smooth.
(b) The coefficient of friction between the particle and the sloping face of the wedge is $\frac{1}{3}$.

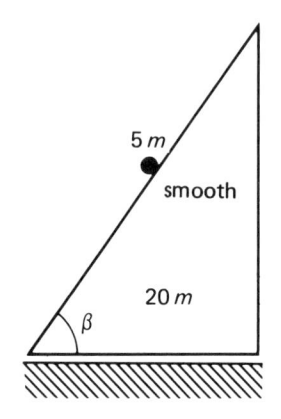

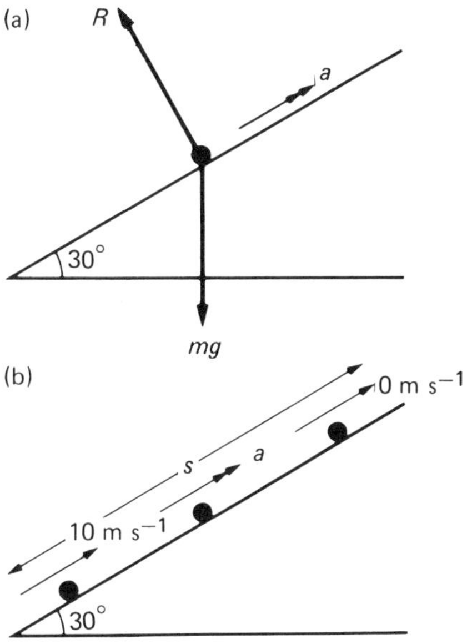

8.3:3 Use of constant acceleration equations

Worked example 5 A particle is projected with speed 10 m s^{-1} up a smooth plane of inclination $30°$ (fig. 8.3:6). Find the distance it travels before coming to rest.

Let the mass of the particle be m.

$$\nearrow) \qquad -mg \sin 30° = ma$$
$$\Rightarrow \qquad a = -g \sin 30° = -\tfrac{1}{2}g = -5 \text{ m s}^{-2}$$

Let the distance travelled be s. $u = 10 \text{ m s}^{-1}$, $a = -5 \text{ m s}^{-2}$, $v = 0 \text{ m s}^{-1}$.

$$v^2 = u^2 + 2as \Rightarrow 0 = 100 - 2 \cdot 5s$$
i.e. $\qquad s = 10 \text{ m}$

So the particle travels a distance 10 m up the plane before coming to rest.

Fig. 8.3:6

Exercise 8.3:3

• **1** A particle of mass 4 kg moves under the action of a resultant force of magnitude 36 N. How long does the particle take to accelerate:

(i) from rest to 12 m s^{-1}?
(ii) from rest to 24 m s^{-1}?

How far does the particle move in each case?

2 A car of mass 500 kg accelerates from rest along a level road. The resistance to the motion of the car is 400 N. The car takes 5 s to accelerate from rest to 8 m s^{-1}. Find the driving force.

Find the distance moved by the car as it accelerates:

(i) from rest to 4 m s^{-1}
(ii) from 4 m s^{-1} to 8 m s^{-1}
(iii) from 8 m s^{-1} to 12 m s^{-1}

3 A particle of mass 2 kg is travelling with a velocity of 10 m s^{-1}. The particle is brought to rest in a distance of 20 m by a constant force of magnitude P. Find P.

4 The velocity of a particle of mass 0.6 kg is reduced from 14 m s^{-1} to 4 m s^{-1} in 1.5 s by a constant force of magnitude P. Find P.

→ **5** A particle of mass 4 kg travelling with a velocity of 60 m s^{-1} is brought to rest by a constant resisting force of 12 N. Find the time taken for its velocity to reach:

(i) 40 m s^{-1} (ii) 20 m s^{-1} (iii) 0 m s^{-1}

Find the total distance it moves.

6 A car of mass 600 kg moving at 10 m s^{-1} can stop within 10 m. Find the braking force (assumed constant). Find its stopping distance when it is moving at 20 m s^{-1}.

7 A bullet of mass 0.05 kg travelling horizontally at 400 m s^{-1} hits a fixed wooden target and penetrates to a distance of 0.1 m. Find the average resisting force of the wood.

8 A nail of mass 0.02 kg, driven with initial speed 40 m s^{-1} into a fixed piece of wood, penetrates a distance 0.01 m. Find the average resistance of the wood.

• **9** A particle is projected with speed 10 m s^{-1} across a rough table. The coefficient of friction between the particle and the table is $\tfrac{1}{2}$. How far does the particle travel before:

(i) its speed is halved?
(ii) it comes to rest?

10 A stone slides across an ice rink. The initial speed of the stone is 8 m s^{-1} and it moves 24 m before coming to rest. Calculate the coefficient of friction between the stone and the ice.

⑤ **11** A particle is projected with speed $10\sqrt{2} \text{ m s}^{-1}$ up a smooth plane of inclination $45°$. Find the time taken for it come to rest and the distance it has travelled.

⑤ **12** A particle slides down from rest down a plane of inclination 30°. Find its speed and the distance it has travelled after 3 s if:

(i) the plane is smooth
(ii) the coefficient of friction between the particle and the plane is $1/2\sqrt{3}$

⑤ **13** A particle slides from rest down a plane of inclination α. Find the time taken to reach a speed v if:

(i) the plane is smooth
(ii) the coefficient of friction between the particle and the plane is μ

⑤ **14** A particle projected with speed 4 m s^{-1} up a rough plane of inclination $\tan^{-1}\frac{4}{3}$ takes $\frac{2}{3}$ s to come to rest. Find the coefficient of friction between the particle and the plane.

⑤ → **15** A particle projected with speed $2\sqrt{(ga)}$ up a rough plane of inclination 30° comes to rest after travelling a distance $3a$. Find the coefficient of friction between the particle and the plane.

⑤ • **16** A particle is projected from a point A with speed 5 m s^{-1} up a rough plane of inclination $\tan^{-1}\frac{3}{4}$. The coefficient of friction between the particle and the plane is $\frac{1}{2}$.

(i) Find the retardation of the particle as it moves up the plane. Hence find how far it is from A when it comes to rest.

(ii) Show that the acceleration of the particle as it moves down the plane is 2 m s^{-2}. Hence find the speed of the particle when it passes through A again.

⑤ **17** A particle is projected from a point A with speed u up a smooth plane of inclination α. Find in terms of u and α the distance from A at which the particle comes to rest. Show that the particle is travelling with speed u when it passes through A again.

⑤ R **18** A particle is projected from a point A with speed u up a rough plane of inclination $\tan^{-1}\frac{3}{4}$. The coefficient of friction between the particle and the plane is $\frac{1}{4}$.

(i) Show that, while the particle is moving up the plane, its deceleration is $\frac{4}{5}g$. Hence show that the particle comes to instantaneous rest after moving a distance $5u^2/8g$.

(ii) Show that, while the particle is moving back down the plane, its acceleration is $\frac{2}{5}g$. Hence find its speed when it passes through A again. Find its distance from A when it is again travelling with speed u.

⑤ → **19** A particle is projected from a point A with speed u up a rough plane of inclination $\tan^{-1}\frac{1}{2}$. The coefficient of friction between the particle and the plane is $\frac{1}{8}$. Find the speed of the particle when it passes through A again.

⑤ **20** A particle is projected from a point A with speed 10 m s^{-1} up a rough plane of inclination 30°. The coefficient of friction between the particle and the plane is $1/2\sqrt{3}$. Find its distance from A when it is again travelling with speed 10 m s^{-1}.

• **21** A particle of mass 0.25 kg moves along Ox, starting from rest at O. The resultant force on the particle after t s is $6t^2\mathbf{i}$. Find its displacement from O after 4 s.

22 A particle of mass 3 kg moves along Ox, starting from rest at O. The resultant force on the particle after t s is 36 cos $2t\mathbf{i}$. Find its velocity and its displacement from O after:

(i) $\frac{1}{8}\pi$ s   (ii) $\frac{1}{4}\pi$ s   (iii) $\frac{3}{8}\pi$ s

23 A car of mass 1000 kg moves from rest on a straight level road. The engine exerts a driving force P N which varies with time t s. The resistance to motion is 200 N. From the table below, and using Simpson's rule (section 4.3), find approximately:

(i) the speeds of the car when $t = 10$ and $t = 20$
(ii) the distance travelled in the first 20 s

t	0	5	10	15	20
P	400	500	580	640	660

• **24** A particle P of mass m falls vertically from rest against an air resistance which is mkv when the speed of P is v (k constant).

(i) Using N2, show that the acceleration of P is $g - kv$.
(ii) Writing the acceleration as dv/dt show that at time t
$$v = \frac{g}{k}(1 - e^{-kt})$$

(iii) Deduce the terminal velocity of P.

R **25** A particle of mass m is projected vertically upwards with speed u against an air resistance which is mkv when the speed of the particle is v. Find the time taken for:

(i) the speed of the particle to be halved
(ii) the particle to reach its greatest height

• **26** A particle of mass m moving along Ox is subject to a force mkx^2 acting towards O. Given that the particle starts from rest when $x = 3a$, find its velocity when it first reaches O.

27 A particle of mass m moving along Ox is subject to a force mk/x^2 acting towards O. When $x = a$, the particle has a velocity u away from O. Find the velocity of the particle when $x = 2a$. Under what circumstances does the particle eventually come to rest?

→ 28 A particle of mass m slides from rest at a point O down a plane inclined at 30° to the horizontal. The resistance to the motion of the particle is mkx where x is the displacement of the particle from O. Find the velocity of the particle when $x = g/2k$. Find the maximum displacement of the particle from O.

+ 29 One end of a light extensible string of natural length l and modulus $3mg$ is attached to a fixed point A, the other end to a particle P of mass m, which hangs in equilibrium at a point O vertically below A. Find the length OA. The particle is pulled a further distance l below O and is released.

(i) Show that while the string is taut and when $OP = x$,

$$v\frac{dv}{dx} = -\frac{3g}{l}x$$

where v is the velocity of P at this instant.

(ii) Hence find the velocity of P when it reaches O.
(iii) Find the velocity of P when the string goes slack.
(iv) Hence find the greatest height above O reached by P.

30 A particle of mass 5 kg moves from rest along Ox under the action of a resultant force of magnitude F. The table shows the values of F against the displacement x of the particle from O. Using Simpson's rule (see section 4.3), find the speed of the particle when its displacement from O is:

(i) 5 m   (ii) 10 m

$F(N)$	6.0	7.4	8.0	8.3	8.5
$x(m)$	0	2.5	5	7.5	10

8.3:4 Connected particles

Worked example 6 Two particles A and B of masses 2 kg and 4 kg are connected to the ends of a light inextensible string. The particles are placed in line on a rough table with the string taut and a force of 30 N is applied to A in the direction BA. If the coefficient of friction between each particle and the table is $\frac{1}{4}$, find the acceleration of the particles and the tension in the string.

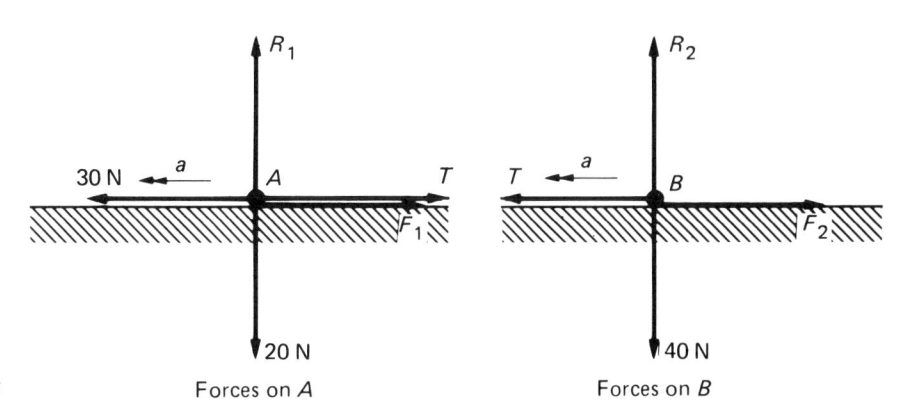

Fig. 8.3:7   Forces on A Forces on B

The forces acting on the particles are shown in fig. 8.3:7. Note that the acceleration of each particle is the same since the string is inelastic.

Consider each particle separately, using N2 (resolved horizontally and vertically):

Particle A

→) $\qquad 30 - T - F_1 = 2a$

↑) $\qquad R_1 - 20 = 0 \quad \Rightarrow R_1 = 20$ N

458 *Connected particles*

Particle B

$\rightarrow)$ $\quad T - F_2 = 4a$

$\uparrow)$ $\quad R_2 - 40 = 0 \quad \Rightarrow R_2 = 40$ N

Since the particles are moving,

$$F = \mu R_1 = \tfrac{1}{4} \cdot 20 = 5 \text{ N}$$
$$F_2 = \mu R_2 = \tfrac{1}{4} \cdot 40 = 10 \text{ N}$$

So

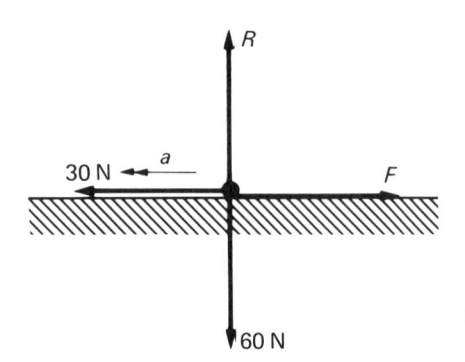

$$\left.\begin{matrix} 25 - T = 2a \\ T - 10 = 4a \end{matrix}\right\} \Rightarrow a = \tfrac{5}{2} \text{ m s}^{-2} \quad \text{and} \quad T = 20 \text{ N}$$

Note: Consider the system of the two particles and the string. Since the string stays taut, we can think of the system as being a rigid body; in fact, since it is moving linearly we can think of it as a particle, as shown in fig. 8.3:**8**. (Since the tension in the string is an *internal* force of the system, it does not appear in this diagram.)

$\rightarrow)$ $\quad 30 - F = 6a$

$\uparrow)$ $\quad R - 60 = 0$

$\Rightarrow$ $\quad R = 60$ N

Fig. 8.3:**8** Since the system is moving $F = \mu R = \tfrac{1}{4} \cdot 60 = 15$ N.

So $\quad 6a = 30 - 15 = 15 \quad \Rightarrow \quad a = \tfrac{5}{2} \text{ m s}^{-2}$

(This is an easier way of finding the acceleration of the particles; but notice that to find the tension in the string we now need to use N2 for one or other of the particles.)

Worked example 7 Two particles A and B of masses 6 kg and 4 kg are connected by a light inextensible string which passes over a smooth fixed pulley (fig. 8.3:**9**). The particles are released from rest with the hanging parts of the string vertical. Find the acceleration of the particles and the tension in the string.

Since the string is inelastic, the magnitude of the acceleration of each particle is the same (though their accelerations are in opposite directions).
Consider each particle separately and use N2:

$A\downarrow)$ $\quad 60 - T = 6a$

$B\uparrow)$ $\quad T - 40 = 4a$

$\Rightarrow$ $\quad a = 2 \text{ m s}^{-2} \quad \text{and} \quad T = 48$ N

Forces on *A*

Forces on *B*

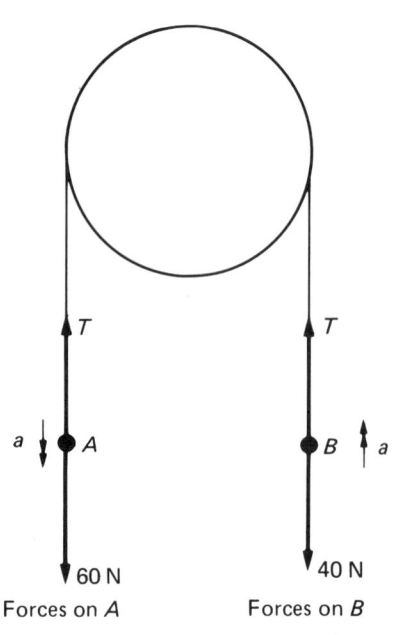

Fig. 8.3:**9**

Note: To use N2 for the *system* is dangerous here since different parts of the system have different accelerations (of the same magnitude but in different directions).

Forces on a particle 459

Worked example 8 A particle A of mass 10 kg rests on a rough plane of inclination $\tan^{-1}\frac{3}{4}$. It is connected by a light inextensively string passing over a smooth pulley at the top of the plane to a particle B of mass 2 kg which hangs freely. The coefficient of friction between A and the plane is $\frac{1}{4}$. Find the acceleration of the particles and the tension in the string.

Considering the relative masses of the particles, we can be sure that A moves down the plane (so the frictional reaction of the plane on A acts up the plane as shown in fig. 8.3:**10**).

$\tan \alpha = \frac{3}{4} \Rightarrow \sin \alpha = \frac{3}{5}, \cos \alpha = \frac{4}{5}.$

Resolving the forces on A

$A\nwarrow)$ $\qquad\qquad\qquad\qquad\qquad R - 100 \cos \alpha = 0 \qquad \Rightarrow \quad R = 80$ N

$A\swarrow)$ $\qquad\qquad\qquad\qquad\qquad 100 \sin \alpha - T - F = 10a$

Resolving the forces on B

$B\uparrow)$ $\qquad\qquad\qquad\qquad\qquad\qquad\qquad\qquad T - 20 = 2a$

Since A is moving, $\qquad F = \mu R = \frac{1}{4} \cdot 80 = 20$ N

So $\qquad \left.\begin{array}{r} 40 - T = 10a \\ T - 20 = 2a \end{array}\right\} \Rightarrow a = \frac{5}{3}$ m s^{-2} and $T = \frac{70}{3}$ N

Fig. 8.3:**10**

Worked example 9 Two particles A and B of masses 5 kg and 3 kg are connected by a light inextensible string passing over a smooth fixed pulley. The system is released from rest with the hanging parts of the string vertical, and A 0.4 m above B. As the particles are passing each other, the string breaks. Find the greatest height reached by B above its initial position.

There are two parts to the motion (fig. 8.3:**11**):

(i) before the string breaks: particles connected
(ii) after the string breaks: particles moving freely under gravity

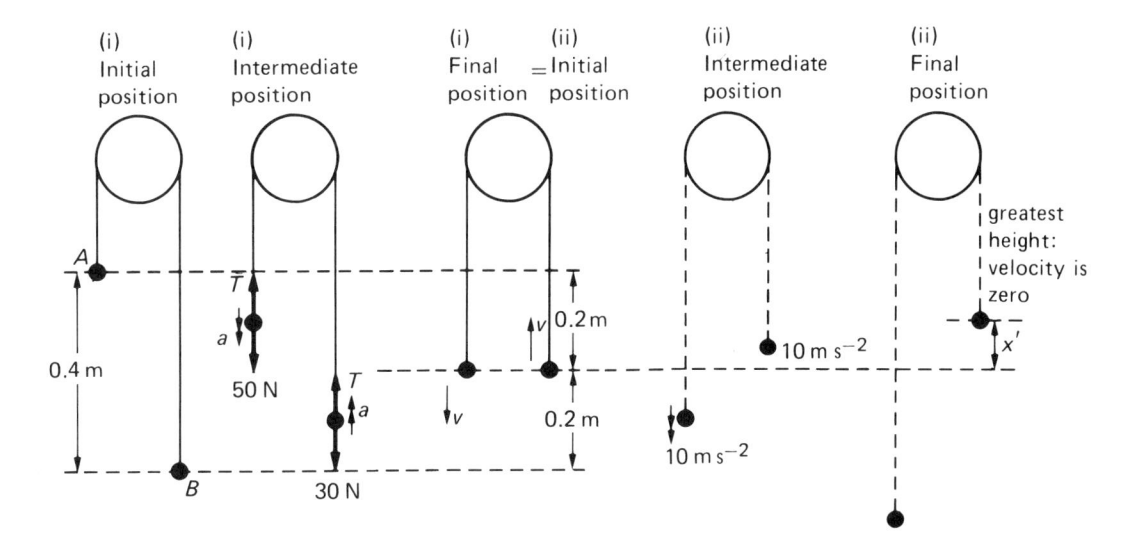

Fig. 8.3:**11**

(i) Before the string breaks:
$A\downarrow)$ $\quad 50 - T = 5a$
$B\uparrow)$ $\quad T - 30 = 3a$ $\Rightarrow a = 2.5 \text{ m s}^{-2}$

Let the velocity of the particles when they are passing one another be v. At this instant they have each travelled 0.2 m, i.e. $u = 0 \text{ m s}^{-1}$, $a = 2.5 \text{ m s}^{-2}$, $x = 0.2 \text{ m}$.

$$v^2 = u^2 + 2ax \Rightarrow 2 \cdot (2.5) \cdot (0.2) = 1 \Rightarrow v = 1 \text{ m s}^{-1}$$

(ii) After the string breaks, both particles have an acceleration due to gravity alone (i.e. 10 m s^{-2} downwards). In particular B continues at first to move up, but is decelerates, stops and then accelerates downwards. Suppose B moves a distance x' from when the string breaks to when it reaches its greatest height.
$u = 1 \text{ m s}^{-1}$, $a = -10 \text{ m s}^{-2}$, $v = 0 \text{ m s}^{-1}$.

$$v^2 = u^2 + 2ax \Rightarrow 0 = 1 + 2(-10)x'$$

i.e. $\quad x' = 0.05 \text{ m}$

So height of B above its initial position is $0.2 \text{ m} + 0.05 \text{ m} = 0.25 \text{ m}$.

Exercise 8.3:4

Questions 1 and 2: Two particles A and B are masses 3 kg and 5 kg are connected to the ends of a light inextensible string. The particles are placed in line on a table with the string taut. Forces are applied as shown (in the line AB). Find the acceleration of the particles and the tension in the string.

⑥ **1** (i) The table is smooth.
   (ii) The table is rough: $\mu = \frac{1}{5}$.

   24 N ← A — B (on hatched table)

⑥ → **2** The table is rough: $\mu = \frac{2}{5}$.

   60 N ← A — B → 24 N   (on hatched table)

⑥ **3** Two particles A and B of masses m and $2m$ are connected to the ends of a light inextensible string. The particles are placed in line on a table with the string taut. A force of $5mg$ is applied to A in the direction BA. Find the acceleration of the particles and the tension in the string given that:

   (i) the table is smooth
   (ii) the table is rough and the coefficient of friction between each particle and the table is $\frac{1}{4}$

⑥ **4** A car of mass 500 kg tows a trailer of mass 400 kg along a horizontal road. The resistance to motion of the car is 600 N, that of the trailer is 300 N. The driving force of the car is 1800 N. Find the acceleration of the car and the trailer and the tension in the towbar.

Questions 5 and 6: Two particles A and B of masses m and $2m$ are connected to the ends of a light inextensible string. The particles are placed on the line of greatest slope of an inclined plane, with the string taut. Forces are applied as shown. Find the acceleration of the particles and the tension in the string.

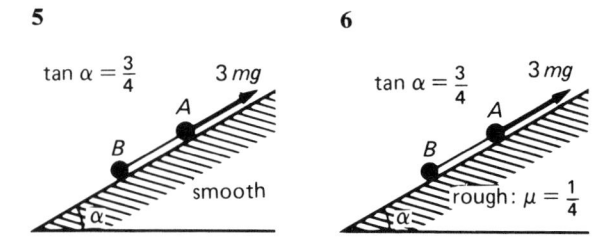

5 $\tan \alpha = \frac{3}{4}$, force $3mg$ up the slope on A; B below A; smooth.

6 $\tan \alpha = \frac{3}{4}$, force $3mg$ up the slope on A; B below A; rough: $\mu = \frac{1}{4}$.

→ **7** A car of mass 600 kg tows a trailer of mass 300 kg. The resistance to motion of the car is 800 N, that of the trailer is 500 N. Find the acceleration of the car and the trailer, and the tension in the towbar, if the driving force of the car is 2250 N and the car and trailer are moving:

   (i) along a horizontal road
   (ii) up a slope of 1 in 10
   (iii) down a slope of 1 in 10

Questions 8 and 9: Three particles *A*, *B* and *C* rest on a table, connected by light inextensible strings *AB* and *BC*. The particles are placed in line with the strings taut. Forces are applied as shown. By considering the forces on *B* decide whether or not the tensions in the strings are the same. Find the acceleration of the particles and the tensions in the strings.

- **8** (i) The table is smooth.
  (ii) The table is rough: $\mu = \tfrac{1}{4}$.

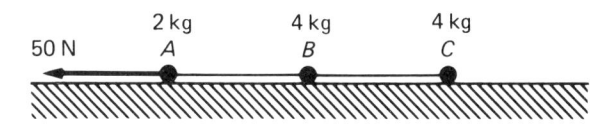

→ **9** The table is rough: $\mu = \tfrac{1}{2}$.

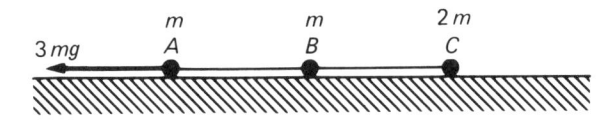

Questions 10–15: Two particles are connected by a light inextensible string passing over a smooth fixed pulley. Find the acceleration of the particles, the tension in the string, and the force exerted by the string on the pulley.

⑦ **10** 7 kg, 3 kg

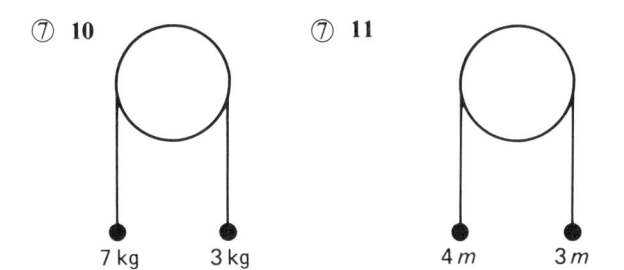

⑦ **11** 4*m*, 3*m*

⑦ **12** *m*, *M*; *M* > *m*

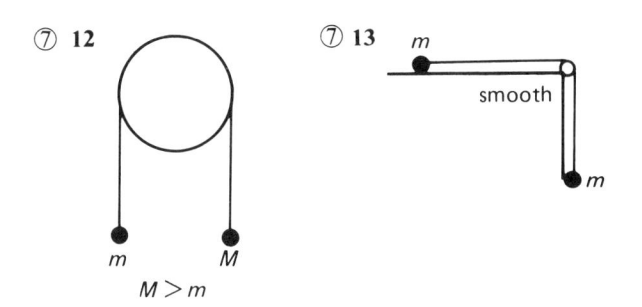

⑦ **13** *m*, *m*; smooth

⑦ **14** *m*, *m*; rough: $\mu = \tfrac{1}{2}$

⑦ **15** 2*m*, 3*m*; rough: $\mu = \tfrac{1}{4}$

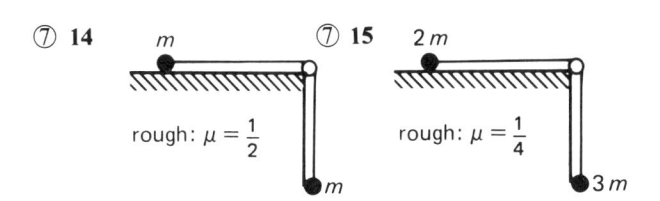

Questions 16–19: Two particles are connected by a light inextensible string passing over a smooth fixed pulley.

⑦ **16** The acceleration of the particles is $\tfrac{1}{2}g$. Find the two possible values of *k*.

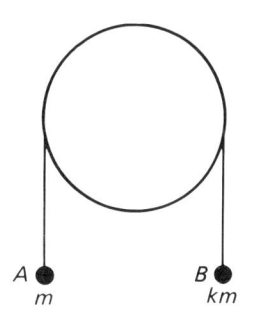

⑦ **17** The acceleration of the particles is $\tfrac{1}{5}g$. Find *k*.

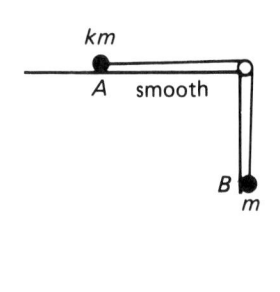

⑦ R **18** (i) The system is in equilibrium. Find the coefficient of friction between *A* and the plane.

(ii) An extra mass 2*m* is added to *B*. Find the acceleration of the particles.

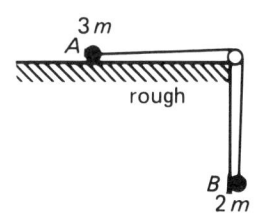

⑦ **19** (i) If *M* = *m*, show that the system is in equilibrium.

(ii) If *M* = 2*m*, find the acceleration of the particles.

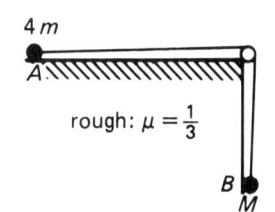

Questions 20 and 21: Particles *A*, *B* and *C* are connected by light inextensible strings *AB* and *BC* passing over smooth fixed pulleys. Find the acceleration of the particles, the tensions in the strings and the forces exerted by the strings on the pulleys.

⑦ **20** (i) The plane is smooth.
(ii) The coefficient of friction between *B* and the plane $\tfrac{1}{5}$.

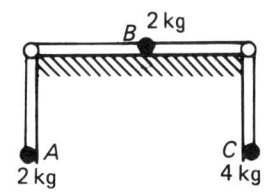

→ ⑦ **21** (i) The plane is smooth.
(ii) The coefficient of friction between *B* and the plane is $\tfrac{1}{6}$.

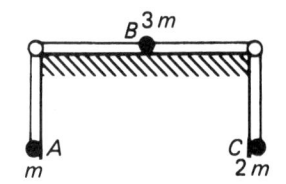

462 Connected particles

Questions 22–31: Two particles are connected by a light inextensible string passing over a smooth fixed pulley. In each case find the quantities indicated, the tension in the string and the force exerted by the string on the pulley.

⑧ **22** Find the acceleration of the particles.

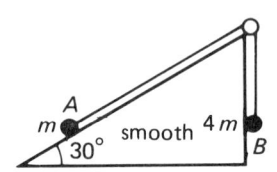

⑧ **23** Find the acceleration of the particles.

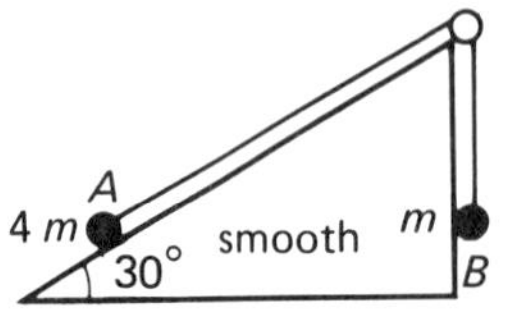

⑧ → **24** (i) The system is in equilibrium. Find θ.

(ii) An extra mass $2m$ is added to B. Find the acceleration of the particles.

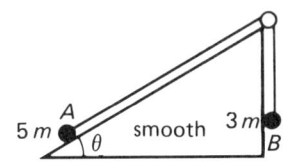

⑧ **25** (i) The system is in equilibrium. Find θ.

(ii) An extra mass m is added to B. Find the acceleration of the particles.

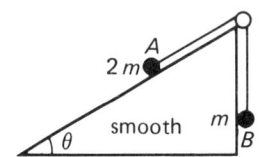

⑧ **26** Find the acceleration of the particles when:

(i) $k = 2$
(ii) $k = 12$

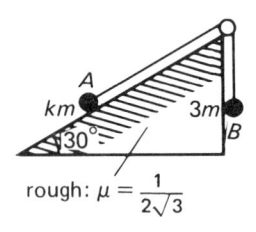

⑧ **27** (i) Find the range of values of k for which the system is in equilibrium ($\tan \alpha = \frac{3}{4}$).

(ii) Find the acceleration of the particles when $k = 7$.

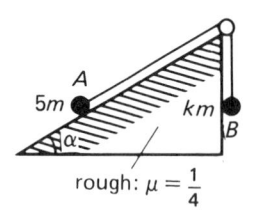

⑧ **28** Find the acceleration of the particles.

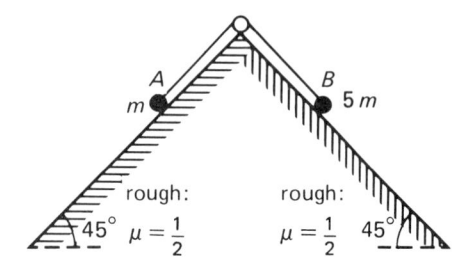

⑧ **29** (i) The system is in equilibrium. Find k.

(ii) Find the acceleration of the particles when $k = 1$.

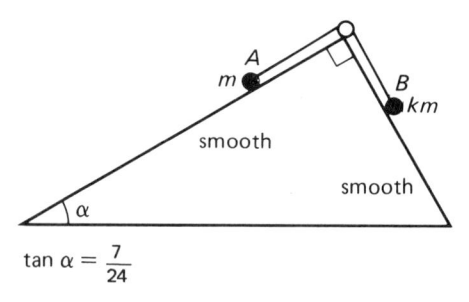

⑧ **R 30** (i) The system is in equilibrium. Find k.

(ii) Find the acceleration of the particles when $k = 2$.

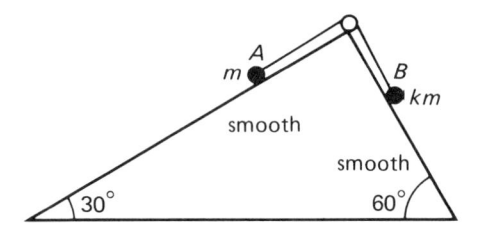

⑧ **31** The coefficient of friction between each particle and the plane is μ.

(i) Show that A moves up the plane if $\mu < \frac{1}{2}$.
(ii) Find the acceleration of the particles if $\mu = \frac{1}{5}$.

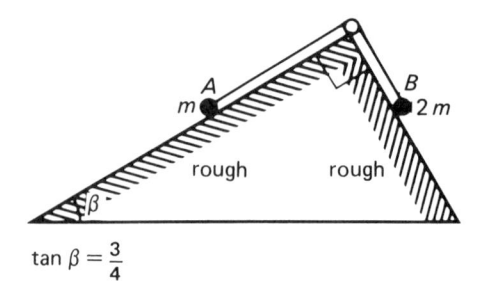

Questions 32–35: A wedge is fixed to a horizontal plane. Two particles are connected by a light inextensible string passing over a smooth pulley fixed to the upper edge of the wedge. Find the acceleration of the particles, the tension in the string and the reactions of the particles on the sloping faces of the wedge.

Draw a diagram showing the forces acting on the wedge (and the pulley—don't forget the force of the string on the pulley). Hence find the magnitude and direction of the force of the plane on the wedge.

32

33

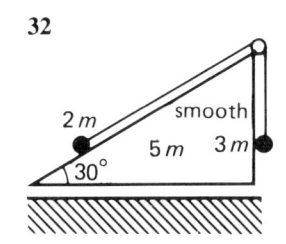

→ **34** (i) The sloping faces of the wedge are smooth.

(ii) The coefficient of friction at both contacts is $\frac{1}{2}$ (tan $\alpha = \frac{3}{4}$).

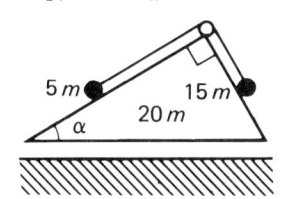

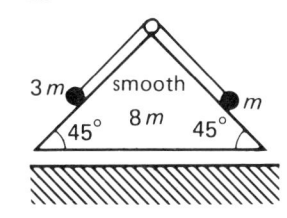

35 (i) The sloping faces of the wedge are smooth.

(ii) The coefficient of friction at both contacts is $\frac{1}{4}$.

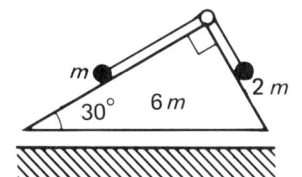

• **36** Two scale pans A and B of masses m and $3m$ are connected by a light inextensible string passing over a smooth fixed pulley. Find their acceleration.

A particle of mass $4m$ is placed in pan A. Find the new acceleration of the scale pans and the reaction between the pan A and the particle.

37 A scale pan A of mass $4m$, containing a particle P of mass $4m$, is connected by a light inextensible string passing over a smooth fixed pulley to a scale pan B of mass $4m$, containing a particle Q of mass $3m$. Find the acceleration of the scale pans, the tension in the string and the reactions between A and P and between B and Q.

→ **38** A particle P of mass $5m$ is connected by a light inextensible string passing over a smooth fixed pulley to a light scale pan containing a particle Q of mass $3m$. Find the acceleration of P and the reaction between Q and the scale pan.

Questions 39–48: Two particles A and B are connected by a light inextensible string passing over a smooth fixed pulley. The system starts from rest.

⑨ **39** Find the speed of A on reaching the pulley.

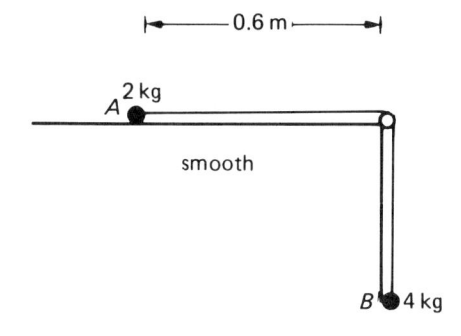

⑨ **40** Find the speed of B on reaching the pulley.

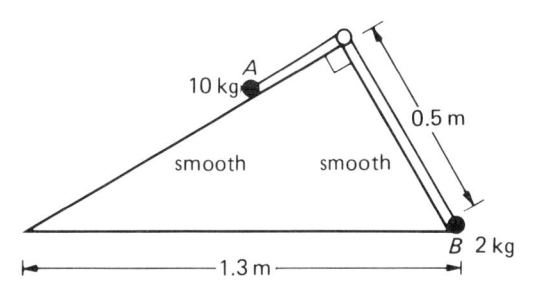

⑨ **41** Find the velocity of A (and B) when B hits the plane. Hence find the greatest height above the plane reached by A.

⑨ **42** Find the velocity of A (and B) when B hits the plane. Hence find the total distance moved by A before it first comes to rest.

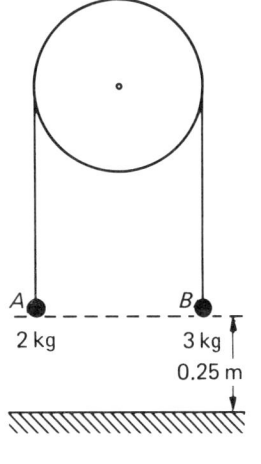

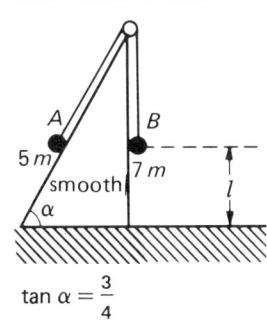

$\tan \alpha = \frac{3}{4}$

⑨ **43** Find the velocity of A (and B) when B hits the plane. Hence find the total distance moved by A before the string next becomes taut, and the total time for which the system has then been moving.

⑨ **44** B hits the ground after 2 s. Find the total time for which the system has been moving when the string next becomes taut.

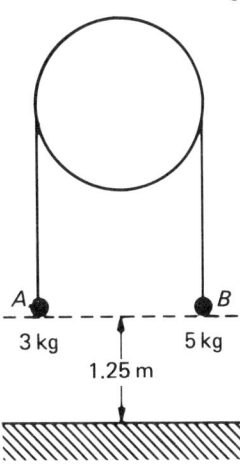

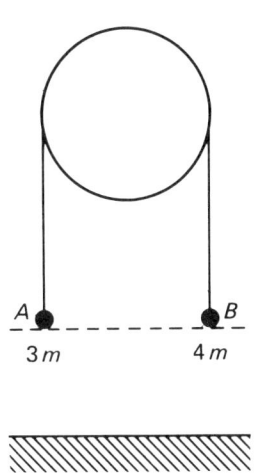

⑨ **45** *B* hits the plane after 3 s. Find the greater height above the plane reached by *A*.

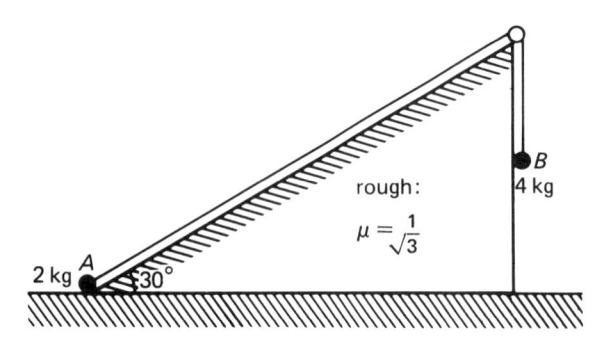

⑨ **46** As the particles are passing each other, the string breaks. Find the greatest height reached by *B* above its initial position.

⑨ →**47** As the particles are passing each other, the string breaks. Find the height of *B* above *A* when *B* reaches its highest point.

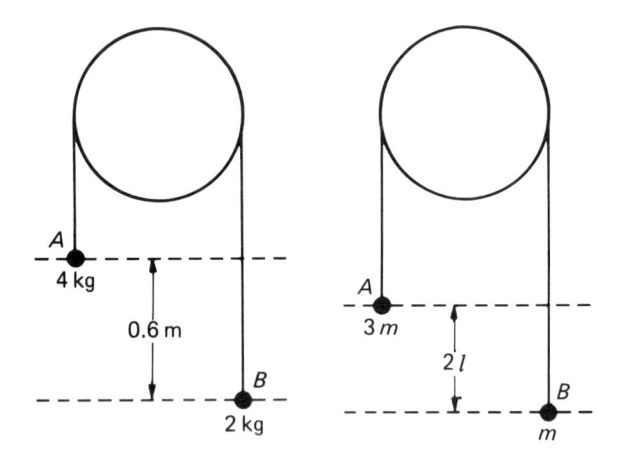

⑨ **48** As the particles are passing each other, the string breaks. *A* hits the plane when moving at 4 m s^{-1}. Find:

(i) the greatest height above the plane reached by *B*
(ii) the height of *B* above the plane when *A* hits the plane

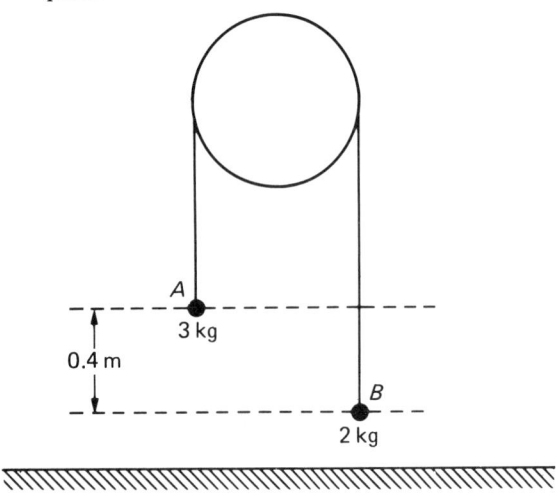

⑨ **49** Two particles *A* and *B* of masses 5 kg and 3 kg are connected by a light inextensible string passing over a smooth fixed pulley. The system is released from rest with the hanging parts of the string vertical and *A* 0.1 m vertically above *B*. As the particles are passing each other, the string breaks. *A* hits the plane after a further 1 s. Find the height of *B* above the plane and its velocity at this instant.

⑨ **50** Two particles *A* and *B* of masses *m* and 3*m* are connected by a light inextensible string of length *l*. *A* is placed on a horizontal table at a perpendicular distance *l* from its edge. The coefficient of friction between *A* and the table is $\frac{1}{3}$. *B*, which is at the edge of the table, is pushed gently over the edge. Find:

(i) the speed of *A* when it reaches the table edge
(ii) the time taken by *A* to reach the edge

⑨ **R 51** Two particles *A* and *B* of masses 3 kg and 1 kg are connected by a light inextensible string passing over a smooth fixed pulley. The system is released from rest with the hanging parts of the string vertical and *A* and *B* at the same level. The string breaks after $\frac{1}{2}$ s and the string is sufficiently long so that *B* does not hit the pulley. Show that the vertical distance between *A* and *B* when *B* reaches its highest point is 2.5 m.

52 A bead of mas *m* can slide on a fixed, smooth, straight, vertical wire. It is attached to one end of a light inextensible string which passes over a smooth fixed pulley. The other end of the string is attached to a particle of mass 2*m*. Find the acceleration of:

(i) the particle   (ii) the bead

in terms of θ.

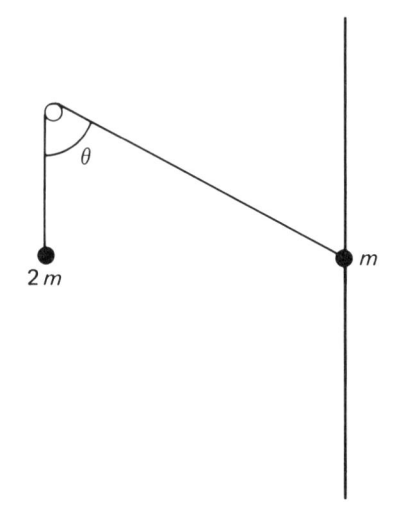

8.3:5 Circular motion

In section 7.4:2 we found that, if a particle moves in a circle of radius r with speed v and angular velocity ω, then

$$v = r\omega$$

and that if v is constant, its acceleration is v^2/r (or $r\omega^2$) towards the centre of the circle.

Worked example 10 A particle P of mass m is attached by a light inelastic string of length l to a fixed point on a smooth horizontal table. If the string breaks when the tension in it exceeds $4mg$, find the greatest possible angular velocity of the particle.

Let the angular velocity be ω. The particle moves under the action of the forces shown in fig. 8.3:**12**. Resolving the forces into components (N2):

$$\leftarrow) \qquad T = ml\omega^2 \qquad (3)$$

$$\uparrow) \qquad R - mg = 0 \qquad (4)$$

From equation (3), when $T = 4mg$,

$$\omega^2 = 4g/l \Rightarrow \omega = 2\sqrt{(g/l)}$$

So the greatest possible angular velocity is $2\sqrt{(g/l)}$.

Worked example 11 A light inelastic string is fixed at end A. At the other end is attached a particle P which rotates in a horizontal circle with angular velocity 2 rad s^{-1}. Find the vertical depth of P below A.

The particle moves under the action of the forces shown in fig. 8.3:**13a**.

$$\leftarrow) \qquad T \sin \theta = mr\omega^2 = 4mr \qquad (5)$$

$$\uparrow) \qquad T \cos \theta - mg = 0 \qquad (6)$$

Note that $r = l \sin \theta$ (fig. 8.3:**13b**); so from equation (5):

$$T \sin \theta = 4ml \sin \theta \Rightarrow T = 4ml \qquad (7)$$

Note also that $h = l \cos \theta$ (fig. 8.3:**13b**); so from equation (6):

$$Th/l - mg = 0 \Rightarrow h = mgl/T$$

So, using equation (7),

$$h = \frac{mgl}{4ml} = \frac{g}{4} = 2.5 \text{ m}$$

Worked example 12 A light inelastic string AC of length $2l$ has a particle of mass m attached at its midpoint B. The ends A and C are fixed to two points in a vertical line such that A is a distance l above C. The particle describes a horizontal circle with constant angular velocity ω. Find the least value of ω so that be strings shall be taut.

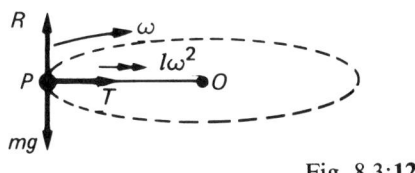

Fig. 8.3:12

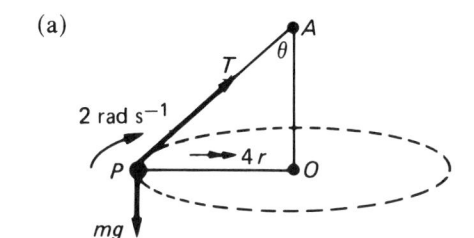

Fig. 8.3:13

Circular motion

Fig. 8.3:**14**

$AB = AC = BC = l$; so $O\hat{A}B = O\hat{C}B = 60°$ and $r = l \sin 60°$ (fig. 8.3:**14**).

←) $\quad (T_1 + T_2) \sin 60° = mr\omega^2 = ml\omega^2 \sin 60°$

⇒ $\quad T_1 + T_2 = ml\omega^2$ (8)

↑) $\quad T_1 \cos 60° - T_2 \cos 60° - mg = 0$

⇒ $\quad T_1 - T_2 = 2mg$ (9)

Clearly AB will remain taut for any value of ω; but BC will become slack if ω is small. In order that BC shall be taut, $T_2 \geq 0$.

Equation (8)-equation (9):

$$2T_2 = ml\omega^2 - 2mg$$

So, when $T_2 \geq 0$, $\quad ml\omega^2 \geq 2mg$

i.e. $\quad \omega^2 \geq 2g/l$

$\quad \omega \geq \sqrt{(2g/l)}$

So the least possible value of ω for both strings to be taut is $\sqrt{(2g/l)}$.

Worked example 13 A rough horizontal plate rotates with constant angular velocity 2 rad s^{-1} about a fixed vertical axis. (We say that a rigid body rotates with angular velocity ω about a fixed axis if the angular velocity of each point of the body about the axis is ω.) A particle of mass 0.5 kg lies on the plate at a distance 1.5 m from this axis. The particle remains at rest relative to the plate. Show that the coefficient of friction between the particle and the plate must be not less than 0.6.

Since the particle remains at rest relative to the plate, the particle rotates in a horizontal circle of radius 1.5 m with angular velocity 2 rad s^{-1}. Its acceleration towards the centre of the circle is $1.5 \times 2^2 = 6$ m s^{-2}. The force towards the centre (see fig. 8.3:**15**) is the frictional component of the reaction between the plate and the particle. Note that there is no frictional component in the direction of the tangent to the circle. (Why not?)

←) $\quad F = 0.5 \cdot 6 = 3$ N

↑) $\quad R - 5 = 0 \Rightarrow R = 5$ N

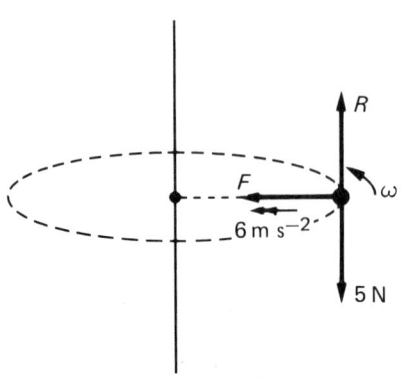

Now, although the particle is moving, it is at rest *relative to* the plate; so

Fig. 8.3:**15** $\quad F \leq \mu R$ (see section 8.1:3). So $3 \leq 5\mu$ i.e. $\mu \geq 0.6$.

Exercise 8.3:5

1 Find the magnitude of the acceleration of a particle moving in a circle:

(i) of radius 2 m with constant angular velocity 3 rad s^{-1}
(ii) of radius 0.3 m with constant speed 4 m s^{-1}
(iii) of radius 0.5 m with constant angular velocity 20 rev min^{-1}

If the mass of the particle is 5 kg, find the magnitude of the resultant force acting on it in each case. What is the direction of the resultant force?

2 A satellite of mass m moves in a circular orbit of radius r, centre O (the centre of the Earth), under the action of a force of magnitude $\mu m / r^2$ (μ constant).

(i) What is the direction of the force?
(ii) Find the angular velocity of the satellite about O and hence show that the period T of its orbit is given by

$$\mu T^2 = 4\pi^2 r^3$$

(iii) Given that $\mu = 4 \times 10^{14}$ N m^2 kg^{-1}, find T if the radius of the orbit is:
(a) 6.4×10^6 m   (b) 8×10^6 m.

Questions 3-6: A particle of mass m is attached to one end of a light inextensible string of length l. The other end is fixed to a point O on a smooth horizontal plane. The particle moves in a circle centre O with constant angular velocity.

3 $m = 2$ kg, $l = 0.5$ m. If the speed of the particle is 5 m s^{-1}, find the tension in the string and the reaction with the plane.

4 If the speed of the particle is $\sqrt{(5gl)}$, find the tension in the string (in terms of m and g).

5 $m = 5$ kg, $l = 2$ m. If the string breaks when the tension in it exceeds 90 N, find the greatest angular velocity at which the particle can travel.

6 If the string breaks when the tension in it exceeds $3mg$, find the greatest angular velocity at which the particle can travel.

Questions 7-10: A particle of mass m is attached by a light elastic string of natural length l and modulus of elasticity λ to a fixed point O on a smooth horizontal plane. The particle moves in a circle centre O with constant angular velocity.

7 $m = 4$ kg, $l = 1$ m, $\lambda = 100$ N. If the angular velocity of the particle is 2 rad s^{-1}, find the extension of the string.

8 $m = 0.5$ kg, $l = 0.6$ m, $\lambda = 30$ N. If the angular velocity of the particle is 5 rad s^{-1}, find the extension of the string.

9 $\lambda = 2mg$. If the speed of the particle is $\sqrt{(5gl/8)}$, find the extension of the string.

R 10 $\lambda = 6mg$. If the extension of the string is $\frac{1}{2}l$, find the speed of the particle.

11 A particle of mass m is attached to the midpoint B of an inextensible string ABC of length $2l$ and another particle of mass $2m$ is attached to the end C. The particles revolve with constant angular velocity $\sqrt{(3g/l)}$ round the end A, which is fixed, on a smooth horizontal table, so that ABC is always a straight line. Find the tensions in AB and BC.

12 A particle of mass $4m$ is attached to the midpoint B of an inextensible string ABC; another particle of mass $3m$ is attached to the end C. The particles revolve with constant angular velocity round the end A, which is fixed, on a smooth horizontal table, so that ABC is always a straight line. Show that the tensions in AB and BC are in the ratio $5:3$.

Questions 13-20: One end of a light inextensible string of length l is attached to a fixed point A, the other end is attached to a particle of mass m which moves in a horizontal circle centre O (so that O is vertically below A) with constant angular velocity.

13 $l = 5$ m, $m = 3$ kg. The radius of the circle is 4 m. Find the angular velocity of the particle and the tension in the string.

14 $l = 1$ m, $m = 4$ kg. O is 0.5 m vertically below A. Find the angular velocity of the particle and the tension in the string.

15 $l = 0.4$ m, $m = 2$ kg. The speed of the particle is 2 m s^{-1}. Find the angle between the string and the vertical, and the tension in the string.

16 The angular velocity of the particle is $\sqrt{(2g/l)}$. Find:

(i) the angle between the string and the vertical
(ii) the speed of the particle (in terms of g and l)

17 The angular velocity of the particle is ω. Find the depth h of O below A. Notice that h is independent of l. Why cannot the string be horizontal?

18 The angle between the string and the vertical is θ. Show that the speed of the particle is $\sqrt{(gl \sin\theta \tan\theta)}$.

468 Circular motion

⑪ **19** If the tension in the string is not to exceed $2mg$, what is the maximum possible speed of the particle? At this speed, what is the depth of O below A?

⑪ R **20** The angle between the string and the vertical is $\tan^{-1}\tfrac{5}{12}$. Find:
 (i) the tension in the string
 (ii) the speed of the particle

21 A particle of mass m is attached to one end of a light inelastic string, the other end of which is fixed at a height h above a smooth horizontal table. The particle moves in a circle on the table with the string taut and with constant angular velocity ω about the centre of the circle. Find the reaction between the particle and the table when:
 (i) $\omega = \sqrt{(g/4h)}$
 (ii) $\omega = \sqrt{(g/2h)}$
 (iii) $\omega = \sqrt{(g/h)}$
What if $\omega > \sqrt{(g/h)}$?

Questions 22–25: A particle of mass m is attached to a fixed point by a light elastic string of natural length l and modulus of elasticity λ. The particle moves in a horizontal circle with constant angular velocity.

• **22** $m = 4$ kg, $l = 2$ m, $\lambda = 60$ N. The angular velocity of the particle is 2 rad s^{-1}. Find the extension of the string and the angle it makes with the vertical.

→ **23** $\lambda = 2mg$. The angular velocity of the particle is $\sqrt{(3g/4l)}$. Find the extension of the string and the angle it makes with the vertical.

24 $\lambda = 2mg$. The angle between the string and the vertical is $60°$. Find the extension of the string and the angular velocity of the particle.

R **25** $\lambda = 3mg$. The angle between the string and the vertical is $\tan^{-1}\tfrac{4}{3}$. Find the extension of the string and the angular velocity of the particle.

26 Particles P and Q of masses m and $2m$ respectively are connected by a light inextensible string of length $2l$ which passes through a smooth fixed ring. P moves in a horizontal circle and Q is at rest at a distance $\tfrac{1}{2}l$ below the ring. Find the angular velocity of P and the radius of the circle in which it moves.

27 Particles P and Q of masses m and km respectively are connected by a light inextensible string of length $3l$ which passes through a smooth fixed ring. P moves in a horizontal circle, centre Q, and Q is at rest at a distance l below the ring. Show that $k = 2$ and find the speed of P.

Questions 28–32: A particle, moving on the smooth inside surface of a fixed spherical bowl of radius l, centre A, describes a horizontal circle centre O (so that O is vertically below A) with constant angular velocity.

• **28** $l = 2$ m. O is at a depth $\tfrac{8}{5}$ m below A. Show that the speed of the particle is 3 m s^{-1}.

29 $l = 13a$, O is at a depth $12a$ below A. Show that the speed of the particle is $\tfrac{5}{2}\sqrt{(\tfrac{1}{3}ga)}$, and find the force exerted by the bowl on the particle.

R **30** $l = 0.6$ m. The speed of the particle is 4 m s^{-1}. Find the depth of O below A.

→ **31** The speed of the particle is $4\sqrt{(\tfrac{1}{15}gl)}$. Find the depth of O below A.

32 The angular velocity of the particle is ω. Find the depth of O below A. Explain why the particle cannot describe a circle level with A.

Questions 33–37: A particle of mass m is attached to a point P on a light inextensible string AB. The ends A and B are attached to two points at a distance h apart in a vertical line, A being above B. The particle moves in a horizontal circle with constant angular velocity ω.

⑫ **33** $AP = PB = l$, $h = l\sqrt{2}$.
 (i) Find the least value of ω for both parts of the string to be taut.
 (ii) If $\omega^2 = 5\sqrt{2}g/l$, find the tensions in AP and BP.

⑫ → **34** $AP = 5a$, $PB = 4a$, $h = 3a$.
 (i) Show that, if neither part of the string is slack, then $\omega \geq \sqrt{(g/3a)}$.
 (ii) If $\omega = 2\sqrt{(g/3a)}$, find the tensions in AP and BP.

⑫ **35** $AP = l\sqrt{3}$, $PB = l$, $h = 2l$.
 (i) Find the least value of ω for both parts of the string to be taut.
 (ii) If the tension in AP is $2\sqrt{3}mg$, find the tension in PB and the angular velocity of the particle.

⑫ **36** $AP = l\sqrt{3}$, $PB = l$; $h = l$ (so P is below the level of B).
 (i) Find the least value of ω for both parts of the string to be taut.
 (ii) Find the value of ω for which the tensions in AP and PB are equal.

⑫ **37** $AP = 4a$, $PB = 3a$, $h = 5a$.
 (i) Show that, if the tensions in AP and PB are T_1 and T_2, then
$$3T_1 + 4T_2 = 12ma\omega^2 \qquad (10)$$
$$4T_1 - 3T_2 = 5mg \qquad (11)$$
 (ii) If neither part of the string is slack, show that $\omega^2 \geq \tfrac{5}{16}ga$.

(iii) If either part of the string will break when subjected to a tension exceeding $6mg$, show that $\omega^2 \leq 55g/16a$. (Hint: it is not obvious which part of the string will break first as ω increases. In order to find out, put $T_1 = 6mg$ in (11) and find T_2.)

(iv) If either part of the string will break when subjected to a tension exceeding $4mg$, show that $\omega^2 \leq 20g/9a$.

• **38** A smooth ring P of mass m is threaded onto a light inelastic string of length $8l$ whose ends are fixed to two points A and B distant $4l$ apart in a vertical line. The ring describes horizontal circles with centre B.

(i) Show that $AP = 5l$ and $BP = 3l$.

(ii) Show that the tension in the string is $\frac{5}{4}mg$ and hence show that the angular velocity of the ring is $\sqrt{(2g/3l)}$.

(Note: since the ring is smooth, the tension in each part of the string is the same.)

→ **39** One end of a light inextensible string of length $3l$ is attached at a fixed point A and the other end is attached at a fixed point B, where A is at a distance $l\sqrt{3}$ vertically above B. A smooth ring of mass m is threaded on the string and rotates in a horizontal circle, centre B, with constant angular velocity. Show that the tension in the string is $2mg/\sqrt{3}$ and find the angular velocity of the ring.

40 One end of a light inextensible string of length $7l$ is attached at a fixed point A and the other end is attached at a fixed point B, where A is at a distance $5l$ vertically above B. A smooth ring P of mass m is threaded on the string and rotates in a horizontal circle about AB with constant angular velocity. If $AP = 4l$, find the tension in the string and the angular velocity of the ring.

R 41 A light inextensible string of length $3a$ has one end fixed at a point A and the other fixed at a point B which is vertically below A and at a distance $2a$ from it. A small ring P of mass m is threaded on the string.

(i) If R is fixed to the midpoint of the string and moves in a horizontal circle with speed $\sqrt{(6ga)}$, find the tensions in the parts AP and BP of the string.

(ii) If P is free to move on the string and moves in a horizontal circle centre B with the string taut, find the speed of P.

(13) **42** A rough horizontal plate is free to rotate about a fixed vertical axis through its centre O. A particle P is placed on the plate at a distance 0.6 m from O. The coefficient of friction between the particle and the plate is 0.3. Find the maximum possible angular velocity of the plate if the particle remains at rest relative to the plate.

(13) **43** A rough horizontal plate rotates with angular velocity $\sqrt{(g/2a)}$ about a fixed vertical axis through its centre O. A particle P is placed on the plate so that $OP = a$. If the particle is about to slip, find the coefficient of friction between the particle and the plate.

(13) → **44** A rough horizontal plate rotates with constant angular velocity ω about a fixed vertical axis. A particle lies on the plate at a distance r from this axis. If the coefficient of friction between the particle and the plate is μ, and the particle remains at rest relative to the plate, show that $\omega^2 \leq \mu g/r$.

(13) **45** A rough horizontal plate rotates with constant angular velocity $\sqrt{(g/3a)}$ about a fixed vertical axis. A particle P of mass m lies on the plate at a distance $\frac{3}{4}a$ from the axis. Find the least possible value of the coefficient of friction between the particle and the plate if the particle is to make horizontal circles about the axis without slipping.

A particle Q of mass $2m$ lies on the plate at a distance $2a$ from the axis. If the angular velocity is suddenly doubled, is it possible for Q to make horizontal circles without slipping? What about P?

(13) **R 46** A rough horizontal plate rotates with constant angular velocity ω about a fixed vertical axis. A particle of mass m connected to the axis by a horizontal light elastic string, of natural length a and modulus of elasticity $3mg$, remains at rest relative to the plate and at a distance $\frac{5}{4}a$ from the axis. Show that, if the coefficient of friction between the particle and the plate is $\frac{1}{3}$, then

$$\frac{g}{3a} \leq \omega^2 \leq \frac{13g}{15a}$$

(13) → **47** A car is turning a horizontal corner of radius 100 m. The maximum speed at which the car can turn the corner without skidding is 20 m s^{-1}. Show that the coefficient of friction between the tyres and the road surface is $\frac{2}{5}$.

(13) **R 48** A car of mass m is turning a corner of radius r. The coefficient of friction between the wheels and the horizontal road surface is μ. What is the maximum speed at which the car can turn the corner without skidding?

8.4 Work and energy; power

8.4:1 Work done by a force

Suppose that the point of application of a constant force **F** acting on an object moves in a straight line from A to B through a displacement **s** (so that $\overrightarrow{AB} = \mathbf{s}$, as shown in fig. 8.4:1).

Definition
The **work done** by **F**, $W_\mathbf{F}$ is given by
$$W_\mathbf{F} = \mathbf{F} \cdot \mathbf{s} = Fs \cos\theta \qquad (1)$$

Fig. 8.4:1

(Note: **F** is not necessarily the *only* force acting on the object. In fact, if **F** is inclined to **s** there *must* be other forces on the object—why?)

In particular,
(a) If **F** is in the same direction as **s** (fig. 8.4:2), $\theta = 0°$, so
$$W_\mathbf{F} = Fs \qquad (2)$$

Fig. 8.4:2

(b) If **F** is in the opposite direction to **s** (fig. 8.4:3), $\theta = 180°$, so
$$W_\mathbf{F} = -Fs \qquad (3)$$

Fig. 8.4:3

(c) If **F** is perpendicular to **s** (fig. 8.4:4), $\theta = 90°$, so
$$W_\mathbf{F} = 0 \qquad (4)$$

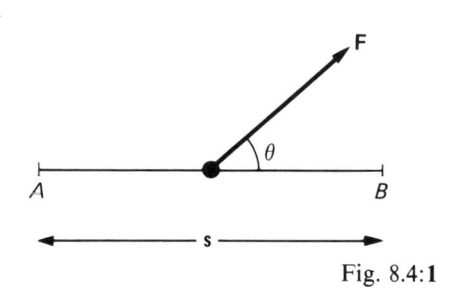

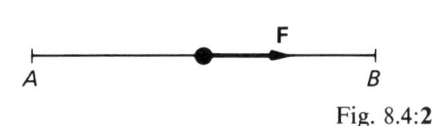

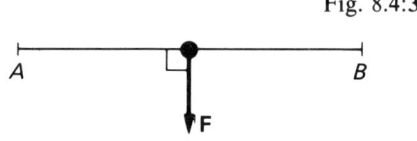

Fig. 8.4:4

Note: if the point of application of **F** does not move, $\mathbf{s} = \mathbf{0}$, so $W_\mathbf{F} = 0$.
Work is a scalar quantity. It is measured in units of newton metre, called **joule** (J).

Worked example 1 A bead of mass 2 kg, threaded onto a rough fixed wire inclined at 30° to the horizontal, moves 4 m along the wire under the action of the forces shown in fig. 8.4:5. Find the work done by each force.

Work done by 25 N force (pulling force) $= 25 \cdot 4 = 100$ J
Work done by 20 N force (weight) $= -20 \cdot 4 \cdot \cos 60° = -40$ J
Work done by 10 N force (frictional reaction) $= -10 \cdot 4 = -40$ J
Work done by $10\sqrt{3}$ N force (normal reaction) $= 0$ J

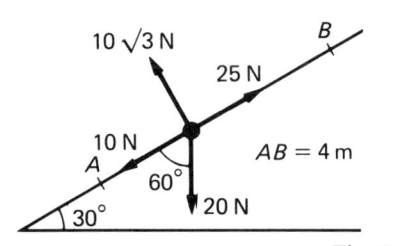

Fig. 8.4:5

Look at worked example 1 again.
The work done on the bead by the frictional reaction is -40 J. We say that the work done by the bead *against* the frictional reaction is $+40$ J.

Definition
If the work done by a force **F** acting on an object is W, then **the work done by the object against F** is $-W$.

This is a particularly useful idea when, as above, W is itself negative.

Theorem 1: The work done in stretching a light elastic string of natural length l_0 and modulus of elasticity λ to an extension x is $\frac{1}{2}\lambda x^2/l_0$ (5)

Forces on a particle 471

Proof
Imagine, for convenience, one end of the string attached to a fixed point X. The string is stretched gradually by a force applied at the other end P, equal in magnitude to the tension in the string (fig. 8.4:**6**).

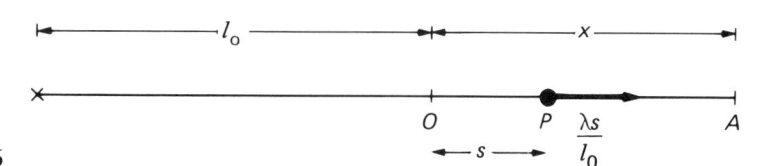

Fig. 8.4:**6**

By Hooke's Law, the tension in the string when its extension is s is $\lambda s/l_0$. So the force needed to stretch it is $\lambda s/l_0$.

Notice that this force is *not constant*; it increases as the point of application P moves from O to A. So it is not at first possible to use the definition of work done by a constant force. However,

$$\text{average force} = \frac{1}{2}\lambda x/l_0$$

$$\text{distance moved by force} = x$$

$\Rightarrow$ $\quad$ work done by force $= \frac{1}{2}(\lambda x/l_0)x = \lambda x^2/2l_0$

(As P moves from O to A, the work done by the tension is $-\lambda x^2/2l_0$.)

Note: the theorem also applies to a light spring compressed to a compression x.

Corollary: The work done in stretching the string from an extension x_1 to an extension x_2 is

$$\tfrac{1}{2}\lambda(x_2^2 - x_1^2)/l_0 \qquad (6)$$

Worked example 2 A light elastic spring has natural length 0.5 m and modulus of elasticity 40 N. Find the work done in:

(i) compressing the spring to a length 0.2 m;
(ii) stretching the spring to a length 0.8 m;
(iii) stretching the spring from a length 0.8 m to a length 1.1 m.

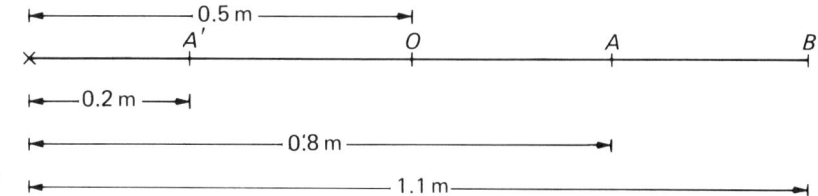

Fig. 8.4:**7**

The positions of the spring are shown in fig. 8.4:**7**.

(i) *O to A'*: from natural length to compression 0.3 m. By Theorem 1,

$$\text{work done} = \frac{\lambda}{2l_0} \cdot (0.3)^2 = \frac{40}{2(0.5)}(0.3)^2 = 3.6 \text{ J}$$

472 *Work done by a force*

(ii) *O to A*: from natural length to extension 0.3 m. By Theorem 1,

$$\text{work done} = 3.6 \text{ J}$$

(iii) *A to B*: from extension 0.3 m to extension 0.6 m. By expression (6),

$$\text{work done} = \frac{\lambda}{2l_0}\{(0.6)^2 - (0.3)^2\}$$

$$= \frac{40}{2 \cdot (0.5)}(0.27)$$

$$= 10.8 \text{ J}$$

General definition of work done by a force

We have defined the work done by a constant force whose point of application moves in a straight line. Consider, more generally, a force **F**, which need not be constant, whose point of application moves along a curve as shown in fig. 8.4:8.

As the point of application of **F** moves between two neighbouring points P and P' on the curve, where $PP' = \delta \mathbf{r}$, we can consider **F** to be approximately constant and PP' approximately straight, so, using our original definition,

$$\text{work done by } \mathbf{F} \text{ in moving from } P \text{ to } P' \approx \mathbf{F} \cdot \delta \mathbf{r}$$

So   work done by **F** in moving from A to B $= \lim_{\delta \mathbf{r} \to 0} \sum_{A}^{B} \mathbf{F} \cdot \delta \mathbf{r}$ (7)

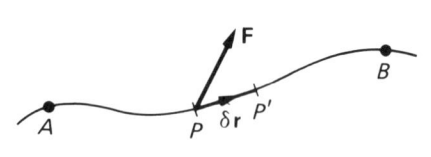

Fig. 8.4:**8**

This is, intentionally, only a rough explanation, and the formula we have obtained leads to the integral of a scalar product:

$$\int \mathbf{F} \cdot d\mathbf{r} \qquad (8)$$

(see section 6.1:3).

However, note two particular cases:

(i) if **F** is always perpendicular to the direction of motion, $\mathbf{F} \cdot \delta \mathbf{r} = 0$; so the work done by **F** is zero (9)

(ii) if **F** has variable magnitude F but constant direction and if its point of application moves in this direction (fig. 8.4:**9**), then work done by **F** in moving from A to B is

$$\lim_{\delta s \to 0} \sum_{A}^{B} F \, \delta s = \int_{s_1}^{s_2} F \, ds \qquad (10)$$

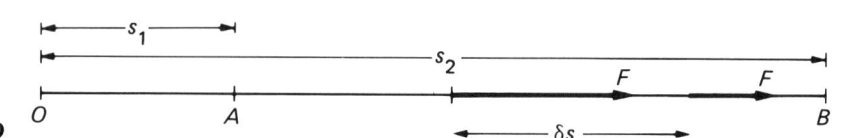

Fig. 8.4:**9**

Exercise 8.4:1

Questions 1–4: A particle moves in a straight line from A to B under the action of a system of forces. Find the work done by each force.

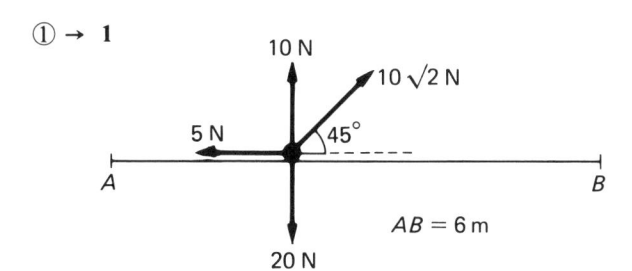

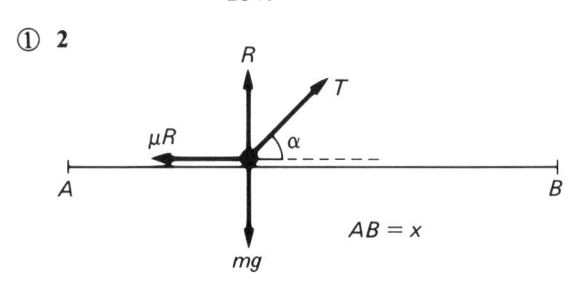

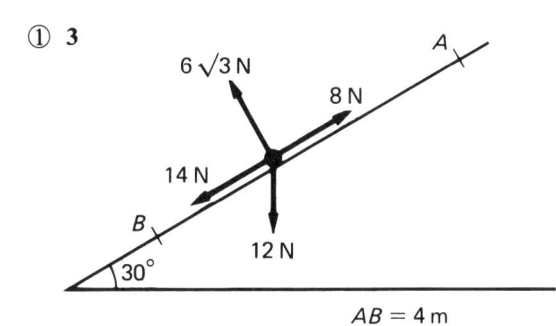

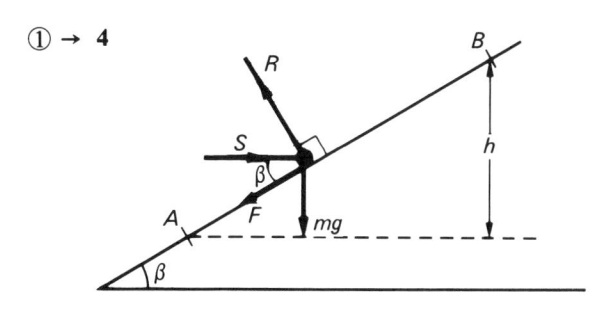

1

(diagram: forces 10 N up, 20 N down, 5 N left, $10\sqrt{2}$ N at 45° above AB; $AB = 6$ m)

2

(diagram: forces R up, mg down, μR left, T at angle α above AB; $AB = x$)

3

(diagram on incline at 30°: $6\sqrt{3}$ N, 8 N along slope up toward A, 14 N down-slope, 12 N; $AB = 4$ m)

4

(diagram: particle moving up incline from A to B with forces R, S, F, mg; height h, angle β)

5 A particle moves from the point A to the point B under the action of a system of forces. Using the definition $\mathbf{F} \cdot \mathbf{s}$ of work, find the work done by each of the forces.

- (i) $\mathbf{F}_1 = 5\mathbf{i} + \mathbf{j}$, $\mathbf{F}_2 = 3\mathbf{i} - \mathbf{j}$, $A(1, 0)$, $B(5, 0)$
- (ii) $\mathbf{F}_1 = 2\mathbf{i} + 2\mathbf{j}$, $\mathbf{F}_2 = 6\mathbf{i}$, $\mathbf{F}_3 = -3\mathbf{i} + 5\mathbf{j}$, $\mathbf{F}_4 = -4\mathbf{j}$, $A(-1, 2)$, $B(4, 5)$

Verify in each case that the total work done by the forces is equal to the work done by the resultant force on the particle. What rule relating to scalar products ensures that this is always true?

Sketch diagrams showing A and B and the forces acting on the particle. Check in each case that the acceleration of the particle is parallel to $\overrightarrow{AB}$.

6 An elastic string of natural length 0.2 m has modulus of elasticity 20 N. What is the work done in stretching it:

(i) to a length 0.4 m?
(ii) from a length 0.4 m to a length 0.6 m?
(iii) from a length 0.6 m to a length 0.8 m?

Explain the meaning of your results.

7 An elastic string of natural length l has modulus of elasticity λ. What is the work done in stretching it:

(i) to a length $\frac{4}{3}l$?
(ii) to a length $\frac{5}{3}l$?
(iii) to a length $2l$?

8 An elastic spring of natural length $3l$ has modulus of elasticity $4mg$. Find the work done in:

(i) stretching it to a length $4l$
(ii) compressing it to a length $2l$

9 The work done in stretching an elastic string of natural length 0.6 m to a length 0.8 m is 3 J. Find the work done in stretching it:

(i) to a length 1 m
(ii) to a length 1.4 m
(iii) from a length 1 m to a length 1.4 m

10 An elastic string of natural length 0.3 m and modulus of elasticity 15 N is stretched until the tension in the string is 10 N. What is the extension of the string? How much work has been done?

11 An elastic spring of natural length l and modulus of elasticity $8mg$ is compressed until the thrust in the spring is $2mg$. What is the compression of the spring? How much work has been done?

12 An elastic string of natural length l and modulus of elasticity $5mg$ breaks if the tension in it is bigger than $6mg$. Find the work done in stretching it to its breaking point.

13 An elastic string of natural length l is stretched to a length $l + x$. The tension in the string is then T. Show that the work done in stretching the string was $\frac{1}{2}Tx$.

R 14 The work done in compressing a spring of natural length $3l$ to a length l is the same as the work done in stretching a string of natural length $2l$ to a length $3l$. Show that the moduli of elasticity of the spring and the string are in the ratio $3:8$.

15 A bead threaded onto a smooth wire moves along the wire. Explain why, whatever the shape of the wire, the work done by the reaction of the wire on the bead is zero.

R 16 A particle moves along Ox from the origin to the point $(2, 0)$ under the action of a force $-kx^2$. Using equation (10), find the work done by the force.

17 The force acting on a particle of mass m at a height x above the Earth's surface is $mgR^2/(R+x)^2$ where R is the radius of the Earth. Show that the work done by the weight in moving the particle to the Earth's surface from a point at a height h above it is

$$\frac{mgh}{1+(h/R)}$$

Find an approximation for this when h is small compared to R.

• → **18** Show, using equation (10), that the work done in stretching a light elastic string of natural length l and modulus of elasticity λ from an extension x_1 to an extension x_2 is

$$\frac{\lambda}{2l}(x_2^2 - x_1^2)$$

19 A particle of mass m, moving in a straight line, accelerates from a velocity u to a velocity v under the action of a resultant force of (not necessarily constant) magnitude F. Using equation (10), show that the work done by the resultant force is $\frac{1}{2}mv^2 - \frac{1}{2}mu^2$.

• → **20** (i) A particle moves along a curve from A to B under the action of a *constant* force $\mathbf{F}$. Using equation (7), show that the work done by $\mathbf{F}$ is $\mathbf{F} \cdot \overrightarrow{AB}$. Notice that in this case the work done is independent of the path of the point of application of the force.

(ii) A particle of mass m moves along a curve from a point A to a point B at a distance h vertically below A. Show that, independent of the path of the particle, the work done by its weight is mgh.

8.4:2 Energy

> **Definition**
> The **energy** of a body is the amount of work it can do.

A body can have chemical, thermal, electrical, and mechanical energy. We are interested here only in its mechanical energy, which is of three types:

(i) kinetic energy (KE);
(ii) gravitational potential energy (PE);
(iii) elastic potential energy (EPE).

The units of energy are those of work, i.e. joule (J).

> **Definition**
> The **KE** of a body is the energy it possesses due to its motion, i.e. the work done by the body in coming to rest. **(11)**

> **Theorem 2:** The KE of a particle of mass m moving with speed u is $\frac{1}{2}mu^2$. **(12)**

Proof
Suppose that the particle can be brought to rest by a constant force $\mathbf{F}$ of magnitude F acting in the opposite direction to its velocity; suppose also that it travels through a displacement s before coming to rest. Let the acceleration of the particle be a (fig. 8.4:10).

Forces on a particle 475

Fig. 8.4:**10**

$\rightarrow$) $\qquad -F = ma \Rightarrow a = -F/m$

F constant $\Rightarrow a$ constant; so we can use the constant acceleration equation $v^2 = u^2 + 2ax$:

$$0 = u^2 - 2 \cdot (F/m) \cdot s$$
$$\Rightarrow \qquad Fs = \tfrac{1}{2}mu^2$$

Work done by **F** $= -Fs$, so work done by particle against **F** is Fs, i.e. $\tfrac{1}{2}mu^2$.

Note: it would have made sense to *define* the KE of a particle as $\tfrac{1}{2}mu^2$ (thus making this proof unnecessary); but to make the relation between energy and work clearer we have defined KE in terms of work.

> **Definition**
> The **PE** of a body with respect to a fixed horizontal reference plane is the work done by the weight of the body in moving it to the plane **(13)**

> **Theorem 3:** If a particle of mass m is at a distance h vertically above a horizontal plane, its PE with respect to the plane is mgh. **(14)**

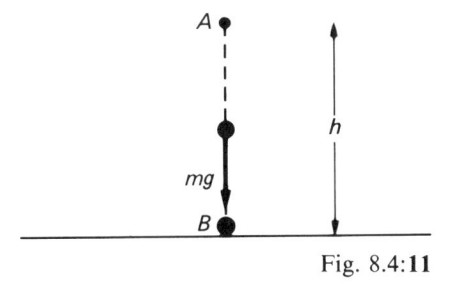

Fig. 8.4:**11**

Proof
When the particle moves from A to B (fig. 8.4:**11**), the work done by its weight is mgh; so the PE of the particle, when it is at A, is mgh.

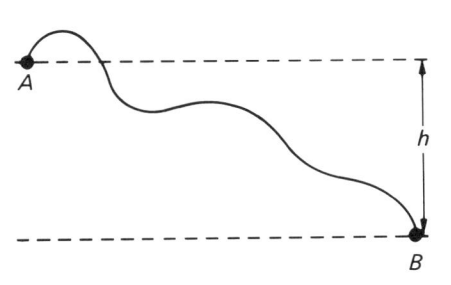

Note: when the particle is at a point distant h vertically below the plane, its PE with respect to the plane is $-mgh$.

We are, on the whole, only going to be interested in *changes* in PE, so the choice of reference plane is unimportant. In particular, if A and B are two points, with A distant h vertically above B (fig. 8.4:**12**) then, as the particle moves from A to B, loss in PE $= mgh$ and as the particle moves from B to A, gain in PE $= mgh$

Fig. 8.4:**12**

> **Definition**
> The **EPE** of a stretched light elastic string is the work done in stretching it. **(15)**

We could have used a more general definition of the EPE of a (general) body, e.g. as the work done by the body in returning to some reference shape; but since the only elastic bodies we will usually consider are strings, this is unnecessary.

476 *Work–energy principle*

> **Theorem 4:** The EPE of a light elastic string of natural length l and modulus of elasticity λ, stretched to an extension x, is $\frac{1}{2}\lambda x^2/l$. **(16)**

Proof
The proof follows from the definition of EPE and Theorem 1.

Exercise 8.4:2

1 What is the KE of a particle of mass 3 kg moving:

(i) with speed 5 m s^{-1}?
(ii) with velocity $3\mathbf{i}+2\mathbf{j}$?
(iii) with velocity $-3\mathbf{i}+2\mathbf{j}$?

2 At time t the position vector of a particle of mass 6 kg is $\mathbf{r}$. Find its KE at the given times:

(i) $\mathbf{r}=4t\mathbf{i}-\frac{1}{3}t^2\mathbf{j}$; $t=0$, $t=1$
(ii) $\mathbf{r}=\frac{1}{t}\mathbf{i}+\frac{1}{2t^2}\mathbf{j}$; $t=2$
→ (iii) $\mathbf{r}=3\cos t\mathbf{i}-\sin t\mathbf{j}$; $t=\frac{1}{4}\pi$
(iv) $\mathbf{r}=e^{2t}\mathbf{i}-e^{-t}\mathbf{j}$; $t=\ln 2$

3 A particle of mass m is moving with velocity $\mathbf{v}$. Show that its KE is $\frac{1}{2}m\mathbf{v}\cdot\mathbf{v}$.

4 Find the increase in PE of a particle of mass 4 kg moving from:

(i) $(2,-1)$ to $(2,5)$
(ii) $(-1,-1)$ to $(7,5)$
(iii) $(-3,8)$ to $(2,3)$

(Ox horizontal, Oy vertically upwards.)

• **5** Find the EPE of a spring of natural length 0.5 m and modulus of elasticity 30 N when it is:

(i) stretched to 0.6 m
(ii) compressed to 0.3 m

6 An elastic string of natural length 0.3 m and modulus of elasticity 12 N is stretched from 0.4 m to 0.5 m. What is its increase in EPE?

R **7** The EPE stored in an elastic string of natural length 1 m that is stretched to 1.2 m is 16 J. Find the energy stored in the string when it is stretched to 1.5 m.

→ **8** The work done in stretching an elastic string of natural length 0.75 m to 1.25 m is 10 J. Find the EPE stored in the string when it is stretched to:

(i) 1 m   (ii) 1.75 m

8.4:3 Work–energy principle

> **Definition**
> The **total mechanical energy** of a body is the sum of its KE, PE and EPE. **(17)**

The definitions of KE, PE and EPE imply that when a body does work W against a set of forces it loses energy W; and, conversely, when a set of forces does work W on a body, the body gains energy W. This is the **Work–Energy Principle**:

> work done by the contact forces acting on a body
> $\qquad\qquad$ = gain in total mechanical energy of the body
>
> i.e. $\sum W_\mathbf{F} = \Delta(\text{KE}) + \Delta(\text{PE}) + \Delta(\text{EPE})$ **(18)**

where $\Delta(\text{KE})$ means increase in KE, i.e. final KE − initial KE; etc. Contact forces do not include the weight; the work done by the weight is $-\Delta(\text{PE})$ and appears on the other side of the equation.

Forces on a particle 477

An immediate consequence of the Work–Energy Principle is the **Conservation of Mechanical Energy Principle**:

> If the work done by the contact forces on a body is zero, the gain in total mechanical energy of the body is zero (i.e. the total mechanical energy of the body is conserved). **(19)**

This is simply a particular case of equation (18); it says that if LHS = 0, then RHS = 0.

Note: The more general **Conservation of Energy Principle** says: the total amount of energy (of all forms) in the universe is fixed, i.e. energy cannot be created or destroyed. (Consequently, in any system the gain in energy of the system is equal to the energy entering the system.)

Worked example 3 Consider worked example 1 again: suppose the particle passes through A with speed 4 m s^{-1}. What is its speed on reaching B?

As the particle moves from A to B (fig. 8.4:**13**),

work done by contact forces

$\quad$ = work done by pulling force + work done by frictional reaction

$\quad$ = 100 J − 40 J

$\quad$ = 60 J

Fig. 8.4:**13**   Let the speed of the particle on reaching B be v.

Gain in KE = $\tfrac{1}{2} \cdot 2 \cdot v^2 - \tfrac{1}{2} \cdot 2 \cdot 4^2 = v^2 - 16$
Gain in PE = $20 \cdot 4 \sin 30° = 40$ J

By the Work–Energy Principle (18),

$$60 = v^2 - 16 + 40 \Rightarrow v = 6 \text{ m s}^{-1}$$

Note: (i) The gain in PE, 40 J = −work done by the weight. We could have used an alternative version of (18):

work done by all the forces on the particle

$\quad$ = increase in KE of particle

(see 8.4:1Q23).

(ii) We could instead have found the acceleration using N2, and then used the constant acceleration equations.

Worked example 4 Water is being raised by a pump from a tank 12 m below the ground and delivered at 4 m s^{-1} at ground level through a pipe of cross-sectional area 0.05 m^2. Find the work done per second by the pump. (1 m^3 of water has mass 1000 kg.)

In one second, a column of water of length 4 m leaves the top of pipe (fig. 8.4:**14**).

$\quad$ volume of water delivered per second = $4 \times 0.05 \text{ m}^3 = 0.2 \text{ m}^3$

Fig. 8.4:**14**   So mass of water delivered per second = $0.2 \times 1000 \text{ kg} = 200 \text{ kg}$

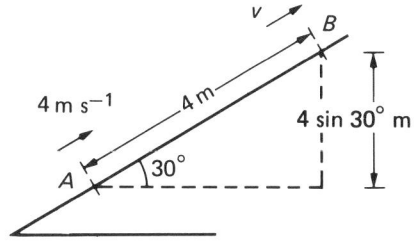

478 *Work–energy principle*

As the 200 kg of water leaves the top of the pipe it is replaced in the pipe by an equal mass of water from the tank, 12 m below. Effectively, the pump raises 200 kg of water (which was at rest in the tank) through 12 m and gives it a speed of 4 m s^{-1}.

Gain in KE of water per second = $\tfrac{1}{2} \cdot 200 \cdot 4^2$ J = 1600 J
Gain in PE of water per second = $200 \cdot 10 \cdot 12$ J = 24000 J

By the Work–Energy Principle (18),

$$\text{work done per second by the pump} = 1600 + 24000 = 25600 \text{ J}$$

Worked example 5 A smooth bead is threaded onto a smooth circular wire, centre O and radius a, which is fixed in a vertical plane. If the bead is projected from the lowest point A of the wire with velocity $\sqrt{(5ga)}$, find its speed when it passes through the point P, where OP makes an angle θ with the upward vertical.

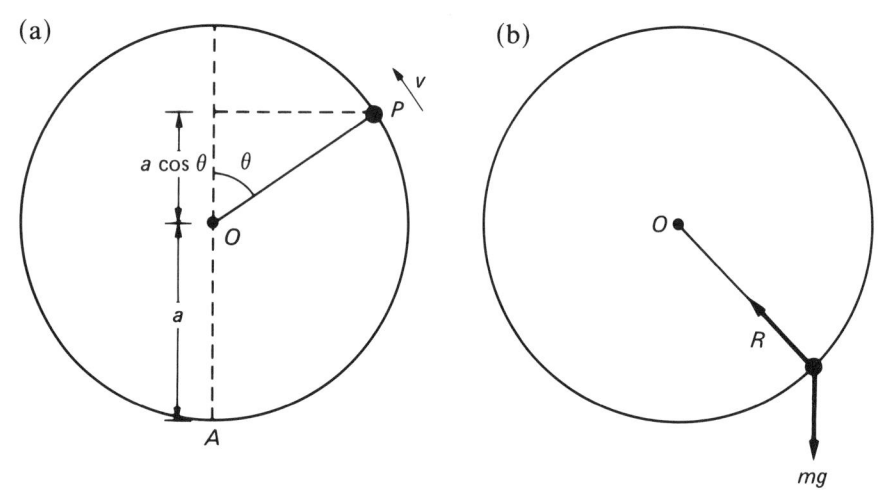

Fig. 8.4:**15**

Let the speed of the bead as it passes through P be v (fig. 8.4:**15a**). As the particle moves from A to P,

$$\text{gain in KE} = \tfrac{1}{2} mv^2 - \tfrac{1}{2} m \cdot 5ga$$

$$\text{gain in PE} = mga(1 + \cos \theta)$$

There is a contact force R, the reaction of the wire on the bead, acting on the bead, but, by statement (9), (8.4:**1**), it does no work since it is always perpendicular to the direction of motion (fig. 8.4:**15b**). So by the Conservation of Energy Principle (19),

$$\tfrac{1}{2} mv^2 - \tfrac{1}{2} mga - mga(1 + \cos \theta) = 0$$

$$\Rightarrow \quad v^2 = 5ga - 2ga(1 + \cos \theta) = ga(3 - 2 \cos \theta)$$

Worked example 6 An elastic string of natural length l and modulus of elasticity $3mg$ has a particle P of mass m attached at one end. The other end of the string is attached to a fixed point A and the string hangs vertically.

(i) Find the extension of the string in the equilibrium position.

The particle is pulled down to B so that $AB = 2l$ and is then released. Find:

(ii) the distance from A of the particle when it first comes to rest;
(iii) the maximum velocity of the particle during the motion.

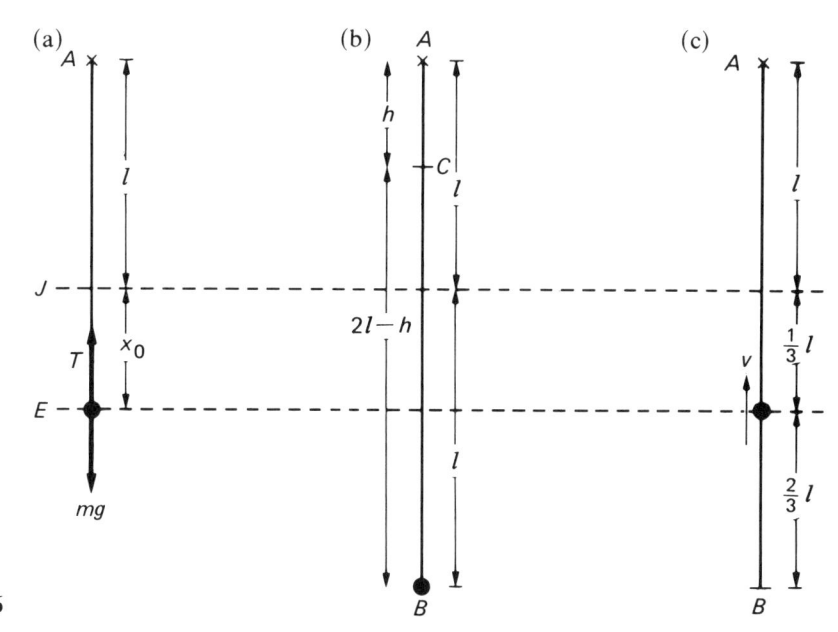

Fig. 8.4:**16**

(i) Let the extension of the string in the equilibrium position E be x_0 (fig. 8.4:**16a**).

$\uparrow)$ $\qquad T - mg = 0$

From Hooke's Law: $\qquad T = 3mgx_0/l$

So $\qquad x_0 = \tfrac{1}{3}l$

Consider the system of particle and string. The only contact force on the system is the reaction at A. This does no work, since it does not move. So the work done by the contact forces on the system is zero.

(ii) Initial KE of particle = final KE of particle = 0. Suppose the particle comes to rest at C and let $AC = h$ (fig. 8.4:**16b**).

$$\text{gain in PE of particle} = mg(2l - h)$$
$$\text{gain in EPE of string} = -\tfrac{1}{2}\lambda l^2/l = -\tfrac{3}{2}mgl$$

By the Conservation of Mechanical Energy Principle (19):

$$0 = mg(2l - h) - \tfrac{3}{2}mgl \Rightarrow h = \tfrac{1}{2}l$$

(iii) Let the maximum velocity during the motion be v. Velocity maximum $\Rightarrow$ acceleration zero $\Rightarrow$ particle in equilibrium position E (fig. 8.4:**16c**).

$$\text{gain in PE of particle} = mg \cdot \tfrac{2}{3}l$$
$$\text{initial EPE of string} = (\lambda/2l) \cdot (l)^2 \quad \text{(see Theorem 4)}$$
$$\text{final EPE of string} = (\lambda/2l) \cdot (\tfrac{1}{3}l)^2$$
$$\Rightarrow \text{gain in EPE of string} = (\lambda/2l)\{(\tfrac{1}{3}l)^2 - l^2\} = -\tfrac{4}{3}mgl$$

(c.f. Corollary (6).)

By the Conservation of Mechanical Energy Principle (19),

$$0 = \tfrac{1}{2}mv^2 + mg \cdot \tfrac{2}{3}l - \tfrac{4}{3}mgl$$
$$\Rightarrow \qquad v = \sqrt{(\tfrac{4}{3}gl)}$$

Exercise 8.4:3

③ **1** A car of mass 800 kg accelerates along a level road from a speed of 5 m s^{-1}. The driving force is 1500 N and the resistance to motion 300 N. Using the Work-Energy Principle, find its speed when it has travelled 40 m.

③→ **2** A particle of mass 3 kg, initially at rest is pulled a distance 4 m across a horizontal plane by a horizontal force of 10 N. Using the Work-Energy Principle, find its final speed if:

(i) the plane is smooth
(ii) the coefficient of friction is $\frac{1}{5}$

Assume that the particle starts from rest.

③ **3** A bullet of mass 0.05 kg travelling horizontally at 400 m s^{-1} hits a fixed wooden target and penetrates to a distance of 0.1 m. Using the Work-Energy Principle, find the average resisting force of the wood.

③ **4** Using the Work-Energy Principle, express the braking distance d of a car in terms of its speed v, its mass m and its braking force R. (Assume that R is constant at all speeds.) Sketch a graph of d against v.
If $m = 750$ kg and $R = 500$ N, find d when:

(i) $v = 15$ m s^{-1} (ii) $v = 30$ m s^{-1}

③ **5** A particle of mass 2 kg slides from rest down a rough plane of inclination 30°. The frictional resistance to motion is 5 N throughout. Using the Work-Energy Principle,

(i) find the speed of the particle when it has travelled a distance 8 m
(ii) find the distance the particle has travelled when its speed is 10 m s^{-1}

③ **6** A particle of mass m is projected with speed u up a rough plane of inclination $\tan^{-1}\frac{3}{4}$. The frictional resistance to motion is $\frac{1}{5}mg$ throughout. Find the distance the particle has travelled when:

(i) its speed is first $\frac{1}{2}u$
(ii) it comes to instantaneous rest

③ **7** A particle of mass 5 kg is pulled at constant speed for a distance 20 m up a smooth slope of inclination $\sin^{-1}\frac{1}{4}$. Find the work done by the pulling force.

③→ **8** A car of mass 1000 kg accelerates up a slope of inclination 1 in 20. The driving force of the engine is 2000 N and the resistance to motion is 700 N. Find the distance travelled by the car as its speed increases

(i) from 2 m s^{-1} to 10 m s^{-1}
(ii) from 10 m s^{-1} to 18 m s^{-1}

The engine cuts out when the car is travelling at 18 m s^{-1}. Find the distance moved by the car before it comes to rest.

③ **9** A particle is projected with speed $\sqrt{(3ga)}$ up a rough plane of inclination $\tan^{-1}\frac{3}{4}$. Find the coefficient of friction between the particle and the plane if after travelling a distance a the speed of the particle is $\sqrt{(ga)}$.

• **10** A force, acting vertically upwards on a particle of mass 0.2 kg, moves the particle vertically from rest to a height 1.2 m above its starting point and gives it a speed of 8 m s^{-1}. Find the work done by the force.

→ **11** A force acting vertically upwards on a body of mass 10 kg does 920 J of work in raising the body from rest to a height of 6 m above its starting point and giving it a speed v m s^{-1}. Find v.

12 Every ten seconds, a machine for getting rid of compressed rubbish picks up a block of average mass 60 kg, moves it vertically from rest to a height 10 m above its starting point and ejects it with a speed of 6 m s^{-1}. Find the work done by the machine in one hour.

Questions 13–15: A particle moves in a vertical plane; **i** and **j** are unit vectors acting horizontally and vertically upwards, respectively.

•→ **13** A particle of mass 0.5 kg starts from rest at the point $A(2, 1)$ and moves to the point $B(-4, 5)$ under the action of its weight and a force $\mathbf{F} = -3\mathbf{i} + 7\mathbf{j}$. Find:

(i) the work done by $\mathbf{F}$
(ii) the increase in PE of the particle

and hence, using the Work-Energy Principle, find:

(iii) the speed of the particle on reaching B

14 A particle of mass 2 kg starts from rest at the point $A(-1, 3)$ and moves to the point $B(2, -1)$ under the action of its weight and forces $\mathbf{F}_1 = 5\mathbf{i} + 8\mathbf{j}$, $\mathbf{F}_2 = 14\mathbf{i}$ and $\mathbf{F}_3 = -7\mathbf{i} - 4\mathbf{j}$. Find:

(i) the total work done by $\mathbf{F}_1$, $\mathbf{F}_2$ and $\mathbf{F}_3$
(ii) the increase in PE of the particle, and hence
(iii) the speed of the particle on reaching B

Find the total work done by all the forces on the particle (including the weight) and show that it equals the increase in KE of the particle.

15 A particle of mass 2.4 kg starts with speed 5 m s^{-1} at the point $A(-2, 3)$ and moves to the point $B(4, 0)$ under the action of its weight and a force $\mathbf{F} = 8\mathbf{i} + 16\mathbf{j}$. Show that the work done by $\mathbf{F}$ is zero and hence find the speed of the particle on reaching B.

R **16** A ring of mass 2 kg is threaded onto a rough straight wire parallel to the vector $3\mathbf{i} + 4\mathbf{j}$. The ring has an initial velocity $18\mathbf{i} + 24\mathbf{j}$ and comes to rest after moving 20 m. Find the frictional force acting on the ring.

Questions 17–22: 1 m³ of water has mass 1000 kg.

④ **17** Water is being raised by a pump from a tank 6 m below the ground and delivered at 5 m s⁻¹ at ground level through a pipe of cross-sectional area 0.12 m². Find the work done per second by the pump.

④ **18** A pump raises water from a depth of 20 m and discharges it through a pipe of cross-sectional area 0.08 m² at a velocity 10 m s⁻¹. Find the work done per second by the pump.

④ **19** A pump raises water from a depth of h m and discharges it through a pipe of cross-sectional area A m² at a velocity v m s⁻¹. Find the work done per second by the pump.

④ → **20** A pump raises water from a depth of 10 m and discharges it through a pipe of diameter 0.1 m at a velocity 8 m s⁻¹. Find the work done per second by the pump.

④ R **21** Water is pumped at the rate of 100 litres s⁻¹ from a tank 2 m below the ground and delivered at ground level through a pipe of cross-sectional area 0.005 m². Find the work done per second by the pump. (1 litre of water has a volume of 0.001 m³.)

④ **22** Water is pumped at the rate of 80 litres s⁻¹ from a tank 10 m below the ground and delivered at ground level at speed v m s⁻¹. If the work done per second by the pump is 20 000 J, find v.

• **23** A particle is projected vertically upwards with speed 10 m s⁻¹. Using the Conservation of Mechanical Energy Principle, find:

(i) the height reached by the particle
(ii) its speed when it is 4 m above the point of projection

→ **24** A particle is projected vertically upwards with speed $\sqrt{(3ga)}$ (a constant). Find:

(i) the height reached by the particle
(ii) its speed when it is at a height a above the point of projection

25 A particle is projected vertically upwards with speed 25 m s⁻¹ from a point A at a height 10 m above ground level. Using the Conservation of Mechanical Energy Principle, find:

(i) the height reached by the particle
(ii) the speed of the particle as it passes through A again
(iii) the speed of the particle as it hits the ground

26 A particle is projected with speed u at an angle α to the horizontal from a point on ground level. Find the height reached by the particle when its velocity is:

(i) $\tfrac{2}{3}u$ (ii) $\tfrac{1}{2}u\sqrt{3}$

Questions 27–31: A smooth bead is threaded onto a smooth circular wire, centre O and radius a, which is fixed in a vertical plane. The bead is projected from the lowest point A of the wire with speed u.

⑤ **27** $u = \sqrt{(7ga)}$. Find the speed of the bead when it passes through the point P where OP makes an angle θ with the upward vertical.

⑤ → **28** $u = \sqrt{(3ga)}$. Find the speed of the bead when it passes through the point P, where OP makes an angle θ with the upward vertical. Hence find:

(i) the greatest height above A reached by the bead
(ii) the speed of the bead as it passes through the point B on the wire level with O

⑤ **29** $u = \sqrt{5ga}$. Find its speed as it passes through

(i) the highest point C of the wire
(ii) the point A again
(iii) the points B and D on the wire level with O

⑤ **30** Find u if the bead just reaches:

(i) the highest point of the wire
(ii) the point on the wire level with O

⑤ **31** Describe the motion of the bead when:

(i) $u = \sqrt{(ga)}$ (ii) $u = \sqrt{(\tfrac{7}{2}ga)}$
(iii) $u = \sqrt{(6ga)}$

⑤ R **32** One end of a light inelastic string is fixed to a point A and the other end is attached to a particle. The particle is held with the string horizontal and is then released. If the length of the string is l, find the speed of the particle when:

(i) the string makes an angle of 60° with the vertical for the first time
(ii) the string is vertical
(iii) the string is again horizontal

• **33** A bead is threaded onto a smooth wire fixed in a vertical plane, as shown in the diagram. The bead is projected along the wire from A with speed 5 m s⁻¹. Find its speed when it passes through:

(i) B (ii) C (iii) D

Explain why the bead does not reach the point E.

Draw a diagram showing the point at which the bead first comes to rest, and describe its subsequent motion.

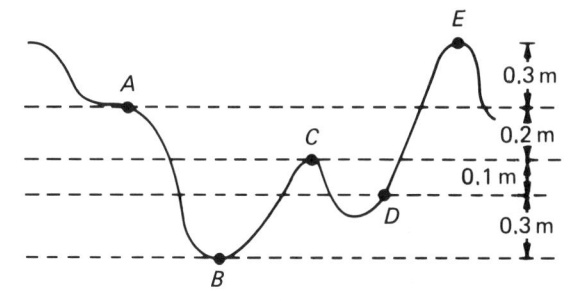

Questions 34–40: An elastic string of natural length l and modulus of elasticity λ has a particle P of mass m attached to one end. The other end is attached to a fixed point A and the string hangs vertically.

34 $\lambda = 2mg$. The particle is pulled down so that $AP = \frac{5}{2}l$, and is then released. Find the distance from A of the particle when it first comes to rest.

35 $\lambda = 4mg$.

(i) Find the extension of the string in the equilibrium position.

The particle is pulled down so that $AP = 2l$, and is then released.

(ii) Show that the particle just reaches A.

(iii) Find the maximum velocity of the particle during the motion.

36 When the particle is in the equilibrium position, $AP = \frac{5}{3}l$. Find λ. Hence find the distance through which the particle must be pulled down vertically from its equilibrium position so that it will just reach A after release.

37 $\lambda = \frac{3}{4}mg$. The particle is held at A and is released. Find:

(i) the length AP when the particle is in its lowest position

(ii) the maximum velocity of the particle during the motion

38 The particle is held at A and is released. The particle first comes to rest when it has fallen a distance $2l$. Find λ. Describe the subsequent motion of the particle.

R 39 The particle is held at A and is released. The particle first comes to rest when it has fallen a distance $3l$

(i) Find λ.

(ii) Find the acceleration of the particle when it reaches its lowest point.

(iii) Find the extension of the string at the instant when its acceleration is $\frac{1}{2}g$ upwards.

(iv) Hence, using the Conservation of Mechanical Energy Principle, find the speed of the particle at this instant.

(v) Find the speed of the particle when its acceleration is $\frac{1}{2}g$ downwards.

40 (i) The particle is lowered slowly (i.e. with negligible velocity) from the position J, where the string is at its natural length, to its equilibrium position E. Find: (a) the loss in PE of the particle; (b) the work done by the tension.

Are these the same? If not, explain the discrepancy.

(ii) The particle is dropped from J. Use the results of (i) to find its velocity when it reaches E.

Questions 41 and 42: One end of an elastic *spring* of natural length l_0 and modulus of elasticity λ is fixed to a point A on a smooth horizontal table, the other end to a particle P of mass m.

41 $l_0 = l$, $\lambda = 3mg$. The particle is pulled away from A until $AP = \frac{3}{2}l$, and is then released.

(i) Find the velocity of the particle when the spring reaches its natural length.

(ii) Find the distance from A at which the particle first comes to instantaneous rest.

(iii) Describe the motion of the particle.

42 $l_0 = 4l$, $\lambda = 2mg$. The particle is pulled away from A until $AP = 6l$, and is then released. Find the velocity of the particle when the length of the spring is:

(i) $6l$ (ii) $5l$ (iii) $4l$ (iv) $3l$ (v) $2l$

Illustrate your results on a diagram.

Questions 43–45: One end of a light elastic *string* of natural length l and modulus of elasticity λ is fixed to a point A on a smooth horizontal table, the other end to a particle P of mass m.

R 43 $\lambda = \frac{5}{2}mg$. The particle is projected from A with speed $\sqrt{(4gl)}$. Find its distance from A when:

(i) its speed is $\sqrt{(gl)}$
(ii) it first comes to instantaneous rest

44 The particle is projected from A with speed u. Find its distance from A when its speed is:

(i) $\frac{2}{3}u$ (ii) $\frac{1}{3}u$ (iii) zero

45 $\lambda = 4mg$. The particle is pulled away from A until $AP = 2l$, and is then released. Find its velocity as it passes through A. Describe the motion of the particle.

46 One end of a light elastic string of natural length $3a$ and modulus of elasticity $2mg$ is fixed to a point A on a horizontal table, the other end to a particle P of mass m. The particle rests at a point O with the string just taut. It is then pulled away from A in the direction AO until $AP = 5a$ and is released. Find its velocity on reaching:

(i) O (ii) A

assuming that: (a) the table is smooth; (b) the coefficient of friction between the particle and the table is $\frac{1}{6}$.

47 A smooth ring of mass m can slide on a smooth vertical wire. It is attached by an elastic string of natural length l and modulus of elasticity $mg\sqrt{2}$ to a point at a distance l from the wire. The ring is released from rest with the string horizontal. Prove that when the ring first comes to instantaneous rest, the length of the string is $3l$.

Questions 48 and 49: An elastic string of natural length l and modulus of elasticity λ has a particle P of mass m attached to one end. The other end is attached to a fixed point A on a smooth slope of inclination α.

48 $\lambda = 3mg$, $\alpha = 30°$. Find the extension of the string when the particle rests in equilibrium on the slope.

The particle is pulled a further distance x down the slope and is released. If the particle just reaches A, find x.

49 $\lambda = mg$, $\alpha = \sin^{-1}\frac{1}{4}$. The particle is projected from A with speed $\sqrt{(2gl)}$ up a line of greatest slope of the plane. Find the distance travelled by the particle before it first comes to rest.

50 An elastic string of natural length l and modulus of elasticity $2mg$ has a particle P of mass m attached to one end. The other end is attached to a fixed point A on a rough slope of inclination $\tan^{-1}\frac{3}{4}$. The coefficient of friction is $\frac{1}{4}$. Find the distance from A of the particle when it first comes to instantaneous rest if:

(i) the particle is released from rest at A
(ii) the particle is projected from A with speed $2\sqrt{(gl)}$ down a line of greatest slope
(iii) the particle is projected from A with speed $2\sqrt{(gl)}$ up a line of greatest slope

• **51** A smooth bead of mass m is threaded onto a smooth circular wire, centre O, radius a, which is fixed in a vertical plane. The bead is attached to one end of a light elastic string of natural length a and modulus of elasticity $4mg$. The other end of the string is attached to C, the highest point of the wire. The bead is slightly disturbed from rest at C. Using the Conservation of Mechanical Energy Principle, find its velocity when:

(i) the string becomes taut for the first time
(ii) the bead is at the point B on the wire level with O
(iii) the string makes an angle $30°$ with the downward vertical

Show that the bead comes to rest at the lowest point A of the wire.

+ → **52** A smooth bead P, of mass m, is threaded onto a smooth circular wire, centre O radius a, which is fixed in a vertical plane. The bead is attached to one end of a light elastic string of natural length a and modulus of elasticity mg. The other end of the string is attached to A, the lowest point of the wire. The bead is slightly disturbed from rest at C, the highest point of the wire. Find its velocity when:

(i) $C\hat{O}P = \theta$ (θ acute)
(ii) the bead is level with O
(iii) the string first becomes slack
(iv) the bead reaches A

53 A particle of mass 8 kg moves along Ox from the origin under the action of a force $-3x^2$. Find the work done by the force in moving the particle to the point $A(4, 0)$. Given that the initial speed of the particle is 5 m s^{-1}, use the Work–Energy Principle to find its speed on reaching A.

54 A particle of mass m starts from O with velocity an and moves along Ox under the action of a force $mn^2 x$ towards O. Use the Work–Energy Principle to find the velocity of the particle when its displacement from O is:

(i) $\frac{1}{3}a$ (ii) $\frac{2}{3}a$ (iii) a

What is the maximum displacement of the particle from O?

8.4:4 Power

Definition
The **power** H generated by a force $\mathbf{F}$ is the rate at which the force is doing work:

$$H = \frac{dW}{dt}$$

Power is measured in joule per second, called watt (W).

Note: we shall be considering motion in a straight line only so we shall use the notation of section 7.1.

Theorem 5: If the point of application of a force $\mathbf{F}$ of magnitude F moves with velocity v, then the power H generated by the force is given by

$$H = Fv \tag{20}$$

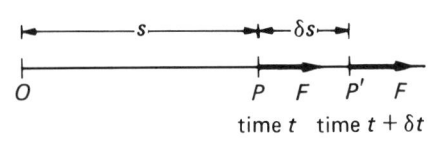

484 Power

Fig. 8.4:17

Proof
Suppose that the point of application of **F** moves through a small displacement δs from P to P' in time δt (fig. 8.4:17).

Work done by **F** in moving from P to $P' = \delta W \approx F \delta s$

$$\Rightarrow \quad \frac{\delta W}{\delta t} = F \frac{\delta s}{\delta t}$$

Taking the limit as $\delta t \to 0$,

$$\frac{dW}{dt} = F \frac{ds}{dt} \quad \text{i.e.} \quad H = Fv$$

(F is not necessarily constant.)

Note: we talk about the work done by a force *during an interval of time* but the power generated by a force *at an instant*.

Worked example 7 A car of mass 1600 kg moves along a level road against a constant resistance of 1800 N. Find its acceleration at the instant when its speed is 6 m s^{-1} if its engine is working at a rate of 18 kW.

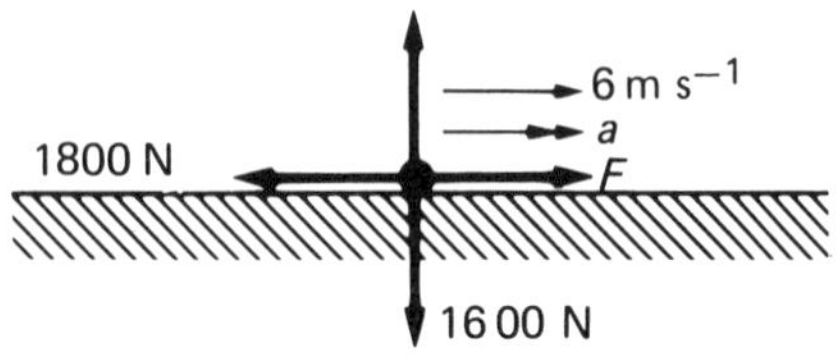

Let the driving force be F, the acceleration a (fig. 8.4:18).

$\rightarrow)$ $\qquad F - 1800 = 1600a$

$H = Fv$ gives $\qquad 18\,000 = F \cdot 6$

So $\qquad a = \tfrac{3}{4}$ m s^{-2}

Fig. 8.4:18

Note: the power generated by the driving force of a car is often loosely called the power of the engine or the power of the car.

Worked example 8 A car of mass 1600 kg whose maximum power is 20 kW moves against a constant resistance of 1800 N. Find the maximum speed of the car

(i) up   (ii) down

a slope of 1 in 16.

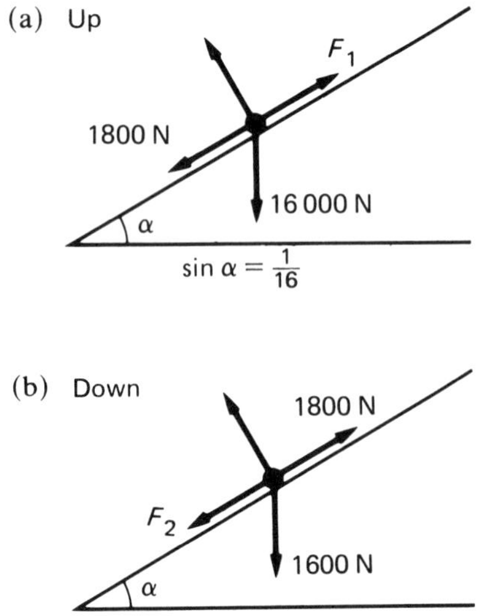

(a) Up

(i) Let the maximum speed of the car up the slope be v_1; let the driving force of the engine be F_1 (fig. 8.4:19a).

speed maximum $\Rightarrow$ acceleration zero

Equating forces parallel to the plane:

$$F_1 - 1800 - 16\,000 \sin \alpha = 0$$

$H = Fv$ gives $\qquad 20\,000 = F_1 v_1$

So $\qquad v_1 = \tfrac{50}{7}$ m s^{-1}

(b) Down

(ii) Let the maximum speed of the car down the slope be v_2; let the driving force of the engine be F_2, different from F_1 (fig. 8.4:19b).

$\nearrow)$ $\qquad F_2 + 16\,000 \sin \alpha - 1800 = 0$

$H = Fv \Rightarrow \qquad 20\,000 = F_2 v_2$

So $\qquad v_2 = 25$ m s^{-1}

Fig. 8.4:19

Exercise 8.4:4

1 Comment on this proof of Theorem 5:

work W done by $\mathbf{F} = Fs$

power H generated by $\mathbf{F} = \dfrac{dW}{dt} = \dfrac{d}{dt}(Fs)$

$= F\dfrac{ds}{dt} = Fv$

Questions 2–6: A car of mass 1000 kg moves along a level road against a constant resistance of 400 N.

2 Find the power exerted by the engine at the instant when its speed is 10 m s^{-1} and its acceleration is 0.5 m s^{-2}.

3 Find the acceleration of the car at the instant when its speed is 12 m s^{-1} and the engine is working at 18 kW.

4 Find the power exerted by the engine when the car moves at a constant speed of 20 m s^{-1}.

5 If the car moves with constant speed v when the power exerted by the engine is 16 kW, find v.

6 If the velocity of the car is maximum, what can you say about its acceleration?
    Show that, if the engine of the car works at a maximum rate of 24 kW, the maximum possible velocity of the car is 60 m s^{-1}.

7 A car of mass m whose engine works at a constant rate H moves along a level road against a constant resistance R.
 (i) Find the acceleration of the car when its speed is v.
 (ii) Find the maximum possible velocity of the car.

R 8 A car of mass 800 kg, whose engine works at a constant rate 15 kW, moves along a level road.
 (i) If its maximum speed is 50 m s^{-1} find the resistance to motion.
 (ii) Assuming the resistance to motion is constant, find the acceleration of the car when the velocity is 30 m s^{-1}.

9 A car of mass m, whose engine works at a constant rate H, moves along a level road.
 (i) If its maximum speed is v, find the resistance to motion.
 (ii) Assuming the resistance to motion is constant, find the acceleration of the car when the velocity is:
 (a) $\tfrac{3}{4}v$; (b) $\tfrac{1}{2}v$; (c) $\tfrac{1}{4}v$.

10 A car of mass 1000 kg tows a caravan of mass 800 kg along a level road. The resistance to motion of each vehicle is 0.5 N per kg. Find:
 (i) the acceleration of car and caravan
 (ii) the tension in the towrope
when they are travelling at 20 m s^{-1} and the engine of the car is working at 50 kW.

11 A car of mass M kg tows a caravan of mass m kg along a level road. The resistance to the motion of each vehicle is k N per kg. Find the tension in the towrope when they are travelling at $v \text{ m s}^{-1}$ and the engine of the car is working at H W.

12 A car of mass 800 kg has a maximum power of 25 kW. The constant resistance to the motion of the car is 1200 N. Find the maximum speed of the car:

(i) up (ii) down

a slope of inclination 1 in 10.

13 A car of mass 750 kg has a maximum power of 20 kW. The constant resistance to the motion of the car is 600 N. Find the maximum speed of the car up a slope of inclination:

(i) 1 in 15 (ii) 1 in 10 (iii) 1 in 5

14 A car of mass m has a maximum power H. The constant resistance to the motion of the car is R. Find the maximum speed of the car:

(i) up (ii) down

a slope of inclination 1 in n.

R 15 A car of mass 1600 kg has maximum power 30 kW. The maximum speed of the car up a slope of inclination 1 in 16 is 12 m s^{-1}. Find its maximum speed down the same slope, assuming the resistance to motion is constant.

16 A car of mass M has maximum power H. The maximum speed of the car up a slope of inclination α is v. Find its maximum speed down the same slope, assuming the resistance to motion is constant.

Questions 17–20: A car of mass 1200 kg moves up a slope of inclination 1 in 20. The constant resistance to the motion of the car is 800 N.

17 Find the power exerted by the engine at the instant when the velocity of the car is 15 m s^{-1} and its acceleration is 0.2 m s^{-2}.

18 Find the acceleration of the car when its velocity is 5 m s^{-1} and its engine is working at 15 kW.

19 When the engine is working at 20 kW, the car can travel with constant speed v. Find v.

⑧ **20** If the maximum possible velocity of the car is 40 m s^{-1}, find the maximum power exerted by the engine.

⑧ **21** A train of mass 50 000 kg, whose engine works at a constant rate of 1800 kW, moves up a slope of inclination α against a constant resistance 10 000 N.

(i) If the maximum possible velocity of the train is 45 m s^{-1}, find sin α.

(ii) Find the acceleration of the train when its velocity is 20 m s^{-1}.

⑧ **22** A train of mass 30 000 kg, whose engine works at a constant rate 750 kW, moves up a slope of inclination 1 in 50 against a constant resistance 15 000 N.

(i) Find its maximum speed.
(ii) Find its acceleration when the speed is 15 m s^{-1}.

Find the maximum speed of the train when it is moving down the same slope.

⑧→ **23** A car of mass 2000 kg moves up a slope of inclination 1 in 25 against a constant resistance 3600 N. If the engine is working at 100 kW, find the maximum velocity of the car.

The engine is now shut off and the car moves down a slope of inclination θ with constant speed. Find sin θ.

⑧ **24** The constant resistance to motion of a car of mass 1000 kg is 800 N.

(i) The car moves along a level road. Find the power exerted by the engine at the instant when the velocity of the car is 30 m s^{-1} and its acceleration is 0.2 m s^{-2}.

(ii) If the maximum power of the engine is 45 kW, find the maximum speed of the car down a slope of inclination 1 in 200.

⑧ **25** A car of mass m is moving up a slope of inclination α against a constant resistance R. At the instant when the velocity of the car is v and its acceleration is a, its engine is working at a rate H. Show that

$$H = (R + mg \sin \alpha + ma)v$$

Find a similar expression for H when the car moves down the same slope with velocity v and acceleration a.

⑧→ **26** A car of mass 800 kg tows a caravan of mass 1200 kg up a slope of inclination 1 in 16. The resistance to the motion of each vehicle is 0.75 N per kg. Find the acceleration of the vehicles and the tension in the towrope if the engine of the car works at a constant rate of 20 kW and the vehicles are travelling at:

(i) 10 m s^{-1} (ii) 5 m s^{-1}

⑧ R **27** A car of mass $3m$ tows a caravan of mass $5m$ up a slope of inclination 1 in 8. The resistance to the motion of each vehicle is $2mg$. Find the acceleration of the vehicles and the tension in the towrope if the engine works at a constant rate of $6mgV$ and the vehicles are travelling with speed V.

⑧ **28** A car of mass 1000 kg whose engine works at a constant rate H moves against constant resistance R. The maximum speeds of the car up and down a slope of inclination 1 in 20 are 25 m s^{-1} and 30 m s^{-1} respectively. Find H and R. Hence find:

(i) the maximum speed of the car on level ground;
(ii) the acceleration of the car when travelling on level ground at a speed 15 m s^{-1}.

⑧ **29** A car whose engine works at a constant rate of 25 kW moves against a constant resistance 400 N. Its maximum speed on level ground is twice its maximum speed up a slope of inclination 1 in 8. Find the mass of the car. Hence find its acceleration when it is moving on level ground at 15 m s^{-1}.

⑧ **30** A car of mass 500 kg moves up a slope of inclination 1 in 10 against a constant resistance of 600 N.

(i) Find its maximum possible velocity when its engine is working at 15 kW.

(ii) When the car is travelling at this speed the power of the engine is suddenly increased to 20 kW. Find the immediate acceleration of the car.

⑧→ **31** A car of mass m whose engine works at a constant rate H moves against a constant resistance R. Its maximum speeds up and down a slope of inclination 1 in n are v and $2v$.

(i) Find R in terms of m, n and g, and find H in terms of m, n, v and g.

(ii) Show that the maximum speed on level ground is $\frac{4}{3}v$.

(iii) Find its acceleration when it moves up the slope at the instant when its speed is $\frac{2}{3}v$.

⑧ R **32** A car of mass m whose engine works at a constant rate H moves against a constant resistance. Its maximum speed on a level road is v. Find its maximum speed:

(i) directly up (ii) directly down

a slope of inclination 1 in n. If the maximum speed down the road is twice the maximum speed up the road, show that $n = 3mgv/H$.

The car moves along a level road with speed $\frac{1}{2}v$. Show that its acceleration is H/mv.

33 A car of mass 800 kg, whose engine works at a constant rate of 25 kW, moves along a level road against a resistance kv^2 where v is its speed and k is a constant.

(i) If its maximum speed is 25 m s^{-1}, find k.

(ii) Find the resistance to motion when its speed is 20 m s^{-1}, and hence find its acceleration at this instant.

34 A car of mass 1500 kg, whose engine works at a constant rate of 30 kW, moves along a level road against a resistance kv where v is its speed and k is a constant. If its maximum speed is 40 m s^{-1}, find k and hence find its acceleration when its speed is 20 m s^{-1}.

35 A car of mass m, whose engine works at a constant rate H, moves along a level road against a resistance proportional to the square of its speed. If its maximum speed is v, find its acceleration when its speed is $\frac{1}{2}v$.

36 A car of mass 500 kg has a maximum power 60 kW. When the engine is working at 40 kW, it moves along a level road at a constant speed of 25 m s^{-1}. If the resistance to motion is proportional to the square of the speed, find its maximum possible speed along this road.

37 A car of mass 750 kg whose engine works at a constant rate of 45 kW moves against a resistance kv where v is its speed and k is a constant.

(i) Its maximum speed on level ground is 40 m s^{-1}. Find k.

(ii) Find its maximum speed up a slope of inclination 1 in 15.

38 A car of mass 1000 kg moves against a resistance kv^2 where v is its speed and k is a constant.

(i) It can maintain a constant speed 30 m s^{-1} when moving up a slope of inclination 1 in 20 with the engine working at 60 kW. Find k.

(ii) Find the acceleration of the car when it is moving down the same slope with the engine working at 40 kW at the instant when its speed is 20 m s^{-1}.

39 A car of mass 1000 kg moves against a resistance proportional to its speed. Its maximum speeds up and down a slope of inclination 1 in 8 are 40 m s^{-1} and 60 m s^{-1}. Find the maximum power of the car. Hence find its maximum speed on level ground.

40 A car of mass 1000 kg has a maximum power of 15 kW. The resistance to motion is proportional to the square of its speed. Given that its maximum speed on level ground is 30 m s^{-1}, find the rate at which the engine is working when the car is moving at a constant speed of 10 m s^{-1} up a road of inclination 1 in 25.

R 41 A lorry of mass 10 000 kg has a maximum speed of 8 m s^{-1} up a slope of 1 in 10 against a resistance of 1200 N. Find the power of the engine. If the resistance to motion is proportional to the square of its speed, find its maximum speed on the level.

42 A car of mass 750 kg moves along a level road. The constant power exerted by the engine is 15 kW. The resistance to motion is negligible.

(i) Show that if the velocity of the car after time t is v then

$$\frac{dv}{dt} = \frac{20}{v}$$

(ii) If the initial velocity of the car is 10 m s^{-1}, use the methods of section 4.4:2 to find its velocity at time t. Sketch a graph of v against t.

(iii) Find the time taken before the velocity is 30 m s^{-1}.

(iv) As the velocity of the car increases, does the driving force increase or decrease?

43 A car of mass m moves along a level road. The constant power exerted by the engine is H. The resistance to motion is negligible. The initial velocity of the car is u.

(i) Find the velocity of the car at time t.

(ii) Find the work done by the driving force between times 0 and t.

(iii) Write down the initial KE of the car and its KE at time t.

What do you conclude?

44 A car of mass 800 kg moves along a level road. The constant power exerted by the engine is 20 kW. The constant resistance to motion is 400 N. If the initial velocity of the car is 25 m s^{-1} and the velocity at time t is v, find a differential equation connecting v and t, and hence show that $t = 50 - 2v - 100 \ln 25/(50 - v)$. Find the time taken for the velocity to reach 30 m s^{-1}. Find the work done by the driving force and the increase in K.E. of the car in this time. Explain the discrepancy.

45 A car of mass m moves along a level road against a constant resistance R. The constant power exerted by the engine is H. If the initial velocity of the car is u, find the time taken before its velocity is $2u$.

46 A car of mass 1000 kg moves along a level road. The constant power exerted by the engine is 15 kW. The constant resistance to motion is 500 N. The initial velocity of the car is 5 m s^{-1}. Find the time taken for the velocity to reach 15 m s^{-1}.

The engine is now shut off. Find the time taken by the car to come to rest.

47 A car of mass 500 kg moves along a level road. The resistance to motion is $20v$ N when the velocity of the car is v m s^{-1}. If the engine works at a constant rate of 30 kW, find the time taken by the car to accelerate:

(i) from 10 m s^{-1} to 20 m s^{-1}
(ii) from 20 m s^{-1} to 30 m s^{-1}

What is the maximum speed of the car?

→ **48** A particle of mass 2 kg moves in a straight line. For speeds less than 6 m s^{-1} the resultant force on the particle is 3 N. For speeds more than 6 m s^{-1} the resultant force does work at the rate 18 W. The particle accelerates from rest.

(i) Show that the particle takes 4 s to reach a speed of 6 m s^{-1} and has then moved a distance 12 m.

(ii) Show that, subsequently, when the velocity of the particle is v, its acceleration is $9/v$. Hence, using the methods of section 4.4:2, show that the particle takes a further 6 s to reach a velocity of 12 m s^{-1} and has then moved a further distance 56 m.

R 49 A car of mass m moves on a level road. The resultant force acting on the car when its speed is less than u is kmg (k constant). When the speed of the car is greater than u, the power exerted by the resultant force is $kmgu$. Find the time taken for the car to accelerate from rest to a speed $2u$, and the distance travelled in this time.

8.5 Momentum

8.5:1 Impulse–Momentum Principle

Remember: The linear momentum of an object of mass m moving with velocity **v** is $m\mathbf{v}$. **(1)**

Definition
The impulse of a force **F** acting in the time interval t_1 to t_2 is

$$\int_{t_1}^{t_2} \mathbf{F}\, dt \qquad (2)$$

Impulse is a vector quantity. Its units are those of force × time, i.e. N s.

Note: if **F** is constant, the impulse of **F** in the time interval t_1 to t_2 is

$$[\mathbf{F}t]_{t_1}^{t_2} = \mathbf{F}(t_2 - t_1)$$

= force × time for which it acts

The Impulse–Momentum Principle (IMP)

The impulse of the resultant force on a body
= change in linear momentum of the body. **(3)**

The Impulse-Momentum Principle is a logical consequence of Newton's Second Law (N2).

Suppose that a resultant force **F** acts on a body of mass m in the time interval t_1 to t_2, changing its velocity from $\mathbf{v}_1$ to $\mathbf{v}_2$ (fig. 8.5:1). Let the acceleration of the body be **a**. By N2,

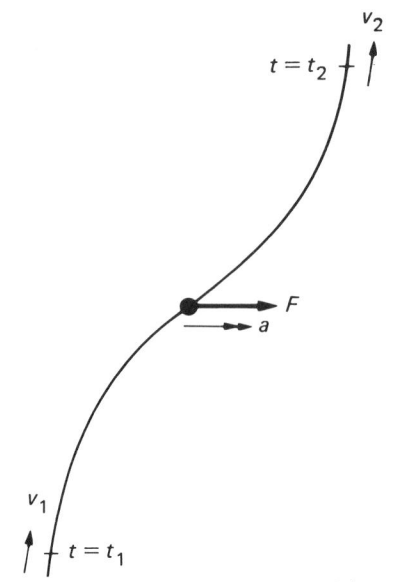

Fig. 8.5:1 $\qquad\qquad\qquad \mathbf{F} = m\mathbf{a} \qquad$ **(4)**

Forces on a particle 489

Integrate both sides of equation (4) with respect to t:

$$\int_{t_1}^{t_2} \mathbf{F}\, dt = \int_{t_1}^{t_2} m\mathbf{a}\, dt = [m\mathbf{v}]_{t_1}^{t_2} = m\mathbf{v}_2 - m\mathbf{v}_1$$

An immediate consequence of the Impulse-Momentum Principle is

---
The Conservation of Momentum Principle.
If the impulse of the resultant force on a body is zero, the change in linear momentum of the body is zero (i.e. the linear momentum of the body is conserved). **(5)**

This is simply a particular case of the Impulse-Momentum Principle (3); it says if LHS = 0, RHS = 0.

Note: we are going, for the moment, to consider bodies moving in a straight line; we will use the notation of section 7.1, so that

$$\text{momentum} = mv \quad \textbf{(1a)}$$

and

$$\text{impulse} = \int_{t_1}^{t_2} F\, dt \quad \textbf{(2a)}$$

and, if F is constant,

$$\text{impulse} = F(t_2 - t_1) \quad \textbf{(2b)}$$

Worked example 1 A constant resultant force $\mathbf{F}$ of magnitude F acting on a particle of mass 2 kg increases its speed from 4 m s^{-1} to 12 m s^{-1} in 4 s (fig. 8.5:2). Find F.

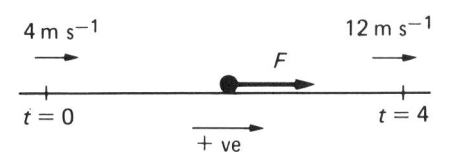

By equation (2b),

$$\text{impulse of force} = F \cdot 4$$

$$\text{change in momentum of particle} = 2 \cdot 12 - 2 \cdot 4 = 16$$

Fig. 8.5:2   By the Impulse-Momentum Principle (3),

$$F \cdot 4 = 16 \quad \Rightarrow \quad F = 4 \text{ N}$$

Note: we could have solved this problem using N2 and the constant acceleration equations. Considering the method by which we proved the Impulse-Momentum Principle, this is hardly surprising.

Worked example 2 A pipe of cross-sectional area 0.08 m^2 discharges water horizontally at a speed of 10 m s^{-1}. The water immediately strikes a wall and does not rebound (see fig. 8.5:3). Find the force F exerted by the wall on the water.

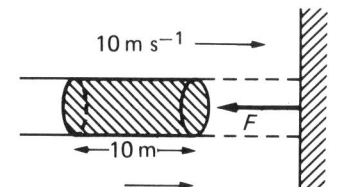

Volume of water discharged per second $= 10 \times 0.08 = 0.8\text{ m}^3$
$\Rightarrow$ Mass of water discharged per second $= 0.8 \times 1000 = 800$ kg
Initial momentum of water $= 800 \times 10 = 8000$ N s
Final momentum of water $= 0$ N s
$\Rightarrow$ Change in momentum of water per second $= -8000$ N s

So   impulse of force of wall on water per second $= -F \cdot 1 = -F$

Fig. 8.5:3   By Impulse-Momentum Principle,

$$-F = -8000\text{ N} \quad \Rightarrow \quad F = 8000\text{ N}$$

Exercise 8.5:1

- **1** Show that the units of linear momentum and impulse are in fact the same.

① **2** A particle of mass 3 kg, whose initial speed is 2 m s^{-1}, moves under the action of a force of magnitude 18 N in the direction of motion. Find the speed of the particle after 5 s.

① **3** Find the time taken for a force of magnitude 3 N to reduce the speed of a particle of mass 4 kg from 15 m s^{-1} to 10 m s^{-1}.

① → **4** A particle of mass 5 kg moves in a straight line under the action of a constant force of magnitude F. Initially it is moving with speed 10 m s^{-1}. Four seconds later it is moving in the opposite direction with the same speed. Find F.

① **5** A bullet of mass 0.05 kg travelling horizontally at 400 m s^{-1} hits a fixed wooden target. It comes to rest 0.002 s later. Using the Impulse–Momentum Principle, find the average resisting force of the wood.

- **6** A force $6t^2$ acts on a particle in the time interval $t = 0$ s to $t = 2$ s. Using definition (2a), find the impulse of the force. The force increases the velocity of the particle from 4 m s^{-1} to 6 m s^{-1}. Find the mass of the particle.

7 A force $12 \sin 2t$ acts on a particle of mass 4 kg. Initially the particle is moving with velocity -6 m s^{-1}. After how long is it:

(i) instantaneously at rest for the first time?
(ii) moving with velocity $+6$ m s^{-1}, for the first time?

Find the speed of the particle after $\frac{1}{3}\pi$ s.

Questions 8–11: A particle of mass 0.25 kg moves in a straight line under the action of a resultant force of magnitude F. Its initial velocity is 3 m s^{-1}. Find its velocity after 1 s for the given value of F.

8 $F = t^3$

9 $F = -t^3$

→ **10** $F = \pi \cos \pi t$

R **11** $F = \pi \sin \pi t$

12 A particle of mass 4 kg moves in a straight line under the action of a resultant force F whose magnitude varies with time t. The table gives the values of F against t.

F (N)	9.2	6.6	5.2	4.7	4.4
t (s)	0	5	10	15	20

Using Simpson's rule (see section 4.3), find the impulse of F in the time intervals:

(i) $t = 0$ s to $t = 10$ s
(ii) $t = 0$ s to $t = 20$ s

Given that the speed of the particle after 10 s is 20 m s^{-1}, find its initial speed and its speed after 20 s.

- **13** A particle moves under the action of a force $24 \cos 3t\mathbf{i} - 6 \sin 3t\mathbf{j}$. Find the change in its momentum in the first $\frac{1}{9}\pi$ seconds of motion.

14 A particle moves from rest under the action of two forces $2\mathbf{i} - 2t\mathbf{j}$ and $-4\mathbf{i} + 3t^2\mathbf{j}$. Find the momentum of the particle at time t.

② → **15** A pipe of cross-sectional area 0.05 m^2 discharges water horizontally at a speed of 3 m s^{-1}. The water immediately strikes a wall and does not rebound. Find the force exerted by the wall on the water.

② **16** A pipe of cross-sectional area A m^2 discharges water horizontally at a speed of v m s^{-1}. The water immediately strikes a wall. Find the force exerted by the wall on the water if:

(i) the water does not rebound from the wall;
(ii) the water rebounds with speed u m s^{-1}.

② **17** A pipe discharges water at a rate of 500 litre s^{-1} with a speed of 5 m s^{-1}. The The water immediately strikes a wall and does not rebound. Find the force exerted by the wall on the water.

② R **18** A pipe discharges water with a speed of 12 m s^{-1}. The water immediately strikes a wall and does not rebound. If the force exerted by the water on the wall is 360 N, find the volume of water leaving the pipe per second. Hence find the diameter of the pipe.

② **19** A cricket sight screen measures 10 m by 6 m. The wind blows at right angles to the screen and the air is brought to rest on impact with the screen.

(i) If the wind speed is 16 m s^{-1}, calculate the force acting on the screen.

(ii) If the screen blows over when the force on it is more than 100 kN, find the maximum possible wind speed at which it stays upright.

(The mass of 1 m^3 of air is 1.25 kg.)

② **20** A sailing dinghy is 'running', i.e. sailing with the wind directly behind and the sail perpendicular to the direction of the wind. The dinghy moves with constant speed v against a resistance of 150 N. The wind, which is blowing at 10 m s^{-1}, has its speed reduced to v when it hits the sail. Find v, given that the sail area is:

(i) 6 m^2 (ii) 9 m^2

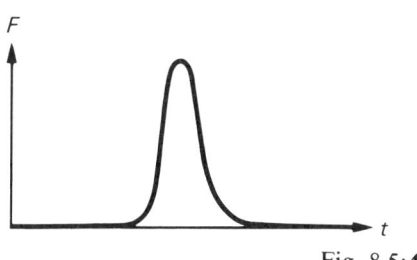

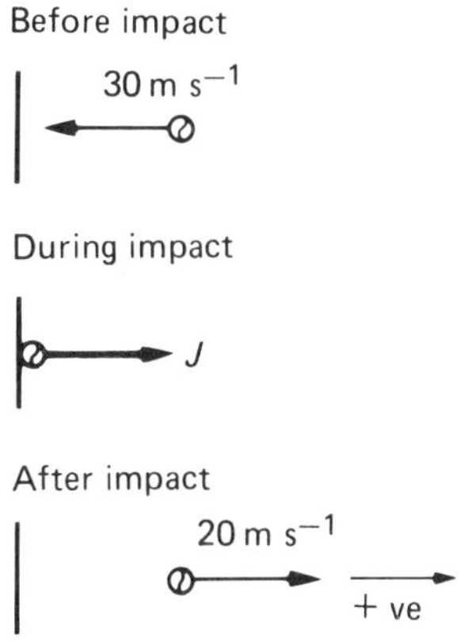

8.5:2 Impulsive forces

An **impulsive force** is a force of large (average) magnitude which acts for a very short time (as in fig. 8.5:4), e.g. the force of the racket on the ball during a tennis shot.

We are not interested in the acceleration of the ball at each instant that the racket and the ball are in contact, only in the change in velocity of the ball. Thus we are not interested in the force itself, only in the impulse J of the force. We call J the *impulse exerted by the racket on the ball*.

Worked example 3 A tennis player is facing a serve. The ball approaches his racket with speed 30 m s^{-1}. He hits the ball with the face of the racket perpendicular to its path. The ball leaves his racket with speed 20 m s^{-1}. If the mass of the ball is 0.1 kg, find the impulse exerted by the racket on the ball. Assuming that the racket and the ball are in contact for 0.04 s, find the average force F exerted by the racket on the ball (fig. 8.5:5).

Initial velocity $v_1 = -30$ m s^{-1}; final velocity $v_2 = 20$ m s^{-1}. Therefore

$$\text{impulse } J = mv_2 - mv_1 = (0.1)(20) - (0.1)(-30) = 5 \text{ N s}$$

The force exerted by the racket on the ball is not constant, but

$$\text{impulse} = \text{average force} \times \text{time for which it acts}$$
$$5 = F \times 0.04$$
$$\Rightarrow \quad F = 100 \text{ N}$$

Important note: we neglect the impulse exerted by the weight of the ball during contact since it is very small compared to the impulse exerted by the racket (see question 3).

Collisions

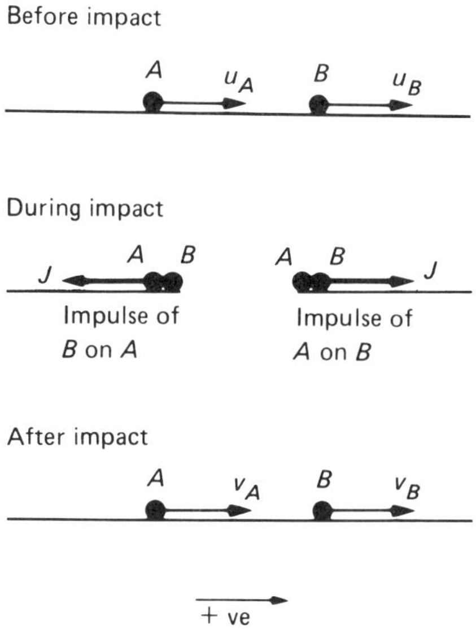

Suppose that two smooth bodies A and B of masses m_A and m_B, travelling with velocities u_A and u_B in the same straight line, collide directly. ('Directly' means that their direction of motion is along their common normal. We usually think of the bodies as being spheres: in this case the common normal is the line joining their centres; so the direction of motion is along the line joining their centres.) Suppose that their velocities after impact are v_A and v_B (see fig. 8.5:6).

During the impact, A and B exert equal and opposite forces on each other (N3); and so they exert equal and opposite impulses on each other. (The fact that the bodies are smooth ensures that the forces, and consequently the impulses, act along the common normal.)

Apply the Impulse–Momentum Principle to each particle

For A: $\quad -J = m_A v_A - m_A u_A \quad$ (6)

For B: $\quad J = m_B v_B - m_B u_B \quad$ (7)

(6)+(7) $\quad 0 = m_A v_A - m_A u_A + m_B v_B - m_B u_B$

i.e. $\quad m_A u_A + m_B u_B = m_A v_A + m_B v_B \quad$ (8)

Total momentum of bodies before collision

= total momentum of bodies after collision

Note: The total impulse exerted on the system is zero, so, by the Conservation of Momentum Principle (5), the change in momentum of the system is zero. This directly implies equation (8).

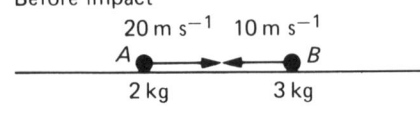

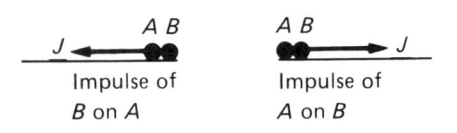

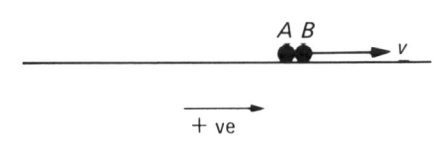

Worked example 4 Two smooth spheres A and B, of masses 2 kg and 3 kg, collide head-on when their speeds are 20 m s^{-1} and 10 m s^{-1} (see fig. 8.5:7). If the impact is inelastic, find:

(i) their combined speed after impact
(ii) the impulse given to B

The initial velocities are $u_A = 20$ m s^{-1} and $u_B = -10$ m s^{-1}.

'The impact is inelastic' means that the particles both move with the same velocity after impact; let this velocity be v, i.e. $v_A = v_B = v$.

(i) Applying the Conservation of Momentum Principle to the system (equation (8)):

$$2 \cdot 20 + 3 \cdot (-10) = 2v + 3v \Rightarrow v = 2 \text{ m s}^{-1}$$

(ii) Applying the Impulse–Momentum Principle to B (equation (7)):

$$J = 3v - 3 \cdot (-10) = 36 \text{ N s}$$

Impulsive tensions

Suppose that two bodies A and B are attached to the ends of a light string. When the string jerks there is a large tension J (an **impulsive tension**) in it for a short time. So the string exerts impulses of equal magnitude on A and B (fig. 8.5:8).

Worked example 5 A particle A of mass 4 kg rests on a table. It is attached to a light inextensible string which passes over a smooth fixed pulley and carries a particle B of mass 2 kg at the other end. If B is raised vertically through a distance 0.2 m and is then dropped, find the speed with which A begins to rise.

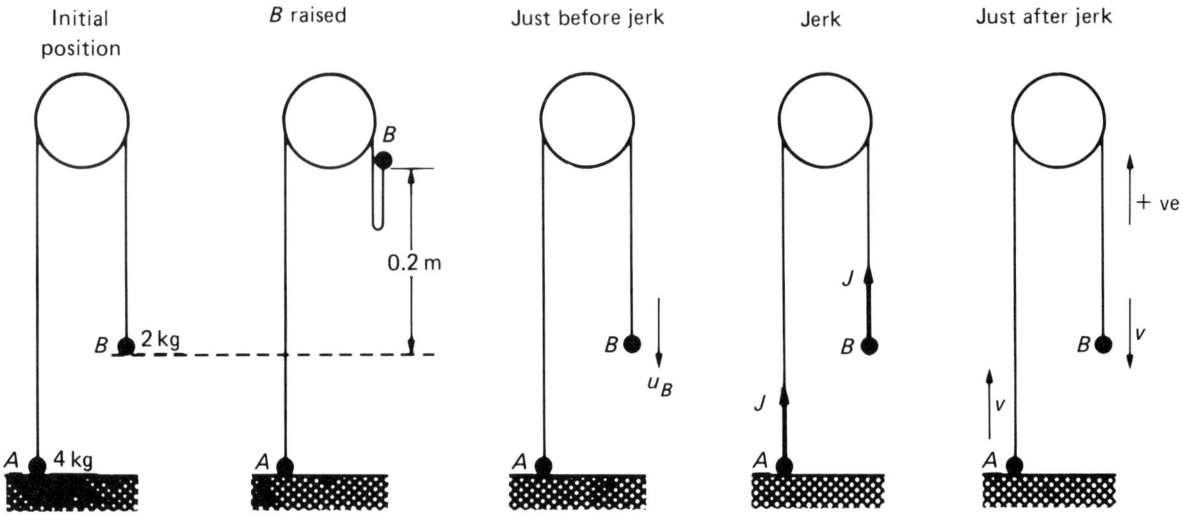

Suppose that just before the jerk B is moving with speed u_B. Using the constant acceleration equation $v^2 = u^2 + 2ax$:

$$u_B^2 = 2 \cdot 10 \cdot 0.2 \implies u_B = 2 \text{ m s}^{-1}$$

Suppose that the common speed of the particles after the jerk is v. Applying the Impulse-Momentum Principle to each particle (taking upwards as positive):

For A: $\qquad J = 4v - 4 \cdot 0$

For B: $\qquad J = -2v - 2 \cdot (-2)$

$\implies \qquad v = \tfrac{2}{3} \text{ m s}^{-1}$

Exercise 8.5:2

1 A sphere of mass 2 kg moving with speed 5 m s^{-1} strikes a wall normally (i.e. at right angles) and bounces off with speed 4 m s^{-1}. Find the impulse exerted by the wall on the sphere.

2 A cricket ball of mass 0.2 kg reaches a batsman with horizontal speed 15 m s^{-1} and is hit straight back horizontally with speed 25 m s^{-1}. Find the impulse of the bat on the ball. Given that the bat and the ball are in contact for 0.05 s, find the average force exerted by the bat on the ball.

3 A sphere of mass 4 kg moving with speed 30 m s^{-1} strikes a wall normally (i.e. at right angles) and bounces back with speed 15 m s^{-1}. Find the impulse of the wall on the sphere. Given that the wall and the sphere are in contact for 0.03 s, find the average force exerted by the wall on the sphere.

Find the impulse of the weight of the sphere during the period of contact (see note on worked example 3).

4 A particle of mass 6 kg is moving with speed 5 m s^{-1} when it receives an impulse of 42 N s in the direction of motion. Find:

(i) the speed of the particle immediately after the impulse
(ii) the increase in KE of the particle

5 A particle of mass 0.2 kg is moving with speed 80 m s^{-1} when it receives an impulse of 50 N s. Find the change in KE of the particle if the impulse acts:

(i) in the same direction as the motion
(ii) in the opposite direction to the motion

R 6 A particle of mass m is moving with speed u when it receives an impulse I in the direction of motion. Find the speed of the particle immediately after the impulse, and show that the increase in its KE is $I(I+2mu)/2m$.

7 A particle of mass m is moving with speed $3u$ when it receives an impulse of magnitude mu in the opposite direction to its direction of motion. Find:

(i) the speed of the particle immediately after the impulse
(ii) the decrease in KE of the particle

8 A bead of mass 2 kg, threaded onto a straight, smooth, horizontal, fixed wire, is moving with speed 20 m s^{-1}. Find the possible values of the magnitude of the impulse which, applied to the particle along the wire, would change the speed of the particle to 30 m s^{-1}.

9 A stream of particles of mass 0.0004 kg, all 0.05 m apart, are travelling in a straight line with a velocity of 16 m s^{-1} perpendicular to a flat screen. As each hits the screen it is instantaneously brought to rest. Find the average force needed to hold the screen in place.

Questions 10–15: Two spheres A and B of masses m_A and m_B collide when moving with velocities u_A and u_B. The impact is inelastic, and their combined velocity after impact is v. Find the quantity indicated, and in each case find also:

(i) the magnitude of the impulse exerted by A on B (and of the impulse exerted by B on A)
Hint: use equation (7).

(ii) the loss in KE caused by the impact (i.e. the total KE of the spheres before impact minus their total KE after impact)

10 $m_A = 5$ kg, $m_B = 2$ kg, $u_A = 11$ m s^{-1}, $u_B = -10$ m s^{-1}. Find v.

11 $m_A = 3$ kg, $m_B = 7$ kg, $u_A = 12$ m s^{-1}, $u_B = 2$ m s^{-1}. Find v.

12 $m_A = 2m$, $m_B = m$, $u_A = 2u$, $u_B = -5u$. Find v.

13 $m_A = 2m$, $m_B = m$, $u_A = 2u$, $u_B = 0$. Find v.

14 $m_A = m$, $m_B = km$, $u_A = 3u$, $u_B = 2u$, $v = \tfrac{12}{5}u$. Find k.

15 $m_A = 4m$, $m_B = 3m$, $u_A = u$, $u_B = ku$, $v = -2u$. Find k.

④ **16** Two spheres of masses 3 kg and 5 kg collide and coalesce (stick together) when they are moving in the same straight line with speeds 7 m s^{-1} and 2 m s^{-1}. Find their combined speed after impact if they were originally moving:

(i) in the same direction
(ii) in opposite directions

④ **17** Two spheres of equal mass collide and coalesce when they are moving in the same straight line with speeds $3u$ and $2u$. Find their combined speed after impact if they were originally moving:

(i) in the same direction
(ii) in opposite directions

④ **18** A sphere of mass m moving with speed u strikes a sphere of mass $3m$ at rest and coalesces with it. Find the ratio of the final KE to the initial KE.

④ • **19** A bullet of mass 0.03 kg is fired horizontally with a velocity of 300 m s^{-1} directly into a stationary wooden block of mass 1.47 kg, which is free to move on a smooth table, and becomes embedded in the block. Find the common speed of the bullet and the block after the impact, and the loss in KE caused by the impact.

④ **20** A bullet is fired horizontally with a velocity of 600 m s^{-1} directly into a wooden block of mass 0.245 kg, resting on a smooth table, and becomes embedded in the block. The block begins to move with a velocity of 12 m s^{-1}. Find the mass of the bullet and the loss in KE caused by the impact.

21 A particle of mass $3m$ travelling horizontally with speed u collides and coalesces with a particle of mass m hanging at rest at the end of a light inextensible string of length l. Find the KE of the combined particle after the impact. Hence, using the Conversvation of Mechanical Energy Principle (section 8.4:3 equation (19)) show that, if the string rotates through an angle 60° before first coming to rest, $u = \frac{1}{3}\sqrt{(8gl)}$.

R **22** Two beads, of masses $3m$ and $2m$, are threaded onto a smooth fixed vertical circular wire, centre O radius a. They are released simultaneously from the opposite ends of the horizontal diameter of the wire (i.e. from the two points level with O). When they reach the lowest point of the wire they collide and coalesce. Find:

(i) the depth below O of the point where they first come to rest
(ii) the loss in KE caused by the impact

• **23** A pile-driver of mass 3000 kg falls from rest through a height 5 m onto a pile of mass 2000 kg and does not rebound.

(i) Find the speed of the pile-driver just before impact.

(ii) Find the common speed of the pile and the pile-driver just after impact.

(iii) Find the combined KE of the pile and the pile-driver just after impact.

Given that the pile is driven 0.5 m into the ground, use the WEP to find the resistance to penetration of the ground. (Assume the resistance is constant.)

24 A pile-driver of mass 2000 kg falls from rest through a height 20 m onto a pile of mass 1000 kg and does not rebound. The pile is driven 0.2 m into the ground. Find:

(i) the common speed of the pile and the pile-driver just after impact
(ii) the average resistance to penetration of the ground

→ **25** A pile-driver of mass 4000 kg falls from rest through a height 7.2 m onto a pile of mass 1000 kg and does not rebound. The pile is driven 0.4 m into the ground. Find the average resistance to penetration of the ground.

• **26** A bullet of mass 0.04 kg is fixed horizontally with a velocity of 100 m s^{-1} into a block of wood of mass 8 kg resting on a smooth table. The bullet goes right through the block and emerges with velocity 40 m s^{-1}.

(i) Find the speed of the block after the impact.
(ii) Find the impulse exerted by the block on the bullet.

27 A bullet of mass m is fired horizontally with a velocity of $3u$ into a block of wood of mass $7m$ resting on a smooth table. The bullet goes right through the block and emerges with velocity u.

(i) Find the speed of the block after the impact.
(ii) Find the impulse exerted by the block on the bullet.

→ **28** A bullet of mass m is fired horizontally with a velocity of $4u$ into a block of wood of mass $15m$ resting on a smooth table. Find the velocity of the block after the impact if:

(i) the bullet goes right through the block and emerges with velocity $3u$
(ii) the bullet becomes embedded in the block

+ **29** A bullet of mass 0.1 kg is fired horizontally with a velocity 200 m s^{-1} into the centre of a vertical face of a stationary cube of mass 0.4 kg and side 0.2 m. The bullet goes right through the cube and emerges with velocity 120 m s^{-1}.

(i) Find the velocity of the block after the impact.
(ii) Using the WEP, find the work done on the cube and the work done on the bullet.

Suppose that the cube moves a distance x before the bullet emerges.

(iii) How far does the bullet move before it emerges?
(iv) If the force exerted by the cube on the bullet (and by the bullet on the cube) is of constant magnitude F, find F and x.

- **30** A gun of mass 50 kg fires a bullet of mass 0.04 kg and recoils horizontally. If the bullet leaves the gun with speed 300 m s^{-1}, find:

  (i) the speed at which the gun begins to recoil
  (ii) the KE imparted to the bullet and the gun

  (Note: when the gun is fired, the explosion exerts equal forces in all directions; so the resultant force on the system of bullet and gun is zero; thus the resultant impulse on the system is zero.)

31 A gun of mass 20 kg fires a bullet of mass 0.02 kg and recoils horizontally. If the bullet leaves the gun with speed 400 m s^{-1}, find:

  (i) the speed at which the gun begins to recoil
  (ii) the KE imparted to the bullet and the gun

32 A gun of mass $24m$ fires a bullet of mass m and recoils horizontally. If the bullet leaves the gun with speed u, find:

  (i) the speed at which the gun begins to recoil
  (ii) the KE imparted to the bullet and the gun

R 33 A gun of mass km fires a bullet of mass m and recoils horizontally. The total KE imparted to the bullet and the gun is E. Find the speed at which the bullet leaves the gun (in terms of k, m and E).

- **34** A gun of mass 1000 kg fires a shell of mass 20 kg and recoils horizontally. If the shell leaves the gun with speed 500 m s^{-1},

  (i) find the speed at which the gun begins to recoil
  (ii) using the IMP, find the constant force needed to bring the gun to rest in 2 s

→ **35** A gun of mass km fires a bullet of mass m and recoils horizontally. If the bullet leaves the gun with speed u, find:

  (i) the speed at which the gun begins to recoil
  (ii) the KE imparted to the gun
  (iii) the constant force needed to bring the gun to rest in time T
  (iv) the constant force needed to bring the gun to rest in a distance x

- **36** A particle of mass 12 kg, moving in a straight line with velocity 20 s^{-1}, explodes, splitting into two fragments—A of mass 5 kg and B of mass 7 kg—which continue to move along the line in the same direction as the original particle. If the velocity of B is 30 m s^{-1}, find the velocity of A and the KE generated by the explosion.

→ **37** A particle of mass $5m$, moving in a straight line with velocity u, explodes, splitting into two fragments: A of mass $2m$ and B of mass $3m$. B moves with velocity $3u$ in the same direction as the original particle. Find the velocity of A and the KE generated by the explosion.

38 A particle of mass $4m$, moving in a straight line with velocity u, explodes, splitting into two fragments—A of mass m and B of mass $3m$—which move in the same straight line as before, but in opposite directions. If the KE generated by the explosion is $5mu^2$, find the velocities of A and B.

Questions 39–41: Two particles A and B of masses m_A and m_B are connected by a light inextensible string. The particles rest on a smooth table with the string taut. An impulse of magnitude I is applied to A in the direction BA. Find:

  (i) the speed with which the particles begin to move
  (ii) the jerk in the string
  (iii) the KE generated by the impulse

- **39** $m_A = 4$ kg, $m_B = 12$ kg, $I = 8$ N s.

40 $m_A = m$, $m_B = m$, $I = 2mu$.

→ **41** $m_A = 5m$, $m_B = 3m$, $I = 2mu$.

Questions 42–45: A particle A of mass m_A rests on a table. It is attached to a light inextensible string which passes over a smooth fixed pulley and carries a particle B of mass m_B at the other end. B is raised vertically through a distance h and is then dropped.

⑤ → **42** $m_A = 3$ kg, $m_B = 2$ kg, $h = 0.8$ m.

  (i) Find the speed of B just before the string becomes taut.
  (ii) Find the impulsive tension in the string.
  (iii) Find the speed with which A begins to rise.

⑤ **43** $m_A = 4$ kg, $m_B = 1$ kg, $h = 1.25$ m.

  (i) Find the speed with which A begins to rise.
  (ii) Find the acceleration of the particles while the string is taut.
  (iii) Find the time taken after the jerk before the system comes instantaneously to rest.

⑤ **44** $m_A = 6$ kg, $m_B = 2$ kg, $h = 5$ m.

  (i) Find the speed with which A begins to rise.
  (ii) Find the time that elapses before A strikes the table again.
  (iii) Find the speed with which A strikes the table.
  (iv) Assuming that A does not rebound, find the impulse exerted by A on the table.

⑤ **R 45** $m_A = 2m$, $m_B = m$.

  (i) Find (in terms of h and g) the speed with which A begins to rise.
  (ii) Find the time that elapses before A strikes the table again.
  (iii) Find the speed with which A strikes the table.

Questions 46 and 47: A particle A of mass m_A rests on a table. It is attached to a light inextensible string of length l which passes over a smooth pulley fixed at a height h above the table, and carries a particle B of mass m_B at the other end. B is raised to the height of the pulley and then dropped.

⑤ → 46 $m_A = 2$ kg, $m_B = 1$ kg, $l = 4.05$ m, $h = 2.25$ m.

(i) Find the speed with which A begins to rise.
(ii) Find the (common) speed of the particles when B reaches the plane.
(iii) Assuming B does not rebound, find the impulse exerted by B on the plane.
(iv) Is there an impulse tension in the string when B hits the plane?

⑤ 47 $m_A = 3m$, $m_B = m$, $l = 5a$, $h = 3a$.

(i) Find the speed with which A begin to rise.
(ii) Show that loss of KE due to the jerk in the string is $3mga$.
(iii) Show that B never reaches the table. Find the total time taken for B to reach its lowest point.

Questions 48-50: Two particles A and B of masses m_A and m_B are connected by a light inextensible string passing over a smooth fixed pulley. Initially the particles are held so that they are both at a height h above a fixed horizontal plane and the string is taut. The system is released from rest.

⑤ 48 $m_A = 2$ kg, $m_B = 1$ kg, $h = 2.4$ m.

(i) Find the velocity with which A strikes the plane.
(ii) Assuming that A does not rebound, find the impulse exerted on A by the plane.
(iii) Find the greatest height reached by B.
(iv) What is the velocity of B just before the string becomes taut again?
(v) Find the velocity with which A leaves the plane.

⑤ 49 $m_A = 3$ kg, $m_B = 2$ kg, $h = 2.25$ m.

(i) Find the time taken before A strikes the plane.
(ii) Find the further time taken before the string is again taut.
(iii) Find the speed with which A leaves the plane.
(iv) Find the further time taken before A strikes the plane for the second time.

⑤ → 50 $m_A = 2m$, $m_B = m$, $h = 3l$. Show that:

(i) A strikes the plane after time $3\sqrt{(2l/g)}$
(ii) A leaves the plane again after a further time $2\sqrt{(2l/g)}$
(iii) the time between the first and second jerks is $\tfrac{8}{3}\sqrt{(2l/g)}$

Questions 51-54: Two particles A and B of masses m_A and m_B are connected by a light inextensible string passing over a smooth fixed pulley. The system is released from rest with the string taut and the straight parts of the string vertical. After t s, B picks up a particle of mass m_C which was at rest. Find:

(i) the impulsive tension in the string
(ii) the (common) speed of the particles just after the jerk
(iii) the time taken after the jerk for the system to come instantaneously to rest
(iv) the total distance travelled by B

• 51 $m_A = 8$ kg, $m_B = 2$ kg, $m_C = 10$ kg, $t = 1$.

R 52 $m_A = 1.3$ kg, $m_B = 1.2$ kg, $m_C = 1.0$ kg, $t = 2$.

53 $m_A = 3m$, $m_B = 2m$, $m_C = 5m$, $t = 2.5$.

→ 54 $m_A = 3m$, $m_B = m$, $m_C = 3m$, $t = \tau$.

8.5:3 Direct elastic impact

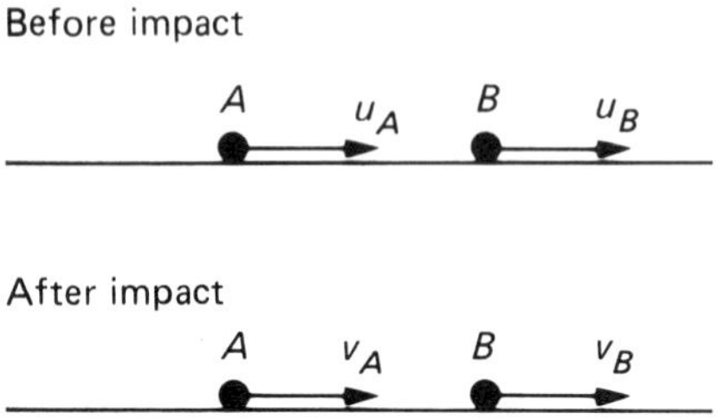

Newton's Law of Restitution

Suppose that two smooth bodies A and B, travelling with velocities u_A and u_B in the same straight line, collide, and that their velocities after impact are v_A and v_B (fig. 8.5:10). Then

$$v_B - v_A = e(u_A - u_B) \tag{9}$$

Before impact

A u_A B u_B

After impact

A v_A B v_B

Fig. 8.5:10

where e, the **coefficient of restitution** between A and B, is a constant depending on the materials of which A and B are made.

Inelastic collision: $v_A = v_B$; so $e = 0$
Elastic collision: $0 < e \leq 1$, i.e. A and B collide and bounce
Perfectly elastic collision: $e = 1$

Forces on a particle 497

Note: Newton's Law of Restitution

(i) applies to the impact of two freely moving bodies *and* to the impact of one moving body and one constrained body (e.g. ball and tennis racket, ball and wall).

(ii) is independent of the masses of the bodies.

(iii) is an experimental law—it cannot be proved only verified.

Worked example 6 A smooth sphere of mass 0.5 kg moving with speed 6 m s^{-1} strikes a wall normally (i.e. at right angles) and rebounds with speed 4 m s^{-1} (fig. 8.5:11). Find:

(i) the coefficient of restitution between the sphere and the wall
(ii) the impulse exerted by the wall on the sphere

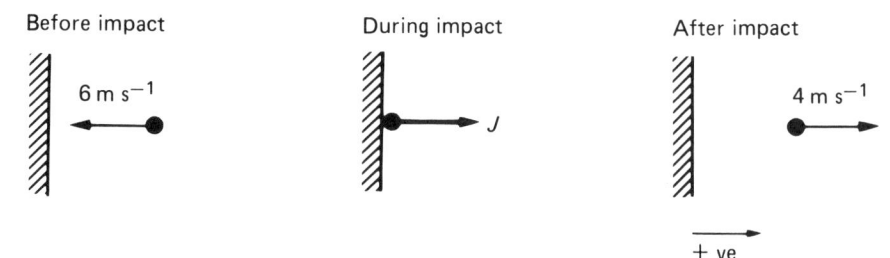

Fig. 8.5:11

Initial velocity of wall, $u_A = 0$ m s^{-1} = v_A, final velocity of wall.
Initial velocity of sphere, $u_B = -6$ m s^{-1}; final velocity $v_B = 4$ m s^{-1}.

(i) NLR $\Rightarrow 4 - 0 = e(0 - (-6))$ i.e. $e = \frac{2}{3}$
(ii) Applying the Impulse-Momentum Principle to the sphere:

$$J = (0.5)(4) - (0.5)(-6) = 5 \text{ N s}$$

(c.f. worked example 3).

Worked example 7 Two smooth spheres A and B of masses 3 kg and 5 kg collide directly head on with speeds 12 m s^{-1} and 2 m s^{-1}. If the coefficient of restitution between them is $\frac{1}{7}$, find:

(i) their speeds after impact
(ii) the overall loss in KE due to the impact

Initial velocities are $u_A = 12$ m s^{-1}, $u_B = -2$ m s^{-1}. Let the final velocities be v_A and v_B (fig. 8.5:12).

(i) Applying the Conservation of Momentum Principle to the system (equation (8)):

$$3 \cdot 12 + 5 \cdot (-2) = 3v_A + 5v_B$$

From Newton's Law of Restitution (equation (9)):

$$v_B - v_A = \tfrac{1}{7}(12 - (-2))$$

$$\Rightarrow \quad v_A = 2 \text{ m s}^{-1} \quad \text{and} \quad v_B = 4 \text{ m s}^{-1}$$

(ii) Total KE of spheres before impact = $\tfrac{1}{2} \cdot 3 \cdot 12^2 + \tfrac{1}{2} \cdot 5 \cdot 2^2 = 226$ J

Total KE of spheres after impact = $\tfrac{1}{2} \cdot 3 \cdot 2^2 + \tfrac{1}{2} \cdot 5 \cdot 4^2 = 46$ J

$\Rightarrow$ overall loss in KE due to impact = $226 - 46 = 180$ J

Fig. 8.5:12

Exercise 8.5:3

1 A smooth sphere of mass 4 kg moving with speed 5 m s^{-1} strikes a wall normally and rebounds with speed 3 m s^{-1}. Find:

(i) the coefficient of restitution between the sphere and the wall
(ii) the impulse exerted by the wall on the sphere

2 A smooth sphere of mass 0.5 kg moving with speed 12 m s^{-1} strikes a wall normally. Find the coefficient of restitution between the sphere and the wall if the impulse exerted by the sphere is:

(i) 6 N s   (ii) 8 N s   (iii) 10 N s   (iv) 12 N s

3 A smooth sphere of mass 1.2 kg moving with speed 4 m s^{-1} strikes a wall normally. The coefficient of restitution between the sphere and the wall is $\frac{3}{8}$. Find:

(i) the speed of the sphere after impact
(ii) the impulse exerted by the wall on the sphere
(iii) the loss in KE caused by the impact

4 A smooth sphere of mass m moving with speed u strikes a wall normally. The coefficient of restitution between the sphere and the wall is e. Find:

(i) the speed of the sphere after impact
(ii) the impulse exerted by the wall on the sphere
(iii) the loss in KE caused by the impact

What is the answer to (iii) if the collision is perfectly elastic?

Questions 5–7: A smooth sphere falls from rest at a height 0.8 m above a horizontal plane. The coefficient of restitution between the sphere and the plane is $\frac{1}{2}$.

5 Find the height to which the sphere rises after its first bounce.

6 Find the height to which the sphere rises after:

(i) its second bounce
(ii) its third bounce
(iii) its nth bounce

Hence find the total distance it moves before coming to rest. (See section 5.1:4.)

7 Find:

(i) the time taken for the sphere to reach the ground for the first time
(ii) the time that elapses between the first and second bounces

Find the total time that elapses before the sphere comes to rest.

8 A smooth sphere falls from rest at a height h above a horizontal plane. The coefficient of restitution between the sphere and the plane is e.

(i) Find the total distance moved by the particle before it comes to rest.
(ii) Find the total time that elapses before the particle comes to rest.

9 A ball is dropped from 20 m above the ground and comes to rest 12 s later. Find the coefficient of restitution between the ball and the ground.

10 A particle is projected from a point O on level ground with initial velocity u at an angle α to the horizontal and hits a vertical wall when travelling horizontally. Show that the time taken to reach the wall is $(u \sin \alpha)/g$. By considering the vertical component of the motion only, show that the time taken to reach the ground after hitting the wall is also $(u \sin \alpha)/g$.

11 A particle is projected from a point O on level ground and hits a vertical wall when travelling horizontally. If the horizontal distance from O to the wall is 40 m and the coefficient of restitution between the particle and the wall is $\frac{2}{5}$, find the distance from O at which the particle strikes the ground.

12 A particle is projected from a point O on level ground and hits a vertical wall 20 m from O when travelling horizontally. If the speed of projection is 25 m s^{-1}, find the two possible angles of projection. If the coefficient of restitution between the particle and the wall is $\frac{1}{4}$, find the distance from O at which the particle strikes the ground.

13 One end of a light inextensible string of length a is fixed to a point A on a wall; the other end is attached to a particle P of mass m. The particle is pulled away from the wall until the string is taut and horizontal, and is then released. The coefficient of restitution between the wall and the particle is $\frac{1}{2}$. Find:

(i) the angle between the string and the vertical when the particle next comes to instantaneous rest
(ii) the loss in KE caused by the impact
(iii) the loss in PE between the two instants when the particle is instantaneously at rest

14 A smooth sphere strikes a plane *obliquely* and rebounds. Explain, using the Impulse–Momentum Principle, why the component of the velocity of the sphere parallel to the wall is unaltered.

The initial velocity of the sphere is u and makes an angle α with the plane. Assuming that Newton's Law of Restitution applies perpendicular to the plane, show that, if the coefficient of restitution between the sphere and the plane is e, the sphere rebounds with speed $v = u\sqrt{(\cos^2 \alpha + e^2 \sin^2 \alpha)}$ at an angle θ to the plane where $\tan \theta = e \tan \alpha$. Interpret this result in the cases $e = 0$ and $e = 1$.

Forces on a particle 499

Questions 15–24: Two smooth spheres A and B of masses m_A and m_B, moving with velocities u_A and u_B in the same straight line, collide; their velocities after impact are v_A and v_B. The coefficient of restitution between the spheres is e. Find the quantity indicated, and in each case find also:

(i) the magnitude of the impulse exerted by A on B (and of the impulse exerted by B on A)
(ii) the loss in KE caused by the impact

⑦ → **15** $m_A = 2$ kg, $m_B = 3$ kg, $u_A = 20$ m s^{-1}, $u_B = 12$ m s^{-1}, $e = \frac{1}{4}$. Find v_A and v_B.

⑦ **16** $m_A = 10$ kg, $m_B = 5$ kg, $u_A = 16$ m s^{-1}, $u_B = -4$ m s^{-1}, $e = \frac{1}{2}$. Find v_A and v_B.

⑦ **17** $m_A = 2m$, $m_B = m$, $u_A = u$, $u_B = 0$, $e = \frac{1}{3}$. Find v_A and v_B.

⑦ **18** $m_A = 2m$, $m_B = m$, $u_A = 2u$, $u_B = u$, $v_B = 2u$. Find v_A and e.

⑦ **19** $m_A = 3$ kg, $u_A = 5$ m s^{-1}, $u_B = 0$ m s^{-1}, $v_A = 0$ s^{-1}, $e = \frac{1}{2}$. Find m_B and v_B.

⑦ **20** $m_A = 4$ kg, $u_A = 12$ m s^{-1}, $u_B = 6$ m s^{-1}, $v_A = 5$ m s^{-1}, $v_B = 8$ m s^{-1}. Find e and m_B.

⑦ → **21** $m_A = m$, $m_B = km$, $u_A = 2u$, $u_B = u$, $v_A = \frac{2}{3}u$, $v_B = \frac{5}{3}u$. Find e and k.

⑦ R **22** $m_A = m$, $u_A = 6u$, $u_B = 2u$, $v_A = 3u$, $v_B = 5u$. Show that $m_B = m$ and find e.

⑦ **23** $m_A = m$, $m_B = 3m$, $u_A = u$, $u_B = 0$. Find v_A and v_B in terms of u and e.

⑦ **24** $m_A = m$, $m_B = 3m$, $u_A = u$, $u_B = -2u$. Find v_A and v_B in terms of u and e.

⑦ **25** Two spheres of masses 6 kg and 4 kg collide when they are moving in the same straight line with speeds 5 m s^{-1} and 4 m s^{-1}. The coefficient of restitution between them is $\frac{1}{2}$. Find their speeds after impact if they collide when moving:

(i) in the same direction
(ii) in opposite directions

⑦ **26** Two spheres of masses m and $3m$ collide when they are moving in the same straight line with speeds $2u$ and u. The coefficient of restitution between them is $\frac{1}{5}$. Find their speeds after impact if they collide when moving:

(i) in the same direction
(ii) in opposite directions

⑦ **27** Two perfectly elastic spheres of equal mass collide directly when they are moving in opposite directions. Show that they exchange velocities.

⑦ **28** Two spheres of equal mass, moving in opposite directions with speeds ku and u $(k > 1)$, collide directly. The coefficient of restitution between them is $\frac{1}{2}$. As a result of the impact one of the spheres is brought to rest. Find k.

⑦ **29** A sphere A of mass m, travelling with speed αu, collides directly with another sphere B of mass βm, travelling with speed u. If the impact brings A to rest show that

$$e = \frac{\alpha + \beta}{\alpha\beta - \beta}$$

Hence show that $\beta > \alpha/(\alpha - 2)$.

⑦ **30** The spheres of equal mass collide head-on when moving with speeds u and $2u$. Half the original KE is lost on impact. Show that $e = \frac{2}{3}$.

⑦ **31** Two smooth spheres A and B of masses m_A and m_B collide when moving with velocities u_A and u_B. If the coefficient of restitution between them is e, find their velocities after impact and hence show that the loss in KE caused by the impact is

$$\frac{1}{2}\left(\frac{m_A m_B}{m_A + m_B}\right)(u_A - u_B)^2(1 - e^2)$$

What happens if the collision is perfectly elastic?

32 Estimate the value of e for the collision of:

(i) two billiard balls
(ii) a car and a lampost
(iii) two cars
(iv) a boxer and the ropes surrounding the ring
(v) a raindrop and a roof
(vi) a hailstone and a roof
(vii) an umbrella and a sewing machine

: **33** Explain carefully what happens to the shape of two spheres when they collide, distinguishing the cases:

(i) $e = 0$ (ii) $0 < e < 1$ (iii) $e = 1$

What happens to the energy lost in the collision when $e \neq 1$? How do you reconcile this with the Work–Energy Principle?

Questions 34–39: Three smooth spheres A, B and C of masses m_A, m_B and m_C are lying in a straight line on a smooth horizontal plane. A is projected with speed u to collide directly with B which goes on to collide directly with C. The coefficient of restitution between each pair of spheres is e.

• **34** $m_A = 3m$, $m_B = 2m$, $m_C = m$, $e = \frac{3}{4}$.

(i) Show that after the first impact, the velocities of A and B are $\frac{3}{10}u$ and $\frac{21}{20}u$ respectively.

(ii) Show that after the second impact the velocities of A, B and C are $\frac{3}{10}u$, $\frac{7}{16}u$ and $\frac{49}{40}u$ respectively.

(Note that since the plane is smooth, the velocities of the spheres are constant between impacts.) Explain why there will be no further impacts.

35 $m_A = 3m$, $m_B = 2m$, $m_C = m$, $e = 1$. Find the velocities of the spheres after the second impact. Explain why there will be no further impacts.

36 $m_A = 3m$, $m_B = 2m$, $m_C = 2m$, $e = \frac{1}{4}$. Find the velocities of the spheres after the second impact. Are there any further impacts?

R 37 $m_A = 2m$, $m_B = m$, $m_C = 3m$, $e = 1$. Show that after the second collision the KE of B is twice the KE of A. Find the velocities of the spheres after the third collision. Explain why there will be no further collisions.

→ **38** $m_A = m$, $m_B = 2m$, $m_C = 4m$. Find the velocities of the spheres after the second impact (in terms of u and e). If $e \geqslant \frac{1}{2}$, show that only two collisions takes place.

39 $m_A = m$, $m_B = m$, $m_C = m$. Find the velocities of the spheres after the second impact (in terms of u and e). Show that the total loss of KE is

$$\tfrac{1}{16}mu^2(5 + 2e - 4e^2 - 2e^2 - e^4)$$

What happens if $e = 1$?

Questions 40-46: Two spheres A and B of masses m_A and m_B lie at rest on a smooth floor in a line perpendicular to a wall, with B nearer to the wall than A. A is projected with speed u to collide directly with B. B goes on to strike the wall, and then after rebounding, collides directly with A again. The coefficient of restitution between A and B is e and between B and the wall is e'.

• **40** $m_A = 2m$, $m_B = m$, $e = 1$, $e' = \frac{1}{4}$. Find the speeds of A and B after their first impact. Find the speed of B after it strikes the wall. Hence find the speeds of A and B after their second impact.

41 $m_A = 3m$, $m_B = m$, $e = \frac{1}{3}$, $e' = \frac{1}{2}$. Find the speeds of A and B after their second impact.

42 $m_A = 2m$, $m_B = 3m$, $e = \frac{1}{2}$, $e' = \frac{2}{3}$. Find the speeds of A and B after their second impact.

→ **43** $m_A = m$, $m_B = 3m$, $e = \frac{1}{4}$. B is brought to rest by its second impact with A. Show that $e' = \frac{1}{11}$.

R 44 $m_A = 2m$, $m_B = 3m$, $e' = \frac{1}{3}$. B is brought to rest by its second impact with A. Find e.

45 $m_A = 2m$, $m_B = 3m$, $e' = e$. B is brought to rest by its second impact with A. Show that $2e^2 - e + 1 = 0$.

46 $m_A = 2m$, $m_B = m$, $e' = 1$. Find the speeds of A and B after their second impact (in terms of u and e).

47 A ball A is dropped from the top of a tower of height 20 m. Simultaneously, another ball B of the same mass is thrown vertically upwards from the foot of the tower with speed 20 m s^{-1}. The balls collide directly. Find their speeds just after the collision, assuming that the collision is perfectly elastic. Hence show that A just reaches the height of the top of the tower again, and find the total time that elapses before B hits the ground again.

48 Two beads A and B of masses $3m$ and m are threaded onto a smooth fixed vertical circular hoop of radius a. B rests at the lowest point of the hoop. A is slightly disturbed from rest at the highest point of the hoop. If the coefficient of restitution between the beads is $\frac{1}{2}$, find the heights above the lowest point of the hoop to which they rise after their first collision.

49 Two beads A and B of masses $3m$ and $2m$ are threaded onto a smooth fixed vertical circular hoop of radius a. They are released simultaneously from the opposite ends of the horizontal diameter of the hoop. The coefficient of restitution between them is $\frac{1}{2}$. Show that their first impact brings A to rest, and find their speeds after their second impact.

Miscellaneous exercise 8

1 A particle, placed on the inner surface of a fixed, rough, hollow sphere of internal radius $2a$, rests in limiting equilibrium at a depth $\sqrt{(3a)}$ below the centre of the sphere. Show that the coefficient of friction between the particle and the sphere is $1/\sqrt{3}$.

2 The ends of a light string are attached to two fixed points A and D at the same level. Particles of masses $2m$ and $4m$ are attached to the string at points B and C respectively. When the system is in equilibrium, $D\hat{A}B = 30°$ and $A\hat{B}C = 60°$. Find the tensions in the sections AB, BC and CD of the string.

3 Two beads P and Q of masses $3m$ and $2m$ respectively are threaded onto a rough, fixed, horizontal wire. The coefficient of friction between each bead and the wire is $\frac{1}{12}$. The beads are connected by a light inelastic string of length $2a$ which passes through a smooth ring of mass m. Show that the greatest possible distance apart of P and Q is $\frac{10}{13}a$.

4 A particle of mass m is placed on a rough plane of inclination α. The angle of friction between the particle and the plane is λ. Show that if $\lambda < \alpha$, the particle cannot rest in equilibrium. P, Q and R are the magnitudes of the least possible forces needed to keep the particle in equilibrium when the force is, respectively, parallel to the slope (P), horizontal (Q), and inclined at an angle α to the normal to the plane (R). Show that:

(i) $P = mg \cos \alpha (\tan \alpha - \tan \lambda)$
(ii) $Q = mg \tan(\alpha - \lambda)$
(iii) $R = mg(\tan \alpha - \tan \lambda)/(\tan \lambda + \tan \alpha)$

5 A particle P of mass m moves along a horizontal straight line. P is projected from a point O on the line with speed u. Find the speed of P when it is a distance a from O, given that P moves under the action of a resisting force of magnitude:

(i) $\tfrac{1}{2} mv^2/a$ (ii) $\tfrac{1}{2} muv/a$ (iii) $\tfrac{1}{2} mu^2 x/a^2$

where v and x are its velocity and displacement from O at time t.

6 A particle of mass m is projected vertically upwards with speed u. It moves against an air resistance which is mkv^2 when the speed of the particle is v.

(i) Show that while the particle is travelling upwards,

$$v \frac{dv}{dx} = -(g + kv^2)$$

where x is the displacement from the point O of projection. Hence show that the greatest height reached by the particle is

$$\frac{1}{2k} \ln(1 + ku^2/g)$$

(ii) Write down a differential equation governing the motion of the particle while it is moving downwards. Hence find its speed as it passes through O again.

7 A parachutist of mass m falls from rest freely under gravity for a time T and then opens her parachute. When it is open she experiences an upward resistance mgv^2/λ^2, where v is the speed and λ is a constant.

(i) If $T = \lambda/2g$, find the total distance she has travelled when her speed is $\tfrac{3}{4}\lambda$.

(ii) Show that, whatever the value of T, her speed approaches a limiting value λ, provided she falls for a sufficiently long time.

8 The position vector of a particle P, of mass 3 kg, at time t is $\{(\cos 2t)\mathbf{i} + (\sin 2t)\mathbf{j}\}$ m.

(i) Find the cartesian equation of the path of P and describe the path.

(ii) Show that the speed of P is constant and find the KE of P.

(iii) Find the magnitude of the force acting on P at time $t = \tfrac{1}{8}\pi$.

(iv) Find the distance of closest approach of P to the point with position vector $(\mathbf{i}+\mathbf{j})$ m.

9 A particle of mass m moves so that its position vector $\mathbf{r}$ at time t is

$$\mathbf{r} = \mathbf{a} \cos \omega t + \mathbf{b} \sin \omega t$$

where $\mathbf{a}$ and $\mathbf{b}$ are constant vectors and ω is a constant. Show that the force acting on the particle is $-m\omega^2 \mathbf{r}$.

(i) If $\mathbf{a}$ is perpendicular to $\mathbf{b}$, show that the KE of the particle is $\tfrac{1}{2} m(|\mathbf{a}|^2 + |\mathbf{b}|^2) \omega^2$.

(ii) If, further, $|\mathbf{a}| = |\mathbf{b}|$, show that $\mathbf{r}$ has constant magnitude and that $\mathbf{r}$ is perpendicular to the velocity of the particle. Describe the motion of the particle.

10 A particle is projected with speed u up a rough plane of inclination α from a point A on the plane. The coefficient of friction between the particle and the plane is μ.

(i) Show that the particle travels a distance

$$\frac{u^2}{2g(\sin \alpha + \mu \cos \alpha)}$$

up the plane before coming to instantaneous rest.

(ii) Show that, when the particle passes through A again, its speed is

$$u \left\{ \frac{\sin \alpha - \mu \cos \alpha}{\sin \alpha + \mu \cos \alpha} \right\}^{1/2}$$

What happens if $\tan \alpha \leq \mu$?

11 A smooth plane and a rough plane, both inclined at 30° to the horizontal, intersect in a fixed horizontal ridge. A particle A of mass m is held on the smooth plane by a light string which passes over a smooth pulley C on the ridge, and is attached to a particle B of mass $4m$ which rests on the rough plane. The plane ABC is perpendicular to the ridge. The system is released from rest with the string taut. Given that the acceleration of each particle is of magnitude $\tfrac{1}{8}g$, find:

(i) the tension in the string
(ii) the coefficient of friction between B and the rough plane
(iii) the magnitude and direction of the force exerted by the string on the pulley

12 A particle A of mass $2m$ resting on a horizontal table is attached to particles B of mass $3m$ and C of mass m by light inextensible strings hanging over fixed smooth pulleys, as shown. The plane ABC is vertical, the strings are taut and pass over the pulleys at right-angles to the edges of the table. The system starts from rest. Find the acceleration of the particles and the tensions in the strings before A reaches a pulley:

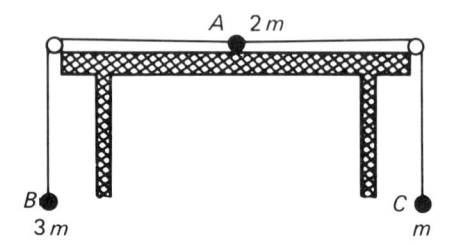

(i) when the table is smooth
(ii) when the table is rough and the coefficient of friction between particle A and the table is $\frac{1}{4}$

13 Two particles A and B, of mass $2m$ and $3m$ respectively, are connected by a light inextensible string of length l. A is placed on a horizontal table at a perpendicular distance l from its edge. The coefficient of friction between A and the table is $\frac{1}{2}$. B, which is at the edge of the table, is pushed gently over the edge. The motion of the particles takes place in a vertical plane perpendicular to the edge. Find:

(i) the acceleration of the particles and the tension in the string while A is moving on the table
(ii) the speed of A when it reaches the edge
(iii) the time taken by A to reach the edge

14 Two particles A and B, of masses 6 kg and 2 kg respectively, are connected by a light inextensible string which passes over a fixed smooth pulley. The system is released from rest with A and B at the same level, the string taut and the hanging parts of the string vertical. The string breaks after $\frac{1}{2}$ s. Assuming that the string is sufficiently long so that B does not hit the pulley, show that the vertical distance between A and B when B reaches its highest point is 2.5 m.

15 Two particles A and B, each of mass m, are attached to the ends of a light inextensible string which passes over a smooth pulley C fixed at the top of a smooth plane inclined at an angle $\tan^{-1}\frac{3}{4}$ to the horizontal. The system is released from rest with the string taut, the section AC of the string lying along a line of greatest slope of the plane and the section CB hanging vertically. Find the acceleration of the particles and the tension in the string.

B moves a distance l before colliding with an inelastic horizontal table. Find the total distance A moves up the plane, assuming that it does not reach C.

16 A wedge of mass $4m$ is placed on a rough table, and one of its sloping faces makes an angle $\tan^{-1}\frac{1}{2}$ with the horizontal. When a particle of mass m slides down this sloping face, the wedge does not move. Find the least possible value of the coefficient of friction between the wedge and the table:

(i) if the sloping face is smooth
(ii) if the coefficient of friction between the sloping face and the particle is $\frac{1}{4}$

17 A man slides down a smooth roof inclined at 30° to the horizontal, starting 1.6 m from the edge. If the edge of the roof is 7 m above ground level, find the horizontal distance from the edge of the roof to the point at which the man lands on the ground. Find also his speed and the direction of his motion as he lands.

18 The attractive force F between two objects of masses m_1 and m_2 is given by $F = Gm_1m_2/r^2$ where G is a constant and r is the distance between the objects.

(i) The Earth may be assumed to be moving round the Sun in a circular orbit of radius 1.5×10^8 m. Given that $G = 6.67 \times 10 \times 10^{-11}$ m^3 kg^{-1} s^{-2}, estimate the mass of the Sun.

(ii) God (or whoever is responsible for these things) brings the Earth to rest (relative to the Sun), so that it then accelerates directly towards the Sun. Find the speed of the Earth when it is halfway to the Sun.

19 (i) A particle P of mass m on, or above, the Earth's surface is attracted towards O, the centre of the Earth, by a force of magnitude mk/r^2, where $r = OP$ and k is constant. Find k, given that the radius of the Earth is 6.4×10^6 m and that at the surface of the Earth the acceleration due to gravity is 10 m s^{-2}.

(ii) A satellite moves with constant angular speed in a circle centre O. The time taken by the satellite to complete one orbit is 3 hours. Find the radius of the orbit.

20 The Wall of Death at a fun-fair consists of a circular cylinder of radius 5 m which can be made to rotate with its axis vertical. Inside the cylinder is a floor which can be lowered while the cylinder is rotating. A man enters through a door in the wall of the cylinder and stands with his back to the wall. The cylinder is made to rotate more and more quickly, and the floor is then lowered.

(i) Explain why, if the angular velocity of the cylinder is large enough, the man will not slide down the wall.

(ii) If the coefficient of friction between the man's body and the wall of the cylinder is $\frac{1}{4}$, find the least possible angular velocity of the cylinder for which the man will not slide down.

21 Two particles are placed, at points A and B, on a rough horizontal plane. O is a point on the plane such that $OA = l$ and $OB = 2l$. The coefficient of friction between each particle and the plane is $\frac{1}{2}$. The plane rotates about a vertical line through O with constant angular velocity ω rad s^{-1}. Discuss the motion of each particle if:

(i) $\omega^2 = g/4l$ (ii) $\omega^2 = g/3l$ (iii) $\omega^2 = g/2l$

22 A particle, moving with constant speed on the smooth inside surface of a fixed spherical bowl of radius l, describes a horizontal circle at a depth $\frac{1}{2}l$ below the centre of the sphere. Find:

(i) the angular speed of the particle
(ii) the speed of the particle
(iii) the magnitude of the reaction between the particle and the sphere

23 The end A of a light inextensible string of length l is attached to a fixed point. A particle P of mass m is attached to the other end of the string. P moves with constant speed v in the horizontal circle with the string inclined at an angle θ to the vertical.

(i) Show that $v^2 = gl \sin \theta \tan \theta$.
(ii) If the string breaks when the tension exceeds $5mg$, find the largest possible value of v.

24 A particle P of mass m is suspended from a fixed point O by a light elastic string of natural length l. In equilibrium, P hangs at a depth $\frac{6}{5}l$ below O. Show that the modulus of elasticity of the string is $5mg$.

The particle P is set in motion and describes a horizontal circle with constant speed and with the string inclined at an angle $\tan^{-1} \frac{4}{3}$ to the vertical. Find:

(i) the tension in the string
(ii) the extension of the string
(iii) the radius of the circle
(iv) the angular speed of the particle

25 A ring of mass m can slide on a fixed, smooth, vertical rod. One end of a light inextensible string of length $2l$ is fastened to the ring, its other end to a point of the rod. A particle of mass $2m$ attached to the midpoint of the string moves in a horizontal circle with constant angular speed ω.

(i) Show that $\omega^2 \geq 2g/l$.
(ii) Find ω if the tension in the lower half of the string is $2mg$.

+ **26** A smooth ring P of mass m is threaded onto a light inextensible string, the ends of which are attached to two points A and B in the same vertical line, with A above B. The ring is moving in a horizontal circle with constant angular velocity ω. $P\widehat{A}B = \alpha$, $P\widehat{B}A = \beta$ and $PA = l$. Show that

$$\omega^2 = \frac{g(\sin \alpha + \sin \beta)}{l \sin \alpha (\cos \alpha - \cos \beta)}$$

Show also that $\beta > \alpha$.

(i) Given that $\alpha = 60°$ and $\beta = 90°$, find the tension in the string.
(ii) Given that $A\widehat{P}B = 90°$ and that the string breaks if the tension in it exceeds $2mg$, show that $\beta - \alpha \geq 2 \sin^{-1}(1/2\sqrt{2})$.

27 One end of a light inextensible string is attached to a particle P of mass 3 kg and the other end to a fixed point A. P moves with speed 4 m s^{-1} in a horizontal circle whose radius is 1.2 m and whose centre is vertically below A. Find the length of the string.

The point A is 4.1 m above level ground. If the string breaks, find:

(i) the time take for P to reach the ground
(ii) the angle which the velocity of P makes with the horizontal when it strikes the ground

28 (i) Find the work done in stretching a light elastic string of natural length 0.8 m and modulus of elasticity 100 N
  (a) from its natural length to a length 1 m;
  (b) from a length 1 m to a length 1.2 m.

(ii) At time t the position vector of a particle of mass 2 kg is $\{(4 \cos 2t)\mathbf{i} + (3 \sin 2t)\mathbf{j}\}$ m. Find the change in KE of the particle between $t = 0$ and $t = \frac{1}{3}\pi$.

29 A particle P of mass 4 kg is moving along the x-axis. At time t seconds the displacement x of P from the origin O is given by $x = 5 + 4 \cos 2t$. Find its KE, V, and its acceleration, f, in terms of x. Find the displacement of P from O when:

(i) V is maximum
(ii) f is maximum

Sketch graphs of V against x and f against x. Describe the motion.

30 The gravitational attraction on a body of mass m at a distance x from the centre of the Earth is mgR^2/x^2, where R is the radius of the Earth. Find the work done by this force if the body travels to the Earth's surface from a great distance. Hence find the minimum speed with which a body from outer space would approach the Earth. (Take $R = 6.4 \times 10^6$ m.)

31 When a car of mass m moves on a straight horizontal road, the engine works at a constant rate mgu and the total resistance to motion is constant. The maximum speed possible under these conditions is u. The car starts with speed $\frac{1}{4}u$. Show that, if it attains a speed v after travelling a distance x,

$$\frac{1}{gu}\frac{dv}{dx} = \frac{1}{v^2} - \frac{1}{uv}$$

Hence find the distance moved by the car in acquiring a speed of $\frac{1}{2}u$.

32 A girder of mass m is raised vertically from rest by a crane which works at a rate $\lambda u m g$, where λ and u are constant.

(i) Show that, if the speed of the girder is v when it has been raised through a distance x,

$$v^2 \frac{dv}{dx} = g(\lambda u - v)$$

(ii) Hence show that when its speed is u it has risen through a height

$$\frac{u^2}{2g}\{2\lambda^2 \ln \lambda/(\lambda - 1) - 2\lambda - 1\}$$

(iii) By considering the increase in the total mechanical energy of the girder, find the time taken to reach this height.

33 A ring of mass 4 kg is acted on by forces of $(4\mathbf{i}+4\mathbf{j}-4\mathbf{k})$ N and $(16\mathbf{i}-8\mathbf{j})$ N, where $\mathbf{k}$ is in the direction of the upward vertical. The ring moves on a smooth rail from the point $A(-1, 5, 1)$ to the point $B(6, 2, 5)$. List the forces acting on the particle and the work done by each of them.

Given that the ring starts from rest at A, find its speed when it passes through B.

+ **34** Two trapezes are supported by ropes 5 m long attached to parallel horizontal rails which are fixed 10 m apart at the same height above the ground, as shown in the diagram. Two trapeze artists P and Q simultaneously climb on to the trapezes when the ropes are horizontal, and swing down, their paths lying in the same vertical plane.

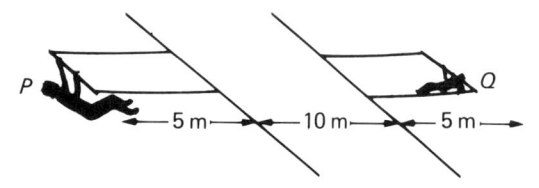

(i) If both P and Q leave go of the trapezes when the ropes are vertical, show that they will meet at a point 6.25 m below the level of the rails.

(ii) If both P and Q leave go of the trapezes when the ropes have moved through an angle $(\frac{1}{2}\pi + \tan^{-1}\frac{3}{4})$, show that they will meet at a point approximately 2.9 m below the level of the rails.

(iii) What happens if Q leaves go when the ropes supporting her trapeze are vertical, while P leaves go when the ropes supporting his trapeze have moved through an angle $(\frac{1}{2}\pi + \tan^{-1}\frac{3}{4})$?

List the assumptions you are making in solving this problem.

35 A particle of mass m is attached to a fixed point O by a light elastic string of natural length l and modulus of elasticity $\frac{3}{2}mg$. The particle is held at O and released from rest. Find the distance through which it falls before coming to instantaneous rest, and the maximum speed it attains during the fall.

36 An elastic string of natural length l and modulus of elasticity $3mg$ has a particle of mass m attached at one end. The other end is attached to a fixed point A on a smooth slope of inclination $\tan^{-1}\frac{3}{4}$. The particle rests in equilibrium at a point O on the plane. Find OA.

The particle is then pulled a further distance l down the slope and is released. Find its velocity as it passes through:

(i) O (ii) A

37 A car of mass 1500 kg moves against a resistance of 750 N. The engine of the car works at a constant rate of 20 kW. Find the maximum speed of the car:

(i) on the level
(ii) directly up a road of inclination $\sin^{-1}\frac{1}{12}$

Find the acceleration of the car when it is travelling at 20 m s^{-1} on a level road.

38 A car of mass m, whose engine works at a constant rate H, moves against a constant resistance R. The maximum speed of the car on level ground is u.

(i) Show that $R = H/u$.

(ii) Find (in terms of m, u, H, n and g) the accelerations f_1 and f_2 of the car when it is moving at speed $\frac{1}{2}u$ directly up (f_1) and directly down (f_2) a road of inclination $\sin^{-1} n$.

(iii) If $f_2 = 2f_1$, show that $n = H/3mgu$. Hence show that the maximum possible speed of the car up the road is $\frac{3}{4}u$ and find its maximum speed down the road.

39 A car of mass m, whose engine works at a constant rate, moves against a constant resistance. The car can attain a maximum speed of u on level ground and a maximum speed of $\frac{1}{2}u$ travelling directly up a slope of inclination $\sin^{-1}\frac{1}{20}$. Find the maximum speed of the car up a slope of inclination $\sin^{-1}\frac{1}{10}$.

The car is towing a trailer of mass $\frac{1}{2}m$ at speed $\frac{1}{4}u$ on level ground. Assuming that the resistance to the motion of the trailer can be neglected, find:

(i) the acceleration of the vehicles
(ii) the tension in the towbar

40 A train of mass m, whose engine works at a constant rate of $\frac{1}{5}mgu$, moves up a straight track which is inclined at $\sin^{-1}\frac{1}{20}$ to the horizontal. When the speed is u, the acceleration is $\frac{1}{8}g$. Find the resistance to motion at this speed. Given that the resistance is proportional to the speed of the train, find the greatest speed of the train:

(i) up the track
(ii) down the track

+ 41 The engine of a car develops constant power. The resistance to the motion of the car is proportional to the square of its speed. The maximum speeds of the car on a level road and up a certain hill are u and $\frac{1}{2}u$. Show that the maximum speed down the same hill is xu, where $4x^3 - 7x - 4 = 0$.

Show graphically that this equation has a root between $x = 1$ and $x = 2$. Find this root, correct to 2 dec. pls.

42 A car of mass 1000 kg has an engine whose maximum rate of working is 10 kW. The resistance to the motion of the car is proportional to the square of its speed. The maximum speed of the car on a level road is 20 m s^{-1}. The car is moving at a constant speed of 10 m s^{-1} up a road of inclination $\sin^{-1}\frac{1}{20}$.

(i) Find the rate at which the engine is working.
(ii) If the engine is shut off, find the distance that the car moves before coming to rest.

43 A pump raises water from a depth of 5 m and discharges it horizontally through a pipe of cross-sectional area 0.01 m^2 at a velocity of 8 m s^{-1}. Find the effective power of the pump.

The water impinges directly with the same velocity on a wall. Assuming that none of the water bounces back, find the force exerted by the water on the wall.

+ 44 A ball is dropped from a height h above a table, and comes to rest after time $k\sqrt{(2h/g)}$ (k constant). Find:

(i) the coefficient of restitution between the ball and the table
(ii) the total distance covered by the ball

45 A sphere A, of mass m and moving with speed $4u$, collides directly with a sphere B of the same radius and mass $4m$ moving in the same direction with speed u. As a result of the impact, A is brought to rest. Find the coefficient of restitution between the spheres.

Find the speeds of A and B after impact if they collide when moving with the same speeds as before but in opposite directions.

46 The M4. Fog and ice. 5.30 on a Friday evening. Traffic leaving London.

(i) A Rolls Royce, of mass 1000 kg, moving with speed 20 m s^{-1}, collides directly with a Ford Granada, of mass 750 kg, moving with speed 10 m s^{-1} in the same direction. If the coefficient of restitution between the cars is $\frac{1}{20}$, find their speeds immediately after the collision. Assuming that the engines cut out on collision and that the coefficient of friction between each car and the road is $\frac{1}{20}$, find the times taken by the cars to come to rest.

(ii) The police are called by a passing lorry driver. After an astonishingly short time, which does the Force credit, a police car of mass 800 kg and moving with speed 60 m s^{-1}, arrives on the scene, colliding directly with the stationary Rolls Royce. If the coefficient of restitution is again $\frac{1}{20}$, find the tragic loss in KE caused by the collision. Find the distance moved by the Rolls Royce before it comes to rest, assuming that it does not collide with the Ford again (or with anything else).

List the assumptions you have made in answering this question.

47 A sphere A of mass $2m$, moving with speed $2u$ on a smooth horizontal plane, collides directly with a sphere B of equal radius and of mass m which is moving with speed u in the opposite direction. Given that the coefficient of restitution between the spheres is $\frac{1}{2}$, find:

(i) their speeds after the collision
(ii) the magnitude of the instantaneous impulses
(iii) the loss in KE caused by the collision

After a short interval the sphere A is given a horizontal impulse of magnitude $7mu$ so that it again collides directly with sphere B. Find the speeds of A and B after the second impact.

48 Three smooth spheres A, B and C, of equal radii and masses $3m$, $2m$ and m respectively, lie at rest on a horizontal plane with their centres in a straight line. A is projected with speed u to collide directly with B which then collides directly with C. The coefficient of restitution between each pair of spheres is $\frac{2}{3}$. Find the speed of each sphere after these collisions. Explain why there are no further collisions.

49 Three smooth spheres A, B and C, of equal radii and masses m, km and k^2m respectively (k constant), lie at rest on a horizontal plane with their centres in a straight line. A is projected with speed u to collide directly with B which then collides directly with C. The coefficient of restitution between each pair of spheres is e. Find the speed of each sphere after these collisions.

Given that $k < 1$, show that there is a third collision if $e < k$.

50 Three smooth spheres A, B and C, of equal radii and masses m, $2m$ and $2m$ respectively, lie at rest on a horizontal plane with their centres in a straight line. A and B are simultaneously projected with speed u towards C which is stationary, so that B collides with C and subsequently with A. The coefficient of restitution between each pair of spheres is e. Find the speeds of the spheres after the second collision. show that there is a third collision if $e < (2 - \sqrt{3})$.

51 A sphere A of mass m is moving with speed u on a smooth horizontal floor when it collides directly with a stationary sphere B of the same radius and of mass km. The coefficient of restitution between the spheres is $\frac{2}{3}$. If the speeds of A and B after the collision

are the same, find k, and show that the KE lost is $\frac{2}{9}mu^2$. Sphere B then strikes a wall normally and rebounds. Show that A and B do not collide again.

52 A small smooth sphere A, moving on a horizontal floor, strikes an identical sphere B lying at rest on the table at a distance 1 m from a wall. The impact is along the line of centres and perpendicular to the wall. The coefficient of restitution between A and B, and between B and the wall, is $\frac{1}{2}$. Show that the next impact between A and B will take place at a distance 0.4 m from the wall.

53 A smooth groove in the shape of a circle of radius a is cut in a horizontal table. Two smooth spheres A and B of equal radii and masses km and m respectively are placed in the groove at opposite ends of a diameter. A is given an impulse and travels round the groove, striking B after a time T. The coefficient of restitution between A and B is e. Find the speeds of A and B after the first collision, and the time τ that elapses between the first and second collisions. Show that B makes at least one complete revolution between the first and second collisions, provided $e \leq k$. Explain why τ is independent of k.

54 A piece of wood of mass $7m$ is suspended from a fixed point O by a light inelastic string of length l. The wood is at rest vertically below O when it is struck by a bullet of mass m moving horizontally with speed u. The bullet remains embedded in the wood. The maximum angle made by the string with the vertical in subsequent motion is 2α where $\alpha < \frac{1}{4}\pi$. Show that $u = 16\sqrt{(gl)} \sin \alpha$.

55 Two beads A and B, of masses $3m$ and $2m$ respectively, are threaded onto a smooth fixed vertical circular hoop of radius a. Bead B rests at the lowest point of the hoop. Bead A is projected from the highest point of the hoop with speed $\sqrt{(2ga)}$. If the coefficient of restitution between A and B is $\frac{1}{4}$, find:

(i) the loss in KE caused by their first collision
(ii) the heights above the lowest point of the hoop to which they rise after this collision

56 One end of a light elastic string of natural length l and modulus of elasticity $2mg$ is attached to a fixed point A, the other to a particle P of mass m, which hangs freely. A bead Q of mass $2m$, which is threaded onto the string, is dropped from A. It strikes P and coalesces with it. Find:

(i) the loss in KE due to the impact
(ii) the length of the string when the combined particle first comes momentarily to rest

57 A particle P of mass m is projected from a point O on level ground with speed $u\sqrt{5}$ at an angle of elevation $\tan^{-1} 2$. Find:

(i) the time taken by P to reach the highest point B of its path
(ii) the vertical height of B above O
(iii) the the speed of P as it passes through B

As P passes through B, it collides directly with a particle Q, of mass $2m$, which was at rest before the impact. The coefficient of restitution between the particles is $\frac{1}{5}$. Show that:

(iv) the loss in KE due to the impact is $\frac{8}{25}mu^2$
(v) the distance between the points at which A and B strike the ground again is $2u^2/5g$

58 Two particles A and B, of mass $2m$ and $3m$ respectively, are attached to each other by a light inelastic string of length $5l$ which passes over a smooth fixed pulley. Initially the particles are at rest vertically below the peg with the string taut and both are at a height $2l$ above a horizontal plane. The particles are released. The coefficient of restitution between B and the plane is $\frac{1}{2}$.

(i) Show that, after B has hit the plane, A does not rise to the level of the pulley.

(ii) Show that B bounces on the plane a second time before the string becomes taut again. Find the time between the first and second bounces.

59 Two particles A and B, of mass $2m$ and $3m$ respectively, are connected by a light inelastic string which passes over a smooth fixed pulley. The system is released from rest with the string taut and the hanging parts vertical. After time T particle A picks up a stationary particle C of mass $3m$.

(i) Find the common speed of the particles just after C has been picked up.

(ii) Show that the loss in KE of the system due to the impulse is $3mg^2T^2/80$.

(iii) Find the common acceleration of the particles after C has been picked up.

(iv) Find the total time that elapses before the system comes to instantaneous rest.

60 A particle A of mass m moves on a smooth horizontal table and is connected to a fixed point O of the table by a light inelastic string of length l. The particle moves with constant speed $\sqrt{(2gl)}$ in a circle centre O. Find the tension in the string.

While moving in this way, A collides directly with a stationary particle B of mass $2m$, the coefficient of restitution being $\frac{1}{8}$. Find:

(i) the speeds of A and B immediately after the impact
(ii) the tension in the string after the impact

9 Forces on a rigid body

9.1 Systems of coplanar forces

9.1:1 Turning effect of a force

Consider a set of coplanar forces acting on a *particle P* (fig. 9.1:1a). The forces must be concurrent (at P); so the forces have no turning effect.

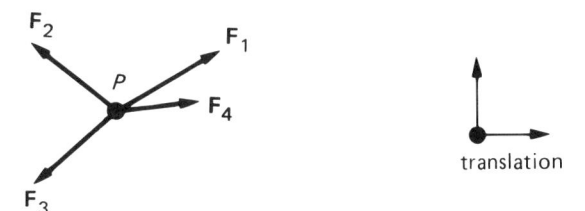

Fig. 9.1:1a   Forces on a particle

The forces can generate movement consisting of two independent factors: translation in two independent directions. In particular, if the resultant of the forces is zero, the particle is in equilibrium.

Now consider a set of coplanar forces acting on a rigid body (fig. 9.1.1b). The forces need not intersect. So the forces may have a turning effect, causing the object to rotate.

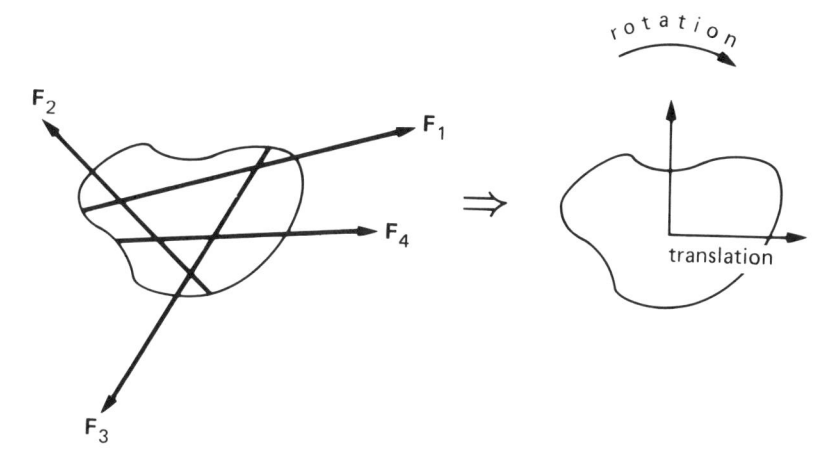

Fig. 9.1:1b   Forces on a rigid body

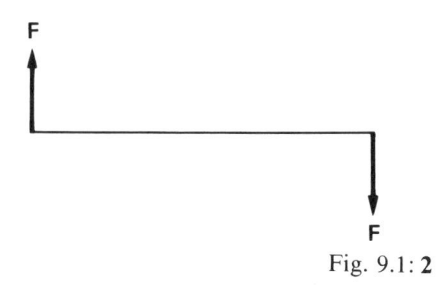

Fig. 9.1: 2

The forces can generate movement consisting of three independent factors: translation in two independent directions (2 factors) and rotation (1 factor). In particular, it is possible that the resultant of the forces is zero, but the object is not in equilibrium; e.g. consider a rod with two forces, of the same magnitude but in opposite directions, acting, one at either end, as in fig. 9.1:2. The resultant force is zero but the rod is not in equilibrium; in fact it will rotate.

So we must concern ourselves with the **line of action** of a force as well as with its magnitude and direction.

508 *Turning effect of a force*

Suppose that a force **F** acts on a rigid body, and that Λ is a line perpendicular to **F**, as in fig. 9.1:3.

Definition The **moment**, or **torque**, $M_\Lambda(\mathbf{F})$ of **F** about Λ is Fd, where F is the magnitude of **F** and d is the shortest distance from Λ to **F**.
(1)

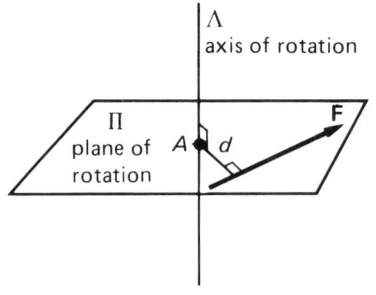

Fig. 9.1:3

The plane Π containing **F** and perpendicular to Λ is the **plane of rotation**. Λ is the **axis of rotation**.

Even though we shall consider objects which are three-dimensional, we shall consider only systems of *coplanar* forces; so

(i) The plane Π of the forces will always be the plane of rotation;
(ii) Λ will always be perpendicular to Π; so instead of talking about the moment of a force **F** about Λ, we often talk loosely about the moment of **F** about A, the point of intersection of Λ and Π.

Moments are measured in units of newton metre, N m (not in joules—a moment is different from work).

Zero moment
The line of action of **F** passes through $\Lambda \Leftrightarrow d = 0 \Leftrightarrow$ the moment of **F** about Λ is zero.

Sense of a moment
Conventionally we call an anticlockwise moment positive, a clockwise moment negative.

Worked example 1 Forces of 2 N, 3 N and 4 N act as shown in fig. 9.1:4 on a rod AB, of length $4a$. Find the moment of each of the forces about A (strictly speaking, about an axis through A perpendicular to the plane of the forces).

Moment of 2 N force about $A = +2a$ N m
Moment of 3 N force about $A = +3 \cdot (4a \sin 30°)$ N m $= +6a$ N m
Moment of 4 N force about $A = -4 \cdot (2a \sin 45°)$ N m $= -4\sqrt{2}a$ N m

Definition The **resultant moment** $\sum M_\Lambda(\mathbf{F})$ of a system of coplanar forces about an axis Λ is the sum of the moments of the individual forces about Λ (taking into account the sense of each moment).

For example, in worked example 1 the resultant moment of the forces is

$$(2a + 6a - 4a\sqrt{2}) \text{ N m} = 4a(2-\sqrt{2}) \text{ N m}$$

When finding the moment of a force **F** about an axis Λ, it is sometimes easier to replace **F** by its components in two perpendicular directions and then find the moment of each component:

Moment of **F** about Λ
  $=$ sum of moments of its components about Λ   (2)

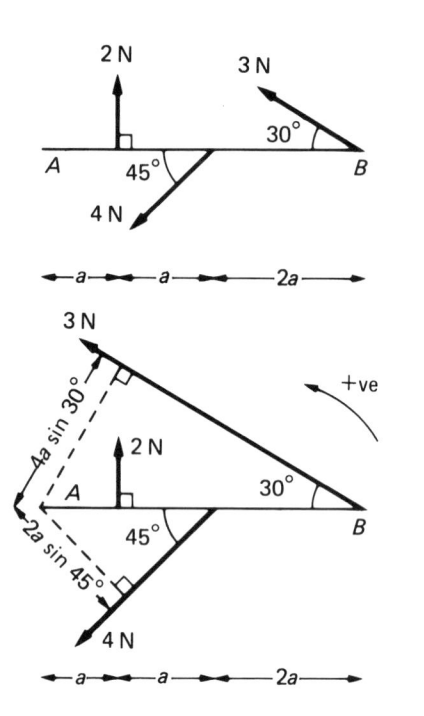

Fig. 9.1:4

Forces on a rigid body 509

(See question 8. This is actually a particular case of the Principle of Moments (see section 9.1:4).)

Worked example 2 Forces $F_1 = 2i + 3j$, $F_2 = -i - 3j$ and $F_3 = 4i - j$ act through the points with position vectors $i + 4j$, $2i - j$ and $-3i$ respectively (see fig. 9.1:5). Find the resultant moment of the forces about O (strictly speaking, about Oz).

Moment of F_1 about $O = 3 \times 1 - 2 \times 4 \quad = -5$

Moment of F_2 about $O = -3 \times 2 - 1 \times 1 = -7$

Moment of F_3 about $O = 1 \times 3 \quad\quad\quad = +3$

Fig. 9.1:5   So resultant moment of forces about $O = -5 - 7 + 3 = -9$

Exercise 9.1:1

① **1** AB is a line segment. Find the resultant moment:

(i) about A (ii) about B

of the systems of (coplanar) forces shown.

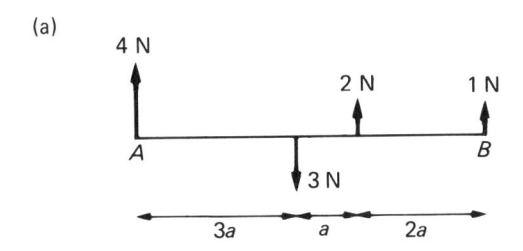

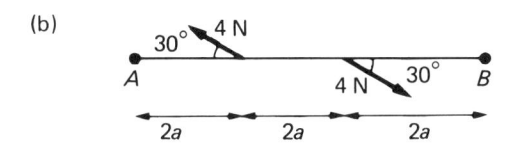

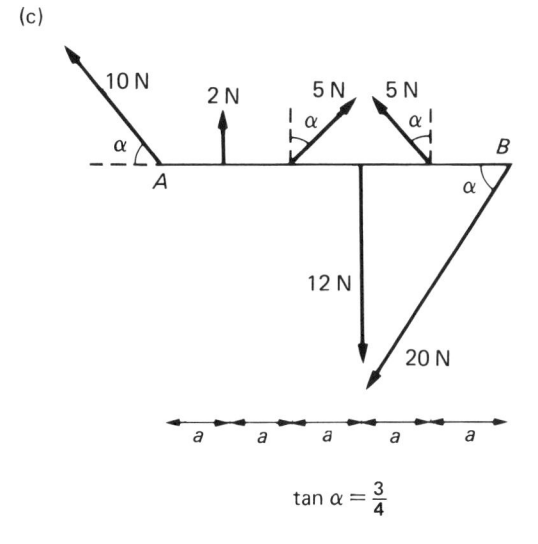

(a)

(b)

(c) $\tan \alpha = \frac{3}{4}$

① **2** $ABCD$ is a square of side $2a$. Find the resultant moment:

(i) about A (ii) about C

of the systems of forces shown.

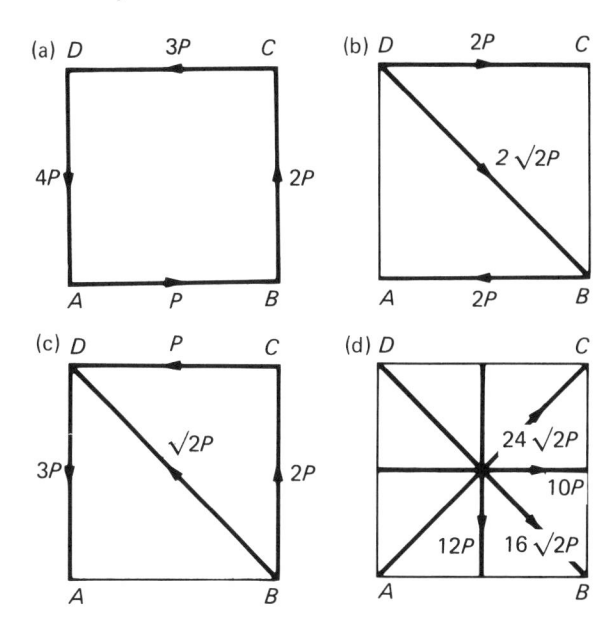

(a) (b) (c) (d)

① **3** Find the resultant moment about A of the systems of forces shown.

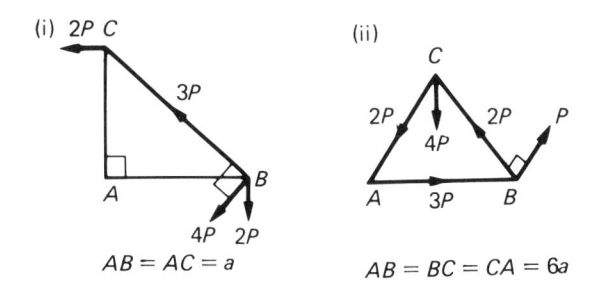

(i) $AB = AC = a$

(ii) $AB = BC = CA = 6a$

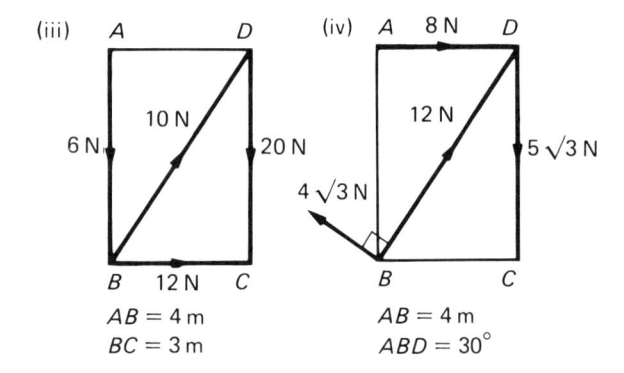

(iii) Rectangle ABCD with A top-left, D top-right, B bottom-left, C bottom-right. Forces: 6 N at B pointing down along BA (shown on left side), 10 N along BD (diagonal from B to D), 20 N along CD (upward on right side), 12 N along BC.
AB = 4 m
BC = 3 m

(iv) Rectangle ABCD with A top-left, D top-right. Forces: 8 N along AD, 12 N along BD (diagonal), 5√3 N along CD, 4√3 N at B at angle (ABD = 30°, shown perpendicular to BD).
AB = 4 m
ABD = 30°

① • **4** *ABCDEF* is a regular hexagon of side $2a$. Find the shortest distance from A to:

(i) CB (produced)   (ii) CD
(iii) CE (iv) CF

Find the shortest distance from O, the centre of the hexagon, to each side.

① **5** *ABCDEF* is a regular hexagon of side $2a$. Find the resultant moment:

(i) about A
(ii) about O, the centre of the hexagon

of the systems of forces shown.

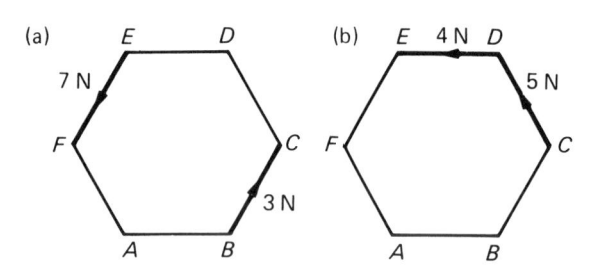

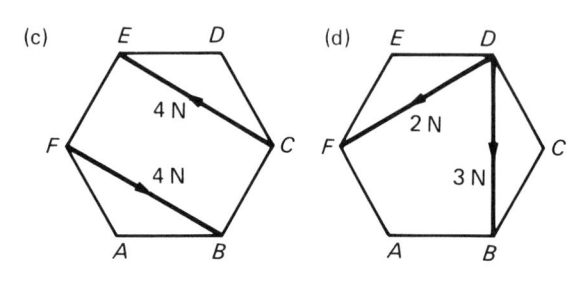

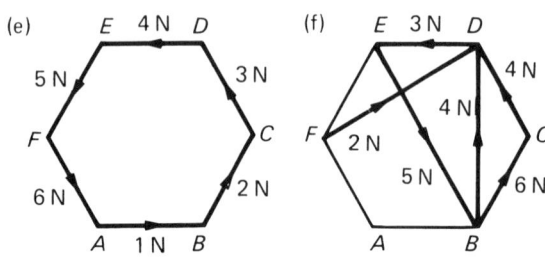

(a) Hexagon ABCDEF with 7 N along FE, 3 N along CB.

(b) Hexagon with 4 N along ED, 5 N along DC.

(c) Hexagon with 4 N along AE and 4 N along AC (or similar diagonals).

(d) Hexagon with 2 N along ED and 3 N along CB (down from D).

(e) Hexagon with 5 N along FE, 4 N along ED, 3 N along DB, 2 N along BC, 1 N along AB, 6 N along AF.

(f) Hexagon with 3 N along ED, 4 N along DC, 4 N along DB, 2 N along AC, 5 N along AD, 6 N along CB.

• → **6** A force $6\mathbf{i}+5\mathbf{j}$ acts through the point $\mathbf{i}-\mathbf{j}$. Find a vector equation and the cartesian equation of its line of action.

7 Find the cartesian equation of the line of action of the force $\mathbf{F}$ acting through the point A

(i) $\mathbf{F}=4\mathbf{i}+5\mathbf{j}$, $\mathbf{a}=4\mathbf{j}$
(ii) $\mathbf{F}=8\mathbf{i}-4\mathbf{j}$, $A(1,1)$
(iii) $\mathbf{F}=28\mathbf{j}$, $A(3,-2)$

• **8** O is a point; Ox and Oy are mutually perpendicular axes through O. A force $\mathbf{F}=p\mathbf{i}+q\mathbf{j}$ acts through the point with position vector $a\mathbf{i}+b\mathbf{j}$.

(i) Find the sum of the moments of the components $p\mathbf{i}$ and $q\mathbf{j}$ of $\mathbf{F}$ about O.
(ii) Find the cartesian equation of the line of action of $\mathbf{F}$.
(iii) Show that the shortest distance from O to this line is

$$|(qa-pb)/\sqrt{(p^2+q^2)}|$$

(iv) Find the magnitude of $\mathbf{F}$ and hence find the moment of $\mathbf{F}$ about O.

Verify that the moment of $\mathbf{F}$ about O equals the sum of the moments of its components about O.

② **9** The force $\mathbf{F}$ acts through the point with position vector $\mathbf{a}$. Find the moment of $\mathbf{F}$ about the origin.

(i) $\mathbf{F}=4\mathbf{j}$, $\mathbf{a}=2\mathbf{i}$
(ii) $\mathbf{F}=5\mathbf{i}$, $\mathbf{a}=3\mathbf{j}$
(iii) $\mathbf{F}=3\mathbf{i}+2\mathbf{j}$, $\mathbf{a}=\mathbf{i}+4\mathbf{j}$
(iv) $\mathbf{F}=-7\mathbf{i}+5\mathbf{j}$, $\mathbf{a}=\mathbf{i}+4\mathbf{j}$
(v) $\mathbf{F}=-5\mathbf{i}-\mathbf{j}$, $\mathbf{a}=-2\mathbf{i}+\mathbf{j}$
(vi) $\mathbf{F}=5\mathbf{i}+\mathbf{j}$, $\mathbf{a}=-2\mathbf{i}+\mathbf{j}$
(vii) $\mathbf{F}=6\mathbf{i}-3\mathbf{j}$, $\mathbf{a}=-2\mathbf{i}+\mathbf{j}$

What do you deduce in (vii)?

② **10** Find the moment about the origin of the force $p\mathbf{i}+q\mathbf{j}$ acting through the point with position vector $a\mathbf{i}+b\mathbf{j}$.

② **11** Find the resultant moment about the origin of the system of forces:

R (i) $\mathbf{F}_1=\mathbf{i}+\mathbf{j}$, $\mathbf{F}_2=-2\mathbf{i}-5\mathbf{j}$, $\mathbf{F}_3=4\mathbf{i}$; acting at the points with position vectors $3\mathbf{i}$, $2\mathbf{i}+\mathbf{j}$, $-\mathbf{j}$.
→ (ii) $\mathbf{F}_1=5\mathbf{i}+2\mathbf{j}$, $\mathbf{F}_2=-3\mathbf{i}-5\mathbf{j}$, $\mathbf{F}_3=-2\mathbf{i}+\mathbf{j}$; acting at the points with position vectors $4\mathbf{i}-\mathbf{j}$, $\mathbf{i}+2\mathbf{j}$, $-3\mathbf{i}+\mathbf{j}$.

• → **12** A and B are the points with position vectors $-\mathbf{i}+5\mathbf{j}$ and $-3\mathbf{i}+2\mathbf{j}$. Find $\overline{AB}$. Hence find the moment about A of the force $2\mathbf{i}+\mathbf{j}$ acting through the point B.

13 A force $\mathbf{F}=-\mathbf{i}+4\mathbf{j}$ acts through the point $2\mathbf{i}+3\mathbf{j}$. Find the moment of $\mathbf{F}$ about the point with position vector:

(i) $\mathbf{i}$ (ii) $\mathbf{j}$ (iii) $-3\mathbf{i}-\mathbf{j}$ (iv) $4\mathbf{i}-5\mathbf{j}$

What do you deduce in (iv)?

14 Find the resultant moment about:
  (i) the origin
  (ii) the point with position vector $\mathbf{i}+\mathbf{j}$

of forces $\mathbf{F}_1 = 4\mathbf{i}+7\mathbf{j}$, $\mathbf{F}_2 = -\mathbf{i}-2\mathbf{j}$ and $\mathbf{F}_3 = 6\mathbf{j}$ acting through the points $-4\mathbf{i}+3\mathbf{j}$, $\mathbf{i}+\mathbf{j}$ and $5\mathbf{i}$ respectively.

15 A and B are the points $(5, 2)$ and $(1, -2)$. A force $\mathbf{F}$ of magnitude $6\sqrt{2}$ N acts along $\overrightarrow{AB}$. Find the moment of $\mathbf{F}$ about the origin.

16 A and B are the points $3\mathbf{i}+4\mathbf{j}$ and $-4\mathbf{i}+3\mathbf{j}$. Forces $\mathbf{F}_1$, $\mathbf{F}_2$ and $\mathbf{F}_3$ of magnitude 10 N, $10\sqrt{2}$ N and 15 N act along $\overrightarrow{OA}$, $\overrightarrow{AB}$ and $\overrightarrow{BO}$ respectively. Find:

  (i) the moment of $\mathbf{F}_2$ about O
  (ii) the moment of $\mathbf{F}_3$ about A
  (iii) the moment of $\mathbf{F}_1$ about B

17 A force $\mathbf{F}$ of magnitude $3\sqrt{5}$ N acts along the line $2x - y + 4 = 0$ in the direction of increasing x and y. Find the moment of $\mathbf{F}$ about the origin.

• **18** By resolving each force parallel and perpendicular to AB, find the resultant moment of the following systems of forces about A.

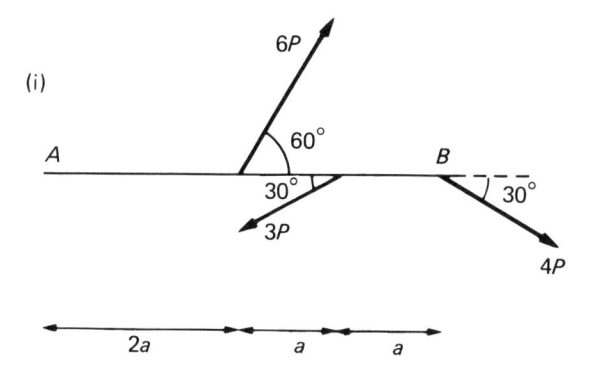

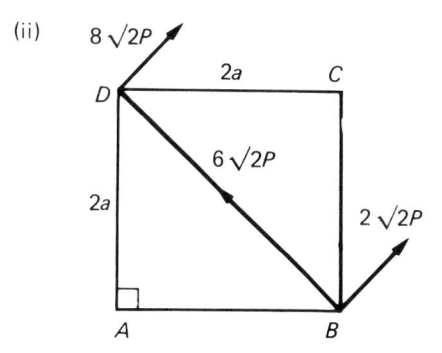

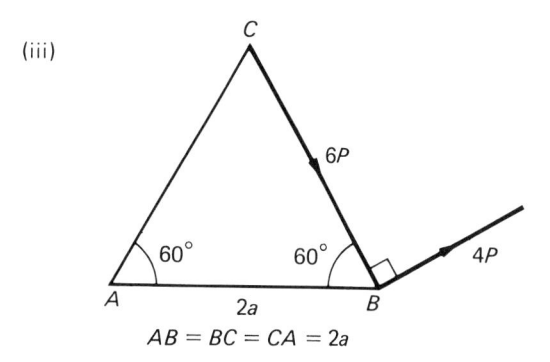

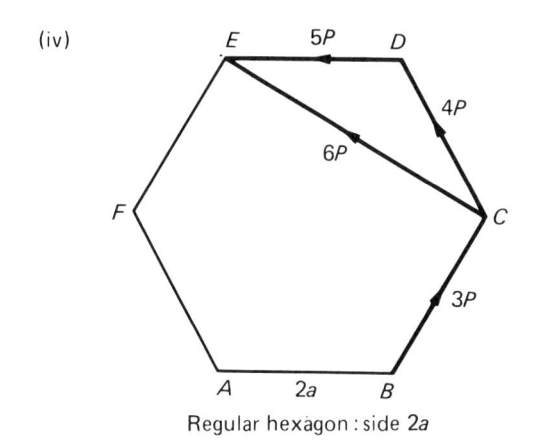

(i)

(ii)

(iii) $AB = BC = CA = 2a$

(iv) Regular hexagon: side $2a$

Do you prefer this method of finding moments to the one on p. 508? If so, keep using it. If not, write to your MP.

9.1:2 Couples

Definition A **couple** is a pair of parallel forces of the same magnitude acting in opposite directions.

A couple acting on a body rotates it but does not translate it.

Theorem 1: The resultant of a couple is zero; the resultant moment of a couple is the same about any point (and is not zero).

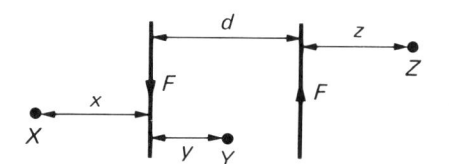

512 *Systems reducing to a single resultant force*

Fig. 9.1:6

Proof
Consider the couple shown in fig. 9.1:**6**.

The resultant of the couple = $\mathbf{F} + (-\mathbf{F}) = \mathbf{0}$.

resultant moment of the couple:
(anticlockwise positive, clockwise negative.)

About X:	$F(d+x) - Fx = +Fd$
About Y:	$F(d-y) + Fy = +Fd$
About Z:	$-Fz + F(d+z) = +Fd$

So the resultant moment of the couple about any point (often called the **magnitude** of the couple) is Fd. The distance d is called the **arm** of the couple.

Exercise 9.1:2

• → 1 Forces $\mathbf{F}_1 = \mathbf{i} + \mathbf{j}$ and $\mathbf{F}_2$ act at points $3\mathbf{j}$ and $4\mathbf{i}$. $\mathbf{F}_1$ and $\mathbf{F}_2$ form a couple. Find $\mathbf{F}_2$ and the magnitude of the couple.

2 Forces $\mathbf{F}_1 = 3\mathbf{i} - 2\mathbf{j}$ and $\mathbf{F}_2$ act at points $\mathbf{i} + 3\mathbf{j}$ and $-4\mathbf{i} + 4\mathbf{j}$. $\mathbf{F}_1$ and $\mathbf{F}_2$ form a couple. Find $\mathbf{F}_2$ and the magnitude of the couple.

3 Forces $\mathbf{F}_1$ of magnitude $4\sqrt{2}$ N and $\mathbf{F}_2$ act along the lines $y = x$ (in the sense of increasing x and y) and $y = x + 2$ respectively. $\mathbf{F}_1$ and $\mathbf{F}_2$ form a couple. Find its magnitude.

9.1:3 Equivalent systems of forces

Suppose that a system S of coplanar forces $\mathbf{F}_1, \mathbf{F}_2, \ldots, \mathbf{F}_n$ acts on a rigid body. We want to find the simplest possible equivalent system S' of forces, i.e. the simplest system which would have the same effect (translational and rotational) on the body as the system S.
We say: S **reduces** to the system S'.
There are three possibilities:

1 The system reduces to a single resultant force.
2 The system reduces to a couple.
3 The system is in equilibrium.

9.1:4 Systems reducing to a single resultant force

Consider a system S of coplanar forces $\mathbf{F}_1, \mathbf{F}_2, \mathbf{F}_3, \ldots, \mathbf{F}_n$.

$\sum \mathbf{F} \neq \mathbf{0} \Leftrightarrow$ the system reduces to a single resultant force $\mathbf{R}$

(i) $\mathbf{R}$ has the same translational effect as $\{\mathbf{F}\} \Leftrightarrow$

$$\mathbf{R} = \sum \mathbf{F}$$

Forces on a rigid body 513

(ii) **R** has the same turning effect about every point X as $\{\mathbf{F}\} \Leftrightarrow$

The moment of the resultant **R** of the forces about every point X equals the resultant moment of the forces about X,

i.e. $$M_X(\mathbf{R}) = \sum M_X(\mathbf{F}) \quad (3)$$

Equation (3) is known as the **Principle of Moments**.
Thus, when a system of forces reduces to a single resultant force **R**, we find the magnitude and direction of **R** using relationship (i):
if each force $\mathbf{F} = F_x\mathbf{i} + F_y\mathbf{j}$ and $\mathbf{R} = R_x\mathbf{i} + R_y\mathbf{j}$

$$R_x = \sum F_x \quad (4)$$
$$R_y = \sum F_y \quad (5)$$
---

and we find the line of action of **R** using relationship (ii) for some convenient point A—this is called **taking moments about** A:

$$M_A(\mathbf{R}) = \sum M_A(\mathbf{F}) \quad (6)$$

Notice that
if each member of a system of forces acts through a point C, the resultant of the system acts through C too. (7)

For, if each force **F** acts through C, $\sum M_C(\mathbf{F}) = 0$ so, by equation (6), $M_C(\mathbf{R}) = 0$, so **R** also acts through C.

Worked example 3 Forces $\mathbf{F}_1 = 3\mathbf{i} + 3\mathbf{j}$ and $\mathbf{F}_2 = 2\mathbf{i} - 4\mathbf{j}$ act through the points A and B with position vectors $\mathbf{a} = -2\mathbf{i} - 3\mathbf{j}$ and $\mathbf{b} = 3\mathbf{i} - \mathbf{j}$ respectively. Show that the forces intersect and find a vector equation of the line of action of their resultant.

Use the methods of section 6.2:3.
The vector equation of the line of action of $\mathbf{F}_1$ is

$$\mathbf{r} = \mathbf{a} + \lambda \mathbf{F}_1 = -2\mathbf{i} - 3\mathbf{j} + \lambda(3\mathbf{i} + 3\mathbf{j}) \quad (8)$$

and the vector equation of the line of action of $\mathbf{F}_2$ is

$$\mathbf{r} = \mathbf{b} + \mu \mathbf{F}_2 = 3\mathbf{i} - \mathbf{j} + \mu(2\mathbf{i} - 4\mathbf{j}) \quad (9)$$

These lines intersect where

$$-2\mathbf{i} - 3\mathbf{j} + \lambda(3\mathbf{i} + 3\mathbf{j}) = 3\mathbf{i} - \mathbf{j} + \mu(2\mathbf{i} - 4\mathbf{j})$$

i.e. $\quad -2 + \lambda = 3 + 2\mu \quad$ and $\quad -3 + 3\lambda = -1 - 4\mu$

$\Rightarrow \quad \lambda = \tfrac{4}{3}, \quad \mu = -\tfrac{1}{2}$

514 *Systems reducing to a single resultant force*

Substituting $\lambda = \frac{4}{3}$ in equation (8) gives

$$\mathbf{r} = 2\mathbf{i} + \mathbf{j}$$

So the forces intersect at the point C with position vector $2\mathbf{i} + \mathbf{j}$ (fig. 9.1:**7a**).

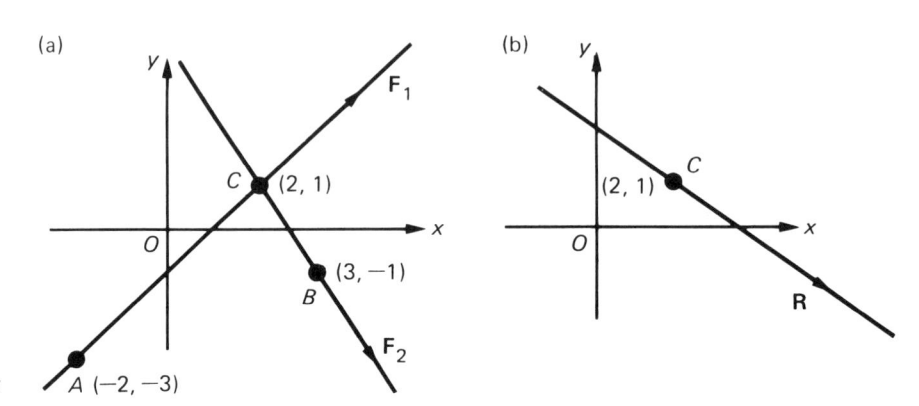

Fig. 9.1:7

Resultant of $\mathbf{F}_1$ and $\mathbf{F}_2$:

$$\mathbf{R} = \mathbf{F}_1 + \mathbf{F}_2 = (3\mathbf{i} + 3\mathbf{j}) + (2\mathbf{i} - 4\mathbf{j}) = 5\mathbf{i} - \mathbf{j}$$

Line of action of $\mathbf{R}$: since $\mathbf{F}_1$ and $\mathbf{F}_2$ act through C, so does $\mathbf{R}$ (by statement (7)); so a vector equation of the line of action of $\mathbf{R}$ (fig. 9.1:**7b**) is

$$\mathbf{r} = \mathbf{c} + \nu\mathbf{R} = 2\mathbf{i} + \mathbf{j} + \nu(5\mathbf{i} - \mathbf{j})$$

Worked example 4   Forces $2P$, $3P$, $3P$, $2P$ and $2\sqrt{2}P$ act along $\overrightarrow{AB}$, $\overrightarrow{BC}$, $\overrightarrow{CD}$, $\overrightarrow{AD}$ and $\overrightarrow{DB}$ where $ABCD$ is a square of side $2a$. Find the magnitude and direction of their resultant and the distance from A of the point where it cuts AB.

Suppose the resultant $\mathbf{R} = R_x\mathbf{i} + R_y\mathbf{j}$ (**i**, **j** as shown in fig. 9.1:**8**), and suppose that the line of action of $\mathbf{R}$ cuts AB at E where $AE = x$. Resolving the forces into components:

$\rightarrow$) $R_x = \sum F_x = 2P - 3P + 2\sqrt{2}P \cos 45° = P$ (by equation (**4**))

$\uparrow$) $R_y = \sum F_y = 3P + 2P - 2\sqrt{2}P \sin 45° = 3P$ (by equation (**5**))

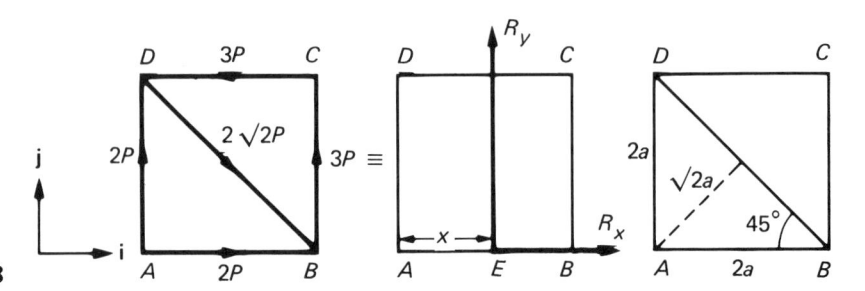

Fig. 9.1:8

Forces on a rigid body 515

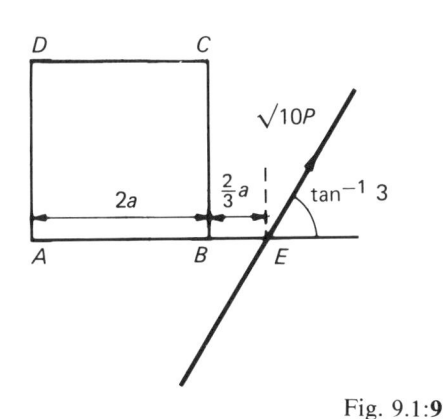

By the Principle of Moments,

$$A\circlearrowleft \qquad R_y \cdot x = 3P \cdot 2a + 3P \cdot 2a - 2\sqrt{2}P \cdot \sqrt{2}a = 8Pa$$
$$\Rightarrow \qquad 3Px = 8Pa \quad \text{i.e.} \quad x = \tfrac{8}{3}a$$

Resultant $\mathbf{R} = P\mathbf{i} + 3P\mathbf{j}$, has magnitude $\sqrt{10}P$. Its line of action makes an angle of $\tan^{-1} 3$ with AB, and cuts AB at E where $AE = \tfrac{8}{3}a$ (fig. 9.1:9).

Fig. 9.1:9

Worked example 5 Forces $\mathbf{F}_1 = 4\mathbf{i} - 3\mathbf{j}$, $\mathbf{F}_2 = 6\mathbf{i} - 7\mathbf{j}$ and $\mathbf{F}_3 = -2\mathbf{i} + 4\mathbf{j}$ act through the points with position vectors $2\mathbf{i} - \mathbf{j}$, $-3\mathbf{i} + \mathbf{j}$ and $4\mathbf{j}$ respectively. Find the resultant of this system and the cartesian equation of its line of action.

Let the resultant $\mathbf{R} = R_x \mathbf{i} + R_y \mathbf{j}$, and suppose that it passes through the point $P(x, y)$ as in fig. 9.1:10).

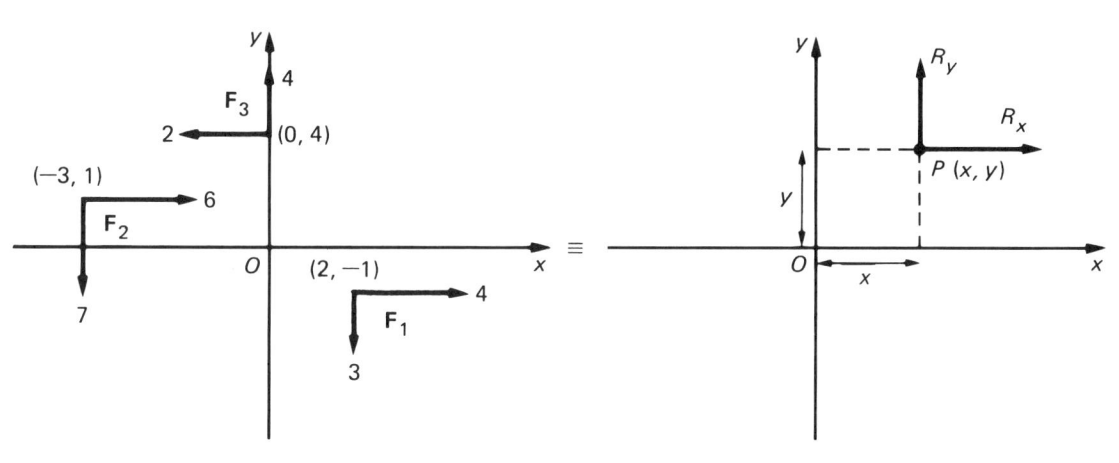

Fig. 9.1:10

$$\mathbf{R} = R_x \mathbf{i} + R_y \mathbf{j} = \mathbf{F}_1 + \mathbf{F}_2 + \mathbf{F}_3$$
$$= (4\mathbf{i} - 3\mathbf{j}) + (6\mathbf{i} - 7\mathbf{j}) + (-2\mathbf{i} + 4\mathbf{j})$$
$$= 8\mathbf{i} - 6\mathbf{j}$$
$$\Rightarrow \qquad R_x = 8, \; R_y = -6$$

By the Principle of Moments, $M_O(\mathbf{R}) = \sum M_O(\mathbf{F})$, i.e.

$$R_y \cdot x - R_x \cdot y = (-3 \times 2 + 4 \times 1) + (7 \times 3 - 6 \times 1) + (2 \times 4)$$
$$= 21$$
i.e. $\qquad -6x - 8y = 21$
i.e. $\qquad 6x + 8y + 21 = 0$

This is the condition that a general point (x, y) lies on the line of action of $\mathbf{R}$; so it is the cartesian equation of this line.

Exercise 9.1:4

1 Find a vector equation of the line of action of the resultant of the following sets of forces. Sketch diagrams showing the forces and their resultant.

- (i) $F_1 = 3i + j$ and $F_2 = -2i + 2j$; acting through the origin.
- (ii) $F_1 = 7i - j$ and $F_2 = -3i + 5j$; acting through the point $3i - j$.
- → (iii) $F_1 = 2i$, $F_2 = 3i - 3j$ and $F_3 = 5i - 2j$; acting through the point $i + j$.

③ **2** Show that the following systems of forces intersect. Find the position vector of their point of intersection. Hence find a vector equation of the line of action of their resultant. Sketch diagrams showing the forces and their resultant.

- (i) $F_1 = 4i + 4j$, $F_2 = 3i + j$; acting through the points $-i - 3j$, $4i$.
- (ii) $F_1 = -6i + 3j$, $F_2 = 2i - 8j$; acting through the origin and the point $3i + 2j$.
- (iii) $F_1 = 3i - 2j$ and $F_2 = 2j$; acting through the points $-2i + 2j$, $i + 5j$.
- → (iv) $F_1 = 10i + 5j$, $F_2 = 4i - 6j$, $F_3 = -9i + 3j$; acting through the points $i + 4j$, $3i + 3j$, $8i + 5j$.
- **R** (v) $F_1 = -8i$, $F_2 = 5j$, $F_3 = 7i - 4j$; acting through the points $3j$, $i + 2j$, $-6i + 7j$.

③ **3** Forces $F_1 = 3i + 2j$ and $F_2 = pi + 4j$ act through the points $-i + qj$ and $-3i + 5j$. A vector equation of the line of action of their resultant is

$$r = 2i + j + \lambda(i - 3j)$$

Find p and q.

③ **4** Find, in the form $r = a + \lambda d$, a vector equation of the line of action of the resultant of each of the following systems of (intersecting) forces.

- (i) $F_1 = i + 3j$, $F_2 = 2i + j$; acting through the point with position vector $i - j$.
- (ii) $F_1 = 3i + 5j$, $F_2 = -i - 4j$, $F_3 = -6i$; acting through the point with position vector $2i + 3j$.

③ **R 5** Forces $F_1 = i - 3j$ and $F_2 = 2i + 5j$ act through the points $4i - 5j$ and $-4j$ respectively. Show that F_1 and F_2 intersect and find a vector equation of the line of action of their resultant.

③ **6** Find a vector equation of the line of action of the resultant of each of the following systems of (intersecting) forces.

- (i) $F_1 = 6i$, $F_2 = 4i + 8j$; acting through the points $2i + 3j$, $-2i + j$.
- (ii) $F_1 = i + 2j$, $F_2 = -3i + 2j$, $F_3 = 4i - 4j$; acting through the points $-i + j$, $6i - j$, $8i - 5j$.

③ → **7** Forces $F_1 = 3i - 3j$, $F_2 = 2i + 4j$ and $F_3 = 5i - 2j$ act through the points $3j$, the origin and $-4i + 4j$.

(i) Find the resultant moment of the forces about the origin O.

(ii) Show that the forces intersect and find their point of intersection.

(iii) Find the resultant of the forces.

(iv) Verify that the moment of the resultant about O equals the resultant moment of the forces about O.

Questions 8–13: Find the magnitude and direction of the resultant of the following systems of forces and the distance from A of the point where its line of action cuts AB (produced if necessary).

④ **8** $AB = 2a$.

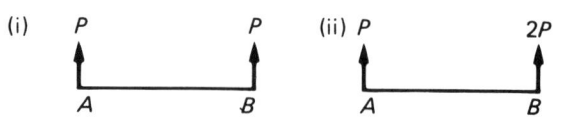

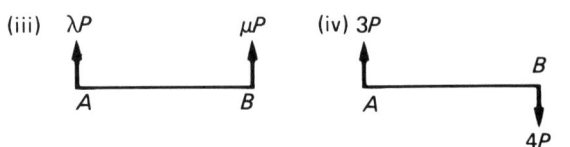

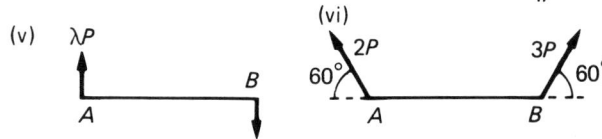

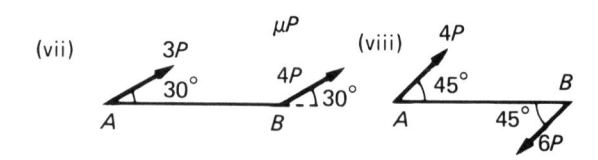

(i)
```
    P           P
    ↑           ↑
    |_____|
    A           B
```

(ii)
```
    P          2P
    ↑           ↑
    |_____|
    A           B
```

(iii)
```
   λP          μP
    ↑           ↑
    |_____|
    A           B
```

(iv)
```
   3P
    ↑           B
    |           |
    A           ↓
                4P
```

(v)
```
   λP
    ↑           
    |_____
    A           B
                ↓
                μP
```

(vi)
```
      2P       3P
      ↖ 60°    60° ↗
    A_____B
```

(vii)
```
   3P      4P
    ↗30°    ↖30°
   A         B
```

(viii)
```
    4P          B
    ↖45°     45°↗
    A          6P
```

④ **9** $ABCD$ is a square of side $2a$.

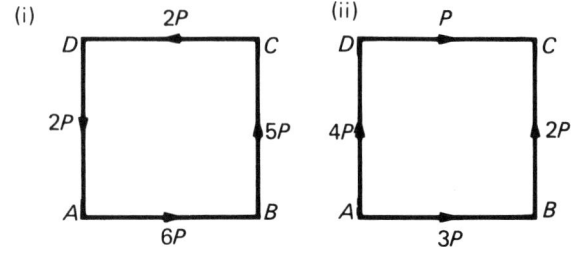

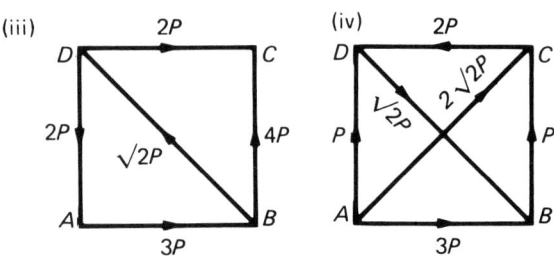

(i)
```
    D    2P    C
    ┌────→────┐
  2P↑         ↓5P
    │         │
    A─────────B
        6P
```

(ii)
```
    D    P     C
    ┌────→────┐
  4P↑         ↑2P
    │         │
    A─────────B
        3P
```

(iii)
```
    D    2P    C
    ┌────→────┐
  2P↑  √2P    ↓4P
    │    ╲    │
    A─────────B
        3P
```

(iv)
```
    D    2P    C
    ┌────→────┐
    │ ╲ 2√2P  │
  P↑  √2P╲   ↑P
    │   ╲    │
    A─────────B
        3P
```

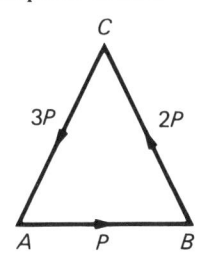

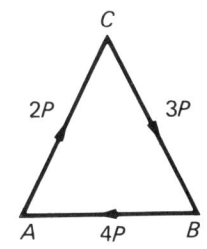

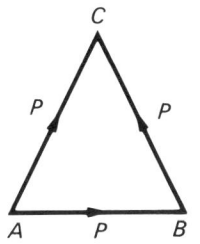

10 ABC is an equilateral triangle of side $2a$; D is the midpoint of AB.

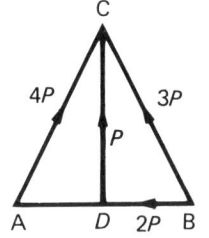

(i) Triangle with $3P$ along AC, $2P$ along CB, P along AB.

(ii) Triangle with $2P$ along CA, $3P$ along CB, $4P$ along BA.

(iii) Triangle with P along AC, P along CB, P along AB.

(iv) Triangle with $4P$ along AC, $3P$ along CB, P along CD (downward), $2P$ along DB.

11 $AB = 4a$, $BC = 3a$.

(i) Right triangle: $2P$ along AB (downward), $5P$ along AC, $2P$ along BC.

(ii) Rectangle $ABCD$: $6P$ along AB (downward), $7P$ along BD, $3P$ along AC, $2P$ along BC.

(iii) Rectangle $ABCD$: $5P$ along DA, $2P$ along AB (downward), $4P$ along CD (upward), P along BD, $4P$ along AC, $3P$ along BC.

(iv) Right triangle with altitude: $6P$ along AB (downward), $10P$ along AC, $4P$ along foot-to-A direction, $8P$ along BC.

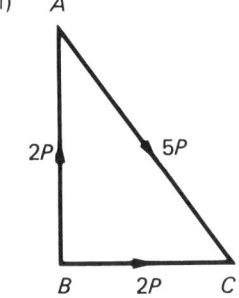

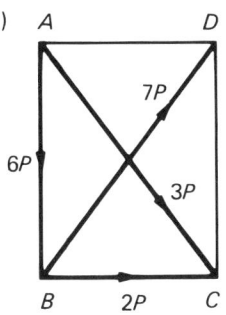

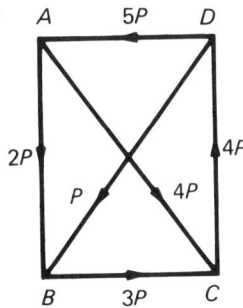

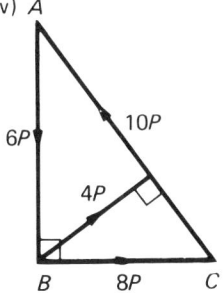

④ **12** $ABCDEF$ is a regular hexagon of side $2a$.

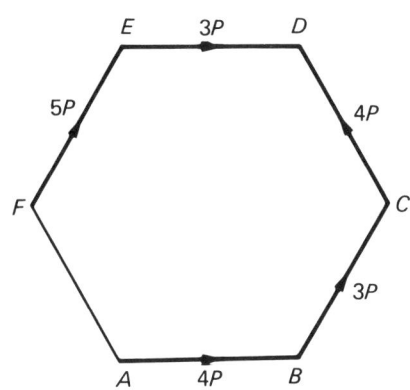

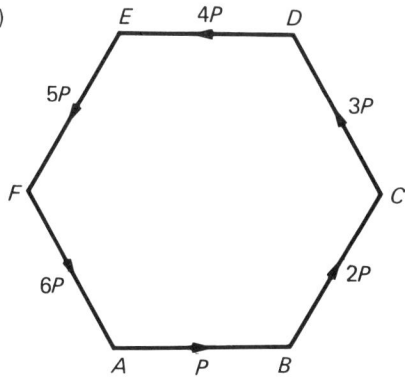

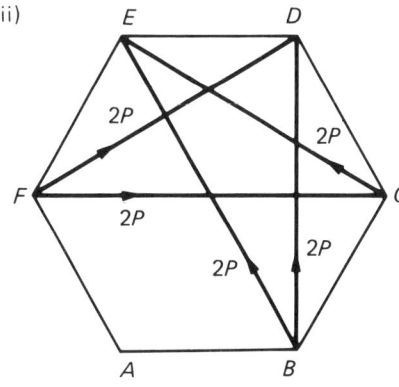

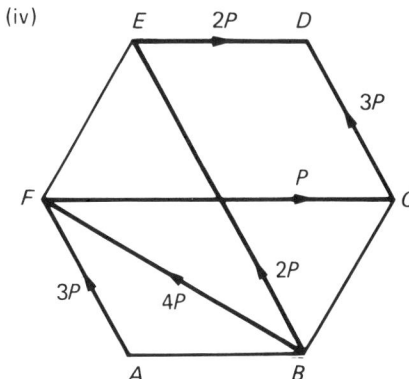

(i) Hexagon: $3P$ along ED, $5P$ along FE, $4P$ along DC, $3P$ along CB, $4P$ along AB.

(ii) Hexagon: $4P$ along ED, $5P$ along FE, $3P$ along DC, $6P$ along AF, $2P$ along CB, P along AB.

(iii) Hexagon with diagonals: $2P$ along EC, $2P$ along DF, $2P$ along FC, $2P$ along EB, $2P$ along DB, $2P$ along AB direction.

(iv) Hexagon: $2P$ along ED, $3P$ along DC, P along FC, $2P$ along CB, $3P$ along AF, $4P$ along FB, $2P$ along EB.

④ **13** $ABCE$ is a square and CDE is an equilateral triangle of side $2a$.

(i) $6P$ along ED, $4P$ along DC, $2P$ along CE, $5P\sqrt{3}$ along EA (downward), $3P$ along CB (downward), $5P$ along AB.

(ii) P along ED, $3P$ along DC, $4P$ along EA (downward), $4P$ along CB (downward), $2P$ along BA.

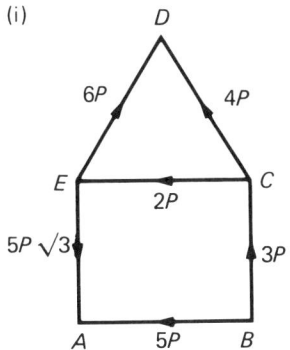

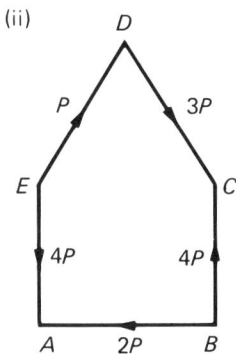

14 Find the resultant of each of the following systems of (parallel) forces, and find the cartesian equation of its line of action.

(i) $\mathbf{F}_1 = 2\mathbf{j}$, $\mathbf{F}_2 = 4\mathbf{j}$; acting through the origin and the point $3\mathbf{i}$.
(ii) $\mathbf{F}_1 = 3\mathbf{i}+\mathbf{j}$, $\mathbf{F}_2 = 3\mathbf{i}+\mathbf{j}$; acting through the points $2\mathbf{i}+\mathbf{j}$, $6\mathbf{i}+3\mathbf{j}$.
(iii) $\mathbf{F}_1 = \mathbf{i}-2\mathbf{j}$, $\mathbf{F}_2 = -2\mathbf{i}+4\mathbf{j}$; acting through the points $\mathbf{i}$, $4\mathbf{i}$.
(iv) $\mathbf{F}_1 = \mathbf{i}+\mathbf{j}$, $\mathbf{F}_2 = 3\mathbf{i}+3\mathbf{j}$, $\mathbf{F}_3 = -4\mathbf{i}-4\mathbf{j}$; acting through the points $2\mathbf{i}-\mathbf{j}$, $4\mathbf{j}$, $-2\mathbf{i}-2\mathbf{j}$.

15 Forces $\mathbf{F}_1 = m(p\mathbf{i}+q\mathbf{j})$ and $\mathbf{F}_2 = n(p\mathbf{i}+q\mathbf{j})$ act through the points $x_1\mathbf{i}+y_1\mathbf{j}$ and $x_2\mathbf{i}+y_2\mathbf{j}$. Find their resultant and the cartesian equation of its line of action.

16 Find the resultant of each of the following systems of forces, and find the cartesian equation of its line of action.

(i) $\mathbf{F}_1 = 6\mathbf{j}$, $\mathbf{F}_2 = 4\mathbf{i}-5\mathbf{j}$, $\mathbf{F}_3 = -2\mathbf{i}+2\mathbf{j}$; acting through the points $4\mathbf{i}$, $\mathbf{i}$, $-3\mathbf{i}$.
(ii) $\mathbf{F}_1 = 2\mathbf{i}+5\mathbf{j}$, $\mathbf{F}_2 = 7\mathbf{i}-\mathbf{j}$, $\mathbf{F}_3 = -8\mathbf{i}-5\mathbf{j}$; acting through the points $\mathbf{i}$, $-2\mathbf{i}+\mathbf{j}$, $\mathbf{i}+2\mathbf{j}$.
(iii) $\mathbf{F}_1 = \mathbf{i}+2\mathbf{j}$, $\mathbf{F}_2 = 4\mathbf{i}-\mathbf{j}$, $\mathbf{F}_3 = -2\mathbf{i}+3\mathbf{j}$; acting through the points $5\mathbf{i}-2\mathbf{j}$, $3\mathbf{i}$, $-4\mathbf{i}+2\mathbf{j}$.
(iv) $\mathbf{F}_1 = -4\mathbf{i}-3\mathbf{j}$, $\mathbf{F}_2 = 3\mathbf{i}-\mathbf{j}$, $\mathbf{F}_3 = -7\mathbf{i}+7\mathbf{j}$, $\mathbf{F}_4 = -2\mathbf{i}+2\mathbf{j}$; acting through the points $\mathbf{i}+\mathbf{j}$, $2\mathbf{j}$, $-\mathbf{i}$, $-3\mathbf{j}$.

17 A, B and C are the points $(2, 0)$, $(8, 8)$ and $(0, 2)$. Forces of magnitudes 20 N, 30 N and $8\sqrt{2}$ N act along $\overrightarrow{AB}$, $\overrightarrow{BC}$ and $\overrightarrow{CA}$ respectively. Find the cartesian equation of the line of action of their resultant.

18 Forces of magnitudes $2\sqrt{5}$ N, $3\sqrt{2}$ N and 4 N act in the direction of x increasing along the lines $x-2y=1$, $x+y=4$ and $y=-1$. Find the cartesian equation of the line of action of their resultant.

19 Forces 3 N, 2 N, p N and 4 N act along the sides $\overrightarrow{AB}$, $\overrightarrow{BC}$, $\overrightarrow{CD}$ and $\overrightarrow{DA}$ of a square $ABCD$. The resultant of the forces is parallel to $\overrightarrow{AD}$. Find p.

20 Forces 2 N, 1 N, 3 N, 2 N and p N act along $\overrightarrow{AB}$, $\overrightarrow{BC}$, $\overrightarrow{DC}$, $\overrightarrow{AD}$ and $\overrightarrow{BD}$, where $ABCD$ is a square. The resultant of the forces is parallel to $\overrightarrow{AC}$. Find p. (Hint: put $\overrightarrow{AC}$ in the $\mathbf{i}$ direction.)

21 Forces 2 N, 3 N, 4 N, 3 N, 2 N and p N act along the sides $\overrightarrow{AB}$, $\overrightarrow{BC}$, $\overrightarrow{CD}$, $\overrightarrow{DE}$, $\overrightarrow{FE}$ and $\overrightarrow{AF}$ of a regular hexagon. The resultant of the forces is parallel to $\overrightarrow{AD}$. Find p.

22 Forces 3 N, 4 N, 2 N, 1 N, p N and q N act along $\overrightarrow{AB}$, $\overrightarrow{BC}$, $\overrightarrow{CD}$, $\overrightarrow{DA}$, $\overrightarrow{AC}$ and $\overrightarrow{BD}$, where $ABCD$ is a rectangle with $AB = 4a$, $BC = 3a$. Find p and q if the resultant of the forces acts:

(i) through B parallel to AD
(ii) through B, parallel to AC

23 Forces 2 N, p N, 4 N, 3 N, q N and $\sqrt{2}$ N act along $\overrightarrow{AB}$, $\overrightarrow{BC}$, $\overrightarrow{DC}$, $\overrightarrow{DA}$, $\overrightarrow{CA}$ and $\overrightarrow{BD}$ where $ABCD$ is a square. Find p and q if the resultant of the forces acts:

(i) through B, parallel to AC
(ii) along ED where E is the point on BA produced such that $EA = 2AB$.

24 A system of forces acts in the xy plane. Its resultant moments about the points O, $A(a, 0)$ and $B(0, a)$ are $-3G$, $-G$ and $+2G$. Find the magnitude of the resultant of the system and the cartesian equation of its line of action. (Hint: let the resultant $\mathbf{R} = R_x\mathbf{i}+R_y\mathbf{j}$; suppose its line of action passes through the point (x, y).)

25 A system of forces acts in the xy plane. Its resultant moments about the points O, $A(2, 1)$ and $B(0, 2)$ are $+3$, $+3$ and 0. Show that the resultant of the system is parallel to $\overrightarrow{AO}$ and find its magnitude.

26 A system of forces acts in the xy plane. Its resultant moments about the points O, $A(4a, 0)$ and $B(0, 3a)$ are $-Pa$, $15Pa$ and $-5Pa$. Find the cartesian equation of the line of action of its resultant.

27 A system of forces acts in the xy plane. Its resultant moments about the points $A(2a, 0)$, $B(2a, 2a)$ and $C(0, 2a)$ are $10Pa$, $-5Pa$, $15Pa$. Find the cartesian equation of the line of action of its resultant.

28 A system of coplanar forces has clockwise moments $5M$, M and $2M$ respectively about the points $(a, 0)$, $(0, a)$ and (a, a) in the plane. Find the magnitude of the resultant of the system and the equation of its line of action.

29 (i) A force $\mathbf{F}$ has equal moments about two points A and B. Show that the line of action of $\mathbf{F}$ is parallel to AB.

(ii) A force of unit magnitude has moment $+3$ Nm about each of the points $(-1, 2)$ and $(3, 5)$. Find the cartesian equation of the line of action of $\mathbf{F}$.

30 ABC is an equilateral triangle of side a. A system of forces acting in the plane of $\triangle ABC$ has resultant moments $+2Pa$, $+3Pa$, $-Pa$ about A, B and C where $+$ indicates the sense ABC. Find the magnitude and direction of the resultant of the system and the distance from A at which its line of action cuts AB.

31 ABC is an equilateral triangle. A system of forces acting in the plane of $\triangle ABC$ has resultant moments $2G$, $-2G$ and G about A, B and C.

(i) Find the point at which the line of action of the resultant of the system cuts AB.
(ii) Find the resultant moment of the system about the centroid of $\triangle ABC$.

- **32** Show that if $\sum \mathbf{F} \neq \mathbf{0}$ a system of forces $\{\mathbf{F}\}$ can be reduced to a single force *through a given point* A, combined with a couple. (Hint: use the fact that the system can be reduced to a single force whose line of action can be found.)

  How would you find the magnitude G of the couple?

33 When the following systems of forces are reduced to a single resultant force through A combined with a couple, find the magnitude and direction of the force and the magnitude of the couple.

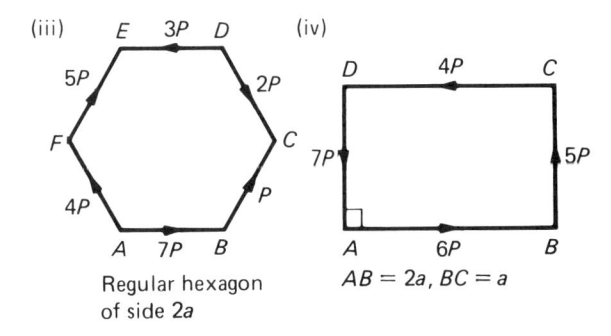

(i) [diagram: forces P, P at distance a, a, and P downward, with points A, B]

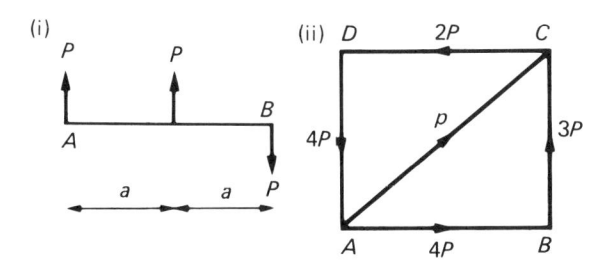

(ii) [diagram: square $ABCD$ with forces $4P$ on AD, $2P$ on DC, $3P$ on CB, $4P$ on AB, and P along diagonal]

(iii) Regular hexagon of side $2a$ [diagram: forces $5P$, $3P$, $2P$, P, $7P$, $4P$ on sides]

(iv) $AB = 2a$, $BC = a$ [diagram: rectangle with forces $4P$, $5P$, $6P$, $7P$]

34 The following systems of forces are reduced to a single resultant force $\mathbf{R}$ through the origin, combined with a couple of magnitude G. Find $\mathbf{R}$ and G:

- → (i) $\mathbf{F}_1 = 2\mathbf{i}$, $\mathbf{F}_2 = 3\mathbf{j}$, $\mathbf{F}_3 = 2\mathbf{i} + \mathbf{j}$; acting through the points $\mathbf{i} + \mathbf{j}$, $3\mathbf{i} + 2\mathbf{j}$, $4\mathbf{j}$.
- R (ii) $\mathbf{F}_1 = 5\mathbf{i} + 3\mathbf{j}$, $\mathbf{F}_2 = -2\mathbf{i} + 4\mathbf{j}$, $\mathbf{F}_3 = 3\mathbf{i} + 3\mathbf{j}$; acting through the points $2\mathbf{i} - \mathbf{j}$, $3\mathbf{i} + 5\mathbf{j}$, $\mathbf{i} - 4\mathbf{j}$.

R **35** Forces $\mathbf{F}_1 = -2\mathbf{i} + 3\mathbf{j}$, $\mathbf{F}_2 = 9\mathbf{i} + \mathbf{j}$ and $\mathbf{F}_3 = 5\mathbf{i} - 12\mathbf{j}$ act at the points $\mathbf{i} + 3\mathbf{j}$, $-2\mathbf{i} + \mathbf{j}$, $\mathbf{i}$. Find the cartesian equation of the line of action of the resultant of each of the following systems:

  (i) $\mathbf{F}_1, \mathbf{F}_2, \mathbf{F}_3$
  (ii) $\mathbf{F}_1, \mathbf{F}_2, \mathbf{F}_3$ and a couple of magnitude $+8$
  (iii) $\mathbf{F}_1, \mathbf{F}_2, \mathbf{F}_3$ and a couple of magnitude -16

9.1:5 Systems reducing to a couple

(a) $\sum \mathbf{F} = \mathbf{0} \Leftrightarrow$ the system has no translational effect
(b) $\sum M_X(\mathbf{F})$ is the same for every point $X \Leftrightarrow$ the system has an equal turning effect about every point X.

So (see 9.1:6):

> $\sum \mathbf{F} = \mathbf{0}$ and $\sum M_X(\mathbf{F})$ is the same for every point $X \Leftrightarrow$ the system reduces to a couple (10)

A rigid body moving in a plane has only three degrees of freedom, so conditions (a) and (b) give only three independent equations. Thus we can show that a given system of forces reduces to a couple by showing

> (i) $\sum F_x = 0$
> (ii) $\sum F_y = 0$ (11)
> (iii) $\sum M_A(\mathbf{F}) \neq 0$ for some point A

(There are other possibilities: see exercise 9.1:5 question 2.)
Note that $\sum M_A(\mathbf{F})$ is the magnitude of the couple.

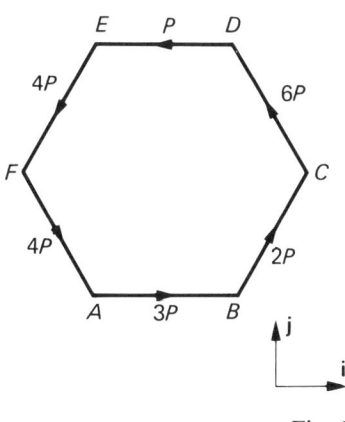

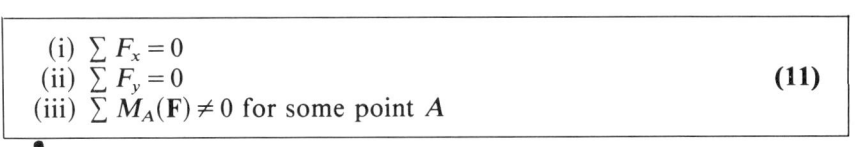

Fig. 9.1:11a [diagram: regular hexagon $ABCDEF$ with forces $3P$ on AB, $2P$ on BC, $6P$ on CD, P on DE, $4P$ on EF, $4P$ on FA]

Worked example 6 Forces $3P, 2P, 6P, P, 4P$ and $4P$ act along the sides $\overrightarrow{AB}, \overrightarrow{BC}, \overrightarrow{CD}, \overrightarrow{DE}, \overrightarrow{EF}$ and $\overrightarrow{FA}$ of a regular hexagon $ABCDEF$, side $2a$, (fig. 9.1:11a). Prove that they reduce to a couple and find its magnitude G.

520 *Systems reducing to a couple*

$\rightarrow) \quad \sum F_x = 3P + 2P\cos 60° - 6P\cos 60° - P - 4P\cos 60° + 4P\cos 60° = 0$

$\uparrow) \quad \sum F_y = 2P\sin 60° + 6P\sin 60° - 4P\sin 60° - 4P\sin 60° = 0$

$\circlearrowleft) \quad G = \sum M_O(\mathbf{F}) = (3P + 2P + 6P + P + 4P + 4P)a\sqrt{3} = 20Pa\sqrt{3}$

Note: (i) The centre O of the hexagon is the easiest point about which to take moments, since the perpendicular distance from O to each force is the same, i.e. $a\sqrt{3}$ (fig. 9.1:**11b**).
(ii) We could instead have shown that the resultant moment of the forces about three non-collinear points is the same—this would have been slightly more laborious. See question 2.

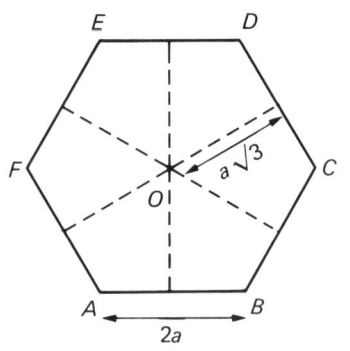

Fig. 9.1:**11b**

Conversely, if we know that a system of forces reduces to a couple we can use three independent equations taken from (a) and (b), usually

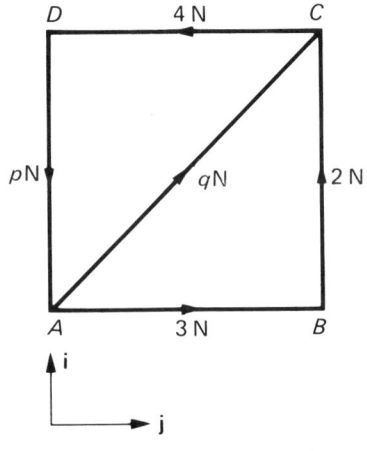

$$\left. \begin{array}{l} \text{(i)} \ \sum F_x = 0 \\ \text{(ii)} \ \sum F_y = 0 \\ \text{(iii)} \ \sum M_A(\mathbf{F}) = G \end{array} \right\} \quad (12)$$

Worked example 7 Forces 3 N, 2 N, 4 N, P N and Q N act along $\overrightarrow{AB}$, $\overrightarrow{BC}$, $\overrightarrow{CD}$, $\overrightarrow{DA}$ and $\overrightarrow{AC}$ where $ABCD$ is a square of side $2a$ (fig. 9.1:**12**). Find P and Q given that the system of forces reduces to a couple G and find the magnitude of G.

$\rightarrow) \quad \sum F_x = 0 \Rightarrow 3 - 4 + Q\cos 45° = 0 \Rightarrow -1 + Q/\sqrt{2} = 0 \Rightarrow Q = \sqrt{2}$

$\uparrow) \quad \sum F_y = 0 \Rightarrow 2 - P + Q\sin 45° = 0 \Rightarrow 2 - P + Q/\sqrt{2} = 0 \Rightarrow P = 3$

Fig. 9.1:**12** $A\circlearrowleft) \quad G = \sum M_A(\mathbf{F}) = 2 \cdot 2a + 4 \cdot 2a = 12a$

Exercise 9.1:5

⑥ **1** Show that each of the following systems of forces reduces to a couple, and find its magnitude:

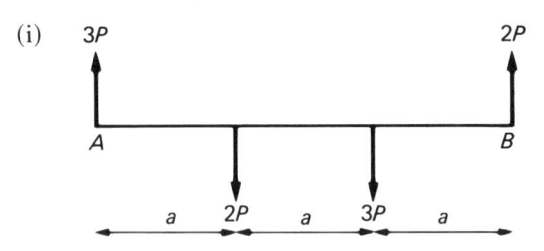

(i)

(ii) *ABC* equilateral triangle side $2a$

(iii) *ABC* right angled triangle; $AB = 3a$, $AC = 4a$

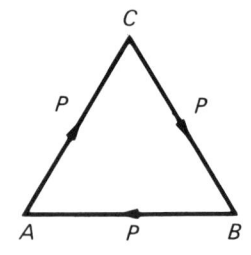

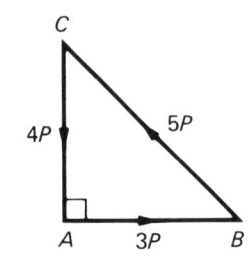

→ (iv) *ABCD* square side $2a$

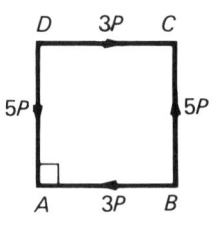

(v) *ABCD* square side $2a$

(vi) *ABCD* rectangle; $AB = 2a$, $BC = a$

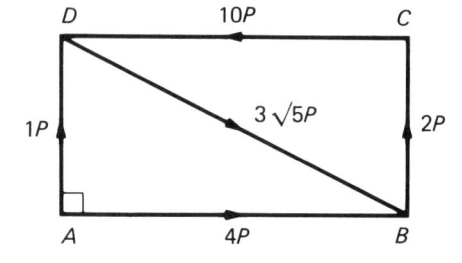

→ (vii) *ABCDEF* regular hexagon side $2a$

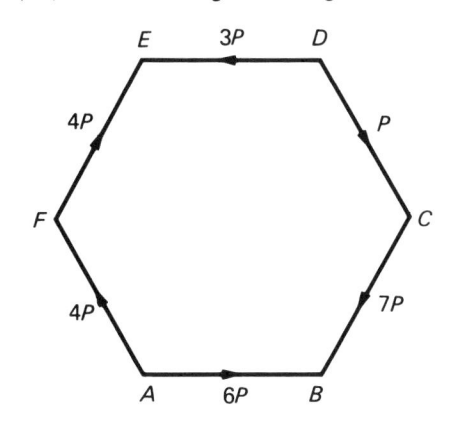

R (viii) *ABCDEF* regular hexagon side $2a$

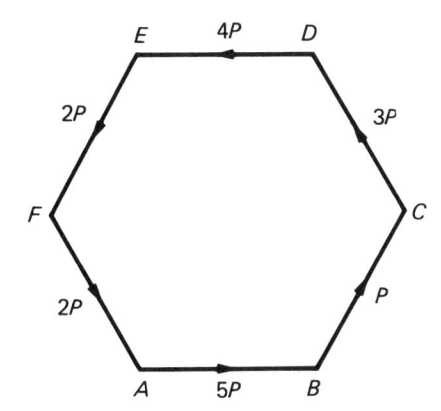

⑥ **2** Consider two of the other ways of showing that a given system of forces reduces to a couple:

(i) $\sum M_A(\mathbf{F}) = \sum M_B(\mathbf{F}) = \sum M_C(\mathbf{F}) \neq 0$
for three non-collinear points *A*, *B* and *C*. Why 'non-collinear'? What is the most you could say about a system of forces for which

$$\sum M_A(\mathbf{F}) = \sum M_B(\mathbf{F}) = \sum M_C(\mathbf{F}) \neq 0$$

for three *collinear* points *A*, *B* and *C*?
(ii) $\sum F_x = 0$, $\sum M_A(\mathbf{F}) = \sum M_B(\mathbf{F}) \neq 0$
for two points *A* and *B* such that *AB* is not parallel to *Oy*. Why '*AB* is not parallel to *Oy*'?

Do one part of question 1 using these methods.

⑥ **3** Show that each of the following systems of forces reduces to a couple, and find its magnitude.

•→ (i) $\mathbf{F}_1 = \mathbf{i} + 3\mathbf{j}$, $\mathbf{F}_2 = -2\mathbf{i} - \mathbf{j}$, $\mathbf{F}_3 = \mathbf{i} - 2\mathbf{j}$; acting through the points $2\mathbf{i} + 5\mathbf{j}$, $4\mathbf{j}$, $-\mathbf{i} + \mathbf{j}$. (Hint: show that $\sum \mathbf{F} = \mathbf{0}$, and find $\sum M_O(\mathbf{F})$.)
(ii) $\mathbf{F}_1 = 4\mathbf{i} + 3\mathbf{j}$, $\mathbf{F}_2 = \mathbf{i} - 4\mathbf{j}$, $\mathbf{F}_3 = -3\mathbf{i} + \mathbf{j}$; acting through the points $\mathbf{i} + \mathbf{j}$, $-3\mathbf{i} + 2\mathbf{j}$, $-3\mathbf{i} + 2\mathbf{j}$.
R (iii) $\mathbf{F}_1 = 2\mathbf{i}$, $\mathbf{F}_2 = -3\mathbf{j}$, $\mathbf{F}_3 = -4\mathbf{i} + 5\mathbf{j}$, $\mathbf{F}_4 = 2\mathbf{i} - 2\mathbf{j}$; acting through the points $2\mathbf{i} + 2\mathbf{j}$, $\mathbf{i} + \mathbf{j}$, $-\mathbf{i} + 2\mathbf{j}$, $-2\mathbf{i} - \mathbf{j}$.

⑦ **4** Find *p* and *q* in each of the following if the system of forces reduces to a couple, and find the magnitude of the couple:

(i) *ABCD* square side $2a$.

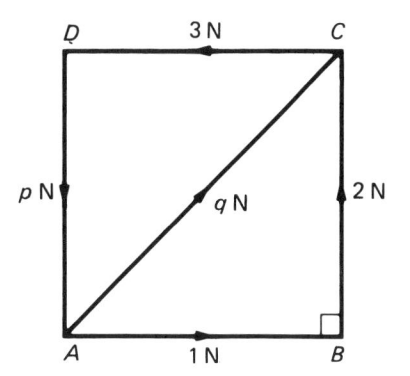

(ii) *ABC* equilateral triangle side $2a$.

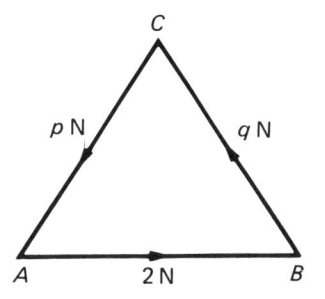

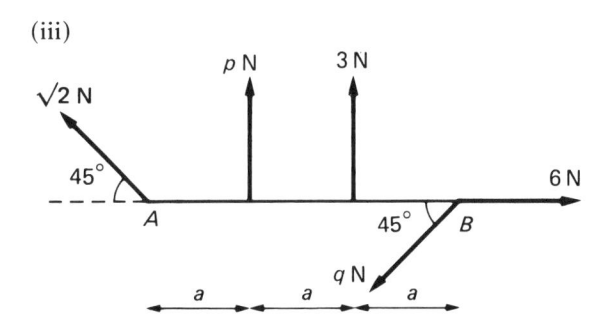

(iii)

(iv) *ABCD* circle radius a.

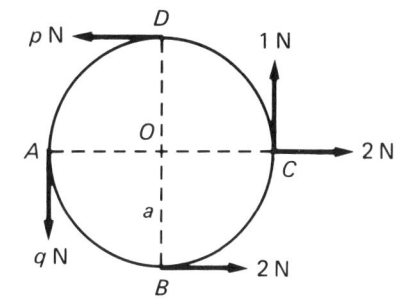

R (vi)

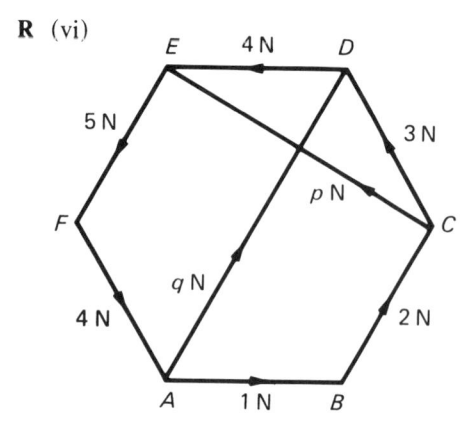

A diagram shows a hexagon ABCDEF with forces: ED = 4 N, EF = 5 N, DC = 3 N, FA = 4 N, CB = 2 N, AB = 1 N, and internal forces p N along DA (or similar diagonal) and q N along another diagonal.

⑦→ **5** Forces $\mathbf{F}_1 = 2\mathbf{i} + 2\mathbf{j}$, $\mathbf{F}_2 = -5\mathbf{i} - 6\mathbf{j}$ and $\mathbf{F}_3$ act at the points $4\mathbf{i} - 3\mathbf{j}$, $2\mathbf{i} + 5\mathbf{j}$ and $4\mathbf{j}$. The forces reduce to a couple. Find $\mathbf{F}_3$ and the magnitude of the couple.

⑦ **6** Find the magnitude and direction of the force, acting at the point A, which will reduce the given system of forces to a couple, and find the magnitude of the couple:

(i) Square ABCD: DC = 2 N, diagonal DB = $2\sqrt{2}$ N, CB = 1 N, AB = 1 N.

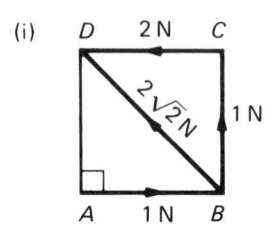

(ii) Hexagon ABCDEF: ED = 2 N, FE = 3 N, DC = 4 N, FA = 2 N, CB = 3 N, AB = 1 N.

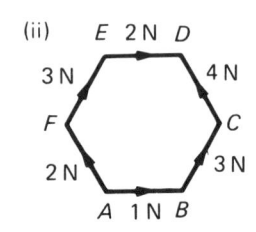

⑦ **7** A, B and C are the points (0, 2), (4, 5) and (4, 0). Forces $\mathbf{F}_1$ and $\mathbf{F}_2$ of magnitudes 10 and 2, act along $\overrightarrow{AB}$ and $\overrightarrow{CB}$ respectively. A force $\mathbf{F}_3$ acts through O; $\mathbf{F}_1$, $\mathbf{F}_2$ and $\mathbf{F}_3$ reduce to a couple. Find $\mathbf{F}_3$ and the magnitude of the couple.

⑦ **8** Find the magnitude, direction and line of action of the force which reduces the given system to a couple of magnitude $+2Pa$:

(i) Square ABCD: AB = 2P, BC = 3P.

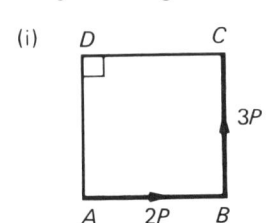

(ii) Hexagon ABCDEF: DC = 3P, CB = 2P, AB = P.

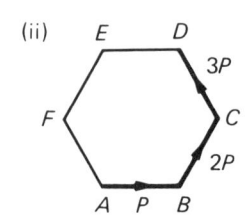

9 Show, using the Principle of Moments, that it is impossible to have a system of forces for which $\sum \mathbf{F} = \mathbf{0}$ and $\sum M_X(\mathbf{F})$ is not the same for every point X.

9.1:6 Systems in equilibrium

(a) $\sum \mathbf{F} = \mathbf{0} \Leftrightarrow$ the system has no translational effect
(b) $\sum M_X(\mathbf{F}) = 0$ for all points $X \Leftrightarrow$ the system has no turning effect

So

> $\sum \mathbf{F} = \mathbf{0}$, $\sum M_X(\mathbf{F}) = 0$ for all points $X \Leftrightarrow$ the system is in equilibrium (13)

A rigid body moving in a plane has only three degrees of freedom; so (a) and (b) give only three independent equations. Thus we can show that a given system of forces is in equilibrium by showing

> (i) $\sum F_x = 0$
> (ii) $\sum F_y = 0$
> (iii) $\sum M_A(\mathbf{F}) = 0$ for some point A (14)

(There are other possibilities; see exercise 9.1:6 question 7)
Note that if the forces intersect at a point A then $\sum M_A(\mathbf{F}) = 0$ and it is enough to show that $\sum F_x = 0$ and $\sum F_y = 0$, i.e.

> A system of intersecting forces is in equilibrium if $\sum \mathbf{F} = \mathbf{0}$. (15)

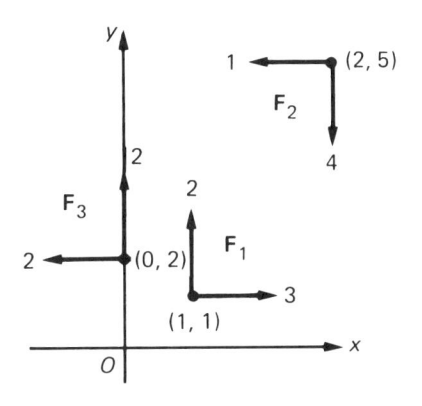

Forces on a rigid body 523

Worked example 8 Forces $\mathbf{F}_1 = 3\mathbf{i} + 2\mathbf{j}$, $\mathbf{F}_2 = -\mathbf{i} - 4\mathbf{j}$ and $\mathbf{F}_3 = -2\mathbf{i} + 2\mathbf{j}$ act through the points with position vectors $\mathbf{i} + \mathbf{j}$, $2\mathbf{i} + 5\mathbf{j}$ and $2\mathbf{j}$ respectively (fig. 9.1:13). Show that the forces are in equilibrium.

$$\sum \mathbf{F} = (3\mathbf{i} + 2\mathbf{j}) + (-\mathbf{i} - 4\mathbf{j}) + (-2\mathbf{i} + 2\mathbf{j}) = \mathbf{0}$$

$$\sum M_O(\mathbf{F}) = (-3 \times 1 + 2 \times 1) + (1 \times 5 - 2 \times 4) + (2 \times 2) = 0$$

So the forces are in equilibrium.

Conversely, if we know that a system of forces is in equilibrium, then we can use three independent equations taken from (a) and (b), usually:

(i) $\sum F_x = 0$
(ii) $\sum F_y = 0$
(iii) $\sum M_A(\mathbf{F}) = 0$ for some point A

Fig. 9.1:13

Exercise 9.1:6

1 Which of the following systems of intersecting forces are in equilibrium?

(i) $\mathbf{F}_1 = 3\mathbf{i} + 5\mathbf{j}$, $\mathbf{F}_2 = -3\mathbf{i} - 5\mathbf{j}$
(ii) $\mathbf{F}_1 = 2\mathbf{i} - \mathbf{j}$, $\mathbf{F}_2 = \mathbf{i} - 2\mathbf{j}$
(iii) $\mathbf{F}_1 = 4\mathbf{i} + 5\mathbf{j}$, $\mathbf{F}_2 = -3\mathbf{i}$, $\mathbf{F}_3 = -\mathbf{i} - 4\mathbf{j}$
(iv) $\mathbf{F}_1 = -6\mathbf{i} + 2\mathbf{j}$, $\mathbf{F}_2 = 5\mathbf{i} + \mathbf{j}$, $\mathbf{F}_3 = \mathbf{i} - 3\mathbf{j}$
(v) $\mathbf{F}_1 = 10\mathbf{i} + 6\mathbf{j}$, $\mathbf{F}_2 = -8\mathbf{i} + 9\mathbf{j}$, $\mathbf{F}_3 = -10\mathbf{j}$, $\mathbf{F}_4 = -2\mathbf{i} - 5\mathbf{j}$

2 The concurrent forces $\mathbf{F}_1 = 3\mathbf{i} + 7\mathbf{j}$, $\mathbf{F}_2 = p\mathbf{i} - 4\mathbf{j}$ and $\mathbf{F}_3 = -5\mathbf{i} + q\mathbf{j}$ are in equilibrium. Find p and q.

3 Forces $\mathbf{F}_1 = 4\mathbf{i} + 5\mathbf{j}$ and $\mathbf{F}_2 = -4\mathbf{i} - 5\mathbf{j}$ act through the points $2\mathbf{i} + \mathbf{j}$ and $\mathbf{b}$, respectively. Explain why the forces are in equilibrium if $\mathbf{b} = 2\mathbf{i} + \mathbf{j}$ but not if $\mathbf{b} = 2\mathbf{i} + 3\mathbf{j}$.

R 4 Forces $\mathbf{F}_1 = -8\mathbf{i} - 5\mathbf{j}$, $\mathbf{F}_2 = \mathbf{i} + \mathbf{j}$, $\mathbf{F}_3 = 2\mathbf{i} + 3\mathbf{j}$ and $\mathbf{F}_4 = 6\mathbf{i} - 2\mathbf{j}$ act at the point $\mathbf{i} + 2\mathbf{j}$. A fifth force $\mathbf{F}_5$ is added to the system which is then in equilibrium. Find a vector equation of the line of action of $\mathbf{F}_5$.

Questions 5 and 6: Show that each of the systems of forces is in equilibrium.

5 (i)

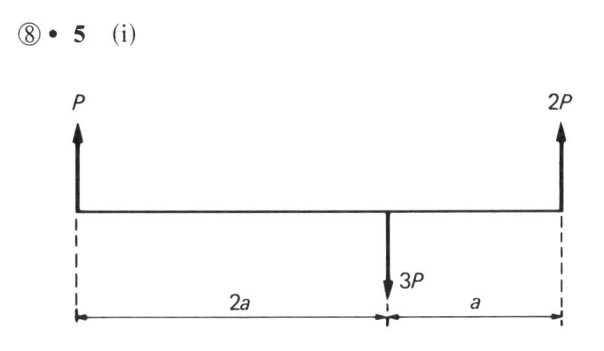

(ii)

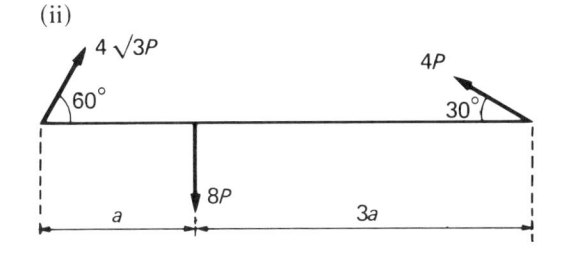

(iii) *ABCD* square

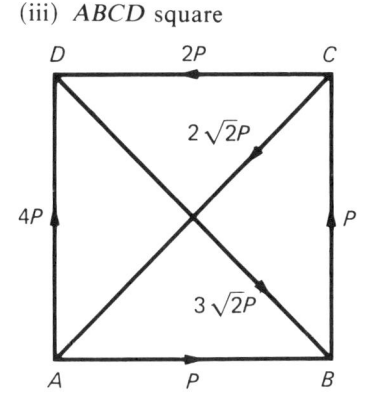

(iv) *ABCDEF* regular hexagon

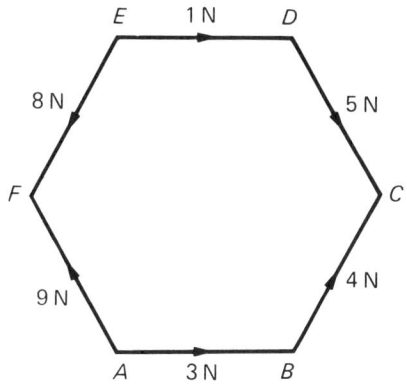

6 (i) $F_1 = 4i-j$, $F_2 = i-5j$, $F_3 = -5i+6j$; acting at the points with position vectors $2i+j$, $3i-j$, $-4j$.

(ii) $F_1 = i+j$, $F_2 = 3i+2j$, $F_3 = -4i-5j$, $F_4 = 2j$; acting at the points with position vectors $i-j$, $-2i-3j$, $3i$, $4i+3j$.

7 Do one part of question 5 and one part of question 6 using these alternative conditions for equilibrium:

(i) $\sum M_A(\mathbf{F}) = \sum M_B(\mathbf{F}) = \sum M_C(\mathbf{F}) = 0$ for three non-collinear points A, B and C.

(ii) $\sum F_x = 0$, $\sum M_A(\mathbf{F}) = \sum M_B(\mathbf{F}) = 0$ for two points A and B such that AB is not parallel to Oy.

Questions 8–10: Show that the forces are in equilibrium. Show also that the forces intersect and find the position vector of their point of intersection.

8 $F_1 = 2i-j$, $F_2 = -3i+2j$, $F_3 = i-j$; acting through the points i, $2i-3j$, $-2i-4j$.

9 $F_1 = -4i-4j$, $F_2 = i+4j$, $F_3 = 3i$; acting through the points $i+2j$, $-5j$, $3i+3j$.

10 $F_1 = 6i+8j$, $F_2 = -9i-3j$, $F_3 = 3i-5j$; acting through the points $i-4j$, $-2i-2j$, $i+5j$.

11 (i) Show that if three coplanar forces are in equilibrium, they are either parallel or concurrent. (Hint: suppose that not all the forces are parallel, in particular that two of them meet in a point A; then show that the third force must also pass through A.)

(ii) Give an example of a system of:
(a) three parallel forces;
(b) three concurrent forces;
which is in equilibrium.

(iii) Give a counter-example to show that the result of (i) cannot be extended to four forces in equilibrium.

(iv) What can you say about a system of two forces in equilibrium?

12 (i) Forces $F_1 = 4i-2j$ and $F_2 = -4i+2j$ act at the points $3i-4j$ and $i-2j$, respectively. Show that the forces are:
(a) in equilibrium; (b) colinear.

(ii) Forces $F_1 = -i+2j$, $F_2 = -3i+6j$ and $F_3 = 4i-8j$ act at the points $3i-2j$, $4i$ and $2i+3j$. Show that the forces are:
(a) in equilibrium; (b) intersect.

(iii) Forces $F_1 = i+3j$, $F_2 = -2j$, $F_3 = 4i-j$ and $F_4 = -5i$ act at the points $3i$, $2i+3j$, $3i-2j$ and $-i+2j$. Show that the forces are in equilibrium.
Find the cartesian equations of the line Λ_1 of action of the resultant of F_1 and F_2, and of the line Λ_2 of action of the resultant of F_3 and F_4. What do you deduce?

13 Find the values of p and q for which the given system of forces is in equilibrium.

(i) *ABCD* square.

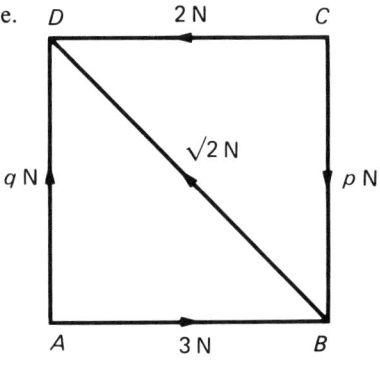

(ii) *ABCDEF* regular hexagon.

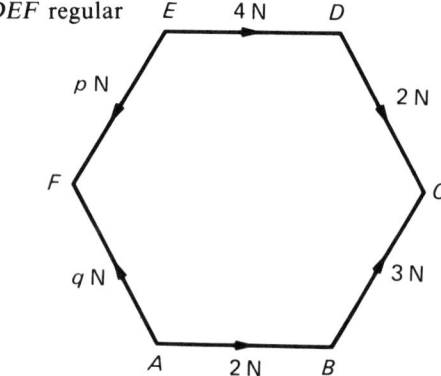

14 Find the values of p, q and r for which the given system of forces is in equilibrium.

(i) *ABCD* square.

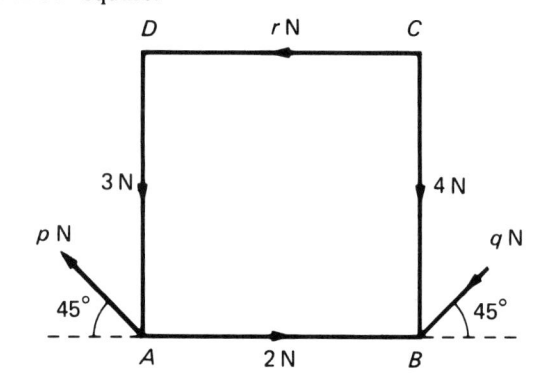

(ii) *ABCDEF* regular hexagon.

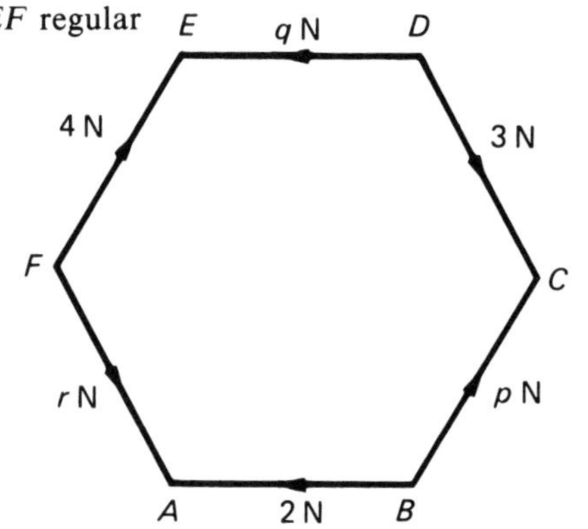

(iii) $AB = AC$.

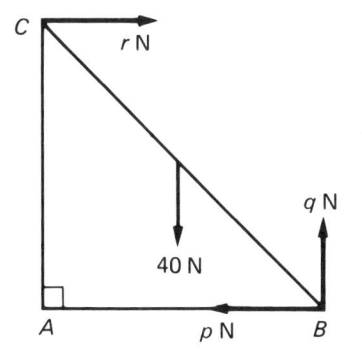

(iv) $AB = BC$.

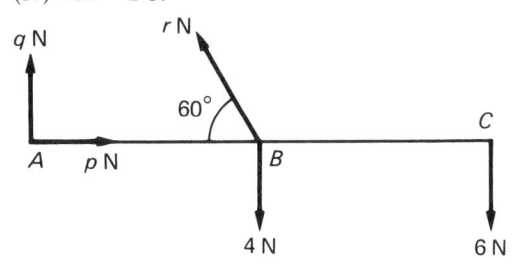

15 The resultant of a system of forces $\{F\}$ is R. Show that the system $\{F, -R\}$ is in equilibrium.

16 Find the magnitude, direction and line of action of the force which reduces the given system of forces to equilibrium.

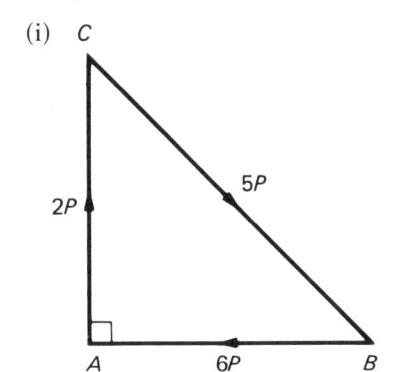

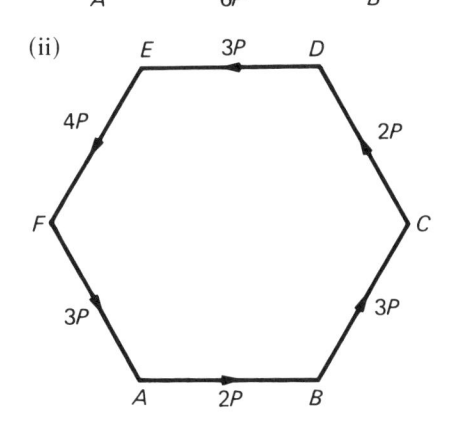

(i)

(ii)

→ (iii) $F_1 = 3i - 5j$, $F_2 = i + j$; acting through the points $2i - j$, $-3i + 2j$.

17 Suppose the given system is reduced to equilibrium by the addition of a force **F** through A and a couple of magnitude G. Find the magnitude and direction of **F** and the value of G.

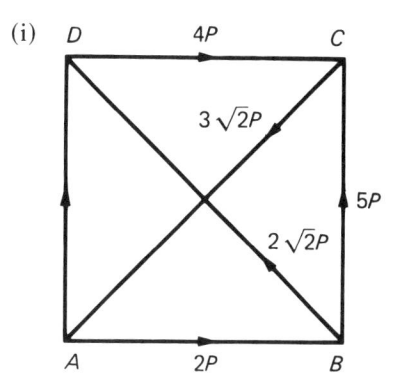

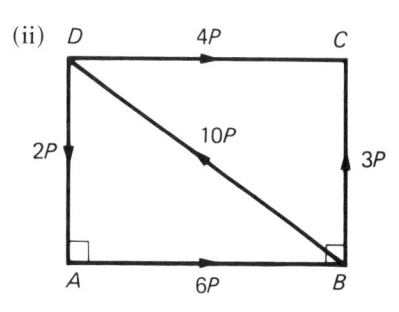

(i)

(ii)

R (iii) $F_1 = 2i - j$, $F_2 = 4i + 5j$, $F_3 = -3i + 2j$; acting through the points $i + j$, $2i - j$, $4i + j$; A is the origin.

18 Explain why none of the following systems of forces can be in equilibrium, whatever the values of p and q.

Which of the systems can reduce to a couple?

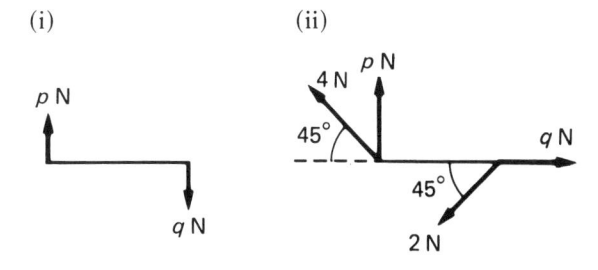

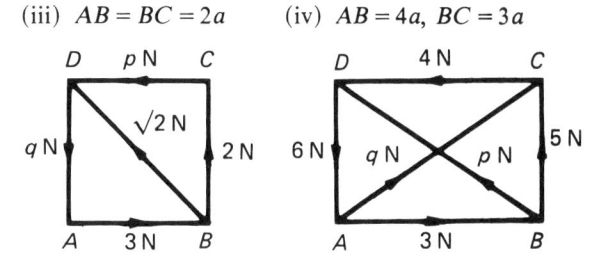

(i)

(ii)

(iii) $AB = BC = 2a$

(iv) $AB = 4a$, $BC = 3a$

(v) *ABC* equilateral triangle.

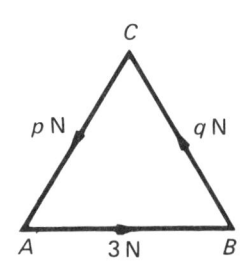

(vi) *ABCDEF* regular hexagon.

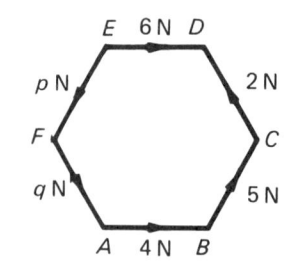

19 Forces $F_1 = 3i + j$, $F_2 = j$ and $F_3 = pi + qj$ act through the points $2i - j$, $i + j$ and $6i + rj$. Find p, q and r if the forces are in equilibrium.

→ **20** Forces $F_1 = 4i + j$, $F_2 = -i - 3j$ and $F_3 = pi + qj$ act through the points $3i + 2j$, $4j$ and $ri - 3j$. Find p, q and r if the forces are in equilibrium. Show that in this case the forces intersect, and find the position vector of their point of intersection.

21 Forces $F_1 = 2i + j$, $F_2 = -6i - 3j$ and $F_3 = pi + qj$ act at the points A, B and C with position vectors $i + j$, $3i + 5j$ and $4i + rj$. Find p, q and r if the forces are in equilibrium. What can you say about F_1, F_2 and F_3?
Show that B divides AC in the ratio $2:1$.

22 Forces $F_1 = 3i + 5j$, $F_2 = pi + 2j$, $F_3 = 4i$ and $F_4 = -i + qj$ act at the points $i + j$, $-i + 2j$, $ri - j$ and $i - j$. Find p, q and r if the forces are in equilibrium.

Show by drawing a diagram that the forces do not intersect.

23 Forces $F_1 = 4i + 5j$ and $F_2 = -3i - 2j$ act at the points $-2i$ and $-4i + j$. Show that the forces intersect and find the position vector of their point of intersection.
The forces F_1, F_2 and F_3 are in equilibrium. Find a vector equation of the line of action of F_3.

24 Forces $F_1 = 2i - j$, $F_2 = 3i + j$ and $F_3 = -4i + 2j$ act at the points $3i - j$, j and $3i$, respectively. The forces F_1, F_2, F_3 and F_4 are in equilibrium. Find a vector equation of the line of action of F_4.
What is the resultant of F_1, F_2 and F_3?

R **25** The position vectors of the vertices B and C of $\triangle ABC$ are $2i + 5j$ and $8i - 4j$. Two forces $F_1 = 4i + 4j$ and $F_2 = -3i + 2j$ act along AB and AC. Find the position vector of A.
If F_1, F_2 and F_3 are in equilibrium, find a vector equation of the line of action of F_3.

: → **26** What can you say about a system of forces for which the following conditions hold?

(i) $\sum F = 0$ and $\sum M_A(F) \neq 0$ for some point A
(ii) $\sum F = 0$
(iii) $\sum M_A(F) = \sum M_B(F) \neq 0$ for two points A and B
(iv) $\sum M_A(F) = \sum M_B(F) = 0$ for two points A and B
(v) $\sum M_A(F) = 0$ for some point A
(vi) $\sum M_A(F) = 0$ for some point A, and $\sum F_y = 0$

9.2 Centres of gravity

9.2:1 Centre of gravity, centre of mass, centroid

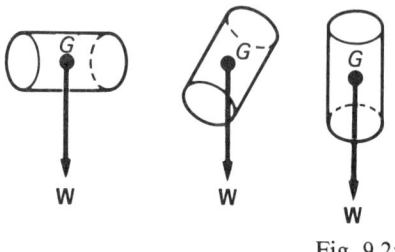

A rigid body is made up of a number of particles joined rigidly together by forces of attraction. The weights of these particles form a system of parallel forces whose resultant is the weight of the whole body. The line of action of this resultant always passes through a fixed point G of the body, *whatever the orientation of the body*, as shown in fig. 9.2:1. This fixed point G is called the **centre of gravity** (CG) of the body.

The point G need not necessarily lie *in* the body (the walking stick in fig. 9.2:2), but it is always in some fixed position *relative* to the body, whatever the orientation of the body. (We shall prove this later.)

Note: this is only true for a *rigid* body. The centre of gravity of a non-rigid body (e.g. a person) changes as the body takes different shapes.

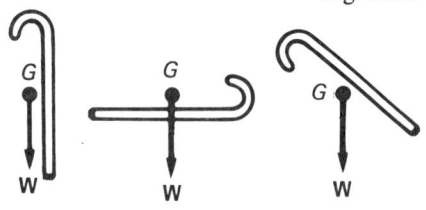

Fig. 9.2:1

Fig. 9.2:2

The **centre of mass** (CM) of a body is the point in the body about which the mass is equally distributed.

A **uniform body** is a body whose density is constant: any two parts of the body with equal volumes have equal masses. Thus the mass of a uniform body will be equally distributed about any line of symmetry, i.e. the CM of a uniform body lies on each line of symmetry of the body. For example, the CM of a uniform rod lies at its midpoint, the CM of a uniform sphere lies at its centre.

Note on rigid bodies: in the real world all objects are three-dimensional. In Applied Maths we often consider two kinds of idealised objects:

(i) two-dimensional objects, i.e. objects with no thickness:
   (a) laminae, e.g. a character in a cartoon who has bumped into a wall;
   (b) hollow objects, e.g. a spherical shell.

(ii) one-dimensional objects, i.e. lines with mass. The lines may be straight (rods) or bent (bent rods).

Worked example 1 Show that the CM of a uniform triangular lamina ABC lies at its centroid.

Think of the lamina ABC as consisting of a large number of thin rods parallel to BC (fig. 9.2:3). The CM of each of these rods lies at its midpoint. So the CM of the lamina lies on the line joining these midpoints, i.e. the median AD.

Similarly, we can show that the CM lies on the other two medians BE and CF; so it lies at the intersection of the medians, i.e. the centroid of the triangle (see 6.1:3 Theorem 3).

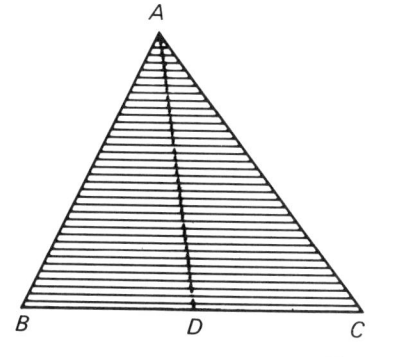

Fig. 9.2:3

The **centroid** of a body is the geometric centre of the body, e.g. the point of intersection of the diagonals of a rectangular lamina.

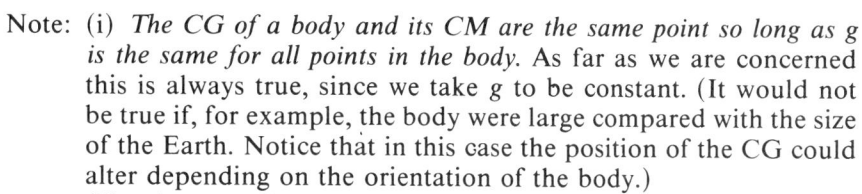

Note: (i) *The CG of a body and its CM are the same point so long as g is the same for all points in the body.* As far as we are concerned this is always true, since we take g to be constant. (It would not be true if, for example, the body were large compared with the size of the Earth. Notice that in this case the position of the CG could alter depending on the orientation of the body.)

(ii) *The CM of a body and its centroid are the same point so long as the body is uniform.* Consider however a rod AB which gets denser towards end B (fig. 9.2:4). The centroid X of the rod is at its midpoint; but its CM Y is somewhere between X and B.

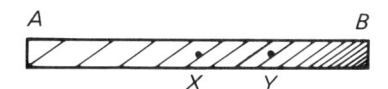

Fig. 9.2:4

Exercise 9.2:1

1 Explain why the forces of attraction which hold together a rigid body form a system which is in equilibrium. (Note: it is therefore not necessary to consider these forces when finding the CG of a body.)

① → **2** Using the result of worked example 1, describe accurately the position of the CM of a uniform lamina in the shape of:

(i) an isosceles triangle ABC with $AB = BC$, $AD = h$ where D is the midpoint of BC
(ii) an equilateral triangle of side $2a$
(iii) a right-angled triangle ABC with $A\widehat{C}B = 90°$, $BC = a$, $AC = b$.

① **3** A uniform lamina is in the shape of a triangle ABC with $AB = AC = 5a$, $BC = 8a$. Find the distance of its CM from BC.

① R **4** A uniform lamina is in the shape of a trapezium $ABCD$ with AB parallel to DC. Show, using the method of worked example 1, that its CM lies on the line joining the midpoints of AB and DC.

① **5** Show that the CM of a uniform lamina in the shape of a parallelogram is at the point of intersection of its diagonals.

① → **6** Show that the CM of a uniform hollow cone of height h with no base is on its axis of symmetry, $\frac{2}{3}h$ from its vertex A. (Hint: think of the surface as consisting of a large number of very thin identical isosceles triangles each with one vertex at A.)

① + **7** (i) Show that the CM G of a uniform solid tetrahedron lies at the point of intersection of the lines joining the vertices to the centroids of the opposite faces. (Hint: begin by choosing one face as base and dividing the tetrahedron into triangular laminae parallel to this base.)

(ii) If the position vectors of the vertices A, B, C and D of the tetrahedron are **a**, **b**, **c** and **d**, what is the position vector of G? (See 6.1:3 Q25.)

(iii) If H is the centroid of the face BCD, in what ratio does G divide AH?

① + **8** A pyramid is a figure whose base is a polygon and whose other faces are triangles meeting in a common vertex. Find the position of the CM G of a uniform solid pyramid. (Hint: using the method of worked example 1, show that G lies on the line joining the vertex V to O, the centroid of the base. Then divide the pyramid into tetrahedra, each having the same height as the pyramid.)

By letting the number of sides of the polygon tend to infinity, find the position of the CM of a uniform solid cone.

9.2:2 Centre of gravity of a set of particles in a plane

Theorem 1: The CG of a set of particles of masses $m_1, m_2, m_3, \ldots$, at points $(x_1, y_1), (x_2, y_2), (x_3, y_3), \ldots$ in the xy plane is

$$\left(\frac{\sum m_i x_i}{\sum m_i}, \frac{\sum m_i y_i}{\sum m_i} \right) \qquad (1)$$

Proof

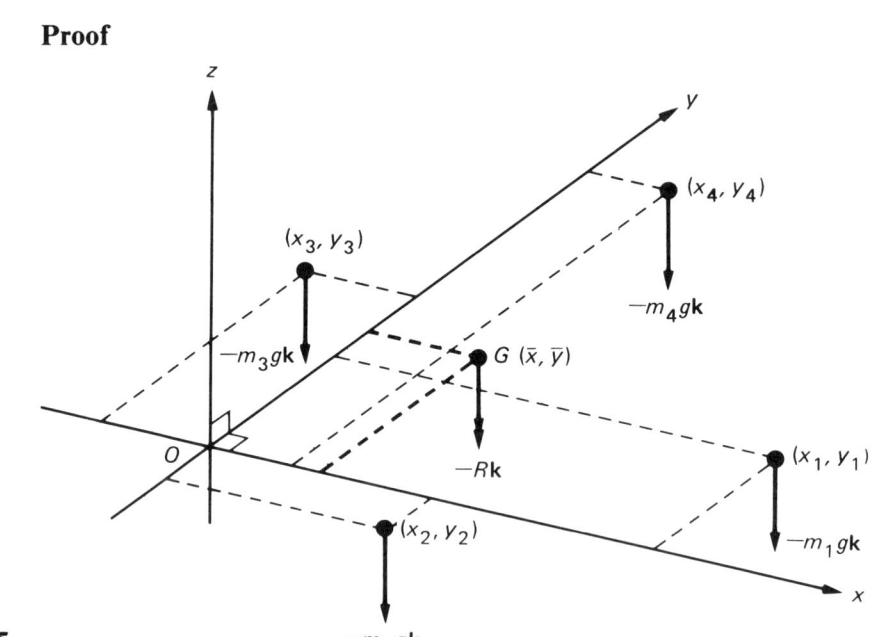

Fig. 9.2:**5**

We need to think of the particles as being rigidly joined by light rods (for otherwise the system need not have constant shape, in which case the CG is not fixed.)

Think of the xy plane as being horizontal. The weights $-m_i g \mathbf{k}$ of the particles form a system of parallel forces. Suppose their resultant is $\mathbf{R}$:

$$\mathbf{R} = -\sum m_i g \mathbf{k}$$

Magnitude of $\mathbf{R}$, $\qquad R = \sum m_i g$

Suppose the line of action of $\mathbf{R}$ cuts the plane at $G(\bar{x}, \bar{y})$; then G is the CG of the set of particles. By the Principle of Moments,

moment of $\mathbf{R}$ about Oy

= resultant moment of weights of particles about Oy

i.e.
$$R\bar{x} = m_1 g x_1 + m_2 g x_2 + m_3 g x_3 + \cdots$$
$$(\sum m_i g)\bar{x} = \sum m_i g x_i$$

So
$$\bar{x} = \frac{\sum m_i g x_i}{\sum m_i g} = \frac{\sum m_i x_i}{\sum m_i}$$

Note that the forces are not coplanar (unless the particles are in a straight line); but that we can still find their resultant moment so long as we are careful to take moments about an *axis*.

Similarly, using the Principle of Moments about the x-axis, we can show that

$$\bar{y} = \frac{\sum m_i y_i}{\sum m_i}$$

> **Corollary:** The position vector of the CG of a set of particles of masses $m_1, m_2, m_3, \ldots$ at the points with position vectors $\mathbf{r}_1, \mathbf{r}_2, \mathbf{r}_3, \ldots$ in the xy plane is
>
> $$\frac{\sum m_i \mathbf{r}_i}{\sum m_i} \tag{2}$$

Proof
See question 6.

Note: in the proof of the theorem we took g constant and found the CG to be independent of g, i.e. dependent only on the masses and relative positions of the particles. This shows that, so long as g is constant:

(i) the CG and the CM are the same point
(ii) the CG is the same whatever the orientation of the plane, since it is independent of the relative direction of the plane and the force of gravity

Worked example 2 Find the position vector $\bar{\mathbf{r}}$ of the CG of particles of masses 3, 2, 5 and 1 at the points $2\mathbf{i} - \mathbf{j}$, $3\mathbf{i} + 5\mathbf{j}$, $-2\mathbf{i} - \mathbf{j}$ and $\mathbf{i} - 3\mathbf{j}$.

$$\sum m_i = 3 + 2 + 5 + 1 = 11$$
$$\sum m_i \mathbf{r}_i = 3(2\mathbf{i} - \mathbf{j}) + 2(3\mathbf{i} + 5\mathbf{j}) + 5(-2\mathbf{i} - \mathbf{j}) + (\mathbf{i} - 3\mathbf{j}) = 3\mathbf{i} - \mathbf{j}$$

So, by the corollary, $\bar{\mathbf{r}} = \tfrac{1}{11}(3\mathbf{i} - \mathbf{j}) = \tfrac{3}{11}\mathbf{i} - \tfrac{1}{11}\mathbf{j}$

Exercise 9.2:2

1 AB is a line segment of length a. Find the distance from A of the CG of two particles placed at A and B if their masses are:

- (i) m and m
  (ii) $4m$ and $4m$
  (iii) $2m$ and $3m$
  (iv) $3m$ and $2m$
→ (v) λm and μm
  (vi) m_1 and m_2

(It is important to notice that the CG of two particles placed at points A and B always lies on the line AB.)

R 2 A and B are the points $(-5, -4)$ and $(1, 4)$. Find the coordinates of the CG of two particles placed at A and B if their masses are:

(i) m and m (ii) $2m$ and $3m$ (iii) $4m$ and m

3 A and B are points with position vectors $\mathbf{a}$ and $\mathbf{b}$. Find the position vector of the CG of particles of masses λm and μm placed at A and B.

4 AB is a line segment of length a and M is the midpoint of AB. Find the distance from A of the CG of three particles placed at A, M and B if their masses are:

(i) m, m, m (ii) $m, m, 2m$ (iii) $m, 2m, 2m$

5 Find the coordinates of the CG of three particles of masses m, $2m$ and $3m$ placed at the points:

(i) $(-1, 0)$, $(5, 0)$, $(8, 0)$
→ (ii) $(1, 0)$, $(2, 3)$, $(0, 2)$
(iii) $(-5, 2)$, $(4, -7)$, $(-1, 4)$

Sketch diagrams to illustrate your results.

6 Using the result of Theorem 1, prove the corollary. (Hint: let $\mathbf{r}_i = x_i\mathbf{i} + y_i\mathbf{j}$.)

② → **7** A, B and C are the points $2\mathbf{i}+\mathbf{j}$, $3\mathbf{i}+5\mathbf{j}$ and $-2\mathbf{i}+3\mathbf{j}$.

(i) Find the position vector of the centroid of $\triangle ABC$.
(ii) Particles of masses m, $2m$ and $3m$ are placed at A, B and C. Find the position vector of their CG.

② **8** A, B and C are the points $2\mathbf{i}-\mathbf{j}$, $a\mathbf{i}+2\mathbf{j}$ and $-4\mathbf{i}+b\mathbf{j}$.

(i) If the position vector of the centroid of $\triangle ABC$ is $-\mathbf{i}+\mathbf{j}$, find a and b.
(ii) Particles of masses $2m$, pm and qm are placed at the points A, B and C. If the position vector of the CG of the particles is $-2\mathbf{i}+\frac{4}{3}\mathbf{j}$, find p and q.

② **R 9** Particles of masses $3m$, pm and qm are placed at $(4, 1)$, $(2, -4)$ and $(-4, 1)$. If their CG is the origin, find p and q.

② **10** D, E and F are the midpoints of the sides BC, CA and AB of an isosceles triangle ABC in which $AB = AC = 5a$ and $BC = 6a$. Find the distance from BC of the CG of particles of masses:

(i) m, m, m placed at A, B, C
(ii) $4m$, $3m$, $3m$ placed at A, B, C
(iii) m, m, m placed at D, E, F
→ (iv) $4m$, $3m$, $3m$ placed at D, E, F

② **11** A, B and C are points with position vectors $\mathbf{a}$, $\mathbf{b}$ and $\mathbf{c}$. Find the position vector of the CG of three particles placed at A, B and C if:

(i) they have masses m, $2m$, m
(ii) they are of equal mass

What do you notice about your answer to (ii)?

② **12** A, B and C are points with position vectors $\mathbf{a}$, $\mathbf{b}$ and $\mathbf{c}$. Find the position vector of the CG of two particles both of mass m placed at A and B. Hence find the position vector of the CG of three particles, all of mass m, placed at A, B and C.

② **13** D, E and F are the midpoints of the sides BC, CA, AB of $\triangle ABC$. Show that:

(i) the CG of three particles of masses m, $3m$ and $5m$ placed at A, B and C is the same as the CG of three particles of masses $4m$, $3m$ and $2m$ placed at D, E and F
(ii) the CG of three particles of equal mass at A, B and C is the same as the CG of three particles of equal mass placed at D, E and F.

What is a simpler way of expressing the result of (ii)?

② **14** Find the coordinates of the CG of four particles of masses $2m$, $2m$, $3m$ and m placed at the points $(4, 1)$, $(-1, 2)$, $(3, 5)$ and $(1, 3)$.

② **15** Find the distances from AB and AD of the CG of four particles of masses m, m, $2m$ and $4m$ placed at the vertices of a square $ABCD$ of side a.

② **R 16** A, B, C and D are the points $\mathbf{i}$, $3\mathbf{j}$, $2\mathbf{i}+5\mathbf{j}$ and $3\mathbf{i}+2\mathbf{j}$. Prove that $ABCD$ is a parallelogram.
Find the position vector of:

(i) the CG of four particles of equal mass placed at A, B, C and D
(ii) the point K of intersection of the diagonals of $ABCD$
(iii) the CG of five particles of masses m, m, $2m$, $3m$ and $4m$ placed at A, B, C, D and K.

② **17** Find the coordinates of the CG of a set of n particles $P_1, P_2, \ldots, P_k, \ldots, P_n$ when:

(i) the particle P_k, of mass $(\frac{1}{2})^k$, is placed at the point $((\frac{3}{2})^k, 0)$
(ii) the particle P_k, of mass k, is placed at the point $(k+1, k^2)$

Discuss in each case the position of the CG as $n \to \infty$.

9.2:3 Centre of gravity of a composite body and a remainder

Suppose that a rigid body consists of several parts whose masses and CGs we can find; then we can use the formulae of section 9.2:2 to find the CG of the whole body.

Worked example 3 A uniform lamina $OABCD$ consists of a square $OACD$ and a right-angled triangle ABC, as shown in fig. 9.2:6. $OA = AB = 2a$. Find its centre of gravity.

Forces on a rigid body 531

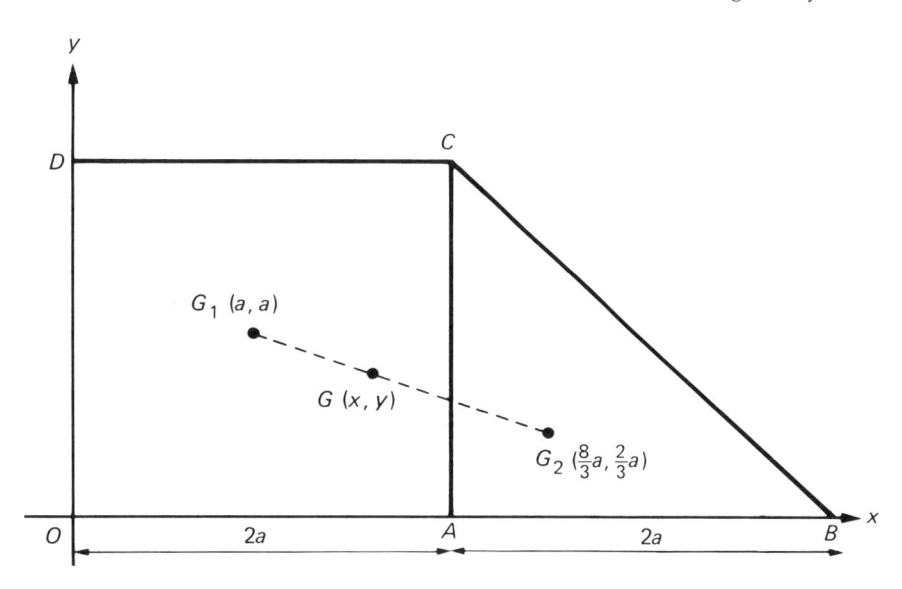

Fig. 9.2:6

Take axes Ox and Oy as shown. Let the mass per unit area of the lamina be m. Let the CG of lamina be $G(\bar{x}, \bar{y})$.

	Mass	CG
Square	$4ma^2$	$G_1(a, a)$: centre of square
Triangle	$2ma^2$	$G_2(\frac{8}{3}a, \frac{2}{3}a)$: centroid of triangle (see 9.1:1 we1)
Complete lamina	$6ma^2$	$G(\bar{x}, \bar{y})$

By Theorem 1,

$$\bar{x} = \frac{4ma^2 \cdot a + 2ma^2 \cdot \frac{8}{3}a}{6ma^2} = \frac{14a}{9}$$

$$\bar{y} = \frac{4ma^2 \cdot a + 2ma^2 \cdot \frac{2}{3}a}{6ma^2} = \frac{8}{9}a$$

So G is the point $(14a/9, 8a/9)$. Note that G lies on $G_1 G_2$. (Why?)

Worked example 4 A lamina consists of a uniform circular disc of radius $2a$ with a circular section of radius a removed, as shown in fig. 9.2:7. Find its centre of gravity.

The centre of gravity of the complete disc is at its centre O. Take axes Ox and Oy, as shown in the figure. The centre of gravity of the lamina remaining is on the x-axis, by symmetry; suppose it is the point $G_2(\bar{x}, 0)$. Let the mass per unit area of the lamina be m.

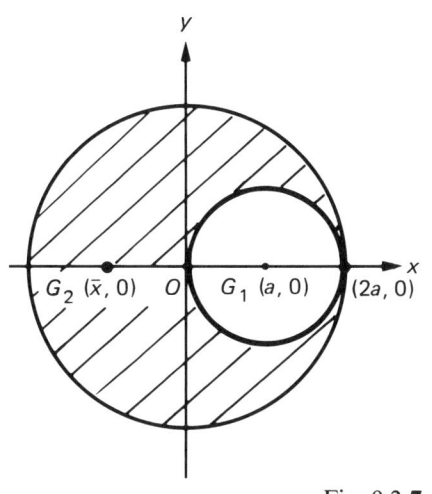

Fig. 9.2:7

	Mass	CG
Disc	$4m\pi a^2$	$O(0, 0)$
Small circular section	$m\pi a^2$	$G_1(a, 0)$: centre of section
Remainder	$3m\pi a^2$	$G_2(\bar{x}, 0)$

Think of the disc as being a composite body consisting of the small circular section and the remainder. By Theorem 1:

$$O = \frac{m\pi a^2 \cdot a + 3m\pi a^2 \bar{x}}{4m\pi a^2}$$

$$\Rightarrow \quad \bar{x} = -\tfrac{1}{3}a$$

So G_2 is the point $(-\tfrac{1}{3}a, 0)$.

Exercise 9.2:3

③ **1** Using a sensible choice of coordinate axes, find the positions of the CGs of these uniform laminae.

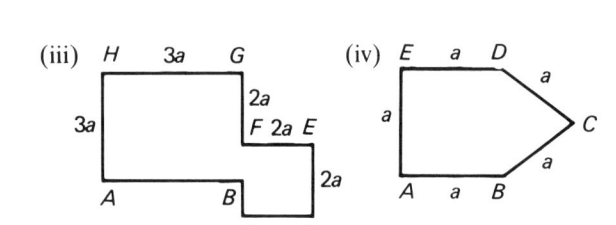

(i) (ii) 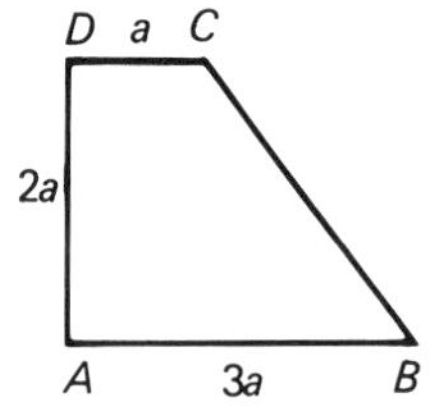 (iii) (iv) (v) (vi)

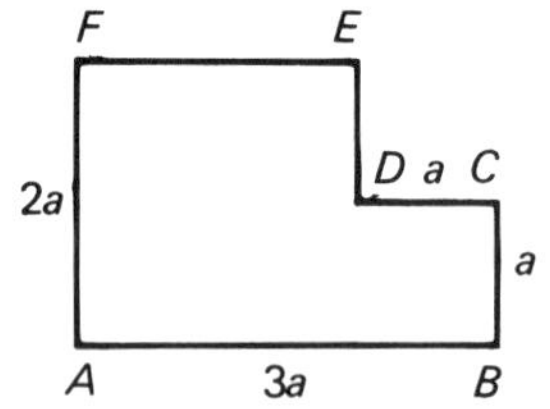

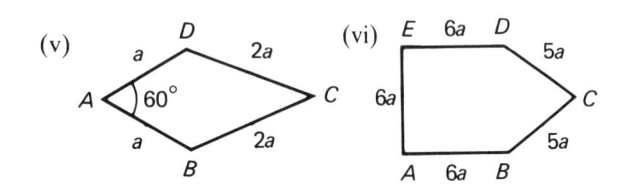

③ **2** $ABCDE$ is a uniform lamina. Find x if its CG lies on BD.

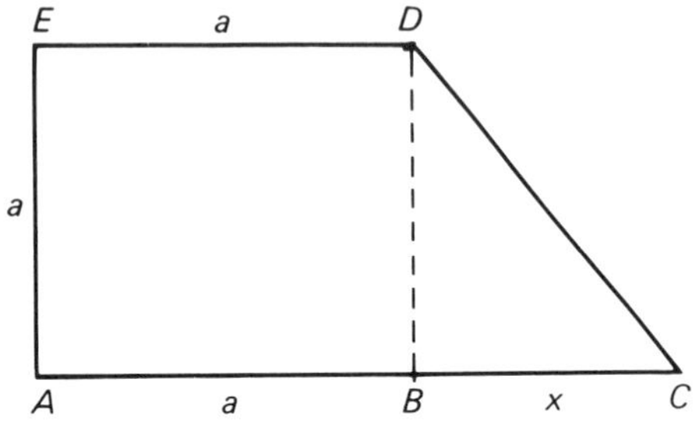

③ **3** Show that the CG of a uniform lamina in the shape of a trapezium $ABCD$ with AB parallel to DC, $AB = a$, $DC = b$, is at the same point as the CG of four particles of masses $(a+b)m, am, (a+b)m, bm$ placed at A, B, C and D respectively.

Find the CGs of these uniform laminae:

(i) (ii)

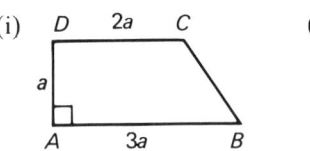

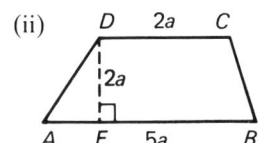

③ **4** Find the position of the centre of gravity of each of these frameworks, made from uniform rods of the same density.

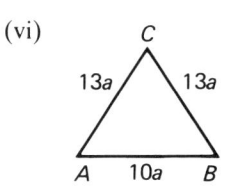

(i) (ii) 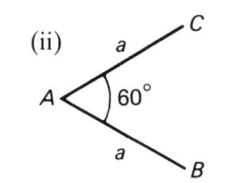 (iii) 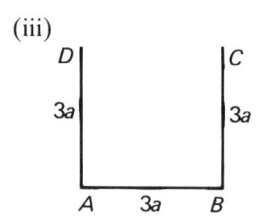 (iv) 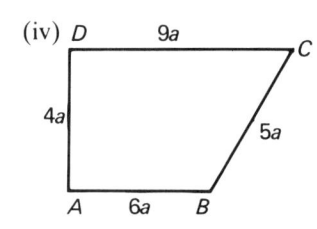 (v) 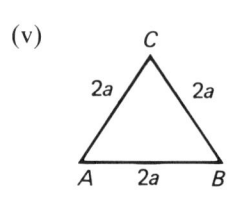 (vi) (vii) 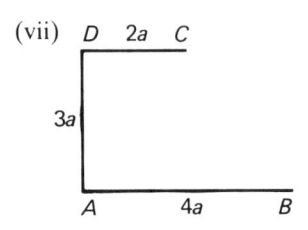 (viii)

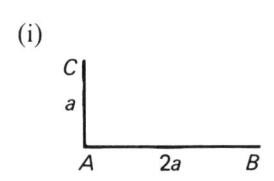

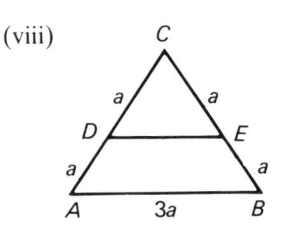

5 Find the position of the centre of gravity of each of these frameworks, made from rods which are uniform but of differing densities (ρ_{AB} means the mass per unit length of AB).

(i) $\rho_{AB} = \rho_{CD} = m$
$\rho_{AD} = 2m$

(ii) $\rho_{AB} = 3m$, $\rho_{BC} = 2m$,
$\rho_{AC} = m$

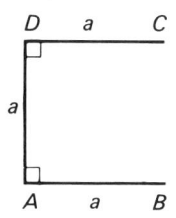

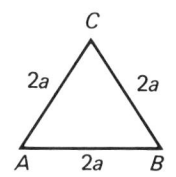

(iii) $\rho_{AB} : \rho_{BC} : \rho_{CA}$
$= 2 : 3 : 2$

(iv) $\rho_{AB} : \rho_{BC} : \rho_{CD} : \rho_{DA}$
$= 1 : 2 : 4 : 3$

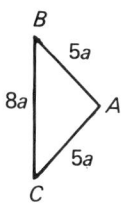

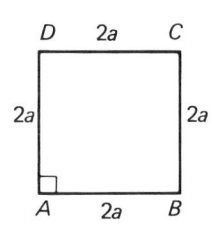

6 A cylindrical can, made of thin material and open at the top, is of height $2a$; the radius of its plane base is a. The mass per unit area of the uniform material making up the base is three times the mass per unit area of the uniform material making up the curved surface. Find the CG of the can.

7 A hollow cone with a base is made of two materials. The base of the cone is made of uniform material whose density is twice that of the uniform material making up the slanting surface. If the height of the cone is $4a$ and the radius of its base is $3a$, find its CG (see 9.2:1 Q6).

8 A hollow cylinder with base and lid is made of two uniform materials. The base and curved surface of the cylinder are made of material whose density is $\frac{2}{3}$ that of the material making up the lid. The cylinder has height $3a$. Find its CG if its base radius is:

(i) $\frac{1}{2}a$ (ii) a (iii) $2a$

9 Find the positions of the CGs of these uniform laminae.

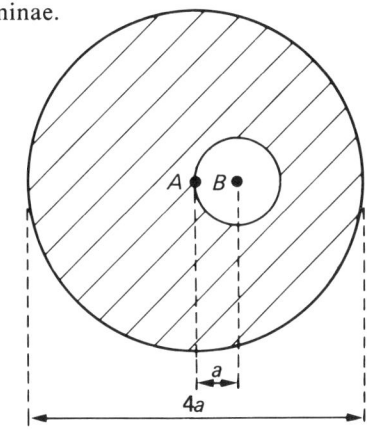

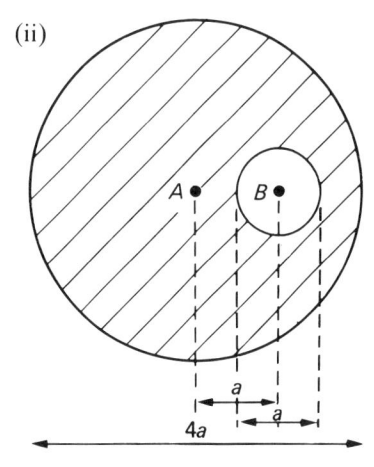

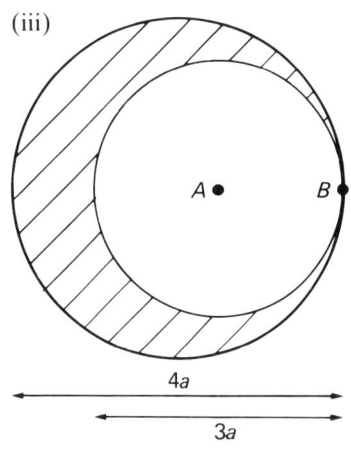

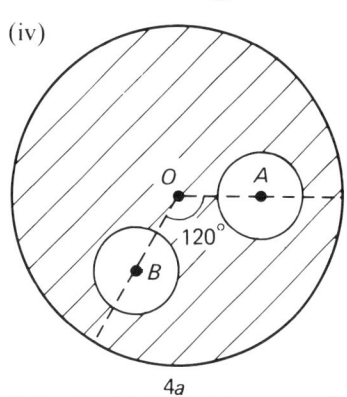

(i)

(ii)

(iii)

(iv)

$OA = OB = a$; radius of small circles $= a/2$

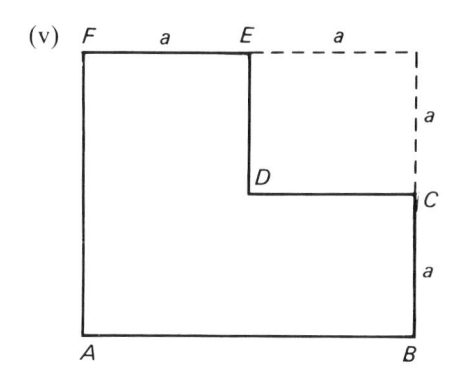

(v)

(vi) [figure: L-shaped lamina with F, E (½a, a along top), height 2a, inner point D, C, with a below C, to B, A at bottom-left]

(vii) [figure: trapezoidal shape D, C (a), with 2a on left side AD, B, A at bottom, a labels]

(viii) [figure: pentagonal lamina with E, D (a), dashed top a, height 3/2 a to C, then ½a down to B, A at bottom-left]

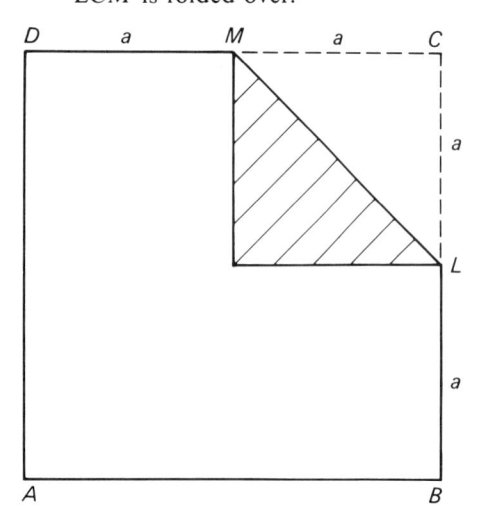

→ (ii) *ABCD* is a square sheet of cardboard. The section *LCMO* is cut out and pasted onto the section *AKON*.

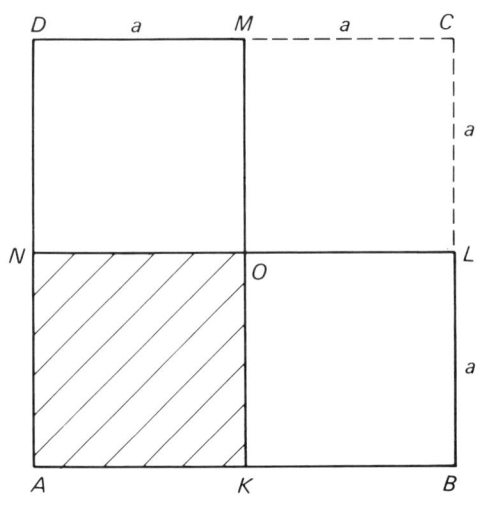

[figure: square ABCD with M midpoint of DC (a, a), N midpoint of AD, L midpoint of BC, O centre, K midpoint of AB; region ANOK shaded]

(iii) *ABCD* is a rectangular sheet of cardboard. The square section *PQRS* is cut out and pasted into the corner *A*.

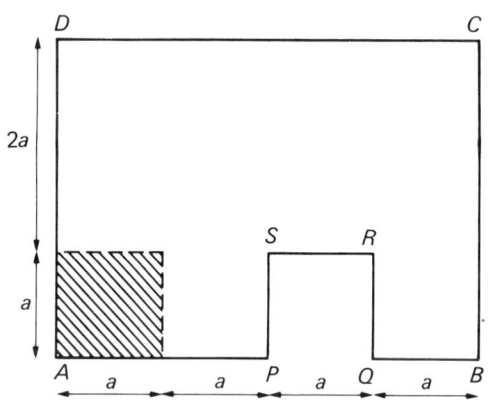

[figure: rectangle ABCD with height 2a, with a shaded square of side a at corner A; PQRS square of side a along AB with AP = a, PQ = a, QB = a]

(iv) *ABC* is a sheet of cardboard: $B\hat{A}C = 90°$, *L* and *M* are the midpoints of *AB* and *BC*. Corner *LBM* is folded over (so that the vertex *B* is at *A*).

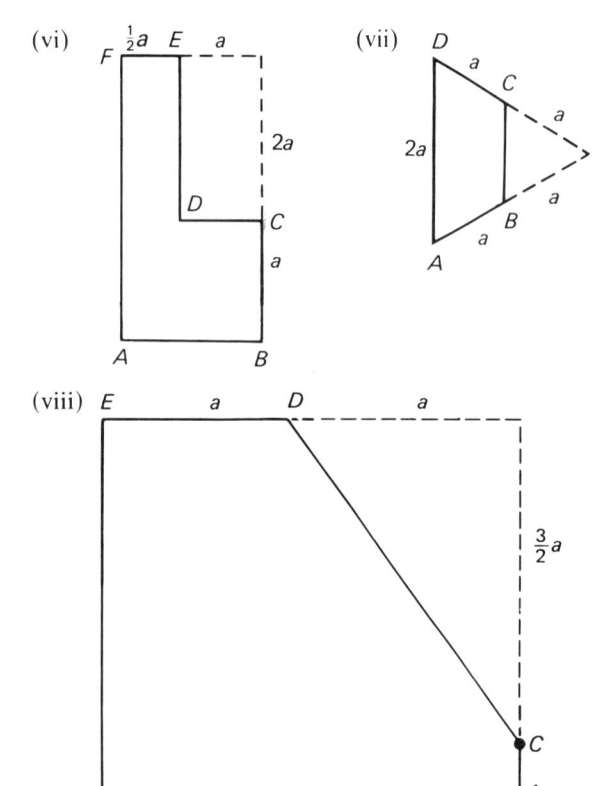

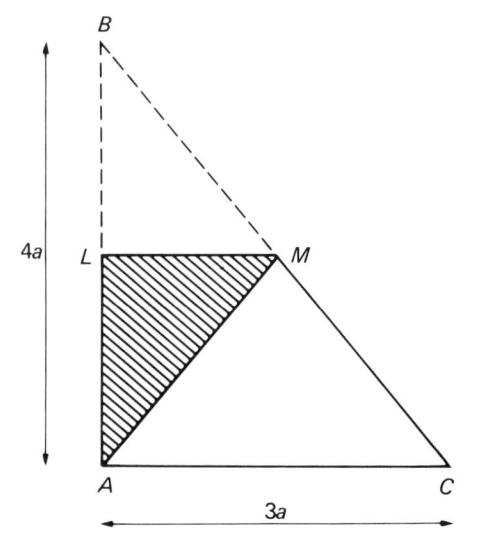

[figure: right triangle with AB = 4a vertical, AC = 3a horizontal, L midpoint of AB, M midpoint of BC, triangle ALM shaded]

10 A cylindrical hole of radius a and length a is bored centrally from one end of a uniform solid cylinder of radius $2a$ and length $3a$. Find the CG of the remainder.

11 A hole of length $2a$, whose cross-section is a square of side a, is cut centrally from one end of a uniform solid cylinder of radius a and length $4a$. Find the CG of the remainder.

12 A cube of side a is cut from the corner of a uniform cube of side $2a$. Find the CG of the remainder.

13 Find the positions of the CGs of these laminae.

- (i) *ABCD* is a square sheet of cardboard. The corner *LCM* is folded over.

[figure: square ABCD with M on DC (DM = a, MC = a), L on BC (LC = a, LB = a); triangle LCM folded over, shaded]

9.2:4 Use of integration

When the CG of a body cannot be found completely by considerations of symmetry, we divide the body into a very large number of very small parts whose masses and CGs are known; we can then think of the body as a composite body and use Theorem 1:

$$\bar{x} = \frac{\sum m_i x_i}{\sum m_i}, \qquad \bar{y} = \frac{\sum m_i y_i}{\sum m_i}$$

Note that $\sum m_i$ is the mass M of the body; so we can write these results as

$$M\bar{x} = \sum m_i x_i, \qquad M\bar{y} = \sum m_i y_i \qquad (3)$$

The summing on the RHS will be done by integration.

Theorem 2: The CG $(\bar{x}, \bar{y})$ of the uniform lamina bounded by the curve $y = f(x)$, the lines $x = a$, $x = b$ and the x-axis is given by

$$A\bar{x} = \int_a^b xy\, dx, \qquad A\bar{y} = \tfrac{1}{2}\int_a^b y^2\, dx \qquad (4)$$

where A is the area of the lamina.

Proof

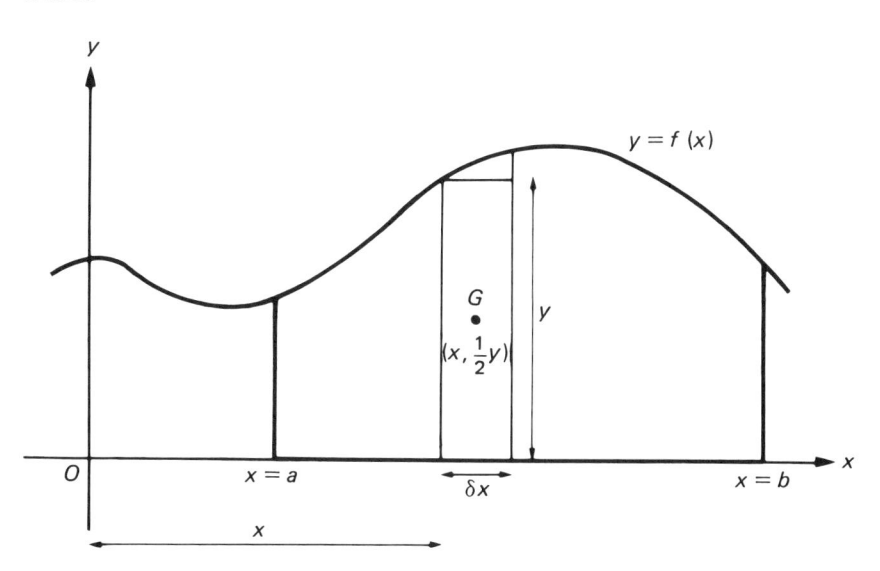

Fig. 9.2:8

Let the mass of the lamina be M, its CG at $G(\bar{x}, \bar{y})$. Let the mass per unit area of the lamina be m.

We divide the lamina into strips, as shown in fig. 9.2:8, of width δx, height y. The area of each strip $\approx y\delta x$; so the mass of each strip $\approx my\delta x$.

$$M = \lim_{\delta x \to 0} \sum_{x=a}^{x=b} my\delta x = m\int_a^b y\, dx \quad \text{(see 4.1:4 equation 15)}$$

Note that $M = mA$.

	Mass	CG
Strip	$my\delta x$	$G_x(x, \tfrac{1}{2}y)$: midpoint of strip
Lamina	mA	$G(\bar{x}, \bar{y})$

By equations (3):

$$mA\bar{x} = \lim_{\delta x \to 0} \sum_{x=a}^{x=b} (my\delta x)x = m\int_a^b xy\,dx$$

$$mA\bar{y} = \lim_{\delta x \to 0} \sum_{x=a}^{x=b} (my\delta x)\cdot \tfrac{1}{2}y = \tfrac{1}{2}m\int_a^b y^2\,dx$$

Since the point (x, y) lies on the curve $y = f(x)$, we evaluate these integrals by putting $y = f(x)$.

Worked example 5 Find the CG of the uniform lamina bounded by the curve $y = x^2$, the line $x = 2$ and the x-axis.

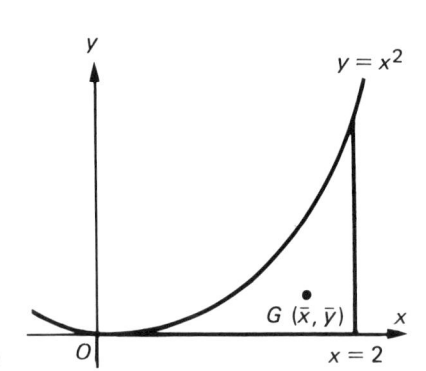

Fig. 9.2:9

$$A = \int_0^2 y\,dx = \int_0^2 x^2\,dx = [\tfrac{1}{3}x^3]_0^2 = \tfrac{8}{3}$$

By equation (4):

$$\tfrac{8}{3}\bar{x} = \int_0^2 xy\,dx = \int_0^2 x^3\,dx = [\tfrac{1}{4}x^4]_0^2 = 4$$

$$\tfrac{8}{3}\bar{y} = \int_0^2 \tfrac{1}{2}y^2\,dx = \int_0^2 \tfrac{1}{2}x^4\,dx = [\tfrac{1}{10}x^5]_0^2 = \tfrac{16}{5}$$

$$\Rightarrow \qquad \bar{x} = \tfrac{3}{2} \quad \text{and} \quad \bar{y} = \tfrac{6}{5}$$

So the CG is the point $(\tfrac{3}{2}, \tfrac{6}{5})$.

Forces on a rigid body

Theorem 3: The CG $(\bar{x}, \bar{y})$ of the uniform solid formed by rotating the section of the curve $y = f(x)$ between $x = a$ and $x = b$ completely about the x-axis is given by:

$$V\bar{x} = \pi \int_a^b xy^2 \, dx \qquad \bar{y} = 0 \qquad (5)$$

where V is the volume of the solid.

Proof
See question 9.

Worked example 6 Find the CG of the uniform solid formed by rotating the section of the curve $y^2 = 4x$ between $x = 0$ and $x = 2$ completely about the x-axis.

The solid is shown in fig. 9.2:**10**.

$$V = \pi \int_0^2 y^2 \, dx = \pi \int_0^2 4x \, dx = \pi [2x^2]_0^2 = 8\pi$$

By Theorem 3:

$$8\pi\bar{x} = \pi \int_0^2 xy^2 \, dx = \pi \int_0^2 4x^2 \, dx = \pi[\tfrac{4}{3}x^3]_0^2 = 32\pi/3$$

$$\Rightarrow \qquad \bar{x} = \tfrac{4}{3}$$

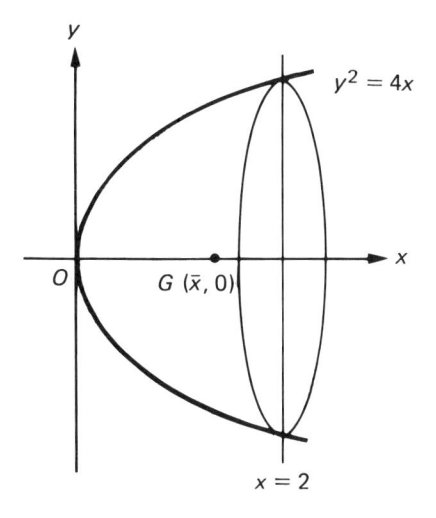

Fig. 9.2:**10**

Worked example 7 Find the CG of a uniform, solid, right circular cone of base radius r height h.

The CG of the cone lies on its axis of symmetry. Let this axis be the x-axis and the take the y-axis through the vertex parallel to a diameter of the base.

The cone is formed by rotating the line $y = rx/h$ completely about the x-axis as shown in fig. 9.2:**11** (see 4.1:4 Q2).

$$V = \pi \int_0^h y^2 \, dx = \pi \int_0^h (rx/h)^2 \, dx$$

$$= \pi \left[\frac{r^2 x^3}{3h^2}\right]_0^h = \tfrac{1}{3}\pi r^2 h$$

By Theorem 3:

$$\tfrac{1}{3}\pi r^2 h \bar{x} = \pi \int_0^h xy^2 \, dx = \pi \int_0^h \left(\frac{rx}{h}\right)^2 \cdot x \, dx$$

$$= \pi \left[\frac{r^2 x^4}{4h^2}\right]_0^h = \tfrac{1}{4}\pi r^2 h^2$$

$$\Rightarrow \qquad \bar{x} = \tfrac{3}{4}h$$

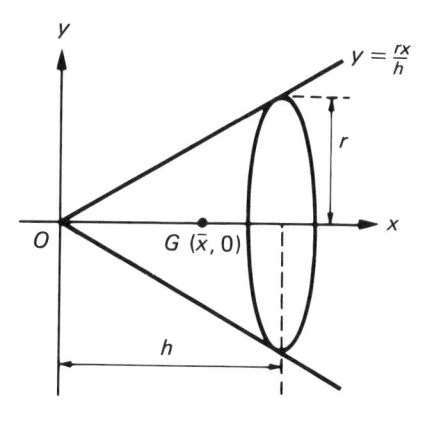

Fig. 9.2:**11**

So the CG lies on the axis of symmetry of a distance $\tfrac{3}{4}h$ from the vertex.

Exercise 9.2:4

⑤ **1** Find the coordinates of the CGs of the uniform lamina bounded by:

(i) $y = 4 - 3x$, x-axis, y-axis
(ii) $y = x^2$, $x = 1$, x-axis
(iii) $y = x(x-2)$, x-axis
→ (iv) $y = e^x$, $x = 0$, $x = \ln 2$, x-axis
(v) $y = \sin x$ between $x = 0$ and $x = \frac{1}{2}\pi$, x-axis, $x = \frac{1}{2}\pi$
(vi) $y = (x+1)/x$, $x = 1$, $x = 2$, $y = 1$

⑤ **2** Find by integration the coordinates of the CG of the uniform triangular lamina bounded by the line $y = 1 - x$ the x-axis and the y-axis. Check that the CG lies at the centroid of the triangle.

⑤ **3** Find the coordinates of the CG of the uniform lamina bounded by the curve $y = 1/x^3$, the lines $x = 1$ and $x = x = n$ and the x-axis. What is the limiting position of this CG as $n \to \infty$?

⑤ → **4** (i) Find the coordinates of the CG of the uniform
• lamina L bounded by $y = 2x^{1/2}$, $x = 9$ and the x-axis.

(ii) Show that the x-coordinate of the CG of the uniform lamina L' bounded by $y^2 - 4x$ and $x = 9$ is the same as that of L. What is its y-coordinate?

(iii) Find the coordinates of the CG of the uniform lamina bounded by $y^2 = 4x$ and $x = 1$.

⑤ **5** Find the coordinates of the CG of the uniform lamina bounded by:

(i) $y^2 = x$, $x = 9$
(ii) $y^2 = 9x$, $x = 1$, $x = 4$
(iii) $y^2 = 1 - x$, y-axis
(iv) $y^2 = x$, $x + y = 2$, $x - y = 2$
(v) $y^2 = x$, $y = \frac{1}{2}x$
R (vi) $y^2 = 4x$, $y = 2x - 4$

⑤ → **6** Show that the CG of a uniform semicircular lamina
• of radius a is on its axis of symmetry at a distance $4a/3\pi$ from its straight edge. (Hint: take the x-axis along the axis of symmetry, the y-axis along the straight edge; the lamina is bounded by the circle $x^2 + y^2 = a^2$.)

⑤ + **7** Using the substitution $x = a \sin \theta$, find

$$\int_0^a \sqrt{(a^2 - x^2)}\, dx$$

Hence find the CG of the uniform lamina bounded by the circle $x^2 + y^2 = 4$ and the ellipse $4x^2 + y^2 = 16$ and lying in the first quadrant.

• **8** (i) Find the coordinates of the CG of the lamina bounded by the curve $y = f(x)$, $y = c$, $y = d$ and the y-axis (see 4.1:4 Q15).

(ii) Find the coordinates of the CG of the uniform lamina bounded by:
→ (a) $y = x^2$, $y = 1$, $y = 9$
(b) $y = e^x$, $y = e$, y-axis

9 Prove Theorem 3. (Hint: divide the solid into discs of width δx by cuts perpendicular to the x-axis—see 4.1:4 Theorem 3).

⑥ **10** Find the coordinates of the CGs of the solids of revolution formed by rotating the following curves about the x-axis:

→ (i) $y^2 = x$ between $x = 0$ and $x = 9$
(ii) $y = 1 - x$ between $x = 0$ and $x = 1$
(iii) $y = 1/x^2$ between $x = 1$ and $x = 2$
(iv) $y^2 = 1/(1 + x^2)$ between $x = 0$ and $x = 1$
R (v) $y = e^x$ between $x = \ln 2$ and $x = \ln 3$
(vi) $y = \sec x$ between $x = 0$ and $x = \frac{1}{3}\pi$
(vii) $y = \sin x$ between $x = 0$ and $x = \frac{1}{2}\pi$

⑥ **11** Find the coordinates of the CG of the solid of revolution formed by rotating the arc of the curve $y = 1/x^2$ between $x = 1$ and $x = n$ completely about the x-axis. What is the limiting position of the CG as $n \to \infty$?

⑦ → • **12** Find the coordinates of the CG of the solid of revolution formed by rotating the circle $x^2 + y^2 = a^2$ between $x = 0$ and $x = a$ about the x-axis. Hence show that the CG of a uniform solid hemisphere of radius a lies on its axis of symmetry at a distance $\frac{3}{8}a$ from its plane face.

⑦ **13** A cap of height $\frac{1}{2}a$ is cut from a uniform solid hemisphere of radius a. Find the position of the CG of the remaining solid. (Hint: use integration between suitable limits.)

⑦ R **14** A uniform solid hemisphere of radius $3a$ is divided into three sections of equal thickness by two cuts parallel to its plane face. Find the position of the CG of each section.

⑦ **15** A cap of height $2a$ is cut from a uniform solid cone of height $4a$ and base radius a. Find the position of the CG of the remaining solid (called a *frustum* of the cone). (Hint: use integration, as in worked example 7, between suitable limits.)

- **16** (i) Find the coordinates of the CG of the solid of revolution formed when the arc of the curve $y = f(x)$ between $y = c$ and $y = d$ is rotated completely about the y-axis.

    (ii) Find the coordinates of the CG of the solid of revolution formed by rotating the given curve about the y-axis:
    - (a) $y = x^2$ between $y = 0$ and $y = 3$
    - (b) $y = x^2 + 4$ between $y = 4$ and $y = 8$
    - (c) $y = x^4$ between $y = 0$ and $y = 16$
    - (d) $y = \ln x$ between $y = 0$ and $y = 1$

- **17** A uniform wire is bent into the shape of an arc of a circle of radius a subtending an angle 2α at its centre O. Find the position of the CG of the wire. (Hint: Take axes Ox and Oy as shown in the diagram. Divide the wire into small arcs each subtending an angle $\delta\theta$ at O.)

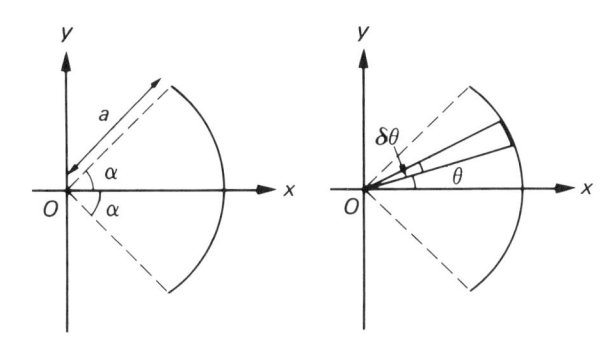

18 Using the result of question 17, find the position of the CG of a semicircular wire of radius a.

- **19** A uniform lamina is in the shape of a sector of a circle of radius a subtending an angle 2α at its centre O. Find the position of its CG. (Hint: divide the lamina into small sectors each subtending an angle $\delta\theta$ at O; each sector is approximately a triangle.)

    Explain how the solution of this question can be simplified by using the result of question 17.

20 Using the result of question 19, find the position of the CG of a uniform semicircular lamina. (See question 6.)

- **21** Find the position of the CG of a uniform hemispherical shell of radius a. (Hint: divide the hemisphere into rings as shown in the diagram. Each ring is approximately a cylinder—find its radius and its width. Note: it is tempting to take the width of the ring as δx, as we did for a solid hemisphere, but this gives a bad approximation to the surface area, particularly towards the point A.)

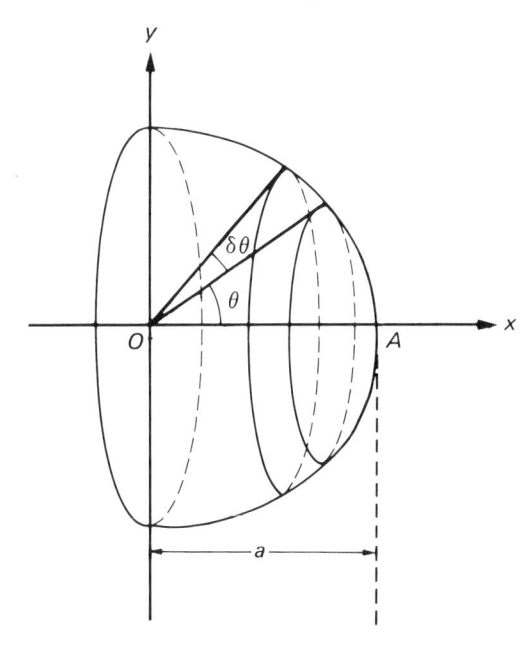

R 22 A section of width h is cut from a uniform hemispherical shell of radius a. Find the position of its centre of gravity.

23 Complete this table:

Body (uniform)	Position of CG
Triangular lamina	
Parallelogram-shaped lamina	
Trapezium-shaped lamina	
Semicircular lamina, radius a	
Sector of circle, radius a, subtending an angle 2α	
Arc of circle, radius a, subtending an angle 2α	
Solid hemisphere, radius a	
Solid cone, height h, base radius r	
Solid pyramid	
Hemispherical shell, radius a	
Conical shell, height h, base radius r	

24 Find the positions of the CGs of the following uniform bodies.

- (i) Solid cylinder, height $2a$ and radius a, glued to solid hemisphere, radius a.

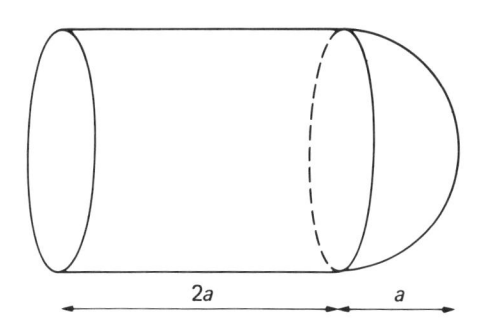

(ii) Solid cone, height $2a$ and base radius a, glued to solid hemisphere, radius a.

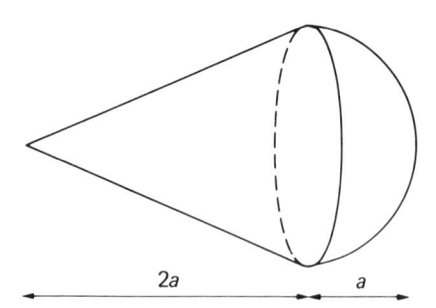

(iii) Solid cone, height $3a$ and base radius a, glued to solid cone, height $2a$ and base radius a.

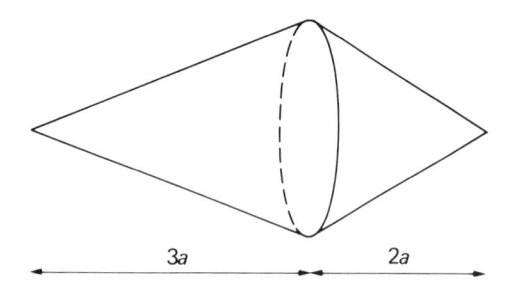

(iv) Solid cone, height $2a$ and base radius a, glued to solid cylinder, height $4a$ and base radius a, glued at its other end to a solid cone, height a and base radius a.

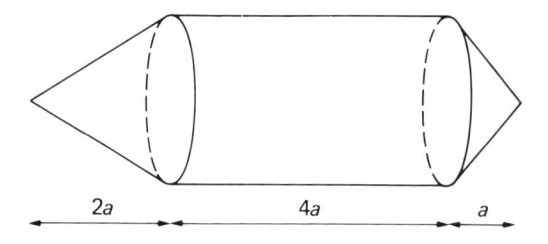

25 Repeat question 25, replacing the solid bodies with hollow ones. (Assume that the resulting object is also hollow.)

R 26 A uniform body consists of a solid cone of height x and base radius a joined by its base to the plane face of a solid hemisphere of radius a. Find x if the CG is at the centre of the common face of the cone and the hemisphere.

→ 27 A uniform wire is bent to form the arc and two radii of a circle of radius a. The arc subtends an angle 2α at the centre of the circle. Find the CG of the wire.

28 A uniform wire is bent to form a semicircular arc of radius a and its diameter. Find its CG.

29 The diagram shows a frustum of a uniform solid cone of height $6a$, base radius $2a$.

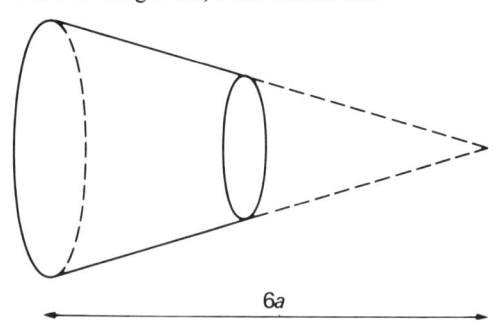

(i) If the radii of the plane faces of the frustum are $2a$ and a, find:
  (a) the height of the frustum (use similar triangles);
  (b) the volume of the frustum;
  (c) using the method of worked example 4, the CG of the frustum.
(ii) The frustum is joined by its smaller plane face to the plane face of a uniform solid hemisphere of radius a made of the same material as the frustum. Find the CG of the composite solid.

30 A trendy wooden soup-bowl consists of a uniform solid hemisphere of radius a with a hemispherical cavity of radius $\frac{1}{2}a$ hollowed out from it. Find the position of its CG.

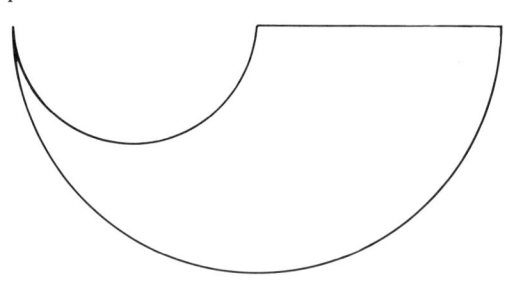

• → 31 The density of a non-uniform rod AB, of length $2a$, at a point P on the rod is proportional to the distance AP. Find the distance of the CG of the rod from A.

R 32 The density of a non-uniform rod AB, of length $2a$, at a point P on the rod is proportional to the square of the distance AP. Find the distance of the CG of the rod from A.

33 The mass per unit length of a non-uniform rod AB, of length $2a$, at a point distant x from A is $(1+x/2a)m$. Find the distance of the CG of the rod from A.

34 The mass per unit area of a semicircular lamina of radius a is mx/a where x is the distance from the centre of the straight edge. By dividing the lamina into semicircular rings and using the result of question 18 find the distance of the CG of the lamina from its straight edge.

9.3 Rigid bodies in equilibrium

9.3:1 Equilibrium problems

In section 9.1:6 we saw that if a rigid body is in equilibrium under a set of coplanar forces $\{\mathbf{F}\}$ then
(a) $\sum \mathbf{F} = \mathbf{0}$
(b) $\sum M_X(\mathbf{F}) = 0$ for all points X

from which we can choose three independent equations, usually:

(i) $\sum F_x = 0$
(ii) $\sum F_y = 0$
(iii) $\sum M_A(\mathbf{F}) = 0$ for some point A

i.e. we resolve in two perpendicular directions and take moments about one point.

Worked example 1 The end A of a uniform rod AB of weight W and length $2a$ stands on rough horizontal ground. The point C on the rod such that $AC = \frac{3}{2}a$ rests on a smooth horizontal cylindrical peg. The rod makes an angle of 45° with the horizontal. Find the least possible value of the coefficient of friction μ between the rod and the ground.

Since the peg is smooth, the reaction at C is normal to the rod (and to the peg), as shown in fig. 9.3:**1**.

$\rightarrow)$ $\sum F_x = 0 \Rightarrow N \cos 45° - F = 0$

$\uparrow)$ $\sum F_y = 0 \Rightarrow R + N \sin 45° - W = 0$

$A\,\rangle$ $\sum M_A(\mathbf{F}) = 0 \Rightarrow Wa \cos 45° - N \cdot \frac{3}{2}a = 0$

$\Rightarrow \qquad N = \frac{1}{3}\sqrt{2}\,W, \quad F = \frac{1}{3}W, \quad R = \frac{2}{3}W$

Note: we have taken moments about A since two of the unknown forces act there; in this case, since there are only three unknown forces in the system, we immediately find the third, N.

Fig. 9.3:**1**

Since the rod is in equilibrium, $F \leq \mu R$ (see section 8.1:3), i.e.

$$\mu \geq \frac{F}{R} = \frac{\frac{1}{3}W}{\frac{2}{3}W} = \frac{1}{2}$$

(Note: since the rod is not necessarily in *limiting* equilibrium, we cannot write $F = \mu R$.)

Worked example 2 A uniform rod AB of length $2a$ and weight W is smoothly hinged to a fixed point A on a wall. A load of weight $2W$ is attached to the end B. The rod is kept horizontal by a string of length $2a$ attached to its midpoint G and to a point C on the wall vertically above A (fig. 9.3:**2**). Find the tension in the string and the magnitude and direction of the reaction at the pivot.

Note: 'smoothly' hinged means that the hinge exerts no frictional turning effect on the rod—it does *not* mean that the force of the hinge on the rod is a normal force (i.e. that $Y = 0$).

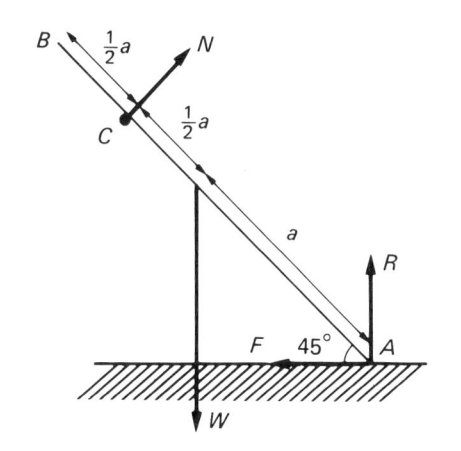

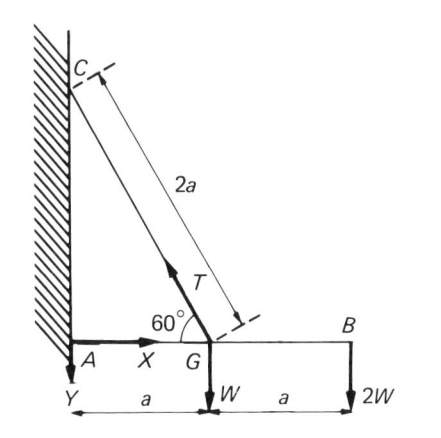

Fig. 9.3:**2**

542 *Equilibrium problems*

From fig. 9.3:**2**, $\cos CGA = a/2a = \frac{1}{2} \Rightarrow CGA = 60°$.

$\rightarrow$) $\quad \Sigma F_x = 0 \Rightarrow X - T \cos 60° = 0$

$\uparrow$) $\quad \Sigma F_y = 0 \Rightarrow T \sin 60° - 3W - Y = 0$

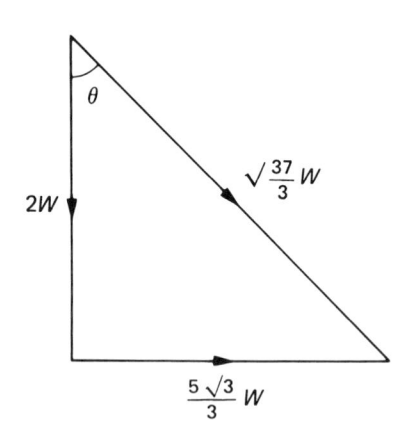

$A\rangle \quad \Sigma M_A(F) = 0 \Rightarrow Ta \sin 60° - Wa - 2W \cdot 2a = 0$

$\Rightarrow \quad\quad\quad T = 10\sqrt{3}\,W/3, \quad X = 5\sqrt{3}\,W3 \quad Y = 2W$

To find the reaction at A:

$$\text{Magnitude} = \sqrt{(X^2 + Y^2)} = \sqrt{\left(\frac{5\sqrt{3}}{3}W\right)^2 + (2W)^2} = \sqrt{\tfrac{37}{3}}\,W$$

Direction: the reaction makes an angle θ with the downward vertical (fig. 9.3:3), where

$$\tan \theta = \frac{5\sqrt{3}\,W/3}{2W} = \frac{5\sqrt{3}}{6}$$

Fig. 9.3:**3**

Exercise 9.3:1

Questions 1–9: A uniform rod AB of length $2a$ and weight W rests in equilibrium.

① **1** Find the least possible value of μ.

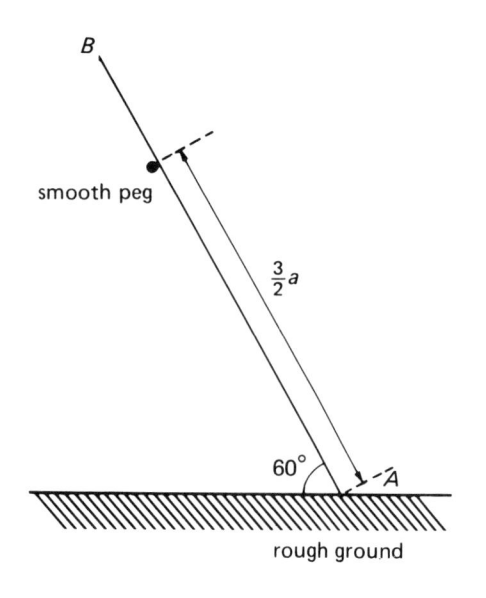

① → **2** Find the least possible value of μ.

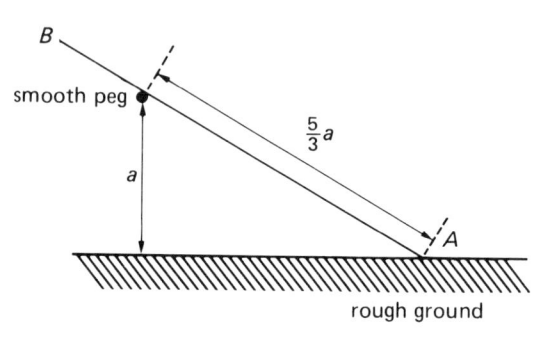

① **3** Tan $\alpha = \frac{1}{2}$. Find the magnitude and direction of the force at A which keeps the rod in equilibrium.

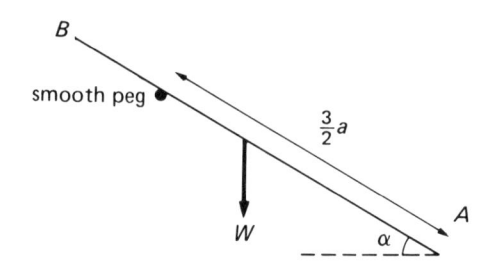

① **4** The rod is in equilibrium under the action of a force F at A; $\tan \alpha = \frac{3}{4}$. (Find F and the magnitude and direction of the reaction at the peg.

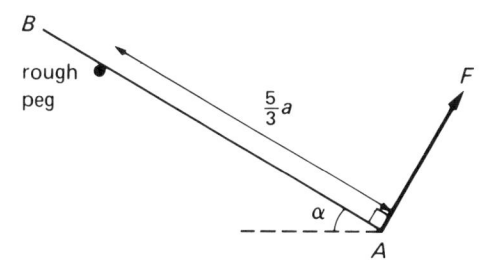

① **5** Find the least possible value of μ.

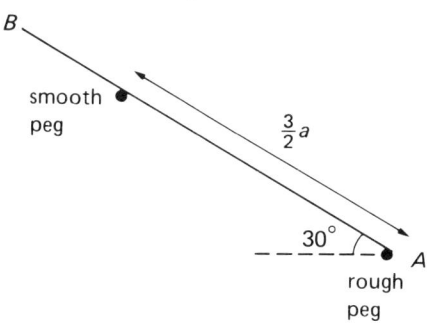

①R **6** $\tan \alpha = \frac{3}{4}$. The rod is just about to slip. Find x and the reactions at the ground and the peg.

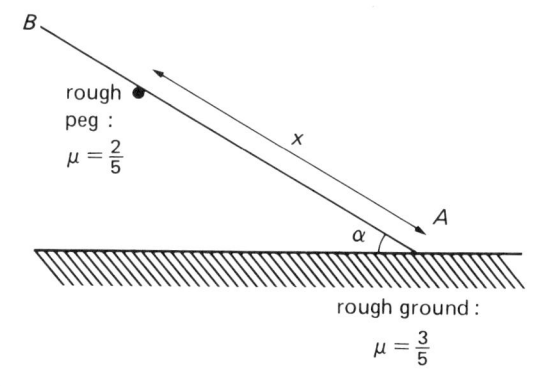

① **7** The rod is in limiting equilibrium. Find μ and the reactions of the sphere and the ground on the rod.

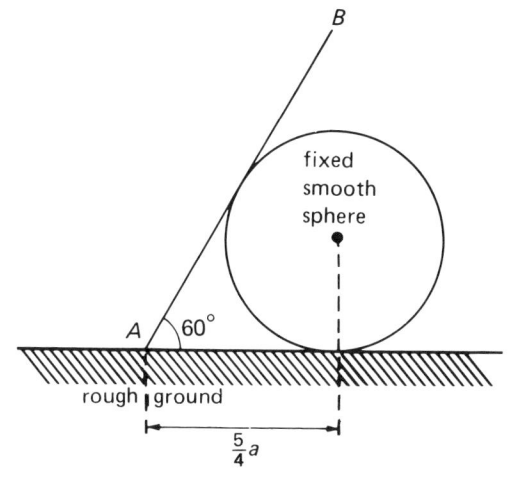

①→ **8** The rod is in limiting equilibrium. Find μ and the reactions of the hemisphere and the ground on the rod.

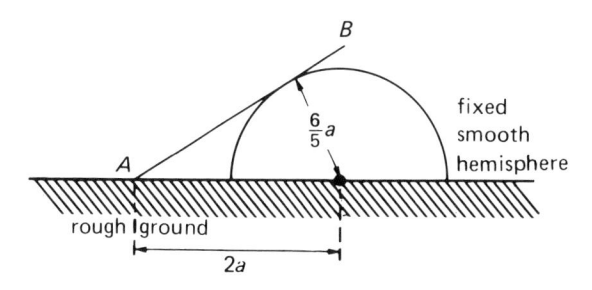

① **9** Show that $\cos \theta = \frac{1}{8}(1 + \sqrt{33})$.

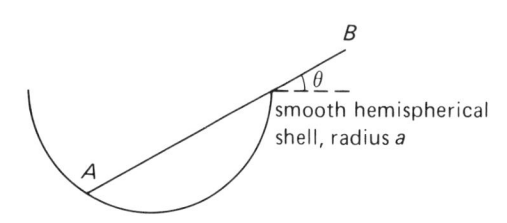

Questions 10–22: A ladder AB of length $2a$ and weight W, uniform unless otherwise stated, with centre of gravity G, rests with one end A on horizontal ground and its other end B against a wall. The angle between the ladder and the horizontal is θ.

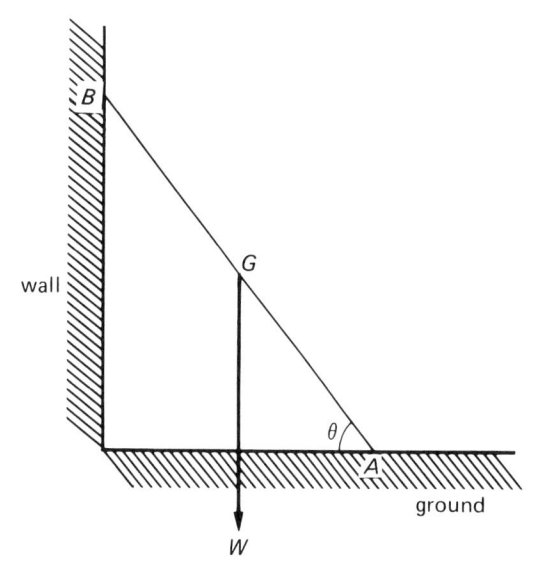

①→ **10** The wall is smooth, the ground rough: $\mu = \frac{1}{3}$. The ladder is in limiting equilibrium. Find θ and the magnitude and direction of the reaction at A.

① **11** The wall is smooth, the ground rough. Find the least possible value of μ, in terms of θ.

① **12** The wall is smooth, the ground rough: $\mu = \frac{3}{4}$. The ladder is non-uniform: $AG = \frac{3}{2}a$. The ladder is in limiting equilibrium. Find θ and the magnitudes of the reactions at A and B.

544 *Equilibrium problems*

① **13** The wall is smooth, the ground rough, the weather fine. The ladder is non-uniform: $AG = \tfrac{7}{4}a$; $\theta = 60°$. Show that the least possible value of μ is $7\sqrt{3}/24$.

①→ **14** The wall is smooth, the ground rough: $\mu = \tfrac{1}{3}$; $\tan\theta = 2$. Find how far up the ladder a man of weight $2W$ can climb without it slipping.

① **15** The wall is smooth, the ground rough: $\mu = \tfrac{1}{2}$; $\tan\theta = \tfrac{3}{4}$. Find how far up the ladder a man of weight W can climb without it slipping.

① **16** The wall is smooth, the ground rough: $\mu = \tfrac{2}{5}$. A man of weight $4W$ can climb a distance $\tfrac{3}{2}a$ up the ladder before it begins to slip. Find $\tan\theta$.

① R **17** The wall is smooth, the ground rough: $\mu = \tfrac{1}{4}$; $\tan\theta = \tfrac{24}{7}$. A man climbing the ladder can reach a height $\tfrac{12}{7}a$ above the ground before the ladder begins to slip. Find the weight of the man.

① **18** The wall and the ground are rough. The coefficient of friction at both surfaces is $\tfrac{1}{5}$. The ladder is in limiting equilibrium. Find θ and the reactions at A and B.

①→ **19** The wall and the ground are rough. The coefficient of friction at both surfaces is μ. Find the least possible value of μ in terms of θ.

① **20** The wall and the ground are rough. The coefficient of friction at both surfaces is μ. The ladder is non-uniform: $AG = \tfrac{5}{4}a$; $\tan\theta = 2$. If the ladder is in limiting equilibrium, find μ and the reactions at A and B.

① **21** The wall and the ground are rough. The coefficient of friction between the ladder and the wall is μ_1, between the ladder and the ground μ_2. If the ladder is just about to slip, show that

$$\tan\theta = \frac{1-\mu_1\mu_2}{2\mu_2}$$

① **22** Prove that it is impossible for a ladder to rest on smooth horizontal ground against a rough vertical wall.

Questions 23–32: A rod of length $2a$ and weight W, uniform unless stated otherwise, with centre of gravity G, rests in equilibrium.

① • **23** The rod is kept in limiting equilibrium by a horizontal force P at B. Find P and μ, and the reaction at A.

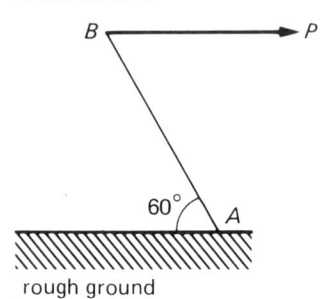

① **24** The rod is kept in limiting equilibrium by the force P. Find P and μ.

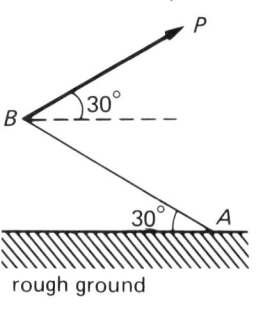

①→ **25** The rod is kept in equilibrium by the vertical force P. Find the least possible value of μ.

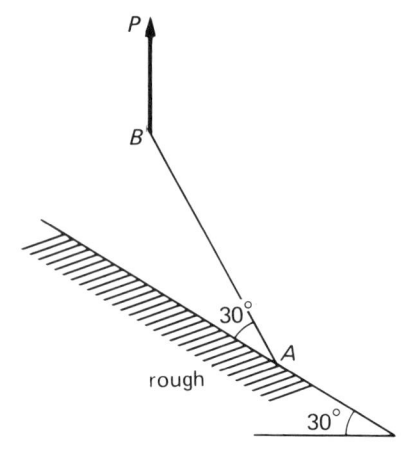

① **26** Find the magnitude and direction of the force at B which keeps the rod in limiting equilibrium.

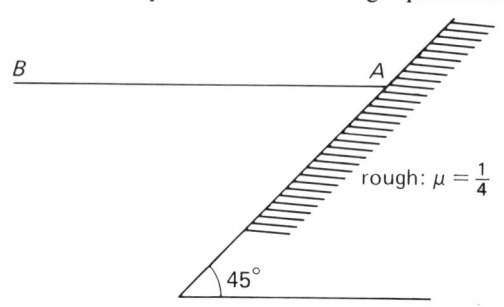

① **27** The rod is non-uniform. Find the position of its centre of gravity.

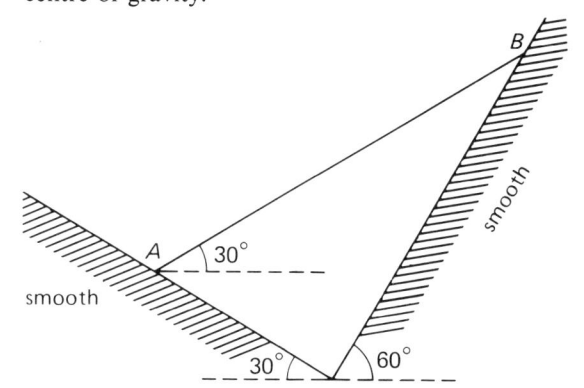

Forces on a rigid body

① → **28** The rod is non-uniform: $AG = \frac{3}{2}a$. Show that $2\tan\alpha \tan\beta = 1$.

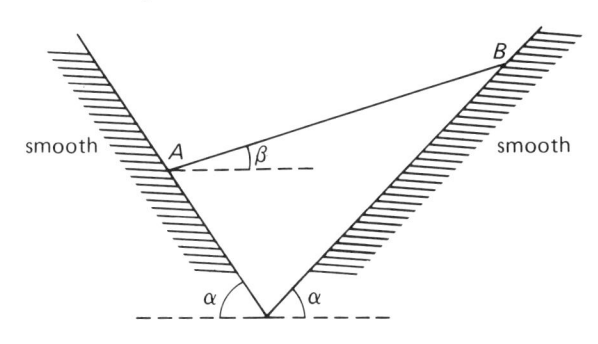

① **29** The rod is in limiting equilibrium. Find the coefficient of friction between the rod and the ground.

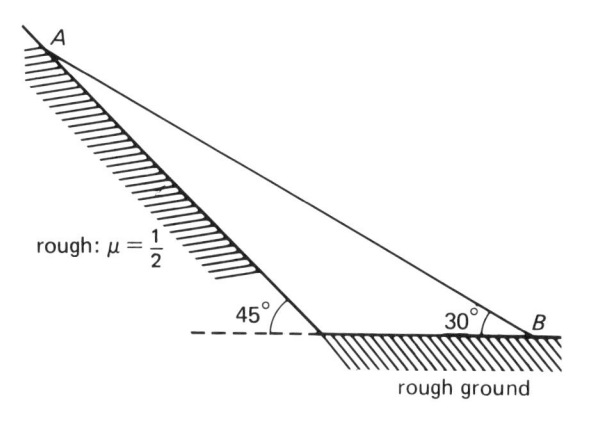

+ ① **30** The rod is non-uniform. If it is in limiting equilibrium, find the possible positions of its centre of gravity, $\alpha = 45°$.

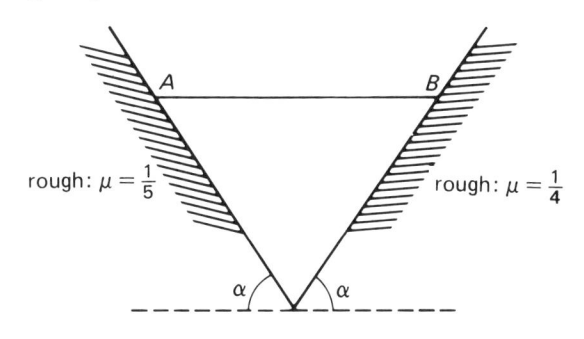

① R **31** The end B is attached by an inelastic string to a point C on the wall vertically above A. Find the least possible value of μ.

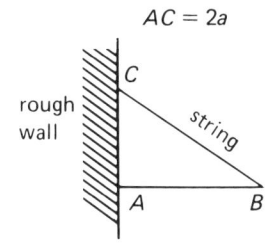

$AC = 2a$

① → **32** The rod is kept in equilibrium by a string attached to a point C on the rod and to a point D on the wall vertically above A. $AD = AC = \frac{1}{2}a$. Find θ and the tension in the string.

If the string is elastic and its natural length is $\frac{1}{3}a$ find its modulus of elasticity.

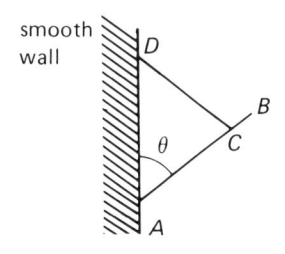

Questions 33 and 34: A solid hemisphere of radius a and weight W rests in equilibrium (see 9.2:4 Q12).

+ **33** The hemisphere is held in limiting equilibrium by an inelastic string of length a; $\tan\alpha = \frac{4}{5}$. Find μ and the tension in the string.

→ **34** The hemisphere is in limiting equilibrium. Find θ and the reactions at A and B.

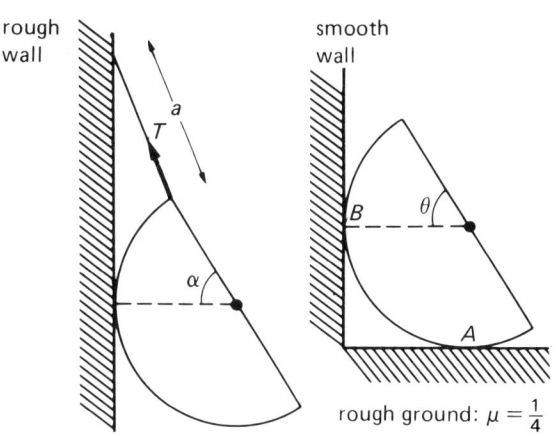

rough ground: $\mu = \frac{1}{4}$

Questions 35–38: A disc of radius a and weight W rests in equilibrium in a vertical plane.

35 The disc is in limiting equilibrium. Find T and show that the reaction of the wall on the disc has no frictional component.

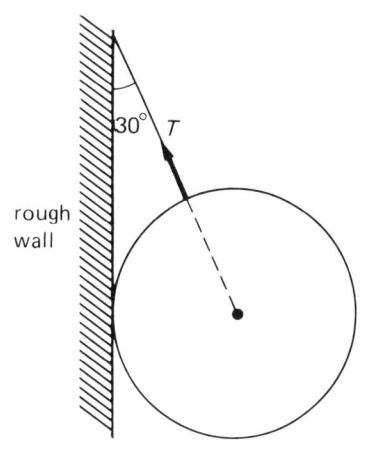

Equilibrium problems

36 Find the reactions at A and B.

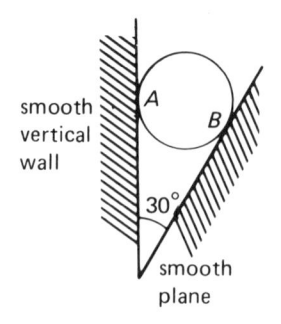

37 The disc is just about to rotate. Find P.

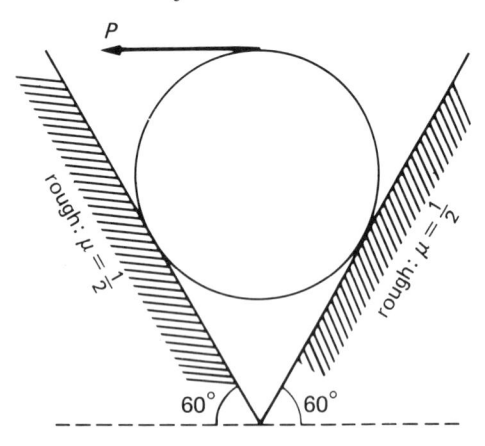

→ **38** The disc is just about to rotate; $\tan \alpha = \frac{4}{3}$. Find P.

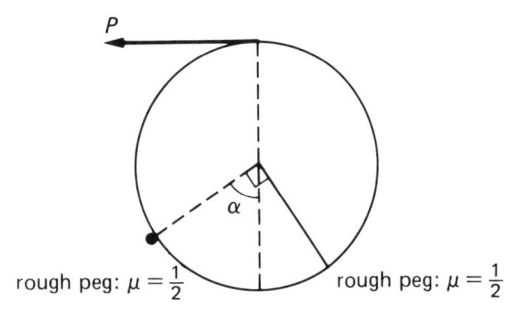

Questions 39–45. A rod AB of length $2a$ and weight W, uniform unless otherwise stated, with centre of gravity G, is smoothly pivoted and rests in equilibrium.

39 $AC = a$. A load of $3W$ is hung at B. Find the tension in the string and the magnitude and direction of the reaction at A.

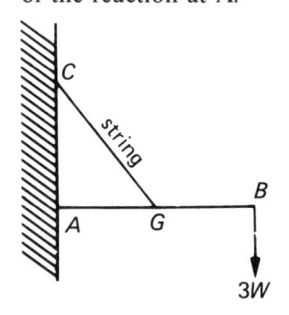

② **40** $AC = a$, $AD = \frac{3}{2}a$. A load of $2W$ is hung at B. Find the tension in the string and the magnitude and direction of the reaction at A.

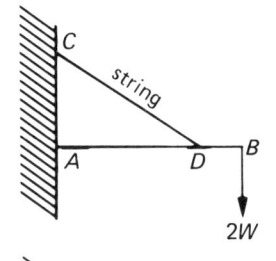

② → **41** $AC = a$. The rod is non-uniform: $AG = \frac{3}{2}a$. Find the tension in the string and the magnitude and direction of the reaction at A.

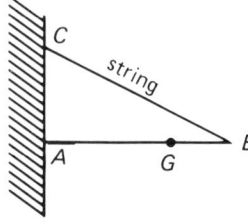

② **42** The rod is kept in equilibrium by a horizontal force $\frac{1}{2}W$ at B. Find θ.

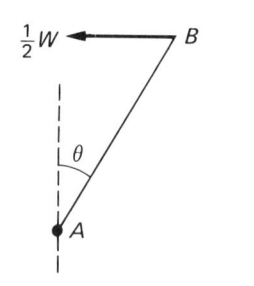

② **43** The rod is kept in equilibrium by a force F at B at right angles to the rod. Find F and the reaction at A.

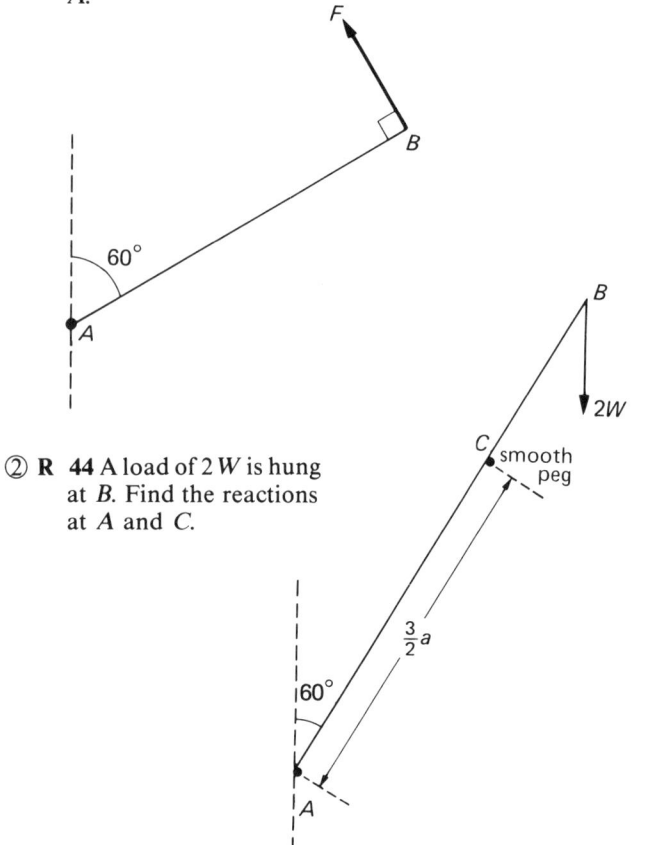

② **R 44** A load of $2W$ is hung at B. Find the reactions at A and C.

② **45** Find the reactions at *A* and *C*.

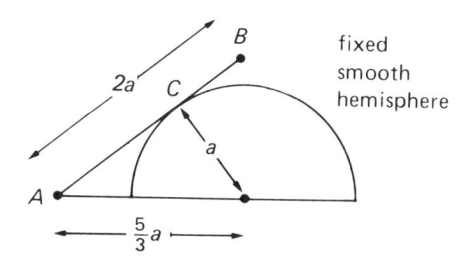

Questions 46–48: A uniform lamina of weight *W* is smoothly pivoted at *A* and rests in equilibrium.

② **46** *ABCD* is a square lamina, side $2a$; $ED = 2a = EA$. Find the tension in the string and the magnitude and direction of the reaction at *A*.

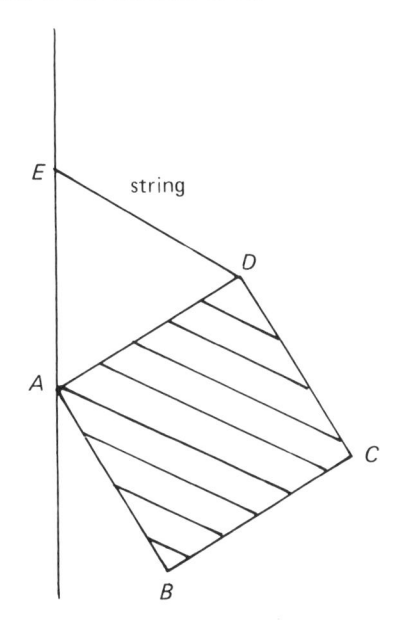

② → **47** *ABCD* is a rectangular lamina; $AB = 3a$, $BC = 4a$, *AC* is horizontal. Find *P*, and the magnitude and direction of the reaction at *A*.

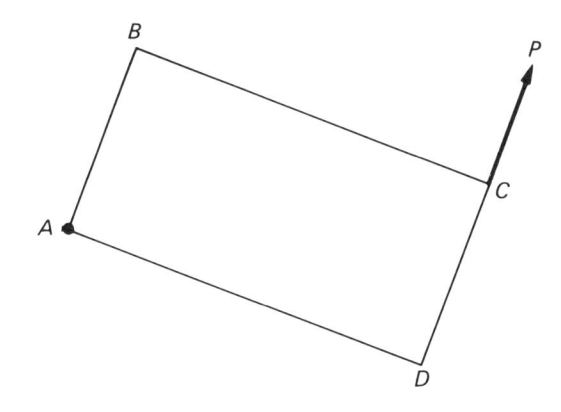

② **R 48** *ABCDEF* is a regular hexagonal lamina, side $2a$. A particle of weight $2W$ is hung from *C*. *AC* is horizontal. Find *P*, and the magnitude and direction of the reaction at *A*.

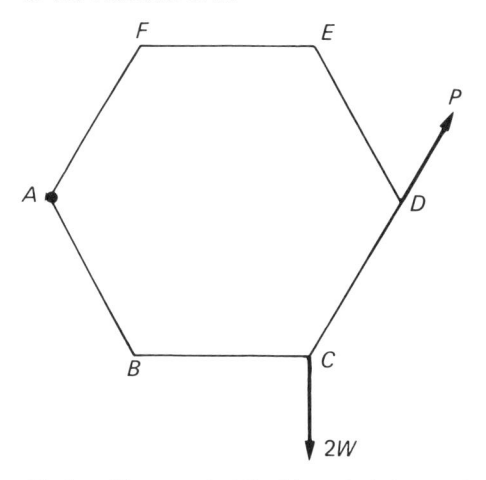

• → **49** A uniform rod *AB* of length 0.6 m and mass 2 kg is smoothly pivoted to a fixed point *A*. A light elastic string *BC* of modulus of elasticity 20 N has one end attached to *B* and its other end *C* fixed to a point at the same level as *A*, where $AC = 1$ m. When the system is in equilibrium $BC = 0.8$ m. Find:

(i) the tension in the string
(ii) the natural length of the string
(iii) the magnitude and direction of the resultant force exerted on the rod at *A*.

50 A uniform rod *AB*, of weight *W* and length $2a$, is smoothly pivoted to a wall at *A*. The point *C* on the wall is at a distance $2a$ vertically above *A*. The ends of a light elastic string of natural length *a* and modulus of elasticity $\frac{1}{2}W$ are attached to *B* and *C*. The rod rests in equilibrium at an angle 2α to the vertical. Show that:

(i) the tension in the string is $W \sin \alpha$
(ii) the extension of the string is $a(4 \sin \alpha - 1)$

Hence show that $\alpha = 30°$.

When a load of weight kW is hung from *B*, the rod rests in equilibrium and is horizontal. Find *k*.

• → **51** A uniform rod *AB* of length $2a$ and weight *W* is smoothly pivoted at *A* to a fixed point. It is held in equilibrium at an angle $\tan^{-1} \frac{3}{4}$ to the downward vertical by a force of magnitude *P* applied at *B*. Find *P* if the force is:

(i) horizontal   (ii) at right-angles to the rod
(iii) vertical

What is the least possible value of *P*?

52 A uniform rod *AB* of length $2a$ and weight *W* rests at an angle 60° to the horizontal with its end *A* on rough horizontal ground. Find the magnitude and direction of the least force which, applied at *B*, keeps the rod in equilibrium.

If the rod is in limiting equilibrium when this force is applied, find the coefficient of friction between the rod and the ground.

9.3:2 Three-force equilibrium problems

Theorem 1: If a rigid body is in equilibrium under three coplanar forces F_1, F_2 and F_3, then the forces are either parallel or intersect.

Proof
Suppose that not all the forces are parallel. Then in particular two of them, $\mathbf{F}_1$ and $\mathbf{F}_2$ say, meet at a point A; so $M_A(\mathbf{F}_1) = M_A(\mathbf{F}_2) = 0$. Since the body is in equilibrium, $\sum M_A(\mathbf{F}) = 0$, so $M_A(\mathbf{F}_3) = 0$, i.e. $\mathbf{F}_3$ also acts through A.

The use of the theorem in equilibrium problems saves having to take moments, since we have effectively taken moments about the point of intersection of the forces.

Worked example 3 A uniform rod AB of length $2a$ and weight W is smoothly pivoted to a fixed point A on a wall. C is a point on the wall at a distance a vertically above A (fig. 9.3:4). Find the magnitude and direction of the reaction at A if the rod is kept horizontal by a string attached to C and to the end B of the rod.

There are three forces acting on the rod, T, W and the reaction P at A. Suppose T and W meet at D, then, by Theorem 1, P acts through D too.
Since $\triangle DGA = \triangle DGB$, $D\hat{A}G = D\hat{B}G = \alpha$ say; so

$$\tan \alpha = \frac{CA}{AB} = \frac{a}{2a} = \frac{1}{2}$$

$\rightarrow)$ $\quad P \cos \alpha - T \cos \alpha = 0$

$\uparrow)$ $\quad P \sin \alpha + T \sin \alpha - W = 0$

$\Rightarrow \quad P = \tfrac{1}{2} W$

So the reaction at A has magnitude $\tfrac{1}{2} W$ and makes an angle $\tan^{-1} \tfrac{1}{2}$ with the horizontal.
Note: we could have solved this problem using the methods of section 9.3:1.

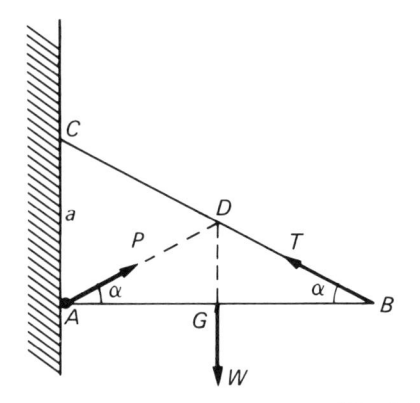

Fig. 9.3:4

Exercise 9.3:2

③ *Questions 1–6:* A uniform rod AB of length $2a$ and weight W, smoothly pivoted at A, is held in equilibrium by an inelastic string. Using Theorem 1, find the tension in the string and the magnitude and direction of the reaction at A.

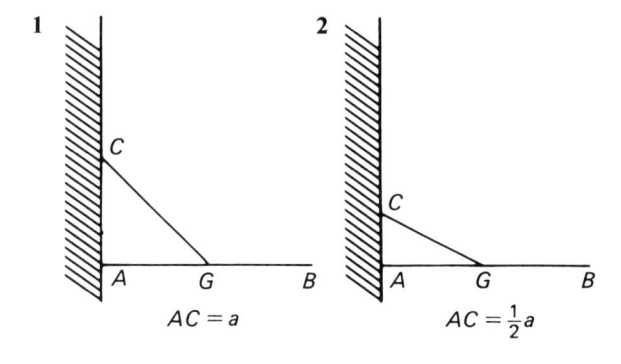

1. $AC = a$

2. $AC = \tfrac{1}{2} a$

Forces on a rigid body 549

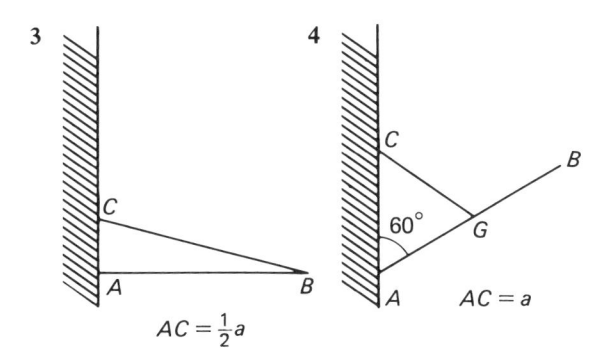

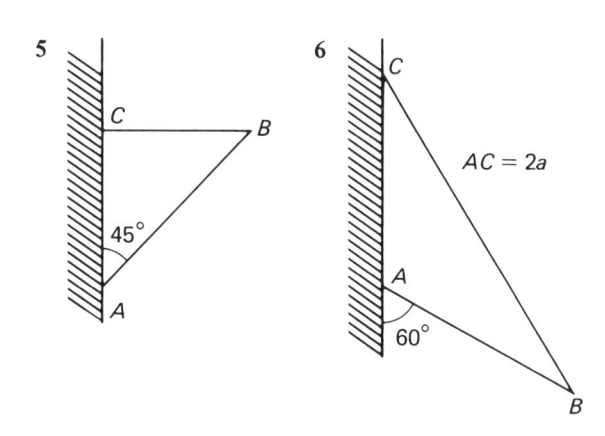

3. $AC = \frac{1}{2}a$

4. $AC = a$, $60°$

5. $45°$

6. $AC = 2a$, $60°$

③→ **7** Using Theorem 1, redo some of the three-force problems of exercise 9.3:1, for example questions 1, 5, 11, 24, 31, 35, 41 and 47. Stop if you do not like the method.

③ **8** Redo exercise 9.3:1 questions 12, 32 and 47 using Theorem 1 and the cotangent rule (see 3.3:1 Q26).

③ **9** A uniform rectangular lamina *ABCD* of weight *W*, with $AB = 8a$ and $AD = 6a$, is smoothly pivoted to a fixed point at *A*. It is kept in equilibrium in a vertical plane with *AC* horizontal and *B* as the lowest point by a force **P**. Find $|\mathbf{P}|$ if:

 (i) **P** acts along *EF*, where *E* and *F* are the midpoints of *AB* and *CD*
 (ii) **P** acts along *BC*
 (iii) the force exerted by the pivot acts in the direction *BA*

③ **R 10** The end *A* of a uniform rod *AB* of length $2a$ and weight *W* is in contact with a wall. A light inelastic string connects the other end *B* to a point *C* on the wall, a distance $2a$ vertically above *A*. The rod makes an angle 2α ($0° < \alpha < 45°$) with the downward vertical.

 (i) If the wall is smooth, show that equilibrium is impossible.
 (ii) If the coefficient of friction between the rod and the wall is $1/\sqrt{3}$, show that the rod can rest in equilibrium if $\alpha < 30°$.

9.3:3 Suspension

Consider a body with centre of gravity *G*, freely suspended from a point *A*, as in fig. 9.3:5. There are only two forces acting on the body, the weight *W* and the tension in the string. So

(i) $T = W$
(ii) *T* and *W* are collinear, i.e. *AG* is vertical.

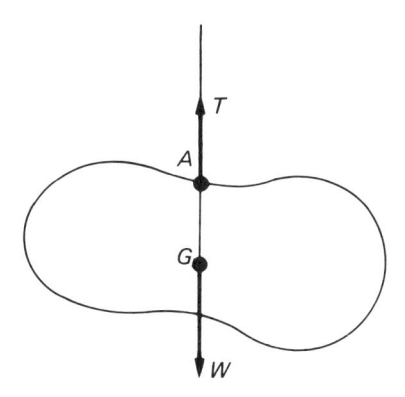

Fig. 9.3:5

Worked example 4 A uniform solid hemisphere of radius *a* is suspended from a point *A* on the rim of its plane face. Find the angle α between *OA* and the vertical, where *O* is the centre of the plane face.

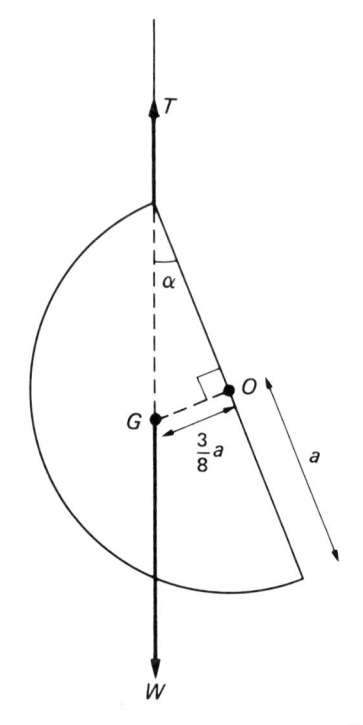

The hemisphere is shown in fig. 9.3:6. (Note: we have not attempted a 3D representation since we are only interested in the plane OAG—it is difficult enough drawing a good 2D representation...)

$$OG = \tfrac{3}{8}a \quad \text{(see 9.2:4 Q12)}$$

$$\tan \alpha = \frac{OG}{OA} = \frac{\tfrac{3}{8}a}{a} = \tfrac{3}{8} \Rightarrow \alpha = 20.55°$$

Worked example 5 A uniform solid hemisphere of weight W and radius a is suspended by two vertical strings, one fastened at O, the centre of its plane face, one at a point A on the rim of its plane face. If OA makes an angle of 60° with the vertical, find the tensions in the strings.

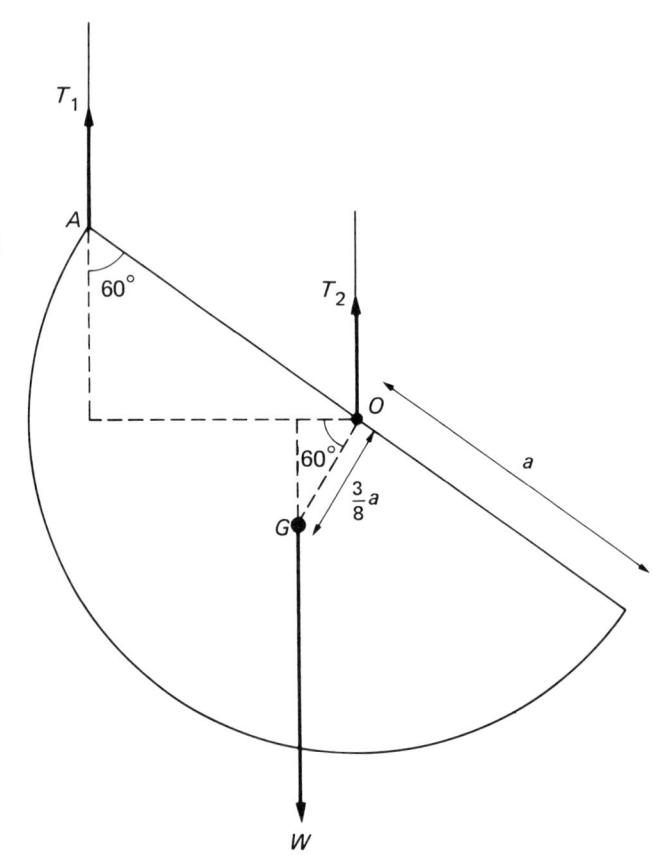

Fig. 9.3:6

Fig. 9.3:7

The hemisphere is shown in fig. 9.3:7. (Note: why is it not possible for O to be above the level of A?)

The body is in equilibrium, so:

$\uparrow)$ $\qquad\qquad\qquad T_1 + T_2 - W = 0$

$O\,\rotatebox{90}{)}$ $\qquad\qquad W \cdot \tfrac{3}{8}a \cos 60° - T_1 a \sin 60° = 0$

$\Rightarrow \qquad\qquad T_1 = \tfrac{1}{8}\sqrt{3}\,W, \quad T_2 = \tfrac{1}{8}(8 - \sqrt{3})\,W$

Exercise 9.3:3

1 A uniform rectangular lamina $ABCD$ is suspended from its vertex A. If $AB = 2a$ and $BC = a$, find the angle between AB and the horizontal.

2 A uniform solid cone of height $\sqrt{3}a$ and base radius a is suspended by a string attached to a point on the circumference of its plane face. Find the angle between the axis of the cone and the vertical (see 9.2:4 WE7).

3 A uniform triangular lamina ABC is suspended from its vertex A. Find the angle between AB and the horizontal if:

(i) $A\widehat{B}C = 90°$, $AB = a$, $BC = a$
(ii) $A\widehat{B}C = 90°$, $AB = 4a$, $BC = 3a$
(iii) $AB = 8a$, $AC = BC = 5a$

4 A uniform wire, bent into the shape of a semicircle, is suspended from one of its endpoints. Find the angle between its axis of symmetry and the vertical. (See 9.2:4Q18.)

Questions 5 and 6: A uniform plank of length $2a$ and weight W, suspended by a string, rests in equilibrium.

5 A load of weight X is hung from C, where $CB = \frac{2}{5}a$. Find X if the plank is horizontal.

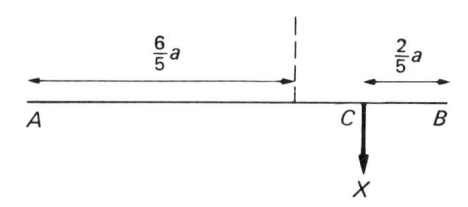

6 A load of weight $\frac{1}{4}W$ is hung from B. Show that the plank can rest in equilibrium whatever the value of θ.

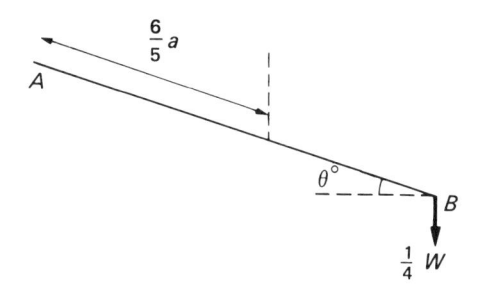

7 A particle of weight $\frac{1}{2}W$ is attached to one vertex A of a uniform square lamina $ABCD$ of weight W, and the lamina is suspended from the vertex B. Find the angle between AB and the horizontal.

8 Two particles of weights W and $\frac{1}{2}W$ are attached to points P and Q at the opposite ends of a diameter of the plane face of a uniform solid hemisphere of weight W. Find the inclination of PQ to the vertical when the hemisphere is suspended from:

(i) P
(ii) Q
(iii) O, the centre of its plane face

R 9 A particle of weight W is attached to the end B of a uniform wire of weight W bent into the shape of a semicircle, and the wire is suspended from the other end A. Find the inclination of AB to the vertical.

10 A uniform solid cone of height $2a$, base radius a and weight W, is suspended by two vertical strings, one attached to its vertex A, the other to a point B on the circumference of its plane face. If AB is horizontal, find the tensions in the strings.

11 A uniform solid hemisphere of weight W is suspended by two vertical strings attached to points P and Q at the ends of a diameter PQ of its plane face. If PQ makes an angle of $30°$ with the horizontal, find the tensions in the strings.

12 $ABCDE$ is a uniform lamina.

$AB = DE = EA = 8a$; $BC = CD = 5a$.
$B\widehat{A}E = A\widehat{E}D = 90°$.

(i) The lamina is suspended from A. Find the angle between the axis of symmetry and the vertical.
(ii) The lamina is suspended by vertical strings attached at A and C. If AB is horizontal, find the ratio of the tensions in the strings.

R 13 The uniform solid shown consists of two cones glued together.

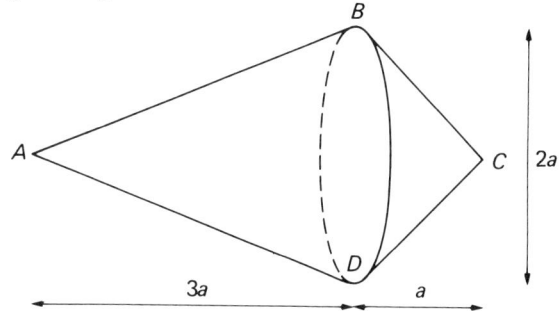

(i) When the solid is suspended from B, find the angle between AC and the vertical.
(ii) The solid is suspended by two vertical strings attached at A and B. Find the ratio of the tensions in the strings if:
   (a) AC is horizontal; (b) AB is horizontal.

Questions 14–17: A plank of length $2a$ and weight W, uniform unless otherwise stated, with centre of gravity G, is suspended by two strings.

14 Either of the strings will snap if the tension in it exceeds $2W$. Find the section of the rod from which a load of weight $2W$ can be hung:

(i) if the plank is uniform
(ii) if $AG = \tfrac{3}{4}a$

15 Either of the strings will snap if the tension in it exceeds $\tfrac{3}{2}W$. Find the section of the rod from which a load of weight W can be hung.

16 Find X if the tensions in the strings are equal.

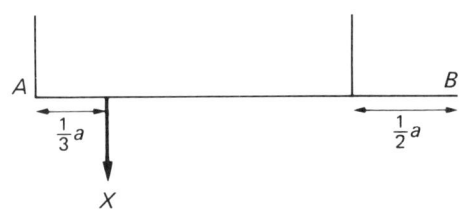

R 17 The plank is non-uniform: $AG = \tfrac{2}{3}a$. AC and BD are elastic strings of natural length a and modulus of elasticity W. Find:

(i) the tensions in the strings
(ii) their extensions
(iii) the distance CD

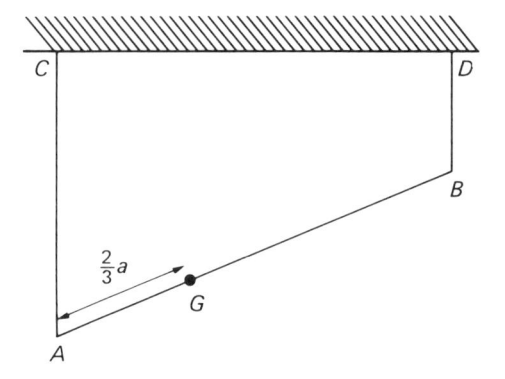

9.3:4 Toppling and sliding

Consider a body resting on a plane. Suppose that A and B are the two outermost points of the body in contact with the plane. Assume that the plane is rough enough to prevent sliding.

(a) If the line of action of the weight falls outside AB (fig. 9.3:**8a**) the body will topple. (Since there is a positive anticlockwise moment about A the body rotates, i.e. topples about A.)
(b) If the body is in equilibrium, the line of action of the weight must fall between A and B (fig. 9.3:**8b**).

In particular, if the body rests in equilibrium with a single point A in contact with the plane, the weight must act through A, i.e. AG must be vertical (fig. 9.3:**8c**).

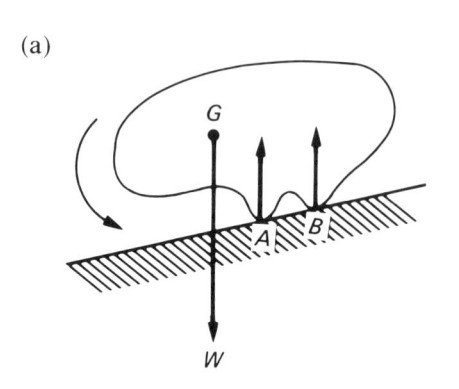

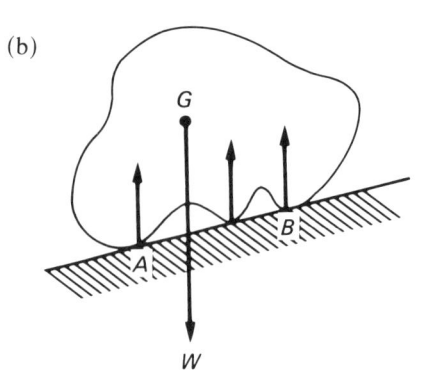

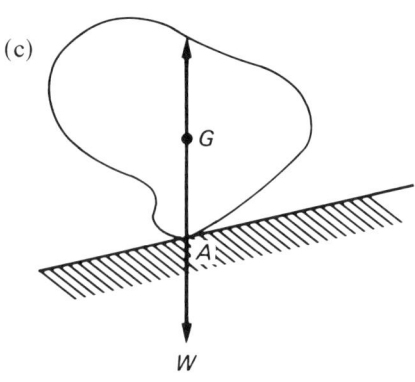

Fig. 9.3:**8**

Forces on a rigid body 553

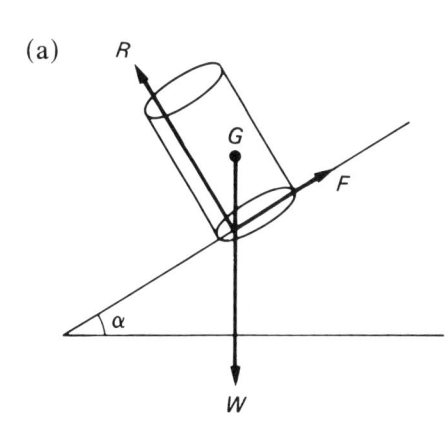

Worked example 6 A solid cylinder of radius a and height $3a$ is placed with one plane face in contact with a rough plane. The inclination of the plane is slowly increased from zero. Show that equilibrium will be broken by sliding if $\mu < \frac{2}{3}$.

We look at the conditions for sliding and toppling separately, and then compare them.

Sliding: suppose the cylinder is about to slide when the inclination of the plane is α, as in fig. 9.3:**9a**. (Note: F, R and W intersect. Why?)

Resolving the forces perpendicular to and down the plane:

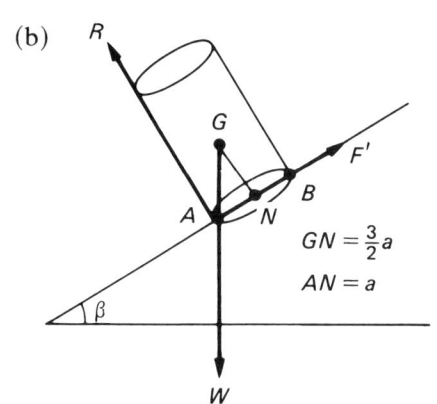

$$\nwarrow) \qquad R - W \cos \alpha = 0$$
$$\swarrow) \qquad W \sin \alpha - F = 0$$
$$\Rightarrow \qquad \tan \alpha = F/R$$

Since the cylinder is about to slide, $F = \mu R \Rightarrow \tan \alpha = \mu$.

Toppling: suppose the cylinder is about to topple when the inclination of the plane is β, as in fig. 9.3:**9b**. The line of action of the weight must pass through A. (Note: the line of action of R also passes through A. Why?)

So
$$\tan \beta = \frac{a}{3a/2} = \frac{2}{3}$$

Fig. 9.3:9   If sliding is to occur before toppling, $\alpha < \beta$,

$$\Rightarrow \qquad \tan \alpha < \tan \beta$$
i.e. $\qquad \mu < \frac{2}{3}$

(a) labels: R, G, F, α, W

(b) labels: R, G, F', B, A, N, $GN = \frac{3}{2}a$, $AN = a$, β, W

Exercise 9.3:4

Questions 1–4: The uniform lamina $ABCD$ is placed in a vertical plane with its edge AD on a horizontal plane. By considering the distance of its centre of gravity from AB, find the least value of x for which the lamina can remain in equilibrium (without toppling about D). (C.f. 9.2:3 Q2.)

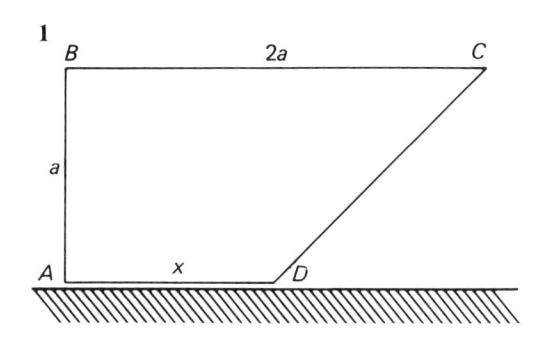

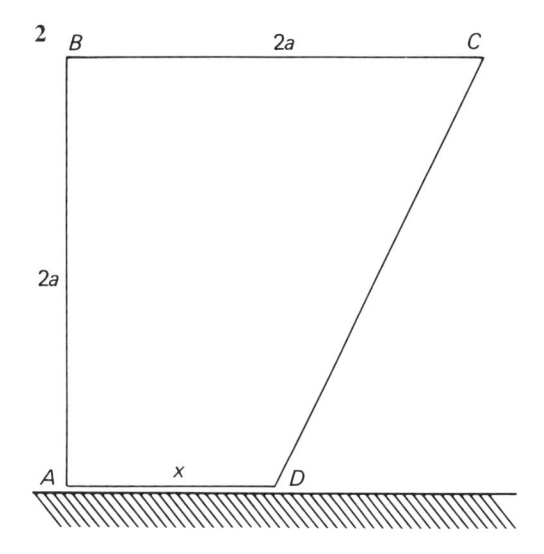

1. Trapezium with B top-left, C top-right, $BC = 2a$, $AB = a$, AD on ground with x shown from A to D.

2. Trapezium with B top-left, C top-right, $BC = 2a$, $AB = 2a$, AD on ground with x shown from A to D.

3

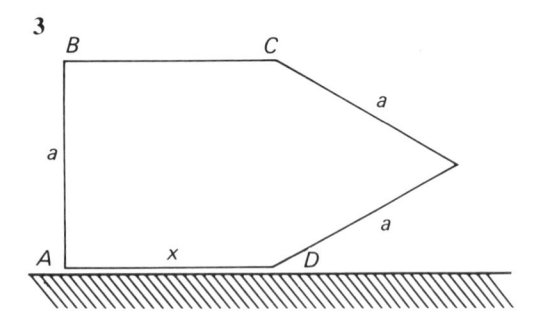

[Figure: Pentagon ABCD on ground with AB = a (vertical), BC = a (horizontal top), CD sloping to apex then to D, with two sides of length a forming point; AD along ground with A at left, D at right, distance x from A]

4

[Figure: L-shaped lamina on ground; AB = a (left vertical side), BC horizontal top, right side down $\frac{1}{2}a$, then horizontal length a, then down to D; AD along ground, distance x from A to D]

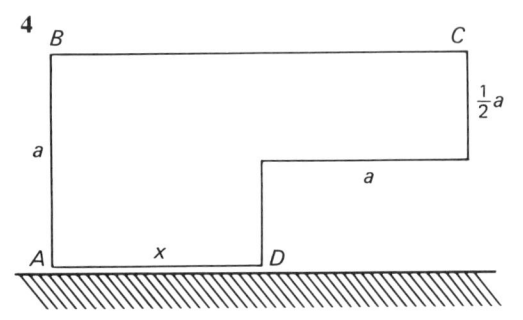

5 A uniform solid body consists of a cylinder of height $2a$ and base radius a joined by one of its ends to the base of a cone of height $\frac{4}{3}a$ and base radius a. Show that if the body is placed with the curved surface of the cone in contact with a horizontal plane, it will topple.

R 6 A uniform solid body consists of a solid cylinder of height x and radius a joined by one of its ends to the base of a solid cone of height $2a$ and base radius a. Find x:

(i) if the body is about to topple when it rests with a generator of the cone in contact with a rough horizontal surface
(ii) if the body is about to topple when it rests on a rough horizontal surface with its axis of symmetry horizontal

7 A particle of weight W is attached to a point on the circumference of the plane face of a uniform solid hemisphere of weight W. The hemisphere rests with its curved surface in contact with a smooth horizontal plane. Find the angle between its plane face and the horizontal.

→ **8** A uniform solid consists of a hemisphere of radius r and a cone of base radius r and height λr glued together so that their plane faces coincide. If the solid can rest in equilibrium with any point of the curved surface of the hemisphere in contact with a horizontal plane, show that $\lambda = \sqrt{3}$.

9 A uniform hollow body consists of a hemispherical shell of radius r and a conical shell of base radius r and height λr glued together so that their rims coincide. If the body can rest in equilibrium with any point of the curved surface of the hemisphere in contact with a horizontal plane, find λ.

Questions 10-14: A plank of length $2a$ and weight W, uniform unless otherwise stated, with centre of gravity, rests on (or under) smooth pegs P and Q.

• **10** P and Q are distant $\frac{3}{2}a$ apart. Find where the plank should be put in order that the heaviest possible person can walk from one end to the other without tipping it:

(i) if the plank is uniform
(ii) if $AG = \frac{3}{4}a$

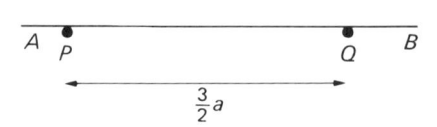

[Figure: horizontal plank A to B with pegs P and Q, distance PQ = $\frac{3}{2}a$]

11 If the peg Q can withstand forces up to $5W$, find the greatest vertical force that can be applied at B:

(i) upwards
(ii) downwards

if equilibrium is not to be broken.

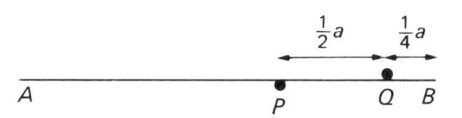

[Figure: plank A to B with pegs P and Q; distance from midpoint region: $\frac{1}{2}a$ and $\frac{1}{4}a$ marked]

→ **12** Loads of $2W$ and $3W$ are hung at A and B. The plank is just about to tilt about Q. Find x.

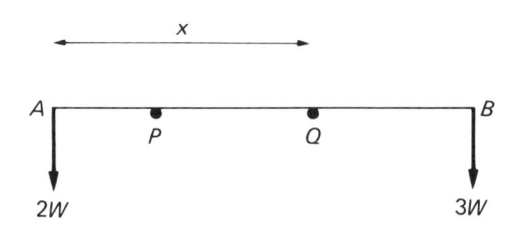

[Figure: plank with pegs P and Q, distance x marked from A; $2W$ hung at A, $3W$ hung at B]

13 The plank is non-uniform.

(i) If the plank starts to tilt when a load of weight $\frac{3}{2}W$ is hung from A, find AG.
(ii) Find the greatest weight that can be hung from B without disturbing equilibrium.

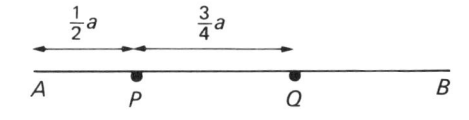

[Figure: plank A to B with peg P at distance $\frac{1}{2}a$ from A, and Q at distance $\frac{3}{4}a$ from P]

14 A load of weight $2W$ is hung from C where $AC = \frac{4}{5}a$. Find the greatest vertical downward force that can be applied at A if equilibrium is not to be broken.

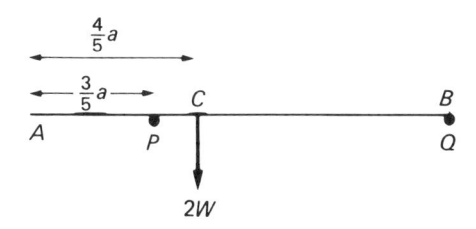

Questions 15–17: A plank of length $2a$ and weight W rests temptingly on the edge of a cliff.

• **15** $AE = \frac{3}{2}a$.
  (i) How far can a man of weight $3W$ walk without tipping the plank?
  (ii) What weight should be placed at A to allow him to walk to B?

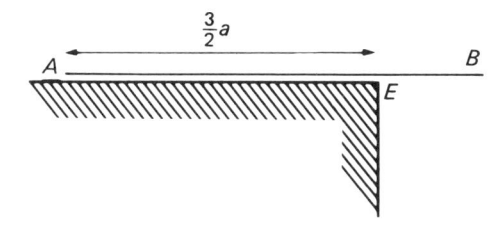

→ **16** $AE = \frac{4}{3}a$.
  (i) When a weight W is placed at A, a man can walk to within $\frac{1}{3}a$ of B without tipping the plank. Find the weight of the man.
  (ii) What weight should be placed at A to allow him to walk to B?

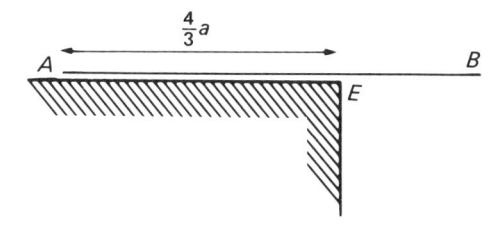

17 $AE = \frac{5}{4}a$. What weight should be placed at A to allow a man of weight $2W$ to walk to B?

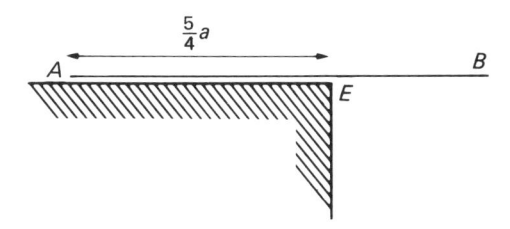

⑥ **18** A solid cylinder of radius $2a$ and height $5a$ is placed with one plane face in contact with a rough inclined plane. The inclination of the plane is slowly increased. Show that equilibrium will be broken by toppling if $\mu > \frac{4}{5}$.

⑥ → **19** A particle of weight $2W$ is attached to the centre of one of the ends of a solid cylinder of weight W, radius a and height $2a$. The other end of the cylinder is placed on a rough plane. The inclination of the plane is slowly increased. Show that equilibrium will be broken by sliding if $\mu < \frac{3}{5}$.

⑥ **20** A uniform solid cone of base radius r and height $4r$ is placed with its plane surface in contact with a rough plane; $\mu = \frac{1}{2}$. The inclination of the plane is slowly increased. Will equilibrium be broken by toppling or sliding?

⑥ **21** A uniform solid hemisphere rests in equilibrium with its curved surface in contact with a rough plane inclined at an angle α to the horizontal. Show that $\sin \alpha \leq \frac{3}{8}$.

⑥ • **22** A uniform square lamina $ABCD$ of side $2a$ rests in a vertical plane with its edge CD in contact with rough horizontal ground, $\mu = \frac{3}{4}$. Will equilibrium be broken by sliding or toppling when a gradually increasing horizontal force P is applied to the lamina (in its own plane):
  (i) at A;
  (ii) at E the midpoint of AD?

⑥ → **23** A uniform solid cone of base radius r and height $5r$ rests with its base on rough horizontal ground; $\mu = \frac{1}{2}$. Will equilibrium be broken by sliding or toppling when a gradually increasing horizontal force is applied:
  (i) at the vertex of the cone;
  (ii) at a point halfway up its slanting surface?

⑥ **R** **24** A uniform cube rests on a rough plane inclined at $\tan^{-1}\frac{1}{2}$ to the horizontal. A gradually increasing horizontal force P is applied, as shown in the diagram. Show that equilibrium will be broken by sliding if $\mu < \frac{8}{11}$.

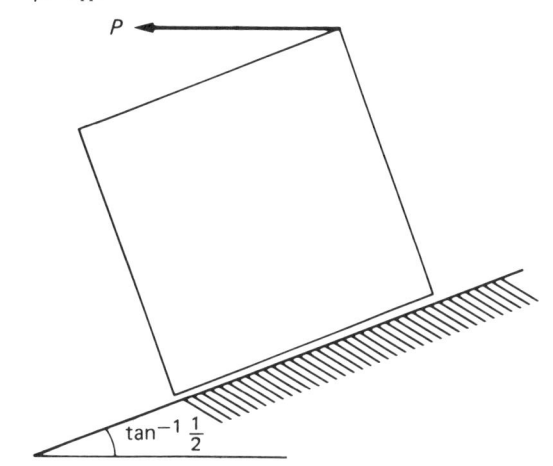

25 A uniform sphere of radius a rests on a step, as shown in the diagram. The coefficient of friction between the sphere and the ground at A is $\frac{3}{4}$, that between the sphere and the step at B is μ. A gradually increasing horizontal force P is applied at C. Show that equilibrium will be broken by sliding if $\mu < \frac{1}{3}$.

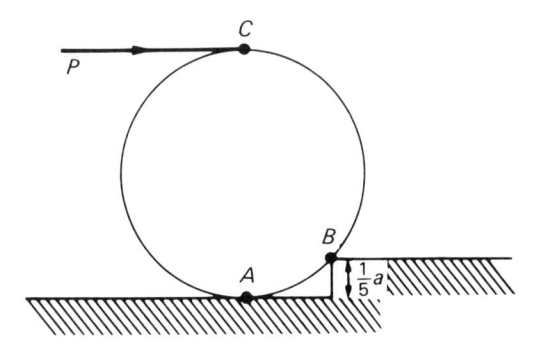

26 A uniform cube of side $8a$ rests on a step, as shown in the diagram. The coefficient of friction between the cube and the step, at both points of contact, is $\frac{1}{3}$. A gradually increasing horizontal force P is applied at A. Will equilibrium be broken by sliding or toppling?

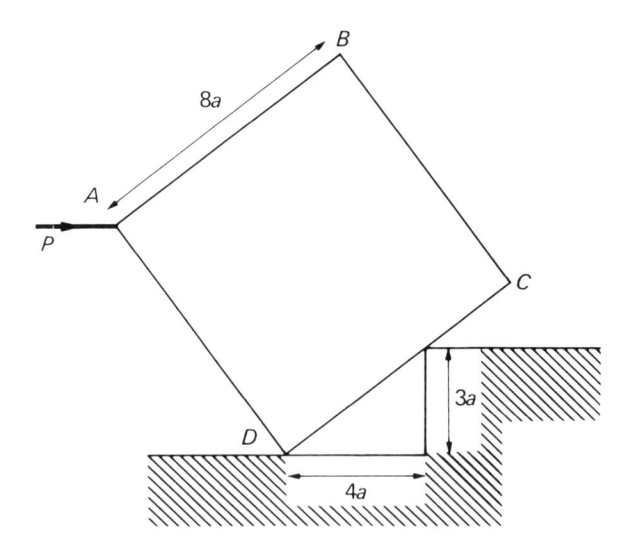

Miscellaneous exercise 9

1 $ABCD$ is a square of side $2a$. Forces of magnitude 4 N, 5 N, 2 N and 2 N act along AB, BC, DC and DA respectively in the directions indicated by the order of the letters. Find the magnitude and direction of the resultant of the forces, and show that its line of action passes through B.

A force of magnitude $6\sqrt{2}$ N acting along BD is added to the system. Show that the resultant of the enlarged system acts along BC.

2 ABC is a triangle in which $AB = AC = 5$ m and $BC = 6$ m. D is the midpoint of BC. Forces of magnitude p N, 15 N, 10 N and 4 N act along AB, BC, CA and DA respectively in the directions indicated by the order of the letters.

(i) Show that if $p = 15$, the system of forces reduces to a couple and find its magnitude.
(ii) If $p = 10$, find the magnitude and direction of the resultant of the forces and the distance from B of the point where its line of action cuts BC.

3 $ABCDEF$ is a regular hexagon. Forces of magnitude 2 N, 3 N, 4 N, 5 N, p N and q N act along AB, BC, CD, ED, EF, and AF respectively in the directions indicated by the order of the letters. Find p and q:

(i) if the system reduces to a couple
(ii) if the system reduces to a single resultant force acting along BE

Explain why

(iii) the system cannot be in equilibrium
(iv) the system cannot reduce to a single resultant force along CF

4 The forces $\mathbf{F}_1 = 4\mathbf{i} + \mathbf{j}$ and $\mathbf{F}_2 = 3\mathbf{i} - 3\mathbf{j}$ act through the points with position vectors $2\mathbf{i} + 2\mathbf{j}$ and $-3\mathbf{i} + 2\mathbf{j}$ respectively. Find the position vector of the point of intersection of the forces. Hence find a vector equation of the line of action of the resultant of $\mathbf{F}_1$ and $\mathbf{F}_2$. Find:

(i) the magnitude of the resultant moment of the forces about O
(ii) the angle between the lines of action of the forces

5 The forces $\mathbf{F}_1 = 6\mathbf{i} - \mathbf{j}$, $\mathbf{F}_2 = 4\mathbf{i} - 8\mathbf{j}$ and $\mathbf{F}_3 = -\mathbf{i} - 3\mathbf{j}$ act through the points with position vectors $-2\mathbf{i} + \mathbf{j}$, $\mathbf{i} + 3\mathbf{j}$ and $2\mathbf{i}$ respectively.

(i) Find the magnitude of the resultant of the forces.
(ii) Find the sum of the moments of the forces about O, and hence show that the equation of the line of action of the resultant is $4x + 3y = 6$.

6 The forces $\mathbf{F}_1 = 4\mathbf{i} - 3\mathbf{j}$, $\mathbf{F}_2 = -5\mathbf{i} + \mathbf{j}$ and $\mathbf{F}_3 = a\mathbf{i} + b\mathbf{j}$ act through the points with position vectors $2\mathbf{i} - 3\mathbf{j}$, $4\mathbf{i} - \mathbf{j}$ and $c\mathbf{i} - \mathbf{j}$ respectively. Find a, b and c if this system of forces:

(i) reduces to a couple of magnitude $+10$
(ii) is in equilibrium

7 Forces $2\mathbf{i} - \mathbf{j}$, $\mathbf{i} + 5\mathbf{j}$ and $-3\mathbf{i} - 4\mathbf{j}$ act at the points with position vectors $-\mathbf{i} + 3\mathbf{j}$, $2\mathbf{i} + 4\mathbf{j}$ and $\mathbf{i} + \mathbf{j}$ respectively. Show that the system of forces is in equilibrium. The force $-3\mathbf{i} - 4\mathbf{j}$ is now reversed in direction, its line of action being unchanged. Find the magnitude and direction of the resultant of the new system and the cartesian equation of its line of action.

8 A and B are the points $(3, 0)$ and $(0, 4)$.

(i) The resultant moments of a system of forces about O, A and B are $+6$ N m, -12 N m and -9 N m respectively. Find the cartesian equation of the line of action of the resultant of the system.
(ii) A force of magnitude 12 N acts along OA and a force of magnitude 15 N acts along AB. Find the cartesian equation of the line of action of the resultant of these forces.

9 A uniform ladder, of weight W and length $2l$, rests against a smooth vertical wall with its foot on rough horizontal ground. The ladder lies in a vertical plane perpendicular to the plane of the wall and is inclined at an angle $\tan^{-1} 2$ to the horizontal. Show that the ladder will rest in equilibrium if $\mu \geq \frac{1}{4}$, where μ is the coefficient of friction between the ladder and the ground.
$\mu = \frac{3}{8}$. A building contractor of weight W climbs the ladder. Show that he can reach the top without the ladder slipping. The contractor sends one of his labourers, of weight $2W$, up the ladder. Find how far the labourer can climb up the ladder before it slips.

10 A uniform rod AB, of weight W and length $2l$, rests with its end A on rough horizontal ground and its end B against a smooth vertical wall. The rod lies in a vertical plane perpendicular to the plane of the wall and is inclined at an angle α to the horizontal. Show that the rod will rest in equilibrium if $\mu \geq \frac{1}{2} \cot \alpha$, where μ is the coefficient of friction between the rod and the ground. Show that, if x is the distance from A to the highest point of the rod at which a load of weight kW can be attached without equilibrium being disturbed, then

$$kx = 2(k+1)l\mu \tan \alpha - l$$

Show that:

(i) if $\mu = \frac{1}{4} \cot \alpha$, $k \geq 1$
(ii) if $\mu = 2 \cot \alpha$, equilibrium will not be disturbed whatever the value of k

11 A uniform rod AB of weight W and length $2a$ rests with its end A on a rough horizontal plane, and a point C of the rod, where $AC = \lambda a$, in contact with a smooth peg. The rod makes an angle of $60°$ with the horizontal. Show that the force exerted by the peg on the rod is $\frac{1}{2} W / \lambda$. Deduce that, if the coefficient of friction between the rod and the plane is $1/2\sqrt{3}$, then $\lambda \geq \frac{7}{4}$.

12 A smooth rail is fixed in a horizontal position parallel to a rough vertical wall and at a distance l from the wall. A uniform rod AB of length $4l$ is in equilibrium resting on the fixed rail with one end A in contact with the wall. If the angle between the rod and the horizontal is $\tan^{-1} \frac{3}{4}$, find:

(i) the force exerted by the rail on the rod
(ii) the least possible value of the coefficient of friction between the rod and the wall

13 A uniform rod AB of length $2a$ and weight W, is hinged to a vertical wall at A and is supported in a horizontal position by a string attached to B and to a point C on the wall vertically above A. The string makes an angle α with the horizontal.

(i) When a load of weight W is hung from B, the reaction of the hinge at A is at right-angles to BC. Show that $\alpha = 60°$.
(ii) When a load of weight W is hung from a point D of the rod, the reaction of the hinge at A makes an angle $45°$ with the horizontal. Show that $AD = \frac{3}{2}a$.

14 A uniform rod AB, of weight W and length $2l$, is smoothly hinged to a vertical wall at A. The point C on the wall is at a distance $2l$ vertically above A. The ends of a light elastic string, of natural length l and modulus of elasticity $2W$, are attached to B and C so that the rod rests in equilibrium. Find:

(i) the angle between the rod and the vertical
(ii) the tension in the string

When a load of weight kW is hung from the rod at B, the rod rests in equilibrium with $AB = BC$. Show that $k = \frac{3}{2}$.

15 A uniform rod AB of weight W and length $2l$, is smoothly hinged at A to a fixed straight horizontal wire. The end B is attached by means of a light inelastic string of length $2l$ to a small ring of weight W, which can slide on the wire. The coefficient of friction between the ring and the wire is μ. The rod is in equilibrium in the vertical plane through the wire and makes an angle of $30°$ with the horizontal. Show that:

(i) the tension in the string is $\frac{1}{2} W$
(ii) $\mu \geq \frac{1}{5}\sqrt{3}$

16 A uniform square lamina $ABCD$ of weight $4W$ and side $2a$ is smoothly pivoted to a fixed point at A. It is kept in equilibrium in a vertical plane with AB making an angle α ($<90°$) with the upward vertical by a force of magnitude $\frac{1}{4}W\sqrt{6}$ acting along CB. Find α and the magnitude and direction of the reaction at A.

17 A uniform hexagonal lamina $ABCDEF$, of weight W and side $2l$, is smoothly hinged to a fixed point at A. It is kept in equilibrium in a vertical plane with AC horizontal and the vertex B above AC by a light inelastic string of length $2l$ whose ends are attached to B and to a fixed point G distant $2l$ vertically above A.

(i) Find the tension in the string and the magnitude and direction of the reaction at A.
(ii) If the string breaks when the tension in it exceeds $3W$, find the weight of the greatest load that can be attached to the lamina at D without disturbing equilibrium. When this load is attached, find the magnitude of the reaction at A, and show that its line of action passes through the midpoint of BC.

18 (i) Find the coordinates of the centroid of the uniform lamina bounded by the curve $y = e^{2x}$, the x-axis, the y-axis and the line $x = \ln 2$.

(ii) A cap of depth $\frac{1}{2}a$ is cut from a uniform solid sphere of radius a. Show that the volume of the cap is $\frac{5}{24}\pi a^3$ and find the distance of its centre of gravity from its plane face.

19 (i) Sketch on the same axes the curve $y = 3x^2$ and the line $4x + y = 20$, and find the coordinates of their point of intersection. Hence find the coordinates of the centre of gravity of the uniform lamina bounded by the curve, the line and the x-axis.

(ii) Sketch the curve
$$y = 1/\sqrt{(4 - x^2)} \text{ for } -2 < x < 2.$$

The region bounded by this curve, the x-axis, the y-axis and the line $x = 1$ is rotated completely about the x-axis. Find the coordinates of the centre of gravity of the uniform solid formed.

20 A uniform lamina is bounded by that part of the parabola $y^2 = 16x$ which lies in the first quadrant, by the axis of the parabola and by the line $x = 1$. Find the coordinates of the centroid of the lamina.

Find the angle of inclination to the vertical of the axis of the parabola when the lamina is suspended from:

(i) the vertex $(0, 0)$ of the parabola
(ii) the point $(1, 4)$

21 Show that the centre of gravity of a uniform solid hemisphere of radius a is at a distance $\frac{3}{8}a$ from its plane face.

(i) The hemisphere is suspended by two vertical strings, one fastened to A, the other to B, where AB is a diameter of its plane face. The angle between AB and the horizontal is $\tan^{-1}\frac{4}{3}$. Show that the tensions in the strings are in the ratio $3:1$.

(ii) The hemisphere rests with the point A on rough horizontal ground and the point B in contact with a smooth vertical wall. When the angle between AB and the horizontal is $\tan^{-1}\frac{4}{3}$, the hemisphere is in limiting equilibrium. Show that the coefficient of friction between the hemisphere and the ground is $\frac{3}{16}$.

22 A uniform lamina $ABCD$ of weight W is in the shape of a trapezium. $AB = 5a$, $CD = 2a$, $DA = 6a$ and $\hat{A} = \hat{D} = 90°$. Find the distances of the centre of gravity of the lamina from AD and AB.

(i) The lamina is suspended from B. Find the angle between AB and the vertical.
(ii) A particle of weight kW is attached to the lamina at B and the lamina is suspended from C. If AB is horizontal, find k.

23 A uniform thin sheet of cardboard is in the form of a square $ABCD$. L and M are the midpoints of BC and CD, and O is the point of intersection of the diagonals. The corner LCM is folded over and stuck to the region LOM, with which it coincides.

(i) Show that, if the object is placed vertically with its edge LD on a horizontal table, it will not quite topple.
(ii) The object is suspended from D. Find the angle between AD and the vertical.

24 A thin uniform wire of length $16a$ and weight W is bent to form the sides AB, BC and CA of a triangle in which $AB = AC = 5a$.

(i) The vertex C of the object is pivoted to a fixed point and the object hangs freely. Find the angle between BC and the vertical.
(ii) When the vertex of the object is pivoted to a fixed point and a force of magnitude P acts along BC, the object is in equilibrium with AC horizontal. Show that $P = \frac{11}{20}W$.

25 Show that the centroid of a uniform solid cone of height $6r$ and base radius r is at a distance $\frac{3}{2}r$ from its plane face.

A toy nuclear warhead is made by sticking the plane face of this cone to one plane face of a cylinder of base radius r and height x, so that the faces coincide. The centre of gravity of this novelty item lies at the centre of the coincident faces. Show that $x = r\sqrt{6}$.

The plaything is suspended by two vertical strings, one attached to the vertex A of the cone and one to a point B on the circumferences of the coincident plane faces. If AB is horizontal, find the ratio of the tensions in the strings.

26 Show that the centre of gravity of a uniform wire in the shape of the arc of a circle of radius a and subtending an angle of 2α at the centre is at distance $(a \sin \alpha)/\alpha$ from the centre.

A thin uniform wire of length $(2+\tfrac{1}{2}\pi)a$ and weight W is bent to form an arc AB and two radii OA and OB of a circle radius a and centre O. Find the distance of the CG of the object from O. The object is smoothly pivoted at O and is kept in equilibrium with OB horizontal by a horizontal force of magnitude P acting at A. Show that $P = 4aW/(4+\pi)$.

27 Show that the centre of gravity of a uniform semi-circular lamina of radius a is at a distance of $\tfrac{4}{3}a/\pi$ from its straight edge.

A semicircular section of radius $\tfrac{1}{2}a$ is cut from this lamina, as shown in the diagram. The resulting lamina rests in equilibrium in a vertical plane with its curved edge AB on a horizontal table. Find the angle between its straight edge AO and the horizontal.

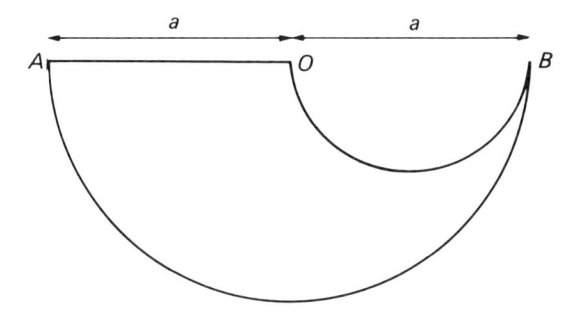

28 A uniform lamina is in the shape of a triangle ABC with $AB = AC = 13a$, $BC = 10a$. Show by integration that the centre of gravity of the lamina lies on the median AD at a distance $8a$ from A.

(i) The lamina, suspended by vertical strings attached to A and to C, rests in equilibrium with AC horizontal. Find the ratio of the tensions in the strings.
(ii) The lamina rests with its vertex A against a smooth wall and its edge BC resting, parallel to the wall, on rough horizontal ground. When the lamina is in limiting equilibrium, the plane of the lamina makes an angle α, where $\tan \alpha = \tfrac{1}{2}$, with the vertical. Show that the coefficient of friction between BC and the ground is $\tfrac{1}{6}$.

Show that if the angle between the lamina and the vertical is 2α, the lamina can just be kept in equilibrium by applying a couple of magnitude $2Wa$ where W is the weight of the lamina.

Examination questions

1.1 The function f is defined by
$$f: x \mapsto \frac{x+3}{x-1}, \ x \in \mathbb{R}, \ x \neq 1.$$
Find (i) the range of f (ii) $ff(x)$ (iii) $f^{-1}(x)$ $(L(B))$

1.2 (i) Find the local maximum and minimum values of
$$\frac{x}{(x+8)(x+2)}.$$
(ii) The equation of a curve is given by
$$6y^2 + 5xy + x^2 + 1 = 0.$$
Using implicit differentiation, or otherwise, find an equation giving $\frac{dy}{dx}$ in terms of x and y. Find the local maximum and minimum values of y. (In both (i) and (ii) distinguish carefully between maximum and minimum values.) (O&C)

1.3 A right circular cylinder is to be cut from a right circular solid cone, of height H and base radius R. The axis of the cylinder lies along the axis of the cone. The circumference of one end of the cylinder is in contact with the curved surface of the cone and the other end of the cylinder lies on the base of the cone. Show that V, the volume of the cylinder, is given by
$$V = \frac{\pi H x^2 (R - x)}{R}$$
where x is the radius of the cylinder.
Show also that, as x varies, the maximum possible value of V is $\frac{4\pi R^2 H}{27}$. $(L(B))$.

1.4 A bowl is filled with water and is of such a shape that, when the depth of the water is h cm, the volume of water is $2\pi h^2$ cm^3. Initially the bowl is empty and it is then filled with water at a steady rate of 3 cm^3 s^{-1}. Determine the rate at which h is increasing after 6 s, giving your answer to 3 significant figures. (O)

2.1 (i) Given that $f(x) = x^3 + kx^2 - 2x + 1$ and that when $f(x)$ is divided by $(x - k)$ the remainder is k, find the possible values of k.
(ii) When the polynomial $p(x)$ is divided by $(x - 1)$ the remainder is 5 and when $p(x)$ is divided by $(x - 2)$ the remainder is 7. Given that $p(x)$ may be written in the form
$$(x-1)(x-2)q(x) + Ax + B,$$
where $q(x)$ is a polynomial and A and B are numbers, find the remainder when $p(x)$ is divided by $(x-1)(x-2)$. (C)

2.2 The quadratic function f is defined by
$$f: x \mapsto 2x^2 + 4x + 5,$$
where $x \in \mathbb{R}$.
(i) Find the set of values of x for which $f(x) > 3x^2$.
(ii) Find the set of values of k for which the equation $f(x) = kx$ has no real roots.
(iii) By considering the identity
$$2x^2 + 4x + 5 \equiv 2(x + A)^2 + B,$$
where A and B constants, find the greatest value of $\frac{1}{f(x)}$.
(iv) The roots of the equation $f(x) = 0$ are α and β. Find an equation with numerical coefficients whose roots are $\alpha^2 + 2$ and $\beta^2 + 2$. (AEB)

2.3 Using the same axes sketch the curves
$$y = \frac{1}{x-1}, \quad y = \frac{x}{x+3}$$
giving the equations of the asymptotes. Hence, or otherwise, find the set of values of x for which $\frac{1}{x-1} > \frac{x}{x+3}$. $(L(B))$

2.4 The Republic of Anarchia first issued a retail price index (r.p.i.) in 1970, setting its initial value at 100. Its value t years later was found to be given by the formula $100 + 5t^2$. The government also announces the annual rate of inflation, R, calculated as
$$R = \frac{(\text{r.p.i. now}) - (\text{r.p.i. a year ago})}{\text{r.p.i. now}}.$$
(i) Show that, for $t \geq 1$, the annual rate of inflation is given by the formula
$$R = \frac{2t - 1}{20 + t^2}.$$
(ii) Find an expression for dR/dt. [Treat t as a continuous variable.]
(iii) Show that $dR/dt > 0$ for $1 \leq t < 5$, and $dR/dt < 0$ for $t > 5$.
(iv) What was the greatest value of the annual rate of inflation between 1970 and the present day? (SMP)

2.5 Two functions f and g each have domain D given by $\{x : x \in \mathbb{R}, x > -1\}$, and codomain $\mathbb{R}$. The rules for the functions are $f: x \mapsto \ln(x+1)$, $g: x \mapsto x^2 + 2x$. State the range of f, and define the inverse function f^{-1}.
Determine whether or not g is one-one, giving a reason for your answer, and state the range of g.
Give the rule for the composite function $f \circ g$, and state its range. (Find x, given that $(f^{-1} \circ g)(x) = e^3 - 1$.) (C)

2.6 Given that $(1+x)y = \ln x$, show that, when y is stationary

$$\ln x = \frac{(1+x)}{x}$$

Show graphically, or otherwise, that this latter equation has only one real root, and prove that this root lies between 3.5 and 3.8.

By taking 3.5 as a first approximation to this root and applying the Newton-Raphson process once to the equation $\ln x - (1+x)/x = 0$, find a second approximation to this root, giving your answer to 3 significant figures. Hence find an approximation to the corresponding stationary value of y. (L(B))

2.7 The population p, in millions, of a small country was recorded in the January of various years and the results are shown in the table below.

Year	1968	1974	1980	1985
p	12.3	13.4	15.1	17.1

Given that $p = 10 + ab^t$ where t is the time measured in years from January 1965 and a and b are constants, express $\log_{10}(p - 10)$ as a linear function of t. Draw a suitable straight line graph for $0 \leq t \leq 25$.

Use your graph to estimate
(i) the values of a and b
(ii) the year in which the population will reach 19.4 million. (AEB)

3.1 The function g is defined by

$$g(x) = 7\cos^2 x + \sin^2 x - 8 \sin x \cos x.$$

(i) Show that $g(x)$ may be expressed in the form

$$a + b \cos(2x + \alpha),$$

where $\tan \alpha = \frac{4}{3}$ and a, b are constants to be determined. Find the greatest and least values of $g(x)$.
(ii) Find the values of x in the range $0 \leq x \leq \pi$ for which $g(x) = 0$ and sketch the graph of $g(x)$ in this range.
(iii) Assuming that $x_1 = \frac{1}{2}\pi$ is an approximate solution of the equation $3g(x) = 2x$, find a closer approximation by one application of Newton's method (O & C)

3.2 Use identities for $\cos(C + D)$ and $\cos(C - D)$ to prove that

$$\cos A + \cos B = 2 \cos \frac{A+B}{2} \cos \frac{A-B}{2}.$$

Hence find, in terms of π, the general solution of the equation

$$\cos 5\theta + \cos \theta = \cos 3\theta.$$

Using both the identity for $\cos A + \cos B$, and the corresponding identity for $\sin A - \sin B$, show that

$$\sin 5\alpha - \sin \alpha = 2 \sin \alpha (\cos 4\alpha + \cos 2\alpha).$$

The triangle PQR has angle $QPR = \alpha$ (which is not zero), angle $PQR = 5\alpha$ and $RP = 3RQ$. Show that $\sin 5\alpha = 3 \sin \alpha$ and deduce that

$$\cos 4\alpha + \cos 2\alpha = 1.$$

By solving a quadratic equation in $\cos 2\alpha$, or otherwise, find the value of α, giving your answer to the nearest one tenth of a degree. (AEB)

3.3 (i) Given that $\sin x + \cos y = \sqrt{3}$, $\cos x + \sin y = 1$, show that $\sin(x + y) = 1$. Hence find the general solution of the given simultaneous equations.
(ii) The roots of the equation $3 \sin 2x + 6 \sin^2 x - 4 = 0$ in the interval $[0, \pi]$ are x_1 and x_2. Without solving the equation, evaluate $\tan(x_1 + x_2)$. (L(B))

3.4 Three spheres are placed on a horizontal plane in such a way that each sphere is touching the other two. The radii of the spheres are 2 cm, 3 cm and 4cm; their centres are A, B, C, respectively; the points of contact with the horizontal plane are p, Q, R, respectively.
(i) Show that $\cos A\hat{C}B = \frac{5}{7}$.
(ii) Show that $PQ = \sqrt{24}$ cm, and find the lengths of QR and RP.
(iii) Show that $\sin P\hat{R}Q = \sqrt{(\frac{47}{96})}$. (C (C))

3.5 A sector of a circle of radius r cm is bounded by the radii OP, OQ and the arc PQ. The angle POQ is θ radians. Given that r and θ vary in such a way that the area of the sector POQ has a constant value of 100 cm^2, obtain a relation between r and θ. Verify that when $r = 10$, $\theta = 2$.

Given that the radius increases at a constant rate of 0.5 cm s^{-1}, find the rate at which the angle POQ is decreasing when the radius is 10 cm. (JMB)

3.6 Show that $\dfrac{d}{dr}[\sin^{-1}(a/r)] = -\dfrac{a}{r\sqrt{(r^2 - a^2)}}$.

A rectangular field has sides of length 10 m and 20 m. A goat is tethered to a corner of the field by an inelastic rope of length r m, where $10 < r < 20$. Show that the goat has access to an area A m^2 of the field, where

$$A = 5\sqrt{(r^2 - 100)} + \tfrac{1}{2}r^2 \sin^{-1}(10/r).$$

Show that $\dfrac{dA}{dr} = r \sin^{-1}(10/r)$.

Apply the Newton-Raphson procedure once to the equation $A - 100 = 0$, with a starting value of $r = 10$, to show that the goat has access to one half of the area of the field when r is approximately equal to 11.4. (L(B))

4.1 (i) Sketch the graph of $f(x) = \dfrac{1}{4 - x^2}$.

(ii) By using partial fractions, find $\displaystyle\int \frac{dx}{4 - x^2}$, $|x| < 2$.

(iii) Show that $\displaystyle\int_3^4 \frac{dx}{4 - x^2} = -\tfrac{1}{4} \ln \tfrac{5}{3}$

and by reference to the graph of f explain the negative sign in the answer.

(iv) Given $|x| < 2$, use the substitution $x = 2 \sin \theta$ to find

$$\int \frac{dx}{4 - x^2} \text{ in terms of } \theta.$$

(You may quote $\int \sec \theta \, d\theta = \ln(\sec \theta + \tan \theta) + C$.)

(v) Show that $\sec\theta = \dfrac{2}{\sqrt{(4-x^2)}}$, and hence that your solutions to parts (ii) and (iv) are equivalent to each other. (SMP)

4.2 Let $y = \sin^{-1}(\sqrt{x})$, where $0 \leq x \leq 1$. Express $\dfrac{dy}{dx}$ in terms of x, and show that $\dfrac{dy}{dx} \geq 1$ for $0 < x < 1$.

Sketch the graph of y.
By considering your sketch, show that
$$\int_0^1 \sin^{-1}(\sqrt{x})\,dx + \int_0^{\frac{1}{2}\pi} \sin^2 y\,dy = \tfrac{1}{2}\pi$$
and hence or otherwise evaluate $\int_0^1 \sin^{-1}(\sqrt{x})\,dx$. (C)

4.3 Evaluate $\int_0^{\pi} e^x \sin x\,dx$ approximately
(i) by using the trapezium rule with ordinates at $x = 0, \tfrac{1}{4}\pi, \tfrac{1}{2}\pi, \tfrac{3}{4}\pi, \pi$
(ii) by using Simpson's rule with the same five ordinates.
Sketch the graph of $y = e^x \sin x$ for $0 \leq x \leq \pi$.
Without further calculation, but with reference to your sketch, explain why both the trapezium rule and Simpson's rule (with three ordinates in each case) would give good approximations to $\int_0^{\frac{1}{2}\pi} e^x \sin x\,dx$. Explain also why the approximation to $\int_0^{\pi} e^x \sin x\,dx$ calculated by Simpson's rule in (ii) above is considerably more accurate than the trapezium rule approximation calculated in (i). (C)

4.4 Given that $f(x) \equiv \dfrac{1}{x(1+x^2)}$, find $\int f(x)\,dx$
(i) by first expressing $f(x)$ in partial fractions,
(ii) by using the substitution $x = \tan\theta$.
Hence, or otherwise, show that, for $x > 0$, the solution of the differential equation
$$\dfrac{dy}{dx} = yf(x)$$
with $y = 1/\sqrt{2}$ when $x = 1$, may be expressed in both of the forms $y = \sin(\tan^{-1} x)$ and $y = x/(1+x^2)^{1/2}$. (L(B))

4.5 Given that k is a positive constant, find the general solution of the differential equation $\dfrac{d\theta}{dt} = k(1000 - \theta)$. (Your solution should involve a new arbitrary constant.)

A block initially at 40°C is put into an oven at time $t = 0$. The oven is kept at a constant temperature of 1000°C. At any subsequent instant the rate of increase of the temperature θ of the block is proportional to the difference in temperature that exists between the block and the oven. If after 1 minute the temperature of the block has risen to 160°C, show that at time t minutes ($t > 0$) the block has temperature θ given by $\theta = 1000 - 960(\tfrac{7}{8})^t$. (SMP)

4.6 An article to be chemically treated is placed at time $t = 0$ in a bath of water in which a kilograms of a chemical have been dissolved. When there are x kilograms of chemical in solution the article absorbs the chemical at a rate $k(x-b)$ kilograms per second, where k and b are positive constants and $b < a$. Write down a differential equation expressing the relation between x and t, and find x in terms of a, b, k and t.

The treatment is complete when the article has absorbed c kilograms of the chemical, where $c < a - b$. Show that the time, T_1 seconds, taken for the completion of the treatment is given by
$$T_1 = \dfrac{1}{k}\ln\dfrac{a-b}{a-b-c}.$$

In an improved process which starts from the same initial state, the bath is fed continuously with the chemical just sufficiently to maintain the amount of chemical in the water at a kilograms throughout the process. Find the time, T_2 seconds, taken to complete the treatment by this method, and show that $kT_2 = 1 - e^{-kT_1}$. (JMB)

5.1 The first, second and third terms of an arithmetic series are p, q, p^2 respectively, where $p < 0$. The first, second and third terms of a geometric series are p, p^2, q respectively.
(i) Show that $p = -\tfrac{1}{2}$ and find the value of q.
(ii) Find the sum to infinity of the geometric series.
(iii) Find the seventeenth term of the arithmetic series. (AEB)

5.2 (i) A geometric series has first term 1 and common ratio r. Given that twice the sum of the first and fourth terms is equal to three times the sum of the second and third terms, and that $r \neq -1$, find the two possible values of r.

Taking that value of r for which $|r| < 1$, state the sum to infinity of the series.
(ii) Expand $(1+2x)^{\frac{1}{2}}$ in a series of ascending powers of x up to and including the term in x^3, simplifying the coefficients. State the set of values of x for which the expansion is valid. (C)

5.3 Show that $n(n+1)(n+2) - (n-1)n(n+1) = 3n(n+1)$.
Hence or otherwise, find a formula for $\sum_1^n r(r+1)$. Use this result to prove that $\sum_1^n r^2 = \tfrac{1}{6}n(n+1)(2n+1)$. Hence or otherwise, find a formula for the sum to n terms of
$$4^2 + 7^2 + 10^2 + 13^2 + \cdots. \text{ (O)}$$

5.4 (i) Given that $(1.01)^{30} = 1.3478$ to 5 significant figures, show that the magnitude of the error in using the first three terms of the binomial expansion of $(1+x)^{30}$ to estimate the value of $(1.01)^{30}$ is less than 0.01. Find also the numerical value of the coefficient of x^5 and x^{25} in this expansion.
(ii) A test consists of 30 questions and each question has only two possible answers, just one of which is correct. Calculate the number of different ways in which it is possible to answer (a) exactly 5 questions correctly (b) exactly 25 questions correctly. (L(D))

5.5 Write down the expansion of $(1+x^2)^{\frac{1}{2}}$, in ascending powers of x up to the term in x^4. Hence, by integrating the series term by term, find an approximate value of $\int_0^{1/2} \sqrt{(1+x^2)}\,dx$, giving your answer to 3 decimal places.
Show that Simpson's rule with two equal intervals leads to the same approximation for $\int_0^{1/2} \sqrt{(1+x^2)}\,dx$. (AEB)

5.6 Alf repeatedly attempts to throw a six with a fair six-sided die. Find the probability that he first throws a six on his third attempt.

Ben has a similar die and also attempts to throw a six. Find the probability that:
(i) Alf and Ben take the same number of attempts to throw a six,
(ii) Alf throws a six in fewer attempts than Ben. (Answers may be given as fractions in their lowest terms.) (O&C)

5.7 In a school 45% of the pupils are girls and 55% are boys. One-fifth of the girls are left-handed, and one-tenth of the boys are left-handed. What is the probability that a pupil selected at random will be left-handed?

A pupil sends an anonymous letter to the head. If it could be proved that it was written by a left-handed person, what would be the conditional probability that the writer is a boy? (SMP)

5.8 (i) A group of 8 people consists of 6 adults and 2 children. Calculate the number of ways in which 5 people may be selected from these 8 in each of the following cases:
(a) if order of selection is unimportant (b) if the 5 people are selected in a definite order (c) if order of selection is unimportant and if at most one child may be selected.
(ii) Two events X and Y are not independent, and their respective probabilities of occurring are given by $P(X) = 0.6$ and $P(Y) = 0.7$. It is also given that $P(X \cup Y) = 0.95$. Calculate (a) $P(X \cap Y)$, (b) $P(Y|X)$, (c) $P(X \cup Y')$. (C)

5.9 Three archers A, B and C shoot arrows at a target. Their probabilities of hitting the target are p_a, p_b and p_c respectively. They each shoot an arrow in turn, first A followed by B and then C, continuing in rotation until an arrow hits the target, at which point the contest ceases. Show that the probability that A hits the target first is

$$p_a / \{1 - (1-p_a)(1-p_b)(1-p_c)\}.$$

Find the probability that
(i) B hits the target first
(ii) C hits the target first.
It is known that these three probabilities are the same. It is also known that the probability that none of the first three arrows hits the target is $\frac{1}{2}$. Find p_a, p_b and p_c. (Answers may be given as fractions in their lowest terms.) (O&C)

6.1 A circle has centre (α, β). Show that if (α, β) lies on the line $2x = 3y + 7$ then the equation of the circle may be written in the form

$$x^2 + y^2 - (3\beta + 7)x - 2\beta y = k.$$

If the circle also passes through the points $(1, 4)$ and $(2, -3)$, find its equation and show that the circle touches the y-axis.
Calculate the length of a tangent from the point $(12, 6)$ to the circle. (AEB)

6.2 A curve is defined by the equations $x = t^2 - 1$, $y = t^3 - t$, where t is a parameter.
Sketch the curve for all real values of t. Find the area of the region enclosed by the loop of the curve. (L(B))

6.3 The line $y = mx + 5a$ cuts the parabola, given by the parametric equations $x = a(t^2+1)$, $y = 2a(2t+1)$, in the distinct points P and Q. Show that the parameters of P and Q are the roots of the equation $mt^2 - 4t + (m+3) = 0$.

Deduce the range of possible values of m. Hence or otherwise find the equations of the tangents to the parabola from the point $(0, 5a)$. (AEB)

6.4 A curve is defined by the parametric equations

$$x = t, \ y = \frac{1}{t}, \ t \neq 0.$$

Sketch the curve, and find the equation of the tangent to the curve at the point $\left(t, \frac{1}{t}\right)$.

The points P and Q on the curve are given by $t = p$ and $t = q$, respectively, and $p \neq q$. The tangents at P and Q meet at R. Show that the coordinates of R are

$$\left(\frac{2pq}{p+q}, \frac{2}{p+q}\right)$$

(i) Given that p and q vary so that $pq = 2$, state the equation of the line on which R moves.
(ii) Given that p and q vary so that $p^2 + q^2 = 1$, show that the point R moves on the curve $y(2x+y) = k$, where k is a constant. State the value of k. (JMB)

6.5 The position vectors with respect to a point O of four distinct points A, B, D, E are $\mathbf{a}$, $\mathbf{b}$, $p\mathbf{a}$, $q\mathbf{b}$ respectively, where $\mathbf{a}$, $\mathbf{b}$ are non-parallel vectors and p, q are unequal scalars. Show that any point on AB has position vector given by $(1-t)\mathbf{a} + t\mathbf{b}$, where t is a scalar.
The lines AB, DE intersect at L. Show that

$$\overrightarrow{OL} = \frac{1}{p-q}[p(1-q)\mathbf{a} + q(p-1)\mathbf{b}].$$

Two further points C, F have position vectors $\mathbf{c}$ and $r\mathbf{c}$ respectively, where $\mathbf{c}$ is not parallel to $\mathbf{a}$ or to $\mathbf{b}$ and r is a scalar not equal to p or to q. The lines CB, FE intersect at N and CA, FD intersect at N. Write down $\overrightarrow{OM}$ and $\overrightarrow{ON}$ in terms of $\mathbf{a}$, $\mathbf{b}$, $\mathbf{c}$, p, q, r.
Taking the values $p = 4$, $q = 3$, $r = 2$, (i) show that L, M, N are collinear points (ii) if $\mathbf{a}$, $\mathbf{b}$ are the perpendicular unit vectors $\mathbf{i}$, $\mathbf{j}$ respectively and $\mathbf{c} = \mathbf{i} + 6\mathbf{j}$, find the acute angle between LMN and OAD. (O & C)

6.6 $\overrightarrow{OA}$ ($= \mathbf{a}$) and $\overrightarrow{OB}$ ($= \mathbf{b}$) are the position vectors of two points A and B referred to an origin O. Describe geometrically the position of the point C where

$$\overrightarrow{OC} = \mathbf{c} = \frac{\alpha \mathbf{a} + \beta \mathbf{b}}{\alpha + \beta},$$

where α and β are scalars.
A, B, C, D are four points in general positions. P, Q, R, S lie respectively on AD, AC, CB, DB where

$$\frac{AP}{PD} = \frac{AQ}{QC} = \frac{1-m}{m} \ \text{and} \ \frac{CR}{RB} = \frac{DS}{SB} = \frac{1-n}{n}.$$

Show that, if $PQRS$ is a parallelogram, then $m + n = 1$.

If D is taken as origin, and
$$\overrightarrow{DA} = 3\mathbf{i}, \overrightarrow{DB} = 3\mathbf{k} - \mathbf{i}, \overrightarrow{DC} = 3\mathbf{j},$$
and if
$$\frac{AP}{PD} = \frac{AQ}{QC} = \frac{BR}{RC} = \frac{BS}{SD} = \frac{1-m}{m}$$
write down the position vectors of P, Q, R and S in terms of $\mathbf{i}$, $\mathbf{j}$ and $\mathbf{k}$ and m. Show that
$$\overrightarrow{RP} = 4m\mathbf{i} + (3m-3)\mathbf{j} + (-3m)\mathbf{k}$$
and obtain a similar expression for $\overrightarrow{SQ}$. (O)

6.7 The lines L and M have the equations
$$\mathbf{r} = (3+s)\mathbf{i} + (2+3s)\mathbf{j} + (4-5s)\mathbf{k} \text{ and}$$
$$\mathbf{r} = (-3+t)\mathbf{i} + (4-2t)\mathbf{j} + (6+2t)\mathbf{k}$$
respectively. The plane π has the equation
$$\mathbf{r} \cdot (2\mathbf{i} - \mathbf{k}) = 16$$
(i) Verify that the point A with coordinates $(1, -4, 14)$ lies on L and on M but not on π.
(ii) Find the position vector of the point of intersection B of L and π.
(iii) Show that M and π have no common point.
(iv) Find the cosine of the angle between the vectors
$$\mathbf{i} + 3\mathbf{j} - 5\mathbf{k} \text{ and } 2\mathbf{i} - \mathbf{k}$$
Hence find, to the nearest degree, the angle between L and π. (JMB)

6.8 Given that $z_1 = 1 - i$ and $z_2 = -1 + i\sqrt{3}$, mark on an Argand diagram the points P_1 and P_2 which represent z_1 and z_2, respectively.
Find $|z_1|$ and $|z_2|$ and write down $|z_1 z_2|$ in surd form.
Find also arg z_1 and arg z_2 and write down arg $z_1 z_2$, giving each argument (in terms of π) between $-\pi$ and π.
Use the given forms of z_1 and z_2 to find $z_1 z_2$ in the form $a + ib$. Deduce that $\cos \frac{5\pi}{12} = \frac{\sqrt{3}-1}{2\sqrt{2}}$. (JMB)

6.9 (i) Prove that $-1 + 2i$ is a root of $3x^3 + 5x^2 + 13x - 5 = 0$, and hence solve the equation completely.
(ii) Express $1 + \sqrt{3}i$ in the form $r(\cos\theta + i\sin\theta)$ where $r > 0$. Hence or otherwise express $(1 + \sqrt{3}i)^9$ in the form $a + ib$ where a and b are real.
(iii) Find the real part of the complex number
$$\frac{1 + \cos\varphi - i\sin\varphi}{1 - \cos\varphi - i\sin\varphi},$$
simplifying your answer as far as possible. (O)

7.1 A particle moves along a straight line ABC. The particle starts from rest at A and moves from A to B with constant acceleration $2f$. It then moves from B to C with acceleration f and reaches C with speed V. The times taken in the motions from A to B and from B to C are each equal to T. Find T in terms of V and f. Show that $AB = \frac{2}{5}BC$. (JMB)

7.2 A bead moves on a very long straight horizontal wire which is lightly greased. When a current is passed through the wire to melt the grease the retardation of the bead is $ba^2 e^{-\frac{1}{2}at}$, where a and b are positive constants and t is the time that has elapsed since the current was switched on. At time $t = 0$, when the current is switched on, the bead is projected with a speed $u (> 2ab)$ from a point O on the wire. Determine
(i) the subsequent speed of the bead at time t, and show that this is always greater than $u - 2ab$
(ii) the displacement of the bead from O at time t.
If, instead, the retardation of the bead is equal to av, where v is its speed, and the bead is again projected with a speed u from O at $t = 0$, determine
(iii) the speed of the bead at time t
(iv) the time taken for the speed to reduce to $\frac{1}{2}u$. (AEB)

7.3 Show in a sketch the vector $\mathbf{i}\cos\theta - \mathbf{j}\sin\theta$, where $0 < \theta < \frac{\pi}{2}$ and $\mathbf{i}$ and $\mathbf{j}$ are unit vectors parallel to the x- and y-axes respectively. Deduce that this is a unit vector.
Show on your diagram a unit vector perpendicular to this and express it in the form $a\mathbf{i} + b\mathbf{j}$.
A particle P describes a circle of radius $2r$ about the origin with constant angular speed 3ω and at time $t = 0$ P is at the point $(2r, 0)$ and is moving in the direction of decreasing y. Express $\overrightarrow{OP}$, at time t, in the form $a\mathbf{i} + b\mathbf{j}$. State the magnitudes of both the velocity and acceleration of P in the form $a\mathbf{i} + b\mathbf{j}$.
A second particle Q has position vector given by $\overrightarrow{OQ} = r(\mathbf{i}\sin\omega t + \mathbf{j}\cos\omega t)$. Obtain, in its simplest possible form, an expression for PQ^2. (AEB)

7.4 A cyclist A is travelling with a constant velocity of 10 kmh^{-1} due east and a cyclist B has a constant velocity of 8 kmh^{-1} in a direction arctan $(4/3)$ east of north. At noon, B is 0.6 km due south of A. Taking the position of B at noon as the origin and $\mathbf{i}$ and $\mathbf{j}$ as unit vectors due east and due north respectively, obtain expressions for the position vectors of A and B at time t hours after noon. Hence find, at time t hours after noon
(i) the position vector of A relative to B
(ii) the velocity of A relative to B.
Show that the cyclists are nearest together at time 4.8 minutes after noon and that, at this time, the distance between them is 360 m. (L(B))

7.5 At time t two points P and Q have position vectors $\mathbf{p}$ and $\mathbf{q}$ respectively, where
$$\mathbf{p} = 2a\mathbf{i} + (a\cos\omega t)\mathbf{j} + (a\sin\omega t)\mathbf{k}$$
$$\mathbf{q} = (a\sin\omega t)\mathbf{i} - (a\cos\omega t)\mathbf{j} + 3a\mathbf{k}$$
and a, ω are constants. Find $\mathbf{r}$, the position vector of P relative to Q, and $\mathbf{v}$, the velocity of P relative to Q. Find also the values of t for which $\mathbf{r}$ and $\mathbf{v}$ are perpendicular.
Determine the smallest and greatest distances between P and Q. (L(B))

7.6 A particle P projected from a point O on level ground strikes the ground again at a distance of 120 m from O after a time 6 s. Find the horizontal and vertical components of the initial velocity of P. The particle passes through a point Q whose horizontal displacement from O is 30 m.

Find
(i) the height of Q above the ground
(ii) the tangent of the angle between the horizontal and the direction of motion of P when it is at Q and also the speed of P at this instant
(iii) the horizontal displacement from O of the point at which P is next at the level of Q. (AEB)

7.7 A particle is projected with speed V m s^{-1} at an angle of elevation θ from a point O on a horizontal plane, and it moves freely under gravity. Denoting the horizontal and vertical displacements of the particle from O at any subsequent time during the flight by x m and y m, show that

$$y = x\tan\theta - \frac{gx^2}{2V^2}(1+\tan^2\theta).$$

Given that $V=50$ and $x=200$, and taking the acceleration due to gravity to be 10 m s^{-2}, show that

$$y = 45 - 80(\tan\theta - \tfrac{5}{4})^2.$$

The particle has to pass over a vertical wall 25 m high at a horizontal distance 200 m from O. Deduce
(i) the greatest distance by which the particle can clear the top of the wall
(ii) the possible values of $\tan\theta$ if the particle *just* clears the top of the wall. (C)

7.8 A particle is projected at an angle of elevation α from a point A on horizontal ground. When travelling upwards at an angle β to the horizontal, the particle passes through a point B. The line AB makes an angle θ with the horizontal. Show that $2\tan\theta = \tan\alpha + \tan\beta$.
The point B is at a horizontal distance of 30 m from point A and is at a height of 20 m above ground. At B, the particle is travelling upwards at an angle $\arctan(1/3)$ to the horizontal. Find the angle of projection and the initial speed of the particle. (L(C))

8.1 A car (of total mass m) is travelling at steady speed u, when the vehicle ahead stops suddenly, and the car must pull up within a distance d to avoid a collision. With a brake fully applied, the car is subject to a total resistance $h+kv^2$ per unit mass (where v is the speed and h, k are constants). Briefly explain how the differential equation

$$mv\frac{dv}{dx} = -m(h+kv^2) \quad (1)$$

is relevant to the situation.
Integrate (1) and hence show that a collision is inevitable unless $u < u_0$ where $u_0^2 = \frac{h}{k}(e^{2kd}-1)$.
Given that $u=20$ and that the values of h, k and d are such that $h=4k$ and $e^{2kd}=50$, show that a collision will occur immediately and that the numerical value of the car's speed immediately before the collision is approximately 2. (Standard units are used throughout this question.) (SMP)

8.2 A smooth plane and a rough plane, both inclined at 45° to the horizontal, intersect in a fixed horizontal ridge. A particle P of mass m is held on the smooth plane by a light string which passes over a small smooth pully A on the ridge, and is attached to a particle Q of mass $3m$ which rests on the rough plane. The plane containing P, Q and A is perpendicular to the ridge. The system is released from rest with the string taut. Given that the acceleration of each particle is of magnitude $g/(5\sqrt{2})$, find
(i) the tension in the string
(ii) the coefficient of friction between Q and the rough plane
(iii) the magnitude and direction of the force exerted by the string on the pulley. (L(D))

8.3 A particle P of mass $8m$ rests on a smooth horizontal rectangular table and is attached by light inelastic strings to particles Q and R of mass $2m$ and $6m$ respectively. The strings pass over light smooth pulleys on opposite edges of the table so that Q and R can hang freely with the strings perpendicular to the table edges. The system is released from rest. Obtain the equations of motion of each of the particles and hence, or otherwise, determine the magnitude of their common acceleration and the tensions in the strings.
After falling a distance x from rest the particle R strikes an inelastic floor and it is brought to rest. Determine the further distance y that Q ascends before momentarily coming to rest. [It is to be assumed that the lengths of the strings are such that P remains on the table and Q does not reach it.] (AEB)

8.4 A light elastic string, of natural length l and modulus $3mg$, has one end A attached to a fixed point O on a smooth horizontal table. A particle of mass m is attached to the other end of the string. The particle moves in a horizontal circle, centre O, so that the string rotates with constant angular velocity $\sqrt{(2g/l)}$. Find the radius of the circle.
The end A of the string is now raised vertically a distance $\tfrac{1}{2}l$ and held fixed. The particle P moves on the table in a circle, with OP having constant angular velocity $\sqrt{(g/l)}$. Show that the radius of the circle is $l/\sqrt{2}$ and that the tension in the string is $\tfrac{3}{2}mg$. Find the magnitude of the reaction between the particle and the table. (C(C))

8.5 Prove that when a particle moves with constant angular speed ω in a circle of radius r the acceleration towards the centre of the circle is $\omega^2 r$.
A particle P of mass m is attached to one end of a light elastic string of natural length a and whose other end is fixed. In the equilibrium position with the string vertical the extension is $\tfrac{1}{4}a$. The string is then moved out of the vertical position and P is projected so that it describes a horizontal circle with constant angular speed and period T. The length of the string during this motion is $a+x$. Write down the equations of motion of P and show that the motion is only possible for $x > \tfrac{1}{4}a$.
Obtain expressions in terms of x, a and g for the angular speed and hence find T. (AEB)

8.6 Show that the potential energy of a light elastic string of modulus λ and natural length l, extended to a length $l+y$, is

$$\frac{\lambda y^2}{2l}.$$

One end of a light elastic string of natural length l and modulus $4mg$ is attached to a fixed point O. The other end is attached to a particle P of mass m. The particle is projected vertically downwards from O with speed $\sqrt{(4gl)}$. By using the principle of conservation of energy, or otherwise, find the speed of P when P is at depth x below O, where $x > l$, and show that the greatest depth below O attained by P is $5l/2$. Find also the maximum value of the speed.

Show that the particle subsequently rises to a maximum height $3l/2$ above O. (JMB)

8.7 (i) A small bead of mass m is threaded on a smooth wire in the form of a circle of radius a. The wire is fixed in a vertical plane and a light elastic string of natural length a and modulus $2mg$ joins the bead to the highest point of the wire. The bead is released from rest in a position where the string is just taut, and in the subsequent motion it passes through the lowest point of the wire with speed U. Using the principle of conservation of energy, or otherwise, find U in terms of g and a.
(ii) A car of mass M is travelling up a straight road inclined at an angle α to the horizontal. The engine of the car is working at a rate P. At an instant when the speed is v and the acceleration is f the resistance to motion due to frictional forces is kMg, where k is a positive constant. Express P in terms of M, g, v, f, k and α. (C(C))

8.8 A motor-cyclist and machine have total mass 100 kg. The frictional resistance to motion is $(a + bv)$ N, where a and b are constants and v m s^{-1} is the speed. When the engine is working at a rate of 2100 W, the maximum speed on the level is 15 m s^{-1} and the maximum speed up a hill of inclination arc sin $(\frac{1}{10})$ is 10 m s^{-1}. Taking the acceleration due to gravity to be 10 m s^{-2}, find a and b. What is the maximum speed *down* a hill of inclination arcsin $(\frac{1}{20})$ if the engine works at the same rate?

At the instant when the speed is 15 m s^{-1}, the motor-cyclist begins to *descend* a hill of inclination arcsin $(\frac{1}{20})$. Assuming that the engine still works at 2100 W, show that the differential equation giving the speed v m s^{-1} at time t s after beginning the descent is $50v\dfrac{dv}{dt} = 3(350 - v^2)$.
Solve this equation to give v^2 in terms of t.
Show that, at time $t = 10$, the speed is about 94% of the maximum speed down this hill. (O&C)

8.9 Three equal smooth spheres A, B and C lie, in that order, at rest on a smooth horizontal table, with their centres collinear. Sphere A is projected with speed u to strike B directly. Sphere B, in turn, collides directly with sphere C. The coefficient of restitution for any impact between the spheres is e. Show that the speeds of A and B after the first collision are $\dfrac{u(1-e)}{2}$ and $\dfrac{u(1+e)}{2}$ respectively. Hence determine the speeds of B and C after the second collision. Determine the condition that e must satisfy in order that there is not a third collision.

When $e = \frac{1}{2}$ show that the speeds of B and C after the second collision are $\dfrac{3u}{16}$ and $\dfrac{9u}{16}$ respectively and prove that the third collision is the last one. (AEB)

8.10 (i) On a smooth horizontal table, n identical particles lie at rest in a straight line. Each particle has mass m and the distance between each pair of adjacent particles is d. The particle at one end of the line is projected with speed u directly towards the others. In each collision that occurs the two particles which are involved coalesce to form a single particle. Find expressions for the time that elapses between the instant of projection and the last collision, and for the total loss of kinetic energy in the $(n-1)$ collisions.
(ii) A particle of mass $(m_1 + m_2)$ is at rest when an internal explosion causes it to split into two particles, of masses m_1 and m_2 moving with speeds u_1 and u_2 respectively. Assuming that the total kinetic energy of the two particles after the explosion is E, find expressions for u_1 and u_2 in terms of m_1, m_2 and E. Find also an expression for the magnitude of the impulse on the particle of mass m_1 when the explosion occurs. (C)

8.11 A particle P, of mass m, is projected from a point O with speed u at an angle of inclination α to the horizontal. Find the time taken for P to reach the highest point A of its path.

When P is at A it collides directly, and coalesces, with a particle Q, of mass $3m$. Just before impact Q was moving horizontally with the same speed as P but in the opposite direction. Find the speed of the composite particle immediately after impact and show that the loss of kinetic energy is $\frac{3}{2}mu^2\cos^2\alpha$.

Given that the composite particle meets the horizontal plane through O at a point B, express OB in terms of u, α and g. (L(B))

9.1 Forces $\mathbf{F}_1$ and $\mathbf{F}_2$ act at points whose position vectors, relative to a fixed origin, are $\mathbf{r}_1$ and $\mathbf{r}_2$ respectively, where

$$\mathbf{F}_1 = (4\mathbf{i} + 3\mathbf{j})\text{N} \qquad \mathbf{r}_1 = (2\mathbf{i} + \mathbf{j})\text{m}$$
$$\mathbf{F}_2 = (6\mathbf{i} - 3\mathbf{j})\text{N} \qquad \mathbf{r}_2 = (-3\mathbf{j})\text{m}$$

Determine the resultant of these forces.
Show that the position vector of the point of intersection of the lines of action of the forces $\mathbf{F}_1$ and $\mathbf{F}_2$ is $(-2\mathbf{i} - 2\mathbf{j})$m and hence write down a vector equation of the line of action of the resultant of these forces.
A third force $\mathbf{F}_3 = (-6\mathbf{i} - 5\mathbf{j})$N acting at the point with position vector $\mathbf{r}_3 = (-2\mathbf{i} - 2\mathbf{j})$m is now added to the system. When the system of forces, $\mathbf{F}_1, \mathbf{F}_2, \mathbf{F}_3$, is applied to a lamina it is found that equilibrium can be maintained by applying a force $\mathbf{F}_4$ at the point with position vector $(-\mathbf{i} + 2\mathbf{j})$m together with a couple G. Find $\mathbf{F}_4$ and G, stating whether G is clockwise or anti-clockwise. (AEB)

9.2 A uniform rod AB has weight W and length $2a$. It rests in limiting equilibrium with A in contact with a rough vertical wall, B being attached to one end of a light string whose other end is fixed at a point O of the wall vertically above A. The rod is horizontal and $A\hat{O}B = 60°$.
Find the coefficient of friction between the wall and the rod.
The string is elastic, having modulus $3W$. Find its natural length. (O&C)

9.3 A *light* aluminium ladder AB of length $6a$ rests with one end A in contact with rough horizontal ground; a point C of the ladder is in contact with the top of a vertical wall of height $4a$ (so that the ladder projects over the wall), the foot A of the ladder being at a distance $3a$ from the wall. The vertical plane containing the ladder is perpendicular to the wall. A man of weight W stands on the ladder at a distance x from A, the system being in equilibrium. Assuming that the contact at C is smooth, show (by taking moments about A, or otherwise) that the force exerted by the wall on the ladder is $\dfrac{3}{25}\dfrac{Wx}{a}$.

Find also R, the normal component of reaction, and F, the frictional force, at A in terms of W, a and x.

The man can reach C without the ladder slipping; find the least possible value for the coefficient of friction at A. (O&C)

9.4 A particle of mass M is attached to the end A, and a particle of mass $3M$ is attached to the end B, of a light rod AB of length $2a$. The rod hangs from a point O to which it is attached by light inextensible strings OA, OB, each of length $a \sec \alpha$. Prove that in equilibrium the inclination of the rod to the horizontal is $\tan^{-1}(\tfrac{1}{2}\cot \alpha)$, and find the tension in the string OA. (O)

9.5 A toy top is constructed by joining, at their circular rims, the bases of a solid uniform hemisphere of base radius a and a solid uniform right circular cone of base radius a and height h. The density of the hemisphere is b times that of the cone.

(i) Show that the centre of gravity of the top is at a distance
$$\frac{3h^2 + b(3a^2 + 8ah)}{4(h + 2ab)}$$
from the vertex of the cone.

(ii) The top, when suspended from a point on the rim of the base of the cone, rests in equilibrium with the axis of the cone inclined at an acute angle, α, to the downward vertical. Find $\tan \alpha$.

(iii) For the case $h = 4a$ determine the range of values of b such that the top cannot rest in equilibrium with the slant surface of the cone in contact with a smooth horizontal plane. (AEB)

9.6 Prove that the centre of mass of a uniform right circular cone of semi-vertical angle α and height h is on the axis of the cone at a distance $\tfrac{3}{4}h$ from the vertex.

Such a cone is supported with its base area in contact with a smooth vertical wall by means of a light inelastic string joining the vertex of the cone to a point on the wall vertically above the centre of the base. Find the maximum possible length of the string. (O)

9.7 The curved surface of a uniform solid hemisphere is in contact with a horizontal floor and a vertical wall. The plane surface of the hemisphere makes an angle α (>0) with the horizontal. Show that equilibrium is
(i) impossible if the floor is smooth,
(ii) possible if the floor is rough and the wall is smooth provided that $8\mu \geq 3 \sin \alpha$, where μ is the coefficient of friction between the floor and the hemisphere.

Given that the floor and the wall are rough and the coefficient of friction between the hemisphere and both the wall and the floor is $\tfrac{1}{8}$, find the value of $\sin \alpha$ when equilibrium is limiting at both points of contact and the hemisphere is on the point of sliding down the wall. (L(B))

Answers

Chapter 1

Exercise 1.1:2

1 (i) T (ii) F (iii) F (iv) F (v) F (vi) F
2 (i) F (ii) F (iii) T
4 (i) (a) $\phi, \{a\}, \{b\}, \{c\}, \{a,b\}, \{a,c\}, \{b,c\}, A$
(b) $\phi, \{a\}, \{b\}, \{c\}, \{d\}, \{a,b\}, \{a,c\}, \{a,d\}, \{b,c\}, \{b,d\}, \{c,d\}, \{a,b,c\}, \{a,b,d\}, \{a,c,d\}, \{b,c,d\}, A$ (ii) 2^k
5 (i) 2, 4, 6, 8, 10 (ii) 3, 6, 9 (iii) 1, 4, 9 (iv) 7, 8, 9, 10 (v) 1, 2, 3 (vi) ϕ; 5, 3, 3, 4, 3, 0; 1, 3, 5, 7, 9; 1, 2, 4, 5, 7, 8, 10; 2, 3, 5, 6, 7, 8, 10; 1, 2, 3, 4, 5, 6; 4, 5, 6, 7, 8, 9, 10; ε
6 (i) (a) $\{2, 5\}$ (b) $\{2\}$ (c) $\{2, 5, 6\}$ (d) $\{2, 5, 6, 7, \pi,$ Ian Botham, Elizabeth Taylor's third marriage, World War III, famine in Ethiopia, the rings of Saturn, 8, 9, etc., etc.$\}$ (ii) (a) ϕ (b) ε (c) A
7 (i) $\{a, b, d, e\}$ (ii) $\{b, d\}$ (iii) $\{c, e\}$ (iv) $\{a, c\}$ (v) $\{e\}$ (vi) $\{a\}$ (vii) $\{c\}$ (viii) $\{a, c, e\}$ (ix) $\{c\}$
8 (i) $\{2, 3, 4, 5, 6, 7, 8, 9\}$ (ii) $\{4, 7\}$ (iii) $\{1, 2, 6, 8, 10\}$ (iv) $\{1, 3, 5, 9, 10\}$ (v) $\{1, 2, 4, 6, 7, 8, 10\}$ (vi) $\{1, 3, 4, 5, 7, 9, 10\}$ (vii) $\{1, 10\}$ (viii) $\{1, 2, 3, 5, 6, 8, 9, 10\}$ (ix) $\{1, 10\}$
9 (i) ε (ii) $\{e\}$ (iii) $\{a, b, c, d, f, g\}$ (iv) ϕ (v) B (vi) $\{d\}$ (vii) $\{a, b, c, e, g\}$ (viii) $\{a, c, d, f\}$
10 (i) (a) ϕ (b) ϕ
13 (i) A (ii) ϕ (iii) ε (iv) A (v) ϕ (vi) ε (vii) $n(\varepsilon)$
20 (i) $A \backslash B$ (ii) B' (iii) $(A \cup B) \cup (A' \cap B')$ × (iv) $(A \cap B) \cup (A' \cap B' \cap C)$×(v) $(A \cap B \cap C') \cup (A' \cap B' \cap C)$ (vi) $(A \cap B' \cap C) \cup (A' \cap B' \cap C) \cup (A' \cap B' \cap C)$ ∤ (viii) $(A \cap B \cap C) \cup (B \cup C)'$ ×
21 (i) $A \cup B$ (ii) $A \cap B$ (iii) $A \cap B$ (iv) ϕ (v) B' (vi) A
22 (i) $A \cup B \cup C$ (ii) $A \cup B \cup C'$ (iii) $A \cap B \cap C$ (iv) $A' \cap B' \cap C$ (v) $A \cup (B \cap C)$ (vi) $A \cap B \cap C'$

Exercise 1.1:3

1 (i) (a) $\frac{17}{20}$ (b) $\frac{5}{8}$ (c) $\frac{19}{40}$ (d) $\frac{15}{4}$ (e) $\frac{54}{25}$ (ii) (a) 1.25 (b) 1.625 (c) 2.175
2 (i) $\frac{5}{9}$ (ii) $\frac{27}{11}$
3 (i) (a) $\frac{2}{9}$ (b) $\frac{7}{9}$ (c) $\frac{58}{11}$ (d) $\frac{13}{45}$ (e) $\frac{51}{37}$ (g) $\frac{508}{111}$ (ii) (a) 1.6 (b) 0.4 (c) 1.142857 (d) 1.297
8 (i) (a) 2 (b) 0 (ii) (a) 1 (b) 1
11 (i) $\{2, 3, 4, 6, 7, 8\}$ (ii) $\{-2, -1, 0, 1, 2, 3\}$ (iii) $\{-1, 0, 1, 2\}$ (iv) $\{1, 2, 3\}$
13 (i) (0, 2) (ii) (−1, 5) (iii) (3, 5) (iv) (0, 6) (v) (−1, 6)
14 (i) $(-\infty, -2) \cup (3, \infty)$ (ii) $(-\infty, -4) \cup (1, \infty)$ (iii) $[-2, 1]$ (iv) $(-\infty, -4) \cup (3, \infty)$ (v) $[-4, 3]$

15 (ii) (a) $(-\infty, -5) \cup [4, \infty)$ (b) $(-\infty, 2] \cup (3, \infty)$ (c) $(2, 4)$ (d) $(-\infty, 5) \cup (6, \infty)$ (e) $[-5, 6]$ (g) $(-\infty, 2] \cup [4, \infty)$
16 (i) [4, 5) (ii) (2, 6) (iii) [5, 6)
(iv) (2, 4)
17 (i) $(2, 3] \cup [4, 5)$ (ii) (1, 6) (iii) $(-\infty, 1] \cup [6, \infty)$ (iv) (3, 4)

Exercise 1.1:4

1 (i) e (ii) i (iii) e (iv) i (v) i (vi) e (vii) e (viii) i
3 (i) $(x-1)(x-2)$ (ii) $(x+4)(x-3)$ (iii) $(5-x)(2+x)$ (iv) $(3x+1)(x+1)$ (v) $(3x+5)(2x-1)$ (vi) $(3x+8)(5x+3)$ (vii) $3x^2(x-2)$ (viii) $x(x-1)(x+5)$
4 (i) $(a-b)(a+b)$; no; $(a-b)(a+b)(a^2+b^2)$ (ii) (a) $(x-1)(x+1)$ (b) $(1-x)(1+x)$ (c) $(1-x)(7+x)$ (iii) (a) 400 (b) −621
5 (i) $(a-b)(a^2+ab+b^2)$ (ii) $(x-1) \times (x^2-x+1)$; $(x-1)(x^2+x+1)$ (iv) $(x-1) \times (x+1)(x^2-x+1)(x^2+x+1)$ (v) (a) $(x+3)(x^2-3x+9)$ (b) $(2x-1)(4x^2+2x+1)$ (c) $(\sqrt{2}x+\sqrt{3})(2x^2-\sqrt{6}x+3)$
6 (i) (a) when $a=0$ or $b=0$ (b) when $a=0$, $b=0$ or $a=-b$ (ii) $(1-x)(1+x+x^2+x^3)$; $(1-x)(1+x+x^2+x^3+x^4)$ (iii) $1-x^2 \equiv (1+x)(1-x)$; $(1+x^3) \equiv (1+x)(1-x+x^2)$; $1-x^4 \equiv (1+x)(1-x+x^2-x^3)$; $1-x^5 \equiv (1+x)(1-x+x^2-x^3+x^4)$ (iv) $\frac{31}{16}$ (ii) $\frac{21}{32}$
7 (ii) $a^4+4a^3b+6a^2b^2+4ab^3+b^4$; $a^5+5a^4b+10a^3b^2+10a^2b^3+5ab^4+b^5$ (vi) $a^6+6a^5b+15a^4b^2+20a^3b^3+15a^2b^4+6ab^5+b^6$; $a^7+7a^6b+21a^5b^2+35a^4b^3+35a^3b^4+21a^2b^5+7ab^6+b^7$ (vii) (a) $1+3x+3x^2+x^3$ (b) $1-3x+3x^2-x^3$ (c) $1+4x+6x^2+4x^3+x^4$ (d) $16-32x+24x^2-8x^3+x^4$ (viii) (i) 1.030301 (ii) 92.3521
8 (i) 10 (ii) $\frac{5}{2}$ (iii) −3, 3 (iv) $-\sqrt{5}, \sqrt{5}$
9 (i) −3, 0 (ii) $-\frac{5}{2}, 2$ (iii) 1, 3, 5
10 (i) 1, 2 (ii) $0, \frac{4}{5}$ (iii) $-\frac{1}{3}, 4$ (iv) $-\frac{8}{3}, -\frac{3}{5}$ (v) $0, \frac{1}{2}$ (iv) −3, 0, 3
11 1.59, 4.41
12 (i) −0.27, −3.73 (ii) −0.08, −5.92 (iii) −0.62, 1.62
13 1, 7 14 (i) 1, 5 (ii) −5, −3 (iii) −2, 5
15 (ii) 10 16 (i) 21 (ii) −20 (iii) −5, 5
18 (i) $x>7$ (ii) $x<-5$ (iii) $x \geq \frac{4}{3}$ (iv) $x<-3$ (v) $x \geq 5$ (vi) $x < \frac{4}{7}$ (vii) $-1<x<2$ (viii) $9 \leq x \leq 15$
19 (i) $x<1$ or $x>3$ (ii) $1<x<3$
20 (i) $x<-4$ or $x>2$ (ii) $-\frac{5}{3}<x<\frac{1}{2}$ (iii) $-1<x<7$ (iv) $x<-\frac{1}{2}$ or $x>3$ (v) $x<\alpha$ or $x>\beta$
21 (i) $x<1$ or $x>7$ (ii) $-\frac{1}{2}<x<3$
22 (i) $x>10$ (ii) $x>1$

Exercise 1.1:5

1 (i) $2\sqrt{3}$ (ii) $3\sqrt{2}$ (iii) $2\sqrt{5}$ (iv) $3\sqrt{3}$ (v) $4\sqrt{2}$ (vi) $3\sqrt{5}$ (vii) $5\sqrt{2}$ (viii) $6\sqrt{2}$ (ix) $4\sqrt{6}$
(x) $6\sqrt{6}$ (xi) $11\sqrt{10}$ (xii) $30\sqrt{21}$ (xiii) $a^2b^3\sqrt{bc}$
2 (i) $2\sqrt{2}$ (ii) $4\sqrt{5}$ (iii) $20\sqrt{2}$ (iv) $40\sqrt{5}$
3 (i) $5\sqrt{2}$ (ii) 0 (iii) $\sqrt{5}$
4 (i) $\frac{\sqrt{3}}{3}$ (ii) $\sqrt{2}$ (iii) $2\sqrt{7}$ (iv) $\frac{\sqrt{2}}{4}$ (v) $5\sqrt{2}$ (vi) $\frac{3\sqrt{6}}{8}$ (vii) $\frac{\sqrt{15}}{3}$ (viii) $\sqrt{a}$ (ix) $a\sqrt{ab}$
5 (i) $\frac{9\sqrt{2}}{4}$ (ii) $\frac{3\sqrt{6}}{10}$ (iii) $\frac{5\sqrt{30}}{12}$ (iv) $\frac{7\sqrt{6}}{27}$ (v) $\frac{a\sqrt{abc}}{b^3}$
6 (i) 2 (ii) $11-5\sqrt{5}$ (iii) $5\sqrt{2}+6$ (iv) $7-2\sqrt{10}$ (v) $a-b$ (vi) 1 (vii) $a+b+2\sqrt{ab}$ (viii) $11+6\sqrt{3}$ (ix) $9\sqrt{3}-11\sqrt{2}$
7 (i) $\frac{3+\sqrt{2}}{5}$ (ii) $\sqrt{2}-1$ (iii) $2+\sqrt{3}$ (iv) $5(2\sqrt{5}+3)$ (v) $\frac{\sqrt{5}+\sqrt{2}}{3}$ (vi) $\frac{3\sqrt{2}+2\sqrt{3}}{6}$ (vii) $\frac{\sqrt{a}+\sqrt{b}}{a-b}$
8 (i) $\frac{4}{5}\sqrt{2}-1$ (ii) $15-4\sqrt{14}$ (iii) $\frac{8}{3}-\frac{1}{3}\sqrt{10}$
9 4 10 0.447; 0.309 11 3.732; 0.268 13 $1+\sqrt{3}$
14 $\pm(\sqrt{5}-\sqrt{2})$
15 4 16 16 17 2 18 (i) 5 (ii) 4

Exercise 1.1:6

1 (i) 1, 1 (ii) $\frac{1}{8}, \frac{1}{16}, \frac{1}{81}, \frac{1}{128}$ (iii) 2, 16, 2, 4, 10, 2 (iv) $\frac{1}{10}, \frac{5}{7}, \frac{1}{2}$ (v) 8, 27, 4, 32, 81, 16, $\frac{16}{81}$ (vi) $\frac{2}{3}, \frac{7}{8}, \frac{8}{2}, \frac{2}{7}, \frac{9}{8}, \frac{1}{625}$ (vii) $\frac{5}{11}, 64, \frac{4}{9}, \frac{27}{8}$ (viii) $\frac{1}{8}, \frac{1}{2}, 25, 4$
2 (i) $\frac{1+x}{x}$ (ii) $\frac{1-x}{1+x}$ (iii) $\frac{1}{(1+x)^2}$ (iv) $\frac{1}{x^2(1+x^2)^{1/2}}$ (v) $\frac{2}{(1+x)^{3/2}(1-x)^{3/2}}$
3 (i) $\frac{3}{2}$ (ii) $\frac{3}{2}$ (iii) $\frac{7}{2}$ (iv) $-\frac{1}{2}$ (v) -4
4 5 5 (i) 3 (ii) 3 (iii) 7 (ii) −1
6 (i) −2, 4 (ii) $\frac{1}{2}, 1$ (iii) $-\frac{1}{2}, 3$
7 (i) 0, 1 (ii) 1, 3 (iii) 1, 2 (iv) 1, −1 (v) −1, 2 (vi) $\frac{1}{2}, -\frac{1}{2}$
9 (i) $\frac{3}{2}$ (ii) −1 (iii) −2

Exercise 1.1:7

2 (i) 2, 3, 1, 1, 0 (ii) −1, −2, −5 (iii) $\frac{1}{2}, \frac{1}{3}, \frac{3}{2}, \frac{2}{3}, \frac{1}{4}, \frac{7}{4}$ (iv) −2, 4
4 (i) (a) 1 (b) 3 (c) −2 (d) 1 (ii) (a) 1 (b) 0 (c) $\log_a x$ (d) $-\log_a(x+1)$
5 (i) (a) lg 0.3 (b) lg 50
6 (i) 0.602 (ii) 1.556 (iii) 0.699
7 (i) 2, 7, 27, 2 (ii) (i) x (ii) x (iii) (a) 0 (b) 2 (c) −1 (d) −3 (e) $\frac{1}{2}$ (f) $\frac{5}{2}$
8 $2\log_a x$; $-\log_a x$; $\frac{1}{2}\log_a x$; $-\frac{3}{2}\log_a x$
9 (i) $\log_a x + \log_a y + \log_a z$ (ii) $\log_a x + \log_a y - \log_a z$ (iii) $2\log_a x - 3\log_a y - \log_a z$ (iv) $\frac{1}{2}\log_a x - \frac{1}{2}\log_a y$ (v) $1 + \log_a x$ (vi) $\log_a x + \log_a y - 2$

Answers 569

10 (i) 3 (ii) $\frac{3}{2}$ (iii) $\frac{2}{3}$ (iv) $\frac{1}{3}$ (v) $\frac{y}{z}$
11 (i) 1.26 (ii) 0.83 (iii) 1.29 (iv) −1.20
 (v) 1.57
12 (i) 2 (ii) 2
13 (i) (a) $\log_c b$ (b) 1 (ii) (a) 3 (b) 6 (c) 1
 (d) $\frac{3}{2}$ (iii) (a) 6 (b) xy
14 (ii) −1 (c) −1; no 15 (ii) 4
16 (i) 1.21 (ii) 0.77 (iii) −1.46 (iv) 1.41
 (v) 1.44 (vi) −1.24, 1.24
17 (i) 1, 0.631 (ii) 2, 1.322 (iii) 1.183, 0.712
 (iv) −0.415
18 (i) 2.710 (ii) 8.599 (iii) 1.484 (iv) 6.213
19 100
20 (i) $\frac{1}{2}$ (ii) 4 (iii) 1.585 (iv) 8.690 (v) 0.177
21 $\frac{2}{3}$
22 (i) 1, 3 (ii) (a) 1, −1 (b) 3 (iii) (a) 0.77
 (b) 4
23 (i) $\frac{1}{100}$, 100 (ii) 4, 22.63 (iii) 2, 16 (iv) $\frac{1}{81}$, 3
 (v) 10, 100
24 20, 5 25 80, 2 26 (i) 256, $\frac{1}{2}$ (ii) 27, $\frac{1}{9}$
27 4, 16
28 (i) 2, 8 (ii) 8, 4 (iii) 4, 2; 9, 3
29 (i) $1000x^2$ (ii) $\frac{9}{\sqrt{x}}$ (iii) $\sqrt{x}$ (iv) x^2

Exercise 1.2:1

2 (i) $\sqrt{52}$ (ii) 10 (iii) $2\sqrt{5}$ (iv) $2\sqrt{10}$ (v) $3\sqrt{2}$
 (vi) $\sqrt{74}$ (vii) $2\sqrt{26}$ (viii) $\sqrt{97}$ (ix) 17 (x) 6
 (xi) $\sqrt{a^2+b^2}$ (xii) $2\sqrt{a^2+4b^2}$
6 (i) (2, 3) (ii) (6, 5) (iii) (2, 2) (iv) (5, 4)
 (v) $(\frac{7}{2}, \frac{5}{2})$ (vi) $(\frac{1}{2}, \frac{1}{2})$ (vii) (−1, 5) (viii) $(-1, \frac{5}{2})$
 (ix) $(-\frac{3}{2}, -3)$ (x) (1, 3) (xi) $(\frac{a}{2}, \frac{b}{2})$
 (xii) $(2a, 3b)$
7 $\sqrt{29}$ 8 10 9 (7, 8) 10 (−2, 4)
11 (i) $\frac{3}{2}$ (ii) $\frac{4}{3}$ (iii) 2 (iv) $\frac{1}{3}$ (v) −1 (vi) $\frac{5}{7}$
 (vii) −5 (viii) $-\frac{9}{4}$ (ix) $-\frac{8}{15}$ (x) 0 (xi) $\frac{b}{a}$
 (xii) $\frac{2b}{a}$
12 undefined 14 45°
15 (i) 0 (ii) 1 (iii) −1 (iv) $\sqrt{3}$ (v) $-\sqrt{3}$
17 yes; no 18 $(-b, a)$; $\frac{b}{a}, -\frac{a}{b}$
21 (i) parallel (ii) perpendicular (iii) neither
22 (i) perpendicular (ii) neither (iii) parallel
27 (i) 5 (ii) 15 (iii) 6 (iv) 16
28 (i) 25 (ii) 10 (iii) 7 (iv) 30
29 (i) (2, −2), $\sqrt{40}$, 20 (ii) (4, −1), $\sqrt{68}$, 34
30 square 31 rhombus; 40 32 6 33 3
34 8, 25 35 $\frac{1}{3}, \frac{1}{3}$; collinear 36 3 37 1
40 5 41 (0, −7) 42 −6 43 (16, 0)

Exercise 1.2:2

1 (i) $y = 3x − 6$; $3x − y − 6 = 0$ (ii) $y = \frac{1}{4}x + \frac{7}{4}$; $x − 4y + 7 = 0$ (iii) $y = −2x + 6$; $2x + y − 6 = 0$ (iv) $y = −\frac{2}{3}x$; $2x + 3y = 0$
 (v) $y = \frac{b}{a}x + b$; $bx − ay + ab = 0$
 (vi) $y = −5$; $y + 5 = 0$ (vii) $y = mx$; $mx − y = 0$
 (viii) $y = mx + c$; $mx − y + c = 0$
 (ix) $y = mx + (y_1 − mx_1)$; $mx − y + y_1 − mx_1 = 0$
2 (i) $y = −\frac{2}{3}x$ (ii) $y = mx$
3 (i) $y = \frac{1}{3}$ (ii) $y = \frac{1}{3}x − 1$ (iii) $y = \frac{1}{3}x + \frac{17}{3}$
4 (i) $y = x − 7$ (ii) $y = −x − 3$

5 $\frac{y+1}{x-4}$; $3x + 5y = 7$
6 (i) $y = \frac{3}{2}x$ (ii) $y = \frac{4}{3}x − 3$ (iii) $y = 2x − 2$
 (iv) $y = \frac{1}{3}x + \frac{7}{3}$ (v) $y = −x + 6$ (vi) $y = \frac{5}{7}x + \frac{1}{7}$
 (vii) $y = −5x$ (viii) $y = −\frac{9}{4}x + \frac{1}{4}$
 (ix) $y = −\frac{8}{15}x − \frac{19}{5}$ (x) $y = 3$ (xi) $y = \frac{bx}{a}$
 (xii) $y = \frac{2bx}{a} − b$
7 (i) $x = 2$ (ii) $x = −5$
8 (i) 3 (ii) $\frac{5}{4}$ (iii) $\frac{1}{3}$ (iv) −2 (v) 0
 (vi) $-\frac{a}{B}$ (vii) $-\frac{3}{2}$ (viii) $-\frac{b}{a}$
10 (i) neither (ii) perpendicular (iii) parallel
11 (i) neither (ii) perpendicular
 (iii) perpendicular
13 $y = x − 1$; $y = −x + 3$
14 $x + 3y + 15 = 0$; $3x − y − 5 = 0$
15 $5x − 2y + 13 = 0$; $2x + 5y − 18 = 0$
16 $4x + y = 0$; $x − 4y = 0$
17 (i) $2x + 3y = 0$ (ii) $2x + 3y + 8 = 0$
18 (i) $6x − y = 0$ (ii) $6x − y − 23 = 0$
19 (i) (2, 3); $2x − 3y + 5 = 0$ (ii) (0, 1); $x + y − 1 = 0$ (iii) (4, −1); $3x + 2y − 10 = 0$
 (iv) $(-\frac{1}{2}, \frac{5}{2})$; $x + y − 2 = 0$
 (v) $(-2a, 2b)$; $3ax − by + 6a^2 + 2b^2 = 0$
20 (i) 1; (2, 0), (0, −2) (ii) 2; $(\frac{3}{2}, 0)$, (0, −3)
 (iii) $\frac{1}{4}$; (−4, 0), (0, 1) (iv) −3; (−2, 0), (0, −6) (v) $-\frac{1}{2}$; (−2, 0), (0, −1)
 (vi) $\frac{3}{4}$; $(\frac{5}{3}, 0)$, $(0, -\frac{5}{4})$ (vii) $-\frac{5}{4}$; (4, 0), (5, 0)
 (viii) $-\frac{1}{3}$; (0, 0), (0, 0)
21 (i) (2, 3) (ii) (−1, 4) (iii) (6, 0) (iv) $(\frac{41}{9}, -\frac{1}{9})$
 (v) (6, 7)
23 (i) (3, −2) (ii) (4, 0) (iii) none
24 (i) (3, 4), (6, −1), (−2, −1) (ii) (2, 6), (4, −2), (−1, 0) (iii) (3, 6), (4, 4), (−3, −1)
25 $4x − 3y − 6 = 0$; $(\frac{6}{5}, -\frac{2}{5})$; 3
26 (i) $4\sqrt{2}$ (ii) $2\sqrt{10}$ (iii) $\frac{3\sqrt{5}}{10}$ (iv) 1
27 (i) 20 (ii) $\frac{8\sqrt{5}}{5}$ 28 (2, 2); 12, 6; 18
29 (i) 30 (ii) 10 (iii) 48
30 −2, 2; $(4n+2, 2n+3)$, $n \in \mathbb{Z}$, $2x + y − 2 = 0$; $89m$; $2x + y − 10n − 2 = 0$, $n \in \mathbb{Z}$

Exercise 1.2:3

1 $\{p, q, r\}$; b, e 2 $\{p, q\}$; c, d, e 3 $\{p, q, r, s, t\}$; b
4 $\{p, q, r, s, t\}$; a 5 $\{p, r, t\}$; none
6 {leg, ear, mouth}
7 {Babs, Dotty} 8 $\{a, b, d\}$ 9 $\{a, b, c, d, e\}$
10 $\{0, 2, 4, 6\}$ 11 $\{-2, 1, 4, 7\}$
14 (i) [−1, 4] (ii) −2, 0, 1, 2, 7 (iii) −5, 10; [−5, 10] (iv) $1, \frac{2}{3}, 3, \frac{5}{6}$
15 (i) 1, 5, −7 (ii) −2 (iii) $-\frac{1}{4}$, 1, −4
16 (i) 2, 14, 14, $\frac{59}{16}$, $\frac{59}{16}$ (ii) 0; 3, −3 (iii) [2, 50]
17 (i) 0, −1, 0 (v) 0, 2; $\frac{1}{2}$, $\frac{3}{2}$
18 (i) −1, 0 (ii) −2, 1 (iii) −3, 2 (iv) −4, 3
19 (a) (i) 0, −3, 5, −4 (ii) 0, 4
 (b) (i) −4, −3, −3, 0 (ii) −0, 2
 (c) (i) 0, −3, −5, −8 (ii) 0, 4
 (d) (i) 0, −3, 3, 0 (ii) −2, 0, 2
20 (i) $2, \frac{4}{3}, \frac{9}{7}$ (ii) 0, 2, 6
21 $2, 5, 0, -\frac{1}{2}$
22 (i) (0, 1), (0, 2) (ii) [0, 1], [2, 5]
 (iii) [−1, 2], [−7, 5] (iv) [−6, 3], [−1, 2]
 (v) $\mathbb{R}, \mathbb{R}$
23 (i) [0, 1] (ii) [0, 25] (iii) [0, 9] (iv) [0, 36]
24 (i) [0, ∞) (ii) [2, ∞) (iii) [−4, ∞)
 (iv) (−∞, 1] (v) (−∞, −3]
25 [9, ∞)

26 (i) [−4, ∞) (ii) [−4, 0) (iii) [−4, 12]
27 (i) $[\frac{1}{2}, \frac{2}{3}]$ (ii) $[\frac{5}{6}, 1]$ (iii) [−5, −2]
28 (i) $\{-1, 1\}$ (ii) $\mathbb{Z}$ (iii) $\mathbb{Z}^+ \cup \{0\}$
29 (i) neither; neither; neither, one-one; into; into
 (ii) (a) one-one (b) into (c) into
 (d) neither
30 (i) (a) [2, ∞) (b) (−∞, 0] (c) (−∞, 3]
 (d) [−1, 1] (e) $(-\infty, -3] \cup [3, \infty)$
 (f) $\left(-\infty, -\frac{\sqrt{5}}{2}\right] \cup \left[\frac{\sqrt{5}}{2}, \infty\right)$
 (g) $(-\infty, -2] \cup [1, \infty)$ (h) [0, 3]
 (ii) (a) $\{x \in \mathbb{R}: x \neq 4\}$ (b) $\{x \in \mathbb{R}: x \neq 3\}$
 (c) $\{x \in \mathbb{R}: x \neq \frac{5}{3}\}$ (d) $\{x \in \mathbb{R}: x \neq 1, x \neq 2\}$
 (e) $\{x \in \mathbb{R}: x \neq -\frac{5}{2}, x \neq -\frac{4}{5}\}$ (f) (1, ∞)
31 (i) (a) $\mathbb{R}, \mathbb{R}$ (b) $\mathbb{R}, [0, \infty)$ (c) $\mathbb{R}, [-8, \infty)$
 (d) $[0, \infty), [0, \infty)$ (e) $(-\infty, 0], [0, \infty)$
 (f) $(-\infty, 4], [0, \infty)$ (g) [−2, 2], [0, 2]
 (h) $(-\infty, -\frac{4}{3}] \cup [\frac{4}{3}, \infty), [0, \infty)$ (i) [1, 3], [0, 1]
 (j) $(-\infty, 0) \cup [\frac{2}{3}, \infty), [0, \infty)$
 (ii) (a) $\{x \in \mathbb{R}: x \neq 0\}$; $\{f(x) \in \mathbb{R}: f(x) \neq 0\}$
 (b) $\{x \in \mathbb{R}: x \neq 0\}$, (0, ∞)
 (c) $\{x \in \mathbb{R}: x \neq 1\}$, $\{f(x) \in \mathbb{R}: f(x) \neq 0\}$
 (d) $\{x \in \mathbb{R}: x \neq -\frac{2}{3}\}$, $\{f(x) \in \mathbb{R}: f(x) \neq 0\}$
32 $\{x \in \mathbb{R}: x \neq -\frac{1}{2}\}$; $\{f(x) \in \mathbb{R}: f(x) \neq \frac{3}{2}\}$
34 (i) no (ii) yes
35 (i) $0, \frac{1}{2}, 2$ (ii) 3 (iv) $-2, -\frac{5}{4}, -\frac{3}{4}$
36 (i) $0, \frac{1}{5}, 2$ (iii) 0, −2 & 4
37 (i) $0, 0, 6\sqrt{2}, 6\sqrt{2}$ (iii) ±5, ±$3\sqrt{5}$
38 (i) $14, 3t + 2, 6x + 2, -12x + 2, 3x^2 + 2,$
 $3\sqrt{x} + 2, 9x + 5, 3s + 3t + 2$
 (ii) $16, t^2, 4x^2, 16x^2, x^4, x, (3x+1)^2, (s+t)^2$
 (iii) $20, t(t+1), 2x(2x+1), 4x(4x-1),$
 $x^2(x^2+1), \sqrt{x}(\sqrt{x}+1), (3x+1)(3x+4),$
 $(s+t)(s+t+1)$
 (iv) $64, t^3, 8x^3, -64x^3, x^6, x^{3/2}, (3x+1)^3, (s+t)^3$
 (v) $2, \frac{t+2}{t-1}, \frac{2x+2}{2x-1}, \frac{2-4x}{4x+1}, \frac{x^2+2}{x^2-1}, \frac{\sqrt{x}+2}{\sqrt{x}-1},$
 $\frac{x+1}{x}, \frac{s+t+2}{s+t-1}$
39 (i) $-\frac{1}{2}$ (ii) $-\frac{4}{3}$ (iii) $-\frac{2}{3}$, 1
40 (i) −1, 0, 1 (ii) $0, -\frac{1}{8}$ (iii) −2
41 (i) 2 (ii) 0, 1 (iii) 3, −2 (iv) $0, \frac{1}{2}, -\frac{1}{2}$ (v) 3
42 (i) 6 (ii) 1.5625 (iii) $\mathbb{R}^+$
43 (i) 12.57 (ii) 2.52 (iii) $\mathbb{R}^+$
44 (i) $\frac{35}{4}m$, $15m$, $\frac{75}{4}m$, $20m$, $\frac{75}{4}m$, $15m$, $\frac{35}{4}m$, 0
 (ii) $20m$ (iii) $t = \frac{4}{3}, \frac{8}{3}$
45 (i) $\sqrt{2x^3+3}$, $2y+3$; $\sqrt{5}$, 5
 (ii) $15x − 2$, $15y + 18$; 13, 33
 (iii) $(2x−5)(2x−1)$, $4y^2 + 12y + 2$; −3, 18
 (iv) $\sqrt{x^4+1}$, $(y+1)^2$; $\sqrt{2}$, 4
 (v) $\frac{9-2x}{x-2}, \frac{y}{y-5}$; $-7, -\frac{1}{4}$
 (vi) x, y; 1, undefined
46 (i) $1 + 9x^2$; 1, 10 (ii) $x^2 + 2x + 2$; 2, 5
 (iii) $x + 2$; 2, 3
48 −3, 2
49 (i) [3, 63] (ii) $[\sqrt{2}, \sqrt{7}]$
50 (i) $\frac{1}{1+x^2}, \frac{-x}{1+x}$ (ii) $\sqrt{1+9x^2}$, $3 + 3x$
51 (i) $x + 2$, $x + 3$ (ii) $9x$, $27x$
 (iii) $25x − 16$, $84 − 125$ (iv) x^4, x^8
 (v) $8x^4 − 8x^2 + 1$,
 $128x^8 − 256x^6 + 32x^4 − 32x^2 + 1$
 (vi) x^9, x^{27} (vii) $x, \frac{1}{x}$ (viii) $x, \frac{x}{x-1}$
52 (i) (a) $x, x^3 − x^4$ (b) $\frac{x^2-x+2}{x^2-1}, \frac{1}{x+1}$
53 (i) x, x (ii) x, x
54 (i) (a) $\frac{x}{2}$ (b) $3x + 6$ (c) $\frac{1}{7}x − 4$ (d) $-\frac{4}{3}x − 5$

570 Answers

(ii) (a) $x^{1/3}$ (b) $x^{1/3}+1$
(iii) (a) $\frac{1}{x}$ (b) $\frac{5}{x}$ (c) $\frac{3+2x}{x}$ (d) $\frac{1-x}{5x}$
(e) $\frac{4x}{x-1}$ (f) $\frac{x+2}{1-x}$ (g) $\frac{4x+5}{2-7x}$

56 (i) $[0, \infty), \sqrt{x-1}, [1, \infty)$
(ii) $[-3, \infty), \sqrt{x-3}, [0, \infty)$
(iii) $[0, \infty), \sqrt{\frac{1-x}{x}}, (0, 1]$

Exercise 1.2:4

4 (i) 0; 0, 3 (ii) -6; -3, 2 (iii) 0; 0, 4
(iv) 2; -2, -1 (v) -5; $-\frac{5}{2}$, 1 (vi) 8; $\frac{4}{3}$, 2
6 (i) -6; 1, 2, 3 (ii) -6; -3, 1, 2 (iii) 0; -1, 0, 1
10 (i) (a) $[1, \infty)$ (b) $[-2, \infty)$ (c) $[\frac{5}{2}, \infty)$
(d) $(-\infty, 1]$
(ii) (a) $(-\infty, -3] \cup [2, \infty)$ (b) $[-1, 1]$
(c) $[-2, 1]$ (d) $(-\infty, 1] \cup [2, \infty)$
11 (i) $(-\infty, -1) \cup [1, \infty)$ (ii) $(-\infty, 4] \cup (6, \infty)$
(iii) $\mathbb{R}$ (iv) $[0, \infty)$
22 (i) odd (ii) neither (iii) neither (iv) even
(v) even (vi) odd
23 (i) odd (ii) even (iii) neither (iv) neither
(v) even (vi) odd (vii) even (viii) odd
25 1
32 (i) (3, 2) (ii) (2, 3) (iii) $(-4, -7)$
(iv) (0, 0), (1, 1) (v) (2, 6), $(-3, 6)$
(vi) (3, 10), $(-3, 4)$ (vii) $(1, -12)$, $(3, -6)$
(viii) (2, 5), $(-5, -16)$ (ix) $(-1, 4)$
(x) $(-1, 4), (\frac{5}{2}, -\frac{5}{4})$ (xi) (10, 3)

Exercise 1.2:5

1 (i) 0, 1, 3; discontinuous
(ii) -1, -1, 0; discontinuous
(iii) 1, 0, 1; discontinuous
(iv) 4, 4, 4; continuous
2 (i) 0, -1, 1; no (ii) yes; no
3 (i) yes (ii) no (iii) yes (iv) no (v) yes
(vi) yes
4 yes **5** no

Exercise 1.2:6

2 (i) $x = 4$ (ii) $x = -3$ (iii) $x = \frac{5}{3}$ (iv) $x = 2$
(v) $x = 1, x = -2$
3 $\frac{5}{2}$ **4** (i) $y = 2$ (ii) $y = \frac{3}{4}$ (iii) $y = -5$ (iv) $y = 2$
(v) $y = 0$ (vi) $y = 0$
5 (i) $y = 3$ (ii) $y = -\frac{4}{3}$ (iii) $y = 0$ (iv) $y = \frac{a}{f}$
6 (i) $(0, -\frac{1}{2})$; $x = 2$; $y = 0$
(ii) $(0, \frac{1}{3})$; $x = -\frac{3}{2}$; $y = 0$
(iii) $(-4, 0), (0, -2)$; $x = 2$; $y = 1$
(iv) $(\frac{5}{2}, 0), (0, -1)$; $x = -\frac{5}{3}$; $y = \frac{2}{3}$
(v) $(-3, 0)$; $x = 0$, $y = 1$
(vi) none; $x = 0$, $y = 0$
10 (i) $(-1, -\frac{1}{2})$ (ii) (4, 1), (5, 2)
(iii) (1, 1), $(-2, 2)$ (iv) $(\sqrt{2}, \sqrt{2}+1)$,
$(-\sqrt{2}, 1-\sqrt{2})$ (v) $(-3, -3), (3, 1)$
(vi) $(-\frac{3}{5}, -\frac{1}{4})$

Exercise 1.2:7

1 (i) -4 (ii) $x > -4$ (iii) $x < -4$
2 (i) $\frac{5}{2}$ (ii) $x < \frac{5}{2}$ (iii) $x > \frac{5}{2}$
3 (i) -3, 2 (ii) $x < -3$ or $x > 2$ (iii) $-3 < x < 2$
4 (i) 1, $\frac{7}{2}$ (ii) $x < 1$ or $x > \frac{7}{2}$ (iii) $1 < x < \frac{7}{2}$
5 (i) $-2, \frac{4}{3}$ (ii) $-2 < x < \frac{4}{3}$ (iii) $x < -2$ or $x > \frac{4}{3}$
6 (i) $-\frac{5}{3}, 0$ (ii) $x < -\frac{5}{3}$ or $x > 0$ (iii) $-\frac{5}{3} < x < 0$
7 (i) $-\frac{3}{2}, \frac{3}{2}$ (ii) $-\frac{3}{2} < x < \frac{3}{2}$ (iii) $x < -\frac{3}{2}$ or $x > \frac{3}{2}$
8 (i) 1, 3, 5 (ii) $1 < x < 3$ or $x > 5$
(iii) $x < 1$ or $3 < x < 5$
9 (i) $-3, \frac{1}{2}, \frac{5}{3}$ (ii) $x < -3$ or $\frac{1}{2} < x < \frac{5}{3}$
(iii) $-3 < x < \frac{1}{2}$ or $x > \frac{5}{3}$
10 (i) $-2, 0$ (ii) $-2 < x < 0$ or $x > 2$
(iii) $x < -2$ or $0 < x < 2$
11 (i) 3 (ii) $x < 0$ or $x > 3$ (iii) $0 < x < 3$
12 (i) -1 (ii) $x < -1$ or $x > 2$ (iii) $-1 < x < 2$
13 (i) 2, 10 (ii) $x < 2$ or $x > 10$ (iii) $2 < x < 10$
14 (i) $-2, 2, 5$ (ii) $x < -2$ or $-1 \le x < 2$ or $x > 5$
(iii) $-2 < x < -1$ or $2 < x < 5$
15 $x < -1$ or $x > \frac{5}{3}$
16 (i) 2 (ii) $x > 2$ (iii) $x < 2$
17 (i) 3 (ii) $x < 3$ (iii) $x > 3$
18 (i) $-2, 3$ (ii) $x < -2$ or $x > 3$ (iii) $-2 < x < 3$
19 (i) $-2, 3$ (ii) $x < -2$ or $x > 3$ (iii) $-2 < x < 3$
20 (i) $-5, 1$ (ii) $-5 < x < 1$ (iii) $x < -5$ or $x > 1$
21 (i) $-1, 2$ (ii) $x < -1$ or $x > 2$ (iii) $-1 < x < 2$
22 (i) 1 (ii) $x > 1$ (iii) $x < 1$
23 (i) $-6, 1$ (ii) $x < -6$ or $-5 < x < 1$
(iii) $-6 < x < -5$ or $x > 1$
24 (i) 2, 4 (ii) $x < 2$ or $3 < x < 4$
(iii) $2 < x < 3$ or $x > 4$
25 (i) 0, 5 (ii) $0 < x < 1$ or $x > 5$
(iii) $x < 0$ or $1 < x < 5$
26 (i) 1, 4 (ii) $0 < x < 1$ or $2 < x < 4$
(iii) $x < 0$ or $1 < x < 2$ or $x > 4$
27 (i) $x < -2$ or $x > 3$ (ii) $2 < x < 3$ or $x > 4$
28 $-3 < x < 4$

Exercise 1.3:1

3 $2x$ **4** $3 - 2x$
5 (i) $10x$ (ii) $2 - 2x$ (iii) $-\frac{1}{3x^2}$ (iv) $-\frac{2}{x^3}$
6 (i) $x^3 + 3x^2\delta x + 3x(\delta x)^2 + (\delta x)^3$; $3x^2$ (ii) $4x^3$

Exercise 1.3:2

1 (i) $7x^6$ (ii) $\frac{5}{2}x^{3/2}$ (iii) $\frac{\sqrt[3]{x}}{3x}$ (iv) $\frac{3}{2}\sqrt{x}$ (v) $-\frac{3}{x^4}$
(vi) $-\frac{1}{2x\sqrt{x}}$ (vii) $-\frac{3}{2x^2\sqrt{x}}$ (viii) $-\frac{1}{3x\sqrt[3]{x}}$
2 (i) $60x''$ (ii) $12x^{1/3}$ (iii) 7 (iv) 0 (v) 7
(vi) $2x^3 + \frac{1}{2}x^2$
(vii) $\frac{1}{\sqrt{x}} - \frac{1}{x\sqrt{x}}$ (viii) $-\frac{9}{x^4} + \frac{2}{x^7}$
3 (i) $4x(1 + x^2)$ (ii) $x^3(5x + 8)$ (iii) $3x(5x + 1)$
(iv) $\frac{2x^2 - 6}{x^3}$ (v) $-\frac{6x + 2}{x^3}$ (vi) $\frac{5x - 1}{2x\sqrt{x}}$
4 m
7 (i) $-4, -2, 0, 2, 4$ (ii) $-7, -5, -3, -1, 1$
(iii) $3, -6, -9, -6, -3$
12 (i) no; no (ii) no; no (iii) yes; yes
13 (i) no (ii) yes (iii) no
14 2, 0, 0, 2
18 (i) $10 - 10t$ (ii) $\frac{\pi}{\sqrt{lg}}$ (iii) $1 - \frac{4}{\theta^3}$
19 (i) (a) $\frac{1}{2}(t+2)g\,wk^{-1}$ (b) $\frac{3}{2}g\,wk^{-1}$
(c) $2g\,wk^{-1}$ (d) $5g\,wk^{-1}$
(ii) (a) $\frac{2(t+2)}{4+(t+2)^2}wk^{-1}$ (b) $\frac{6}{3}wk^{-1}$
(c) $\frac{2}{5}wk^{-1}$ (d) $\frac{5}{21}wk^{-1}$
20 (i) 8000, 4000, 6000, 8000, 4000, -1500
(ii) (a) £3000 per month (b) 0 (c) $-$£9000 per month
(iii) $t = 0, 3$
(iv) (a) $1 < t < 3$ (b) $0 \le t < 1, t > 3$

Exercise 1.3:3

1 (i) 5, $-\frac{1}{5}$ (ii) $\frac{3}{2}, -\frac{2}{3}$ (iii) $-5, \frac{1}{5}$ (iv) $-6, \frac{1}{6}$
(v) 0, undefined
2 (i) -12 (ii) -6 (iii) 0 (iv) 6 (v) 12
3 (i) (1, 2) (ii) $(\frac{1}{2}, \frac{1}{8})$, $(-\frac{1}{2}, -\frac{1}{8})$
(iii) $(-\frac{2}{3}, \frac{37}{27})$, $(2, -5)$
(iv) $(-1, -5)$ (v) $(0, 5), (2, 1)$
4 (i) $(3, -5)$ (ii) $(2, -8)$ (iii) $(1, -9)$
(iv) $(0, -8)$ (v) $(-1, -5)$
5 $y = 4x - 4$
6 (i) $6x - y - 3 = 0$, $x + 6y - 19 = 0$
(ii) $6x - y - 2 = 0$, $x + 6y + 49 = 0$
(iii) $3x - 4y + 1 = 0$, $8x + 6y - 31 = 0$
(iv) $5x - 4y + 20 = 0$, $4x + 5y - 66 = 0$
(v) $2x + 3y - 9 = 0$, $3x - 2y - 21 = 0$
(vi) $x - y - 1 = 0$, $x + y - 1 = 0$
7 $2x - y + 1 = 0$, $x + 2y - 7 = 0$
8 156 **9** $\frac{15}{2}$
10 $4x$; (3, 18); $y = 12x - 18$
11 $y = 2x - 9$
12 $y = 4x + 1$, $y = 4x + 3$
13 $y = x + 16$, $y = x - 16$
14 $y = (1 + 2a)x - a^2$; $-1, 5$; $y = -x - 1$, $y = 11x - 25$
15 $y = -2x + 2$, $y = 6x + 10$
16 (i) $(-1, -4), (3, 0)$ (ii) $(0, 0), (2, 2)$
(iii) (1, 2)
17 $(-\frac{9}{4}, \frac{81}{16})$ **18** $(3, \frac{7}{2})$ **19** $(-\frac{1}{8}, -8)$ **20** $(6, \frac{20}{3})$
21 $\frac{1}{2\sqrt{2}}$

Exercise 1.3:4

1 (i) $5x^4, 20x^3$ (ii) $-\frac{2}{x^3}, \frac{6}{x^4}$ (iii) $\frac{4}{3}x^{1/3}, \frac{4}{9}x^{-2/3}$
(iv) $4x^2 - 9x^2 + 4$, $12x^2 - 18x$
(v) $3\sqrt{x} + \frac{3}{2\sqrt{x}}, \frac{3}{2\sqrt{x}} - \frac{3}{4x\sqrt{x}}$
(vi) $2x + \frac{1}{2x\sqrt{x}}, 2 - \frac{3}{4x^2\sqrt{x}}$
2 $15x^4 - 4x^3 + 4x + 8$; $60x^3 - 12x^2 + 4$;
$180x^2 - 24x$; $360x - 24$; 360

Exercise 1.3:5

1 (i) -9; $(3, -9)$ (ii) 9; $(-2, 9)$ (iii) $\pm\frac{16}{9}$;
$(-\frac{2}{3}, -\frac{16}{9}), (\frac{2}{3}, \frac{16}{9})$ (iv) $0, -27$; $(-1, 0)$,
$(2, -27)$ (v) ± 2; $(-1, -2), (1, 2)$
(vi) 0; (1, 0)
(vii) $-3, 0, -128$; $(-1, -3), (0, 0), (4, -128)$
2 (i) $(3, -9)$ minimum (ii) $(-2, 9)$ maximum
(iii) $(-\frac{2}{3}, -\frac{16}{9})$ minimum, $(\frac{2}{3}, \frac{16}{9})$ maximum
(iv) $(-1, 0)$ maximum, $(2, -27)$ minimum
(v) $(-1, -2)$ maximum, $(1, 2)$ minimum
(vi) $(1, 0)$ point of inflexion
(vii) $(-1, -3)$ minimum, $(0, 0)$
maximum, $(4, -128)$ minimum
3 (i) $(0, 0)$ maximum, $(\frac{2}{3}, -\frac{4}{27})$ minimum
(ii) $(-2\sqrt{2}, 32\sqrt{2})$ maximum, $(2\sqrt{2}, 32\sqrt{2})$
minimum
(iii) $(\frac{1}{3}, \frac{175}{27})$ maximum, $(3, -3)$ minimum
(iv) $(2, 10)$ point of inflexion
(v) $(2, 32)$ maximum, $(6, 0)$ minimum
(vi) $(-1, -1)$ minimum, $(0, 0)$ point of
inflexion
5 (3, 9) **6** (1, 3) **7** 5, 9; $\infty, -\infty$
9 (i) $(0, 0)$ turning point (ii) $(2, -11)$
(iii) $(-1, -25), (3, -173)$
(iv) $(2, 0)$ turning point, $(1, -1)$
10 (i) (a) $(0, 0), (1, -1)$; $(0, 0), (\frac{2}{3}, -\frac{16}{27})$
(b) $(1, -5), (-\frac{1}{2}, \frac{7}{4})$; $(\frac{1}{4}, -\frac{13}{8})$

(ii) (a) (3, 6), (−3, −6); none (b) (2, 12); (−2.52, 0)
11 $2x^2 - 4x + 10$; (1, 3)
12 (i) $\frac{1}{\sqrt{5}}$ (ii) $\frac{6}{\sqrt{5}}$
13 (i) $(-\frac{1}{2}, \frac{1}{2}), (\frac{1}{\sqrt{2}}, \frac{1}{2})$ (ii) $\sqrt{17}$
14 0.128 m^3 16 18 cm^3 17 $400 \text{ m}^2, 800 \text{ m}^2$
18 $225 \text{ m}^2, 450 \text{ m}^2$; regular 20-sided polygon
19 $2a \text{ m} \times 2a \text{ m} \times a \text{ m}$
22 (i) 255.8 cm^3 (ii) $1:2$
23 (i) 3 l (ii) 289.6 l
24 £2500 25 $\frac{p}{2b}$
26 3 27 5 m 28 l

Exercise 1.3:6

1 (i) $24x(3x^2+1)^3$ (ii) $\frac{3x^2}{2\sqrt{x^3+4}}$
2 (i) $-12x(1-x^2)^5$ (ii) $5(2x-1)(x^2-x)^4$
(iii) $3\left(1-\frac{1}{x^2}\right)\left(x+\frac{1}{x}\right)^2$ (iv) $\frac{-6x}{(3x^2-1)^2}$
(v) $\frac{-6x}{(x^2+2)^4}$ (vi) $\frac{1}{\sqrt{x}(1-\sqrt{x})^3}$ (vii) $\frac{x}{\sqrt{1+x^2}}$
(viii) $\frac{3}{2\sqrt{1+3x}}$ (ix) $\frac{-2x}{3(1-x^2)^{2/3}}$
(x) $\frac{1}{2(1-x)^{3/2}}$
(xi) $\frac{-3x^2}{(3+2x^3)^{3/2}}$ (xii) $\frac{3(\sqrt{x}+1)^2}{2\sqrt{x}}$
3 (i) $an(ax+b)^{n-1}$ (ii) $2anx(ax^2+b)^{n-1}$
(iii) $\frac{an(a\sqrt{x}+b)^{n-1}}{2\sqrt{x}}$
4 (i) $4x(1+x^2)$ (ii) $6x(1+x^2)^2$
6 $\frac{2+\frac{1}{\sqrt{1+x}}}{4\sqrt{x+\sqrt{1+x}}}$
7 (i) (a) $x = \sqrt{y}$ (b) $\frac{1}{2\sqrt{y}}$; yes (c) $-\frac{1}{4y\sqrt{y}}$; no
(ii) (a) $x = y^{1/3} - 1$ (b) $\frac{1}{3}y^{-2/3}$; yes
(c) $-\frac{2}{9}y^{-5/3}$; no
8 $\sqrt{2}, 0$
9 $\frac{2x}{5} + \frac{2\sqrt{x^2-2ax+2a^2}}{3}$
10 $15\sqrt{10}$ s 11 no 12 $3.14 \text{ cm}^3 \text{ s}^{-1}$
13 0.26 m s^{-1}
14 $201.1 \text{ cm}^2 \text{ s}^{-1}$ 15 $1.6 \text{ cm}^2 \text{ s}^{-1}$ 17 9.7 hours
18 $0.38 \text{ m}^2 \text{ s}^{-1}$; 0.012 m s^{-1}
19 (i) $1.2 \text{ m}^3 \text{ s}^{-1}$ (ii) $0.05 \text{ m}^3 \text{ s}^{-1}$
20 $\frac{2x}{3}$; $\frac{8}{3}$ m s^{-1}
21 0.13 m s^{-1}
22 0.09 m s^{-1}
23 0.04 cm s^{-1}
24 (i) 0.028 cm s^{-1} (ii) 0.020 cm s^{-1} (iii) 0.013 cm s^{-1}
25 (i) 0.84 cm s^{-1} (ii) 0.53 cm s^{-1}
26 (i) 0.1 m s^{-1} (ii) 0.05 m s^{-1}
27 10 days
29 (i) $\frac{1}{3}$ cm s^{-1} (ii) $\frac{2}{9}$ cm s^{-1}
30 $\frac{3a}{20}$ cm s^{-1}

Exercise 1.3:7

1 (i) $(1+x)^2(1+4x)$ (ii) $(3x+1)^3(18x^2+2x)$
(iii) $(x+1)(x-2)^2(5x-1)$
(iv) $(x+1)(4x^2+2x-10)$
(v) $\frac{5x^3+1}{\sqrt{2x^3+1}}$ (vi) $\frac{2x-x^3}{(1+x^2)^{3/2}}$ (vii) $\frac{2}{(x+1)^2}$
(viii) $\frac{1+3x^2}{2\sqrt{x}(1-x^2)^2}$
2 (i) $2x(1+2x^2)$ (ii) $2x(1+x^2)^2(1+4x^2)$
3 $\frac{2x^2+3x+2}{2\sqrt{1+x}(1+2x)^2}$
4 $11x - 4y - 9 = 0$, $4x + 11y - 78 = 0$
5 (i) $(\frac{2}{3}, \frac{32}{7})$, $(2, 0)$ (ii) $(-3, 0), (-1, 16), (1, 0)$
6 (i) $0.29, 0$ (ii) $0, -1.05$
8 (i) $\frac{1-2x^2}{(2x^2+1)^2}$ (ii) $\frac{2}{(x+1)^2}$ (iii) $\frac{-4x}{(1+x^2)^2}$
(iv) $\frac{x+1}{(3x+1)^3}$ (v) $\frac{-6(x+1)}{(x-4)^3}$ (vi) $\frac{1}{\sqrt{x}(1-\sqrt{x})^2}$
(vii) $\frac{-1}{\sqrt{1-x^2}(1+x)}$ (vii) $\frac{1}{\sqrt{2x-x^2}(2-x)}$
(ix) $\frac{a^2}{(a^2-x^2)^{3/2}}$
9 (i) $\frac{1-x^2}{(x^2+1)^2}$ (ii) $\frac{1}{(2x-1)^2}$
13 $7x - 8y - 6 = 0$, $8x + 7y - 23 = 0$
14 $y = x - 3$, $y = -x + 1$
15 $\frac{1-x^2}{(1+x^2)^2}$; $\frac{1}{2}, -\frac{1}{2}$
16 (i) $(0, 0), (2, 4)$ (ii) $(3, 0)$
(iii) $\left(-\sqrt{3}, \frac{3\sqrt{3}}{2}\right), (0, 0), \left(\sqrt{3}, -\frac{3\sqrt{3}}{2}\right)$
17 (i) $\frac{a}{(a+bI)^2}$ (ii) $\frac{1}{a}, 0$

Exercise 1.3:8

1 (i) $-\frac{x}{y}$ (ii) $\frac{18-4x}{9y}$ (iii) $\frac{3x^2}{5-3y^2}$
(iv) $\frac{3x^2-8x-11}{3y^2-3}$ (v) $\frac{2\sqrt{y}}{2\sqrt{y}+1}$
2 $\frac{-2xy^2}{3x^2y+2}$
3 (i) $-\frac{2y}{x}$ (ii) $\frac{3-4y}{2x}$ (iii) $\frac{3x^2+y^2-5y}{5x-2xy}$
(iv) $\frac{3x-4y}{4x-3y}$
4 (i) 2 (ii) 4 (iii) $-\frac{10}{13}$ (iv) $-\frac{3}{4}$
5 -1; $y = -x + 2$
6 $(4, -3)$, $(4, 3)$; $4x + 3y - 7 = 0$, $4x - 3y - 7 = 0$
7 $(7, 0)$
8 (i) $y = 3x - 2$, $y = -3x + 6$ (ii) $(0, 0), (0, 4)$
9 $y = 3x + \frac{1}{3}$; $(\frac{1}{9}, \frac{2}{3})$; $y = -\frac{1}{3}x + \frac{19}{27}$
10 $\frac{2(1-2y)^2 - 8x^2}{(1-2y)^3}$
11 (i) $\frac{x}{y}$; $\frac{y^2-x^2}{y^3}$ (ii) $\frac{3x-1}{2y}$; $\frac{6y^2-(3x-1)^2}{4y^3}$
(iii) $\frac{3x^2}{3y^2+1}$; $\frac{6x(3y^2+1)^2 - 9x^4y}{(3y^2+1)^3}$
12 (i) $\frac{3}{4}, \frac{25}{64}$ (ii) 3, -10 13 maximum
14 (i) $\frac{m}{n}x^{\frac{m-n}{n}}$ (ii) $-\frac{m}{n}x^{-\frac{m+n}{n}}$

Exercise 1.3:9

1 (i) 2.0025 (ii) 1.9975
2 (i) 4.0025 (ii) 4.99 (iii) 2.0008 (iv) 5.97 (v) 8.15
3 (i) 0.12 (ii) 0 (iii) 0.12 (iv) 0.48 (v) 12
4 25.1 cm^2; 100.5 cm^2 5 0.11 cm

6 (i) 6% (ii) $\frac{3}{2}$% (iii) $-\frac{3}{2}$% (iv) -6%
7 (i) αp% (ii) αp% 8 4%
9 (i) $\frac{1}{2}$% (ii) 10%
10 5.48 days 11 1% 12 4% 13 $\frac{1}{2}p$%
14 -2.8%; -1.43%
15 (i) 9% (ii) 3% (iii) 1%
16 (i) 4% decrease (ii) $\frac{4}{3}$%

Miscellaneous Exercise 1

1 (i) $\frac{1}{4}$, 16 (ii) $\frac{3}{4}$ (iii) $-3, 2$
2 (i) (a) 9 (b) 1.530 (ii) (a) 1.079 (b) 1.398 (c) 1.477
3 (i) 1 (ii) $\frac{1}{2}\sqrt{2}, \frac{1}{2}$
4 $[-5, 3]$; $\sqrt{4-x}$
5 $3+2x^2$; $[0, 1]$; $[3, 5]$; $\sqrt{\frac{x-3}{2}}$
6 (i) $\frac{x^2}{x^2-1}, \frac{x^2}{(x-1)^2}$; 1, 1 (ii) $\frac{x}{x-1}$
7 (i) $\frac{3x+1}{8-x}$ (ii) $\frac{3x-1}{x-2}$; $\{x : x \in \mathbb{R}, x \neq 2\}$
8 (i) x (iv) $2 \pm \sqrt{2}$
9 (i) $-\frac{4}{9}$ (ii) -1 (iii) 1
12 (i) $(-2, 0), (0, 0), (2, 0)$; $x > 2$ or $-2 < x < 0$
(ii) $(-1, -5), (2, 4)$; $-1 < x < 2$
13 (i) (a) $\frac{1}{1-2x\sqrt{1-4x^2}}$ (b) $\frac{2y^2-2xy^3}{3x^2y^2-4xy-3}$
(ii) (a) $\frac{3}{4}$ (b) $-\frac{1}{32}$
14 (i) $(1, -2), (-1, 2)$ (ii) $(2, 4), (-2, -4)$
15 $(1, 4), (3, 0)$; $(2, 2)$; $(2, 2), (4, 4)$; $\frac{64}{9}$
16 6.83 cm, 3.41 cm
17 $0.5 \text{ m}^2 \text{ s}^{-1}$
18 (ii) $\frac{br}{\sqrt{l^2-b^2}}$ m s^{-1}
19 0.036 cm s^{-1} 20 1% 21 3%
22 (i) 6 pints (ii) (a) $\frac{3}{2}$ IEU (b) -7 IEU
(iii) 29 pints would do the trick

Chapter 2

Exercise 2.1:1

1 (i) $(x+3)(x-1)$; $-3, 1$
(ii) $(4+x)(1-x)$; $-4, 1$
(iii) $(2x+7)(x-1)$; $-\frac{7}{2}, 1$
(iv) $(5x+12)(4x-3)$; $-\frac{12}{5}, \frac{3}{4}$
(v) $(2x+3)(2x-3)$; $-\frac{3}{2}, \frac{3}{2}$
(vi) $(\sqrt{3}x+\sqrt{5})(\sqrt{3}x-\sqrt{5})$; $-\sqrt{\frac{5}{3}}, +\sqrt{\frac{5}{3}}$
2 (i) $-\frac{5}{3}, 0$ (ii) (a) $-\frac{1}{5}, 0$ (b) $0, \frac{4}{3}$
3 (i) $-\frac{1}{2}, 1$ (ii) $-\frac{1}{3}, \frac{5}{3}$
4 (i) $(x-2)^2$; 2 (ii) $(x-p)^2$; p
(iii) $(3x+4)^2$; $-\frac{4}{3}$
5 $-3, 7$ $(x+3)(x-7)$ 6 $\frac{-5 \pm \sqrt{17}}{4}$ 7 $\frac{1 \pm \sqrt{7}}{3}$
8 (i) $-5, 1$ $(x+5)(x-1)$
(ii) $-2, 4$ $(x+2)(x-4)$
(iii) $-1, 7$ $(x+1)(x-7)$
(iv) $-\frac{3}{2}, -1$ $(2x+3)(x+1)$
(v) $-5, \frac{1}{3}$ $(x+5)(3x-1)$
(vi) $-\frac{1}{2}, \frac{2}{3}$ $(2x+1)(3x-2)$
9 (i) $-1 \pm \sqrt{2}$ (ii) $2 \pm \sqrt{3}$
(iii) $-4 \pm \sqrt{21}$ (iv) $\frac{1}{2}(-1 \pm \sqrt{5})$
(v) $\frac{1}{2}(-3 \pm \sqrt{5})$ (vi) $\frac{1}{4}(3 \pm \sqrt{17})$
(vii) $\frac{1}{6}(-7 \pm \sqrt{37})$ (viii) $\frac{1}{10}(-5 \pm \sqrt{5})$
(ix) $p \pm \sqrt{p^2-q}$ (x) $\frac{1}{2}(-p \pm \sqrt{p^2-4p})$
10 $\frac{-b \pm \sqrt{b^2-4ac}}{2a}$

11 (i) $1+\sqrt{3}$ (ii) $\frac{1}{2}(-5\pm\sqrt{5})$ (iii) $\frac{1}{2}(1\pm\sqrt{13})$
(iv) $\frac{1}{10}(-3\pm\sqrt{29})$ (v) $\frac{1}{2}(2\pm\sqrt{5})$
12 $2\pm\sqrt{2}$ (ii) $1, 3$ (iii) 2
13 (i) $4, 3$; $-3, -4$ (ii) $1.62, -0.62$; $-0.62, 1.62$
14 $\frac{1}{2}(\sqrt{5}-1)$
15 (a) (i) 4, distinct real, $(x-2)(x-4)$
(ii) 0, equal, $(x-3)^2$ (iii) -4, non-real
(b) (i) 12, distinct real (ii) 0, equal
(iii) -12, non-real
(c) (i) 1, distinct real (ii) 0, equal
(iii) -15, non-real
(d) (i) 16, distinct real (ii) 0, equal
(iii) -16, non-real
16 (i) 64 (ii) $-8, 8$ (iii) $-2, 1$ (iv) $1, 9$
(v) $-14, 2$
18 (i) $k\leq 8$ (ii) $k\leq\frac{61}{12}$ (iii) $k\leq-\frac{2}{3}$ (iv) $k\geq-\frac{1}{16}$
19 $k\leq 1$; $-1, 3$; -1
20 (i) real (ii) $k\leq-\frac{7}{24}$

Exercise 2.1:2

1 (i) -9, min; $-1, 5$ (ii) $\frac{25}{4}$, max; $-3, 2$
(iii) $-\frac{1}{4}$, min; $2, 3$ (iv) $\frac{81}{8}$, max; $-4, \frac{1}{2}$
(v) 1, max; $-1, 1$ (vi) -9, min; $-\frac{3}{2}, \frac{3}{2}$
(vii) -1, min; $-2, 0$ (viii) -4, min; $0, \frac{4}{3}$
(ix) $-\frac{5}{4}$, min; $\frac{1}{2}(3\pm\sqrt{5})$
(x) $-\frac{7}{2}$, min; $\frac{1}{2}(-1\pm\sqrt{7})$
2 (i) 0, min; 2 (ii) 0, min; -1 (iii) 0, max; 1
3 (i) 9, min; none (ii) -1, max; none
(iii) $\frac{7}{8}$, min; none (iv) -1; max; none
(v) $-\frac{1}{12}$, min; none
9 $2, -4, -1$ **10** $-3x^2+5x+2$
11 $-2x^2-8x+10$; $-50x^2+40x+10$
13 4 m **14** (i) more than 20 (ii) 80 (iii) 10
15 (ii) 4 m (iii) 9.1 m
16 (i) 45 m (ii) 4 s (iii) 6s

Exercise 2.1:3

1 (i) $-4<x<1$; $x<-4$ or $x>1$
(ii) $x<2$ or $x>5$; $2<x<5$
(iii) $-3<x<\frac{1}{2}$; $x<-3$ or $x>\frac{1}{2}$
(iv) $x<-\frac{3}{2}$ or $x>\frac{1}{5}$; $-\frac{3}{2}<x<\frac{1}{5}$
(v) $x<-\frac{5}{2}$ or $x>\frac{5}{2}$; $-\frac{5}{2}<x<\frac{5}{2}$
(vi) $x<0$ or $x>\frac{7}{3}$; $0<x<\frac{7}{3}$
2 (i) $x<-1$ or $x>3$ (ii) $x<1$ or $x>2$
(iii) $-5\leq x\leq\frac{1}{2}$ (iv) $-\frac{4}{3}<x<2$
(v) $0\leq x\leq\frac{3}{4}$ (vi) $-\frac{5}{3}<x<\frac{1}{2}$
(vii) $x<-1$ or $x>7$ (viii) $-\frac{3}{5}<x<1$
3 (i) $x<-3$ or $x>3$ (ii) $-\sqrt{3}<x<\sqrt{3}$
(iii) $x<-\frac{2}{\sqrt{5}}$ or $x>\frac{2}{\sqrt{5}}$
4 (i) $x<1-\sqrt{2}$ or $x>1+\sqrt{2}$
(ii) $x<\frac{1}{2}(3-\sqrt{5})$ or $x>\frac{1}{2}(3+\sqrt{5})$
(iii) $2-\sqrt{7}<x<2+\sqrt{7}$
(iv) $x<\frac{1}{4}(-1-\sqrt{41})$ or $x>\frac{1}{4}(-1+\sqrt{41})$
5 $-\frac{1}{2}<x<1$
6 (i) $-2<x<0$ or $1<x<3$
(ii) $-1<x<1$ or $2<x<4$
(iii) $-5<x<-3$ or $-1<x<1$
(iv) $-3<x<-2$ or $2<x<3$
(v) $3<x<4$ (vi) $-7<x<-\frac{3}{2}$ or $1<x<2$
(vii) $1<x<3$
7 $-3<x<-2$ or $\frac{1}{3}<x<\frac{4}{3}$
8 $-1<x<-\frac{1}{2}$ or $1<x<\frac{5}{2}$
9 $(x+2)^2+1$
10 $2(x-2)^2+6$
11 $-3\{(x-\frac{5}{6})^2+\frac{11}{36}\}$; $-\frac{11}{12}$
13 (i) $x\in\mathbb{R}, x\neq 3$ (ii) $x\in\mathbb{R}$

15 (i) $-6, 6$; $k<-6$ or $k>6$; $-6<k<6$
(ii) $-1, 7$; $k<-1$ or $k>7$; $-1<k<7$
(iii) $-3, \frac{5}{9}$; $k<-3$ or $k>\frac{5}{9}$; $-3<k<\frac{5}{9}$
(iv) $-2, \frac{2}{3}$; $-2<k<\frac{2}{3}$; $k<-2$ or $k>\frac{2}{3}$
(v) $-3, 2$; $-3<k<2$; $k<-3$ or $k>2$
16 (i) $5, -3$ (ii) $x<-3$ or $x>5$ (iii) $-3<x<5$
17 $-\frac{3}{2}<k<\frac{1}{2}$
18 (i) $k>1$ (ii) $k<-\frac{1}{3}$
19 $-7<k<1$
20 (i) $p<-1$ (ii) $p>\frac{1}{3}$
23 $y=-x+3$, $y=3x+3$
24 $y=-6x-9$, $y=2x-1$
25 $m\leq-7$ or $m\geq 1$; $y=-7x$, $y=x$
26 (i) 2 (ii) 1 (iii) 0; 9; $y=-4x+16$
27 $4y=3x+25$, $3y=-4x+25$

Exercise 2.1:4

1 (ii) (a) $-1\pm\sqrt{2}$; $(x+1+\sqrt{2})(x+1-\sqrt{2})$
(b) $-1\pm\frac{1}{2}\sqrt{10}$; $2(x+1+\frac{1}{2}\sqrt{10})(x+1-\frac{1}{2}\sqrt{10})$
(iii) (a) $(x+2+\sqrt{6})(x+2-\sqrt{6})$
(b) $(x+\frac{1}{2}+\frac{1}{2}\sqrt{17})(x+\frac{1}{2}-\frac{1}{2}\sqrt{17})$
(c) $5(x-\frac{2}{5}+\frac{1}{5}\sqrt{14})(x-\frac{2}{5}-\frac{1}{5}\sqrt{14})$
2 (i) $4, 7$ (ii) $\frac{4}{5}, \frac{3}{5}$ (iii) $\frac{2}{3}, 2$
3 (i) $1, 3$; $4, 3$ (ii) $-3, \frac{5}{2}$; $-\frac{1}{2}, -\frac{15}{2}$
(iii) $-1\pm\frac{1}{3}\sqrt{6}$; $-2, \frac{1}{3}$ (iv) $p+\sqrt{p^2-q}$; $2p, -q$
5 (i) $\frac{7}{12}$ (ii) 33 (iii) -126 (iv) $\frac{1}{12}$ (v) $-\frac{27}{2}$ (vi) $\frac{7}{4}$
6 (i) 72 (ii) 5 (iii) $-\frac{9}{8}$ (iv) -32 (v) 26 (vi) 24
7 (i) $\dfrac{p^2-2q}{q^2}$ (ii) p^2-4q (iii) p^2q-2q^3 (iv) $\dfrac{p}{q^2}$
(v) $\dfrac{3q+p^3}{q}$ (vi) $\dfrac{p+4}{2p+q+4}$
8 (i) $-\dfrac{b}{c}$ (ii) $\dfrac{b^2-2ac}{c^2}$ (iii) $-\dfrac{bc}{a^2}$ (iv) $\dfrac{3abc-b^3}{a^3}$
(v) $\dfrac{3abc-b^3}{ac^2}$ (vi) $\dfrac{ab+b^2-2ac}{a^2-ab+ac}$
9 (i) $x^2-7x+1=0$ (ii) $x^2-x-1=0$
(iii) $x^2+3x+1=0$ (iv) $x^2-5x+1=0$
(v) $x^2+3x+1=0$ (vi) $x^2-14x+45=0$
10 (i) $4x^2-12x+1=0$ (ii) $2x^2-12x+17=0$
(iii) $x^2-4x+2=0$ (iv) $x^2-6x+1=0$
(v) $8x^2-8x+1=0$ (vi) $x^2-6x+8=0$
11 (i) $a^2x^2+(2ac-b^2)x+c^2=0$
(ii) $ax^2+3bx+9c=0$
(iii) $cx^2+bx+a=0$
(iv) $c^2x^2+(2ac-b^2)x+a^2=0$
(v) $acx^2+(2ac-b^2)x+ax=0$
(vi) $a^3x^2+(b^3-3abc)x+a^2c=0$
12 (i) $4x^2-37x+9=0$ (ii) $-2, \frac{1}{3}$
13 6 **14** $\frac{11}{12}$ **16** 6
17 $-\frac{2}{3}, 2$ **18** $\frac{5}{3}, -8$
19 (i) $k\leq-2$ or $k\geq 1$ (ii) $k\leq-2$
20 (i) $k<1$ or $k>9$ (ii) $0<k<1$ or $k>9$
21 (iii) $2k^2x-3kx+1=x^2$
22 (i) $cx^2+bx+a=0$
23 $0, 0$; $-1, 0$; $1, -2$

Exercise 2.2:1

1 (i) $3; 1$ (ii) $5; 0$ (iii) $1; 0$ (iv) $0; 0$ (v) $4; 3$
(vi) $3; -1$ (vii) $4; 1$
2 (i) $3x^5-2x^4-4x^3+2x^2+x$
(ii) $4x^5-4x^4+9x^3-22x^2+17x-4$
(iii) x^4-1 (iv) x^4-1
3 (i) $2x^2-x-2$; 9 (ii) $2x^2-2x-1$; 4
(iii) x^3-x; x (iv) x^2+x+1; $-2x+6$
(v) $\frac{1}{2}(x-5)$; $\frac{1}{2}(25x-3)$ (vi) $x+2$; $-x+3$
4 $(x-1)(x+1)(x^2+1)$
5 $(x+2)(2x+3)(x-1)(x-2)$
6 (i) $2x+2$ (ii) -2 (iii) 0

Exercise 2.2:2

1 (i) 9 (ii) 4 (iii) $-\frac{1724}{125}$
2 2 **3** $-2x+9$ **4** $\frac{1}{2}(1-12x)$
5 $2-x$ **6** $1, 0, 1$ **7** $-2, -1, 16$ **9** $3, 2$

Exercise 2.2:3

2 $x-4$ **3** 3
4 (i) $1, 2, 3$ (ii) $-1, -1, 2$ (iii) $\frac{1}{2}, 4, -2$
(iv) $3, \pm\sqrt{2}$ (v) $-1, 2\pm\sqrt{3}$ (vi) $-2, -1\pm\sqrt{2}$
5 $(2x+1), (x-4)$
6 (i) $(x-1)^2(x+1)^2$; $-1, 1$
(ii) $(x+2)(x+1)(2x-5)(x-3)$; $-2, -1, \frac{5}{2}, 3$
(iii) $(x+2)^2(x+1)(3x-1)(x-1)$;
$-2, -1, \frac{1}{3}, 1$
7 $4, -1$; $(x+2), (x+3)$
8 (i) 0 (ii) 3 (iii) -2 (iv) -4
9 (i) $(x-1)(x^2+x+1)$ (ii) $(x+1)(x^2-x+1)$
(iii) $(a-b)(a^2+ab+b^2)$,
$(a+b)(a^2-ab+b^2)$;
$(3x+2)(9x^2-6x+4)$
10 (i) $(x-1)(x+1)(x^2+1)$
(ii) $(x-1)(x+2)(x^2+x+1)$
(iii) $(x^2+2)(x^2+x+1)$
11 $1, -37$ **12** $-9, 7, 6$; $-\frac{1}{2}, 2, 3$
13 (i) $\pm 1, \pm 2$ (ii) $\pm 1, \pm\sqrt{2}$ (iii) $\pm\sqrt{2}$
(iv) $\pm\frac{1}{2}, \pm\sqrt{5}$
14 $-6, -5, 2, 3$
15 (i) $-\frac{2}{3}, 0, \frac{1}{6}, 1$
(ii) $-2, -1, \frac{2}{3}, \frac{5}{3}$

Exercise 2.2:4

1 (i) $x<-1$ or $0<x<1$; $-1<x<0$ or $x>1$
(ii) $x<-\frac{4}{5}$ or $\frac{2}{3}<x<2$; $-\frac{4}{5}<x<\frac{2}{3}$ or $x>2$
(iii) $1<x<3$ or $x>5$; $x<1$ or $3<x<5$
(iv) $x<-2$ or $-1<x<1$ or $x>4$;
$-2<x<-1$ or $1<x<4$
3 (i) $-2, 1, 2$; $x<-2$ or $1<x<2$; $-2<x<1$
or $x>2$ (ii) $-3, -\frac{1}{2}, 1$; $x<-3$ or $-\frac{1}{2}<x<1$;
$-3<x<-\frac{1}{2}$ or $x>1$ (iii) $-2, -1, 1, 2$;
$-2<x<-1$ or $1<x<2$; $x<-2$ or
$-1<x<1$ or $x>2$
4 (i) $1<x<2$ or $x>3$ (ii) $-4<x<\frac{3}{2}$ or $x>2$
(iii) $x\leq-3$ or $-1\leq x\leq 1$ (iv) $x<-1$ or
$0<x<5$
(v) $-2\leq x\leq-1$ or $x\leq 3$ (vi) $-1<x<2-\sqrt{3}$
or $x>2+\sqrt{3}$ (vii) $-\sqrt{3}\leq x\leq-1$ or
$1\leq x\leq\sqrt{5}$
(viii) $x<-2$ or $-\frac{3}{2}<x<0$ or $x>2$
5 $(-3, -45), (\frac{1}{2}, \frac{1}{2}), (2, 20)$; $-3<x<\frac{1}{2}$ or $x>2$
6 $(-3, 15), (-1, -3), (2, 0)$;
$x<-3$ or $-1<x<2$
7 $-1; 1$; (a) $x<1$ (b) $x>1$
(ii) none; 0; (a) $x<0$ (b) $x>0$
(iii) none; -1; (a) $x<-1$ (b) $x>-1$
(iv) none; 3; (a) $x>3$ (b) $x<3$
(v) $47, 15$; -4; (a) $x<-4$ (b) $x>-4$
(vi) $-1; -1, 1$; (a) $-1<x<1$ (b) $x<-1$ or
$x>1$
8 (i) $x>3$ (ii) $x\leq\frac{5}{2}$ (iii) $x\geq 2$ (iv) $x>2$
(v) $x<-1$ (vi) $x\leq-2$ or $x\geq 2$
9 $(-2, 8)$; $x\geq-2$
10 3 (2 of which are equal); $1; 3$
13 (i) $x<-2$; $x>-2$, $x\neq 1$; $x\geq-2$
(ii) $x<0$ or $x>2$; $0<x<2$; $0\leq x\leq 2$
(iii) $2<x<3$; $x<2$ or $x>3$, $x\neq-1$;
$x\leq 2$ or $x\geq 3$
(iv) no real values of x; $x\in\mathbb{R}$, $x\neq-1$,
$x\neq-2$; $x\in\mathbb{R}$
(v) $x<4$, $x\neq 1$; $x>4$; $x=1$ or $x\geq 4$

15 (i) $x > \frac{1}{2}$ (ii) $1 \le x \le 2$ (iii) $x \in \mathbb{R}$ (iv) $x \le \frac{1}{2}$ or $x = 2$
16 $(-2, -8)$ **17** $(4, 52)$ **18** $(1, 2)$
19 $(-1, 6)$; $x > -1, x \ne 3$; $x \ge 1$
20 $-4, 0$; $4, 0$
21 (i) $-1, 2, \frac{5}{2}$; $-1 < x < 2$ or $x > \frac{5}{2}$ (ii) 3; $x > 3$
(iii) $-\frac{1}{2}, 1, 1$; $-\frac{1}{2} < x < 1$ or $x > 1$

Exercise 2.3:1

5 $(-\infty, 3) \cup (3, \infty)$; $\frac{x+2}{3-x}$; $(-\infty, 3) \cup (3, \infty)$, $(-\infty, -1) \cup (-1, \infty)$
6 $a = -q$
8 $(3, -4)$
11 (i) $(0, 1)$ (ii) $(-1, \frac{1}{4})$
12 (i) $(-1, -\frac{1}{2}), (1, \frac{1}{2})$ (ii) $(-1, -\frac{1}{2}), (3, \frac{1}{6})$
(iii) $(-\frac{2}{3}, \frac{9}{4}), (6, -\frac{1}{4})$
13 $(1, 1)$ **15** $(-2, -\frac{1}{2}), (1, 1)$
22 (i) $[-\frac{3}{4}\sqrt{2}, \frac{3}{4}\sqrt{2}]$ (ii) $(\frac{1}{4}, 1]$ (iii) $[-1, \frac{1}{9}]$
(iv) $[-\frac{3}{4}, \infty)$
23 (i) $-\frac{1}{2}, \frac{1}{2}$ (iii) $[-\frac{1}{2}, \frac{1}{2}]$
24 (i) $-\frac{1}{4}, \frac{1}{4}$
25 $-\frac{1}{8}$
26 $(-1, 1)$; $\frac{1+\sqrt{1-x^2}}{x}$; $(-1, 1)$
27 $0.79, 2.21$
28 (i) $[-9, \infty)$ (ii) $(-\infty, -1]$

Exercise 2.3:2

1 (i) $-1 < x < 2$ (ii) $x < 1$ or $x > 3$
(iii) $x < \frac{4}{3}$ or $x > \frac{11}{2}$
(iv) $-4 < x < \frac{1}{2}$ or $x > 5$
(v) $x < -3$ or $-2 < x < 2$
(vi) $x < -\frac{4}{3}$ or $-\frac{1}{2} < x < 0$
(vii) $x < -4$ or $2 < x < 3$
(viii) $-2 < x < \frac{1}{2}$ or $x > 2$
2 $-2 < x < -1$ or $1 < x < 4$
3 (i) $x < -1$ or $x > 2$ (ii) $x < -1$ or $1 < x < 2$
(iii) $\frac{2}{3} < x < 1$ or $\frac{3}{2} < x < 4$
(iv) $-1 < x < \frac{1}{3}$ or $1 < x < 2$
4 $\frac{1}{2} < x < 1$ **5** $1 < x < 3$
6 $-2 < x < 1$ or $x > 3$
7 $x < -1$ or $x > 2$
8 $(0, 3], [1, 4)$: $-1 - \frac{1}{2}\sqrt{2} < x < -1 + \frac{1}{2}\sqrt{2}$
9 $x < -3$ or $-\frac{11}{4} < x < 0$ or $1 < x < 2$
10 (ii) $(0, 1), (3, 4)$ (iv) $x < 0$ or $0 < x < 3$

Exercise 2.3:3

1 (i) $\frac{1}{x-2} - \frac{1}{x-1}$ (ii) $\frac{1}{4x} - \frac{1}{4(x+4)}$
(iii) $\frac{1}{x-1} - \frac{1}{x+1}$ (iv) $\frac{5}{3(x-2)} - \frac{2}{3(x+1)}$
(v) $\frac{2}{x-2} + \frac{1}{x+2}$ (vi) $\frac{1}{19(2x+3)} + \frac{8}{19(3x-5)}$
2 (i) $\frac{1}{2}, -4, \frac{9}{2}$
(ii) (a) $\frac{1}{35(x+2)} - \frac{4}{45(2x-1)} + \frac{1}{63(x-5)}$
(b) $\frac{1}{8(x-2)} + \frac{1}{4x} - \frac{3}{8(x+2)}$
4 (i) $\frac{1}{x+1} + \frac{1-x}{x^2+4}$ (ii) $\frac{1}{11(x+2)} - \frac{2x-4}{11(2x^2+3)}$
(iii) $\frac{3}{x-2} + \frac{1-3x}{x^2+x+1}$ (iv) $\frac{x}{x^2+1} - \frac{x}{x^2+4}$
(v) $\frac{x+2}{x^2+1} - \frac{x-1}{x^2+4}$

6 (i) $1 + \frac{4}{x-2} - \frac{1}{x-1}$
(ii) $x + 1 + \frac{1}{3(x+1)} + \frac{8}{3(x-2)}$
(iii) $1 + \frac{1}{x-1} - \frac{1}{x+1}$ (iv) $1 + \frac{1}{x} - \frac{x+1}{x^2+1}$
7 $\frac{1}{2(x+1)} + \frac{1}{2(x-3)}$

Exercise 2.4:1

6 2 **7** 4 **8** 6

Exercise 2.4:2

1 (i) $1, 9$ (ii) $-\frac{2}{5}, 2$ (iii) $0, 1$ (iv) no solution
2 $2, 5$
3 (i) -1 (ii) $-\frac{1}{3}, 3$ (iii) $\frac{1}{2}, 1$ (iv) $-\frac{1}{3}, 7$ (v) $\frac{4}{3}, 4$
(vi) $\frac{5}{2}, 7$
4 $-7, 1$ **5** $\pm\sqrt{2}, -1 \pm \sqrt{3}$
6 (i) $x < 1$ or $x > 7$ (ii) $-1 < x < 4$
(iii) $x < \frac{2}{3}$ or $x > \frac{4}{3}$ (iv) $-\frac{5}{4} \le x \le \frac{9}{4}$
7 $x < \frac{1}{3}$ or $x > 3$
8 (i) $x > -\frac{5}{2}$ (ii) $x < -$ or $x > \frac{7}{3}$
(iii) $-\frac{3}{2} < x < \frac{5}{8}$ (iv) $x \le -\frac{4}{3}$ or $x \ge 0$
(v) $x < 1$ or $x > 3$
9 $0 < x < 2$
10 (i) $-\frac{\sqrt{5}}{2} < x < \frac{\sqrt{5}}{2}$
(ii) $x < -\sqrt{5}$ or $-\frac{1}{\sqrt{3}} < x < \frac{1}{\sqrt{3}}$ or $x > \sqrt{5}$
11 $-2\sqrt{2} < x < -2$ or $2 < x < 2\sqrt{2}$
12 (i) $x < -1$ or $x > 1$
(ii) $-3 < x < -1$ or $1 < x < 3$
13 (i) $1 < x < 9$ (ii) $\frac{1}{2} < x < 4$
14 $x < -\frac{1}{2}$ **17** $-1 \le x \le 5$; $[-\frac{5}{2}, \infty)$; 1; $\frac{1}{2}$

Exercise 2.5:1

1 $(0, \infty)$ **2** $(0, 1)$
6 (i) $759, 351, 199$ (ii) 3.96 days, 5.68 days, 6.68 days
7 (i) £1154, £1331, £1536
(ii) 2.75 years, 4.25 years, 7.27 years
8 1.29 years, 2.29 years, 3.11 years
9 (i) 0.95% (ii) 5.83% (iii) 76.23%; 6.12 years
10 (i) 11.90 years (ii) 18.85 years
11 1698 yrs **12** (i) 1803 yrs (ii) 1.23%
13 9.65% **14** $1.06:1$; $322.5, 341.7, 383.6$
15 1.033; (i) 55.4 m s^{-1}
(ii) 145.4 m s^{-1}; 200 m s^{-1}

Exercise 2.5:2

4 $(0, 1), (2, e^4)$
5 (i) $3e^{3x}$ (ii) $-e^{-x}$ (iii) $-\frac{3}{2}e^{-3/2x}$
(iv) $x(x+2)e^x$ (v) $(1-5x)e^{-5x}$
6 (i) $3x^2 e^{x^3}$ (ii) $\frac{1}{2\sqrt{x}}e^{\sqrt{x}}$ (iii) $\frac{2}{x^2}e^{-2/x}$
(iv) $\frac{1}{(x^x+1)^2}$
7 (i) $a e^{ax+b}$ (ii) $f'(x) e^{f(x)}$
13 $(0, 0), \left(2, \frac{4}{e^2}\right)$; $y = \frac{1}{e}x$
14 $1, -6, 9$
15 26 months; never
16 (i) 1 (ii) 129

Exercise 2.5:4

1 (i) 1 (ii) 2 (iii) -1 (iv) $\frac{1}{2}$
2 (i) 2 (ii) 8 (iii) 2 (iv) 4 (v) $\frac{1}{4}$ (vi) $\frac{5}{2}$ (viii) $\frac{1}{\sqrt{2}}$
2 (i) x^2 (ii) xy (iii) $\frac{1}{x}$ (iv) $\frac{1}{\sqrt{x}}$
8 (i) $\frac{1}{x}$ (ii) $\frac{1}{x}$ (iii) $\frac{3}{2+3x}$ (iv) $\frac{3x^2}{1+x^2}$ (v) $\frac{2 e^{2x}}{1+e^{2x}}$
(vi) $-\frac{1}{x(\ln x)^2}$ (vii) $\frac{1}{2x\sqrt{\ln x}}$ (viii) $1 + \ln x$
(ix) $e^x\left(\frac{1}{x} + \ln x\right)$ (x) $\frac{\ln 3x^2}{x(\ln 3x)^2}$
9 (i) $\frac{a}{ax+b}$ (ii) $\frac{f'(x)}{f(x)}$
10 (i) $\frac{2}{x}$ (ii) $\frac{1}{2x}$ (iii) $-\frac{1}{x}$ (iv) $\frac{2}{1-x^2}$
(v) $\frac{2}{x} - \frac{x}{1-x^2}$ (vi) $\frac{2x}{1-x^4}$
11 $\frac{1}{x}$ **12** (i) $\frac{1}{x \ln a}$ (ii) $\frac{2}{x \ln 10}$
13 (i) $10^x \ln 10$ (ii) $x^x(1 + \ln x)$
18 (i) $\ln 2$ (ii) $\frac{1}{2}\ln 3$ (iii) $0, \frac{1}{2}\ln 3$
19 (i) $(0, -2)$ max, $(\ln 2, 2\ln 2 - 6)$ min
(ii) $(\ln \frac{2}{3}, \frac{20}{9})$ max, $(\ln \frac{4}{3}, \frac{16}{9})$ min
20 $(\ln 3, 3)$
21 $1, -1$; $-1 < f(x) < 1$; $f^{-1}(x) = \ln\sqrt{\frac{1-x}{1+x}}$; $-1 < x < 1$
22 $(1, \infty)$; $0, \ln(2 \pm \sqrt{3})$; $\ln\{x + \sqrt{x^2 - 1}\}$

Exercise 2.5:5

1 (i) $3^x \ln 3$ (ii) $8^x \ln 8$ **2** $\ln a$
3 (i) $3(10^{3x} \ln 10)$ (ii) $5(2^{5x+1} \ln 2)$
(iii) $2x(3^{x^2} \ln 3)$
4 (i) $2x^{2x}(1+\ln x)$ (ii) $3x^x(1 + \ln x)$
(iii) $(5x)^x(1 + \ln 5x)$
5 (i) $(x+1)^x\left\{\frac{x}{x+1} + \ln(x+1)\right\}$
(ii) $(1+2x)^{3x-1}\left\{\frac{6x-2}{1+2x} + 3\ln(1-2x)\right\}$
(iii) $(3x-1)^{1+2x}\left\{\frac{3+6x}{3x-1} + 2\ln(3x-1)\right\}$
7 (a) (i) $a, \frac{a}{at+b}$ (b) $a e^{at+b}, a$
(c) $(2at+b) e^{at^2+bt+c}, 2at+b$
8 -1 **9** (ii) $\frac{3}{2}\%$

Exercise 2.6:1

1 (i) $6, 2$; 1 (ii) none; 1 (iii) $3, -1$; 3
(iv) $2 \pm 4\sqrt{2}$; 3 (v) $4, 13$; 1 (vi) $-\frac{11}{16}, -\frac{38}{27}$; 1
(vii) $-\frac{7}{4}, 2$
(viii) $-4, 1, -31$; 4 (ix) $7, -1$; 3 (x) $11, 3$; 1
(xi) $1, \frac{1}{4}$; 1
2 (i) 1 (ii) 3 (2 equal) (iii) 3 (iv) 3
(v) 3 (2 equal) (vi) 1
3 (i) (a) 4 (b) 4 (c) 2 (d) none
(ii) (a) $k \le 5$ (b) $5 < k \le 32$ (c) $k > 32$
4 $(-1, 3), (\frac{5}{3}, -\frac{175}{27})$; $-\frac{175}{27} < x < 3$
5 $(-1, 0), (1, 64), (4, -125)$; $0 < k < 64$
7 (i) -3 (ii) 1 (iii) $-1, 0, 2$ (iv) $-3, 0, 2$ (v) 0
(vi) 0 (vii) $-1, 0$ (viii) $-2, -1, 0, 2$
(ix) $-2, 0, 1$ (x) -2 (xi) -1

Exercise 2.6:2

1. (i) 1.909, 1.904 (ii) 3.037, 3.037
   (iii) 3.222, 3.196 (iv) 2.458, 2.457
   (v) 4.765, 4.744 (vi) 2.032, 2.031
   (vii) 1.407, 1.395
2. 2.714, 2.702, 2.7016; 2.7016 (to 4 d.pl)
3. (i) 4.02 (ii) (a) 4.12 (b) 1.97
4. 0.0321, 0.0322, 0.0323 5. 0.2; 0.1999
7. 2.833; 2.806
8. (i) 1.913, 1.905, 1.904 (ii) 2.844, 2.811, 2.804
   (iii) 1.057, 1.070, 1.072
   (iv) 2.031, 2.0315, 2.03148
   (v) −1.167, −1.022, −1.150
9. (i) 0.714, 0.653
   (ii) $x = \frac{2^2}{x^2 + x + 2}$: 0.500, 0.717
10. (i) 1.148 (ii) $x = \ln(x+2)$: 1.131

Exercise 2.7:1

1. 0.4, 0.12 2. 1.2, 6.4 3. 0.05, 0.3 4. 40, 60
5. 0.7, 0.45 6. 4, 1.5 7. 24, 32
8. (i) x^2, y, a, b (ii) $x^2, \frac{y}{x}, a, b$ (iii) $\sqrt{x}, y, a, b$
   (iv) $x, x^2 - y, a+b, -ab$ (v) $x, y - x^2, a, b$
9. (i) $\frac{1}{x}, y, a, b$ (ii) $\frac{1}{x^2}, \frac{y}{x}, b, a$ (iii) $\frac{1}{x}, \frac{y}{\sqrt{x}}, b, a$
   (iv) $x, \frac{1}{y}, \frac{1}{a}, \frac{b}{a}$ (v) $x, \frac{1}{y}, a, b$ (vi) $x, \frac{1}{y} - x^2, a, b$
10. $\frac{1}{y}, \frac{1}{x}, -1, \frac{1}{a}$
11. (i) x, y^2, a, b (ii) $x, \frac{y^2}{x}, \frac{1}{a}, \frac{b}{a}$
    (iii) $y, y^2 - x, a, -b$
12. (i) $\log x, \log y, b, -\log a$
    (ii) $\log(1+x), \log y, b, \log a$
13. (i) $x, \log y, \log b, \log a$
    (ii) $x, \log y, -2\log b, \log a$
    (iii) $x, \log y, \log a, b \log a$
14. (i) $x, 10^y, a, b$ (ii) $x, \frac{10^y}{x}, a, b$
    (iii) $\lg x, y, a, a \lg b$ (iv) $x, \frac{y^2}{\lg x}, \frac{1}{a}, \frac{b}{a}$
15. (i) ? (ii) $\frac{y}{10^x}, \frac{y^2}{10^x}, a, b$
    (iii) $x, \lg\left(\frac{y}{x}\right), -\frac{1}{a}, \frac{\lg b}{a}$
    (iv) $x, \frac{2^y}{x}, \frac{1}{a}, -\frac{b}{a}$
16. 0.8 17. 10 18. 2.3
19. 3, −2 20. 5, 3 21. 0.5, 0.3 22. 10, 40
23. −3, 8 24. 0.3, 2.5 25. 50, −0.7
26. 30, 1.2 27. 0.6, 0.9 28. 4.2, 2.5
29. 0.3, 1.2 30. 2, 0.1 31. −1.5, 20 32. 15, 50
33. (3, 1.11), (5, 0.71); 5, −0.8
34. (2, 3.6), (5, 24.2); 0.4, 0.3
35. (1.62, 0.61), (2.24, 0.84); 0.2, −3.0
36. 0.06; 0.4 37. 15.6, 3.1
38. (i) 40, 1.4 (ii) 300 (iii) 40% (iv) 2.06 days
39. (i) 1, $\frac{3}{2}$ (ii) 29.5 yrs
40. 21, 0.55; 6 months if he's lucky
41. (i) 2, 0.5 (ii) 1.90 s (iii) 0.25 m

Miscellaneous Exercise 2

1. $(\frac{1}{2}, -\frac{3}{4}); (2, -3), (-1, -3)$
2. (i) 0, 0; 1, −2 (ii) −5

3. (i) $-1 \le x \le 3$ (ii) $1 - \sqrt{2} \le x \le 1 + \sqrt{2}$
   (iii) $x = 1$ (iv) Φ
4. (i) $-\frac{1}{2} \le k \le 4$ (ii) $(2k-7)x - 4x + k = 0$
5. (i) $-\frac{1}{2} \le k \le \frac{9}{2}$ (ii) $k < -\frac{1}{2}$
6. (i) $-\frac{3}{2}, 2$ (ii) $\frac{1}{2}, 4$ (iii) $k \le \frac{49}{8}$
7. (ii) $k < \frac{7}{5}$ or $k > 7$
8. (i) $-25; \frac{3}{5}, -\frac{7}{2}$ (iii) $y < -\frac{1}{25}$ or $y > 0$ (iv) 4
11. $\frac{1}{3}(7 - 2x)$ 12. −40
13. −7, −6; $(x+2)(x+1)(x-3)$; $-2 \le x \le -1$ or $x \ge 3$
14. (i) $(-2, 0), (-1, 3), (3, 15)$; $-2 < x < -1$) or $x > 3$
    (ii) $y = 8x - 16$; $(-4, -48)$ (iii) 2.46
15. (i) $(x+5)(x-2)(x-3)$; $-5 \le x \le 2$ or $x \ge 3$
    (ii) $(x+5)(x-2)^2$; $x \ge -5$
16. (i) $x < -2$ (ii) $x < -2$ or $1 < x < 2$
    (iii) $x < -2$ or $x > 1$ (iv) $x > 1$
17. (i) $-1 < x < 4$ (ii) $-2 < x < 1$ or $x > 2$
    (iii) $x < -1$ or $0 < x < \frac{1}{2}$ or $x > 5$
    (iv) $-2 < x < 0$
18. (ii) (a) $-2 < x < \frac{5}{3}$ (b) $\frac{1}{5} < x < \frac{5}{3}$
19. $1 < x < 2$ or $3 < x < 4$
20. $-\frac{1}{3} < x < 1$ 21. (i) $1 < x < 3$ (ii) $x > \frac{3}{4}$
23. (i) $-7 < x < -5$ or $0 < x < 1$
    (ii) $(3, 2), (\frac{3}{2}, \frac{1}{2})$; $x < \frac{3}{2}$ or $2 < x < 3$
24. (i) $(0, -1)$; $y = \frac{1}{2}$; $[-1, \frac{1}{2})$ (ii) $[0, \frac{1}{3}]$
25. (i) $[1, \infty)$ (ii) $[-\frac{1}{4}\sqrt{2}, \frac{1}{4}\sqrt{2}]$ (iii) $(\frac{1}{2}, 1]$
26. (i) $\frac{8}{5} < x < 4$ (ii) $x - 4, 5x - 8, 4 - x$
    (iii) $\frac{9}{5} < x < 3$
27. (i) 3, −1 (ii) $x < \frac{1}{3}$ or $x > 3$ (iii) 2
28. (i) 0.01 (ii) 0
29. $x + y = 1$ 30. 2
31. (i) $\frac{1}{x(1-x^2)\ln 2}$ (ii) $(1 - 2x^2 \ln 2)2^{1-x^2}$
32. (i) $\frac{86}{27}$ (ii) 0, 1
33. (i) −1, 1; (−1, 1) (ii) $\ln \frac{1+x}{1-x}$; (−1, 1)
34. 2.04 35. $(-\frac{1}{2}, \frac{1}{2}e)$; (0, 1)
36. (i) 2.073, 2.071 (ii) 2.079, 2.070
37. (i) 2.926 (ii) 2.935
39. (i) £3.99 (ii) £2, 20,000 (iii) 2.77; less
41. 2.08, 2.06 42. 8.7
43. 15, 2.3; (i) 0.51 (ii) 1.66

Chapter 3

Exercise 3.1:1

2. 3.3 3. $\frac{4}{5}, \frac{3}{4}$
4. (i) $\frac{12}{13}, \frac{5}{13}$ (ii) $\frac{1}{\sqrt{5}}, \frac{2}{\sqrt{5}}$ (iii) $\frac{3}{\sqrt{10}}, \frac{1}{\sqrt{10}}$
5. (i) $\frac{4}{5}, \frac{4}{3}$ (ii) $\frac{15}{17}, \frac{15}{8}$ (iii) $\frac{\sqrt{15}}{4}, \sqrt{15}$
6. $\frac{1}{2}, \sqrt{3}; \frac{1}{\sqrt{3}}$ 7. $\frac{1}{\sqrt{2}}, \frac{1}{\sqrt{2}}, 1$ 8. $\frac{\sqrt{3}}{2}, \frac{1}{2}, \sqrt{3}, \frac{1}{2}, \frac{\sqrt{3}}{2}, \frac{1}{\sqrt{3}}$ 9. $(45°, \frac{1}{2}\sqrt{2})$
11. (ii) $\sqrt{3}, \sqrt{2}, \frac{2}{3}\sqrt{3}$ (iii) $\frac{15}{17}, \frac{17}{8}$
12. (i) 0, 1, 0 (ii) 1, 0 (iii) 0, −1, 0 (iv) −1, 0
15. $\frac{1}{2}, -\frac{\sqrt{3}}{2}, -\frac{1}{\sqrt{3}}$ 16. $-\frac{\sqrt{3}}{2}, -\frac{1}{2}, \sqrt{3}$ 17. $-\frac{1}{\sqrt{2}}, \frac{1}{\sqrt{2}}, -1$
18. (i) $\sin \theta, -\cos \theta, -\tan \theta$ (ii) $-\sin \theta, -\cos \theta, \tan \theta$ (iii) $-\sin \theta, \cos \theta, -\tan \theta$
19. $\frac{1}{\sqrt{2}}, \sqrt{3}, 0, -\frac{1}{\sqrt{3}}, 0, -\frac{\sqrt{3}}{2}, 0, 1$

20. $\frac{\sqrt{3}}{2}, -1, 1, \frac{\sqrt{3}}{2}$, 21 $\frac{1}{\sqrt{2}}, \frac{1}{2}, \sqrt{3}$
22. (i) $\cos \theta$ (ii) $-\sin \theta$; -0.34
23. (i) $-\cos \theta$ (ii) $-\sin \theta$, $\frac{1}{\tan \theta}$

Exercise 3.1:2

8. (i) 720° (ii) 240° (iii) 1080°
10. (i) 5, 240°; 5, −5 (ii) $\frac{3}{5}$, 216°; $\frac{3}{5}, -\frac{3}{5}$
    (iii) 4,360°; 4, −4
12. (i) 5, −5 (ii) 4, −2 (iii) 2, −4 (iv) 6, 2
18. (i) 1, 120° (ii) 3, 720° (iii) 5; 540°
21. (i) $-\cos \theta$ (ii) $4 \sin 5\theta$ (iii) $1 + 2 \cos \theta$
    (iv) $5(\cos 4\theta + 30°)$

Exercise 3.1:3

3. (i) 360° (ii) 108°
8. $\frac{1}{\sqrt{3}}, \frac{2}{\sqrt{3}}, \sqrt{2}, 0, -2, -\frac{2}{\sqrt{3}}, -\sqrt{3}, -\sqrt{2}$
9. (i) 1 (ii) cosec 2θ (iii) $\tan \theta$ (iv) $\sec \frac{\theta}{2}$ (v) 1
10. $\{\theta° : \theta \in \mathbb{R}, \theta \ne 180n°\}$, $(-\infty, -1) \cup (1, \infty)$, 360°, odd; $\{\theta° : \theta \in \mathbb{R}, \theta \ne 180n°\}$, $\mathbb{R}$, 180°, odd

Exercise 3.2:1

1. (i) 60°, −30°, 0 (ii) 30°, 120°, 90°
   (iii) 45°, −60°, 0
2. (i) 17.46° (ii) 41.41° (iii) 26.57° (iv) 78.69°
   (v) −41.81° (vi) 25.84° (vii) −56.31°
3. (i) 23.58°, −23.58° (ii) 78.46°, 101.54°
   (iii) 45°, 135° (iv) 71.57°, −71.57°
4. (i) 60°, 120° + 360n° (ii) 11.5°, 168.5° + 360n°
   (iii) −23.6°, 203.6° + 360n° (iv) 90° + 360n°
5. (i) ±75.5° + 360n° (ii) ±120° + 360n°
   (iii) ±45.6° + 360n° (iv) 180° + 360n°
6. (i) 30° + 180n° (ii) 71.6° + 180n°
   (iii) −18.4° + 180n° (iv) −45° + 180n°
7. 40.5°, 139.5° + 360n°
8. (i) ±70.5° + 360n° (ii) ±109.5° + 360n°
9. (i) 63.4° + 360n° (ii) 26.6° + 360n°
10. (i) 44.4°, 135.6° + 360n°; 44.4°, 135.6°
    (ii) ±126.9° + 360n°; 126.9°, 233.1°
    (iii) −63.4° + 180n°; 116.6°, 296.6°
    (iv) −5.7°, 185.7° + 360n°; 185.7°, 354.3°
    (v) −90° + 360n°; 270°
11. (i) −15°, 105° + 180n°; 105°, 165°, 295°, 345°
    (ii) ±15° + 120n°; 15°, 105°, 135°, 225°, 255°, 345° (iii) 120° + 360n°; 120°
    (iv) 10.6° + 36n°; 10.6°, 46.6°, 82.6°, 118.6°, 154.6°, 190.6°, 226.6°, 262.6°, 298.6°, 334.6°
    (v) −75°, 525° + 900n°; none
    (vi) 83.1°, 203.1° + 360n°; 83.1°, 156.9°
    (vii) 17.9°, 62.1° + 120n°; 17.9°, 62.1°, 137.9°, 182.1°, 257.9°, 302.1° (viii) ±30° + 90n°; 30°, 60°, 120°, 150°, 210°, 240°, 300°, 330°
12. (i) −285°, −165°, 75°, 195° (ii) −195°, −75°, 165°, 285°
13. (i) −160°, −10°, 20°, 170° (ii) −155°, −125°, −65°, −35°, 25°, 55°, 115°, 145°
    (iii) −120°, −30°, 0, 90°, 120°
14. (i) $0 \le \theta \le 30°$ or $150° \le \theta \le 360°$
    (ii) $0 \le \theta \le 26.6°$ or $153.4° \le \theta \le 206.6°$ or $333.4° \le \theta \le 360°$

Answers 575

16 (i) $\pm 45°$, $\pm 135° + 360n°$ (ii) 0, $\pm 30°$, $180° + 360n°$ (iii) 0, $30°$, $150°$, $180° + 360n°$ (iv) 0, $\pm 104.5°$, $180° + 360n°$ (v) $\pm 120°$, $180° + 360n°$ (vi) $-45°$, $63.4° + 180n°$ (vii) $45°$, $71.6° + 180n°$ (viii) $90°$, $\pm 120° + 360n°$
17 (i) $90n°$ (ii) $60n°$, $90°$ (iii) $45° + 90n°$, $180n°$ (iv) $120n°$, $180° + 360n°$ (v) $45n°$ (vi) $45n°$ (vii) $7.5° + 90n°$, $45° + 180n°$
18 $-45° + 180n°$, $22.5° + 90n°$
19 (i) $30° + 120n°$, $90° + 360n°$ (ii) $18° + 72n°$ (iii) $12.9° + 51.4n°$, $-90° + 360n°$ (iv) $-30° + 120n°$, $90° + 360n°$ (v) $30° + 60n°$ (vi) $60° + 120n°$ (vii) $52.5° + 180n°$ (viii) $40° + 90n°$, $-80° + 180n°$
20 $\left(-300°, \frac{\sqrt{3}}{2}\right)$, $(-180°, 0)$, $\left(-60°, \frac{-\sqrt{3}}{2}\right)$, $(0, 0)$, $\left(60°, \frac{\sqrt{3}}{2}\right)$, $(180°, 0)$, $\left(300°, \frac{-\sqrt{3}}{2}\right)$
21 (i) $\frac{360n°}{k-1}$, $\frac{180° + 360n°}{k+1}$ (ii) $72n°$; 6
22 (i) $\pm 60° + 360n°$, $\pm 60° + 360n°$ (ii) $-31.7° + 180n°$, $58.3° + 180n°$ (iii) $360n°$, $\pm 120° + 360n°$

Exercise 3.2:2

2 (i) $-90°$, $30°$, $150° + 360n°$ (ii) $\pm 41.4° + 360n°$ (iii) $30°$, $41.8°$, $138.2°$, $150° + 360n°$ (iv) $\pm 70.5°$, $\pm 120° + 360n°$ (v) $19.5°$, $30°$, $150°$, $160.5° + 360n°$ (vi) $17.6°$, $162.4° + 360n°$
3 0, $26.6° + 180n°$
4 (i) $\pm 60° + 360n°$ (ii) $45°$, $76.0° + 180n°$ (iii) $-14.0°$, $45° + 180n°$ (iv) 0, $\pm 131.8° + 360n°$
5 (i) $30°$, $90°$, $150° + 360n°$ (ii) $16.6°$, $23.6°$, $156.4°$, $163.4° + 360n°$ (iii) $41.8°$, $138.2° + 360n°$
6 $41.8°$, $138.2° + 360n°$
7 (i) $30°$, $150° + 360n°$ (ii) $-63.4°$, $26.6° + 180n°$ (iii) $-53.1°$, $90°$, $233.1° + 360n°$ (vi) $-45°$, $53.1° + 180n°$
8 $38.2°$, $141.8° + 360n°$
11 $\frac{5}{3}$, $\frac{4}{3}$ **12** $\pm 65.5° + 360n°$
13 (i) $\frac{x^2}{4} + y^2 = 1$ (ii) $(x-2)^2 + (y-1)^2 = 1$ (iii) $\frac{y^2}{16} - \frac{x^2}{9} = 1$ (iv) $\frac{a^2}{x^2} + \frac{y^2}{b^2} = 1$
14 $\frac{9}{x^2} - \frac{16}{y^2} = 1$
15 (i) $\frac{25}{x^2} - \frac{y^2}{4} = 1$ (ii) $\frac{225}{x^2} = \left(y - \frac{3}{4x}\right)^2$
16 (i) $x^2 = y^2(1-y^2)$ (iii) $x^2 = 1 - \frac{2}{y}$
17 (i) $x^2 + y^2 = 2$ (ii) $xy = 1$ (iii) $x^2 = (1-y)^2 + (1-y)^4$ (iv) $xy = 1$
18 $\pm\frac{12}{13}$, $\pm\frac{5}{13}$; $-\frac{12}{13}$, $-\frac{5}{13}$
19 (i) $\frac{1}{2}$, $-\frac{2}{\sqrt{3}}$ (ii) $\frac{24}{25}$, $-\frac{7}{24}$ (iii) $-\frac{2}{\sqrt{5}}$, $\sqrt{5}$ (iv) $\frac{-2\sqrt{2}}{3}$, $-2\sqrt{2}$
20 $\pm\frac{3}{\sqrt{10}}$, $\pm\frac{1}{\sqrt{10}}$; $\frac{3}{\sqrt{10}}$, $\frac{1}{\sqrt{10}}$
21 $\pm\frac{\sqrt{7}}{3}$, $\pm\sqrt{3}$ **22** $71.6°$; 25.8m   **23** $45°$, $71.6°$

Exercise 3.2:3

3 $-\sin\theta$, $\cos\theta$, $-\tan\theta$
4 (i) $\frac{1}{\sqrt{2}}(\sin\theta + \cos\theta)$ (ii) $\frac{1}{2}(\sqrt{3}\cos\theta + \sin\theta)$ (iii) $\frac{\sqrt{3} - \tan\theta}{1 + \sqrt{3}\tan\theta}$
5 (i) $\sin\theta$ (ii) $\cos\theta$ (iii) $-\cos 3\theta \cos\theta$ (iv) $8\cos 2\theta$
6 $\tan(\theta - 45°)$; $-45° + 180n°$
7 (i) $\frac{\sqrt{2}}{4}(\sqrt{3}-1)$ (ii) $2+\sqrt{3}$ (iii) $\sqrt{3}-2$ (iv) $\frac{\sqrt{2}}{4}(1-\sqrt{3})$ (v) $\frac{\sqrt{2}}{4}(\sqrt{3}-1)$
8 2   **9** (i) $60° + 180n°$ (ii) $13.2° + 180n°$ (iii) $52.5° + 180n°$ (iv) $45° + 180n°$
10 $75° + 180n°$ **11** $\frac{\cos\beta - \sin\alpha}{\sin\beta + \cos\alpha}$
12 $\pm\frac{117}{125}$, $\pm\frac{3}{5}$; $-\frac{44}{125}$, $\frac{4}{5}$
13 $-\frac{33}{65}$; $\frac{16}{65}$ **14** $\frac{168}{625}$; 0   **15** $\pm\frac{19}{4\sqrt{5}}$, $\pm\frac{11}{8\sqrt{5}}$
16 $\pm\frac{1}{5\sqrt{2}}$; $\pm\frac{1}{\sqrt{2}}$; $45°$ **17** -0.6, 0.8

Exercise 3.2:4

1 (i) $\frac{1}{2}$ (ii) $-\frac{\sqrt{3}}{2}$ (iii) -1
2 (i) $-\frac{1}{2}$ (ii) $\frac{7}{25}$ **3** $-\frac{24}{25}$, $-\frac{7}{25}$; $\frac{336}{625}$ **4** $-\frac{28560}{28561}$
5 (i) $\pm\frac{24}{25}$, $-\frac{7}{25}$ (ii) $\pm\frac{336}{625}$, $\frac{527}{625}$ (iii) $\pm\frac{4}{5}$, $-\frac{3}{5}$
6 $\frac{24}{7}$, $-\frac{336}{527}$ **7** $\frac{24}{7}$, $\pm\frac{120}{119}$
8 $-3 \pm \sqrt{10}$ **9** $\sqrt{2}+1$
10 $\frac{\cot^2\theta - 1}{2\cot\theta}$; $\sqrt{2}+1$
11 $\frac{1}{3}$; $\frac{1}{\sqrt{10}}$, $\frac{3}{\sqrt{10}}$; $\frac{3}{5}$, $\frac{4}{5}$
12 $\frac{1}{2}(1-\cos 2\theta)$, $\frac{1}{2}(1+\cos 2\theta)$; $\frac{1}{2}\sqrt{2-\sqrt{3}}$, $\frac{1}{2}\sqrt{2+\sqrt{3}}$
13 (i) $30°$, $90°$, $150° + 360n°$ (ii) $\pm 48.2°$, $\pm 60° + 360n°$ (iii) $\pm 120°$, $\pm 126.9° + 360n°$ (iv) $30°$, $\pm 90°$, $150° + 360n°$ (v) $-30°$, $\pm 90°$, $210° + 360n°$ 49(vi) $17.8°$, $90°$, $162.2° + 360n°$ (vii) $\pm 72.8° + 360n°$ (viii) $45°$, $90° + 180n°$ (ix) $\pm 39.2 + 180n°$ (x) $-75°$, $-15° + 180n°$
14 0, $45°$, $\pm 53.1°$, $135°$, $180° + 360n°$
15 $21.5°$, $158.5° + 360n°$
16 $-90°$, $30°$, $150° + 360n°$
17 $\pm 60° + 180n°$
20 (i) $\tan^2\theta$ (ii) $\tan\theta$ (iii) 2 (iv) $\tan\theta$
21 (i) $\frac{\sqrt{3}}{2}$ (ii) $-\frac{\sqrt{3}}{2}$
22 $2\cos^2\theta - 1$; $2\cos^2 2\theta - 1$; $2\cos^2 4\theta - 1$
23 (i) $x = 2y^2 - 1$ (ii) $x = \frac{4}{4-y^2}$ (iii) $x = 1 - \frac{2}{y^2}$ (iv) $y = \frac{x^2}{x^2-2}$ (v) $x^2 + y^2 = 2$ (vi) $x^2 = 4(y^2 - y^4)$
24 $3\sin\theta - 4\sin^3\theta$ **25** -3.411, 1.185, 2.227
26 $\frac{1}{4}(\sqrt{5}+1)$ **27** $\pm 30° + 180n°$; $2 - \sqrt{3}$
28 0, $\pm 16.8° + 180n°$
29 (i) $\pm 60°$, $\pm 90°$, $\pm 120° + 360n°$ (ii) 0, $\pm 90°$, $\pm 120° + 360n°$ (iii) $-48.6°$, $30°$, $150°$, $228.6° + 360n°$ (iv) 0, ± 41.4, $180° + 360n°$
30 $231.3°$, $308.7°$

Exercise 3.2:5

2 $-\frac{12}{13}$, $\frac{5}{13}$; $-\frac{120}{169}$, $-\frac{119}{169}$
3 $\frac{24}{7}$ **4** $-\frac{2}{3}$, $\frac{3}{2}$
5 (i) $2+\sqrt{3}$ (ii) $\tan\left(45° - \frac{\theta}{2}\right)$, $2-\sqrt{3}$
6 (i) $\frac{1}{t}$ (ii) t^2 (iii) $\frac{1-t^2}{2t^2}$ (iv) $\frac{(t-1)^2}{3-t^2}$ (v) t
7 $\frac{2+t^2}{t}$
8 (i) $-143.1°$, $-90° + 360n°$ (ii) $-145.6°$, $26.1° + 360n°$ (iii) $-36.9°$, $53.1° + 360n°$ (iv) $-62.3°$, $17.1° + 360n°$ (v) $-90°$, $\pm 45° + 360n°$
9 $53.1° + 360n°$ (ii) $-143.1°$, $180° + 360n°$
10 $90°$, $143.1° + 360n°$
11 (ii) $(0, 1)$, $(-\frac{4}{5}, -\frac{3}{5})$ (iii) $-143.1°$, $90° + 360n°$
12 0, $106.3° + 360n°$
13 $(1, 0)$, $(2, \sqrt{3})$; 0, $60° + 360n°$
14 No
15 (i) $-58.3°$, $31.7° + 180n°$ (ii) $\pm 30° + 180n°$ (iii) 0, $-45° + 180n°$
16 $67.4° + 360n°$
19 (i) $-126.9°$, $90° + 360n°$ (ii) $71.5 + 360n°$

Exercise 3.2:6

1 (i) 2, $30°$ (ii) 13, $22.6°$ (iii) $\sqrt{5}$, $26.6°$ (iv) 5, $36.9°$
2 $5\cos(\theta - 53.1°)$
3 $\sqrt{a^2+b^2}$, $\frac{b}{a}$; $\sqrt{a^2+b^2}$, $\frac{a}{b}$; Yes, No
4 (i) $\sqrt{10}\cos(3\theta - 18.4°)$ (ii) $17\cos(2\theta + 61.9°)$
6 $5\cos(\theta + 36.9°)$; $16.3°$, $-90° + 360n°$
7 (i) $-15°$, $105° + 360n°$ (ii) $-1.9°$, $-121.9° + 360n°$ (iii) $-83.1°$, $-23.1° + 360n°$ (iv) $-36.9°$, $90° + 360n°$ (v) $37.5°$, $127.5° + 180n°$ (vi) $-18°$, $40.5° + 72n°$
8 (i) $113.1°$, $353.1°$ (ii) $90°$, $330°$ (iii) $233.1°$, $339.4°$ (iv) $270°$, $323.1°$ (v) $43.3°$, $163.3°$, $223.3°$, $343.3°$
9 $\alpha + 30°$, $90° - \alpha + 360n°$
10 $4\sin 2\theta + 3\cos 2\theta + 3$; $5\sin(2\theta - 36.9°) + 3$; $23.4°$, $96.6° + 180n°$
11 (i) $-73.8°$, $55.4° + 180n°$ (ii) $45°$, $90° + 180n°$
13 (i) 2, $30°$; -2, $210°$
14 (i) 33, 7 (ii) $\frac{1}{7}$, $\frac{1}{33}$
15 (i) 25, -25 (ii) $\sqrt{13}$, $-\sqrt{13}$ (iii) 22, 2 (iv) $\frac{1}{2}$, $\frac{1}{22}$ (v) 1, $\frac{1}{5}$
16 $26.6°$, $-153.4° + 360n°$
17 $\sqrt{a^2+b^2}$, $-\sqrt{a^2+b^2}$
19 0, $126.9° + 360n°$
20 (i) $103.3° \le \theta \le 330.5°$ (ii) $0 \le \theta \le 127.1$ or $173.8° \le \theta \le 307.1°$ or $353.8° \le \theta \le 360°$
21 $\sqrt{13}\sin(\theta - 33.7°)$; $33.7°$, $213.7° + 360n°$
23 $\sqrt{17}\cos(\theta - 14.0°)$ (iii) $-29.3°$, $57.3° + 360n°$
24 $36.9°$, $180° + 360n°$
26 $5\cos(2\theta + 53.1°) + 3$; 8, -2
27 $\sqrt{5}$, $63.4°$, -2; $-18.4°$, $45° + 180n°$

Exercise 3.2:7

2 (i) $\sin 3\theta + \sin\theta$ (ii) $\sin 6\theta - \sin 4\theta$ (iii) $\cos 7\theta + \cos\theta$ (iv) $\cos 2\theta - \cos 6\theta$ (v) $\cos 6\theta - \cos 2\theta$
3 (i) $\frac{1}{2}$ (ii) $\frac{1}{2}(\sqrt{3}-\sqrt{2})$
4 $\sin 3\theta - \sin\theta$
5 $\sin(2\theta + 30°) - \sin 30°$; $30° + 180n°$

576 Answers

6 $\sin(2\theta + \alpha) - \sin \alpha$; $45° - \frac{1}{2}\alpha + 180n°$
7 $\frac{1}{2}\{\cos(X+Y) + \cos(X-Y)\}$,
$\frac{1}{2}\{\cos(X-Y) - \cos(X+Y)\}$;
$-45°$, $15° + 180n°$
8 $-33.8°$, $63.8° + 180n°$
9 (i) $2\sin 2\theta \cos \theta$ (ii) $2\cos 3\theta \cos \theta$
(iii) $\cos \theta$ (iv) $-\sqrt{2}\sin \theta$
10 (i) $\frac{1}{\sqrt{2}}$ (ii) $\sqrt{3}$
15 (i) $120n°$, $180n°$ (ii) $60n°$, $\pm 30° + 180n°$
(iii) $\pm 30° + 120n°$, $\pm 30° + 180n°$
(iv) $30° + 120n°$, $180n°$ (v) $72n°$, $90n°$
16 (i) $30°$, $45°$, $135°$, $150°$ (ii) 0, $90°$, $180°$, $270°$, $360°$ (iii) $13.2°$, $226.8°$ (iv) $120°$, $240°$, $300°$
17 $90° - \alpha$, $270° - \alpha$
18 0, $\pm 60°$, $180° + 360n°$
19 $2\cos(2n-1)\theta \cos \theta$; $4\cos^3\theta - 3\cos\theta$,
$8\cos^4\theta - 8\cos^2\theta + 1$, $16\cos^5\theta - 20\cos^3\theta + 5\cos\theta$

Exercise 3.3:1

1 (i) $b = 6.76$, $c = 8$, $C = 65°$ (ii) $a = 4.90$, $b = 6.69$, $A = 45°$ (iii) $a = 4.04$, $b = 1.60$, $B = 20°$
2 (i) $b = 5.41$, $B = 72.9°$, $C = 62.1°$;
$b = 1.66$, $B = 17.1°$, $C = 117.9°$
(ii) $b = 9.20$, $B = 113.1°$, $C = 36.9°$;
$b = 1.20$, $B = 6.9°$, $C = 143.1°$
(iii) $a = 5.58$, $A = 71.7°$, $B = 58.3°$;
$a = 0.85$, $A = 8.34°$, $B = 121.7°$
3 (i) $b = 6.65$, $B = 109.9°$, $C = 25.1°$
(ii) $b = 5.21$, $B = 34.3°$, $C = 25.7°$
5 (i) $A = 22.3°$, $B = 49.5°$, $C = 108.2°$
(ii) $A = 47.2°$, $B = 54.7°$, $C = 78.1°$
(iii) $A = 29.5°$, $B = 38.1°$, $C = 112.4°$
6 $98.2°$ **7** $49.5°$
8 (i) $c = 2.65$, $A = 40.9°$, $B = 79.1°$ (ii) $c = 9.76$, $A = 98.8°$, $B = 41.2°$ (iii) $c = 8.06$, $A = 38.3°$, $C = 93.7°$ (iv) $a = 25.70$, $B = 23.9°$, $C = 36.1°$
9 $2r\sin\frac{\alpha}{2}$
11 (i) $c = 5.99$, $A = 56.4°$, $C = 93.6°$; $c = 2.67$, $A = 123.6°$, $C = 26.4°$ (ii) impossible
(iii) $c = 4.56$, $A = 19.5°$, $C = 130.5°$
(iv) $c = 1.10$, $A = 19.5°$, $C = 10.5°$
12 3.6 km, 73.9° **13** 8.7 km, 116.2°
14 190.8 km, 147.0° **15** 31.2 kmh^{-1}, 95.4°
16 60°, 13 **19** 5 **21** 2.87 cm
23 234.6°, 52.8°, 72.8°; 3.38 **25** 90°, 63.4°

Exercise 3.3:2

1 (i) 8.67 (ii) 3.56 (iii) 9.68 (iv) 12.97 (v) 17.02
2 5 **3** 4.11, 6.45 **4** (i) 12 (ii) 18.0

Exercise 3.4:1

1 (i) (a) π (b) $\frac{\pi}{2}$ (c) $\frac{\pi}{6}$ (d) $\frac{\pi}{9}$ (e) $\frac{\pi}{10}$ (f) $\frac{3\pi}{4}$
(g) $\frac{5\pi}{6}$ (h) $\frac{3\pi}{2}$ (i) $\frac{5\pi}{3}$ (j) $\frac{5\pi}{4}$ (k) $\frac{5\pi}{2}$ (l) 4π
(ii) (a) 0.0175 (b) 0.7191 (c) 3.7477
2 (i) (a) 90° (b) 135° (c) 120° (d) 36° (e) 210° (f) 330° (g) 15° (h) 12° (i) 54° (j) 900°
(ii) (a) 57.30° (b) 8.02° (c) 250.96°
3 (i) 1 (ii) $\frac{1}{\sqrt{2}}$ (iii) $\sqrt{3}$ (iv) $-\frac{1}{2}$ (v) -2 (vi) -1
(vii) $\sqrt{2}$ (viii) 0 (ix) $\frac{\sqrt{3}}{2}$

4 (i) 0.977 (ii) -0.452 (iii) 1.601
5 (i) $\sin x$ (ii) $\cos x$ (iii) $\cot x$ (iv) $\sin x$
(v) $-\cos x$ (vi) $\cot x$ (vii) $-\cos x$
6 (i) $\pm\frac{5\pi}{6} + 2\pi n$; $\frac{5\pi}{6}$, $\frac{7\pi}{6}$
(ii) $0.08 + \frac{\pi n}{3}$; 0.08, 1.13, 2.18, 3.22
(iii) $\frac{\pi}{8} + \frac{\pi n}{2}$, $-\frac{\pi}{4} + \pi n$; $\frac{\pi}{8}$, $\frac{5\pi}{8}$, $\frac{3\pi}{4}$, $\frac{9\pi}{8}$, $\frac{13\pi}{8}$, $\frac{7\pi}{4}$ (iv) $\pm\frac{\pi}{3} + 2\pi n$; $\frac{\pi}{3}$, $\frac{5\pi}{3}$ (v) $\frac{\pi}{4} + \pi n$; $\pm\frac{\pi}{2} + 2\pi n$, $\frac{\pi}{4}$, $\frac{\pi}{2}$, $\frac{5\pi}{4}$, $\frac{3\pi}{2}$ (vi) -2.44, $1.26 + 2\pi n$; 1.26, 3.84 (vii) $\pm\frac{\pi}{3} + \frac{4\pi n}{3}$, $\frac{\pi}{2} + 2\pi n$; $\frac{\pi}{3}$, $\frac{\pi}{2}$, π, $\frac{5\pi}{3}$

8 4 **13** 245.7
15 (i) $\frac{1}{2}r^2(\alpha - \sin\alpha)$ (ii) $\frac{1}{2}r^2(2\pi - \alpha + \sin\alpha)$
16 $\frac{2\pi r}{i}$ **18** 27.94 cm
19 $r^2(2\theta - \sin 2\theta)$ **20** $\frac{10\pi + 3\sqrt{3}}{2\pi - 3\sqrt{3}}$
23 76.53 m **24** 5.8 m^2, 36.9 km
25 (i) 58.3s (ii) 67.2s; $\frac{2\pi}{3}$ **26** 206.8 m^2

Exercise 3.4:2

1 (i) $\frac{\theta}{2}$ (ii) $\frac{1}{2}$ (iii) 4 (iv) 2 (v) $\frac{1}{8}$ (vi) $\frac{1}{2}$
(vii) $\frac{1-\theta}{1+\theta}$
2 $\frac{1-\cos\theta}{\cos\theta - \cos 2\theta}$; $\frac{1}{3}$ **3** $\frac{1}{10}$ **4** $\frac{1}{9}$ **5** $\frac{a}{b}$
6 (i) $\frac{a^2}{b^2}$ (ii) 0
8 (i) $\frac{2}{3}$ (ii) $\frac{1}{4}$ (iii) 2 (iv) $\frac{2}{5}$ (v) $\frac{3}{8}$ (vi) $\frac{1}{4}$
9 2 **10** (i) 2 (ii) $\frac{1}{4}$
11 (i) $\frac{1}{4}$ (ii) $-\sin\alpha$

Exercise 3.4:3

3 (i) $2\cos 2x$ (ii) $-3\sin 3x$ (iii) $\frac{3}{5}\cos\frac{3x}{5}$
(iv) $2\sin x\cos x$ (v) $-4\cos^3 x\sin x$
(vi) $-10\cos 5x\sin 5x$ (vii) $\frac{5}{3}\sin^4\frac{x}{3}\cos\frac{x}{3}$
(viii) $\frac{\cos x}{2\sqrt{\sin x}}$ (ix) $\frac{\tan x}{2\sqrt{\cos x}}$
(x) $2(\sin 2x + \cos x)(2\cos 2x - \sin x)$
(xi) $\frac{\cos\sqrt{x}}{2\sqrt{x}}$ (xii) $-\sin(1+x)$
(xiii) $\frac{\pi}{180}\cos x°$
4 (i) $a\cos(ax+b)$ (ii) $-a\sin(ax+b)$
(iii) $n\sin^{n-1}x\cos x$ (iv) $-n\cos^{n-1}x\sin x$
(vi) $an\sin^{n-1}(ax+b)\cos(ax+b)$
(vii) $-an\cos^{n-1}(ax+b)\sin(ax+b)$
5 (i) $\sin 3x + 3x\cos 3x$
(ii) $2\cos 2x\cos 3x - 3\sin 2x\sin 3x$
(iii) $\sin x\cos x(2\cos^2 x - 3\sin^2 x)$
(iv) $-\frac{4x\sin 4x + \cos 4x}{x^2}$
(v) $\frac{x\{(2+x)\sin x + (2-x)\cos x\}}{1 + 2\sin x\cos x}$

9 (ii) $-\csc x\cot x$, $\sec x\tan x$
10 (i) $3\sec^2 3x$ (ii) $\frac{1}{2}\sec^2\frac{1}{2}x$ (iii) $\frac{\pi}{180}\sec^2 x°$
(iv) $4\sec 4x\tan 4x$ (v) $-\frac{2}{3}\csc^2\frac{2x}{3}$
(vi) $-5\csc 5x\cot 5x$ (viii) $\frac{\sec^2\sqrt{x}}{2\sqrt{x}}$
(viii) $2x\sec(x^2+1)\tan(x^2+1)$
(ix) $\frac{-2}{(1-x)^2}\csc^2\frac{1+x}{1-x}$ (x) $2\tan x\sec^2 x$
(xi) $\frac{-\csc^2 x}{2\sqrt{\cot x}}$ (xii) $3\sec^3 x\tan x$
(xiii) $-9\cot^2 3x\csc^2 3x$
(xiv) $-15\csc^3 5x\cot 5x$
11 (i) $-\csc x\cot x$
(ii) $-4\cos 2x\csc^2 4x - 2\sin 2x\cot 4x$
(iii) $x\sec^2 x + \tan x$
(iv) $2x\tan x(x\sec^2 x + \tan x)$
(v) $\frac{\sec x\tan x + \tan^2 x - 1}{(1+\sin x)^2}$
13 (i) (a) $e^x(\cos x - \sin x)$
(b) $e^{-x/3}(3\cos 3x - \frac{1}{3}\sin 3x)$
(c) $\frac{\sec^2 x - \tan x}{e^x}$
(ii) (a) $-\sin x\, e^{\cos x}$
(b) $\sec x(\sec x + \tan x)e^{\tan x + \sec x}$
(c) $(x\cos x + 1)e^{\sin x}$
(d) $6\sec^3 2x\tan 2x\, e^{\sec^3 2x}$
15 5, 0.93
16 (i) (a) $-\tan x$ (b) $3\cot 3x$ (c) $-2\cot x$
(d) $\frac{\cos x}{2(1+\sin x)}$ (e) $\frac{\cos x}{1+\sin x} - \frac{2\cos 2x}{1+\sin 2x}$
(f) $2\sec 2x$
(ii) (a) $x^{\sin x}\left(\frac{\sin x}{x} + \cos x\ln x\right)$
(b) $\sin x^{\cos x}(\cos x\cot x - (\sin x)(\ln\sin x))$
(c) $\cos x^{\tan x}(-\tan^2 x + (\sec^2 x)(\ln\cos x))$
18 -1, 3 max, $-\frac{3}{2}$ min; 0.38, 2.76
19 $\left(\sqrt{3} - \frac{\pi}{3}\right)$ max, $\left(\frac{\pi}{3} - \sqrt{3}\right)$ min
20 2, $2\sqrt{2}$, $-2\sqrt{2}$, -2, $-2\sqrt{2}$, $2\sqrt{3}$, 2; $\frac{\pi}{2}$, $\frac{3\pi}{2}$
23 $\sqrt{5}-1:2$ **24** $3\pi\sqrt{6}$ m^2 **25** $\frac{3}{4}r^2\sqrt{3}$ **26** $\frac{1}{16}$ m^2
27 13.7 km **28** 17.7s **30** 18.2m

Exercise 3.5:1

1 $[0,\pi]$; $[-1,1]$
2 $\left[-\frac{\pi}{2},\frac{\pi}{2}\right]$; $y = \pm\frac{\pi}{2}$; $\mathbb{R}$
3 (i) 0 (ii) $\frac{\pi}{6}$ (iii) $-\frac{\pi}{6}$ (iv) $-\frac{\pi}{4}$ (v) $\frac{\pi}{6}$
(vi) $\frac{\pi}{3}$ (vii) $\frac{\pi}{2}$
6 $\sqrt{1-x^2}$ **7** $\frac{\pi}{2} - |x|$ **8** (i) $\frac{1-x^2}{1+x^2}$ (ii) $\frac{2x}{1-x^2}$
11 (i) $\frac{\pi}{4}$ (ii) $\frac{\pi}{2}$ (iii) $\frac{\pi}{4}$ (iv) $\frac{\pi}{4}$
12 (i) $\frac{2}{9}$ (ii) $\frac{\sqrt{21}}{14}$ (iii) ± 4 (iv) $\frac{1}{\sqrt{3}}$
13 (i) F (ii) T (iii) F (iv) T (v) T

Exercise 3.5:2

2 $-\frac{1}{x\sqrt{x^2-1}}$, $-\frac{1}{1+x^2}$

Answers 577

3 (i) $\frac{2}{\sqrt{1-4x^2}}$ (ii) $-\frac{1}{1-2x+2x^2}$
(iii) $\frac{-\sec^2 x}{\sqrt{1-\tan^2 x}}$ (iv) $\frac{1}{1+x^2}\sin(\cot^{-1} x)$
6 7.8

Miscellaneous Exercise 3

1 (i) $50.3° < x < 120°$ or $189.6° < x < 360°$
 (ii) $2\sin(3x+60°)$
2 (i) (a) π (b) 6π (c) 12π (d) the L.C.M. of a and b (ii) 2π
3 (ii) (a) π (b) π (c) π (d) π
5 (i) $-57.5°, 36.4°, 143.6°, 227.5° + 360n°$
 (ii) $\pm 109.5° + 360n°$ (iii) $60n°$
6 (i) $-23.1°, 96.9° + 360n°$ (ii) $30°, 150° + 360n°, \pm 45° + 180n°$ (iii) $-45°, 33.7° + 180n°$
7 (i) $-53.1°, 36.9° + 360n°$ (ii) $7.5° + 180n°$
 (iii) $\pm 70.5°, 180° + 360n°$
8 (i) $\frac{1}{2}, -2$ (ii) $\frac{17}{32}$ (iii) $\pm \mu \sqrt{1-\lambda^2} \pm \lambda\sqrt{1-\mu^2}$
9 (i) $2 + \sqrt{3}$ (ii) $\pm 35.3° + 180n°$
10 (ii) $\pm 90°, \pm 120° + 360n°$ (iii) $0 < x < 90°$ or $120° < x < 240°$ or $270° < x < 360°$
11 $10\cos(x + 36.9°)$; (i) $10, -10$
 (ii) $-96.9°, 23.1° + 360n°$
12 $25\cos(x-16.3°)$; (ii) $\frac{1}{5}, \frac{1}{55}$
 (iii) $-50.2°, 82.7° + 360n°$
14 $-61.9°, 90° + 360n°$
15 (ii) $30° - \alpha, 150° - \alpha + 360n°$
16 (i) $\pm 45° + 180n°$; $\pm 22\frac{1}{2}° + 90n°$
17 (i) $22\frac{1}{2}° + 90n°$ (ii) $45°, 90° + 180n°$
 (iii) $51.3°, 128.7° + 360n°$
 (iv) $\pm 45° + 180n°$; $-30°, 210° + 360n°$
18 (ii) $\pm 30°, \pm 60° + 180n°$
19 21.65 km, 14.82 km
20 (i) 1 am Saturday (ii) 4.14 am Saturday, 5 am Tuesday
21 2.34 22 (i) 1.17 (ii) 1.18
23 $\frac{1}{12}$ 24 (i) 2 (ii) 5 (iii) 1 (iv) 4
25 1 26 (i) $\sqrt{3}, -\sqrt{3}$ (ii) $20.8°, 122.3° + 360n°$
27 (i) $0, \pi, \frac{3\pi}{2}, 2\pi$ (ii) $\left(\frac{\pi}{2}, -4\right), \left(\frac{7\pi}{6}, \frac{1}{2}\right)$,
 $\left(\frac{3\pi}{2}, 0\right), \left(\frac{11\pi}{6}, \frac{1}{2}\right)$; 3
28 (i) 1.46 pm, 1.46 am, 13 m, 12 h
 (ii) (a) -1.57 mh^{-1} (b) ± 2.39 mh^{-1}; 2.62 mh^{-1}
29 121.89 30 1.114, 1.114
31 (i) $\mathbb{R}$ (ii) $g(x) = \begin{cases} \frac{1}{2}x & x \leq 0 \\ \frac{1}{2}\tan^{-1} x & x > 0 \end{cases}$; $\mathbb{R}$
 (iii) (a) $\frac{\pi}{8}$ (b) 1
32 (i) $[\frac{1}{2}, 1]$ (ii) $\sin^{-1}\frac{1-x}{x}$; $[\frac{1}{2}, 1]$, $\left[0, \frac{\pi}{2}\right]$
 (iv) 0.63

Chapter 4

Exercise 4.1:2

1 (a) (i) $\frac{1}{8}x^8 + c$ (ii) $\frac{-1}{3x^3} + c$ (iii) $\frac{2}{3}x\sqrt{x} + c$
 (iv) $\frac{2}{5}x^2\sqrt{x} + c$ (v) $-\frac{2}{3x\sqrt{x}} + c$
2 (i) $\frac{4x^3}{3} - \frac{5x^4}{4} + c$ (ii) $\frac{t^4}{4} - \frac{t^3}{3} + c$

(iii) $2x^3 + \frac{x^2}{2} - 2x + c$
(iv) $\frac{x^5}{5} + \frac{10x^3}{3} + 25x + c$ (v) $-\frac{5}{u} - \frac{2}{u^2} + c$
(vi) $-\frac{7}{x} - x + c$ (vii) $\frac{2}{3}p\sqrt{p} + p + c$
(viii) $\frac{2}{3}x\sqrt{x} + 2\sqrt{x} + c$ (ix) $-\frac{4}{\sqrt{s}} + \frac{3}{s} + c$
3 (i) $\frac{1}{2}ax^2 + bx$ (ii) $\frac{1}{3}ax^3 + \frac{1}{2}bx^2 + cx$
 (iii) $\frac{2}{5}ax\sqrt{x} + bx$ (iv) $\frac{2}{3}ax\sqrt{x} + 2b\sqrt{x} + c$
4 (i) $3 - \frac{1}{x}$ (ii) $2x^2\sqrt{x} - 4$ (iii) $x^3 + x^4 - 2$
6 (i) $-2x^2 + 3$ (ii) $-2x^2 + 1$ (iii) $-2x^2$
7 $x^2 + 3x - 2; -\frac{17}{4}$
8 $y = 2x^2 - \frac{1}{x}$
9 (i) $y = 8x + 5$ (ii) $\frac{31}{27}$, 1
10 $f(x) = \begin{cases} 2x - 2 & x < 0 \\ x^2 - 1 & x \geq 1 \end{cases}$
11 $f(x) = \begin{cases} 6x + 12 & x \leq -2 \\ 2x - x^2 + 8 & -2 < x < 4 \\ 6x - 24 & x \geq 4 \end{cases}$;
$-12, 8, 0, 24$; yes

12 (i) $12(3x+5)^3$; $\frac{1}{12}(3x+5)^4 + c$ (ii) $\frac{x}{\sqrt{1-x^2}}$;
$\sqrt{1-x^2} + c$ (iii) $3x\sqrt{4+x^2}$; $\frac{1}{3}(4+x^2)^{3/2} + c$
13 $-\frac{1}{(x-1)^2}$; $\frac{1}{1-x} + c$; $x - \frac{1}{1-x} + c$
14 $-\frac{2}{(2x+1)^2}$; $-\frac{1}{2(2x+1)} + c$; $\frac{1}{2}x + \frac{1}{4(2x+1)} + c$
15 e^{ax}; $\frac{1}{a}\mathrm{e}^{ax} + c$ (i) $\frac{1}{5}\mathrm{e}^{5x} + c$ (ii) $\frac{2}{3}\mathrm{e}^{3x/2} + c$
 (iii) $-\mathrm{e}^{-x} + c$ (iv) $-\frac{5}{4}\mathrm{e}^{-4x/5} + c$
16 (i) $3x^2 \mathrm{e}^{x^3}$; $\frac{1}{3}\mathrm{e}^{x^3} + c$ (ii) $\cos x \mathrm{e}^{\sin x}$, $\mathrm{e}^{\sin x} + c$
17 $(x+1)\mathrm{e}^x$; $(x-1)\mathrm{e}^x + c$
18 (i) $\frac{1}{4}(\mathrm{e}^{4x} + 3)$ (ii) $\frac{1}{2} - \mathrm{e}^{-x}$ (iii) $2\mathrm{e}^{x/2}$
19 $a^x \ln a$; $\frac{a^x}{\ln a} + c$ (i) $\frac{10^x}{\ln 10} + c$ (ii) $\frac{3^{2x}}{2\ln 3} + c$
 (iii) $2x(5^{x^2})$; $\frac{1}{2}(5^{x^2})$
20 (i) $-2\sin 2x$; $-\frac{1}{2}\cos 2x + c$
 (ii) $5\cos 5x$; $\frac{1}{5}\sin 5x + c$; $\frac{1}{a}\sin ax + c$;
$-\frac{1}{a}\cos ax + c$ (a) $2\sin\frac{x}{2} + c$ (b) $-\frac{3}{5}\cos\frac{5x}{3} + c$
21 $\sec^2 x$; $\tan x + c$; $\tan x - x + c$; $-\cot x + c$;
$-(\cot x + x) + c$
22 (i) $4\sin^3 x \cos x$; $\frac{1}{4}\sin^4 x + c$
 (ii) $-3\cos^2 x \sin x$; $-\frac{1}{3}\cos^3 x + c$
 (iii) $4\tan^3 x \sec^2 x$; $\frac{1}{4}\tan^4 x + c$

Exercise 4.1:3

1 (i) 3 (ii) $\frac{43}{3}$ (iii) $\frac{4}{3}(4-\sqrt{2})$ (iv) 0
 (v) $\frac{6}{5}(25\sqrt{5} - 4\sqrt{2})$
2 (i) $2\ln 2$ (ii) $2\ln 2$ (iii) $\ln\frac{3}{2}$ (iv) 1
 (v) $1 + \ln 2$
3 (i) $\mathrm{e} - 1$ (ii) 1 (iii) 2
4 (i) $\frac{1}{2}$ (ii) 2 (iii) $\frac{1}{2}(\sqrt{2} - 1)$ (iv) 1
5 (i) $\frac{15}{4}$ (ii) $\frac{2}{3}$ (iii) $\frac{62}{5}$ (iv) $\frac{4}{3}$ (v) 9 (vi) $\ln 3$
 (vii) 4 (viii) 1 (ix) 1 (x) $\frac{1}{\sqrt{2}}$
6 $\frac{32}{3}$ 7 (i) $\frac{32}{3}$ (ii) $\frac{27}{4}$ 8 (i) $\frac{9}{2}$ (ii) $\frac{9}{2}$
9 $\frac{4}{3}$ 10 (i) $\frac{27}{4}$ (ii) $\frac{79}{6}$ 11 $\frac{13}{3}$
12 (i) $\frac{5}{2}$ (ii) $\frac{1}{2}$ (iii) $\frac{1}{3}(2a^2 + ab + b^2)$
13 (i) $\frac{31}{6}$ (ii) $\frac{1}{3}(b^3 - 2c^3 + d^3) - \frac{1}{2}(a+c) \times (b^2 - 2c^2 + d^2) + ac(b - 2c^3 + d)$ 14 36
17 (i) $-\frac{15}{4}$ (ii) $-\frac{22}{3}$ (iii) $-\frac{16}{3}$ (iv) -160 (v) $-\frac{1}{2}$
18 (i) $-\frac{32}{3}$ (ii) $\frac{2}{3}$
19 (i) $-\frac{32}{3}$ (ii) $-\frac{4}{3}$ (iii) $-\frac{16}{3}$ (iv) $\frac{32}{3}$

20 (i) $\frac{26}{3}$ (ii) 13 (iii) $\frac{117}{3}$ (iv) $\frac{25}{2}$
21 (i) $\frac{20}{3}$ (ii) $\frac{28}{3}$ (iii) $\frac{49}{2}$
22 (i) $\frac{79}{16}\pi - 1$ (ii) $\frac{25}{12}\pi - \frac{3}{2}$
23 (i) $\frac{81}{4}$ (ii) -4 (iii) $\frac{65}{4}, \frac{65}{4}, \frac{97}{4}$
24 (i) $-\frac{2}{3}$ (ii) $\frac{4}{3}$ (iii) $\frac{2}{3}; \frac{2}{3}, 2$
25 (i) $\frac{65}{6}$ (ii) 24 (iii) $\frac{49}{6}$ (iv) 8 (v) $\frac{37}{12}$ (vi) $\frac{1}{2}$
 (vii) $\frac{3}{2}$ (viii) 2
27 $\frac{104}{3}, \frac{208}{3}$ 28 48
33 $0; 0$ (iii) $0; \sin a; 2\sin a$
34 (iii) (a) $\int_{-1}^{3} x^3 \, \mathrm{d}x$ (b) $\int_0^2 x^3 \, \mathrm{d}x$
35 (i) $\frac{33}{4}, \frac{33}{4}, \frac{33}{4}$ (ii) 0 36 $\frac{13}{4}; \frac{2}{3}$ 37 $\frac{32}{3}; \frac{4}{3}$
38 (i) $\frac{1}{6}$ (ii) $\frac{1}{6}$ 40 $(\frac{1}{2}, 0); \frac{1}{4}$ 41 $\frac{1}{12}$
42 $\frac{9}{4}; 18$ 45 $\frac{1}{4}\pi^2 - 2$ 46 $(\frac{1}{2}\pi - 1, 1); \frac{1}{2}(\pi - 3)$
48 (i) $\frac{1}{48}$ (ii) $\frac{1}{12}$ (iii) $\frac{27}{64}$ (iv) $\frac{61}{12}$ (v) 0.06
49 36 50 (i) $\frac{4}{27}$ (ii) $\frac{4}{3}$ (iii) $\frac{1}{3}$ (iv) $\frac{1}{3}$
51 (i) $\frac{4}{3}\sqrt{2}$ (ii) $\frac{1}{6}$ (iii) $\frac{9}{2}$ (iv) $\frac{1}{3}$ (v) $\frac{64}{3}$
52 (i) $\frac{8}{3}$ (ii) $\frac{21}{4}$ (iii) $\frac{93}{4}$
53 180,000 m^2; 720,000 m^2

Exercise 4.1:4

1 (i) $\frac{26}{3}\pi$ (ii) $\frac{1}{5}\pi$ (iii) $\frac{1}{6}\pi$ (iv) 40π (v) 108π
 (vi) $\frac{16}{15}\pi$ (vii) 3π 3 5250π cm^3 5 $\frac{5}{24}\pi r^3$
6 2304π cm^3 7 $\frac{3}{5}a$ 8 0.27 m^3
10 (i) $\frac{2}{15}\pi$ (ii) $\frac{243}{5}\pi$ (iii) $\frac{1}{5}\pi a^5$ (iv) $\frac{2}{3}\pi$ (v) $\frac{96}{5}\pi$
11 $\frac{2}{3}; \frac{2}{5}\pi$ 12 $\frac{1}{30}\pi$ 13 $\frac{16}{15}\pi$
14 (i) $\frac{32}{3}; 8\pi$ (ii) $\frac{40}{3}; \frac{112\pi}{3}$ (iii) $\frac{5}{3}; \frac{20\pi}{7}$
15 (ii) $x = y^{1/3}$; (a) $\frac{45}{4}$ (b) $\frac{93}{5}\pi$
16 (i) $\frac{7}{3}; \frac{31}{6}\pi$ (ii) 36; $\frac{81}{2}\pi$ (iii) $\frac{8}{5}(8-\sqrt{2})$; 60π
 (iv) $\frac{56}{3}, 6\pi$
17 (i) 8π (ii) 8π (iii) $\frac{100\pi}{3}$ (iv) $\frac{444\pi}{5}$
18 $\frac{6\pi}{5}$ 19 $\frac{4\pi ab^2}{3}, \frac{4\pi a^2 b}{3}$ 20 $\frac{64\pi}{15}$ 21 (ii) $\frac{968}{5}\pi$
22 (i) $\frac{1}{3}\pi r^2 h$ (ii) $\frac{1}{3}a^2 h$ (iii) $\frac{1}{6}a^3$
23 (i) $\int_0^h A(x)\,\mathrm{d}x$ (ii) 1800 cm^3 (iii) 0.07 m^3
24 (i) $(\frac{3}{4}, \frac{3}{10})$ (ii) $(\frac{12}{5}, \frac{3}{4})$ (iii) $(\frac{1}{2}, \frac{1}{10})$
26 (i) $(\frac{3}{4}, 0)$ (ii) $(\frac{8}{3}, 0)$ (iii) $(\frac{1}{2}, 0)$ 28 $\frac{85}{12}$

Exercise 4.2:1

1 (i) (a) $\frac{1}{21}(3x+2)^7 + c$ (b) $\frac{1}{3}(2x+1)^{3/2} + c$
 (c) $-\frac{1}{5(5x-4)} + c$
 (ii) (a) $\frac{2}{135}(3x-1)^{3/2}(9x+2) + c$
 (b) $\frac{2}{3}\sqrt{x-2}(x+4) + c$
 (c) $-\frac{1}{5}\sqrt{3-2x}(x+3) + c$
 (d) $\frac{2}{105}(x+1)^{3/2}(15x^2 - 12x + 8) + c$
 (e) $\frac{1}{72}(x-1)^8(8x+1) + c$
 (f) $4x - 17\ln|x+3| + c$
 (g) $\frac{1}{16}(4x + 17\ln|4x-5|) + c$
 (h) $\frac{1}{4}(6x - \ln|2x+1|) + c$
 (i) $\frac{1}{4}(x^2 + 2x + 4\ln|2x+1|) + c$
 (j) $\ln|x+2| + \frac{2}{x+2} + c$
2 (i) $\frac{2}{5}$ (ii) $\frac{8}{9}$ (iii) $\frac{16}{15}$ (iv) $\frac{184}{105}$ (v) $\frac{22}{3}$ (vi) 1.87
 (vii) 1.11 (viii) $\frac{35}{3}$
3 (i) $\frac{1}{a(\alpha+1)}(ax+b)^{\alpha+1} + c$; (a) $\frac{1}{21}(3x+2)^7 + c$
 (b) $\frac{1}{65}(5x+1)^{13} + c$ (c) $-\frac{1}{24}(1-4x)^6 + c$
 (d) $\frac{1}{3}(2x-1)^{3/2} + c$ (e) $-\frac{1}{x+1} + c$
 (f) $-\frac{1}{3(3x-1)} + c$ (g) $\frac{1}{4(5-2x)^2} + c$
 (ii) (a) $\ln|x+2| + c$ (b) $\ln|4x+5| + c$
 (c) $-\frac{2}{3}\ln|6-3x| + c$
 (iii) (a) $-\frac{1}{6(3x+1)^2} + c$ (b) $-\frac{1}{3(3x+1)} + c$
 (c) $\frac{1}{3}\ln|3x+1| + c$ (d) $\frac{2}{3}\sqrt{3x+1} + c$

578 Answers

4 $\frac{1}{a}e^{ax+b}+c$; (i) $\frac{1}{5}e^{5x}+c$ (ii) $\frac{1}{3}e^{3x-2}+c$
(iii) $\frac{2}{3}e^{3x/2}+c$ (iv) $-\frac{1}{4}e^{-4x}+c$
(v) $-\frac{1}{5}e^{1-5x}+c$

5 $-\frac{1}{a}\cos(ax+b)+c$, $\frac{1}{a}\sin(ax+b)+c$;
(i) $-\frac{1}{2}\cos 2x+c$ (ii) $\frac{1}{5}\sin 5x+c$
(iii) $-3\cos\frac{x}{3}+c$ (iv) $10\sin\frac{x}{2}+c$
(v) $-\frac{1}{\pi}\cos\pi x+c$ (vi) $-\frac{1}{3}\sin(5-3x)+c$

6 (i) $\frac{1}{4}\tan 4x$ (ii) $-\frac{1}{5}\operatorname{cosec} 5x$ (iii) $2\sec\frac{x}{2}$

7 (i) $\frac{1}{2}(e^3-1)$ (ii) 12 (iii) $\frac{2}{\pi}$ (iv) $\frac{121}{5}$ (v) $\frac{1}{5}$
(vi) ln 5 (vii) $-\frac{1}{2}\ln 3$ (viii) $-\frac{4}{3}\ln\frac{5}{2}$

8 (i) $2\sqrt{2}\sin\frac{1}{2}x+c$ (ii) $-2\sqrt{2}\cos\frac{1}{2}x+c$

9 $2\cos^2\frac{3x}{2}-1$; (i) $\frac{2\sqrt{2}}{3}\cos\frac{3x}{2}+c$
(ii) $\frac{1}{3}\tan\frac{3x}{2}+c$

10 (i) $\frac{\sqrt{6}}{2}$ (ii) $\frac{2}{3}(\sqrt{2}-1)$ (iii) 1

11 $-2\sqrt{2}\sin\left(\frac{\pi}{4}-\frac{x}{2}\right)+c$

12 (i) $\frac{1}{2}(1-\cos 2x)$; $\frac{1}{4}(2x-\sin 2x)+c$
(ii) $\frac{1}{4}(2x+\sin 2x)+c$

13 $\frac{1}{2}(1+\cos 4x)$; (i) $\frac{1}{8}(4x+\sin 4x)+c$
(ii) $\frac{3\pi-8}{32}$ (iii) $\frac{4\pi-5\sqrt{3}}{8}$

14 (i) $\frac{1}{8}(4x-\sin 4x)+c$ (ii) $\frac{1}{32}(4x-\sin 4x)+c$
(iii) $\frac{1}{12}(6x-\sin 6x)+c$

14 $\sin 4x + \sin x$; $-\frac{1}{8}\cos 4x - \frac{1}{2}\cos x$

16 (i) $-\frac{1}{8}\cos 4x - \frac{1}{4}\cos 2x$ (ii) $\frac{1}{6}\sin 3x + \frac{1}{2}\sin x$
(iii) $\frac{1}{4}\sin 2x - \frac{1}{12}\sin 6x$
(iv) $-\frac{1}{20}\cos 5x - \frac{1}{6}\cos 3x - \frac{1}{4}\cos x$

17 (i) $-\frac{1}{4}$ (ii) 0.127 (iii) $-\frac{14}{15}$ **18** $\frac{1}{2}\ln\frac{5}{3}$; $\frac{1}{15}\pi$

19 $\frac{\pi}{18}(4\pi-3\sqrt{3})$ **20** $\frac{1}{12}(2\pi+3\sqrt{3})$

21 4π; $\frac{9\pi^2}{2}$ **22** (i) $\frac{1}{2}$ (ii) $\frac{9}{4}\pi$

Exercise 4.2:2

1 (i) $\frac{1}{3}(x^2-1)^{3/2}+c$ (ii) $\frac{2}{5}\sqrt{(x^3+1)^5}+c$
(iii) $\frac{1}{3}\ln|x^3+1|+c$ (iv) $\frac{1}{2}e^{x^2}+c$
(v) $e^{\tan x}+c$ (vi) $-\frac{1}{4}\cos^4 x+c$
(vii) $\frac{1}{5}\tan^5 x+c$ (viii) $\frac{2}{5}\sqrt{x^3+3x+4}+c$
(ix) $-\frac{2}{3}(\cos x)^{3/2}+c$ (x) $-2\cos\sqrt{x}$

2 (i) $\frac{1}{n+1}\sin^{n+1}x+c$ (ii) $-\frac{1}{n+1}\cos^{n+1}x+c$
(iii) $\frac{1}{n+1}\tan^{n+1}x+c$ (iv) $\frac{1}{n}\sec^n x+c$

3 (i) 6 (ii) $\sqrt{5}-\sqrt{2}$ (iii) $\frac{1}{160}$ (iv) $\frac{15}{4}$ (v) $\frac{1}{2}(\ln 2)^2$
(vi) $\frac{2}{3}(3\sqrt{3}-1)$ (vii) $\frac{4\sqrt{2}}{3}$ (viii) $\frac{4}{3}(3\sqrt{3}-2\sqrt{2})$

4 (i) $e^{\tan x}+c$ (ii) $-e^{1/x}+c$ (iii) $2e^{\sqrt{x}}+c$

5 (i) $\frac{1}{2}e(e^3-1)$ (ii) $\frac{\sqrt{e-1}}{2e}$

6 (i) $\ln|x^2+1|+c$ (ii) $\frac{1}{3}\ln|x^3+1|+c$
(iii) $\frac{1}{2}\ln|x^2-4x+7|+c$
(iv) $\ln|1+\tan x|+c$ (v) $-\ln|2-\sin x|+c$
(vi) $\frac{1}{3}\ln|e^{3x}+1|+c$ (vii) $\ln|\ln x|+c$

7 (i) $\frac{1}{2}\ln\frac{3}{2}$ (ii) $\frac{1}{2}\ln\frac{18}{7}$ (iii) $\ln\frac{6}{5}$ (iv) $\frac{1}{2}\ln 2$
(v) $\ln\frac{3}{2}$ (vi) $\ln\frac{16}{9}$

8 (i) $-\frac{1}{4(x^2+2)^2}+c$ (ii) $-\frac{1}{2(x^2+2)}+c$
(iii) $\frac{1}{2}\ln(x^2+2)+c$ (iv) $\sqrt{x^2+2}+c$

9 (i) $\ln|\sin x|$; (a) $\ln 2$ (b) $-\frac{1}{2}\ln 2$
(ii) $-\ln(\operatorname{cosec} x+\cot x)$; $\ln\frac{1+\sqrt{2}}{\sqrt{3}}$

12 $\ln|1+e^x|+c$; $x-\ln|1+e^x|+c$
13 $\frac{1}{3}\{x-\ln|2+e^x|\}+c$
14 $\frac{2}{5},\frac{2}{5},\frac{1}{5}\{x+2\ln|\cos x+2\sin x|\}+c$
15 (i) (a) $x-\frac{1}{2}\ln|1+e^{2x}|+c$
(b) $\frac{1}{2}\{x+\ln|\cos x+\sin x|\}+c$
(c) $\frac{1}{12}\{7x-\ln|2e^x+3e^{3x}|\}+c$
(d) $\frac{1}{13}\{12x-5\ln|2\cos x+3\sin x|\}+c$
(ii) $\frac{1}{50}\{3\pi-8\ln\frac{4}{3}\}$

16 (i) $\frac{2}{3}$ (ii) $\frac{1}{8}$ (iii) 0 (iv) $\frac{2}{3}(8-3\sqrt{3})$ (v) $\frac{7}{18}$
17 $-\frac{1}{4}\cos^4 x+c$, $-\frac{1}{6}\cos^6 x+c$; $\frac{5}{384}$
18 $\frac{1}{3}\sin^3 x+c$; $\frac{1}{5}\sin^5 x+c$; $\sin x-\frac{1}{3}\sin^3 x+c$;
$\sin x - \frac{2}{3}\sin^3 x + \frac{1}{5}\sin^5 x + c$

19 (i) $\frac{3\sqrt{3}}{8}$ (ii) $\frac{203}{480}$ (iii) $\frac{2}{15}$ (iv) $\frac{2}{35}$

20 $\frac{1}{3}\tan^3 x + c$; $\tan x + \frac{1}{3}\tan^3 x$; π
21 $\frac{1}{2}\sec^2 x + c$; $\frac{1}{2}\sec^2 x + \ln|\cos x|+c$;
$\frac{1}{4}\sec^4 x - \sec^2 x - \ln|\cos x|+c$

22 (i) $\ln\dfrac{1}{|2-\tan x|}$ (ii) $\ln\dfrac{1}{|1-\tan^2 x|}$

23 (i) $\ln 2$ (ii) $\dfrac{e-1}{2e}$ **24** $1-\ln 2$

Exercise 4.2:3

1 (i) (a) 1 (b) $\dfrac{\pi}{3}$ (ii) $\frac{1}{2}(2\theta+\sin 2\theta)+c$; $\frac{1}{2}\pi$

2 (i) $\dfrac{1}{\sqrt{3}}$ (ii) $\dfrac{\pi}{6}$ (iii) $\sqrt{3}+\dfrac{2\pi}{3}$

3 (i) $\frac{1}{3}$ (ii) $\dfrac{2}{3\sqrt{3}}$

4 $\frac{1}{32}(4\theta - \sin 4\theta)+c$; $\frac{1}{192}(4\pi-3\sqrt{3})$

5 (i) $\ln(1+\sqrt{2})$ (ii) $\dfrac{\pi}{12}$ (iii) $\dfrac{1}{9\sqrt{2}}$

6 (i) $\ln\dfrac{2+\sqrt{3}}{1+\sqrt{2}}$ (ii) $\dfrac{\pi}{12}$ (iii) $\ln\dfrac{1+\sqrt{2}}{\sqrt{3}}$

7 (i) $\dfrac{\pi}{12}$ (ii) $\ln\dfrac{2+\sqrt{3}}{\sqrt{3}}$

8 (i) 0.39 (ii) 0.06 (iii) 0.05

9 $\frac{1}{2}(2\theta-\sin 2\theta)+c$; $\dfrac{\pi}{4}-\frac{1}{2}$

11 (i) $\dfrac{x}{4\sqrt{4-x^2}}+c$ (ii) $\dfrac{x}{9\sqrt{x^2+9}}+c$
(iii) $\frac{1}{3}\sec^{-1}x+c$

12 (i) $\frac{1}{3}\sin^{-1}\dfrac{3x}{2}+c$ (ii) $\frac{1}{5}\tan^{-1}\dfrac{x}{5}+c$

13 $\sin^{-1}x + x\sqrt{1-x^2}+c$

15 $\dfrac{1}{\sqrt{a^2-x^2}}$; $\dfrac{a}{a^2+x^2}$ (i) $\sin^{-1}\dfrac{x}{a}+c$
(ii) $\dfrac{1}{a}\tan^{-1}\dfrac{x}{a}+c$

16 (i) $\sin^{-1}\dfrac{x}{3}+c$ (ii) $\frac{1}{4}\tan^{-1}\dfrac{x}{4}+c$
(iii) $\dfrac{1}{\sqrt{3}}\tan^{-1}\dfrac{z}{\sqrt{3}}+c$

17 (i) $\dfrac{\pi}{3}$ (ii) $\dfrac{\pi}{10}$

18 $\ln|x^2+4|+c$; $\ln|x^2+4|+\frac{3}{2}\tan^{-1}\dfrac{x}{2}+c$;
(i) $\frac{1}{2}\ln|x^2+5|+\sqrt{5}\tan^{-1}\dfrac{x}{\sqrt{5}}+c$
(ii) $\frac{3}{2}\ln|3+x^2|+\dfrac{2}{\sqrt{3}}\tan^{-1}\dfrac{x}{\sqrt{3}}+c$

19 (i) $\frac{1}{3}\sin^{-1}\dfrac{3x}{2}+c$ (ii) $\frac{1}{5}\sin^{-1}\dfrac{x}{5}+c$
(iii) $\frac{1}{12}\tan^{-1}\dfrac{4x}{3}+c$ (iv) $\dfrac{1}{2\sqrt{5}}\tan^{-1}\dfrac{2x}{\sqrt{5}}+c$

20 (i) $\dfrac{\pi}{6}$ (ii) 0.05 (iii) 0.27 **21** $\frac{1}{4}(\pi-3)$

Exercise 4.2:4

1 (i) $\frac{1}{3}\ln\frac{5}{2}$ (ii) 1 (iii) $\sqrt{3}-1$ (iv) $2-\sqrt{3}$ (v) $\dfrac{\sqrt{3}}{2}$

2 $\ln|\tan\frac{1}{2}x|$

3 $-\cot\dfrac{x}{2}+c$; $-x-\cot\dfrac{x}{2}+c$;
$\ln|1-\cos x|+2x+2\cot\dfrac{x}{2}+c$

4 $\dfrac{1}{\sqrt{3}}$ **5** $\frac{1}{3}\tan^{-1}3t+c$; 0.11

Exercise 4.2:5

1 (i) $x+2\ln|x-1|+c$
(ii) $3x-6\ln|x+2|+c$
(iii) $-4x-16\ln|4-x|+c$
(iv) $3x-5\ln|x+1|+c$
(v) $\frac{1}{4}\{2x-9\ln|2x+1|+c$
(vi) $\frac{1}{9}\{12x+7\ln|3x+4|+c$
(vii) $-\frac{1}{36}\{30x+5\ln|1-6x|\}+c$

2 $\frac{1}{9}\{15x+\ln|3x+1|\}+c$
3 (i) $\frac{1}{2}x^2+x+\ln|x-1|+c$
(ii) $\frac{1}{2}x^2+2x-\frac{5}{3}\ln|3x-1|+c$
(iii) $\frac{1}{8}\{2x^2+2x-5\ln|2x+1|+c$
(iv) $\frac{1}{3}x^3-x^2+3x-5\ln|x+2|+c$

4 (i) $\frac{2}{9}(3-4\ln 4)$ (ii) $1+4\ln\frac{2}{3}$
(iii) $\frac{1}{4}(4+5\ln 5)$ (iv) $10\ln 2$

5 (i) $\ln|x+1|+\dfrac{1}{x+1}+c$
(ii) $\ln|x+3|+\dfrac{5}{x+3}+c$
(iii) $\frac{1}{2}(x^2+4x-5)+3\ln|x-1|-\dfrac{1}{x-1}+c$

6 (i) $\frac{1}{6}$ (ii) $\ln\frac{3}{2}-\frac{1}{6}$ (iii) $\frac{7}{6}-2\ln\frac{3}{2}$

7 (i) $\ln\left|\dfrac{x+1}{x+2}\right|+c$ (ii) $\ln\left|\dfrac{x-1}{x+1}\right|+c$
(iii) $\ln|x^2-1|+c$ (iv) $\ln|x^2-3x|+c$
(v) $\frac{1}{5}\{\ln|x-3|+4\ln|x-2|\}+c$
(vi) $\frac{1}{10}\{4\ln|x-1|+\ln|2x+3|\}+c$

8 (i) $\ln\left|\dfrac{x}{\sqrt{x^2+1}}\right|+c$ (ii) $\frac{1}{4}\ln\left|\dfrac{(x-2)^2}{x^2+2}\right|+c$
(iii) $\ln|x+1|-\tan^{-1}\frac{1}{2}x+c$
(iv) $\ln\left|\dfrac{(x-1)^2}{x^2+x+3}\right|+c$

9 (i) $\frac{1}{2}x^2-3x+8\ln|x+2|-\ln|x+1|+c$
(ii) $\frac{1}{2}x^2-2x+\frac{1}{3}\{4\ln|2x-1|-\ln|x+1|\}+c$
10 (i) $\frac{1}{2}\ln\frac{25}{21}$ (ii) $\frac{1}{4}\ln\frac{3}{2}$ (iii) $\frac{1}{2}\ln\frac{32}{27}$ (iv) $\frac{1}{2}\ln\frac{40}{17}$
(v) $2\ln 2+\tan^{-1}4-\tan^{-1}3$ (vi) $\frac{1}{2}(1+\ln\frac{5}{16})$
11 (i) $\frac{1}{3}\ln(4+3\sqrt{3})$ (ii) $\frac{1}{4}\ln(7+4\sqrt{3})$
(iii) $\frac{1}{5}\ln(8+5\sqrt{2})$ (iv) $\ln\frac{1}{3}(5+2\sqrt{3})$

12 $\ln\dfrac{1+\tan\dfrac{z}{2}}{1-\tan\dfrac{x}{2}}+c$

13 (i) $\frac{1}{3}\ln(7+4\sqrt{3})$ (ii) $\tan^{-1}\frac{1}{3}$
14 (i) $\dfrac{1}{2a}\ln\left|\dfrac{x-a}{x+a}\right|$ (ii) $-\frac{1}{2}\ln|a^2-x^2|$

Answers 579

(iii) $\frac{1}{a}\arctan\frac{x}{a}$ (iv) $\frac{1}{2}\ln|a^2+x^2|$
15 (i) $\frac{1}{2}\ln 3$ (ii) $\tan^{-1}\frac{1}{2}$ 16 $1-\ln 2; \frac{3}{2}-2\ln 2$

Exercise 4.2:6

1 $e^x(x-1)+c$
2 (i) $\sin x - x\cos x + c$ (ii) $\frac{1}{9}e^{3x}(3x-1)+c$
   (iii) $\frac{1}{25}(5x\sin 5x - \cos 5x)+c$
   (iv) $x\tan x + \ln|\cos x|+c$
3 $\frac{2^x}{(\ln 2)^2}(x\ln 2 - 1)+c$
4 (i) $e^x(x^2-2x+2)+c$
   (ii) $-\frac{1}{4}e^{-2x}(2x^2+2x+1)+c$
   (iii) $(2-x^2)\cos x + 2x\sin x + c$
   (iv) $\frac{1}{27}\{(9x^2-2)\sin 3x + 6x\cos 3x\}+c$
5 (i) $\frac{1}{2}\pi - 1$ (ii) $\frac{1}{4}$ (iii) $\frac{1}{2}(1-\ln 2)$ (iv) $\frac{1}{4}(\pi + 2\ln 2)$ (v) $\frac{1}{4}\pi^2 - 2$ (vi) $\frac{1}{125}(17e^5-2)$
6 $\frac{1}{4}(\pi + 2\ln 2); \frac{1}{32}(8\pi - \pi^2 + 16\ln 2)$
8 (i) $\frac{1}{2}(e^2+1)$ (ii) $e-2$
9 (i) $10-\frac{9}{\ln 10}$ (ii) $50-\frac{99}{4\ln 10}$
10 (i) $x\tan^{-1}x - \frac{1}{2}\ln|1+x^2|+c$
   (ii) $x\sin^{-1}x - \sqrt{1-x^2}+c$
11 $\frac{1}{2}e^x(\sin x - \cos x)$
12 (i) $\frac{1}{2}e^x(\cos x + \sin x)$
   (ii) $\frac{1}{5}e^{2x}(2\sin x - \cos x)$
   (iii) $\frac{1}{17}e^{1/2x}(\sin 2x - 4\cos 2x)$
   (iv) $-\frac{1}{25}e^{-3x}(3\sin 4x + 4\cos 4x)$
13 $\frac{1}{a^2+b^2}e^{ax}(a\cos bx + b\sin bx)$;
   $\frac{1}{a^2+b^2}e^{ax}(a\sin bx - b\cos bx)$
14 (i) $\frac{1}{2}\{\sec\theta\tan\theta + \ln|\sec\theta+\tan\theta|\}+c$
   (ii) $\frac{1}{2}\{\sqrt{2}+\ln(1+\sqrt{2})\}$
15 $(0.65, 1)$ 16 $(0.36, 0.21)$ 17 $(2.10, 0.36)$

Exercise 4.3:1

1 (i) 1.117 (ii) 1.104 (iii) 0.6111 (iv) 0.6240
  (v) 0.3882 (vi) 0.4173 (vii) 2.106
  (viii) 0.6819 (ix) 3.092 (x) 7.369
2 (i) 14.539 (ii) 4.3558 (iii) 1.3686
3 (i) (a) 0.5437 (b) 0.5195 (c) 0.5197
  (ii) (a) 1.944 (b) 1.978 (c) 1.992
  (iii) (a) 2.467 (b) 3.041 (c) 3.182
4 (i) (a) 5.93 (b) 14.7
  (ii) (a) 7.00 (b) 7.11 (c) 6.10
  (iii) (a) 0.844 (b) 0.325 (c) 5.71
5 (i) 0.26 (ii) 0.25
6 (i) $\frac{52}{3}$
7 (i) 1.100 (ii) 1.105 (iii) 0.7407 (iv) 0.7141
  (v) 0.3884 (vi) 0.4543 (vii) 2.285
  (viii) 0.6805 (ix) 3.104 (x) 7.341
8 (i) (a) 0.4837 (b) 0.5114 (c) 0.5198
  (ii) (a) 1.97 (b) 1.990 (c) 1.996
  (iii) (a) 3.290 (b) 3.232 (c) 3.229
9 (i) (a) 5.92 (b) 14.7
  (ii) (a) 7.07 (b) 7.19 (c) 6.25
  (iii) (a) 0.850 (b) 0.319 (c) 5.84
10 51.9 11 68 540 m³
12 (i) (a) 22 (b) 64.33 (ii) 64.33
13 (i) (a) 0.3413 (b) 0.3468 (ii) 0.3465
14 (i) (a) 8.827 (b) 8.400 (ii) 8.389
15 (i) (a) 1.853 (b) 1.887 (ii) 1.886
16 (i) (a) 0.3552 (b) 0.3473 (ii) 0.3465
17 (i) (a) 4.016 (b) 3.911 (ii) 3.909
18 (i) (a) 3.131 (b) 3.142 (ii) 3.142

Exercise 4.4:1

2 (i) $\frac{dy}{dx}=\frac{2y}{x}$ (ii) $\frac{dy}{dx}=-\frac{y}{x}$ (iii) $\frac{dy}{dx}=y\tan x$
  (iv) $2y\frac{dy}{dx}-1$ (v) $\frac{dy}{dx}=\frac{x}{y}$
  (vi) $\frac{dy}{dx}+y\tan 2x = 0$ (vii) $2\frac{dy}{dx}=3(1-y^2)$
  (viii) $(\cos x)\frac{dy}{dx}=1-y\sin x$
  (ix) $\frac{dy}{dx}=2x(1-y)$
4 (i) $y=3e^{x^2}$ (ii) $y=2e^{x^2-1}$

Exercise 4.4:2

1 (i) $y=Ae^x$ (ii) $y=Ae^{-2x}$ (iii) $y=\frac{1}{x+A}$
  (iv) $y=Ax$ (v) $x^2+y^2=A$
2 (i) $y=3e^{4x}$ (ii) $y=\frac{1}{3}(2e^{3x}-1)$
  (iii) $y=2-e^{-x}$
3 (i) $x^2+y^2=25$ (ii) $x^2-y^2=1$
  (iii) $y=xe^{x^2}$ (iv) $y^2=2(x-1)$
  (v) $y=\frac{2}{5-2x-x^2}$ (vi) $y=\ln x$
  (vii) $y=\frac{e^x-1}{e^x+1}$ (viii) $y=\frac{e^{2x/(1+x)}-1}{e^{2x/(1+x)}+1}$
4 (i) $y=\frac{1}{1-e^{-1/x}}$ (ii) $y=e^{\sqrt{x^2+1}}$
  (iii) $y=\frac{1}{2}\sec x$ (iv) $y=\frac{4\sin^2 x - 1}{4\sin^2 x + 1}$
  (v) $y=\frac{x^2-1}{x^2+3}$ (vi) $y=\frac{2}{\sqrt{1+\cos 2x}}$
  (vii) $y=\tan\frac{1}{x}$ (viii) $y=\sin^{-1}\frac{x}{2}$
5 (i) $y=e^{1-x^2}$; $y=-2x+3$
  (ii) $y=xe^{1-x^2}$; $y=-x+2$
6 $4x+3y=0$ 7 $y=xe^{-x}$; $x+e^2y=3$
8 $x=\frac{Ne^{Nkt}}{N-1+e^{Nkt}}$
9 (i) $y=2\{e^{-x}+x-1\}$ (ii) $y=e^{(1-x)^2}-x^2$
10 (i) $y=\frac{1}{3}Ax^3+Ax+B$
   (ii) $y=\frac{x}{A}-\frac{1}{A^2}\ln|1-Ax|+B$
11 (i) $y=Ae^{Bx}$ 12 $y=3e^{x^2}$

Exercise 4.4:3

1 (iii) 47.55 min 3 2000 years
5 (i) 61.67° (ii) 11.12 min
6 3.6° 9 10.2 cm 10 $\frac{13}{3}T$ 11 $2r_0+k$
12 85.7 s 15 $3\ln 3$ 16 $T_0\left\{1-\left(\frac{\gamma-1}{\gamma}\right)\frac{gx}{kT_0}\right\}$

Miscellaneous Exercise 4

11 (i) π (ii) $\frac{2}{3}$ (iii) $\frac{1}{2}\ln\frac{3}{2}$
12 (i) $\frac{1}{3}(2\sqrt{2}-1)$ (ii) $1-\frac{\pi}{4}$ (iii) $\ln\frac{32}{9}$
13 (i) $\frac{100}{27}$ (ii) 0.58 (iii) $\frac{1}{4}$
14 (i) 4.57 (ii) $\frac{3\sqrt{3}+4\pi}{48}$ (iii) $\frac{2\pi}{3}$
15 (i) 0.745 (ii) 0.519
17 (i) $-\frac{4}{5}$ (ii) $\frac{1}{5}$ (iii) $2(\sqrt{2}-1)$
   (iv) $-\frac{3\pi+2}{18}$

19 (i) $\frac{3\sqrt{3}+2\pi}{24}$ (ii) $\frac{\pi+3\sqrt{3}-6}{12}$ (iii) $\frac{1}{18}\sqrt{2}$
21 (i) $\frac{\pi}{4}$ (iii) 0
22 $\frac{1}{5}\ln 6$ 23 $2\ln 2 - 1$; $2(\ln 2)^2 - 4\ln 2 + 2$
24 $\frac{2}{3}$; (i) $\frac{8}{3}\pi$ (ii) $\frac{8}{15}\pi$ 25 (i) $\frac{4}{3}$ (ii) $\frac{8}{3}$ (iii) $1-\frac{\pi}{4}$
26 (i) $\ln 2$ (ii) $\frac{1}{2}\ln 3$ 27 (i) $2-2\ln 3$ (ii) $\ln\frac{16}{9}$
28 (i) $4\ln 2 + \frac{3}{2}$ (ii) $2\ln\frac{8}{3}+\frac{17}{8}$
29 $\frac{1}{2}(9-12\ln 2)$ 30 $\frac{3-e}{e}$
31 (i) $\frac{1}{4}\pi - \frac{1}{2}$ (ii) $\frac{1}{4}\pi - \frac{1}{2}\ln 2$ (iii) $\frac{1}{4}\pi$
32 (i) 5 (ii) $\frac{1}{2}\pi(2\ln 2 - 1)$ (iii) $6\ln 3 - 4$
33 3.13 34 (i) 36.095 (ii) 29.021 (iii) 26.800
35 1.491, 0.614
36 (i) (a) 0.127 (b) 0.132 (ii) 0.424
37 (ii) 1.44
38 (i) $2\frac{dy}{dx}+3y\tan 3x = 0$ (ii) $y=xe^x$; $-\frac{1}{e}$
39 $y=2\sin x$ 40 $y=\frac{x}{4-x}$
42 $y=\frac{2}{1-4x^2}$; $x=\pm\frac{1}{2}$, $y=0$
43 $y=\frac{\tan x}{2+\tan x}$; $1, \frac{1}{3}$
44 $\tan x$; (i) $\frac{1}{2}2$ (ii) $y=e^x\sec x$
45 $y=\frac{1}{1+x^2}$; $\frac{1}{2}\pi$
46 (i) $\cos^{-1}(1-2\sin x)$ (ii) $\tan^{-1}(\ln x)$
47 $y=2xe^{-x^2}$; $\frac{e-1}{e^2}$

50 $-5°C$ 51 3 weeks

Chapter 5

Exercise 5.1:1

4 (i) 4, 7, 10, 151, $3n-2$, $3n+1$, $3n+4$, $6n+1$
  (ii) $\frac{1}{3}$, $\frac{1}{15}$, $\frac{1}{35}$, $\frac{1}{9999}$, $\frac{1}{(2n-1)(3n-3)}$,
  $\frac{1}{(2n+1)(2n-1)}$, $\frac{1}{(2n+3)(2n+1)}$,
  $\frac{1}{(4n+1)(4n-1)}$
  (iii) $\frac{1}{6}$, $\frac{3}{20}$, $\frac{2}{15}$, $\frac{51}{2756}$, $\frac{n}{(n+1)(n+2)}$,
  $\frac{n+1}{(n+2)(n+3)}$, $\frac{n+2}{(n+3)(n+4)}$, $\frac{2n+1}{(2n+2)(2n+3)}$
  (iv) 2, 4, 8, 2^{50}, 2^{n-1}, 2^n, 2^{n+1}, 2^{2n}
  (v) 5, 1, $\frac{1}{5}$, 5^{-48}, 5^{3-n}, 5^{2-n}, 5^{1-n}, 5^{2-2n}
  (vi) $-\frac{1}{3}$, $\frac{1}{27}$, $-\frac{1}{243}$, $\frac{1}{3^{99}}$, $\frac{(-1)^{n-1}}{3^{2n-3}}$, $\frac{(-1)^n}{3^{2n-1}}$,
  $\frac{(-1)^{n+1}}{3^{2n+1}}$, $\frac{1}{3^{4n-1}}$
  (vii) -1, 1, -1, 1, $\cos(n-1)\pi$, $\cos n\pi$,
  $\cos(n+1)\pi$, 1
5 (i) ∞ (ii) $-\infty$ (iii) ∞ (iv) 1 (v) 0
9 1, 2, 6, 24, 120, 720, 5040
10 (i) 7! (ii) 8! (iii) $(n+1)!$ (iv) 6! (v) 5!
   (vi) 42 (vii) 120 (viii) $(n+1)!$ $(n+1)(n-1)!$
   (ix) $n+1$ (x) 5(8!) (xi) $(n^2+n-1)(n-1)!$

11 (i) $\frac{15!}{10!}$ (ii) $\frac{10!}{6!4!}$ (iii) $\frac{n!}{(n-4)!}$ (iv) $\frac{n!}{(n-5)!5!}$

Exercise 5.1:2

1 (i) 1, 3, 6, 10, 15, 21 (ii) 3, 4, 3, 0, −5, −12
(iii) $\frac{3}{2}, \frac{17}{6}, \frac{49}{12}, \frac{317}{60}, \frac{377}{60}, \frac{3119}{420}$
(iv) 1, 3, 7, 15, 31, 63
(v) −1, 0, −1, 0, −1, 0
(vi) 1, 4, 10, 20, 35, 56
3 (i) $1+2+3+4+5+6+7$; 28
(ii) $2+4+8+16+32$; 62
(iii) $1+3+5+7+9+11+13+15+17+19$; 100
(iv) $9+16+25+36+49+64$; 199
4 (i) $\sum_{r=1}^{10} r$ (ii) $\sum_{r=5}^{8}\frac{7}{r}$ (iii) $\sum_{r=1}^{7} 2r-1$ (iv) $\sum_{r=1}^{8} 2^{3-r}$
(v) $\sum_{r=m}^{2m} r^2$ (v) $\sum_{r=1}^{2m}(-1)^{r+1}x^3$ (vi) $\sum_{r=1}^{50}(2r)^2$
(vii) $\sum_{r=1}^{100}\frac{1}{r(r+2)}$
5 (i) $10+20+40+\cdots+5.2^n$
(ii) $\frac{1}{6}+\frac{1}{12}+\frac{1}{20}+\cdots+\frac{1}{(n+1)(n+2)}$
(iii) $n^2+(n+1)^2+(n+2)^2+\cdots+(2n)^2$
(iv) $-\frac{1}{25^3}+\frac{1}{26^3}-\frac{1}{27^3}+\cdots-\frac{1}{75^3}$

Exercise 5.1:3

1 (i) 13, 17, 21 (ii) −1, −3, −5 (iii) $\frac{2}{3}$, 0, $-\frac{2}{3}$
(iv) $4x-3$, $5x-4$, $6x-5$
(v) $2y-x$, $3y-2x$, $4y-3x$
(vi) $\cos^2\theta+1$, $\cos^2\theta+2$, $\cos^3\theta+3$
2 (i) 13, $3r-2$ (ii) 0, $\frac{1}{3}(5r-20)$
(iii) $-\frac{11}{2}, \frac{1}{2}(1-3r)$ (iv) $7x$, $(2r-1)x$
(v) ln 24, $\ln(3.2^{r-1})$
3 (i) 2, $\frac{7}{3}, \frac{8}{3}, 3, \frac{10}{3}, \frac{11}{3}$
(ii) $\frac{1}{2}, -\frac{3}{2}, -\frac{7}{2}, -\frac{11}{2}, -\frac{15}{2}, -\frac{19}{2}$
(iii) $x-3$, −2, $-1-x$, $-2x$, $1-3x$, $2-4x$
4 (i) 109 (ii) 95 (iii) 26 (iv) n (v) $n+1$
(vi) $m-2$
5 (i) −2, 3 (ii) 1, 4 (iii) $\frac{9}{2}, -\frac{1}{2}$ (iv) $\frac{15x}{4}, -\frac{x}{4}$
6 22 **7** −32
8 (i) 21 (ii) −3 (iii) $4x$ (iv) y
9 $\frac{19}{3}, \frac{11}{3}$ **10** 8, 18
11 (a) (i) 5, 2 (ii) 4, −3 (iii) $p+q, p$
12 (i) 5050 (ii) 3240
13 (i) 710 (ii) −950 (iii) $15(9-7x)$
(iv) $101g2+451g3$
14 (i) 16524 (ii) −5050 (iii) −1312 (iv) −2120
15 (i) 210 (ii) −117 (iii) $\frac{1}{6}n(5-n)$
(iv) $\frac{1}{2}n\{2p+(n+1)q\}$
16 $\frac{1}{2}n(n+1)$
17 (i) $m(2m+1)$ (ii) $\frac{1}{2}m(3m+1)$ (iii) $-m$
(iv) $m(3m+1)$
18 (i) 45150 (ii) 15150 (iii) 30000
19 78750 **20** (i) 68629 (ii) 64000 (iii) 54739
21 62499 **22** $60n^2$
23 35 **24** 3, −2 **25** $\frac{15}{4}$ **26** 540
27 (i) 447 (ii) 366 (iii) 261 (iv) 283
28 (i) $n \geq 2236$ (ii) $n \geq 116$
29 (i) 102 (ii) 51
30 (i) 20 (ii) 50 (iii) 18 (iv) 25

Exercise 5.1:4

1 (i) 27, 81, 243 (ii) $\frac{1}{2}, \frac{1}{4}, \frac{1}{8}$ (iii) $-\frac{16}{9}, \frac{64}{27}, -\frac{256}{81}$
(iv) $2\sqrt{2}$, 4, $4\sqrt{2}$ (v) 0.72, 0.432, 0.2592

(vi) $-24x^4$, $48x^5$, $96x^6$ (vii) $\frac{8}{x}, \frac{16}{x^2}, \frac{32}{x^3}$
(viii) −3, 3, −3
(ix) $\sin^2\theta\cos^2\theta$, $\sin^3\theta\cos^4\theta$, $\sin^4\theta\cos^6\theta$
2 (i) $\frac{3}{2}, \frac{24}{2^r}$ (ii) $\frac{10}{3}, \frac{6}{5}(-\frac{5}{3})^{r-1}$ (iii) $\mathbf{x}^5$, $(-1)^{r-1}x^r$
(iv) e^{-5x}, $e^{(3-2r)x}$
(v) $-\sin^3\theta$, $-\sin r2(r-1)\theta$
3 (i) $\frac{1}{4^{10}}$ (ii) $-\frac{1}{4^{10}}$ (iii) -2.7^{11} (iv) $32x^6$
(v) e^{43x}
4 (i) AP (ii) neither (iii) GP (iv) both
(v) neither (vi) neither (vii) neither
(viii) GP (ix) neither
5 (i) $\frac{1}{9}, \frac{1}{4}, \frac{3}{8}, \frac{9}{16}, \frac{27}{32}$
(ii) 10, −2, 0.4, −0.08, 0.016, −0.0032
(iii) x^2, $2x$, 4, $\frac{8}{x}, \frac{16}{x^2}, \frac{32}{x^3}$
6 (i) 13 (ii) 21 (iii) 33 (iv) m
7 (i) $\frac{5}{3}$, 2 (ii) $\frac{2}{81}$, 3; $\frac{2}{81}$, −3 (iii) 1458, $\frac{1}{3}$; 1458, $-\frac{1}{3}$ (iv) $x^{11}, \frac{1}{x^2}$; $-x^{11}, -\frac{1}{x^2}$
8 1.07 m **9** 43 **10** (i) 5 (ii) 8 (iii) 10 **11** 15
12 (i) 6, $\frac{27}{2}$ (ii) $-\frac{1}{2}$, 8 (iii) 10,500; −3, $-\frac{8}{3}$ (iv) $\frac{7}{2}$, 32
13 (i) 16 (ii) 30 (iii) $\frac{2}{3}$ (iv) x^4y^2
14 (i) 15, 12 (ii) 13, 5 (iii) $\frac{145}{8}$, 3
21 4, $\frac{16}{3}$ **22** 4, 9 **23** 4924
24 (i) (a) 3, 3 (ii) $\frac{5}{8}, \frac{1}{8}$ (c) $\frac{9}{2}, \frac{9}{2}$ (d) pq, q
25 2046
26 (i) 121.49 (ii) 4.8650 (iii) $\frac{x^{33}-1}{x^{10}(x^3-1)}$
27 (i) 1275 (ii) 0.74897 (iii) $\frac{e^{20x}-1}{e^x(e^{2x}-1)}$
34 (i) 3, −2 (ii) 81, $\frac{2}{3}$ (iii) $\frac{4}{63}$, 2; $-\frac{1}{21}$, −2
35 2
36 (i) $\frac{1-(n+1)x^n+nx^{n+1}}{(1-x)^2}$ (ii) $\frac{1-x^{n+1}}{1-x}$
(iii) $2+(n-1)2^{n+1}$
37 (i) 16 (ii) 11 (iii) 24 (iv) impossible

Exercise 5.1:6

1 (i) 8 (ii) $\frac{9}{5}$ (iii) $\frac{25}{3}$ (iv) 10 (v) $\frac{10}{19}$
2 (i) converges, $\frac{16}{3}$ (ii) converges, $\frac{27}{5}$
(iii) diverges (iv) diverges (v) converges, $\frac{10}{3}$
(vi) diverges
3 (i) $\frac{2}{3}$ (ii) $\frac{4}{5}$ **4** $\frac{4}{9}$
5 (i) $\frac{1}{9}$ (ii) $\frac{4}{33}$ (iii) $\frac{4}{333}$ (iv) $\frac{125}{999}$
6 (i) $\frac{2}{3}$ (ii) $\frac{64}{99}$ (iii) $\frac{128}{999}$
7 20 m **8** 28 s
12 (i) 37 (ii) 23 (iii) 12 (iv) 11
13 (i) 23 (ii) 7
14 (i) (a) $\frac{1}{1-x}$ (b) $\frac{1}{(1-x)^2}$ (c) $-\ln(1-x)$
(ii) (a) 4 (b) ln 2 (c) $\ln\frac{3}{2}$
15 $\frac{x(x^{2n}-1)}{x^2-1}$
16 (i) $|x|<1$; $\frac{1}{1+x}$ (ii) $|x|<\frac{1}{2}$; $\frac{1}{1-2x}$
(iii) $-3<x<-1$; $\frac{-1}{1+x}$ (iv) $x>\frac{1}{2}$; $\left(\frac{x+1}{x}\right)^2$
(v) $x<\frac{1}{2}$; $\frac{1-x}{1-2x}$ (vi) $-1<x<2$; $\frac{2+x}{2+x-x^2}$
(vii) $x<0$; $\frac{1}{1-e^{2x}}$ (viii) $x>0$; $\frac{1}{e^x-1}$
(ix) $-\frac{\pi}{6}+2n\pi<x<\frac{\pi}{6}+2n\pi$ or

$\frac{5\pi}{6}+2n\pi<x<\frac{7\pi}{6}+2n\pi$; $\frac{1}{1-2\sin x}$
(x) $-\frac{\pi}{4}+n\pi<x<\frac{\pi}{4}+n\pi$; $\frac{\cos x}{1-\tan x}$
17 $x<-\frac{1}{2}$; $-\frac{3}{4}$
18 $-\frac{1}{3}<x<3$; 2
19 $\frac{1}{2}<x<1$; $\frac{2}{3}$
20 (i) $|x|<1$ (ii) $|x|<\frac{1}{3}$; $\frac{1}{1-x}, \frac{1}{1-3x}$, $\frac{5-13x}{(1-x)(1-3x)}$
21 $\theta\neq(2n-1)\pi$; $\frac{1}{1-\sin\theta}$; $\frac{1}{(1-\sin\theta)^2}$
22 $x<0$; $\frac{1}{1-e^x}$; (i) $\frac{e^x}{(1-e^x)^2}$ (ii) $x-\ln(1-e^x)$

Exercise 5.1:7

1 $2r$ **2** $4r^3$; $\frac{1}{4}n^2(n+1)^2$
3 $2r-1$; n^2
4 $3r^2-3r+1$; n^3
5 $\frac{1}{3}n(n+1)(n+2)$
6 $4r(r+2)$; $\frac{1}{4}n(n+1)(n+2)(n+3)$; $\frac{1}{2}n(n+1)$, $\frac{1}{6}n(n+1)(2n+1)$
7 (a) $\frac{1}{5}(r+1)(r+2)(r+3)(r+4)$; $\frac{1}{5}n(n+1)(n+2)(n+3)(n+4)$
(b) $\frac{1}{k+2}n(n+1)(n+2)\ldots(n+k+1)$
8 2, 2, −2
9 $\frac{1}{2}, \frac{1}{2}, -\frac{1}{2}, -\frac{1}{2}$
10 $\frac{1}{4}, -\frac{1}{2}, 0$
11 $r(r!)$; $(n+1)!-1$
12 $\frac{1}{2}\sec\theta\{\sin(2n+1)\theta-\sin\theta\}$
13 $\frac{1}{2}\csc\theta\{1-\cos 2n\theta\}$
14 (i) $\frac{1}{2}n(n+1)$, $\frac{1}{6}n(n+1)(2n+1)$, $\frac{1}{4}n^2(n+1)^2$
(ii) (a) $\frac{1}{6}n(n+1)(2n+13)$
(b) $\frac{1}{6}n(n+1)(4n-1)$
(c) $\frac{1}{12}n(n+1)(3n^2+7n+2)$
(d) $\frac{1}{6}n(2n^2+3n-5)$
(e) $\frac{1}{6}n(4n^2+n-21)$
15 $\frac{1}{6}n(n+1)(n+2)$
16 $\frac{1}{12}n(n+1)(n+2)(3n+13)$; $\frac{1}{12}n(n+1)(n+2)(3n+5)$
17 $\frac{1}{6}n(n+1)(2n+1)$; $\frac{1}{3}n(2n+1)(4n+1)$; $\frac{1}{6}n(2n+1)(7n+1)$; 295425
18 $\frac{1}{4}n^2(15n^2+18n+3)$; 42075
19 (i) $\frac{1}{6}(n+2)(n+3)(2n+5)$
(ii) $\frac{1}{6}n(2n^2+15n+37)$
(iii) $\frac{1}{3}n(3n+1)(6n+1)$ (iv) $\frac{3}{2}n(n+1)(2n+1)$
20 (i) $\frac{1}{3}n(n^2+3n-16)$; $\frac{1}{3}n(7n^3+19n-16)$
(ii) 32845
21 $\frac{1}{6}n(n+1)(n+2)$
22 (i) (a) $\frac{1}{3}n(2n+1)(4n+1)$
(b) $\frac{2}{3}n(n+1)(2n+1)$
(c) $\frac{1}{3}n(2n+1)(2n-1)$ (iii) 166650
23 $n^2(2n^2-1)$; 12497500 **24** (iii) 4949
26 1804511 **27** 16594300 **29** 45
30 (i) $r(3r+5)$ (ii) $r(3r-1)$ (iii) $3r^2-3r+1$
32 $1-\frac{1}{n+1}$
33 $\frac{1}{4}-\frac{1}{2(n+1)(n+2)}$
34 $1-\frac{1}{(n+1)!}$ **35** $\frac{3}{2}-\frac{n+3}{(n+1)(n+2)}$
36 (ii) $\frac{5}{2}-\frac{2n+5}{(n+1)(n+2)}$ (iii) $\frac{5}{2}$
37 (ii) $\frac{1}{12}-\frac{1}{4(2n+1)(2n+3)}$ (iii) $\frac{1}{12}$ (iv) 25
38 (ii) $\frac{1}{24}-\frac{1}{6(3n+1)(3n+4)}$ (iii) $\frac{1}{24}$ (iv) 22

39 (i) $1-\dfrac{1}{n+1}$; 1 (ii) $\dfrac{1}{2}\left\{1-\dfrac{1}{2n+1}\right\}$; $\dfrac{1}{2}$
(iii) $1-\dfrac{1}{4n+1}$; 1 (iv) $\dfrac{1}{2}\left\{\dfrac{3}{2}-\dfrac{1}{n+1}-\dfrac{1}{n+2}\right\}$; $\dfrac{3}{4}$
(v) $\dfrac{1}{3}\left\{\dfrac{11}{6}-\dfrac{1}{n+1}-\dfrac{1}{n+2}-\dfrac{1}{n+3}\right\}$; $\dfrac{11}{18}$
(vi) $\dfrac{1}{4}-\dfrac{1}{2(n+1)(n+2)}$; $\dfrac{1}{4}$
(vii) $\dfrac{3}{2}-\dfrac{n+3}{(n+1)(n+2)}$; $\dfrac{3}{2}$
(viii) $\dfrac{1}{12}-\dfrac{1}{4(2n+1)(2n+3)}$; 12
(ix) $\dfrac{5}{24}-\dfrac{3}{16(2n+1)}-\dfrac{1}{16(2n+3)}$; $\dfrac{5}{24}$
(x) $\dfrac{1}{8}\left\{\dfrac{7}{12}+\dfrac{1}{n}-\dfrac{1}{n+1}-\dfrac{1}{n+2}+\dfrac{1}{n+3}\right\}$; $\dfrac{7}{96}$
(xi) $\dfrac{3}{4}-\dfrac{4n+3}{2(n+1)(n+2)^2}$; $\dfrac{3}{4}$
40 (i) $\dfrac{1}{6}-\dfrac{n+2}{(n+3)(n+4)}$ (ii) $\dfrac{1}{6}$

Exercise 5.3:1

1 (i) 15 (ii) 17 (iii) 11
2 (i) 6 (ii) 4
3 (i) 5040 (ii) 40320
4 120 **5** 5040 **6** 3628800
7 (i) 12 (ii) 120 **8** (i) 3024 (ii) 6720
9 360 **10** 64 **11** 1920
12 (i) 3360 (ii) 3360 **13** 15120
14 (i) 3603600 (ii) 3.41×10^9 **15** 120960
16 (i) 50400 (ii) 25200
17 (i) 8! (ii) $\dfrac{8!}{2}$
18 (i) 12! (ii) $\dfrac{12!}{2}$ (iii) $\dfrac{12!}{3!}$ (iv) $\dfrac{12!}{2}$
19 18 **20** 12
21 (i) 1440 (ii) 1080
22 (i) 10080 (ii) 30240 (iii) 8640
23 480 **24** 360 **25** 48
26 10 **27** 42 **28** 260; 88
29 (i) 48 (ii) 240 (iii) 4320 (iv) 720
30 40320 (i) 10080 (ii) 30 240
31 480 **32** (i) 14 400 (ii) 43200
33 (i) 72 (ii) 3 628 800 (iii) $(r!)^{r+1}p_q$
34 (i) 21772800 (ii) 1.22×10^8
35 (i) 240 (ii) 480
36 (i) 5040 (ii) 720
37 1344 **38** 3 **39** 210
40 (i) 48 (ii) 48 (iii) 72
41 (i) $(n-1)!$ (ii) $(n-3)\{(n-2)!\}$
42 (i) 120 (ii) 3125
43 (i) 1470 (ii) 3584
44 (i) 336 (ii) 512
45 (i) 6 (ii) 64 (iii) 4; 32 (iv) 2; 32
46 (i) 604 800 (ii) 10 000 000

Exercise 5.3:2

1 2 598 960 **2** 20 825 **3** 2002
4 (i) 8008 (ii) 6006 **5** 62 322
6 (i) $^{12}C_5$ (ii) $\dfrac{^{12}C_6}{2}$
7 (i) $^{16}C_6$ (ii) $\dfrac{^{16}C_8}{2}$
8 (i) 30 (ii) 20 **9** (i) 56 (ii) 70
10 (i) 27 720 (ii) 5775
11 (i) 6 (ii) 36 **12** 1749

13 (i) 256 (ii) 512
14 84 (i) 30 (ii) 40
15 (i) 35 (ii) 315 (iii) 756 (iv) 588 (v) 126; 1820 i.e. $^{16}C_4$
16 (i) 786 (ii) 786 (iii) 780
17 (i) 165 (ii) 285
18 (i) 791 (ii) 690 (iii) 66 (iv) 440
19 (i) 6 (ii) 45 (iii) 60 (iv) 15
20 (i) 10 (ii) 120 (iii) 80
21 (i) 60 (ii) 160 (iii) 32
22 (i) 14 850 (ii) 7560
23 (i) 38 760 (ii) 6300 (iii) 30 375
24 9450
25 (i) 35 (ii) 425 (iii) 304 (iv) 270
26 (i) 715 (ii) 11 154 (iii) 57 798 (iv) 118 807 (v) 82 251
27 (i) 6084 (ii) 13 182 (iii) 28 561
28 (i) 189 189 (ii) 69 667 (iii) 11 869
29 (i) 192 (ii) 495 (iii) 91 390
30 (i) 255 775 (ii) 140 686
31 (i) 211 820 (ii) 58 905
32 240 **33** 2100
34 1680 (i) 720; 360 (ii) 720; 180
35 (i) 31 (ii) 255
37 15 **38** 63 **39** 15
40 (i) 8 (ii) 32 (iii) 2^n **41** 59 049
42 (i) 36 (ii) 216 (iii) 6^n **43** k^n
44 (i) 1 (ii) 3 (ii) 3 (iv) 1
45 (i) 1 (ii) 5 (ii) 10 (iii) 10 (iv) 5 (v) 1; 32 i.e. 2^5
46 ^nC_r **47** (i) 255 (ii) 247
48 (i) 1 048 576 (ii) 2 621 440 (iii) 2 949 120 (iv) 9 765 624 (v) 9 765 584
49 (i) 220 (ii) 243
50 (i) 1 (ii) 10 (iii) 40 (iv) 80 (v) 80 (vi) 32; 243
51 (i) 1 (ii) 10 (iii) 55 (iv) 210
52 (i) 216 (ii) (a) 1 (b) 15 (c) 75 (d) 125
53 28 **54** 2970
55 (i) 784, 140 (ii) 29 241, 2109
56 (i) 12 (ii) rs (iii) 21 (iv) $rs+st+tr$
57 (i) 45 (ii) 210 **58** 132 **59** 66
60 (i) 5 (ii) 54 (iii) $\tfrac{1}{2}n(n-3)$
61 (i) 3, 7 (ii) 6, 11 (iii) 10, 16 (iv) $\tfrac{1}{2}n(n-1)$, $\tfrac{1}{2}(n^2+n+2)$

Exercise 5.3:3

1 (i) 360 (ii) 20 (iii) 90 720 (iv) 60 (v) 332 640
2 (i) 60 (ii) 20 (iii) 10
3 (i) 4^5 (ii) 3^5 (iii) 2^5
4 (i) 420 (ii) 84
5 (i) 60 (ii) 90 (iii) 729
6 (i) 60 (ii) 30 (iii) 15
7 (i) 126 (ii) ^nC_r (iii) ^nC_r (iv) ^nC_r; ^nC_r
8 11 440
9 (i) 5^4 (ii) 5^4 (iii) 5^4
10 (i) 120 (ii) 120 (iii) 1800
11 (i) 120 (ii) 2160 (iii) 1440 (iv) 151 200 (v) 1440
12 (i) 180 (ii) 9648
13 (i) 2520 (ii) 2520 (iii) 720
14 (i) 1200 (ii) 211 680 (iii) 1200 (iv) 360 (v) 423 360
15 (i) 2400 (ii) 720 (iii) 3600
16 (i) 14; 72 (ii) 50; 1020 (iii) 10; 42 (iv) 30; 500 (v) 41; 626 (vi) 41; 2250
17 (i) 11 (ii) 11 **18** (i) 10 (ii) 11
19 (i) 108 (ii) 444 (iii) 1620
20 (i) 72 (ii) 1230 **21** (i) 130 (ii) 170
22 (i) 60 (ii) 60 (iii) 38 (iv) 19
23 (i) 24 (ii) 25 (iii) 60 (iv) 52
24 (i) (a) n^2 (b) $n(n^2-1)$ (ii) (a) n^3 (b) $n(n^3-1)$ (c) $n(n^4-5n+4)$

25 (i) n^r (ii) $n(n^r-1)$ (ii) $n\{n^{r+1}-(r+2)n^2+(r+1)n\}$
26 (i) 59 (ii) 159
27 59; (i) 12 (ii) 12 (iii) 48
28 (i) 159 (ii) 160 **29** 71
30 (i) 14 (ii) 29 (iii) 35
31 (i) 16 (ii) 32 (iii) 2^n
32 (i) 27 (ii) 81 (iii) 3^n

Exercise 5.4:1

2 (i) $1+3x+3x^2+x^3$
 (ii) $1+4x+6x^2+4x^3+x^4$
 (iii) $1+\dfrac{5x}{2}+\dfrac{5x^2}{2}+\dfrac{5x^3}{4}+\dfrac{5x^4}{16}+\dfrac{x^5}{32}$
 (iv) $1-\dfrac{8x}{3}+\dfrac{8x^2}{3}-\dfrac{32x^3}{27}+\dfrac{16x^4}{81}$
 (v) $64-144x+108x^2-27x^3$
3 (i) $8x^3+12x^2y+6xy^2+y^3$
 (ii) $x^3-3x^2y+3xy^2-y^3$
 (iii) $8x^3-36x^2y+54xy^2-27y^3$
 (iv) $256x^4+128x^3y+24x^2y^2+2xy^3+\dfrac{y^4}{16}$
 (v) $729x^6-1458x^4y+3645x^4y^2-540x^3y^3+135x^2y^4-18xy^5+y^6$
4 (i) $x^3+3x+\dfrac{3}{x}+\dfrac{1}{x^3}$
 (ii) $x^4-4x^2+6-\dfrac{4}{x^2}+\dfrac{1}{x^4}$
 (iii) $x^{10}+5x^7+10x^4+10x+\dfrac{5}{x^2}+\dfrac{1}{x^5}$
5 (i) $1+8x+28x^2$ (ii) $1+12x+60x^2$
 (iii) $1-\dfrac{20x}{3}+\dfrac{190x^2}{9}$ (iv) $32-240x+720x^2$
 (v) $\dfrac{59\,049}{1024}-\dfrac{98\,415}{128}x+\dfrac{295\,245}{64}x^2$
6 (i) -5280; 101 376 (ii) $\dfrac{15309}{16}$ (iii) -448
 (iv) 120; 120 (v) 8505
7 (i) -4 (ii) 6 (iii) 12 (iv) 15 (v) $^{3n}C_n$
8 3:5 **9** (i) $-\tfrac{1}{3}$ or 3 (ii) $\tfrac{2}{3}$ or $\tfrac{3}{2}$
10 (i) 5 **11** 11
12 (i) $\tfrac{3}{2}$, $\tfrac{27}{2}$, 4 (ii) -3, 270, 5
13 $1+6x+21x^2+50x^3$
14 (i) $1-5x+15x^2-30x^3$
 (ii) $1+4x+18x^2+40x^3+91x^4$
 (iii) $1+3x+\dfrac{23x^2}{4}$
 (iv) $32+80x+160x^2+200x^3$
 (v) $1+nx+\dfrac{n(n+1)}{2}x^2$
15 (i) $1-2x+2x^3-x^4$
 (ii) $3+10x+10x^2-5x^4-2x^5$
 (iii) $8+20x+10x^2-5x^3-5x^4-x^5$
16 (i) $1-11x+36x^2$
 (ii) $1+21x+200x^2+1140x^3$
 (iii) $1+(n-1)x+\dfrac{n(n-3)}{2}x^2+\dfrac{n(n-1)(n-5)}{6}x^3$
17 1452 **18** (i) 92 378 (ii) $-\dfrac{4257}{256}$ **19** $-4, 2$
21 (i) $(2^r+1)^nC_x$ (ii) $2^nC_r+^nC_{r-1}$
 (iii) $2^{r-1}(2^nC_r-^nC_{r-1})$
 (iv) $^nC_r+^nC_{r-1}+^nC_{r-2}$
22 (i) $^nC_r(-3)^r$ (ii) $(-1)^r\{2^nC_r-3^nC_{r-1}\}$
 (iii) $(-2)^{r-2}\{4^nC_r-2^nC_{r-1}-^nC_{r-2}\}$
23 (i) 2^n (ii) $n2^{n-1}$
24 0 **25** (i) 56 (ii) 149
26 0.406 (i) 1.867 (ii) 5.742
27 (i) $x^3+3x^2\delta x+3x(\delta x)^2+(\delta x)^3$
 (ii) $x^n+nx^{n-1}\delta x+\dfrac{n(n-1)}{2}x^{n-2}(\delta x)^2$

582 Answers

$$+\frac{n(n-1)(n-2)}{6}x^{n-3}(\delta x)^3$$

28 $1+6x+15x^2+20x^3+15x^4+6x^5+x^6$; 1.00602

29 $1+5x+\frac{45x^2}{2}+15x^3$; 1.0511

30 $1+27x+324x^2+21168x^3$; 1.0273

31 $1-6x+15x^2-20x^3+15x^4$; 0.941480

32 (i) 95 099 · 004 99 (ii) 108 243 216

33 $1+7x+21x^2+35x^3+21x^4$ (i) 5846 (ii) 629

Exercise 5.4:2

1 (i) $\frac{1}{1-x}$; $|x|<1$ (ii) $\frac{1}{1-3x}$; $|x|<\frac{1}{3}$
  (iii) $\frac{1}{1+x}$; $|x|<1$ (iv) $\frac{1}{1-x^2}$; $|x|<1$
  (v) $\frac{1}{(1-x)^2}$; $|x|<1$ (vi) $\frac{2+x}{1-x^2}$; $|x|<1$
  (vii) $\frac{1+2x}{1+x}+\frac{4x}{(1+x)(1-x)^2}$; $|x|<1$

2 (i) $1+4x+6x^2+4x^3+x^4$;
  $1+5x+10x^2+10x^3+5x^4+x^5$
  (ii) 1, nx, $\frac{n(n-1)}{2}x^2$, $\frac{n(n-1)(n-2)}{6}x^3$, nx^{n-1}, x^n, 0, 0.

3 (i) $\frac{1}{1-x}$; $|x|<1$ (ii) $1-x+x^2-x^3+\ldots$; $|x|<1$

4 (i) $1+\frac{2}{3}x-\frac{1}{9}x^2+\frac{4}{81}x^3$; $|x|<1$
  (ii) $1-4x+10x^2-20x^3$; $|x|<1$
  (iii) $1-2x+3x^2-4x^3$; $|x|<1$
  (iv) $1-\frac{1}{3}x-\frac{1}{9}x^2-\frac{5}{81}x^3$; $|x|<1$
  (v) $1-3x+\frac{3}{2}x^2+\frac{1}{2}x^3$; $|x|<\frac{1}{2}$
  (vi) $1-\frac{3}{2}x+\frac{27}{8}x^2-\frac{135}{16}x^3$; $|x|<\frac{1}{3}$
  (vii) $1-\frac{1}{3}x-\frac{1}{18}x^2-\frac{5}{27}x^3$; $|x|<\frac{3}{2}$

5 (i) $\frac{1}{2}+\frac{1}{4}x+\frac{1}{8}x^2+\frac{1}{16}x^3$; $|x|<2$
  (ii) $\frac{1}{2}-\frac{1}{16}x+\frac{3}{256}x^2-\frac{5}{2048}x^3$; $|x|<4$
  (iii) $3+\frac{2}{3}x-\frac{2}{27}x^2+\frac{4}{243}x^3$; $|x|<\frac{9}{4}$
  (iv) $\frac{1}{9}-\frac{2}{27}x+\frac{2}{27}x^2-\frac{2}{243}x^3$; $|x|<3$
  (v) $-1-x-x^2-x^3$; $|x|<1$

6 (i) $5x-5x^2+35x^3-65x^4$; $|x|<\frac{1}{3}$
  (ii) $3-3x+9x^2-15x^3$; $|x|<\frac{1}{2}$
  (iii) $-2-x-2x^2-2x^3$; $|x|<1$
  (iv) $\frac{5}{2}-\frac{23}{4}x+\frac{145}{8}x^2-\frac{863}{16}x^3$; $|x|<\frac{1}{3}$
  (v) $\frac{1}{4}-\frac{5}{16}x+\frac{21}{64}x^2-\frac{85}{64}x^3$; $|x|<\frac{1}{3}$
  (vi) $-x+2x^2-x^3-x^5$; $|x|<1$
  (vii) $1+4x+11x^2+26x^3$; $|x|<\frac{1}{5}$

7 (i) $1+\frac{5}{2}x+\frac{7}{8}x^2-\frac{3}{16}x^3$; $|x|<1$
  (ii) $1+2x+2x^2+2x^3$; $|x|<1$
  (iii) $1-\frac{2x}{3}+\frac{2x^2}{9}-\frac{2x^3}{27}$; $|x|<3$
  (iv) $1-\frac{3x}{2}+\frac{7x^2}{8}-\frac{11x^3}{16}$; $|x|<1$
  (v) $1+4x+7x^2+10x^3$; $|x|<1$

8 (i) $1-\frac{3}{2}x+\frac{11}{8}x^2-\frac{23}{16}x^3$; $|x|<1$
  (ii) $1+3x+9x^2+49x^3$; $|x|<\frac{1}{4}$
  (iii) $1+2x+2x^2+4x^3$; $|x|<\frac{1}{2}$

9 $-2, 2, 2$ **10** (i) $-\frac{1}{2}, -\frac{1}{8}$ (ii) $-3, -\frac{1}{3}$

11 $-\frac{1}{2}, -2, \frac{1}{2}$; $|x|<2$

12 (i) $1+2x+3x^2+4x^3+5x^4$
  (ii) $1+3x+6x^2+10x^3+15x^4$

13 $1+\frac{1}{3}x-\frac{1}{9}x^2+\frac{5}{81}x^3$; 1.040 04; 2.080

14 $1-\frac{1}{3}x-\frac{1}{9}x^2-\frac{5}{81}x^3$; 0.999 666 555 5; 3.332 222

15 $1-\frac{1}{2}x^2-\frac{1}{8}x^4-\frac{1}{16}x^6$; 0.994 987 4; 3.316 62

16 $1+\frac{1}{2}x+\frac{3}{8}x^2+\frac{5}{16}x^3$; 1.414 21

17 $1-x-\frac{1}{2}x^2$; 3.162

18 $1-2x+3x^2-4x^3$; $1+\frac{1}{2}x-\frac{1}{8}x^2+\frac{1}{16}x^3$;
  (i) 0.994 027 (ii) 2.004 994

19 $1+x+\frac{1}{2}x^2$

20 $1+2x+2x^2$; 7.141

21 $1+\frac{3}{2}x+\frac{3}{8}x^2$

22 $1+\frac{5}{2}x-\frac{61}{8}x^2$; $2\frac{1}{2}\%$

23 $1-\frac{1}{2}x+\frac{3}{8}x^4-\frac{5}{16}x^6$; 0.390 09

24 (i) 0.231 82 (ii) $\frac{1}{2}\tan^{-1}\frac{1}{2}$; 0.463 64

25 (i) 0.47832 (ii) $\dfrac{3\sqrt{3}+2\pi}{24}$

Exercise 5.5:1

1 (i) $\frac{3}{7}$ (ii) $\frac{4}{7}$; 1

2 (i) (a) $\frac{4}{9}$ (b) $\frac{5}{9}$; 1 (ii) $\frac{2}{9}$ **3** $\frac{5}{9}$ **4** (i) $\frac{3}{10}$ (ii) $\frac{3}{5}$ (iii) 0

5 (i) $\frac{1}{2}$ (ii) $\frac{1}{3}$ (iii) $\frac{1}{6}$ (iv) $\frac{2}{3}$ (v) $\frac{1}{2}$

6 (i) $\frac{1}{13}$ (ii) $\frac{1}{4}$ (iii) $\frac{1}{26}$ (iv) $\frac{3}{13}$ (v) $\frac{1}{2}$

7 (i) $\frac{1}{52}$ (ii) $\frac{4}{13}$ (iii) $\frac{3}{52}$ (iv) $\frac{3}{13}$ (v) $\frac{9}{13}$

8 (i) $\frac{1}{3}$ (ii) $\frac{1}{6}$ (iii) $\frac{1}{2}$; 1

9 (i) $\frac{3}{10}$ (ii) $\frac{1}{5}$ (iii) $\frac{3}{5}$ (iv) $\frac{2}{5}$

10 (i) $\frac{1}{500}$ (ii) $\frac{1}{50}$ (iii) $\frac{49}{50}$

11 (i) $\frac{1}{2}$ (ii) $\frac{1}{4}$ (iii) $\frac{7}{8}$

12 (i) $\frac{1}{16}$ (ii) $\frac{3}{4}$ (iii) $\frac{5}{16}$ (iv) $\frac{1}{4}$ (v) $\frac{1}{16}$

13 (i) $\frac{5}{16}$ (ii) $\frac{1}{2}$ (iii) $\frac{5}{16}$

14 (i) $\frac{1}{4}$ (ii) $\frac{3}{4}$

15 (i) $\frac{3}{8}$ (ii) $\frac{1}{2}$ (iii) $\frac{1}{2}$ (iv) $\frac{3}{8}$

16 (i) $0, 0, \frac{1}{36}, \frac{2}{36}, \frac{3}{36}, \frac{4}{36}, \frac{5}{36}, \frac{6}{36}, \frac{5}{36}, \frac{4}{36}, \frac{3}{36}, \frac{2}{36}, \frac{1}{36}$
  (ii) 1 (iii) 7

17 (i) $\frac{1}{6}$ (ii) $\frac{5}{18}$

18 (i) $\frac{5}{12}$ (ii) $\frac{5}{12}$ (iii) $\frac{1}{6}$

19 (i) $\frac{1}{6}$ (ii) $\frac{5}{18}$ (iii) $\frac{2}{9}$ (iv) $\frac{1}{6}$ (v) $\frac{1}{9}$ (vi) $\frac{1}{18}$

20 (i) $\frac{1}{36}$ (ii) $\frac{10}{36}$ (iii) $\frac{10}{36}$ (iv) $\frac{25}{36}$

21 (i) $\frac{2}{3}$ (ii) $\frac{2}{3}$ (iii) $\frac{2}{3}$ (iv) $\frac{2}{3}$

22 (i) (a) $\frac{2}{5}$ (b) $\frac{1}{5}$ (c) $\frac{16}{25}$
  (ii) (a) $\frac{2}{25}$ (b) $\frac{3}{25}$ (c) $\frac{6}{25}$
  (iii) (a) $\frac{7}{25}$ (b) $\frac{16}{25}$ (c) $\frac{9}{25}$ (d) $\frac{6}{25}$

23 (i) $\frac{1}{10}$ (ii) $\frac{9}{50}$ (iii) $\frac{3}{25}$ (iv) $\frac{9}{20}$ (v) $\frac{11}{20}$

24 216; (i) $\frac{13}{54}$ (ii) $\frac{22}{27}$ (iii) $\frac{1}{2}$

25 64; (i) $\frac{1}{16}$ (ii) $\frac{27}{32}$ (iii) $\frac{9}{16}$ (iv) $\frac{9}{32}$

26 (i) $\frac{1}{216}$ (ii) $\frac{1}{36}$ (iii) $\frac{7}{72}$

27 2704 (i) 16; $\frac{25}{169}$ (ii) 384; $\frac{24}{169}$ (iii) 2304; $\frac{144}{169}$

28 (i) $\frac{1}{64}$ (ii) $\frac{3}{64}$ (iii) $\frac{3}{8}$ (iv) $\frac{64}{2197}$

29 (i) $\frac{81}{256}$ (ii) $\frac{1}{256}$ (iii) $\frac{1}{128}$ (iv) $\frac{3}{32}$

30 (i) $\dfrac{ns(t-s)^{n-1}}{t^n}$ (ii) $\dfrac{{}^nC_2s^2(t-s)^{n-2}}{t^n}$
  (iii) $\dfrac{s^n}{t^n}$ (iv) $\dfrac{{}^nC_rs^r(t-s)^{n-r}}{t}$

31 (i) $\frac{1}{21}$ (ii) $\frac{5}{14}$ (iii) $\frac{10}{21}$ (iv) $\frac{5}{42}$; 1

32 (i) $\frac{1}{126}$ (ii) $\frac{20}{63}$ (iii) $\frac{5}{6}$

33 (i) $\frac{4}{45}$ (ii) $\frac{8}{15}$, $\frac{1}{2}$

34 $\frac{7}{248}$ **35** $\frac{39}{105}$

36 (i) $\frac{1}{462}$ (ii) $\frac{100}{231}$ (iii) $\frac{25}{77}$

37 (i) $\frac{11}{850}$ (ii) $\frac{117}{850}$ (iii) $\frac{39}{850}$ (iv) $\frac{39}{204}$

39 (i) 6.3×10^{-12} (ii) 0.019 (iii) 0.304

40 (i) $\frac{3}{5}$ (ii) $\frac{54}{125}$ **41** (i) $\frac{5}{9}$ (ii) $\frac{1}{2}$

42 (i) 0.001 (ii) 0.08e

43 (i) 0.060 (ii) 0.063

44 (i) 0.040 (ii) 0.008

45 (i) $\frac{1}{56}$ (ii) $\frac{15}{56}$ (iii) $\frac{15}{28}$ (iv) $\frac{5}{28}$

46 (i) $\frac{1}{20}$ (ii) $\frac{9}{20}$ (iii) $\frac{1}{20}$, $\frac{3}{8}$, $\frac{3}{8}$

47 (i) $\frac{1}{20}$ (ii) $\frac{1}{2}$ **48** (i) $\frac{9}{14}$ (ii) $\frac{5}{6}$

49 $\frac{2}{55}$ **50** (i) $\frac{1}{105}$ (ii) $\frac{10}{21}$

51 (i) $\frac{1}{15}$ (ii) $\frac{2}{5}$ (iii) $\frac{8}{15}$

Exercise 5.2:2

1 (i) $\frac{1}{12}$ (ii) $\frac{11}{12}$

2 0.41, 0.59

3 0.28, 0.999996

4 $\frac{125}{126}$, $\frac{105}{126}$ **5** $\frac{9}{14}$

6 (i) 0.264 (ii) 0.495 (iii) 0.242

7 (i) (a) $\frac{91}{216}$ (b) $\frac{215}{216}$
  (ii) (a) $\frac{4083}{4096}$ (b) $\frac{4083}{4096}$

8 (i) $\frac{1}{7}$ (ii) $\frac{6}{7}$ (iii) $\frac{2}{7}$

9 (i) $\frac{1}{49}$ (ii) $\frac{30}{49}$ (iii) $\frac{4}{49}$

10 $1-\dfrac{{}^{365}p_m}{(365)^n}$; 23

12 (i) $\frac{1}{6}$ (ii) $\frac{1}{3}$

13 (i) $\frac{5}{8}$ (ii) $\frac{19}{30}$

14 (i) $\frac{4}{13}$ (ii) $\frac{9}{13}$

15 (i) $\frac{8}{13}$ (ii) $\frac{9}{13}$

16 (i) $\frac{5}{8}$ (ii) $\frac{5}{8}$, $\frac{1}{2}$ (iii) $\frac{3}{8}$ (iv) $\frac{3}{4}$ (v) $\frac{1}{8}$ (vi) $\frac{1}{4}$

17 (i) $\frac{1}{3}$ (ii) $\frac{2}{3}$ (iii) $\frac{1}{12}$ (iv) $\frac{5}{12}$

18 (i) $\frac{3}{10}$ (ii) $\frac{1}{5}$ **19** (i) $\frac{3}{20}$ (ii) $\frac{11}{30}$ **20** $\frac{1}{20}$

21 (i) $\frac{1}{4}$, $\frac{1}{7}$ (ii) $\{x\in\mathbb{Z}: x \text{ is divisible by } 28\}$, $\frac{1}{28}$
  (iii) $\frac{5}{14}$

22 (i) $\frac{8}{15}$ (ii) $\frac{2}{3}$

Exercise 5.5:3

1 (i) Yes; $\frac{1}{2}, \frac{1}{2}, 1$ (ii) No; $\frac{1}{6}, \frac{1}{9}, \frac{5}{18}$

2 (i) Yes; $\frac{1}{4}, \frac{1}{4}, \frac{1}{13}$ (ii) No; $\frac{1}{4}, \frac{4}{13}, \frac{1}{13}$

3 (i) No; $\frac{1}{12}, \frac{1}{6}, \frac{2}{9}$ (ii) Yes; $\frac{1}{12}, \frac{7}{9}, \frac{7}{36}$
  (ii) Yes; $\frac{1}{9}, \frac{1}{9}, \frac{2}{9}$

4 (i) No (ii) Yes (iii) No

5 A and B, B and C

6 (i) Yes (ii) No

7 A and B, B and C; No

8 (i) A and B (ii) A and C

13 (i) $\frac{1}{2}$ (ii) $\frac{4}{9}$ (iii) $\frac{2}{3}$ (iv) $\frac{6}{25}$

14 (i) $\frac{2}{15}$ (ii) $\frac{3}{5}$ (iii) $\frac{4}{13}$, $\frac{4}{15}$

15 (i) (a) $1-p-p'+pp'$ (b) $p'(1-p)$

16 $\frac{3}{4}$ **17** $\frac{11}{20}$

18 (i) $\frac{1}{12}$ (ii) $\frac{1}{2}$ (iii) $\frac{1}{2}$ (iv) $\frac{1}{4}$

19 (i) $\frac{7}{16}$ (ii) $\frac{55}{784}$

20 (i) $\frac{7}{12}$ (ii) $\frac{1}{12}$ (iii) $\frac{7}{8}$

21 (i) $\frac{36}{125}$ (ii) $\frac{54}{125}$

22 (i) $\frac{27}{256}$ (ii) $\frac{255}{256}$ **23** $\frac{5}{72}$

24 (i) $\frac{75}{512}$ (ii) $\frac{27}{64}$

25 $\frac{1}{21}$ (i) (a) $\frac{10}{441}$ (b) $\frac{8}{63}$ (c) $\frac{169}{441}$ (ii) $\frac{275}{343}$

26 (i) (a) $\frac{1}{16}$ (b) 0.240
  (ii) (a) $\frac{1}{4}$ (b) 0.412
  (iii) (a) $\frac{3}{8}$ (b) 0.265
  (iv) (a) $\frac{1}{4}$ (b) 0.076
  (v) (a) $\frac{1}{16}$ (b) 0.008

27 (i) $\frac{16}{81}$ (ii) $\frac{4}{9}$ (iii) $\frac{41}{81}$

28 (i) $\frac{16}{81}$ (ii) $\frac{4}{9}$ (iii) $\frac{29}{81}$

29 (i) 0.246 (ii) 0.167

30 (i) 0.007 (ii) 0.323

31 (i) 1×10^{-7} (ii) 0.322

32 (i) 0.803 (ii) 0.347

33 (i) 0.386 (ii) 0.003

34 0.088; (a) 0.631 (b) 5×10^{-6}

35 (i) 1 (ii) q^n, npq^{n-1}, ${}^nC_2p^2q^{n-2}$, p^n
  (iii) ${}^nC_rp^rq^{n-r}$

36 (i) 0.098 (ii) 0.004 (iii) 0.0002

37 11 **38** 5 **39** 17

40 (i) 0.311 (ii) 0.001 (iii) 0.029 (iv) 0.117
  (v) 7×10^{-4}

41 (i) (a) 0.035 (b) 0.095 (ii) (a) 0.16 (b) 0.18

42 (i) 0.015 (ii) 0.155

43 (i) 0.077 (ii) 0.125 (iii) 0.246

44 0.896

45 (i) $\frac{1}{2}$ (ii) $\frac{1}{8}$ (iii) $\frac{1}{32}$ (iv) $\frac{1}{128}$; $\frac{2}{3}$

46 (i) $\frac{6}{11}$ (ii) $\frac{5}{11}$

47 $\dfrac{p}{p+p'-pp'}$

48 (i) $\frac{4}{7}$ (ii) $\frac{2}{7}$ (iii) $\frac{1}{7}$

49 (i) (a) $\frac{25}{39}$ (b) $\dfrac{1}{3-3p+p^2}$
  (ii) (a) $\frac{10}{39}$ (b) $\dfrac{1-p}{3-3p+p^2}$
  (iii) (a) $\frac{4}{39}$ (b) $\dfrac{(1-p)^2}{3-3p+p^2}$ No

50 0.096

51 (i) 0.079 (ii) 0.033

52 (i) 0.148 (ii) 0.066 (iii) 0.026

53 (i) $\dfrac{1}{2^r}$ (ii) $\dfrac{2}{3^r}$

Answers 583

54 $p(1-p)^{r-1}$
55 (i) 0.031 (ii) 0.104 (iii) 0.071
57 (i) 0.027 (ii) 0.051 (iii) 0.209

Exercise 5.5:4

1 (i) $\frac{1}{6}, \frac{2}{5}$ (ii) $\frac{2}{11}, \frac{2}{5}$ (iii) $\frac{7}{11}, \frac{7}{12}$
2 (i) 0, 0 (ii) $\frac{1}{2}, \frac{1}{3}$ (iii) $\frac{1}{2}, \frac{1}{2}$
5 (i) $\frac{3}{4}$ (ii) $\frac{1}{2}$ (iii) $\frac{7}{12}$ (iv) $\frac{5}{12}$ (v) $\frac{5}{8}$ (vi) $\frac{5}{6}$
6 (i) $\frac{3}{10}$ (ii) $\frac{1}{8}$ (iii) $\frac{1}{4}$ (iv) $\frac{1}{5}$ (v) $\frac{2}{5}$ (vi) $\frac{1}{3}$
7 $\frac{2}{5}, \frac{1}{3}$ **8** (i) $\frac{1}{5}$ (ii) $\frac{4}{15}$
9 (i) $\frac{5}{8}$ (ii) $\frac{15}{28}$ (iii) No (iv) No
10 (i) $\frac{2}{3}$ (ii) $\frac{3}{5}$ (iii) $\frac{14}{17}$
11 0.167; (i) 0.125 (ii) 0.667 (iii) 0.035 (iv) 0.087
12 It doesn't matter—the probability of improvement in each case is $\frac{9}{16}$.
13 (i) (a) $\frac{3}{7}$ (b) $\frac{1}{2}$ (ii) (a) $\frac{3}{7}$ (b) $\frac{1}{2}$
14 (i) $\frac{1}{3}$ (ii) $\frac{1}{2}$
15 (i) 0.4616 (ii) 0.6918
16 (i) $\frac{131}{336}$ (ii) $\frac{1}{6}$ (iii) $\frac{35}{131}$ (iv) $\frac{21}{149}$
17 (i) $\frac{7}{24}$ (ii) $\frac{3}{7}$
18 $\frac{3}{13}$ **19** (i) $\frac{113}{360}$ (ii) $\frac{48}{113}$
20 (i) $\frac{27}{43}$ (ii) $\frac{1}{17}$
21 (i) $\frac{25}{48}$ (ii) $\frac{18}{25}$ (a) $\frac{18}{25}$ (b) $\frac{1}{5}$ (c) $\frac{2}{25}$ (iii) (a) $\frac{6}{23}$ (b) $\frac{3}{23}$ (c) $\frac{14}{23}$
22 (i) $\frac{2}{5}, \frac{1}{5}, \frac{1}{5}$ (ii) $\frac{3}{5}, \frac{2}{5}$ (iii) $\frac{2}{3}, \frac{1}{2}$
23 (i) $\frac{27}{125}$ (ii) $\frac{1}{10}$
24 (i) $\frac{3}{8}$ (ii) $\frac{27}{64}$
25 (i) $\frac{15}{32}$ (ii) $\frac{45}{96}$
26 (i) 0.0002 (ii) 0.013
27 (i) $\frac{3}{44}$ (ii) $\frac{3}{11}$
29 $\frac{3}{56}$ **30** (i) 0.119 (ii) $\dfrac{3 \times {}^{17}C_{r-1}}{{}^{20}C_r}$
31 (i) 0.148 (ii) 0.088 (iii) 0.030
32 (i) $\frac{3}{10}$ (ii) $\frac{3}{10}$ (iii) $\frac{1}{15}$ (iv) $\frac{8}{15}$ (v) $\frac{2}{9}$
33 (i) $\dfrac{s}{s+t}$ (ii) $\dfrac{s(s-1)}{(s+t)(s+t-1)}$ (iii) $\dfrac{t}{s+t-1}$ (iv) $\dfrac{s-1}{s+t-1}$
34 (i) 0.659 (ii) 0.341 (iii) 0.040 (iv) 0.208 (v) 0.659; 0.060

Miscellaneous Exercise 5

1 (i) 4, 9, 16 (ii) 6, 4, $\frac{8}{3}$; $9(\frac{2}{3})^{n-1}$; 27
2 $2n+3$; 99 **3** $x>2$; 13
4 (i) 1920 (ii) 1287
5 (i) $-1, 4$; $-1, \frac{3}{2}$ (ii) $7+4\sqrt{3}$
6 $x > \ln 2$; $\dfrac{e^{3x}}{1-4e^{-2x}}$ (i) 48.6 (ii) 1
7 (ii) $\lambda\mu$ (iii) $-\frac{1}{3}, \frac{1}{3}, \frac{1}{9}$
8 (i) (a) $-2<x<0$; $-\dfrac{1}{x}$ (b) $x<-2$ or $x>0$; $\dfrac{1+x}{x}$
(c) $-\frac{1}{3}<x<1$; $\dfrac{1+x}{1-x}$ (ii) $\frac{2}{3}$, 144
9 (i) $\frac{1}{2}$ **10** the largest integer less than or equal to $\log_z\left(\dfrac{y}{y-x(z-1)}\right)$, where $z = 1 + \dfrac{p}{100}$
11 (i) $\frac{1}{3}n(n+1)(n+2)$ (ii) $1 - \dfrac{1}{(n+1)^2}$
16 (i) $\frac{1}{4} - \dfrac{1}{2(n+1)(n+2)}$ (ii) $\frac{1}{4}$; 70
17 (i) $\frac{1}{6}n(2n^2+15n+31)$ (ii) $\frac{1}{2}n(n^2+5n+5)$ (iii) $\frac{1}{6}(21n^2+16n+1)$ (iv) $-n(2n+1)$ (v) $-8n^2$
18 (i) 24; 5200 (ii) (a) $\frac{1}{20}$ (b) 1
19 (i) -536 (ii) 3008 (iii) 2.37 (iv) $\frac{16}{105}$
23 (i) 3.3×10^8; 1.8×10^8 (ii) 8 990 000; 541 296

24 90 720; (i) 66 (ii) 5070
25 (i) 462 (ii) 792 (iii) 2047 (iv) 5775 (v) 5544
26 (i) 30? (ii) 167960
27 (i) 210, 210 (iii) (a) 210 (b) 210 (iv) (a) 210 (b) 210 (v) 1023
28 (i) 90; 86 (ii) $(2n)(3^{n-1})$
29 $-4, \frac{1}{2}, -4$; $|x| < \frac{1}{4}$
30 $-1, 2$, $|x|<\frac{1}{2}$; $-7, -10$, $|x| < \frac{1}{10}$
31 $1 + x + x^2 + 3x^3$; $|x| < \frac{1}{2}\sqrt{2}$
32 $\frac{3}{2} + \frac{3}{4}x + \frac{9}{8}x^2 + \frac{15}{16}x^3$; $|x|<1$
33 (iii) $\dfrac{3}{3-2r}$ (iv) $\dfrac{1}{1+x^2}$ (v) $\dfrac{2+x}{1-x^2}$
34 (i) 0.0234 (ii) 0.0235
35 (i) 5.099 (ii) 0.319
36 $1 - \frac{1}{2}y + \frac{3}{8}y^2$; $|y|<1$; (i) 0.009 685 (ii) 0.009 975
37 (i) $2 + \dfrac{x^2}{4} - \dfrac{x^4}{128}$, $|x|<2$
(ii) $x + \dfrac{2}{x} - \dfrac{2}{x^3}$, $|x|>2$ (a) 10.198
38 (i) 0.22 (ii) 0.73
39 $\frac{1}{4}$ **40** (i) $\frac{1}{2}$ (ii) $\frac{11}{108}$ (iii) $\frac{17}{108}$
41 (i) $\frac{7}{12}$ (ii) $\frac{1}{4}$ (iii) $\frac{7}{18}$ (iv) $\frac{5}{12}$ (v) $\frac{1}{6}$
42 (i) $\frac{1}{6}$ (ii) $\frac{1}{36}$ (iii) 0 (iv) $\frac{1}{7776}$ (v) $\frac{1}{12}$ (vi) $\frac{5}{216}$ (vii) $\frac{5}{7776}$
43 (i) $\frac{3}{10}$ (ii) $\frac{5}{9}$ (iii) $\frac{43}{45}$ (iv) $\frac{1}{6}$ (v) $\frac{28}{45}$
44 (i) $\frac{3}{44}$ (ii) $\frac{3}{22}$ (iii) $\frac{5}{12}$
45 (i) $\frac{1}{126}$ (ii) $\frac{5}{126}$ (iii) $\frac{10}{21}$
46 (i) 66 (ii) (a) $\frac{1}{2}$ (b) $\frac{2}{5}$
48 (i) 0.011 (ii) 0.135 (iii) 0.105 (iv) 0.00005 (v) 0.010 (vi) 0.676
49 $\frac{5}{24}, \frac{1}{24}; \frac{1}{12}$
50 (i) 0.4 (ii) 0.2 (iii) 0.8
51 (i) (a) $\frac{1}{221}$ (b) $\frac{32}{221}$ (ii) $\frac{25}{3}$ (iii) $\frac{25}{169}$
52 (i) $\frac{5}{8}$ (ii) (a) $\frac{1}{10}$ (b) 0.65
53 (i) 0.165 (ii) 0.184 (iii) 0.026 (iv) 0.110 (v) 0.006
54 $\frac{19}{48}$ (i) $\frac{9}{19}$ (ii) $\frac{9}{11}$ (iii) $\frac{9}{23}$
55 0.0002, 0.41, 0.59 (i) 0.00002 (ii) 0.504 (iii) 0.75
56 (i) $\frac{9}{25}, \frac{12}{25}, \frac{4}{25}$ (ii) $\frac{4}{729}$
57 $\dfrac{1}{1-x}$ (i) $-\dfrac{1}{(1-x)^2}$ (ii) $\dfrac{2}{(1-x)^3}$, $\frac{7}{27}$

Chapter 6

Exercise 6.1:2

3 (i) $-\mathbf{a}$ (ii) $3\mathbf{a}$ (iii) $-2\mathbf{a}$ (iv) $-3\mathbf{a}$
4 (i) 12, NE (ii) 2, NE (iii) 4, SW (iv) 6, SW
6 (i) $\overrightarrow{AC}$ (ii) $\overrightarrow{AB}$ (iii) $\overrightarrow{AD}$ (iv) $\overrightarrow{BC}$ (v) $\overrightarrow{AC}$
7 (i) $\overrightarrow{AB}$ (ii) $\overrightarrow{AD}$ (iii) $\overrightarrow{AD}$ (iv) $\overrightarrow{CA}$
8 (i) $\overrightarrow{AD}$ (ii) $\overrightarrow{AE}$ (iii) $2\overrightarrow{AD}$ (iv) $3\overrightarrow{AC}$
9 (i) $\overrightarrow{AC}$ (ii) $\overrightarrow{AB}$ (iii) $2\overrightarrow{AC}$ (iv) $\overrightarrow{CD}$ (v) $\overrightarrow{AD}$ (vi) $2\overrightarrow{AC}$
10 (i) $\mathbf{u}+\mathbf{v}$ (ii) $\frac{1}{2}(\mathbf{u}+\mathbf{v})$ (iii) $\frac{1}{2}(\mathbf{v}-\mathbf{u})$ (iv) $\frac{1}{2}(\mathbf{u}-\mathbf{v})$
12 (i) $\overrightarrow{AB}$ (ii) $\overrightarrow{AC}$ (iii) $\overrightarrow{AO}$ (iv) $\overrightarrow{AE}$ (v) $\overrightarrow{CE}$ (vi) $\overrightarrow{AD}$ (vii) $\overrightarrow{AC}$
13 (i) $\mathbf{u}+\mathbf{v}$ (ii) $\mathbf{v}-\mathbf{u}$ (iii) $-\mathbf{u}$ (iv) $\mathbf{v}-2\mathbf{u}$
14 (i) $\mathbf{b}-\mathbf{a}$ (ii) $-\mathbf{b}$ (iii) $\mathbf{a}-2\mathbf{b}$ (iv) $\mathbf{b}-2\mathbf{a}$ (v) $2\mathbf{a}-2\mathbf{b}$
15 (i) $\mathbf{u}+\mathbf{v}$ (ii) $\mathbf{u}+\mathbf{v}+\mathbf{w}$ (iii) $\mathbf{v}-\mathbf{u}$ (iv) $\mathbf{u}-\mathbf{v}-\mathbf{w}$ (v) $-\mathbf{u}-\mathbf{v}$
16 (i) $\mathbf{0}$ (ii) $\mathbf{0}$ (iii) $\mathbf{0}$
17 (i) $\sqrt{2}$, 45° (ii) 4, 30° (iii) 13, 67.4°
18 4.47 m, N63.4°E
19 24 **20** (i) $\sqrt{3}$, 30° (ii) 11.36, 37.6°
21 (i) 7.21 m, N43.9°E (ii) 5.29 m, N10.9°W
22 (i) 4.36, 36.6° to $\mathbf{u}$ (ii) 2.65, 79.1° to $\mathbf{u}$

23 (i) 8, 0° to $\mathbf{u}$ (ii) 7.54, 25.2° to $\mathbf{u}$ (iii) 2.83, 118.0° to $\mathbf{u}$ (iv) 2, 180° to $\mathbf{u}$
24 75.5° **25** 158.2° **26** 25, 25 **28** 7.81, 4.58 **29** 125.1°
30 (i) $2\sqrt{2}$ m, NW (ii) 8 m, due N (iii) 10.67 m, N 81.0°E
31 (i) 6.16 m (ii) 7.09 m
32 (i) 3.61, N56.3°E (ii) 3.61, S56.3°E (iii) 2.83, NE (iv) 2.83, NE
33 (i) 3.90, S7.4°W (ii) 7.80, S7.4°W (iii) 3.90, N7.4°E
35 (i) 6 (ii) 5.20 (iii) 3 (iv) 0 (v) -3 (vi) -5.20 (vii) -6
36 (i) 8.66 (ii) 4 (iii) 17.32 (iv) 21.32 (v) -5.32 (vi) 25.98 (vii) 0
37 (i) 0 (ii) a^2 (iii) 0 (iv) $-\frac{1}{2}a^2$
38 (i) $\frac{1}{2}a^2$ (ii) $-\frac{1}{2}a^2$ (iii) a^2 (iv) 0 (v) $\frac{3}{2}a^2$

Exercise 6.1:3

3 $\mathbf{b}-\mathbf{a}$, $2\mathbf{a}$, $-\mathbf{a}-\mathbf{b}$
4 $\mathbf{b}-\mathbf{a}$, $\mathbf{b}-\mathbf{a}$
7 (i) Yes (ii) No (iii) Yes
9 (i) $\mathbf{v}$ (ii) $-\frac{1}{2}\mathbf{u}-\mathbf{v}$ (iii) $-\frac{1}{2}\mathbf{u}-\mathbf{v}$ (iv) $-\frac{1}{2}\mathbf{u}+\mathbf{v}$ (v) $-\frac{1}{2}\mathbf{u}+\mathbf{v}$
19 $\frac{1}{2}\mathbf{a}$, $\frac{1}{2}(\mathbf{c}-\mathbf{b})$, $\frac{1}{2}(\mathbf{c}-\mathbf{a})$
20 (i) $\frac{1}{2}(\mathbf{u}+\mathbf{v})$, $\frac{1}{2}\mathbf{u}+\frac{1}{2}\mathbf{v}+\mathbf{w}$ (ii) $\frac{1}{2}(\mathbf{u}+\mathbf{v}+\mathbf{w})$, $\frac{1}{2}(\mathbf{u}+\mathbf{v}+\mathbf{w})$, $\frac{1}{2}(\mathbf{u}+\mathbf{v}+\mathbf{w})$, $\frac{1}{2}(\mathbf{u}+\mathbf{v}+\mathbf{w})$ (iii) $\frac{1}{2}(\mathbf{u}+\mathbf{v}+\mathbf{w})$
29 (iii) $2, -2$; $2\mathbf{b}-\mathbf{a}$, $3\mathbf{a}-2\mathbf{b}$
30 (i) $\mathbf{r}=2\mathbf{p}+\mathbf{q}-\lambda(\mathbf{p}+\mathbf{q})$ (ii) $\mathbf{r}=3\mathbf{p}+2\mathbf{q}+\mu(4\mathbf{p}+3\mathbf{q})$
32 2, 1 **33** $7\mathbf{a}-3\mathbf{b}$
34 (i) $\frac{3}{4}(\mathbf{a}+\mathbf{b})$ (ii) $\frac{2}{3}(2\mathbf{a}+\mathbf{b})$
35 (i) $\frac{1}{3}(\mathbf{a}+2\mathbf{b})$ (ii) $\frac{1}{3}(2\mathbf{a}+\mathbf{b})$; yes
36 (i) $\frac{1}{4}(\mathbf{a}+3\mathbf{b})$ (ii) $\frac{1}{4}(3\mathbf{a}+\mathbf{b})$
37 $\mathbf{r}=\mathbf{a}+\lambda(\mathbf{b}-\mathbf{a})$; $\frac{1}{4}(\mathbf{a}+3\mathbf{b})$
38 (i) $\mathbf{r}=\frac{1}{2}\mathbf{a}+\lambda(\mathbf{b}+\mathbf{c}-\mathbf{a})$ (ii) $\frac{1}{4}(\mathbf{a}+\mathbf{b}+\mathbf{c})$
39 $\mathbf{r}=(1-\mu)\mathbf{b}+\frac{1}{2}\mu(\mathbf{a}+\mathbf{c})$, $\mathbf{r}=(1-\nu)\mathbf{c}+\frac{1}{2}\nu(\mathbf{a}+\mathbf{b})$; $\frac{1}{3}(\mathbf{a}+\mathbf{b}+\mathbf{c})$
40 $\frac{1}{2}(\mathbf{a}+\mathbf{b})$, $\frac{1}{3}(\mathbf{b}+2\mathbf{c})$, $2\mathbf{d}-\mathbf{c}$; $\mathbf{r}=\mathbf{a}+\lambda(\mathbf{b}-\mathbf{a})$, $\mathbf{r}=\frac{1}{2}(\mathbf{a}+\mathbf{b})+\mu(3\mathbf{a}+\mathbf{b}-12\mathbf{c})$; $\frac{1}{2}(\mathbf{a}+\mathbf{b})$

Exercise 6.2:1

1 (i) $4\mathbf{u}-\mathbf{v}$ (ii) $2\mathbf{u}+5\mathbf{v}$ (iii) $9\mathbf{u}-5\mathbf{v}$
2 $\frac{1}{7}(3\mathbf{a}+\mathbf{b})$, $\frac{1}{7}(\mathbf{a}-2\mathbf{b})$
3 (i) 3, due E (ii) 2, due W (iii) 1, due S (iv) $\sqrt{2}$, NE (v) $\sqrt{2}$, SE (vi) 5, N36.9°E (vii) 5, N36.9°W
4 (i) 3, SW (ii) $3\sqrt{2}$, due S (iii) $3\sqrt{2}$, due E (iv) $\sqrt{5}$, S18.4°E
5 $\frac{1}{2}\mathbf{a}$, $\mathbf{a}+\mathbf{b}$, $\frac{1}{2}(\mathbf{a}+\mathbf{b})$, $2\mathbf{b}$, $2\mathbf{a}+\frac{1}{2}\mathbf{b}$, $2\mathbf{a}+\mathbf{b}$; $\frac{1}{2}\mathbf{a}+\mathbf{b}$, $-\mathbf{a}-\frac{1}{2}\mathbf{b}$, $-2\mathbf{a}+\mathbf{b}$
6 (i) (a) $5\mathbf{i}$ (b) $2\sqrt{2}\mathbf{i}+2\sqrt{2}\mathbf{j}$ (c) $\mathbf{i}+\sqrt{3}\mathbf{j}$ (d) $-6.43\mathbf{i}-7.66\mathbf{j}$ (ii) $v\cos\alpha\mathbf{i}+v\sin\alpha\mathbf{j}$ (iii) $3\sqrt{3}\mathbf{i}+3\mathbf{j}$, $-4\sqrt{2}\mathbf{i}+4\sqrt{2}\mathbf{j}$, $-8\mathbf{i}-6\mathbf{j}$, $6\mathbf{i}-6\sqrt{3}\mathbf{j}$
7 ± 6
8 (i) $\frac{1}{5}(3\mathbf{i}-4\mathbf{j})$ (ii) $\dfrac{1}{\sqrt{5}}(-\mathbf{i}+2\mathbf{j})$ (iii) $\dfrac{1}{\sqrt{2}}(-\mathbf{i}-\mathbf{j})$
9 $\pm\frac{1}{2}\sqrt{3}$ **10** $\pm\frac{3}{5}$
11 (i) $6\mathbf{i}+8\mathbf{j}$ (ii) $36\mathbf{i}-15\mathbf{j}$ (iii) $-8\mathbf{i}+4\mathbf{j}$ (iv) $-2\sqrt{3}\mathbf{i}-2\mathbf{j}$
12 (i) and (iv)
13 (i) 3, along Ox (ii) 6, $-90°$ with Ox (iii) $\sqrt{2}$, 63.4° with Ox (iv) 13, $-22.6°$ with Ox (v) 2, 120° with Ox (vi) $\sqrt{17}$, 166.0° with Ox (vii) $\sqrt{2}$, $-135°$ with Ox (viii) $\sqrt{13}$, $-123.7°$ with Ox

14 (i) $9\mathbf{i}+\mathbf{j}$ (ii) $-5\mathbf{j}$ **15** $8\mathbf{i}-4\mathbf{j}$
16 (i) $\mathbf{i}+6\mathbf{j}$ (ii) $7\mathbf{i}+8\mathbf{j}$ (iii) $5\mathbf{i}-4\mathbf{j}$ (iv) $\tfrac{9}{2}\mathbf{i}-7\mathbf{j}$
17 (i) $3\mathbf{i}+3\mathbf{j}$ (ii) $\mathbf{0}$
18 (i) $-4\mathbf{i}-2\mathbf{j}$ (ii) $-2\sqrt{3}\mathbf{i}-10\mathbf{j}$
19 (i) $3\sqrt{2}$, $-45°$ with Ox
(ii) 25.08, 113.5° with Ox
20 (i) 5, 36.9° with Ox (ii) $20\sqrt{2}$, 98.1° with Ox
21 (ii) $\tfrac{5}{3}, -\tfrac{11}{3}$ **22** $3\mathbf{u}-2\mathbf{v}$
23 $\mathbf{u}-2\mathbf{v}$ **24** $2\sqrt{5}, 4\sqrt{2}$
25 (i) 22 (ii) -5 (iii) 4 (iv) 0
26 (i) 5 (ii) 3 (iii) 8
28 (i) and (iii)
29 -2
31 $\lambda(2\mathbf{i}-\mathbf{j}); \pm\dfrac{1}{\sqrt{5}}(2\mathbf{i}-\mathbf{j})$
32 (i) $\dfrac{1}{\sqrt{2}}$ (ii) $-\dfrac{1}{\sqrt{10}}$ (iii) -1 (iv) 0
33 (i) $\tfrac{1}{2}\mathbf{i}\pm\tfrac{1}{2}\sqrt{3}\mathbf{j}$ (ii) $0.060\mathbf{i}+0.998\mathbf{j}, 0.835\mathbf{i}-0.551\mathbf{j}$

Exercise 6.2:2

1 (i) $\mathbf{i}+2\mathbf{j}$ (ii) $3\mathbf{i}-\mathbf{j}$ (iii) $-\tfrac{1}{2}\mathbf{i}+4\mathbf{j}$ (iv) $4\mathbf{i}$ (v) $-\mathbf{j}$
2 (i) $(3, 7)$ (ii) $(-3, -2)$ (iii) $(0, 5)$ (iv) $(-2, 0)$
3 (i) $3\mathbf{i}+4\mathbf{j}, -3\mathbf{i}-4\mathbf{j}, 5$ (ii) $4\mathbf{i}-4\mathbf{j}, -4\mathbf{i}+4\mathbf{j}, 4\sqrt{2}$
(iii) $-3\mathbf{i}-\mathbf{j}, 3\mathbf{i}+\mathbf{j}, \sqrt{10}$ (iv) $\mathbf{i}+7\mathbf{j}, -\mathbf{i}-7\mathbf{j}, 5\sqrt{2}$
(v) $-9\mathbf{j}, 9\mathbf{j}, 9$
4 $(x_2-x_1)\mathbf{i}+(y_2-y_1)\mathbf{j}, \sqrt{(x_1-x_2)^2+(y_2-y_1)^2}$
6 $-3\mathbf{i}-2\mathbf{j}, \mathbf{i}-6\mathbf{j}$
9 $(10, 4), (5, 5)$
10 $\sqrt{130-8x+x^2}, \sqrt{82-2x+x^2}; (3, 0)$
11 (i) $(-3, 0)$ (ii) $(0, -3)$
13 $2x+y-5=0$
14 (i) 8.13° (ii) 45° (iii) 98.1°
16 5 **17** 5.10, 7.21, 7
20 $\pm(b\mathbf{i}-a\mathbf{j}); (7, 4), (2, 7); (1, -6), (-4, -3)$
21 $(-4, -4), (2, -3); (0, 9), (-6, 8)$
22 $2\mathbf{i}-6\mathbf{j}; (2, 2); (3, -1), (1, 5)$
23 $(2, 7), (-4, 1)$
24 $(9, 0), (5, 8), (-3, 4)$
25 $x^2+y^2-x-y-8=0$
26 (i) (a) $\mathbf{i}-\mathbf{j}$ (b) $3\mathbf{i}-2\mathbf{j}$ (c) $4\mathbf{i}-\tfrac{5}{2}\mathbf{j}$ (d) $-2\mathbf{i}+\tfrac{1}{2}\mathbf{j}$
(ii) (a) $43\mathbf{i}-22\mathbf{j}$ (b) $13\mathbf{i}-7\mathbf{j}$ (c) $10\mathbf{i}-\tfrac{11}{2}\mathbf{j}$
27 (i) (a) $-16\mathbf{i}+12\mathbf{j}$ (b) $-8\mathbf{i}+8\mathbf{j}$ (c) $-6\mathbf{i}+7\mathbf{j}$
(ii) (a) $14\mathbf{i}-3\mathbf{j}$ (b) $6\mathbf{i}+\mathbf{j}$ (c) $4\mathbf{i}+2\mathbf{j}$
28 (i) $\tfrac{4}{5}\mathbf{i}+\tfrac{8}{5}\mathbf{j}$ (ii) $8\mathbf{i}+4\mathbf{j}$
29 $\left(\dfrac{nx_1+mx_2}{m+n}, \dfrac{ny_1+my_2}{m+n}\right); \left(\dfrac{x_1+x_2}{2}, \dfrac{y_1+y_2}{2}\right)$
30 (i) $-5\mathbf{i}+11\mathbf{j}$ (ii) $13\mathbf{i}-13\mathbf{j}$
31 $-9\mathbf{i}-3\mathbf{j}, 3\mathbf{i}+3\mathbf{j}$
32 $\tfrac{11}{18}(9\mathbf{i}+7\mathbf{j})$
33 $\tfrac{1}{4}(7\mathbf{i}+23\mathbf{j})$
34 (i) yes (ii) yes (iii) no
35 -5 **36** $\tfrac{6}{5}$
37 $\mathbf{a}+3(\mathbf{b}+\mathbf{a}); 10\mathbf{i}-7\mathbf{j}$
38 $x-4y+5=0$
39 (i) $(2, 3)$ (ii) $(4, 2)$ (iii) $(\tfrac{5}{3}, \tfrac{5}{3})$ (iv) $(0, 1)$;
$\left(\dfrac{x_1+x_2+x_3}{3}, \dfrac{y_1+y_2+y_3}{3}\right)$
40 $2\mathbf{i}-\mathbf{j}$
41 $(\tfrac{7}{3}, \tfrac{7}{3}); (\tfrac{7}{3}, \tfrac{7}{3})$

Exercise 6.2:3

1 (i) $\mathbf{r}=\mathbf{i}+\mathbf{j}+\lambda(\mathbf{i}+2\mathbf{j}), 2x-y-1=0$
(ii) $\mathbf{r}=\lambda(3\mathbf{i}-\mathbf{j}), x+3y=0$
(iii) $\mathbf{r}=4\mathbf{i}-\mathbf{j}+\lambda(\mathbf{i}-\mathbf{j}), x+y-3=0$
(iv) $\mathbf{r}=3\mathbf{j}+\lambda(3\mathbf{i}+5\mathbf{j}), 5x-3y+9=0$
2 (i) $3x+2y-11=0, 3x+2y-1=0$
(ii) $\mathbf{r}=2\mathbf{i}+\mathbf{j}+\lambda(-3\mathbf{i}+4\mathbf{j})$
3 (i) $\mathbf{r}=\mathbf{i}+3\mathbf{j}+\lambda(\mathbf{i}-\mathbf{j}), x+y-4=0$
(ii) $\mathbf{r}=\mathbf{i}+3\mathbf{j}+\lambda(\mathbf{i}+\mathbf{j}), x-y+2=0$

4 $\mathbf{r}=\dfrac{C}{B}\mathbf{j}+\lambda(B\mathbf{i}-A\mathbf{j}), Ax+By=C$
5 (i) $\mathbf{i}+2\mathbf{j}$ (ii) $4\mathbf{i}+2\mathbf{j}$ (iii) $\mathbf{i}-2\mathbf{j}$;
$\mathbf{r}=\mathbf{i}+2\mathbf{j}+\lambda(\mathbf{i}-2\mathbf{j}); 2x+y-4=0$
6 (i) $\mathbf{r}=2\mathbf{i}+\lambda(\mathbf{i}-5\mathbf{j}); 5x+y=10$
(ii) $\mathbf{r}=\tfrac{3}{2}a\mathbf{i}+2b\mathbf{j}+\lambda(2b\mathbf{i}-a\mathbf{j});$
$2ax+4by=3a^2+8b^2$
7 (i) and (ii)
8 (i) $\mathbf{r}=x_1\mathbf{i}+y_1\mathbf{j}+\lambda\{(x_2-x_1)\mathbf{i}+(y_2-y_1)\mathbf{j}\}$
(ii) $\mathbf{r}=b\mathbf{j}+\lambda(a\mathbf{i}-b\mathbf{j})$
9 (i) $\mathbf{r}=\mathbf{j}+\lambda(\mathbf{i}+\mathbf{j})$ (ii) $\mathbf{r}=-3\mathbf{j}+\lambda(\mathbf{i}-2\mathbf{j})$
(iii) $\mathbf{r}=3\mathbf{j}+\lambda(4\mathbf{i}-3\mathbf{j})$ (iv) $\mathbf{r}=-2\mathbf{i}+\lambda(3\mathbf{i}+\mathbf{j})$
(v) $\mathbf{r}=5\mathbf{i}+\lambda\mathbf{j}$ (vi) $\mathbf{r}=-\mathbf{j}+\lambda\mathbf{i}$
10 (i) $\mathbf{r}=c\mathbf{j}+\lambda(\mathbf{i}+m\mathbf{j})$ (ii) $\mathbf{r}=\dfrac{C}{B}\mathbf{j}+\lambda(A\mathbf{i}+B\mathbf{j})$
11 (i) $\mathbf{i}+4\mathbf{j}$ (ii) $2\mathbf{i}-\mathbf{j}$ (iii) $5\mathbf{i}+3\mathbf{j}$
12 (i) $\tfrac{2}{3}$ (ii) -1 (iii) 0 (iv) $\dfrac{b}{a}$
13 $-1, -2; -2\mathbf{i}-3\mathbf{j}$
14 (i) $-2\mathbf{i}+5\mathbf{j}$ (ii) $x+y-3=0, 4x+y+3=0$
16 $(2, 2); 2:1$
17 $\mathbf{r}=\lambda\mathbf{i}, \mathbf{r}=4\mathbf{i}-2\mathbf{j}+\mu\mathbf{j}; 4\mathbf{i}$

Exercise 6.2:4

1 (i) $\mathbf{r}\cdot(3\mathbf{i}-2\mathbf{j})=1, 3x-2y=1$
(ii) $\mathbf{r}\cdot(\mathbf{i}+4\mathbf{j})=3, x+4y=3$
(iii) $\mathbf{r}\cdot(3\mathbf{i}+4\mathbf{j})=5, 3x+4y=5$
(iv) $\mathbf{r}\cdot(3\mathbf{i}-5\mathbf{j})=15, 3x-5y=15$
2 (i) $3\mathbf{i}+2\mathbf{j}$ (ii) $\mathbf{i}-3\mathbf{j}$ (iii) $4\mathbf{i}-\mathbf{j}$ (iv) $3\mathbf{i}+7\mathbf{j}$
(v) $\mathbf{j}$
3 (i) $\mathbf{r}=\mathbf{j}+\lambda(2\mathbf{i}-3\mathbf{j}); \mathbf{r}\cdot(3\mathbf{i}+2\mathbf{j})=2$
(ii) $\mathbf{r}=\mu(\mathbf{i}+\mathbf{j}); \mathbf{r}\cdot(\mathbf{i}-\mathbf{j})=0$
4 $2\mathbf{i}+\mathbf{j}$
5 $\mathbf{r}\cdot(\cos\alpha\mathbf{i}+\sin\alpha\mathbf{j})=p; x\cos\alpha+y\sin\alpha=p$
6 (i) 2 (ii) 30°; $\tfrac{1}{2}\sqrt{3}\mathbf{i}+\tfrac{1}{2}\mathbf{j}$
7 (i) $y=mx+c$ (ii) $y-y_1=m(x-x_1)$
(iii) $\dfrac{y-y_1}{x-x_1}=\dfrac{y_2-y_1}{x_2-x_1}$ (iv) $x\cos\alpha+y\sin\alpha=p$
(v) $\dfrac{x}{a}+\dfrac{y}{b}=1$
8 (i) 2 (ii) $\sqrt{2}$ (iii) $\sqrt{5}$ (iv) 0
9 $\mathbf{r}\cdot(2\mathbf{i}-\mathbf{j})=5; \tfrac{2}{5}\sqrt{5}$
10 (i) $\tfrac{1}{2}\sqrt{10}$ (ii) $\tfrac{1}{2}\sqrt{10}$
11 $\left|\dfrac{mh-k+c}{\sqrt{1+m^2}}\right|$
12 (i) 3 (ii) $\tfrac{4}{5}$ (iii) $\sqrt{17}$
13 (i) 0.56 (ii) 0.56 (iii) 0.56 (iv) 0.56
14 (i) same (ii) opposite
15 5, 2; 5
16 (i) $\tfrac{13}{2}$ (ii) 10
(iii) $\tfrac{1}{2}|(x_2-x_1)(y_3-y_1)-(x_3-x_1)(y_2-y_1)|$
18 $9X^2+16Y^2-24XY-26X-82Y+91=0$
20 (i) $x=-3, y=4$ (ii) $3x+11y-19=0,$
$11x-3y-17=0$
(iii) $2x+4y+8=0, 12x-6y-2=0$
21 $x+y=1$
22 $\tfrac{3}{10}\sqrt{10}, \tfrac{4}{5}\sqrt{10}; \tfrac{1}{2}\sqrt{10}$
23 (i) $\tfrac{2}{3}$ (ii) $\tfrac{3}{5}\sqrt{5}$
24 $3x-4y-3=0, 3x-4y+7=0$
25 (i) 36.9° (ii) 60.3° (iii) 90° (iv) 0°
26 (i) 63.4° (ii) 26.6° (iii) 45°
29 (i) 45° (ii) 8.1° **30** 26.6°

Exercise 6.2:5

1 $x^2+y^2-8x-2y+13=0$
2 (i) $x^2+y^2-6x-4y+12=0$
(ii) $x^2+y^2=16$
(iii) $x^2+y^2+6x-8y=0$

3 (i) $x+y-6=0$ (ii) $2x-4y-5=0$ (iii) $x=1$
4 (i) $x^2+y^2-12x-10y+53=0$
(ii) $3x^2+3y^2-8x+16y+20=0$
(iii) $3x^2+3y^2-46x-12y+103=0$
5 (i) $8x+6y-5=0$
(ii) $3x^2+3y^2-26x-12y+35=0$
(iii) $8x^2+8y^2-56x-22y+85=0$
7 (i) $x^2+y^2-2x-6y-3=0$
(ii) $x^2+y^2-6x+5=0$
9 $\dfrac{x^2}{4}+\dfrac{y^2}{3}=1$
10 $x^2+y^2-2xy-8x-12y+46=0$
11 (i) $x^2+y^2-2xy-14x-14y+35=0$
(ii) $x^2+4y^2-4xy-22x+6y-4=0$
(iii) $y^2-4x=0$
12 $(\sqrt{5}-\sqrt{2})x+(\sqrt{5}+2\sqrt{2})y-(\sqrt{5}+2\sqrt{2})=0,$
$(\sqrt{5}+\sqrt{2})x+(\sqrt{5}-2\sqrt{2})y-(\sqrt{5}-2\sqrt{2})=0$
13 (i) $x-7y-2=0, 7x+y=0$
(ii) $4y-1=0, x=0$
(iii) $x+y-3=0, x-y+1=0$
(iv) $(2\sqrt{2}-1)x-(\sqrt{2}+3)y-(\sqrt{2}+1)=0,$
$(2\sqrt{2}+1)x+(3-\sqrt{2})y+(1-\sqrt{2})=0$

Exercise 6.2:6

1 (i) $(3, 2), 1$ (ii) $(-2, 4), \sqrt{3}$ (iii) $(4, -4), 2$
(iv) $(-3, 5), \sqrt{3}$ (v) $(-3, -4), \sqrt{5}$
(vi) $(-3, 0), \sqrt{5}$ (vii) $(0, -2), \sqrt{2}$
(viii) $(0, 0), \sqrt{6}$
2 (i) $(x+1)^2+(y-4)^2=4,$
$x^2+y^2+2x-8y+13=0$
(ii) $(x-7)^2+(y+6)^2=36,$
$x^2+y^2-14x+12y+49=0$
(iii) $(x-4)^2+(y-2)^2=20,$
$x^2+y^2-8x-4y=0$
(iv) $x^2+(y+1)^2=16, x^2+y^2+2y-15=0$
(v) $(x-5)^2+y^2=25, x^2+y^2-10x=0$
3 (i) (a) $k=r$ (b) $h=r$ (c) $h^2+k^2=r^2$
(ii) $(\pm r, \pm r)$ (iii) ± 3
4 (i) $(4, 1), 3$ (ii) $(-3, 7), 5$ (iii) $(-2, -2), 3$
(iv) $(1, -5), 5$ (v) $(\tfrac{1}{2}, 0), \tfrac{5}{2}$ (vi) $(-\tfrac{3}{2}, -\tfrac{5}{2}), \tfrac{5}{2}\sqrt{2}$
(vii) $(4, 3), 10$ (viii) $(-\tfrac{1}{4}, -\tfrac{3}{4}), \tfrac{3}{4}\sqrt{2}$
(ix) $(-\tfrac{2}{5}, 1), \tfrac{7}{5}$
5 $(5, 3), 4\sqrt{2}$
6 $(-5, \tfrac{1}{2}), \tfrac{1}{2}\sqrt{43}; (-2, 2), (-8, -1)$
7 (i) $c=h^2$ (ii) $c=k^2$ (iii) $c=0$
8 (i) $(5, 6), 4; 1, 6$
9 (i) $3-\sqrt{5}$ (ii) $3-\sqrt{2}$
10 (i) $2\sqrt{5}-2$ **11** (i) 3 (ii) 5 (iii) 5
12 (i) (a) $4\sqrt{2}-2$ (b) $4\sqrt{2}+2$ (ii) (a) $2-\sqrt{2}$
(b) $2+\sqrt{2}$
13 (i) $(-2, 0), (0, 4)$ (ii) $(5, 0), (1, 4)$
(iii) $(-2, 1), (4, 3)$
14 $2\sqrt{17}$ **15** $2\sqrt{5}$
17 (i) $x^2+y^2-2x-24=0$
(ii) $x^2+y^2-3x-5y+6=0$
(iii) $x^2+y^2-8x-14y+40=0$
(iv) $x^2+y^2-2x-14y=0$
(v) $x^2+y^2-4x-2y-45=0$
(vi) $x^2+y^2-7x-y+4=0$
(vii) $x^2+y^2-4ax-2ay=0$
19 (i) $x^2+y^2-8x-2y+8=0$
(ii) $x^2+y^2-6x-4y=0$
(iii) $x^2+y^2-x-10y+18=0$
(iv) $x^2+y^2+4x-5y+1=0$
(v) $(x-x_1)(x-x_2)+(y-y_1)(y-y_2)=0$
21 (i) $x^2+y^2-10x-6y+24=0$
(ii) $x^2+y^2-2x-2y-23=0$
22 (i) $x^2+y^2-4x-8y=0$
(ii) $x^2+y^2-6x-2y-10=0$
(iii) $x^2+y^2-8y-1=0$
23 (i) $x-2y+11=0$ (ii) $x+5y+25=0$
(iii) $3x-y-8=0$ (iv) $2x-y+10=0$
(v) $(h-x_1)(x-x_1)+(k-y_1)(y-y_1)=0$

(vi) $2x - y = 0$ (vii) $x + 7y = 0$
(viii) $hx + gy = 0$
25 (i) $x + 3y - 10 = 0$
 (ii) $x - y - 6 = 0$, $x + y - 12 = 0$
 (iii) $3x - 2y + 1 = 0$, $3x + 2y - 25 = 0$
26 (i) $y = 5$, $y = -3$; $x = 1$, $x = 9$
 (ii) $y = 10$, $y = -4$; $x = -9$, $x = 5$
 (iii) $y = \pm 4$; $x = 0$, $x = 8$
27 $(-1, 3), 2$ **28** (i) and (iii) **29** (i) and (ii)
30 $(-1, 5)$ **31** $(-6, 3)$ **32** $-11, 9$
33 (i) $x - y + 1 = 0$, $(0, 1)$; $x - y + 9 = 0$, $(-4, 5)$
 (ii) $x - 3y + 5 = 0$, $(7, 4)$;
 $x - 3y + 25 = 0$, $(5, 10)$
 (iii) $2x - y = 0$, $(1, 2)$; $2x - y - 15 = 0$, $(7, -1)$
 (iv) $4x - 3y + 7 = 0$, $(-1, 1)$;
 $4x - 3y - 18 = 0$, $(3, -2)$
34 $0, \frac{3}{4}$; $y = 0$, $3x - 4y = 0$
35 (i) $3x + 4y = 0$ (ii) $x - 3y = 0$, $3x - y = 0$
 (iii) $y = 0$, $15x + 8y = 0$
36 $(3m + 2 + c)^2 = 13(1 + m^2)$; $3x - 2y + 8 = 0$,
 $2x + 3y + 1 = 0$
37 $(mc - 2)^2 = c^2(1 + m^2)$; $3x - 4y = 0$; $x = 0$
38 $y = \pm x + 1$
39 (i) $3x - y = \pm \frac{10}{3}$ (ii) $\pm x + 2y = 10$
40 (ii) $(2, 1), 3$; $\sqrt{34}$; 5
41 (i) 3 (ii) 6 (iii) 7
42 (ii) $8X - 2Y - 8 = 0$
43 (i) $5x^2 + 5y^2 - 29x - 20y + 20 = 0$
 (ii) $x^2 + y^2 - 5y = 0$
 (iii) $x^2 + y^2 - 8x - 10y + 16 = 0$
 (iv) $x^2 + y^2 - 2x + 5y - 24 = 0$
44 (i) $x^2 + y^2 - 10x - 8y + 33 = 0$
 (ii) $x^2 + y^2 - 8x - 6y + 19 = 0$
45 (i) $x^2 + y^2 + 8x - 10y + 16 = 0$
 (ii) $x^2 + y^2 + 2y - 1 = 0$, $x^2 + y^2 - 14y - 1 = 0$
46 $x^2 + y^2 - 6x - 6y + 9 = 0$
47 $x^2 + y^2 - 24x - 12y + 144 = 0$,
 $x^2 + y^2 - 12x - 6y + 36 = 0$,
 $x^2 + y^2 - 12x + 24y + 36 = 0$,
 $x^2 + y^2 - 6x - 12y + 9 = 0$
48 $x^2 + y^2 - 6x - 6y + 17 = 0$
49 (i) $x^2 + y^2 - 4x - 4y + 4 = 0$,
 $9x^2 + 9y^2 + 12x - 12y + 4 = 0$
 (ii) $x^2 + y^2 - 4x + 2y + 4 = 0$,
 $x^2 + y^2 + 6x - 12y + 9 = 0$
51 (i) $5x + 3y + 4 = 0$ (ii) $x - y = 0$
 (iii) $4x - y + 2 = 0$
52 $2x + y - 7 = 0$; $(1, 5), (3, 1)$
53 (i) $(2, 4), (4, 2)$ (ii) $(0, 2), (6, 0)$
 (iii) $(2, 1), (-4, -3)$
54 $2(h_2 - h_1)x + 2(k_2 - k_1)y + (c_1 - c_2) = 0$
56 (ii) $(3, 2)$
57 $(2, 1)$; $x - y - 1 = 0$
58 $4x + 3y - 28 = 0$, $(4, 4)$
59 (i) $3x + 4y - 6 = 0$, $(\frac{6}{5}, \frac{8}{5})$
 (ii) $3x + 4y + 14 = 0$, $(\frac{21}{5}, \frac{28}{5})$
 (iii) $4x + y = 0$, $(0, 0)$
 (iv) $4x + 3y - 50 = 0$, $(8, 6)$
 (v) $3x - 4y - 27 = 0$, $(\frac{17}{5}, \frac{21}{5})$
 (vi) $3x + 2y - 5 = 0$, $(-1, 2)$
60 $6, 20$ **61** (i) $1, 7$ (ii) $4, 6$ **62** $1, 6$

Exercise 6.2:7

1 (i) $x^2 + y^2 + 2xy - 14x - 6y + 41 = 0$
 (ii) $x^2 + 9y^2 + 6xy + 20x - 40y + 50 = 0$
 (iii) $4x^2 + y^2 - 4xy - 10x - 20y + 25 = 0$
 (iv) $y = x^2$ (v) $y^2 = 12x$
3 (i) 1 (ii) $\frac{3}{2}$ (iii) $\frac{1}{4}$
4 (ii) (a) $(1, 0), 1$ (b) $(2, 1), 1$ (c) $(-\frac{3}{2}, -2), \frac{1}{2}$
 (d) $(\frac{2}{3}, -3), \frac{3}{2}$
5 (i) $(x - h)^2 = 4a(y - k)$
 (ii) (a) $(-2, -1), \frac{1}{2}$ (b) $(-\frac{3}{4}, \frac{15}{8}), \frac{1}{8}$
6 (i) $x + y + a = 0$, $x - y - 3a = 0$
 (ii) $x - 3y + 9a = 0$, $3x + y - 33a = 0$

7 $y = \pm(x + 2)$; $(-2, 0)$
9 (i) $(\frac{9}{2}, 2)$ (ii) 36
10 $(\frac{121}{9}, -\frac{22}{3})$; $\frac{220}{3}$

Exercise 6.2:8

1 (i) $2mc = 1$ (ii) $x - 4y + 8 = 0$ (iii) $(8, 4)$
2 $(8, 8)$
3 (i) $2x + y + 2 = 0$ (ii) $x - y + 4 = 0$
 (iii) $x - y - 4 = 0$, $2x - y + 2 = 0$
4 $x - 2y + 8 = 0$, $2x + y + 1 = 0$
5 (i) $y = -\dfrac{1}{m}x - am$
 (ii) $x - 3y + 9a = 0$, $x - 2y + 4a = 0$
6 (i) $x = -1$ (ii) $x = -\frac{1}{2}$ (iii) $x = \frac{1}{3}$
7 (i) $y = 3x \pm 5$; $(-\frac{3}{2}, \frac{1}{2}), (\frac{3}{2}, -\frac{1}{2})$
 (ii) $y = \pm 3x - 10$; $(-3, -1), (3, -1)$
8 $4m^2 - 3m = 0$; $y = 0$, $3x - 4y = 0$
10 $y = \pm(x + 3)$
11 $\dfrac{4 - 2mc}{m^2}$; $2, -3$
12 $\frac{1}{5}(m - 3)$; $-2, 1$

Exercise 6.3:1

9 (i) $y^2 = 8x$ (ii) $y^2 + 6y + 17 = 8x$ (iii) $xy = 9$
 (iv) $x(1 - y) = 1$ (v) $y^2 - x^3 - x^2 = 0$
10 (i) $x^2 + y^2 = 4$ (ii) $(x - 1)^2 + (y - 3)^2 = 4$
 (iii) $\frac{1}{9}x^2 + \frac{1}{25}y^2 = 1$ (iv) $x^2 + \frac{1}{4}y^2 = 1$
 (v) $y^2 = 4x^2(1 - x^2)$
11 (i) $x = 3 + t$, $y = 2 - t$; $x + y - 5 = 0$
 (ii) $x = t^2$, $y = t^3$; $y^2 = x^3$
 (iii) $x = 3 + 2\cos\theta$, $y = -4 + 2\sin\theta$;
 $(x - 3)^2 + (y + 4)^2 = 4$
 (iv) $x = 3\cos\theta$, $y = -1 + \sin\theta$;
 $\frac{1}{9}x^2 + (y + 1)^2 = 1$
12 $x^2 + y^2 = r^2$
13 $(x - h)^2 + (y - k)^2 = r^2$; $r, (h, k)$
14 (i) $4, 0$ (ii) $6, 4$ **15** $y^2 = 4ax$
17 (i) $\frac{1}{9}x^2 + \frac{1}{4}y^2 = 1$ (ii) $\frac{1}{4}x^2 + \frac{1}{9}y^2 = 1$
18 $\dfrac{x^2}{a^2} + \dfrac{y^2}{b^2} = 1$
19 $xy = c^2$; $x = 0$, $y = 0$
21 (i) $\frac{1}{9}(x - 5)^2 + \frac{1}{4}(y - 4)^2 = 1$ (ii) $x - 3 = (y + 1)^2$
 (iii) $(x + 2)(y - 1) = 1$
22 $\tan\phi = \tan^3\theta$; $\theta = \dfrac{n\pi}{4}$
24 (i) $x = \dfrac{1}{t^2 - 1}$, $y = \dfrac{t}{t^2 - 1}$
 (ii) $x = t^2 + 4$, $y = t(t^2 + 4)$
25 (i) $2, -2$ (ii) $3, -3$ (iii) $\frac{1}{2}, -\frac{1}{2}$ (iv) $-2, -\frac{1}{2}$
 (v) $\frac{1}{3}$ (vi) $0, \pi, -\dfrac{\pi}{2}, -\dfrac{\pi}{4}$
26 (i) $x + 2y - 6 = 0$ (ii) $\sqrt{3}x - 2y - 2 = 0$
 (iii) $2x + y - 4a = 0$
 (iv) $(p^2 + pq + q^2)x - (p + q)y - p^2q^2 = 0$
27 (i) $2x - (p + q)y + 2apq = 0$ (ii) $(-apq, 0)$
 (iii) $(\frac{1}{3}a(p^2 + q^2), \frac{2}{3}a(p + q))$
30 (i) $(4, \pm 8)$ (ii) $(-3, 0), (0, -2)$
 (iii) $(0, 0), (1, -1)$
 (iv) $(a, 2a), (16a, -8a)$ (v) $(-2, 0), (\frac{10}{3}, \frac{8}{3})$
 (vi) $(-3, -4), (5, 0)$
31 (i) $(2, \pm 4)$ (ii) $(2, 1), (1, -1), (2, -3)$
 (iii) $(1, 4), (4, 1), (-1, -4), (-4, -1)$
32 (i) $(1, 4)$ (ii) $(3, 3)$ (iii) $(2, 2)$
33 (i) $(5, -2)$ (ii) $(-3, 3)$ (iii) $(\frac{19}{2}, 3)$

Exercise 6.3:2

1 (i) $\dfrac{1}{t}, -\dfrac{1}{2at^3}$ (ii) $-\dfrac{1}{t^2}, \dfrac{2}{ct^3}$ (iii) $t, \dfrac{1}{6t}$

(iv) $\dfrac{t^2 + 1}{t^2 - 1}, -4\left(\dfrac{t}{t^2 - 1}\right)^3$ (v) $\dfrac{2t}{1 - t^2}, 2\left(\dfrac{1 + t^2}{1 - t^2}\right)^3$
 (vi) $-\frac{1}{3}\cot\theta, -\frac{1}{9}\csc^2\theta$
 (vii) $\csc\theta, -\cot^3\theta$
 (viii) $-\tan\theta, \dfrac{1}{3a\cos^4\theta\sin\theta}$
2 (i) $x + t^2y - 2ct = 0$,
 $t^3x - ty + c - ct^4 = 0$;
 $x + y - 2c = 0$, $x - y = 0$;
 $x + 9y - 6c = 0$, $27x - 3y - 80c = 0$
 (ii) $x - ty + at^2 = 0$, $tx + y - at^3 - 2at = 0$;
 $x - y + a = 0$, $x + y - 3a = 0$;
 $x + 3y + 9a = 0$, $3x - y33a = 0$
 (iii) $x - ty + at^2 - 2a = 0$,
 $tx + y - at^3 - 4at = 0$;
 $x - 2y + 2a = 0$, $2x + y - 16a = 0$
 (iv) $x + t^2y - 6t = 0$, $t^3x - ty + 3 - 3t^4 = 0$;
 $x + 9y - 18 = 0$, $9x - y - 80 = 0$;
 $9x + y + 18 = 0$, $x - 9y - 80 = 0$
 (v) $3tx - 2y - at^3 = 0$,
 $2x + 3ty - 2at^2 - 3at^4 = 0$;
 $12x + 16y - a = 0$, $32x - 24y - 11a = 0$
 (vi) $x\cos\theta - 2y\sin\theta - 4 = 0$,
 $2x\sin\theta - y\cos\theta - 6\sin\theta\cos\theta = 0$;
 $\sqrt{3}x + 2y - 8 = 0$, $2x - \sqrt{3}y - 3\sqrt{3} = 0$;
 $y = -2$, $x = 0$
 (vii) $x\sin\theta + y\cos\theta - a\sin\theta\cos\theta = 0$,
 $x\cos\theta - y\sin\theta + a\sin^2\theta - a\cos^2\theta = 0$;
 $x + y + \frac{1}{2}a\sqrt{2} = 0$, $x - y = 0$
3 (i) $(\frac{2}{3}, \frac{4}{3}c)$ (ii) $(11, -\frac{17}{3})$
6 $(apq, a(p + q))$ **8** (ii) $x - ty - a = 0$
11 $x\cos\theta + y\sin\theta - (1 + \cos\theta) = 0$
13 (ii) $(a(\cos\phi - \sin\phi), b(\cos\phi + \sin\phi))$
 (iv) $\sqrt{(a^2 + b^2)}\cos 2\phi, \sqrt{a^2 + b^2}$
18 $x - ty + at^2 = 0$; (i) $x - 3y + 6a = 0$, $(9a, 6a)$
 (ii) $x - 2y + 4a = 0$, $(4a, 4a)$;
 $x - 4y + 16a = 0$, $(16a, 8a)$
 (iii) $4x - 2y + a = 0$, $x + 2y + 4a = 0$
19 (i) $x - 2y + 4a = 0$, $x - 3y + 9a = 0$
 (ii) $2x + y - 12a = 0$, $x - y - 3a = 0$
20 (i) $x + 16y - 8c = 0$, $x + 4y + 4c = 0$
 (ii) $x + y - 2c = 0$, $x + 9y - 6c = 0$
22 $\dfrac{1}{t}$; $(\frac{5}{4}, 1)$; $\frac{9}{10}\sqrt{5}$ **23** $\frac{1}{15}$
27 (i) $2x - y + \frac{1}{2}a = 0$
 (ii) $x - 2y + 4a = 0$, $x - y + a = 0$
 (iii) $x - 16y + 256a = 0$, $x - y + a = 0$
28 $c^2 + 36m = 0$;
 (i) $x + y - 6 = 0$, $x + y + 6 = 0$
 (ii) $4x + y - 12 = 0$, $16x + 25y + 120 = 0$
29 $2x - (p + q)y - 2apq = 0$; $x - py - ap^2 = 0$
30 $x + pqy - c(p + q) = 0$; $x + p^2y - 2cp = 0$;
 $x + 4y - 4c = 0$, $4c + y + 4x = 0$
31 $4t^4 - r^2t^2 + 4 = 0$; $2\sqrt{2}$; $(2, 2), (-2, -2)$
32 (i) $\left(a\left(t + \dfrac{2}{t}\right)^2, -2a\left(t + \dfrac{2}{t}\right)\right)$
 (ii) $(-2t, -8t^3)$ (iii) $(a, -a)$
 (iv) $(-\frac{1}{8}t^3 - \frac{1}{4}t^2 + 2, \frac{1}{4}t - 1)$
 (v) $\left(-\dfrac{1}{2t}, 4t^2\right)$
33 $x - 2y + 4 = 0$
34 $(\frac{1}{4}t^2, -\frac{1}{8}t^3)$
35 $y = tx - t^3$; $\left(\dfrac{3t^2}{4}, -\dfrac{t^3}{4}\right)$
36 $\left(\dfrac{8}{9}, \dfrac{16\sqrt{2}}{27}\right), \left(\dfrac{2}{9}, -\dfrac{2\sqrt{2}}{27}\right)$

Exercise 6.3:3

1 (i) $\left(t, \dfrac{1}{t}\right)$; $xy = 1$ (ii) $(x - 4)(y - 4) = 1$

2 (i) $\dfrac{x^2}{64}+\dfrac{y^2}{36}=1$ (ii) $\dfrac{9x^2}{256}+\dfrac{y^2}{16}=1$

3 $y^2=2ax-a$

4 $(x-2)y=9$ **5** $y^2=ax-a$

6 $x^3+xy+ay=0$

7 $(x^2+y^2)^2=4c^2xy$

8 $2y^2=16x^3+24x^2+9x$

9 $c^4-x^4=2c^2xy$

10 $\dfrac{25x^2}{64}+\dfrac{9y^2}{64}=1$

11 $5(x^2-25y^2)=9x$

13 $x(x^2+y^2)+ay^2=0$

14 (i) $y=x^2+2$ (ii) $x^2-y^2+16=0$
(iii) $y^2=4x-16$

15 $(apq, a(p+q))$ (i) $y^2=2ax+2a^2$
(ii) $y^2=2ax+4a^2$
(iii) $|\frac{1}{2}a^2(p-q)^3|$; $y^2=4ax+4a^2$

16 $(x^2-y^2)^2+4c^2xy=0$

17 $y^2(x+2a)+4a^3=0$

18 $y^2+8x^2+4ax=0$

19 $y^2-4ax=k^2(x+a)^2$

20 $\left(\dfrac{a\cos\alpha\cos\beta(\sin\alpha-\sin\beta)}{\sin(\alpha-\beta)}, \dfrac{a\sin\alpha\sin\beta(\cos\alpha-\cos\alpha)}{\sin(\alpha-\beta)}\right)$

Exercise 6.3:4

2 (i) $\frac{4}{3}a^2$ (ii) $2\pi a^3$

3 πab

4 $\frac{4}{3}\pi ab^2, \frac{4}{3}\pi a^2 b$

7 7.82 **8** $\frac{77}{5}$ **9** $\frac{8}{15}$ **10** $s=t=\pm\sqrt{2}$

Exercise 6.4:1

1 (i) 7; $\frac{2}{7}\mathbf{i}+\frac{6}{7}\mathbf{j}+\frac{3}{7}\mathbf{k}$
(ii) 9; $-\frac{1}{9}\mathbf{i}+\frac{8}{9}\mathbf{j}-\frac{4}{9}\mathbf{k}$
(iii) $\sqrt{14}$; $-\dfrac{1}{\sqrt{14}}\mathbf{i}-\dfrac{2}{\sqrt{14}}\mathbf{j}-\dfrac{3}{\sqrt{14}}\mathbf{k}$
(iv) 5; $-\frac{4}{5}\mathbf{j}+\frac{3}{5}\mathbf{k}$

2 (i) ± 4 (ii) $\pm\frac{2}{3}$

3 (i) 9 (ii) $\frac{2}{3}\mathbf{i}-\frac{2}{3}\mathbf{j}+\frac{1}{3}\mathbf{k}$
(iii) $16\mathbf{i}-16\mathbf{j}+8\mathbf{k}$

4 (iii) $\frac{2}{7}, \frac{6}{7}, \frac{3}{7}$ (iv) $\pm\dfrac{1}{\sqrt{2}}$; 45°, 135°
(v) $\dfrac{1}{\sqrt{3}}\mathbf{i}+\dfrac{1}{\sqrt{3}}\mathbf{j}+\dfrac{1}{\sqrt{3}}\mathbf{k}$

5 (i) & (iii)

6 (i) $4\mathbf{i}+3\mathbf{j}$, $-2\mathbf{i}+\mathbf{j}+2\mathbf{k}$, $-3\mathbf{i}+4\mathbf{j}+5\mathbf{k}$
(ii) $\frac{4}{5}\mathbf{i}+\frac{3}{5}\mathbf{j}$, $-\frac{2}{3}\mathbf{i}+\frac{1}{3}\mathbf{j}+\frac{2}{3}\mathbf{k}$

8 2, 3, −1

9 (i) 9 (ii) 1 (iii) 0 **10** (i) 1 (ii) 3

12 $\frac{7}{9}, \frac{4}{9}$; $\dfrac{1}{\sqrt{3}}\mathbf{i}+\dfrac{1}{\sqrt{3}}\mathbf{j}+\dfrac{1}{\sqrt{3}}\mathbf{k}$

13 $\dfrac{2}{\sqrt{17}}\mathbf{i}-\dfrac{3}{\sqrt{17}}\mathbf{j}+\dfrac{2}{\sqrt{17}}\mathbf{k}$

14 (i) $\dfrac{4}{\sqrt{66}}$ (ii) $-\frac{1}{2}$ (iii) 0 (iv) −1

Exercise 6.4:2

1 (i) $-2\mathbf{i}-6\mathbf{j}+3\mathbf{k}$ (ii) $2\mathbf{i}+6\mathbf{j}-3\mathbf{k}$ (iii) 7

2 $-\frac{4}{21}$ **3** $2\mathbf{j}-\mathbf{k}$, $6\mathbf{i}+\mathbf{j}+2\mathbf{k}$; 7.16

4 162.5°, 9.75 **5** 2

6 (i) and (iii)

7 $\mathbf{a}+3(\mathbf{b}-\mathbf{a})$; $10\mathbf{i}-7\mathbf{j}-4\mathbf{k}$

8 (i) $-\frac{1}{3}\mathbf{j}-\mathbf{k}$ (ii) $3\mathbf{i}+\frac{11}{3}\mathbf{j}-4\mathbf{k}$

9 (i) $2\mathbf{i}+\frac{3}{2}\mathbf{k}$ (ii) $\frac{8}{3}\mathbf{i}+\frac{2}{3}\mathbf{j}+\frac{11}{3}\mathbf{k}$
(iii) $9\mathbf{i}+7\mathbf{j}+5\mathbf{k}$

10 (i) $\frac{14}{5}\mathbf{i}+2\mathbf{j}-\frac{8}{5}\mathbf{k}$ (ii) $10\mathbf{i}+2\mathbf{j}+8\mathbf{k}$

11 $-\mathbf{i}-11\mathbf{j}+9\mathbf{k}$, $7\mathbf{i}+13\mathbf{j}-11\mathbf{k}$

12 (i) $\frac{3}{2}\mathbf{i}-\frac{1}{2}\mathbf{j}+\frac{3}{2}\mathbf{k}$, $2\mathbf{i}+\frac{3}{2}\mathbf{j}+\frac{1}{2}\mathbf{k}$, $\frac{7}{2}\mathbf{i}+\mathbf{j}$, $3\mathbf{i}-\mathbf{j}+\mathbf{k}$
(ii) $\frac{5}{2}\mathbf{i}+\frac{1}{4}\mathbf{j}+\frac{3}{4}\mathbf{k}$

13 (i) $2\mathbf{i}+3\mathbf{j}-\mathbf{k}$ (ii) $-\frac{8}{3}\mathbf{i}+\frac{5}{3}\mathbf{j}$ (iii) $\mathbf{0}$

14 $\frac{7}{3}\mathbf{i}+2\mathbf{j}+\frac{14}{3}\mathbf{k}$, $\frac{7}{3}\mathbf{i}+2\mathbf{j}+\frac{14}{3}\mathbf{k}$, $\frac{7}{3}\mathbf{i}+2\mathbf{j}+\frac{14}{3}\mathbf{k}$

15 $2a\mathbf{i}+2a\mathbf{j}+2a\mathbf{k}$, $a\mathbf{i}+a\mathbf{j}+2a\mathbf{k}$, $2a\mathbf{i}+a\mathbf{j}+a\mathbf{k}$
(i) 19.5° (ii) 35.3°

Exercise 6.4:3

1 (i) $\mathbf{r}=-\mathbf{i}+4\mathbf{j}+5\mathbf{k}+\lambda(2\mathbf{i}+3\mathbf{j}+2\mathbf{k})$,
$\dfrac{x+1}{2}=\dfrac{y-4}{3}=\dfrac{x-5}{2}$
(ii) $\mathbf{r}=\mathbf{i}+2\mathbf{j}+3\mathbf{k}+\lambda(\mathbf{i}+2\mathbf{j}-5\mathbf{k})$,
$\dfrac{x-1}{1}=\dfrac{y-2}{2}=\dfrac{z-3}{-5}$
(iii) $\mathbf{r}=\lambda(4\mathbf{i}+5\mathbf{j}-\mathbf{k})$, $\dfrac{x}{4}=\dfrac{y}{5}=\dfrac{z}{-1}$
(iv) $\mathbf{r}=3\mathbf{i}+2\mathbf{j}+\lambda(5\mathbf{i}-2\mathbf{j}-3\mathbf{k})$,
$\dfrac{x-3}{5}=\dfrac{y-2}{-2}=\dfrac{z}{-3}$
(v) $\mathbf{r}=\mathbf{i}+3\mathbf{j}+4\mathbf{k}+\lambda(\mathbf{i}-\mathbf{j}-\mathbf{k})$,
$\dfrac{x-1}{1}=\dfrac{y-3}{-1}=\dfrac{z-4}{-1}$

5 (i) only

6 $-2\mathbf{i}-\mathbf{j}+4\mathbf{k}$, $3\mathbf{i}-5\mathbf{j}-\mathbf{k}$;
$\mathbf{r}=3\mathbf{i}-5\mathbf{j}-\mathbf{k}+\lambda(-2\mathbf{i}-\mathbf{j}+4\mathbf{k})$

7 (i) $\mathbf{r}=3\mathbf{i}-2\mathbf{j}+6\mathbf{k}+\lambda(4\mathbf{i}-5\mathbf{j}+3\mathbf{k})$
(ii) $\mathbf{r}=\lambda(\mathbf{i}+2\mathbf{j}+3\mathbf{k})$
(iii) $\mathbf{r}=-\frac{1}{3}\mathbf{i}+\frac{1}{2}\mathbf{j}+\lambda(4\mathbf{i}-3\mathbf{j}+2\mathbf{k})$

8 $\mathbf{r}=-\mathbf{j}+2\mathbf{k}+\lambda(\mathbf{i}+2\mathbf{j}-3\mathbf{k})$

9 (i) $\mathbf{r}=-14\mathbf{j}-20\mathbf{k}+\lambda(\mathbf{i}+4\mathbf{j}+5\mathbf{k})$
(ii) $\mathbf{r}=\frac{3}{2}\mathbf{i}-3\mathbf{k}+\lambda(\mathbf{i}-2\mathbf{j}-2\mathbf{k})$
(iii) $\mathbf{r}=\mathbf{j}-4\mathbf{k}+\lambda(\mathbf{i}-\mathbf{j}+2\mathbf{k})$
(iv) $\mathbf{r}=\frac{7}{2}\mathbf{i}+\frac{5}{2}\mathbf{k}+\lambda(\mathbf{i}+4\mathbf{j}+5\mathbf{k})$

10 $\dfrac{x-1}{2}=\dfrac{y+2}{-3}=\dfrac{z-4}{4}$

11 (i) $4\mathbf{j}-2\mathbf{k}$ (ii) $4\mathbf{i}+14\mathbf{k}$ (iii) $\frac{1}{2}\mathbf{i}+\frac{7}{2}\mathbf{j}$

12 (iii) $10\mathbf{i}-9\mathbf{j}-3\mathbf{k}$, $-6\mathbf{i}+7\mathbf{j}+5\mathbf{k}$

13 $17\mathbf{i}-3\mathbf{j}-7\mathbf{k}$, $-19\mathbf{i}+9\mathbf{j}+11\mathbf{k}$

14 $\mathbf{r}=\mathbf{i}+2\mathbf{j}+\mathbf{k}+\lambda(\mathbf{i}+3\mathbf{k})$

15 (i) $\mathbf{r}=3\mathbf{i}+2\mathbf{j}+\mathbf{k}+\lambda(4\mathbf{i}+3\mathbf{j})$,
$\dfrac{x-3}{4}=\dfrac{y-2}{3}$, $z=1$
(ii) $\mathbf{r}=2\mathbf{i}-\mathbf{j}+5\mathbf{k}+\lambda(\mathbf{j}+\mathbf{k})$,
$x=2$, $\dfrac{y+1}{1}=\dfrac{z-5}{1}$

16 (i) $\mathbf{r}=5\mathbf{j}+2\mathbf{k}+\lambda(3\mathbf{i}-\mathbf{j})$
(ii) $\mathbf{r}=\mathbf{i}-3\mathbf{k}+\lambda\mathbf{j}$

17 (i) 6 (ii) 7 (iii) 15 (iv) $5\sqrt{3}$

19 0, 1, 2

20 $\mathbf{i}+2\mathbf{j}+3\mathbf{k}$; $\mathbf{r}=\mathbf{j}+2\mathbf{k}+\mu(\mathbf{i}+\mathbf{j}+\mathbf{k})$; $2\mathbf{i}+3\mathbf{j}+4\mathbf{k}$

21 (i) $5\mathbf{i}+7\mathbf{j}+\mathbf{k}$ (ii) $5\mathbf{i}+3\mathbf{j}+3\mathbf{k}$

22 (i) skew (ii) parallel (iii) skew
(iv) intersecting; $-\mathbf{i}+\mathbf{j}-\mathbf{k}$ (v) parallel
(vi) intersecting; $\mathbf{i}-\mathbf{j}-3\mathbf{k}$ (vii) parallel
(viii) intersecting; $2\mathbf{i}-5\mathbf{j}+6\mathbf{k}$

24 5; $\mathbf{i}+2\mathbf{j}+\mathbf{k}$

25 0; $7\mathbf{i}-2\mathbf{j}-3\mathbf{k}$

26 $(2, \frac{1}{3}, \frac{7}{3})$

27 (i) 21.0° (ii) 90° (iii) 36.7° (iv) 0°

Exercise 6.4:4

1 (i) $\mathbf{r}=2\mathbf{i}+\mathbf{j}+3\mathbf{k}+\lambda(\mathbf{i}+\mathbf{j}+3\mathbf{k})+\mu(\mathbf{j}+5\mathbf{k})$,
$2x-5y+z=2$
(ii) $\mathbf{r}=\lambda(\mathbf{i}+\mathbf{j})+\mu(\mathbf{j}+\mathbf{k})$, $x-y+z=0$
(iii) $\mathbf{r}=3\mathbf{i}+4\mathbf{j}-\mathbf{k}+\lambda(\mathbf{i}+4\mathbf{j}-2\mathbf{k})$
$+\mu(4\mathbf{i}+2\mathbf{j}-\mathbf{k})$, $y+2z=2$
(iv) $\mathbf{r}=-2\mathbf{i}+4\mathbf{j}+5\mathbf{k}+\lambda\mathbf{i}+\mu\mathbf{k}$, $y=4$
(v) $\mathbf{r}=\lambda(\mathbf{i}+2\mathbf{j}+3\mathbf{k})+\mu(3\mathbf{i}-2\mathbf{j}+4\mathbf{k})$,
$14x+5y-8z=0$

2 $\mathbf{r}=3\mathbf{i}-2\mathbf{j}+2\mathbf{k}+\lambda(2\mathbf{i}-\mathbf{j}+\mathbf{k})+\mu(\mathbf{i}+\mathbf{j}+2\mathbf{k})$

3 $\mathbf{r}=4\mathbf{i}+\mathbf{j}+\lambda(5\mathbf{i}-\mathbf{j}+2\mathbf{k})+\mu(2\mathbf{i}-5\mathbf{j}-3\mathbf{k})$

4 $\mathbf{r}=2\mathbf{i}-\mathbf{j}+\lambda(-3\mathbf{i}+\mathbf{j}-\mathbf{k})+\mu(\mathbf{i}+\mathbf{j}-3\mathbf{k})$

6 $\mathbf{r}=4\mathbf{i}-\mathbf{j}+3\mathbf{k}$

7 $\mathbf{r}=2\mathbf{i}+\lambda(4\mathbf{i}+2\mathbf{j}-\mathbf{k})$

8 $\mathbf{r}=5\mathbf{i}+3\mathbf{j}-\mathbf{k}+\lambda(3\mathbf{i}+4\mathbf{j}-\mathbf{k})$

9 $\mathbf{r}=2\mathbf{i}-\mathbf{j}+\nu(\mathbf{i}+\mathbf{j}-\mathbf{k})$; $-\mathbf{i}-4\mathbf{j}+3\mathbf{k}$; $3\sqrt{3}$

Exercise 6.4:5

1 (i) $\mathbf{r}\,.\,(4\mathbf{i}+\mathbf{j}+3\mathbf{k})=2$, $4x+y+3z=2$
(ii) $\mathbf{r}\,.\,(\mathbf{i}+\mathbf{j}+\mathbf{k})=0$, $x+y+z=0$
(iii) $\mathbf{r}\,.\,(2\mathbf{i}-\mathbf{k})=1$, $2x-z=1$
(iv) $\mathbf{r}\,.\,(4\mathbf{i}+\mathbf{j}-4\mathbf{k})=-10$, $4x+y-4z=-10$
(v) $\mathbf{r}\,.\,(2\mathbf{i}-\mathbf{j}-\mathbf{k})=-7$, $2x-y-z=-7$
(vi) $\mathbf{r}\,.\,(2\mathbf{i}-\mathbf{j}+3\mathbf{k})=14$, $2x-y+3z=14$

2 (i) $3\mathbf{i}+5\mathbf{j}-2\mathbf{k}$ (ii) $\mathbf{i}+\mathbf{j}-\mathbf{k}$ (iii) $\mathbf{i}-2\mathbf{j}$ (iv) $\mathbf{k}$

3 (i) $\mathbf{r}\,.\,(3\mathbf{i}-4\mathbf{j}-5\mathbf{k})=-2$ (ii) $\mathbf{r}\,.\,(\mathbf{i}+2\mathbf{k})=3$
(iii) $\mathbf{r}\,.\,\mathbf{i}=2$

4 $\mathbf{r}=2\mathbf{i}-4\mathbf{j}-3\mathbf{k}+\lambda(2\mathbf{i}+5\mathbf{j}-2\mathbf{k})$

5 $x-3y-2z=-2$; $\mathbf{r}\,.\,(\mathbf{i}-3\mathbf{j}-2\mathbf{k})=-2$

6 $\mathbf{r}\,.\,(2\mathbf{i}-5\mathbf{j}+\mathbf{k})=2$

7 $\mathbf{r}\,.\,(\mathbf{j}+2\mathbf{k})=2$

8 $6\mathbf{i}+\mathbf{j}-4\mathbf{k}$; $\mathbf{r}\,.\,(6\mathbf{i}+\mathbf{j}-4\mathbf{k})=11$, $6x+y-4z=$
11; $\mathbf{t}=\mathbf{i}+\mathbf{j}-\mathbf{k}+\lambda(\mathbf{i}+2\mathbf{j}+2\mathbf{k})+\mu(2\mathbf{i}+3\mathbf{k})$

9 (i) $\mathbf{r}\,.\,(\mathbf{i}-\mathbf{j}+\mathbf{k})=0$, $x-y+z=0$
(ii) $\mathbf{r}\,.\,(\mathbf{i}-\mathbf{k})=-3$, $x-z=-3$
(iii) $\mathbf{r}\,.\,(2\mathbf{i}+3\mathbf{j}-3\mathbf{k})=0$, $2x+3y-3z=0$
(iv) $\mathbf{r}\,.\,(5\mathbf{i}-2\mathbf{j}-13\mathbf{k})=-8$, $5x-2y-13z=-8$
(v) $\mathbf{r}\,.\,(4\mathbf{i}-3\mathbf{j}+3\mathbf{k})=2$, $4x-3y+3z=2$

10 $4\mathbf{i}$, $2\mathbf{j}$, $\mathbf{i}+\mathbf{k}$; $\mathbf{r}=4\mathbf{i}+\lambda(2\mathbf{i}-\mathbf{j})+\mu(3\mathbf{i}-\mathbf{k})$

11 (i) $\frac{4}{3}$ (ii) 1 (iii) $\sqrt{5}$ (iv) 0 (v) $2\sqrt{6}$ (vi) 0

13 (i) $\frac{8}{3}$ (ii) $\frac{5}{3}$ (iii) 0 (iv) $\frac{17}{3}$; same, same

15 4.17

16 $x-4y+3z=-9$, $3x-z=-1$

17 (i) $\frac{11}{6}$; $\frac{11}{6}$; $\frac{8}{3}$, $\frac{5}{6}$ (ii) $\frac{17}{6}$; $\frac{17}{6}$; $\frac{1}{6}$

18 (i) $\dfrac{3}{\sqrt{11}}$ (ii) $\frac{1}{2}$ (iii) $\dfrac{7\sqrt{2}}{15}$

19 $\mathbf{r}\,.\,(3\mathbf{i}+6\mathbf{j}-6\mathbf{k})=28$, $\mathbf{r}\,.\,(3\mathbf{i}+6\mathbf{j}-6\mathbf{k})=-8$

20 $2x+3y-6z=12$, $2x+3y-6z=-2$

21 (i) $4\mathbf{i}-\mathbf{j}+\mathbf{k}$ (ii) $-2\mathbf{j}+3\mathbf{k}$ (iii) $3\mathbf{i}+2\mathbf{j}+4\mathbf{k}$

22 3, $-\frac{1}{3}$ **23** $\frac{4}{3}$

25 $\mathbf{r}=-7\mathbf{i}-\mathbf{j}+12\mathbf{k}+\lambda(4\mathbf{i}-\mathbf{j}-5\mathbf{k})$; $\mathbf{i}-3\mathbf{j}+2\mathbf{k}$;
(i) 12.96 (ii) $9\mathbf{i}-5\mathbf{j}-8\mathbf{k}$

26 $\mathbf{i}-5\mathbf{j}-6\mathbf{k}$

27 (i) $-\mathbf{i}-2\mathbf{j}+\mathbf{k}$ (ii) $\frac{1}{9}\mathbf{i}+\frac{11}{9}\mathbf{j}-\frac{2}{9}\mathbf{k}$
(iii) $\mathbf{r}=-\mathbf{i}-2\mathbf{j}+\mathbf{k}+\mu(10\mathbf{i}+29\mathbf{j}-11\mathbf{k})$

28 $\mathbf{r}=3\mathbf{i}-2\mathbf{j}+\mathbf{k}+\lambda(\mathbf{i}-\mathbf{j})$

29 $\mathbf{i}-3\mathbf{j}+\mathbf{k}$; $\mathbf{r}\,.\,(\mathbf{i}-3\mathbf{j}+\mathbf{k})=6$; $\mathbf{i}+\mathbf{j}-4\mathbf{k}$; 66.0°

30 (i) $\mathbf{r}=\mathbf{i}+\lambda(\mathbf{i}-\mathbf{j}+2\mathbf{k})$
(ii) $\mathbf{r}=2\mathbf{i}+\lambda(3\mathbf{i}-2\mathbf{j}-3\mathbf{k})$
(iii) $\mathbf{r}=\mathbf{i}+\mathbf{j}+\lambda(2\mathbf{i}+\mathbf{j}-\mathbf{k})$
(iv) $\mathbf{r}=\frac{6}{7}\mathbf{j}-\frac{11}{7}\mathbf{k}+\lambda(7\mathbf{i}-9\mathbf{j}+20\mathbf{k})$

31 $\mathbf{i}+\mathbf{j}+4\mathbf{k}$; 37.6°

32 (i) 5.2° (ii) 54.7° (iii) 90°

33 (i) 70.5° (ii) 90° (iii) 27.0° (iv) 60°

34 $|\mathbf{r}|=a$, $x^2+y^2+z^2=a^2$

Exercise 6.5:1

1 (i) $\pm 2i$ (ii) $\pm 9i$ (iii) $\pm\sqrt{3}i$ (iv) $\pm 2\sqrt{5}i$
(v) $\pm\frac{5}{3}i$ (vi) $\pm\frac{1}{2}\sqrt{7}i$

2 $\pm bi$; (i) $x^2+9=0$ (ii) $x^2+12=0$

3 (i) $-1\pm i$ (ii) $3\pm 2i$ (iii) $-\frac{1}{2}\pm\dfrac{\sqrt{3}}{2}i$
(iv) $-\frac{3}{4}\pm\dfrac{\sqrt{7}}{4}i$ (iv) $\frac{3}{2}\pm 2i$

Answers 587

4 $a \pm bi$; (i) $x^2 - 6x + 10 = 0$ (ii) $x^2 - 4x + 7 = 0$ (iii) $x^2 + 6x + 25 = 0$
5 (i) $\frac{3}{2}, 2i$ (ii) $-5, \sqrt{7}i$ (iii) $5, 0$ (iv) $0, -9i$ (v) $5, -4i$
6 (i) true (ii) false (iii) false

Exercise 6.5:2

1 (i) $5 + 12i$ (ii) $11 + i$ (iii) $1 + 2i$
(iv) $-3 + 5i$ (v) $\sqrt{3} - 4\sqrt{5}i$ (vi) $3\sqrt{3} + 2\sqrt{5}i$
(vii) $9a + 6ai$ (viii) $-a + 8ai$ (ix) $2p$ (x) $2qi$
(xi) $(p+r) + (q+s)i$ (xii) $(p-r) + (q-s)i$
2 (i) $3 + 11i$ (ii) $28 + 6i$ (iii) $-2 + 5i$ (iv) 4
(v) $p^2 + q^2$ (vi) $8 + 6i$ (vii) $18 + 26i$
(viii) -10 (ix) $(p^2 + q^2)i$
(x) $(pr - qs) + (ps + qr)i$
3 (i) $-i$ (ii) 1 (iii) i (iv) -1 (v) i
4 (i) 1 (ii) i (iii) -1 (iv) $-i$
5 (i) $-3 + 4i$, $-11 - 2i$, $-7 - 24i$
(ii) $2 - 2\sqrt{3}i$, $8i$, $-8 - 8\sqrt{3}i$
6 $3 + 4i$; $\pm(2+i)$ **7** (i) i (ii) -1
8 (i) 1 (ii) $\cos 2\theta + i \sin 2\theta$
(iii) $\cos(\theta + \phi) + i \sin(\theta + \phi)$
9 (i) $3, -2$ (ii) $2, 3$
11 (i) $\frac{9}{10} + \frac{7}{10}i$ (ii) $-\frac{3}{5} + \frac{3}{10}i$ (iii) $\frac{3}{26} + \frac{15}{26}i$
(iv) $1 + \sqrt{3}i$ (v) $2 - 5i$ (vi) $\frac{x^2 - y^2}{x^2 + y^2} - \frac{2xy}{x^2 + y^2}i$
(vii) $\frac{3}{25} - \frac{4}{25}i$ (viii) $-\frac{3}{2} + i$
(ix) $\frac{pr + qs}{r^2 + s^2} + \frac{qr - ps}{r^2 + s^2}i$
12 (i) $-i$ (ii) -1 (iii) i (iv) i
13 $\frac{2}{5} + \frac{1}{5}i$; $\frac{3}{10} + \frac{1}{10}i$; $\frac{4}{5} + \frac{3}{5}i$
14 (i) $(x^2 - y^2) + 2xyi$ (ii) $\frac{x}{x^2 + y^2} - \frac{y}{x^2 + y^2}i$
(iii) $\frac{x^2 + y^2 - x}{x^2 + y^2} + \frac{y}{x^2 + y^2}i$
(iv) $\frac{x(x-1) + y^2}{(x-1)^2 + y^2} - \frac{y}{(x-1)^2 + y^2}i$
15 $-8 + 6i$; $\frac{1}{10} - \frac{3}{10}i$; $-\frac{2}{25} - \frac{3}{50}i$; $-\frac{2}{25} - \frac{3}{50}i$
16 $5 - 2i$, $-1 - 4i$, $9 - 7i$; $\frac{5}{13} + \frac{11}{13}i$; $8 + 6i$;
$-5 - 12i$; $-\frac{112}{169} + \frac{66}{169}i$
17 $\frac{8}{25} + \frac{31}{25}i$; $\frac{8}{41} - \frac{31}{41}i$
18 (i) $\cos \theta + i \sin \theta$ (ii) $\frac{1}{2} - \frac{i \sin \theta}{2(1 - \cos \theta)}$
(iii) $\frac{1}{2} + \frac{i \sin \theta}{2(1 + \cos \theta)}$
(iv) $\cos(\theta - \phi) + i \sin(\theta - \phi)$
19 (i) $2 \cos \theta$ (ii) $2i \sin \theta$ (iii) $2i \sin 2\theta$
20 (i) $\frac{5}{2} + \frac{5}{2}i$ (ii) $\frac{5}{13} + \frac{12}{13}i$ (iii) $-\frac{11}{2} + i$
21 (i) $\frac{2}{5}i$ (ii) $\frac{2x}{x^2 + 9}$ (iii) $\frac{2x}{x^2 + y^2}$
22 (i) $\frac{1}{10} + \frac{3}{10}i$ (ii) $-i$
23 (i) $\frac{11}{13} - \frac{10}{13}i$ (ii) $1 - i$ (iii) $\frac{40}{29} + \frac{45}{29}i$ (iv) $\frac{4}{5} - \frac{3}{5}i$
24 (i) $\frac{9}{13} + \frac{7}{13}i$ (ii) $\frac{103}{68} + \frac{21}{68}i$
25 $\pm \sqrt{\dfrac{L - CR_1^2}{CL(L - CR_2^2)}}$
26 $x + \dfrac{x}{x^2 + y^2}$, $\left(y - \dfrac{y}{x^2 + y^2}\right)i$
27 (i) 25, $-\frac{7}{25} + \frac{24}{25}i$ (ii) $\frac{1}{13}$, $-\frac{5}{13} - \frac{12}{13}i$
(iii) 1, $\cos 2\theta + i \sin 2\theta$
29 (i) $-3, 5$ (ii) $3, 2$ (iii) $2, 1$ (iv) $1, 2$
30 $\frac{4}{13}, -\frac{7}{13}$
31 (i) $\frac{7}{29} + \frac{26}{29}i$ (ii) $1 - i$
32 (i) $3, 5$ (ii) $1, -1$
33 (i) $3 + 2i$ (ii) $3 - i$, $3 + 3i$
34 (i) $\pm(3 + 2i)$ (ii) $\pm(2 - i)$ (iii) $\pm(1 + i)$
(iv) $\pm \dfrac{1}{\sqrt{2}}(\sqrt{3} + i)$ (v) $\pm(4 - 3i)$ (vi) $\pm(1 + 3i)$
35 $\pm \left\{\sqrt{\frac{1}{2}(a + \sqrt{a^2 + b^2})} + \dfrac{b}{\sqrt{2(a + \sqrt{a^2 + b^2})}}\right\}$

(i) $\pm\left(3 + \dfrac{\sqrt{3}}{3}i\right)$ (ii) $\pm\left(3 - \dfrac{\sqrt{3}}{3}i\right)$
36 (i) $\pm(2+i)$, $\pm(2-i)$ (ii) $\pm(2+2i)$, $\pm(2-2i)$
(iii) $\pm\left(\dfrac{\sqrt{3}}{2} + \dfrac{1}{2}i\right)$, $\pm\left(\dfrac{\sqrt{3}}{2} - \dfrac{1}{2}i\right)$

Exercise 6.5:3

1 $-4, 29$ **2** $-2, 4$
3 (i) $z^2 + 4 = 0$ (ii) $z^2 - 6z + 10 = 0$
(iii) $z^2 + 4z + 7 = 0$
(iv) $z^2 - 2az + (a^2 + b^2) = 0$
4 $-\dfrac{1}{2} \pm \dfrac{\sqrt{39}}{2}i$
5 $-1 \pm 3i$; $(z + 1 - 3i)(z + 1 + 3i)$
6 $\dfrac{1}{2} \pm \dfrac{\sqrt{7}}{2}i$; $\left(z - \dfrac{1}{2} - \dfrac{\sqrt{7}}{2}i\right)\left(z - \dfrac{1}{2} + \dfrac{\sqrt{7}}{2}i\right)$
7 (i) $(z - 4i)(z + 4i)$ (ii) $(z + 2 - i)(z + 2 - i)$
(iii) $(z + 1 + i)(z - 2 + 3i)$
(iv) $(2z - 1 - 3i)(2z - 1 + 3i)$
8 (i) $1 + i$, 2; -4 (ii) $2 - 3i$, -1; 13
(iii) $1 - i$, $\frac{3}{2}$; -6 (iv) $1 + 2i$, -2; 10
9 (i) 2 (ii) 15 (iii) -15
10 $1 + i, 1$ **11** $2 - i, \frac{1}{2}$
12 $1 - i, -3$; $1, -4$
13 $3 + 2i, -1$; $-5, 13$
14 (ii) $1 - 2i, -1$; 5
15 (i) $-2 + i, 1$; $1, 5$ (ii) $1 - 2i, 2$; $9, -10$
(iii) $3 + i, -1$; $4, 10$ (iv) $1 - 3i, \frac{1}{3}$; $11, -10$
(v) $\frac{3}{2} - i, -\frac{2}{3}$; $15, 26$ (vi) $4 + 3i, 1$; $-9, 33$
(vii) $2 + i, -3$; $-1, 15$
16 (i) $-1, \pm i$ (ii) $2, 1 \pm i$ (iii) $-2, \dfrac{1}{2} \pm \dfrac{\sqrt{3}}{2}i$
(iv) $3, -\dfrac{1}{4} \pm \dfrac{\sqrt{7}}{4}i$
17 (i) 1; $-\dfrac{1}{2} \pm \dfrac{\sqrt{3}}{2}i$
18 (i) $-1, \dfrac{1}{2} \pm \dfrac{\sqrt{3}}{2}i$
19 (i) $z^3 - 5z^2 + 8z - 6 = 0$ (ii) $z^3 - 2z^2 + z - 2 = 0$
(iii) $z^3 - 5z^2 + 19z + 25 = 0$ (iv) $z^3 - 8 = 0$
20 (i) $(x-2)(x+1-\sqrt{3}i)(x+1+\sqrt{3}i)$
(ii) $(x-1)(x-i)(x+i)$
23 $\pm 1, \pm i$
24 (i) $1 + i, 3, -1$ (ii) $1 \pm i, 2 - 3i$
(iii) $1 \pm i, -1 - 2i$
25 $3 + i, \pm \frac{1}{2}$
26 $2 \pm i, 1 - 2i$
27 -1; $\pm 2i, 1 - i$
28 (i) $i, 2$ (ii) $1 - 2i, i$ (iii) $1 - i, -1 - i, i$

Exercise 6.5:4

8 (i) $(3, 4)$, $3\mathbf{i} + 4\mathbf{j}$ (ii) $(-2, -5)$, $-2\mathbf{i} - 5\mathbf{j}$
(iii) $(-3, 0)$, $-3\mathbf{i}$ (iv) $(0, 2)$, $2\mathbf{j}$
(v) $(-7, 3)$, $-7\mathbf{i} + 3\mathbf{j}$
9 (i) $4 + 6\mathbf{i}$ (ii) $-6 + 4\mathbf{i}$ (iii) -7

Exercise 6.5:5

10 (i) 10 (ii) 13
11 (i) 5 (ii) $53.1°$
12 (i) $75°$ (ii) $65.1°$ (iii) $39.9°$
14 $3 + 4i$
15 $-7, -1 - \frac{1}{2}i$
16 (i) $7 + 2i$, $1 + 6i$ (ii) $7 + 5i$, $-1 - i$
(iii) $3 - 2i$, $-5 + 2i$
17 $10 - 8i$, $-9 - i$, $8 + 10i$
18 $1 - 3i, -2$
19 $7 - i, 6 + 6i, -2i$
20 $-5i, -4 + 4i, -2i$
21 $\frac{1}{2}(z_1 + z_3) \pm \frac{1}{2}i(z_3 - z_1)$

22 (i) $-1, -i$; $1 + 2i, 2 + i$
(ii) $-7 + 4i, -4 - 2i$; $5 + 10i, 8 + 4i$
24 $-\frac{3}{2} + \frac{1}{2}i, \frac{9}{2} - \frac{3}{2}i$
25 (i) $(-1 + i) + 4\sqrt{3}i$ (ii) $\pm(2\sqrt{3} + 3\sqrt{3}i)$
(iii) $(1 + \frac{5}{2}i) \pm \dfrac{\sqrt{3}}{2}(3 + 4i)$
(iv) $(-1 + \sqrt{2}i) \pm \sqrt{3}(\sqrt{2} - i)$
26 (i) 10 (ii) $\frac{15}{2}$ (iii) $\frac{25}{2}$

Exercise 6.5:6

1 (i) $5, 143.1°$ (ii) (a) $\sqrt{5}, -26.6°$ (b) $\sqrt{5}, 153.4°$
2 (i) $\sqrt{2}, 45°$ (ii) $2, 60°$ (iii) $8, -30°$
(iv) $3\sqrt{2}, 135°$ (v) $5, -143.1°$ (vi) $\sqrt{13}, -33.7°$
3 (i) (a) $2, 90°$ (b) $1, 180°$ (c) $5, 180°$
(d) $\sqrt{3}, -90°$ (e) $6, 0°$
(ii) (a) $\lambda^2, 0°$ (b) $\lambda^2, 90°$ (c) $\lambda^2, 180°$
(d) $\lambda^2, -90°$
5 (i) $\sqrt{2}, 45°$; $2, 30°$; $2\sqrt{2}, 75°$; $\frac{1}{2}\sqrt{2}, 15°$
(ii) $4, -60°$; $5, -126.9°$; $20, 179.1°$; $\frac{4}{5}, 66.9°$
7 (i) $5\sqrt{2}, 45°$ (ii) $\sqrt{2}, -45°$ (iii) $\frac{13}{5}, 14.3°$
8 $\sqrt{13}, -33.7°$; $2\sqrt{5}, 116.6°$;
$\sqrt{5}, 63.4°$, $\sqrt{61}, -50.2°$

Exercise 6.5:7

1 (i) $4 \operatorname{cis} 30°$ (ii) $\sqrt{2} \operatorname{cis}(-45°)$
(iii) $\sqrt{5} \operatorname{cis}(-153.4°)$
(iv) $\operatorname{cis} 120°$ (v) $5 \operatorname{cis} 180°$ (vi) $6 \operatorname{cis} 90°$
(vii) $5 \operatorname{cis}(-126.9°)$ (viii) $\operatorname{cis}(-90°)$
(ix) $\sqrt{10} \operatorname{cis}(-71.6°)$ (x) $\operatorname{cis} 143.1°$
2 (i) $\sqrt{13} \operatorname{cis} 33.7°$ (ii) $\sqrt{13} \operatorname{cis}(-33.7°)$
(iii) $\sqrt{13} \operatorname{cis} 146.3°$ (iv) $\sqrt{13} \operatorname{cis}(-146.3°)$
3 (i) $\operatorname{cis}(-\alpha)$ (ii) $\operatorname{cis}(90° - \alpha)$ (iii) $\operatorname{cis}(\alpha - 90°)$
4 (i) $2 + 2\sqrt{3}i$ (ii) $1.88 + 0.68i$ (iii) $5i$ (iv) -3
(v) $-3 - 3i$ (vi) $-2.30 - 1.93i$
(vii) $-0.35 + 1.97i$ (viii) $-\frac{1}{4} + \frac{1}{4}\sqrt{3}i$
5 (i) 8 (ii) $2\sqrt{3} - 2i$ (iii) $-\sqrt{2} + \sqrt{2}i$
(iv) $-\dfrac{1}{2} - \dfrac{\sqrt{3}}{2}i$ (v) -3
6 (ii) $1, i, -1, -i, -\dfrac{1}{2} + \dfrac{\sqrt{3}}{2}i, -\frac{1}{2}\sqrt{2} + \frac{1}{2}\sqrt{2}i$,
$0.31 - 0.95i, -\dfrac{\sqrt{3}}{2} + \dfrac{1}{2}i$
8 (i) (a) $\operatorname{cis}(\pm 120°)$ (b) $2\sqrt{2} \operatorname{cis}(\pm 135°)$
(c) $\frac{1}{2}\sqrt{2} \operatorname{cis} 45°$, $\frac{1}{2}\sqrt{2} \operatorname{cis}(-135°)$
(ii) (a) $\operatorname{cis}(\pm 90°)$ (b) $\operatorname{cis} 0, \operatorname{cis}(\pm 120°)$
(iii) $\operatorname{cis} 0, \operatorname{cis}(\pm 90°), \operatorname{cis} 180°$
9 $\operatorname{cis}(\pm \alpha)$; (i) $z^2 - \sqrt{3}z + 1 = 0$
(ii) $z^3 - (2 + \sqrt{2})z^2 + (1 + 2\sqrt{2})z - 2 = 0$
(iii) $z^5 - 1 = 0$

Exercise 6.5:8

2 (i) $\sqrt{2} \operatorname{cis} 45°$, $2 \operatorname{cis} 30°$ $2\sqrt{2} \operatorname{cis} 75°$, $\frac{1}{2}\sqrt{2} \operatorname{cis} 15°$
(ii) $\operatorname{cis} 60°$, $\operatorname{cis} 150°$, $\operatorname{cis}(-150°)$, $\operatorname{cis}(-90°)$
(iii) $5 \operatorname{cis} 36.9°$, $5 \operatorname{cis}(-53.1°)$, $25 \operatorname{cis}(-16.2°)$, $\operatorname{cis} 90°$
(iv) $\sqrt{10} \operatorname{cis}(-18.4°)$, $\sqrt{5} \operatorname{cis} 63.4°$, $5\sqrt{2} \operatorname{cis} 45°$, $\sqrt{2} \operatorname{cis}(-81.8°)$
3 $\sqrt{2}, 45°$; $2, -60°$; (i) $2\sqrt{2} \operatorname{cis}(-15°)$
(ii) $\frac{1}{2}\sqrt{2} \operatorname{cis} 105°$ (iii) $\sqrt{2} \operatorname{cis}(-105°)$
(iv) $2 \operatorname{cis}(90°)$ (v) $8 \operatorname{cis} 180°$ (vi) $\frac{1}{8} \operatorname{cis}(-30°)$
4 $4, -45°$; $2, 120°$; (i) $8 \operatorname{cis} 75°$
(ii) $2 \operatorname{cis}(-165°)$ (iii) $64 \operatorname{cis}(-135°)$
(iv) $32 \operatorname{cis}(-120°)$ (v) $16 \operatorname{cis}(60°)$
(vi) $\frac{1}{2} \operatorname{cis} 15°$

5 cis 0°, 3 cis 90°, 4 cis 180°, 3 cis 48.2°
(i) $\frac{1}{3}$ cis $(-48.2°)$ (ii) 9 cis 138.2°
(iii) cis $(-41.8°)$
(iv) 12 cis $(-131.8°)$ (v) $\frac{4}{3}$ cis 131.8°
6 (i) $r, -\theta$ (ii) $r^2, 2\theta$ (iii) $\frac{1}{r}, -\theta$ (iv) $r, 180° + \theta$
7 (i) $\sqrt{2}$ cis 135° (ii) 4 cis 30° (iii) 2 cis 0°
8 (i) $|z_1 z_2| = |z_1||z_2|$; $\left|\frac{z_1}{z_2}\right| = \frac{|z_1|}{|z_2|}$;
arg $(z_1 z_2)$ = arg z_1 + arg z_2;
arg $\frac{z_1}{z_2}$ = arg z_1 - arg z_2
(ii) (a) $\frac{5}{2}\sqrt{2}$ (b) 1 (c) 60° (d) $-165°$
9 (i) $2\sqrt{10}$ (ii) 10 (iii) $\frac{1}{5}\sqrt{10}$
10 (i) cis $\frac{7}{12}\pi$ (ii) cis π (iii) cis $\frac{4}{3}\pi$ (iv) cis 0
11 cis $(-\theta)$
12 (i) $2\cos\frac{1}{2}\theta$ cis $\frac{1}{2}\theta$ (ii) $2\sin\frac{1}{2}\theta$ cis $(\frac{1}{2}\theta + \frac{1}{2}\pi)$
(iii) $\tan\frac{1}{2}\theta$ cis $\frac{1}{2}\pi$ (iv) cosec θ cis $\frac{1}{2}\pi$
13 (i) $\frac{1}{2} - \frac{1}{2}i \tan\frac{\theta}{2}$ (ii) sec θ (iii) $-i \tan\theta$
20 (i) 27 cis $(-\frac{4}{5}\pi)^*$ (ii) 32 cis $\frac{\pi}{3}$
21 2 cis $\frac{\pi}{6}$; (i) $8i$ (ii) $-128 - 128\sqrt{3}i$
22 (i) -4 (ii) $8 + 8i$ (iii) $-32i$
23 (i) $16 - 16\sqrt{3}i$ (ii) $512 + 512\sqrt{3}i$
(iii) $-128 + 128\sqrt{3}i$
24 $-4 - 4i$
25 (i) $(2\sqrt{2} - 2) + (2\sqrt{2} - 2)i$, $(24 - 16\sqrt{2})i$,
$-(160 - 112\sqrt{2}) + (160 - 112\sqrt{2})i$
26 64
29 (ii) (a) $\frac{1}{8}i$ (b) $-\frac{1}{512} - \frac{1}{512}\sqrt{3}i$
30 (i) -1 (ii) $-2 - 2\sqrt{3}i$
33 $\frac{1}{2} + \frac{1}{2}\sqrt{3}i$, i, $-\frac{1}{2} + \frac{1}{2}\sqrt{3}i$
34 (i) $\sqrt{3} + i, 2 + 2\sqrt{3}i, 8i, \frac{1}{4}\sqrt{3} - \frac{1}{4}i, \frac{1}{8} - \frac{1}{8}\sqrt{3}i$
(ii) $1 - i, -2i, -2 - 2i, \frac{1}{2} + \frac{1}{2}i, \frac{1}{2}i$
37 cis 0, cis $\frac{2\pi}{3}$, cis $\frac{4\pi}{3}$
39 (i) 53.1° (ii) 45° (iii) 8.1° (iv) 8.1°
41 (ii) $-4 + 2i, 1 + 2i$
45 $q, p \cos\gamma + (p \sin\gamma)i$
46 (i) 90°, 26.6°, 63.4°; 5
(ii) 59.0°, 61.9°, 59.0°; $\frac{15}{2}$
48 $\frac{1}{2}(z_1 + z_2), \frac{1}{2}(z_2 + z_3), \frac{1}{2}(z_3 + z_4), \frac{1}{2}(z_4 + z_1)$

Miscellaneous Exercise 6

1 (i) $25a^2$ (ii) (a) $-\frac{5}{2}a\mathbf{i} + 5a\mathbf{j}$
(b) $-\frac{2}{3}a\mathbf{i} + \frac{14}{3}a\mathbf{j}, -\frac{10}{3}a\mathbf{i} + \frac{16}{3}a\mathbf{j}$
(iii) $14a\mathbf{i} + 2a\mathbf{j}$
2 $\mathbf{r} \cdot (-2\mathbf{i} + \mathbf{j}) = 2$ (i) $\frac{2}{5}\sqrt{5}$ (ii) $2\sqrt{5}$; $(-\frac{4}{5}, \frac{2}{5})$
3 $4\mathbf{i} - \mathbf{j}$; 45° (i) $\sqrt{13}$ (ii) $6\mathbf{i} + 2\mathbf{j}$
4 (4, 8)
6 (i) $x^2 + y^2 - 4x - 8y = 0$
(ii) $x^2 + y^2 - 8x - 10y + 16 = 0$
7 $2x - y + 7 = 0$, $(-2, 3)$; $2x - y - 3 = 0$; 41.17
8 $8\mathbf{j}, 8\mathbf{i} + 4\mathbf{j}$ (ii) 37.31 (iii) 1.107
9 $4x - y + 12 = 0$, $4x - y - 22 = 0$;
$4x + y - 28 = 0$,
$4x + y + 6 = 0$ (i) 153 (ii) 16
10 (i) 2.80 (ii) 5.57 (iii) 6.46 (iv) 7
12 (i) $-\frac{1}{2}$ (ii) $\frac{1}{48}$ **13** $\frac{1}{6}$ **14** ± 4
15 (i) $y = \pm\frac{16}{9}\sqrt{3}$ (ii) $\frac{256}{15}$
16 (i) $|t|$ (ii) $2a^2|t|(1 + t^2)$
17 $(4a, \pm 4a)$
18 $(2a + at^2, 0)$; $x^2 + y^2 - 12ax + 16a^2 = 0$; $20a^2$
19 $9, (9a, 6a)$; $18a^2$
20 $y^2 = x - 1$
21 $(\frac{1}{2}at^2 + a, \frac{1}{2}at)$
22 $a(1 - \cos\theta); \frac{1}{2}a^2|\tan\theta|(1 - \cos\theta)$
24 (i) $\left(-\frac{2}{t^3}, -2t^3\right)$ (ii) 8

25 $\left(\frac{2ct}{1+t^4}, \frac{2ct^3}{1+t^4}\right)$
26 (i) $(\frac{1}{4}t^2, -\frac{1}{8}t^3)$ (ii) $\frac{3t^5}{32}(4 + 9t^2)$
27 $(a, -a), (4a, 8a); \frac{27}{10}a^2$
28 $x \pm 2y = 0$; $\frac{8}{3}a^2$
31 (i) -3 (ii) $\frac{16}{3}$ (iii) $\frac{4}{3}$
32 (i) $-2\mathbf{j} + 3\mathbf{k}$ (ii) $\frac{1}{\sqrt{15}}$ (iii) $\sqrt{14}$
33 $0, 5\mathbf{i} + 3\mathbf{j} - \mathbf{k}$; $\mathbf{r} \cdot (\mathbf{j} + \mathbf{k}) = 2; \frac{1}{\sqrt{2}}$
34 $5, 2\mathbf{i} + 3\mathbf{j} - \mathbf{k}$;
(i) $\mathbf{r} = 2\mathbf{i} + 3\mathbf{j} - \mathbf{k} + s(-\mathbf{i} + \mathbf{j} + \mathbf{k}) + t(\mathbf{j} + \mathbf{k})$
(ii) $\mathbf{r} \cdot (\mathbf{j} - \mathbf{k}) = 4$
35 (i) $\frac{1}{3}\sqrt{6}$ (ii) $-\mathbf{j} + 2\mathbf{k}$ (iii) 60°
36 $\mathbf{r} = -2\mathbf{j} + 2\mathbf{k} + \lambda(3\mathbf{i} + \mathbf{j} - 2\mathbf{k})$;
$\mathbf{r} = 3\mathbf{i} - \mathbf{j} + \mu(2\mathbf{i} + 4\mathbf{j} + \mathbf{k})$; $\sqrt{21}$
37 (i) $3\mathbf{i} + \mathbf{j} - \mathbf{k}$ (ii) $\mathbf{r} \cdot (3\mathbf{i} + \mathbf{j} - \mathbf{k}) = 6$
(iii) $\frac{6}{11}\sqrt{11}$ (iv) $4\mathbf{i} - 3\mathbf{j} + 3\mathbf{k}$ (v) 52.0°
38 $\mathbf{r} = 3\mathbf{i} + 4\mathbf{j} - 4\mathbf{k} + \lambda(\mathbf{i} - 3\mathbf{j} + 2\mathbf{k})$; $4\mathbf{i} - 2\mathbf{j}$;
(i) $2\sqrt{14}$ (ii) $7\mathbf{i} - 8\mathbf{j} + 4\mathbf{k}$; $\sqrt{266}$
39 (ii) $\mathbf{r} = \mathbf{i} + \mathbf{k} + \lambda(\mathbf{i} + 3\mathbf{j} - 2\mathbf{k})$,
$\mathbf{r} = \mathbf{i} + 2\mathbf{j} + 4\mathbf{k} + \mu(4\mathbf{i} - 3\mathbf{j} + 2\mathbf{k})$; skew; 63.5°
(iii) $\mathbf{r} = \mathbf{i} + \mathbf{k} + \lambda(\mathbf{i} + 3\mathbf{j} - 2\mathbf{k}) + \mu(2\mathbf{j} + 3\mathbf{k})$;
$13x - 3y + 2z = 15$; $13\mathbf{i} - 3\mathbf{j} + 2\mathbf{k}$;
$\mathbf{r} \cdot (13\mathbf{i} - 3\mathbf{j} + 2\mathbf{k}) = 15$; 1.11 (iv) $\mathbf{i} - 4\mathbf{j} - 5\mathbf{k}$
40 (i) $\frac{1}{2}(\mathbf{i} + \mathbf{j} + \mathbf{k} + \mathbf{l})$; 60° (ii) $\mathbf{i} - \mathbf{l}$; 35.3° (iv) 3; 1
41 (i) $z_1 + z_2 - z_3$
43 (i) $2 - i$ (ii) $1 - 2i$ (iii) $-\sqrt{2}(1 + i)$
44 (i) $\pm(2 + i)$ (ii) $\pm 2 \pm i$
45 (i) $-2 + i$ (ii) $\pm(4 + i)$ (ii) $1 - i, 1; 4, -2$
(iv) (a) $\frac{3}{2} + \frac{3}{2}i$ (b) $4 + 4i$
46 $2 - 3i, 2$; -26 (ii) 2 cis $\left(\pm\frac{2\pi}{3}\right)$; $3\sqrt{3}$
47 (i) $2, \frac{1}{6}\pi$ (ii) $2, \frac{1}{3}\pi$ (iii) $4, \frac{1}{2}\pi$ (iv) $4, \frac{1}{3}\pi$
49 $1 \pm \sqrt{3}i$; $4, 0$; $\sqrt{5} + 2, \sqrt{5} - 2$
50 (ii) $-1 - 2i, 1 + 6i; \frac{1}{2}\pi$

Chapter 7

Exercise 7.1:1

1 $4 - 2t, -2$
2 $3t^2 - 24t, 6t - 24$
3 $10 \cos 2t - 20 \sin 2t$
4 $-e^{-t} + 8e^{-2t}, e^{-t} - 16e^{-2t}$
5 $\frac{20}{1+t}, -\frac{20}{(1+t)^2}$
6 (i) 3, 8, 13, 18, 23 (iii) 0
(iv) $5T + 3, 5T + 8, 5T + 5\alpha + 3$ (a) 5 (b) 5α
7 (ii) -20 ms^{-1}; 20 ms^{-1} (iii) 60 m (iv) 3s
8 (i) 0 (ii) q
9 (i) $\frac{1}{2}$ kmh^{-1} (ii) $\frac{5}{36}$ ms^{-1}; 70 h
10 (i) 0, 7, 12, 15, 16, 15, 12; 0, 8; (a) $0 < t < 8$
(b) $t > 8$
(ii) $8 - 2t$; 4; 16 (maximum); -2;
(a) $0 < t < 4$ (b) $t > 4$; 6, 2, 2, 6
11 (i) 2, 6 (ii) (a) 36 (b) 12 (iii) 24
12 (i) 4 s (ii) 5 m, 25 m, 35 m (iii) 40 ms^{-1}
13 (i) 20 m; 2 s (ii) 20 m s^{-1}; 20 m s^{-1} (iii) 2 s
14 (i) 0 (ii) $2pt + q$; q
15 (i) 50 m (ii) 30 s (iii) $\frac{45}{4}$ m s^{-1}
16 (i) $3t^2 - 12t + 9$, $6t - 12$ (ii) 1, 3; 8, 4 (iii) 2
(iv) 3
17 (i) 12 m from 0 (ii) 12 m
(iii) $(-36 \sin 3t)$ m s^{-1}
(iv) 0 (v) 36 m s^{-1}
18 (i) $-20 \sin 2t$ (ii) $-4x$ (iii) 20
19 (i) $R \sin\varepsilon$ (ii) $v = \pm n\sqrt{R^2 - x^2}$ (iii) Rn
(iv) $-n^2 x$

20 (i) $3 \cos t - 4 \sin t, -3 \sin t - 4 \cos t$
(ii) $\tan^{-1}\frac{3}{4}$ (iii) -5, 11
21 (i) $2 \cos 2t - 2 \sin t, -4 \sin 2t - 2 \cos t$
(ii) $\frac{\pi}{6}$ (iii) $-3\sqrt{3}, \frac{3\sqrt{3}}{2}$
22 (i) $6\lambda - 4x$ (a) 2λ (b) λ
23 (i) $0.02(4 - \frac{1}{2}e^{-2t}); v \to 0.08$
(ii) $0.02 e^{-2t}; a \to 0$
24 (i) $\frac{\pi}{2}$

Exercise 7.1:2

1 (i) $12t + 4; 6t^2 + 4t + 3$ (ii) $6t^2 + 4; 2t^3 + 4t + 3$
(iii) $4t^3 + 4; t^4 + 4t + 3$
(iv) $8t\sqrt{t} + 4; \frac{16}{5}t^2\sqrt{t} + 4t + 3$
2 $\frac{1}{2}\lambda t^2, \frac{1}{6}\lambda t^3$
3 $5 - \frac{2}{t^2}, 5t - \frac{2}{t} + 1$
4 $3 + 4t - 6t^2, 3t + 2t^2 - 2t^3$
5 $-12 \sin 3t, 4 \cos 3t$
6 $\frac{2}{t}, 2 \ln t$
7 $4 - 4e^{-3t}, 4t + \frac{4}{3}e^{-3t} - \frac{4}{3}$
8 $\lambda\omega \cos\omega t, \lambda(1 + \sin\omega t)$
9 $\frac{3\lambda}{2} - \frac{\lambda}{t+2}, \frac{3\lambda t}{2} - \lambda \ln(t + 2) + \lambda \ln 2$
10 $-\lambda\omega e^{-\omega t}, \lambda(1 + e^{-\omega t})$
11 2; 16, 48
12 1, 5; (i) 7, 7 (ii) 2, 12 (iii) -27, 27 (iv) -20, 34
13 (i) $8t - 12$ (ii) $\frac{3}{2}$ (iii) 34 (iv) 5
14 (i) $3t^2 - 12$ (ii) 2; 12 (iii) 3; 9 (iv) $\sqrt{12}$; 24
15 2, 4; $\frac{20}{3}, \frac{16}{3}$
16 (i) $\frac{2}{3}$, 2 (ii) $\frac{5}{12}$
17 $\frac{3}{10}$
19 0.54 **20** 16
21 (i) $8\sqrt{2}$ (ii) $\frac{256}{5}$
22 (i) $7 - \frac{5}{t^2}$ (ii) $\frac{9}{2}$
23 (i) 45 km min^{-1} (ii) 144 km min^{-1}
(iii) 144 km min^{-1}; 1036.8 km
24 (i) $4(1 + 3t - \cos 3t)$ (ii) $4\left(1 + \frac{3\pi}{2}\right)$
25 (i) 4 (ii) 6
26 (i) $a = -4x$ (ii) $\sqrt{5}$ (iii) $\frac{1}{2}\tan^{-1} 2$; $\frac{1}{2}\tan^{-1} 2 + \frac{\pi}{2}$; $\frac{1}{2}\tan^{-1} 2 + (n-1)\frac{\pi}{2}$ (iv) $2\sqrt{5}$
27 (i) $t + \frac{2}{t}$; $2\sqrt{2}$ (ii) $\frac{5}{2} + 2 \ln\frac{3}{2}$
28 $9 - \ln 4$
29 (i) $(3 - 3 e^{-2t})$ m s^{-1}
30 6.9 s, 96.6 m; 13.9 m s^{-1}
31 (i) $2 e^{2t} - 5 e^t + 2$; ln 2 minimum; $-6 + 2 \ln 2$
(ii) $\frac{7}{2} - (\ln 2)^2$
32 (i) (a) 39.17 m (b) 39.18 m (ii) (a) 27.78 m
(b) 88.83 m
33 (i) 0.304 m s^{-1}, 1.732 m s^{-1}, 4.914 m s^{-1},
10.341 m s^{-1} (ii) 24.240 m
34 (i) $3t^2$, t^3 (iii) $20 - 4t, -24 + 20t - 2t^2$; (a) 5
(b) $5 + \sqrt{13}$
35 58 m s^{-1}, 64 m

Exercise 7.1:3

2 (i) 5 (ii) 10; 10
3 (i) 4 (ii) $\frac{9}{2}$
4 (ii) $\sqrt{3} V, 2V, \sqrt{5} V$
6 70 m **7** 2 m s^{-2}; 3 m s^{-1}; 28 m

Answers 589

8 (i) $(-\frac{5}{3}, 0)$ (ii) $(\frac{27}{8}, 0)$
9 $\frac{14}{9}$ m s^{-1}
10 (i) 1 m s^{-2} (ii) $-\frac{1}{15}$ m s^{-2}
11 (i) 15 m s^{-1}, $\frac{225}{2}$ m (ii) $15T$ m
(iii) 225 m; 111 s
12 (i) 6 m s^{-1} (ii) 15 m
13 40 m **14** (i) 6 s (ii) 8 s
15 (i) 380 m (ii) 5 m s^{-2}
16 160 s
17 (i) 4 m s^{-1} (ii) 0.8 m s^{-2}, 1 m s^{-2}
18 (i) 20 m s^{-1}, 50 s
(ii) $8\sqrt{10}$ m s^{-1}, $14\sqrt{10}$ s
20 (i) $\sqrt{\dfrac{(f+f')s}{ff'}}$ (ii) $\dfrac{2ff's + f'V^2 + fV^2}{2Vff'}$
21 (i) 8 m s^{-1}, 12 m s^{-1}, 0 (ii) 72 m
22 (i) $8fT^2$ (ii) $\frac{19}{6}fT^2$

Exercise 7.1:4

1 (i) $10t$ m s^{-1} (ii) $5t^2$ m (iii) $10\sqrt{10}$ m s^{-1}
(iv) $\sqrt{10}$ s
2 (i) 20 m (ii) 2 s (iii) 20 m s^{-1} (iv) 4 s
3 $\frac{1}{5}$ s, 1 s; $\frac{4}{5}$ s **4** 1.54 s
5 (i) 3 s (ii) 1.8 s (iii) 1 s
6 (i) 5 m (ii) 10 m s^{-1} (iii) $\sqrt{120}$ m s^{-1}
(iv) 2.1 s (v) 1.6 s
8 320 m; 80 m s^{-1}
9 100 m; 5 m s^{-1}
10 125 m
11 320 m
12 $\dfrac{5}{2\{1+\sqrt{1-k}\}-k}$
13 (i) (a) 10 m s^{-1} (b) 4 m s^{-1}

Exercise 7.1:5

5 (i) $6\,e^{-3t}$, $2-2\,e^{-3t}$ (iii) $x=2$
6 (i) $\frac{3}{2}(1-e^{-2t})$, $\frac{3}{4}(2t+e^{-2t}-1)$ (iii) $\frac{3}{2}$
7 $\frac{1}{5}\ln 4$
8 (i) 2 (ii) $\dfrac{2(e^{4t}-1)}{e^{4t}+1}$
9 (ii) $\frac{5}{2}, \frac{13}{4}, \frac{29}{8}, \frac{61}{16}$ (iii) $v \to 4$
11 (i) $\dfrac{1}{2k}\ln\dfrac{16b^2+16u^2}{16b^2+u^2}$
(ii) $\dfrac{1}{2k}\ln\dfrac{b^2+u^2}{b^2}$
12 (ii) $\frac{1}{2}(1+3\ln\frac{3}{2})$
13 (i) $\frac{1}{5}\left\{3-v+\ln\dfrac{1+v}{4}\right\}$
14 $\dfrac{u^2(3+2u)}{6k}$
15 (i) $\dfrac{u}{k}\left\{\dfrac{1}{\sqrt{e}}-\dfrac{1}{e}\right\}$ (ii) $\dfrac{u}{k}\left(1-\dfrac{1}{e}\right)$; $\dfrac{u^2}{k}\left\{1-\dfrac{1}{e}\right\}$
16 (iv) (á) $\dfrac{g}{k^2}(\ln\frac{3}{2}-\frac{1}{3})$ (b) $\dfrac{g}{h^2}(\ln 3 - \frac{2}{3})$
(c) infinity
17 (i) $\dfrac{1}{k}\ln\dfrac{2g+2kU}{2g+kU}$
(ii) $\dfrac{1}{k}\ln\dfrac{g+kU}{g}$; $\dfrac{u}{k}+\dfrac{g}{k^2}\ln\dfrac{g+kU}{g}$
18 $\dfrac{1}{2k}\ln\left(1+\dfrac{ku^2}{g}\right)$
19 (i) $\dfrac{1}{\sqrt{gk}}\tan^{-1}\sqrt{\dfrac{k}{g}}U$

(ii) $\dfrac{1}{\sqrt{gk}}\left\{\tan^{-1}\sqrt{\dfrac{k}{g}}U + \frac{1}{2}\ln\dfrac{1+U\sqrt{\dfrac{k}{g+kU^2}}}{1-U\sqrt{\dfrac{k}{g+kU^2}}}\right\}$

20 57.9 m s^{-1}, 356.6 m
22 (i) $\sqrt{2x^3}+1$
23 2.89
24 $\sqrt{\dfrac{2kA^3}{3}}$
25 0
26 (a) An (b) An^2
27 (i) $6\sqrt{3}, \dfrac{\pi}{9}$ (ii) $12, \dfrac{\pi}{6}$ (iii) $6\sqrt{3}, \dfrac{2\pi}{9}$ (iv) $0, \dfrac{\pi}{3}$
29 $\dfrac{2500}{(5+x)^3}$; 21.8 h
30 (ii) (a) 3.54 s (b) 5 s (iii) (a) 17.06 s
(b) 22.06 s

Exercise 7.1:6

1 5 s, 75 m **2** 6 s, 48 m **3** 6 s, 60 m
4 3 s, 18 m **5** 9 s, 36 m **6** 3, $3V$
7 (i) $\frac{4}{3}$ m s^{-2} (ii) 3200 m
8 (i) 8 s (ii) $4\sqrt{2}$ m s^{-1}
9 20 m **10** $\frac{15}{4}$ m
11 (i) $\dfrac{5u}{4g}$ (ii) $\dfrac{3u}{4g}$
12 $\dfrac{u}{g}+\dfrac{\tau}{2}, \dfrac{u^2}{2g}-\dfrac{g\tau^2}{8}$
14 (i) 60 m (ii) 30 m s^{-1}, 10 m s^{-1}
16 2 **17** $-8, 1$ **20** $\frac{3}{2}$ m

Exercise 7.2:1

1 $6t\mathbf{i}+3t^2\mathbf{j}$, $6\mathbf{i}+6t\mathbf{j}$; $\mathbf{0}$, $6\mathbf{i}$; $6\mathbf{i}+3\mathbf{j}$, $6\mathbf{i}+6\mathbf{j}$; $12\mathbf{i}+12\mathbf{j}$, $6\mathbf{i}+12\mathbf{j}$
2 $40\mathbf{i}-10t\mathbf{j}$, $-10\mathbf{j}$; $40\mathbf{i}-10\mathbf{j}$, $-10\mathbf{j}$; $40\mathbf{i}-30\mathbf{j}$, $-10\mathbf{j}$
3 $-\dfrac{2}{t^2}\mathbf{i}+\dfrac{2}{t^3}\mathbf{j}, \dfrac{4}{t^3}\mathbf{i}-\dfrac{6}{t^4}\mathbf{j}$; $-2\mathbf{i}+2\mathbf{j}$, $4\mathbf{i}-6\mathbf{j}$
4 $-5\sin t\mathbf{i}-3\cos t\mathbf{j}$, $-5\cos t\mathbf{i}+3\sin t\mathbf{j}$; $-5\mathbf{i}$, $3\mathbf{j}$; $3\mathbf{j}$, $5\mathbf{i}$
5 $-6\sin 3t\mathbf{i}+6\cos 3t\mathbf{j}$, $-18\cos 3t\mathbf{i}-18\sin 3t\mathbf{j}$; $6\mathbf{j}$, $-18\mathbf{i}$; $-3\sqrt{3}\mathbf{i}+3\mathbf{j}$, $-9\mathbf{i}-9\sqrt{3}\mathbf{j}$
6 $\sin 2t\mathbf{i}+(\cos t+\sin 2t)\mathbf{j}$, $2\cos 2t\mathbf{i}+(-\sin t+2\cos 2t)\mathbf{j}$; $\dfrac{\sqrt{3}}{2}\mathbf{i}+\sqrt{3}\mathbf{j}$, $-\mathbf{i}-\left(1+\dfrac{\sqrt{3}}{2}\right)\mathbf{j}$
7 $-e^{-t}\mathbf{j}+2e^{2t}\mathbf{j}$; $e^{-t}\mathbf{i}+4e^{2t}\mathbf{j}$; $-\mathbf{i}+2\mathbf{j}$, $\mathbf{i}+4\mathbf{j}$; $-\dfrac{1}{e}\mathbf{i}+2e^2\mathbf{j}$, $\dfrac{1}{e}\mathbf{i}+4e^2\mathbf{j}$; $-\frac{1}{3}\mathbf{i}+18\mathbf{j}$, $\frac{1}{3}\mathbf{i}+36\mathbf{j}$
8 $\dfrac{1}{t}\mathbf{i}+\mathbf{j}$, $-\dfrac{1}{t^2}\mathbf{i}$; $\frac{1}{3}\mathbf{i}+\mathbf{j}$, $-\frac{1}{9}\mathbf{j}$
9 $(1-\sin t)\mathbf{i}+\cos t\mathbf{j}$, $-\cos t\mathbf{i}-\sin t\mathbf{j}$; $2n\pi+\dfrac{\pi}{2}$
10 (i) $y=x-x^2$ (ii) $4x=(y-1)^2$ (iii) $xy=1$
(iv) $x^2+y^2=1$ (v) $x^2+y^2=25$
(vi) $(x-4)^2+(y+3)^2=1$
11 (i) $\dfrac{x^2}{25}+\dfrac{y^2}{16}=1$ (ii) $y^2=4x^2(1-x^2)$
(iii) $x^2-y^2=1$ (iv) $y=1-2x^2$ (v) $y^2=x^3$
(vi) $y^2=x^3$
12 3ω
14 $y=x^2-2x$
17 (i) $10\sqrt{t^2-4t+5}$; $10\sqrt{5}, 10, 10\sqrt{5}$ (ii) 6; 6, 6
(iii) $|\sec t|\sqrt{\sec^2 t+\tan^2 t}$; 1, $2\sqrt{7}$
18 (i) $\tan^{-1}\frac{4}{3}$, $\tan^{-1}\frac{2}{3}$, $0°$, $\tan^{-1}(-\frac{2}{3})$, $\tan^{-1}(-\frac{4}{3})$, $\tan^{-1}(-2)$ with Ox (ii) $\tan^{-1}(-\frac{16}{3})$ with Ox
19 (i) $\tan^{-1} 3$ with Ox (ii) $\tan^{-1} 7$ with Ox
21 $\frac{5}{2}$, $25\mathbf{i}+\frac{75}{4}\mathbf{j}$
22 $\mathbf{i}$, $\frac{13}{9}\mathbf{i}-\frac{28}{27}\mathbf{j}$
23 $\tan^{-1} 2+n\pi$
24 $2\sqrt{2}\,e^{2t}$, $8e^{2t}$
25 (i) 4 s; 2 m s^{-1}, 2 m s^{-1} (ii) $\sqrt{13}$ m s^{-1}, 0
28 $x^2+4y^2=4a^2$
29 (i) $2\mathbf{i}-3\mathbf{j}-10t\mathbf{k}$, $-10\mathbf{k}$
(ii) $3t^2\mathbf{i}+t\mathbf{j}+3t^2\mathbf{k}$, $6t\mathbf{i}+\mathbf{j}+6t\mathbf{k}$
(iii) $4\cos t\mathbf{i}-4\sin t\mathbf{j}+8\mathbf{k}$, $-4\sin t\mathbf{i}-4\cos t\mathbf{j}$
30 $-3\mathbf{i}+2\mathbf{j}+2\mathbf{k}, 0$; $\dfrac{x-4}{-3}=\dfrac{y}{2}=\dfrac{z-3}{2}$ **31** 7
32 $-an\sin nt\mathbf{i}+an\cos nt\mathbf{j}+2a\mathbf{k}$, $-an^2\cos nt\mathbf{i}-an^2\sin nt\mathbf{j}$

Exercise 7.2:2

1 $3t^2\mathbf{i}+\mathbf{j}$, $t^3\mathbf{i}+t\mathbf{j}$
2 $2t\mathbf{i}+t\mathbf{j}$, $(t^2+3)\mathbf{i}+(\frac{1}{2}t^2+1)\mathbf{j}$
3 $\mathbf{i}+(1-4t)\mathbf{j}$, $(1+t)\mathbf{i}+(t-2t^2)\mathbf{j}$
4 $(4t^3+1)\mathbf{i}+(3t^2+2)\mathbf{j}$, $(t^4+t)\mathbf{i}+(t^3+2t)\mathbf{j}$
5 $10t\sqrt{t}\mathbf{i}+24\sqrt{t}\mathbf{j}$, $4t^2\sqrt{t}\mathbf{i}+(16\sqrt{t}-12)\mathbf{j}$
6 $(5+4t-6t^2)\mathbf{i}+(14-12t^3)\mathbf{j}$, $(-8+5t+2t^2-2t^3)\mathbf{i}+(-16+14t-3t^4)$
7 $-3\cos t\mathbf{i}+4\sin t\mathbf{j}$, $(8-3\sin t)\mathbf{i}+(1-4\cos t)\mathbf{j}$
8 $18\mathbf{i}-54\mathbf{j}$ **9** $9\mathbf{i}+7\mathbf{j}$
10 (i) $2t\mathbf{i}+(3t^2-1)\mathbf{j}$, $t^2+(t^3-t)\mathbf{j}$
(iii) $2\mathbf{i}+(4-2t)\mathbf{j}$, $(2t-1)\mathbf{i}+(4t-t^2-3)\mathbf{j}$
11 (i) $2t^3\mathbf{i}+4t\mathbf{j}$, $\frac{1}{2}t^4\mathbf{i}+(2+2t^2)\mathbf{j}$
(ii) $16\mathbf{i}+8\mathbf{j}$, $8\mathbf{i}+10\mathbf{j}$
(iii) $16\mathbf{i}+(3t^2-4)\mathbf{j}$, $(16t-24)\mathbf{i}+(t^3-4t+10)\mathbf{j}$
(iv) $16\mathbf{i}+23\mathbf{j}$, $24\mathbf{i}+25\mathbf{j}$
12 $(3-t)\mathbf{i}+(1+2t)\mathbf{j}$; $2x+y=7$
13 $6\mathbf{i}+8\mathbf{j}$, 1 s
14 (i) $y=0$ (ii) $2x+y=2$
(iii) $\dfrac{x+1}{1}=\dfrac{y-2}{1}=\dfrac{z+3}{4}$
15 (i) $(3-2t)\mathbf{i}+(1+t)\mathbf{j}$
(ii) $(-2+4t)\mathbf{i}+(11-10t)\mathbf{j}$
17 $y=\frac{4}{3}x-\dfrac{5x^2}{144}$ **18** $3x=y^2+2y+6$
19 $(1+3t+t^2)\mathbf{i}+(1+4t)\mathbf{j}$ **20** $2x-y=3$ **21** $6\mathbf{i}$

Exercise 7.2:3

1 $(2+3\sqrt{3})\mathbf{i}+(3+2\sqrt{3})\mathbf{j}$, $(-6-2\sqrt{3})\mathbf{i}+(2+6\sqrt{3})\mathbf{j}$
2 (i) $6t^2\mathbf{i}-(1+2t)\mathbf{j}$, $12t\mathbf{i}$ (ii) $6\mathbf{i}-3\mathbf{j}$, $12\mathbf{i}$; $3\sqrt{5}$
3 $-2\mathbf{i}-\mathbf{j}$, $\mathbf{j}$
4 (i) $-t^2\mathbf{i}+(2-t^2)\mathbf{j}$; $2\sqrt{5}$
(ii) $(1-\frac{1}{3}t^3)\mathbf{i}+(2t-\frac{1}{3}t^3)\mathbf{j}$; $\frac{1}{3}\sqrt{41}$
5 (i) t s (ii) $(2+t)\mathbf{i}+(-14+t)\mathbf{j}$
(iii) $\dfrac{|t|}{2}\sqrt{t^2-48t+800}$
6 $2t\mathbf{i}+3\mathbf{j}$ **7** $(1+2t+t^2)\mathbf{i}+(-1+2t-t^3)\mathbf{j}$
8 $y=\frac{3}{4}x^2+1$ **9** (i) $\mathbf{0}$ (ii) $\mathbf{0}$
10 (i) $\mathbf{0}$ (ii) $5\mathbf{i}-15\mathbf{j}$ (iii) $5t\mathbf{i}-15t\mathbf{j}$
11 35 km h^{-1}, S21.8°E
12 7 m s^{-1}, due S **13** 5 m s^{-1}, S36.9°E
14 5 m s^{-1}, S81.9°W **15** 5 m s^{-1}, S66.9°W
16 3.6 m s^{-1}, S16.1°E **17** 5.59 m s^{-1}, S1.2°E
18 6.97 m s^{-1}, S34.3°W **19** 1.56 m s^{-1}, S79.0°E
20 62.45 m s^{-1}, S76.1°E
21 15.73 m s^{-1}, from N67.9°W

22 10 m s^{-1}, $36.9°$ from vertical
23 $78.7°$ 24 87.01 km h^{-1}, N80.6°W
25 33.71 km h^{-1}, N80.2°E 26 10.39 m s^{-1}
27 8.66 km h^{-1} 28 $10\mathbf{i}+9\mathbf{j}$
29 28.03 km h^{-1}, N45.5°E
30 22.9 km h^{-1}, from N10.9°E
31 10.39 knots, S60°E
32 S30°W 33 from NW
34 N38.6°E, 19.89 km h^{-1}
35 N27.1°W, 26.3 min
36 $39.8°$ to AB, 100.1 s
37 4.35 h 38 1.07 h
39 (i) 125 s (ii) 100 s
40 (i) 420 s (ii) 336 s
41 95.2 s; 92.3 s; 125 s
42 $t=1$, $a=5$; $\mathbf{i}+4\mathbf{j}$
43 $t=3$, $a=-3$; $-11\mathbf{i}+2\mathbf{j}$
44 $t=2$, $a=8$; $7\mathbf{i}+8\mathbf{j}$
45 $t=4$, $a=3$; $10\mathbf{i}+17\mathbf{j}$
46 $t=3$, $a=\tfrac{4}{5}$; $3\mathbf{j}$
47 $t=2$; $5\mathbf{j}$
48 $t=3$, 5; $\mathbf{i}-2\mathbf{j}$
49 $4\mathbf{i}+3\mathbf{j}$, $t=2$, $9\mathbf{i}+10\mathbf{j}$
50 $\tfrac{5}{2}\mathbf{i}+6\mathbf{j}$, $t=10$, $-5\mathbf{i}+20\mathbf{j}$
51 424.3 km h^{-1}, 44 s
52 N49.5°E, 15.9 s
53 $t=2$, $3\sqrt{5}$; $7\mathbf{i}+11\mathbf{j}$, $10\mathbf{i}+17\mathbf{j}$
54 $t=1$, $2\sqrt{2}$; $2\mathbf{i}+\mathbf{j}$, $3\mathbf{j}$
55 $t=6$, $\sqrt{2}$; $-9\mathbf{i}+12\mathbf{j}$, $-10\mathbf{i}+13\mathbf{j}$
56 $t=4$, $4\sqrt{2}$; $20\mathbf{i}+12\mathbf{j}$, $24\mathbf{i}+16\mathbf{j}$
57 3, $\sqrt{13}$ 58 1 59 4
60 (i) $\mathbf{i}+3\mathbf{j}$ (ii) $(t-8)\mathbf{i}+(3t-14)\mathbf{j}$ (iii) $\sqrt{10}$
   (iv) 6
61 (i) $\sqrt{10}$ (ii) $\sqrt{6}$
62 $-9\mathbf{i}+12\mathbf{j}$; $2\sqrt{5}$
63 8.9 min, 7.53 km; 13.2 min
64 (i) 9.73 km, 6.14 h (ii) 5.03 h
65 $(\sin 2t-2)\mathbf{i}-2\cos 2t\mathbf{j}+(3-\sin 2t)\mathbf{k}$,
   $2\cos 2t\mathbf{i}+4\sin 2t\mathbf{j}-2\cos 2t\mathbf{k}$; $\sqrt{5}$, 5

Exercise 7.3:1

1 (i) $3\mathbf{i}+(4-10t)\mathbf{j}$, $3t\mathbf{i}+(4t-5t^2)\mathbf{j}$
  (ii) $40\mathbf{i}+(35-10t)\mathbf{j}$, $40t\mathbf{i}+(3t-5t^2)\mathbf{j}$
  (iii) $20\sqrt{3}\mathbf{i}+(20-10t)\mathbf{j}$, $20\sqrt{3}t\mathbf{i}+(20t-5t^2)\mathbf{j}$
  (iv) $12\mathbf{i}+(9-10t)\mathbf{j}$, $12t\mathbf{i}+(9t-5t^2)\mathbf{j}$
  (v) $6\mathbf{i}+(18-10t)\mathbf{j}$, $6t\mathbf{i}+(18t-5t^2)\mathbf{j}$
  (vi) $4U\mathbf{i}+(U-gt)\mathbf{j}$, $4Ut\mathbf{i}+(Ut-\tfrac{1}{2}gt^2)\mathbf{j}$
  (vii) $8\mathbf{i}-(8\sqrt{3}+10t)\mathbf{j}$, $8t\mathbf{i}-(8\sqrt{3}t+5t^2)\mathbf{j}$
  (viii) $16\mathbf{i}-(8+10t)\mathbf{j}$, $16t\mathbf{i}-(8t+5t^2)\mathbf{j}$
  (ix) $20\mathbf{i}-10t\mathbf{j}$, $20t\mathbf{i}-5t^2\mathbf{j}$
2 (i) $12\sqrt{3}\mathbf{i}+(12-10t)\mathbf{j}$, $12\sqrt{3}t\mathbf{i}+(12t-5t^2)\mathbf{j}$
  (ii) $12\sqrt{3}\mathbf{i}-(12+10t)\mathbf{j}$, $12\sqrt{3}t\mathbf{i}-(12t+5t^2)\mathbf{j}$
3 $16\mathbf{i}+(12-10t)\mathbf{j}$, $16t\mathbf{i}+(12t-5t^2)\mathbf{j}$; $y = \tfrac{3}{4}x - \tfrac{5}{256}x^2$
4 $U\mathbf{i}+(2U-gt)\mathbf{j}$, $Ut\mathbf{i}+(2Ut-\tfrac{1}{2}gt^2)\mathbf{j}$;
  $y = x - \tfrac{1}{2}\dfrac{gx^2}{U^2}$
5 (i) $(\tfrac{24}{5}, \tfrac{9}{5})$ (ii) $(\tfrac{48}{5}, 0)$ (iii) $(24, -27)$
6 (i) $\tfrac{9}{10}$ s (ii) $\tfrac{81}{20}$ m (iii) $\tfrac{9}{5}$ s (iv) $\tfrac{108}{5}$ m
7 (i) $\tfrac{3}{2}$ s (ii) $\tfrac{45}{4}$ m (iii) 3 s (iv) $45\sqrt{3}$ m
8 (i) $\tfrac{4}{5}$ s (ii) $\tfrac{16}{5}$ m (iii) $\tfrac{8}{5}$ s (iv) $\tfrac{64}{5}$ m
9 (i) 3s (ii) 45 m (iii) 6s (iv) 120 m
10 (i) $\dfrac{U}{g}$ (ii) $\dfrac{U^2}{2g}$ (iii) $\dfrac{2U}{g}$ (iv) $\dfrac{6U^2}{g}$
11 (i) $\dfrac{V}{g}$ (ii) $\dfrac{V^2}{2g}$ (iii) $\dfrac{2V}{g}$ (iv) $\dfrac{2UV}{g}$
12 (i) $\dfrac{u \sin\alpha}{g}$ (ii) $\dfrac{u^2 \sin^2\alpha}{2g}$ (iii) $\dfrac{2u\sin\alpha}{g}$
   (iv) $\dfrac{2u^2 \sin\alpha\cos\alpha}{g}$

13 (i) h, $2\sqrt{\dfrac{2h}{g}}$, 0 (ii) $\tfrac{3}{4}h$, $2\sqrt{\dfrac{3h}{2g}}$, $\sqrt{3}h$
   (iii) $\tfrac{1}{2}h$, $2\sqrt{\dfrac{h}{g}}$, $2h$ (iv) $\tfrac{1}{4}h$, $2\sqrt{\dfrac{h}{2g}}$, $\sqrt{3}h$
14 9.60 s, 13.15 m
15 $\dfrac{9U^2}{2g}$, $\dfrac{6U^2}{g}$; $y = 3x - \dfrac{gx^2}{2U^2}$
16 $\tfrac{144}{5}$ m, 48 m; $y = \tfrac{12}{5}x - \tfrac{1}{20}x^2$
17 2s, 16 m 18 6s, 60 m
19 $5\sqrt{2}$ s, 200 m
21 (i) 21771 m (ii) 120.6 m
22 (i) 3.6 s, 88.4 m (ii) 2 s, 48 m
24 35 m 25 45 m
26 $\dfrac{d}{2u^2}(gd - 2u^2 \tan\alpha)$
27 15 m s^{-1}, $15\sqrt{3}$ m
28 20 m s^{-1}, $\tan^{-1}\tfrac{3}{4}$
29 16 m s^{-1}, $45°$
30 $10\sqrt{5} \text{ m s}^{-1}$, $\tan^{-1}(-\tfrac{1}{2})$
31 $\tfrac{35}{4}$ m, 15 m, 20 m, 15 m, $\tfrac{35}{4}$ m
32 $4h$, $\dfrac{9h}{2}$, $4h$
35 2 s, 2.8 s; 0.8 s
36 3.6 s 37 $2\sqrt{4\sin^2\alpha - 2}$; $60°$
38 2s; 4 m 45 $\dfrac{\sqrt{3}V^2}{g}$
47 $60\mathbf{i}+45\mathbf{j}$
48 36.9 m s^{-1} at $12.5°$ to the horizontal
49 $\tan^{-1}\tfrac{3}{4}$, 1.09 s, $56.5\mathbf{i}+26.7\mathbf{j}$
50 $2\sqrt{2}$ s, 20 m 51 1.5 m

Exercise 7.3:2

1 (ii) 160 m, $45°$
  (iii) 80 m, 80 m, $80\sqrt{3}$ m, $80\sqrt{3}$ m
2 (ii) 80 m, $90°$ (iii) $30°$, $45°$, $60°$
3 $30°$ 4 $\tan^{-1}\tfrac{2}{3}$
5 $\tan^{-1}2$ 6 $20\sqrt{5} \text{ m s}^{-1}$, 50 m
8 $20\sqrt{15} \text{ m s}^{-1}$ 9 38.7 m s^{-1}; 20.1 m
10 $30°$, 2 s, 5 m; $60°$, $2\sqrt{3}$ s, 15 m
11 $21.41°$, 2.14 s, 5.99 m, $68.59°$, 5.59 s, 39.01 m
12 $15°$, 2.07 s, 5.36 m, $75°$, 7.73 s, 74.64 m
13 $26.57°$, 4.47 s, 25.01 m; $63.43°$, 8.94 s, 99.99 m
14 $45°$ 16 (ii) $45°$ 19 20 m s^{-1}
20 40 m s^{-1} 21 $12\sqrt{5} \text{ m s}^{-1}$
22 15 m s^{-1}, $\tfrac{45}{4}$ m
24 $15°$, $75°$ 25 $26.57°$, $71.57°$
26 $63.43°$, $86.82°$ 27 $40.09°$, $76.48°$
28 $0°$, $26.57°$ 29 $67.5°$, $-22.5°$
30 $0°$, $78.70°$ 31 $45°$, $71.57°$ 32 $35.26°$
36 (i) 40 m, 78.4 m (ii) 160 m, 44.8 m
37 $0°$, $78.70°$; 11.71 s
38 $10\sqrt{10} \text{ m s}^{-1}$, $\tan^{-1}\tfrac{1}{3}$
39 $6h$ 40 $25\sqrt{5} \text{ m s}^{-1}$, $\tan^{-1}\tfrac{1}{2}$, 250 m
41 $4h$
43 (i) $\sqrt{u^2 - 2ugt \sin\alpha + g^2 t^2}$
   (ii) at time $t = \dfrac{u \sin\alpha}{g}$
44 $5\sqrt{5} \text{ m s}^{-1}$ 45 25 m
46 $\dfrac{U}{g}(3 \pm \sqrt{3})$
48 $\tan^{-1}\tfrac{1}{2}$, $0°$, $\tan^{-1}(-\tfrac{1}{2})$, $\tan^{-1}(-\tfrac{3}{4})$ with Ox
49 $\tan^{-1}\tfrac{2}{5}$, $0°$, $\tan^{-1}(-\tfrac{2}{5})$, $\tan^{-1}(-\tfrac{4}{5})$,
   $\tan^{-1}(-\tfrac{6}{5})$ with Ox
50 $\dfrac{u\sin\alpha - gt}{u\cos\alpha}$ (ii) $\tan\alpha - \dfrac{gx}{u^2 \cos^2\alpha}$
51 $49.34°$ with Ox
52 $71.57°$
53 horizontal
54 80 m s^{-1} (i) $2\sqrt{2}$ s (ii) $4\sqrt{2}$ s (iii) $6\sqrt{2}$ s

55 $\tfrac{160}{3}$ m 56 $\tfrac{15}{4}$ 57 $\tfrac{25}{3}$ 58 $\tfrac{2}{3}$
59 11.55 s 60 $\dfrac{U^2 + V^2}{gV}$
61 $\dfrac{u}{g \sin\alpha}$ 62 $51.34°$
63 (i) $\tfrac{3}{2}$, 2 (ii) $\tan^{-1}(-\tfrac{15}{16})$, $\tan^{-1}(-\tfrac{7}{4})$
64 (i) 2, 5 (ii) $\tan^{-1}\tfrac{4}{.7}$, $\tan^{-1}(-\tfrac{17}{.7})$
65 18.44 m s^{-1}, 7.04 m
66 23.45 m s^{-1}, $68.33°$, 11.37 m

Exercise 7.4:1

1 $5-2t$; $t=\tfrac{5}{2}$
2 $t = \ln\tfrac{3}{2}$; No; -2 rad s^{-1}, -10 rad s^{-2}
3 $\tfrac{67}{4}$ rad 4 $\dfrac{2\pi}{\omega}$ s 5 π s
8 (i) $\dfrac{-1}{1+t^2}$ (ii) $\dfrac{-(t^4 + 2t)}{t^6 + t^4 - 2t^3 + 1}$
9 (i) 2 (ii) 1
11 $\dfrac{-11}{5t^2 - 10t + 7}$; $-\mathbf{i} + 5\mathbf{j}$
12 (i) $\tfrac{1}{5}16\sqrt{5} \text{ m s}^{-1}$, $\tfrac{8}{25} \text{ rad s}^{-1}$
13 $\tfrac{8}{25} \text{ rad s}^{-1}$ 14 $\dfrac{4\pi}{5}$ s
16 6 rad s^{-1} 17 8.5 m s^{-1}
18 $-\dfrac{8\pi}{25} \text{ rad s}^{-2}$ 19 $2\pi \text{ m s}^{-1}$
20 (i) $\tfrac{1}{8}\pi$ (ii) $\tfrac{3}{8}\pi$ (iii) $\tfrac{5}{8}\pi$
21 (i) 4 s (ii) $4\sqrt{2}$ s (iii) $4\sqrt{3}$ s

Exercise 7.4:2

1 (i) $x^2 + y^2 = 25$
  (ii) $-10\sin 2t\mathbf{i} + 10\cos 2t\mathbf{j}$; 2
  (iii) $-20\cos 2t\mathbf{i} - 20\sin 2t\mathbf{j}$
2 $3\cos 2t\mathbf{i} + 3\sin 2t\mathbf{j}$; $-3\sqrt{3}\mathbf{i} + 3\mathbf{j}$, $-6\mathbf{i} - 6\sqrt{3}\mathbf{j}$;
  6; π
3 (i) $2\cos 5t\mathbf{i} + 2\sin 5t\mathbf{j}$
  (ii) $-2\sin 5t\mathbf{i} + 2\cos 5t\mathbf{j}$
4 $\{3 + 2\cos(\tfrac{1}{4}\pi t)\}\mathbf{i} + \{4 + 2\sin(\tfrac{1}{4}\pi t)\}\mathbf{j}$
  $\dfrac{\pi}{2} \text{ m s}^{-1}$, $\dfrac{\pi^2}{8} \text{ m s}^{-2}$; $\tfrac{1}{2}\pi + \tfrac{1}{4}\pi t$, $\tfrac{1}{4}\pi t$ with Ox
5 (i) $2\cos 3t\mathbf{i} + 2\sin 3t\mathbf{j}$, $-3\sin 3t\mathbf{i} + 3\cos 3t\mathbf{j}$;
   $(-3\sin 3t - 2\cos 3t)\mathbf{i} + (3\cos 3t - 2\sin 3t)\mathbf{j}$
  (ii) $(-9\cos 3t + 6\sin 3t)\mathbf{i}$
     $+(-9\sin 3t - 6\cos 3t)\mathbf{j}$; $\sqrt{117} \text{ m s}^{-1}$
6 (i) $2\omega\{(2\cos 2\omega t + \sin \omega t)\mathbf{i}$
   $-(2\sin 2\omega t + \cos \omega t)\mathbf{j}\}$ (ii) $\dfrac{\pi}{2\omega}$
   (iii) $2\omega^2\{(-4\sin 2\omega t + \cos \omega t)\mathbf{i}$
   $+(-4\cos 2\omega t + \sin \omega t)\mathbf{j}\}$
7 (i) $3 \times 10^4 \text{ m s}^{-1}$ (ii) $6 \times 10^{-3} \text{ m s}^{-2}$
8 0.5 9 (i) $\tfrac{1}{4} \text{ rad s}^{-1}$ (ii) 6.25 m s^{-2} (iii) 8π s

Miscellaneous Exercise 7

1 43 s 2 (i) $\tfrac{112}{27}$ (ii) 1, 2; 1 m s^{-1}
3 $1.76a$ 5 $\dfrac{7u}{6g}$, $\dfrac{15u^2}{16g}$
7 (i) ue^{-kx}; $\dfrac{1}{k} \ln 2$
8 $\tfrac{1}{4} \text{ m s}^{-1}$; 20 min
9 (i) 16.46 m (ii) 20 m
10 $\sqrt{\dfrac{k}{a}}$, $\sqrt{\dfrac{2k}{a}}$
11 32, 0.015; 32 m s^{-1}; 6.2 m s^{-2}
12 (i) $y = -x^3$ (ii) $\sqrt{10}$ (iii) $\tfrac{3}{2}$ (iv) $-6\mathbf{j}$

Answers 591

13 (iii) $n\pi$ 14 $-a\omega^2 \mathbf{j}$
15 $\mathbf{i}+\mathbf{j}, \mathbf{i}-\mathbf{j}; x^2-y^2=4$
16 (i) $64\mathbf{i}+12\mathbf{j}, 30\mathbf{i}+15\mathbf{j}$
    (ii) $64\mathbf{i}+18\mathbf{j}, 94\mathbf{i}+30\mathbf{j}$
17 (i) $\frac{3}{2}\mathbf{i}+2\mathbf{j}$ (ii) $\frac{5}{2}\sqrt{3}$ m s^{-1}
18 (i) $5\sqrt{2}$ m s^{-1} (ii) $\sqrt{10}$ m s^{-1}
21 (i) $(a-3)\mathbf{i}-2\mathbf{j}$ (ii) $4\mathbf{i}+8\mathbf{j}+\{(a-3)\mathbf{i}-2\mathbf{j}\}t$; 2, $12\mathbf{i}-7\mathbf{j}; 2\sqrt{2}$
22 6.78 m
23 $2u$, from N60°W; $2.65u$, from S19°W
24 (i) $\frac{1}{2}a$ (ii) $\dfrac{2a\pi}{3u}$
25 $\dfrac{u}{g \sin \alpha}; \dfrac{2u \sin \alpha}{g}$ 26 (i) $\dfrac{u^2}{g}$
27 35 s
28 (i) $\dfrac{u}{\sqrt{2}}, \dfrac{u}{\sqrt{2}} - gt$ (ii) $\dfrac{ut}{\sqrt{2}}, \dfrac{ut}{\sqrt{2}} - \frac{1}{2}gt^2$
    (iii) $\tan^{-1} \frac{1}{2}$ (iv) $10\sqrt{5}$ m s^{-1} (v) 3 s
29 (i) $\sqrt{6gh}$ (ii) $2\sqrt{\dfrac{2h(4 \sin^2 \alpha - 1)}{g}}$; 0.55
30 (i) 39.19 m (ii) 40 m
31 (i) 38.14 m s^{-1} (ii) $30.4° < \alpha < 73.7°$
32 (i) $2\sqrt{\dfrac{5a}{g}}$ (ii) $16a$ (iii) $2\sqrt{5ga}, \frac{3}{4}$
33 $\tan^{-1} 3; 5a$
34 (ii) 53.6°, 38.23 m; 81.4°, 11.77 m
36 (i) 40.25 m s^{-1}, 26.6° (ii) 24.08 m s^{-1}, 36.06 m s^{-1}
37 2
39 (i) $2\pi, \pi^2; 2\sqrt{\pi^2+16}$

Chapter 8

Exercise 8.1:1

1 (i) $F \cos \alpha, F \sin \alpha$ (ii) (a) $3\mathbf{i}, 3\sqrt{3}\mathbf{i}+3\mathbf{j},$
    $2\mathbf{i}+2\sqrt{3}\mathbf{j}, 3\mathbf{j}, -2\sqrt{2}\mathbf{i}+2\sqrt{2}\mathbf{j}, -2\sqrt{3}\mathbf{i}-2\mathbf{j},$
    $-3\sqrt{2}\mathbf{i}-3\sqrt{2}\mathbf{j}, 2\mathbf{i}-2\sqrt{3}\mathbf{j}$ (b) $8\mathbf{i}+6\mathbf{j}, 6\mathbf{i}+8\mathbf{j},$
    $6\mathbf{i}+8\mathbf{j}, -8\mathbf{i}+6\mathbf{j}, -8\mathbf{i}-6\mathbf{j}, 6\mathbf{i}-8\mathbf{j}$ (c) $-3\mathbf{i},$
    $-3\sqrt{3}\mathbf{i}-3\mathbf{j}, -4\mathbf{j}, 2\sqrt{3}\mathbf{i}-2\mathbf{j}, 4\mathbf{i}+4\sqrt{3}\mathbf{j}, 2\mathbf{j}$
2 (i) horizontal: 15 N, 0, −10 N, 0;
    vertical: $15\sqrt{3}$ N, 25 N, 0, −40 N
    (ii) horizontal: 12 N, −3 N, 0;
    vertical: 9 N, 4 N, −13 N
    (iii) horizontal: −5 N, 0, 7 N, 20 N, 0;
    vertical: 5 N, 15 N, −7 N, 0, −3 N
3 (i) up the plane: 5 N, 0, $-10\sqrt{3}$ N
    perpendicular to the plane: 0, 10, −10 N
    (ii) up the plane: 4 N, 0, 12 N, −18 N
    perpendicular to the plane: 0, 33 N, −9 N, −24 N
4 (i) $5\sqrt{2}$ N, 45° with Ox (ii) 5 N, 143.1° with Ox
    (ii) 5 N, 143.1° with Ox
    (iii) 5 N, −36.9° with Ox (iv) $\sqrt{5}$ N, −153.4° with Ox
5 12, 22.6°; −12, 157.4°
6 $4\sqrt{3}, 120°; -4\sqrt{3}, -120°$
7 (i) $\dfrac{1}{\sqrt{2}}\mathbf{i} + \dfrac{1}{\sqrt{2}}\mathbf{j}$ (ii) $-\frac{12}{13}\mathbf{i}+\frac{5}{13}\mathbf{j}$
   (iii) $-\dfrac{1}{\sqrt{10}}\mathbf{i} - \dfrac{3}{\sqrt{10}}\mathbf{j}$ (iv) $-\mathbf{i}$
8 $\frac{4}{5}\mathbf{i}+\frac{3}{5}\mathbf{j}; 12\mathbf{i}+9\mathbf{j}$
9 (i) $5\mathbf{i}-5\mathbf{j}$ (ii) $-4\sqrt{3}\mathbf{i}+4\mathbf{j}$ (iii) $10\mathbf{i}+24\mathbf{j}$
    (iv) $6\mathbf{i}-2\mathbf{j}$ (v) $\dfrac{5}{\sqrt{2}}\mathbf{i}+\dfrac{5}{\sqrt{2}}\mathbf{j}$ (vi) $-30\mathbf{i}-16\mathbf{j}$
10 $2\sqrt{2}$ N, 45° to each force
11 10 N, 36.9° to 8 N force
12 $2\sqrt{3}$ N, 30° to each force
13 2 N, 60° to each force
14 12 N, down slope
15 7 N, down slope
16 16 N horizontally
17 12.04 N, 48.4° to 12 N force
18 6.24 N, 16.1° to 3 N force
19 5.10 N, 11.3° to 4 N force
20 2.12 N, 13.6° to dotted direction
21 (i) 5 N; in direction of 3 N force (ii) 4.36 N; 23.4° to 3 N force (iii) 3.61 N; 33.7° to 3 N force (iv) 2.65 N; 40.9° to 3 N force (v) 1 N; in direction of 3 N force
22 (i) 0 (ii) 71.8° (iii) 104.5° (iv) 133.4° (v) 180°
23 (i) 8.60 N; 54.5° to AB
    (ii) 7.07 N; 81.9° to AB
24 (i) 6.92 N, 30° to AB
    (ii) 2.65 N, 19.1° to AB
    (iii) 10.97 N, 35.1° to AB
    (iv) 1.73 N, parallel to BA
25 (i) 14.04 N, 4.1° to BA
    (ii) 4.80 N, parallel to AB
26 (i) 8 N, 60° to AB
    (ii) 7.66 N, 13.1° to AB
    (iii) 8.35 N, 63.4° to AB
    (iv) 6 N, 60° to BA
27 (i) $5\mathbf{i}+4\mathbf{j}$ (ii) $6\mathbf{i}+3\mathbf{j}$
28 −3, −2 29 1, −4 30 −9
31 (i) $10\mathbf{i}+20\mathbf{j}$ (ii) $11\mathbf{i}+\mathbf{j}$ (iii) $22\mathbf{i}-6\mathbf{j}$
32 20 N, $6\sqrt{5}$ N 33 $12\mathbf{i}+15\mathbf{j}$
34 $5\mathbf{i}+9.8\mathbf{j}$ 35 $2\mathbf{i}+6\mathbf{j}$

Exercise 8.1:2

1 0.4 m s^{-2} 2 12.5 m s^{-2}
3 $\frac{3}{4}g$ 4 $6\mathbf{i}+3\mathbf{j}$
5 (i) $3\mathbf{i}-2\mathbf{j}$ (ii) $-\frac{1}{2}\mathbf{i}$ (iii) $-\frac{1}{2}\mathbf{i}+\frac{5}{2}\mathbf{j}$
6 $-\mathbf{i}+6\mathbf{j}$ 7 $-\mathbf{i}-5\mathbf{j}$
8 −9, 4 9 7; 2 kg
10 $13\mathbf{i}-\frac{31}{2}\mathbf{j}$
11 $4\mathbf{i}-\dfrac{4}{t^2}\mathbf{j}, 4t\mathbf{i}+\left(\dfrac{2}{t}+2\right)\mathbf{j}; y=\dfrac{8}{x}+2$
12 (i) 2 (ii) $3\mathbf{i}+4\mathbf{j}, 2$
13 (i) $30t\mathbf{i}-30\mathbf{j}$ (ii) $30\mathbf{i}$
14 $-\frac{3}{2}\sqrt{3}\mathbf{i}-\frac{5}{2}\mathbf{j}; 6\mathbf{i}-10\sqrt{3}\mathbf{j}$
15 $40\mathbf{i}-15\mathbf{j}$
16 $80\mathbf{i}+80t\mathbf{j}-30t^2\mathbf{k}, 80\mathbf{j}-60t\mathbf{k}$
17 $\mathbf{0}$; constant
18 (i) $3\mathbf{i}+2\mathbf{j}$ (ii) $6\mathbf{i}+3\mathbf{j}$
20 2, −1
21 (i) $\frac{7}{3}, \frac{1}{3}$ (ii) $-\frac{5}{3}, \frac{13}{3}$

Exercise 8.1:3

4 0.48 N 6 mg

Exercise 8.1:5

1 (i) (a) 3 m (b) 4 m (c) 6 m
   (ii) (a) 2 N (b) 4 N (c) 6 N (iii) 7 m
2 (i) (a) 0.225 m (b) 0.12 m (ii) (a) 20 N, tension
   (b) 0 (c) 12 N, thrust (d) 20 N, thrust
3 (i) (a) $\frac{5}{4}l$ (b) $\frac{3}{2}l$ (c) $\frac{7}{4}l$
   (ii) (a) $\frac{3}{4}l$ (b) $\frac{1}{2}l$ (c) $\frac{1}{4}l$
4 (i) 2.4 m (ii) 1.2 m (iii) 0.96 m
5 (i) $3l$ (ii) $2l$ (iii) $\frac{3}{2}l$ (iv) $\frac{4}{3}l$
6 5 N 7 0.36 m, 90 N
8 (i) (a) 2.2 m (b) 2.8 m (c) 3.4 m
   (ii) (a) 1.3 m (b) 32.5 N

9 0.45 m
10 $\frac{4}{7}a$
12 $\dfrac{l}{\lambda}$

Exercise 8.2:1

1 (i) 400 N (ii) 500 N (iii) $10m$ N
2 2 kg
3 600 N
4 (i) 800 N (ii) 800 N (iii) 800 N
5 $\frac{1}{3}40\sqrt{3}$ N, $\frac{1}{3}80\sqrt{3}$ N
6 71.6°, 94.87 N
7 $2mg, \sqrt{5}mg$
8 $8\sqrt{3}$, 120 N
9 58.56 N, 71.73 N
10 $\dfrac{mg \sin \phi}{\sin (\theta + \phi)}$ 11 $\frac{3}{2}mg$
12 $\dfrac{3mg}{4 \sin \theta + 3 \cos \theta}, \tan^{-1} \frac{4}{3}$
13 $10\sqrt{3}$ N 14 $\frac{25}{6}$ N
15 64 N, 8 kg
16 22.6°, 3.9 kg
17 27.36 N, 21.28 N
18 $\dfrac{mg \cos \beta}{\sin (\alpha + \beta)}, \dfrac{mg \cos \alpha}{\sin (\alpha + \beta)};$
   $mg \sin \alpha, mg \cos \alpha$
19 33.09 N, 13.09 N
20 (i) 130 N, 90 N (ii) 170 N, 60 N
   (iii) 210 N, 30 N; 200 N
21 $\dfrac{2mg}{\sqrt{3}+1}, \dfrac{mg(\sqrt{3}-1)}{\sqrt{3}+1}$
22 $\frac{1}{2}mg, \frac{5}{8}mg$
23 $\dfrac{2\sqrt{5}mg}{5+\sqrt{5}}, \dfrac{\sqrt{5}mg}{5+\sqrt{5}}$
24 $\dfrac{17mg}{31}, \dfrac{25mg}{31}$
26 $50\sqrt{3}$ N 27 $mg \tan \alpha$
28 10 N 29 $10\sqrt{3}$ N
30 $\sqrt{2}mg, mg$ 31 $\sqrt{3}mg, mg$
32 $\dfrac{mg}{\sqrt{3}}, \dfrac{mg}{\sqrt{3}}$
33 $mg(2 \sin \theta - \text{cosec } \theta), mg \cot \theta$
34 $\dfrac{mg}{\cos \theta}$

Exercise 8.2:2

1 20 N, $20\sqrt{5}$ N
2 $\mu mg, mg\sqrt{1+\mu^2}$
3 $\frac{3}{5}$, 58.31 N
4 $\frac{1}{20}$, 2.002 N
5 (i) 10 N (ii) 25 N (iii) 25 N
6 20 N; (i) 30.4 N (ii) 31.6 N (iii) 36.1 N (iv) 36.1 N
7 1500 N 8 R
9 15 N 10 $\frac{1}{2}$
11 $\dfrac{\sqrt{3}}{5}$ 12 $\dfrac{2 \cos \theta}{1-2 \sin \theta}$; 2
14 (ii) $mg \sin \lambda; \lambda$
16 $\frac{1}{4}\sqrt{2}$ 17 $\frac{1}{3}$
18 $\dfrac{1}{\sqrt{5}}$ 19 0.22
20 35 N, 5 N
21 28.50 N, 6.14 N
22 $\dfrac{1}{\sqrt{3}}$; 40 N, vertically upwards
23 $\frac{3}{4}$; 100 N, vertically upwards

24 26.6°; 50 N vertically upwards
25 $\tan \alpha$; mg vertically upwards
26 $\tan \alpha$ **27** 18.4°
28 (i) 2000 N (ii) 4000 N
29 (i) $R - Mg \sin \alpha$ (ii) $R + Mg \sin \alpha$
30 $\frac{1}{2}$ **31** $160\sqrt{2}$ N
32 $\frac{1}{3}(280\sqrt{3})$ N **33** 15.0°
34 $\frac{1}{7}(60\sqrt{3})$ N **35** $\dfrac{mg(\sin\alpha + \mu\cos\alpha)}{\mu\sin\beta + \cos\beta}$
36 (i) 10 N (ii) 5 N **37** 16.88 N
38 $4\sqrt{2}$ kg **39** $3\sqrt{5}$ N
40 $\dfrac{mg(\sin\alpha - \mu\cos\alpha)}{\cos\beta - \mu\sin\beta}$
41 $\frac{36}{17}$ N $\leq S \leq$ 108 N
42 $\dfrac{mg(\sin\alpha - \mu\cos\alpha)}{\cos\alpha + \mu\sin\alpha} \leq S$
$\leq \dfrac{mg(\sin\alpha + \mu\cos\alpha)}{\cos\alpha - \mu\sin\alpha}$
46 (i) $mg \sin(\alpha + \lambda)$ (ii) $mg \sin(\alpha - \lambda)$
47 $\frac{3}{4}$; mg, vertically upwards
48 $\dfrac{3a}{\sqrt{10}}$; mg, vertically upwards
50 (i) 36.9°; 40 N, vertically downwards
  (ii) 240 N, vertically upwards
51 38.4 N; 210.6 N, 71.7° to vertical

Exercise 8.2:3

1 0.25 m **2** 1 m
3 80 N **4** $\dfrac{mgl_0}{\lambda}$
5 0.6 m, 60 N
6 2.48 m **7** $\frac{6}{7}$
8 0.8 m **9** $2mg$
10 53.1° **11** $\dfrac{mgl_0 \sin\alpha}{\lambda}$
12 $\dfrac{7a}{4}$ **13** $\dfrac{20a}{7}$
14 $\dfrac{9mg}{5}$
15 50 N, $50\sqrt{3}$ N: 3 m, $\sqrt{3}$ N: 3 m, $\sqrt{3}$ m
16 48 N **17** $50\sqrt{3}$ N
18 0.62 **19** 10 mg
20 4.42 **21** $\frac{1000}{13}$ N, $\frac{3000}{13}$ N
22 $\sqrt{3}mg$, $2mg$, a
23 $\tan^{-1}\frac{2}{3}$; $2\sqrt{13}mg$
24 $\dfrac{mg}{\sqrt{3}-1}$, $\dfrac{mg(\sqrt{3}-1)}{2}$

Exercise 8.2:4

1 2; $2\sqrt{2}mg$, 45° to the vertical
2 30°; $mg\sqrt{3}$, 30° to the vertical
3 3; 5.37mg, 26.6° to the vertical
4 $\dfrac{1}{\sqrt{3}}$;
5 $30\sqrt{3}$ N, 30° to the vertical
6 $\tan^{-1}\frac{1}{2}$
7 $\frac{7}{26} \leq k \leq \frac{23}{14}$
9 0.06 $\leq k \leq$ 1.51
11 60°, 40 N **12** 60°
13 $\sqrt{3}$; mg, $\sqrt{3}mg$; 2m
14 $\frac{21}{24}$, $\frac{11}{16}$
16 49.1°; $2mg$, $\sqrt{3}mg$, $\sqrt{7}mg$
17 $\sqrt{3}$ **18** $\tan^{-1} 3$, $\tan^{-1} 2$
19 60°; $\sqrt{3}mg$, mg, $\sqrt{3}mg$

Exercise 8.3:1

1 (i) 5 m s^{-2} upwards (ii) 0
  (iii) 5 m s^{-2} downwards
  (iv) $3g$ upwards
  (v) 0
  (vi) $\frac{3}{4}g$ downwards
2 (i) 8 m s^{-2} (ii) 15 N
3 (i) 1050 N (ii) 900 N
  (iii) 600 N (iv) 450 N
4 (i) 108 N (ii) 60 N
  (iii) 0 (iv) 0
5 $\frac{3}{2}$ m s^{-2}; 40 N, vertically downwards
6 15 N; 50 N, vertically downwards
7 $\frac{1}{3}g$; $3mg$, vertically downwards
8 $\dfrac{5\sqrt{3}}{8}$ m s^{-2}; 75 N, vertically downwards
9 5 N; 2 N, vertically downwards
10 $\dfrac{T\cos\theta}{m}$; $mg - T\sin\theta$, vertically downwards
11 (i) $g\sqrt{3}$ (ii) $g\sqrt{2}$ (iii) $\frac{1}{3}g\sqrt{3}$
12 (i) 2.68 m s^{-2} (ii) 5.7°
13, 14 2 m s^{-2}; 15 N, vertically downwards
  2 m s^{-2}; 45 N, vertically downwards
  3 m s^{-2}; 50.39 N, vertically downwards
15 $5\sqrt{2}$ m s^{-2}, 45° to each force
16 $\frac{5}{2}$ m s^{-2}, 53.1° to 6 N force
17 4.36 m s^{-2}, 36.6° to 8 N force
18 1.58 m s^{-2}, 26.6° to $5\sqrt{2}$ N force
19 5 m s^{-2}; $15\sqrt{3}$ N perpendicular to the plane
20 $5\sqrt{2}$ m s^{-2}; $20\sqrt{2}$ N, perpendicular to the plane
21 53.1°; 12 N, perpendicular to the plane
22 $g \sin\alpha$; $mg \cos\alpha$, perpendicular to the plane
24 $2\sqrt{2}$ m s^{-2} **25** 2 m s^{-2}
26 $\frac{14}{5}$ m s^{-2} **27** $\dfrac{9\sqrt{3}}{2}$ m s^{-2}
28 (i) 16 N, 12 N (ii) 24 N, 12 N (iii) 32 N, 12 N
29 (i) 5 m s^{-2} (ii) 0 N
30 (i) g upwards, $4mg$ (ii) $\frac{1}{5}g$ upwards, $\frac{12}{5}mg$
  (iii) 0, $2mg$ (iv) $\frac{1}{5}g$ downwards, $\frac{8}{5}mg$
31 (i) 5 m s^{-2} upwards, 1200 N (ii) 0, 800 N
  (iii) 5 m s^{-2} downwards, 400 N
32 (i) 5 m s^{-2} (ii) 30 N (iii) 90 N

Exercise 8.3:2

1 2 m s^{-2}; 43.08 N, 31.8° to vertical
2 $\frac{1}{4}$; 41.23 N, 14.0° to vertical
3 $\frac{3}{4}g$; $\dfrac{\sqrt{5}}{2}mg$, to the vertical
4 7 m s^{-2}; 12.37 N, 14.0° to the vertical
5 $\dfrac{28\sqrt{2}}{3}$ N; $\dfrac{16\sqrt{5}}{3}$ N, 26.6° to the vertical
6 $\dfrac{T}{m}(\cos\theta - \mu\sin\alpha) - \mu g$;
  $(mg - T\sin\theta)\sqrt{1 + \mu^2}$, $\tan^{-1}\mu$ to the vertical
9 (i) 5 m s^{-2} (ii) 3 m s^{-2}
  (iii) 1 m s^{-2} (iv) 0
  (v) 0
10 (i) 15 m s^{-2} (ii) 7.32 m s^{-2} (iii) 0
  (iv) 0; 50.5°
11 0.37g
12 $\frac{5}{2}$ m s^{-2}, 45° to each force
13 8.68 m s^{-2}, 26.6° to 40 N force
14 9.01 m s^{-2}, 13.9° to 10 N force
15 $\frac{1}{2}$ m s^{-2}, in direction of central 10 N force
16 (i) 0 (ii) $\frac{5}{4}$ m s^{-2} (iii) $\frac{5}{2}$ m s^{-2}
17 4200 N; (i) $\frac{5}{3}$ m s^{-2} (ii) $\frac{10}{3}$ m s^{-2}

18 (i) $\dfrac{7Mg}{20}$ (ii) $\dfrac{33Mg}{80}$ (iii) $\dfrac{23Mg}{80}$
19 (i) $R + Ma$ (ii) $R + Ma + Mg\sin\alpha$
  (iii) $R + Ma - Mg\sin\alpha$
20 $\frac{5}{3}$ m s^{-2}; 36.06 N, 73.9° to the plane
21 $\frac{1}{3}$; 50.60 N, 71.6° to the plane
22 19.35 N; 82.9° to the plane
23 2 m s^{-2}; 44.72 N, 63.4° to the plane
24 (i) 0 (ii) 0 (iii) 2.99 m s^{-2}
  (iv) 5.77 m s^{-2}
25 $g(\sin\alpha - \mu\cos\alpha)$; (i) $\tan\alpha$ (ii) $g\sin\alpha$
26 (i) $\frac{2}{3}g$ (ii) $\frac{3}{4}g$ (iii) g
27 (i) -9.66 m s^{-2} (ii) 7.66 m s^{-2}
28 (i) -10 m s^{-2} (ii) 6 m s^{-2}
29 (i) -12.07 m s^{-2} (ii) 2.07 m s^{-2}
30 3.95 m s^{-2} up the plane
31 1.60 m s^{-2} down the plane
32 5.6 m s^{-2} up the plane
33 0.08
34 (i) $\frac{3}{5}g$ (ii) $\frac{4}{5}mg$ (iii) 6.67mg, 85.9° to the horizontal
35 (i) 10 m s^{-2} (ii) 34.64 N
  (iii) 131.15 N, 82.4° to the horizontal
36 (i) 0.62g (ii) $\dfrac{\sqrt{5}mg}{4}$
  (iii) 5.52mg, 82.3° to the horizontal
37 (a) (i) $\frac{4}{5}g$ (ii) $3mg$
  (iii) 21.93mg, 83.7° to the horizontal
  (b) (i) $\frac{3}{5}g$ (ii) 3.16mg
  (iii) 22.67mg, 85.4° to the horizontal

Exercise 8.3:3

1 (i) $\frac{4}{3}$ s, 8 m (ii) $\frac{8}{3}$ s, 32 m
2 1200 N; (i) 5 m (ii) 15 m (iii) 25 m
3 5 N **4** 4 N
5 (i) $\frac{20}{3}$ s (ii) $\frac{40}{3}$ s (iii) 20 s; 600 m
6 3000 N; 40 m
7 40 000 N **8** 1600 N
9 (i) 7.5 m (ii) 10 m
10 $\frac{2}{15}$ **11** 2 s, $10\sqrt{2}$ m
12 (i) 15 m s^{-1} (ii) 7.5 m s^{-1}
13 (i) $\dfrac{v}{g\sin\alpha}$ (ii) $\dfrac{v}{g(\sin\alpha - \mu\cos\alpha)}$
14 $\frac{1}{3}$ **15** $\dfrac{1}{3\sqrt{3}}$
16 (i) 10 m s^{-2}; 1.25 m (ii) 2.24 m s^{-2}
17 $\dfrac{u^2}{g\sin\alpha}$ **18** (ii) $\dfrac{u}{\sqrt{2}}$; $\dfrac{5u^2}{8g}$
19 $u\sqrt{\frac{3}{5}}$ **20** $\frac{40}{3}$ m
21 512i
22 (i) $3\sqrt{2}i$, $\left(3 - \dfrac{3}{\sqrt{2}}\right)i$ (ii) $6i$, $3i$
  (iii) $3\sqrt{2}i$, $\left(3 + \dfrac{3}{\sqrt{2}}\right)i$
23 (i) 2.97 m s^{-2}, 7.30 m s^{-2} (ii) 63.9 m
24 (iii) $\dfrac{g}{k}$
25 (i) $\dfrac{1}{k}\ln\dfrac{2g + 2ku}{2g + ku}$ (ii) $\dfrac{1}{k}\ln\dfrac{g + ku}{g}$
26 $3\sqrt{2ka^3}$
27 $\sqrt{u^2 - \dfrac{k}{a}}$; $u^2 < \dfrac{2k}{a}$
28 $\dfrac{g^2}{4k^3}$; $\dfrac{g}{k}$
29 $\frac{1}{3}l$; (ii) $\sqrt{3gl}$ (iii) $\dfrac{2\sqrt{2}}{3}gl$ (iv) $\dfrac{7l}{9}$
30 (i) 3.81 m s^{-1} (ii) 5.58 m s^{-1}

Exercise 8.3:4

1. (i) 3 m s^{-2}, 15 N
   (ii) 1 m s^{-2}, 15 N
2. $\frac{1}{2} \text{ m s}^{-2}$, $46\frac{1}{2}$ N
3. (i) $\frac{5g}{3}$, $\frac{10mg}{3}$ (ii) $\frac{17g}{12}$, $\frac{10mg}{3}$
4. 1 m s^{-2}, 700 N
5. $\frac{2g}{5}$, $2mg$
6. $\frac{g}{5}$, $2mg$
7. (i) 1.06 m s^{-2}, 816.67 N
   (ii) 0.06 m s^{-2}, 816.67 N
   (iii) 2.06 m s^{-2}, 816.67 N
8. (i) 5 m s^{-2}, 40 N, 20 N
   (ii) 2.5 m s^{-2}, 40 N, 20 N
9. $\frac{1}{4}g$, $\frac{9mg}{4}$, $\frac{3mg}{2}$
10. 4 m s^{-2}, 42 N, 84 N
11. $\frac{1}{7}g$, $\frac{24}{7}mg$, $\frac{48}{7}mg$
12. $\frac{M-m}{M+m}g$, $\frac{2Mm}{M+m}g$, $\frac{4Mm}{M+m}g$
13. $\frac{1}{2}g$, $\frac{1}{2}mg$, $\frac{\sqrt{2}}{2}mg$
14. $\frac{1}{4}g$, $\frac{3}{4}mg$, $\frac{3\sqrt{2}}{4}mg$
15. $\frac{1}{2}g$, $\frac{3}{2}mg$, $\frac{3\sqrt{2}}{2}mg$
16. $\frac{1}{3}$, 3 **17** 4
18. (i) $\frac{2}{3}$ (ii) $\frac{2}{7}g$
20. (i) $\frac{5}{2} \text{ m s}^{-2}$; 25 N, 30 N; $25\sqrt{2}$ N, $30\sqrt{2}$ N
    (ii) $\frac{8}{5} \text{ m s}^{-2}$, $\frac{116}{5}$ N, $\frac{168}{5}$ N; $\frac{116\sqrt{2}}{5}$ N, $\frac{168\sqrt{2}}{5}$ N
21. (i) $\frac{1}{6}g$; $\frac{7}{6}mg$, $\frac{5}{3}mg$; $\frac{7\sqrt{2}}{6}mg$, $\frac{5\sqrt{2}}{3}mg$
    (ii) $\frac{1}{12}g$; $\frac{13}{12}mg$, $\frac{11}{6}mg$; $\frac{13\sqrt{2}}{12}mg$, $\frac{11\sqrt{2}}{6}mg$
22. $\frac{7}{10}g$, $\frac{6}{5}mg$, $\frac{6\sqrt{3}}{5}mg$
23. $\frac{1}{5}g$, $\frac{4}{5}mg$, $\frac{4\sqrt{3}}{5}mg$
24. (i) $36.9°$, $3mg$, $5.37mg$
    (ii) $\frac{1}{5}g$, $4mg$, $7.16mg$
25. (i) $30°$, mg, $\sqrt{3}mg$
    (ii) $\frac{1}{4}g$, $\frac{3}{2}mg$, $\frac{3\sqrt{3}}{2}mg$
26. (i) $\frac{3}{10}g$, $\frac{21}{10}mg$, $\frac{21\sqrt{3}}{10}mg$
    (ii) 0, $3mg$, $3\sqrt{3}mg$
27. (i) $2 \le k \le 4$ (ii) $\frac{1}{4}g$, $\frac{21}{4}mg$, $9.39mg$
28. $\frac{1}{6\sqrt{2}}g$, $\frac{5}{3\sqrt{2}}mg$, $\frac{5}{3}mg$
29. (i) $\frac{7}{24}g$, $\frac{7}{25}mg$, $\frac{7\sqrt{2}}{25}mg$
    (ii) $\frac{17}{50}g$, $\frac{31}{50}mg$, $\frac{31\sqrt{2}}{50}mg$
30. (i) $\frac{1}{\sqrt{3}}$, $\frac{1}{2}mg$, $\frac{\sqrt{2}}{2}mg$
    (ii) $0.41g$, $0.91mg$, $1.29mg$
31. (ii) $\frac{1}{5}g$, $\frac{24}{25}mg$, $\frac{24\sqrt{2}}{25}mg$
32. $\frac{2g}{5}$, $\frac{mg}{5}$; $mg\sqrt{3}$; $9.23mg$, $85.7°$ to the horizontal
33. $\frac{1}{2\sqrt{2}}g$, $\frac{3}{2\sqrt{2}}mg$; $\frac{3}{\sqrt{2}}mg$, $\frac{1}{\sqrt{2}}mg$; $11.54mg$, $85.0°$ to the horizontal
34. (i) $\frac{9}{20}g$, $\frac{21}{4}mg$; $4mg$, $9mg$; $36.42mg$, $80.8°$ to the horizontal
    (ii) $\frac{1}{8}g$, $\frac{45}{8}mg$; $4.47mg$, $10.62mg$; $38.93mg$, $86.9°$ to the horizontal
35. (i) $0.41g$, $0.91mg$; $0.87mg$, mg, $8.53mg$, $84.8°$ to the horizontal
    (ii) $0.26g$, $0.97mg$; $0.89mg$, $1.03mg$, $8.70mg$, $86.9°$ to the horizontal
36. $\frac{1}{2}g$; $\frac{1}{4}g$, $3mg$
37. $\frac{1}{15}g$, $\frac{112}{15}mg$; $\frac{56}{15}mg$, $\frac{16}{5}mg$
38. $\frac{1}{4}g$, $\frac{15}{4}mg$
39. 2.83 m s^{-1} **40** 1.29 m s^{-1}
41. 1 m s^{-1}; 0.3 m
42. $\sqrt{\frac{2gl}{3}}$; $\frac{14l}{9}$
43. 2.5 m s^{-1}; 1.875 m, 1.5 s
44. $\frac{24}{7}$ s **45** 10 m
46. 0.4 m **47** $2l$
48. (i) 0.8 m (ii) 0.56 m
49. 1 m, 9.5 m s^{-1}
50. (i) $\sqrt{\frac{4gl}{3}}$ (ii) $\sqrt{\frac{3l}{g}}$
52. (i) $\frac{2\cos\theta - 1}{3\cos\theta}g$ (ii) $\frac{2\cos\theta - 1}{3}g$

Exercise 8.3:5

1. (i) 18 m s^{-2}; 90 N (ii) 53.33 m s^{-2}; 266.67 N
   (iii) 2.19 m s^{-2}; 10.97 N; towards the centre of the circle
2. (iii) (a) 84.8 min (b) 118.5 min
3. 100 N, 20 N **4** $5mg$
5. 3 rad s^{-1} **6** $\sqrt{\frac{3g}{l}}$
7. 0.19 m **8** 0.2 m
9. $\frac{1}{4}l$ **10** $\sqrt{\frac{2g}{l}}$
11. $15mg$, $12mg$ **13** 1.83 rad s^{-1}, 50 N
14. 4.47 rad s^{-1}, 80 N
15. $45°$, $20\sqrt{2}$ N
16. (i) $60°$ (ii) $\sqrt{\frac{3gl}{2}}$
17. $\frac{g}{\omega^2}$
19. $\sqrt{\frac{3gl}{2}}$; $\frac{l}{2}$
20. (i) $\frac{13mg}{12}$ (ii) $\frac{5}{2}\sqrt{\frac{gl}{39}}$
21. (i) $\frac{3mg}{4}$ (ii) $\frac{mg}{2}$ (iii) 0
22. 1 m, $48.2°$ **23** $\frac{3}{5}l$, $33.6°$
24. l, $\sqrt{\frac{g}{l}}$ **25** $\frac{5l}{9}$, $\sqrt{\frac{15g}{14l}}$
26. $\sqrt{\frac{4g}{3l}}$, $\frac{3\sqrt{3}l}{4}$ **27** $\sqrt{2\sqrt{3}gl}$
29. $\frac{13mg}{12}$ **30** 0.2 m **31** $\frac{3}{5}l$
33. (i) $\sqrt{\frac{\sqrt{2}g}{l}}$ (ii) $3mg\sqrt{2}$, $2mg\sqrt{2}$
34. $\frac{5mg}{3}$, $4mg$
35. (i) $\sqrt{\frac{2g}{3l}}$ (ii) $4mg$, $\sqrt{\frac{6g}{l}}$
36. (i) $\sqrt{\frac{2g}{3l}}$ (ii) $\sqrt{\frac{2g}{l\sqrt{3}}}$
39. $\sqrt{\frac{\sqrt{3}g}{l}}$

41. (i) mg, $2.85mg$ (ii) $\frac{1}{2}\sqrt{5ga}$
42. 2.23 rad s^{-1}
43. $\frac{1}{2}$ **45** $\frac{1}{4}$ **48** $\sqrt{\mu gr}$

Exercise 8.4:1

1. 60 J, 0, -30J, 0
2. $Tx \cos\alpha$, 0, $-\mu Rx$, 0
3. -32 J, 0, 56 J, 24 J
4. 0, $Sh \cot\beta$, $-Fh \csc\beta$, $-mgh$
5. (i) 20 J, 12 J (ii) 16 J, 30 J, 0, -12 J
6. (i) 2 J (ii) 6 J (iii) 24 J
7. (i) $\frac{1}{18}\lambda l$ (ii) $\frac{2}{9}\lambda l$ (iii) $\frac{1}{2}\lambda l$
8. (i) $\frac{1}{5}mgl$ (ii) $\frac{2}{5}mgl$
9. (i) 12 J (ii) 48 J (iii) 36 J
10. 0.2 m, 1 J **11** $\frac{l}{4}$, $4mgl$
12. $\frac{18}{5}mgl$ **16** $-\frac{8k}{3}$ **17** mgh

Exercise 8.4:2

1. (i) 37.5 J (ii) 19.5 J (iii) 19.5 J
2. (i) 12 J, 12.17 J (ii) 0.84 J (iii) 6.71 J
   (iv) 12.09 J
4. (i) 240 J (ii) 240 J (iii) -200 J
5. (i) 0.3 J (ii) 1.2 J
6. 0.6 J
7. 50 J
8. (i) 2.5 J (ii) 40 J

Exercise 8.4:3

1. 12.04 m s^{-1}
2. (i) 5.16 m s^{-1} (ii) 3.27 m s^{-1}
3. 40 000 N
4. $\frac{mv^2}{2R}$ (i) 168.75 m (ii) 675 m
5. (i) 6.32 m s^{-1} (ii) 20 m
6. (i) $\frac{15u^2}{16g}$ (ii) $\frac{5u^2}{4g}$
7. 250 J
8. (i) 60 m (ii) 140 m; 270 m
9. $\frac{1}{2}$ **10** 8.8 J
11. 8 m s^{-1}
12. 2.55×10^6 J
13. (i) 46 J (ii) 20 J (iii) 10.20 m s^{-1}
14. (i) 20 J (ii) -80 J (iii) 10 m s^{-1}; 100 J
15. 9.22 m s^{-1}
16. $-27\mathbf{i} - 36\mathbf{j}$
17. 43 500 J
18. 200 000 J
19. $1000Av(\frac{1}{2}v^2 + 10h)$
20. 8294 J
21. 22 000 J **22** 17.32 m s^{-1}
23. (i) 5 m (ii) 4.47 m s^{-1}
24. (i) $\frac{3}{2}a$ (ii) $\sqrt{ga}$
25. (i) 41.25 m above ground level
    (ii) 25 m s^{-1} (iii) 28.72 m s^{-1}
26. (i) $\frac{5u^2}{9g}$ (ii) $\frac{u^2}{8g}$
27. $\sqrt{ga(5 - 2\cos\theta)}$
28. $\sqrt{ga(1 - 2\cos\theta)}$ (i) $\frac{3}{2}a$ (ii) $\sqrt{ga}$
29. (i) $\sqrt{ga}$ (ii) $\sqrt{5ga}$ (ii) $\sqrt{3ga}$
30. (i) $2\sqrt{ga}$ (ii) $\sqrt{2ga}$
32. (i) $\sqrt{gl}$ (ii) $\sqrt{2gl}$ (iii) 0
33. (i) 6.08 m s^{-1} (ii) 5.39 m s^{-1} (iii) 5.57 m s^{-1}
34. $\frac{l}{4}$ **35** (i) $\frac{l}{4}$ (iii) $\frac{3}{2}\sqrt{gl}$
36. $\frac{3}{2}mg$; $\frac{4}{3}l$
37. (i) $\frac{7 + 2\sqrt{10}}{3}l$ (ii) $\sqrt{\frac{10gl}{3}}$

38 $4mg$
39 (i) $\frac{3mg}{2}$ (ii) $2g$ (iii) l (iv) $\sqrt{\frac{5gl}{2}}$ (v) $\sqrt{\frac{5gl}{2}}$
40 (i) (a) $\frac{m^2g^2l}{\lambda}$ (b) $\frac{m^2g^2l}{2\lambda}$
(ii) $\sqrt{2gl\left(1+\frac{mg}{2\lambda}\right)}$
41 (i) $\frac{1}{2}\sqrt{3gl}$ (ii) $\frac{l}{2}$
42 (i) 0 (ii) $\sqrt{2gl}$ (iii) $2\sqrt{2gl}$ (iv) $\sqrt{2gl}$ (v) 0
43 (i) $2.10l$ (ii) $2.26l$
44 (i) $l+\frac{u}{3}\sqrt{\frac{5ml}{\lambda}}$ (ii) $l+\frac{2u}{3}\sqrt{\frac{2ml}{\lambda}}$
(iii) $l+u\sqrt{\frac{ml}{\lambda}}$
45 $2\sqrt{gl}$
46 (i) (a) $\sqrt{\frac{8ga}{3}}$ (b) $\sqrt{2ga}$ (ii) (a) $\sqrt{\frac{8ga}{3}}$
(b) $\sqrt{ga}$
48 $\frac{1}{6}l$; $\frac{\sqrt{13}}{6}l$ **49** $2a$
50 (i) $2.13l$ (ii) $2.94l$ (iii) $1.92l$
51 (i) $\sqrt{ga}$ (ii) $1.15\sqrt{ga}$ (iii) $0.93\sqrt{ga}$
52 (i) $\sqrt{ga\left(4\cos\frac{\theta}{2}-4\cos\theta\right)}$ (ii) $\sqrt{2\sqrt{2}ga}$
(iii) $2\sqrt{ga}$ (iv) $\sqrt{5ga}$
53 3 m s^{-1}
54 (i) $\frac{2\sqrt{2}}{3}$ an (ii) $\frac{\sqrt{5}}{3}$ an (iii) 0

Exercise 8.4:4

2 9 kW **3** 1.1 m s^{-2}
4 8 kW **5** 40 m s^{-1}
7 (i) $\frac{H-Rv}{mv}$ (ii) $\frac{H}{R}$
8 (i) 300 N (ii) 0.25 m s^{-2}
9 (i) $\frac{H}{v}$ (ii) (a) $\frac{H}{3mv}$ (b) $\frac{H}{mv}$ (c) $\frac{3H}{mv}$
10 (i) $\frac{8}{9}$ m s^{-2} (ii) $\frac{10\,000}{9}$ N
11 $\frac{Hm}{(M+m)v}$
12 (i) 12.5 m s^{-1} (ii) 62.5 m s^{-1}
13 (i) 18.18 m s^{-1} (ii) 14.81 m s^{-1}
(iii) 9.52 m s^{-1}
14 (i) $\frac{Hn}{Rn+Mg}$ (ii) $\frac{Hn}{Rn-Mg}$
15 60 m s^{-1} **16** $\frac{Hv}{H-2Mgv\sin\alpha}$
17 24.6 kW **18** $\frac{4}{3}$ m s^{-2}
19 14.29 m s^{-1} **20** 56 kW
21 (i) $\frac{3}{50}$ (ii) 1 m s^{-2}
22 (i) 35.71 m s^{-1} (ii) 0.97 m s^{-2}; 83.33 m s^{-1}
23 (i) 22.73 m s^{-1} (ii) $\frac{9}{50}$
24 (i) 30 kW (ii) 60 m s^{-1}
25 $(R-mg\sin\alpha+ma)v$
26 (i) $-\frac{3}{8}$ m s^{-2}, 1200 N (ii) $\frac{5}{8}$ m s^{-2}, 2400 N
27 $\frac{1}{8}g$; $\frac{13}{4}mg$
28 150 kW, 5500 N;
(i) 27.27 m s^{-1} (ii) 4.5 m s^{-2}
29 320 kg; 3.96 m s^{-2}
30 (i) 13.64 m s^{-1} (ii) 0.73 m s^{-2}
31 (i) $\frac{3mg}{n}$; $\frac{4mgv}{n}$ (iii) $\frac{2g}{n}$
32 (i) $\frac{Hnv}{Hn+mgv}$ (ii) $\frac{Hnv}{Hn-mgv}$

33 (i) 1.6 (ii) 640 N, 0.76 m s^{-2}
34 $\frac{75}{4}$, $\frac{3}{4}$ m s^{-2}
35 $\frac{7H}{4mv}$
36 28.62 m s^{-1}
37 (i) 28.125 (ii) 32.09 m s^{-1}
38 (i) 1.68 (ii) 1.82 m s^{-2}
39 150 kW; 48.98 m s^{-1}
40 4556 W
41 89.6 kW; 16.84 m s^{-1}
42 (ii) $\sqrt{40t+100}$ (iii) 20 s
43 (i) $\sqrt{u^2+\frac{2Ht}{m}}$ (ii) Ht (iii) $\frac{1}{2}mu^2$, $\frac{1}{2}mu^2+Ht$
44 12.31 s; 2.46×10^5 J, 1.1×10^5 J
45 $\frac{mH}{R^2}\ln\frac{H-Ru}{H-2Ru}-\frac{mu}{R}$
46 10.65 s, 30 s
47 (i) 3.0 s (ii) 7.58 s; 38.73 m s^{-1}
49 $\frac{5u}{2kg}$, $\frac{17u^2}{6kg}$

Exercise 8.5:1

2 32 m s^{-1} **3** $\frac{20}{3}$ s
4 25 N **5** 40 000 N
6 16 Ns; 8 kg
7 (i) $\frac{\pi}{4}$ s (ii) $\frac{\pi}{2}$ s (iii) 3 m s^{-1}
8 4 m s^{-1} **9** 2 m s^{-1}
10 3 m s^{-1} **11** 11 m s^{-1}
12 (i) 68 Ns (ii) 115.3 Ns; 3 m s^{-1}, 31.8 m s^{-1}
13 $4\sqrt{3}\mathbf{i}-\mathbf{j}$
14 $2t\mathbf{i}+(t^3-t^2)\mathbf{j}$
15 450 N
16 (i) $1000Av^2$ (ii) $1000Av(v+u)$
17 2500 N
18 0.003 m^3; 0.018 m
19 (i) 19.2 kN (ii) 36.51 m s^{-1}
20 (i) 5.53 m s^{-1} (ii) 6.35 m s^{-1}

Exercise 8.5:2

1 18 Ns **2** 8 Ns, 160 N
3 180 Ns, 6000 N, 1.2 Ns
4 (i) 12 m s^{-1} (ii) 357 J
5 (i) 10 250 J (ii) 2250 J
6 $u+\frac{I}{m}$ **7** (i) $2u$ (ii) $\frac{5}{2}mu^2$
8 20 Ns, 100 Ns
9 2.05 N
10 5 m s^{-1}; (i) 30 Ns (ii) 315 J
11 2.2 m s^{-1}; (i) 29.4 Ns (ii) 205.8 J
12 $-\frac{1}{3}u$; (i) $\frac{14}{3}mu$ (ii) $\frac{49}{3}mu^2$
13 $\frac{4}{3}u$; (i) $\frac{4}{3}mu$ (ii) $\frac{4}{3}mu^2$
14 $\frac{3}{2}$; (i) $\frac{3}{5}mu$ (ii) $\frac{39}{10}mu^2$
15 -6; (i) $12mu$ (ii) $15mu^2$
16 (i) $\frac{13}{8}$ m s^{-1} (ii) $\frac{9}{8}$ m s^{-1}
17 (i) $\frac{5u}{2}$ (ii) $\frac{u}{2}$ **18** 1:4
19 6 m s^{-1}, 1323 J
20 0.005 kg, 882 J
21 $\frac{3}{5}mu^2$ **22** (i) $\frac{12}{25}a$ (ii) $\frac{12}{5}mg$
23 (i) 10 m s^{-1} (ii) 6 m s^{-1} (iii) 90 kJ; 230 kN
24 (i) $\frac{40}{3}$ m s^{-1} (ii) 454 kN
25 125 kN
26 (i) 0.3 m s^{-1} (ii) 2.4 Ns
27 (i) $\frac{2u}{7}$ (ii) $2mu$
28 (i) $\frac{1}{15}u$ (ii) $\frac{4}{15}u$

29 (i) 20 m s^{-1} (ii) 80 J, 1280 J (iii) $(0.2+x)m$
(iv) 6000 N, 0.013 m
30 (i) 0.24 m s^{-1} (ii) 1801.4 J
31 (i) 0.4 m s^{-1} (ii) 1601.6 J
32 (i) $\frac{1}{24}u$ (ii) $\frac{25}{48}mu^2$
33 $\sqrt{\frac{2kE}{m(k+1)}}$
34 (i) 10 m s^{-1} (ii) 5000 N
35 (i) $\frac{u}{k}$ (ii) $\frac{mu^2}{2k}$ (iii) $\frac{mu}{T}$ (iv) $\frac{mu^2}{2kx}$
36 6 m s^{-1}, 840 J
37 $-2u$, $15mu^2$
38 $-2u$, $2u$
39 (i) $\frac{1}{2}$ m s^{-1} (ii) 6 Ns (iii) 2 J
40 (i) u (ii) mu (iii) mu^2
41 (i) $\frac{1}{4}u$ (ii) $\frac{3}{4}mu$ (iii) $\frac{1}{4}mu^2$
42 (i) 4 m s^{-1} (ii) 4.8 Ns (iii) 1.6 m s^{-1}
43 (i) 1 m s^{-1} (ii) 6 m s^{-2} (iii) $\frac{1}{6}$ s
44 (i) 2.5 m s^{-1} (ii) 1 s (iii) 2.5 m s^{-1} (iv) 15 Ns
45 (i) $\frac{1}{3}\sqrt{2gh}$ (ii) $2\sqrt{\frac{2h}{g}}$ (iii) $\frac{1}{3}\sqrt{2gh}$
46 (i) 2 m s^{-1} (ii) 2.65 m s^{-1} (iii) 2.65 Ns
47 (i) $\frac{1}{2}\sqrt{ga}$ (iii) $\sqrt{\frac{a}{g}}$
48 (i) 4 m s^{-1} (ii) 8 Ns (iii) 5.6 m above plane
(iv) 4 m s^{-1} (v) 1.33 m s^{-1}
49 (i) 1.5 s (ii) 0.6 s (iii) 1.2 m s^{-1} (iv) 1.2 s
51 (i) 24 Ns (ii) 3 m s^{-1} (iii) 1.5 s (iv) 5.25 m
52 (i) 0.30 Ns (ii) 0.57 m s^{-1}
(iii) 0.22 s (iv) 0.86 m
53 (i) $\frac{3}{4}mg$ (ii) $\frac{1}{4}g$ (iii) 1.25 s (iv) $\frac{15}{16}g$
54 (i) $\frac{9}{7}mg\tau$ (ii) $\frac{2}{7}g\tau$ (iii) 2τ (iv) $\frac{23}{28}g\tau^2$

Exercise 8.5:3

1 (i) $\frac{3}{5}$ (ii) 32 Ns
2 (i) 0 (ii) $\frac{1}{3}$ (iii) $\frac{2}{3}$ (iv) 1
3 (i) 0.75 m s^{-1} (ii) 5.7 Ns (iii) 9.26 J
4 (i) eu (ii) $mu(1+e)$ (iii) $\frac{1}{2}mu^2(1-e^2)$; 0
5 0.2 m
6 (i) 0.05 m (ii) 0.0125 m (iii) $(0.8)(\frac{1}{4})^n$; 1.33 m
7 (i) 0.4 s (ii) 0.4 s; 1.2 s
8 (i) $\frac{1-e+2e^2}{1-e}h$ (ii) $\frac{1+e}{1-e}\sqrt{\frac{2h}{g}}$
9 $\frac{5}{7}$ **11** 24 m
12 19.9°, 70.1°; 15 m
13 (i) 41.4° (ii) $\frac{3}{4}mga$ (iii) $\frac{3}{4}mga$
14 14 m s^{-1}, 16 m s^{-1}; (i) 12 Ns (ii) 36 J
15 8 m s^{-1}; (i) 8 Ns (ii) 32 J
16 6 m s^{-1}, 16 m s^{-1}; (i) 100 Ns (ii) 500 J
17 $\frac{5}{9}u$, $\frac{8}{9}u$; (i) $\frac{8}{27}mu$ (ii) $\frac{8}{27}mu^2$
18 $\frac{3}{4}u$, $\frac{1}{2}$; (i) mu (ii) $\frac{1}{4}mu^2$
19 6 kg, 2.5 m s^{-1}; (i) 15 Ns (ii) 18.75 J
20 $\frac{1}{2}$, 14 kg; (i) 28 Ns (ii) 42 J
21 1, 2; (i) $\frac{4}{3}mu$ (ii) 0
22 $\frac{1}{2}$; (i) $3mu$ (ii) $3mu^2$
23 $\frac{1}{4}(1-3e)u$, $\frac{1}{4}(1+e)u$ (i) $\frac{3}{4}(1+e)mu$
(ii) $\frac{1}{16}mu^2(7-5e)(1+e)$
24 $-\frac{1}{4}(5+9e)u$, $\frac{1}{4}(-5+3e)u$ (i) $\frac{9}{4}(1+e)mu$
(ii) $\frac{27}{8}(1-e^2)mu^2$
25 (i) 4.4 m s^{-1}, 4.9 m s^{-1} (ii) 0.4 m s^{-1}, 4.1 m s^{-1}
26 (i) $\frac{13}{10}u$, $\frac{3}{10}u$ (ii) $\frac{7}{10}u$, $\frac{1}{10}u$ **28** 3
35 $\frac{5}{4}u$, $\frac{6}{4}u$, $\frac{8}{4}u$
36 $\frac{5}{4}u$, $\frac{9}{32}u$, $\frac{15}{32}u$; yes
37 $-\frac{1}{3}u$, $\frac{2}{3}u$, $\frac{2}{3}u$
38 $\frac{1}{3}(1-2e)u$, $\frac{1}{9}(1-2e)(1+e)u$, $\frac{1}{9}(1+e)^2u$
39 $\frac{1}{2}(1-e)u$, $\frac{1}{4}(1-e^2)u$, $\frac{1}{4}(1+e)^2u$
40 $\frac{1}{3}u$, $\frac{1}{18}u$, $\frac{1}{9}u$
41 $\frac{5}{18}u$, $\frac{5}{9}u$
42 $\frac{7}{20}u$, $\frac{1}{10}u$
44 $\frac{3}{7}$

46 $\frac{1}{9}(2-8e-e^2)u$, $\frac{2}{9}(1+e)^2 u$
47 10 m s^{-1}, 10 m s^{-1}; 2 s
48 $\frac{1}{2}\sqrt{10}u$, $\frac{3}{2}\sqrt{2}a$
49 $\frac{2}{5}\sqrt{2ga}$, $\frac{1}{10}\sqrt{2ga}$

Miscellaneous Exercise 8

2 $3mg$, $\sqrt{7}mg$, $3\sqrt{3}mg$
5 (i) $\dfrac{u}{\sqrt{e}}$ (ii) $\frac{1}{2}u$ (iii) $\frac{1}{2}\sqrt{3}u$
6 (ii) $u\sqrt{\dfrac{1}{g+ku^2}}$
7 (i) $\dfrac{\lambda^2}{8g}(1+4\ln\frac{12}{17})$
8 (i) $x^2 + y^2 = 1$ (ii) 6 J (iii) 12 N
 iv) $(\sqrt{2}-1)m$
11 (i) $\frac{5}{8}mg$ (ii) $\frac{7}{48}\sqrt{3}$ (iii) $\frac{5}{8}mg$
12 (i) $\frac{1}{3}g$, $2mg$, $\frac{4}{3}mg$ (ii) $\frac{1}{4}g$, $\frac{9}{4}mg$, $\frac{5}{4}mg$
13 (i) $\frac{2}{5}g$, $\frac{9}{5}mg$ (ii) $2\sqrt{\dfrac{gl}{5}}$ (iii) $\sqrt{\dfrac{5l}{g}}$
15 $\frac{1}{5}g$, $\frac{6}{5}mg$; $\frac{4}{3}l$
16 (i) $\frac{1}{12}$ (ii) $\frac{2}{49}$
17 $2\sqrt{3}$ m; $2\sqrt{39}$ m s^{-1}, $\tan^{-1} 2\sqrt{3}$ with the horizontal
18 (i) 2.02×10^{24} kg (ii) 1340 m s^{-1}
19 (i) 4.1×10^{14} m^3 s^{-2} (ii) 1.1×10^7 m
20 $2\sqrt{2}$ rad s^{-1}
22 (i) $\sqrt{\dfrac{2g}{l}}$ (ii) $\sqrt{\dfrac{3gl}{2}}$ (iii) $2mg$
23 (ii) $\sqrt{\dfrac{24gl}{5}}$
24 (i) $\dfrac{5mg}{3}$ (ii) $\frac{1}{3}l$ (iii) $\frac{16}{15}l$ (iv) $\sqrt{\dfrac{5g}{4l}}$
25 (ii) $\sqrt{\dfrac{4g}{l}}$
26 (i) $2mg$
27 1.5 m; (i) 0.8 s (ii) $63.4°$
28 (i) (a) 2.5 J (b) 7.5 J (ii) 21 J
29 $8(10x - x^2 - 24)$, $-4x + 20$ (i) 5 (ii) 1 or 9
30 6.4×10^7 J; 1.3×10^4 m s^{-1}
31 $\dfrac{u^2}{8g}(8\ln 2 - 5)$
32 (iii) $\dfrac{u}{g}\left(\lambda \ln \dfrac{\lambda}{\lambda - 1} - 1\right)$
33 5 m s^{-1} 35 $3l$, $\sqrt{\dfrac{8gl}{3}}$
36 $\frac{6l}{5}$ (i) $\sqrt{3gl}$ (ii) $\sqrt{\dfrac{42gl}{5}}$
37 (i) 26.67 m s^{-1} (ii) 10 m s^{-1}; 0.17 m s^{-2}
38 (ii) $\dfrac{H}{mu} \mp gn$ (iii) $\frac{3}{2}u$
39 $\frac{1}{3}u$ (i) $\frac{1}{10}g$ (ii) $\frac{1}{20}mg$
40 $\dfrac{mg}{40}$ (i) $2u$ (ii) $4u$
42 (i) 6.25 kW (ii) 89.26 m
43 6.56 kW; 640 N
44 (i) $\dfrac{\lambda - 1}{\lambda + 1}$ (ii) $\dfrac{(\lambda^2+1)h}{\lambda}$
45 $\frac{2}{3}$; $\dfrac{8u}{3}$, $\dfrac{2u}{3}$
46 (i) 15.5 m s^{-1}, 16 m s^{-1}; 31 s, 32 s
 (ii) 7.98×10^5 J, 625 m
47 (i) $\dfrac{u}{2}$, $2u$ (ii) $3mu$ (iii) $\frac{9}{4}mu^2$; $3u$, $4u$
48 $\frac{1}{3}u$, $\frac{4}{9}u$, $\frac{10}{9}u$
49 $\dfrac{1-ke}{1+k}u$, $\dfrac{(1+ke)(1+e)}{(1+k)^2}u$, $\dfrac{(1+e)^2}{(1+k)^2}u$

50 $\frac{1}{3}u(2-e-e^2)$, $\frac{1}{6}u(4-e-e^2)$, $\frac{1}{2}u(1+e)$
51 4
53 $\dfrac{\pi(k-e)a}{(k+1)T}$, $\dfrac{\pi k(1+e)a}{(k+1)T}$; $\dfrac{2T}{e}$
55 (i) $\frac{27}{8}mga$ (ii) $\frac{3a}{4}$, $\frac{27a}{16}$
56 (i) mgl (ii) $\frac{1}{2}l(5+2\sqrt{2})$
58 (ii) $\sqrt{\dfrac{4l}{5g}}$
59 (i) $\frac{1}{8}gT$ (iii) $\frac{1}{4}g$ (iv) $\frac{3}{2}T$
60 $2mg$; (i) $\frac{3}{4}\sqrt{\dfrac{gl}{2}}$, $\frac{1}{2}\sqrt{\dfrac{gl}{2}}$ (ii) $\frac{1}{8}mg$

Chapter 9

Exercise 9.1:1

1 (a) (i) $5a$ Nm (ii) $-19a$ Nm
 (b) (i) $-4a$ Nm (ii) $-4a$ Nm
 (c) (i) $-82a$ Nm (ii) $-38a$ Nm
2 (a) (i) $10Pa$ (ii) $10Pa$
 (b) (i) $8Pa$ (ii) $4Pa$
 (c) (i) $-54Pa$ (ii) $54Pa$
3 (i) $-\frac{1}{2}\sqrt{2}Pa$ (ii) $(6\sqrt{3}-9)Pa$ (iii) 12 Nm
 (iv) -20 Nm
4 (i) $a\sqrt{3}$ (ii) $2a\sqrt{3}$ (iii) $3a$ (iv); $a\sqrt{3}$; $a\sqrt{3}$
5 (a) (i) $10\sqrt{3}a$ Nm (ii) $10\sqrt{3}a$ Nm
 (b) (i) $18\sqrt{3}a$ Nm (ii) $9\sqrt{3}a$ Nm
 (c) (i) $8a$ Nm (ii) $8a$ Nm
 (d) (i) $-2a$ Nm (ii) $-a$ Nm
 (e) (i) $21\sqrt{3}a$ Nm (ii) $21\sqrt{3}a$ Nm
 (f) (i) $(4+15\sqrt{3})a$ Nm
  (ii) $(-2+13\sqrt{3})a$ Nm
6 $\mathbf{r} = \mathbf{i} - \mathbf{j} + \lambda(6\mathbf{i} + 5\mathbf{j})$, $5x - 6y - 11 = 0$
7 (i) $5x - 4y + 16 = 0$ (ii) $x + 2y - 3 = 0$
 (iii) $x = 3$
8 (i) $aq - bp$ (ii) $qx - py - (aq - bp) = 0$
 (iv) $\sqrt{p^2 + q^2}$; $aq - bp$
9 (i) 8 (ii) -15 (iii) -10 (iv) 33 (v) 7 (vi) -7
 (vii) 0
10 $aq - bp$ 11 (i) -1 (ii) 13 12 4
13 (i) 7 (ii) 10 (iii) 24 (iv) 0
14 (i) -11 (ii) -19 15 -18 Nm
16 (i) 50 Nm (ii) -75 Nm (iii) 50 Nm
17 12 Nm
18 (i) $\frac{1}{2}(12\sqrt{3} - 25)Pa$ (ii) 0 (iii) $(4 - 6\sqrt{3})Pa$
 (iv) $(18 + 21\sqrt{3})Pa$

Exercise 9.1:2

1 -7 2 -7 3 8

Exercise 9.1:4

1 (i) $\mathbf{r} = \lambda(\mathbf{i} + 3\mathbf{j})$ (ii) $\mathbf{r} = 3\mathbf{i} - \mathbf{j} + \lambda(\mathbf{i} + \mathbf{j})$
 (iii) $\mathbf{r} = \mathbf{i} + \mathbf{j} + \lambda(2\mathbf{i} - \mathbf{j})$
2 (i) $\mathbf{r} = \mathbf{i} - \mathbf{j} + \lambda(7\mathbf{i} + 5\mathbf{j})$
 (ii) $\mathbf{r} = 4\mathbf{i} - 2\mathbf{j} + \lambda(4\mathbf{i} + 5\mathbf{j})$
 (iii) $\mathbf{r} = \lambda \mathbf{i}$ (iv) $\mathbf{r} = 5\mathbf{i} + 6\mathbf{j} + \lambda(5\mathbf{i} + 2\mathbf{j})$
 (v) $\mathbf{r} = \mathbf{i} + 3\mathbf{j} + \lambda(\mathbf{i} - \mathbf{j})$
3 -5, -1
4 (i) $\mathbf{r} = \mathbf{i} - \mathbf{j} + \lambda(3\mathbf{i} + 4\mathbf{j})$
 (ii) $\mathbf{r} = 2\mathbf{i} + 3\mathbf{j} + \lambda(4\mathbf{i} - \mathbf{j})$
5 $\mathbf{r} = 2\mathbf{i} + \mathbf{j} + \lambda(3\mathbf{i} + 2\mathbf{j})$
6 (i) $\mathbf{r} = -\mathbf{i} + 3\mathbf{j} + \lambda(5\mathbf{i} + 4\mathbf{j})$ (ii) $\mathbf{r} = 3\mathbf{j} + \lambda \mathbf{i}$
7 (i) -21 (ii) $\mathbf{i} + 2\mathbf{j}$ (iii) $10\mathbf{i} - \mathbf{j}$
8 (i) $2P$, perpendicular to AB; a
 (ii) $3P$, perpendicular to AB; $\frac{4}{3}a$
 (iii) $(\lambda + \mu)P$, perpendicular to AB; $\dfrac{2\mu}{\lambda + \mu}a$
 (iv) P, perpendicular to AB; $8a$

 (v) $(\lambda - \mu)P$, perpendicular to AB;
  $\left|\dfrac{2\mu}{\lambda - \mu}\right|a$
 (vi) $4.36P$, $83.4°$ to AB, $\frac{6}{5}a$
 (vii) $7P$, $30°$ to AB, $\frac{8}{9}a$
 (viii) $2P$, $45°$ to AB, $6a$
9 (i) $5P$, $36.9°$ to AB, $\frac{14}{3}a$
 (ii) $7.21P$, $56.3°$ to AB, $\frac{1}{3}a$
 (iii) $5P$, $36.9°$ to AB, $2a$
 (iv) $5P$, $36.9°$ to AB, $\frac{4}{3}a$
10 (i) $\sqrt{3}P$, $30°$ to BA, $4a$
 (ii) $\sqrt{3}P$, $30°$ to BA, $3a$
 (iii) $\sqrt{3}P$, perpendicular to AB, a
 (iv) $7.09P$, $77.8°$ to BA, $0.89a$
11 (i) $5.39P$, $68.2°$ to AB, $\frac{8}{5}a$
 (ii) $8.48P$, $70.7°$ to AB, $3.1a$
 (iii) $2.01P$, $5.7°$ to AB, $108a$
 (iv) $6.81P$, $49.8°$ to BA, $8.62a$
12 (i) $13.75P$, $40.9°$ to AB, 0
 (ii) $6P$, $60°$ to BA, $7a$
 (iii) $5.82P$, $80.1°$ to AB, $1.05a$
 (iv) $9.98P$, $63.4°$ to BA, $1.03a$
13 (i) $6.71P$, $26.6°$ to BA, $4.98a$
 (ii) $\sqrt{3}P$, perpendicular to AB, $0.69a$
14 (i) $6\mathbf{j}$, $x = 2$
 (ii) $6\mathbf{i} + 2\mathbf{j}$, $x - 3y + 2 = 0$
 (iii) $-\mathbf{i} + 2\mathbf{j}$, $2x + y - 14 = 0$
 (iv) $2\mathbf{i} + 2\mathbf{j}$, $2x - 2y + 1 = 0$
15 $(m+n)(p\mathbf{i} + q\mathbf{j})$, $q(m+n)x - p(m+n)y + (pmv_1 + pny_2 - qmx_1 - qnx_2) = 0$
16 (i) $2\mathbf{i} + 3\mathbf{j}$, $3x - 2y - 13 = 0$
 (ii) $\mathbf{i} - \mathbf{j}$, $x + y + 1l = 0$
 (iii) $3\mathbf{i} + 4\mathbf{j}$, $4x - 3y - 1 = 0$
 (iv) $-10\mathbf{i} + 5\mathbf{j}$, $5x + 10y + 18 = 0$
17 $5x - 2y + 32 = 0$
18 $x + 11y - 6 = 0$
19 3 20 $\sqrt{2}$ 21 3
22 (i) $\frac{25}{6}$, $\frac{65}{12}$ (ii) $\frac{25}{6}$, $\frac{65}{7}$
23 (i) 7, $\sqrt{2}$ (ii) $\frac{11}{3}$, $\frac{17}{3}\sqrt{2}$
24 $\dfrac{\sqrt{29}G}{a}$, $2x + 5y - 3 = 0$
25 $\frac{4}{3}\sqrt{5}$
26 $12x + 4y = 3a$
27 $4x + 3y = 12a$
28 $\dfrac{\sqrt{10}m}{a}$, $x - 3y - 4a = 0$
29 $3x - 4y - 15 = 0$
30 $4.16P$, $13.9°$ to BA, $2a$
31 (i) midpoint of AB (ii) $\frac{1}{3}G$
33 (i) P, perpendicular to AB, $-Pa$
 (ii) $3P$, along AB, $10P$
 (iii) $8.60P$, $49.1°$ to AB, $-2\sqrt{3}Pa$
 (iv) $2.83P$, $45°$ to AB, $14Pa$
34 (i) $4\mathbf{i} + 4\mathbf{j}$, -1 (ii) $6\mathbf{i} + 10\mathbf{j}$, 48
35 (i) $4x + 6y - 7 = 0$ (ii) $4x + 6y - 3 = 0$
 (iii) $4x + 6y - 15 = 0$

Exercise 9.1:5

1 (i) $-2Pa$ (ii) $-\sqrt{3}Pa$ (iii) $12Pa$ (iv) $4Pa$
 (v) $18Pa$ (vi) $8Pa$ (vii) $-7\sqrt{3}Pa$
 viii) $17\sqrt{3}Pa$
3 (i) 10 (ii) 12 (iii) 2
4 (i) 4, $2\sqrt{2}$; $10a$ Nm (ii) 2, 2; $2\sqrt{3}a$ Nm
 (iii) 1, $5\sqrt{2}$; $-8a$ Nm (iv) 4, 1; $8a$ Nm
 (v) 1, 7; $(1 + 19\sqrt{3})a$ Nm
5 $3\mathbf{i} + 4\mathbf{j}$, 15
6 (i) 4.24 N, $45°$ to AB, $10a$ Nm
 (ii) 10.82 N, $73.9°$ to BA, $6.93a$ Nm
7 $-8\mathbf{i} - 11\mathbf{j}$, 4
8 (i) 3.61 N, $56.3°$ to BA, cuts AB at E where $AE = \frac{4}{3}a$

596   Answers

(ii) 4.36 N, 83.4° to *BA*, cuts *AB* produced at *G* where $AG = 2.74a$

Exercise 9.1:6

1 (i), (iv), (v)   2 2, −3
4 $r = i + 2j + \lambda(-i + 3j)$
8 $-13i + 7j$ 9 $2i + 3j$ 10 $4i$
12 (iii) $y = x - 5$, $y = x - 5$
13 3, 2   (ii) 9, 8
14 (i) $4\sqrt{2}, 3\sqrt{2}, 1$ (ii) 2, 4, 3
    (iii) 20, 40, 20 (iv) $\frac{16}{3}\sqrt{3}, 6, \frac{32}{3}\sqrt{3}$
16 (i) $3.61P$, 56.3° to *AB*, cuts *AB* produced at *D* where $AD = 6a$
    (ii) $2P$, 60° to *AB*, cuts *BA* produced at *G* where $AG = 17a$
    (iii) $4\sqrt{2}$, 135° to $0x$, $x + y = 3$
17 (i) $11.05P$, 84.8° to *BA*; $-6Pa$
    (ii) $7.28P$, 74.1° to *BA*; $-24Pa$
    (iii) 6.71, $-116.6°$ to $0x$; -22
19 $-3, -3, 2$
20 $-3, 2, 5$; $-i + j$
21 $4, 2, 7$
22 $-6, -7, 2$
23 $2i + 5j$; $r = 2i + 5j + \lambda(i + 3j)$
24 $r = 2i - j + \lambda(i + 2j)$
25 $-i + 2j$; $r = -i + 2j + \lambda(i + 6j)$

Exercise 9.2:1

2 (i) At the point *G* on *AD* such that $AG = \frac{2}{3}h$
   (ii) At the point *G* on *AD* such that $AG = \frac{2}{\sqrt{3}}a$, where *D* is the midpoint of *BC*
   (iii) At the point *G* inside $\triangle ABC$ distant $\frac{1}{3}a$ from *AC*, $\frac{1}{3}b$ from *BC*
3 a
7 (ii) $\frac{1}{4}(a + b + c + d)$ (iii) $3:1$
8 At the point *G* on *VO* such that $VG : GO = 3:1$

Exercise 9.2:2

1 (i) $\frac{1}{2}a$ (ii) $\frac{1}{2}a$ (iii) $\frac{3}{5}a$ (iv) $\frac{2}{5}a$
   (v) $\frac{\mu a}{\lambda + \mu}$ (vi) $\frac{m_2 a}{m_1 + m_2}$
2 (i) $(-2, 0)$ (ii) $(-\frac{7}{5}, \frac{4}{5})$ (iii) $(-\frac{19}{5}, -\frac{12}{5})$
3 $\frac{\lambda a + \mu b}{\lambda + \mu}$
4 (i) $\frac{1}{2}a$ (ii) $\frac{5}{8}a$ (iii) $\frac{3}{5}a$
5 (i) $(\frac{11}{2}, 0)$ (ii) $(\frac{5}{6}, 2)$ (iii) $(0, 0)$
7 (i) $i + 3j$ (ii) $\frac{1}{3}i + \frac{10}{3}j$
8 (i) $-1, 2$ (ii) $2, 5$ 9 $4, 5$
10 (i) $\frac{4}{3}a$ (ii) $\frac{8}{5}a$ (iii) $\frac{3}{4}a$ (iv) $\frac{6}{5}a$
11 (i) $\frac{1}{4}(a + 2b + c)$ (ii) $\frac{1}{3}(a + b + c)$
12 $\frac{1}{2}(a + b)$; $\frac{1}{3}(a + b + c)$
14 $(2, 3)$ 15 $\frac{3}{4}a, \frac{3}{8}a$
16 (i) $\frac{3}{2}i + \frac{5}{2}j$ (ii) $\frac{3}{2}i + \frac{5}{2}j$ (iii) $\frac{20}{11}i + \frac{29}{11}j$
17 (i) $\left(\frac{3\{1-(\frac{3}{4})^n\}}{\{1-(\frac{1}{2})^n\}}, 0\right)$
    (ii) $(\frac{1}{3}(n+2), \frac{1}{2}n(n+1))$

Exercise 9.2:3

1 (i) $\frac{24}{14}a$ from *A*, *F* $\frac{13}{14}a$ from *AB*
   (ii) $\frac{13}{12}a$ from *AD*, $\frac{7}{6}a$ from *AB*
   (iii) $\frac{59}{26}a$ from *AH*, $\frac{27}{26}a$ from *AB*
   (iv) $0.56a$ from *AE*, $\frac{1}{2}a$ from *AB*
   (v) $0.36a$ from *BD*, on *AC*
   (vi) $0.38a$ from *AB*, on axis of symmetry
2 $\sqrt{3}a$
3 (i) $\frac{19}{15}a$ from *AD*, $\frac{7}{15}a$ from *AB*
   (ii) $\frac{5}{7}a$ from *ED*, $\frac{6}{7}a$ from *AB*
4 (i) $\frac{2}{3}a$ from *AC*, $\frac{1}{6}a$ from *AB*
   (ii) on axis of symmetry, $\frac{1}{4}\sqrt{3}a$ from *A*
   (iii) $\frac{3}{4}a$ from *AD*, a from *AB*
   (iv) $4a$ from *AD*, $\frac{9}{4}a$ from *AB*
   (v) on *CD*, where *D* is the midpoint of *AB*, $\frac{1}{3}\sqrt{3}a$ from *AB* (vi) on *CD*, where *D* is the midpoint of *AB*, $\frac{13}{3}a$ from *AB* (vii) $\frac{10}{9}a$ from *AD*, $\frac{7}{6}a$ from *AB* (viii) on *CD*, where *D* is the midpoint of *AB*, from *AB* $0.43a$
5 (i) $\frac{1}{4}a$ from *AD*, $\frac{1}{2}a$ from *AB*
   (ii) $\frac{1}{12}a$ from *CD*, where *D* is the midpoint of *AB*, $\frac{1}{4}\sqrt{3}a$ from *AB*
   (iii) on *AD*, where *D* is the midpoint of *BC*, $\frac{15}{22}a$ from *BC*
   (iv) $\frac{1}{2}a$ from *AD*, $\frac{13}{10}a$ from *AB*
6 On axis of symmetry, $\frac{4}{7}a$ from base
7 On axis of symmetry, $\frac{20}{33}a$ from base
8 On axis of symmetry, (i) $\frac{45}{29}a$ from base
   (ii) $\frac{27}{17}a$ from base (iii) $\frac{18}{11}a$ from base
9 (i) on *AO* produced, $\frac{1}{30}a$ from *O*
   (ii) on *AO* produced, $\frac{1}{15}a$ from *O*
   (iii) on *AO* produced, $\frac{9}{14}a$ from *O*
   (iv) on axis of symmetry (angle bisector of $A\hat{O}B$), $\frac{1}{14}a$ from *O*
   (v) $\frac{5}{6}a$ from *AF*, $\frac{5}{6}a$ from *AB*
   (vi) $\frac{17}{10}a$ from *AF*, $\frac{7}{6}a$ from *AB*
   (vii) on axis of symmetry, $0.38a$ from *AD*
   (viii) $\frac{11}{13}a$ from *AE*, $\frac{23}{26}a$ from *AB*
10 on axis of symmetry, $\frac{35}{22}a$ from end
11 on axis of symmetry, $2.19a$ from end
12 $\frac{13}{14}a$ from each large face
13 (i) $\frac{23}{24}a$ from *AD*, $\frac{23}{24}a$ from *AB*
    (ii) $\frac{3}{4}a$ from *AD*, $\frac{3}{4}a$ from *AB*
    (iii) $\frac{11}{6}a$ from *AD*, $\frac{3}{2}a$ from *AB*
    (iv) a from *AB*, a from *AC*

Exercise 9.2:4

1 (i) $(\frac{4}{9}, \frac{4}{3})$ (ii) $(\frac{3}{4}, \frac{3}{10})$ (iii) $(1, -\frac{2}{5})$
   (iv) $(2 \ln 2 - 1, \frac{3}{4})$
   (v) $(1, \frac{1}{8}\pi)$ (vi) $\left(\frac{1}{\ln 2}, 1 + \frac{1}{4 \ln 2}\right)$ 2 $(\frac{1}{3}, \frac{1}{3})$
3 $\left(\frac{2n}{n+1}, \frac{n^4 + n^3 + n^2 + 1}{5n^3(n+1)}\right)$; $(2, \frac{1}{5})$
4 (i) $(\frac{27}{5}, \frac{9}{4})$ (ii) 0 (iii) $(\frac{3}{5}, 0)$
5 (i) $(\frac{27}{7}, 0)$ (ii) $(\frac{93}{35}, 0)$ (iii) $(\frac{2}{5}, 0)$
   (iv) $(\frac{32}{35}, 0)$ (v) $(\frac{6}{5}, 1)$ (vi) $(\frac{8}{5}, 1)$
7 $\frac{1}{4}\pi a^2$; $\left(\frac{8}{3\pi}, \frac{8}{\pi}\right)$
8 (i) $\left(\frac{\int_c^d \frac{1}{2}x^2 \, dy}{\int_c^d x \, dy}, \frac{\int_c^d xy \, dy}{\int_c^d x \, dy}\right)$
   (ii) (a) $(0, 5.58)$ (b) $(0.36, 2.10)$
10 (i) $(6, 0)$ (ii) $(\frac{1}{4}, 0)$ (iii) $(\frac{9}{7}, 0)$ (iv) $(0.44, 0)$
    (v) $(0.92, 0)$ (vi) $(0.64, 0)$ (vii) $1.10, 0)$
11 $\left(\frac{3n(n+1)}{2(n^2+n+1)}, 0\right)$; $(\frac{3}{2}, 0)$
13 $\frac{7}{40}a$ from plane face
14 $\frac{51}{108}a$ from larger plane face; $\frac{37}{80}a$ from larger plane face; $\frac{11}{32}a$ from plane face
15 $\frac{11}{14}a$ from larger plane face
16 (i) $\left(0, \frac{\int_c^d x^2 y \, dy}{\int_c^d x^2 \, dy}\right)$ (ii) (a) $(0, 2)$ (b) $(0, \frac{20}{3})$
    (c) $(0, \frac{48}{e})$ (d) $(0, e - 1)$

17 On the axis of symmetry, $\dfrac{a \sin \alpha}{\alpha}$ from *O*
18 On the axis of symmetry $\dfrac{2a}{\pi}$ from the centre
19 On the axis of symmetry, $\dfrac{2a \sin \alpha}{3\alpha}$ from *O*
20 On the axis of symmetry, $\dfrac{4a}{3\pi}$ from the straight edge
21 On the axis of symmetry $\frac{1}{2}a$ from the bounding plane
22 On the axis of symmetry, $\frac{1}{2}h$ from the bounding plane
24 (i) On the axis of symmetry, $\frac{21}{32}a$ from common face
    (ii) On the axis of symmetry $\frac{1}{16}a$ from common face
    (iii) On the axis of symmetry, $\frac{3}{20}a$ from common face
    (iv) On the axis of symmetry, $\frac{229}{60}a$ from the end
25 (i) On the axis of symmetry, $\frac{1}{2}a$ from common face
    (ii) On the axis of symmetry, $0.12a$ from common face
    (iii) On the axis of symmetry $0.31a$ from common face
    (iv) On the axis of symmetry $3.78a$ from the end
26 $\sqrt{3}a$
27 On the axis of symmetry, $\dfrac{a(\cos + 2 \sin \alpha)}{2(1 + \alpha)}$ from centre
28 On the axis of symmetry, $\dfrac{2a}{2 + \pi}$ from centre
29 (i) (a) $3a$ (b) $7\pi a^3$ (c) $\frac{33}{28}a$ from larger plane face
    (ii) $\frac{63}{46}a$ from larger plane face of frustrum
30 $\frac{7}{16}a$ from axis of symmetry, $\frac{1}{6}a$ from plane face of original hemisphere
31 $\frac{4}{3}a$ 32 $\frac{3}{2}a$ 33 $\frac{10}{9}a$ 34 $\dfrac{3a}{2\pi}$

Exercise 9.3:1

1 $\frac{1}{5}\sqrt{3}$ 2 $\frac{36}{77}$
3 $\frac{1}{15}\sqrt{65}W$, $60.3°$ to the horizontal
4 $\frac{8}{25}W$; $0.77W$, $51.3°$ to the rod
5 $\dfrac{1}{\sqrt{3}}$
6 $1.21a$, $0.36W$, $59.0°$ to the horizontal; $0.71W$, $68.2°$ to the rod
7 $\dfrac{\sqrt{3}}{4}$; $\frac{2}{3}W$, perpendicular to the rod; $0.87W$, $66.6°$ to the horizontal
8 $\frac{1}{2}$; $\frac{1}{2}W$, perpendicular to the rod; $0.67W$, $63.4°$ to the horizontal
10 $33.6°$; $1.05W$, $18.4°$ to the vertical
11 $\frac{1}{2} \cot \theta$ 12 $45°$; $\frac{5}{4}W$, $\frac{3}{4}W$
14 $\frac{1}{4}a$ 15 $\frac{1}{2}a$ 16 $\frac{7}{4}$ 17 $10W$
18 $67.4°$; $0.98W$, $11.3°$ to the vertical; $0.20W$, $11.3°$ to the horizontal
19 $\sqrt{\tan^1 \theta + \sec \theta} - \tan \theta$
20 0.63; $0.85W$, $32.0°$ to the vertical; $0.53W$, $32.0°$ to the horizontal
23 $\dfrac{1}{2\sqrt{3}}W$, $\dfrac{1}{2\sqrt{3}}$; $1.04W$
24 $\frac{1}{2}W$, $\dfrac{1}{\sqrt{3}}$ 25 $\dfrac{1}{\sqrt{3}}$
26 $0.97W$, $59.0°$ to the horizontal

27 At the endpoint of AB
29 0.24
30 $\frac{4}{3}a$ from A or $\frac{4}{7}a$ from B
31 1 32 60°, $2W$; $4W$
33 0.12, 1.02 W
34 41.8°; 1.03 W, 0.26 W
35 $\frac{2W}{\sqrt{3}}$ 36 $\sqrt{3}W$, $2W$
37 0.54 W 38 $\frac{3}{5}W$
39 9.90 W, 7.62 W, 23.2° below AB
40 6.01 W; 5.01 W, 3.8° below AB
41 1.68 W, 1.52 W, 9.5° above AB
42 26.6°
43 0.43 W; 0.66 W, 70.9° above horizontal
44 1.53 W, 19.1° above horizontal; 2.89 W, perpendicular to AB
45 0.63 W; 55.3° above horizontal; 0.6 W, perpendicular to AB
46 0.79 W, 0.91 W, 41.6° above horizontal
47 $\frac{5}{8}W$; $\frac{5}{8}W$, 53.1° above horizontal
48 2.31 W, 1.53 W, 40.9° above horizontal
49 (i) 12 N (ii) 0.5 m (iii) 16 N, 53.1° above horizontal
50 0.15
51 (i) $\frac{3}{8}W$ (ii) $\frac{3}{10}W$ (iii) $\frac{1}{2}W$; $\frac{3}{10}W$
52 $\frac{1}{4}W$, perpendicular to AB; 0.25

Exercise 9.3:2

1 $\sqrt{2}W$; W, along AB
2 $\sqrt{5}W$; $2W$, along AB
3 2.06 W; 2.06 W, 14.0° above AB
4 W; W along AB
5 $\frac{1}{2}W$; 1.12 W, 26.6° to the vertical
6 0.87 W; $\frac{1}{2}W$, 30° above horizontal
9 (i) $\frac{5}{4}W$ (ii) $\frac{5}{8}W$ (iii) $\frac{4}{5}W$

Exercise 9.3:3

1 63.4° 2 66.6°
3 (i) 63.4° (ii) 69.4° (iii) 76.0°
4 57.5° 5 $\frac{1}{2}W$ 7 63.4°
8 (i) 10.6° (ii) 7.1° (iii) 36.9°
9 12.0° 10 $\frac{2}{5}W, \frac{3}{5}W$
11 0.61 W, 0.39 W
12 (i) 39.9° (ii) 1.3:1
13 (i) 63.4° (ii) 1:5 (iii) 1:3
14 (i) between $\frac{1}{2}a$ & $\frac{3}{2}a$ from A
   (ii) between $\frac{5}{8}a$ & $\frac{13}{8}a$ from A
15 between $\frac{3}{8}a$ & $\frac{43}{24}a$ from A
16 $\frac{3}{4}W$
17 (i) $\frac{2}{3}W, \frac{1}{3}W$ (ii) $\frac{2}{3}a, \frac{1}{3}a$ (iii) 1.97 a

Exercise 9.3:4

1 0.73 a 2 0.73 a
3 $\frac{1}{2}a$ 4 0.71 a
6 (i) 1.21 a (ii) 0.82 a
7 69.4° 9 1.59
10 (i) $AP = \frac{1}{4}a$ (ii) $AP = \frac{3}{16}a$
11 (i) $3W$ (ii) $\frac{1}{3}W$
12 $\frac{7}{6}a$
13 (i) $\frac{1}{12}a$ (ii) $\frac{1}{9}W$ 14 $\frac{4}{3}W$
15 (i) $\frac{1}{8}a$ from E (ii) $\frac{2}{3}W$
16 (i) $5W$ (ii) $\frac{9}{4}W$
17 W 20 sliding

22 (i) toppling (ii) sliding
23 (i) toppling (ii) toppling
26 toppling

Miscellaneous Exercise 9

1 $3\sqrt{5}$ N, $\tan^{-1}\frac{1}{2}$ to AB produced
2 (i) 60 N m (ii) 25 N, parallel to BA; 4 m
3 (i) 10, 3 (ii) 10, 14
4 $-2\mathbf{i} + \mathbf{j}$; $\mathbf{r} = -2\mathbf{i} + \mathbf{j} + \lambda(\mathbf{i} - 2\mathbf{j})$
   (i) 3 (ii) 59.0°
5 (i) 15 (ii) -30
6 (i) 1, 2, 2 (ii) 1, 2, -3
7 10, $\tan^{-1}\frac{4}{3}$ to Ox; $4x - 3y - 1 = 0$
8 (i) $2x + y - 2 = 0$ (ii) $4x - y - 12 = 0$
9 $\frac{7}{4}l$ 12 (i) $\frac{32}{25}W$ (ii) $\frac{1}{32}$
14 (i) 33.2° (ii) $\frac{2}{7}W$
16 15°, 1.03 W, 54.9° to the horizontal; or 75°, 0.44 W, 68.8° to the horizontal
17 (i) W, W, along AB (ii) W; $\sqrt{7}W$
18 (i) (0.42, 1.25) (ii) $\frac{7}{40}a$
19 (i) (1.43, 3.88) (ii) (0.52, 0)
20 $(\frac{3}{5}, \frac{3}{2})$; (i) 68.2° (ii) 80.9°
22 $\frac{13a}{7}, \frac{18a}{7}$. (i) 39.3° (ii) $\frac{1}{21}$
23 42.6° 24 22.6°
25 36:1 26 $\frac{4a\sqrt{2}}{4+\pi}$
27 18.6° 28 (i) 96:73

Examination questions

1.1 (i) $\{f(x) \in \mathbb{R}, f(x) \neq 1\}$ (ii) x (iii) $\frac{x+3}{x-1}$
1.2 (i) $\frac{1}{18}$(max), $\frac{1}{2}$(min)
   (ii) $-\frac{2x+5y}{5x+12y}$; 2(max), -2(min)
1.4 0.141 cms^{-1}
2.1 (i) 1, $\frac{1}{2}(-1 \pm \sqrt{3})$ (ii) $2x + 3$
2.2 (i) $-1 < x < 5$ (ii) $4 - 2\sqrt{10} < k < 4 + 2\sqrt{10}$
   (iii) $\frac{1}{3}$ (iv) $4x^2 - 12x + 33 = 0$
2.3 $-3 < x < -1$ or $1 < x < 3$
2.4 (ii) $\frac{2(4+t)(5-t)}{(20+t^2)^2}$ (iv) 20%
2.5 $\mathbb{R}$, $e^x - 1$; yes, $[-1, \infty)$; $2\ln(x+1)$, $\mathbb{R}$, -3, 1
2.6 3.59; 0.278
2.7 $\lg(p - 10) = t \lg b + \lg a$; (i) 1.9, 1.07 (ii) 1989
3.1 (i) 9, -1 (ii) 0.785, 1.429 (iii) 1.584
3.2 $\pm 30° + 120n°$, $\pm 30° + 180n°$, 19.3°
3.3 (i) $\frac{1}{3}\pi + 2n\pi$, $\frac{1}{6}\pi + 2n\pi$ (ii) -1
3.4 (ii) $\sqrt{48}$ cm, $\sqrt{32}$ cm
3.5 $\theta = \frac{200}{\tau^2}$; 0.2 s^{-1}
4.1 (ii) $\frac{1}{4}\ln\frac{2+x}{2-x} + c$
   (iv) $\frac{1}{2}\ln(\sec\theta + \tan\theta) + c$
4.2 $\frac{1}{2\sqrt{x(1-x)}}$; $\frac{1}{4}\pi$
4.3 (i) 10.86 (ii) 11.96
4.4 (i) $\frac{1}{2}\ln\frac{x^2}{1+x^2}$ (ii) $\ln\sin(\tan^{-1} x)$

4.5 $\theta = 1000 - Ae^{-kt}$
4.6 $\frac{dx}{dt} = -k(x-b)$;
   $b + (a-b)e^{-kt}$; $\frac{c}{k(a-b)}$
5.1 (i) $-\frac{1}{8}$ (ii) $-\frac{1}{3}$ (iii) $\frac{11}{2}$
5.2 (i) $\frac{1}{2}$, 2; 2
   (ii) $1 + \frac{2}{3}x - \frac{4}{9}x^2 + \frac{40}{81}x^3$, $|x| < \frac{1}{2}$
5.3 $\frac{1}{3}n(n+1)(n+2)$; $\frac{1}{6}n(6n^2 + 15n + 11)$
5.4 (i) 142 506, 142 506
   (ii) 142 506, 142 506
5.5 $1 + \frac{1}{2}x^2 - \frac{1}{8}x^4$; 0.520
5.6 $\frac{25}{216}$; (i) $\frac{1}{11}$ (ii) $\frac{5}{11}$
5.7 $\frac{31}{200}$; $\frac{9}{31}$
5.8 (i) (a) 56 (b) 6720 (c) 36
   (ii) (a) 0.35 (b) 0.57 (c) 0.65
5.9 (i) $(1-p_a)p_b/\{1-(1-p_a)(1-p_b)(1-p_c)\}$
   (ii) $(1-p_a)(1-p_b)p_c/\{1-(1-p_a)(1-p_b)(1-p_c)\}$; $\frac{1}{6}, \frac{1}{3}, \frac{1}{2}$
6.1 $x^2 + y^2 - 10x - 2y + 1 = 0$; 7
6.2 $\frac{8}{15}$
6.3 $-4 < m < 1$; $y = x + 5a$, $y = -4x + 5a$
6.4 (i) $x - 2y = 0$ (ii) 4
6.5 $\frac{1}{q-r}\{q(1-r)\mathbf{b} + r(q-1)\mathbf{c}\}$,
   $\frac{1}{r-p}\{r(1-p)\mathbf{c} + p(r-1)\mathbf{a}\}$;
   (ii) 45°
6.6 $3m\mathbf{i}$, $3m\mathbf{i} + (3-3m)\mathbf{j}$,
   $(-m)\mathbf{i} + (3-3m)\mathbf{j} + 3m\mathbf{k}$, $(-m)\mathbf{i} + 3m\mathbf{k}$;
   $4m\mathbf{i} + (3-3m)\mathbf{j} + (-3m)\mathbf{k}$
6.7 (ii) $5\mathbf{i} + 8\mathbf{j} - 6\mathbf{k}$ (iv) $\frac{1}{2}\sqrt{7}$, 58°
6.8 $\sqrt{2}$, 2, $2\sqrt{2}$; $-\frac{1}{4}\pi$, $\frac{2}{3}\pi$, $\frac{5}{12}\pi$;
   $(-1 + \sqrt{3}) + (1 + \sqrt{3})i$
6.9 (i) $\frac{1}{3}$, $-1 \pm 2i$
   (ii) $2(\cos\frac{\pi}{3} + i\sin\frac{\pi}{3})$; -1024
   (iii) $\frac{\sin\phi(1 + \cos\phi)}{1 - \cos\phi}$
7.1 $\frac{V}{3f}$
7.2 (i) $u - 2ab(1 - e^{-\frac{1}{2}at})$
   (ii) $(u - 2ab)t + 4b(1 - e^{-\frac{1}{2}at})$
   (iii) ue^{-at} (iv) $\frac{1}{a}\ln 2$
7.3 $\pm(\sin\theta\mathbf{i} + \cos\theta\mathbf{j})$;
   $2r(\cos 3\omega t\mathbf{i} - \sin 3\omega t\mathbf{j})$;
   $6\omega r$ perpendicular to OP, $18\omega^2 r$ parallel to OP, $-6\omega r(\sin 3\omega t\mathbf{i} + \cos 3\omega t\mathbf{j})$,
   $-18\omega^2 r(\cos 3\omega t\mathbf{i} - \sin 3\omega t\mathbf{j})$;
   $5r^2 + 4r^2 \sin 2\omega t$
7.4 $(10t\mathbf{i} + 0.6\mathbf{j})$ km, $(6.4t\mathbf{i} + 4.8t\mathbf{j})$ km;
   (i) $\{3.6t\mathbf{i} + (0.6 - 4.8t)\mathbf{j}\}$ km
   (ii) $(3.6\mathbf{i} - 4.8\mathbf{j})$ kmh^{-1}
7.5 $(2a - a\sin\omega t)\mathbf{i} + (2a\cos\omega t)\mathbf{j} + (a\sin\omega t - 3a)\mathbf{k}$;
   $-a\omega\cos\omega t\mathbf{i} - 2a\omega\sin\omega t\mathbf{j} + a\omega\cos\omega t\mathbf{k}$;
   $\frac{1}{2}(2n+1)\pi$; $\sqrt{5}a$, $5a$
7.6 20 ms^{-1}, 30 ms^{-1}; (i) 33.8 m (ii) 0.75, 25 ms^{-1} (iii) 90 m
7.7 (i) 20 m (ii) $\frac{3}{4}$, $\frac{7}{4}$
7.8 45°, 30 ms^{-1}
8.2 (i) $\frac{3}{5}\sqrt{2}$ mg (ii) $\frac{2}{5}$ (iii) $\frac{6}{5}mg$ vertically downwards
8.3 0.25 g, 4.5 mg, 2.5 mg; 1.25 x
8.4 $3l$; $\frac{1}{2}mg$
8.5 $\sqrt{\frac{4gx}{a(a+x)}}$; $\pi\sqrt{\frac{a(a+x)}{gx}}$
8.6 $\sqrt{\frac{2gx}{l}(5l - 2x)}$; $\frac{5}{2}\sqrt{gl}$
8.7 (i) $\sqrt{ga}$ (ii) $Mv(kg + g\sin\alpha + f)$

8.8 50, 6; $5\sqrt{14}$ ms^{-1}; $v^2 = 25(14 - 5e^{-\frac{3}{25}t})$
8.9 $\frac{1}{4}(1-e^2)u$; $\frac{1}{4}(1+e)^2 u$; $e = 1$
8.10 (i) $\dfrac{n(n-1)d}{2u}$; $\dfrac{(n-1)mu^2}{2n}$

(ii) $\sqrt{\dfrac{2Em_2}{m_1(m_1+m_2)}}$, $\sqrt{\dfrac{2Em_1}{m_2(m_1+m_2)}}$; $\sqrt{\dfrac{2Em_1m_2}{m_1+m_2}}$

8.11 $\dfrac{u\sin\alpha}{g}$; $\frac{1}{2}u\cos\alpha$; $\dfrac{u^2\sin 2\alpha}{4g}$

9.1 10**i** N, **r** = $(-2\mathbf{i} - 2\mathbf{j}) + \lambda\mathbf{i}$; $(-4\mathbf{i} + 5\mathbf{j})$ N, 21 Nm, clockwise

9.2 $\dfrac{1}{\sqrt{3}}$; $a\sqrt{3}$

9.3 $W\left(1 - \dfrac{9x}{125a}\right)$, $\dfrac{12Wx}{125a}$; $\frac{3}{4}$

9.4 $\dfrac{2Mg}{\sqrt{1 + 3\sin^2\alpha}}$

9.5 (ii) $\dfrac{4ah + 8a^2b}{h^2 - 3a^2b}$ (iii) $b > 20$

9.6 $\frac{1}{3}h\sqrt{9 + 16\tan^2\alpha}$

9.7 $\frac{24}{65}$

Index

Abscissa, 21
Acceleration, 369, 386
  angular, 412
  as a function of displacement, 382
  as a function of time, 372, 392
  as a function of velocity, 380
  constant, 375
  due to gravity, 378
  relative, 394
Addition of vectors, 281
Air resistance, 456
Algebraic area, 174, 373
Altitude of a triangle, 152, 289
Amplitude, 126
Angle
  between a line and a plane, 340
  between two lines, 303, 334
  between two planes, 340
  between two vectors, 294, 331
  bisectors of two lines, 304
  of friction, 428
  small angles, 156
Angular acceleration, 412
Angular motion, 367
Angular speed, 412
Angular velocity, 411
Approximate integration, 200, 256
Approximate solution of equations, 110
Approximations for trigonometric functions, 156
Arbitrary constant, 168
Arc, length of, 154
Area
  algebraic, 174, 373
  numerical, 174, 373
  of a triangle, 27, 152
  of a curved surface of cone, 155
  of sector, 154
  under a curve, 173, 180, 182, 327
Argand Diagram, 352
Argument of a complex number, 355
Arithmetic mean, 219
Arithmetic progression, 216
Arrangement, 236, 246
Associative, 282, 284
Astroid, 320
Asymptote, 43
Axis of a parabola, 313
Axis of rotation, 508

Base of a logarithm, 17
Basis, 291, 328
Bearing, 151
Bernouilli trial, 268
Bijective function, 33
Binomial series, 252
Binomial theorem, 249

Cartesian components of a vector, 292, 329
Cartesion coordinates, 21
Cartesian equation(s),
  of a curve, 36
  of a line, 25, 302, 332
  of a plane, 337, 338

Centre of gravity, 526
  of a composite body, 530
  of a lamina, 535
  of a set of particles, 528
  of a solid of revolution, 537
  use of integration to find, 535
Centre of mass, 526
Centroid, 183, 288, 527
Centroid of a triangle, 153
Chain rule for differentiation, 65
Change of shape, 367
Choice, 241, 246
Chord (of a curve), 50
Circle, 306
  common chord of two circles, 312
  common tangent of two circles, 312
  motion in a, 414, 465
  tangent to a, 308
  touching circles, 312
Circular motion, 414, 465
Circumcentre of a triangle, 152
Circumcircle of a triangle, 152
Codomain, 28
Coefficient of friction, 428
Coefficient of restitution, 496
Collision, 384, 396, 491
Combination, 241, 246
Common chord of two circles, 312
Common difference, 216
Common ratio, 220
Common tangent of two circles, 312
Commutative, 282, 284
Complement of a set, 3, 262
Completing the square, 74
Complex number, 8, 75, 343
  addition and subtraction, 345, 353
  argument of a, 355
  complex roots of polynomial equations, 349
  conjugate of a, 345
  imaginary part of a, 344
  modulus of a, 355
  modulus-argument (polar) form of a, 357
  multiplication and division, 345, 358
  real part of a, 344
  square roots of a, 347
  vector representation of a, 352
  zero, 345
Components of a vector, 291
  Cartesian, 292, 329
Composition of functions, 30, 63
Compound angle formulae, 139
Compression in a spring, 434
Conditional probability, 270
Cone
  area of curved surface of a, 155
  volume of a, 181
Conjugate of a complex number, 345
Connected particles, 445, 457
Conservation of Energy Principle, 477
Conservation of Mechanical Energy Principle, 477
Conservation of Momentum Principle, 489
Constant acceleration equations, 375, 455
Contact forces, 427
Continuity, 40

Continuous function, 41
Convergence of a geometric series, 225
Convergence of a series, 223
Cosine formula, 149
Cotangent formula, 152
Counter example, 3
Couple, 511
  arm of a, 512
  moment of a, 512
  system of forces equivalent to a, 519
Cubic equation, 87, 114
Cubic function, 37, 88
Cycloid, 320

Decreasing function, 40
Deduction, 2
Definite integral, 173
Degrees of freedom, 35, 338, 507
De Moivre's Theorem, 359
De Morgan's Laws, 5
Dependent trial, 271
Dependent variable, 28
Derivative, 52
  higher, 57
Differences, method of, 229
Differential coefficient, 58
Differential equation, 203
  natural occurrence of 207
  with separable variables, 205
Differentiation, 52
  chain rule for, 63
  implicit function rule for, 68
  logarithmic, 109
  of a vector, 387
  of e^x, 104
  of inverse trigonometric functions, 164
  of ln x, 107
  of trigonometric functions, 159
  of x^a, 53
  parametric, 321
  product rule for, 66
  quotient rule for, 67
Direct impact,
  elastic, 496
  inelastic, 492
Direction cosines, 330
Directrix of a parabola, 313
Discontinuity, 41
Discontinuous function, 41
Discriminant, 74
Displacement, 280, 368
  relative, 394
Distance, 368
  between two points, 21, 296
  of a point from a circle, 307
  of a point from a line, 302, 333
  of a point from a plane, 339
Distributive, 284
Divergence of a geometric series, 226
Divergence of a series, 224
Division of a line segment, 287, 297, 331
Divisor, 85
Domain, 28
Dot product, 284

Double angle formulae, 141
Dummy variable, 178

Elastic
  impact, 496
  limit, 434
  motion, 367
  potential energy, 475
  string, 437
Elements of a set, 3
Ellipse, 50, 319
Empty set, 3
Energy, 474
  Conservation of Energy Principle, 477
  Conservation of Mechanical Energy Principle, 477
  elastic potential, 475
  gravitational potential, 475
  kinetic, 474
  total mechanical, 476
  Work-Energy Principle, 476
Equal vectors, 281
Equally likely outcomes, 256
Equation, 9
  approximate solution, 110
  complex roots of, 349
  cubic, 87, 114
  modulus, 100
  number of real roots, 110
  of a circle, 307
  of a curve, 36
  of a parabola, 313
  of a plane, 336, 338
  of a straight line, 25, 288, 299, 301, 332
  polynomial, 87
  quadratic, 10, 74
  solution of an, 10, 46
  trigonometric, 131
Equilibrium, 420, 435
  limiting, 428
  particle in, 435
  rigid body in, 541
  system of forces in, 522
  three forces in, 548
Equivalent operators, 185
Equivalent systems of forces, 512
Euler's equation, 366
Even function, 38
Event, 256
  independent events, 265
  mutually exclusive events, 264
Exponential function, 105

Factor formulae, 147
Factor theorem, 87
Factorial, 215
Fibonacci sequence, 215
First order differential equation, 204
  with separable variables, 205
Focal length of a parabola, 313
Focus of a parabola, 313
Force, 420
  contact, 427
  equivalent systems of forces, 512
  forces at a distance, 427
  impulse of, 488
  impulsive, 489
  line of action of 507
  moment of, 508
  resultant, 420
  turning effect of, 507
Free vector, 281
Friction, 428
Function, 28
  bijective, 33
  composition of functions, 30, 63

Function
  continuous, 41
  cubic, 37, 88
  decreasing, 40
  discontinuous, 41
  even, 38
  exponential, 105
  implicit, 49
  increasing, 40
  injective (into), 33
  inverse, 30
  linear, 37
  logarithmic 106
  modulus, 99
  odd, 38
  of a function rule for differentiation, 63
  one-one, 30, 33
  onto, 33
  periodic, 39
  polynomial, 88
  power, 102
  quadratic, 37, 76
  rational, 91
  surjective, 33
  symmetrical, 84
  trigonometric, 125
Fundamental Theorem of Algebra, 350

Geometric mean, 222
  progression, 220
Gradient of a curve, 50
  of a line, 22
Graph, 36
Gravitation, Newton's Law of, 429
Gravitational potential energy, 475
Gravity, acceleration due to, 378
Guinness, overindulgence in, 73

Half-line, 208
Higher derivatives, 57
Hinge, 541
Hooke's Law, 434
Hyperbola, rectangular, 320

Identity, 9
  trigonometric, 131
Image, 28
Imaginary axis, 352
Imaginary number, 8, 343
Imaginary part of a complex number, 344
Impact, 491
Implicit, 49
  rule of differentiation, 68
Impulse, 488
  -Momentum Principle, 488
  instantaneous, 491
Impulsive force, 491
Impulsive tension, 492
Increasing function, 40
Indefinite integral, 173
Independent events 265
Independent trials, 265
Independent variable, 28
Indices, 15
Induction, method of, 234
Inelastic impact, 492
Inelastic string, 431
Inequality, 9
  involving rational functions, 94
  modulus, 100
  polynomial, 88
  quadratic, 79
  solution of, 10, 46
Inextensible string, 431
Inflexion, point of, 59, 60
Initial condition, 168, 204

Injective (into) function, 33
Integer, 6
Integral, 169
  as the limit of a sum, 180
  curve, 168, 204
  definite, 173
  indefinite, 173
Integrand, 169
Integration, 169
  approximate, 200, 256
  by parts, 197
  by substitution, 184
  of a vector, 391
  of e^x, 170
  of ln x, 170
  of trigonometric functions, 170
  of x^α, 169
  reverse of differentiation, 168
  use of partial fractions, 195
Intersecting forces, 507, 513, 522, 548
Intersection
  of a line and a plane, 339
  of two lines, 26, 301, 333
  of two planes, 340
  of two sets, 4, 262
Invariant element, 34
Inverse function, 30
Inverse trigonometric functions, 162
  differentiation of, 164
  use in integration, 192
Irrational number, 7
Irreducible polynomial, 87

Kinetic Energy, 474

Lamina, 527
Learning curve, 106
Length of arc, 154
Light object, 425
Limit of a function, 40
Limit of a sequence, 223
Limiting equilibrium, 428
Limiting friction, 428
Line
  angle between plane and, 340
  angle between two lines, 303, 334
  Cartesian equation of, 25, 302, 332
  distance of point from, 302, 333
  intersection of plane and, 339
  intersection of two lines, 26, 301, 333
  vector equation of, 288, 299, 301, 332
Line of action
  of a force, 507
  of a resultant force, 513
Linear
  function, 37
  momentum, 425
  motion, 367
Local maximum and minimum, 59, 61
Localised vector, 281
Locus, 305
  secondary, 324
Logarithm, 17
  base of, 17
  change of base of, 18
  laws of logarithms, 17
  natural, 106
Logarithmic differentiation, 109
Logarithmic function, 106
Logically equivalent statements, 1

Macdonalds, 106, 247
Magnitude of a couple, 512
Magnitude of a vector, 280, 292
Mapping, 28
Mass, 425

Index

Mathematical model, 368
Maximum value, 59
Mechanical Energy, 476
   conservation of, 477
Median of a triangle, 23, 152, 288
Method of differences, 229
Method of induction, 234
Midpoint of two points, 21, 287
Minimum value, 59
Modulus
   -argument form of a complex number, 357
   equation, 100
   function, 99
   inequality, 100
   of a complex number, 355
   of a real number, 12, 99
   of elasticity, 434
Moment
   of a couple, 512
   of a force, 508
   Principle of Moments, 513
   resultant, 508
Momentum, 425
   conservation of, 489
   Impulse-Momentum Principle, 488
Motion
   angular, 367
   elastic, 367
   in a circle, 414, 465
   in a straight line, 368
   in a plane, 386
   in three dimensions, 391
   linear, 367
   Newton's Laws of, 420
   relative, 394
Mutually exclusive events, 264
Mutually exclusive sets, 5

Natural Logarithm, 106
Natural number, 6
Negation of a statement, 1
Newton-Raphson method, 112
Newton's
   First Law, 420
   Law of Cooling, 208
   Law of Restitution, 496
   Second Law, 426
   Third Law, 430
   Universal Law of Gravitation, 429
Normal
   reaction, 428
   to a curve, 56, 320
   to a parabola, 314
   to a plane, 338
Number
   complex, 8, 75, 343
   imaginary, 343
   irrational, 7
   natural, 6
   rational, 6
   real, 6
   telephone, 276
Numerical area 174, 373

Odd function, 38
One-one function, 30, 33
Onto function, 33
Order of a differential equation, 204
Ordinate, 21
Outcome, 256

Parabola, 37, 313, 407
   axis of, 313
   directrix of, 313
   focal length of, 313
   focus of, 313

Parabola
   of safety, 410
   path of a projectile, 407
   standard, 314
   vertex of a, 313
Parallel lines, 23
Parameter, 288, 317
Parametric
   coordinates, 317
   differentiation, 321
   equations of a curve, 317
   equations of a line, 300, 332
   equations of a plane, 337
   equations of the path of a particle, 388
Partial fractions, 96
   use in integration, 195
   use in series expansion, 253
Particle, 368
Parts, integration by, 197
Pascal's triangle, 11, 249
Path of a particle, 386
   Cartesian equation of, 388
Perfectly elastic impact, 496
Period of a function, 39
Period of circular motion, 415
Periodic function, 39
Permutation, 236, 246
Perpendicular lines, 23
Phase difference, 127
Pivot, 541
Plane
   angle between line and, 340
   angle between two planes, 340
   distance from a point to, 339
   distance from the origin to, 339
   equation of, 336, 338
   intersection of line and, 339
   intersection of two planes, 340
   motion in, 386
   normal to, 338
   of rotation, 508
Point of inflexion, 59
Polar form of a complex number, 357
Polygon of vectors, 282
Polynomial, 85
   equation, 87, 349
   function, 88
   inequality, 88
   irreducible, 87
Position vector, 287, 296, 331
Possibility space, 256
Potential energy
   elastic, 475
   gravitational, 475
Power, 483
Power function, 102
Pre-image, 29
Probability, 256
   conditional, 270
   tree, 265
Product of two functions, 35, 66, 197
Product rule for differentiation, 66
Proper subset, 3
Projectiles, 401
   equation of path, 407
   greatest height, 402, 406
   horizontal range, 402, 406
   maximum range, 406
   time of flight, 402, 406
Pulley, 431, 445, 458

Quadratic equation, 10, 74
Quadratic function, 37, 76
Quadratic inequality, 79
Quotient of two polynomials, 85
Quotient rule for differentiation, 67

Radian, 153
Range
   of a function, 28
   of a projectile, 402, 406
   of a rational function, 92
Rate of change, 64
Rational function, 91
   integration of, 195
   range of, 92
Rational number, 6
Reaction, frictional, 428
Reaction, normal, 428
Real
   axis, 352
   line, 6
   number, 6
   part of a complex number, 344
   roots of equations, 110
Recurrence relation, 215
Rectangular hyperbola, 320
Reduction to linear form, 115
Relative
   acceleration, 394
   displacement, 394
   growth rate, 55
   notion, 394
   velocity, 394
Remainder, 85
Remainder Theorem, 86
Resolving into components, 295, 436
Resultant
   force, 420
   moment, 508
   vector, 281, 293, 329
Restitution, coefficient of, 496
Restitution, Newton's Law of, 496
Rigid body, 368
   centre of gravity of 526
   equilibrium, 541
Roots of equations, 10
   complex, 349
   quadratic, 82, 349
   polynomial, 87, 349

Sample space, 256
Scalar, 281
   product, 284, 294, 329
Secondary locus, 324
Sector area of, 154
Segment, 154
Selection, 241, 246
Semi-cubical parabola, 50, 319
Sequence, 214
   arithmetic, 216
   geometric, 220
   limit of, 223
Series, 216
   binomial, 253
   convergence of, 223
Set, 3
Shortest distance
   of a point from a circle, 307
   of a point from a line, 302, 333
   of a point from a plane, 339
   of the origin from a plane, 339
Sigma notation, 216
Simpson's Rule, 201
Sine formula, 149
Skew lines, 333
Sliding and toppling, 552
Small angles, 156
Small changes, 70
Smoothly contact, 429
Smooth hinge, 541
Solution of a triangle, 149
Soufflé, 209

Speed, 369, 388
 angular, 412
 terminal, 40, 381
Sphere, vector equation of, 343
Sphere, volume of, 182
Spring, 434
Square roots of a complex number, 347
Standard parabola, 314
Stationary point, 59
Stationary value, 59
String, elastic, 434
String, inelastic, 431
Subset, 3
Substitution, integration by, 184
Subtraction of vectors, 282
Sum
 integration as the limit of, 180
 of a geometric series, 220
 of an arithmetic series, 217
 of two functions, 35, 53, 170
 to infinity of a geometric series, 226
 to infinity of a series, 224
Surd, 13
Surjective function, 33
Suspension, 549
Symmetrical function, 84

Taking moments, 513
Tangent
 to a circle, 308
 to a curve, 50, 56, 315, 320
 to a parabola, 314, 323
Target, 28
Tension
 impulsive, 489
 in an elastic spring, 434
 in a rod, 431
 in a string, 431
Terminal speed, 40, 381
t formulae, 143
 use in integration, 194
Three forces in equilibrium, 548
Thrust in a light rod, 431
Thrust in a spring, 434
Tied vector, 281

Time constant, 209
Total Mechanical Energy, 476
Torque, 508
Touching circles, 312
Translation, 367
Trapezium Rule, 200
Tree diagram, 200, 265
Trial, 256
 Bernouilli, 268
Triangle
 altitude of, 152, 289
 area of, 27, 152
 centroid of, 153, 288
 circumcentre of, 152
 circumcircle of, 152
 law of addition of vectors, 280, 281
 median of, 23, 152, 288
 solution of, 149
Trigonometric equations, 131
Trigonometric identities, 131
Trigonometric functions, 125
 approximations for, 156
 differentiation of, 159
 integration of, 170
 inverse, 162
Turning effect of a force, 508
Turning point, 59

Uniform body, 526
Union of two sets, 4, 262
Unit vector, 281, 292, 329
Universal Set, 3
Universal Law of Gravitation, 429

Vector, 280
 addition of vectors, 281, 293
 angle between two vectors, 294, 331
 Cartesian components of 292, 329
 components of, 291
 differentiation of, 387
 direction cosines of, 330
 dot product, 284
 equal vectors, 281
 equation of a curve, 318, 388
 equation of a line, 288, 299, 301, 332, 393

Vector
 equation of a plane, 336, 338
 equation of a sphere, 343
 free, 281
 integration of, 391
 localised, 281
 magnitude of a, 292
 polygon of vectors, 282
 position, 287, 296, 331
 representation of a complex number, 352
 resolving into components, 295
 resultant, 281, 293, 329
 scalar product, 284, 294, 329
 subtraction of vectors, 282
 tied, 281
 triangle law of addition of vectors, 280, 281
 unit, 281, 292, 329
 zero, 282
Velocity, 369, 386
 angular, 411
 relative, 394
Venn Diagram, 3
Vertical motion under gravity, 378
Volume
 of a cone, 181
 of a pyramid, 183
 of a solid of revolution, 180, 182
 of a sphere, 182

Weight, 427
Wok, 182
Work
 done by a force, 470, 472
 done by an object against a force, 470
 done in stretching an elastic string, 470
 Energy Principle, 476

$X = \phi(x)$ method, 113

Zero
 complex number, 345
 moment, 508
 vector, 282